COMMON GRAPHS AND MODELS

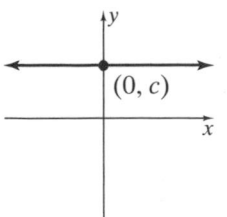

Horizontal Line;
Zero Slope
$y = c$

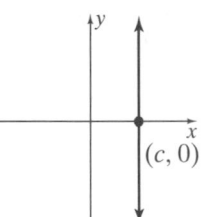

Vertical Line;
Undefined Slope
$x = c$

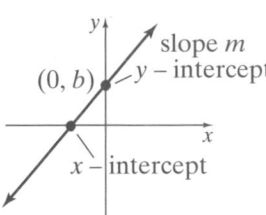

Linear Function;
Positive Slope
$y = mx + b; m > 0$

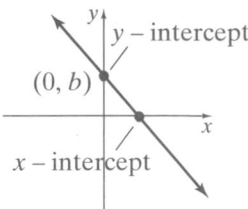

Linear Function;
Negative Slope
$y = mx + b; m < 0$

$y = x$

$y = x^2$

$y = x^3$

$y = |x|$

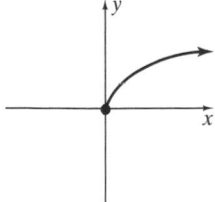

$y = \sqrt{x}; x \geq 0$

Quadratic Function
$y = ax^2 + bx + c; a \neq 0$
Parabola opens upward if $a > 0$
Parabola opens downward if $a < 0$

vertex has
x – coordinate: $-\dfrac{b}{2a}$.

Quadratic Function
$y = a(x - h)^2 + k; a \neq 0$
Parabola opens upward if $a > 0$
Parabola opens downward if $a < 0$

vertex is
(h, k).
axis of
symmetry,
$x = h$.

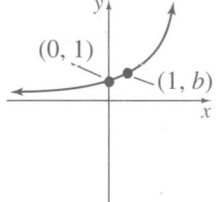

Exponential
Function
$y = b^x$ for $b > 1$

SYSTEMS OF LINEAR EQUATIONS

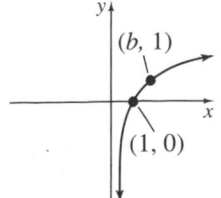

Logarithmic Function
$y = \log_b x$ for $b > 1$

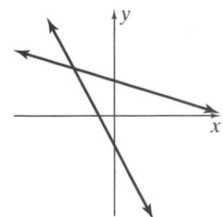

Independent and
consistent; one solution

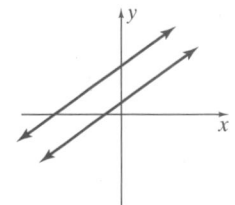

Independent and
inconsistent, no solution

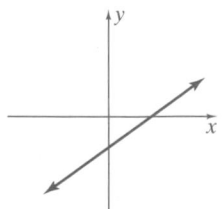

Dependent and consistent;
infinitely many solutions

ANNOTATED INSTRUCTOR'S EDITION

Intermediate Algebra
A Graphing Approach

THIRD EDITION

K. Elayn Martin-Gay
University of New Orleans

Margaret Greene
Florida Community College at Jacksonville

This Annotated Instructor's Edition is exactly like your student's text, but it contains Teaching Tips, Classroom Examples, and all exercise answers, including Vocabulary Checks. Where possible the answer is displayed on the same page as the exercise. Exercise answers that require graphical solutions and answers to Spotlight on Decision Making and Chapter Projects can be found in the Graphing Answer Section.

The Student Edition of the text contains answers to selected exercises. Located in the back of the Student Edition are answers to odd-numbered exercises and review exercises, all chapter test exercises, and all cumulative review exercises including text references. It does not contain answers to even-numbered exercises, Spotlight on Decision Making, Chapter Projects, Vocabulary Checks, group activities, or instructor's notes.

PEARSON

Prentice
Hall

Upper Saddle River, New Jersey 07458

DEDICATION

To my husband Jack and our family Kevin, Elizabeth, Dan, Cara, and Matt, and my dear friend, Eleanor King

Editor in Chief: Christine Hoag
Senior Acquisitions Editor: Paul Murphy
Project Manager: Mary Beckwith
Assistant Editor: Christine Simoneau
Project Management: Elm Street Publishing Services, Inc.
Senior Managing Editor: Linda Mihatov Behrens
Executive Managing Editor: Kathleen Schiaparelli
Assistant Vice President of Production and Manufacturing: David W. Riccardi
Assistant Manufacturing Manager/Buyer: Michael Bell
Manufacturing Manager: Trudy Pisciotti
Art Editor: Tom Benfatti
Executive Marketing Manager: Eilish Collins Main
Marketing Assistant: Annett Uebel
Director of Marketing: Patrice Jones
Marketing Project Manager: Barbara Herbst
Development Editor: Elka Block
Editor in Chief, Development: Carol Trueheart
Art Director/Cover Designer: Maureen Eide
Cover Art: Geoffrey Cassar
Assistant to the Art Director: Dina Curro
Interior Designer: Donna Wickes
Creative Director: Carole Anson
Director of Creative Services: Paul Belfanti
Media Project Manager, Developmental Math: Audra J. Walsh

Composition: Pearson formatting:
Manager, Electronic Composition: Jim Sullivan; Assistant Formatting Manager: Allyson Graesser; Electronic Production Specialists: Karen Noferi, Joanne Del Ben, Karen Stephens, Jacqueline Ambrosius, Julita Nazario, Vickie Croghan and Progressive Information Technologies
Director, Image Resource Center: Melinda Reo
Manager, Rights and Permissions: Zina Arabia
Interior Image Specialist: Beth Brenzel
Image Permission Coordinator: Craig Jones
Photo Researcher: Melinda Alexander
Art Studios: Artworks, Scientific Illustrators, Laserwords
Artworks
Managing Editor, AV Production & Management: Patricia Burns
Production Manager: Ronda Whitson
Production Technologies Manager: Matthew Haas
Project Coordinator: Dan Missildine
Art Supervisor: Kathryn Anderson
Illustrators: Audrey Simonetti, Daniel Knopsnyder, Mark Landis, Nathan Storck, Ryan Currier, Stacy Smith, Scott Wieber
Quality Supervisor: Pamela Taylor
Quality Assurance: Cathy Shelly, Tim Nguyen, Ken Mooney

©2005 Pearson Education, Inc.
Pearson Prentice Hall
Pearson Education, Inc.
Upper Saddle River, New Jersey 07458

10 9 8 7 6 5 4 3 2 1

ISBN 0-13-146901-0
ISBN 0-13-146900-2 (student edition)

Pearson Education Ltd., London
Pearson Education Australia Pty. Limited, Sydney
Pearson Education Singapore Pte. Ltd.
Pearson Education North Asia, Ltd., Hong Kong
Pearson Education Canada, Ltd., Toronto
Pearson Educacion de Mexico, S.A., de C.V.
Pearson Education, Japan, Tokyo
Pearson Education Malaysia, Pte. Ltd.

CONTENTS

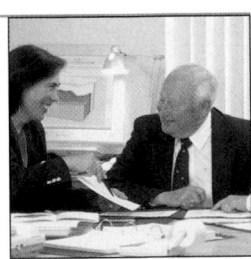

PREFACE

ABOUT THE BOOK

Intermediate Algebra: A Graphing Approach, Third Edition was written to provide a **solid foundation in algebra.** Specific care has been taken to ensure that students have the most **up-to-date and relevant** text preparation for their next mathematics course, as well as to help students to succeed in nonmathematical courses that require a grasp of algebraic fundamentals. I have tried to achieve this by writing a user-friendly text that is keyed to objectives and contains many worked-out examples. The basic concepts of graphing and functions are introduced early, and problem solving techniques, real-life and real-data applications, data interpretation, appropriate use of technology, mental mathematics, number sense, critical thinking, decision-making, and geometric concepts are emphasized and integrated throughout the book.

The new edition includes an increased emphasis on study and test preparation skills. In addition, the third edition now includes a new resource, the Chapter Test Prep Video CD. With this CD/Video, students have instant access to video solutions for each of the chapter test questions contained in the text. It is designed to help them study efficiently.

The many factors that contributed to the success of the previous editions have been retained. In preparing this edition, I considered the comments and suggestions of colleagues throughout the country, students, and many users of the prior editions. The AMATYC Crossroads in Mathematics: Standards for Introductory College Mathematics before Calculus and the MAA and NCTM standards (plus Addenda), together with advances in technology, also influenced the writing of this text.

Intermediate Algebra: A Graphing Approach, Third Edition is **part of a series of texts** that can include *Basic College Mathematics, Second Edition, Prealgebra, Fourth Edition, Beginning Algebra, Fourth Edition.* Also, available are *Intermediate Algebra, Fourth Edition*, and *Beginning and Intermediate Algebra, Third Edition*, a combined algebra text. Throughout the series, pedagogical features are designed to develop student proficiency in algebra and problem solving, and to prepare students for future courses.

NEW FEATURES AND CHANGES IN THE NEW EDITION

The following new features have been added to the fourth edition.

INCREASED EMPHASIS ON STUDY SKILLS

New! Study Skills Reminders Integrated throughout the text to help students hone their study skills and serve as a point-of-use support resource, reinforcing the skills covered in Section 1.1. Also, beginning in Section 1.5, and continued throughout the text, there are special Study Skills Reminders that help students organize an outline on how to recognize and how to solve various equations and inequalities. See the Study Skills Reminders in Sections 1.5, 3.2, 3.3, and 3.4 for examples.

New! Integrated Reviews   serve as mid-chapter reviews and help students assimilate new skills and concepts they have learned separately over several sections. The reviews provide students with another opportunity to practice with *mixed* exercises as they master the topics.

ENHANCED SECTION EXERCISE SETS

Mixed Practice exercises. Exercise sets have been organized to include mixed practice exercises where appropriate. They give students the chance to assimilate the concepts and skills covered in separate objectives. Students have the opportunity to practice the kind of decision making they will encounter on tests.

New! Concept Extension exercises have been added to the end of the section exercise sets. They extend the concepts and require students to combine several skills or concepts. They expose students to the way math ideas build upon each other and offer the opportunity for additional challenges.

MORE OPPORTUNITIES FOR STUDENTS TO CHECK THEIR UNDERSTANDING

New! Concept Checks are special exercises found in most sections following key examples. Working these will help students check their grasp of the concept being developed before moving to the next example.

INCREASED EMPHASIS ON IMPROVING TEST PREPARATION

New! Chapter Test Prep Video CD packaged with each text and presented by Elayn Martin-Gay provides students with a resource to take and correct sample tests as they prepare for exams. Step-by-Step solutions are presented for every Chapter Test exercise contained in the text. Easy video navigation allows students to instantly access the solutions to the exact exercises they need help with.

KEY CONTINUING FEATURES

The following key features have been retained from previous editions.

Readability and Connections I have tried to make the writing style as clear as possible while still retaining the mathematical integrity of the content. When a new topic is presented, an effort has been made to **relate the new ideas to those that students may already know**. Constant reinforcement and connections within problem solving strategies, data interpretation, geometry, patterns, graphs, and situations from everyday life can help students gradually master both new and old information.

Problem Solving Process This is formally introduced in Chapter 2 with a **four-step process that is integrated throughout the text**. The four steps are Understand, Translate, Solve, and Interpret. The repeated use of these steps throughout the text in a variety of examples shows their wide applicability. Reinforcing the steps can increase students' confidence in tackling problems.

Applications and Connections Every effort was made to include as many accessible, interesting, and relevant real-life applications as possible throughout the text in both worked-out examples and exercise sets. The applications **strengthen students' understanding of mathematics in the real world** and help to motivate students. They show connections to a wide range of fields including agriculture, allied health, art, astronomy, automotive ownership, aviation, biology, business, chemistry, communication, computer technology, construction, consumer affairs, demographics, earth science, education, entertainment, environmental issues, finance and economics, food service, geography, government, history, hobbies, labor and career issues, life science, medicine, music, nutrition, physics, political science, population, recreation, sports, technology, transportation, travel, weather, and important related mathematical areas such as geometry and statistics. (See the Index of Applications on page xxiv.) Many of the applications are based on **recent and interesting real-life data**. Sources for data include newspapers, magazines, government publications, publicly held companies, special interest groups, research organizations, and reference books. Opportunities for obtaining your own real data are also included.

Helpful Hints   Helpful Hints, contain practical advice on applying mathematical concepts. These are found throughout the text and **strategically placed** where students are most likely to need immediate reinforcement. They are highlighted in a box for quick reference and, as appropriate, an indicator line is used to precisely identify the particular part of a problem or concept being discussed. For instance, see page 98.

Visual Reinforcement of Concepts   The text contains numerous graphics, models, and illustrations to visually clarify and reinforce concepts. These include **new and updated** bar graphs, circle graphs in two and three dimensions, line graphs, calculator screens, application illustrations, photographs, and geometric figures.

Discover the Concept   These explorations help students recognize patterns or discover a concept on their own immediately before the concept is introduced.

Technology Notes   These notes contain specific suggestions for problem solving with technology.

Real World Chapter Openers   The chapter openers focus on how math is used in a specific career, and reference a "Spotlight on Decision Making" feature within the chapter for further exploration of the **career and the relevance of algebra**.

Student Resource Icons   At the beginning of each exercise set, videotape, tutorial software CD-Rom, Student Solutions Manual, Study Guide, and tutor center icons are displayed. These icons help reinforce that these learning aids are available should students wish to use them to review concepts and skills at their own pace. These items have **direct correlation to the text** and emphasize the text's methods of solution.

Chapter Highlights   Found at the end of each chapter, the Chapter Highlights contain key definitions, concepts, *and* examples to **help students understand and retain** what they have learned.

Chapter Project   This feature occurs at the end of each chapter, often serving as a chapter wrap-up. For **individual or group completion**, the multi-part Chapter Project, usually hands-on or data based, allows students to problem solve, make interpretations, and to think and write about algebra.

EXERCISE SETS

Each text section ends with an exercise set. The exercises are carefully graded in level of difficulty. In the beginning, **exercises are keyed** to at least one worked example in the text. Once a student has gained confidence, **the latter exercises are not keyed to examples**. Exercises and examples marked with a video icon ($\blacksquare$) have been worked out step-by-step by the author in the videos that accompany this text.

Throughout the text exercises there is an emphasis on data and graphical interpretation via tables, charts, and graphs. The ability to interpret data and read and create a variety of types of graphs is developed gradually so students become comfortable with it. Similarly, throughout the text there is integration of geometric concepts, such as perimeter and area. Exercises and examples marked with a geometry icon ($\triangle$) have been identified for convenience.

Each exercise set contains one or more of the following features.

Mental Math   These problems are found at the beginning of many exercise sets. They are mental warm-ups that **reinforce concepts** found in the accompanying section and increase students' confidence before they tackle an exercise set. By relying on their own mental skills, students increase not only their confidence in themselves, but also their number sense and estimation ability. This edition includes a greater number of Mental Math exercises.

Writing Exercises These exercises are found in almost every exercise set and are marked with the icon (✎). They require students to **assimilate information** and provide a written response to explain concepts or justify their thinking. Guidelines recommended by the American Mathematical Association of Two Year Colleges (AMATYC) and other professional groups recommend incorporating writing in mathematics courses to reinforce concepts.

Mixed Practice exercises. Exercise sets have been organized to include mixed practice exercises where appropriate. They give students the chance to assimilate the concepts and skills covered in separate objectives. Students have the opportunity to practice the kind of decision making they will encounter on tests.

New! Concept Extension exercises have been added to the end of the section exercise sets. They extend the concepts and require students to combine several skills or concepts. They expose students to the way math ideas build upon each other and offer the opportunity for additional challenges.

Data and Graphical Interpretation Throughout the text there is an emphasis on data interpretation in exercises via tables, bar charts, line graphs, or circle graphs. The ability to interpret data and read and create a variety of graphs is **developed gradually** so students become comfortable with it.

Review and Preview These exercises occur in each exercise set (except for those in Chapter 1). These problems are **keyed to earlier sections** and review concepts learned earlier in the text that are needed in the next section or in the next chapter. These exercises show the **links between earlier topics and later material**.

Vocabulary Checks Vocabulary checks, provide an opportunity for students to become more familiar with the use of mathematical terms as they strengthen verbal skills. They are found at the end of each chapter.

Chapter Review and Chapter Test The end of each chapter contains a review of topics introduced in the chapter. The review problems are keyed to sections. The Chapter Test is not keyed to sections. The Chapter Test Prep Video CD provides solutions by the author to every Chapter Test exercise in the text.

Cumulative Review Each chapter after the first contains a **cumulative review of all chapters beginning with the first** up through the chapter at hand. The odd problems contained in the cumulative reviews are actually earlier worked examples in the text. The even problems are keyed to sections where students can go to review the material in an exercise.

KEY CONTENT CHANGES IN THE FOURTH EDITION

Overview Throughout the new edition, you will find numerous new real-life applications in examples and exercises. See Sections 1.9, 2.6, and 8.7 for examples. Some sections have internal reorganization to better clarify and enhance the presentation.

 Specific discussions have been enhanced in the third edition. Some of the changes include:

- In every section where a new type of equation or inequalities is solved, **a special study skill box has been added. Students are encouraged to develop a cumulative outline** in their own words that will help them learn to recognize and solve these equations and inequalities. This is designed to help students throughout intermediate algebra—as well as to be better prepared for college algebra.

- **Increased coverage on factoring trinomials by grouping**. See Section 5.6.

- **Increased discussion about slope as a rate of change**. See Sections 2.4 and 2.5.

- **Improved diagrams for student understanding**. For example, see Sections 9.1 and 9.2.

Exercise Sets. All applications and data have been thoroughly updated. In addition, the following changes have been made:

- All exercise sets are better organized. Most now include a section of exercises called **"Mixed Practice"**. These exercises are designed to help prepare students for any testing of concepts.
- Each exercise set now includes **Concept Extension** exercises. These exercises allow instructors to pick and choose more difficult and/or more conceptual types of exercises for students.
- Overall, the range of the exercise sets has extended so that there are now more choices for exercises at increased level of difficulty. See Sections 1.5, 2.3, 3.4, 4.1, 4.4, 5.3, 5.7, 6.4, 6.8, 7.4, 7.6, 9.3.
- Overall, the types of the exercises available for students has increased. This is intended to help them better prepare for standardized testing. For example, see the new multiple choice exercises in Sections 3.2, 4.2, 6.1, 7.1, 7.2.
- More real-life data exercises have been included. See Sections 1.6, 1.8, 2.1, 4.3.
- Increased Mental Math exercises: These can be worked as a class or alone before the exercise set. These exercises not only increase student's mental computation skills, but prepares a student for the upcoming exercise set. See Sections 2.1, 3.5, 6.2, 6.8, 7.1, 7.2, 9.1, 9.5, 10.2.

INSTRUCTOR AND STUDENT RESOURCES

The third edition is supported by a comprehensive resource for instructors and students.

INSTRUCTOR RESOURCES—PRINT

Annotated Instructor's Edition (ISBN 0-13-146901-0)

- Answers to exercises on the same text page or in Graphing Answer Section
- Graphing Answer Section contains answers to exercises requiring graphical solutions, chapter projects, and Spotlight on Decision Making exercises
- Teaching Tips throughout the text placed at key points in the margin, found in places where students historically need extra help together with ideas on how to help students through these concepts, as well as placed appropriately to provide ideas for expanding upon a certain concept, other ways to present a concept, or ideas for classroom activities
- New Classroom Examples have been added. Each Classroom Example parallels the text example for an added resource during lecture

Instructor's Solutions Manual (ISBN 0-13-126854-6)

- Detailed step-by-step solutions to even-numbered section exercises
- Solutions to every Spotlight on Decision Making exercise
- Solutions to every Calculator Exploration exercise
- Solutions to every Chapter Test and Chapter Review exercise
- Solution methods reflect those emphasized in the textbook

Instructor's Resource Manual with Tests (ISBN 0-13-126853-8)

- Notes to the Instructor that include new suggested assignments for each exercise set

- Eight Chapter Tests per chapter (5 free response, 3 multiple choice)
- Two Cumulative Review Tests (one free response, one multiple choice)
- Eight Final Exams (4 free response, 4 multiple choice)
- Twenty additional exercises per section for added test exercises or worksheets, if needed
- Group Activities (on average of two per chapter; providing short group activities in a convenient ready-to-use handout format)
- Answers to all items

INSTRUCTOR RESOURCES—MEDIA

TestGen with QuizMaster   enables instructors to build, edit, print, and administer tests using a computerized bank of questions developed to cover all the objectives of the text. Instructors can modify test bank questions or add new questions by using the built-in question editor, which allows users to create graphs, import graphics, and insert math notation, variable numbers, or text. Tests can be printed or administered online via the Internet or another network. TestGen comes packaged with QuizMaster, which allows students to take tests on a local area network. The software is available on a dual-platform Windows/Macintosh CD-ROM.

"Instructor to Instructor" Videos   authored by Elayn Martin-Gay these videos offer topical and teaching technique instruction to new instructors and adjuncts to enhance effective classroom communication, and provide seasoned faculty with additional teaching ideas and approaches. They also provide suggestions for presenting the topics in class, alternative approaches, time saving strategies, classroom activities, and much more.

New MyMathLab® (instructor)   is a series of text-specific, easily customizable online courses for Prentice Hall mathematics textbooks. MyMathLab is powered by CourseCompass™—Pearson Education's online teaching and learning environment—and by MathXL®—our online homework, tutorial, and assessment system. MyMathLab gives you the tools you need to deliver all or a portion of your course online, whether your students are in a lab setting or working from home.

MyMathLab provides a rich and flexible set of course materials, featuring free-response exercises that are algorithmically generated for unlimited practice and mastery. Students can also use online tools such as video lectures, animations, and a multimedia textbook to independently improve their understanding and performance. Instructors can use MyMathLab's homework and test managers to select and assign online exercises correlated directly to the textbook, and they can also import TestGen tests into MyMathLab for added flexibility. MyMathLab's online gradebook—designed specifically for mathematics—automatically tracks students' homework and test results and gives the instructor control over how to calculate final grades.

MyMathLab is available to qualified adopters. For more information, visit our website at www.mymathlab.com or contact your Prentice Hall sales representative for a product demonstration.

MathXL®   is a powerful online homework, tutorial, and assessment system that accompanies your Prentice Hall mathematics textbook. With MathXL, instructors can create, edit, and assign online homework and tests using algorithmically generated exercises correlated at the objective level to your textbook. All student work is tracked in MathXL's online gradebook. Students can take chapter tests in MathXL and receive personalized study plans based on their test results. The study plan diagnoses weaknesses and links students directly to tutorial exercises for the objectives they need to study and retest. Students can also access supplemental animations and video clips directly from selected exercises. MathXL is available to qualified adopters. For more information, visit our website at www.mathxl.com, or contact your Prentice Hall sales representative for a product demonstration.

MathXL® Tutorials on CD   This interactive tutorial CD-ROM provides algorithmically generated practice exercises that are correlated at the objective level to the exercises in the textbook. Every practice exercise is accompanied by an example and a guided solution designed to involve students in the solution process. Selected exercises may also include a video clip to help students visualize concepts. The software tracks student activity and scores and can generate printed summaries of students' progress.

STUDENT RESOURCES—PRINT

Student Solutions Manual (ISBN 0-13-126864-3)

- Detailed step-by-step solutions to odd-numbered section exercises
- Solutions to every (odd and even) Mental Math exercise
- Solutions to odd-numbered Calculator Exploration exercises
- Solutions to every (odd and even) exercise found in the Chapter Reviews and Chapter Tests
- Solution methods reflect those emphasized in the textbook
- Ask your bookstore about ordering

Graphing Calculator Guide (ISBN 0-13-126852-X)

STUDENT RESOURCES—MEDIA

New! Chapter Test Prep Video CD   Provides a step-by-step video solution to each problem in the textbook Chapter Test, presented by Elayn Martin-Gay.

MyMathLab® (student)   is a complete online course designed to help students succeed in learning and understanding mathematics. MyMathLab contains an online version of your textbook with links to multimedia resources—such as video clips, practice exercises, and animations—that are correlated to the examples and exercises in the text. MyMathLab also provides students with online homework and tests and generates a personalized study plan based on their test results. The study plan links directly to unlimited tutorial exercises for the areas students need to study and re-test, so they can practice until they have mastered the skills and concepts in the textbook. All of the online homework, tests, and tutorial work students do is tracked in their MyMathLab gradebook.

MathXL®   is a powerful online homework, tutorial, and assessment system that accompanies your Prentice Hall mathematics textbook. With MathXL, instructors can create, edit, and assign online homework and tests using algorithmically generated exercises correlated at the objective level to your textbook. All student work is tracked in MathXL's online gradebook. Students can take chapter tests in MathXL and receive personalized study plans based on their test results. The study plan diagnoses weaknesses and links students directly to tutorial exercises for the objectives they need to study and retest. Students can also access supplemental animations and video clips directly from selected exercises. MathXL is available to qualified adopters. For more information, visit our website at www.mathxl.com, or contact your Prentice Hall sales representative for a product demonstration.

MathXL® Tutorials on CD   This interactive tutorial CD-ROM provides algorithmically generated practice exercises that are correlated at the objective level to the exercises in the textbook. Every practice exercise is accompanied by an example and a guided solution designed to involve students in the solution process. Selected exercises may also include a video clip to help students visualize concepts. The software tracks student activity and scores and can generate printed summaries of students' progress.

Lecture Series Videos   Digitized on CD-ROM (0-13-126862-7) and on VHS Tape (0-13-126860-0)

- Keyed to each section of the text
- Step-by-step solutions to exercises from each section of the text
- In-text exercises marked with a video icon appear on the videos
- Digitized videos offer convenient anytime access to video tutorial support when shrinkwrapped with the text

PH Tutor Center (0-13-064604-0)

- Free tutorial support via phone, fax or email
- Available Sunday—Thursday 5pm e.s.t. to midnight—5 days a week, 7 hours a day
- Staffed by developmental math faculty
- Accessed through a registration number that may be bundled with a new text or purchased separately with a used book.
- See www.prenhall.com/tutorcenter for FAQ

ACKNOWLEDGMENTS

First, as usual, I would like to thank my husband, Clayton, for his constant encouragement. I would also like to thank my children, Eric and Bryan, for continuing to eat my burnt meals. Thankfully, they have started to cook a little themselves.

I would also like to thank my extended family for their invaluable help and wonderful sense of humor. Their contributions are too numerous to list. They are Rod and Karen Pasch; Peter, Michael, Christopher, Matthew, and Jessica Callac; Stuart, Earline, Melissa, Mandy, Bailey, and Ethan Martin; Mark, Sabrina, and Madison Martin; Leo and Barbara Miller; and Jewett Gay.

We would like to thank the following reviewers for their input and suggestions:

Jeffrey Berg, *Arapahoe Community College*
June Decker, *Three Rivers Community College*
Kathryn Gunderson, *Three Rivers Community College*
Margaret Harris, *Columbus State Community College*
Celeste Harnandez, *Richland College*
Steven Hundert, *College of Southern Maryland*
Erica M. Johnson, *Arapahoe Community College*
Kandace Kling, *Portland Community College*
David Kratzer, *University of New Mexico–Los Alamos Brand*
Vicki Lackey, *Columbus State Community College*
John Martin, *Santa Rosa Junior College*
Rick Pal, *Valencia Community College*
Richard Plagge, *Highline Community College*
Michael Reynolds, *Valencia Community College*
Owen Thomas, *Columbus State Community College*

There were many people who helped us develop this text and we will attempt to thank some of them here. Lauri Semarne was invaluable for contributing to the overall accuracy of this text. Elka Block, Chris Callac, and Miriam Daunis were invaluable for their many suggestions and contributions during the development and writing of this third edition. Brandi Nelson provided guidance throughout the production process.

Sadly, executive editor of this project, Karin Wagner, passed away this year. She will be dearly remembered by us and the rest of the staff at Prentice Hall for her integrity, wisdom, commitment, and especially her sense of humor and infectious laugh.

A very special thank you to our project manager, Mary Beckwith, for taking over during a difficult period for all of us.

Lastly, my thanks to the staff at Prentice Hall for all their support: Linda Behrens, Mike Bell, Patty Burns, Tom Benfatti, Paul Belfanti, Maureen Eide, Eilish Main, Patrice Jones, Chris Hoag, Paul Corey, Tim Bozik, Jim Sullivan, Allyson Graesser, Karen Noferi, Joanne Del Ben, Karen Stephens, Jacqueline Ambrosius, Julita Nazario, and Vickie Croghan.

K. Elayn Martin-Gay
Margaret (Peg) Greene

ABOUT THE AUTHORS

K. Elayn Martin-Gay has taught mathematics at the University of New Orleans for 25 years. Her numerous teaching awards include the local University Alumni Association's Award for Excellence in Teaching, and Outstanding Developmental Educator at University of New Orleans, presented by the Louisiana Association of Developmental Educators.

Prior to writing textbooks, K. Elayn Martin-Gay developed an acclaimed series of lecture videos to support developmental mathematics students in their quest for success. These highly successful videos originally served as the foundation material for her texts. Today the tapes specifically support each book in the Martin-Gay series.

Elayn is the author of over ten published textbooks as well as multimedia interactive mathematics, all specializing in developmental mathematics courses such as basic mathematics, prealgebra, beginning and intermediate algebra. She has provided author participation across the broadest range of materials: textbook, videos, tutorial software, and Interactive Math courseware. All the components are designed to work together. This offers an opportunity of various combinations for an integrated teaching and learning package offering great consistency and comfort for the student.

Margaret (Peg) Green has taught mathematics at Florida Community College at Jacksonville for more than 18 years. Peg has also been an instructor of the College Instructors Training Network, an Ohio State Short course program for teaching college instructors how to enhance their teaching by using technology in the mathematics classroom. A recipient of excellence in teaching awards, Peg's experience also includes: being a member of the NASA/MAATYC Coalition: I and II; teaching a telecourse; teaching online courses as well as training instructors to teach via the online environment; developing a video series on using the graphing calculator; conducting workshops; authoring numerous publications; and teaching at the Southeast Center for Cooperative Learning, certified through the University of Minnesota.

Highlights of *Intermediate Algebra, A Graphing Approach, Third Edition*

Every Student Can Succeed

Intermediate Algebra, A Graphing Approach, Third Edition has been written and designed to help you succeed in this course. Special care has been taken to ensure students have the most up-to-date and relevant text features, and as many real-world applications as possible to provide you with a solid foundation in algebra and prepare you for future courses.

Good study skills are essential for success in mathematics. This edition provides an increased emphasis on study and test preparation skills. Take a few minutes to examine the features and resources that have been incorporated into *Intermediate Algebra, A Graphing Approach, Third Edition* to help students excel.

CHAPTER 2

Graphs and Functions

2.1 Introduction to Graphing
2.2 Introduction to Functions
2.3 Graphing Linear Functions
2.4 The Slope of a Line
2.5 Equations of Lines
   Integrated Review—Linear Equations in Two Variables
2.6 Interpreting Data: Linear Models

The Centers for Disease Control and Prevention (CDC), located in Atlanta, Georgia, was founded by the United States government as the primary federal agency devoted to public health. Public health officials there work constantly to provide information about current health issues. Prevention of diseases, injury, and disability is another focus of the public health specialists at the CDC.

The CDC employs approximately 8500 people in Atlanta, Washington, D.C., and at its many field sites nationwide. Positions at the CDC include medical officers, public health advisors, epidemiologists, behavioral scientists, statisticians, and information and technology specialists.

PAGE 115

◀ Real-World Chapter Openers

Real-world chapter openers focus on how algebraic concepts relate to the world around you. They also reference a **Spotlight on Decision Making** feature within the chapter for further exploration.

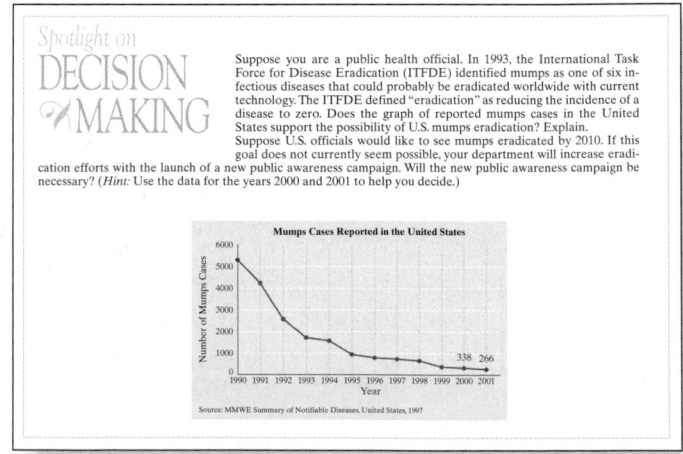

PAGE 187

Spotlight on Decision Making ▲

These unique applications encourage students to develop decision making and problem solving abilities. Primarily workplace or career-related situations are highlighted in each feature.

Become a Confident Problem Solver!

A goal of this text is to help you develop problem solving abilities.

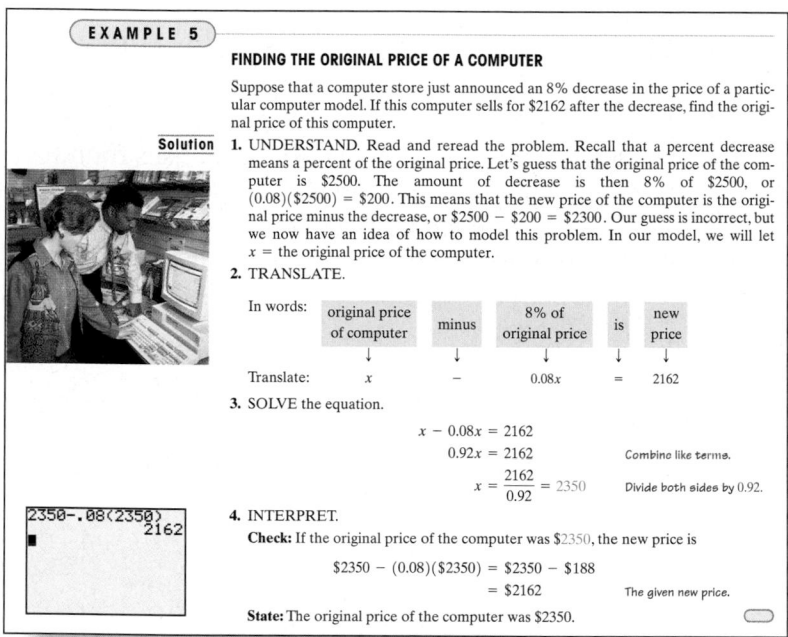

EXAMPLE 5

FINDING THE ORIGINAL PRICE OF A COMPUTER

Suppose that a computer store just announced an 8% decrease in the price of a particular computer model. If this computer sells for $2162 after the decrease, find the original price of this computer.

Solution

1. UNDERSTAND. Read and reread the problem. Recall that a percent decrease means a percent of the original price. Let's guess that the original price of the computer is $2500. The amount of decrease is then 8% of $2500, or $(0.08)(\$2500) = \$200$. This means that the new price of the computer is the original price minus the decrease, or $\$2500 - \$200 = \$2300$. Our guess is incorrect, but we now have an idea of how to model this problem. In our model, we will let x = the original price of the computer.

2. TRANSLATE.

In words:

original price of computer	minus	8% of original price	is	new price
↓	↓	↓	↓	↓

Translate:

| x | $-$ | $0.08x$ | $=$ | 2162 |

3. SOLVE the equation.

$$x - 0.08x = 2162$$
$$0.92x = 2162 \qquad \text{Combine like terms.}$$
$$x = \frac{2162}{0.92} = 2350 \qquad \text{Divide both sides by } 0.92.$$

4. INTERPRET.

Check: If the original price of the computer was $2350, the new price is

$$\$2350 - (0.08)(\$2350) = \$2350 - \$188$$
$$= \$2162 \qquad \text{The given new price.}$$

State: The original price of the computer was $2350.

`2350-.08(2350)`
`               2162`

PAGE 58

◀ General Strategy for Problem Solving

Save time by having a plan. The organization of this text can help you. Note the outlined problem-solving steps: *Understand, Translate, Solve,* and *Interpret.* Problem solving is introduced early, emphasized, and integrated throughout the chapters. The problem-solving procedure is illustrated step-by-step in the in-text examples.

Geometry ▶

Geometric concepts are integrated throughout the text in examples and exercises and are identified with a triangle icon. The text includes dedicated appendices on geometry.

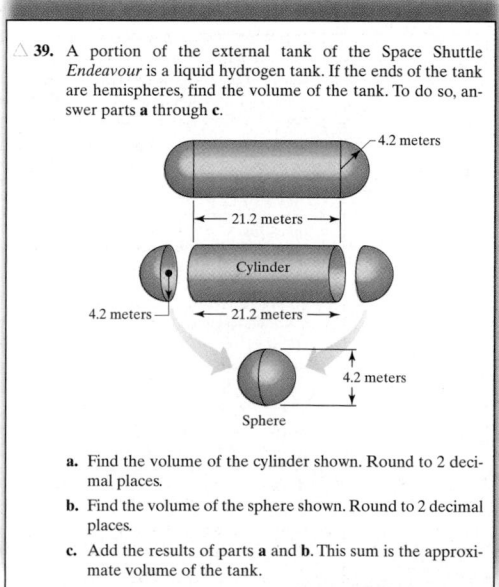

△ **39.** A portion of the external tank of the Space Shuttle *Endeavour* is a liquid hydrogen tank. If the ends of the tank are hemispheres, find the volume of the tank. To do so, answer parts **a** through **c**.

4.2 meters

21.2 meters

Cylinder

4.2 meters ─ ── 21.2 meters ──

4.2 meters

Sphere

a. Find the volume of the cylinder shown. Round to 2 decimal places.

b. Find the volume of the sphere shown. Round to 2 decimal places.

c. Add the results of parts **a** and **b**. This sum is the approximate volume of the tank.

PAGE 87

Learn Better-Master and Apply Skills and Concepts

K. Elayn Martin-Gay provides thorough explanations of concepts and enlivens the content by integrating successful and innovative instructional tools. These features have been included to enhance your understanding of mathematical concepts.

✔ **CONCEPT CHECK**

Explain why a function can contain both the ordered pairs $(1, 3)$ and $(2, 3)$ but not both $(3, 1)$ and $(3, 2)$.

We will call an equation such as $y = 2x + 1$ a **relation** since this equation defines a set of ordered pair solutions.

PAGE 137

◀ New Concept Checks

Concept Checks are special exercises found in most sections following key examples. Work these to help measure your grasp of the concept being explained before moving to the next example.

Mixed Practice Exercises ▶

The exercise sets have been reorganized and include mixed practice exercises. These exercises give students the chance to assimilate the concepts and skills that have been covered in separate objectives.

MIXED PRACTICE

Graph each linear equation.

35. $x + 2y = 8$		**36.** $x - 3y = 3$
37. $3x + 5y = 7$		**38.** $3x - 2y = 5$
39. $x + 8y = 8$		**40.** $x - 3y = 9$
41. $5 = 6x - y$		**42.** $4 = x - 3y$
43. $-x + 10y = 11$		**44.** $-x + 9 = -y$
45. $y = \frac{3}{2}$		**46.** $x = \frac{3}{2}$
47. $2x + 3y = 6$		**48.** $4x + y = 5$
49. $x + 3 = 0$		**50.** $y - 6 = 0$
51. $f(x) = \frac{3}{4}x + 2$		**52.** $f(x) = \frac{4}{3}x + 2$
53. $f(x) = x$		**54.** $f(x) = -x$
55. $f(x) = \frac{1}{2}x$		**56.** $f(x) = -2x$

PAGE 161

Concept Extensions

31. In the United States, eating at a restaurant, or "dining in," is decreasing while "take-out," that is picking up and eating restaurant food at a different location, is increasing. The data below represents the annual restaurant meals per person in each category. Write a linear regression equation for "dining in" and one for "take-out." Use these equations and their graphs to approximate the year that the two types of dining occurred the same number of times. (*Source: Wall Street Journal Almanac*) Let $x = 0$ represent the number of years since 1980.

Year	1984	1988	1990	1992	1997	2000	2003
Dining In	69	68	64	63	64	69	70
Take-Out	43	53	55	57	66	73	76

Source: Wall Street Journal Almanac

32. **GROUP ACTIVITY.** Divide into groups of 5–6 students to see if a relationship exists between the length of a person's forearm (in cm) and their height (in cm).

 a. Complete a table recording the data of each student in the group by measuring the forearm length in cm in L1 and the height in cm in L2.

 b. Make a scatter plot of the data. What type of graph does the data seem to best fit?

 c. Find the regression equation that best fits the data.

 d. Interpret the slope and the y-intercept of the regression equation.

33. **GROUP ACTIVITY.** Complete a similar table comparing the circumference of a student's neck to the circumference of his or her wrist.

◀ New Concept Extensions

Concept Extension exercises have also been added to the exercise sets. They extend the concepts and require students to combine several skills or concepts. These exercises expose students to the way math ideas build upon each other.

PAGE 202

Study Better-Build Confidence and Develop Study Skills

Several features of this text can be helpful in building your confidence and mathematical competence. They will also help improve your study skills.

Tips for Success ▶

Coverage of study skills in Section 1.1 reinforces this important component to success in this course.

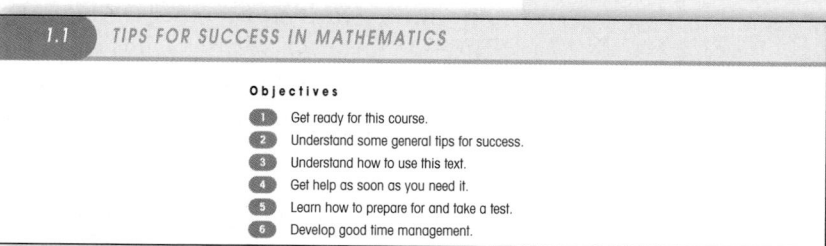

PAGE 2

◀ Study Skills Reminders

Study Skills Reminders are integrated throughout the text to reinforce Section 1.1 and encourage the development of strong study skills. Beginning in Section 1.5 a special type of reminder is included. It helps students develop a cumulative outline that will help them learn to recognize and solve equations and inequalities. This is designed to help students throughout intermediate algebra--as well as to be better prepared for college.

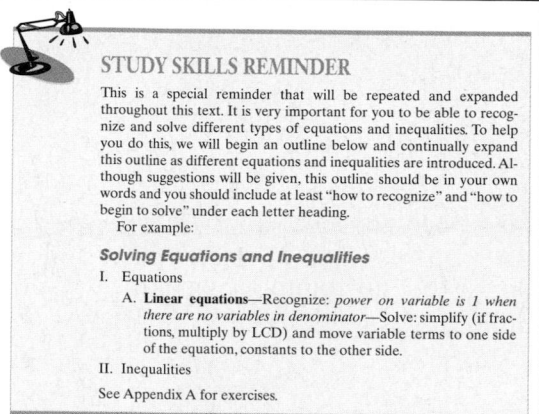

PAGE 52

Mental Math ▶

Mental Math warm-up exercises reinforce concepts found in the accompanying section and can increase your confidence before beginning an exercise set.

PAGE 175

◀ Helpful Hints

Found throughout the text, these features contain practical advice on applying mathematical concepts. They are strategically placed where students are most likely to need immediate reinforcement.

PAGE 141

Vocabulary Checks ▶

Vocabulary Checks help students strengthen verbal skills and answer questions about a chapter's content by filling in the blank with the correct word from the vocabulary list.

2 CHAPTER VOCABULARY CHECK

Fill in each blank with one of the words or phrases listed below.

relation	standard	slope–intercept	range	point–slope
line	slope	x	parallel	perpendicular
function	domain	y	linear function	

1. A _____ is a set of ordered pairs.
2. The graph of every linear equation in two variables is a ___.
3. The equation $y - 8 = -5(x + 1)$ is written in _____ form.
4. _____ form of linear equation in two variables is $Ax + By = C$.
5. The ___ of a relation is the set of all second components of the ordered pairs of the relation.
6. _____ lines have the same slope and different y-intercepts.
7. _____ form of a linear equation in two variables is $y = mx + b$.
8. A _____ is a relation in which each first component in the ordered pairs corresponds to exactly one second component.

PAGE 204

Test Better-Test Yourself and Check Your Understanding

Good exercise sets and an abundance of worked-out examples are essential for building confidence. The exercises in this text are intended to help you build skills and understand concepts as well as motivate and challenge you. In addition, features such as *Integrated Reviews*, *Chapter Highlights*, *Chapter Reviews*, *Chapter Tests*, and *Cumulative Reviews* are found in each chapter to help you study and organize your notes. A new resource—*Chapter Test Prep Video*—will help students study more effectively than ever before.

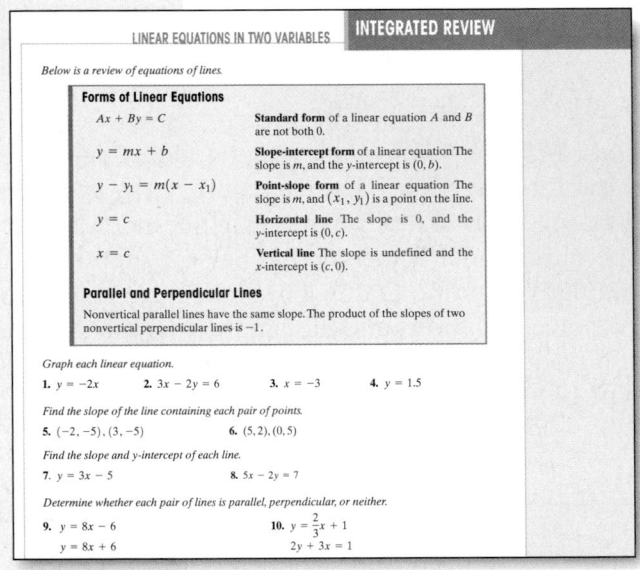

PAGE 191

◀ New Integrated Reviews

Integrated Reviews serve as mid-chapter reviews and help you learn the new skills you have been studying over several sections. This allows students to practice making decisions before taking a test.

Chapter Highlights ▶

Each chapter ends with *Chapter Highlights*. They contain key definitions, concepts, and examples to help you understand and retain what you have learned in the chapter.

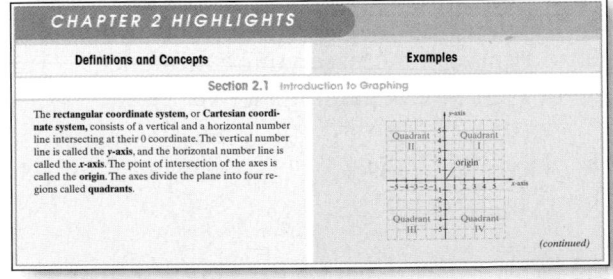

PAGE 204

◀ Chapter Tests

Take the *Chapter Test* at the end of each chapter to prepare for a class exam. With the new *Chapter Test Prep Video* (packaged with your text), you can instantly view the solution to the exact exercises you're working as part of your studying. See the next page for more details.

PAGE 214

Begin Studying More Effectively Today!

New Chapter Test Prep Video CD

Good study skills are essential for success in mathematics. Use the student resources that accompany *Intermediate Algebra, A Graphing Approach, Third Edition* to make the most of your valuable study time.

The *Chapter Test Prep Video CD*, presented by K. Elayn Martin-Gay, provides students with instant access to step-by-step video solutions for each of the Chapter Test questions in the text. To make the most of this resource when studying for a test, follow these three steps:

1. Take the Chapter Test at the end of each chapter.
2. Check your answers in the back of the text.
3. Use the *Chapter Test Prep Video* to review **every step** of the worked-out solution for those specific questions you answered incorrectly or didn't understand on the test.

The Video CD's easy navigation is designed to help students study efficiently. Students select the exact test questions they missed or don't understand and instantly view the solution worked out by the author.

Previous and Next buttons allow easy navigation within tests.

The Chapter Test menu allows students to select the test questions they wish to view.

Rewind, Pause, and Play buttons let students view the solutions at their own pace.

The Video CD includes an Introduction with instructions on how to use the *Chapter Test Prep Video CD* as well as test-taking study skills.

To begin studying and preparing for your test, see the Video CD and Note to Students in the back of your text. The Video CD is also available through your campus bookstore.

Get Motivated!

The third edition strongly emphasizes visualization. Graphing is introduced early and intuitively. Knowing how to read and use graphs is a valuable skill in the workplace as well as in this and other courses. This edition integrates the use of the graphing calculator and emphasizes interpreting data and exploration.

Discover the Concept ▼

Explorations help students recognize patterns or discover a concept on their own immediately before the concept is introduced.

Discover the Concept

How do you think the graph of $y = x^3$ is affected when we place a negative sign in front of the x^3? Test your prediction by graphing $y = x^3$ and $y = -x^3$ separately in a standard window. State in words how you can obtain the graph of $y = -x^3$ from the graph of $y = x^3$.

PAGE 127

Technology Notes ▶

Found in the margin and text, contain specific suggestions for problem solving with technology. They also contain notes on extra features students might find available on their graphing calculator.

Technology Note

Although the graphs of the equations shown below to the left are perpendicular lines, they do not appear to be so when graphed using a standard viewing window.

This is because a graphing calculator screen is rectangular, so the tick marks along the *x*-axis are farther apart than the tick marks along the *y*-axis. The graphs of the same lines at the right do appear to be perpendicular because the viewing window has been selected so that there is equal spacing between tick marks on both axes. We say that the lines to the right are graphed using a **square viewing window**. Some graphing utilities can automatically provide a square setting. (This technology note shows the necessity of being able to algebraically determine whether lines are perpendicular. A graphing utility can then be used to partially verify the result.)

PAGE 172

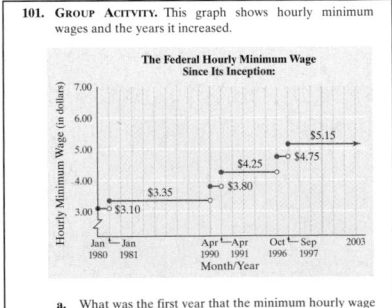

101. GROUP ACTIVITY. This graph shows hourly minimum wages and the years it increased.

a. What was the first year that the minimum hourly wage rose above $4.00?
b. What was the first year that the minimum hourly wage rose above $5.00?
c. Why do you think that this graph is shaped the way it is?
d. The federal hourly minimum wage started in 1938 at $0.25. How much has it increased by in 2003?

PAGE 134

◀ Visualization of Topics

Many illustrations, models, photographs, tables, and graphs provide visual reinforcement of concepts and opportunities for data interpretation. ▼

4   In fact, the equation $3x - y = 12$ has an infinite number of ordered pair solutions. Since it is impossible to list all solutions, we visualize them by graphing them.
A few more ordered pairs that satisfy $3x - y = 12$ are $(4, 0), (3, -3), (5, 3)$, and $(1, -9)$. These ordered pair solutions along with the ordered pair solutions from Example 3 are plotted on the following graph. The graph of $3x - y = 12$ is the single line containing these points. Every ordered pair solution of the equation corresponds to a point on this line, and every point on this line corresponds to an ordered pair solution.

Using a table on the graphing utility to verify the ordered pairs solutions.

x	y	$3x - y = 12$
5	3	$3 \cdot 5 - 3 = 12$
4	0	$3 \cdot 4 - 0 = 12$
3	-3	$3 \cdot 3 - (-3) = 12$
2	-6	$3 \cdot 2 - (-6) = 12$
1	-9	$3 \cdot 1 - (-9) = 12$
0	-12	$3 \cdot 0 - (-12) = 12$

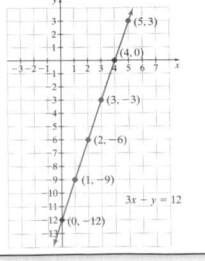

PAGE 121

Resources for Student Success

Intermediate Algebra, A Graphing Approach, Third Edition is supported by comprehensive resource packages for both instructors and students. Please refer to the Preface for detailed descriptions.

Highlights of the packages include the following:

Chapter Test Prep Video CD—included in every student text and annotated instructor's edition. It provides step-by-step worked out solutions to every question contained on the chapter tests in the text. Hosted by K. Elayn Martin-Gay.

MyMathLab®—a series of text-specific, easily customizable online courses for Prentice Hall mathematics texts.

MathXL®—a powerful online homework, tutorial, and assessment system.

MathXL® tutorials on CD—provides algorithmically generated practice exercises that are correlated at the objective level to the exercises in the text.

CD Lecture Series—text-specific videos available on CD. They are hosted by K. Elayn Martin-Gay and cover each objective in every chapter section of the text.

Prentice Hall Tutor Center—provides text-specific tutoring via phone, fax, and email.

Student Solutions Manual and Student Graphing Calculator Guide

FOR INSTRUCTORS

Annotated Instructor's Edition—now includes one additional classroom example for each example in the text, Teaching Tips, and answers to all text exercises.

Instructor's Solutions Manual and Instructor's Resource Manual with Tests

TestGen—enables instructors to build, edit, print, and administer tests using a computerized bank of questions developed to cover all objectives of the text.

Instructor-to-Instructor Videos—provide suggestions for presenting topics, time-saving tips, alternative strategies, and much more.

Real Numbers, Algebraic Expressions, and Equations

Mathematics is a tool for solving problems in such diverse fields as transportation, engineering, economics, medicine, business, and biology. We solve problems using mathematics by modeling real-world phenomena with mathematical equations or inequalities. Our ability to solve problems using mathematics, then, depends in part on our ability to solve equations and inequalities. In this chapter, we review operations on and properties of real numbers. We then solve linear equations and problems that can be modeled by linear equations.

T he travel industry is the third largest retail industry in the United States, accounting for the employment of over 16 million people. One segment of this ever-growing population is made up of travel agents. Travel agents are professionals trained to search for and book airline fares, rail tickets, hotel reservations, car rentals, cruise packages, tours and much more. They also address topics important to travelers, including special events, climate conditions, and travel and safety regulations for domestic and international travel.

Successful travel agents need good communication and business skills. Because the job involves searching for competitive prices, basic computer skills are also a must as well as mathematics and number sense.

In the Spotlight on Decision Making feature on page 27, you will have the opportunity to compare tour packages as a travel agent.

NUM	BONUS	TOTAL 3
15	100	▪▪▪▪▪
60	25	
25	75	
40	50	
------	------	
TOTAL(1)=		

Link: www.astanet.com *(American Society of Travel Agents)*
Source of text: (same)

1.1   *TIPS FOR SUCCESS IN MATHEMATICS*

Objectives

1. Get ready for this course.
2. Understand some general tips for success.
3. Understand how to use this text.
4. Get help as soon as you need it.
5. Learn how to prepare for and take a test.
6. Develop good time management.

Before reading this section, remember that your instructor is your best source for information. Please see your instructor for any additional help or information.

1   Getting Ready for This Course

Now that you have decided to take this course, remember that a *positive attitude* will make all the difference in the world. Your belief that you can succeed is just as important as your commitment to this course. Make sure that you are ready for this course by having the time and positive attitude that it takes to succeed.

Next make sure that you have scheduled your math course at a time that will give you the best chance for success. For example, if you are also working, you may want to check with your employer to make sure that your work hours will not conflict with your course schedule. Also, schedule your class during a time of day when you are more attentive and do your best work.

On the day of your first class period, double-check your schedule and allow yourself extra time to arrive in case of traffic problems or difficulty locating your classroom. Make sure that you bring at least your textbook, paper, and a writing instrument. Are you required to have a lab manual, graph paper, calculator, or other supplies besides this text? If so, also bring this material with you.

2   General Tips for Success

Below are some general tips that will increase your chance for success in a mathematics class. Many of these tips will also help you in other courses you may be taking.

Exchange names and phone numbers with at least one other person in class. This contact person can be a great help if you miss an assignment or want to discuss math concepts or exercises that you find difficult.

Choose to attend all class periods and be on time. If possible, sit near the front of the classroom. This way, you will see and hear the presentation better. It may also be easier for you to participate in classroom activities.

Do your homework. You've probably heard the phrase "practice makes perfect" in relation to music and sports. It also applies to mathematics. You will find that the more time you spend solving mathematics problems, the easier the process becomes. Be sure to schedule enough time to complete your assignments before the next class period.

Check your work. Review the steps you made while working a problem. Learn to check your answers in the original problems. You may also compare your answers with the answers to selected exercises section in the back of the book. If you have made a mistake, try to figure out what went wrong. Then correct your mistake. If you can't find what went wrong, don't erase your work or throw it away. Bring your work to your instructor, a tutor in a math lab, or a classmate. It is easier for someone to find where you had trouble if they look at your original work.

Learn from your mistakes and be patient with yourself. Everyone, even your instructor, makes mistakes. Use your errors to learn and to become a better math student. The key is finding and understanding your errors.

Was your mistake a careless one, or did you make it because you can't read your own math writing? If so, try to work more slowly or write more neatly and make a conscious effort to carefully check your work.

Did you make a mistake because you don't understand a concept? Take the time to review the concept or ask questions to better understand it.

Did you skip too many steps? Skipping steps or trying to do too many steps mentally may lead to preventable mistakes.

Know how to get help if you need it. It's OK to ask for help. In fact, it's a good idea to ask for help whenever there is something that you don't understand. Make sure you know when your instructor has office hours and how to find his or her office. Find out whether math tutoring services are available on your campus. Check out the hours, location, and requirements of the tutoring service. Videotapes and software are available with this text. Learn how to access these resources.

Organize your class materials, including homework assignments, graded quizzes and tests, and notes from your class or lab. All of these items will make valuable references throughout your course especially when studying for upcoming tests and the final exam. Make sure that you can locate these materials when you need them.

Read your textbook before class. Reading a mathematics textbook is unlike leisure reading such as reading a book or newspaper. Your pace will be much slower. It is helpful to have a pencil and paper with you when you read. Try to work out examples on your own as you encounter them in your text. You may also write down any questions that you want to ask in class. When you read a mathematics textbook, some of the information in a section may be unclear. But after you hear a lecture or watch a videotape on that section, you will understand it much more easily than if you had not read your text beforehand.

Don't be afraid to ask questions. Instructors are not mind readers. Many times we do not know a concept is unclear until a student asks a question. You are not the only person in class with questions. Other students are normally grateful that someone has spoken up.

Hand in assignments on time. This way you can be sure that you will not lose points for being late. Show every step of a problem and be neat and organized. Also be sure that you understand which problems are assigned for homework. You can always double-check this assignment with another student in your class.

3   **Using This Text**

There are many helpful resources that are available to you in this text. It is important that you become familiar with and use these resources. This should increase your chances for success in this course.

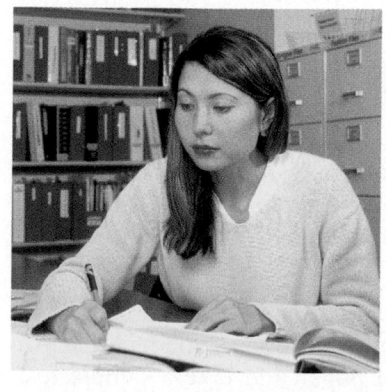

- The main section of exercises in each exercise set is referenced by examples. Use this referencing if you have trouble completing an assignment from the exercise set.

- If you need extra help in a particular section, look at the beginning of the section to see what videotapes and software are available.

- Make sure that you understand the meaning of the icons that are beside many exercises. The video icon tells you that the corresponding exercise may be viewed on the videotape that corresponds to that section. The pencil icon ✎ tells you that this exercise is a writing exercise in which you should answer in complete sentences. The △ icon tells you that the exercise involves geometry.

- Integrated Reviews in each chapter offer you a chance to practice—in one place—the many concepts that you have learned separately over several sections.

- There are many opportunities at the end of each chapter to help you understand the concepts of the chapter.

    Chapter Highlights contain chapter summaries and examples.

    Chapter Reviews contain review problems organized by section.

    Chapter Tests are sample tests to help you prepare for an exam.

    Cumulative Reviews are reviews consisting of material from the beginning of the book to the end of that particular chapter.

See the preface at the beginning of this text for a more thorough explanation of the features of this text.

4 Getting Help

If you have trouble completing assignments or understanding the mathematics, get help as soon as you need it! This tip is presented as an objective on its own because it is so important. In mathematics, usually the material presented in one section builds on your understanding of the previous section. What does this mean? It means that if you don't understand the concepts covered during a class period, there is a good chance that you will not understand the concepts covered during the next class period. If this happens to you, get help as soon as you can.

Where can you get help? Many suggestions have been made in this section on where to get help, and now it is up to you to do it. Try your instructor, a tutoring center, or math lab, or you may want to form a study group with fellow classmates. If you do decide to see your instructor or go to a tutoring center, make sure that you have a neat notebook and be ready with your questions.

5 Preparing for and Taking a Test

Make sure that you allow yourself plenty of time to prepare for a test. If you think that you are a little "math anxious," it may be that you are not preparing for a test in a way that will ensure success. The way that you prepare for a test in mathematics is important. To prepare for a test,

1. Review your previous homework assignments.

2. Review any notes from class and section-level quizzes you may have taken. (If this is a final exam, also review chapter tests you have taken.)

3. Review concepts and definitions by reading the Highlights at the end of each chapter.

4. Practice working exercises by completing the Chapter Review found at the end of each chapter. (If this is a final exam, go through a Cumulative Review. There is one found at the end of each chapter (except Chapter 1). Choose the review found at the end of the latest chapter that you have covered in your course.) *Don't stop here!*

5. It is important that you place yourself in conditions similar to test conditions to find out how you will perform. In other words, as soon as you feel that you know the material, get a few blank sheets of paper and take a sample test.

There is a Chapter Test available at the end of each chapter. During this sample test, do not use your notes or your textbook. Once you complete the Chapter Test, check your answers in the back of the book. If any answer is incorrect, there is a CD available with each exercise of each chapter test worked. Use this CD or your instructor to correct your sample test. Your instructor may also provide you with a review sheet. If you are not satisfied with the results, study the areas that you are weak in and try again.

6. Get a good night's sleep before the exam.

7. On the day of the actual test, allow yourself plenty of time to arrive at where you will be taking your test.

When taking your test,

1. Read the directions on the test carefully.

2. Read each problem carefully as you take the test. Make sure that you answer the question asked.

3. Watch your time and pace yourself so that you can attempt each problem on your test.

4. If you have time, check your work and answers.

5. Do not turn your test in early. If you have extra time, spend it double-checking your work.

6 Managing Your Time

As a college student, you know the demands that classes, homework, work, and family place on your time. Some days you probably wonder how you'll ever get everything done. One key to managing your time is developing a schedule. Here are some hints for making a schedule:

1. Make a list of all of your weekly commitments for the term. Include classes, work, regular meetings, extracurricular activities, etc. You may also find it helpful to list such things as laundry, regular workouts, grocery shopping, etc.

2. Next, estimate the time needed for each item on the list. Also make a note of how often you will need to do each item. Don't forget to include time estimates for reading, studying, and homework you do outside of your classes. You may want to ask your instructor for help estimating the time needed.

3. In the following Exercise Set, you are asked to block out a typical week on the schedule grid given. Start with items with fixed time slots like classes and work.

4. Next, include the items on your list with flexible time slots. Think carefully about how best to schedule some items such as study time.

5. Don't fill up every time slot on the schedule. Remember that you need to allow time for eating, sleeping, and relaxing! You should also allow a little extra time in case some items take longer than planned.

6. If you find that your weekly schedule is too full for you to handle, you may need to make some changes in your workload, classload, or in other areas of your life. You may want to talk to your advisor, manager or supervisor at work, or someone in your college's academic counseling center for help with such decisions.

Note: In this chapter, we begin a feature called Study Skills Reminder. The purpose of this feature is to remind you of some of the information given in this section and to further expand on some topics in this section.

EXERCISE SET 1.1

STUDY   CD/   PH MATH   MathXL®Tutorials   MathXL®   MyMathLab®
GUIDE/SSM   VIDEO   TUTOR CENTER   ON CD

1. What is your instructor's name?

2. What are your instructor's office location and office hours?

3. What is the best way to contact your instructor?

4. What does the ⟍ icon mean?

5. What does the 🔒 icon mean?

6. What does the △ icon mean?

7. Where are answers located in this text?

8. What Exercise Set answers are available to you in the answers section?

9. What Chapter Review, Chapter Test, and Cumulative Review answers are available to you in the answer section?

10. Are there worked-out solutions to exercises in this text?

11. If the answer to Exercise 10 is yes, what worked-out solutions are available to you in this text?

12. Go to the Highlights section at the end of this chapter. Describe how this section may be helpful to you when preparing for a test.

13. Do you have the name and contact information of at least one other student in class?

14. Will your instructor allow you to use a calculator in this class?

15. Are videotapes, CDs, and/or tutorial software available to you? If so, where?

16. Is there a tutoring service available? If so, what are its hours?

17. Have you attempted this course before? If so, write down ways that you might improve your chances of success during this next attempt.

18. List some steps that you can take if you begin having trouble understanding the material or completing an assignment.

19. Read or reread objective ⑥ and fill out the schedule grid below.

	Monday	*Tuesday*	*Wednesday*	*Thursday*	*Friday*	*Saturday*	*Sunday*
7:00 a.m.							
8:00 a.m.							
9:00 a.m.							
10:00 a.m.							
11:00 a.m.							
12:00 a.m.							
1:00 p.m.							
2:00 p.m.							
3:00 p.m.							
4:00 p.m.							
5:00 p.m.							
6:00 p.m.							
7:00 p.m.							
8:00 p.m.							
9:00 p.m.							

20. Study your filled-out grid from Exercise 19. Decide whether you have the time necessary to successfully complete this course and any other courses you may be registered for.

1.2   ALGEBRAIC EXPRESSIONS AND SETS OF NUMBERS

Objectives

1   Identify and evaluate algebraic expressions.

2   Identify natural numbers, whole numbers, integers, and rational and irrational real numbers.

3   Find the absolute value of a number.

4   Find the opposite of a number.

5   Write phrases as algebraic expressions.

TECHNOLOGY NOTE

Throughout this text, we assume that students have access to a graphing utility. Technology notes such as this one will appear often to alert students to possible commands that may be available or to alert students to special considerations that they need to watch for when using a graphing utility.

1   Recall that letters that represent numbers are called **variables**. An **algebraic expression** is formed by numbers and variables connected by the operations of addition, subtraction, multiplication, division, raising to powers, and/or taking roots. For example,

$$2x + 3, \quad \frac{x + 5}{6} - \frac{z^2}{y^2}, \quad \text{and} \quad \sqrt{y} - 1.6$$

are algebraic expressions or, more simply, expressions.

Algebraic expressions occur often during problem solving. For example, the B747-400 aircraft costs $8158 per hour to operate. The algebraic expression $8158t$ gives the total cost to operate the aircraft for t hours. (*Source: The World Almanac*, 2003) To find the cost to operate the aircraft for 5.2 hours, for example, we replace the variable t with 5.2 and perform the indicated operation. This process is called **evaluating** an expression, and the result is called the **value** of the expression for the given replacement value.

In our example, when $t = 5.2$ hours,

$$8158t = 8158(5.2) = 42{,}421.60$$

Helpful Hint

Recall that $8158t$ means $8158 \cdot t$.

Thus, it costs $42,421.60 to operate the B747-400 aircraft for 5.2 hours.

FINDING THE AREA OF A TILE

The research department of a flooring company is considering a new flooring design that contains parallelograms. The area of a parallelogram with base b and height h is bh. Find the area of a parallelogram with base 10 centimeters and height 8.2 centimeters.

Solution We replace b with 10 and h with 8.2 in the algebraic expression bh.

$$bh = 10 \cdot 8.2 = 82$$

The area is 82 square centimeters.

Algebraic expressions simplify to different values depending on replacement values.

CLASSROOM EXAMPLE
The area of a rectangle with length l and width w is lw. Find the area of a rectangle with length 5 meters and width 3.2 meters.
answer: 16 sq. m

EXAMPLE 2

Evaluate: $3x - y$ when $x = 15$ and $y = 4$.

Solution We replace x with 15 and y with 4 in the expression.

$$3x - y = 3 \cdot 15 - 4 = 45 - 4 = 41$$

CLASSROOM EXAMPLE
Evaluate $b - 5a$ when $a = 4$ and $b = 35$.
answer: 15

When evaluating an expression to solve a problem, we often need to think about the kind of number that is appropriate for the solution. For example, if we are asked to determine the maximum number of parking spaces for a parking lot to be constructed, an answer of $98\frac{1}{10}$ is not appropriate because $\frac{1}{10}$ of a parking space is not realistic.

2 Let's review some common sets of numbers and their graphs on a number line. To construct a number line, we draw a line and label a point 0 with which we associate the number 0. This point is called the **origin**. Choose a point to the right of 0 and label it 1. The distance from 0 to 1 is called the **unit distance** and can be used to locate more points. The **positive numbers** lie to the right of the origin, and the **negative numbers** lie to the left of the origin. The number 0 is neither positive nor negative.

 CONCEPT CHECK

Concept Check Answer:
a number that is 0 or positive

Use the definitions of positive numbers, negative numbers, and zero to describe the meaning of *nonnegative numbers*.

A number is **graphed** on a number line by shading the point on the number line that corresponds to the number. Some common sets of numbers and their graphs include:

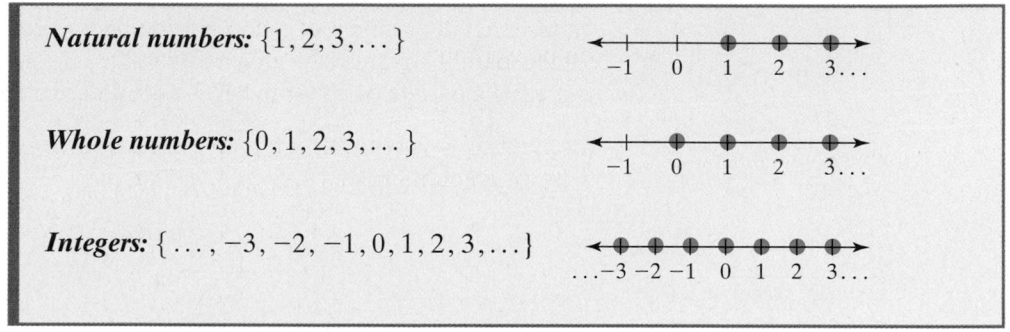

The symbol ... is used in each set above. This symbol, consisting of three dots, is called an **ellipsis** and means to continue in the same pattern.

The members of a set are called its **elements**. When the elements of a set are listed, such as those displayed in the previous paragraph, the set is written in **roster** form. A set can also be written in **set builder notation**, which describes the members of a set but does not list them. The following set is written in set builder notation.

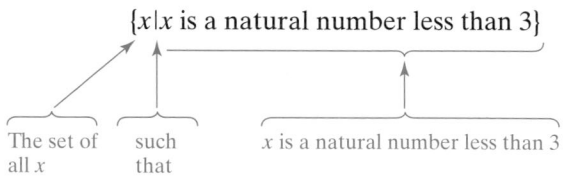

This same set written in roster form is $\{1, 2\}$.

A set that contains *no* elements is called the **empty set** (or **null set**) symbolized by $\{\ \}$ or $\varnothing$. The set

$$\{x \mid x \text{ is a month with 32 days}\} \text{ is } \varnothing \text{ or } \{\ \}$$

because no month has 32 days. The set has no elements.

CLASSROOM EXAMPLE
Write each set in roster form.
$\{x \mid x \text{ is a whole number between 0 and 4}\}$
$\{x \mid x \text{ is a natural number greater than 80}\}$
answer: a. $\{1, 2, 3\}$ **b.** $\{81, 82, 83, \ldots\}$

▶ **Helpful Hint**
Use $\{\ \}$ or $\varnothing$ to write the empty set. $\{\varnothing\}$ is **not** the empty set because it has one element: $\varnothing$.

EXAMPLE 3

List the elements in each set.

a. $\{x \mid x \text{ is a whole number between 1 and 6}\}$

b. $\{x \mid x \text{ is a natural number greater than 100}\}$

Solution **a.** $\{2, 3, 4, 5\}$ **b.** $\{101, 102, 103, \ldots\}$

The symbol $\in$ is used to denote that an element is in a particular set. The symbol $\in$ is read as "is an element of." For example, the true statement

3 is an element of $\{1, 2, 3, 4, 5\}$

can be written in symbols as

$$3 \in \{1, 2, 3, 4, 5\}$$

The symbol $\notin$ is read as "is not an element of." In symbols, we write the true statement "p is not an element of $\{a, 5, g, j, q\}$" as

$$p \notin \{a, 5, g, j, q\}$$

EXAMPLE 4

Determine whether each statement is true or false.

a. $3 \in \{x \mid x \text{ is a natural number}\}$ **b.** $7 \notin \{1, 2, 3\}$

Solution   **a.** True, since 3 is a natural number and therefore an element of the set.

b. True, since 7 is not an element of the set $\{1, 2, 3\}$.

We can use set builder notation to describe three other common sets of numbers.

Identifying Numbers

Real numbers: $\{x \mid x \text{ corresponds to a point on the number line}\}$

Rational numbers: $\left\{ \dfrac{a}{b} \,\middle|\, a \text{ and } b \text{ are integers and } b \neq 0 \right\}$

Irrational numbers: $\{x \mid x \text{ is a real number and } x \text{ is not a rational number}\}$

> **Helpful Hint**
>
> Notice from the definition that all real numbers are either rational or irrational.

Every rational number can be written as a decimal that either repeats or terminates. For example,

Rational Numbers

$$\frac{1}{2} = 0.5 \qquad\qquad \frac{5}{4} = 1.25$$

$$\frac{2}{3} = 0.6666666\ldots = 0.\overline{6} \qquad \frac{1}{11} = 0.090909\ldots = 0.\overline{09}$$

An irrational number written as a decimal neither terminates nor repeats. For example, π and $\sqrt{2}$ are irrational numbers. Their decimal form neither terminates nor repeats. Decimal approximations of each are below:

Irrational Numbers

$$\pi \approx 3.141592\ldots \qquad \sqrt{2} \approx 1.414213\ldots$$

Notice that every integer is also a rational number since each integer can be written as the quotient of itself and 1:

$$3 = \frac{3}{1}, \quad 0 = \frac{0}{1}, \quad -8 = \frac{-8}{1}$$

Not every rational number, however, is an integer. The rational number $\frac{2}{3}$, for example, is not an integer. Some square roots are rational numbers and some are irrational numbers. For example, $\sqrt{2}$, $\sqrt{3}$ and $\sqrt{7}$ are irrational numbers while $\sqrt{25}$ is a rational number because $\sqrt{25} = 5 = \frac{5}{1}$. The set of rational numbers together with the set of irrational numbers make up the set of real numbers. To help you make the distinction between rational and irrational numbers, here are a few examples of each.

Real Numbers		
Rational Numbers		*Irrational Numbers*
Numbers	*Equivalent Quotient of Integers,* $\dfrac{a}{b}$ *(Decimal Form Terminates or Repeats)*	*(Decimal Form neither Terminates nor Repeats)*
$-\dfrac{2}{3}$	$\dfrac{-2}{3}$ or $\dfrac{2}{-3}$	$\sqrt{5}$
$\sqrt{36}$	$\dfrac{6}{1}$	$\dfrac{\sqrt{6}}{7}$
5	$\dfrac{5}{1}$	$-\sqrt{13}$
0	$\dfrac{0}{1}$	π
1.2	$\dfrac{12}{10}$	$\dfrac{2}{\sqrt{3}}$
$3\dfrac{7}{8}$	$\dfrac{31}{8}$	

Some rational and irrational numbers are graphed below.

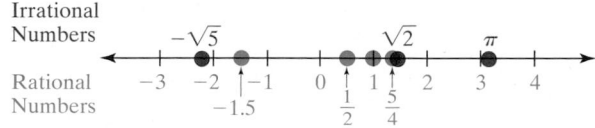

Earlier we mentioned that every integer is also a rational number. In other words, all the elements of the set of integers are also elements of the set of rational numbers. When this happens, we say that the set of integers, set Z, is a subset of the set of rational numbers, set Q. In symbols,

$$\underbrace{Z \subseteq Q}_{\text{is a subset of}}$$

The natural numbers, whole numbers, integers, rational numbers, and irrational numbers are each a subset of the set of real numbers. The relationships among these sets of numbers are shown in the following diagram.

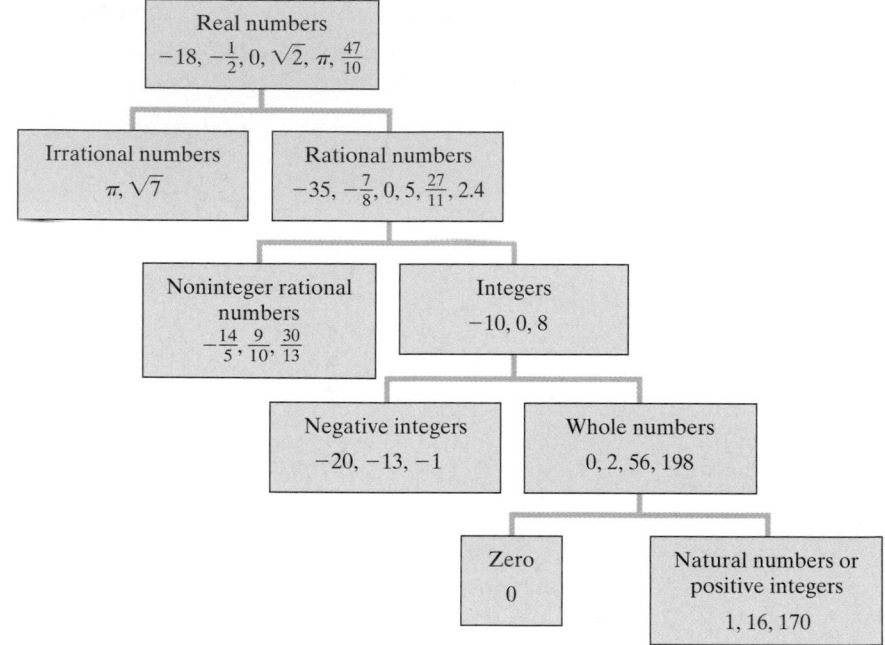

EXAMPLE 5

Determine whether the following statements are true or false.

a. 3 is a real number.

b. $\dfrac{1}{5}$ is an irrational number.

c. Every rational number is an integer.

d. $\{1, 5\} \subseteq \{2, 3, 4, 5\}$

Solution

a. True. Every whole number is a real number.

b. False. The number $\dfrac{1}{5}$ is a rational number, since it is in the form $\dfrac{a}{b}$ with a and b integers and $b \neq 0$.

c. False. The number $\dfrac{2}{3}$, for example, is a rational number, but it is not an integer.

d. False since the element 1 in the first set is not an element of the second set.

3   The number line can also be used to visualize distance, which leads to the concept of absolute value. The **absolute value** of a real number a, written as $|a|$, is the distance between a and 0 on the number line. Since distance is always positive or zero, $|a|$ is always positive or zero.

Using the number line, we see that

$$|4| = 4 \qquad \text{and also} \qquad |-4| = 4.$$

Why? Because both 4 and -4 are a distance of 4 units from 0.

An equivalent definition of the absolute value of a real number a is given next.

Absolute Value

The absolute value of a, written as $|a|$, is

$$|a| = \begin{cases} a \text{ if } a \text{ is } 0 \text{ or a positive number} \\ -a \text{ if } a \text{ is a negative number} \end{cases}$$

the opposite of

EXAMPLE 6

Find each absolute value.

a. $|3|$ **b.** $|-5|$ **c.** $-|2|$ **d.** $-|-8|$ **e.** $|0|$

CLASSROOM EXAMPLE

Find each absolute value.
a. $|7|$ **b.** $|-1|$ **c.** $|-9|$ **d.** $-|5|$
e. $-|-3|$
answer: **a.** 7 **b.** 1 **c.** 9 **d.** -5 **e.** -3

Solution   **a.** $|3| = 3$ since 3 is located 3 units from 0 on the number line.

b. $|-5| = 5$ since -5 is 5 units from 0 on the number line.

c. $-|2| = -2$. The negative sign outside the absolute value bars means to take the opposite of the absolute value of 2.

d. $-|-8| = -8$. Since $|-8|$ is 8, $-|-8| = -8$.

e. $|0| = 0$ since 0 is located 0 units from 0 on the number line.

TECHNOLOGY NOTE

To find absolute value using a graphing utility, look for a key named ABS or check in a math menu. Check on the use of parentheses with absolute value for your graphing utility.

The screens below show a check using a graphing utility. Many calculators have a subtraction key and a separate negative key. Make sure that you are using the negative key for this exercise.

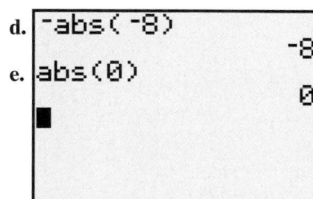

✔ CONCEPT CHECK

Explain how you know that $|14| = -14$ is a false statement.

TEACHING TIP

Remind students that the negative key on the calculator may be different from the minus key.

4   The number line can also help us visualize opposites. Two numbers that are the same distance from 0 on the number line but are on opposite sides of 0 are called **opposites**.

See the definition illustrated on the number lines below.

The opposite of 6 is -6

The opposite of $\frac{2}{3}$ is $-\frac{2}{3}$.

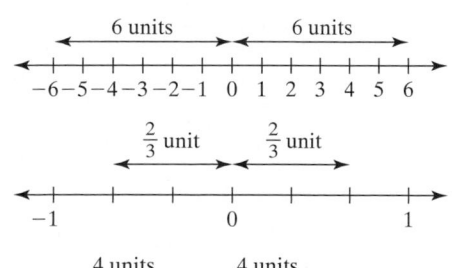

Helpful Hint

The opposite of 0 is 0.

Concept Check Answer:
$|14| = 14$ since the absolute value of a number is the distance between the number and 0 and distance cannot be negative.

The opposite of -4 is 4.

> ### Opposite
> The opposite of a number a is the number $-a$.

Above we state that the opposite of a number a is $-a$. This means that the opposite of -4 is $-(-4)$. But from the number line above, the opposite of -4 is 4. This means that $-(-4) = 4$, and in general, we have the following property.

> ### Double Negative Property
> For every real number a, $-(-a) = a$.

EXAMPLE 7

Write the opposite of each.

a. 8   **b.** $\frac{1}{5}$ **c.** -9.6

Solution   **a.** The opposite of 8 is -8.

b. The opposite of $\frac{1}{5}$ is $-\frac{1}{5}$.

c. The opposite of -9.6 is $-(-9.6) = 9.6$.

5   Often, solving problems involves translating a phrase to an algebraic expression. The following is a partial list of key words and phrases and their usual direct translations.

ADDITION	SUBTRACTION	MULTIPLICATION	DIVISION
sum	difference of	product	quotient
plus	minus	times	divide
added to	subtracted from	multiply	into
more than	less than	twice	ratio
increased by	decreased by	of	
total	less		

EXAMPLE 8

Translate each phrase to an algebraic expression. Use the variable x to represent each unknown number.

a. Eight times a number
b. Three more than eight times a number
c. The quotient of a number and -7
d. One and six-tenths subtracted from twice a number

Solution   **a.** $8 \cdot x$ or $8x$ **b.** $8x + 3$ **c.** $x \div -7$ or $\frac{x}{-7}$ **d.** $2x - 1.6$

Technology Note

Recall that most calculators have a subtraction key and a separate negative key. If this is so on your calculator, notice the difference in their displays. The subtraction sign is usually a longer horizontal dash and the negative sign is a shorter, raised horizontal dash.

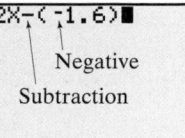

Spotlight on
DECISION
MAKING

Suppose you work for an auto insurance company. Your company has just announced a new partnership with an automobile club that will allow club members to receive a 5% discount on their auto insurance. In addition, any auto club members who are safe drivers (rated 1 or 2 on the safety scale) will receive an additional 10% safe-driver discount. Your supervisor has asked you to compile a master mailing list of all current insurance clients who already belong to the auto club so they may be notified of their eligibility for the 5% discount. You must also identify the subset of auto club members who should receive a separate mailing about the additional 10% discount.

Using the given list of current insurance clients, decide who should be included in the master mailing list for notification of the 5% discount. Which of these clients should also be sent information about the 10% safe-driver discount?

Current Client Database

Client Name	Age (years)	Safety Rating (1–5)	Airbags (0 = no, 1 = yes)	Annual Mileage (miles)	Auto Club (0 = no, 1 = yes)
Alvarez, Wendy	29	1	1	5000	1
Brown, Keisha	19	2	0	5000	0
Cardoni, Anthony	43	1	0	10,000	0
Darden, Clay	35	3	1	7500	1
Evans, Gabriella	26	2	1	5000	1
Fonteneau, Monique	38	4	1	7500	0
Greenberg, Ira	49	1	0	5000	1
Hakkinen, Mika	31	1	1	15,000	0
Issacson, Maude	55	2	0	2000	1
Jones, Harold	47	1	0	7500	1
Khalosef, Avi	52	3	1	5000	1
Lee, Feng	33	2	1	10,000	0
Martinez, Ricardo	25	4	0	5000	1
Nunn, Destiny	21	4	1	5000	0

TEACHING TIP
A Group Activity for this section is available in the Instructor's Resource Manual.

EXERCISE SET 1.2

STUDY GUIDE/SSM   CD/VIDEO   PH MATH TUTOR CENTER   MathXL®Tutorials ON CD   MathXL®   MyMathLab®

Find the value of each algebraic expression at the given replacement values. See Examples 1 and 2.

1. $5x$ when $x = 7$ 35

2. $3y$ when $y = 45$ 135

3. $9.8z$ when $z = 3.1$ 30.38

4. $7.1a$ when $a = 1.5$ 10.65

5. ab when $a = \dfrac{1}{2}$ and $b = \dfrac{3}{4}$ $\dfrac{3}{8}$

6. yz when $y = \dfrac{2}{3}$ and $z = \dfrac{1}{5}$ $\dfrac{2}{15}$

7. $3x + y$ when $x = 6$ and $y = 4$ 22

8. $2a - b$ when $a = 12$ and $b = 7$ 17

9. The aircraft B737-400 flies an average speed of 400 miles per hour. The expression $400t$ gives the distance traveled by the aircraft in t hours. Find the distance traveled by the B737-400 in 5 hours.   2000 mi

10. The algebraic expression $1.5x$ gives the total length of shelf space needed in inches for x encyclopedias. Find the length of shelf space needed for a set of 30 encyclopedias.   45 in.

△ **11.** Employees at Wal-Mart constantly reorganize and reshelve merchandise. In doing so, they calculate floor space needed for displays. The algebraic expression $l \cdot w$ gives the floor space needed in square units for a display that measures length l units and width w units. Calculate the floor space needed for a display whose length is 5.1 feet and whose width is 4 feet.   20.4 sq. ft

12. The algebraic expression $\frac{x}{5}$ can be used to calculate the distance in miles that you are from a flash of lightning, where x is the number of seconds between the time you see a flash of lightning and the time you hear the thunder. Calculate the distance that you are from the flash of lightning if you hear the thunder 2 seconds after you see the lightning. $\frac{2}{5}$ of a mile

13. The B737-400 aircraft costs \$2948 dollars per hour to operate. The algebraic expression $2948t$ gives the total cost to operate the aircraft for t hours. Find the total cost to operate the B737-400 for 3.6 hours.   \$10,612.80

14. Flying the SR-71A jet, Capt. Elden W. Joersz, USAF, set a record speed of 2193.16 miles per hour. At this speed, the algebraic expression $2193.16t$ gives the total distance flown in t hours. Find the distance flown by the SR-71A in 1.7 hours.   3728.372 mi

17. $\{11, 12, 13, 14, 15, 16\}$

List the elements in each set. See Example 3.

 15. $\{x \mid x$ is a natural number less than 6$\}$ $\{1, 2, 3, 4, 5\}$

16. $\{x \mid x$ is a natural number greater than 6$\}$ $\{7, 8, 9, \ldots\}$

17. $\{x \mid x$ is a natural number between 10 and 17$\}$

18. $\{x \mid x$ is an odd natural number$\}$ $\{1, 3, 5, \ldots\}$

19. $\{x \mid x$ is a whole number that is not a natural number$\}$ $\{0\}$

20. $\{x \mid x$ is a natural number less than 1$\}$ $\varnothing$

21. $\{x \mid x$ is an even whole number less than 9$\}$ $\{0, 2, 4, 6, 8\}$

22. $\{x \mid x$ is an odd whole number less than 9$\}$ $\{1, 3, 5, 7\}$

26.

28.

Graph each set on a number line.

23. $\{0, 2, 4, 6\}$

24. $\{-1, -2, -3\}$

25. $\left\{\dfrac{1}{2}, \dfrac{2}{3}\right\}$

26. $\{1, 3, 5, 7\}$

27. $\{-2, -6, -10\}$

28. $\left\{\dfrac{1}{4}, \dfrac{1}{3}\right\}$

✎ **29.** In your own words, explain why the empty set is a subset of every set.   answers may vary

✎ **30.** In your own words, explain why every set is a subset of itself.   answers may vary

List the elements of the set $\left\{3, 0, \sqrt{7}, \sqrt{36}, \dfrac{2}{5}, -134\right\}$ *that are also elements of the given set. See Example 4.*

31. Whole numbers $\{3, 0, \sqrt{36}\}$ **32.** Integers

33. Natural numbers $\{3, \sqrt{36}\}$ **34.** Rational numbers

35. Irrational numbers $\{\sqrt{7}\}$ **36.** Real numbers

Place $\in$ *or* $\notin$ *in the space provided to make each statement true. See Example 4.*

32. $\{3, 0, \sqrt{36}, -134\}$

37. $-11 \in \{x \mid x$ is an integer$\}$ **38.** $-6 \notin \{2, 4, 6, \ldots\}$

39. $0 \notin \{x \mid x$ is a positive integer$\}$ **40.** $12 \in \{1, 2, 3, \ldots\}$

41. $12 \notin \{1, 3, 5, \ldots\}$ **34.** $\left\{3, 0, \sqrt{36}, \dfrac{2}{5}, -134\right\}$

42. $\dfrac{1}{2} \notin \{x \mid x$ is an irrational number$\}$

43. $0 \notin \{1, 2, 3, \ldots\}$ **36.** $\left\{3, 0, \sqrt{7}, \sqrt{36}, \dfrac{2}{5}, -134\right\}$

44. $0 \notin \{x \mid x$ is a natural number$\}$

Determine whether each statement is true or false. See Examples 4 and 5. Use the following sets of numbers.

$$N = \text{set of natural numbers}$$
$$Z = \text{set of integers}$$
$$I = \text{set of irrational numbers}$$
$$Q = \text{set of rational numbers}$$
$$\mathbb{R} = \text{set of real numbers}$$

45. $Z \subseteq \mathbb{R}$ true   **46.** $\mathbb{R} \subseteq N$ false

47. $-1 \in Z$ true   **48.** $\dfrac{1}{2} \in Q$ true

49. $0 \in N$ false   **50.** $Z \subseteq Q$ true

51. $\sqrt{5} \notin I$ false   **52.** $\pi \notin \mathbb{R}$ false

53. $N \subseteq Z$ true   **54.** $I \subseteq N$ false

55. $\mathbb{R} \subseteq Q$ false   **56.** $N \subseteq Q$ true

✎ **57.** In your own words, explain why every natural number is also a rational number but not every rational number is a natural number.   answers may vary

58. In your own words, explain why every irrational number is a real number but not every real number is an irrational number. answers may vary

Find each absolute value. See Example 6.

59. $-|2|$ -2

60. $|8|$ 8

61. $|-4|$ 4

62. $|-6|$ 6

63. $|0|$ 0

64. $|-1|$ 1

65. $-|-3|$ -3

66. $-|-11|$ -11

67. Explain why $-(-2)$ and $-|-2|$ simplify to different numbers. answers may vary

68. The boxed definition of absolute value states that $|a| = -a$ if a is a negative number. Explain why $|a|$ is always nonnegative, even though $|a| = -a$ for negative values of a. answers may vary

Write the opposite of each number. See Example 7.

69. -6.2 6.2

70. -7.8 7.8

71. $\dfrac{4}{7}$ $-\dfrac{4}{7}$

72. $\dfrac{9}{5}$ $-\dfrac{9}{5}$

73. $-\dfrac{2}{3}$ $\dfrac{2}{3}$

74. $-\dfrac{14}{3}$ $\dfrac{14}{3}$

75. 0 0

76. 10.3 -10.3

Write each phrase as an algebraic expression. Use the variable x to represent each unknown number. See Example 8.

77. Twice a number. $2x$

78. Six times a number. $6x$

79. Five more than twice a number. $2x + 5$

80. One more than six times a number. $6x + 1$

81. Ten less than a number. $x - 10$

82. A number minus seven. $x - 7$

83. The sum of a number and two. $x + 2$

84. The difference of twenty-five and a number. $25 - x$

85. A number divided by eleven.

86. The quotient of twice a number and thirteen.

87. Twelve added to three times a number. $3x + 12$

88. Four subtracted from a number. $x - 4$

89. Seventeen subtracted from a number. $x - 17$

90. Four subtracted from three times a number. $3x - 4$

91. Twice the sum of a number and three. $2(x + 3)$

Concept Extensions   **85.** $\dfrac{x}{11}$ **86.** $\dfrac{2x}{13}$

Write each phrase as an algebraic expression. Use the variable x to represent each unknown number.

92. The quotient of four and the sum of a number and one.

93. The quotient of five and the difference of four and a number.

92. $\dfrac{4}{x + 1}$ **93.** $\dfrac{5}{4 - x}$

94. Eight times the difference of a number and 9. $8(x - 9)$

95. The following bar graph shows the top five countries with the projected number of tourists visiting in 2020.

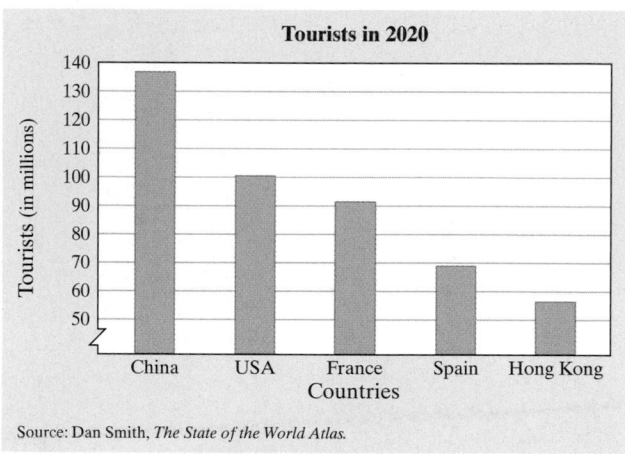

Source: Dan Smith, *The State of the World Atlas.*

Use the height of each bar to estimate the millions of tourists for each country. (Use whole numbers.)

China	137
USA	100
France	91
Spain	69
Hong Kong	56

96. In your own words, explain how the graphing utility screen suggests that $\sqrt{25}$ is a rational number and $\sqrt{10}$ is an irrational number.   answers may vary

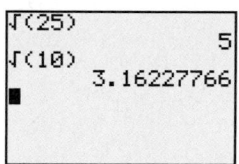

97. In your own words, explain how the graphing utility screen suggests that $\sqrt{\dfrac{4}{9}}$ is a rational number and $\sqrt{7}$ is an irrational number.   answers may vary

1.3 OPERATIONS ON REAL NUMBERS

Objectives

1 Add and subtract real numbers.

2 Multiply and divide real numbers.

3 Simplify expressions containing exponents.

4 Find roots of numbers.

5 Use the order of operations.

6 Evaluate algebraic expressions.

1 When solving problems, we often have to add real numbers. For example, if the New Orleans Saints lose 5 yards in one play, then lose another 7 yards in the next play, their total loss may be described by $-5 + (-7)$.

The addition of two real numbers may be summarized by the following.

Adding Real Numbers

1. To add two numbers with the *same sign*, add their absolute values and attach their common sign.
2. To add two numbers with *different signs*, subtract the smaller absolute value from the larger absolute value and attach the sign of the number with the larger absolute value.

TEACHING TIP

Before discussing the rules for adding real numbers, have students work in groups to categorize the types of numbers that can be added the same way and then describe their method for adding each category. If necessary, get them started by saying. "One category could be adding a positive number to a positive number." Then ask groups how many categories they found. Have groups share their results.

For example, to add $-5 + (-7)$, first add their absolute values.

$$|-5| = 5, |-7| = 7, \quad \text{and} \quad 5 + 7 = 12$$

Next, attach their common negative sign.

$$-5 + (-7) = -12$$

(This represents a total loss of 12 yards for the New Orleans Saints in the example above.)

To find $(-4) + 3$, first subtract their absolute values.

$$|-4| = 4, |3| = 3, \quad \text{and} \quad 4 - 3 = 1$$

Next, attach the sign of the number with the larger absolute value.

$$(-4) + 3 = -1$$

EXAMPLE 1

Add.

a. $-3 + (-11)$ **b.** $3 + (-7)$ **c.** $-10 + 15$

d. $-8.3 + (-1.9)$ **e.** $-\dfrac{1}{4} + \dfrac{1}{2}$ **f.** $-\dfrac{2}{3} + \dfrac{3}{7}$

Solution

a. $-3 + (-11) = -(3 + 11) = -14$

b. $3 + (-7) = -4$

c. $-10 + 15 = 5$

d. $-8.3 + (-1.9) = -10.2$

e. $-\dfrac{1}{4} + \dfrac{1}{2} = -\dfrac{1}{4} + \dfrac{2}{4} = \dfrac{1}{4}$

f. $-\dfrac{2}{3} + \dfrac{3}{7} = -\dfrac{14}{21} + \dfrac{9}{21} = -\dfrac{5}{21}$

The screens below show a check using a calculator.

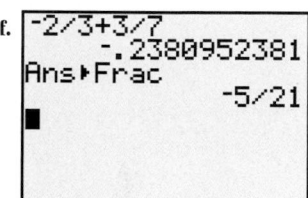

TECHNOLOGY NOTE

To evaluate expressions with a calculator, we sometimes need to insert parentheses that may not be shown in the expression. This is especially true when entering a fraction whose numerator or denominator contains more than one term. (See the screen below.)

A calculator check for Example 9c and d. Notice the use of parentheses.

Subtraction of two real numbers may be defined in terms of addition.

Subtracting Real Numbers

If a and b are real numbers,

$$a - b = a + (-b)$$

In other words, to subtract a real number, we add its opposite.

EXAMPLE 2

Subtract.

a. $2 - 8$ **b.** $-8 - (-1)$ **c.** $-11 - 5$ **d.** $10.7 - (-9.8)$

e. $\dfrac{2}{3} - \dfrac{1}{2}$ **f.** $1 - 0.06$ **g.** Subtract 7 from 4.

Solution

Add the opposite

a. $2 - 8 = 2 + (-8) = -6$

b. $-8 - (-1) = -8 + (1) = -7$

c. $-11 - 5 = -11 + (-5) = -16$

d. $10.7 - (-9.8) = 10.7 + 9.8 = 20.5$

e. $\dfrac{2}{3} - \dfrac{1}{2} = \dfrac{2 \cdot 2}{3 \cdot 2} - \dfrac{1 \cdot 3}{2 \cdot 3} = \dfrac{4}{6} + \left(-\dfrac{3}{6}\right) = \dfrac{1}{6}$

f. $1 - 0.06 = 1 + (-0.06) = 0.94$ **g.** $4 - 7 = 4 + (-7) = -3$

To add or subtract three or more real numbers, add or subtract from left to right.

EXAMPLE 3

Simplify the following expressions.

a. $11 + 2 - 7$ **b.** $-5 - 4 + 2$

Solution   **a.** $11 + 2 - 7 = 13 - 7 = 6$ **b.** $-5 - 4 + 2 = -9 + 2 = -7$

2 In order to discover sign patterns when you multiply real numbers, recall that multiplication by a positive integer is the same as repeated addition. For example,

$$3(2) = 2 + 2 + 2 = 6$$

$$3(-2) = (-2) + (-2) + (-2) = -6$$

Notice here that $3(-2) = -6$. This illustrates that the product of two numbers with different signs is negative. We summarize sign patterns for multiplying any two real numbers as follows.

Multiplying Two Real Numbers

The product of two numbers with the *same* sign is positive.
The product of two numbers with *different* signs is negative.

Also recall that the product of zero and any real number is zero.

$$0 \cdot a = 0$$

EXAMPLE 4

Multiply.

a. $(-8)(-1)$ **b.** $(-2)\dfrac{1}{6}$ **c.** $3(-3)$ **d.** $0(11)$

e. $\left(\dfrac{1}{5}\right)\left(-\dfrac{10}{11}\right)$ **f.** $(7)(1)(-2)(-3)$ **g.** $8(-2)(0)$

Solution

a. Since the signs of the two numbers are the same, the product is positive. Thus $(-8)(-1) = +8$, or 8.

b. Since the signs of the two numbers are different or unlike, the product is negative. Thus $(-2)\dfrac{1}{6} = -\dfrac{2}{6} = -\dfrac{1}{3}$.

c. $3(-3) = -9$

d. $0(11) = 0$

e. $\left(\dfrac{1}{5}\right)\left(-\dfrac{10}{11}\right) = -\dfrac{10}{55} = -\dfrac{2}{11}$

f. To multiply three or more real numbers, you may multiply from left to right.

$$(7)(1)(-2)(-3) = 7(-2)(-3)$$
$$= -14(-3)$$
$$= 42$$

g. Since zero is a factor, the product is zero.

$$(8)(-2)(0) = 0$$

The screen below shows options for checking parts b and e where fractions are involved.

> ### Helpful Hint
> The following sign patterns may be helpful when we are multiplying.
>
> **1.** An odd number of negative factors gives a negative product.
> **2.** An even number of negative factors gives a positive product.

Recall that $\frac{8}{4} = 2$ because $2 \cdot 4 = 8$. Likewise, $\frac{8}{-4} = -2$ because $(-2)(-4) = 8$. Also, $\frac{-8}{4} = -2$ because $(-2)4 = -8$, and $\frac{-8}{-4} = 2$ because $2(-4) = -8$. From these examples, we can see that the sign patterns for division are the same as for multiplication.

> ### Dividing Two Real Numbers
> The quotient of two numbers with the *same* sign is positive.
> The quotient of two numbers with *different* signs is negative.

TECHNOLOGY NOTE

A calculator will give some type of error message when asked to divide by 0. For example:

Also recall that division by a nonzero real number b is the same as multiplication by $\frac{1}{b}$. In other words,

$$\frac{a}{b} = a \cdot \frac{1}{b}$$

This means that to simplify $\frac{a}{b}$, we can divide by b or multiply by $\frac{1}{b}$. The nonzero numbers b and $\frac{1}{b}$ are called **reciprocals**. Notice that b *must* be a nonzero number. We do not define division by 0. For example, $5 \div 0$, or $\frac{5}{0}$, is undefined. To see why, recall that if $5 \div 0 = n$, a number, then $n \cdot 0 = 5$. This is not possible since $n \cdot 0 = 0$ for any number n, and never 5. Thus far we have learned that we cannot divide 5 or any other nonzero number by 0.

Can we divide 0 by 0? By the same reasoning, if $0 \div 0 = n$, a number, then $n \cdot 0 = 0$. This is true for any number n so that the quotient $0 \div 0$ would not be a single number. To avoid this, we say that

> Division by 0 is undefined.

EXAMPLE 5

Divide.

a. $\dfrac{20}{-4}$ **b.** $\dfrac{-9}{-3}$ **c.** $-\dfrac{3}{8} \div 3$ **d.** $\dfrac{-40}{10}$ **e.** $\dfrac{-1}{10} \div \dfrac{-2}{5}$ **f.** $\dfrac{8}{0}$

Solution   **a.** Since the signs are different or unlike, the quotient is negative and $\dfrac{20}{-4} = -5$.

b. Since the signs are the same, the quotient is positive and $\dfrac{-9}{-3} = 3$.

c. $-\dfrac{3}{8} \div 3 = -\dfrac{3}{8} \cdot \dfrac{1}{3} = -\dfrac{1}{8}$

d. $\dfrac{-40}{10} = -4$

e. $\dfrac{-1}{10} \div \dfrac{-2}{5} = -\dfrac{1}{10} \cdot -\dfrac{5}{2} = \dfrac{1}{4}$

f. $\dfrac{8}{0}$ is undefined.

With sign rules for division, we can understand why the positioning of the negative sign in a fraction does not change the value of the fraction. For example,

$$\frac{-12}{3} = -4, \quad \frac{12}{-3} = -4, \quad \text{and} \quad -\frac{12}{3} = -4$$

Since all the fractions equal -4, we can say that

$$\frac{-12}{3} = \frac{12}{-3} = -\frac{12}{3}$$

In general, the following holds true.

If a and b are real numbers and $b \neq 0$, then $\dfrac{a}{-b} = \dfrac{-a}{b} = -\dfrac{a}{b}$.

---

3   Recall that when two numbers are multiplied, they are called **factors**. For example, in $3 \cdot 5 = 15$, the 3 and 5 are called factors.

A natural number *exponent* is a shorthand notation for repeated multiplication of the same factor. This repeated factor is called the **base**, and the number of times it is used as a factor is indicated by the **exponent**. For example,

$$\text{base} \; 4^3 = \underbrace{4 \cdot 4 \cdot 4}_{4 \text{ is a factor 3 times}} \overset{\text{exponent}}{} = 64$$

Also,

$$\text{base} \; 2^5 = \underbrace{2 \cdot 2 \cdot 2 \cdot 2 \cdot 2}_{2 \text{ is a factor 5 times}} \overset{\text{exponent}}{} = 32$$

Exponents

If a is a real number and n is a natural number, then the **nth power of a**, or **a raised to the nth power**, written as a^n, is the product of n factors, each of which is a.

$$\text{base} \; a^n = \underbrace{a \cdot a \cdot a \cdot a \cdot \ldots \cdot a}_{a \text{ is a factor } n \text{ times}} \overset{\text{exponent}}{}$$

It is not necessary to write an exponent of 1. For example, 3 is assumed to be 3^1.

EXAMPLE 6

Simplify each expression.

a. 3^2 **b.** $\left(\dfrac{1}{2}\right)^4$ **c.** -5^2

d. $(-5)^2$ **e.** -5^3 **f.** $(-5)^3$

Solution   **a.** $3^2 = 3 \cdot 3 = 9$ **b.** $\left(\dfrac{1}{2}\right)^4 = \left(\dfrac{1}{2}\right)\left(\dfrac{1}{2}\right)\left(\dfrac{1}{2}\right)\left(\dfrac{1}{2}\right) = \dfrac{1}{16}$

c. $-5^2 = -(5 \cdot 5) = -25$ **d.** $(-5)^2 = (-5)(-5) = 25$

e. $-5^3 = -(5 \cdot 5 \cdot 5) = -125$ **f.** $(-5)^3 = (-5)(-5)(-5) = -125$

The screens below show a check using a calculator.

> **Helpful Hint**
>
> Be very careful when simplifying expressions such as -5^2 and $(-5)^2$.
>
> $$-5^2 = -(5 \cdot 5) = -25 \quad \text{and} \quad (-5)^2 = (-5)(-5) = 25$$
>
> Without parentheses, the base to square is 5, not -5.

4   The opposite of squaring a number is taking the **square root** of a number. For example, since the square of 4, or 4^2, is 16, we say that a square root of 16 is 4. The notation $\sqrt{a}$ is used to denote the **positive, or principal, square root** of a nonnegative number a. We then have in symbols that $\sqrt{16} = 4$. The negative square root of 16 is written $-\sqrt{16} = -4$. The square root of a negative number, such as $\sqrt{-16}$ is not a real number. Why? There is no real number, that when squared gives a negative number.

EXAMPLE 7

Find the square roots.

a. $\sqrt{9}$ **b.** $\sqrt{25}$ **c.** $\sqrt{\dfrac{1}{4}}$ **d.** $-\sqrt{36}$ **e.** $\sqrt{-36}$

Solution   **a.** $\sqrt{9} = 3$ since 3 is positive and $3^2 = 9$.

b. $\sqrt{25} = 5$ since $5^2 = 25$.   **c.** $\sqrt{\dfrac{1}{4}} = \dfrac{1}{2}$ since $\left(\dfrac{1}{2}\right)^2 = \dfrac{1}{4}$.

d. $-\sqrt{36} = -6$ **e.** $\sqrt{-36}$ is not a real number.

TECHNOLOGY NOTE

Some graphing utilities contain a square root key as well as other root options under a menu such as a math menu.

We can find roots other than square roots. Since 2 cubed, written as 2^3, is 8, we say that the **cube root** of 8 is 2. This is written as

$$\sqrt[3]{8} = 2.$$

Also, since $3^4 = 81$ and 3 is positive,

$$\sqrt[4]{81} = 3.$$

EXAMPLE 8

Find the roots.

a. $\sqrt[3]{27}$ **b.** $\sqrt[5]{1}$ **c.** $\sqrt[4]{16}$

Solution   **a.** $\sqrt[3]{27} = 3$ since $3^3 = 27$.

b. $\sqrt[5]{1} = 1$ since $1^5 = 1$.

c. $\sqrt[4]{16} = 2$ since 2 is positive and $2^4 = 16$.

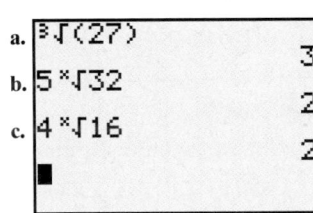

CLASSROOM EXAMPLE
Find each root.
a. $\sqrt[3]{8}$ b. $\sqrt[4]{1}$ c. $\sqrt[5]{32}$
answer:
a. 2   b. 1   c. 2

Of course, as mentioned in Section 1.2, not all roots simplify to rational numbers. We study radicals further in Chapter 7.

5   Expressions containing more than one operation are written to follow a particular agreed-upon **order of operations**. For example, when we write $3 + 2 \cdot 10$, we mean to multiply first, and then add.

Order of Operations

Simplify expressions using the order that follows. If grouping symbols such as parentheses are present, simplify expressions within those first, starting with the innermost set. If fraction bars are present, simplify the numerator and denominator separately.

1. Raise to powers or take roots in order from left to right.
2. Multiply or divide in order from left to right.
3. Add or subtract in order from left to right.

EXAMPLE 9

Simplify.

a. $3 + 2 \cdot 10$ **b.** $2(1 - 4)^2$ **c.** $\dfrac{|-2|^3 + 1}{-7 - \sqrt{4}}$ **d.** $\dfrac{(6 + 2) - (-4)}{2 - (-3)}$

CLASSROOM EXAMPLE
Simplify.
a. $15 - 2 \cdot 5$ b. $5(2 - 6)^2$
c. $\dfrac{|-3|^2 + 5}{\sqrt{9} - 10}$ d. $\dfrac{(8 - 3) - (-6)}{4 - (-1)}$
answer:
a. 5   b. 80   c. -2 d. $\dfrac{11}{5}$

Solution   **a.** First multiply; then add.

$3 + 2 \cdot 10 = 3 + 20 = 23$

b. $2(1 - 4)^2 = 2(-3)^2$ Simplify inside grouping symbols first.

$= 2(9)$ Write $(-3)^2$ as 9.

$= 18$ Multiply.

TECHNOLOGY NOTE

To evaluate expressions with a calculator, we sometimes need to insert parentheses that may not be shown in the expression. This is especially true when entering a fraction whose numerator or denominator contains more than one term. (See the screen below.)

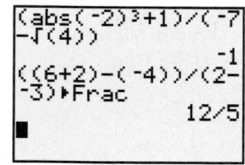

A calculator check for Example 9c and d. Notice the use of parentheses.

c. Simplify the numerator and the denominator separately; then divide.

$$\frac{|-2|^3 + 1}{-7 - \sqrt{4}} = \frac{2^3 + 1}{-7 - 2}$$ Write $|-2|$ as 2 and $\sqrt{4}$ as 2.

$$= \frac{8 + 1}{-9}$$ Write 2^3 as 8.

$$= \frac{9}{-9} = -1$$ Simplify the numerator, then divide.

d.
$$\frac{(6 + 2) - (-4)}{2 - (-3)} = \frac{8 - (-4)}{2 - (-3)}$$ Simplify inside grouping symbols first.

$$= \frac{8 + 4}{2 + 3}$$ Write subtractions as equivalent additions.

$$= \frac{12}{5}$$ Add in both the numerator and denominator.

Besides parentheses, other symbols used for grouping expressions are brackets [ ] and braces { }. These other grouping symbols are commonly used when we group expressions that already contain parentheses.

EXAMPLE 10

Simplify: $3 - [(4 - 6) + 2(5 - 9)]$

Solution

$$3 - [(4 - 6) + 2(5 - 9)] = 3 - [-2 + 2(-4)]$$ Simplify within the innermost sets of parentheses.

$$= 3 - [-2 + (-8)]$$

$$= 3 - [-10]$$

$$= 13$$

CLASSROOM EXAMPLE

Simplify: $7 - [2(1 - 3) + 5(10 - 12)]$
answer: 21

TEACHING TIP

Before beginning objective 2, tell students that the concepts they learn and the skills they develop in this section will be useful when they begin using algebra to solve equations.

> **Helpful Hint**
>
> When grouping symbols occur within grouping symbols, remember to perform operations on the innermost set first.

✔ **CONCEPT CHECK**

True or false? If two different people use the order of operations to simplify a numerical expression and neither makes a calculation error, it is not possible that they each obtain a different result. Explain.

6   Recall from Section 1.2 that an algebraic expression is formed by numbers and variables connected by the operations of addition, subtraction, multiplication, division, raising to powers, and/or taking roots. Also, if numbers are substituted for the variables in an algebraic expression and the operations performed, the result is called **the value of the expression** for the given replacement values. This entire process is called **evaluating an expression**.

EXAMPLE 11

Concept Check Answer:
true; answers may vary

Evaluate each algebraic expression when $x = 2$, $y = -1$, and $z = -3$.

a. $z - y$ **b.** z^2 **c.** $\dfrac{2x + y}{z}$ **d.** $-x^2 - 4x$

Solution   **a.** $z - y = -3 - (-1) = -3 + 1 = -2$ **b.** $z^2 = (-3)^2 = 9$

c. $\dfrac{2x + y}{z} = \dfrac{2(2) + (-1)}{-3} = \dfrac{4 + (-1)}{-3} = \dfrac{3}{-3} = -1$

d. $-x^2 - 4x = -2^2 - 4(2) = -4 - 8 = -12$

How can we use a calculator to evaluate the expressions in Example 11? One way to evaluate Example 11 d with a calculator is to just enter the numerical expression and evaluate.

Another way to evaluate $-x^2 - 4x$ when $x = 2$ is to store the value 2 in the variable x and then enter and evaluate the algebraic expression.

The value of $-x^2 - 4x$ when $x = 2$ is -12.

Thus far, we have seen that we can evaluate an expression or perform a calculation mentally, using paper and pencil, or using a calculator. In general, how do we decide which method to use?

The following example shows that all three methods are useful.

EXAMPLE 12

Evaluate the expression $3x^2 + 4$ for the following values of x:

a. $x = 2$ **b.** $x = 12$ **c.** $x = 3.91$

Solution

a. Never underestimate the power and speed of mental calculation. Replace x with 2 and mentally calculate x^2 to be 4, multiply by 3 to get 12, and add 4 to get 16. The value of $3x^2 + 4$ when $x = 2$ is 16.

b. Replace x with 12 and mentally calculate the square of 12 to be 144. Next, you might use paper and pencil to evaluate $3 \cdot 144$, which equals 432, and then add 4 to get 436. The value of $3x^2 + 4$ when $x = 12$ is 436.

c. Since the value of x is a decimal, using a calculator is the most appropriate choice. First, let's mentally approximate the value of the expression. $3.91 \approx 4$, and the value of $3x^2 + 4$ at 4 is 52. The screen at the left shows that the value of $3x^2 + 4$ when $x = 3.91$ is 49.8643. This is close to our approximation of 52 and thus is reasonable.

> **Helpful Hint**
>
> Sometimes variables such as x_1 and x_2 will be used in this book. The small 1 and 2 are called **subscripts**. The variable x_1 can be read as "x sub 1," and the variable x_2 can be read as "x sub 2." The important thing to remember is that they are two different variables. For example, if $x_1 = -5$ and $x_2 = 7$, then
>
> $$x_1 - x_2 = -5 - 7 = -12.$$

EXAMPLE 13

The algebraic expression $\dfrac{5(x - 32)}{9}$ represents the equivalent temperature in degrees

CLASSROOM EXAMPLE

Use the algebraic expression given in Example 13 to complete the following table.

Degrees Fahrenheit	x	−13	0	41
Degrees Celsius	$\frac{5(x-32)}{9}$			

answer: $-25; -\frac{160}{9}$ or $-17\frac{7}{9}; 5$

Celsius when x is the temperature in degrees Fahrenheit. Complete the following table by evaluating this expression at the given values of x.

Degrees Fahrenheit	x	−4	10	32
Degrees Celsius	$\frac{5(x-32)}{9}$			

Solution

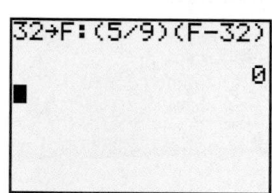

To complete the table, evaluate $\frac{5(x-32)}{9}$ at each given replacement value.

When $x = -4$,

$$\frac{5(x-32)}{9} = \frac{5(-4-32)}{9} = \frac{5(-36)}{9} = -20$$

When $x = 10$,

$$\frac{5(x-32)}{9} = \frac{5(10-32)}{9} = \frac{5(-22)}{9} = -\frac{110}{9}$$

When $x = 32$,

$$\frac{5(x-32)}{9} = \frac{5(32-32)}{9} = \frac{5 \cdot 0}{9} = 0$$

The completed table is

Degrees Fahrenheit	x	−4	10	32
Degrees Celsius	$\frac{5(x-32)}{9}$	−20	$-\frac{110}{9}$	0

Thus, −4°F is equivalent to −20°C, 10°F is equivalent to $-\frac{110}{9}$°C, and 32°F is equivalent to 0°C.

Spotlight on
DECISION
MAKING

Suppose you are a travel agent. A tour company is offering bonuses to travel agents booking clients on selected tour packages for a limited time. However, prior to participating in this bonus program, you must select only one type of tour package for which you will receive the bonus. Information about the selected tour packages is shown in the table.

Based on client inquiries during the past week, you estimate that you could probably interest 30 clients in a cruise to Alaska, 60 in a Bermuda package, 50 in a trip to Cancun, and 40 in a Hawaii package. However, you also estimate that in each case, only half of the clients would book the trip if it cost over $1000 per person.

Which one of the tour packages would you choose for participating in the tour company's bonus program? Why?

Selected Tour Packages

Destination	Cost per person	Bonus per person booked
Alaska cruise	$2029	$100
Bermuda	$ 699	$ 25
Cancun, Mexico	$1349	$ 75
Hawaii	$ 840	$ 50

MENTAL MATH

Choose the fraction(s) equivalent to the given fraction. (There may sometimes be more than one correct choice.)

1. $-\dfrac{1}{7}$ **a.** $\dfrac{-1}{-7}$ **b.** $\dfrac{-1}{7}$ **c.** $\dfrac{1}{-7}$ **d.** $\dfrac{1}{7}$ b, c

2. $\dfrac{-x}{y}$ **a.** $\dfrac{x}{-y}$ **b.** $-\dfrac{x}{y}$ **c.** $\dfrac{x}{y}$ **d.** $\dfrac{-x}{-y}$ a, b

3. $\dfrac{5}{-(x+y)}$ **a.** $\dfrac{5}{(x+y)}$ **b.** $\dfrac{-5}{(x+y)}$ **c.** $\dfrac{-5}{-(x+y)}$ **d.** $-\dfrac{5}{(x+y)}$ b, d

4. $-\dfrac{(y+z)}{3y}$ **a.** $\dfrac{-(y+z)}{3y}$ **b.** $\dfrac{-(y+z)}{-3y}$ **c.** $\dfrac{(y+z)}{3y}$ **d.** $\dfrac{(y+z)}{-3y}$ a, d

5. $\dfrac{-9x}{-2y}$ **a.** $\dfrac{-9x}{2y}$ **b.** $\dfrac{9x}{2y}$ **c.** $\dfrac{9x}{-2y}$ **d.** $-\dfrac{9x}{2y}$ b

6. $\dfrac{-a}{-b}$ **a.** $\dfrac{a}{b}$ **b.** $\dfrac{a}{-b}$ **c.** $\dfrac{-a}{b}$ **d.** $-\dfrac{a}{b}$ a

EXERCISE SET 1.3

STUDY GUIDE/SSM   CD/VIDEO   PH MATH TUTOR CENTER   MathXL®Tutorials ON CD   MathXL®   MyMathLab®

Find each sum or difference. See Examples 1 through 3.

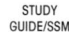 **1.** $-3 + 8$ 5   **2.** $-5 + (-9)$ -14

 3. $-14 + (-10)$ -24 **4.** $12 + (-7)$ 5

5. $-4.3 - 6.7$ -11 **6.** $-8.2 - (-6.6)$ -1.6

7. $13 - 17$ -4 **8.** $15 - (-1)$ 16

9. $\dfrac{11}{15} - \left(-\dfrac{3}{5}\right)$ $\dfrac{4}{3}$ **10.** $\dfrac{7}{10} - \dfrac{4}{5}$ $-\dfrac{1}{10}$

11. $19 - 10 - 11$ -2 **12.** $-13 - 4 + 9$ -8

Find each product or quotient. See Examples 4 and 5.

13. $(-5)(12)$ -60 **14.** $6(-3)$ -18

15. $(-8)(-10)$ 80   **16.** $7(0)$ 0

17. $\dfrac{-12}{-4}$ 3   **18.** $\dfrac{60}{-6}$ -10

19. $\dfrac{0}{-2}$ 0   **20.** $\dfrac{-2}{0}$ undefined

21. $(-4)(-2)(-1)$ -8 **22.** $5(-3)(-2)$ 30

23. $\dfrac{-6}{7} \div 2$ $-\dfrac{3}{7}$ **24.** $\dfrac{-9}{13} \div (-3)$ $\dfrac{3}{13}$

25. $\left(-\dfrac{2}{7}\right)\left(-\dfrac{1}{6}\right)$ $\dfrac{1}{21}$ **26.** $\dfrac{5}{9}\left(-\dfrac{3}{5}\right)$ $-\dfrac{1}{3}$

Evaluate. See Example 6.

27. -7^2 -49 **28.** $(-7)^2$ 49

29. $(-6)^2$ 36   **30.** -6^2 -36

31. $(-2)^3$ -8 **32.** -2^3 -8

Find the following roots. See Examples 7 and 8.

33. $\sqrt{49}$ 7   **34.** $\sqrt{81}$ 9

35. $-\sqrt{\dfrac{1}{9}}$ $-\dfrac{1}{3}$ **36.** $-\sqrt{\dfrac{1}{25}}$ $-\dfrac{1}{5}$

37. $\sqrt[3]{64}$ 4   **38.** $\sqrt[5]{32}$ 2

39. $\sqrt[4]{81}$ 3   **40.** $\sqrt[3]{1}$ 1

41. $\sqrt{-100}$ not a real number   **42.** $\sqrt{-25}$ not a real number

MIXED PRACTICE

Simplify each expression. See Examples 9 and 10.

43. $3(5-7)^4$ 48   **44.** $7(3-8)^2$ 175

45. $-3^2 + 2^3$ -1 **46.** $-5^2 - 2^4$ -41

47. $\dfrac{3-(-12)}{-5}$ -3 **48.** $\dfrac{-4-(-8)}{-4}$ -1

49. $|3.6 - 7.2| + |3.6 + 7.2|$ 14.4

50. $|8.6 - 1.9| - |2.1 + 5.3|$ -0.7

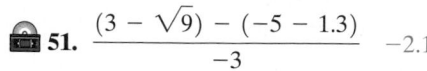**51.** $\dfrac{(3 - \sqrt{9}) - (-5 - 1.3)}{-3}$ -2.1

52. $\dfrac{-\sqrt{16} - (6 - 2.4)}{-2}$ 3.8

53. $\dfrac{|3 - 9| - |-5|}{-3}$ $-\dfrac{1}{3}$

54. $\dfrac{|-14| - |2 - 7|}{-15}$ $-\dfrac{3}{5}$

55. $(-3)^2 + 2^3$ 17 **56.** $(-15)^2 - 2^4$ 209

57. $4[8 - (2 - 4)]$ 40 **58.** $3[11 - (1 - 3)]$ 39

59. $2 - [(7 - 6) + (9 - 19)]$ 11

60. $8 - [(4 - 7) + (8 - 1)]$ 4

61. $\dfrac{(-9 + 6)(-1^2)}{-2 - 2}$ $-\dfrac{3}{4}$

62. $\dfrac{(-1 - 2)(-3^2)}{-6 - 3}$ -3

63. $(\sqrt[3]{8})(-4) - (\sqrt{9})(-5)$ 7

64. $(\sqrt[3]{27})(-5) - (\sqrt{25})(-3)$ 0

65. $25 - [(3 - 5) + (14 - 18)]^2$ -11

66. $10 - [(4 - 5)^2 + (12 - 14)]^4$ 9

67. $\dfrac{\frac{1}{3} \cdot 9 - 7}{3 + \frac{1}{2} \cdot 4}$ $-\dfrac{4}{5}$ **68.** $\dfrac{\frac{1}{5} \cdot 20 - 6}{10 + \frac{1}{4} \cdot 12}$ $-\dfrac{2}{13}$

69. $\dfrac{3(-2 + 1)}{5} - \dfrac{-7(2 - 4)}{1 - (-2)}$ $-\dfrac{79}{15}$

70. $\dfrac{-1 - 2}{2(-3) + 10} - \dfrac{2(-5)}{-1(8) + 1}$ $-\dfrac{61}{28}$

71. $\dfrac{\frac{-3}{10}}{\frac{42}{50}}$ $-\dfrac{5}{14}$ **72.** $\dfrac{\frac{-5}{21}}{\frac{-6}{42}}$ $\dfrac{5}{3}$

Find the value of each expression when $x = -2$, $y = -5$, and $z = 3$. See Examples 11 and 12.

73. $x^2 + z^2$ 13 **74.** $y^2 - z^2$ 16

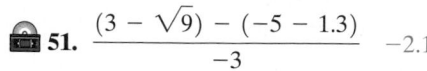**75.** $-5(-x + 3y)$ 65 **76.** $-7(-y - 4z)$ 49

77. $\dfrac{3z - y}{2x - z}$ -2 **78.** $\dfrac{5x - z}{-2y + z}$ -1

Find the value of each expression when $x = 1.4$ and $y = -6.2$. If necessary, round the result to the nearest hundredth. See Examples 11 and 12.

79. $3x - 2y$ 16.6 **80.** $5y^2 + x$ 193.6

81. $\dfrac{|x - y|}{2y}$ -0.61 **82.** $\dfrac{\sqrt{x - y}}{2x}$ 0.98

83. $-3(x^2 + y^2)$ -121.2 **84.** $1.6(3x^2 + 1.8)$ 12.29

State the expression being evaluated, the values of the variables, and the value of the expression.

85.   **86.**

87.   **88.**

89. Consider the expressions $(1/2)x$ and $1/2x$.
   a. Use your calculator to store 2 in x, and then enter and evaluate each expression. $(1/2)x = 1$; $1/2x = 1$
   b. Are the results of part a the same or different? Explain why.   same: answers may vary

90. Consider the expression $12/(x - 2)$ and $12/x - 2$.
   a. Use your calculator to store 4 in x, and then enter and evaluate each expression. $12/(x - 2) = 6$; $12/x - 2 = 1$
   b. Are the results of part a the same or different? Explain why.   different: answers may vary

See Example 13.

91. The algebraic expression $8 + 2y$ represents the perimeter of a rectangle with width 4 and length y.

   a. Complete the table that follows by evaluating this expression at the given values of y.

Length	y	5	7	10	100
Perimeter	$8 + 2y$	18	22	28	208

   b. Use the results of the table in **a** to answer the following question. As the width of a rectangle remains the same and the length increases, does the perimeter increase or decrease? Explain how you arrived at your answer.   increase

85. $x = -2.7$; $y = 0.9$; $x^2 + 2y = 9.09$ **86.** $x = 5.1$; $y = -3$; $2x - y^2 = 1.2$ **87.** $A = 4$; $B = -2.5$; $A^2 + B^2 = 22.25$
88. $B = 8$; $H = 4.5$; $\frac{1}{2}BH = 18$

△ **92.** The algebraic expression πr^2 represents the area of a circle with radius r.

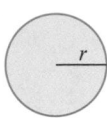

a. Complete the table below by evaluating this expression at given values of r. (Use 3.14 for π.)

Radius	r	2	3	7	10
Area	πr^2	12.56	28.26	153.86	314

✎ **b.** As the radius of a circle increases, does its area increase or decrease? Explain your answer.   increase

93. The algebraic expression $\dfrac{100x + 5000}{x}$ represents the cost per bookshelf (in dollars) of producing x bookshelves.

a. Complete the table below.

Number of Bookshelves	x	10	100	1000
Cost per Bookshelf	$\dfrac{100x + 5000}{x}$	600	150	105

b. As the number of bookshelves manufactured increases, does the cost per bookshelf increase or decrease? Why do you think that this is so?   decrease

94. If c is degrees Celsius, the algebraic expression $1.8c + 32$ represents the equivalent temperature in degrees Fahrenheit.
a. Complete the table below.

Degrees Celsius	c	-10	0	50
Degrees Fahrenheit	$1.8c + 32$	14	32	122

✎ **b.** As degrees Celsius increase, do degrees Fahrenheit increase or decrease?   increase

Concept Extensions

Each circle below represents a whole, or 1. Determine the unknown fractional part of each circle.

95. 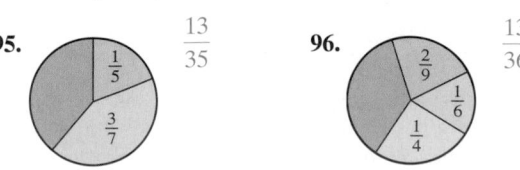 $\dfrac{13}{35}$ **96.** $\dfrac{13}{36}$

97. Most of Mauna Kea, a volcano on Hawaii, lies below sea level. If this volcano begins at 5998 meters below sea level and then rises 10,203 meters, find the height of the volcano above sea level.   4205 meters

98. The highest point on land on Earth is the top of Mt. Everest in the Himalayas, at an elevation of 29,028 feet above sea level. The lowest point on land is the Dead Sea, between Israel and Jordan, at 1319 feet below sea level. Find the difference in elevations.   30,347 feet

99. The following bar graph shows the U.S. life expectancy at birth for females born in the years shown. Use the graph to fill in the table below by calculating the *increase* in life expectancy over each ten-year period shown.

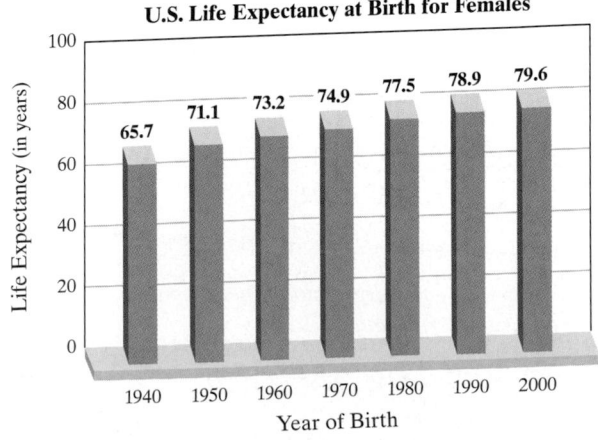

Source: Social Security Administration

Year	Increase in Life Expectancy (in years) from 10 Years Earlier
1950	5.4
1960	2.1
1970	1.7
1980	2.6
1990	1.4
2000	0.7

Investment Risk Over Time

Average Annual Compound Return (%)

181.5

34.9

15.6

12.9

−13.2   −2.1   2.1

−64.0

1 Year   5 Years   15 Years   25 Years

Years

Source: *Investment Digest,* VALIC

Insert parentheses so that each expression simplifies to the given number.

100. $2 + 7 \cdot 1 + 3$; 36   **101.** $6 - 5 \cdot 2 + 2$; -6

102. Explain why -3^2 and $(-3)^2$ simplify to different numbers. answers may vary

103. Explain why -3^3 and $(-3)^3$ simplify to the same number. answers may vary

Approximate each square root. For exercises 108 and 109, simplify the expression. Round answers to four decimal places.

104. $\sqrt{10}$ 3.1623   **105.** $\sqrt{273}$ 16.5227

106. $\sqrt{7.9}$ 2.8107   **107.** $\sqrt{19.6}$ 4.4272

108. $\dfrac{-1.682 - 17.895}{(-7.102)(-4.691)}$ **109.** $\dfrac{(-5.161)(3.222)}{7.955 - 19.676}$ 1.4187

Investment firms often advertise their gains and losses in the form of bar graphs such as the one that follows. This graph shows investment risk over time for the S&P 500 Index by showing average annual compound returns for 1 year, 5 years, 15 years, and 25 years. For example, after one year, the annual compound return in percent for an investor is anywhere from a gain of 181.5% to a loss of 64%. Use this graph to answer the questions below.

110. A person investing in the S&P 500 Index may expect at most an average annual gain of what percent after 15 years?   15.6%

111. A person investing in the S&P 500 Index may expect to lose at most an average per year of what percent after 5 years? 13.2%

112. Find the difference in percent of the highest average annual return and the lowest average annual return after 15 years. 17.7%

113. Find the difference in percent of the highest average annual return and the lowest average annual return after 25 years. 10.8%

114. Do you think that the type of investment shown in the figure is recommended for short-term investments or long-term investments? Explain your answer.   Long-term. Short-term is very volatile.

A fair game is one in which each team or player has the same chance of winning. Suppose that a game consists of three players taking turns spinning a spinner. If the spinner lands on yellow, player 1 gets a point. If the spinner lands on red, player 2 gets a point, and if the spinner lands on blue, player 3 gets a point. After 12 spins, the player with the most points wins.

a.   *b.*

c.   *d.*

115. Which spinner would lead to a fair game?   b

116. If you are player 2 and want to win the game, which spinner would you choose?   a

117. If you are player 1 and want to lose the game, which spinner would you choose?   d

118. Is it possible for the game to end in a three-way tie? If so, list the possible ending scores.   Yes. Each player has 4 points.

119. Is it possible for the game to end in a two-way tie? If so, list the possible ending scores.   Yes. Two players have 6 points each (the third player has 0 points) or two players have 5 points each (the third has 2 points)

100. $(2 + 7) \cdot (1 + 3)$ **101.** $6 - (5 \cdot 2 + 2)$ **108.** -0.5876

1.4 PROPERTIES OF REAL NUMBERS

Objectives

① Use operation and order symbols to write mathematical sentences.

② Identify identity numbers and inverses.

③ Identify and use the commutative, associative, and distributive properties.

④ Write algebraic expressions.

⑤ Simplify algebraic expressions.

TEACHING TIP

Start reinforcing the differences between an equation and an expression. An equation has "=" which an expression does not.

① In Section 1.2, we used the symbol = to mean "is equal to." All of the following key words and phrases also imply equality.

Equality

equals	is/was	represents	is the same as
gives	yields	amounts to	is equal to

EXAMPLE 1

Write each sentence using mathematical symbols.

a. The sum of x and 5 is 20.

b. Two times the sum of 3 and y amounts to 4.

c. Subtract 8 from x, and the difference is the same as twice x.

d. The quotient of z and 9 is 3 times the difference of z and 5.

Solution

a.

the sum of x and 5	is	20
↓	↓	↓
$x + 5$	$=$	20

b. $2(3 + y) = 4$

c. $x - 8 = 2x$

d. $\dfrac{z}{9} = 3(z - 5)$

If we want to write in symbols that two numbers are not equal, we can use the symbol $\neq$, which means "**is not equal to.**" For example,

$$3 \neq 2$$

Graphing two numbers on a number line gives us a way to compare two numbers. For two real numbers a and b, we say **a is less than b** if on the number line a lies to the left of b. Also, if b is to the right of a on the number line, then **b is greater than a**. The symbol $<$ means "**is less than.**" Since a is less than b, we write

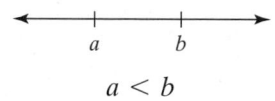

$$a < b$$

The symbol $>$ means "**is greater than.**" Since b is greater than a, we write

$$b > a$$

> ### Helpful Hint
> Notice that if $a < b$, then $b > a$.

EXAMPLE 2

Insert $<, >,$ or $=$ between each pair of numbers to form a true statement.

a. $-1 \quad -2$ **b.** $\dfrac{12}{4} \quad 3$ **c.** $-5 \quad 0$ **d.** $-3.5 \quad -3.05$

Solution   **a.** $-1 > -2$ since -1 lies to the right of -2 on the number line.

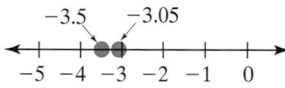

b. $\dfrac{12}{4} = 3$.

c. $-5 < 0$ since -5 lies to the left of 0 on the number line.

d. $-3.5 < -3.05$ since -3.5 lies to the left of -3.05 on the number line.

> ### Helpful Hint
> When inserting the $>$ or $<$ symbol, think of the symbols as arrowheads that "point" toward the smaller number when the statement is true.

In addition to $<$ and $>$, there are the inequality symbols $\leq$ and $\geq$. The symbol

$$\leq \text{ means "is less than or equal to"}$$

and the symbol

$$\geq \text{ means "is greater than or equal to"}$$

For example, the following are true statements.

$10 \leq 10$	since	$10 = 10$
$-8 \leq 13$	since	$-8 < 13$
$-5 \geq -5$	since	$-5 = -5$
$-7 \geq -9$	since	$-7 > -9$

EXAMPLE 3

Write each sentence using mathematical symbols.

a. The sum of 5 and y is greater than or equal to 7.

b. 11 is not equal to z.

c. 20 is less than the difference of 5 and twice x.

Solution   **a.** $5 + y \geq 7$ **b.** $11 \neq z$ **c.** $20 < 5 - 2x$

2   Of all the real numbers, two of them stand out as extraordinary: 0 and 1. Zero is the only number that when *added* to any real number, the result is the same real number. Zero is thus called the **additive identity**. Also, one is the only number that when *multiplied* by any real number, the result is the same real number. One is thus called the **multiplicative identity**.

	Addition	**Multiplication**
Identity Properties	The additive identity is 0.	The multiplicative identity is 1.
	$a + 0 = 0 + a = a$	$a \cdot 1 = 1 \cdot a = a$

In Section 1.2, we learned that a and $-a$ are opposites.

Another name for opposite is **additive inverse.** For example, the additive inverse of 3 is -3. Notice that the sum of a number and its opposite is always 0.

In Section 1.3, we learned that, for a nonzero number, b and $\dfrac{1}{b}$ are reciprocals.

Another name for reciprocal is **multiplicative inverse.** For example, the multiplicative inverse of $-\dfrac{2}{3}$ is $-\dfrac{3}{2}$. Notice that the product of a number and its reciprocal is always 1.

	Opposite or Additive Inverse	***Reciprocal or Multiplicative Inverse***
Inverse Properties	For each number a, there is a unique number $-a$ called the **additive inverse** or **opposite** of a such that $$a + (-a) = (-a) + a = 0$$	For each nonzero a, there is a unique number $\dfrac{1}{a}$ called the **multiplicative inverse** or **reciprocal** of a such that $$a \cdot \dfrac{1}{a} = \dfrac{1}{a} \cdot a = 1$$

CLASSROOM EXAMPLE

Find the opposite or additive inverse of each.

a. 10 **b.** $-\dfrac{2}{3}$ **c.** 8.1

answer:

a. -10 **b.** $\dfrac{2}{3}$ **c.** -8.1

EXAMPLE 4

Write the additive inverse, or opposite, of each.

a. 8   **b.** $\frac{1}{5}$ **c.** -9.6

Solution **a.** The opposite of 8 is -8.

b. The opposite of $\frac{1}{5}$ is $-\frac{1}{5}$.

c. The opposite of -9.6 is $-(-9.6) = 9.6$.

 EXAMPLE 5

Write the multiplicative inverse, or reciprocal, of each.

a. 11   **b.** -9 **c.** $\frac{7}{4}$

Solution **a.** The reciprocal of 11 is $\frac{1}{11}$.

b. The reciprocal of -9 is $-\frac{1}{9}$.

c. The reciprocal of $\frac{7}{4}$ is $\frac{4}{7}$ because $\frac{7}{4} \cdot \frac{4}{7} = 1$.

CLASSROOM EXAMPLE

Find the multiplicative inverse, or reciprocal, of each.
a. 19 **b.** $\frac{4}{9}$ **c.** -8
answer:
a. $\frac{1}{19}$ **b.** $\frac{9}{4}$ **c.** $-\frac{1}{8}$

> **Helpful Hint**
> The number 0 has no reciprocal. Why? There is no number that when multiplied by 0 gives a product of 1.

✔ **CONCEPT CHECK**

Can a number's additive inverse and multiplicative inverse ever be the same? Explain.

3   In addition to these special real numbers, all real numbers have certain properties that allow us to write equivalent expressions—that is, expressions that have the same value. These properties will be especially useful in Chapter 2 when we solve equations.

The **commutative properties** state that the order in which two real numbers are added or multiplied does not affect their sum or product.

Concept Check Answer:
no; answers may vary

Commutative Properties

For real numbers a and b,

$$Addition \quad a + b = b + a$$
$$Multiplication \quad a \cdot b = b \cdot a$$

The **associative properties** state that regrouping numbers that are added or multiplied does not affect their sum or product.

Associative Properties

For real numbers a, b, and c,

$$Addition \quad (a + b) + c = a + (b + c)$$
$$Multiplication \quad (a \cdot b) \cdot c = a \cdot (b \cdot c)$$

EXAMPLE 6

CLASSROOM EXAMPLE

Use the commutative property of addition to write an expression equivalent to $9 + 4x$.

answer: $4x + 9$

Use the commutative property of addition to write an expression equivalent to $7x + 5$.

Solution $7x + 5 = 5 + 7x$.

EXAMPLE 7

Use the associative property of multiplication to write an expression equivalent to $4 \cdot (9y)$. Then simplify this equivalent expression.

Solution $4 \cdot (9y) = (4 \cdot 9)y = 36y$.

The **distributive property** states that multiplication distributes over addition.

CLASSROOM EXAMPLE

Use the associative property of multiplication to write an expression equivalent to $5 \cdot (6x)$. Then simplify.

answer: $30x$

Distributive Property

For real numbers a, b, and c,
$$a(b + c) = ab + ac$$

EXAMPLE 8

Use the distributive property to multiply.

a. $3(2x + y)$ **b.** $-(3x - 1)$ **c.** $0.7a(b - 2)$

Solution **a.** $3(2x + y) = 3 \cdot 2x + 3 \cdot y$ Apply the distributive property.

$\qquad\qquad\quad = 6x + 3y$ Apply the associative property of multiplication.

CLASSROOM EXAMPLE

Use the distributive property to multiply.

a. $7(4x - y)$ **b.** $-8(3 + x)$

c. $5x(y - 4)$

answer:

a. $28x - 7y$ **b.** $-24 - 8x$

c. $5xy - 20x$

b. Recall that $-(3x - 1)$ means $-1(3x - 1)$.

$$-1(3x - 1) = -1(3x) + (-1)(-1)$$

$$= -3x + 1$$

c. $0.7a(b - 2) = 0.7a \cdot b - 0.7a \cdot 2 = 0.7ab - 1.4a$

✔ **CONCEPT CHECK**
Is the statement below true? Why or why not?
$6(2a)(3b) = 6(2a) \cdot 6(3b)$

In the following example, a calculator screen illustrates some of the properties.

EXAMPLE 9

TECHNOLOGY NOTE

In the third line of Example 9a the multiplicative inverse is indicated by using the x^{-1} key on the calculator. We can refer to the x^{-1} key as the *reciprocal key*, or the *multiplicative inverse* key.

State the basic property that is being illustrated on each calculator screen:

a.   **b.**   **c.**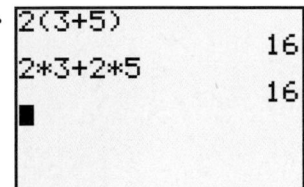

Solution   **a.** Multiplicative inverses   **b.** Double negative property
c. Distributive property

CLASSROOM EXAMPLE
State the property illustrated.
a. $5(6x) = (5 \cdot 6)x$
b. $2 + (a + b) = 2 + (b + a)$
c. $2(a + b) = 2 \cdot a + 2 \cdot b$
answer:
a. Associative Property of Multiplication
b. Commutative Property of Addition
c. Distributive Property

4   As mentioned earlier, an important step in problem solving is to be able to write algebraic expressions from word phrases. Sometimes this involves a direct translation, but often an indicated operation is not directly stated but rather implied.

EXAMPLE 10

Write each as an algebraic expression.

a. A vending machine contains x quarters. Write an expression for the *value* of the quarters.
b. The number of grams of fat in x pieces of bread if each piece of bread contains 2 grams of fat.
c. The cost of x desks if each desk costs $156.
d. Sales tax on a purchase of x dollars if the tax rate is 9%.

Each of these examples implies finding a product.

Solution   **a.** The value of the quarters is found by multiplying the value of a quarter (0.25 dollar) by the number of quarters.

CLASSROOM EXAMPLE
Write as an algebraic expression.
a. Express the *value* of y nickels.
b. The cost of x DVDs if each one costs $21.
answer:
a. $0.05y$ **b.** $21x$

In words:	Value of a quarter	·	Number of quarters	
Translate:	0.25	·	x,	or $0.25x$

b.

In words:	Number of grams of fat in one piece of bread	·	Number of pieces or bread	
Translate:	2	·	x,	or $2x$

Concept Check Answer:
no; $6(2a)(3b) = 12a(3b) = 36ab$

c. In words:

Cost of a desk	·	Number of desks	
Translate: 156	·	x,	or $156x$

d. In words:

Sales tax rate	·	purchase price	
Translate: 0.09	·	x,	or $0.09x$

(Here, we wrote 9% as a decimal, 0.09.)

Two or more unknown numbers in a problem may sometimes be related. If so, try letting a variable represent one unknown number and then represent the other unknown number or numbers as expressions containing the same variable.

EXAMPLE 11

Write each as an algebraic expression.

a. Two numbers have a sum of 20. If one number is x, represent the other number as an expression in x.

b. The older sister is 8 years older than her younger sister. If the age of the younger sister is x, represent the age of the older sister as an expression in x.

△ **c.** Two angles are complementary if the sum of their measures is 90°. If the measure of one angle is x degrees, represent the measure of the other angle as an expression in x.

d. If x is the first of two consecutive integers, represent the second integer as an expression in x.

Solution

a. If two numbers have a sum of 20 and one number is x, the other number is "the rest of 20."

In words:

Twenty	minus	x
Translate: 20	−	x

b. The older sister's age is

In words:

Eight years	added to	younger sister's age
Translate: 8	+	x

c. In words:

Ninety	minus	x
Translate: 90	−	x

d. The next consecutive integer is always one more than the previous integer.

In words:

The first integer	plus	one

Translate: x $+$ 1

5   Often, an expression may be **simplified** by removing grouping symbols and combining any like terms. The **terms** of an expression are the addends of the expression. For example, in the expression $3x^2 + 4x$, the terms are $3x^2$ and $4x$.

Expression	*Terms*
$-2x + y$	$-2x, y$
$3x^2 - \dfrac{y}{5} + 7$	$3x^2, -\dfrac{y}{5}, 7$

Terms with the same variable(s) raised to the same power are called **like terms.** We can add or subtract like terms by using the distributive property. This process is called **combining like terms.**

EXAMPLE 12

Use the distributive property to simplify each expression.

a. $3x - 5x + 4$ **b.** $7yz + yz$ **c.** $4z + 6.1$

Solution   **a.** $3x - 5x + 4 = (3 - 5)x + 4$ *Apply the distributive property.*
$$= -2x + 4$$

b. $7yz + yz = (7 + 1)yz = 8yz$

c. $4z + 6.1$ cannot be simplified further since $4z$ and 6.1 are not like terms.

Let's continue to use properties of real numbers to simplify expressions. Recall that the distributive property can also be used to multiply. For example,

$$-2(x + 3) = -2(x) + (-2)(3) = -2x - 6$$

The associative and commutative properties may sometimes be needed to re-arrange and group like terms when we simplify expressions.

$$-7x^2 + 5 + 3x^2 - 2 = -7x^2 + 3x^2 + 5 - 2$$
$$= (-7 + 3)x^2 + (5 - 2)$$
$$= -4x^2 + 3$$

EXAMPLE 13

Simplify each expression.

a. $3xy - 2xy + 5 - 7 + xy$

b. $7x^2 + 3 - 5(x^2 - 4)$

c. $(2.1x - 5.6) - (-x - 5.3)$

d. $\frac{1}{2}(4a - 6b) - \frac{1}{3}(9a + 12b - 1) + \frac{1}{4}$

Solution

a. $3xy - 2xy + 5 - 7 + xy = 3xy - 2xy + xy + 5 - 7$ Apply the commutative property.

$= (3 - 2 + 1)xy + (5 - 7)$ Apply the distributive property.

$= 2xy - 2$ Simplify.

b. $7x^2 + 3 - 5(x^2 - 4) = 7x^2 + 3 - 5x^2 + 20$ Apply the distributive property.

$= 2x^2 + 23$ Simplify.

c. Think of $-(-x - 5.3)$ as $-1(-x - 5.3)$ and use the distributive property.

$(2.1x - 5.6) - 1(-x - 5.3) = 2.1x - 5.6 + 1x + 5.3$

$= 3.1x - 0.3$ Combine like terms.

d. $\frac{1}{2}(4a - 6b) - \frac{1}{3}(9a + 12b - 1) + \frac{1}{4}$

$= 2a - 3b - 3a - 4b + \frac{1}{3} + \frac{1}{4}$ Use the distributive property.

$= -a - 7b + \frac{7}{12}$ Combine like terms.

CLASSROOM EXAMPLE

Simplify each expression.

a. $7st - 4st + 3 - 11 + st$

b. $4y^2 + 2 - 3(y^2 + 3)$

c. $\frac{1}{3}(9x - 6y) - \frac{1}{2}(8x - 4y + 1) - \frac{3}{4}$

answer:

a. $4st - 8$

b. $y^2 - 7$

c. $-x - \frac{5}{4}$

✔ **CONCEPT CHECK**

Find and correct the error in the following

$x - 4(x - 5) = x - 4x - 20$

$= -3x - 20$

Concept Check Answer:

$x - 4(x - 5) = x - 4x + 20$

$= -3x + 20$

TEACHING TIP

A Group Activity for this section is available in the Instructor's Resource Manual.

Spotlight on
DECISION
& MAKING

Suppose you are a geologist studying Hawaiian volcanoes. When lava from a volcano flows into the ocean, it heats the water around it. You are color-coding a map of ocean water temperatures around a lava flow. If the water temperature is less than or equal to 22°C, the map will be colored dark blue. If the water temperature is greater than 29°C, the map will be colored tan. Otherwise, the map will be colored green. Decide what color on the map should be used for the following temperatures.

a. 25°C   **b.** 18°C   **c.** 29°C   **d.** 22°C   **e.** 31°C

EXERCISE SET 1.4

| STUDY | CD/ | PH MATH | MathXL®Tutorials | MathXL® | MyMathLab® |
| GUIDE/SSM | VIDEO | TUTOR CENTER | ON CD | | |

Insert $<$, $>$, or $=$ in the space provided to form a true statement. See Example 2.

1. $0 > -2$

2. $-5 < 0$

 3. $7.4 = 7.40$

 4. $\dfrac{7}{11} < \dfrac{9}{11}$

5. $-7.9 < -7.09$

6. $-13.07 > -13.7$

MIXED PRACTICE

Write each sentence using mathematical symbols. See Examples 1 and 3.   **10.** $9(4 - m) = 1$

 7. Twice x plus 5 is the same as -14. $2x + 5 = -14$

8. The sum of 10 and x is -12. $10 + x = -12$

9. 3 times the sum of x and 1 amounts to 7. $3(x + 1) = 7$

10. 9 times the difference of 4 and m amounts to 1.

11. The quotient of n and 5 is 4 times n.

12. The quotient of 8 and y is 3 more than y.

13. The difference of z and 2 is the same as the product of z and 2. $z - 2 = 2z$

14. Five added to twice q is the same as 4 more than q.

15. The product of 7 and x is less than or equal to -21.

16. 10 subtracted from the reciprocal of x is greater than 0.

17. The sum of -2 and x is not equal to 10. $-2 + x \neq 10$

18. Twice the quotient of y and 3 is less than or equal to y.

19. Twice the difference of x and 6 is greater than the reciprocal of 11. $2(x - 6) > \frac{1}{11}$

20. Four times the sum of 5 and x is not equal to the opposite of 15. $4(5 + x) \neq -15$

21. 7 subtracted from the product of 5 and y is 6. $5y - 7 = 6$

22. The sum of z and w, divided by 2, is 12.

23. Twice the difference of x and 6 is -27 $2(x - 6) = -27$

24. 5 times the sum of 6 and y is -35. $5(6 + y) = -35$

Fill in the chart. See Examples 4 and 5.

	Number	**Opposite**	**Reciprocal**
25.	5	-5	$\dfrac{1}{5}$
26.	7	-7	$\dfrac{1}{7}$
27.	-8	8	$-\dfrac{1}{8}$
28.	-4	4	$-\dfrac{1}{4}$
29.	$-\dfrac{1}{7}$	$\dfrac{1}{7}$	-7
30.	$\dfrac{1}{11}$	$-\dfrac{1}{11}$	11
31.	0	0	undefined
32.	1	-1	1
33.	$\dfrac{7}{8}$	$-\dfrac{7}{8}$	$\dfrac{8}{7}$
34.	$-\dfrac{23}{5}$	$\dfrac{23}{5}$	$-\dfrac{5}{23}$

35. Name the only real number that has no reciprocal, and explain why this is so.

36. Name the only real number that is its own opposite, and explain why this is so.   zero; answers may vary

Use a commutative property to write an equivalent expression. See Example 6.

37. $7x + y$ $y + 7x$

38. $3a + 2b$ $2b + 3a$

39. $z \cdot w$ $w \cdot z$

40. $r \cdot s$ $s \cdot r$

41. $\dfrac{1}{3} \cdot \dfrac{x}{5}$ $\dfrac{x}{5} \cdot \dfrac{1}{3}$

42. $\dfrac{x}{2} \cdot \dfrac{9}{10}$ $\dfrac{9}{10} \cdot \dfrac{x}{2}$

43. Is subtraction commutative? Explain why or why not.   no; answers may vary

44. Is division commutative? Explain why or why not.   no; answers may vary

11. $\dfrac{n}{5} = 4n$ **12.** $\dfrac{8}{y} = y + 3$ **14.** $5 + 2q = q + 4$ **15.** $7x \leq -21$ **16.** $\dfrac{1}{x} - 10 > 0$ **18.** $2\left(\dfrac{y}{3}\right) \leq y$ **22.** $\dfrac{z + w}{2} = 12$ **35.** Zero. For

every real number x, $0 \cdot x \neq 1$, so 0 has no reciprocal. It is the only real number that has no reciprocal because if $x \neq 0$, then $x \cdot \dfrac{1}{x} = 1$ by definition.

Use an associative property to write an equivalent expression. See Example 7. **47.** $x + (1.2 + y)$ **48.** $(5q + 2r) + s$

 45. $5 \cdot (7x)$ $(5 \cdot 7)x$ **46.** $3 \cdot (10z)$ $(3 \cdot 10)z$

47. $(x + 1.2) + y$ **48.** $5q + (2r + s)$

49. $(14z) \cdot y$ $14(z \cdot y)$ **50.** $(9.2x) \cdot y$ $9.2(x \cdot y)$

51. Evaluate $12 - (5 - 3)$ and $(12 - 5) - 3$. Use these two expressions and discuss whether subtraction is associative.

52. Evaluate $24 \div (6 \div 3)$ and $(24 \div 6) \div 3$. Use these two expressions and discuss whether division is associative.

Use the distributive property to find the product. See Example 8.

 53. $3(x + 5)$ $3x + 15$ **54.** $7(y + 2)$ $7y + 14$

55. $-(2a + b)$ $-2a - b$ **56.** $-(c + 7d)$ $-c - 7d$

57. $2(6x + 5y + 2z)$ **58.** $5(3a + b + 9c)$

59. $-4(x - 2y + 7)$ **60.** $-10(2a - 3b - 4)$

61. $0.5x(6y - 3)$ **62.** $1.2m(9n - 4)$

Complete the statement to illustrate the given property.

63. $3x + 6 = \underline{6 + 3x}$ Commutative property of addition

64. $8 + 0 = \underline{8}$ Additive identity property

65. $\frac{2}{3} + \left(-\frac{2}{3}\right) = \underline{0}$ Additive inverse property

66. $4(x + 3) = \underline{4x + 12}$ Distributive property

67. $7 \cdot 1 = \underline{7}$ Multiplicative identity property

68. $0 \cdot (-5.4) = \underline{0}$ Multiplication property of zero

69. $10(2y) = \underline{(10 \cdot 2)y}$ Associative property

70. $9y + (x + 3z) = \underline{(9y + x) + 3z}$ Associative property

Write the property illustrated on each screen. See Example 9.

71.   **72.**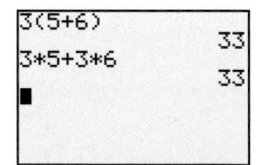

71. Associative property of addition **72.** Distributive property

73.   **74.**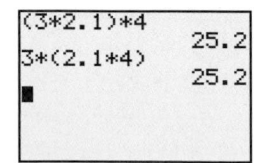

73. Commutative property of multiplication
74. Associative property of multiplication

75. To demonstrate the distributive property geometrically, represent the area of the larger rectangle in two ways: First as length a times width $b + c$, and second as the sum of the areas of the smaller rectangles. $a(b + c) = ab + ac$

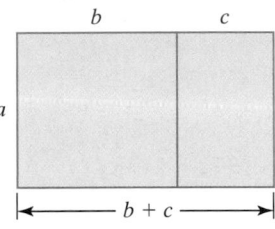

Write each of the following as an algebraic expression. See Examples 10 and 11.

76. Write an expression for the amount of money (in dollars) in n nickels. $0.05n$

77. Write an expression for the amount of money (in dollars) in d dimes. $0.1d$

78. Two numbers have a sum of 25. If one number is x, represent the other number as an expression in x. $25 - x$

79. Two numbers have a sum of 112. If one number is x, represent the other number as an expression in x. $112 - x$

80. Two angles are supplementary if the sum of their measures is 180°. If the measure of one angle is x degrees, represent the measure of the other angle as an expression in x. $180 - x$

81. If the measure of an angle is $5x$ degrees, represent the measure of its complement as an expression in x. $90 - 5x$

82. The cost of x compact discs if each compact disc costs $6.49.

83. The cost of y books if each book costs $35.61. $35.61y$

84. If x is an odd integer, represent the next odd integer as an expression in x. $x + 2$

85. If $2x$ is an even integer, represent the next even integer as an expression in x. $2x + 2$

Simplify each expression. See Examples 8, 12, and 13.

 86. $-9 + 4x + 18 - 10x$ **87.** $5y - 14 + 7y - 20y$

88. $5k - (3k - 10)$ **89.** $-11c - (4 - 2c)$

90. $(3x + 4) - (6x - 1)$ **91.** $(8 - 5y) - (4 + 3y)$

92. $3(xy - 2) + xy + 15 - x^2$ $-x^2 + 4xy + 9$

93. $-4(yz + 3) - 7yz + 1 + y^2$ $y^2 - 11yz - 11$

94. $-(n + 5) + (5n - 3)$ **95.** $-(8 - t) + (2t - 6)$

96. $4(6n^2 - 3) - 3(8n^2 + 4)$ -24

97. $5(2z^3 - 6) + 10(3 - z^3)$ 0

---

51. 10 and 4. Subtraction is not associative. **52.** 12 and $\frac{4}{3}$. Division is not associative. **57.** $12x + 10y + 4z$ **58.** $15a + 5b + 45c$
59. $-4x + 8y - 28$ **60.** $-20a + 30b + 40$ **61.** $3xy - 1.5x$ **62.** $10.8mn - 4.8m$ **82.** $6.49x$ **86.** $-6x + 9$ **87.** $-8y - 14$
88. $2k + 10$ **89.** $-9c - 4$ **90.** $-3x + 5$ **91.** $4 - 8y$ **94.** $4n - 8$ **95.** $3t - 14$

98. $3x - 2(x - 5) + x$ $2x + 10$ **99.** $7n + 3(2n - 6) - 2$

100. $1.5x + 2.3 - 0.7x - 5.9$ $0.8x - 3.6$

101. $6.3y - 9.7 + 2.2y - 11.1$ $8.5y - 20.8$

102. $\dfrac{3}{4}b - \dfrac{1}{2} + \dfrac{1}{6}b - \dfrac{2}{3}$ **103.** $\dfrac{7}{8}a - \dfrac{11}{12} - \dfrac{1}{2}a + \dfrac{5}{6}$

104. $\dfrac{1}{6}(24a - 18b) - \dfrac{1}{7}(7a - 21b - 2) - \dfrac{1}{5}$ $3a + \dfrac{3}{35}$

105. $\dfrac{1}{3}(6x - 33y) - \dfrac{1}{8}(24x - 40y + 1) - \dfrac{1}{3}$ $-x - 6y - \dfrac{11}{24}$

Concept Extensions

Simplify each expression.

106. $-1.2(5.7x - 3.6) + 8.75x$ $1.91x + 4.32$

107. $5.8(-9.6 - 31.2y) - 18.65$ $-180.96y - 74.33$

108. $8.1z + 7.3(z + 5.2) - 6.85$ $15.4z + 31.11$

109. $6.5y - 4.4(1.8x - 3.3) + 10.95$ $6.5y - 7.92x + 25.47$

△ **110. GROUP ACTIVITY.** Do figures with the same surface area always have the same volume? To see, take two $8\frac{1}{2}$-by-11-inch sheets of paper and construct two cylinders using the following figures as a guide. Working with a partner, measure the height and the radius of each resulting cylinder and use the expression $\pi r^2 h$ to approximate each volume to the nearest tenth of a cubic inch. Explain your results.   no

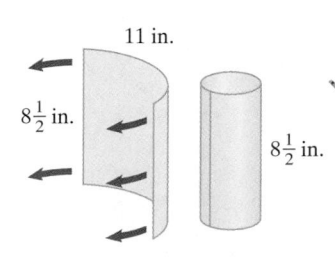

Cylinder 1   Cylinder 2

Volume = 63.2 cu. in.   Volume = 81.8 cu. in.

△ **111.** Use the same idea as in Exercise 110, work with a partner, and discover whether two rectangles with the same perimeter always have the same area. Explain your results.   no

The following graph is called a broken-line graph, or simply a line graph. This particular graph shows the past, present, and future predicted U.S. population over 65. Just as with a bar graph, to find the population over 65 for a particular year, read the height of the corresponding point. To read the height, follow the point horizontally to the left until you reach the vertical axis. Use this graph to answer Exercises 112 through 117.

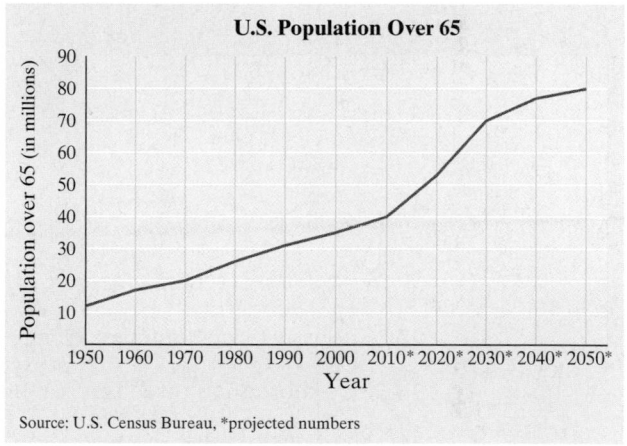

Source: U.S. Census Bureau, *projected numbers

112. Estimate the population over 65 in the year 1970.   20 million

113. Estimate the predicted population over 65 in the year 2050.   80 million

114. Estimate the predicted population over 65 in the year 2030.   70 million

115. Estimate the population over 65 in the year 2000.   35 million

116. Is the population over 65 increasing as time passes or decreasing? Explain how you arrived at your answer.   increasing; answers may vary

117. The percent of Americans over 65 in 1950 was 8.1%. The percent of Americans over 65 in 2050 is expected to be 2.5 times the percent over 65 in 1950. Estimate the percent of Americans expected to be over age 65 in 2050.   20.25%

99. $13n - 20$ **102.** $\dfrac{11}{12}b - \dfrac{7}{6}$ **103.** $\dfrac{3}{8}a - \dfrac{1}{12}$

| INTEGRATED REVIEW | ALGEBRAIC EXPRESSIONS, OPERATIONS ON REAL NUMBERS, AND PROPERTIES |

Fine the value of each expression when $x = -1$, $y = 3$, and $z = -4$.

1. z^2 16   **2.** $-z^2$ -16 **3.** $\dfrac{4x - z}{2y}$ 0   **4.** $x(y - 2z)$ -11

Perform indicated operations.

5. $-7 - (-2)$ -5 **6.** $\dfrac{9}{10} - \dfrac{11}{12}$ $-\dfrac{1}{60}$

7. $\dfrac{-13}{2 - 2}$ undefined   **8.** $(1.2)^2 - (2.1)^2$ -2.97

9. $\sqrt{64} - \sqrt[3]{64}$ 4   **10.** $-5^2 - (-5)^2$ -50

11. $9 + 2[(8 - 10)^2 + (-3)^2]$ 35   **12.** $8 - 6[\sqrt[3]{8}(-2) + \sqrt{4}(-5)]$ 92

Write each phrase as an algebraic expression. Use x to represent each unknown number.

13. Subtract twice a number from -15. $-15 - 2x$

14. Five more than three times a number. $3x + 5$

15. Name the whole number that is not a natural number.   0

16. True or false: A real number is either a rational number or an irrational number, but never both.   True

Use properties of real numbers to simplify each expression.

17. $-5(9x)$ $-45x$ **18.** $(3x - 7) - (4x + 1)$ $-x - 8$

19. $8.6a + 2.3b - a + 4.9b$ $7.6a + 7.2b$ **20.** $\dfrac{2}{3}y - \dfrac{2}{3} + y - \dfrac{1}{9}y + \dfrac{9}{10}$ $\dfrac{14}{9}y + \dfrac{7}{30}$

| **1.5** | *SOLVING LINEAR EQUATIONS ALGEBRAICALLY* |

Objectives

1.   Solve linear equations using properties of equality.
2.   Solve linear equations that can be simplified by combining like terms.
3.   Solve linear equations containing fractions.
4.   Recognize when an equation is an identity and when it has no solution.

1 Linear equations model many real-life problems. For example, we can use a linear equation to calculate the increase in households (in millions) with digital cameras.

With the help of your computer, digital cameras allow you to see your pictures and make copies immediately, send them in e-mail or use them on a Web page. Current projected number of households with these cameras are shown in the graph below.

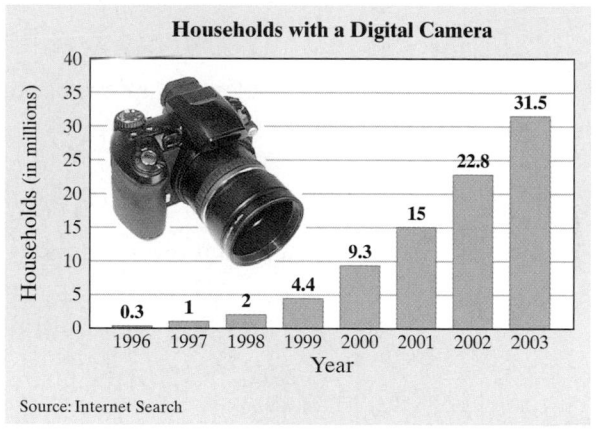

Households with a Digital Camera

Source: Internet Search

To find the increase in households from 2002 to 2003, for example, we can use the equation below.

In words:

Increase in households	is	households in 2003	minus	households in 2002
Translate: x	$=$	31.5	$-$	22.8

Since our variable x (increase in households) is by itself on one side of the equation, we can find the value of x by simplifying the right side.

$$x = 8.7$$

The increase in households with digital cameras from 2002 to 2003 is $8.7 million.

The **equation**, $x = 31.5 - 22.8$, like every other equation, is a statement that two expressions are equal. Oftentimes, the unknown variable is not by itself on one side of the equation. In these cases, we will use properties of equality to write equivalent equations so that a solution many be found. This is called **solving the equation**. In this section, we concentrate on solving equations such as this one, called **linear equations** in one variable. Linear equations are also called **first-degree equations** since the exponent on the variable is 1.

Linear Equations in One Variable

$$3x = -15 \qquad 7 - y = 3y \qquad 4n - 9n + 6 = 0 \qquad z = -2$$

TEACHING TIP

Consider having students read the box explaining the addition property of equality and the multiplication property of equality, write what they think the properties mean, and then have a few of them share their thoughts with the class.

Linear Equations in One Variable

A linear equation in one variable is an equation that can be written in the form

$$ax + b = c$$

where a, b, and c are real numbers and $a \neq 0$.

When a variable in an equation is replaced by a number and the resulting equation is true, then that number is called a **solution** of the equation. For example, 1 is a solution of the equation $3x + 4 = 7$, since $3(1) + 4 = 7$ is a true statement. But 2 is not a solution of this equation, since $3(2) + 4 = 7$ is not a true statement. The **solution set** of an equation is the set of solutions of the equation. For example, the solution set of $3x + 4 = 7$ is $\{1\}$.

To **solve an equation** is to find the solution set of an equation. Equations with the same solution set are called **equivalent equations**. For example,

$$3x + 4 = 7 \qquad 3x = 3 \qquad x = 1$$

are equivalent equations because they all have the same solution set, namely $\{1\}$. To solve an equation in x, we start with the given equation and write a series of simpler equivalent equations until we obtain an equation of the form

$$x = \textbf{number}$$

Two important properties are used to write equivalent equations.

The Addition and Multiplication Properties of Equality

If a, b, and c, are real numbers, then

$$a = b \quad \text{and} \quad a + c = b + c \text{ are equivalent equations.}$$

Also, $a = b$ and $ac = bc$ are equivalent equations as long as $c \neq 0$.

The **addition property of equality** guarantees that the same number may be added to both sides of an equation, and the result is an equivalent equation. The **multiplication property of equality** guarantees that both sides of an equation may be multiplied by the same nonzero number, and the result is an equivalent equation. Because we define subtraction in terms of addition $(a - b = a + (-b))$, and division in terms of multiplication $\left(\dfrac{a}{b} = a \cdot \dfrac{1}{b}\right)$, these properties also guarantee that we may *subtract* the same number from both sides of an equation, or *divide* both sides of an equation by the same nonzero number and the result is an equivalent equation.

For example, to solve $2x + 5 = 9$, use the addition and multiplication properties of equality to isolate x—that is, to write an equivalent equation of the form

$$x = \text{number}$$

EXAMPLE 1

Solve for x: $2x + 5 = 9$.

Solution   First, use the addition property of equality and subtract 5 from both sides. We do this so that our only variable term, $2x$, is by itself on one side of the equation.

CLASSROOM EXAMPLE

Solve: $3x + 6 = 12$.
answer: 2

$$2x + 5 = 9$$
$$2x + 5 - 5 = 9 - 5 \qquad \text{Subtract 5 from both sides.}$$
$$2x = 4 \qquad \text{Simplify.}$$

Now that the variable term is isolated, we can finish solving for x by using the multiplication property of equality and dividing both sides by 2.

$$\frac{2x}{2} = \frac{4}{2} \qquad \text{Divide both sides by 2.}$$
$$x = 2 \qquad \text{Simplify.}$$

Check   To see that 2 is the solution, replace x in the original equation with 2 and see that a true statement results using any method below.

$$2x + 5 = 9 \qquad \text{Original equation}$$

$2(2) + 5 = 9$ Let $x = 2$.

$4 + 5 = 9$

$9 = 9$ True

Let $x = 2$.

The left side of the equation evaluates to 9, the value of the right side.

Store 2 in x and evaluate the left side of the equation.

The left side of the equation evaluates to 9, the value of the right side.

Since we arrive at a true statement, 2 is the solution or the solution set is $\{2\}$.

EXAMPLE 2

Solve: $0.6 = 2 - 3.5c$.

Solution   We use both the addition property and the multiplication property of equality.

$$0.6 = 2 - 3.5c$$

$0.6 - 2 = 2 - 3.5c - 2$ Subtract 2 from both sides.

$-1.4 = -3.5c$ Simplify. The variable term is now isolated.

$\dfrac{-1.4}{-3.5} = \dfrac{-3.5c}{-3.5}$ Divide both sides by -3.5.

$0.4 = c$ Simplify $\dfrac{-1.4}{-3.5}$.

> **Helpful Hint**
> Don't forget that
>
> $0.4 = c$ and $c = 0.4$ are equivalent equations.
>
> We may solve an equation so that the variable is alone on either side of the equation.

Check

$$0.6 = 2 - 3.5c$$

$0.6 \overset{?}{=} 2 - 3.5(0.4)$ Replace c with 0.4.

$0.6 \overset{?}{=} 2 - 1.4$ Multiply.

$0.6 = 0.6$ True.

The solution is 0.4, or the solution set is $\{0.4\}$.

CLASSROOM EXAMPLE
Solve: $4.5 = 3 + 2.5x$.
answer: 0.6

---

2   Often, an equation can be simplified by removing any grouping symbols and combining any like terms.

EXAMPLE 3

Solve: $-6x - 1 + 5x = 3$.

Solution   First, the left side of this equation can be simplified by combining like terms $-6x$ and $5x$. Then use the addition property of equality and add 1 to both sides of the equation.

CLASSROOM EXAMPLE
Solve: $-2x + 2 - 4x = 20$.
answer: -3

$$-6x - 1 + 5x = 3$$

$-x - 1 = 3$ Combine like terms.

$-x - 1 + 1 = 3 + 1$ Add 1 to both sides of the equation.

$-x = 4$ Simplify.

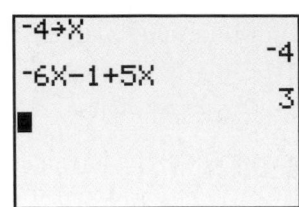

Notice that this equation is not solved for x since we have $-x$ or $-1x$, not x. To solve for x, divide both sides by -1.

$$\frac{-x}{-1} = \frac{4}{-1} \qquad \text{Divide both sides by } -1.$$

$$x = -4 \qquad \text{Simplify.}$$

Check to see that the solution is -4. A calculator check is shown to the left.

If an equation contains parentheses, use the distributive property to remove them.

EXAMPLE 4

Solve: $2(x - 3) = 5x - 9$.

Solution   First, use the distributive property.

CLASSROOM EXAMPLE
Solve: $4(x - 2) = 6x - 10$.
answer: 1

$$2(x - 3) = 5x - 9$$

$$2x - 6 = 5x - 9 \qquad \text{Use the distributive property.}$$

Next, get variable terms on the same side of the equation by subtracting $5x$ from both sides.

$$2x - 6 - 5x = 5x - 9 - 5x \qquad \text{Subtract } 5x \text{ from both sides.}$$

$$-3x - 6 = -9 \qquad \text{Simplify.}$$

$$-3x - 6 + 6 = -9 + 6 \qquad \text{Add 6 to both sides.}$$

$$-3x = -3 \qquad \text{Simplify.}$$

$$\frac{-3x}{-3} = \frac{-3}{-3} \qquad \text{Divide both sides by } -3.$$

$$x = 1$$

Check   Let $x = 1$ in the original equation and check that a true statement results.

$$2(x - 3) = 5x - 9 \qquad \text{Original equation}$$

To use a calculator to check, store 1 in x and evaluate the left side and the right side of the equation.

Left side of equation.
Right side of equation.

Same value, so 1 is the solution.

The solution is 1.

3   If an equation contains fractions, we first clear the equation of fractions by multiplying both sides of the equation by the *least common denominator* (LCD) of all fractions in the equation.

EXAMPLE 5

Solve for y: $\dfrac{y}{3} - \dfrac{y}{4} = \dfrac{1}{6}$.

Solution

First, clear the equation of fractions by multiplying both sides of the equation by 12, the LCD of denominators 3, 4, and 6.

$$\frac{y}{3} - \frac{y}{4} = \frac{1}{6}$$

$$12\left(\frac{y}{3} - \frac{y}{4}\right) = 12\left(\frac{1}{6}\right) \qquad \text{Multiply both sides by the LCD 12.}$$

$$12\left(\frac{y}{3}\right) - 12\left(\frac{y}{4}\right) = 2 \qquad \text{Apply the distributive property.}$$

$$4y - 3y = 2 \qquad \text{Simplify.}$$

$$y = 2 \qquad \text{Simplify.}$$

Check

To check using a calculator see the screen below. The solution is 2.

```
2→Y:Y/3-Y/4
         .1666666667
Ans▶Frac
               1/6
■
```

As a general guideline, the following steps may be used to solve a linear equation in one variable.

Solving A Linear Equation in One Variable

Step 1: Clear the equation of fractions by multiplying both sides of the equation by the least common denominator (LCD) of all denominators in the equation.

Step 2: Use the distributive property to remove grouping symbols such as parentheses.

Step 3: Combine like terms on each side of the equation.

Step 4: Use the addition property of equality to rewrite the equation as an equivalent equation with variable terms on one side and numbers on the other side.

Step 5: Use the multiplication property of equality to isolate the variable.

Step 6: Check the proposed solution in the original equation.

EXAMPLE 6

Solve for x: $\dfrac{x+5}{2} + \dfrac{1}{2} = 2x - \dfrac{x-3}{8}$.

Solution   Multiply both sides of the equation by 8, the LCD of 2 and 8.

$$8\left(\frac{x+5}{2} + \frac{1}{2}\right) = 8\left(2x - \frac{x-3}{8}\right) \qquad \text{Multiply both sides by } 8.$$

$$8\left(\frac{x+5}{2}\right) + 8 \cdot \frac{1}{2} = 8 \cdot 2x - 8\left(\frac{x-3}{8}\right) \qquad \text{Apply the distributive property.}$$

$$4(x+5) + 4 = 16x - (x-3) \qquad \text{Simplify.}$$

$$4x + 20 + 4 = 16x - x + 3 \qquad \begin{array}{l}\text{Use the distributive property} \\ \text{to remove parentheses.}\end{array}$$

$$4x + 24 = 15x + 3 \qquad \text{Combine like terms.}$$

$$-11x + 24 = 3 \qquad \text{Subtract } 15x \text{ from both sides.}$$

$$-11x = -21 \qquad \text{Subtract } 24 \text{ from both sides.}$$

$$\frac{-11x}{-11} = \frac{-21}{-11} \qquad \text{Divide both sides by } -11.$$

$$x = \frac{21}{11} \qquad \text{Simplify.}$$

To check, verify that replacing x with $\dfrac{21}{11}$ makes the original equation true. If you use a calculator to verify, make sure that parentheses are placed about a numerator or denominator that contains more than one term, as shown to the left. The solution is $\dfrac{21}{11}$.

> **Helpful Hint**
>
> When we multiply both sides of an equation by a number, the distributive property tells us that each term of the equation is multiplied by the number.

```
21/11→X
         1.909090909
(X+5)/2+1/2
         3.954545455
2X-(X-3)/8
         3.954545455
■
```

If an equation contains decimals, you may want to first clear the equation of decimals.

EXAMPLE 7

Solve: $0.3x + 0.1 = 0.27x - 0.02$.

Solution   To clear this equation of decimals, we multiply both sides of the equation by 100. Recall that multiplying a number by 100 moves its decimal point two places to the right.

$$100(0.3x + 0.1) = 100(0.27x - 0.02)$$

$$100(0.3x) + 100(0.1) = 100(0.27x) - 100(0.02) \qquad \text{Use the distributive property.}$$

$$30x + 10 = 27x - 2 \qquad \text{Multiply.}$$

$$30x - 27x = -2 - 10 \qquad \begin{array}{l}\text{Subtract } 27x \text{ and } 10 \text{ from} \\ \text{both sides.}\end{array}$$

$$3x = -12 \qquad \text{Simplify.}$$

$$\frac{3x}{3} = \frac{-12}{3} \qquad \text{Divide both sides by } 3.$$

$$x = -4 \qquad \text{Simplify.}$$

Check to see that the solution is -4. A calculator check is shown to the left.

```
-4→X
              -4
.3X+.1
            -1.1
.27X-.02
            -1.1
■
```

✔ **CONCEPT CHECK**

Explain what is wrong with the following:

$$3x - 5 = 16$$
$$3x = 11$$
$$\frac{3x}{3} = \frac{11}{3}$$
$$x = \frac{11}{3}$$

4   So far, each linear equation that we have solved has had a single solution. A linear equation in one variable that has exactly one solution is called a **conditional equation**. We will now look at two other types of equations: contradictions and identities.

An equation in one variable that has no solution is called a **contradiction**, and an equation in one variable that has every number (for which the equation is defined) as a solution is called an **identity**. The next examples show how to recognize contradictions and identities.

EXAMPLE 8

Solve for x: $3x + 5 = 3(x + 2)$.

Solution   First, use the distributive property and remove parentheses.

$$3x + 5 = 3(x + 2)$$
$$3x + 5 = 3x + 6 \qquad \text{Apply the distributive property.}$$
$$3x + 5 - 3x = 3x + 6 - 3x \qquad \text{Subtract } 3x \text{ from both sides.}$$
$$5 = 6$$

> **Helpful Hint**
>
> A solution set of $\{0\}$ and a solution set of $\{\ \}$ are not the same. The solution set $\{0\}$ means 1 solution, 0. The solution set $\{\ \}$ means no solution.

The equation $5 = 6$ is a false statement no matter what value the variable x might have. Thus, the original equation has no solution. Its solution set is written either as $\{\ \}$ or $\emptyset$. This equation is a contradiction.

EXAMPLE 9

Solve for x: $6x - 4 = 2 + 6(x - 1)$.

Solution   First, use the distributive property and remove parentheses.

$$6x - 4 = 2 + 6(x - 1)$$
$$6x - 4 = 2 + 6x - 6 \qquad \text{Apply the distributive property.}$$
$$6x - 4 = 6x - 4 \qquad \text{Combine like terms.}$$

At this point we might notice that both sides of the equation are the same, so replacing x by any real number gives a true statement. Thus the solution set of this equation is the set of real numbers, and the equation is an identity. Continuing to "solve" $6x - 4 = 6x - 4$, we eventually arrive at the same conclusion.

Concept Check Answer:

$$3x - 5 = 16$$
$$3x = 21$$
$$x = 7$$

Therefore the correct solution set is $\{7\}$.

$$6x - 4 + 4 = 6x - 4 + 4 \qquad \text{Add 4 to both sides.}$$
$$6x = 6x \qquad \text{Simplify.}$$
$$6x - 6x = 6x - 6x \qquad \text{Subtract } 6x \text{ from both sides.}$$
$$0 = 0 \qquad \text{Simplify.}$$

Since $0 = 0$ is a true statement for every value of x, all real numbers are solutions. The solution set is the set of all real numbers or, $\mathbb{R}$, $\{x \mid x \text{ is a real number}\}$, and the equation is called an identity.

> **Helpful Hint**
>
> For linear equations, *any* false statement such as $5 = 6, 0 = 1$, or $-2 = 2$ informs us that the original equation has no solution. Also, *any* true statement such as $0 = 0, 2 = 2$, or $-5 = -5$ informs us that the original equation is an identity.

STUDY SKILLS REMINDER

This is a special reminder that will be repeated and expanded throughout this text. It is very important for you to be able to recognize and solve different types of equations and inequalities. To help you do this, we will begin an outline below and continually expand this outline as different equations and inequalities are introduced. Although suggestions will be given, this outline should be in your own words and you should include at least "how to recognize" and "how to begin to solve" under each letter heading.

For example:

Solving Equations and Inequalities

I. Equations

    A. **Linear equations**—Recognize: *power on variable is 1 when there are no variables in denominator*—Solve: simplify (if fractions, multiply by LCD) and move variable terms to one side of the equation, constants to the other side.

II. Inequalities

See Appendix A for exercises.

MENTAL MATH

Simplify each expression by combining like terms.

1. $3x + 5x + 6 + 15$ $8x + 21$

2. $8y + 3y + 7 + 11$ $11y + 18$

3. $5n + n + 3 - 10$ $6n - 7$

4. $m + 2m + 4 - 8$ $3m - 4$

5. $8x - 12x + 5 - 6$ $-4x - 1$

6. $4x - 10x + 13 - 16$ $-6x - 3$

Identify each as an equation or an expression.

7. $\frac{1}{3}x - 5$ expression

8. $2(x - 3) = 7$ equation

9. $\frac{5}{9}x + \frac{1}{3} = \frac{2}{9} - x$ equation

10. $\frac{5}{9}x + \frac{1}{3} - \frac{2}{9} - x$ expression

11. all real numbers   **12.** no solution   **13.** no solution   **14.** all real numbers

Decide which equations have no solution and which equations have all real numbers as solutions.

11. $2x + 3 = 2x + 3$

12. $2x + 1 = 2x + 3$

13. $5x - 2 = 5x - 7$

14. $5x - 3 = 5x - 3$

A Group Activity for this section is available in the Instructor's Resource Manual.

EXERCISE SET 1.5

STUDY   CD/   PH MATH   MathXL®Tutorials   MathXL®   MyMathLab®
GUIDE/SSM   VIDEO   TUTOR CENTER   ON CD

Solve for the variable. See Examples 1 and 2.

1. $-3x = 36$ -12

2. $8x = -40$ -5

3. $x + 2.8 = 1.9$ -0.9

4. $y - 8.6 = -6.3$ 2.3

5. $5x - 4 = 26$ 6

6. $2y - 3 = 11$ 7

7. $-4 = 3x + 11$ -5

8. $-9 = 5x + 11$ -4

9. $-4.1 - 7z = 3.6$ -1.1

10. $10.3 - 6x = -2.3$ 2.1

11. $5y + 12 = 2y - 3$ -5

12. $4x + 14 = 6x + 8$ 3

Solve for the variable. See Examples 3 and 4.

13. $8x - 5x + 3 = x - 7 + 10$ 0

14. $6 + 3x + x = -x + 2 - 26$ -6

15. $5x + 12 = 2(2x + 7)$ 2

16. $2(x + 3) = x + 5$ -1

17. $3(x - 6) = 5x$ -9

18. $6x = 4(5 + x)$ 10

19. $-2(5y - 1) - y = -4(y - 3)$ $-\frac{10}{7}$

20. $-3(2w - 7) - 10 = 9 - 2(5w + 4)$ $-\frac{5}{2}$

21. **a.** Simplify the expression $4(x + 1) + 1$. $4x + 5$

    b. Solve the equation $4(x + 1) + 1 = -7$. -3

    c. Explain the difference between solving an equation for a variable and simplifying an expression.

22. Explain why the multiplication property of equality does not include multiplying both sides of an equation by 0. (*Hint:* Write down a false statement and then multiply both sides by 0. Is the result true or false? What does this mean?)

21. c. answers may vary   **22.** answers may vary

Solve for the variable. See Examples 5 through 7.

23. $\dfrac{x}{2} + \dfrac{2}{3} = \dfrac{3}{4}$ $\dfrac{1}{6}$

24. $\dfrac{x}{2} + \dfrac{x}{3} = \dfrac{5}{2}$ 3

25. $\dfrac{3t}{4} - \dfrac{t}{2} = 1$ 4

26. $\dfrac{4r}{5} - 7 = \dfrac{r}{10}$ 10

27. $\dfrac{n - 3}{4} + \dfrac{n + 5}{7} = \dfrac{5}{14}$ 1

28. $\dfrac{2 + h}{9} + \dfrac{h - 1}{3} = \dfrac{1}{3}$ 1

29. $0.6x - 10 = 1.4x - 14$ 5

30. $0.3x + 2.4 = 0.1x + 4$ 8

Solve the following. See Examples 8 and 9.

31. $4(n + 3) = 2(6 + 2n)$ all real numbers

32. $6(4n + 4) = 8(3 + 3n)$ all real numbers

33. $3(x - 1) + 5 = 3x + 7$ $\varnothing$

34. $5x - (x + 4) = 5 + 4(x - 2)$ $\varnothing$

35. In your own words, explain why the equation $x + 7 = x + 6$ has no solution while the solution set of the equation $x + 7 = x + 7$ contains all real numbers.

    answers may vary

36. In your own words, explain why the equation $x = -x$ has one solution, namely 0, while the solution set of the equation $x = x$ is all real numbers. answers may vary

Each screen shows a calculator check of a proposed solution to an equation. Write the equation and the verified solution.

37.

38.

39.

40.

37. $6x - 5 = 4x - 21$; $x = -8$
38. $7x - 2 = 5x$; $x = 1$
39. $3(N - 3) + 2 = -1 + N$; $N = 3$

MIXED PRACTICE

Solve the following.

41. $6x + 9 = 51$ 7

42. $4x + 11 = 47$ 9

43. $-5x + 1.5 = -19.5$ 4.2

44. $-3x - 4.7 = 11.8$ -5.5

45. $x - 10 = -6x + 4$ 2

46. $4x - 7 = 2x - 7$ 0

47. $3x - 4 - 5x = x + 4 + x$ -2

48. $13x - 15x + 8 = 4x + 2 - 24$ 5

49. $5(y + 4) = 4(y + 5)$ 0

50. $6(y - 4) = 3(y - 8)$ 0

40. $7(C + 2) = 5C + 4$: $C = -5$

51. $-1.2x + 20 = -2.8x + 28$ 5

52. $-0.9x - 7.2 = -0.3x - 12$ 8

53. $6x - 2(x - 3) = 4(x + 1) + 4$ $\varnothing$

54. $10x - 2(x + 4) = 8(x - 2) + 6$ $\varnothing$

55. $\dfrac{3}{8} + \dfrac{b}{3} = \dfrac{5}{12}$ $\dfrac{1}{8}$

56. $\dfrac{a}{2} + \dfrac{7}{4} = 5$ $\dfrac{13}{2}$

57. $z + 3(2 + 4z) = 6(z + 1) + 5z$ 0

58. $4(m - 6) - m = 8(m - 3) - 5m$ all real numbers

59. $\dfrac{3t + 1}{8} = \dfrac{5 + 2t}{7} + 2$ 29

60. $4 - \dfrac{2z + 7}{9} = \dfrac{7 - z}{12}$ 19

61. $\dfrac{m - 4}{3} - \dfrac{3m - 1}{5} = 1$ -8

62. $\dfrac{n + 1}{8} - \dfrac{2 - n}{3} = \dfrac{5}{6}$ 3

63. $5(x - 2) + 2x = 7(x + 4) - 38$ all real numbers

64. $3x + 2(x + 4) = 5(x + 1) + 3$ all real numbers

65. $y + 0.2 = 0.6(y + 3)$ 4

66. $-(w + 0.2) = 0.3(4 - w)$ -2

67. $-(3x - 5) - (2x - 6) + 1 = -5(x - 1) - (3x + 2) + 3$
   67. -2

68. $-4(2x - 3) - (10x + 7) - 2 =$
$-(12x - 5) - (4x + 9) - 1$ 4

69. $2(x - 8) + x = 3(x - 6) + 2$ all real numbers

70. $4(x + 5) = 3(x - 4) + x$ $\varnothing$

71. $\dfrac{3x - 1}{9} + x = \dfrac{3x + 1}{3} + 4$ $\dfrac{40}{3}$

72. $\dfrac{2z + 7}{8} - 2 = z + \dfrac{z - 1}{2}$ $-\dfrac{1}{2}$

73. $1.5(4 - x) = 1.3(2 - x)$ 17

74. $2.4(2x + 3) = -0.1(2x + 3)$ -1.5

75. $-2(b - 4) - (3b - 1) = 5b + 3$ $\dfrac{3}{5}$

76. $4(t - 3) - 3(t - 2) = 2t + 8$ -14

77. $\dfrac{1}{3}(y + 4) + 5 = \dfrac{1}{4}(3y - 1) - 2$ $\dfrac{103}{5}$ or 20.6

78. $\dfrac{1}{5}(2y - 1) - 2 = \dfrac{1}{2}(3y - 5) + 3$ $-\dfrac{27}{11}$

Concept Extensions

Find the value of K such that the equations are equivalent.

79. $3.2x + 4 = 5.4x - 7$ $K = -11$
$3.2x = 5.4x + K$

80. $-7.6y - 10 = -1.1y + 12$ $K = 22$
$-7.6y = -1.1y + K$

81. $\dfrac{x}{6} + 4 = \dfrac{x}{3}$ $K = 24$
$x + K = 2x$

82. $\dfrac{5x}{4} + \dfrac{1}{2} = \dfrac{x}{2}$ $K = 2$
$5x + K = 2x$

Solve the following.

83. $x(x - 6) + 7 = x(x + 1)$ 1

84. $7x^2 + 2x - 3 = 6x(x + 4) + x^2$ $-\dfrac{3}{22}$

85. $3x(x + 5) - 12 = 3x^2 + 10x + 3$ 3

86. $x(x + 1) + 16 = x(x + 5)$ 4

Solve and check.

87. $2.569x = -12.48534$ -4.86

88. $-9.112y = -47.537304$ 5.217

89. $2.86z - 8.1258 = -3.75$ 1.53

90. $1.25x - 20.175 = -8.15$ 9.62

91. Recall from Section 1.3 that a game is fair if each team or player has an equal chance of winning. Cut or tear a sheet of paper into 10 pieces, numbering each piece from 1 to 10. Place the pieces into a bag. Draw 2 pieces from the bag, record their sum, and then return them to the bag. If the sum is 10 or less, player 1 gets a point. If their sum is more than 10, player 2 gets a point. Is this a fair game? Try it and see.   not a fair game

---

1.6   AN INTRODUCTION TO PROBLEM SOLVING

Objectives

1 Write algebraic expressions that can be simplified.

2 Apply the steps for problem solving.

1   In order to prepare for problem solving, we practice writing algebraic expressions that can be simplified.

Our first example involves consecutive integers and perimeter. Recall that *consecutive integers* are integers that follow one another in order. Study the examples of consecutive, even, and odd integers and their representations.

Consecutive Integers:

Consecutive Even Integers:

Consecutive Odd Integers:

EXAMPLE 1

Write the following as algebraic expressions. Then simplify.

a. The sum of three consecutive integers, if x is the first consecutive integer.

b. The perimeter of the triangle with sides of length $x, 5x,$ and $6x - 3$.

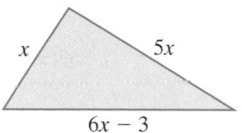

Solution

a. Recall that if x is the first integer, then the next consecutive integer is 1 more, or $x + 1$ and the next consecutive integer is 1 more than $x + 1$, or $x + 2$.

CLASSROOM EXAMPLE

Write as algebraic expressions. Then simplify.
a. The sum of two consecutive even integers, if x is the first even integer.
b. The perimeter of the rectangle with sides of length x and $2x - 1$.

answer: a. $2x + 2$ **b.** $6x - 2$

In words:

first integer	plus	next consecutive integer	plus	next consecutive integer

Translate: x $+$ $(x + 1)$ $+$ $(x + 2)$

Then $x + (x + 1) + (x + 2) = x + x + 1 + x + 2$

$= 3x + 3$ *Simplify by combining like terms.*

b. The perimeter of a triangle is the sum of the lengths of the sides.

In words:

side	+	side	+	side

Translate: x $+$ $5x$ $+$ $(6x - 3)$

Then $x + 5x + (6x - 3) = x + 5x + 6x - 3$

$= 12x - 3$ *Simplify.*

EXAMPLE 2

The three busiest airports in the United States are in the cities of Chicago, Atlanta, and Los Angeles. The airport in Atlanta has 15.3 million more arrivals and departures than the Los Angeles airport. The Chicago airport has 5.8 million more arrivals and departures than the Los Angeles airport. Write the sum of the arrivals and departures from these three cities as a simplified algebraic expression. Let x be the number of arrivals and departures at the Los Angeles airport. (*Source: The World Almanac,* 2003)

Solution

If x = millions of arrivals and departures at the Los Angeles airport, then

$x + 15.3$ = millions of arrivals and departures at the Atlanta airport and

$x + 5.8$ = millions of arrivals and departures at the Chicago airport

Since we want their sum, we have

In words:

arrivals and departures at Los Angeles	+	arrivals and departures at Atlanta	+	arrivals and departures at Chicago.

Translate: x $+$ $(x + 15.3)$ $+$ $(x + 5.8)$

Then $x + (x + 15.3) + (x + 5.8) = x + x + 15.3 + x + 5.8$

$= 3x + 21.1$ *Combine like terms.*

In Exercise 27, we will find the actual number of arrivals and departures at these airports.

2 Our main purpose for studying algebra is to solve problems. The following problem-solving strategy will be used throughout this text and may also be used to solve real-life problems that occur outside the mathematics classroom.

> **Helpful Hint**
>
> You may want to begin this section by studying key words and phrases and their translations in Sections 1.2 Objective 5 and 1.4 Objective 4.

> ### General Strategy for Problem Solving
>
> **1.** UNDERSTAND the problem. During this step, become comfortable with the problem. Some ways of doing this are:
>
> Read and reread the problem.
>
> Choose a variable to represent the unknown.
>
> Construct a drawing.
>
> Propose a solution and check. Pay careful attention to how you check your proposed solution. This will help when writing an equation to model the problem.
>
> **2.** TRANSLATE the problem into an equation.
>
> **3.** SOLVE the equation.
>
> **4.** INTERPRET the results. *Check* the proposed solution in the stated problem and *state* your conclusion.

Let's review this strategy by solving a problem involving unknown numbers.

EXAMPLE 3

FINDING UNKNOWN NUMBERS

Find two numbers such that the second number is 3 more than twice the first number and the sum of the two numbers is 72.

Solution

1. UNDERSTAND the problem. First let's read and reread the problem and then propose a solution. For example, if the first number is 25, then the second number is 3 more than twice 25, or 53. The sum of 25 and 53 is 78, not the required sum, but we have gained some valuable information about the problem. First, we know that the first number is less than 25 since our guess led to a sum greater than the required sum. Also, we have gained some information as to how to model the problem with an equation.

> **Helpful Hint**
>
> The purpose of proposing a solution is not to guess correctly but to gain confidence and to help understand the problem and how to model it.

Next let's assign a variable and use this variable to represent any other unknown quantities. If we let

the first number $= x$, then

the second number $= \underline{2x} + 3$

$\qquad\qquad\uparrow\qquad\uparrow$
$\qquad\qquad\qquad$ 3 more than
$\qquad$ twice the second number

2. TRANSLATE the problem into an equation. To do so, we use the fact that the sum of the numbers is 72. First let's write this relationship in words and then translate to an equation.

In words:

first number	added to	second number	is	72
↓	↓	↓	↓	↓

Translate: $\quad\quad x \quad\quad + \quad\quad (2x + 3) \quad\quad = \quad\quad 72$

3. SOLVE the equation.

$$x + (2x + 3) = 72$$
$$x + 2x + 3 = 72 \quad \text{Remove parentheses.}$$
$$3x + 3 = 72 \quad \text{Combine like terms.}$$
$$3x = 69 \quad \text{Subtract 3 from both sides.}$$
$$x = 23 \quad \text{Divide both sides by 3.}$$

TEACHING TIP
Remind students to be sure to answer the question that was asked. Then point out that the answer to a problem should include units if appropriate and is usually given as a complete sentence.

4. INTERPRET. Here, *we check* our work and *state* the solution. Recall that if the first number $x = 23$, then the second number $2x + 3 = 2 \cdot 23 + 3 = 49$.

Check: Is the second number 3 more than twice the first number? Yes, since 3 more than twice 23 is $46 + 3$, or 49. Also, their sum, $23 + 49 = 72$, is the required sum.

State: The two numbers are 23 and 49.

Many of today's rates and statistics are given as percents. Interest rates, tax rates, nutrition labeling, and percent of households in a given category are just a few examples. Before we practice solving problems containing percents, let's briefly take a moment and review the meaning of percent and how to find a percent of a number.

The word *percent* means "per hundred," and the symbol % is used to denote percent. This means that 23% is 23 per hundred, or $\frac{23}{100}$. Also,

CLASSROOM EXAMPLE
Find 48% of 200.
answer: 96

$$41\% = \frac{41}{100} = 0.41$$

To find a percent of a number, we multiply.

EXAMPLE 4

Find 16% of 25.

Solution To find 16% of 25, we find the product of 16% (written as a decimal) and 25.

$$16\% \cdot 25 = 0.16 \cdot 25$$
$$= 4$$

Thus, 16% of 25 is 4.

Next, we solve a problem containing percent.

✔ **CONCEPT CHECK**

Suppose you are finding 112% of a number x. Which of the following is a correct description of the result? Explain.

a. The result is less than x.
b. The result is equal to x.
c. The result is greater than x.

Concept Check Answer:
c; the result is greater than x

EXAMPLE 5

FINDING THE ORIGINAL PRICE OF A COMPUTER

Suppose that a computer store just announced an 8% decrease in the price of a particular computer model. If this computer sells for $2162 after the decrease, find the original price of this computer.

Solution

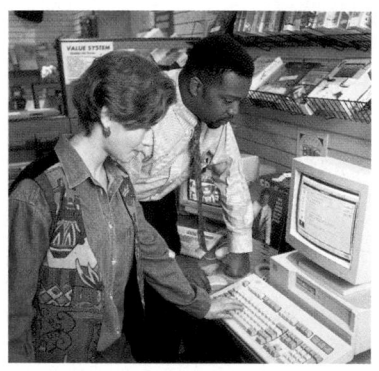

1. UNDERSTAND. Read and reread the problem. Recall that a percent decrease means a percent of the original price. Let's guess that the original price of the computer is $2500. The amount of decrease is then 8% of $2500, or $(0.08)(\$2500) = \$200$. This means that the new price of the computer is the original price minus the decrease, or $\$2500 - \$200 = \$2300$. Our guess is incorrect, but we now have an idea of how to model this problem. In our model, we will let x = the original price of the computer.

2. TRANSLATE.

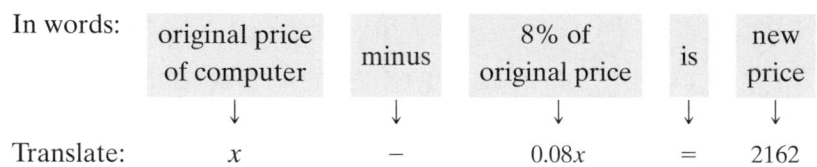

In words:

original price of computer	minus	8% of original price	is	new price
↓	↓	↓	↓	↓
x	$-$	$0.08x$	$=$	2162

Translate:

3. SOLVE the equation.

$$x - 0.08x = 2162$$
$$0.92x = 2162 \qquad \text{Combine like terms.}$$
$$x = \frac{2162}{0.92} = 2350 \qquad \text{Divide both sides by } 0.92.$$

4. INTERPRET.

Check: If the original price of the computer was $2350, the new price is

$$\$2350 - (0.08)(\$2350) = \$2350 - \$188$$
$$= \$2162 \qquad \text{The given new price.}$$

State: The original price of the computer was $2350.

⚠ **EXAMPLE 6**

FINDING THE LENGTHS OF A TRIANGLE'S SIDES

A pennant in the shape of an isosceles triangle is to be constructed for the Slidell High School Athletic Club and sold at a fund-raiser. The company manufacturing the pennant charges according to perimeter, and the athletic club has determined that a perimeter of 149 centimeters should make a nice profit. If each equal side of the triangle is twice the length of the third side, increased by 12 centimeters, find the lengths of the sides of the triangular pennant.

Solution

1. UNDERSTAND. Read and reread the problem. Recall that the perimeter of a triangle is the distance around. Let's guess that the third side of the triangular pennant is 20 centimeters. This means that each equal side is twice 20 centimeters, increased by 12 centimeters, or $2(20) + 12 = 52$ centimeters.

52 centimeters

20 centimeters

52 centimeters

This gives a perimeter of $20 + 52 + 52 = 124$ centimeters. Our guess is incorrect, but we now have a better understanding of how to model this problem.

Now we let the third side of the triangle $= x$

the first side $=$ twice   the third side   increased by 12
                      $\downarrow$ $\downarrow$ $\downarrow$
              $=$ 2 x $+$ 12,
              or $2x + 12$
the second side $= 2x + 12$

2x + 12

x

2x + 12

2. TRANSLATE.

In words: first $+$ second $+$ third $=$ 149
            side   side   side
            $\downarrow$ $\downarrow$ $\downarrow$ $\downarrow$

Translate: $(2x + 12)$ $+$ $(2x + 12)$ $+$ x $=$ 149

3. SOLVE the equation.

$$(2x + 12) + (2x + 12) + x = 149$$
$$2x + 12 + 2x + 12 + x = 149 \quad \text{Remove parentheses.}$$
$$5x + 24 = 149 \quad \text{Combine like terms.}$$
$$5x = 125 \quad \text{Subtract 24 from both sides.}$$
$$x = 25 \quad \text{Divide both sides by 5.}$$

TEACHING TIP
Remind students to be sure to answer the question that was asked. Then point out that the answer to a problem should include units if appropriate and is usually given as a complete sentence.

4. INTERPRET. If the third side is 25 centimeters, then the first side is $2(25) + 12 = 62$ centimeters and the second side is 62 centimeters also.

Check: The first and second sides are each twice 25 centimeters increased by 12 centimeters or 62 centimeters. Also, the perimeter is $25 + 62 + 62 = 149$ centimeters, the required perimeter.

State: The lengths of the sides of the triangle are 25 centimeters, 62 centimeters, and 62 centimeters.

EXAMPLE 7

Kelsey Ohleger was helping her friend Benji Burnstine study for an algebra exam. Kelsey told Benji that her three latest art history quiz scores are three consecutive even integers whose sum is 264. Help Benji find the scores.

Solution   **1.** UNDERSTAND. Read and reread the problem. Since we are looking for consecutive even integers, let

$$x = \text{the first integer. Then}$$

$$x + 2 = \text{the second consecutive even integer}$$

$$x + 4 = \text{the third consecutive even integer.}$$

2. TRANSLATE.

In words:

first integer	+	second even integer	+	third even integer	=	264
↓		↓		↓		↓

Translate: x + $(x + 2)$ + $(x + 4)$ =   264

3. SOLVE.

$$x + (x + 2) + (x + 4) = 264$$

$$3x + 6 = 264 \quad \text{Combine like terms.}$$

$$3x = 258 \quad \text{Subtract 6 from both sides.}$$

$$x = 86 \quad \text{Divide both sides by 3.}$$

4. INTERPRET. If $x = 86$, then $x + 2 = 86 + 2$ or 88, and $x + 4 = 86 + 4$ or 90.

Check: The numbers 86, 88, and 90 are three consecutive even integers. Their sum is 264, the required sum.

State: Kelsey's art history quiz scores are 86, 88, and 90.

CLASSROOM EXAMPLE

Find three consecutive integers whose sum is 378.

answer: 125, 126 and 127

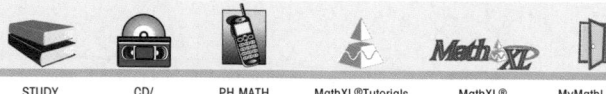

Spotlight on

DECISION ✗ MAKING

Suppose you are a reporter for the *Marston Gazette*, a daily newspaper serving the medium-sized industrial city of Marston. Your editor has assigned you to a feature story about food pantries, soup kitchens, and other efforts to alleviate hunger among the city's homeless, poor, and working poor.

While researching your story assignment, you find that, according to the U.S. Department of Agriculture, 10.2% of all households in the United States do not have access to enough food to meet their basic needs. A survey conducted by Marston Social Services reveals that approximately 4950 of the estimated 39,400 Marston households must cope with hunger.

Armed with these basic facts, you now need to decide what angle to take with your story. Which of the following approaches would you choose? Why? What other information would you want to consider?

a. Marston lags behind nation in fight against hunger.
b. Marston mirrors national hunger picture.
c. Marston makes progress against hunger.

TEACHING TIP

A Group Activity for this section is available in the Instructor's Resource Manual.

EXERCISE SET 1.6

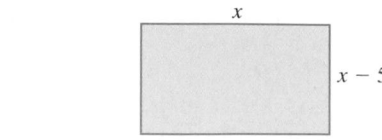

STUDY GUIDE/SSM   CD/VIDEO   PH MATH TUTOR CENTER   MathXL®Tutorials ON CD   MathXL®   MyMathLab®

Write the following as algebraic expressions. Then simplify. See Examples 1 and 2.

1. The perimeter of the square with side length y. $4y$

[square with side labeled y]

2. The perimeter of the rectangle with length x and width $x - 5$. $4x - 10$

[rectangle with length x and width $x - 5$]

3. The sum of three consecutive integers if the first integer is z. $3z + 3$

4. The sum of three consecutive odd integers if the first integer is x. $3x + 6$

5. The total amount of money (in cents) in x nickels and $(x + 3)$ dimes. (*Hint:* the value of a nickel is 5 cents and the value of a dime is 10 cents.) $(15x + 30)$ cents

6. The total amount of money (in cents) in y quarters and $(2y - 1)$ nickels. (Use the hint for Exercise 5.) $(35y - 5)$ cents

△ **7.** A piece of land along Bayou Liberty is to be fenced and subdivided as shown so that each rectangle has the same dimensions. Express the total amount of fencing needed as an algebraic expression in x. $10x + 3$

△ **8.** Write the perimeter of the floor plan shown as an algebraic expression in x. $2x + 14$

Solve. See Example 3.

9. Four times the difference of a number and 2 is the same as 6 times the number, increased by 2. Find the number. -5

10. Twice the sum of a number and 3 is the same as 1 subtracted from the number. Find the number. -7

🔒 **11.** One number is 5 times another number. If the sum of the two numbers is 270, find the numbers.   45, 225

12. One number is 6 less than another number. If the sum of the two numbers is 150, find the numbers.   72, 78

Solve. See Example 4.

13. Find 30% of 260.   78   **14.** Find 70% of 180.   126

15. Find 12% of 16.   1.92   **16.** Find 22% of 12.   2.64

Solve. See Examples 4 through 6.

17. The United States consists of 2271 million acres of land. Approximately 29% of this land is federally owned. Find the number of acres that are not federally owned. (*Source:* U.S. General Services Administration)   1612.41 million acres

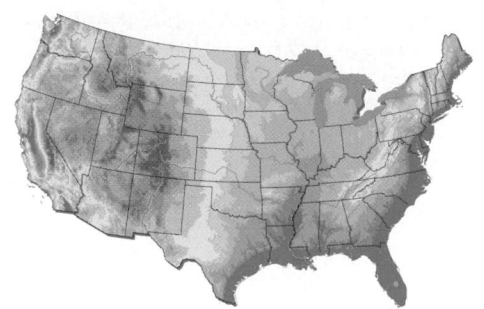

18. The state of Nevada contains the most federally owned acres of land in the United States. If 90% of the state's 70 million acres of land is federally owned, find the number of acres that are not federally owned. (*Source:* U.S. General Services Administration)   7 million acres

19. In 2000, a total of 2342 earthquakes occurred in the United States. Of these, 85% were minor tremors with magnitudes of 3.9 or less on the Richter scale. How many minor earthquakes occurred in the United States in 2000? Round to the nearest whole. (*Source*: U.S. Geological Survey National Earthquake Information Center)   1991 earthquakes

20. Of the 1071 tornadoes that occurred in the United States during 2000, 22.5% occurred during the month of May. How many tornadoes occurred during May 2000? Round to the nearest whole. (*Source:* Storm Prediction Center)   241 tornadoes

21. According to a recent survey, $33\frac{1}{3}$% of online shoppers in the United States say that they spend more than they intended when shopping online. In a group of 1290 online shoppers, how many would you expect don't spend more than they intended when shopping online? (*Source:* Cyber Dialogue, 04/11/2001)   860 shoppers

22. Only 6% of Americans eat a commercially prepared breakfast five times or more per week. As of 2000, Cincinnati, Ohio, had a population of 331,285 people. How many of these people would you expect do not eat a commercially prepared breakfast fives or more times per week? Round to the nearest whole. (*Sources:* National Restaurant Association, U.S. Census Bureau)   311,408 people

The following graph is called a circle graph or a pie chart. The circle represents a whole, or in this case, 100%. This particular graph shows the number of times e-mail users check their e-mail. Use this graph to answer Exercises 23 through 26.

How Often We Check Our E-mail

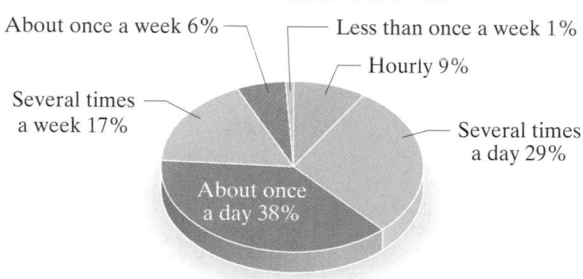

Source: The UCLA Internet Report: "Surveying the Digital Future,"
UCLA Center for Communication Policy

23. What percent of e-mail users check their e-mail several times per week? 17%

24. Among e-mail users, what is the most popular frequency for checking e-mail? once a day

25. If it is estimated that Fort Wayne, Indiana, has 112,500 e-mail users, how many of these would you expect check their e-mail about once a week? 6750 users

26. If it is estimated that the city of New Orleans has 265,000 e-mail users, how many of these would you expect check their e-mail either several times a day or hourly?
100,700 users

MIXED PRACTICE

Solve. See Examples 5 through 7.

27. The airports in Chicago, Atlanta, and Los Angeles have a total of 205.9 million annual arrivals and departures. Use this information and Example 2 in this section to find the number from each individual airport.

28. The perimeter of the triangle in Example 1b is 483 feet. Find the length of each side. 40.5 feet, 202.5 feet, 240 feet

29. America West Airlines has 3 models of Boeing aircraft in their fleet. The 737-300 contains 21 more seats than the 737-200. The 757-200 contains 36 less seats than twice the number of seats in the 737-200. Find the number of seats for each aircraft if the total number of seats for the 3 models is 437. (*Source*: America West Airlines) 737-200: 113 seats; 737-300: 134 seats; 757-200: 190 seats

27. Los Angeles 61.6 million arrivals and departures; Atlanta 76.9 million arrivals and departures; Chicago 67.4 million arrivals and departures

30. The new governor of California will make $4000 less than the governor of New York, and $89,000 more than the governor of Alaska. If the sum of these 3 salaries is $440,000, find the salary of each governor. (*Source: World Almanac*) CA: $175,000; NY: $179,000; AL: $86,000

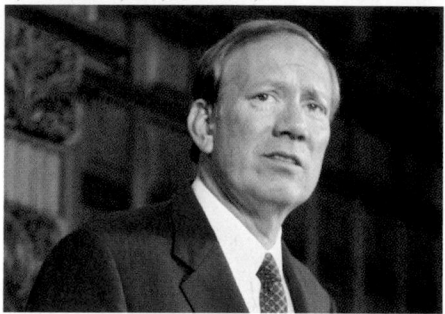

31. $430.00 **33.** Mile High Stadium: 76,125 seats; Heinz Field Stadium: 64,450 seats **34.** Minnesota Twins: $24,350,000; Colorado Rockies: $71,068,000 **35.** AOL: 23,700,000; Earthlink: 4,700,000; MSN: 4,000,000 **36.** Guy's Tower: 469 ft; Queen Mary Hospital: 449 ft; Galter Pairlion: 402 ft

31. A new fax machine was recently purchased for an office in Hopedale for $464.40 including tax. If the tax rate in Hopedale is 8%, find the price of the fax machine before taxes.

32. A premedical student at a local university was complaining that she had just paid $86.11 for her human anatomy book, including tax. Find the price of the book before taxes if the tax rate at this university is 9%. $79.00

33. INVESCO Field at Mile High, home to the Denver Broncos, has 11,675 more seats than Heinz Field, home to the Pittsburgh Steelers. Together, these two stadiums can seat a total of 140,575 NFL fans. How many seats does each stadium have? (*Sources:* Denver Broncos, Pittsburgh Steelers)

34. For the 2001 Major League Baseball season, the opening day payroll for the Minnesota Twins was $46,718,000 less than the opening day payroll for the Colorado Rockies. The total of the opening day payrolls for these two teams was $95,418,000. What was the opening day payroll for each team? (*Source*: Associated Press)

35. In a recent year, the three Internet access companies with the most subscribers were Earthlink, MSN, and AOL. Earthlink had 700,000 more subscribers than MSN and AOL had 3,700,000 more than 5 times the subscribers that MSN had. If the total subscribers for these Internet access companies is 32,400,000, find the number of subscribers for each company. (*Source:* Jupiter Media Metrix)

36. The three tallest hospitals in the world are Guy's Tower in London, Queen Mary Hospital in Hong Kong, and Galter Pairlion in Chicago. These buildings have a total height of 1320 feet. Guy's Tower is 67 feet taller than Galter Pairlion and the Queen Mary Hospital is 47 feet taller than Galter Pairlion. Find the heights of the three hospitals.

37. In 2003, the population of Morocco was 31.2 million. This represented an increase in population of 9.1% from the year 2000. What was the population of Morocco in 2000? (*Source:* Population Reference Bureau) 28.6 million

38. Palm, Inc. is a worldwide supplier of handheld computing devices. On October 25, 2001, shares of Palm stock closed at $2.44 per share. This represents a 97.4% decrease in stock price from the closing price on March 2, 2000, the day when Palm stock first traded on the NASDAQ exchange. Find the closing price on March 2, 2000. (*Source:* Financial Insight Systems, Inc.) $93.85

39. According to government statistics, the number of switchboard operators in the United States is expected to decrease to 185,000 by the year 2008. This represents a decrease of 13.9% from the number of switchboard operators in 1998. (*Source:* U.S. Bureau of Labor Statistics)

   a. Find the number of switchboard operators in 1998. Round to the nearest whole number. 214,866 operators

   b. In your own words, explain why you think that the need for switchboard operators is decreasing. answers may vary

40. The number of deaths caused by tornadoes decreased 59.2% from the 1950s to the 1990s. There were 579 deaths from tornadoes in the 1990s. (*Source:* National Weather Service)

   a. Find the number of deaths caused by tornadoes in the 1950s. Round to the nearest whole number. 1419

   b. In your own words, explain why you think that the number of tornado-related deaths has decreased so much since the 1950s. answers may vary

41. The sum of three consecutive integers is 228. Find the integers. 75, 76, 77

42. The sum of three consecutive odd integers is 327. Find the integers. 107, 109, 111

Recall that the sum of the angle measures of a triangle is 180°.

43. Find the measures of the angles of a triangle if the measure of one angle is twice the measure of a second angle and the third angle measures 3 times the second angle decreased by 12. 64°, 32°, 84°

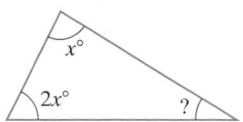

44. Find the angles of a triangle whose two base angles are equal and whose third angle is 10° less than three times a base angle. 38°, 38°, 104°

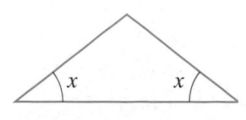

45. The official manual for traffic signs is the *Manual on Uniform Traffic Control Devices* published by the Government Printing Office. The rectangular sign below has a length 12 inches more than twice its height. If the perimeter of the sign is 312 inches, find its dimensions. height: 48 in.; width: 108 in.

46. Two frames are needed with the same outside perimeter: one frame in the shape of a square and one in the shape of an equilateral triangle. Each side of the triangle is 6 centimeters longer than each side of the square. Find the dimensions of each frame. (An equilateral triangle has sides that are the same length.) square: 18 cm on each side; triangle: 24 cm on each side

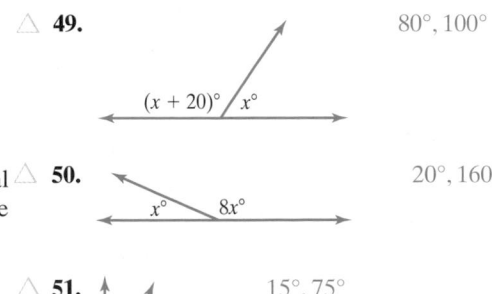

47. length: 14 cm; width: 6 cm

47. In a blueprint of a rectangular room, the length is to be 2 centimeters greater than twice its width. Find the dimensions if the perimeter is to be 40 centimeters.

48. A plant food solution contains 5 cups of water for every 1 cup of concentrate. If the solution contains 78 cups of these two ingredients, find the number of cups of concentrate in the solution. 13 cups

Recall that two angles are complements of each other if their sum is 90°. Two angles are supplements of each other if their sum is 180°. Find the measure of each angle.

49. 80°, 100°

50. 20°, 160°

51. 15°, 75°

△ **52.**

25°, 65°

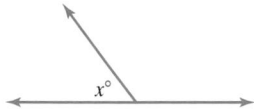

△ **53.** One angle is three times its supplement increased by 20°. Find the measures of the two supplementary angles.   40°, 140°

57. Incandescent: 1500 bulb hours; Fluorescent: 100,000 bulb hours; Halogen: 4000 bulb hours

△ **54.** One angle is twice its complement increased by 30°. Find the measures of the two complementary angles.   20°, 70°

55. The external tank of a NASA Space Shuttle contains the propellants used for the first 8.5 minutes after launch. Its height is 5 times the sum of its width and 1. If the sum of the height and width is 55.4 meters, find the dimensions of this tank. (*Source:* NASA/Kennedy Space Center)   width: 8.4 m; height: 47 m

height

width = *x* meters

56. The blue whale is the largest of whales. Its average weight is 3 times the difference of the average weight of a humpback whale and 5 tons. If the total of the average weights of a blue whale and a humpback whale is 117 tons, find the average weight of each type of whale.   humpback: 33 tons; blue whale: 84 tons

57. Incandescent, fluorescent, and halogen bulbs are lasting longer today than ever before. On average, the number of bulb hours for a fluorescent bulb is 25 times the number of bulb hours for a halogen bulb. The number of bulb hours for an incandescent bulb is 2,500 less than the halogen bulb. If the total number of bulb hours for the three types of bulbs is 105,500, find the number of bulb hours for each type. (*Source:* Popular Science Magazine)

Incandescent bulb	Fluorescent bulb	Halogen bulb
1870	1938	1964

58. Finland, Sweden, and Austria have the greatest percent of their population as cell phone subscribers in the world. In Sweden, 7% more of its population are cell phone subscribers than in Austria. In Finland, 13% more of its population are cell phone subscribers than in Austria. If the sum of their percents is 179%, find the percent of population that are cell phone subscribers for each country. (Currently, 29% of the U.S. population are cell phone subscribers.) (*Source:* EMC World Cellular Database; Cellular Telecommunications Industry Asso.)   Finland: 66%; Sweden: 60%; Austria: 53%

59. In 2002, 54.1 million income tax returns were filed electronically (online filing, professional electronic filing, and Tele-File). This represents a 156.4%. increase over the year 1999. How many income tax returns were filed electronically in 1999? Round to the nearest tenth of a million.

60. The popularity of the Harry Potter series of books continues to increase among readers of all ages. To satisfy their desires for Harry Potter adventures, author J.K. Rowling has been increasing the number of pages in each new Harry Potter adventure. The newest book, *The Order of the Phoenix* has 870 pages. If the number of pages has increased by 182% from the first book to the newest book, how many pages are in the first Harry Potter book, *The Sorcerer's Stone*? (Round to the nearest whole page.) (*Source:* Amazon.com)   309 pages

61. During the 2001 Major League Baseball season, the numbers of home runs hit by Jim Thome of the Cleveland Indians, Rafael Palmeiro of the Texas Rangers, and Richie Sexson of the Milwaukee Brewers were three consecutive odd integers. Of these three players, Thome hit the most home runs, and Sexson hit the fewest home runs. The total number of home runs hit by these three players over the course of the season was 141. How many home runs did each player hit during the 2001 season?(*Source:* Major League Baseball)   Thome: 49; Palmeiro: 47; Sexson: 45

62. During the 2000 Olympic Games in Sydney, Australia, the total number of medals won by Germany, Australia, and the People's Republic of China were three consecutive integers. Of these three countries, Germany won the fewest medals and China won the most. If the sum of the first integer, twice the second integer, and four times the third integer is 409, find the number of medals won by each country. (*Source:* International Olympic Committee)   Germany: 57; Australia: 58; China: 59

Concept Extensions

63. Newsprint is either discarded or recycled. Americans recycle about 27% of all newsprint, but an amount of newsprint equivalent to 30 million trees is discarded every year. About how many trees' worth of newsprint is *recycled* in the United States each year? (*Source:* The Earth Works Group)

64. Find an angle such that its supplement is equal to twice its complement increased by 50°.   50°

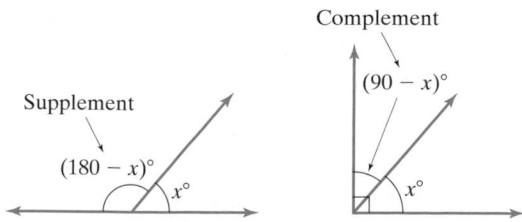

Supplement

$(180 - x)°$

$x°$

Complement

$(90 - x)°$

$x°$

65. The average annual number of cigarettes smoked by an American adult continues to decline. For the years 1991–2000, the equation $y = -64.45x + 2795.5$ approximates this data. Here, x is the number of years after 1990 and y is the average annual number of cigarettes smoked.

   a. If this trend continues, find the year in which the average annual number of cigarettes smoked is 0. To do this, let $y = 0$ and solve for x.   during the year 2033

   b. Predict the average annual number of cigarettes smoked by an American adult in 2005. To do so, let $x = 15$ (Since $2005 - 1990 = 15$) and find y.   1828.75

   c. Use the result of part b to predict the average *daily* number of cigarettes smoked by an American adult in 2005. Round to the nearest whole. Do you think this number represents the average daily number of cigarettes smoked by an adult smoked? Why or why not?

66. Determine whether there are three consecutive integers such that their sum is three times the second integer.   any three consecutive integers

67. Determine whether there are two consecutive odd integers such that 7 times the first exceeds 5 times the second by 54.

To break even in a manufacturing business, income or revenue R must equal the cost of production C. Use this information to answer Exercises 68 through 71.

68. The cost C to produce x number of skateboards is $C = 100 + 20x$. The skateboards are sold wholesale for $24 each, so revenue R is given by $R = 24x$. Find how many skateboards the manufacturer needs to produce and sell to break even. (*Hint:* Set the cost expression equal to the revenue expression and solve for x.)   25 skateboards

69. The revenue R from selling x number of computer boards is given by $R = 60x$, and the cost C of producing them is given by $C = 50x + 5000$. Find how many boards must be sold to break even. Find how much money is needed to produce the break-even number of boards.   500 boards; $30,000

70. In your own words, explain what happens if a company makes and sells fewer products than the break-even number.

71. In your own words, explain what happens if more products than the break-even number are made and sold.

72. In 2002, 54.1 million income tax returns were filed electronically. From 2002 to 2009, the average annual increase in electronic filing is projected to be 6.8%. If this holds true, what will the number of electronic tax returns be for the next 5 years? (IRS) (Round to the nearest tenth of a million.)   2003: 57.8 million; 2004: 61.7 million; 2005: 65.9 million; 2006: 70.4 million; 2007: 75.2 million

59. 21.1 million returns   **63.** 11 million trees   **65. c.** 5; no: this is the average daily number of cigarettes for all American adults—smokers and non smokers   **67.** no such odd integers exist   **70.** company loses money   **71.** company makes a profit

STUDY SKILLS REMINDER

Are You Organized?

Have you ever had trouble finding a completed assignment? When it's time to study for a test, are your notes neat and organized? Have you ever had trouble reading your own mathematics handwriting? (Be honest—I have had trouble reading my own handwriting before.)

When any of these things happen, it's time to get organized. Here are a few suggestions:

Write your notes and complete your homework assignments in a notebook with pockets (spiral or ring binder). Take class notes in this notebook, and then follow the notes with your completed homework assignment. When you receive graded papers or handouts, place them in the notebook pocket so that you will not lose them.

Place a mark (possibly an exclamation point) beside any note(s) that seem especially important to you. Also place a mark (possibly a question mark) beside any note(s) or homework that you are having trouble with. Don't forget to see your instructor, a tutor, or your fellow classmates to help you understand the concepts or exercises you have marked.

Also, if you are having trouble reading your own handwriting, *slow down* and write your mathematics work clearly!

1.7 A NUMERICAL APPROACH: MODELING WITH TABLES

Objectives

1 Create tables.

2 Use tables to solve problems.

1 In this section, we introduce a numerical approach to problem solving. This approach may be used instead of or in addition to an algebraic approach. A numerical approach is invaluable for developing estimation skills and number sense.

In earlier sections, we introduced methods for using a calculator to evaluate an algebraic expression at given values of the variable. Often we organized our results in tables such as the one below.

Time (seconds)	x	2	3	4	5	6
Speed (feet per second)	$32x$	64	96	128	160	192

Notice for this particular table that the minimum x-value is 2, and the x-values increase by 1. Another method for completing a table such as this is the table feature of your calculator.

When generating a **table** on a calculator, we need to give the calculator specific instructions as to a minimum x-value to *start* with and the *increment* or change in the x-values. (For many calculators, the Greek letter delta, Δ, is used in the table setup menu. Here, Δ indicates the change in x.) In this text, we will use the term *table start* to indicate the first x-value to appear in the table and the term *table increment* to indicate the increment or change in x-values.

To display the table above on your calculator, go to the Y= editor and enter the expression $32x$. Next, go to the table setup menu. There, let table start $= 2$ and table increment $= 1$. Then the displayed table should look like the table below.

Table start →
Table increment

2   Next, we generate tables to help us solve problems.

FINDING COMMISSION

Bette Meish is considering a job as a salesperson for a computer software company. She is offered monthly gross pay of $2500 plus 5% commission on sales. An expression that models her gross pay is

In words:	$2500	plus	5% of sales
Translate:	2500	+	0.05x

To help her decide about the job, she considers her monthly pay for sales amounts of $1000 to $9000 in $1000 increments.

Sales, x	1000	2000	3000	4000	5000	6000	7000	8000	9000
Gross pay *2500 + 0.05x*									

a. Complete the table.

b. The company informs her that the average sales per month for its sales force is $6000. Find her annual gross pay if she maintains this average.

c. Look for a pattern in the table. Complete the following sentence. For each additional $1000 in sales, Bette's gross pay is increased by _____.

d. What could Bette expect her annual gross pay to be if she averaged $5000 in sales per month?

e. If Bette decides that she needs to earn a gross pay of $2900 per month how much must she average in sales per month?

Solution   **a.** To complete the table, go to the Y= editor and enter $y_1 = 2500 + 0.05x$. Notice that the sales values in the given table start at 1000 and the change or increment is 1000. In the table setup menu, let table start $= 1000$ and table increment $= 1000$.

From the screens above we can complete the table as follows:

Sales, x	1000	2000	3000	4000	5000	6000	7000	8000	9000
Gross pay $2500 + 0.05x$	2550	2600	2650	2700	2750	2800	2850	2900	2950

b. To find her annual gross pay if she maintains $6000 in sales per month, find 6000 under x, and the corresponding y_1 entry is 2800. Therefore, if she averages $6000 worth of sales per month, she will average $2800 gross pay per month. To find her annual gross pay, find $12(\$2800) = \$33,600$.

c. Observing the table entries, we see that for each additional $1000 sales, Bette's gross pay is increased by $50.

d. To find her annual gross pay if she averages $5000 in sales monthly, scroll to the 5000 entry in the x column. Read the corresponding y_1-value to find a monthly gross pay of $2750. The annual gross pay is $12(\$2750) = \$33,000$.

e. Find the 2900 entry in the y_1(gross pay) column and read the corresponding x-value which is sales. She must average $8000 in sales per month to earn a gross pay of $2900 per month.

EXAMPLE 2

CALCULATING COSTS

Tom Sabo, a licensed electrician, charges $45 per house visit and $30 per hour.

a. Write an equation for Tom's total charge given the number of hours, x, on the job.

b. Use a table to find Tom's total charge for a job that could take from 1 to 6 hours.

Solution   **a.** To model this problem, recall that we are given that x is the number of hours.

In words:	Total Charge	is	$45	plus	$30 per hour
Translate:	Total Charge	=	45	+	$30x$

b. To produce a table to model what Tom charges, go to the Y= editor and enter $y_1 = 45 + 30x$. Notice that the x-values (hours) we are interested in start at 1. In the table setup menu, set table start at 1 and table increment at 1.

From the table, we read that Tom charges $75 for 1 hour, $105 for 2 hours, $135 for 3 hours, and so on, up to $255 for 7 hours.

There are numerous advantages to producing a table rather than evaluating the algebraic expression for each value of x, but one of the most advantageous situations occurs when you want to change the increment slightly. Let's say that Tom decides to calculate his fee every 15 minutes, or in 15-minute increments. In this case, let's start the table at 0 and have the table increment be 0.25, since every 15 minutes is a quarter of an hour.

From the table we see that it would cost us $45 just for having Tom show up. We can look down the table (scrolling if necessary) to find other charges. For example, we see that the charge is $82.50 to hire him for 1 hour and 15 minutes. This is a reasonable number since we found earlier that the charge is $75 for 1 hour and $105 for 2 hours.

Recall that a *mathematical model* is a graph, table, list, equation, or inequality that describes a situation. We call the equations that model Examples 1 and 2 **linear models** because they can be written in the form $y = mx + b$.

	Equation	$y = mx + b$
Example 1	$y = 2500 + 0.05x$	$y = 0.05x + 2500$
Example 2	$y = 45 + 30x$	$y = 30x + 45$

For linear models, as the x-values increase, the y-values or expressions in x always increase, always decrease, or always stay the same. Check the tables in Examples 1 and 2 to see that this is true. We study linear models and the equation $y = mx + b$ further in Chapter 2.

It is possible to enter more than one expression in the Y = editor.

EXAMPLE 3

FINDING DISCOUNTED PRICE, SALES TAX, AND TOTAL COST

Matthew Kramer wants to buy stereo equipment at a store which has a 20% off the original price sale. There is also a 7% sales tax. Each of the items' original price, x, is shown in the table below.

a. Find the discounted price of each item and the total cost if he buys all the items during the sale.

b. Find the total cost if he does not purchase them during the sale.

c. How much does he save if he purchases them during the sale?

Item	*Original Price*	*Discounted Price*	*Tax*	*Total Discounted Cost*
Stereo Receiver	199.99			
DVD/CD Player	179.99			
Pair of Speakers	89.99			
Totals (for Discounted Prices)				

Solution   **a.** To model this problem, recall that we are given that x is the original price. Since we are looking for the discounted price and the sale is 20% off the original price, we have that

In words: | Discounted price | is | 80% | of | original cost |

Translate: Discounted price   =   0.80   · x

We define $y_1 = 0.80x$ to represent the discounted price.

To calculate sales tax, we know that

In words: | Sales tax | is | 7% | of | discounted price |

Translate: Sales tax   =   0.07   · $(0.80x)$

We define $y_2 = 0.07(0.80x)$ to be the tax, and $y_3 = y_1 + y_2$ to be the total cost of each item. To access Y1 and Y2 enter VARS, YVars, Function, and the appropriate Y=.

Notice that there is no constant change in the original price, x, of the three items. To keep from scrolling over a large span of numbers in a table, we introduce the ASK feature. This feature allows us to enter x-values one at a time and see the corresponding y-values. Access the ASK feature of the table and enter the tag prices in the x-column. The corresponding y_1 entry represents the discounted price, y_2 the tax, and y_3 the total cost.

Fill in the corresponding entries on the table and find the sum of each column below.

Item	Original Price	Discounted Price	Tax	Total Discounted Cost
Stereo Receiver	199.99	159.99	11.20	171.19
DVD/CD Player	179.99	143.99	10.08	154.07
Pair of Speakers	89.99	71.99	5.04	77.03
Totals (for Discounted Prices)		375.97	26.32	402.29

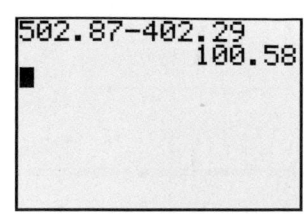

The discounted price with tax is $402.29.

b. To find the total cost if he does not purchase the items during the sale, we can calculate separately.

Total of original prices = $199.99 + $179.99 + $89.99 = $469.97

Tax of total = 0.07($469.97) = $32.90

Total Original Cost (if not discounted) = $469.97 + $32.90 = $502.87

c. His savings = Total Original Cost − Total Discounted Cost
= $502.87 − $402.29 = $100.58

He saves $100.58.

To break even in a manufacturing business, income or revenue R must equal the cost of production C. In other words the break even point, sometimes called the equilibrium point in economics, is where the Revenue = Cost.

EXAMPLE 4

BREAK-EVEN POINT

The cost to produce a certain type of recycled tire is $1400 in fixed costs plus $50 per tire. The tires sell for $85 per tire. The revenue to produce x tires is $R = 85x$. Complete the table below to find the break-even point

No. of Tires	10	20	30	40	50
Cost					
Revenue					

Solution

1. UNDERSTAND. Read and reread the problem. Recall that the break-even point is where the cost and the revenue are equal. Let's first guess what the cost would be if they produced 10 tires. The fixed cost is $1400 and the cost per tire is $50, so the total cost is $1400 + 50(10) = 1400 + 500 = 1900$. The revenue brought in by the sale of 10 tires would be $R = 85(10) = 850$ since each tire sells for $85 each. The cost in this case is $1900 and the revenue is $850, which are not equal. Our guess is not correct, but we have a better understanding of the problem and see that there is a need to produce more tires to break even.

2. TRANSLATE.

In words:

Cost	=	Revenue

Translate: $1400 + 50x = 85x$

3. COMPLETE THE TABLE. Enter the cost in y_1, as $y_1 = 1400 + 50x$, and the revenue as $y_2 = 85x$. Set the table to start at 10 and the increment (Δ table) to be 10.

No. of Tires, x	10	20	30	40	50
Cost (x Tires)	1900	2400	2900	3400	3900
Revenue (x Tires)	850	1700	2550	3400	4250

4. INTERPRET. Since the table entries for 40 tires are the same for both the cost and the revenue, the sale and production of 40 tires produces the break-even point.

Thus far, we have studied linear models only. Another common model is a quadratic model. A quadratic model has the form $y = ax^2 + bx + c, a \neq 0$. For quadratic models, as x-values increase, y-values increase and then decrease, or decrease and then increase. We study quadratic models further in Chapters 5 and 8.

EXAMPLE 5

HEIGHT OF A ROCKET

A small rocket is fired straight up from ground level on a campus parking lot with a velocity of 80 feet per second. Neglecting air resistance, the rocket's height y feet at time x seconds is given by the equation

$$y = -16x^2 + 80x.$$

If you are standing at a window that is 64 feet high,

a. In how many seconds will you see the rocket pass by the window?

b. When will you see the rocket pass by the window again?

c. From firing, how many seconds does it take the rocket to hit the ground?

d. Approximate to the nearest tenth of a second when the rocket reaches its maximum height. Approximate the maximum height to the nearest foot.

64 ft

Solution

Enter $y_1 = -16x^2 + 80x$ in the Y=editor. Using the table feature, let table start = 0 and table increment = 1.

X	Y1	
0	0	
1	64	
2	96	
3	96	
4	64	
5	0	
6	-96	

Y1▤-16X²+80X

a. To find the number of seconds at which the rocket passes the window, recall that the window height is 64 feet and look for a y_1 value of 64. From the table we can see that it will first pass the window 1 second after it is shot off.

b. Scrolling down the y_1 column in the table we see that at 4 seconds the rocket again passes the window on the way down.

c. We see that at 5 seconds the height is again 0, indicating that the rocket has hit the ground.

d. From the table, the rocket appears to reach its maximum height somewhere between 2 and 3 seconds. To get a better approximation, start the table at 2 and set the table increment to be 0.1 to find one-tenth of a second intervals of time.

X	Y₁
2	96
2.1	97.44
2.2	98.56
2.3	99.36
2.4	99.84
2.5	100
2.6	99.84

Y₁=100

The maximum height, 100 feet, occurs 2.5 seconds after the rocket is launched.

Notice that the table above displays a relation between two values. For example, the *x*-value 2 is paired with the *y*-value 96 and so on. We study relations and paired data further in Chapter 2.

MENTAL MATH

For each table below, the y_1 column is generated by evaluating an expression in x for corresponding values in the x column. Match each table with the expression in x that generates the y_1 column.

a. $5x$　　　　　　**b.** $3x + 1$　　　　　　**c.** x^2

d. $x^2 + 1$　　　　**e.** $-x + 5$　　　　　　**f.** $-2x$

1.

x	y_1	
1	-2	
2	-4	f
3	-6	

2.

x	y_1	
1	5	
2	10	a
3	15	

3.

x	y_1	
1	4	
2	7	b
3	10	

4.

x	y_1	
1	4	
2	3	e
3	2	

5.

x	y_1	
1	2	
2	5	d
3	10	

6.

x	y_1	
1	1	
2	4	c
3	9	

EXERCISE SET 1.7

| STUDY GUIDE/SSM | CD/ VIDEO | PH MATH TUTOR CENTER | MathXL®Tutorials ON CD | MathXL® | MyMathLab® |

MIXED PRACTICE

Use the table feature to complete each table for the given expression. If necessary, round results to two decimal places. See Examples 1 through 5.

1.

Side Length of a Cube	x	7	8	9	10	11
Volume	x^3	343	512	729	1000	1331

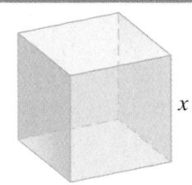

a. From the table, give the volume of a cube whose side measures 9 centimeters. Use correct units.   729 cu. cm

b. From the table, give the volume of a cube whose side measures 11 feet. Use correct units.   1331 cu. ft

c. If a cube has volume 343 cubic inches, what is the length of its side?   7 in.

2.

Radius of Cylinder	x	2	3	4	5	6
Volume (If height is 3 units)	$3\pi x^2$	37.70	84.82	150.80	235.62	1339.29

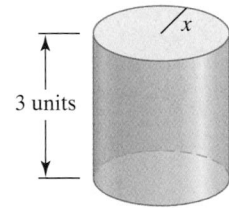

3 units

a. Give the volume of a cylinder whose height is 3 inches and whose radius is 5 inches. Use correct units.   235.62 cu. in.

b. Give the volume of a cylinder whose height is 3 meters and whose radius is 4 meters. Use correct units.   150.80 cu. m

c. If a cylinder with height 3 kilometers has a volume of 150.8 cubic kilometers, approximate the radius of the cylinder.   4 km

3.

Hours Worked	x	39	39.5	40	40.5
Gross Pay (Dollars)	$25 + 7x$	298	301.50	305	308.50

a. Scroll down the table on your calculator to find how many hours worked gives a gross pay of $333.   44 hr

b. Scroll up the table to find how many hours worked gives a gross pay of $270.   35 hr

c. If gross pay is $284, how many hours were worked?   37 hr

4.

Minutes on Phone	x	7	7.5	8	8.5
Total Charge (Dollars)	$1.30 + 0.15x$	2.35	2.43	2.5	2.58

a. Scroll on your calculator to find how long a person with $3.25 can talk on a pay phone.   13 min

b. If a person wants to spend no more than $4 for a single phone call, how long can that person stay on the phone?   18 min

Solve. If necessary, round amounts to two decimal places.

5. Emily Keaton has a job typing term papers for students on her computer. She charges $5.25 per hour for a rough draft and $7 per hour for a final bound manuscript. Her price increases in 15-minute intervals with a 1-hour minimum.   **a.** $y_1 = 5.25x$

a. Let x = number of hours and write an algebraic representation for y_1, the cost of preparing a rough draft.

b. Let x = number of hours and write an algebraic representation for y_2, the cost of preparing a final manuscript.

c. Using a table, complete the following chart for the costs of jobs ranging from 1 to 3 hours.   **5. b.** $y_2 = 7x$

Hours	1	1.25	1.5	1.75	2	2.25	2.5	2.75	3
Rough Draft	5.25	6.56	7.88	9.19	10.50	11.81	13.13	14.44	15.75
Manuscript	7	8.75	10.5	12.25	14	15.75	17.5	19.25	21

d. If Emily can type at a rate of 2 pages per 15 minutes (0.25 of an hour), how much does she charge for a 10-page rough draft? $6.56

e. If a job that consists of typing a bound manuscript takes Emily 10.75 hours, how much does she charge? $75.25

6. The cost of tuition at the local community college is $49.30 per credit hour for in-state students and $182.50 per credit hour for out-of-state students.

a. Write an algebraic representation, y_1, for the cost of in-state tuition dependent on x, the number of credit hours.

b. Write an algebraic representation, y_2, for the cost of out-of-state tuition dependent on x, the number of credit hours. $y_2 = 182.50x$

c. Complete the following table given the number of credit hours.   **6. a.** $y_1 = 49.30x$

10. At the end of 5 years she would make the most money staying with Company B. Company A total is $127,500 and Company B is $129,750

Credit Hours	2	3	4	5	6
In-State Cost (Dollars)	98.6	147.9	197.2	246.5	295.8
Out-of-State Cost (Dollars)	365	547.5	730	912.5	1095

d. What is the total tuition due if Mandy, an in-state student, is taking a 3-hour math course, a 4-hour science course, a 2-hour study skills course, a 3-hour history course, and a 3-hour Spanish course? The science course has an additional $15 lab fee. $754.50

Use the table and the ask feature of your calculator to complete each table for the given expression. If necessary, round to two decimal places. See Example 3.

7.

Radius of Circle	x	2	5	21	94.2
Circumference	$2\pi x$	12.57	31.42	131.95	591.88
Area	πx^2	12.57	78.54	1385.44	27,877.36

a. Find the circumference and the area of a circle whose radius is 30 yards. Use correct units. 188.50 yd: 2827.43 sq. yd

b. Find the circumference and the area of a circle whose radius is 78.5 millimeters. Use correct units. 493.23 mm; 19,359.28 sq. mm

8.

Radius of Sphere	x	3	7.1	43	50
Volume	$\frac{4}{3}\pi x^3$	113.1	1499.2	333,038.14	523,598.78
Surface Area	$4\pi x^2$	113.1	633.47	23,235.22	31,415.93

a. Find the volume and surface area of a sphere whose radius is 1.7 centimeters. Use correct units.

b. Find the volume and surface area of a sphere whose radius is 25 feet. Use correct units. 64,449.85 cu. ft: 7853.98 sq. ft

9. Krista Handefeld is considering a job as a real estate agent. She is offered a monthly salary of $500 plus 4% commission on sales of homes. She would like to know her possible gross pay if the homes in the area range from a sales price of $120,000 to $400,000 and she averages selling one home per month.

a. Complete the table below. 8. a. 20.58 cu. cm: 36.32 sq. cm

Sales (In Thousands)	125	200	250	350	400
Gross Pay	5500	8500	10,500	14,500	16,500

b. If she sells at least one home a month, her gross salary would be between __$5500 and $16,500__ .

10. Elizabeth Gocek is considering two job offers. Company A pays a starting salary of $23,500 with raises of $1000 per year. Company B pays a starting salary of $25,000 with raises of $475 each year.

a. Make a table showing her salary during each year if she plans to stay with the company for 5 years.

Years	1	2	3	4	5
Gross Pay A	23,500	24,500	25,500	26,500	27,500
Gross Pay B	25,000	25,475	25,950	26,425	26,900

b. Which job do you think pays the most money if she stays for five years?

11. Kinsley Water Service charges a basic water fee of $12.96 per month plus $1.10 per thousand gallons of water used.

a. Write an algebraic representation for the total cost, y_1, in terms of the number of thousands of gallons of water used, x. $y_1 = 12.96 + 1.10x$

b. Make a table for the cost each month of 0 to 6 thousand gallons of water used.

Gallons of Water Used (Thousands)	0	1	2	3	4	5	6
Monthly Cost (Dollars)	12.96	14.06	15.16	16.26	17.36	18.46	19.56

12. Jewell Water Service charges a basic water fee of $6.00 per month plus $1.72 per thousand gallons of water used.

a. Write an algebraic representation for the total cost, y_1, in terms of the number of thousands of gallons of water used, x. $y_1 = 6.00 + 1.72x$

b. Make a table for the cost each month of 0 to 6 thousand gallons of water.

Gallons of Water Used (Thousands)	0	1	2	3	4	5	6
Monthly Cost (Dollars)	6	7.72	9.44	11.16	12.88	14.6	16.32

13. A company that has 500 computers has been infected by a computer virus. The virus is infecting the computers at a rate of 8 per minute.

a. Write an algebraic expression representation, y_1, for the number of computers that have not been infected. Let x be the number of minutes that have elapsed since the first computer has been infected. $y_1 = 500 - 8x$

b. Complete the table for the number of computers not infected depending on the time elapsed.

No. of Minutes	0	10	20	30	40	50	60	70
No. of Computers Not Infected	500	420	340	260	180	100	20	−60

c. When will there be 260 computers that are not infected?

d. How long will it take for all the computers to be infected?

14. Mark Cook is the general manager of the Acme Used Car lot. At the beginning of the month, he has 850 cars on the lot and his sales team has been selling the cars at the rate of 25 per day.

    a. Write an algebraic representation for the number of cars on the lot, y_1, in terms of the number of days that have passed, x. (Assume no new cars are delivered.)

    b. Complete the table for the number of cars remaining on the lot given the number of days that have elapsed.

Days Ellapsed	0	10	15	20	25	30
Cars Remaining	850	600	475	350	225	100

16. Rory Baruch considers two different salary plans that his company offers. Plan A starts at $25,000, with raises of $650 per year. Plan B starts at $22,000, with raises of $1000 per year.

    a. If Rory plans to stay with the company less than 10 years, make a chart showing his salary for both plans during years 1 to 9.

    b. Which plan looks more advantageous for Rory? Why do you think so?   answers may vary

Year	1	2	3	4	5	6	7	8	9
Plan A	25,000	25,650	26,300	26,950	27,600	28,250	28,900	29,550	30,200
Plan B	22,000	23,000	24,000	25,000	26,000	27,000	28,000	29,000	30,000

17. The Jackson Electric Authority charges a basic fee of $15.00 per month plus $0.08924 per kilowatt-hour of electricity used.

    a. Write an algebraic representation for the total cost, y_1, in terms of the number of kilowatt-hours used, x.

    b. Complete the table for the cost each month of 0 to 2000 kilowatt-hours.   **17. a.** $y_1 = 15 + 0.08924x$

Kilowatt-Hours	0	500	1000	1500	2000
Monthly Cost (Dollars)	15	59.62	104.24	148.86	193.48

18. The First Coast Electric charges a basic fee of $10.00 per month plus $0.09850 per kilowatt-hour of electricity used.

    a. Write an algebraic representation for the total cost, y_1, in terms of the number of kilowatt-hours used, x.

    b. Complete the table for the cost each month of 0 to 2000 kilowatt-hours.   **18. a.** $y_1 = 10 + 0.09850x$

c. If Mark does not replace any of the sold inventory, when will his lot be completely empty of cars?   34 days

15. Kelsey is considering two job offers. The first job pays a starting salary of $14,500, with raises of $1000 every year. The second job pays a starting salary of $20,000 with raises of $575 each year.

    a. Make a chart showing both salaries during 5, 10, 15, and 20 years.

Year	5	10	15	20
First Job	$18,500	$23,500	$28,500	$33,500
Second Job	$22,300	$25,175	$28,050	$30,925

    b. Which plan do you think is more advantageous for Kelsey? Why do you think so?   answers may vary

13. c. In 30 minutes there will be 260 uninfected computers.

13. d. In 60 minutes there will be 20 computers not infected. Since the rate is 8 per minute it will take another 2.5 minutes to infect the remaining computers. In 62.5 minutes they will all be infected.

14. a. $y_1 = 850 - 25x$

Kilowatt-Hours	0	500	1000	1500	2000
Monthly Cost (Dollars)	10	59.25	108.5	157.75	207

19. Kevin's Barbeque Deluxe is having an all-you-can-eat special on chicken. He barbequed 150 chickens. At 5 pm, the local high school football team unexpectedly came into the restaurant after practice and his customers are now consuming chickens at the rate of 42 chickens per half hour.

    a. Write an algebraic representation, y_1, for the number of chickens remaining where x is the number of minutes past 5 pm. $y_1 = 150 - x(42/30)$

    b. Complete the table.

No. of Minutes	0	30	60	90
No. Chickens	150	108	66	24

    c. When will Kevin run out of chicken if he does not barbeque any additional chickens?   In 107 minutes, or at 6:47 pm.

20. d. It will take 12.5 hrs. If she works nonstop she will have the order finished at 2:30 a.m.

20. At 2 pm, Julie Aires T-Shirt Shop received a rush order for 500 t-shirts to be printed and delivered by 10 pm that day. She averages 40 t-shirts per hour.

   a. Write an algebraic representation for the total number of shirts completed, y_1, in terms of the number of hours that has elapsed since 2 pm, x. $y_1 = 40x$

   b. Complete the table.   **21. a.** $y_1 = 24 - \dfrac{x}{15}$

Hours	0	5	10	15	20
Shirts	0	200	400	600	800

   c. Can Julie complete the order in time?   no

   d. If not, when will she be able to complete the order?

21. Cara D'Alesio bought a new Sports Utility Vehicle (SUV) that has a 24-gallon gas tank. The salesperson told her that the car averages 15 miles per gallon (mpg) in the city and 25 mpg for highway travel. (If needed round answers to one decimal place.)

   a. Write an algebraic representation for the amount of gas left in the tank, y_1 in terms of miles driven, x, for city driving.

   b. Write an algebraic representation for the amount of gas left in the tank, y_2, in terms of miles driven, x, for highway driving.

   c. Complete the table.   **21. b.** $y_2 = 24 - \dfrac{x}{25}$

Miles	0	100	200	300	400	500	600
City	24	17.3	10.7	4	−2.7	−9.3	−16
Highway	24	20	16	12	8	4	0

   d. How much gas is left in the tank after the first 100 miles of city driving?   17.3 gallons

   e. How much gas is left in the tank after the first 100 miles of highway driving?   20 gallons

   f. How many miles can she drive before her tank is empty if she is only driving on a highway?   600 miles

   g. How many miles can she drive before her tank is empty if she is only driving in the city?   360 miles

   h. Write a sentence telling what the negative entries in the table represent.   answers may vary.

22. Jesus Thornbury has an economy car that has a 14-gallon tank. His car averages 24 miles per gallon (mpg) in the city and 30 mpg when traveling on a highway. (If needed, round answers to one decimal place.)

   a. Write an algebraic representation for the amount of gas left in the tank, y_1, in terms of miles driven, x, for city driving.

   b. Write an algebraic representation for the amount of gas left in the tank, y_2, in terms of miles driven, x, for highway driving.

   c. Complete the table.

Miles	0	50	100	150	200	250	300	350	400
City	14	11.9	9.8	7.8	5.7	3.6	1.5	−0.6	−2.7
Highway	14	12.3	10.7	9	7.3	5.7	4	2.3	0.7

   d. How much gas is left in the tank after the first 300 miles of city driving?   1.5 gallons

   e. How much gas is left in the tank after the first 300 miles of highway driving?   4 gallons

   f. How many miles can he drive before his tank is empty if he is only driving on a highway?   420 miles

   g. How many miles can he drive before his tank is empty if he is only driving in the city?   336 miles.

   h. Write a sentence telling what the negative entries in the table represent.   answers may vary

23. Judy Martinez has been shopping for school clothes and has picked out several outfits. She knows that the following Wednesday there will be a Super Wednesday sale and everything in the store will be discounted 35%.

   a. Write an algebraic representation for the sale price, y_1, in terms of the price, x, on the price tag. $y_1 = 0.65x$

   b. Complete the table and find the cost of each item on sale, given the price on the price tag. Round dollar amounts to the nearest cent.

Item	Blouse	Skirt	Shorts	Shoes	Purse	Earrings	Backpack
Price Tag	$29.95	$35.95	$19.25	$39.95	$17.95	$9.95	$25.75
Sale Price	$19.47	$23.37	$12.51	$25.97	$11.67	$6.47	$16.74

   c. If she purchases all the items when on sale, what will her total bill be before taxes? $116.20

   d. What is the total after taxes if a 7% tax is added to the bill? $124.33

   e. How much did she save before taxes by waiting for the sale? $62.55

24. Saumil has been shopping for tools and has picked out several at Sears. Next month, Sears is having a tool sale and all tools in the store will be discounted 30%.

   a. Write an algebraic representation for the sale price, y_1, in terms of the price, x, on the price tag. $y_1 = 0.70x$

   b. Complete the table and find the cost of each item on sale, given the price on the price tag.

Item	Hammer	Drill	Sander	Glue gun	Screwdriver set	Socket wrench set
Price Tag	$9.95	$29.95	$49.75	$19.95	$27.95	$79.95
Sale Price	$6.97	$20.97	$34.83	$13.97	$19.57	$55.97

22. a. $y_1 = 14 - \dfrac{x}{24}$ **22. b.** $y_2 = 14 - \dfrac{x}{30}$

c. If he purchases all the items when on sale, what will his total bill be before taxes? $152.28

d. What is the total after taxes if a 7% tax is added to the bill? $162.94

e. How much did he save before taxes by waiting for the sale? $65.22

25. The cost of producing x calculators is $3500 in fixed costs and $55 per calculator. The calculators sell for $75 each. Find the break-even point by completing the following table.

x Calculators	100	125	150	175	200
Cost	9000	10.375	11.750	13.125	14.500
Revenue	7500	9375	11.250	13.125	15.000

26. The cost of producing x video tapes is $1750 in fixed costs and $21 per tape. The tapes sell for $35 each wholesale. Find the break even point by completing the following table.

x Tapes	25	50	75	100	125
Cost	2275	2800	3325	3850	4375
Revenue	875	1750	2625	3500	4375

27. In Exercise 25, write in words what the entries for the cost and revenue tell you about the profit when 200 calculators are produced and sold.   When 200 calculators are produced and sold, the profit is $500.

28. In Exercise 26, write in words what the entries for the cost and revenue tell you about the profit when 100 video tapes have been produced and sold.   When 100 video tapes are produced and sold, there is $350 profit.

29. **GROUP ACTIVITY.** The owner of Descartes Pizzaria is in the process of changing his pizza prices. He is considering two methods for determining the new price for a pizza. One method is to charge 5¢ per square inch of area of pizza. Another method is to charge 25¢ per inch of circumference. Fill in the table below and compare the costs of pizzas for the given *diameters* using the two methods. Round costs to 2 decimal places. Which price do you think the owner should choose for each size of pizza? Why?

Diameter (Inches)	Area	Circumference	Cost (Area)	Cost (Circumference)
6	28.274	18.85	$1.41	$4.71
12	113.1	37.699	$5.65	$9.42
18	254.17	56.549	$12.72	$14.14
24	452.39	75.398	$22.62	$18.85
30	706.86	94.248	$35.34	$23.56
36	1017.9	113.1	$50.89	$28.27

Diameter

30. A firecracker rocket is fired at ground level with an initial velocity of 100 feet per second. Neglecting air resistance the height of the rocket at time x seconds is given by the equation $y = -16x^2 + 100x$ where y is measured in feet.

a. Find the height of the rocket for the first five seconds in one-second intervals.

Seconds	1	2	3	4	5
Height (Feet)	84	136	156	144	100

b. Explain why the rocket heights increase and then decrease.   The rocket rises, peaks, then falls.

c. If it doesn't explode, determine to the nearest tenth of a second when the rocket hits the ground.   6.3 sec

31. A rocket with an initial velocity of 85 feet per second is launched from the top of a mountain 1500 feet above sea level. Neglecting air resistance, the height of the rocket above sea level at the time x seconds is given by the equation $y = -16x^2 + 85x + 1500$ where y is measured in feet.

a. Find its height for the first 5 seconds in one-second intervals.   1569, 1606, 1611, 1584, 1525 ft   **31. b.** 1613 ft

b. Find the rocket's maximum height to the nearest foot.

32. Dan Newman is going to the county fair. The advertisement in the paper states that there are two different ticket options. The first option is to pay $4.00 for an admission ticket and $0.35 for each ride. The second option is to pay $1.50 for an admission ticket and $0.75 for each ride.

a. For 5 rides, determine the price of each option.

b. After how many rides is option 1 the better buy?

33. Anne Holloway rents a car for $25 per day plus $0.08 per mile. Michelle Goods rents a similar car from a different rental agency for $35 per day plus $0.05 per mile.

a. Find both rental charges for one day if they each drive 300 miles.   Anne: $49.00: Michelle: $50.00

b. For what distance is Anne's rental agreement cheaper than Michelle's for one day?   distances less than $333\frac{1}{3}$ miles

32. a. First Option: $5.75: Second Option; $5.25

32. b. 7 or more

34. Matthew Aires repairs computers and charges $35 per house visit plus $25 per hour.

   a. Write an expression for Matthew's total charge given the number of hours, x, on the job.　$C = 35 + 25x$

   b. Complete the following table for the total charges given the number of hours on the job.

Hours on Job	1.5	2	2.5	3
Cost (Dollars)	72.5	85	97.5	110

35. William Kramer, a lawyer, charges an initial consultation fee of $90 plus $75 per hour after this consultation.

   a. Write an expression for William's total charges given the number of hours, x, on the job after the consultation.

   b. Complete the table below for the total charges given the number of hours *after* the initial consultation.

No. of Hours	2	4	6	8
Total Fee (Dollars)	240	390	540	690

36. An object is *dropped* from the top of the Nations Bank Tower in Atlanta, Georgia, Neglecting air resistance, the height of the object at time x seconds is given by

$$y = -16x^2 + 1050$$

where y is measured in feet. Use a table to determine the following.

   a. What is the object's maximum height? How did you arrive at your answer?　1050 ft

   b. To the nearest tenth of a second, determine when the object hits the ground.　8.1 sec

37. The path of some leaping animals can be described by a quadratic model. A particular leap of a frog can be modeled by the equation

$$y = -0.14x^2 + 1.4x$$

where x is the number of feet horizontally from a starting point and y is the vertical height of the frog in feet. Use a table to determine the following.

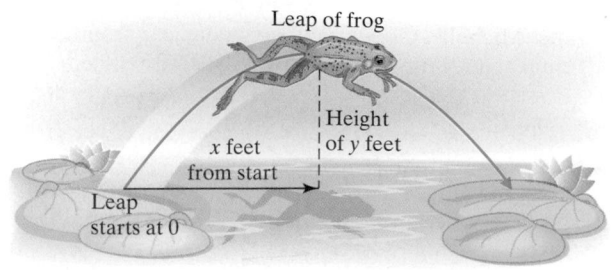

Leap of frog

Height of y feet

x feet from start

Leap starts at 0

   a. Determine to the nearest tenth of a foot the maximum height reached by the frog.　3.5 ft

   b. What is the frog's horizontal distance from the starting point when the maximum height is reached?　5 ft

   c. How far is the frog from the starting point when it lands? (By the way, the longest frog leap on record is 21 feet $5\frac{3}{4}$ inches.)　10 ft

Concept Extensions

38. The high temperature for a particular week in Jackson, Mississippi can be described by the quadratic model

$$y = 1.05x^2 - 8.38x + 89.43$$

where x is the day of the week (Sunday, $x = 1$; Monday, $x = 2$, and so on) and y is the temperature in degrees Fahrenheit.

   a. Find the day of the week with the lowest temperature.

   b. What is the lowest temperature for the week rounded to the nearest tenth?　72.7　　38. a. Wednesday

   c. Do you think that this equation models the temperature for much longer than a week, say 2 weeks? Why or why not? No: answers may vary

39. The sunrise times in Valdivia, Chile, for the months of April through October can be described by the quadratic model

$$y = -0.14x^2 + 1.09x + 5.11$$

where $x = 1$ is the first day of April, $x = 2$ is the first day of May, $x = 3$ is the first day of June, and so on and y is the time of sunrise in hours A.M.

South America

Valdivia

   a. Approximate the time of sunrise on May 1. Give the time in hundredths of an hour and also in hours and minutes.　6.73 or 6:44 A.M.

   b. For what integer value of x is the sunrise the latest? What calendar date does this correspond to?　4: July 1

   c. Give this latest sunrise time in hours and minutes.　7:14 A.M.

   d. Approximate the sunrise time in hours and minutes for September 15.　6:17 A.M.

   e. Do you think that this equation models the sunrise times for all the months of the year? Why or why not? answers may vary

1.8   *FORMULAS AND PROBLEM SOLVING*

Objectives

 Solve a formula for a specified variable.

2   Use formulas to solve problems.

1   Solving problems that we encounter in the real world sometimes requires us to express relationships among measured quantities. A **formula** is an equation that describes a known relationship among quantities such as time, area, and gravity. Some examples of formulas are

Formula	Meaning
$I = PRT$	Interest = principal · rate · time
$A = lw$	Area of a rectangle = length · width
$d = rt$	Distance = rate · time
$C = 2\pi r$	Circumference of a circle = $2 \cdot \pi \cdot$ radius
$V = lwh$	Volume of a rectangular solid = length · width · height

TEACHING TIP

Remind students that "Solve for a variable" means find what the variable equals. Then point out that a variable can equal a number or an expression involving other variables.

Other formulas are listed in the front cover of this text. Notice that the formula for the volume of a rectangular solid $V = lwh$ is solved for V since V is by itself on one side of the equation with no V's on the other side of the equation. Suppose that the volume of a rectangular solid is known as well as its width and its length, and we wish to find its height. One way to find its height is to begin by solving the formula $V = lwh$ for h.

EXAMPLE 1

Solve $V = lwh$ for h.

Solution   To solve $V = lwh$ for h, isolate h on one side of the equation. To do so, divide both sides of the equation by lw.

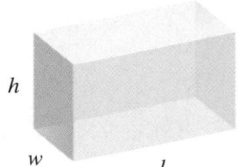

$$V = lwh$$

$$\frac{V}{lw} = \frac{lwh}{lw} \qquad \text{Divide both sides by } lw.$$

$$\frac{V}{lw} = h \qquad \text{Simplify.}$$

Then to find the height of a rectangular solid, divide the volume by the product of its length and its width.

CLASSROOM EXAMPLE

Solve $d = rt$ for r.

answer: $r = \dfrac{d}{t}$

The following steps may be used to solve formulas and equations in general for a specified variable.

> ## Solving Equations for a Specified Variable
>
> **Step 1:** Clear the equation of fractions by multiplying each side of the equation by the least common denominator.
>
> **Step 2:** Use the distributive property to remove grouping symbols such as parentheses.
>
> **Step 3:** Combine like terms on each side of the equation.
>
> **Step 4:** Use the addition property of equality to rewrite the equation as an equivalent equation with terms containing the specified variable on one side and all other terms on the other side.
>
> **Step 5:** Use the distributive property and the multiplication property of equality to isolate the specified variable.

EXAMPLE 2

Solve $3y - 2x = 7$ for y.

Solution   This is a linear equation in two variables. Often an equation such as this is solved for y in order to reveal some properties about the graph of this equation, which we will learn more about in Chapter 3. Since there are no fractions or grouping symbols, we begin with Step 4 and isolate the term containing the specified variable y by adding $2x$ to both sides of the equation.

TEACHING TIP

Ask students what is wrong with the following solution for y.

$$3y + 7 = x - 4y$$
$$3y + 7 - 7 = x - 4y - 7$$
$$3y - 2y = x - 4y - 7 - 2y$$
$$y = x - 6y - 7$$

Be sure they understand that when an equation is solved for a variable, the variable occurs only on one side of the equation.

$$3y - 2x = 7$$
$$3y - 2x + 2x = 7 + 2x \qquad \text{Add } 2x \text{ to both sides.}$$
$$3y = 7 + 2x$$

To solve for y, divide both sides by 3.

$$\frac{3y}{3} = \frac{7 + 2x}{3} \qquad \text{Divide both sides by 3.}$$

$$y = \frac{2x + 7}{3} \qquad \text{or} \qquad y = \frac{2x}{3} + \frac{7}{3}$$

CLASSROOM EXAMPLE

Solve $2y + 5x = 10$ for y.

answer: $y = \dfrac{10 - 5x}{2}$

EXAMPLE 3

Solve $A = \dfrac{1}{2}(B + b)h$ for b.

Solution   Since this formula for finding the area of a trapezoid contains fractions, we begin by multiplying both sides of the equation by the LCD 2.

$$A = \frac{1}{2}(B + b)h$$

$$2 \cdot A = 2 \cdot \frac{1}{2}(B + b)h \qquad \text{Multiply both sides by 2.}$$

$$2A = (B + b)h \qquad \text{Simplify.}$$

Next, use the distributive property and remove parentheses.

$$2A = (B + b)h$$

$$2A = Bh + bh \qquad \text{Apply the distributive property.}$$

$$2A - Bh = bh \qquad \text{Isolate the term containing } b \text{ by subtracting } Bh \text{ from both sides.}$$

$$\frac{2A - Bh}{h} = \frac{bh}{h} \qquad \text{Divide both sides by } h.$$

$$\frac{2A - Bh}{h} = b, \quad \text{or} \quad b = \frac{2A - Bh}{h}$$

CLASSROOM EXAMPLE

Solve $A = \dfrac{1}{2}(B + b)h$ for B.

answer: $B = \dfrac{2A - bh}{h}$

> **Helpful Hint**
> Remember that we may isolate the specified variable on either side of the equation.

EXAMPLE 4

FINDING DEGREES CELSIUS

The formula $C = \dfrac{5}{9}(F - 32)$ converts degrees Fahrenheit to degrees Celsius. Use this formula and the table feature of your calculator to complete the given table. If necessary, round values to the nearest hundredth.

Fahrenheit	−4	10	32	70	100
Celsius					

Solution Let $x =$ degrees Fahrenheit. Then enter $y_1 = \dfrac{5}{9}(x - 32)$ to find the corresponding degrees Celsius. Notice that there is no constant increment change in the Fahrenheit values in the given table. To keep from scrolling over a large span of numbers in a table (not a good use of time), we activate the ask feature in table setup.

Enter each x-value from the table. The corresponding y_1-values are shown to the left. The completed table is

Fahrenheit	x	−4	10	32	70	100
Celsius	$\frac{5}{9}(x - 32)$	−20	−12.22	0	21.11	37.78

2   In this section, we also solve problems that can be modeled by known formulas. We use the same problem-solving steps that were introduced in the previous section.

Formulas are very useful in problem solving. For example, the compound interest formula

$$A = P\left(1 + \frac{r}{n}\right)^{nt}$$

is used by banks to compute the amount A in an account that pays compound interest. The variable P represents the principal or amount invested in the account, r is the annual rate of interest, t is the time in years, and n is the number of times compounded per year.

EXAMPLE 5

FINDING THE AMOUNT IN A SAVINGS ACCOUNT

Karen Estes just received an inheritance of $10,000 and plans to place all the money in a savings account that pays 5% compounded quarterly to help her son go to college in 3 years. How much money will be in the account in 3 years?

Solution

1. UNDERSTAND. Read and reread the problem. The appropriate formula needed to solve this problem is the compound interest formula

$$A = P\left(1 + \frac{r}{n}\right)^{nt}$$

Make sure that you understand the meaning of all the variables in this formula.

A = amount in the account after t years

P = principal or amount invested

t = time in years

r = annual rate of interest

n = number of times compounded per year

CLASSROOM EXAMPLE

If $5000 is invested in an account paying 4% compounded monthly, determine how much money will be in the account in 2 years.

answer: $5415.71

TEACHING TIP

As an extention to Example 4, ask, "How would the answer change if the question was changed to "How much has the account earned in 3 years?"

2. TRANSLATE. Use the compound interest formula and let P = $10,000, r = 5% = 0.05, t = 3 years, and n = 4 since the account is compounded quarterly, or 4 times a year.

Formula: $A = P\left(1 + \frac{r}{n}\right)^{nt}$

Substitute: $A = 10,000\left(1 + \frac{0.05}{4}\right)^{4 \cdot 3}$

3. SOLVE. We simplify the right side of the equation. Here, we can store the known values in the calculator and evaluate the algebraic expression, or we can use our calculator to evaluate the numerical expression.

4. INTERPRET.

   Check: Repeat your calculations to make sure that no error was made. Notice that $11,607.55 is a reasonable amount to have in the account after 3 years.

   State: In 3 years, the account will contain $11,607.55.

Discover the Concept

Finding the Maximum Area. While this discovery can be done by an individual, we suggest working in groups of 2 or 3 if feasible. Each group/person begins with 12 toothpicks.

Suppose that each toothpick represents a 1-foot piece of fencing. Use all the toothpicks (without bending or breaking) to form a rectangular fence. First, form a rectangular fence of width 1 foot, next, a rectangular fence of width 2 feet, then 3 feet and so on. For each rectangular fence constructed, record the width, length area, and perimeter in a table such as the one below.

Width	Length	Area	Perimeter
1			
2			
3			
4			
5			
6			

a. Find the dimensions of the rectangle with the largest area.

b. Compare the areas of the rectangles with their perimeters. What patterns do you notice?

From the discovery above, we see that the rectangle with largest area has dimensions 3 feet by 3 feet and an area of 9 square feet. Notice that although a change in dimensions results in a change in area, the perimeter remains constant.

EXAMPLE 6

FINDING MAXIMUM AREA

Sylvia Daschle has purchased 60 feet of fencing in 1-foot sections. She wants to enclose the largest possible rectangular garden using a portion of a side of her house as a side of the rectangle. Use a table to find the dimensions that will give Sylvia the largest area.

1. UNDERSTAND. Read and reread the problem and then try some dimensions in particular. For example, if the width is 1 foot, the length is $60 - 2(1)$ or 58 feet. The area of the rectangle is $w \cdot l$ or $1(58) = 58$ square feet. Next, let the width be 2 feet and continue until a pattern is found. A table can be useful when looking for a pattern.

Solution

1 foot

No fencing needed

58 feet

1 foot

Width	Length	Area
1	$60 - 2(1)$ or 58	$1 \cdot 58 = 58$
2	$60 - 2(2)$ or 56	$2 \cdot 56 = 112$
3	$60 - 2(3)$ or 54	$3 \cdot 54 = 162$

2. TRANSLATE. Let $x =$ the width of the garden. From the table, we can see that
$60 - 2x =$ the length of the garden

In words: Area   =   width   ·   length   or

Translate: Area   = x · $(60 - 2x)$

3. SOLVE. Since Sylvia has fencing in 1-foot sections, we know that the width and the length are natural numbers. To solve, we can use a table. We can enter the expression for length in y_1 and the area in y_2.

$$y_1 = 60 - 2x \qquad \text{Length}$$
$$y_2 = x(60 - 2x) \qquad \text{Area}$$

Set table start to be 1 and table increment to be 1. Scroll down the y_2 or area column to find the largest area. Notice that equation $y_2 = x(60 - 2x)$ is a quadratic model so for increasing values of x, the y_2 values will increase and then decrease, or decrease and then increase.

X	Y₁	Y₂
1	58	58
2	56	112
3	54	162
4	52	208
5	50	250
6	48	288
7	46	322

Y₁=60-2X

X	Y₁	Y₂
10	40	400
11	38	418
12	36	432
13	34	442
14	32	448
15	30	450
16	28	448

Y₂=450

The largest area is 450 square feet.

4. INTERPRET. *Check* the solution and *state*. From the table, we see that the dimensions that give the largest area are 15 feet by 30 feet. The largest area is 450 square feet.

EXAMPLE 7

FINDING CYCLING TIME

The fastest average speed by a cyclist across the continental United States is 15.4 mph, by Pete Penseyres. If he traveled a total distance of about 3107.5 miles at this speed, find his time cycling. Write the time in days, hours, and minutes. *(Source: The Guinness Book of World Records)*

Solution

1. UNDERSTAND. Read and reread the problem. The appropriate formula needed is the distance formula

$$d = rt \qquad \text{where}$$
$$d = \text{distance traveled} \quad r = \text{rate and} \quad t = \text{time}$$

2. TRANSLATE. Use the distance formula and let $d = 3107.5$ miles and $r = 15.4$ mph.

Check: $d = rt$

State: $3107.5 = 15.4t$

3. SOLVE.

$$\frac{3107.5}{15.4} = \frac{15.4t}{15.4} \qquad \text{Divide both sides by 15.4.}$$

$$201.79 \approx t$$

The time is approximately 201.79 hours. Since there are 24 hours in a day, we divide 201.79 by 24 and find that the time is approximately 8.41 days. Now, let's convert the decimal part of 8.41 days back to hours. To do this, multiply 0.41 by 24 and the result is 9.84 hours. Next, we convert the decimal part of 9.84 hours to minutes by multiplying by 60 since there an 60 minutes in an hour. We have $0.84 \cdot 60 \approx 50$ minutes rounded to the nearest whole. The time is then approximately

8 days, 9 hours, 50 minutes.

4. INTERPRET.

Check: Repeat your calculations to make sure that an error was not made.

State: Pete Penseyres's cycling time was approximately 8 days, 9 hours, 50 minutes.

Spotlight on
DECISION
of MAKING

Suppose you are a dentist. Although fluoride can play an important part in a treatment plan to prevent tooth decay, you know that large doses of fluoride can be lethal. The formula $F = 10qpr$ can be used to calculate the number of milligrams (mg) of fluoride F ingested by a patient who receives q milliliters of a fluoride solution with p percent concentration and molecular weight ratio r. The molecular weight ratios for common fluoride compounds are given in Table 1. Table 2 shows the maximum safe doses of fluoride for children, along with certainly lethal doses.
  Decide whether or not a fluoride treatment of 5 milliliters of an 8 percent SnF_2 solution is safe for a 50-pound child.

Molecular Weight Ratios

Fluoride Compound	Ratio
NaF	$\dfrac{1}{2.2}$
Na_2FPO_3	$\dfrac{1}{7.6}$
SnF_2	$\dfrac{1}{4.1}$

Fluoride Doses For Children

Weight (pounds)	Maximum Safe Dose (mg)	Certainly Lethal Dose (mg)
20	73	291
30	109	436
40	146	582
50	182	727
60	218	873
70	255	1018
80	291	1163
90	327	1309
100	363	1454

(*Source*: Based on data from S.B. Heifetz and H.S Horowitz. "The Amounts of Fluoride in Current Fluoride Therapies: Safety Considerations for Children," *ASDC J. Dent. Child.*, July–Aug. 1984.)

TEACHING TIP
A Group Activity for this section is available in the Instructor's Resource Manual.

MENTAL MATH

Solve each equation for the specified variable. See Examples 1 through 3.

1. $2x + y = 5$; for y $y = 5 - 2x$ **2.** $7x - y = 3$; for y $y = 7x - 3$ **3.** $a - 5b = 8$; for a $a = 5b + 8$

4. $7r + s = 10$; for s $s = 10 - 7r$ **5.** $5j + k - h = 6$; for k $k = h - 5j + 6$ **6.** $w - 4y + z = 0$; for z $z = 4y - w$

5. $y = \dfrac{9x - 16}{4}$ **6.** $y = \dfrac{17 - 2x}{3}$ **7.** $W = \dfrac{P - 2L}{2}$ **8.** $N = \dfrac{3M - A}{2}$ **9.** $A = \dfrac{J + 3}{C}$ **11.** $g = \dfrac{W}{h - 3t^2}$ **13.** $B = \dfrac{T - 2C}{AC}$

EXERCISE SET 1.8

   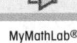

STUDY GUIDE/SSM   CD/VIDEO   PH MATH TUTOR CENTER   MathXL®Tutorials ON CD   MathXL®   MyMathLab®

Solve each equation for the specified variable. See Examples 1–3.

14. $b = \dfrac{A - 5HB}{5H}$ **16.** $h = \dfrac{S - 2\pi r^2}{2\pi r}$

1. $D = rt$; for t $t = \dfrac{D}{r}$ **2.** $W = gh$; for g $g = \dfrac{W}{h}$ **9.** $J = AC - 3$; for A **10.** $y = mx + b$; for x $x = \dfrac{y - b}{m}$

3. $I = PRT$; for R $R = \dfrac{I}{PT}$ △ **4.** $V = lwh$; for l $l = \dfrac{V}{wh}$ **11.** $W = gh - 3gt^2$; for g **12.** $A = Prt + P$; for P $P = \dfrac{A}{rt + 1}$

5. $9x - 4y = 16$; for y **6.** $2x + 3y = 17$; for y **13.** $T = C(2 + AB)$; for B **14.** $A = 5H(b + B)$; for b

△ **7.** $P = 2L + 2W$; for W **8.** $A = 3M - 2N$; for N △ **15.** $C = 2\pi r$; for r $r = \dfrac{C}{2\pi}$ △ **16.** $S = 2\pi r^2 + 2\pi rh$; for h

17. $E = I(r + R)$; for r

18. $A = P(1 + rt)$; for t

19. $s = \dfrac{n}{2}(a + L)$; for L

20. $C = \dfrac{5}{9}(F - 32)$; for F

21. $N = 3st^4 - 5sv$; for v

22. $L = a + (n - 1)d$; for d

△ **23.** $S = 2LW + 2LH + 2WH$; for H $H = \dfrac{S - 2LW}{2L + 2W}$

24. $T = 3vs - 4ws + 5vw$; for v $v = \dfrac{T + 4ws}{3s + 5w}$

In this exercise set, round all dollar amounts to two decimal places. Solve. See Example 5.

25. Complete the table and find the balance A if $3500 is invested at an annual percentage rate of 3% for 10 years and compounded n times a year.

n	1	2	4	12	365
A	$4703.71	$4713.99	$4719.22	$4722.74	$4724.45

26. Complete the table and find the balance A if $5000 is invested at an annual percentage rate of 6% for 15 years and compounded n times a year.

n	1	2	4	12	365
A	$11,982.79	$12,136.31	$12,216.10	$12,270.47	$12,297.11

27. A principal of $6000 is invested in an account paying an annual percentage rate of 4%. Find the amount in the account after 5 years if the account is compounded

  a. semiannually $7313.97

  b. quarterly $7321.14

  c. monthly $7325.98

28. A principal of $25,000 is invested in an account paying an annual percentage rate of 5%. Find the amount in the account after 2 years if the account is compounded

  a. semiannually $27,595.32

  b. quarterly $27,612.15

  c. monthly $27,623.53

17. $r = \dfrac{E - IR}{I}$ **18.** $t = \dfrac{A - P}{Pr}$

19. $L = \dfrac{2s - na}{n}$

MIXED PRACTICE

Solve. See Examples 4 and 6.

29. The day's high temperature in Phoenix, Arizona, was recorded as 104°F. Write 104°F as degrees Celsius. [Use the formula $C = \frac{5}{9}(F - 32)$]   40°C

30. The annual low temperature in Nome, Alaska, was recorded as −15°C. Write −15°C as degrees Fahrenheit. [Use the formula $F = \frac{9}{5}C + 32$]   5°F

31. Omaha, Nebraska, is about 90 miles from Lincoln, Nebraska. Irania must go to the law library in Lincoln to get a document for the law firm she works for. Find how long it takes her to drive round-trip if she averages 50 mph.

32. It took the Selby family $5\frac{1}{2}$ hours round-trip to drive to and from their house to their beach house 154 miles away. Find their average speed.   56 mph

△ **33.** A package of floor tiles contains 24 one-foot-square tiles. Find how many packages should be bought to cover a square ballroom floor whose side measures 64 feet.   171 packages

△ **34.** One-foot-square ceiling tiles are sold in packages of 50. Find how many packages must be bought for a rectangular ceiling 18 feet by 12 feet.   5 packages

△ **35.** If the area of a triangular kite is 18 square feet and its base is 4 feet, find the height of the kite.   9 ft

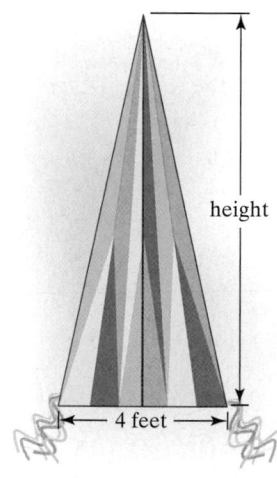

36. Bryan, Eric, Mandy, and Melissa would like to go to Disneyland in 3 years. Their total cost should be $4500. If each invests $1000 in a savings account paying 5.5% interest, compounded semiannually, will they have enough in 3 years?   yes, total of $4707.07

△ **37.** A gallon of latex paint can cover 500 square feet. Find how many gallon containers of paint should be bought to paint two coats on each wall of a rectangular room whose dimensions are 14 feet by 16 feet (assume 8-foot ceilings).   2 gal

△ **38.** A gallon of enamel paint can cover 300 square feet. Find how many gallon containers of paint should be bought to paint three coats on a wall measuring 21 feet by 8 feet.   2 gal

20. $F = \dfrac{9C + 160}{5}$ or $F = \dfrac{9}{5}C + 32$ **21.** $v = \dfrac{3st^4 - N}{5s}$ **22.** $d = \dfrac{L - a}{n - 1}$ **31.** 3.6 hr, or 3 hr and 36 min

△ **39.** A portion of the external tank of the Space Shuttle *Endeavour* is a liquid hydrogen tank. If the ends of the tank are hemispheres, find the volume of the tank. To do so, answer parts **a** through **c**.

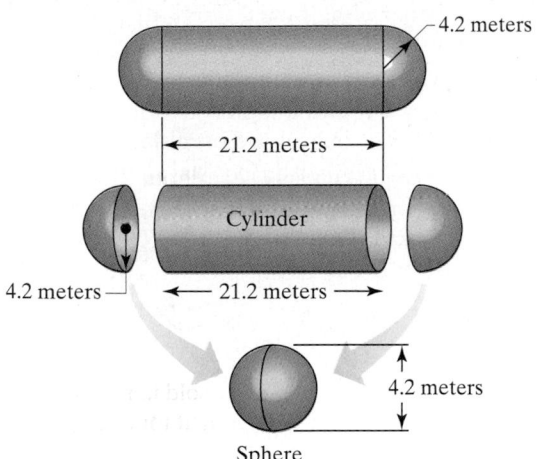

4.2 meters

21.2 meters

Cylinder

4.2 meters — ← 21.2 meters →

4.2 meters

Sphere

a. Find the volume of the cylinder shown. Round to 2 decimal places. 1174.86 cu. m

b. Find the volume of the sphere shown. Round to 2 decimal places. 310.34 cu. m

c. Add the results of parts **a** and **b**. This sum is the approximate volume of the tank. 1485.20 cu. m

40. The Cassini spacecraft mission to Saturn was launched October 15, 1997. It will take more than six and a half years to reach Saturn, arriving in July 2004. During its mission, Cassini will travel a total distance of 2 billion miles in 80.5 months. Find the average speed of the spacecraft in miles per hour. (*Hint:* Convert 80.5 months to hours using 1 month = 30 days and then use the formula $d = rt$.) (*Source:* NASA Jet Propulsion Laboratory)

△ **41.** In 1945, Arthur C. Clarke, a scientist and science-fiction writer, predicted that an artificial satellite placed at a height of 22,248 miles directly above the equator would orbit the globe at the same speed with which the Earth was rotating. This belt along the equator is known as the Clarke belt. Use the formula for circumference of a circle and find the "length" of the Clarke belt. (*Hint:* Recall that the radius of the Earth is approximately 4000 miles. Round to the nearest whole mile.) 164,921 mi

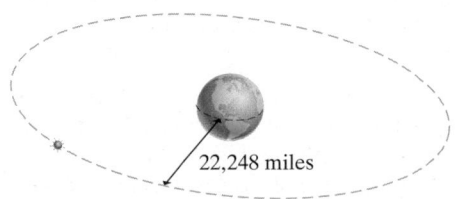

22,248 miles

△ **42.** The Space Shuttle *Endeavour* has a cargo bay that is in the shape of a cylinder whose length is 18.3 meters and whose diameter is 4.6 meters. Find its volume.
96.807π cu. m ≈ 304.12816 cu. m

△ **43.** The deepest hole in the ocean floor is beneath the Pacific Ocean and is called Hole 504B. It is located off the coast of Ecuador. Scientists are drilling it to learn more about the Earth's history. Currently, the hole is in the shape of a cylinder whose volume is approximately 3800 cubic feet and whose length is 1.3 miles. Find the radius of the hole to the nearest hundredth of a foot. (*Hint:* Make sure the same units of measurement are used.) 0.42 ft

44. The deepest man-made hole is called the Kola Superdeep Borehole. It is approximately 8 miles deep and is located near a small Russian town in the Arctic Circle. If it takes 7.5 hours to remove the drill from the bottom of the hole, find the rate that the drill can be retrieved in feet per second. Round to the nearest tenth. (*Hint:* Write 8 miles as feet, 7.5 hours as seconds, then use the formula $d = rt$.) 1.6 ft/sec

△ **45.** Eartha is the world's largest globe. It is located at the headquarters of DeLorme, a mapmaking company in Yarmouth, Maine. Eartha is 41.125 feet in diameter. Find its exact circumference (distance around) and then approximate its circumference using 3.14 for π. (*Source:* DeLorme)
41.125π ft ≈ 129.1325 ft

△ **46.** Eartha is in the shape of a sphere. Its radius is about 20.6 feet. Approximate its volume to the nearest cubic foot. (*Source:* DeLorme) 36,618 cu. ft

47. Find *how much interest* $10,000 earns in 2 years in a certificate of deposit paying 8.5% interest compounded quarterly. $1831.96

48. Find how long it takes Mark to drive 135 miles on I-10 if he merges onto I-10 at 10 A.M. and drives nonstop with his cruise control set on 60 mph. 2.25 hr, or 2 hr and 15 min

40. approximately 34,507 mph

The calorie count of a serving of food can be computed based on its composition of carbohydrate, fat, and protein. The calorie count C for a serving of food can be computed using the formula $C = 4h + 9f + 4p$, where h is the number of grams of carbohydrate contained in the serving, f is the number of grams of fat contained in the serving, and p is the number of grams of protein contained in the serving.

49. Solve this formula for *f*, the number of grams of fat contained in a serving of food.

50. Solve this formula for *h*, the number of grams of carbohydrate contained in a serving of food.

49. $f = \dfrac{C - 4h - 4p}{9}$ **50.** $h = \dfrac{C - 9f - 4p}{4}$

51. A serving of cashews contains 14 grams of fat, 7 grams of carbohydrate, and 6 grams of protein. How many calories are in this serving of cashews?   178 cal

52. A serving of chocolate candies contains 9 grams of fat, 30 grams of carbohydrate, and 2 grams of protein. How many calories are in this serving of chocolate candies?   209 cal

53. A serving of raisins contains 130 calories and 31 grams of carbohydrate. If raisins are a fat-free food, how much protein is provided by this serving of raisins?   1.5 g

54. A serving of yogurt contains 120 calories, 21 grams of carbohydrate, and 5 grams of protein. How much fat is provided by this serving of yogurt? Round to the nearest tenth of a gram.   1.8 g

Concept Extensions

55. Solar system distances are so great that units other than miles or kilometers are often used. For example, the astronomical unit (AU) is the average distance between Earth and the Sun, or 92,900,000 miles. Use this information to convert each planet's distance in miles from the Sun to astronomical units. Round to three decimal places. (*Source:* National Space Science Data Center)

Planet	Miles from the Sun	AU from the Sun	Planet	Miles from the Sun	AU from the Sun
Mercury	36 million	0.388	Saturn	886.1 million	9.538
Venus	67.2 million	0.723	Uranus	1783 million	19.193
Earth	92.9 million	1.00	Neptune	2793 million	30.065
Mars	141.5 million	1.523	Pluto	3670 million	39.505
Jupiter	483.3 million	5.202			

57. $6.80 per person

60. $n_e = \dfrac{N}{R^* \times f_p \times f_l \times f_i \times f_c \times L}$

56. answers may vary   **58.** answers may vary

56. An orbit such as Clarke's belt in Exercise 41 is called a geostationary orbit. In your own words, why do you think that communications satellites are placed in geostationary orbits?

57. How much do you think it costs each American to build a space shuttle? Write down your estimate. The space shuttle *Endeavour* was completed in 1992 and cost approximately $1.7 billion. If the population of the United States in 1992 was 250 million, find the cost per person to build the *Endeavour*. How close was your estimate?

58. If you are investing money in a savings account paying a rate of *r*, which account should you choose—an account compounded 4 times a year or 12 times a year? Explain your choice.

59. To borrow money at a rate of *r*, which loan plan should you choose—one compounding 4 times a year or 12 times a year? Explain your choice.   answers may vary

60. The Drake Equation is a formula used to estimate the number of technological civilizations that might exist in our own Milky Way Galaxy. The Drake Equation is given as $N = R^* \times f_p \times n_e \times f_l \times f_i \times f_c \times L$. Solve the Drake Equation for the variable n_e. (*Note:* Descriptions of the meaning of each variable in this equation, as well as Drake Equation calculators, exist online. For more information, try doing a Web search on "Drake Equation.")

61. On April 1, 1985, *Sports Illustrated* published an April Fool's story by writer George Plimpton. He wrote that the New York Mets had discovered a man who could throw a 168-miles-per-hour fast ball. If the distance from the pitcher's mound to the plate is 60.5 feet, how long would it take for a ball thrown at that rate to travel that distance? (*Hint:* Write the rate 168 miles per hour in feet per second. Then use the formula $d = r \cdot t$.)   0.25 sec

$$168 \text{ miles per hour} = \frac{168 \text{ miles}}{1 \text{ hour}}$$

$$= \frac{\underline{} \text{ feet}}{\underline{} \text{ seconds}}$$

$$= \frac{\underline{} \text{ feet}}{1 \text{ second}}$$

$$= \underline{} \text{ feet per second}.$$

The measure of the chance or likelihood of an event occurring is its **probability**. A formula basic to the study of probability is the formula for the probability of an event when all the outcomes are equally likely. This formula is

$$\text{Probabilit y of an event} = \frac{\begin{array}{c}\text{number of ways that}\\ \text{the event can occur}\end{array}}{\begin{array}{c}\text{number of possible}\\ \text{outcomes}\end{array}}$$

For example, to find the probability that a single spin on the spinner will result in red, notice first that the spinner is divided into 8 parts, so there are 8 possible outcomes. Next, notice that there is only one sector of the spinner colored red, so the number of ways that the spinner can land on red is 1. Then this probability denoted by P(red) is

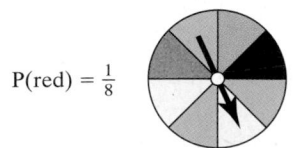

$P(red) = \frac{1}{8}$

Find each probability in simplest form.

62. P(green) $\frac{1}{8}$

63. P(yellow) $\frac{1}{4}$

64. P(black) $\frac{1}{8}$

65. P(blue) $\frac{3}{8}$

66. P(green or blue) $\frac{1}{2}$

67. P(black or yellow) $\frac{3}{8}$

68. P(red, green, or black) $\frac{3}{8}$

69. P(yellow, blue, or black) $\frac{3}{4}$

70. P(white)   0

71. P(red, yellow, green, blue, or black)   1

72. From the previous probability formula, what do you think is always the probability of an event that is impossible occuring?   0

73. What do you think is always the probability of an event that is sure to occur?   1

1.9   *INTERPRETING DATA AND READING BAR, LINE, AND CIRCLE GRAPHS*

Objectives

1 Find the mean, median, and mode of a set of data.

2 Interpret circle graphs.

3 Interpret bar graphs.

4 Interpret line graphs.

1   Many real-world situations involve problem solving based on collecting, graphing, and analyzing data. Computers can quickly organize and analyze the data whether they are collected by means of a survey, a collection device such as the Calculator Based Laboratory that can be linked to the TI calculators, or simply by using numerical methods. Once we collect data we need to organize and interpret the results.

To organize and analyze data, it is sometimes desirable to be able to describe the set of data, or a set of numbers, by a single "middle" number. Three such **measures of central tendency** are the *mean*, the *median*, and the *mode*.

The most common measure of central tendency is the mean (sometimes called the arithmetic mean or the average). The **mean** of a set of data items, denoted by $\overline{x}$, is the sum of the items divided by the number of items.

EXAMPLE 1

ANALYZING A PSYCHOLOGY EXPERIMENT

Seven students in a psychology class conducted an experiment on mazes. Each student was given a pencil and asked to successfully complete the same maze. The time results are below.

Student	Ann	Thanh	Carlos	Jesse	Melinda	Ramzi	Dayni
Time (Seconds)	13.2	11.8	10.7	16.2	15.9	13.8	18.5

a. What was the shortest time and the longest time for the maze to be completed?

b. Find the mean of the times.

c. How many students took more than the mean time? How many students took less than the mean time?

Solution

a. Carlos completed the maze in 10.7 seconds, the shortest time. Dayni completed the maze in 18.5 seconds, the longest time.

b. To find the mean, $\overline{x}$, find the sum of the data items and divide by 7, the number of items.

$$\overline{x} = \frac{13.2 + 11.8 + 10.7 + 16.2 + 15.9 + 13.8 + 18.5}{7} = \frac{100.1}{7} = 14.3 \text{ seconds}$$

c. Three students, Jesse, Melinda, and Dayni, had times longer than the mean time. Four students, Ann, Thanh, Carlos, and Ramzi, had times shorter than the mean time.

Two other measures of central tendency are the median and the mode.

The **median** of an ordered set of numbers is the middle number. If the number of items is even, the median is the mean of the two middle numbers. The **mode** of a set of numbers is the number that occurs most often. It is possible for a data set to have no mode or more than one mode.

EXAMPLE 2

ANALYZING TEMPERATURE

Find the median and the mode of the following set of numbers. These numbers were high temperatures for fourteen consecutive days in a city in Montana.

$$76, 80, 85, 86, 89, 87, 82, 77, 76, 79, 82, 89, 89, 92$$

Solution   First, write the numbers in order.

$$76, 76, 77, 79, 80, 82, \underbrace{82, 85}_{\substack{\text{two} \\ \text{middle numbers}}}, 86, 87, \underbrace{89, 89, 89}_{\text{mode}}, 92$$

Since there are an even number of items, the median is the mean of the two middle numbers.

$$\text{median} = \frac{82 + 85}{2} = 83.5$$

The mode is 89, since 89 occurs most often.

If a data set contains a large number of items, calculating the mean, median, and mode by hand can be tedious. In this section we introduce **lists** to help us to calculate these measures of central tendency. A list is different from a table in many ways. Whereas a *table* may give *x* and *y* values that are related by some equation, it is possible to enter a *list* of single values only.

EXAMPLE 3

FINDING THE MEAN AND THE MEDIAN OF A SET OF DATA

Find the mean and the median of the following set of temperatures for a given week.

Mon	*Tues*	*Wed*	*Thurs*	*Fri*	*Sat*	*Sun*
89	82	76	80	68	70	74

Solution   To find the mean by hand, find the sum of the temperatures and divide by 7.

$$\frac{89 + 82 + 76 + 80 + 68 + 70 + 74}{7} = \frac{539}{7} = 77, \text{ or } 77°F.$$

To find the median by hand, arrange the data in numerical order and find the middle entry.

68 70 74 76 80 82 89

Since there is an odd number of entries, the median is the middle entry, or fourth entry, 76, which represents 76°F.

To find the mean and the median using a graphing utility, access the list in the statistics feature of your calculator and enter the given temperatures in the first list L_1. Next, access a list math menu and select the mean and median entries.

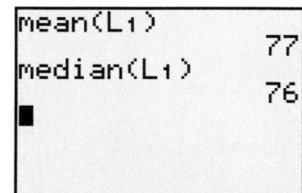

We see that the mean is 77°F and the median is 76°F.

EXAMPLE 4

PRESIDENTS' AGE PROBLEM

Here is a list of the first 43 presidents of the United States and their ages when they took office. Use a graphing utility to find the mean, median, and mode of the ages.

1.	George Washington	57	**6.**	John Quincy Adams	57
2.	John Adams	61	**7.**	Andrew Jackson	61
3.	Thomas Jefferson	57	**8.**	Martin Van Buren	54
4.	James Madison	58	**9.**	William Harrison	68
5.	James Monroe	58	**10.**	John Tyler	51

11. James Polk   49
12. Zachary Taylor   64
13. Millard Fillmore   50
14. Franklin Pierce   48
15. James Buchanan   65
16. Abraham Lincoln   52
17. Andrew Johnson   56
18. Ulysses S. Grant   46
19. Rutherford Hayes   54
20. James Garfield   49
21. Chester Arthur   50
22. Grover Cleveland   47
23. Benjamin Harrison   55
24. Grover Cleveland   55
25. William McKinley   54
26. Theodore Roosevelt   42
27. William Taft   51

28. Woodrow Wilson   56
29. Warren Harding   55
30. Calvin Coolidge   51
31. Herbert Hoover   54
32. Franklin Roosevelt   51
33. Harry Truman   60
34. Dwight Eisenhower   62
35. John Kennedy   43
36. Lyndon Johnson   55
37. Richard Nixon   56
38. Gerald Ford   61
39. Jimmy Carter   52
40. Ronald Reagan   69
41. George Bush   64
42. William Clinton   46
43. George W. Bush   54

Solution   Enter the data in a list, and find the mean, median, and mode of the ages.

   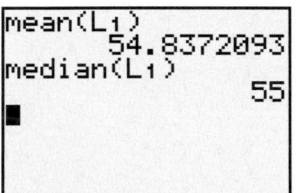

We can see that the mean is about 54.8 years and the median is 55 years. To find the mode, use a sort feature to *arrange the ages in ascending order* and scroll to count the number of multiple entries.

The mode is 54 years.

2   Throughout Chapter 1, we have slowly introduced different types of graphs. In this section, we go one step further and interpret data from graphs. The next type of graph is called a circle graph or pie chart. The circle is divided into sectors that represent different parts of the whole. The circle represents the whole, in this case 100%.

EXAMPLE 5

INTERPRETING DATA FROM A CIRCLE GRAPH

United States astronaut Shannon Lucid spent more hours in space than any other American at the time of her return in September 1996. The circle graph below represents the percent of time spent in space by Lucid, Norm Thagard, Skylab 4 crew, Skylab 3 crew, and Skylab 2 crew. If the total number of hours recorded by the five components is 11,398 hours, approximate the number of hours that each spent in space. (*Source*: Lyndon B. Johnson Space Center.)

Space Marathoners

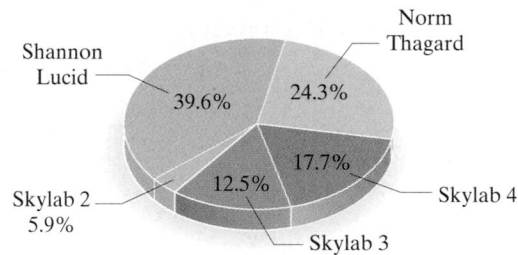

CLASSROOM EXAMPLE

Use the circle graph in Example 5 to answer the following.

a. What percent of the circle graph is the number of hours for the Skylab 2, 3, or 4 crew?

b. Find the total number of hours that Skylab 2, 3 and 4 spent in space.

answer: a. 36.1%   **b.** 4115 rounded to the nearest whole.

Solution   To approximate the number of hours that each spent in space, we find percents of the total hours.

Lucid:	39.6% of 11,398 or (0.396)(11,398) ≈ 4514 hours
Thagard:	24.3% of 11,398 or (0.243)(11,398) ≈ 2770 hours
Skylab 4 crew:	17.7% of 11,398 or (0.177)(11,398) ≈ 2017 hours
Skylab 3 crew:	12.5% of 11,398 or (0.125)(11,398) ≈ 1425 hours
Skylab 2 crew:	5.9% of 11,398 or (0.059)(11,398) ≈ 672 hours

Notice that the percents have a sum of 100% and the hours found, although they are rounded, have a sum of 11,398.

3   Another common type of graph is the bar graph. A bar graph consists of a series of bars arranged vertically or horizontally as shown next.

EXAMPLE 6

INTERPRETING DATA FROM A BAR GRAPH

The bar graph below shows the monthly average temperatures for Albany, New York.

Source: The World Almanac

CLASSROOM EXAMPLE

Use the bar graph in Example 6 to answer the following.

a. Approximate the monthly average temperature for May.

b. Approximate the monthly average temperature for November.

answer: a. 58°   **b.** 39°

a. Approximate the highest monthly average temperature.

b. Approximate the lowest monthly average temperature.

c. Which month shows the greatest increase in temperature?

d. Which month shows the greatest decrease in temperature?

e. Discuss any patterns noticed in this graph.

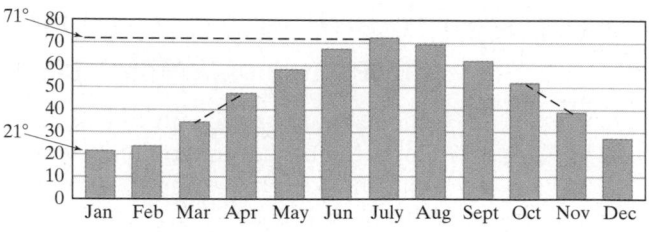

Source: The World Almanac

a. The highest monthly average temperature is approximately 71 °F in July.

b. The lowest monthly average temperature is approximately 21 °F in January.

c. The month of April shows the greatest increase in monthly temperature. (To see this, use a straightedge and see that the temperature is increasing the fastest when the straightedge is the steepest.)

d. The month of November shows the greatest decrease.

e. The graph clearly shows the temperature climbing from January until July and dropping from July to January.

4 A third type of graph is a broken line graph, or simply a line graph.

EXAMPLE 7

INTERPRETING DATA FROM A LINE GRAPH

The broken line graph shows the incidence of cancer among Americans from 1975 to the beginning of 2004.

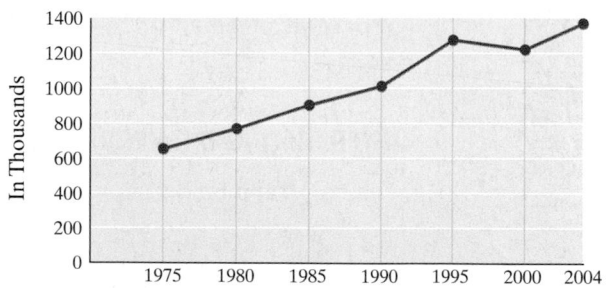

Source: *Cancer Facts and Figures*, American Cancer Society, 2000.

a. Approximate the number of people with cancer in 1980.

b. Approximate the number of people with cancer in 1995.

c. Which year shows the greatest increase in the incidence of cancer?

d. In what year shown did the cancer rate rise above 1,200,000?

e. Discuss any patterns that you notice.

CLASSROOM EXAMPLE

Use the graph in Example 7 to answer the following.

a. Approximate the number of people with cancer in 2000.

b. Which year shown shows a decrease in the incidence of cancer?

c. In what year was the cancer rate about 900,000?

answer: a. 1,230,000 **b.** 2000 **c.** 1985

Solution

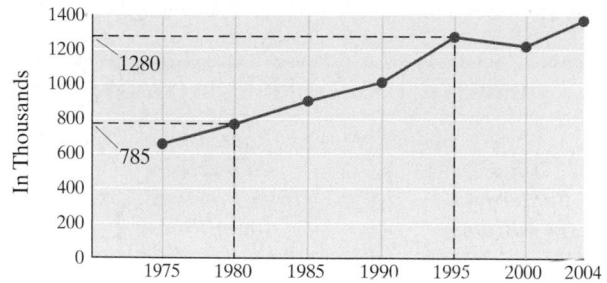

Source. *Cancer Facts and Figures*, American
Cancer Society, 2000.

a. The incidence of cancer in 1980 was approximately 785,000.

b. The incidence of cancer in 1995 was approximately 1,280,000.

c. The greatest increase appears to be in the year 1995.

d. The cancer rate rose above 1,200,000 in the year 1990, and continued to be above 1,200,000.

e. This graph clearly shows that overall, the incidence of cancer is increasing.

EXERCISE SET 1.9

STUDY   CD/   PH MATH   MathXL®Tutorials   MathXL®   MyMathLab®
GUIDE/SSM   VIDEO   TUTOR CENTER   ON CD

Use the graph below for Exercises 1 and 2.

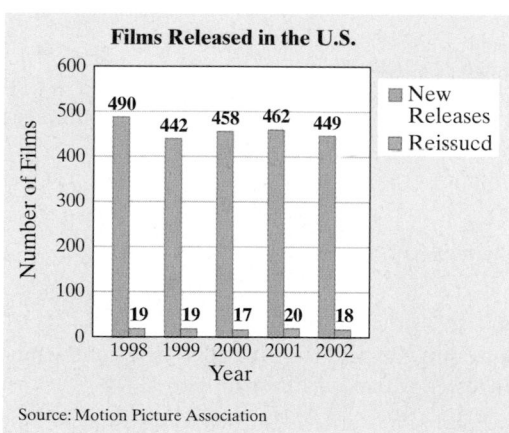

Source: Motion Picture Association

1. Find the mean number of reissued films for the years 1998 through 2002.   18.6

2. Find the mean number of new released films for the years 1998 through 2002.   460.2

Use the table below for Exercises 3 and 4. If necessary, round answers to two decimal places.

MOVIE THEATER AVERAGE ANNUAL ADMISSION PRICE		
Year	*Ave. Annual Admission Price**	*Increase in Admission Price from Previous Year*
1998	$4.69	
1999	$5.08	$0.39
2000	$5.39	$0.31
2001	$5.66	$0.27
2002	$5.81	$0.15

*Price includes all special pricing. Source: Motion Picture Asso.

3. Find the mean annual admission price of a movie for the years 1998 through 2002. $5.33

4. **a.** Complete the table shown by computing the increase of the admission price from the previous year. The first entry has been completed for you.

   b. Find the average change in price per year by finding the mean of the numbers in the increase column. $0.28

Use the table below for Exercises 5 through 8. If necessary, round answers to one decimal place.

DOMESTIC U.S. COMMERCIALLY-OPERATIONAL WIRELESS PROVIDERS

Year	Total Subscribers (in millions)	Increase in Subscribers from Previous Year (in millions)	Number of Cell Sites (in thousands)	Increase in Cell Sites from Previous Year (in thousands)
1993	13.1		12	
1994	19.3	6.2	15	3
1995	28.2	8.9	20	5
1996	38.2	10	25	5
1997	48.7	10.5	39	14
1998	60.8	12.1	58	19
1999	76.3	15.5	74	16
2000	97.0	20.7	96	22
2001	118.4	21.4	114	18
2002	134.6	16.2	131	17

Source: Cellular Telecommunications and Internet Asso.

5. Find the mean number of cell phone subscribers for the years 2000 through 2002.   116.7 million

6. Find the mean number of cell sites for the years 1993 through 2002.   58.4 thousand

7. a. Complete the column entitled *Increase in Cell Sites*. The first few entries have been computed.

   b. Find the average yearly increase in cell sites for the years shown by finding the mean of the numbers in the *Increase in Cell Sites* column.   13.2 thousand

8. a. Complete the column entitled *Increase in Subscribers*. The first few entries have been computed.

   b. Find the average yearly increase in cell phone subscribers for the years shown by finding the mean of the numbers in the *Increase in Subscribers* column.   13.5 million

9. Given the following grades for a particular history exam, find the mode(s) of the set of data.   85

92, 75, 85, 91, 83, 85, 92, 90, 85, 87, 76, 93, 81

10. Given the following temperatures recorded for a given week, find the mode(s) of the set of data.   85

90, 86, 85, 82, 85, 86, 85

Financial planners say that "empty nesters" should save the money that they previously spent on their children in a mutual fund portfolio. A possible mutual fund portfolio for someone 45 to 50 years old is exhibited in the following circle graph. Use this graph for Exercises 11 and 12. See Example 5.

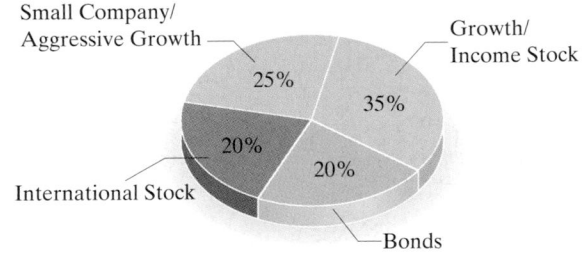

Mutual Fund Portfolio

Small Company/Aggressive Growth — 25%
Growth/Income Stock — 35%
International Stock — 20%
Bonds — 20%

11. If an "empty nester" has $15,000 to invest, how much should be invested in each category?

12. If an "empty nester" has $40,000 to invest, how much should be invested in each category?

13. GROUP ACTIVITY. Form groups of 9 to 10 people. Measure each person's height. Record the heights in L_1 and find the mean, median, and mode of the data.   answers may vary

14. GROUP ACTIVITY. Form groups of 9 to 10 people. Measure each person's length of foot in centimeters. Record the lengths in L_1 and find the mean, median, and mode of the data.   answers may vary

11. Growth/Income: $5250; Small Company/Aggressive Growth: $3750; International: $3000: Bonds: $3000
12. Growth/Income: $14.000: Small Company/Aggressive Growth: $10,000: International: $8000: Bonds: $8000

The ten tallest buildings in the world and in the USA are listed in the table below. Use this table for Exercises 15 through 18.

	World	City	Height (meters)	USA	City	Height (meters)
1.	Taipei 101	Taipei	509	Sears Tower	Chicago	442
2.	Petronas Tower 1	Kuala Lumpar	452	Empire State Building	New York City	381
3.	Petronas Tower 2	Kuala Lumpur	452	Aon Center	Chicago	346
4.	Sears Tower	Chicago	442	John Hancock Center	Chicago	344
5.	Jin Mao Tower	Shanghai	421	Chrysler Building	NYC	319
6.	Two International Finance	Hong Kong	415	Bank of America Plaza	Atlanta	312
7.	Citic Plaza	Guangzhou	391	US Bank Tower	Los Angeles	310
8.	Shun Hing Square	Shenzhen	384	AT & T Center	Chicago	307
9.	Empire State Building	New York City	381	JP Morgan Chase Tower	Houston	305
10.	Central Plaza	Hong Kong	374	Two Prudential Plaza	Chicago	303

15. Find the mean height for the ten tallest buildings in the world.   422.1 m

16. Find the mean height for the ten tallest buildings in the USA.   336.9 m

17. Find the median height for the ten tallest buildings in the world.   418 m

18. Find the median height for the ten tallest buildings in the USA.   315.5 m

During an experiment, the following times (in seconds) were recorded: 7.8, 6.9, 7.5, 4.7, 6.9, 7.0.

19. Find the mean. Round to the nearest tenth.   6.8 sec

20. Find the median.   6.95 sec   **21.** Find the mode.   6.9 sec

In a mathematics class, the following test scores were recorded for a student: 86, 95, 91, 74, 77, 85.

22. Find the mean. Round to the nearest hundredth.   84.67

23. Find the median.   85.5   **24.** Find the mode.   No mode

The following pulse rates were recorded for a group of fifteen students: 78, 80, 66, 68, 71, 64, 82, 71, 70, 65, 70, 75, 77, 86, 72.

25. Find the mean.   73   **26.** Find the median.   71

27. Find the mode.   70 and 71

28. How many rates were higher than the mean?   6

29. How many rates were lower than the mean?   9

30. Have each student in your algebra class take his or her pulse rate. Record the data and find the mean, the median, and the mode.   answers may vary

Find the missing numbers in each set of numbers. (These numbers are not necessarily in numerical order.)

31. __, __, 16, 18, __
The mode is 21.
The median is 20.
21, 21, 20

32. __, __, __, __, 40
The mode is 35.
The median is 37.
The mean is 38.
35, 35, 37, 43

The following bar graph shows the professions that walk the farthest in average miles walked per year. Use this graph for Exercises 33 through 38. See Example 6. (Source: Dr. Scholl's and the American Podiatry Association)

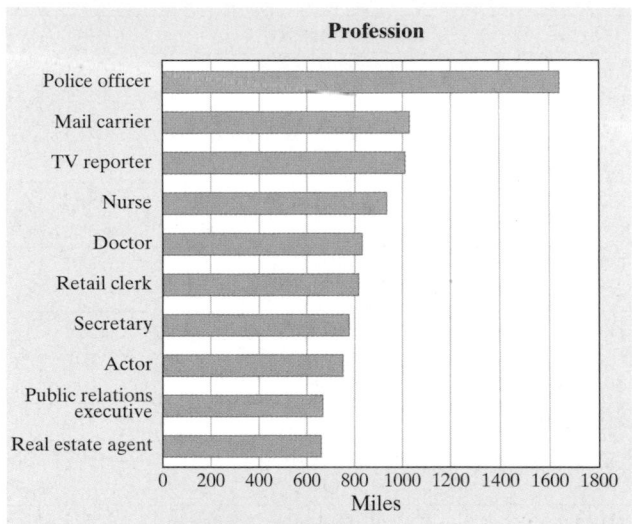

Profession

33. Which profession walks the farthest?   Police officer

34. Which profession does the least amount of walking?   Real estate agent

35. Estimate the average miles walked per year by a nurse.   940 mi

36. Estimate the average miles walked per year by a retail clerk.   810 mi

37. Estimate the number of miles you walk in a week. Multiply this number by 52 to find the average number of miles you walk in a year. How does your distance compare with the distances in the bar graph?   answers may vary

38. Estimate how much farther a police officer walks in a year than a mail carrier.   620 mi

The following graph shows the political party identification of the adult population for the years shown. Use this graph for Exercises 39 through 46. See Example 7.

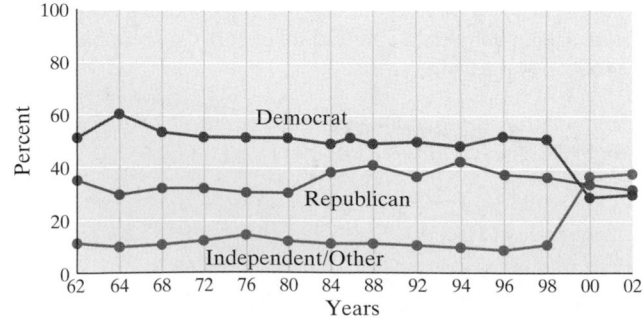

39. Which political party shows the greatest percent of the adult population for most of the years shown?   Democrat

40. Which political party shows the least percent of the adult population for most of the years shown?   Independent

41. What year shows the greatest difference in percent of Democrats and percent of Republicans?   1964

42. What year shows the least difference in percent of Democrats and percent of Republicans?   2002

43. Estimate the percent of adult population identified as Democrat, Republican, and Independent/Other in the year 1992.   Democrat: 50%: Republican: 38%: Independent: 10%

44. Estimate the percent of adult population identified as Democrat, Republican, and Independent/Other in the year 1964.   Democrat: 60%: Republican: 30%: Independent: 10%

45. Discuss any patterns you notice in the graph.   answers may vary

46. Do you think the percents for any given year have a sum of 100%? Why or why not?   answers may vary

According to USA Today, parents spend an average of $363 per child on back-to-school items for a school year, and 58% of this cost will be spent on clothes. The following circle graph shows the distribution method used to pay for the clothes. Use this graph for Exercises 47 through 50. If necessary, round amounts to the nearest hundredth.

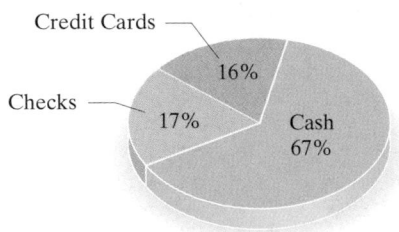

School Clothes Budget

47. What was the average amount of money spent on purchasing clothes? $210.54

48. What was the average amount paid in cash for clothes? $141.06

49. What was the average amount paid by credit card for clothes? $33.69

50. What was the average amount paid by checks for clothes? $35.79

The data below represents the 2015 projected population in millions of the world's 15 largest cities. Use this data to answer Exercises 51 through 54. (Source: The World Almanac.)

28.7	17.6	20.8	18.8	23.4	27.4
14.3	19.4	17.6	13.1	21.2	12.4
10.6	17.0	11.6			

51. Find the mean of the 5 most populous cities.   24.3 million
52. Find the median of the 5 most populous cities.   23.4 million
53. Find the median of the 15 most populous cities.   17.6 million
54. Find the mean of the 15 most populous cities.   18.26 million

Concept Extensions

55. A survey was conducted to determine which age category knows their history best. When asked to name the state where the final battle of the Revolutionary War was fought, these percents of each age group correctly named Virginia. Make a bar graph using the data listed below. (*Source: USA Today* and Opinion Research for Colonial Williamsburg Foundation.)   See graphing answer section.

Age	18–24	25–34	35–44	45–54	55–64	65 +
Percent	44	38	43	43	51	43

56. Many business and government computers were not set up to handle the year 2000. Their computers would falsely read 00 as 1900. A survey was conducted to see how aware companies were of the "century problem." The data listed below was from a *USA Today* and Ofsten Corporation survey of senior executives. Construct a circle graph using the data.   See graphing answer section.

Conversion program under way	28%
Planning conversion	34%
Unaware of the problem	13%
Not a problem for our company	21%
Don't know	4%

57. **GROUP ACTIVITY.** Decide on a topic and survey your algebra class. Some suggested topics are favorite brand of soft drink given 5 choices, favorite leisure activity, number of hours spent each week watching television, or the best time of day to study. Tabulate your data and use it to construct a bar graph or a circle graph. Make sure that your graph is labeled properly.   answers may vary

58. The annual starting salaries for 5 levels of employees at a local factory are $15,000, $22,000, $30,000, $60,000, and $110,000. During a salary dispute, the employees claim that a typical employee makes approximately $30,000 a year (the median). The owners of the factory claim that a typical employee makes almost $50,000 a year (the mean). Which number do you think is a better representation of the situation? Why?   answers may vary

59. Suppose that a set of data contains only numbers between 21.6 and 46.8. Can the mean of this data set be 20.2? Why or why not?   no; answers may vary

60. In a set of data, suppose that the largest number (not the mode) is replaced with a larger number. How does this affect the mean, median, and mode?   The mean is raised; the median and mode are unchanged.

61. In a set of data, suppose that the smallest number (not the mode) is replaced with a smaller number. How does this affect the mean, median, and mode?   The mean is lowered; the median and mode are unchanged.

62. Suppose that a set of data consists of positive integers. Is it possible for the mean not to be an integer? Is it possible for the median not to be an integer? Is it possible for the mode not to be an integer?   yes; yes; no

STUDY SKILLS REMINDER

Are You Preparing for a Test on Chapter 1?

Below I have listed some common trouble areas for students in Chapter 1. After studying for your test—but before taking your test—read these.

▶ Do you remember the meaning of a negative exponent?

$$7^2 = 7 \cdot 7 = 49$$

▶ Don't forget the order of operations and the distributive property.

$$7 - 3(2x - 6y) + 5 = 7 - 6x + 18y + 5 \quad \text{Use the distributive}$$
$$\uparrow \qquad\qquad\qquad \text{property.}$$
$$\text{Notice the sign.}$$
$$= -6x + 18y + 12 \quad \text{Combine like terms.}$$

▶ Don't forget the difference between $(-3)^2$ and -3^2.

$$(-3)^2 = (-3)(-3) = 9$$
$$-3^2 = -(3 \cdot 3) = -9$$

▶ Remember that

$$\frac{0}{8} = 0 \text{ while } \frac{8}{0} \text{ is undefined.}$$

▶ Don't forget the difference between reciprocal and opposite.

The opposite of $-\dfrac{3}{5}$ is $\dfrac{3}{5}$.

The reciprocal of $-\dfrac{3}{5}$ is $-\dfrac{5}{3}$.

Remember: This is simply a checklist of common trouble areas. For a review of Chapter 1, see the Highlights and Chapter Review at the end of Chapter 1.

CHAPTER 1 PROJECT

Analyzing Newspaper Circulation

The number of daily newspapers in business in the United States has declined steadily recently, continuing a trend that started in the mid-1970s. In 1900, there were roughly 2300 daily newspapers in operation. However, by 1998, that number had dropped to only 1509 newspapers in existence. Average overall daily newspaper circulation also continues to decline, from 60.2 million in 1992 to 56.7 million in 1997.

The table on the next page gives data about daily newspaper circulation for New York City–based newspapers for the years 2000 and 2001. In this project, you will have the opportunity to analyze this data. This project may be completed by working in groups or individually.

New York–based Newspaper	2000 Daily Circulation	2001 Daily Circulation	Daily Edition Newsstand Price
Wall Street Journal	1,762,751	1,780,605	$1.00
New York Times	1,097,180	1,109,371	$1.00
New York Daily News	704,463	734,473	$0.50
New York Post	443,951	533,860	$0.50

1. Find the change in circulation from 2000 to 2001 for each newspaper. Did any newspaper gain circulation? If so, which one(s) and by how much?

2. Construct a bar graph showing the change in circulation from 2000 to 2001 for each newspaper. Which newspaper experienced the largest change in circulation?

3. What was the total daily circulation of these New York City–based newspapers in 2000? In 2001? Did total circulation increase or decrease from 2000 to 2001? By how much?

4. Discuss factors that may have contributed to the overall change in daily newspaper circulation.

5. The population of New York City is approximately 7,381,000. Find the number of New York City–based newspapers sold per person in 2001 for New York City. (*Source:* U.S. Bureau of the Census)

6. The population of the New York City metropolitan area is approximately 16,332,000. Find the number of New York City–based newspapers sold per person in 2001 for the New York City metropolitan area. (*Source:* The United Nations)

7. Which of the figures found in Questions 5 and 6 do you think is more meaningful? Why? Why might neither of these figures be capable of describing the full circulation situation?

8. Assuming that each copy was sold from a newsstand in the New York City metropolitan area, use the daily edition newspaper prices given in the table to approximate the total amount spent each day on these New York City–based newspapers in 2001. Find the total amount spent annually on these daily newspapers in 2001.

9. How accurate do you think the figures you found in Question 8 are? Explain your reasoning.

CHAPTER VOCABULARY CHECK

Fill in each blank with one of the words or phrases listed below.

identity   distributive   real   reciprocals
solution   absolute value   opposite   associative
formula   inequality   commutative   whole   contradiction
algebraic expression   exponent   variable   consecutive integers   linear equation in one variable

1. A(n) algebraic expression is formed by numbers and variables connected by the operations of addition, subtraction, multiplication, division, raising to powers, and/or taking roots.
2. The opposite of a number a is $-a$.
3. $3(x - 6) = 3x - 18$ by the distributive property.
4. The absolute value of a number is the distance between that number and 0 on the number line.
5. A(n) exponent is a shorthand notation for repeated multiplication of the same factor.
6. A letter that represents a number is called a variable.
7. The symbols $<$ and $>$ are called inequality symbols.
8. If a is not 0, then a and $1/a$ are called reciprocals.
9. $A + B = B + A$ by the commutative property.
10. $(A + B) + C = A + (B + C)$ by the associative property.
11. The numbers $0, 1, 2, 3, \ldots$ are called whole numbers.
12. If a number corresponds to a point on the number line, we know that number is a real number.
13. An equation in one variable that has no solution is called a(n) contradiction.
14. An equation in one variable that has every number (for which the equation is defined) as a solution is called a(n) identity.
15. The equation $d = rt$ is also called a(n) formula.
16. When a variable in an equation is replaced by a number and the resulting equation is true, then that number is called a(n) solution of the equation.
17. The integers 17, 18, 19 are examples of consecutive integers.
18. The statement $5x - 0.2 = 7$ is an example of a(n) linear equation in one variable.

CHAPTER 1 HIGHLIGHTS

Definitions and Concepts	**Examples**

Section 1.2 Algebraic Expressions and Sets of Numbers

Letters that represent numbers are called **variables.**

An **algebraic expression** is formed by numbers and variables connected by the operations of addition, subtraction, multiplication, division, raising to powers, and/or taking roots.

To **evaluate** an algebraic expression containing variables, substitute the given numbers for the variables and simplify. The result is called the **value** of the expression.

Natural numbers: $\{1, 2, 3, \ldots\}$
Whole numbers: $\{0, 1, 2, 3, \ldots\}$
Integers: $\{\ldots, -3, -2, -1, 0, 1, 2, 3, \ldots\}$
Each listing of three dots above is called an **ellipsis,** which means the pattern continues.
The members of a set are called its **elements.**
Set builder notation describes the elements of a set but does not list them.
Real numbers: $\{x \mid x$ corresponds to a point on the number line.$\}$
Rational numbers: $\{\frac{a}{b} \mid a$ and b are intergers and $b \neq 0\}$.
Irrational numbers: $\{x \mid x$ is a real number and x is not a rational number$\}$.
If 3 is an element of set A, we write $3 \in A$.
If all the elements of set A are also in set B, we say that set A is a **subset** of set B, and we write $A \subseteq B$.
Absolute value:

$$|a| = \begin{cases} a \text{ if } a \text{ is 0 or a positive number} \\ -a \text{ if } a \text{ is a negative number} \end{cases}$$

The opposite of a number a is the number $-a$.

Examples of variables are
$$x, a, m, y$$
Examples of algebraic expressions are

$$7y, -3, \frac{x^2 - 9}{-2} + 14x, \sqrt{3} + \sqrt{m}$$

Evaluate $2.7x$ if $x = 3$.
$$2.7x = 2.7(3)$$
$$= 8.1$$

Given the set $\{-9.6, -5, -\sqrt{2}, 0, \frac{2}{5}, 101\}$ list the elements that belong to the set of

Natural numbers 101

Whole numbers $0, 101$

Integers $-5, 0, 101$

Real numbers $-9.6, -5, -\sqrt{2}, 0, \frac{2}{5}, 101$

Rational numbers $-9.6, -5, 0, \frac{2}{5}, 101$

Irrational numbers $-\sqrt{2}$

List the elements in the set
$\{x \mid x$ is an integer between -2 and $5\}$.

$$\{-1, 0, 1, 2, 3, 4\}$$
$$\{1, 2, 4\} \subseteq \{1, 2, 3, 4\}.$$
$$|3| = 3, |0| = 0, |-7.2| = 7.2$$

```
abs(3)
              3
abs(0)
              0
abs(-7.2)
            7.2
■
```

The opposite of 5 is -5. The opposite of -11 is 11.

Section 1.3 Operations on Real Numbers

Adding real numbers:

1. To add two numbers with the same sign, add their absolute values and attach their common sign.

2. To add two numbers with different signs, subtract the smaller absolute value from the larger absolute value and attach the sign of the number with the larger absolute value.

$$20.8 + (-10.2) = 10.6$$

$$-18 + 6 = -12$$
$$\frac{2}{7} + \frac{1}{7} = \frac{3}{7}$$
$$-5 + (-2.6) = -7.6$$

```
-18+6
            -12
2/7+1/7▶Frac
            3/7
-5+(-2.6)
           -7.6
■
```

(continued)

Definitions and Concepts	**Examples**

Section 1.3   Operations on Real Numbers

Subtracting real numbers:

$$a - b = a + (-b)$$

Multiplying and dividing real numbers:

The product or quotient of two numbers with the same sign is positive.

The product or quotient of two numbers with different signs is negative.

$$18 - 21 = 18 + (-21) = -3$$

$$(-8)(-4) = 32 \qquad \frac{-8}{-4} = 2$$

$$8 \cdot 4 = 32 \qquad \frac{8}{4} = 2$$

$$-17 \cdot 2 = -34 \qquad \frac{-14}{2} = -7$$

$$4(-1.6) = -6.4 \qquad \frac{22}{-2} = -11$$

A natural number **exponent** is a shorthand notation for repeated multiplication of the same factor.

The notation $\sqrt{a}$ is used to denote the **positive**, or **principal, square root** of a nonnegative number a.

$$\sqrt{a} = b \text{ if } b^2 = a \text{ and } b \text{ is positive}.$$

Also,

$$\sqrt[3]{a} = b \text{ if } b^3 = a$$
$$\sqrt[4]{a} = b \text{ if } b^4 = a \text{ and } b \text{ is positive}$$

$$3^4 = 3 \cdot 3 \cdot 3 \cdot 3 = 81$$

$$\sqrt{49} = 7$$
$$\sqrt[3]{64} = 4$$
$$\sqrt[4]{16} = 2$$

Order of Operations

Simplify expressions using the order that follows. If grouping symbols such as parentheses are present, simplify expressions within those first, starting with the innermost set. If fraction bars are present, simplify the numerator and denominator separately.

1. Raise to powers or take roots in order from left to right.

2. Multiply or divide in order from left to right.

3. Add or subtract in order from left to right.

Simplify $\dfrac{42 - 2(3^2 - \sqrt{16})}{-8}$.

$$\frac{42 - 2(3^2 - \sqrt{16})}{-8} = \frac{42 - 2(9 - 4)}{-8}$$

$$= \frac{42 - 2(5)}{-8}$$

$$= \frac{42 - 10}{-8}$$

$$= \frac{32}{-8} = -4$$

To **evaluate** an algebraic expression containing variables, substitute the given numbers for the variables and simplify. The result is called the **value** of the expression.

Evaluate $-2.7x$ if $x = 3$.

$$-2.7x = -2.7(3)$$
$$= -8.1$$

Section 1.4   Properties of Real Numbers

Symbols : $=$ is equal to
$\neq$ is not equal to
$>$ is greater than
$<$ is less than
$\geq$ is greater than or equal to
$\leq$ is less than or equal to

$$-5 = -5$$
$$-5 \neq -3$$
$$1.7 < 1.2$$
$$-1.7 < -1.2$$
$$\frac{5}{3} \geq \frac{5}{3}$$
$$-\frac{1}{2} \leq \frac{1}{2}$$

(continued)

Definitions and Concepts	**Examples**

Section 1.4 Properties of Real Numbers

Identity:
$$a + 0 = a \qquad 0 + a = a$$
$$a \cdot 1 = a \qquad 1 \cdot a = a$$

Inverse:
$$a + (-a) = 0 \qquad -a + a = 0$$
$$a \cdot \frac{1}{a} = 1 \qquad \frac{1}{a} \cdot a = 1, a \neq 0$$

Commutative:
$$a + b = b + a$$
$$a \cdot b = b \cdot a$$

Associative:
$$(a + b) + c = a + (b + c)$$
$$(a \cdot b) \cdot c = a \cdot (b \cdot c)$$

Distributive:
$$a(b + c) = ab + ac$$

$$3 + 0 = 3 \qquad\qquad 0 + 3 = 3$$
$$-1.8 \cdot 1 = -1.8 \qquad 1 \cdot -1.8 = -1.8$$

$$7 + (-7) = 0 \qquad -7 + 7 = 0$$

$$5 \cdot \frac{1}{5} = 1 \qquad \frac{1}{5} \cdot 5 = 1$$

$$x + 7 = 7 + x$$
$$9 \cdot y = y \cdot 9$$

$$(3 + 1) + 10 = 3 + (1 + 10)$$
$$(3 \cdot 1) \cdot 10 = 3(1 \cdot 10)$$

$$6(x + 5) = 6 \cdot x + 6 \cdot 5$$
$$= 6x + 30$$

Section 1.5 Solving Linear Equations Algebraically

An **equation** is a statement that two expressions are equal.

A **linear equation in one variable** is an equation that can be written in the form $ax + b = c$, where a, b, and c are real numbers and a is not 0.

A **solution** of an equation is a value for the variable that makes the equation a true statement.

Equations:
$$5 = 5 \qquad 7x + 2 = -14 \qquad 3(x - 1)^2 = 9x^2 - 6$$
Linear equations:
$$7x + 2 = -14 \qquad x = -3$$
$$5(2y - 7) = -2(8y - 1)$$
Check to see that -1 is a solution of
$$3(x - 1) = 4x - 2.$$

$$3(-1 - 1) = 4(-1) - 2$$
$$3(-2) = -4 - 2$$
$$-6 = -6 \qquad \text{True}$$

Thus, -1 is a solution.

Equivalent equations have the same solution.

The **addition property of equality** guarantees that the same number may be added to (or subtracted from) both sides of an equation, and the result is an equivalent equation.

$x - 12 = 14$ and $x = 26$ are equivalent equations.

Solve for x: $-3x - 2 = 10$.
$$-3x - 2 + 2 = 10 + 2 \qquad \text{Add 2 to both sides.}$$
$$-3x = 12$$

The **multiplication property of equality** guarantees that both sides of an equation may be multiplied by (or divided by) the same nonzero number, and the result is an equivalent equation.

$$\frac{-3x}{-3} = \frac{12}{-3} \qquad \text{Divide both sides by } -3.$$
$$x = -4$$

To solve linear equations in one variable:

Solve for x:
$$x - \frac{x - 2}{6} = \frac{x - 7}{3} + \frac{2}{3}$$

1. Clear the equation of fractions.

1. $6\left(x - \frac{x - 2}{6}\right) = 6\left(\frac{x - 7}{3} + \frac{2}{3}\right)$ Multiply both sides by 6.

$$6x - (x - 2) = 2(x - 7) + 2(2)$$

2. Remove grouping symbols such as parentheses.

2. $6x - x + 2 = 2x - 14 + 4$ Remove grouping symbols.

3. Simplify by combining like terms.

3. $5x + 2 = 2x - 10$

(continued)

Definitions and Concepts	**Examples**

Section 1.5 Solving Linear Equations Algebraically

4. Write variable terms on one side and numbers on the other side using the addition property of equality.

4.
$$5x + 2 - 2 = 2x - 10 - 2 \quad \text{Subtract 2.}$$
$$5x = 2x - 12$$
$$5x - 2x = 2x - 12 - 2x \quad \text{Subtract } 2x.$$
$$3x = -12$$

5. Isolate the variable using the multiplication property of equality.

5.
$$\frac{3x}{3} = \frac{-12}{3} \quad \text{Divide by 3.}$$
$$x = -4$$

6. Check the proposed solution in the original equation.

6.
$$-4 - \frac{-4 - 2}{6} \stackrel{?}{=} \frac{-4 - 7}{3} + \frac{2}{3} \quad \begin{array}{l}\text{Replace } x \text{ with } -4 \\ \text{in the original equation.}\end{array}$$
$$-4 - \frac{-6}{6} \stackrel{?}{=} \frac{-11}{3} + \frac{2}{3}$$
$$-4 - (-1) \stackrel{?}{=} \frac{-9}{3}$$
$$-3 = -3 \quad \text{True}$$

Section 1.6 An Introduction to Problem Solving

Problem-Solving Strategy

Colorado is shaped like a rectangle whose length is about 1.3 times its width. If the perimeter of Colorado is 2070 kilometers, find its dimensions.

1. UNDERSTAND the problem.

1. Read and reread the problem. Guess a solution and check your guess.
Let x = width of Colorado in kilometers. Then
$1.3x$ = length of Colorado in kilometers

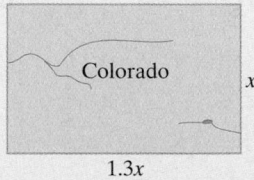

2. TRANSLATE the problem.

2. In words:

twice the length	+	twice the width	=	perimeter
↘		↓		↓

Translate: $2(1.3x) + \quad 2x \quad = \quad 2070$

3. SOLVE the equation.

3.
$$2.6x + 2x = 2070$$
$$4.6x = 2070$$
$$x = 450$$

4. INTERPRET the results.

4. If $x = 450$ kilometers, then $1.3x = 1.3(450) = 585$ kilometers. *Check:* The perimeter of a rectangle whose width is 450 kilometers and length is 585 kilometers is $2(450) + 2(585) = 2070$ kilometers, the required perimeter. *State:* The dimensions of Colorado are 450 kilometers by 585 kilometers

Definitions and Concepts	**Examples**

Section 1.7 A Numerical Approach: Modeling with Tables

To generate a table on a calculator, a **table start** x-value and a **table increment**, or change in x-values, must be given.

Use your calculator to complete the given table.

Hours Worked	x	8	10	12	14	16
Gross Pay	$8 + 6.25x$					

Go to the Y= editor and enter $y_1 = 8 + 6.25x$. In the table setup menu, let table start = 8 and table increment = 2 since the change in x-values in the table is 2. The table generated is shown below.

X	Y1	
8	58	
10	70.5	
12	83	
14	95.5	
16	108	
18	120.5	
20	133	
Y1■8+6.25X		

Section 1.8 Formulas and Problem Solving

An equation that describes a known relationship among quantities is called a **formula.**

Formulas:

$$A = \pi r^2 \text{ (area of a circle)}$$
$$I = PRT \quad (\text{interest} = \text{principal} \cdot \text{rate} \cdot \text{time})$$

To solve a formula for a specified variable, use the steps for solving an equation. Treat the specified variable as the only variable of the equation.

Solve $A = 2HW + 2LW + 2LH$ for H

$A - 2LW = 2HW + 2LH$	Subtract $2LW$.
$A - 2LW = H(2W + 2L)$	Factor out H.
$\dfrac{A - 2LW}{2W + 2L} = \dfrac{H(2W + 2L)}{2W + 2L}$	Divide by $2W + 2L$.
$\dfrac{A - 2LW}{2W + 2L} = H$	Simplify.

Section 1.9 Interpreting Data and Reading Bar, Line, and Circle Graphs

Find the mean, the median, and the mode of the following list of numbers.

$$23, 23, 31, 35, 46, 63$$

The **mean** of a set of data items, denoted by $\overline{x}$, is the sum of the data items divided by the number of items.

$$\overline{x} = \frac{23 + 23 + 31 + 35 + 46 + 63}{6} \approx 36.8$$

The **median** of an ordered set of numbers is the middle number. If the number of items is even, the median is the mean of the two middle numbers.

$$\text{median} = \frac{31 + 35}{2} = 33$$

The **mode** of a set of numbers is the number that occurs most often.

$$\text{mode} = 23$$

1 CHAPTER REVIEW

(1.2) Find the value of each algebraic expression at the given replacement values.

1. $7x$ when $x = 3$ 21

2. st when $s = 1.6$ and $t = 5$ 8

3. The hummingbird has an average wing speed of 90 beats pcr second. The expression $90t$ gives the number of wing beats in t seconds. Calculate the number of wing beats in *1 hour* for the hummingbird.   324,000

List the elements in each set.

4. $\{x \mid x \text{ is an odd integer between } -2 \text{ and } 4\}$ $\{-1, 1, 3\}$

5. $\{x \mid x \text{ is an even integer between } -3 \text{ and } 7\}$ $\{-2, 0, 2, 4, 6\}$

6. $\{x \mid x \text{ is a negative whole number}\}$ $\varnothing$

7. $\{x \mid x \text{ is a natural number that is not a rational number}\}$ $\varnothing$

8. $\{x \mid x \text{ is a whole number greater than } 5\}$ $\{6, 7, 8, \dots\}$

9. $\{x \mid x \text{ is an integer less than } 3\}$ $\{\dots, -1, 0, 1, 2\}$

Determine whether each statement is true or false if $A = \{6, 10, 12\}$, $B = \{5, 9, 11\}$, $C = \{\dots, -3, -2, -1, 0, 1, 2, 3, \dots\}$, $D = \{2, 4, 6, \dots, 16\}$ $E = \{x \mid x \text{ is a rational number}\}$, $F = \{\ \}$, $G = \{x \mid x \text{ is an irrational number}\}$, and $H = \{x \mid x \text{ is a real number}\}$.

10. $10 \in D$ true

11. $B \in 9$ false

12. $\sqrt{169} \notin G$ true

13. $0 \notin F$ true

14. $\pi \in E$ false

15. $\pi \in H$ true

16. $\sqrt{4} \in G$ false

17. $-9 \in E$ true

18. $A \subseteq D$ true

19. $C \nsubseteq B$ true

20. $C \nsubseteq E$ false

21. $F \subseteq H$ true

22. $B \subseteq B$ true

23. $D \subset C$ true

24. $C \subseteq H$ true

25. $G \subseteq H$ true

26. $\{5\} \in B$ false

27. $\{5\} \subseteq B$ true

List the elements of the set
$$\left\{5, -\frac{2}{3}, \frac{8}{2}, \sqrt{9}, 0.3, \sqrt{7}, 1\frac{5}{8}, -1, \pi\right\} \text{ that are also elements of}$$
each given set.

28. Whole numbers $\{5, \frac{8}{2}, \sqrt{9}\}$

29. Natural numbers $\{5, \frac{8}{2}, \sqrt{9}\}$

30. Rational numbers $\{5, -\frac{2}{3}, \frac{8}{2}, \sqrt{9}, 0.3, 1\frac{5}{8}, -1\}$

31. Irrational numbers $\{\sqrt{7}, \pi\}$

32. Real numbers $\{5, -\frac{2}{3}, \frac{8}{2}, \sqrt{9}, 0.3, \sqrt{7}, 1\frac{5}{8}, -1, \pi\}$

33. Integers $\{5, \frac{8}{2}, \sqrt{9}, -1\}$

Find the opposite.

34. $-\frac{3}{4}$ $\frac{3}{4}$

35. 0.6 -0.6

36. 0 0

37. 1 -1

Find the reciprocal.

38. $-\frac{3}{4}$ $-\frac{4}{3}$

39. 0.6 $\frac{1}{0.6}$

40. 0 undefined

41. 1 1

(1.3) Simplify.

42. $-7 + 3$ -4

43. $-10 + (-25)$ -35

44. $5(-0.4)$ -2

45. $(-3.1)(-0.1)$ 0.31

46. $-7 - (-15)$ 8

47. $9 - (-4.3)$ 13.3

48. $(-6)(-4)(0)(-3)$ 0

49. $(-12)(0)(-1)(-5)$ 0

50. $(-24) \div 0$ undefined

51. $0 \div (-45)$ 0

52. $(-36) \div (-9)$ 4

53. $60 \div (-12)$ -5

54. $\left(-\frac{4}{5}\right) - \left(-\frac{2}{3}\right)$ $-\frac{2}{15}$

55. $\left(\frac{5}{4}\right) - \left(-2\frac{3}{4}\right)$ 4

56. Determine the unknown fractional part. $\frac{5}{12}$

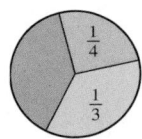

Simplify.

57. $-5 + 7 - 3 - (-10)$ 9

58. $8 - (-3) + (-4) + 6$ 13

59. $3(4 - 5)^4$ 3

60. $6(7 - 10)^2$ 54

61. $\left(-\dfrac{8}{15}\right) \cdot \left(-\dfrac{2}{3}\right)^2$ $-\dfrac{32}{135}$ **62.** $\left(-\dfrac{3}{4}\right)^2 \cdot \left(-\dfrac{10}{21}\right)$ $-\dfrac{15}{56}$

63. $\dfrac{-\dfrac{6}{15}}{\dfrac{8}{25}} - \dfrac{5}{4}$ **64.** $\dfrac{\dfrac{4}{9}}{-\dfrac{8}{45}} - \dfrac{5}{2}$

65. $-\dfrac{3}{8} + 3(2) \div 6\dfrac{5}{8}$ **66.** $5(-2) - (-3) - \dfrac{1}{6} + \dfrac{2}{3} - 6\dfrac{1}{2}$

67. $|2^3 - 3^2| - |5 - 7|$ -1 **68.** $|5^2 - 2^2| + |9 \div (-3)|$ 24

69. $(2^3 - 3^2) - (5 - 7)$ 1 **70.** $(5^2 - 2^4) + [9 \div (-3)]$ 6

71. $\dfrac{(8 - 10)^3 - (-4)^2}{2 + 8(2) \div 4}$ -4 **72.** $\dfrac{(2 + 4)^2 + (-1)^5}{12 \div 2 \cdot 3 - 3}$ $\dfrac{7}{3}$

73. $\dfrac{(4 - 9) + 4 - 9}{10 - 12 \div 4 \cdot 8}$ $\dfrac{5}{7}$ **74.** $\dfrac{3 - 7 - (7 - 3)}{15 + 30 \div 6 \cdot 2}$ $-\dfrac{8}{25}$

75. $\dfrac{\sqrt{25}}{4 + 3 \cdot 7}$ $\dfrac{1}{5}$ **76.** $\dfrac{\sqrt{64}}{24 - 8 \cdot 2}$ 1

Find the value of each expression when $x = 0$, $y = 3$, and $z = -2$.

77. $x^2 - y^2 + z^2$ -5 **78.** $\dfrac{5x + z}{2y}$ $-\dfrac{1}{3}$

79. $\dfrac{-7y - 3z}{-3}$ 5 **80.** $(x - y + z)^2$ 25

△ **81.** The algebraic expression $2\pi r$ represents the circumference of (distance around) a circle of radius r.

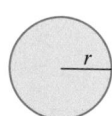

a. Complete the table below by evaluating the expression at given values of r. (Use 3.14 for π)

Radius	r	1	10	100
Circumference	$2\pi r$	6.28	62.8	628

b. As the radius of a circle increases, does the circumference of the circle increase or decrease? increase

(1.4) *Simplify each expression.*

82. $5xy - 7xy + 3 - 2 + xy$ $-xy + 1$

83. $4x + 10x - 19x + 10 - 19$ $-5x - 9$

84. $6x^2 + 2 - 4(x^2 + 1)$ $2x^2 - 2$

85. $-7(2x^2 - 1) - x^2 - 1$ $-15x^2 + 6$

86. $(3.2x - 1.5) - (4.3x - 1.2)$ $-1.1x - 0.3$

87. $(7.6x + 4.7) - (1.9x + 3.6)$ $5.7x + 1.1$

Write each statement using mathematical symbols.

88. Twelve is the product of x and negative 4. $12 = -4x$

89. The sum of n and twice n is negative fifteen. $n + 2n = -15$

90. Four times the sum of y and three is -1. $4(y + 3) = -1$

91. The difference of t and five, multiplied by six is four.
$6(t - 5) = 4$

92. Seven subtracted from z is six. $z - 7 = 6$

93. Ten less than the product of x and nine is five. $9x - 10 = 5$

94. The difference of x and 5 is at least 12. $x - 5 \geq 12$

95. The opposite of four is less than the product of y and seven.
$-4 < 7y$

96. Two-thirds is not equal to twice the sum of n and one-fourth.
$\frac{2}{3} \neq 2\left(n + \frac{1}{4}\right)$

97. The sum of t and six is not more than negative twelve.
$t + 6 \leq -12$

Name the property illustrated.

98. $(M + 5) + P = M + (5 + P)$ associative property of addition

99. $5(3x - 4) = 15x - 20$ distributive property

100. $(-4) + 4 = 0$ additive inverse property

101. $(3 + x) + 7 = 7 + (3 + x)$ commutative property of addition

102. $(XY)Z = (YZ)X$ associative and commutative properties of multiplication

103. $\left(-\dfrac{3}{5}\right) \cdot \left(-\dfrac{5}{3}\right) = 1$ multiplicative inverse property

104. $T \cdot 0 = 0$ multiplication property of zero

105. $(ab)c = a(bc)$ associative property of multiplication

106. $A + 0 = A$ additive identity property

107. $8 \cdot 1 = 8$ multiplicative identity property

Complete the equation using the given property.

108. $5x - 15z = \underline{5(x - 3z)}$ Distributive property
109. $(3 + x) + (7 + y)$
109. $(7 + y) + (3 + x) = $ _____ Commutative property

110. $0 =$ _____ Additive inverse property $2 + (-2)$, for example

111. $1 =$ _____ Multiplicative inverse property $2 \cdot \frac{1}{2}$, for example

112. $[(3.4)(0.7)]5 = \underline{(3.4)[(0.7)5]}$ Associative property

113. $7 = \underline{7 + 0}$ Additive identity property

Insert $<$, $>$, *or* $=$ *to make each statement true.*

114. $-9 \; > \; -12$ **115.** $0 \; > \; -6$

116. $-3 \; < \; -1$ **117.** $7 \; = \; |-7|$

118. $-5 \; < \; -(-5)$

(1.5) *Solve each linear equation.*

119. $4(x - 5) = 2x - 14$ 3

120. $x + 7 = -2(x + 8)$ $-\frac{23}{3}$

121. $3(2y - 1) = -8(6 + y)$ $-\frac{45}{14}$

122. $-(z + 12) = 5(2z - 1)$ $-\frac{7}{11}$

123. $0.3(x - 2) = 1.2$ 6

124. $1.5 = 0.2(c - 0.3)$ 7.8

125. $-4(2 - 3h) = 2(3h - 4) + 6h$ all real numbers

126. $6(m - 1) + 3(2 - m) = 0$ 0

127. $6 - 3(2g + 4) - 4g = 5(1 - 2g)$ $\varnothing$

128. $20 - 5(p + 1) + 3p = -(2p - 15)$ all real numbers

129. $\frac{x}{3} - 4 = x - 2$ -3 **130.** $\frac{9}{4}y = \frac{2}{3}y$ 0

131. $\frac{3n}{8} - 1 = 3 + \frac{n}{6}$ $\frac{96}{5}$ **132.** $\frac{z}{6} + 1 = \frac{z}{2} + 2$ -3

133. $\frac{b - 2}{3} = \frac{b + 2}{5}$ 8 **134.** $\frac{2t - 1}{3} = \frac{3t + 2}{15}$ 1

135. $\frac{x - 2}{5} + \frac{x + 2}{2} = \frac{x + 4}{3}$ 2

136. $\frac{2z - 3}{4} - \frac{4 - z}{2} = \frac{z + 1}{3}$ $\frac{37}{8}$

(1.6) *Solve.*

137. Twice the difference of a number and 3 is the same as 1 added to three times the number. Find the number. -7

138. One number is 5 more than another number. If the sum of the numbers is 285, find the numbers.   140, 145

139. Find 40% of 130.   52

140. Find 1.5% of 8.   0.12

141. In 1995, the Academy Awards (Oscars) had a peak in television viewership. By 2003, the number of Oscar viewers had decreased to 33 million. If this represented a decrease of 40%, find the number of Oscar viewers in 1995. (*Source:* Nielsen Media Research).   55 million viewers

142. Find four consecutive integers such that twice the first subtracted from the sum of the other three integers is sixteen.   10, 11, 12, 13

143. Determine whether there are two consecutive odd integers such that 5 times the first exceeds 3 times the second by 54.   no such odd integers exist

144. The length of a rectangular playing field is 5 meters less than twice its width. If 230 meters of fencing goes around the field, find the dimensions of the field.   width, 40 m; length, 75 m

x

$2x - 5$

145. A car rental company charges $29.95 per day for a compact car plus 15 cents per mile for every mile over 100 miles driven per day. If Mr. Woo's bill for 2 days' use is $83.60 before taxes, find how many miles he drove.   358 mi

146. The cost C of producing x number of scientific calculators is given by $C = 4.50x + 3000$, and the revenue R from selling them is given by $R = 16.50x$. Find the number of calculators that must be sold to break even. (To break even, revenue = cost.)   250 calculators

147. An entrepreneur can sell her musically vibrating plants for $40 each, while her cost C to produce x number of plants is given by $C = 20x + 100$. Find her break-even point. Find her revenue if she sells exactly that number of plants.   5 plants, $200

(1.7) *Use your calculator and the table feature to complete each table. If necessary, round any amounts to 2 decimal places.*

148.

Radius of Cone	x	1	1.5	2	2.5	3
Volume (If Height is 10 Inches)	$\frac{10}{3}\pi x^2$	10.47	23.56	41.89	65.45	94.25

10

x

149.

Width of Rectangle	x	2	4.68	9.5	12.68
Perimeter (If Length is 15 Units)	$30 + 2x$	34	39.36	49	55.36

x

15

150. Gina Saltalamacchia works as a hostess at a restaurant and makes $5.75 per hour. Complete the table and find her gross pay for the hours given.

Hours	5	10	15	20	25	30
Gross Pay (Dollars)	28.75	57.5	86.25	115	143.75	172.5

151. Coast Waterworks, Inc. charges $8.00 for the first 4 thousand gallons of water used and then $1.10 per thousand gallons used over 4 thousand gallons. Cross Gates Water Company charges $12.00 for the first 5 thousand gallons and $1.50 per thousand gallons used over 5 thousand gallons. Complete the table and find the total charges for the gallons used.

Gallons (In Thousands Used)	5	6	7	8
Coast Charge (Dollars)	9.1	10.2	11.3	12.4
Cross Gates Charge (Dollars)	12	13.5	15	16.5

152. A rock is thrown upward from a bridge. Neglecting air resistance, the height of the rock at time x seconds is given by

$$y = -16x^2 + 40x + 500$$

where y is measured in feet.

a. Complete the table to find the height of the rock at the given times.

Seconds	0	1	2	3	4	5
Height in Feet	500	524	516	476	404	300

b. Find the maximum height of the rock to the nearest foot.   525 ft

c. Find to the nearest tenth of a second when the rock strikes the ground.   7.0 sec

(1.8) Solve each equation for the specified variable.

153. $V = lwh; w$ $w = \dfrac{V}{lh}$ **154.** $C = 2\pi r; r$ $r = \dfrac{C}{2\pi}$

155. $5x - 4y = -12; y$ **156.** $5x - 4y = -12; x$

157. $y - y_1 = m(x - x_1); m$ $m = \dfrac{y - y_1}{x - x_1}$

158. $y - y_1 = m(x - x_1); x$ $x = \dfrac{y - y_1 + mx_1}{m}$

159. $E = I(R + r); r$ **160.** $S = vt + gt^2; g$

161. $T = gr + gvt; g$ **162.** $I = Prt + P; P$

△ **163.** $A = \dfrac{h}{2}(B + b); B$ △ **164.** $V = \dfrac{1}{3}\pi r^2 h; h$

165. $R = \dfrac{r_1 + r_2}{2}; r_1$ **166.** $\dfrac{V_1}{T_1} = \dfrac{V_2}{T_2}; T_2$

Solve.

167. A principal of $3000 is invested in an account paying an annual percentage rate of 3%. Find the amount in the account after 7 years if the interest is compounded

a. semiannually $3695.27

b. weekly $3700.81

(Approximate to the nearest cent.)

168. Angie has a photograph in which the length is 2 inches longer than the width. If she increases each dimension by 4 inches, the area is increased by 88 square inches. Find the original dimensions.   8 in. × 10 in.

169. The formula for converting Celsius temperatures to Fahrenheit temperatures is $F = \dfrac{9C + 160}{5}$.

a. Complete the following table to find Fahrenheit temperatures given the Celsius temperatures.

Celsius	-40	-15	10	60
Fahrenheit	-40	5	50	140

b. If the boiling point of water is 100 degrees Celsius, what is this in degrees Fahrenheit?   212°F

c. If the freezing point of water is 0 degrees Celsius, what is this in degrees Fahrenheit?   32°F

170. The formula for finding the distance traveled at the rate of 55 miles per hour for x hours is given by $y = 55x$, where y is measured in miles.

a. Find the distance traveled in 15-minute increments for 0 to 1.5 hours.

Hours	0	0.25	0.50	0.75	1	1.25	1.5
Miles	0	13.75	27.5	41.25	55	68.75	82.5

b. If the distance traveled was 192.5 miles, how many hours did it take to travel this distance at 55 mph?   b. 3.5 hrs

171. One-square-foot floor tiles come 24 to a package. Find how many packages are needed to cover a rectangular floor 18 feet by 21 feet.   16 packages

172. Determine which container holds more ice cream, an 8 inch × 5 inch × 3 inch box or a cylinder with radius 3 inches and height 6 inches.   cylinder

173. Angie left Los Angeles at 11 A.M. and drove nonstop to San Diego, 130 miles away. If she arrived at 1:15 P.M., find her average speed, rounded to the nearest mile per hour.
58 mph

The volume of a cone can be found using the formula $V = \dfrac{1}{3}\pi r^2 h$.

Find the volumes of the cones with the following heights and radii. Round to two decimal places.

174.

Height h	6	6	6	6	6
Radius x	1	1.5	2	2.25	3
Volume	6.28	14.14	25.13	31.81	56.55

175.

Height h	10	10	10	10	10
Radius x	1.5	2.1	2.75	3	3.5
Volume	23.56	46.18	79.19	94.25	128.28

(1.9) For each of the following sets of numbers, find the mean, the median, and the mode(s). If necessary, round the mean to two decimal places.

176. 21, 28, 16, 42, 38   mean: 29; median: 28; mode: none

177. 42, 35, 36, 40, 50   mean: 40.6; median: 40; mode: none

178. 7.6, 8.2, 8.2, 9.6, 5.7, 9.1   mean: 8.07; median: 8.2; mode: 8.2

179. 4.9, 7.1, 6.8, 6.8, 5.3, 4.9   mean: 5.97; median: 6.05; mode: 4.9 and 6.8

180. 0.2, 0.3, 0.5, 0.6, 0.6, 0.9, 0.2, 0.7, 1.1   mean: 0.57; median: 0.6; mode: 0.2 and 0.6

181. 0.6, 0.6, 0.8, 0.4, 0.5, 0.3, 0.7, 0.8, 0.1   mean: 0.53; median: 0.6; mode: 0.6 and 0.8

The circle graph below shows percents of financial aid for college students from federal, state, and institutional sources. Use this graph to answer Exercises 182 and 183. (Source: College Board Trends in Student Aid.)

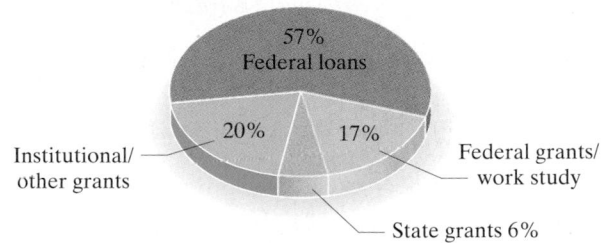

Source: College Board Trends in Student Aid

182. hat type of financial aid do most students receive?
Federal loans

183. If the total financial aid from the categories shown is $50.3 billion for 1995–96, find the amount of money spent in each category. Round to the nearest hundredth of a billion.
$28.67 billion $10.06 billion $8.55 billion $3.02 billion

The bar graph below compares the size of a company (number of employees) with the percent of those employees receiving health benefits. Use this graph for Exercises 184 through 187. (Source: Lewin Group for American Hospital Association.)

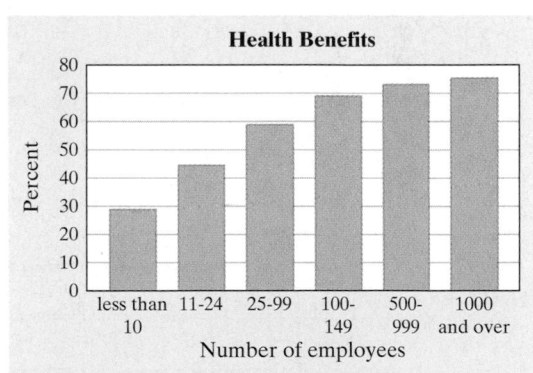

184. Estimate the percent of employees receiving health benefits for companies with 11–24 employees.   44%

185. Which size company has the lowest percent of employees receiving health benefits? Estimate the percent.   Less than 10: 29%

186. Which size company has the highest percent of employees receiving health benefits? Estimate the percent.   1000 and over: 75%

187. Discuss a trend shown by this bar graph.   answers may vary

155. $y = \dfrac{5x + 12}{4}$ **156.** $x = \dfrac{4y - 12}{5}$ **159.** $r = \dfrac{E - IR}{I}$ **160.** $g = \dfrac{S - vt}{t^2}$

161. $g = \dfrac{I}{r + vt}$ **162.** $P = \dfrac{T}{1 + rt}$ **163.** $B = \dfrac{2A - hb}{h}$ **164.** $h = \dfrac{3V}{\pi r^2}$ **165.** $r_1 = 2R - r_2$ **166.** $T_2 - \dfrac{I_1 V_2}{V_1}$

CHAPTER 1 TEST

Remember to use your Chapter Test Prep Video CD to help you study and view solutions to the test questions you need help with.

Determine whether each statement is true or false.

1. $-2.3 > -2.33$ true   **2.** $-6^2 = (-6)^2$ false

3. $-5 - 8 = -(5 - 8)$ false   **4.** $(-2)(-3)(0) - \dfrac{-4}{0}$ false

5. All natural numbers are integers.   true

6. All rational numbers are integers.   false

Simplify.   **12.** $-\dfrac{3}{2}$ **17.** $\dfrac{(6-y)^2}{7} < -2$ **18.** $\dfrac{9z}{|-12|} \neq 10$

7. $5 - 12 \div 3(2)$ -3 **8.** $5^2 - 3^4$ -56

9. $(4-9)^3 - |-4 - 6|^2$ -225

10. $12 + \{6 - [5 - 2(-5)]\}$ 3   **19.** $3\left(\dfrac{n}{5}\right) = -n$

11. $\dfrac{6(7-9)^3 + (-2)}{(-2)(-5)(-5)}$ 1   **12.** $\dfrac{(4 - \sqrt{16}) - (-7 - 20)}{-2(1-4)^2}$

21. $-2 = \dfrac{x}{x+5}$ **23.** associative property of addition

Evaluate each expression when $q = 4$, $r = -2$, and $t = 1$.

13. $q^2 - r^2$ 12   **14.** $\dfrac{5t - 3q}{3r - 1}$ 1

15. The algebraic expression $5.75x$ represents the total cost for x adults to attend the theater.

   a. Complete the table that follows.

   b. As the number of adults increases does the total cost increase or decrease? increase

Adults	x	1	3	10	20
Total Cost	$5.75x$	5.75	17.25	57.50	115.00

Write each statement using mathematical symbols.

16. Twice the sum of x and five is 30. $2(x+5) = 30$

17. The square of the difference of six and y, divided by seven, is less than -2.

18. The product of nine and z, divided by the absolute value of -12, is not equal to 10.

19. Three times the quotient of n and five is the opposite of n.

20. Twenty is equal to 6 subtracted from twice x. $20 = 2x - 6$

21. Negative two is equal to x divided by the sum of x and five.

Name each property illustrated.

22. $6(x - 4) = 6x - 24$ distributive property

23. $(4 + x) + z = 4 + (x + z)$

24. $(-7) + 7 = 0$ additive inverse property

25. $(-18)(0) = 0$ multiplication property of zero

26. Write an expression for the total amount of money (in dollars) in n nickels and d dimes. $0.05n + 0.1d$

Simplify each expression.

27. $-2(3x + 7)$ $-6x - 14$

28. $\dfrac{1}{3}a - \dfrac{3}{8} + \dfrac{1}{6}a - \dfrac{3}{4}$ $\dfrac{1}{2}a - \dfrac{9}{8}$

29. $4y + 10 - 2(y + 10)$ $2y - 10$

30. $(8.3x - 2.9) - (9.6x - 4.8)$ $-1.3x + 1.9$

31. Evaluate the expression $0.2x^3 + 5x^2 - 6.2x + 3$ if $x = -3.1$.   64.3118

32. The circumference of a circle can be found by multiplying π times the diameter or using the formula, $C = \pi d$. The area of a circle is found by multiplying the radius squared times π or using the formula $A = \pi r^2$. Complete the table below to find the circumference and area, given the following diameters. If necessary, round answers to the nearest hundredth.

DIAMETER	d	2	3.8	10	14.9
RADIUS	r	1	1.9	5	7.45
CIRCUMFERENCE	πd	6.28	11.94	31.42	46.81
AREA	πr^2	3.14	11.34	78.54	174.37

Solve each equation.

33. $8x + 14 = 5x + 44$ 10

34. $3(x + 2) = 11 - 2(2 - x)$ 1

35. $3(y - 4) + y = 2(6 + 2y)$ $\varnothing$

36. $7n - 6 + n = 2(4n - 3)$ all real numbers

37. $\dfrac{7w}{4} + 5 = \dfrac{3w}{10} + 1$ $-\dfrac{80}{29}$

Solve each equation for the specified variable.

38. $3x - 4y = 8$; y $y = \dfrac{3x - 8}{4}$

39. $S = gt^2 + gvt$; g $g = \dfrac{S}{t^2 + vt}$

40. $F = \dfrac{9}{5}C + 32$; C $C = \dfrac{5}{9}(F - 32)$

41. Find 12% of 80.   9.6

Solve.   **44.** more than 850 sunglasses

42. In 2006, the number of people employed as database administrators, computer support specialists, and all other computer scientists is expected to be 461,000 in the United States. This represents a 118% increase over the number of people employed in these occupations in 1996. Find the number of database administrators, computer support specialists, and all other computer scientists employed in 1996. (*Source:* U.S. Bureau of Labor Statistics)   211,468 people

△ **43.** A circular dog pen has a circumference of 78.5 feet. Approximate π by 3.14 and estimate how many hunting dogs could be safely kept in the pen if each dog needs at least 60 square feet of room.   approximately 8 dogs

44. The company that makes Photoray sunglasses figures that the cost C to make x number of sunglasses weekly is given by $C = 3910 + 2.8x$, and the weekly revenue R is given by $R = 7.4x$. Use an inequality to find the number of sunglasses that must be made and sold to make a profit. (Revenue must exceed cost in order to make a profit.)

45. Find the amount of money in an account after 10 years if a principal of $2500 is invested at 3.5% interest compounded quarterly. (Round to the nearest cent.) $3542.27

46. Ann-Margaret Tober is deciding whether to accept a part-time sales position at Campo Electronics. She is offered a gross monthly pay of $1500 plus 5% commission on her sales.

  a. Complete the gross monthly pay table for the given amounts of sales.

Sales (Dollars)	8000	9000	10,000	11,000	12,000
Gross Monthly Pay (Dollars)	1900	1950	2000	2050	2100

  b. If she has sales of $11,000 per month, what is her gross annual pay? $24,600

  c. If Ann-Margaret decides she needs a gross monthly pay of $2200, how much must she sell every month? $14,000

47. The fireworks display for a small community is launched from a platform that is 20 feet from ground level. A particular rocket is launched with a velocity of 80 feet per second and explodes 0.5 second after it reaches its maximum height. The height of the rocket at time x seconds is given by

$$y = -16x^2 + 80x + 20$$

where y is measured in feet from the ground.

  a. When is the height of the rocket 116 feet?
  At 2 and 3 seconds

  b. What is the rocket's maximum height?
  120 ft

  c. When does the rocket reach its maximum height?
  At 2.5 sec

  d. When does the rocket explode?   3.0 sec

The following chart shows the number of books a graduate student checked out of the library for a period of 12 weeks while working on a term paper.

Week	1	2	3	4	5	6	7	8	9	10	11	12
Number of Books	12	0	6	5	2	10	3	1	5	7	1	5

48. Find the mean number of books checked out per week.   4.75

49. Find the median of the number of books checked out per week.   5

50. Find the mode of the number of books checked out per week.   5

The graph below shows U.S. consumption of fruits and vegetables per person. Use this graph for Exercises 51 through 54.

51. Approximate the pounds of fresh fruit consumed per person in 1970.   80 lb

52. Approximate the pounds of fresh vegetables consumed per person in 1995.   119 lb

53. Estimate the year in which there was the greatest difference in consumption of fresh fruits and vegetables.   2000

54. Discuss any trends shown in the graph.   answers may vary

Fruit and Vegetable Consumption

The linear equations we explored in Chapter 1 are statements about a single variable. This chapter examines statements about two variables: linear equations in two variables. We focus particularly on graphs of those equations which lead to the notion of relation and to the notion of function, perhaps the single most important and useful concept in all of mathematics.

CHAPTER 2

Graphs and Functions

The Centers for Disease Control and Prevention (CDC), located in Atlanta, Georgia, was founded by the United States government as the primary federal agency devoted to public health. Public health officials there work constantly to provide information about current health issues. Prevention of diseases, injury, and disability is another focus of the public health specialists at the CDC.

The CDC employs approximately 8500 people in Atlanta, Washington, D.C., and at its many field sites nationwide. Positions at the CDC include medical officers, public health advisors, epidemiologists, behavioral scientists, statisticians, and information and technology specialists.

In the Spotlight on Decision Making feature on page 187, you will have the opportunity as a public health official to make a decision about a public awareness campaign concerning the eradication of mumps in the United States.

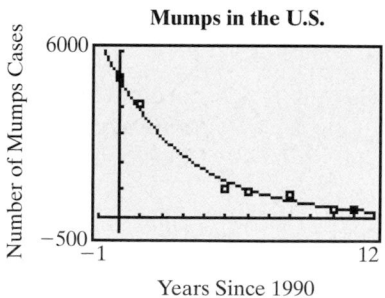

Link: www.cdc.gov *(Centers for Disease Control and Prevention)*
Source of text: (same)

2.1 INTRODUCTION TO GRAPHING

Objectives

1. Plot ordered pairs.
2. Introduce the window setting capability of a graphing utility.
3. Determine whether an ordered pair of numbers is a solution to an equation in two variables.
4. Graph linear equations.
5. Graph nonlinear equations.

1   Graphs are widely used today in newspapers, magazines, and all forms of newsletters. A few examples of graphs are shown here.

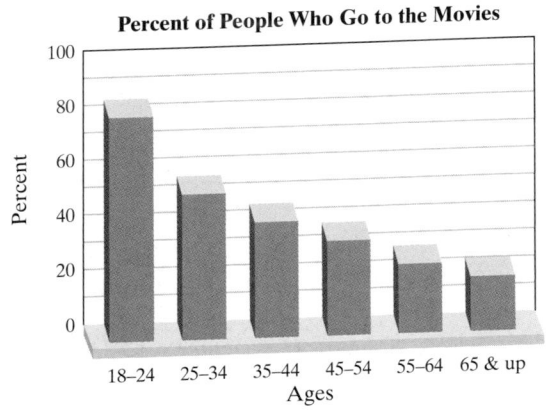

Percent of People Who Go to the Movies

Source: TELENATION/Market Facts, Inc.

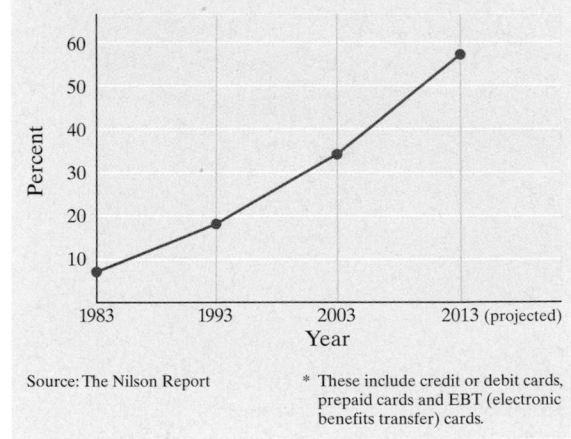

Percent of Sales Completed Using Cards*

Source: The Nilson Report

* These include credit or debit cards, prepaid cards and EBT (electronic benefits transfer) cards.

To review how to read these graphs, we review their origin—the rectangular coordinate system. One way to locate points on a plane is by using a **rectangular coordinate system**, which is also called a **Cartesian coordinate system** after its inventor, René Descartes (1596–1650).

A rectangular coordinate system consists of two number lines that intersect at right angles at their 0 coordinates. We position these axes on paper such that one number line is horizontal and the other number line is then vertical. The horizontal number line is called the **x-axis** (or the axis of the **abscissa**), and the vertical number line is called the **y-axis** (or the axis of the **ordinate**). The point of intersection of these axes is named the **origin.**

Notice in the left figure on the next page that the axes divide the plane into four regions. These regions are called **quadrants.** The top-right region is quadrant I. Quadrants II, III, and IV are numbered counterclockwise from the first quadrant as shown. The x-axis and the y-axis are not in any quadrant.

Each point in the plane can be located, or **plotted**, or graphed by describing its position in terms of distances along each axis from the origin. An **ordered pair**, represented by the notation (x, y), records these distances.

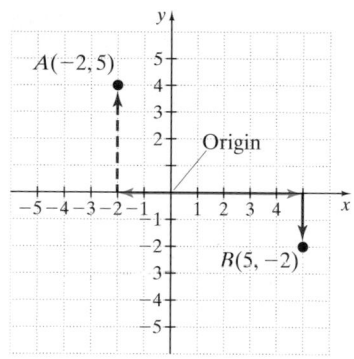

For example, the location of point A in the figure on the right above is described as 2 units to the left of the origin along the x-axis and 5 units upward parallel to the y-axis. Thus, we identify point A with the ordered pair $(-2, 5)$. Notice that the order of these numbers is critical. The x-value -2 is called the **x-coordinate** and is associated with the x-axis. The y-value 5 is called the **y-coordinate** and is associated with the y-axis. Compare the location of point A with the location of point B, which corresponds to the ordered pair $(5, -2)$.

Keep in mind that **each ordered pair corresponds to exactly one point in the real plane and that each point in the plane corresponds to exactly one ordered pair.** Thus, we may refer to the ordered pair (x, y) as the point (x, y).

TEACHING TIP

Before doing Example 1, ask "Why are the points in a rectangular coordinate system called ordered pairs?"

EXAMPLE 1

Plot each ordered pair on a Cartesian coordinate system and name the quadrant in which the point is located.

a. $(2, -1)$ **b.** $(0, 5)$ **c.** $(-3, 5)$ **d.** $(-2, 0)$ **e.** $\left(-\frac{1}{2}, -4\right)$ **f.** $(1.5, \ 1.5)$

Solution   The six points are graphed as shown.

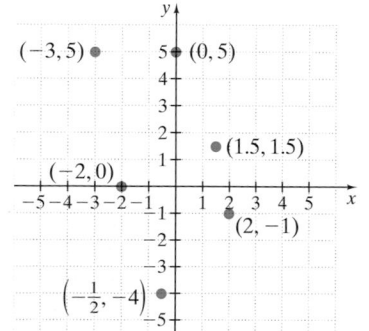

a. $(2, -1)$ lies in quadrant IV.

b. $(0, 5)$ is not in any quadrant.

c. $(-3, 5)$ lies in quadrant II.

d. $(-2, 0)$ is not in any quadrant.

e. $\left(-\frac{1}{2}, -4\right)$ is in quadrant III.

f. $(1.5, 1.5)$ is in quadrant I.

CLASSROOM EXAMPLE

Plot each ordered pair on a rectangular coordinate system and name the quadrant in which the point is located.

a. $(3, -2)$ **b.** $(0, 3)$
c. $(-4, 1)$ **d.** $(-1, 0)$
e. $\left(-2\frac{1}{2}, -3\right)$ **f.** $(3.5, 4.5)$

answer:
a. quadrant IV   **b.** not in any quadrant
c. quadrant II   **d.** not in any quadrant
e. quadrant III   **f.** quadrant I

Concept Check Answer:

c

Notice that the y-coordinate of any point on the x-axis is 0. For example, the point with coordinates $(-2, 0)$ lies on the x-axis. Also, the x-coordinate of any point on the y-axis is 0. For example, the point with coordinates $(0, 5)$ lies on the y-axis. These points that lie on the axes do not lie in any quadrants.

✔ **CONCEPT CHECK**

Which of the following best describes the location of the point $(3, -6)$ in a rectangular coordinate system?

a. 3 units to the left of the y-axis and 6 units above the x-axis
b. 3 units above the x-axis and 6 units to the left of the y-axis
c. 3 units to the right of the y-axis and 6 units below the x-axis
d. 3 units below the x-axis and 6 units to the right of the y-axis

2 The rectangular coordinate system extends infinitely in all directions with the number lines that form it. A graphing utility can be used to display a portion of the system. The portion being viewed is called a **viewing window** or simply a **window**.

Below, the rectangular coordinate system is shown with a corresponding viewing window indicated.

Yscl indicates the number of units per tick mark on the y-axis.
Xscl indicates the number of units per tick mark on the x-axis.

The dimensions of a window are selected by entering the minimum value, maximum value, and scale for both x and y under the window setting.

Several viewing windows can be accessed automatically on most graphing utilities. The screen below illustrates viewing the rectangular coordinate system from -10 to 10 on both the x-axis and the y-axis, with each tick mark representing 1 unit. This particular window setting is called the **standard window** on some graphing utilities.

Standard Window

To refer to the window dimensions throughout this text, we will use the following notation. $[-10, 10, 1]$ by $[-10, 10, 1]$ refers to [Xmin, Xmax, Xscl] by [Ymin, Ymax, Yscl].

If we select the same standard window setting but change the Xmin to -20, the Xmax to 20, and the Xscl to 5, observe to the left below how the window has been changed. Next, select the standard window setting but change the Ymin to -50, the Ymax to 50, and the Yscl to 10 and observe the changes in the window to the right.

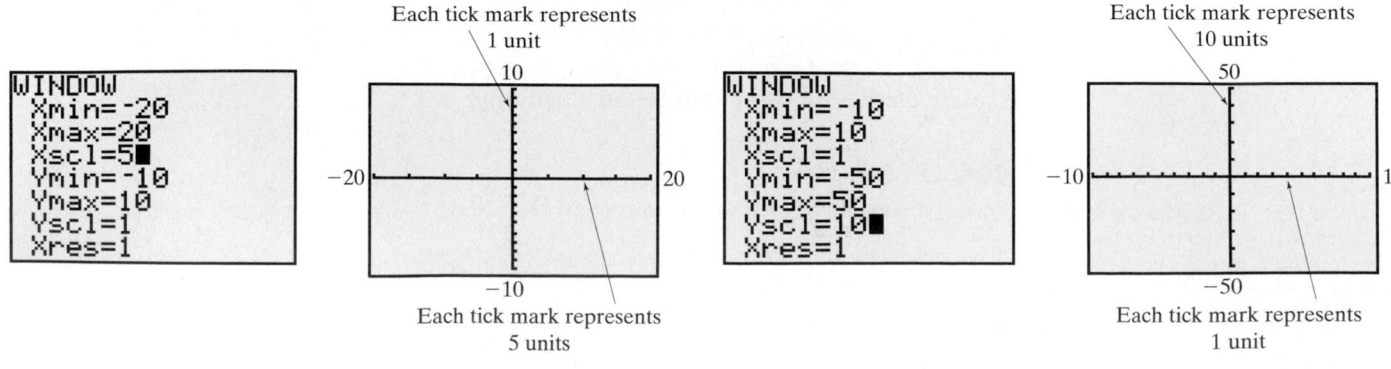

Many graphing utilities have the ability to show placement (via coordinates) and movement about the viewing window by means of a **cursor.**

Discover the Concept

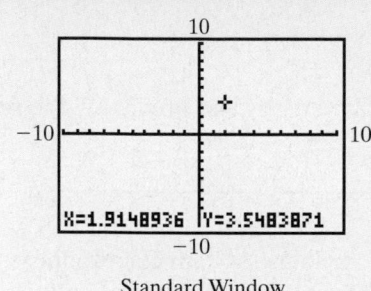

Standard Window

Clear any equations from the Y= editor and access the **standard window.** The screen shows the standard window with the cursor located at the ordered pair (1.9148936, 3.5483871).

Use the arrow keys and move the cursor around the rectangular coordinate system. Using the arrow keys, move to some points in quadrant I. What do you observe about the signs of the *x*-and *y*-coordinates in this quadrant? Repeat this process in quadrants II, III, and IV. What observations can you make about the signs of the coordinates in each quadrant?

From the discovery above, we see that the signs of the coordinates of an ordered pair depend on the quadrant in which it is located.

Quadrant I:	$(+, +)$	Quadrant III:	$(-, -)$
Quadrant II:	$(-, +)$	Quadrant IV:	$(+, -)$
x-axis:	$(x, 0)$	*y*-axis:	$(0, y)$

An **integer window** is another window that you may be able to automatically access on your graphing utility. In an integer window the *x*-coordinates are integers as you move the cursor around the coordinate system.

Access an integer window on your graphing utility and observe how the *x*-and *y*-coordinates are different from the *x*-and *y*-coordinates in a standard window. The screen below shows an integer window with the ordered pair (10, 7) indicated.

TECHNOLOGY NOTE

The settings for an integer window vary among graphing utilities. Look in your graphing utility manual to find the integer setting for your particular graphing utility.

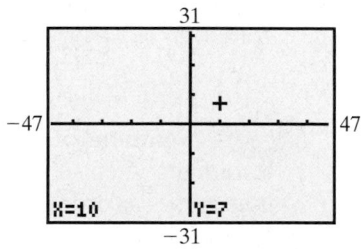

Integer Window

Another common window on many graphing utilities is a **decimal window.** Access this window and then move the cursor around the screen. Notice that the *x*-coordinates are incrementing or changing by 0.1 as you move from one point to another. The screen below shows an example of a decimal window on a graphing utility and its window settings.

Decimal Window

Technology Note

For many graphing utilities, both the standard window and the decimal window are centered at the origin. However, your graphing utility may allow you to center an integer window anywhere in the rectangular coordinate system. The center may depend on where you last left a cursor before selecting the integer setting. Check your graphing utility manual to see what window options are available to you.

Experiment on your own by changing the window settings and predicting the resulting windows.

Helpful Hint

Remember that each point in the rectangular coordinate system corresponds to a unique ordered pair (x, y), and each ordered pair corresponds to a unique point.

EXAMPLE 2

Display a window of the rectangular coordinate system with a setting of $[-20, 40, 5]$ by $[-30, 30, 10]$.

Solution The window setting and the resulting window are shown next.

CLASSROOM EXAMPLE

Display a window of the rectangular coordinate system with a setting of $[-40, 40, 20]$ by $[-10, 50, 10]$

answer:

3   **Solutions** of equations in two variables consist of two numbers that form a true statement when substituted into the equation. A convenient notation for writing these numbers is as ordered pairs. A solution of an equation containing the variables x and y is written as a pair of numbers in the order (x, y). If the equation contains other variables, we will write ordered pair solutions in alphabetical order.

EXAMPLE 3

Determine whether $(0, -12)$, $(1, 9)$, and $(2, -6)$ are solutions of the equation $3x - y = 12$.

Solution To check each ordered pair, replace x with the x-coordinate and y with the y-coordinate and see whether a true statement results.

CLASSROOM EXAMPLE

Determine whether $(0, -6)$, $(1, 4)$, and $(-1, -4)$ are solutions of the equation $2x + y = -6$.

answer: yes, no, yes

Let $x = 0$ and $y = -12$.

$$3x - y = 12$$
$$3(0) - (-12) \stackrel{?}{=} 12$$
$$0 + 12 \stackrel{?}{=} 12$$
$$12 = 12 \quad \text{True}$$

Let $x = 1$ and $y = 9$.

$$3x - y = 12$$
$$3(1) - 9 \stackrel{?}{=} 12$$
$$3 - 9 \stackrel{?}{=} 12$$
$$-6 = 12 \quad \text{False}$$

Let $x = 2$ and $y = -6$.

$$3x - y = 12$$
$$3(2) - (-6) \stackrel{?}{=} 12$$
$$6 + 6 \stackrel{?}{=} 12$$
$$12 = 12 \quad \text{True}$$

TEACHING TIP
Before discussing objective 4, challenge students to find another solution to the equation in Example 3.

Thus, $(1, 9)$ is not a solution of $3x - y = 12$, but both $(0, -12)$ and $(2, -6)$ are solutions.

4 In fact, the equation $3x - y = 12$ has an infinite number of ordered pair solutions. Since it is impossible to list all solutions, we visualize them by graphing them.

A few more ordered pairs that satisfy $3x - y = 12$ are $(4, 0)$, $(3, -3)$, $(5, 3)$, and $(1, -9)$. These ordered pair solutions along with the ordered pair solutions from Example 3 are plotted on the following graph. The graph of $3x - y = 12$ is the single line containing these points. Every ordered pair solution of the equation corresponds to a point on this line, and every point on this line corresponds to an ordered pair solution.

Using a table on the graphing utility to verify the ordered pairs solutions.

x	y	$3x - y = 12$
5	3	$3 \cdot 5 - 3 = 12$
4	0	$3 \cdot 4 - 0 = 12$
3	-3	$3 \cdot 3 - (-3) = 12$
2	-6	$3 \cdot 2 - (-6) = 12$
1	-9	$3 \cdot 1 - (-9) = 12$
0	-12	$3 \cdot 0 - (-12) = 12$

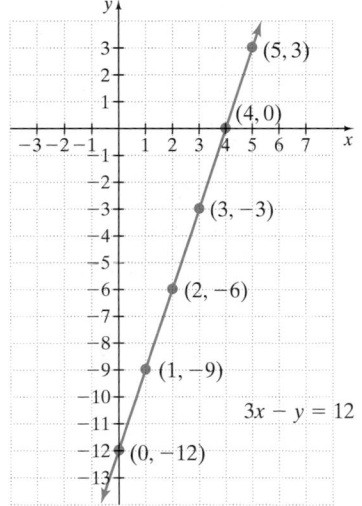

The equation $3x - y = 12$ is called a linear equation in two variables, and **the graph of every linear equation in two variables is a line.**

TEACHING TIP
Ask which of the following are linear equations:
$3x^2 + 4y = 1$
$7x + 3 = 2y$
$7y = 14$
$3x = -27$

> ### Linear Equation in Two Variables
>
> A linear equation in two variables is an equation that can be written in the form
> $$Ax + By = C$$
> where A and B are not both 0. This form is called **standard form**.

Some examples of equations in standard form:

$$3x - y = 12$$
$$-2.1x + 5.6y = 0$$

> **Helpful Hint**
>
> Remember: A linear equation is written in standard form when all of the variable terms are on one side of the equation and the constant is on the other side.

Many real-life applications are modeled by linear equations. Suppose you have a part-time job at a store that sells office products.

Your pay is $1500 plus 10% or $\frac{1}{10}$ of the price of the products you sell. If we let x represent products sold and y represent monthly salary, the linear equation that models your salary is

$$y = 1500 + \frac{1}{10}x$$

(Although this equation is not written in standard form, it is a linear equation. To see this, subtract $\frac{1}{10}x$ from both sides.)

Some ordered pair solutions of this equation are below.

Products Sold	x	0	100	200	300	400	1000
Monthly Salary	y	1500	1510	1520	1530	1540	1600

For example, we say that the ordered pair $(100, 1510)$ is a solution of the equation $y = 1500 + \frac{1}{10}x$ because when x is replaced with 100 and y is replaced with 1510, a true statement results.

$$y = 1500 + \frac{1}{10}x$$

$$1510 \stackrel{?}{=} 1500 + \frac{1}{10}(100) \quad \text{Let } x = 100 \text{ and } y = 1510.$$

$$1510 \stackrel{?}{=} 1500 + 10$$

$$1510 = 1510 \qquad \text{True.}$$

A portion of the graph of $y = 1500 + \frac{1}{10}x$ is shown in the next example.

Since we assume that the smallest amount of product sold is none, or 0, then x must be greater than or equal to 0. Therefore, only the part of the graph that lies in quadrant I is shown. Notice that the graph gives a visual picture of the correspondence between products sold and salary.

> **Helpful Hint**
>
> A line contains an infinite number of points and each point corresponds to an ordered pair that is a solution of its corresponding equation.

EXAMPLE 4

Use the graph of $y = 1500 + \frac{1}{10}x$ to answer the following questions.

a. If the salesperson has $800 of products sold for a particular month, what is the salary for that month?

b. If the salesperson wants to make more than $1600 per month, what must be the total amount of products sold?

Solution
a. Since x is products sold, find 800 along the x-axis and move vertically up until you reach a point on the line. From this point on the line, move horizontally to the left until you reach the y-axis. Its value on the y-axis is 1580, which means if $800 worth of products is sold, the salary for the month is $1580.

CLASSROOM EXAMPLE

Use the graph in Example 4.

a. If the salesperson sells $700 of products for a particular month, what is the salary?

b. Find the total amount of products needed to be sold to make more than $1550 per month.

answer: a. $1570

b. more than $500 of products

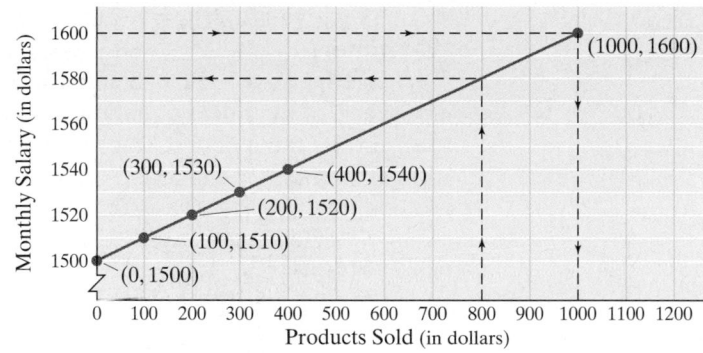

b. Since y is monthly salary, find 1600 along the y-axis and move horizontally to the right until you reach a point on the line. Either read the corresponding x-value from the labeled ordered pair, or move vertically downward until you reach the x-axis. The corresponding x-value is 1000. This means that $1000 worth of products sold gives a salary of $1600 for the month. For the salary to be greater than $1600, products sold must be greater that $1000.

Recall from geometry that a line is determined by two points. This means that to graph a linear equation in two variables, just two solutions are needed. We will find a third solution, just to check our work. To find ordered pair solutions of linear equations in two variables, we can choose an x-value and find its corresponding y-value, or we can choose a y-value and find its corresponding x-value. The number 0 is often a convenient value to choose for x and also for y.

EXAMPLE 5

Graph the equation $y = -2x + 3$. Then use a graphing utility to check.

Solution

This is a linear equation. (In standard form it is $2x + y = 3$.) Find three ordered pair solutions, and plot the ordered pairs. The line through the plotted points is the graph. Since the equation is solved for y, let's choose three x-values. Let's let x be 0, 2, and then -1 to find our three ordered pair solutions.

CLASSROOM EXAMPLE

Graph $y = 4x - 3$.

answer:

Let $x = 0$

$y = -2x + 3$
$y = -2 \cdot 0 + 3$
$y = 3$ Simplify.

Let $x = 2$

$y = -2x + 3$
$y = -2 \cdot 2 + 3$
$y = -1$ Simplify.

Let $x = -1$

$y = -2x + 3$
$y = -2(-1) + 3$
$y = 5$ Simplify.

The three ordered pairs $(0, 3), (2, -1)$ and $(-1, 5)$ are listed in the table and the graph is shown.

x	y
0	3
2	-1
-1	5

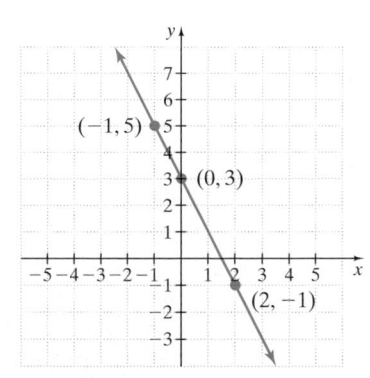

Next we graph $y = -2x + 3$ using a graphing utility. To do so, enter it into the Y= editor.

Define $y_1 = -2x + 3$ and select the integer window. The screen below illustrates the graph of the linear equation. Access a trace feature to locate the ordered pair solutions in the table above. Trace to verify that the point $(0, 3)$ is a solution for the equation as well as $(2, -1)$ and $(-1, 5)$.

Notice that the graph crosses the y-axis at the point $(0, 3)$. This point is called the **y-intercept**. (You may sometimes see just the number 3 called the y-intercept.) This graph also crosses the x-axis at the point $\left(\frac{3}{2}, 0\right)$. This point is called the **x-intercept**. (You may also see just the number $\frac{3}{2}$ called the x-intercept.)

Since every point on the y-axis has an x-value of 0, we can find the y-intercept of a graph by letting $x = 0$ and solving for y. Also, every point on the x-axis has a y-value of 0. To find the x-intercept, we let $y = 0$ and solve for x.

TEACHING TIP

Ask students why 0 is a convenient value to substitute for x or y.

Finding x- and y-Intercepts

To find an x-intercept, let $y = 0$ and solve for x.
To find a y-intercept, let $x = 0$ and solve for y.

We will study intercepts further in Section 2.3.

EXAMPLE 6

Graph $5x - 3y = 10$ in an integer window centered at the origin. Find the y-coordinate that makes the ordered pair $(8, ?)$ a solution for the given equation.

Solution   To enter the equation in the Y= editor, first solve the equation for y.

$$5x - 3y = 10$$
$$5x - 3y - 5x = -5x + 10 \quad \text{Subtract 5x from both sides.}$$
$$-3y = -5x + 10$$
$$y = \frac{-5x + 10}{-3} \quad \text{Divide both sides by } -3.$$

CLASSROOM EXAMPLE

Graph $2x + 3y = -9$ in an integer window centered at the origin. Find the y-coordinate that makes the ordered pair $(3, ?)$ a solution.

answer: $(3, -5)$

Define $y_1 = \frac{-5x + 10}{-3}$ and graph in an integer window centered at $(0, 0)$. Trace to locate the point with x-coordinate 8 and find the corresponding y-value to be 10.

The ordered pair (8, 10) is a solution of the given solution. To check, let $x = 8$ and $y = 10$ in the equation $5x - 3y = 10$ and see that a true statement results.

5   Not all equations in two variables are linear equations, and not all graphs of equations in two variables are lines.

EXAMPLE 7

Graph $y = x^2$.

Solution   This equation is not linear because the x^2 term does not allow us to write it in the form $Ax + By = C$. Its graph is not a line. We begin by finding ordered pair solutions. Because this graph is solved for y, we choose x-values and find corresponding y-values.

TEACHING TIP

Tell students that if they are unsure about the graph of a nonlinear equation, they should find more ordered pair solutions.

CLASSROOM EXAMPLE

Graph $y = -x^2$.
answer:

If $x = -3$, then $y = (-3)^2$, or 9.

If $x = -2$, then $y = (-2)^2$, or 4.

If $x = -1$, then $y = (-1)^2$, or 1.

If $x = 0$, then $y = 0^2$, or 0.

If $x = 1$, then $y = 1^2$, or 1.

If $x = 2$, then $y = 2^2$, or 4.

If $x = 3$, then $y = 3^2$, or 9.

x	y
-3	9
-2	4
-1	1
0	0
1	1
2	4
3	9

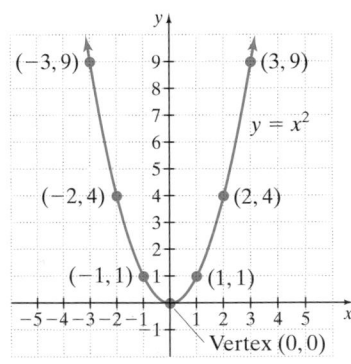

Study the table a moment and look for patterns. Notice that the ordered pair solution $(0, 0)$ contains the smallest y-value because any other x-value squared will give a positive result. This means that the point $(0, 0)$ will be the lowest point on the graph. Also notice that all other y-values correspond to two different x-values. For example, $3^2 = 9$ and also $(-3)^2 = 9$. This means that the graph will be a mirror image of itself across the y-axis. Connect the plotted points with a smooth curve to sketch its graph.

To check, graph $y_1 = x^2$ in an integer window. Compare the graph and table shown with the calculations done by hand.

This curve is given a special name called a **parabola**. It can be shown that the graph of a quadratic equation of the form $y = ax^2 + bx + c$ is a parabola. We will study more about parabolas in later chapters.

The graphing utility finds and plots various ordered pair solutions in the same manner that we find and plot ordered pair solutions. The advantage of a graphing utility is that it can perform this process much faster and with greater accuracy than we can.

Discover the Concept

a. Compare the graphs of $y = x^2$ and $y = -x^2$ by graphing each in a standard window.

b. How can the graph of $y = x^2$ be transformed to look like the graph of $y = -x^2$?

In the above discovery, we see that if the coefficient of x^2 is positive, the parabola opens upward, and if the coefficient is negative, the parabola opens downward. Compare this fact with the table of values below. We defined $y_1 = x^2$ and $y_2 = -x^2$.

For $y = -x^2$, the y-value is always negative or 0 and the parabola opens downward.

For $y = x^2$, the y-value is always positive or 0 and the parabola opens upward.

We say the graph of $y = x^2$ is reflected across the x-axis to obtain the graph of $y = -x^2$.

Another nonlinear equation is a third-degree or cubic equation.

EXAMPLE 8

Use a graphing utility to graph $y = x^3$ and examine a table of values with x equal to the integers from -3 to 3.

Solution   Define $y_1 = x^3$ and graph the equation in a standard window. The figure below shows the graph of $y = x^3$ with the trace cursor on $(2, 8)$. The table of values shows ordered pair solutions for $x = -3, -2, -1, 0, 1, 2,$ and 3.

CLASSROOM EXAMPLE

Review Example 8 to your right.

Notice in the table that when x is negative y is also negative. Thus, these points lie in the third quadrant. When x is positive, y is positive; thus, this portion of the graph lies in the first quadrant. When $x = 0$, then $y = 0$, so the graph crosses the axes at the origin.

Discover the Concept

How do you think the graph of $y = x^3$ is affected when we place a negative sign in front of the x^3? Test your prediction by graphing $y = x^3$ and $y = -x^3$ separately in a standard window. State in words how you can obtain the graph of $y = -x^3$ from the graph of $y = x^3$.

We next graph the basic absolute value equation, $y = |x|$. Begin by finding and plotting ordered pair solutions and analyze any patterns.

EXAMPLE 9

Make a table of values for $y = |x|$. Let x be the integers from -3 to 3. Use the table to graph the equation $y = |x|$. Use a graphing utility to check the table of values and the shape of the graph.

Solution This is not a linear equation, so its graph is not a line. We list x-values and substitute to find corresponding y-values. Then we plot the ordered pair solutions.

CLASSROOM EXAMPLE

Review Example 9 to your right.

	x	y
If $x = -3$, then $y = \lvert -3 \rvert$, or 3.	-3	3
If $x = -2$, then $y = \lvert -2 \rvert$, or 2.	-2	2
If $x = -1$, then $y = \lvert -1 \rvert$, or 1.	-1	1
If $x = 0$, then $y = \lvert 0 \rvert$, or 0.	0	0
If $x = 1$, then $y = \lvert 1 \rvert$, or 1.	1	1
If $x = 2$, then $y = \lvert 2 \rvert$, or 2.	2	2
If $x = 3$, then $y = \lvert 3 \rvert$, or 3.	3	3

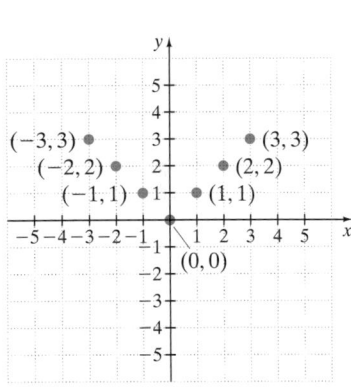

Study the table of values and the plotted points and notice any patterns.

From the plotted ordered pairs, we see that the graph of this absolute value equation is V-shaped. The completed graph is shown next. To check, we show the graph of $y = |x|$ using a standard window alongside.

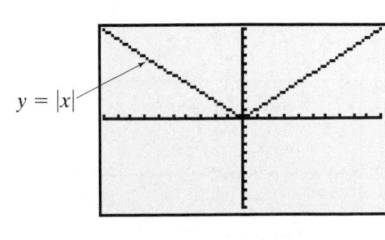

Notice that the basic absolute value graph is V-shaped as compared to the U-shape of a parabola.

We now examine how a graph is affected when a constant is added.

EXAMPLE 10

Graph the equation $y = |x| - 3$ and examine a table of values for x equal to the integers from -3 to 3. Compare the graph and the table to the graph and table for $y = |x|$.

Solution

Graph both $y_1 = |x|$ and $y_2 = |x| - 3$ using a standard window. Notice that compared with the equation $y = |x|$, the equation $y = |x| - 3$ decreases the y-values by 3 units for the same x-values. This means that the graph of $y = |x| - 3$ is the same as the graph of $y = |x|$ lowered by 3 units.

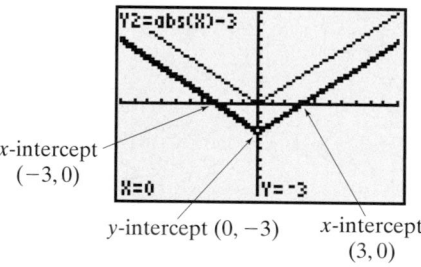

x-intercept $(-3, 0)$

y-intercept $(0, -3)$ x-intercept $(3, 0)$

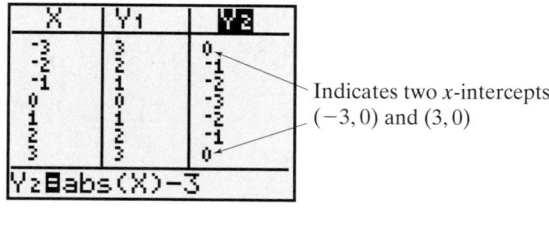

Indicates two x-intercepts $(-3, 0)$ and $(3, 0)$

This graph of $y = |x| - 3$ shows us that it is possible to have more than one x- or y-intercept. Notice that this graph has one y-intercept $(0, -3)$ and two x-intercepts $(-3, 0)$ and $(3, 0)$.

Discover the Concept

a. Predict how the graph of $y = |x| + 5$ can be obtained from the graph of $y = |x|$. Then check your prediction by graphing both equations using a standard window.

b. Predict how the graph of $y = x^2 - 3$ can be obtained from the graph of $y = x^2$. Use a graphing utility to check your prediction. Next, predict how the graph of $y = x^2 + 5$ can be obtained from $y = x^2$.

c. In general, can you predict how the graph of a basic equation with a constant added can be obtained from the graph of the basic equation? Test your prediction with the basic cubic equation $y = x^3$.

If we add a constant K to a basic graph, it is said to cause a **vertical translation** of the graph. That is, if the constant K is positive, the basic graph is moved or translated K units upward. Similarly, if the constant K is negative, the basic graph is moved, or translated, $|K|$ units downward.

To summarize, the graphs of the basic equations we have discussed so far are shown below.

Standard Window

Basic Linear Equation
$y = x$

Absolute Value Equation
$y = |x|$

Quadratic Equation (Parabola)
$y = x^2$

Cubic Equation
$y = x^3$

Spotlight on
DECISION
MAKING

Suppose you are a meteorologist. You are tracking Hurricane Felix. Hurricane position information is issued every 6 hours. The table lists Felix's most recent positions, given in latitude (vertical scale on the hurricane tracking chart) and longitude (horizontal scale on the hurricane tracking chart). Plot the position of the hurricane on the tracking chart and decide whether Felix is a threat to the United States. If so, what part?

National Hurricane Center Reconnaissance Report

		Felix Coordinates	
Date	**Time**	**Latitude**	**Longitude**
9/7	4 PM	23.2	88.0
9/7	10 PM	23.7	88.6
9/8	4 AM	23.4	89.1
9/8	10 AM	22.8	89.9
9/8	4 PM	22.2	91.2
9/8	10 PM	21.7	91.8
9/9	4 AM	21.5	92.4
9/9	10 AM	21.2	93.1

Hurricane Tracking Chart

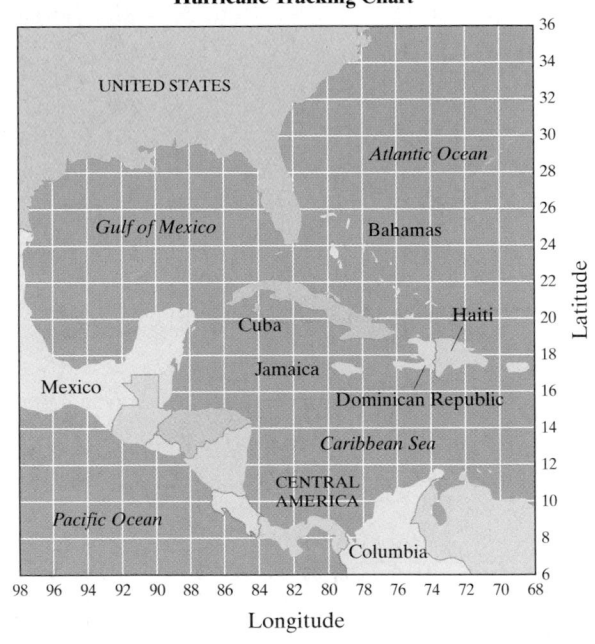

MENTAL MATH

Determine the coordinates of each point on the graph.

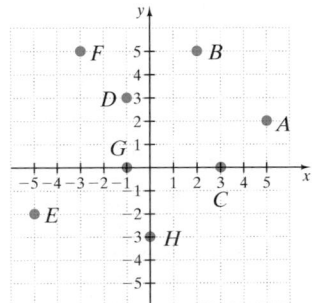

1. Point A $(5, 2)$
2. Point B $(2, 5)$
3. Point C $(3, 0)$
4. Point D $(-1, 3)$
5. Point E $(-5, -2)$
6. Point F $(-3, 5)$
7. Point G $(-1, 0)$
8. Point H $(0, -3)$

Without graphing, visualize the location of each point. Then give its location by quadrant or x- or y-axis.

9. $(2, 3)$ QI
10. $(0, 5)$ y-axis
11. $(-2, 7)$ QII
12. $(-3, 0)$ x-axis
13. $(-1, -4)$ QIII
14. $(4, -2)$ QIV
15. $(0, -100)$ y-axis
16. $(10, 30)$ QI
17. $(-10, -30)$ QIII
18. $(0, 0)$ x- and y-axis
19. $(-87, 0)$ x-axis
20. $(-42, 17)$ QII

EXERCISE SET 2.1

1.–10. See graphing answer section.

STUDY GUIDE/SSM   CD/VIDEO   PH MATH TUTOR CENTER   MathXL®Tutorials ON CD   MathXL®   MyMathLab®

Plot each point and name the quadrant or axis in which the point lies. See Example 1.

1. $(3,2)$ quadrant I
2. $(2,-1)$ quadrant IV
3. $(-5,3)$ quadrant II
4. $(-3,-1)$ quadrant III
5. $\left(5\frac{1}{2},-4\right)$ quadrant IV
6. $\left(-2,6\frac{1}{3}\right)$ quadrant II
7. $(0,3.5)$ y-axis
8. $(-5.2,0)$ x-axis
9. $(-2,-4)$ quadrant III
10. $(-4.2,0)$ x-axis

Given that x is a positive number and that y is a positive number, determine the quadrant or axis in which each point lies.

11. $(x,-y)$ quadrant IV
12. $(-x,y)$ quadrant II
13. $(x,0)$ x-axis
14. $(0,-y)$ y-axis
15. $(-x,-y)$ quadrant III
16. $(0,0)$ origin

Determine the coordinates of the point shown and name the quadrant in which the point lies.

17.

$(2,8)$; quadrant I

18.

$(-3,2)$; quadrant II

19.

$(1,-4)$; quadrant IV

20.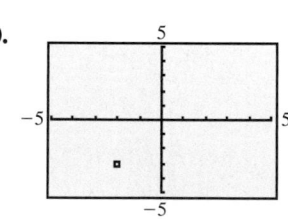

$(-2,-3)$; quadrant III

Determine a window setting so that the three given points will lie in that window. Report your answer in the [Xmin, Xmax, Xscl] by [Ymin, Ymax, Yscl] format. See Example 2.

21. $(-5,2),(3,8),(10,15)$
22. $(-10,12),(0,9),(6,-15)$
23. $(-25,0),(5,7),(20,100)$
24. $(0,50),(25,75),(50,150)$

21. possible answer: $[-10,10,1]$ by $[0,20,1]$
22. possible answer: $[-15,15,1]$ by $[-15,15,1]$
23. possible answer: $[-100,100,10]$ by $[-100,100,10]$
24. possible answer: $[-150,150,10]$ by $[-150,150,10]$

Match the following screens with the window settings listed below. **25.** C **26.** B **27.** D **28.** A

25.

26.

27.

28.

A. WINDOW
Xmin=0
Xmax=100
Xscl=10
Ymin=0
Ymax=50
Yscl=10
Xres=1■

B. WINDOW
Xmin=-3
Xmax=3
Xscl=1
Ymin=-5
Ymax=5
Yscl=1
Xres=1■

C. WINDOW
Xmin=-5
Xmax=5■
Xscl=1
Ymin=-5
Ymax=5
Yscl=1
Xres=1

D. WINDOW
Xmin=-1
Xmax=6
Xscl=1
Ymin=-3
Ymax=10
Yscl=1■
Xres=1

Use a table or a graph to determine whether each ordered pair is a solution of the given equation. See Example 3.

29. $y=3x-5$; $(0,5),(-1,-8)$ no; yes
30. $y=-2x+7$; $(1,5),(-2,3)$ yes; no
31. $-6x+5y=-6$; $(1,0),\left(2,\frac{6}{5}\right)$ yes; yes
32. $5x-3y=9$; $(0,3),\left(\frac{12}{5},-1\right)$ no; no
33. $y=2x^2$; $(1,2),(3,18)$ yes; yes
34. $y=2|x|$; $(-1,2),(0,2)$ yes; no

37. $y = -2x + 10$ **38.** $y = 6x + 2$ **39.** $y = (-7x - 4)/3$ **40.** $y = (5x + 1.2)/11$

MIXED PRACTICE

Determine whether each equation is linear or nonlinear and tell the basic shape of the graph (line, parabola, cubic, V-shaped). See Examples 5 through 10.

35.

Equation	Linear or nonlinear	Shape (Line, Parabola, Cubic, V-shaped)
$y - x = 8$	linear	line
$y = 6x$	linear	line
$y = x^2 + 3$	nonlinear	parabola
$y = 6x - 5$	linear	line
$y = -|x| + 2$	nonlinear	V-shaped
$y = 3x^2$	nonlinear	parabola
$y = -4x + 2$	linear	line
$y = -|x|$	nonlinear	V-shaped
$y = x^3$	nonlinear	cubic

36.

Equation	Linear or nonlinear	Shape (Line, Parabola, Cubic, V-shaped)
$x + y = 3$	linear	line
$y = 4 - x$	linear	line
$y = 2x^2 - 5$	nonlinear	parabola
$y = -8x + 6$	linear	line
$y = |x - 3|$	nonlinear	V-shaped
$y = 7x^2$	nonlinear	parabola
$2x - y = 5$	linear	line
$y = -|x - 1|$	nonlinear	V-shaped
$y = x^3 - 2$	nonlinear	cubic

Rewrite each equation as an equivalent equation that can be entered in the Y= *editor of your graphing utility. See Example 6.*

37. $2x + y = 10$ **38.** $-6x + y = 2$

39. $-7x - 3y = 4$ **40.** $5x - 11y = -1.2$

41. The graph of $y = 2x^2 + 1.2x - 5.6$ has two x-intercepts and one y-intercept. Use your graphing utility and graph this equation using a decimal window. Find the coordinates of the intercepts.   See graphing answer section.

42. The graph of $y = x^2 + 7x - 6$ has two ordered pair solutions whose y-value is 12. Use your graphing utility and graph this equation using an integer window. Find the coordinates of these points.   See graphing answer section.

43. If you trace along the graph of $y = 1.5x - 6$ using an integer window, which of the following coordinates would not be displayed?   d
a. $(0, -6)$ **b.** $(1, -4.5)$
c. $(4, 0)$ **d.** $(1.2, -4.2)$

44. If you trace along the graph of $y = x + 2.3$ using an integer window, which of the following coordinates would not be displayed?   a
a. $(1.7, 4)$ **b.** $(0, 2.3)$
c. $(2, 4.3)$ **d.** $(-3, -0.7)$

Match each equation with its graph without using a graphing utility. Then check your answers by graphing each equation using a standard window.

45. $y = x^2 - 4$ C   **46.** $y = x + 5$ D
47. $y = x^3 + x^2 - 4$ A   **48.** $y = |x| + 3$ B

A.   **B.**

C.   **D.**

Match each equation with its graph without using a graphing utility. Then check your answers by graphing each equation using a standard window.

49. $y = x - 5$ D   **50.** $y = -x^2 + 4$ B
51. $y = -x^3 + x^2 + 2x - 3$ C   **52.** $y = |x + 3|$ A

A.   **B.**

C.   **D.**

Match each graph with its equation without using a graphing utility. Then graph using a standard window to check your answers.

53. A

54. B

55. C

56. D

A. $y = x^2 - 3$
B. $y = x^2 + 3$
C. $y = |x| + 5$
D. $y = |x| - 3$

57.–86. See graphing answer section.

Sketch the graphs of the three equations by hand on the same coordinate plane. Then use your graphing utility to check your results. Describe the patterns you see in these graphs.

57. $y = x$
$y = x + 2$
$y = x - 3$

58. $y = x^2$
$y = x^2 + 1$
$y = x^2 - 3$

59. $y = -x^2$
$y = -x^2 + 1$
$y = -x^2 - 2$

60. $y = -|x|$
$y = -|x| + 2$
$y = -|x| - 4$

Graph each equation by plotting ordered pair solutions by hand. If the equation is not linear, suggested x-values have been given for generating ordered pair solutions. Check the graph using a graphing utility. See Examples 4 through 10.

61. $x + y = 3$

62. $y - x = 8$

63. $y = 4x$

64. $y = 6x$

65. $y = 4x - 2$

66. $y = 6x - 5$

67. $y = |x| + 3$
Let $x = -3, -2, -1, 0, 1, 2, 3$.

68. $y = |x| + 2$
Let $x = -3, -2, -1, 0, 1, 2, 3$.

69. $2x - y = 5$

70. $4x - y = 7$

71. $y = 2x^2$
Let $x = -3, -2, -1, 0, 1, 2, 3$.

72. $y = 3x^2$
Let $x = -3, -2, -1, 0, 1, 2, 3$.

73. $y = x^2 - 3$
Let $x = -3, -2, -1, 0, 1, 2, 3$.

74. $y = x^2 + 3$
Let $x = -3, -2, -1, 0, 1, 2, 3$.

75. $y = -2x$

76. $y = -3x$

77. $y = -2x + 3$

78. $y = -3x + 2$

79. $y = |x + 2|$
Let $x = -4, -3, -2, -1, 0, 1$.

80. $y = |x - 1|$
Let $x = -1, 0, 1, 2, 3, 4$.

81. $y = x^3$
Let $x = -3, -2, -1, 0, 1, 2$.

82. $y = x^3 - 2$
Let $x = -3, -2, -1, 0, 1, 2$.

83. $y = -|x|$
Let $x = -3, -2, -1, 0, 1, 2, 3$.

84. $y = -x^2$
Let $x = -3, -2, -1, 0, 1, 2, 3$.

85. Given the following table of values for ordered pairs satisfying an absolute value equation, plot the points and join them to complete the graph. Then write an equation of the graph. $y = -|x|$

X	Y1	
-3	-3	
-2	-2	
-1	-1	
0	0	
1	-1	
2	-2	
3	-3	
X = -3		

86. Given the following table of values for ordered pairs satisfying a second degree equation, plot the points and join them to complete the graph. Then write an equation of the graph. $y = -x^2$

X	Y1	
-3	-9	
-2	-4	
-1	-1	
0	0	
1	-1	
2	-4	
3	-9	
X = -3		

For Exercises 87 through 90, fill in the blank with the word "line" or "parabola."

△ **87.**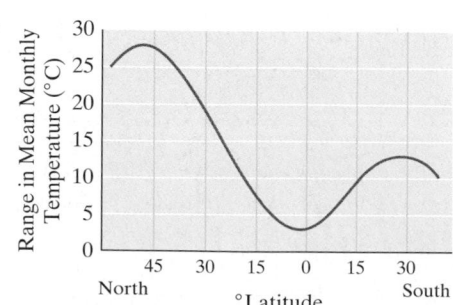

a. The shape of the graph from 55°N-15°N only resembles a __parabola__.

b. The shape of the graph from 30°N to 15°N resembles a __line__.

△ **88.**

a. The shape of the graph from 20°C to 60°C resembles a   parabola.

b. The shape of the graph from 0°C to 20°C resembles a   line .

△ **89.**

a. The shape of the graph from 1650 to 1800 resembles a   line .

b. The shape of the graph from 1900 to 2000 resembles a   parabola.

△ **90.**

a. The shape of the graph from 600 nm to 700 nm resembles a   parabola.

b. The shape of the graph from 550 nm to 625 nm resembles a   line .

Solve See Example 4.   **91.a., 92.a.** See graphing answer section.

△ **91.** The perimeter y of a rectangle whose width is a constant 3 inches and whose length is x inches is given by the equation

$$y = 2x + 6$$

a. Draw a graph of this equation.

b. Read from the graph the perimeter y of a rectangle whose length x is 4 inches.   14 in.

x inches

3 inches

92. The distance y traveled in a train moving at a constant speed of 50 miles per hour is given by the equation

$$y = 50x$$

where x is the time in hours traveled.

a. Draw a graph of this equation.

b. Read from the graph the distance y traveled after 6 hours.   300 mi

REVIEW AND PREVIEW

Solve the following equations. See Section 1.5.

93. $3(x - 2) + 5x = 6x - 16$ -5

94. $5 + 7(x + 1) = 12 + 10x$ 0

95. $3x + \dfrac{2}{5} = \dfrac{1}{10}$ $-\dfrac{1}{10}$ **96.** $\dfrac{1}{6} + 2x = \dfrac{2}{3}$ $\dfrac{1}{4}$

Concept Extensions

For Exercises 97 through 100, match each description with the graph that best illustrates it.

97. Moe worked 40 hours per week until the fall semester started. He quit and didn't work again until he worked 60 hours a week during the holiday season starting mid-December break.   b

98. Kawana worked 40 hours a week for her father during the summer. She slowly cut back her hours to not working at all during the fall semester. During the holiday season in December, she started working again and increased her hours to 60 hours per week.   d

99. Wendy worked from July through February, never quitting. She worked between 10 and 30 hours per week.   c

100. Bartholomew worked from July through February. The rest of the time, he worked between 10 and 40 hours per week. During the holiday season between mid-November and the beginning of January, he worked 40 hours per week.   a

a.

b.

c.

d.

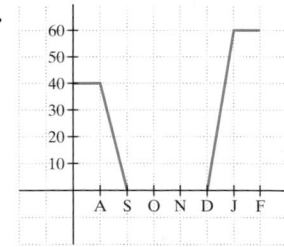

101. **GROUP ACITVITY.** This graph shows hourly minimum wages and the years it increased.

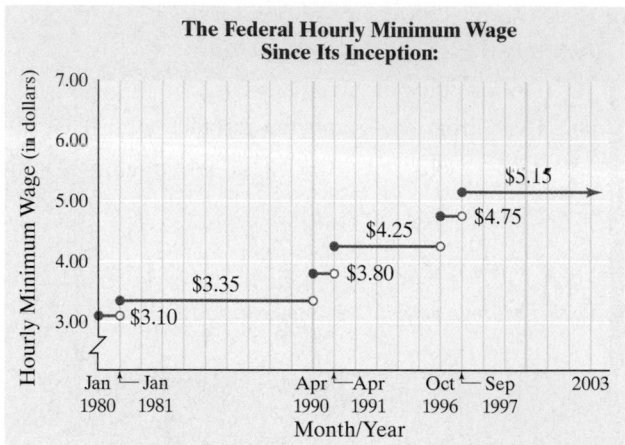

a. What was the first year that the minimum hourly wage rose above $4.00?   1991
b. What was the first year that the minimum hourly wage rose above $5.00?   1997
c. Why do you think that this graph is shaped the way it is?
d. The federal hourly minimum wage started in 1938 at $0.25. How much has it increased by in 2003? $4.90

102. **GROUP ACITVITY.** For income tax purposes, Jason Verges, owner of Copy Services, uses a method called ***straight-line depreciation*** to show the loss in value of a copy machine he recently purchased. Jason assumes that he can use the machine for 7 years. The following graph shows the value of

the machine over the years. Use this graph to answer the following questions.

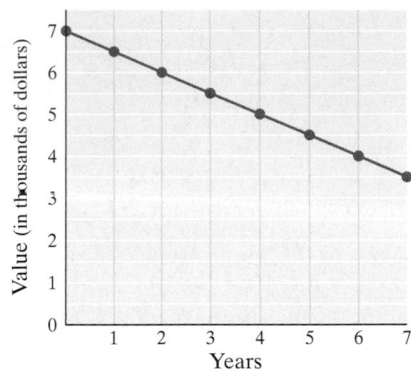

a. What was the purchase price of the copy machine? $7000
b. What is the depreciated value of the machine in 7 years?
c. What loss in value occurred during the first year? $500
d. What loss in value occurred during the second year? $500
e. Why do you think that this method of depreciating is called straight-line depreciation?
f. Why is the line tilted downward?   answers may vary

Write each statement as an equation in two variables. Then graph each equation.   103.–106. See graphing answer section.

103. The *y*-value is 5 more than three times the *x*-value.
104. The *y*-value is −3 decreased by twice the *x*-value.
105. The *y*-value is 2 more than the square of the *x*-value.
106. The *y*-value is 5 decreased by the square of the *x*-value.

101.c. answers may vary   **102.b.** $3500   **102.e.** Depreciation is the same from year to year.   **103.** $y = 3x + 5$ **104.** $y = -3 - 2x$
105. $y = x^2 + 2$ **106.** $y = 5 - x^2$

2.2   INTRODUCTION TO FUNCTIONS

Objectives

1   Define relation, domain, and range.

2   Identify functions.

3   Use the vertical line test for functions.

4   Find the domain and range of a function.

5   Use function notation.

1   Recall our example from the last section about products sold and monthly salary. We modeled the data given by the equation $y = 1500 + \frac{1}{10}x$. This equation describes a relationship between *x*-values and *y*-values. For example, if $x = 100$, then this equation describes how to find the *y*-value related to $x = 100$. In words, the equation $y = 1500 + \frac{1}{10}x$ says that 1500 plus $\frac{1}{10}$ of the *x*-value gives the corresponding

y-value. The x-value of 100 corresponds to the y-value of $1500 + \frac{1}{10} \cdot 100 = 1510$ for this equation, and we have the ordered pair $(100, 1510)$.

There are other ways of describing relations or correspondences between two numbers or, in general, a first set (sometimes called the set of *inputs*) and a second set (sometimes called the set of *outputs*). For example,

First Set: Input	*Correspondence*	*Second Set: Output*
People in a certain city	Each person's age	The set of nonnegative integers

A few examples of ordered pairs from this relation might be (Ana, 4); (Bob, 36); (Trey, 21); and so on.

Below are just a few other ways of describing relations between two sets and the ordered pairs that they generate.

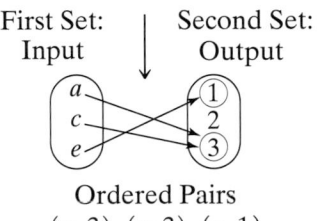

Correspondence

First Set: Input   Second Set: Output

Ordered Pairs
$(a, 3), (c, 3), (e, 1)$

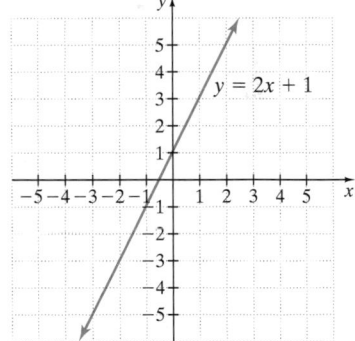

Ordered Pairs
$(-3, -1), (1, 1), (2, 3), (3, -2)$

Some Ordered Pairs
$(1, 3), (0, 1)$, and so on

Relation, Domain, and Range

A **relation** is a set of ordered pairs.
The **domain** of the relation is the set of all first components of the ordered pairs.
The **range** of the relation is the set of all second components of the ordered pairs.

For example, the domain for our relation in the middle of the page above is $\{a, c, e\}$ and the range is $\{1, 3\}$. Notice that the range does not include the element 2 of the second set. This is because no element of the first set is assigned to this element. If a relation is defined in terms of x- and y-values, we will agree that the domain corresponds to x-values and that the range corresponds to y-values that have x-values assigned to them.

EXAMPLE 1

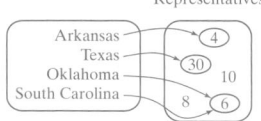
Determine the domain and range of each relation.

a. $\{(2,3),(2,4),(0,-1),(3,-1)\}$

b.

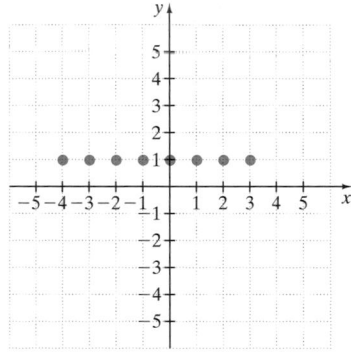

c.

Input:	Output:
Cities	Population (in thousands)

Solution

a. The domain is the set of all first coordinates of the ordered pairs, $\{2,0,3\}$. The range is the set of all second coordinates, $\{3,4,-1\}$.

b. Ordered pairs are not listed here, but are given in graph form. The relation is $\{(-4,1),(-3,1),(-2,1),(-1,1),(0,1),(1,1),(2,1),(3,1)\}$. The domain is $\{-4,-3,-2,-1,0,1,2,3\}$. The range is $\{1\}$.

c. The domain is the first set, $\{$Erie, Escondido, Gary, Miami, Waco$\}$. The range is the numbers in the second set that correspond to elements in the first set $\{104,109,117,359\}$.

2 Now we consider a special kind of relation called a function.

Function

A **function** is a relation in which each first component in the ordered pairs corresponds to *exactly* one second component.

Helpful Hint

A function is a special type of relation, so all functions are relations, but not all relations are functions.

EXAMPLE 2

Which of the following relations are also functions?

a. $\{(-2,5),(2,7),(-3,5),(9,9)\}$

CLASSROOM EXAMPLE

Determine whether each relation is also a function.

a. $\{(-3,7),(1,7),(2,2)\}$

b.

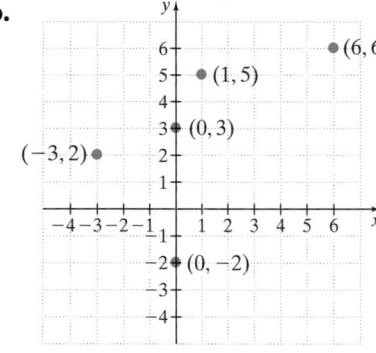

c.

Input	Correspondence	Output
People in a certain state	County/ parish that a person lives in	Counties of that state

answer:

a. function

b. not a function

c. function

b.

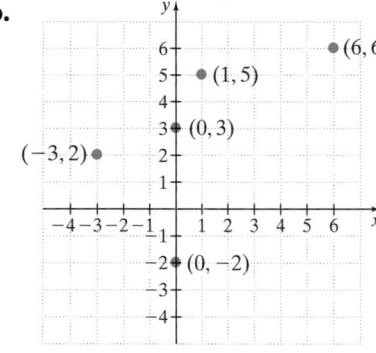

c.

Input	Correspondence	Output
People in a certain city	Each person's age	The set of nonnegative integers

Solution

a. Although the ordered pairs $(-2,5)$ and $(-3,5)$ have the same y-value, each x-value is assigned to only one y-value, so this set of ordered pairs is a function.

b. The x-value 0 is assigned to two y-values, -2 and 3, in this graph so this relation does not define a function.

c. This relation is a function because although two different people may have the same age, each person has only one age. This means that each element in the first set is assigned to only one element in the second set.

✔ **CONCEPT CHECK**

Explain why a function can contain both the ordered pairs $(1,3)$ and $(2,3)$ but not both $(3,1)$ and $(3,2)$.

We will call an equation such as $y = 2x + 1$ a **relation** since this equation defines a set of ordered pair solutions.

EXAMPLE 3

Is the relation $y = 2x + 1$ also a function?

Solution The relation $y = 2x + 1$ is a function if each x-value corresponds to just one y-value. For each x-value substituted in the equation $y = 2x + 1$, the multiplication and addition performed on each gives a single result, so only one y-value will be associated with each x-value. Thus, $y = 2x + 1$ is a function.

CLASSROOM EXAMPLE

Determine whether the relation $y = 3x + 2$ is also a function.

answer: yes

 EXAMPLE 4

Is the relation $x = y^2$ also a function?

Solution In $x = y^2$, if $y = 3$, then $x = 9$. Also, if $y = -3$, then $x = 9$. In other words, the x-value 9 corresponds to two y-values, 3 and -3. Thus, $x = y^2$ is not a function.

Concept Check Answer:
Two different ordered pairs can have the same y-value, but not the same x-value in a function.

3 As we have seen so far, not all relations are functions. Consider the graphs of $y = 2x + 1$ and $x = y^2$ shown next. For the graph of $y = 2x + 1$, notice that each

x-value corresponds to only one y-value. Recall from Example 3 that $y = 2x + 1$ is a function.

 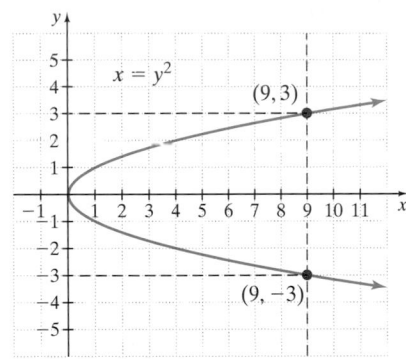

For the graph of $x = y^2$ the x-value 9, for example, corresponds to two y-values, 3 and -3, as shown by the vertical line. Recall from Example 4 that $x = y^2$ is not a function.

Graphs can be used to help determine whether a relation is also a function by the following vertical line test.

Vertical Line Test

If no vertical line can be drawn so that it intersects a graph more than once, the graph is the graph of a function.

EXAMPLE 5

Which of the following graphs are graphs of functions?

a. **b.** **c.**

d. **e.** **f.**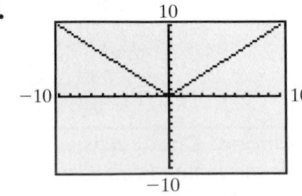

Solution **a.** This graph is the graph of a function since no vertical line will intersect this graph more than once.

b. This graph is also the graph of a function.

c. This graph is not the graph of a function. Note that vertical lines can be drawn that intersect the graph in two points.

d. This graph is the graph of a function.

e. This graph is not the graph of a function. A vertical line can be drawn that intersects this line at every point.

f. This graph is the graph of a function.

Recall that the graph of a linear equation in two variables is a line, and a line that is not vertical will pass the vertical line test. Thus, **all linear equations are functions except those whose graph is a vertical line.**

✔ **CONCEPT CHECK**

Determine which equations represent functions. Explain your answer.

a. $y = 14$ **b.** $x = -5$ **c.** $x + y = 6$

4 Next, we practice finding the domain and range of a relation from its graph.

EXAMPLE 6

Find the domain and range of each relation. Determine whether the relation is also a function.

TEACHING TIP
Classroom Activity

Give students a domain and range to use to make 4 sketches: 2 of a relation that is not a function and 2 of a relation that is a function. When they have finished, have them exchange sketches with a partner to determine which ones are sketches of functions.

a.

b.

c.

d.

e.
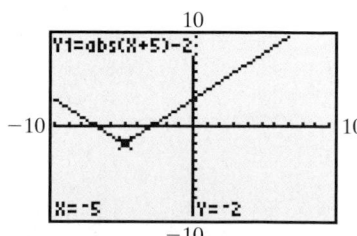

Helpful Hint

In Example 6, Part **a**, notice that the graph contains the endpoints $(-3, 1)$ and $(5, -2)$ whereas the graphs in Parts **c** and **d** contain arrows that indicate that they continue forever.

Concept Check Answer:
a, c

Solution By the vertical line test, graphs **a**, **c**, and **d** are graphs of functions. The domain is the set of values of *x* and the range is the set of values of *y*. We read these values from each graph.

a.

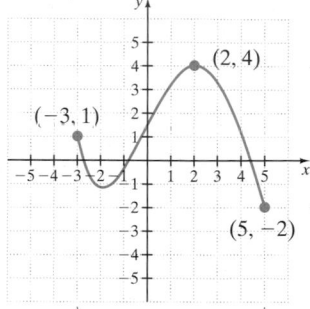

Range: The *y* values graphed are from -2 to 4, or $[-2, 4]$.

Domain:
The *x*-values graphed are from -3 to 5, or $[-3, 5]$.

b.

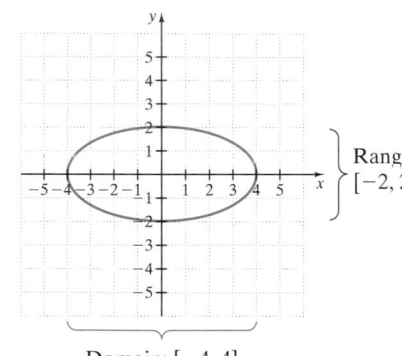

Range: $[-2, 2]$

Domain: $[-4, 4]$

c.

Domain: $(-\infty, \infty)$

Range: $[3, \infty)$

d.

Domain: $(-\infty, \infty)$

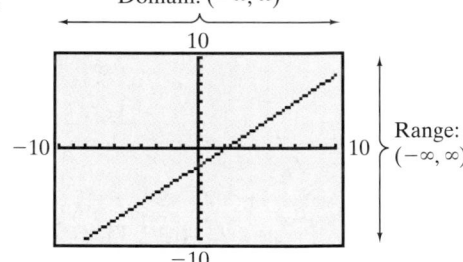

Range: $(-\infty, \infty)$

e.

Domain: $(-\infty, \infty)$

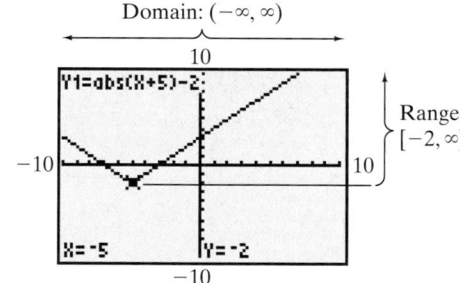

Range: $[-2, \infty)$

5 Many times letters such as *f*, *g*, and *h* are used to name functions. To denote that *y* is a function of *x*, we can write

$$y = f(x)$$

This means that **y is a function of x** or that *y depends on x*. For this reason, *y* is called the **dependent variable** and *x* the **independent variable**. The notation $f(x)$ is read "*f* of *x*" and is called **function notation**.

For example, to use function notation with the function $y = 4x + 3$, we write $f(x) = 4x + 3$. The notation $f(1)$ means to replace *x* with 1 and find the resulting *y* or function value. Since

$$f(x) = 4x + 3$$

then

$$f(1) = 4(1) + 3 = 7$$

This means that when $x = 1$, *y* or $f(x) = 7$. The corresponding ordered pair is $(1, 7)$. Here, the input is 1 and the output is $f(1)$ or 7. Now let's find $f(2), f(0)$, and $f(-1)$.

<table>
<tr><td>

$f(x) = 4x + 3$
$f(2) = 4(2) + 3$
$\quad = 8 + 3$
$\quad = 11$

</td><td>

$f(x) = 4x + 3$
$f(0) = 4(0) + 3$
$\quad = 0 + 3$
$\quad = 3$

</td><td>

$f(x) \;\; = 4(x) + 3$
$f(-1) = 4(-1) + 3$
$\qquad = -4 + 3$
$\qquad = -1$

</td></tr>
</table>

> **Helpful Hint**
>
> Make sure you remember that $f(2) = 11$ corresponds to the ordered pair $(2, 11)$.

Ordered Pairs:

$(2, 11)$ $\qquad\qquad\qquad (0, 3)$ $\qquad\qquad\qquad (-1, -1)$

There are many ways of evaluating function values using a graphing utility. One way is to use the store feature. For example, to find $f(2)$ when $f(x) = 4x + 3$, store the number 2, enter the expression, and calculate. To find $f(0)$ and $f(-1)$ for the same function and save time, have your graphing utility replay the last entry and then edit it.

Replay, then replace 2 with 0 and enter.

Replay, then replace 0 with -1 and enter.

We can also use the graph to evaluate a function at a given value. The graph of $y_1 = 4x + 3$ in an integer window is shown below. Move the cursor to the point with x-coordinate 2 to find $f(2)$. Continue in this manner to find $f(0)$ and $f(-1)$.

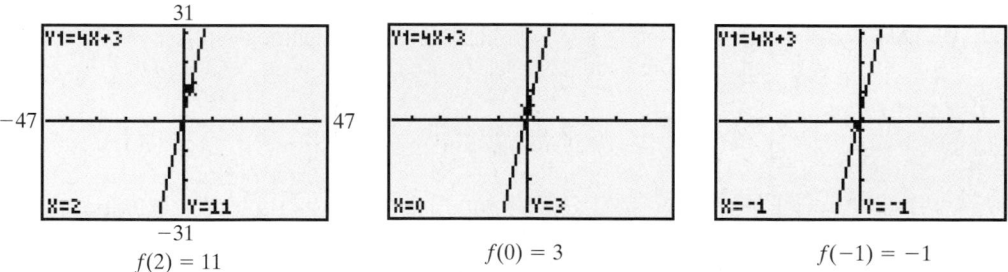

$f(2) = 11$ $\qquad\qquad\qquad f(0) = 3$ $\qquad\qquad\qquad f(-1) = -1$

A third method for finding function values is by using a table.

$f(-1) = -1 \longrightarrow$
$f(0) = 3 \longrightarrow$
$f(2) = 11 \longrightarrow$

The method you use to find function values depends on the particular situation. For instance, in Example 7, we evaluate several different functions for various values of x, so we calculate by hand and use the store feature to check. However, in Example 8 we evaluate the same function, but for different values of x, so we choose to look at the graph of the function. These are just a few methods that can be used to evaluate functions at given values.

> **Helpful Hint**
>
> Note that $f(x)$ is a special symbol in mathematics used to denote a function. The symbol $f(x)$ is read "f of x." It does *not* mean $f \cdot x$ (f times x).

EXAMPLE 7

If $f(x) = 7x^2 - 3x + 1$, $g(x) = 3x - 2$, and $h(x) = x^2$ find the following.

a. $f(1)$ **b.** $g(3)$ **c.** $h(-2)$

```
1→X:7X²-3X+1
                5  ← f(1) = 5
3→X:3X-2
                7  ← g(3) = 7
-2→X:X²
                4  ← h(-2) = 4
■
```

Solution

a. Substitute 1 for x in $f(x) = 7x^2 - 3x + 1$ and simplify.

$$f(x) = 7x^2 - 3x + 1$$
$$f(1) = 7(1)^2 - 3(1) + 1 = 5$$

b. $g(x) = 3x - 2$
$$g(3) = 3(3) - 2 = 7$$

c. $h(x) = x^2$
$$h(-2) = (-2)^2 = 4$$

CLASSROOM EXAMPLE

If $g(x) = 4x + 5$ and
$f(x) = 3x^2 - x + 2$, find:

a. $g(0)$ **b.** $g(-5)$

c. $f(2)$ **d.** $f(-1)$

answer:

a. 5   **b.** -15

c. 12   **d.** 6

✔ **CONCEPT CHECK**

Suppose $y = f(x)$ and we are told that $f(3) = 9$. Which is not true?

a. When $x = 3$, $y = 9$.
b. A possible function is $f(x) = x^2$.
c. A point on the graph of the function is $(3, 9)$.
d. A possible function is $f(x) = 2x + 4$.

TEACHING TIP

Typically students need extra time to master function notation. Continue to review the idea that $f(0) = 2$ corresponds to the ordered pair $(0, 2)$, for example. An example similar to Section 2.2, Exercises 81 through 92 is an excellent way to help students understand this notation.

If it helps, think of a function, f, as a machine that has been programmed with a certain correspondence or rule. An input value (a member of the domain) is then fed into the machine, the machine does the correspondence or rule and the result is the output (a member of the range).

EXAMPLE 8

If $f(x) = 0.5x - 25$, find

a. $f(-10)$ **b.** $f(18)$ **c.** $f(11)$ **d.** $f(0)$

Solution

a. We choose to use a graphical method to evaluate. Define $y_1 = 0.5x - 25$ and graph using an integer window. Move the cursor to find $x = -10$. In the screen on the next page, we see that -10 is paired with $y = -30$; therefore $f(-10) = -30$.

b. Move the cursor to $x = 18$ and see that it is paired with -16 and thus $f(18) = -16$.

c. Move the cursor to $x = 11$ and see that $y = -19.5$, so $f(11) = -19.5$.

Concept Check Answer:
d

d. Move the cursor to $x = 0$ and see that $y = -25$, so $f(0) = -25$.

$f(-10) = -30$

Many formulas that are familiar to you describe functions. For example, we have used the formula for finding the area of a circle, $A = \pi r^2$. The area of the circle is actually a function of the length of the radius. Using this function notation, we write

$$A(r) = \pi r^2$$

$A(r)$ can be read as the area with respect to r. To find the area of the circle whose radius is 3 cm, we write

$$A(r) = \pi r^2$$
$$A(3) = \pi(3)^2 = 9\pi \text{ square centimeters}$$

An approximation to two decimal places is

$$9\pi \text{ sq. cm} \approx 28.27 \text{ sq. cm}$$

EXAMPLE 9

If the area of a circle with respect to its radius can be described by the formula $A(r) = \pi r^2$, find the following:

a. $A(0.5)$ **b.** $A(3.1)$ **c.** $A(3.5)$

Solution We choose to evaluate the function $A(r) = \pi r^2$ by graphing $y_1 = \pi x^2$ in a decimal window. Here, $y_1 =$ area and $x =$ radius. The screens below illustrate entering the function in the Y= editor, and the various ordered pair solutions found by using the trace feature. We round each function value to two decimal places.

a. $A(0.5) \approx 0.79$ b. $A(3.1) \approx 30.19$ c. $A(3.5) \approx 38.48$

EXAMPLE 10

Given the graphs of the functions f and g, find each function value by inspecting the graphs.

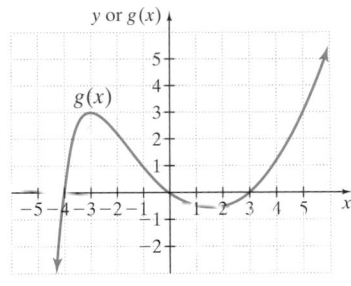

a. $f(4)$ b. $f(-2)$ c. $g(5)$ d. $g(0)$

e. Find all x-values such that $f(x) = 1$.

f. Find all x-values such that $g(x) = 0$.

Solution

a. To find $f(4)$, find the y-value when $x = 4$. We see from the graph that when $x = 4$, y or $f(x) = 2$. Thus, $f(4) = 2$.

b. $f(-2) = 1$ from the ordered pair $(-2, 1)$.

c. $g(5) = 3$ from the ordered pair $(5, 3)$.

d. $g(0) = 0$ from the ordered pair $(0, 0)$.

e. To find x-values such that $f(x) = 1$, I'm looking for any ordered pairs on the graph of f whose $f(x)$ or y-value is 1. They are $(2, 1)$ and $(-2, 1)$. Thus $f(2) = 1$ and $f(-2) = 1$. The x-values are 2 and −2.

f. Find ordered pairs on the graph of g whose $g(x)$ or y-value is 0. They are $(3, 0)(0, 0)$ and $(-4, 0)$. Thus $g(3) = 0$, $g(0) = 0$, and $g(-4) = 0$. The x-values are 3, 0, and −4.

When a store manager is setting the retail price of an article, the price may be dependent on the wholesale price of the article. When this happens, we say that the retail price is a function of the wholesale price.

EXAMPLE 11

FINDING RETAIL PRICES

Elizabeth Lockwood manages the college bookstore and purchases the books from several wholesale companies. She finds that she needs to mark up the wholesale cost by 25%.

a. Write the retail price as a function of the wholesale cost.

b. Find the retail price of the following books given the wholesale cost.

Wholesale Cost	$15.00	$22.75	$38.50	$53.00
Retail Price				

Solution

a. Let $w =$ the wholesale cost of a book. To denote that the retail price is a function of wholesale cost, we define the function $R(w)$.

In words:	Retail price (function of wholesale cost)	is	wholesale cost	plus	25% of wholesale cost
Translate:	$R(w)$	$=$	w	$+$	$0.25w$

b. Here we evaluate $R(w) = w + 0.25w$ for the given values of w.

Completing the table, we have

Wholesale Cost	w	$15.00	$22.75	$38.50	$53.00
Retail Price	$R(w)$	$18.75	$28.44	$48.13	$66.25

Many types of real-world paired data form functions. The broken-line graph below shows the research and development spending by the Pharmaceutical Manufacturers Association.

EXAMPLE 12

The following graph shows the research and development expenditures by the Pharmaceutical Manufacturers Association as a function of time.

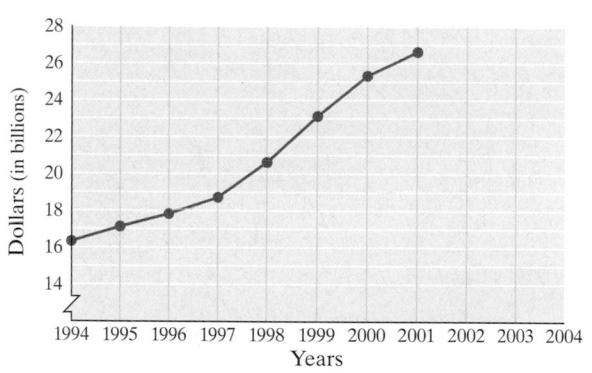

Source: Pharmaceutical Manufacturers Association

a. Approximate the money spent on research and development in 2000.

b. In 1958, research and development expenditures were $200 million. Find the increase in expenditures from 1958 to 2001.

Solution

a. Find the year 2000 and move upward until you reach the graph. From the point on the graph move horizontally to the left until the other axis is reached. In 2000, approximately $25.3 billion was spent.

b. In 2001, approximately $26.6 billion, or $26,600 million, was spent. The increase in spending from 1958 to 2001 is $26,600 − $200 = $26,400 million or $26.4 billion.

CLASSROOM EXAMPLE

Use the graph for Example 12 and approximate the money spent in 1999.
answer: approximately 23.1 billion

Notice that the graph in Example 12 is the graph of a function since for each year there is only one total amount of money spent by the Pharmaceutical Manufacturers Association on research and development. Also notice that the graph resembles the graph of a line. Often, businesses depend on equations that "closely fit" data-defined functions like this one in order to model the data and predict future trends. For example, by a method called **least squares**, the function $f(x) = 1.558x - 3092$ approximates the data shown. For this function, x is the year and $f(x)$ is total money spent. Its graph and the actual data function are shown next.

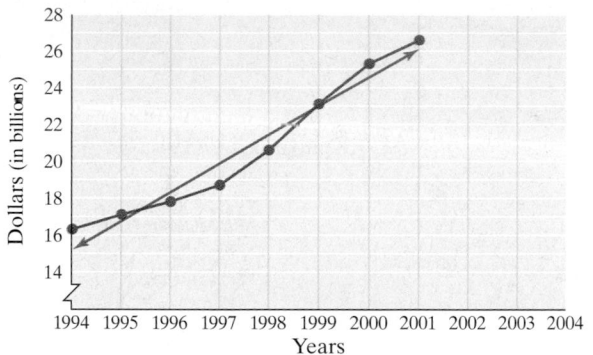

In Section 2.6 we will discuss how to use a graphing calculator to model the data from Example 12.

EXAMPLE 13

Use the function $f(x) = 1.558x - 3092$ to predict the amount of money that will be spent by the Pharmaceutical Manufacturers Association on research and development in 2010.

Solution

To predict the amount of money that will be spent in the year 2010 we use $f(x) = 1.558x - 3092$ and find $f(2010)$.

$$f(x) = 1.558x - 3092$$
$$f(2010) = 1.558(2010) - 3092$$
$$= 39.58$$

We predict that in the year 2010, $39.58 billion dollars will be spent on research and development by the Pharmaceutical Manufacturers Association.

CLASSROOM EXAMPLE

Use $f(x) = 1.558x - 3092$ to approximate the money spent in 2009.
answer: $38.022 billion

1. domain: $\{-1, 0, -2, 5\}$; range: $\{7, 6, 2\}$; function
2. domain: $\{4, -4, 2, 10\}$; range: $\{9, 3, -5\}$; function
3. domain: $\{-2, 6, -7\}$; range: $\{4, -3, -8\}$; not a function
4. domain: $\{6, 5, 7\}$; range: $\{6, -2\}$; not a function
5. domain: $\{1\}$; range: $\{1, 2, 3, 4\}$; not a function
6. domain: $\{1, 2, 3, 4\}$; range: $\{1\}$; function

7. domain: $\left\{\dfrac{3}{2}, 0\right\}$; range: $\left\{\dfrac{1}{2}, -7, \dfrac{4}{5}\right\}$; not a function
8. domain: $\{\pi, 0, -2, 4\}$; range: $\{0, \pi, 4, -2\}$; function
9. domain: $\{-3, 0, 3\}$; range: $\{-3, 0, 3\}$; function
10. domain: $\left\{\dfrac{1}{2}, 0, 0.5\right\}$; range: $\left\{\dfrac{1}{4}, \dfrac{7}{8}, \pi\right\}$; not a function

EXERCISE SET 2.2

STUDY GUIDE/SSM   CD/ VIDEO   PH MATH TUTOR CENTER   MathXL®Tutorials ON CD   MathXL®   MyMathLab®

Find the domain and the range of each relation. Also determine whether the relation is a function. See Examples 1 and 2.

1. $\{(-1, 7), (0, 6), (-2, 2), (5, 6)\}$
2. $\{(4, 9), (-4, 9), (2, 3), (10, -5)\}$
3. $\{(-2, 4), (6, 4), (-2, -3), (-7, -8)\}$
4. $\{(6, 6), (5, 6), (5, -2), (7, 6)\}$
5. $\{(1, 1), (1, 2), (1, 3), (1, 4)\}$

6. $\{(1, 1), (2, 1), (3, 1), (4, 1)\}$
7. $\left\{\left(\dfrac{3}{2}, \dfrac{1}{2}\right), \left(1\dfrac{1}{2}, -7\right), \left(0, \dfrac{4}{5}\right)\right\}$
8. $\{(\pi, 0), (0, \pi), (-2, 4), (4, -2)\}$
9. $\{(-3, -3), (0, 0), (3, 3)\}$
10. $\left\{\left(\dfrac{1}{2}, \dfrac{1}{4}\right), \left(0, \dfrac{7}{8}\right), (0.5, \pi)\right\}$

11.

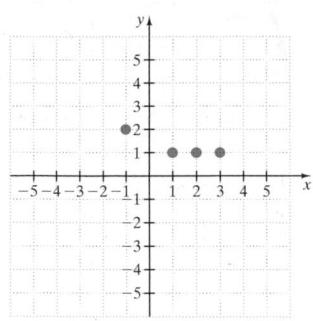

domain: $\{-1, 1, 2, 3\}$; range: $\{2, 1\}$; function

15.   Input: Output:

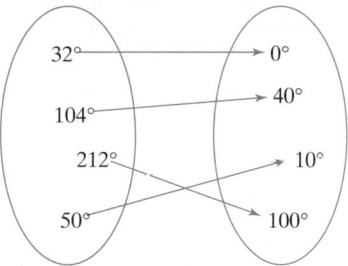

Degrees Fahrenheit   Degrees Celsius

domain: $\{32°, 104°, 212°, 50°\}$; range: $\{0°, 40°, 10°, 100°\}$; function

12.

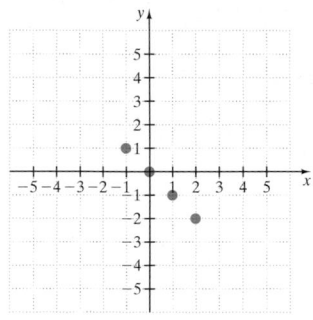

domain: $\{-1, 0, 1, 2\}$; range: $\{1, 0, -1, -2\}$; function

16.   Input: Output:

Words   Number of Letters

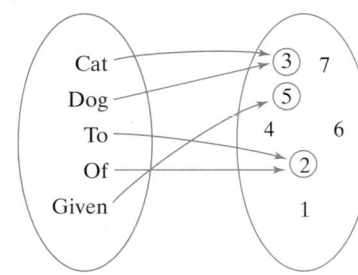

domain: $\{$cat, dog, to, of, given$\}$; range: $\{3, 5, 2\}$; function

13.   Input: Output:

State   Number of Congressional Representatives

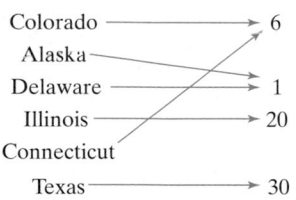

domain: (Colorado, Alaska, Delaware, Illinois, Connecticut, Texas); range: $\{6, 1, 20, 30\}$; function

17.   Input: Output:

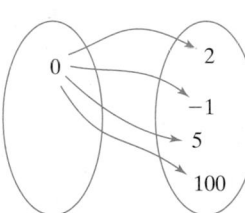

domain: $\{0\}$; range: $\{2, -1, 5, 100\}$; not a function

14.   Input: Output:

Animal   Average Life Span (in years)

domain: (polar bear, cow, chimpanzee, giraffe, gorilla, kangaroo, red fox); range: $\{20, 15, 10, 7\}$; function

18.   Input: Output:

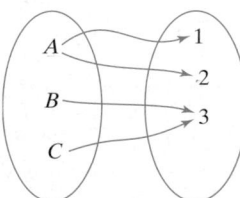

domain: $\{A, B, C\}$; range: $\{1, 2, 3\}$; not a function

In Exercises 19 through 22, determine whether the relation is a function.

First Set: Input	Correspondence	Second Set: Output
19. Class of algebra students	Final grade average	non negative numbers
20. People who live in Cincinnati, Ohio	Birth date	days of the year
21. blue, green, brown	Eye color	People who live in Cincinnati, Ohio
22. Whole numbers from 0 to 4	Number of children	50 Women in a water aerobics class

Use the vertical line test to determine whether each graph is the graph of a function. See Example 5.

23.

24.

25.

26.

27.

28.

29.

function

30.

not a function

31.

not a function

32.

function

Find the domain and the range of each relation. Use the vertical line test to determine whether each graph is the graph of a function. See Example 6.

33.

34.
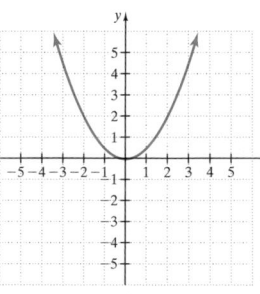

33. domain: $[0, \infty)$; range: $(-\infty, \infty)$; not a function
34. domain: $(-\infty, \infty)$; range: $[0, \infty)$; function

35.

36.
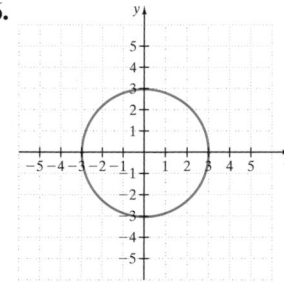

35. domain: $[-1, 1]$; range: $(-\infty, \infty)$; not a function
36. domain: $[-3, 3]$; range: $[-3, 3]$; not a function

37.

38.
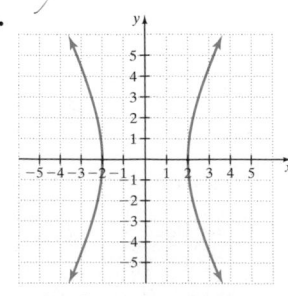

37. domain: $(-\infty, \infty)$; range: $(-\infty, -3] \cup [3, \infty)$; not a function
38. domain: $(-\infty, -2] \cup [2, \infty)$; range: $(-\infty, \infty)$; not a function

19. function   20. function   21. not a function   22. not a function   23. function   24. not a function   25. not a function   26. not a function   27. function   28. not a function

39.

40.

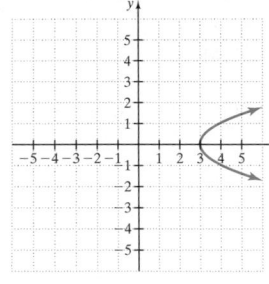

39. domain: $[2, 7]$; range: $[1, 6]$; not a function
40. domain: $[3, \infty)$; range: $(-\infty, \infty)$; not a function

41.

42.

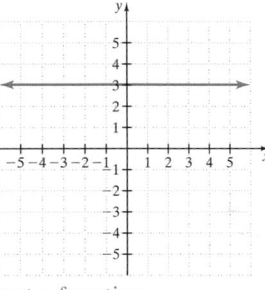

41. domain: $\{-2\}$; range: $(-\infty, \infty)$; not a function
42. domain: $(-\infty, \infty)$; range: $\{3\}$; function

43.

44.

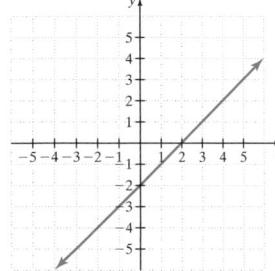

43. domain: $(-\infty, \infty)$; range: $(-\infty, 3)$; function
44. domain: $(-\infty, \infty)$; range: $(-\infty, \infty)$; function

45.

46.

45. domain: $(-\infty, \infty)$; range: $(-\infty, \infty)$; function
46. domain: $(-\infty, \infty)$; range: $[-5, \infty)$; function
47. In your own words define **(a)** function; **(b)** domain; **(c)** range.   **47.–48.** answers may vary
48. Explain the vertical line test and how it is used.

Decide whether each is a function. See Examples 3 and 4.

49. $y = x + 1$ yes
50. $y = x - 1$ yes
51. $x = 2y^2$ no
52. $y = x^2$ yes
53. $y - x = 7$ yes
54. $2x - 3y = 9$ yes
55. $y = \dfrac{1}{x}$ yes
56. $y = \dfrac{1}{x - 3}$ yes

57. $y = 5x - 12$ yes
58. $y = \dfrac{1}{2}x + 4$ yes
59. $x = y^2$ no
60. $x = |y|$ no

If $f(x) = 3x + 3$, $g(x) = 4x^2 - 6x + 3$, and $h(x) = 5x^2 - 7$, find the following. See Example 7.

61. $f(4)$ 15
62. $f(-1)$ 0
63. $h(-3)$ 38
64. $h(0)$ -7
65. $g(2)$ 7
66. $g(1)$ 1
67. $g(0)$ 3
68. $h(-2)$ 13

Given the following functions, find the indicated values. See Examples 7 through 9.

69. $f(x) = \dfrac{1}{2}x$;   **a.** $f(0)$ 0
   b. $f(2)$ 1   **c.** $f(-2)$ -1
70. $g(x) = -\dfrac{1}{3}x$;   **a.** $g(0)$ 0
   b. $g(-1)$ $\frac{1}{3}$ **c.** $g(3)$ -1
71. $g(x) = 2x^2 + 4$;   **a.** $g(-11)$ 246
   b. $g(-1)$ 6   **c.** $g\left(\dfrac{1}{2}\right)$ $\dfrac{9}{2}$
72. $h(x) = -x^2$;   **a.** $h(-5)$ -25
   b. $h\left(-\dfrac{1}{3}\right)$ $-\dfrac{1}{9}$ **c.** $h\left(\dfrac{1}{3}\right)$ $-\dfrac{1}{9}$
73. $f(x) = -5$;   **a.** $f(2)$ -5
   b. $f(0)$ -5 **c.** $f(606)$ -5
74. $h(x) = 7$;   **a.** $h(7)$ 7
   b. $h(542)$ 7   **c.** $h\left(-\dfrac{3}{4}\right)$ 7
75. $f(x) = 1.3x^2 - 2.6x + 5.1$ **a.** $f(2)$ 5.1
   b. $f(-2)$ 15.5   **c.** $f(3.1)$ 9.533
76. $g(x) = 2.7x^2 + 6.8x - 10.2$ **a.** $g(1)$ -0.7
   b. $g(-5)$ 23.3   **c.** $g(7.2)$ 178.728
77. Given the following table of values for $f(x) = |2x - 5| + 8$, find the indicated values.
   a. $f(-5)$ 23   **b.** $f(10)$ 23   **c.** $f(15)$ 33

X	Y1
-5	23
0	13
5	13
10	23
15	33
20	43
25	53

Y1 ◼ abs(2X−5)+8

78. Given the following table of values for $f(x) = -2x^3 + x$ find the indicated values.
   a. $f(-4)$ 124   **b.** $f(2)$ -14 **c.** $f(6)$ -426

X	Y1
-4	124
-2	14
0	0
2	-14
4	-124
6	-426
8	-1016

Y1 ◼ −2X3+X

79. Given the graph of the function $f(x) = x^2 + x + 1$, find the value of $f(2)$.   7

80. Given the graph of the function $f(x) = x^3 + x^2 + 1$, find $f(-2)$. -3

Use the graph of the functions below to answer Exercises 81 through 92. See Example 10.   **81.** $(1, -10)$ **82.** $(-5, -10)$

 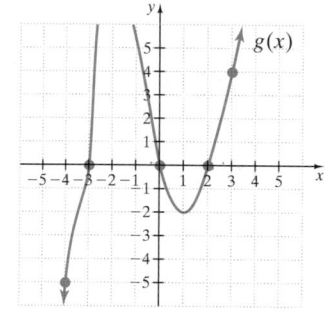

85. $f(-1) = -2$ **86.** $f(-2) = -1$ **87.** $g(2) = 0$ **88.** $g(-4) = -5$

81. If $f(1) = -10$ write the corresponding ordered pair.

82. If $f(-5) = -10$, write the corresponding ordered pair.

83. If $g(4) = 56$, write the corresponding ordered pair. $(4, 56)$

84. If $g(-2) = 8$, write the corresponding ordered pair. $(-2, 8)$

85. Find $f(-1)$.   **86.** Find $f(-2)$.

87. Find $g(2)$.   **88.** Find $g(-4)$.

89. Find all values of x such that $f(x) = -5$. $-4, 0$

90. Find all values of x such that $f(x) = -2$. $-3, -1$

91. Find all values of x such that $g(x) = -5$ -4

92. Find all values of x such that $g(x) = 0$. $-3, 0, 2$

93. What is the greatest number of x-intercepts that a function may have? Explain your answer.   infinite number

94. What is the greatest number of y-intercepts that a function may have? Explain your answer.   1

Solve. See Examples 10 and 11.

95. Julie needs to hire a plumber. The cost of hiring a plumber, C, in dollars is a function of the time spent on the job, t, in hours. If the plumber charges a one-time fee of $20 plus $35 per hour, we can write this in function notation as $C(t) = 20 + 35t$. Complete the table.

Time In Hours	t	1	2	3	4	5
Total Cost	$C(t)$	55	90	125	160	195

96. The cost of hiring a secretary to type a term paper is $10 plus $5.25 per hour. Therefore, the cost, C, in dollars is a function of time, t, in hours, or $C(t) = 10 + 5.25t$. Complete the table.

Time In Hours	t	0.5	1	1.5	2.5	3
Total Cost	$C(t)$	12.63	15.25	17.88	23.13	25.75

97. The height h of a firecracker being shot into the air from ground level is a function of time t and can be represented using the following formula from physics: $h = -16t^2 + vt$ where v is initial velocity. If the velocity is 80 feet per second, the height in terms of time is $h(t) = -16t^2 + 80t$. Find the following:

a. $h(0.5)$ 36   **b.** $h(1)$ 64   **c.** $h(1.5)$ 84

d. $h(2)$ 96   **e.** $h(2.5)$ 100   **f.** $h(5)$ 0

98. The height h of a rocket launched from a 200-foot high building with velocity of 120 feet per second can be expressed as a function of time t using the formula $h = -16t^2 + vt + h_0$, where h_0 stands for the initial height. If $h(t) = -16t^2 + 120t + 200$, find the following:

a. $h(0.2)$ 223.36   **b.** $h(0.6)$ 266.24   **c.** $h(2.25)$ 389

d. $h(3)$ 416   **e.** $h(4)$ 424

Use the graph in Example 12 to answer the following. Also see Example 13.

99. a. Use the graph to approximate the money spent on research and development in 1995. $17.1 billion

b. Recall that the function $f(x) = 1.558x - 3092$ approximates the graph of Example 9. Use this equation to approximate the money spent on research and development in 1995. $16.21 billion

100. a. Use the graph to approximate the money spent on research and development in 1999. $23.1 billion

b. Use the function $f(x) = 1.558x - 3092$ to approximate the money spent on research and development in 1999. $22.442 billion

101. Use the function $f(x) = 1.558x - 3092$ to predict the money that will be spent on research and development in 2008. $36.464 billion

102. Use the function $f(x) = 1.558x - 3092$ to predict the money that will be spent on research and development in 2011. $41.138 billion

103. Since $y = x + 7$ describes a function, rewrite the equation using function notation. $f(x) = x + 7$

104. In your own words, explain how to find the domain of a function given its graph.   answers may vary

The function $A(r) = \pi r^2$ may be used to find the area of a circle if we are given its radius.

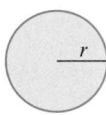

△ **105.** Find the area of a circle whose radius is 5 centimeters. (Do not approximate π.) 25π sq. cm

△ **106.** Find the area of a circular garden whose radius is 8 feet. (Do not approximate π.) 64π sq. ft

The function $V(x) = x^3$ may be used to find the volume of a cube if we are given the length x of a side.

107. Find the volume of a cube whose side is 14 inches. 2744 cu. in.

108. Find the volume of a die whose side is 1.7 centimeters. 4.913 cu. cm

Forensic scientists use the following functions to find the height of a woman if they are given the height of her femur bone f or her tibia bone t in centimeters.

$$H(f) = 2.59f + 47.24$$
$$H(t) = 2.72t + 61.28$$

46 cm Femur — 35 cm Tibia

109. Find the height of a woman whose femur measures 46 centimeters. 166.38 cm

110. Find the height of a woman whose tibia measures 35 centimeters. 156.48 cm

The dosage in milligrams D of Ivermectin, a heartworm preventive, for a dog who weighs x pounds is given by

$$D(x) = \frac{136}{25}x$$

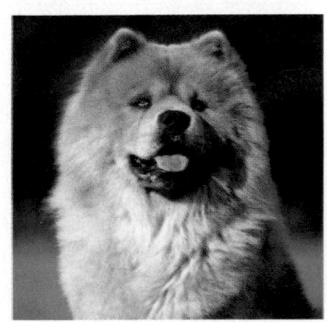

111. Find the proper dosage for a dog that weighs 30 pounds.

112. Find the proper dosage for a dog that weighs 50 pounds.

113. The per capita consumption (in pounds) of all poultry in the United States is approximated by the function $C(x) = 1.69x + 87.54$, where x is the number of years since 1995. (*Source*: Based on actual and estimated data from the Economic Research Service, U.S. Department of Agriculture, 1995–2002)

  a. Find and interpret $C(5)$.

  b. Estimate the per capita consumption of all poultry in the United States in 2002. 99.37 lb

114. The average length of U.S. hospital stays has been decreasing, following the equation $y = -0.09x + 8.02$ where x is the number of years since 1970. (*Source:* National Center for Health Statistics)

  a. What was the length of the average hospital stay in 1995? 5.77 days

  b. If this trend continues, what will the average length be in 2007? 4.69 days

111. 163.2 mg
112. 272 mg
113. a. per capita consumption of poultry was 95.99 lb in 2000

REVIEW AND PREVIEW

Complete the given table and use the table to graph the linear equation. See Section 2.1.

115. $x - y = -5$

x	0	-5	1
y	5	0	6

116. $2x + 3y = 10$

x	0	5	2
y	$\frac{10}{3}$	0	2

117. $7x + 4y = 8$

x	0	$\frac{8}{7}$	$\frac{12}{7}$
y	2	0	-1

118. $5y - x = -15$

x	0	15	-2
y	-3	0	$-\frac{17}{5}$

119. $y = 6x$

x	0	0	-1
y	0	0	-6

120. $y = -2x$

x	0	0	-2
y	0	0	4

△ **121.** Is it possible to find the perimeter of the following geometric figure? If so, find the perimeter.   yes, 170 m

45 meters

40 meters

Concept Extensions

Given the following functions, find the indicated values.

122.–125. See graphing answer section.

122. $f(x) = 2x + 7$;

   a. $f(2)$ 11   **b.** $f(a)$ $2a + 7$

123. $g(x) = -3x + 12$;

   a. $g(s)$ $-3s + 12$ **b.** $g(r)$ $-3r + 12$

124. $h(x) = x^2 + 7$;

   a. $h(3)$ 16   **b.** $h(a)$ $a^2 + 7$

125. $f(x) = x^2 - 12$;

   a. $f(12)$ 132   **b.** $f(a)$ $a^2 - 12$

126. Describe a function whose domain is the set of people in your hometown.   answers may vary

127. Describe a function whose domain is the set of people in your algebra class.   answers may vary

---

2.3 GRAPHING LINEAR FUNCTIONS

Objectives

1 Graph linear functions.

2 Graph linear functions by finding intercepts.

3 Graph vertical and horizontal lines.

4 Use a graph to solve problems.

1 In this section, we identify and graph linear functions. By the vertical line test, we know that all linear equations except those whose graphs are vertical lines are functions. For example, we know from Section 2.1 that $y = 2x$ is a linear equation in two variables. Its graph is shown.

x	$y = 2x$
1	2
0	0
-1	-2

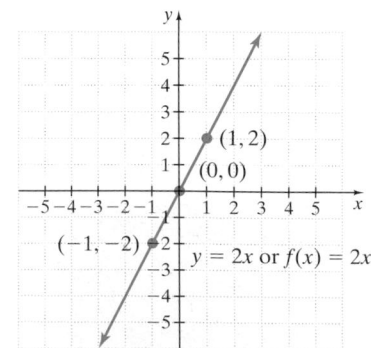

Because this graph passes the vertical line test, we know that $y = 2x$ is a function. If we want to emphasize that this equation describes a function, we may write $y = 2x$ as $f(x) = 2x$.

A graphing utility is a very versatile tool for exploring the graphs of functions.

Discover the Concept

a. On your graphing utility, graph both $f(x) = 2x$ as $y_1 = 2x$, and $g(x) = 2x + 10$ as $y_2 = 2x + 10$ using an integer window.

Integer Window

b. Trace on the graph of $y_1 = 2x$ to find the point whose ordered pair has x-coordinate 3 and find the corresponding y-coordinate. Now, press the down arrow key and find the corresponding y-coordinate on the graph of $y_2 = 2x + 10$. Compare the two y-coordinates for the same x-coordinate.

c. Trace to another point on the graph of y_1 and repeat this process. Compare the two lines, and see if you can state how we could sketch the graph of $g(x) = 2x + 10$ from the graph of $f(x) = 2x$.

d. Predict how the graph of $h(x) = 2x - 15$ could be drawn. Graph $y_3 = 2x - 15$ and see if your prediction is correct.

In the discovery above, we confirmed what we discovered on page 128. The y-values for the graph of $g(x)$ or $y = 2x + 10$ are obtained by adding 10 to the y-value of each corresponding point of the graph of $f(x)$ or $y = 2x$. The graph of $g(x) = 2x + 10$ is the same as the graph of $f(x) = 2x$ shifted upward 10 units.

x	-1	0	1
$f(x) = 2x$	-2	0	2
$g(x) = 2x + 10$	8	10	12

add 10

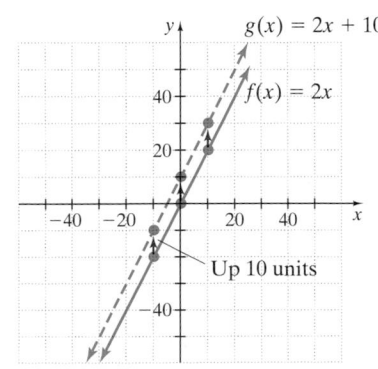

Also, the graph of $h(x)$ or $y = 2x - 15$ is obtained by subtracting 15 from the y-value of each corresponding point of the graph of $f(x)$ or $y = 2x$. Thus, the graph of $h(x) = 2x - 15$ is the same as the graph of $f(x) = 2x$ shifted downward 15 units.

The functions $f(x) = 2x$, $g(x) = 2x + 10$, and $h(x) = 2x - 15$ are called linear functions—"linear" because each graph is a line, and "function" because each graph passes the vertical line test. In general, a **linear function** is a function that can be written in the form $f(x) = mx + b$. For example, $g(x) = 2x + 10$ is in this form, with $m = 2$ and $b = 10$.

EXAMPLE 1

Graph both linear functions $f(x) = 0.5x$ and $g(x) = 0.5x + 8$ using the same integer window.

a. Use the graphs to complete the following table of solution pairs.

x	−6	0	5	13	18
$f(x) = 0.5x$					
$g(x) = 0.5x + 8$					

b. Complete the following sentence. The graph of $g(x) = 0.5x + 8$ can be obtained from the graph of $f(x) = 0.5x$ by _____.

Solution **a.** Define $y_1 = 0.5x$ and $y_2 = 0.5x + 8$ and graph in the integer window. The screens below show the values of $f(x)$ and $g(x)$ when $x = 4$. In other words, $f(4) = 2$ and $g(4) = 10$.

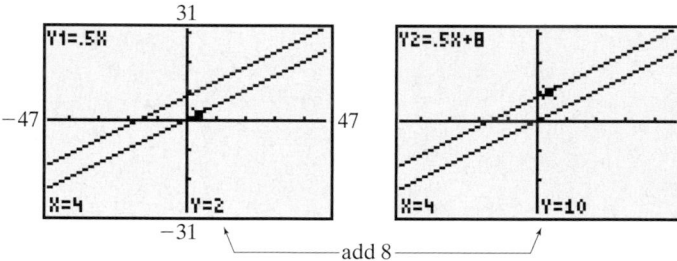

Trace along each graph to complete the table as follows:

x	−6	0	5	13	18
$f(x) = 0.5x$	−3	0	2.5	6.5	9
$g(x) = 0.5x + 8$	5	8	10.5	14.5	17

b. The graph of $g(x) = 0.5x + 8$ can be obtained from the graph of $f(x) = 0.5x$ by shifting it upward 8 units.

EXAMPLE 2

Graph both linear functions $f(x) = -x$ and $g(x) = -x - 16$ using an integer window. Complete the following statement. The graph of $g(x) = -x - 16$ can be obtained from the graph of $f(x) = -x$ by _____.

Solution Graph $y_1 = -x$ and $y_2 = -x - 16$ using an integer window as shown on the next page. We can complete the statement as follows. The graph of $g(x) = -x - 16$ can be obtained from the graph of $f(x)$ by shifting it downward 16 units. To further illustrate this fact, see the table on the next page. And compare the y_1 and y_2 values for the same x-value.

subtract 16

In general, for any function $f(x)$, the graph of $y = f(x) + K$ is the same as the graph of $y = f(x)$ shifted $|K|$ units upward if K is positive and downward if K is negative.

2 The graph of $y = 2x + 10$ is shown in both screens below. Notice that this graph crosses both the x-axis and the y-axis. Recall that a point where a graph crosses the x-axis is called the **x-intercept,** and a point where a graph crosses the y-axis is called the **y-intercept.**

 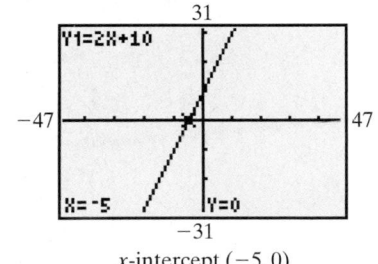

y-intercept $(0, 10)$ x-intercept $(-5, 0)$

By tracing along the graph, we can see that the y-intercept of the graph of $y = 2x + 10$ is $(0, 10)$, or we can say simply that the y-intercept is 10. Also the x-intercept of the graph is $(-5, 0)$ or the x-intercept is -5.

One way to find the y-intercept of the graph of an equation is to let $x = 0$, since a point on the y-axis has an x-coordinate of 0. To find the x-intercept, let $y = 0$, since a point on the x-axis has a y-coordinate of 0.

Finding x-and y-Intercepts

To find an x-intercept, let $y = 0$ or $f(x) = 0$ and solve for x.
To find a y-intercept, let $x = 0$ and solve for y.

Intercepts are usually easy to find and plot since one coordinate is 0.

In the next example, we sketch a linear function by plotting x- and y- intercepts.

EXAMPLE 3

Graph $x - 3y = 6$ by plotting intercepts. Check using a graphing utility.

Solution Let $y = 0$ to find the x-intercept and $x = 0$ to find the y-intercept.

$$\text{If } y = 0 \quad \text{then} \qquad \text{If } x = 0 \quad \text{then}$$
$$x - 3(0) = 6 \qquad\qquad 0 - 3y = 6$$
$$x - 0 = 6 \qquad\qquad -3y = 6$$
$$x = 6 \qquad\qquad\quad y = -2$$

The x-intercept is $(6, 0)$ and the y-intercept is $(0, -2)$. We find a third ordered pair solution to check our work. If we let $y = -1$, then $x = 3$. Plot the points $(6, 0)$, $(0, -2)$, and $(3, -1)$. The graph of $x - 3y = 6$ is the line drawn through these points, as shown.

x	y
6	0
0	-2
3	-1

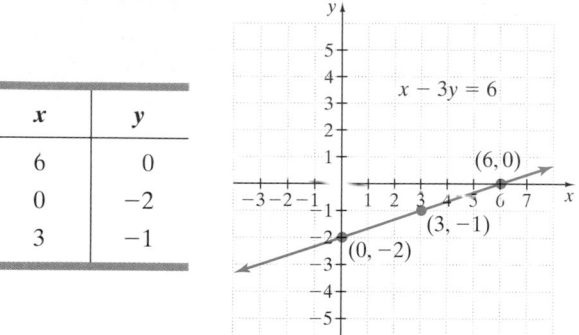

To check using a graphing utility, solve for y to enter the equation in the Y= editor.

$$x - 3y = 6$$
$$-3y = -x + 6$$
$$y = \frac{-x + 6}{-3}$$

Define $y_1 = \dfrac{-x + 6}{-3}$ and graph in an integer window as shown below.

x-intercept $(6, 0)$

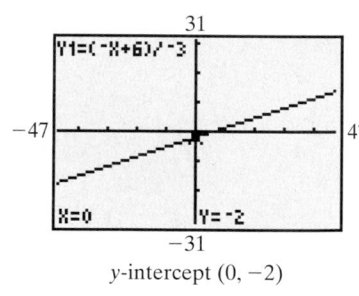

y-intercept $(0, -2)$

TEACHING TIP

Remind students that any function can be written with or without function notation.
Help them complete:
Linear Functions Written:

• with function notation
$-y = 7x - 1$;
$2x + 3y = 4$

• without function notation
$f(x) = -\dfrac{1}{2}x$;
$f(x) = -x + 2$

Notice that the equation $x - 3y = 6$ describes a linear function—"linear" because its graph is a line and "function" because the graph passes the vertical line test.

If we want to emphasize that the equation $x - 3y = 6$ from Example 3 describes a function, remember to first solve the equation for y.

$$x - 3y = 6$$
$$-3y = -x + 6 \qquad \text{Subtract } x \text{ from both sides.}$$
$$\frac{-3y}{-3} = \frac{-x}{-3} + \frac{6}{-3} \qquad \text{Divide both sides by } -3.$$
$$y = \frac{1}{3}x - 2 \qquad \text{Simplify.}$$

Next, let
$$y = f(x).$$
$$f(x) = \frac{1}{3}x - 2$$

> ### Helpful Hint
>
> Any linear equation that describes a function can be written using function notation. To do so,
>
> **1.** solve the equation for y and then
> **2.** replace y with $f(x)$, as we did above.

> ### Helpful Hint
>
> Recall that when generating a graph using a graphing utility we first solve for the variable y. When a function is given in the $f(x)$ notation, it is already solved for y and can be entered in the Y = editor directly.

Discover the Concept

a. Use an integer setting and graph each linear function.

$$y_1 = x + 15 \qquad y_2 = x \qquad y_3 = x - 11$$

b. Decide how the y-intercept of the graph of the equation compares with the equation.

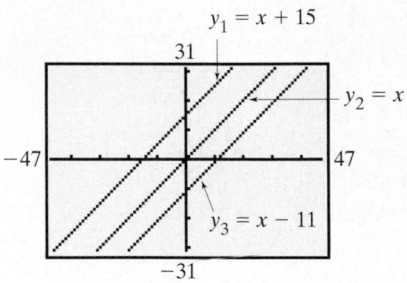

From the discovery above, we found the following:

Equation	y-intercept
$y = x + 15$	$(0, 15)$
$y = x$	$(0, 0)$
$y = x - 11$	$(0, -11)$

In general,

$$y = mx + b \qquad (0, b)$$

Notice that the y-intercept of the graph of the equation of the form $y = mx + b$ is $(0, b)$ each time. This is because we find the y-intercept of the graph of an equation by letting $x = 0$. Thus,

$$y = mx + b$$
$$y = m - 0 + b \quad \text{Let } x = 0.$$
$$y = b$$

The intercept is $(0, b)$.

EXAMPLE 4

Find the y-intercept of the graph of each equation.

a. $f(x) = \dfrac{1}{2}x + \dfrac{3}{7}$ **b.** $y = -2.5x - 3.2$

Solution **a.** The y-intercept of $f(x) - \dfrac{1}{2}x + \dfrac{3}{7}$ is $\left(0, \dfrac{3}{7}\right)$.

b. The y-intercept of $y = -2.5x - 3.2$ is $(0, -3.2)$.

In general, to find the y-intercept of the graph of an equation not in the form $y = mx + b$, let $x = 0$ since any point on the y-axis has an x-coordinate of 0. To find the x-intercept of a line, let $y = 0$ or $f(x) = 0$ since any point on the x-axis has a y-coordinate of 0.

3   The equations $x = c$ and $y = c$, where c is a real number constant, are both linear equations in two variables. Why? Because $x = c$ can be written as $x + 0y = c$ and $y = c$ can be written as $0x + y = c$. We graph these two special linear equations below.

EXAMPLE 5

Graph $x = 2$.

Solution The equation $x = 2$ can be written as $x + 0y = 2$. For any y-value chosen, notice that x is 2. No other value for x satisfies $x + 0y = 2$. Any ordered pair whose x-coordinate is 2 is a solution to $x + 0y = 2$ because 2 added to 0 times any value of y is $2 + 0$, or 2. We will use the ordered pairs $(2, 3), (2, 0)$ and $(2, -3)$ to graph $x = 2$.

x	y
2	3
2	0
2	-3

x-intercept $\rightarrow$ (row with $x=2$, $y=0$)

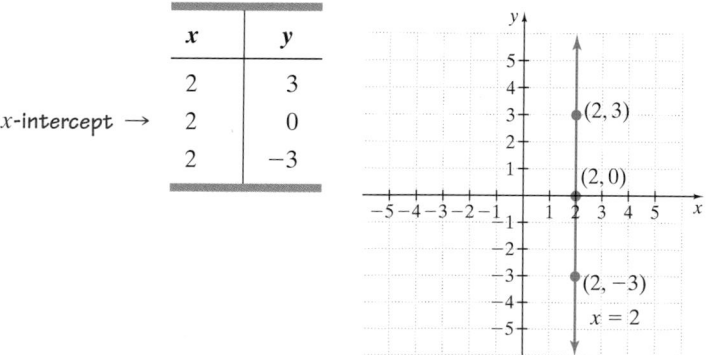

The graph is a vertical line with x-intercept $(2, 0)$. Notice that this graph is not the graph of a function, and it has no y-intercept because x is never 0.

EXAMPLE 6

Graph $y = -3$.

Solution The equation $y = -3$ can be written as $0x + y = -3$. For any x-value chosen, y is -3. If we choose 4, 0, and -2 as x-values, the ordered pair solutions are $(4, -3), (0, -3)$, and $(-2, -3)$. We will use these ordered pairs to graph $y = -3$.

x	y
4	−3
0	−3	← y-intercept
−2	−3

The graph is a horizontal line with y-intercept $(0, -3)$ and no x-intercept. Notice that this graph is the graph of a function. Above to the right is the graph of $y = -3$ using a graphing utility. Trace along the line to see that all ordered pair solutions have a y-coordinate of -3.

From Examples 5 and 6, we have the following generalization.

Graphing Vertical and Horizontal Lines

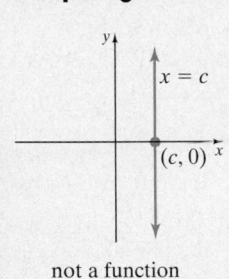

The graph of $x = c$, where c is a real number, is a vertical line with x-intercept $(c, 0)$.

The graph of $y = c$, where c is a real number, is a horizontal line with y-intercept $(0, c)$.

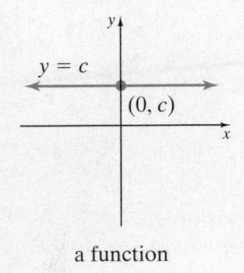

not a function

a function

---

4 We can use the graph of a linear equation to solve problems.

EXAMPLE 7

COST OF RENTING A CAR

The cost of renting a car for a day is given by the linear function $C(x) = 35 + 0.15x$ where $C(x)$ represents the cost in dollars and x is the number of miles driven. Use the graph of the function to complete the table below in dollars, and find the cost $C(x)$ for the given number of miles.

No. of Miles		x	150	200	325	500
Cost	$C(x) = 35 + 0.15x$					

Solution Define $y_1 = 35 + 0.15x$ and graph in a $[0, 600, 100]$ by $[0, 200, 50]$ window. We choose $0 \le x \le 600$ since the values in the table are in this interval. We choose $0 \le y \le 200$ as we are estimating our cost to be less than \$200. The graph below illustrates how to obtain the cost for 150 miles.

$x = 150$ miles $y = 57.5$

Finding the remaining values from the graph, we can complete the table.

No. of Miles	x	150	200	325	500
Cost	$C(x) = 35 + 0.15x$	57.5	65	83.75	110

Notice from the graph that the cost $C(x)$ increases as the number of miles, x, increases.

EXERCISE SET 2.3

| STUDY GUIDE/SSM | CD/ VIDEO | PH MATH TUTOR CENTER | MathXL®Tutorials ON CD | MathXL® | MyMathLab® |

1.–8. See graphing answer section.

Graph each linear function. See Examples 1 and 2.

1. $f(x) = -2x$
2. $f(x) = 2x$
3. $f(x) = -2x + 3$
4. $f(x) = 2x + 6$
5. $f(x) = \dfrac{1}{2}x$
6. $f(x) = \dfrac{1}{3}x$
7. $f(x) = \dfrac{1}{2}x - 4$
8. $f(x) = \dfrac{1}{3}x - 2$

The graph of $f(x) = 5x$ in a standard window follows. Use this graph to match each linear function with its graph. See Examples 1 and 2.

13. $f(x) = x - 3$ **14.** $f(x) = x + 4$

17. $f(x) = \dfrac{1}{2}x + 3$
18. $f(x) = \dfrac{1}{2}x + 4$
19. $f(x) = \dfrac{1}{2}x - 2$

A.

B.

C.

D.

9. $f(x) = 5x - 6$ C
10. $f(x) = 5x - 2$ A
11. $f(x) = 5x + 7$ D
12. $f(x) = 5x + 3$ B

20. $f(x) = -\dfrac{2}{3}x + 2$

Graph each linear function by finding x- and y-intercepts. Then write each equation using function notation. See Examples 3 and 4.

13. $x - y = 3$
14. $x - y = -4$
15. $x = 5y$ $f(x) = \dfrac{x}{5}$
16. $2x = y$ $f(x) = 2x$
17. $-x + 2y = 6$
18. $x - 2y = -8$
19. $2x - 4y = 8$;
20. $2x + 3y = 6$

21. In your own words, explain how to find x- and y-intercepts.

22. Explain why it is a good idea to use three points to graph a linear equation. answers may vary

21. answers may vary **23.–28.** See graphing answer section.

Graph each linear equation. See Examples 5 and 6.

23. $x = -1$
24. $y = 5$
25. $y = 0$
26. $x = 0$
27. $y + 7 = 0$
28. $x - 3 = 0$

Match each equation on the next page with its graph.

A.

B.

C.

D.

29. $y = 2$ C   **30.** $x = -3$ D

31. $x - 2 = 0$ A   **32.** $y + 1 = 0$ B

33. Discuss whether a vertical line ever has a y-intercept.

34. Discuss whether a horizontal line ever has an x-intercept.
The horizontal line $y = 0$ has x-intercepts.

MIXED PRACTICE

Graph each linear equation. **35.–60.** *See graphing answer section.*

35. $x + 2y = 8$ **36.** $x - 3y = 3$

37. $3x + 5y = 7$ **38.** $3x - 2y = 5$

39. $x + 8y = 8$ **40.** $x - 3y = 9$

41. $5 = 6x - y$ **42.** $4 = x - 3y$

43. $-x + 10y = 11$ **44.** $-x + 9 = -y$

45. $y = \frac{3}{2}$ **46.** $x = \frac{3}{2}$

47. $2x + 3y = 6$ **48.** $4x + y = 5$

49. $x + 3 = 0$ **50.** $y - 6 = 0$

51. $f(x) = \frac{3}{4}x + 2$ **52.** $f(x) = \frac{4}{3}x + 2$

53. $f(x) = x$ **54.** $f(x) = -x$

55. $f(x) = \frac{1}{2}x$ **56.** $f(x) = -2x$

57. $f(x) = 4x - \frac{1}{3}$ **58.** $f(x) = -3x + \frac{3}{4}$

59. $x = -3$ **60.** $f(x) = 3$

61. Given the graph of $f(x)$, which of the following statements are true?   b, d

a. $f(7) = 0$ **b.** $f(0) = 7$
c. The x-intercept is $(7, 0)$.   **d.** The y-intercept is $(0, 7)$.

62. Given the graph of $g(x)$, which of the following statements are true?   a, c

a. $g(-2) = 0$ **b.** $g(0) = -2$
c. The x-intercept is $(-2, 0)$.
d. The y-intercept is $(0, -2)$.

33. The vertical line $x = 0$ has y-intercepts.

Solve. See Example 7.

63. Kevin Elliott works in the Human Resource department at a rate of $15.75 per hour for a 40-hour work week. He makes time and a half for overtime. His salary, before any deductions, can be represented by the function $S(x) = 15.75(40) + 1.5(15.75)x$ where x is the number of overtime hours. Graph the function and use it to complete the table below for the salary $S(x)$ in dollars.

No. of Over-time Hours	x	2	3.25	7.5	9.75
Salary in Dollars	$S(x)$	677.25	706.78	807.19	860.34

64. The average hourly earnings of someone who works in retail trade is $8.34. If the normal hourly rate for overtime work (work over 40 hours per week) is 1.5 ($8.34), the salary, before deductions, can be represented by $S(x) = 8.34(40) + 1.5(8.34)x$, where x is the number of overtime hours. Graph the function $S(x)$ and use the graph to complete the table below. (*Source:* U.S. Bureau of Labor Statistics)

No. of Overtime Hours	2	10	12.5	15.75
Salary in Dollars	358.62	458.70	489.98	530.63

REVIEW AND PREVIEW

Simplify. See Section 1.3.

65. $\frac{-6 - 3}{2 - 8}$ $\frac{3}{2}$ **66.** $\frac{4 - 5}{-1 - 0}$ 1

67. $\frac{-8 - (-2)}{-3 - (-2)}$ 6   **68.** $\frac{12 - 3}{10 - 9}$ 9

69. $\frac{0 - 6}{5 - 0}$ $-\frac{6}{5}$ **70.** $\frac{2 - 2}{3 - 5}$ 0

Concept Extensions

Solve.

71.a. $(0, 500)$; if no tables are produced, 500 chairs can be produced.
71.b. $(750, 0)$; if no chairs are produced, 750 tables can be produced.

71. Broyhill Furniture found that it takes 2 hours to manufacture each table for one of its special dining room sets. Each chair takes 3 hours to manufacture. A total of 1500 hours is available to produce tables and chairs of this style. The linear equation that models this situation is $2x + 3y = 1500$, where x represents the number of tables produced and y the number of chairs produced.

a. Complete the ordered pair solution $(0, \quad)$ of this equation. Describe the manufacturing situation this solution corresponds to.

b. Complete the ordered pair solution $(\quad, 0)$ for this equation. Describe the manufacturing situation this solution corresponds to.

c. If 50 tables are produced, find the greatest number of chairs the company can make.   466 chairs

72. While manufacturing two different camera models, Kodak found that the basic model costs $55 to produce, whereas the deluxe model costs $75. The weekly budget for these two models is limited to $33,000 in production costs. The linear equation that models this situation is $55x + 75y = 33,000$, where x represents the number of basic models and y the number of deluxe models.

   a. Complete the ordered pair solution (0,) of this equation. Describe the manufacturing situation this solution corresponds to.

   b. Complete the ordered pair solution (, 0) of this equation. Describe the manufacturing situation this solution corresponds to.

   c. If 350 deluxe models are produced, find the greatest number of basic models that can be made in one week.

73. The cost of renting a car for a day is given by the linear function $C(x) = 0.2x + 24$, where $C(x)$ is in dollars and x is the number of miles driven.

   a. Find the cost of driving the car 200 miles. $64

   b. Graph $C(x) = 0.2x + 24$.   See graphing answer section.

   c. How can you tell from the graph of $C(x)$ that as the number of miles driven increases, the total cost increases also?   The line moves upward from left to right.

74. The cost of renting a piece of machinery is given by the linear function $C(x) = 4x + 10$, where $C(x)$ is in dollars and x is given in hours.

   a. Find the cost of renting the piece of machinery for 8 hours. $42

   b. Graph $C(x) = 4x + 10$.   See graphing answer section.

   c. How can you tell from the graph of $C(x)$ that as the number of hours increases, the total cost increases also?

75. The yearly cost of tuition (in state) and required fees for attending a public two-year college full time can be estimated by the linear function $f(x) = 53.6x + 849.88$, where x is the number of years after 1990 and $f(x)$ is the total cost. (*Source:* U.S. National Center for Education Statistics)

   a. Use this function to approximate the yearly cost of attending a two-year college in the year 2010. [*Hint:* Find $f(20)$.]

   b. Use the given function to predict in what year the yearly cost of tuition and required fees will exceed $2000. [*Hint:* Let $f(x) = 2000$ and solve for x.]   2012

   c. Use this function to approximate the yearly cost of attending a two-year college in the present year. If you attend a two-year college, is this amount greater than or less than the amount that is currently charged by the college that you attend?   answers may vary

76. The yearly cost of tuition (in state) and required fees for attending a public four-year college can be estimated by the linear function $f(x) = 174.4x + 2074.38$, where x is the number of years after 1990 and $f(x)$ is the total cost in dollars. (*Source:* U.S. National Center for Education Statistics)

   a. Use this function to approximate the yearly cost of attending a four-year college in the year 2010. [*Hint:* Find $f(20)$.]

   b. Use the given function to predict in what year the yearly cost of tuition and required fees will exceed $5000. [*Hint:* Let $f(x) = 5000$ and solve for x.]   year 2007

   c. Use this function to approximate the yearly cost of attending a four-year college in the present year. If you attend a four-year college, is this amount greater than or less than the amount that is currently charged by the college that you attend?   answers may vary

77. The graph of $f(x)$ or $y = -4x$ is given below.

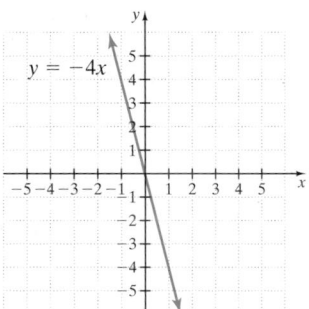

Without actually graphing, describe the shape and location of

   a. $y = -4x + 2$ **b.** $y = -4x - 5$

It is true that for any function $f(x)$, the graph of $f(x) + K$ is the same as the graph of $f(x)$ shifted K units up if K is positive and $|K|$ units down if K is negative.

The graph of $y = |x|$ is

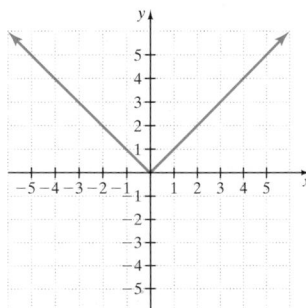

Without actually graphing, match each equation with its graph.

   a. $y = |x| - 1$ **b.** $y = |x| + 1$

   c. $y = |x| - 3$ **d.** $y = |x| + 3$

72.a. $(0, 440)$; if no basic models are produced, 440 deluxe models can be produced.   **72.b.** $(600, 0)$; if no deluxe models are produced, 600 basic models can be produced.   **72.c.** 122 basic models   **74.c.** The line moves upward from left to right.   **75.a.** $1921.88.
76.a. $5562.38   **77.a.** a line parallel to $y = -4x$ but with y-intercept of $(0, 2)$ **77.b.** a line parallel to $y = -4x$ but with y-intercept of $(0, -5)$

78. d   **79.** b   **80.** c   **81.** a

STUDY SKILLS REMINDER

Tips for Studying for an Exam

To prepare for an exam, try the following study techniques.

▶ Start the study process days before your exam.

▶ Make sure that you are current and up-to-date on your assignments.

▶ If there is a topic that you are unsure of, use one of the many resources that are available to you. For example,
See your instructor.
Visit a learning resource center on campus where math tutors are available.
Read the textbook material and examples on the topic.
View a videotape on the topic.

▶ Reread your notes and carefully review the Chapter Highlights at the end of the chapter.

▶ Work the review exercises at the end of the chapter and check your answers. Make sure that you correct any missed exercises. If you have trouble on a topic, use a resource.

▶ Find a quiet place to take the Chapter Test found at the end of the chapter. Do not use any resources when taking this sample test. This way you will have a clear indication of how prepared you are for your exam. Check your answers and use the Chapter Test Prep Video CD to correct any missed exercises.

▶ Get lots of rest the night before the exam. It's hard to show how well you know the material if your brain is foggy from lack of sleep.

Good luck and keep a positive attitude.

2.4 THE SLOPE OF A LINE

Objectives

1. Find the slope of a line given two points on the line.
2. Find the slope of a line given the equation of a line.
3. Interpret the slope–intercept form in an application.
4. Find the slopes of horizontal and vertical lines.
5. Compare the slopes of parallel and perpendicular lines.

1. You may have noticed by now that different lines tilt differently as shown.

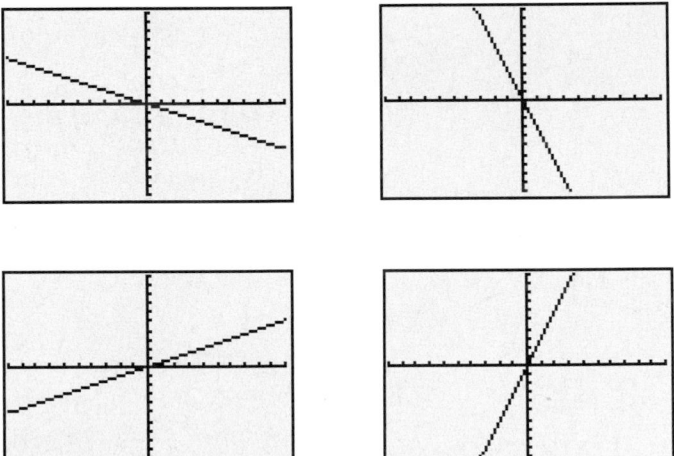

It is very important in many fields to be able to measure and compare the tilt, or **slope**, of lines. For example, a wheelchair ramp with a slope of $\frac{1}{12}$ means that the ramp rises 1 foot for every 12 horizontal feet. A road with a slope or grade of 11% (or $\frac{11}{100}$) means that the road rises 11 feet for every 100 horizontal feet.

1 ft
12 ft

11 ft
100 ft

We measure the slope of a line as a ratio of **vertical change** to **horizontal change**. Slope is usually designated by the letter m.

Suppose that we want to measure the slope of the following line.

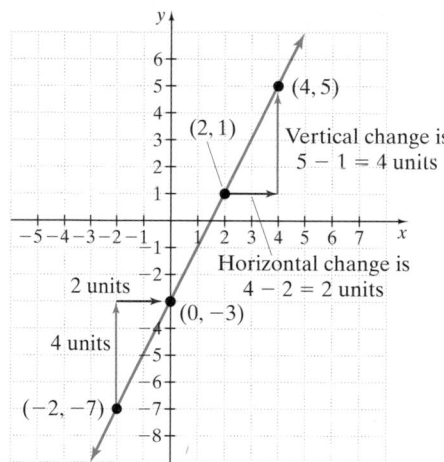

The vertical change between *both* pairs of points on the line is 4 units per horizontal change of 2 units. Then

$$\text{slope } m = \frac{\text{change in } y \text{ (vertical change)}}{\text{change in } x \text{ (horizontal change)}} = \frac{4}{2} = 2$$

We can also think of slope as a *rate of change* between points. A slope of 2 or $\frac{2}{1}$ means that between pairs of points on the line, the rate of change is a vertical change of 2 units per horizontal change of 1 unit.

Consider the line in the box below, which passes through the points (x_1, y_1) and (x_2, y_2). (The notation x_1 is read "x-sub-one.") The vertical change, or *rise*, between these points is the difference of the y-coordinates: $y_2 - y_1$. The horizontal change, or *run*, between the points is the difference of the x-coordinates: $x_2 - x_1$.

Slope of a Line

Given a line passing through points (x_1, y_1) and (x_2, y_2) the **slope** m of the line is

$m = \dfrac{\text{rise}}{\text{run}} = \dfrac{y_2 - y_1}{x_2 - x_1}$, as long as $x_2 \neq x_1$.

✔ **CONCEPT CHECK**

In the definition of slope, we state that $x_2 \neq x_1$. Explain why.

EXAMPLE 1

Find the slope of the line containing the points $(0, 3)$ and $(2, 5)$. Graph the line.

Solution We use the slope formula. It does not matter which point we call (x_1, y_1) and which point we call (x_2, y_2). We'll let $(x_1, y_1) = (0, 3)$ and $(x_2, y_2) = (2, 5)$.

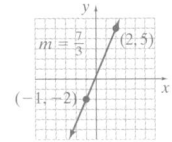

$$m = \frac{y_2 - y_1}{x_2 - x_1}$$

$$= \frac{5 - 3}{2 - 0} = \frac{2}{2} = 1$$

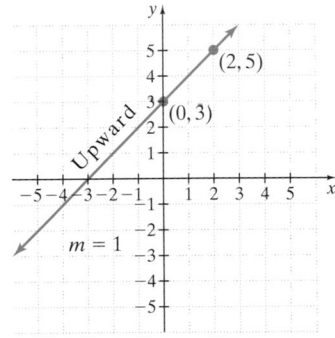

Notice in this example that the slope is positive and that the graph of the line containing $(0, 3)$ and $(2, 5)$ moves upward, or increases, as we go from left to right.

> **Helpful Hint**
>
> The slope of a line is the same no matter which 2 points of a line you choose to calculate slope. The line in Example 1 also contains the point $(-3, 0)$. Below, we calculate the slope of the line using $(0, 3)$ as (x_1, y_1) and $(-3, 0)$ as (x_2, y_2).
>
> $$m = \frac{y_2 - y_1}{x_2 - x_1} = \frac{0 - 3}{-3 - 0} = \frac{-3}{-3} = 1 \quad \text{Same slope as found in Example 1.}$$

EXAMPLE 2

Find the slope of the line containing the points $(5, -4)$ and $(-3, 3)$. Graph the line.

Solution We use the slope formula, and let $(x_1, y_1) = (5, -4)$ and $(x_2, y_2) = (-3, 3)$.

$$m = \frac{y_2 - y_1}{x_2 - x_1}$$

$$= \frac{3 - (-4)}{-3 - 5} = \frac{7}{-8} = -\frac{7}{8}$$

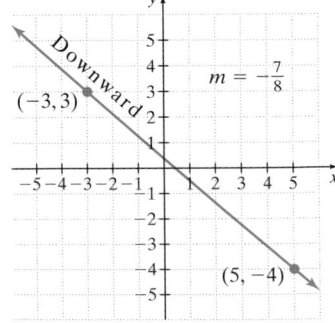

Notice in this example that the slope is negative and that the graph of the line through $(5, -4)$ and $(-3, 3)$ moves downward, or decreases, as we go from left to right.

> **Helpful Hint**
>
> When we are trying to find the slope of a line through two given points, it makes no difference which given point is called (x_1, y_1) and which is called (x_2, y_2). Once an x-coordinate is called x_1, however, make sure its corresponding y-coordinate is called y_1.

✔ **CONCEPT CHECK**

Find and correct the error in the following calculation of slope of the line containing the points $(12, 2)$ and $(4, 7)$.

$$m = \frac{12 - 4}{2 - 7} = \frac{8}{-5} = -\frac{8}{5}$$

To give meaning to the fact that the slope of a line is constant, see the following.

$$y = 2x + 3$$

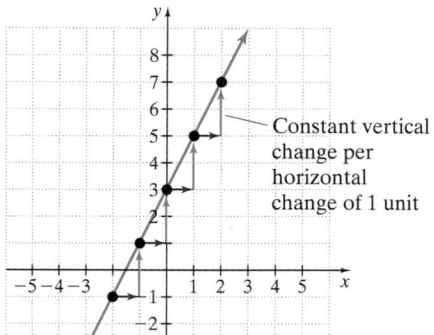

Constant vertical change per horizontal change of 1 unit

$$m = \frac{2}{1} = 2.$$

For linear equations, there is a constant vertical change per constant horizontal change. This means that the slope is the same between any two points of the graph.

Concept Check Answer:
$$m = \frac{2 - 7}{12 - 4} = \frac{-5}{8} = -\frac{5}{8}$$

2   The slope of a line indicates the direction of the line, that is, whether it slants upward or downward from left to right, as well as the amount of tilt to the line. In the following discovery we investigate how the equation itself can indicate direction and amount of tilt.

Discover the Concept

a. Graph $y_1 = x$, $y_2 = 3x$, and $y_3 = 0.5x$ using a standard window. Examine the graphs. What observation can you make with regard to the slant of the lines? What observation can you make about each line with respect to the amount of tilt and the coefficient of x in the equation?

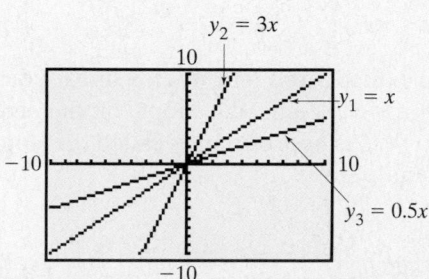

b. Can you change the equations so that the graphs slant downward instead of slanting upward from left to right?

c. Graph each equation separately. Predict the direction of slant and the amount of tilt of the line before seeing its graph on a graphing utility.

i. $y = x + 3$ **ii.** $y = 2x + 3$ **iii.** $y = -2x + 3$ **iv.** $y = 0.5x + 3$

TEACHING TIP
Have students use their graphing
calculator to graph the following
equations.
$y = 5x$
$y = x$
$y - 0.5x$
$y = -0.5x$
$y = -x$
$y = -5x$

On the same coordinate axes, have
them sketch each graph and label
the line with its slope at the right
arrowhead of the line. Then, ask
the students what they notice
about the slope values. Finally have
them sketch an estimate of where
the lines $y = 2x$ and $y = -2x$
would be.

In part (a) of the discovery, we see that each line moves upward from left to right. Also, notice that the greater the positive coefficient of x is, the steeper the tilt of the line. From part (b), we discover that if the coefficient of x is a negative number, the line moves downward from left to right.

As we have seen, the slope of a line is defined by two points on the line. Thus, if we know an equation of a line, we can find its slope.

Discover the Concept

Let's see if we can discover a relationship between an equation written in the form $y = ax + b$ and its slope. We'll use the equations from the previous Discover the Concept box.

a. Find the slope of the graph of each equation below. To do so, use two ordered pair solutions and the slope formula. (We will use the y-intercept as one ordered pair solution.)

$$y = x + 3 \qquad\qquad y = 2x + 3$$
$$y = -2x + 3 \qquad\qquad y = 0.5x + 3$$

Equation	y-intercept	Any second ordered-pair solution	Slope
$y = x + 3$ or $y = 1x + 3$	$(0,3)$	$(4,7)$	$m = \dfrac{7-3}{4-0} = 1$
$y = 2x + 3$	$(0,3)$	$(1,5)$	$m = \dfrac{5-3}{1-0} = 2$
$y = -2x + 3$	$(0,3)$	$(1,1)$	$m = \dfrac{1-3}{1-0} = -2$
$y = 0.5x + 3$	$(0,3)$	$(2,4)$	$m = \dfrac{4-3}{2-0} = 0.5$

b. For the equations in the general form $y = ax + b$, compare the slope of each equation above with a and compare the y-intercept with b.

We discovered that when a linear equation is written in the form $f(x) = ax + b$ or $y = ax + b$, a is the slope of the line and $(0, b)$ is its y-intercept. The form $y = mx + b$ is appropriately called the **slope-intercept form**. (Instead of a, we now use m to denote slope.)

Slope–Intercept Form

When a linear equation in two variables is written in slope–intercept form,

$$\underset{\uparrow}{slope} \quad \underset{\uparrow}{y\text{-intercept is } (0,b)}$$
$$y = mx + b$$

then m is the slope of the line and $(0, b)$ is the y-intercept of the line.

EXAMPLE 3

Find the slope and the *y*-intercept of the line $3x - 4y = 4$. Use a graphing utility to check.

Solution   We write the equation in slope–intercept form by solving for *y*.

$$3x - 4y = 4$$
$$-4y = -3x + 4 \qquad \text{Subtract } 3x \text{ from both sides.}$$
$$\frac{-4y}{-4} = \frac{-3x}{-4} + \frac{4}{-4} \qquad \text{Divide both sides by } -4.$$
$$y = \frac{3}{4}x - 1 \qquad \text{Simplify.}$$

The coefficient of x, $\frac{3}{4}$, is the slope, and the *y*-intercept is $(0, -1)$.

To check, we graph the equation $y_1 = (-3x + 4)/-4$ in a decimal window. Since the slope is positive, the line does indeed slant upward from left to right (as it should) and the *y*-intercept is $(0, -1)$.

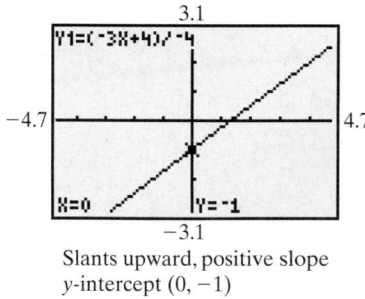

Slants upward, positive slope
y-intercept $(0, -1)$

3   Below is the graph of one-day ticket prices at Disney World for the years shown. Notice that the graph resembles the graph of a line. Recall that businesses often depend on equations that "closely fit" graphs like this one to model the data and to predict future trends. By the **least squares** method, the linear function $f(x) = 1.618x + 32.17$ approximates the data shown, where *x* is the number of years since 1990 and *y* is the ticket price for that year.

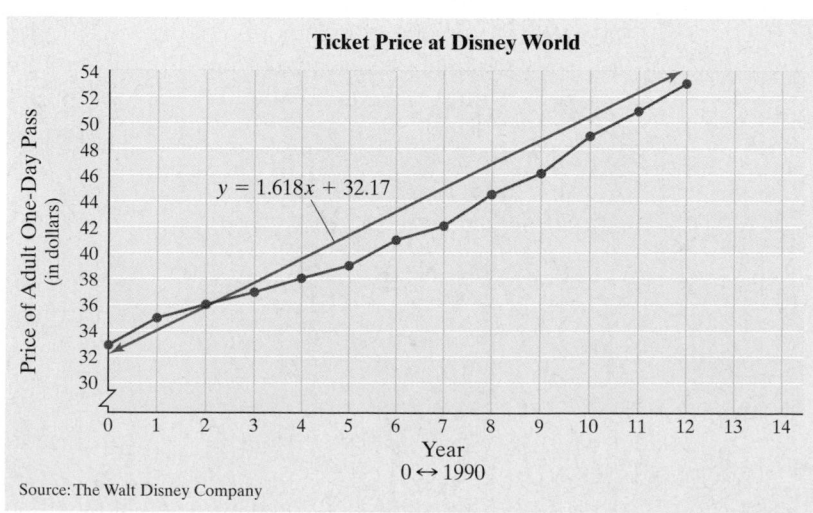

Source: The Walt Disney Company

EXAMPLE 4

PREDICTING FUTURE PRICES

The adult one-day pass price $f(x)$ for Disney World is given by

$$f(x) = 1.618x + 32.17$$

where x is the number of years since 1990.

a. Use this equation to predict the ticket price for the year 2008.

b. What does the slope of this equation mean?

c. What does the y-intercept of this equation mean?

Solution   **a.** To predict the price of a pass in 2008, we need to find $f(18)$. (Since year 1990 corresponds to $x = 0$, year 2008 corresponds to $x = 18$.)

$$f(x) = 1.618x + 32.17$$
$$f(18) = 1.618(18) + 32.17 \quad \text{Let } x = 18.$$
$$= 61.294$$

We predict that in the year 2008 the price of an adult one-day pass to Disney World will be about $61.29.

b. The slope of $f(x) = 1.618x + 32.17$ is 1.618. We can think of this number as $\dfrac{\text{rise}}{\text{run}}$ or $\dfrac{1.618}{1}$. This means that the ticket price increases on the average by $1.618 every 1 year.

c. The y-intercept of $y = 1.618x + 32.17$ is $(0, 32.17)$.

$$\underset{\underset{\text{year}}{\uparrow} \quad \underset{\text{price}}{\nwarrow}}{}$$

This means that at year $x = 0$ or 1990, the ticket price was about $32.17.

4 Next we find the slopes of two special types of lines: vertical lines and horizontal lines.

EXAMPLE 5

Find the slope of the line $x = -5$.

Solution   Recall that the graph of $x = -5$ is a vertical line with x-intercept $(-5, 0)$. To find the slope, we find two ordered pair solutions of $x = -5$. Of course, solutions of $x = -5$ must have an x-value of -5. We will let $(x_1, y_1) = (-5, 0)$ and $(x_2, y_2) = (-5, 4)$. Then

$$m = \frac{y_2 - y_1}{x_2 - x_1}$$
$$= \frac{4 - 0}{-5 - (-5)}$$
$$= \frac{4}{0}$$

Since $\dfrac{4}{0}$ is undefined, we say that the slope of the vertical line $x = -5$ is undefined.

EXAMPLE 6

Find the slope of the line $y = 2$.

Solution

The graph of $y = 2$ is shown. Trace to find two ordered pair solutions. We select $(0, 2)$ and $(1, 2)$. Use these points to find the slope.

$$m = \frac{2 - 2}{1 - 0}$$

$$= \frac{0}{1}$$

$$= 0$$

The slope of the horizontal line $y = 2$ is 0.

From the previous two examples, we have the following generalization.

The slope of any vertical line is undefined.
The slope of any horizontal line is 0.

CLASSROOM EXAMPLE

Find the slope of the line $y = 7$.
answer: slope is 0

Helpful Hint

Slope of 0 and undefined slope are not the same. Vertical lines have undefined slope, whereas horizontal lines have slope of 0.

TEACHING TIP

For the first two lines, point out that we describe the trend in the graph as we read it, like a book, from left to right.

The following four graphs summarize the overall appearance of lines with positive, negative, zero, or undefined slopes.

Increasing line, positive slope

Decreasing line, negative slope

Horizontal line, zero slope

Vertical line, undefined slope

The appearance of a line can give us further information about its slope.

The graphs of $y = \frac{1}{2}x + 1$ and $y = 5x + 1$ are shown to the right. Recall that the graph of $y = \frac{1}{2}x + 1$ has a slope of $\frac{1}{2}$ and that the graph of $y = 5x + 1$ has a slope of 5.

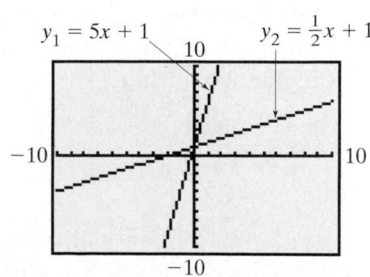

Notice that the line with the slope of 5 is steeper than the line with the slope of $\frac{1}{2}$. This is true in general for positive slopes.

> For a line with positive slope m, as m increases, the line becomes steeper.

To see why this is so, compare the slopes from above.

$\frac{1}{2}$ means a vertical change of 1 unit per horizontal change of 2 units

5 or $\frac{10}{2}$ means a vertical change of 10 units per horizontal change of 2 units

For larger positive slopes, the vertical change is greater for the same horizontal change. Thus, larger positive slopes mean steeper lines.

5   Slopes of lines can help us determine whether lines are parallel or perpendicular. Recall that parallel lines are distinct lines with the same tilt that do not meet, and perpendicular lines are lines that intersect to form right angles.

Technology Note

Although the graphs of the equations shown below to the left are perpendicular lines, they do not appear to be so when graphed using a standard viewing window.

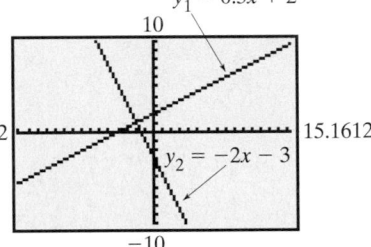

Standard Viewing Window   Square Viewing Window

This is because a graphing calculator screen is rectangular, so the tick marks along the x-axis are farther apart than the tick marks along the y-axis. The graphs of the same lines at the right do appear to be perpendicular because the viewing window has been selected so that there is equal spacing between tick marks on both axes. We say that the lines to the right are graphed using a **square viewing window**. Some graphing utilities can automatically provide a square setting. (This technology note shows the necessity of being able to algebraically determine whether lines are perpendicular. A graphing utility can then be used to partially verify the result.)

Discover the Concept

Graph each pair of equations in a square window, such as an integer window. Do the lines appear to be parallel, perpendicular, or neither? Find the relationship between the slopes of the lines and whether they appear parallel, perpendicular, or neither.

a. $y = 3x - 5$ and $y = 3x + 7$ **b.** $y = 2x + 3$ and $y = 4x - 5$

c. $y = 0.5x - 9$ and $y = 2x - 3$ **d.** $y = -5x - 6$ and $y = -5x - 2$

e. $y = -\frac{1}{3}x - 7$ and $y = 3x + 5$

In the previous discovery, we see that parallel lines have the same tilt and therefore have the same slope.

Parallel Lines

Two nonvertical lines are parallel if they have the same slope and different *y*-intercepts.

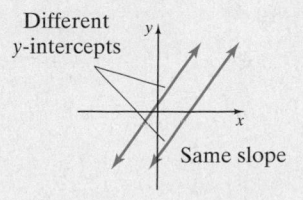

How do the slopes of perpendicular lines compare? (Two lines intersecting at right angles are called **perpendicular lines**.) Suppose that a line has a slope of $\frac{a}{b}$. If the line is rotated 90°, the rise and run are now switched, except that the run is now negative. This means that the new slope is $-\frac{b}{a}$. Notice that

$$\left(\frac{a}{b}\right) \cdot \left(-\frac{b}{a}\right) = -1$$

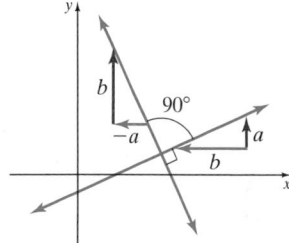

This is how we tell whether two lines are perpendicular.

Perpendicular Lines

Two nonvertical lines are perpendicular if the product of their slopes is -1.

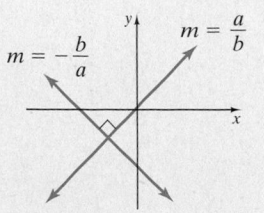

In other words, two nonvertical lines are perpendicular if the slope of one is the negative reciprocal of the slope of the other.

EXAMPLE 7

Given the following pairs of equations, determine whether their graphs are parallel lines, perpendicular lines, or neither. Then use a graphing utility and a square window setting to visualize your results.

a. $3x + 7y = 21$ **b.** $-x + 3y = 2$ **c.** $2x - 3y = 12$

$6x + 14y = 7$ $2x + 6y = 5$ $6x + 4y = 16$

Solution Find the slope of each line by solving each equation for y.

a.

$$3x + 7y = 21 \qquad\qquad 6x + 14y = 7$$
$$7y = -3x + 21 \qquad\qquad 14y = -6x + 7$$
$$\frac{7y}{7} = \frac{-3x}{7} + \frac{21}{7} \qquad\qquad \frac{14y}{14} = \frac{-6x}{14} + \frac{7}{14}$$
$$y = -\frac{3}{7}x + 3 \qquad\qquad y = -\frac{3}{7}x + \frac{1}{2}$$

The slope of both lines is $-\dfrac{3}{7}$. The y-intercept of one line is 3, whereas the

y-intercept of the other line is $\dfrac{1}{2}$. Since these lines have the same slope and

different y-intercepts, the lines are parallel.

b. Solve each equation for y.

$$-x + 3y = 2 \qquad \text{and} \qquad 2x + 6y = 5$$
$$y = \frac{1}{3}x + \frac{2}{3} \qquad\qquad y = -\frac{1}{3}x + \frac{5}{6}$$

The slopes $\dfrac{1}{3}$ and $-\dfrac{1}{3}$ are not equal nor are they negative reciprocals of each other,

so they are simply intersecting lines.

c. Solve each equation for y.

$$2x - 3y = 12 \qquad \text{and} \qquad 6x + 4y = 16$$
$$y = \frac{2}{3}x - 4 \qquad\qquad y = -\frac{3}{2}x + 4$$

The slopes of the lines are $\dfrac{2}{3}$ and $-\dfrac{3}{2}$, which are negative reciprocals of each other,

so the lines are perpendicular.

To visualize the results of parts (a), (b) and (c), see the graphs below using square window settings.

$$y_1 = -\frac{3}{7}x + 3 \qquad y_1 = \frac{1}{3}x + \frac{2}{3} \qquad y_1 = \frac{2}{3}x - 4$$

$$y_2 = -\frac{3}{7}x + \frac{1}{2} \qquad y_2 = -\frac{1}{3}x + \frac{5}{6} \qquad y_2 = -\frac{3}{2}x + 4$$

| Parallel lines | Neither parallel nor perpendicular | Perpendicular lines |

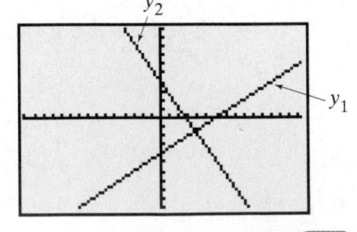

Square Viewing Window

✔ **CONCEPT CHECK**

What is *different* about the equations of two parallel lines?

Spotlight on
DECISION
✕MAKING

Suppose you are the manager of an apartment complex. You have just notified residents of a rent increase. Some residents think that the increase may be unjustified and out of line with recent increases. A group of concerned residents asks you to hold an open meeting to answer questions about the increase. You are preparing a set of overheads to use during the meeting to show the history of rent increases at the apartment complex. Which overhead would you use and why?

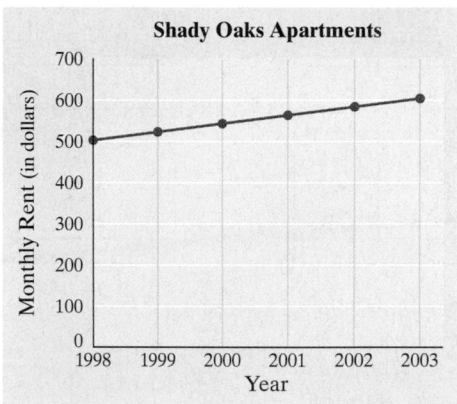

MENTAL MATH

Decide whether a line with the given slope slants upward, downward, horizontally, or vertically from left to right.

1. $m = \dfrac{7}{6}$ upward   **2.** $m = -3$ downward   **3.** $m = 0$ horizontally   **4.** m is undefined vertically

Concept Check Answer:
y-intercepts are different

TEACHING TIP
A Group Activity for this section is available in the Instructor's Resource Manual.

EXERCISE SET 2.4

STUDY GUIDE/SSM   CD/ VIDEO   PH MATH TUTOR CENTER   MathXL®Tutorials ON CD   MathXL®   MyMathLab®

Find the slope of the line that goes through the given points. See Examples 1 and 2.

1. $(3, 2), (8, 11)$ $\frac{9}{5}$ **2.** $(1, 6), (7, 11)$ $\frac{5}{6}$

3. $(-2, 8), (4, 3)$ $-\frac{5}{6}$ **4.** $(3, 7), (-2, 11)$ $-\frac{4}{5}$

🔒 **5.** $(-2, -6), (4, -4)$ $\frac{1}{3}$ **6.** $(-3, -4), (-1, 6)$ 5

7. $(-2, 5), (3, 5)$ 0 **8.** $(4, 2), (4, 0)$ undefined

9. $(-1, 1), (-1, -5)$ undefined **10.** $(-2, -5), (3, -5)$ 0

🔒 **11.** $(-1, 2), (-3, 4)$ -1 **12.** $(3, -2), (-1, -6)$ 1

A linear function is graphed below. Using the trace feature, we found two points on the line. Give the slope of the line.

13. $-\dfrac{2}{3}$

A linear function is graphed below. Using the trace feature we found two points on the line. Give the slope of the line.

14.

  4

Examine the table of values for a linear function. Select two ordered pairs and find the slope of the line joining these points.

15. 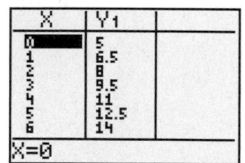 1.5 **16.** 2

17. 0.2 **18.** −4

Two lines are graphed on each set of axes. Decide whether l_1 or l_2 has the greater slope.

19. **20.**

21. **22.**

23. **24.**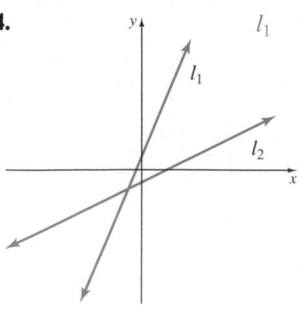

Find the slope and the y-intercept of each line. See Example 3.

25. $f(x) = 5x - 2$ **26.** $f(x) = -2x + 6$

27. $2x + y = 7$ $m = -2, b = 7$ **28.** $-5x + y = 10$

29. $2x - 3y = 10$ **30.** $-3x - 4y = 6$

31. $f(x) = \frac{1}{2}x$ $m = \frac{1}{2}, b = 0$ **32.** $f(x) = -\frac{1}{4}x$

25. $m = 5, b = -2$ **26.** $m = -2, b = 6$ **28.** $m = 5, b = 10$
29. $m = \frac{2}{3}, b = -\frac{10}{3}$ **30.** $m = -\frac{3}{4}, b = -\frac{3}{2}$ **32.** $m = -\frac{1}{4}, b = 0$

Match each graph with its equation.

A **B**

C **D**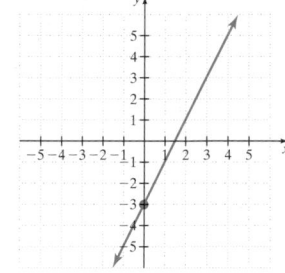

33. $f(x) = 2x + 3$ A **34.** $f(x) = 2x - 3$ D
35. $f(x) = -2x + 3$ B **36.** $f(x) = -2x - 3$ C

43.–44. answers may vary **45.** $m = -1, b = 5$ **46.** $m = 1, b = 2$
47. $m = \frac{6}{5}, b = 6$ **48.** $m = \frac{4}{7}, b = -4$ **49.** $m = 3, b = 9$

Find the slope of each line. See Examples 5 and 6.

37. $x = 1$ undefined **38.** $y = -2$ 0
39. $y = -3$ 0 **40.** $x = 4$ undefined
41. $x + 2 = 0$ undefined **42.** $y - 7 = 0$ 0

43. Explain how merely looking at a line can tell us whether its slope is negative, positive, undefined, or zero.

44. Explain why the graph of $y = b$ is a horizontal line.

50. $m = \frac{1}{2}, b = \frac{7}{2}$ **52.** slope is undefined, no y-intercept
54. $m = \frac{1}{7}, b = 0$

Find the slope and the y-intercept of each line.

45. $f(x) = -x + 5$ **46.** $f(x) = x + 2$
47. $-6x + 5y = 30$ **48.** $4x - 7y = 28$
49. $3x + 9 = y$ **50.** $2y - 7 = x$
51. $y = 4$ $m = 0, b = 4$ **52.** $x = 7$
53. $f(x) = 7x$ $m = 7, b = 0$ **54.** $f(x) = \frac{1}{7}x$

55. $6 + y = 0$ $m = 0, b = -6$ **56.** $x - 7 = 0$

57. $2 - x = 3$ **58.** $2y + 4 = -7$

56.–57. slope is undefined, no y-intercept

Determine whether the lines are parallel, perpendicular, or neither. See Example 7.

△ **59.** $f(x) = -3x + 6$ △ **60.** $f(x) = 5x - 6$

  $g(x) = 3x + 5$ 🔒 $g(x) = 5x + 2$ parallel

△ **61.** $-4x + 2y = 5$ △ **62.** $2x - y = -10$

  $2x - y = 7$ parallel $2x + 4y = 2$

△ **63.** $-2x + 3y = 1$ △ **64.** $x + 4y = 7$

  $3x + 2y = 12$ $2x - 5y = 0$ neither

✏ **65.** Explain whether two lines, both with positive slopes, can be
△ perpendicular.   answers may vary

✏ **66.** Explain why it is reasonable that nonvertical parallel lines
△ have the same slope.   answers may vary

        58. $m = 0, b = -\dfrac{11}{2}$

Determine the slope of each line.

67. $\dfrac{3}{2}$ **68.** -3

69. $-\dfrac{1}{2}$ **70.** 1

            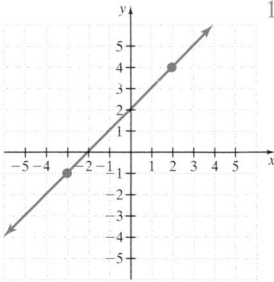

59. neither **62.–63.** perpendicular

Find each slope.

71. Find the pitch, or slope, of the roof shown. $\dfrac{2}{3}$

72. Upon takeoff, a Delta Airlines jet climbs to 3 miles as it passes over 25 miles of land below it. Find the slope of its climb. $\dfrac{3}{25}$

73. Driving down Bald Mountain in Wyoming, Bob Dean finds that he descends 1600 feet in elevation by the time he is 2.5 miles (horizontally) away from the high point on the mountain road. Find the slope of his descent rounded to two decimal places (1 mile = 5280 feet).   approximately -0.12

74. Find the grade, or slope, of the road shown. $\dfrac{3}{20}$

Solve. See Example 4.

75. The annual average income y of an American man over 25 with an associate's degree is approximated by the linear equation $y = 1545.4x + 33,858.4$, where x is the number of years after 1997. (*Source:* Based on data from the U.S. Bureau of the Census)

   a. Predict the average income of an American man with an associate's degree in 2005. $\$46,221.60$

   b. Find and interpret the slope of the equation.

   c. Find and interpret the y-intercept of the equation.

76. The annual average income of an American woman over 25 with a bachelor's degree is given by the linear equation $y = 1280.3x + 28,421.6$, where x is the number of years after 1997. (*Source:* Based on data from the U.S. Bureau of the Census)

   a. Find the average income of an American woman with a bachelor's degree in 2005. $\$38,664$

   b. Find and interpret the slope of the equation.

   c. Find and interpret the y-intercept of the equation.

75.b. $m = 1545.4$; The annual income increases $\$1545.40$ every year. **75.c.** $b = 33,858.4$, At year $x = 0$, or 1997, the annual average income was $\$33,858.40$. **76.b.** $m = 1280.3$; The annual average income increases $\$1280.30$ every year. **76.c.** $b = 28,421.6$; At year $x = 0$, or 1997, the annual average income was $\$28,421.80$.

77. With wireless Internet (WiFi) gaining popularity, the number of public wireless Internet access points (in thousands) is projected to grow from 2002 to 2006 according to the equation

$$-245x + 10y = 59$$

where x is the number of years after 2002.

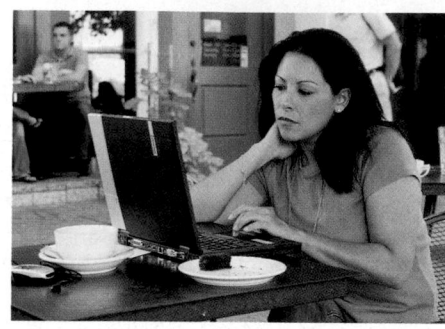

   a. Find the slope and y-intercept of the linear equation.

   b. What does the slope mean in this context?

   c. What does the y-intercept mean in this context?

78. One of the faster growing occupations over the next few years is expected to be paralegal. The number of people y in thousands employed as paralegals in the United States can be estimated by the linear equation $-63x + 10y = 1880$, where x is the number of years after 1996. (*Source:* Based on projections from the U.S. Bureau of Labor Statistics, 2000–2010)

   a. Find the slope and y-intercept of the linear equation.

   b. What does the slope mean in this context?

   c. What does the y-intercept mean in this context?

79. In an earlier section, it was given that the yearly cost of tuition and required fees for attending a public four-year college full-time can be estimated by the linear function

$$f(x) = 174.4x + 2074.38$$

where x is the number of years after 1990 and $f(x)$ is the total cost. (*Source:* U.S. National Center for Education Statistics)

   a. Find and interpret the slope of this equation.

   b. Find and interpret the y-intercept of this equation.

80. If an earlier section, it was given that the yearly cost of tuition and required fees for attending a public two-year college full-time can be estimated by the linear function

$$f(x) = 53.6x + 849.88$$

where x is the number of years after 1990 and $f(x)$ is the total cost. (*Source:* U.S. National Center for Education Statistics)

   a. Find and interpret the slope of this equation.

   b. Find and interpret the y-intercept of this equation.

Solve.

△ **81.** Find the slope of a line parallel to the line

$$f(x) = -\frac{7}{2}x - 6. \quad -\frac{7}{2}$$

△ **82.** Find the slope of a line parallel to the line $f(x) = x$.   1

△ **83.** Find the slope of a line perpendicular to the line

$$f(x) = -\frac{7}{2}x - 6. \quad \frac{2}{7}$$

△ **84.** Find the slope of a line perpendicular to the line $f(x) = x$.

△ **85.** Find the slope of a line parallel to the line $5x - 2y = 6.$ $\frac{5}{2}$

△ **86.** Find the slope of a line parallel to the line $-3x + 4y = 10.$

△ **87.** Find the slope of a line perpendicular to the line $5x - 2y = 6.$ $-\frac{2}{5}$ **84.** -1 **86.** $\frac{3}{4}$

REVIEW AND PREVIEW

Recall the formula

$$\text{Probability of an event} = \frac{\text{number of ways that the event can occur}}{\text{number of possible outcomes}}$$

Suppose these cards are shuffled and one card is turned up. Find the possibility of selecting each letter.

88. $P(R)$ $\frac{1}{11}$ **89.** $P(B)$ $\frac{2}{11}$

90. $P(E)$ 0   **91.** $P(I \text{ or } T)$ $\frac{3}{11}$

92. $P(\text{selecting a letter of the alphabet})$ 1

93. $P(\text{vowel})$ $\frac{4}{11}$

Simplify and solve for y. See Section 1.5.

94. $y - 2 = 5(x + 6)$ $y = 5x + 32$

95. $y - 0 = -3[x - (-10)]$ $y = -3x - 30$

96. $y - (-1) = 2(x - 0)$ $y = 2x - 1$

97. $y - 9 = -8[x - (-4)]$ $y = -8x - 23$

Concept Extensions

98. Each line below has negative slope.

   a. Find the slope of each line. $l_1: -2, l_2: -1, l_3: -\frac{2}{3}$

   b. Use the result of Part **a** to fill in the blank. For lines with negative slopes, the steeper line has the _____ (greater/lesser) slope.   lesser

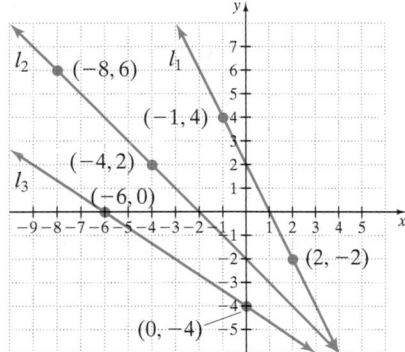

99. The following graph shows the altitude of a seagull in flight over a time period of 30 seconds.

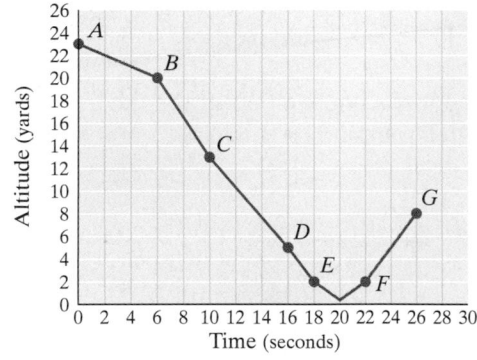

   a. Find the coordinates of point B. $(6, 20)$

   b. Find the coordinates of point C. $(10, 13)$

   c. Find the rate of change of altitude between points B and C. (Recall that the rate of change between points is the slope between points. This rate of change will be in yards per second.) $-\frac{7}{4}$ or -1.75 yd per sec

   d. Find the rate of change of altitude (in yards per second) between points F and G. $\frac{3}{2}$ or 1.5 yd per sec

100. Professional plumbers suggest that a sewer pipe should be sloped 0.25 inch for every foot. Find the recommended slope for a sewer pipe. (*Source: Rules of Thumb* by Tom Parker, 1983, Houghton Mifflin Company) $\frac{1}{48}$

101. Support the result of Exercise 61 by graphing the pair of equations on a graphing calculator.

102. Support the result of Exercise 62 by graphing the pair of equations on a graphing calculator. (*Hint:* Use the window showing $[-15, 15]$ on the x-axis and $[-10, 10]$ on the y-axis.)

103. a. On a single screen, graph $y = \frac{1}{2}x + 1$, $y = x + 1$ and $y = 2x + 1$. Notice the change in slope for each graph.

   b. On a single screen, graph $y = -\frac{1}{2}x + 1$, $y = -x + 1$ and $y = -2x + 1$. Notice the change in slope for each graph.

   c. Determine whether the following statement is true or false for slope m of a given line. As $|m|$ becomes greater, the line becomes steeper.   true

101.–102., 103.a.–103.b. See graphing answer section.

2.5   **EQUATIONS OF LINES**

Objectives

  1 Use the slope–intercept form to write the equation of a line.

  2 Graph a line using its slope and y-intercept.

  3 Use the point–slope form to write the equation of a line.

  4 Write equations of vertical and horizontal lines.

  5 Find equations of parallel and perpendicular lines.

  1 In the last section, we learned that the slope–intercept form of a linear equation is $y = mx + b$. When a linear equation is written in this form, the slope of the line is the same as the coefficient m of x. Also, the y-intercept of the line is $(0, b)$. For example, the slope of the line defined by $y = 2x + 3$ is, 2, and its y-intercept is $(0, 3)$.

We may also use the slope–intercept form to write the equation of a line given its slope and y-intercept. The equation of a line is a linear equation in 2 variables that, if graphed, would produce the line described.

EXAMPLE 1

Write an equation of the line with y-intercept $(0, -3)$ and slope of $\frac{1}{4}$.

Solution We want to write a linear equation in 2 variables that describes the line with y-intercept $(0, -3)$ and has a slope of $\frac{1}{4}$. We are given the slope and the y-intercept. Let $m = \frac{1}{4}$ and $b = -3$, and write the equation in slope–intercept form, $y = mx + b$.

$$y = mx + b$$

$$y = \frac{1}{4}x + (-3) \qquad \text{Let } m = \frac{1}{4} \text{ and } b = -3.$$

$$y = \frac{1}{4}x - 3 \qquad \text{Simplify.}$$

✔ **CONCEPT CHECK**

What is wrong with the following equation of a line with y-intercept $(0, 4)$ and slope 2?

$$y = 4x + 2$$

2   Given the slope and y-intercept of a line, we may graph the line as well as write its equation. Let's graph the line from Example 1.

EXAMPLE 2

Graph $y = \frac{1}{4}x - 3$.

Solution Recall that the slope of the graph of $y = \frac{1}{4}x - 3$ is $\frac{1}{4}$ and the y-intercept is $(0, -3)$. To graph the line, we first plot the y-intercept $(0, -3)$. To find another point on the line, we recall that slope is $\frac{\text{rise}}{\text{run}} = \frac{1}{4}$. Another point may then be plotted by starting at $(0, -3)$, rising 1 unit up, and then running 4 units to the right. We are now at the point $(4, -2)$. The graph is the line through these two points.

Notice that the line does have a y-intercept of $(0, -3)$ and a slope of $\frac{1}{4}$.

EXAMPLE 3

Graph $2x + 3y = 12$.

Solution

First, we solve the equation for y to write it in slope–intercept form. In slope–intercept form, the equation is $y = -\dfrac{2}{3}x + 4$. Next we plot the y-intercept $(0, 4)$. To find another point on the line, we use the slope $-\dfrac{2}{3}$, which can be written as $\dfrac{\text{rise}}{\text{run}} = \dfrac{-2}{3}$. We start at $(0, 4)$ and move down 2 units since the numerator of the slope is -2; then we move 3 units to the right since the denominator of the slope is 3. We arrive at the point $(3, 2)$. The line through these points is the graph, shown below to the left.

TEACHING TIP

Remind students that there are many equivalent ways to write slope.

$$\frac{3}{4} = \frac{-3}{-4} = \frac{6}{8}, \text{ etc.}$$

$$-5 = \frac{-5}{1} = \frac{5}{-1} = \frac{-15}{3} = \frac{15}{-3}, \text{ etc.}$$

Any equivalent way can be used to correctly graph an equation.

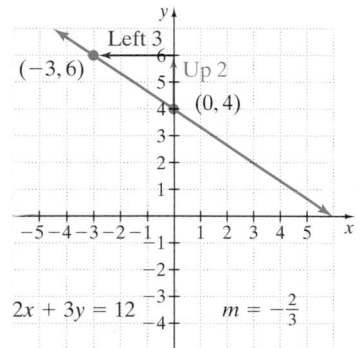

The slope $\dfrac{-2}{3}$ can also be written as $\dfrac{2}{-3}$, so to find another point in Example 3 we could start at $(0, 4)$ and move up 2 units and then 3 units to the left. We would arrive at the point $(-3, 6)$. The line through $(-3, 6)$ and $(0, 4)$ is the same line as shown previously through $(3, 2)$ and $(0, 4)$. See the graph above to the right.

3 When the slope of a line and a point on the line are known, the equation of the line can also be found. To do this, use the slope formula to write the slope of a line that passes through points (x_1, y_1) and (x, y). We have

$$m = \frac{y - y_1}{x - x_1}$$

Multiply both sides of this equation by $x - x_1$ to obtain

$$y - y_1 = m(x - x_1)$$

This form is called the **point–slope form** of the equation of a line.

Point–Slope Form of the Equation of a Line

The point–slope form of the equation of a line is $y - y_1 = m(x - x_1)$, where m is the slope of the line and (x_1, y_1) is a point on the line.

EXAMPLE 4

Find an equation of the line with slope -3 containing the point $(1, -5)$. Write the equation in slope–intercept form $y = mx + b$. Then use a graphing utility to check.

Solution

Because we know the slope and a point of the line, we use the point–slope form with $m = -3$ and $(x_1, y_1) = (1, -5)$.

$$
\begin{aligned}
y - y_1 &= m(x - x_1) & &\text{Point–slope form.} \\
y - (-5) &= -3(x - 1) & &\text{Let } m = -3 \text{ and } (x_1, y_1) = (1, -5). \\
y + 5 &= -3x + 3 & &\text{Apply the distributive property.} \\
y &= -3x - 2 & &\text{Write in slope–intercept form.}
\end{aligned}
$$

In slope–intercept form, the equation is $y = -3x - 2$.

To check, define $y_1 = -3x - 2$ and graph in an integer window. From the equation we see that $m = -3$, so the slope is -3. Next, check on the graph to see that it contains the point $(1, -5)$ as in the screen below to the left.

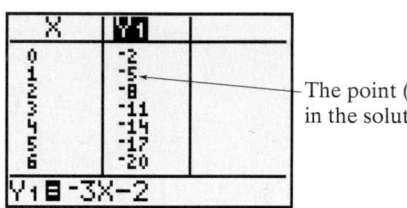

The point $(1, -5)$ is included in the solution pairs.

A second way to see that $(1, -5)$ is an ordered pair solution is to examine a table of values as shown above to the right.

> **Helpful Hint**
>
> Remember, "slope-intercept form" means the equation is "solved for y."

EXAMPLE 5

Find an equation of the line through points $(4, 0)$ and $(-4, -5)$. Write the equation using function notation. Then use a graphing utility to check.

Solution

First, find the slope of the line.

$$
m = \frac{-5 - 0}{-4 - 4} = \frac{-5}{-8} = \frac{5}{8}
$$

Next, make use of the point–slope form. Replace (x_1, y_1) by either $(4, 0)$ or $(-4, -5)$ in the point–slope equation. We will choose the point $(4, 0)$. The line through $(4, 0)$ with slope $\frac{5}{8}$ is

---

CLASSROOM EXAMPLE

Write an equation of the line with slope -2 and containing the point $(2, -4)$. Write the equation in slope-intercept form, $y = mx + b$.

answer: $y = -2x$

CLASSROOM EXAMPLE

Write an equation of the line through points $(3, 0)$ and $(-2, 4)$. Write the equation using function notation.

answer:

$f(x) = -\dfrac{4}{5}x + \dfrac{12}{5}$

$$y - y_1 = m(x - x_1) \qquad \text{Point–slope form.}$$

$$y - 0 = \frac{5}{8}(x - 4) \qquad \text{Let } m = \frac{5}{8} \text{ and } (x_1, y_1) = (4, 0).$$

$$8y = 5(x - 4) \qquad \text{Multiply both sides by } 8.$$

$$8y = 5x - 20 \qquad \text{Apply the distributive property.}$$

To write the equation using function notation, we solve for y.

$$8y = 5x - 20$$

$$y = \frac{5}{8}x - \frac{20}{8} \qquad \text{Divide both sides by } 8.$$

$$f(x) = \frac{5}{8}x - \frac{5}{2} \qquad \text{Write using function notation.}$$

To check, graph $y_1 = \frac{5}{8}x - \frac{5}{2}$ in an integer window. Trace to find $(4, 0)$ and $(-4, -5)$ on the graph of the line. We can also examine a table of ordered pair solutions to check.

> **Helpful Hint**
>
> If two points of a line are given, either one may be used with the point-slope form to write an equation of the line.

The point–slope form of an equation is very useful for solving real-world problems.

EXAMPLE 6

PREDICTING SALES

Southern Star Realty is an established real estate company that has enjoyed constant growth in sales since 1995. In 1997 the company sold 200 houses, and in 2002 the company sold 275 houses. Use these figures to predict the number of houses this company will sell in the year 2011.

Solution   1. UNDERSTAND.   Read and reread the problem. Then let
  $x =$ the number of years after 1995 and
  $y =$ the number of houses sold in the year corresponding to x.
  The information provided then gives the ordered pairs $(2, 200)$ and $(7, 275)$. To better visualize the sales of Southern Star Realty, we graph the linear equation that passes through the points $(2, 200)$ and $(7, 275)$.

2. TRANSLATE. We write a linear equation that passes through the points $(2, 200)$ and $(7, 275)$. To do so, we first find the slope of the line.

$$m = \frac{275 - 200}{7 - 2} = \frac{75}{5} = 15$$

Then, using the point–slope form and the point $(2, 200)$ to write the equation, we have

$$y - y_1 = m(x - x_1)$$

$$y - 200 = 15(x - 2) \quad \text{Let } m = 15 \text{ and } (x_1, y_1) = (2, 200).$$

$$y - 200 = 15x - 30 \quad \text{Multiply.}$$

$$y = 15x + 170 \quad \text{Add } 200 \text{ to both sides.}$$

3. SOLVE. To predict the number of houses sold in the year 2011, we use $y = 15x + 170$ and complete the ordered pair $(16, \quad)$, since $2011 - 1995 = 16$.

$$y = 15(16) + 170 \quad \text{Let } x = 16.$$

$$y = 410$$

4. INTERPRET.
Check: Verify that the point $(16, 410)$ is a point on the line graphed in step 1.
State: Southern Star Realty should expect to sell 410 houses in the year 2011.

4   A few special types of linear equations are linear equations whose graphs are vertical and horizontal lines.

EXAMPLE 7

Find an equation of the horizontal line containing the point $(2, 3)$.

Solution   Recall that a horizontal line has an equation of the form $y = b$. Since the line contains the point $(2, 3)$, the equation is $y = 3$. Below is a graph of $y = 3$ with the point $(2, 3)$ indicated.

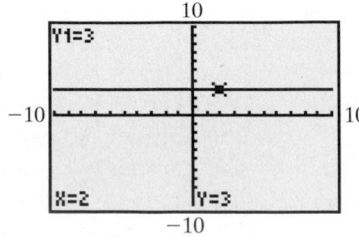

EXAMPLE 8

Find an equation of the line containing the point $(2, 3)$ with undefined slope.

Solution

Since the line has undefined slope, the line must be vertical. A vertical line has an equation of the form $x = c$, and since the line contains the point $(2, 3)$, the equation is $x = 2$.

TECHNOLOGY NOTE

Recall that a vertical line does not represent the graph of a function and cannot be graphed in the Y= editor in function mode. Vertical lines may be drawn using the draw menu. Below is an example of using a draw vertical feature.

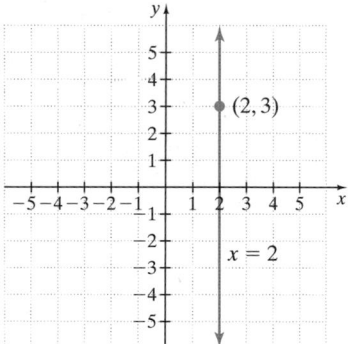

5 Next, we find equations of parallel and perpendicular lines.

EXAMPLE 9

Find an equation of the line containing the point $(4, 4)$ and parallel to the line $2x + 3y = -6$. Write the equation in standard form.

Solution

Because the line we want to find is *parallel* to the line $2x + 3y = -6$, the two lines must have equal slopes. Find the slope of $2x + 3y = -6$ by writing it in the form $y = mx + b$. In other words, solve the equation for y.

$$2x + 3y = -6$$

$$3y = -2x - 6 \qquad \text{Subtract } 2x \text{ from both sides.}$$

$$y = \frac{-2x}{3} - \frac{6}{3} \qquad \text{Divide by 3.}$$

$$y = -\frac{2}{3}x - 2 \qquad \text{Write in slope-intercept form.}$$

The slope of this line is $-\dfrac{2}{3}$. Thus, a line parallel to this line will also have a slope of $-\dfrac{2}{3}$.

The equation we are asked to find describes a line containing the point $(4, 4)$ with a slope of $-\dfrac{2}{3}$. We use the point-slope form.

The graphs of parallel lines $2x + 3y = -6$ and $2x + 3y = 20$.

$$y - y_1 = m(x - x_1)$$

$$y - 4 = -\frac{2}{3}(x - 4) \qquad \text{Let } m = -\frac{2}{3}, \; x_1 = 4, \text{ and } y_1 = 4.$$

$$3(y - 4) = -2(x - 4) \qquad \text{Multiply both sides by 3.}$$

$$3y - 12 = -2x + 8 \qquad \text{Apply the distributive property.}$$

$$2x + 3y = 20 \qquad \text{Write in standard form.}$$

> ▶ **Helpful Hint**
>
> Multiply both sides of the equation $2x + 3y = 20$ by -1 and it becomes $-2x - 3y = -20$. Both equations are in standard form, and their graphs are the same line.

EXAMPLE 10

Write a function that describes the line containing the point $(4, 4)$ and is perpendicular to the line $2x + 3y = -6$.

Solution

In the previous example, we found that the slope of the line $2x + 3y = -6$ is $-\frac{2}{3}$. A line perpendicular to this line will have a slope that is the negative reciprocal of $-\frac{2}{3}$, or $\frac{3}{2}$.

From the point-slope equation, we have

$$y - y_1 = m(x - x_1)$$

$$y - 4 = \frac{3}{2}(x - 4) \qquad \text{Let } x_1 = 4, y_1 = 4 \text{ and } m = \frac{3}{2}.$$

$$2(y - 4) = 3(x - 4) \qquad \text{Multiply both sides by } 2.$$

$$2y - 8 = 3x - 12 \qquad \text{Apply the distributive property.}$$

$$2y = 3x - 4 \qquad \text{Add } 8 \text{ to both sides.}$$

$$y = \frac{3}{2}x - 2 \qquad \text{Divide both sides by } 2.$$

$$f(x) = \frac{3}{2}x - 2 \qquad \text{Write using function notation.}$$

The graphs of perpendicular lines $2x + 3y = -6$ and $f(x) = \frac{3}{2}x - 2$.

Forms of Linear Equations

$Ax + By = C$	**Standard form** of a linear equation A and B are not both 0.
$y = mx + b$	**Slope–intercept form** of a linear equation The slope is m, and the y-intercept is $(0, b)$.
$y - y_1 = m(x - x_1)$	**Point–slope form** of a linear equation The slope is m, and (x_1, y_1) is a point on the line.
$y = c$	**Horizontal line** The slope is 0, and the y-intercept is $(0, c)$.
$x = c$	**Vertical line** The slope is undefined and the x-intercept is $(c, 0)$.

Parallel and Perpendicular Lines

Nonvertical parallel lines have the same slope. The product of the slopes of two nonvertical perpendicular lines is -1.

Spotlight on

DECISION
✕ MAKING

Suppose you are a public health official. In 1993, the International Task Force for Disease Eradication (ITFDE) identified mumps as one of six infectious diseases that could probably be eradicated worldwide with current technology. The ITFDE defined "eradication" as reducing the incidence of a disease to zero. Does the graph of reported mumps cases in the United States support the possibility of U.S. mumps eradication? Explain.

Suppose U.S. officials would like to see mumps eradicated by 2010. If this goal does not currently seem possible, your department will increase eradication efforts with the launch of a new public awareness campaign. Will the new public awareness campaign be necessary? (*Hint:* Use the data for the years 2000 and 2001 to help you decide.)

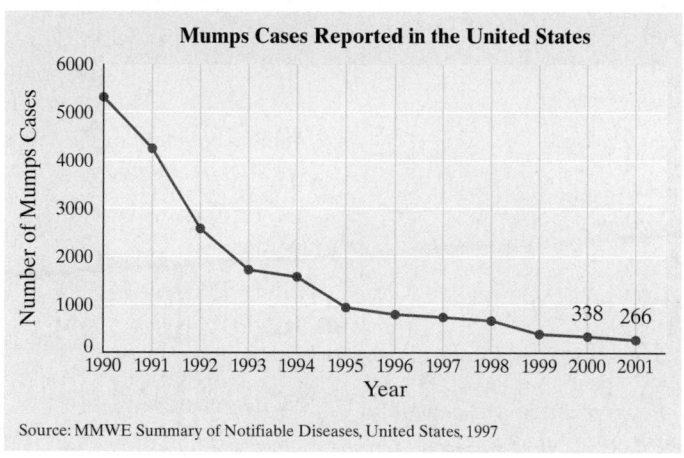

TEACHING TIP

A Group Activity for this section is available in the Instructor's Resource Manual.

MENTAL MATH

State the slope and the *y*-intercept of each line with the given equation.

1. $y = -4x + 12$ $m = -4, b = 12$

2. $y = \frac{2}{3}x - \frac{7}{2}$ $m = \frac{2}{3}, b = -\frac{7}{2}$

3. $y = 5x$ $m = 5, b = 0$

4. $y = -x$ $m = -1, b = 0$

5. $y = \frac{1}{2}x + 6$ $m = \frac{1}{2}, b = 6$

6. $y = -\frac{2}{3}x + 5$ $m = -\frac{2}{3}, b = 5$

Decide whether the lines are parallel, perpendicular, or neither.

7. $y = 12x + 6$
$y = 12x - 2$ parallel

8. $y = -5x + 8$
$y = -5x - 8$ parallel

9. $y = -9x + 3$
$y = \frac{3}{2}x - 7$ neither

10. $y = 2x - 12$
$y = \frac{1}{2}x - 6$ neither

EXERCISE SET 2.5

STUDY GUIDE/SSM CD/VIDEO PH MATH TUTOR CENTER MathXL®Tutorials ON CD MathXL® MyMathLab®

Use the slope–intercept form of the linear equation to write the equation of each line with the given slope and y-intercept. See Example 1.

 1. Slope -1; y-intercept $(0, 1)$ $y = -x + 1$

2. Slope $\frac{1}{2}$; y-intercept $(0, -6)$ $y = \frac{1}{2}x - 6$

3. Slope 2; y-intercept $\left(0, \frac{3}{4}\right)$ $y = 2x + \frac{3}{4}$

4. Slope -3; y-intercept $\left(0, -\frac{1}{5}\right)$ $y = -3x - \frac{1}{5}$

5. Slope $\frac{2}{7}$; y-intercept $(0, 0)$ $y = \frac{2}{7}x$

6. Slope $-\frac{4}{5}$; y-intercept $(0, 0)$ $y = -\frac{4}{5}x$

7.–12. See graphing answer section.

Graph each linear equation. See Examples 2 and 3.

7. $y = 5x$ **8.** $y = 2x + 12$

9. $x + y = 7$ **10.** $3x + y = 9$

11. $-3x + 2y = 3$ **12.** $-2x + 5y = -16$

15. $y = -2x - 1$ **16.** $y = -4x + 4$

Find an equation of the line with the given slope and containing the given point. Write the equation in slope–intercept form. See Example 4. **13.** $y = 3x - 1$ **14.** $y = 4x - 19$

 13. Slope 3; through $(1, 2)$ **14.** Slope 4; through $(5, 1)$

15. Slope -2; through $(1, -3)$ **16.** Slope -4; through $(2, -4)$

17. Slope $\frac{1}{2}$; through $(-6, 2)$ $y = \frac{1}{2}x + 5$

18. Slope $\frac{2}{3}$; through $(-9, 4)$ $y = \frac{2}{3}x + 10$

19. Slope $-\frac{9}{10}$; through $(-3, 0)$ $y = -\frac{9}{10}x - \frac{27}{10}$

20. Slope $-\frac{1}{5}$; through $(4, -6)$ $y = -\frac{1}{5}x - \frac{26}{5}$

Find an equation of each line graphed. Write the equation in standard form.

21.

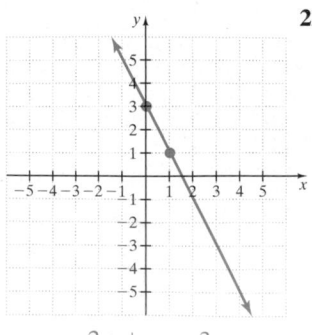

$2x + y = 3$

22.

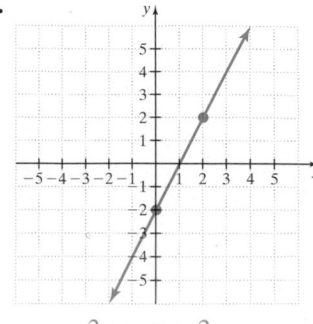

$2x - y = 2$

23.

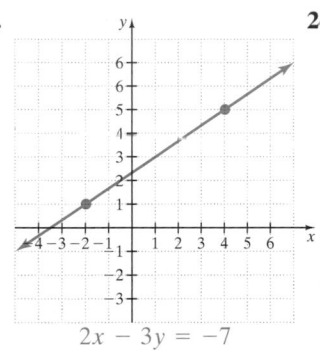

$2x - 3y = -7$

24.

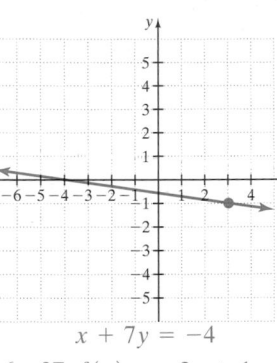

$x + 7y = -4$

25. $f(x) = 3x - 6$ **26.** $f(x) = 2x - 6$ **27.** $f(x) = -2x + 1$

28. $f(x) = -2x + 10$ **29.** $f(x) = -\frac{1}{2}x - 5$ **30.** $f(x) = 2x + 16$

Find an equation of the line passing through the given points. Use function notation to write the equation. See Example 5.

25. $(2, 0), (4, 6)$ **26.** $(3, 0), (7, 8)$

27. $(-2, 5), (-6, 13)$ **28.** $(7, -4), (2, 6)$

29. $(-2, -4), (-4, -3)$ **30.** $(-9, -2), (-3, 10)$

 31. $(-3, -8), (-6, -9)$

32. $(8, -3), (4, -8)$

33. Describe how to check to see if the graph of $2x - 4y = 7$ passes through the points $(1.4, -1.05)$ and $(0, -1.75)$. Then follow your directions and check these points.

31. $f(x) = \frac{1}{3}x - 7$ **32.** $f(x) = \frac{5}{4}x - 13$ **33.** answers may vary

Use the graph of the following function f(x) to find each value.

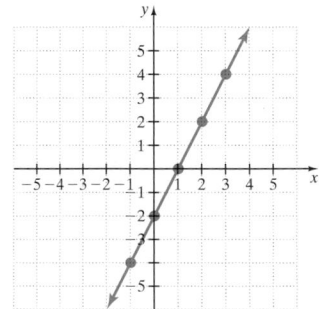

34. $f(1)$ 0 **35.** $f(0)$ -2

36. $f(-1)$ -4 **37.** $f(2)$ 2

38. Find x such that $f(x) = 4$. 3

39. Find x such that $f(x) = -6$. -2

Write an equation of each line. See Examples 7 and 8.

40. Vertical; through $(2, 6)$ $x = 2$

41. Slope 0; through $(-2, -4)$ $y = -4$

42. Horizontal; through $(-3, 1)$ $y = 1$

43. Vertical; through $(4, 7)$ $x = 4$

44. Undefined slope; through $(0, 5)$ $x = 0$

45. Horizontal; through $(0, 5)$ $y = 5$

△ **46.** Answer the following true or false. A vertical line is always perpendicular to a horizontal line. true

49. $f(x) = -3x + 1$ **50.** $f(x) = -\frac{3}{2}x + 2$

Find an equation of each line. Write the equation using function notation. Use a graphing utility to check. See Examples 9 and 10.

△ **47.** Through $(3, 8)$; parallel to $f(x) = 4x - 2$ $f(x) = 4x - 4$

△ **48.** Through $(1, 5)$; parallel to $f(x) = 3x - 4$ $f(x) = 3x + 2$

🔒 **49.** Through $(2, -5)$; perpendicular to $3y = x - 6$

△ **50.** Through $(-4, 8)$; perpendicular to $2x - 3y = 1$

△ **51.** Through $(-2, -3)$; parallel to $3x + 2y = 5$

△ **52.** Through $(-2, -3)$; perpendicular to $3x + 2y = 5$

51. $f(x) = -\frac{3}{2}x - 6$ **52.** $f(x) = \frac{2}{3}x - \frac{5}{3}$ **55.** $f(x) = -x + 7$

MIXED PRACTICE **58.** $f(x) = -4x + \frac{2}{9}$

Find the equation of each line. Write the equation in standard form unless indicated otherwise.

53. Slope 2; through $(-2, 3)$ $2x - y = -7$

54. Slope 3; through $(-4, 2)$ $3x - y = -14$

55. Through $(1, 6)$ and $(5, 2)$; use function notation.

56. Through $(2, 9)$ and $(8, 6)$ $x + 2y = 20$

57. With slope $-\frac{1}{2}$; y-intercept 11 $x + 2y = 22$

58. With slope -4; y-intercept $\frac{2}{9}$; use function notation.

59. Through $(-7, -4)$ and $(0, -6)$ $2x + 7y = -42$

60. Through $(2, -8)$ and $(-4, -3)$ $5x + 6y = -38$

61. Slope $-\frac{4}{3}$; through $(-5, 0)$ $4x + 3y = -20$

62. Slope $-\frac{3}{5}$; through $(4, -1)$ $3x + 5y = 7$

63. Vertical line; through $(-2, -10)$ $x = -2$

64. Horizontal line; through $(1, 0)$ $y = 0$

△ **65.** Through $(6, -2)$; parallel to the line $2x + 4y = 9$

△ **66.** Through $(8, -3)$; parallel to the line $6x + 2y = 5$

67. Slope 0; through $(-9, 12)$ $y = 12$

68. Undefined slope; through $(10, -8)$ $x = 10$

△ **69.** Through $(6, 1)$; parallel to the line $8x - y = 9$

△ **70.** Through $(3, 5)$; perpendicular to the line $2x - y = 8$

△ **71.** Through $(5, -6)$; perpendicular to $y = 9$ $x = 5$

△ **72.** Through $(-3, -5)$; parallel to $y = 9$ $y = -5$

73. Through $(2, -8)$ and $(-6, -5)$; use function notation.

74. Through $(-4, -2)$ and $(-6, 5)$; use function notation.

Solve. See Example 6.

75. Del Monte Fruit Company recently released a new applesauce. By the end of its first year, profits on this product amounted to \$30,000. The anticipated profit for the end of the fourth year is \$66,000. The ratio of change in time to change in profit is constant. Let x be years and P be profit.
   a. Write a linear function $P(x)$ that expresses profit as a function of time. $P(x) = 12{,}000x + 18{,}000$
   b. Use this function to predict the company's profit at the end of the seventh year. \$102,000
   c. Predict when the profit should reach \$126,000.

76. The value of a computer bought in 2000 depreciates, or decreases, as time passes. Two years after the computer was bought, it was worth \$2600; 4 years after it was bought, it was worth \$1000.
   a. If this relationship between number of years past 2000 and value of computer is linear, write an equation describing this relationship. [Use ordered pairs of the form (years past 2000, value of computer).] $y = -800x + 4200$
   b. Use this equation to estimate the value of the computer in the year 2005. \$200

77. The Pool Fun Company has learned that, by pricing a newly released Fun Noodle at \$3, sales will reach 10,000 Fun Noodles per day during the summer. Raising the price to \$5 will cause the sales to fall to 8000 Fun Noodles per day.
   a. Assume that the relationship between sales price and number of Fun Noodles sold is linear and write an equation describing this relationship. $y = -1000x + 13{,}000$
   b. Predict the daily sales of Fun Noodles if the price is \$3.50. 9500 Fun Noodles

78. The value of a building bought in 1990 appreciates, or increases, as time passes. Seven years after the building was bought, it was worth \$165,000; 12 years after it was bought, it was worth \$180,000.
   a. If this relationship between number of years past 1990 and value of building is linear, write an equation describing this relationship. [Use ordered pairs of the form (years past 1980, value of building).] $y = 3000x + 144{,}000$
   b. Use this equation to estimate the value of the building in the year 2010. \$204,000

79. In 2002, the median price of an existing home in the United States was approximately \$147,802. In 1999, the median price of an existing home was \$133,300. Let y be the median price of an existing home in the year x, where $x = 0$ represents 1999. (*Source:* National Association of REALTORS®)
   a. Write a linear equation that models the median existing home price in terms of the year x. [*Hint:* The line must pass through the points $(0, 133{,}300)$ and $(3, 147{,}802)$] $y = 4834x + 133{,}300$
   b. Use this equation to predict the median existing home price for the year 2008. \$176,806

---

65. $x + 2y = 2$ **66.** $3x + y = 21$ **69.** $8x - y = 47$ **70.** $x + 2y = 13$ **73.** $f(x) = -\frac{3}{8}x - \frac{29}{4}$ **74.** $f(x) = -\frac{7}{2}x - 16$
75.c. end of the the ninth yr

c. Interpret the slope of the equation found in part **a.**

80. The number of births (in thousands) in the United States in 2000 was 4060. The number of births (in thousands) in the United States in 1997 was 3895. Let y be the number of births (in thousands) in the year x, where $x = 0$ represents 1997. (*Source:* National Center for Health Statistics)

   a. Write a linear equation that models the number of births (in thousands) in terms of the year x. (See hint for Exercise 79a.) $y = 55x + 3895$

   b. Use this equation to predict the number of births in the United States for the year 2010.   4610 thousand births

   c. Interpret the slope of the equation in part a.

81. The number of people employed in the United States as medical assistants was 757 thousand in 2000. By the year 2010, this number is expected to rise to 1052 thousand. Let y be the number of medical assistants (in thousands) employed in the United States in the year x, where $x = 0$ represents 2000. (*Source:* Bureau of Labor Statistics)

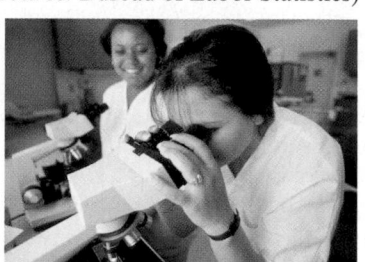

93. $-4x + y = 4$
94. $x - y = -5$
95. $2x + y = -23$
96. $x - 3y = -9$
97. $3x - 2y = -13$
98. $x - 5y = -20$

   a. Write a linear equation that models the number of people (in thousands) employed as medical assistants in the year x. (See hint for Exercise 79a.) $y = 29.5x + 757$

   b. Use this equation to estimate the number of people who will be employed as medical assistants in the year 2004.

82. The number of people employed in the United States as systems analysts was 431 thousand in 2000. By the year 2010, this number is expected to rise to 689 thousand. Let y be the number of systems analysts (in thousands) employed in the United States in the year x, where $x = 0$ represents 2000. (*Source:* Bureau of Labor Statistics)

   a. Write a linear equation that models the number of people (in thousands) employed as systems analysts in the year x. (See hint for Exercise 79a.) $y = 25.8x + 431$

   b. Use this equation to estimate the number of people who will be employed as systems analysts in the year 2008.

Use a graphing calculator with a TRACE feature to see the results of each exercise.   83.–86. See graphing answer section.

83. Exercise 55; graph the function and verify that it passes through $(1, 6)$ and $(5, 2)$.

84. Exercise 56; graph the equation and verify that it passes through $(2, 9)$ and $(8, 6)$.

85. Exercise 61; graph the equation. See that it has a negative slope and passes through $(-5, 0)$.

86. Exercise 62; graph the equation. See that it has a negative slope and passes through $(4, -1)$.

REVIEW AND PREVIEW

Solve. See Section 1.5.   **89.** $\dfrac{7}{2}$ **90.** -12

87. $2x - 7 = 21$ 14

88. $-3x + 1 = 0$ $\dfrac{1}{3}$

89. $5(x - 2) = 3(x - 1)$

90. $-2(x + 1) = -x + 10$

91. $\dfrac{x}{2} + \dfrac{1}{4} = \dfrac{1}{8}$ $-\dfrac{1}{4}$

92. $\dfrac{x}{5} - \dfrac{3}{10} = \dfrac{x}{2} - 1$ $\dfrac{7}{3}$

Concept Extensions

Example:

Find an equation of the perpendicular bisector of the line segment whose endpoints are $(2, 6)$ and $(0, -2)$.

Solution:

A perpendicular bisector is a line that contains the midpoint of the given segment and is perpendicular to the segment.

79.c. Every year, the median price of a home increases by $4834.

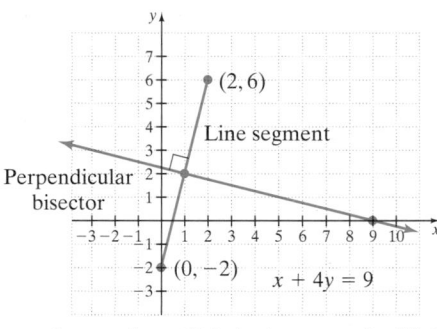

80.c. Every year the number of births increases by 55 thousand.
81.b. 875 thousand people   **82.b.** 637.4 thousand people

Step 1: The midpoint of the segment with endpoints $(2, 6)$ and $(0, -2)$ is $(1, 2)$.

Step 2: The slope of the segment containing points $(2, 6)$ and $(0, -2)$ is 4.

Step 3: A line perpendicular to this line segment will have slope of $-\dfrac{1}{4}$.

Step 4: The equation of the line through the midpoint $(1, 2)$ with a slope of $-\dfrac{1}{4}$ will be the equation of the perpendicular bisector. This equation in standard form is $x + 4y = 9$.

Find an equation of the perpendicular bisector of the line segment whose endpoints are given. See the previous example.

△ **93.** $(3, -1); (-5, 1)$ △ **94.** $(-6, -3); (-8, -1)$
△ **95.** $(-2, 6); (-22, -4)$ △ **96.** $(5, 8); (7, 2)$
△ **97.** $(2, 3); (-4, 7)$ △ **98.** $(-6, 8); (-4, -2)$

Answer true or false.

  99. A vertical line is always perpendicular to a horizontal line. true

  100. A vertical line is always parallel to a vertical line.   true

TEACHING TIP
After reviewing the box for Forms of Linear Equations ask, "How many pieces of information about a line must you know to determine the equation of the line?"

LINEAR EQUATIONS IN TWO VARIABLES

INTEGRATED REVIEW

Below is a review of equations of lines.

Forms of Linear Equations

$Ax + By = C$ — **Standard form** of a linear equation A and B are not both 0.

$y = mx + b$ — **Slope-intercept form** of a linear equation The slope is m, and the y-intercept is $(0, b)$.

$y - y_1 = m(x - x_1)$ — **Point-slope form** of a linear equation The slope is m, and (x_1, y_1) is a point on the line.

$y = c$ — **Horizontal line** The slope is 0, and the y-intercept is $(0, c)$.

$x = c$ — **Vertical line** The slope is undefined and the x-intercept is $(c, 0)$.

Parallel and Perpendicular Lines

Nonvertical parallel lines have the same slope. The product of the slopes of two nonvertical perpendicular lines is -1.

Graph each linear equation.   **1.–4.** See graphing answer section.

1. $y = -2x$ **2.** $3x - 2y = 6$ **3.** $x = -3$ **4.** $y = 1.5$

Find the slope of the line containing each pair of points.   **5.** 0

5. $(-2, -5), (3, -5)$ **6.** $(5, 2), (0, 5)$ $-\dfrac{3}{5}$

Find the slope and y-intercept of each line.

7. $y = 3x - 5$ $m = 3; (0, -5)$ **8.** $5x - 2y = 7$ $m = \dfrac{5}{2}; \left(0, -\dfrac{7}{2}\right)$

Determine whether each pair of lines is parallel, perpendicular, or neither.

9. $y = 8x - 6$
$y = 8x + 6$
parallel

10. $y = \dfrac{2}{3}x + 1$
$2y + 3x = 1$
perpendicular

Find the equation of each line. Write the equation in the form $x = a$, $y = b$, or $y = mx + b$. For Exercises 14 through 17, write using function notation.

11. Through $(1, 6)$ and $(5, 2)$
$y = -x + 7$

12. Vertical line; through $(-2, -10)$
$x = -2$

13. Horizontal line; through $(1, 0)$
$y = 0$

14. Through $(2, -8)$ and $(-6, -5)$
$f(x) = -\dfrac{3}{8}x - \dfrac{29}{4}$

15. Through $(-2, 4)$ with slope -5
$f(x) = -5x - 6$

16. Slope -4; y-intercept $\left(0, \dfrac{1}{3}\right)$
$f(x) = -4x + \dfrac{1}{3}$

17. Slope $\frac{1}{2}$; y-intercept $(0, -1)$

$f(x) = \frac{1}{2}x - 1$

18. Through $\left(\frac{1}{2}, 0\right)$ with slope 3

$y = 3x - \frac{3}{2}$

19. Through $(-1, -5)$; parallel to $3x - y = 5$

$y = 3x - 2$

20. Through $(0, 4)$; perpendicular to $4x - 5y = 10$

$y = -\frac{5}{4}x + 4$

21. Through $(2, -3)$; perpendicular to $4x + y = \frac{2}{3}$

$y = \frac{1}{4}x - \frac{7}{2}$

22. Through $(-1, 0)$; parallel to $5x + 2y = 2$

$y = -\frac{5}{2}x - \frac{5}{2}$

23. Undefined slope; through $(-1, 3)$

$x = -1$

24. $m = 0$; through $(-1, 3)$

$y = 3$

2.6 INTERPRETING DATA: LINEAR MODELS

Objective

1 Use a graphing utility to find a linear equation that models data.

The graph below shows the number of passengers who rode public transportation such as a train, bus, or subway during the years shown.

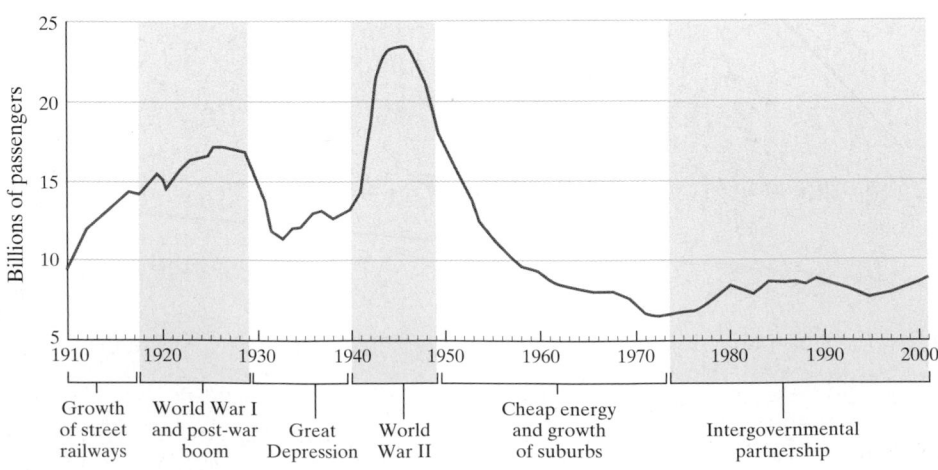

Source: American Public Transit Association

We see that different periods of history affect the number of passengers. For this complex graph there is not one model that best describes the number of passengers. Instead, we can take different periods of time and model each period to study it more closely. Once we get an equation (function) that best models a period of time, we can further study trends by evaluating the function within that period of time, or domain. It is important to remember that if a model is used to predict results outside its domain, this prediction is accurate only if the trend continues in the same manner. For example, if we had modeled the data from 1910–1925 and used it to predict the number of passengers for the next decade, we would have expected the number to continue to increase rapidly. Instead, World War I erupted and our prediction would have been very inaccurate.

The table below shows different periods of time for our graph above and possible types of models.

Period of Time	Approximate Shape of Graph	Type of Model (Equation)
1910–1925	Line	Linear
1929–1945	Upward parabola	Quadratic
1940–1950	Downward parabola	Quadratic

1   **Regression analysis** is the process of fitting a line or a curve to a set of data points. The equation of this line or curve can then be used to further study the graph itself or to predict something in the future. Your calculator has many regression equations to choose from, but in this section we will concentrate on **linear regression**. In other words, we will study data that resembles a line and model it with a linear equation that best fits the data. (In Section 8.7 we will study another regression equation.)

Below is public transit data from the years 1910–1929.

Year $(1910 = 0)$	Billions of Riders
0	9
2	12
6	14
8	13.8
10	15.5
11	15
13	16.2
15	16.8

First we will draw a scatter plot of the ordered pairs and examine it. A scatter plot is a quick way to visually see if there is a relationship, perhaps linear, between the two sets of data.

Plotting the Data

To plot the data, see the steps below.

1. Enter the data. To do so, use the EDIT menu found in the STAT feature. Enter years as L1 and billions of passengers as L2.

2. Use Stat Plot feature to indicate the type of graph as shown in the screens to the right.

3. Find an appropriate window. In this case $0 \leq x \leq 15$, so we choose $[-2, 20, 1]$ for the x-window; and $9 \leq y \leq 16.8$, so we choose $[8, 18, 1]$ for the y-window. You can choose various windows. Just make sure the domain and range are both included in your choice. Press GRAPH to see the individual points plotted.

Fitting a Line to the Data

Since this data resembles a line, let's model it with a linear equation. Your calculator will use a method beyond the scope of this course to find a line that best fits this data. Under the Stat Calc menu select LinReg $(ax + b)$. On the home screen indicate the x-list, y-list, and the Y= position where you want the regression equation to be stored. (See the second screen.)

On the previous page, notice in the bottom left screen that the calculator gives the slope, a, and the y-intercept b, for the linear equation of the form $y = ax + b$. It is up to us to insert these values in this form. Thus, the linear regression equation is $y = 0.4639816634x + 10.26764898$ where x is the number of years since 1910 and y is the number of passengers in the billions.

The r-value shown in the third screen is called the **coefficient correlation**. The values r and r^2 indicate how well the regression equation fits the data. If the diagnostic feature of your calculator is turned on, these values are shown. Check with your instructor to see whether you are to make use of this feature.

EXAMPLE 1

USING A LINEAR REGRESSION EQUATION

Use the equation $y = 0.4639816634x + 10.26764898$ from the above data and estimate the number of passengers in the year 1915.

Solution   Trace on the above graph of the regression equation to evaluate the function at $x = 5$. Another alternative method is to evaluate the function in Y1 for 5 as shown next.

 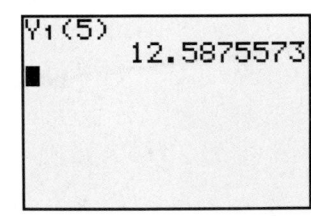

Using either method, the estimate of the number of passengers in 1915 is approximately 12.59 billion.

You may want to use the following steps to find a linear regression model.

Finding a Linear Regression Model

1. Enter the given data into lists.
2. Draw a scatter plot of the ordered pairs to see if the data appears linear.
3. Find the linear regression equation.
4. Graph the linear regression equation and use it to answer questions.

EXAMPLE 2

CIGARETTE USAGE DECLINE

The United States consumption of cigarettes (in billions of cigarettes) is given in the table below.

Years after 1980 (x)	0	5	10	15	20	23
Cigarettes in Billions (y)	632	615	525	487	461	449

Let x = the number of years since 1980.

a. Plot the data points.
b. Use linear regression to fit a line to the data.
c. Predict the cigarette consumption in the year 2012 if the trend continues.

Solution   **a.** Enter the data into L1 and L2. Find an appropriate window and graph as shown. We used the window $[0, 25, 5]$ by $[400, 700, 100]$.

b.

If a and b are rounded to 3 decimal places, the linear regression equation is $y = -8.646x + 633.360$.

c. Since this equation is stored in Y1 we evaluate the equation for the year 2012 as Y1 (32).

We predict the cigarette consumption in 2012 to be approximately 357 billion.

EXAMPLE 3

NUMBER OF INJURED WORKERS DECREASES

The number of cases of injuries and sickness suffered in the workplace continues to decrease according to the Bureau of Labor and Statistics.

Year	1997	1998	1999	2000	2001	2002
Injuries and Sickness per 100 workers	7.1	6.5	6.3	6.1	5.8	5.3

a. Plot the data points. (Let $x = 0$ represent the number of years since 1990.)

b. Find the regression equation that best fits the data. Then indicate whether the line increasing or decreasing?

c. What is the rate at which the number of sick or injured workers is increasing or decreasing?

d. If the trend continues at the same rate, predict the number of sick or injured workers out of 100 to expect in the year 2010.

Solution a. Enter the years in L1 with 1990 represented by $x = 0$ and the corresponding number of cases per 100 workers in L2. Find an appropriate window. We choose $[6, 13, 1]$ by $[5, 8, 1]$.

b.

If a and b are rounded to three decimal places, the linear regression equation is $y = -0.323x + 9.250$. Since the slope is negative (-0.323), the line is decreasing.

c.

The number of injured or sick workers is decreasing at the rate of 0.323 per 100 workers.

d. Since the equation is stored in Y1, we evaluate the equation for the year 2010 by finding Y1(20).

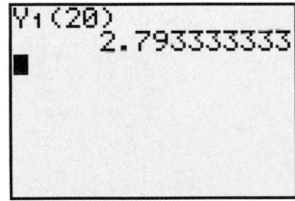

Y1(20) is equal to 2.793 or approximately 2.8 workers per 100 in the year 2010.

EXAMPLE 4

A LINEAR MODEL FOR PREDICTING SELLING PRICE

The following houses were sold in an Orange Park neighborhood in the past month. Listed below are each house's selling price and the number of square feet of living area in the house.

Total Square Feet	2150	2273	2474	2416
Selling Price in Dollars	120,000	134,500	158,900	149,800

a. Plot the data points.

b. Use linear regression to fit a line to the data.

c. Predict the approximate selling price for a house in the same neighborhood that has 2350 square feet of living area.

Solution　**a.** Enter the total number of square feet in a first list, L1, and the corresponding selling price in a second list, L2. Find an appropriate window and graph as shown.

b.

If a and b are rounded to 2 decimal places, the linear regression equation is $y = 117.35x - 132{,}428.80$.

c. Since this equation is stored in Y1, we evaluate the equation for 2350 square feet, or Y1(2350).

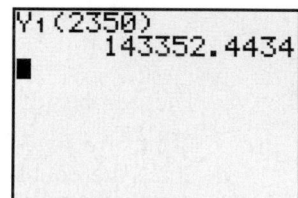

We predict the selling price for a house in the same neighborhood that has 2350 square feet of living area to be $143,352.

EXERCISE SET 2.6

STUDY   CD/   PH MATH   MathXL®Tutorials   MathXL®   MyMathLab®
GUIDE/SSM   VIDEO   TUTOR CENTER   ON CD

Solve. Predictions using linear regression equations may differ slightly depending on different rounding of the coefficients in the equation. See Examples 1 through 4.

1. Use the function from Section 2.2, Example 10, $f(x) = 1.558x - 3092$ to predict the amount of money that will be spent by the Pharmaceutical Manufacturers Association on research and development in 2009. (*Hint: x* corresponds to the year and $f(x)$ represents amount of money in billions of dollars. (See Example 1.)   38.022 billion

2. Use the function from Exercise 1 to predict the year that $34.9 billion was spent on research and development. (*Hint:* 0 corresponds to the year and $f(x)$ represents amount of money in billions of dollars. (See Example 1.)   2007

3. In Section 2.4, Example 5, the adult one-day pass price $f(x)$ for Disney World is given by $f(x) = 1.618x + 32.17$ where x is the number of years since 1990. Find what year the price was approximately $41.   1995

4. Using the function in Exercise 3, predict the price of a one-day pass to Disney World in 2008. $61.29

Solve. Round the values in each linear regression equation to three decimal places.

5. The number of outpatient visits has increased significantly from 1965 to 2002.   **5. c.** 11,575,000 visits per year
   a. Use linear regression to fit a line to the data in the table. $y = 11575.003x + 58440.426$

Years since 1960	5	15	25	35	40	42
No. of Outpatient Visits (in thousands)	125,793	254,844	282,140	483,195	528,134	551,687

   b. Predict the number of outpatient visits in the year 2010 if the trend continues to increase at the same rate. Round to the nearest thousand.   637,191,000 visits
   c. Find the rate at which the number of visits is increasing per year. Round to the nearest thousand.

6. The percent of male smokers 18 years and older has decreased from 1965 to 2002.   **a.** $y = -0.824x + 54.064$
   a. Use linear regression to fit a line to the data in the table.

Years since 1960	5	11	25	30	35	40	42
Percent of Male Smokers	51.9	43.1	32.6	28.4	27	21	19.5

   b. Predict the percent of male smokers in the year 2011 if the trend continues to decrease at the same rate. Round to the nearest tenth of a percent.   12.0%
   c. Predict the percent at which male smokers are decreasing per year.   0.824% per year

7. The percent of female smokers 18 years and older has decreased from 1965 to 2002.   **a.** $y = -0.411x + 36.836$
   a. Use linear regression to fit a line to the data in the table.

Years since 1960	5	14	25	30	35	40	42
Percent of Female Smokers	33.9	32.1	27.9	22.8	22.6	21	19.1

   b. Predict the percent of female smokers in the year 2011 if the trend continues to decrease at the same rate. Round to the nearest tenth of a percent.   15.9%
   c. Predict the percent at which female smokers are decreasing per year.   0.411% per year

8. The number of sentenced male prisoners in state and federal prisons has risen from 1980 to 2000.
   a. Using the data in the table, find the linear regression equation that best fits the data. $y = 47.84x + 262$

Years since 1980	0	5	10	15	20
Number of Male Prisoners (in thousands)	304	459	699	1021	1219

   b. Predict the number of male prisoners in the year 2009 if the trend continues to increase at the same rate. Round to the nearest thousand.   1649 thousand or 1,649,000 male prisoners
   c. Predict the rate at which male prisoners are increasing. Round to the nearest thousand.   48 thousand or 48,000 per year

9. The number of sentenced female prisoners in state and federal prisons has risen from 1980 to 2000.
   a. Using the data in the table, find the regression equation that best fits the data. $y = 3.82x + 6.6$

Years since 1980	0	5	10	15	20
Number of Female Prisoners (in thousands)	12	21	41	64	86

   b. Predict the number of female prisoners in the year 2009 if the trend continues to increase at the same rate. Round to the nearest thousand.
   c. Predict the rate at which female prisoners are increasing. Round to the nearest thousand.   4 thousand or 4000 per year   **b.** 117 thousand or 117,000 female prisoners

10. The number of high school girls softball teams since Title IX was enacted has grown rapidly. From 1972 to 1974, the number of teams rose from 373 to 6000. Since 1974, the growth of number of teams has been more steady. The table below shows the increase between 1974 and 2001.

   a. Plot the data in the table and find the linear regression equation that best fits the data. (Let $x = 0$ represent the number of years since 1970.) $y = 265.844x + 4548.305$

Year	1974	1984	1985	1990	1994	2001
No. of teams	6000	8700	8100	8200	12,000	13,001

   b. Predict the number of girls softball teams in 2008 if this trend continues. Round to the nearest whole.   14,650 teams
   c. Find the rate at which the number of teams is rising annually.   The number of teams is increasing at the rate of approximately 266 per year.

Given the following data for four houses sold in comparable neighborhoods and their corresponding number of square feet, draw a scatter plot and find a linear regression equation representing a relationship between the number of square feet and the selling price of the house.   **11.** $y = 114.453x - 81,378.675$

 11.

Square Feet	1519	2593	3005	3016
Selling Price	$91,238	$220,000	$254,000	$269,000

12.

Square Feet	1754	2151	2587	2671
Selling Price	$91,238	$130,000	$155,000	$172,500

13. Use the equation found in Exercise 11 to predict the selling price of a house with 2200 square feet. $170,418
14. Use the equation found in Exercise 12 to predict the selling price of a house with 2400 square feet. $146,161
   12. $y = 82.173x - 51,054.198$

A salesperson's salary depends on the amount of sales for the week. Given the following amounts of sales and the corresponding weekly salaries, plot the data points and find a linear regression equation that describes the salary in terms of the amount of sales.

15.

Sales (in dollars)	1000	3000	5000	8000
Salary (in dollars)	780	940	1100	1340

$y = 0.08x + 700$

16.

Sales (in dollars)	5000	10,000	25,000	30,000
Salary (in dollars)	550	800	1550	1800

$y = 0.05x + 300$

17. The revenue produced from DVD rentals has increased from 2001 through 2003.   **c.** $0.475 billion per year

Years since 2000	1	2	3
DVD Rental Revenue (in billions of dollars)	0.45	0.92	1.40

   a. Using the data in the table, find the linear regression equation that best fits the data. $y = 0.475x - 0.027$
   b. If this trend continues at the same rate, predict the DVD rental revenue in the year 2010. $4.723 billion
   c. Predict the rate at which the revenue is increasing.

18. The revenue produced from VCR/VHS rentals has decreased from 2001 through 2003.

Years since 2000	1	2	3
VCR/VHS Rental Revenue (in billions of dollars)	2.55	2.02	1.43

   a. Using the data in the table, find the linear regression equation that best fits the data. $y = -0.56x + 3.12$
   b. If this trend continues at the same rate, predict the VCR/VHS rental revenue in the year 2005, then the year 2006. Describe the meaning of your answers.   2005: $0.32 billion; 2006: −$0.24 billions; answers may vary

19. Life expectancy at birth of males was rising in the U.S. between 1900–2001 according to the table below.

Year	1900	1930	1950	1970	1990	1999	2001
Life Expectancy	46.3	57.5	65.6	67.1	71.8	73.9	74.4

   a. Using the data in the table, find the regression equation that best fits this data. (Let $x = 0$ represent the number of years since 1900.) $y = 0.262x + 48.740$
   b. What is the rate at which the life expectancy at birth of males is rising per year?   The life expectancy at birth of males is increasing at the rate of 0.262 years annually.
   c. Predict the life expectancy of a male born in the year 2010 if this trend continues.   77.6 years old

20. Life expectancy at birth of females was rising in the U.S. between 1900 and 2001 according to the table below.

Year	1900	1930	1950	1970	1990	1999	2001
Life Expectancy	48.3	59.8	71.1	74.7	78.8	79.4	79.8

  a. Using the data in the table, find the regression equation that best fits this data. (Let $x = 0$ represent the number of years since 1900.) $y = 0.305x + 51.073$

  b. What is the rate at which the life expectancy at birth of females is rising?   The life expectancy at birth of females is increasing at the rate of 0.305 years annually.

  c. Predict the life expectancy of a female born in the year 2010 if the trend continues.   84.6 years old

21. The number of people receiving care in PPOs has increased between the years of 1992 and 2001 according to the American Association of Health Plans.

Year	1992	1994	1996	2001
Number of PPO Members (in millions)	50.4	79.2	96.1	127.8

  a. Using the data in the table, find the regression equation that best fits the data. (Let $x = 0$ represent the number of years since 1990.) $y = 8.209x + 41.170$

  b. Predict the number of people receiving care in PPOs in the year 2010 if the trend continues to increase at the same rate.   205.36 million members

  c. If this trend continues, predict the year that there will be over 240 million members.   2015

  d. What does the slope of the linear regression equation represent?   The number of PPO members is increasing at the rate of 8.209 million per year.

22. The number of people receiving care in HMOs has increased between the years of 1988 and 2001 according to the American Association of Health Plans.

Year	1988	1990	1992	1994	1996	2001
Number of HMO Members (in millions)	32.7	36.5	41.4	51.1	61.8	73.5

  a. Using the data in the table, find the regression equation that best fits the data. (Let $x = 0$ represent the number of years since 1980.) $y = 3.364x + 4.090$

  b. Predict the number of people receiving care in HMOs in the year 2010 if the trend continues to increase at the same rate.   105.01 million members

  c. If this trend continues, predict the year that there will be over 90 million members.   2006

  d. What does the slope of the linear regression equation represent?   The number of HMO members is increasing at the rate of 3.364 million per year.

23. The average top ticket price for Broadway musicals has increased dramatically between 1975 and 2003.

Year	1975	1985	1998	2000	2003
Average Ticket Price	13.76	45.26	73.03	118.89	130.50

  a. Using the data in the table, find the regression equation that best fits the data. (Let $x = 0$ represent the number of years since 1970.) $y = 3.933x - 11.025$

  b. Predict the average top ticket price for Broadway musicals in the year 2013 if the trend continues to increase at the same rate. $158.09 per ticket

  c. Find the rate at which the cost is rising. $3.93 per year

24. The average top ticket price for Broadway plays has increased dramatically between 1975 and 2003.

Year	1975	1985	1998	2000	2003
Average Ticket Price	10.76	35.29	56.35	91.35	100.05

  a. Using the data in the table, find the regression equation that best fits the data. (Let $x = 0$ represent the number of years since 1970.) $y = 3.008x - 8.012$

  b. Predict the average top ticket price for Broadway plays in the year 2013 if the trend continues to increase at the same rate. $121.33 per ticket

  c. Find the rate at which the cost is rising. $3.01 per year

25. Academy Award Commercials. The Academy Awards is one of the most watched television shows. The table below gives the average cost of a 30-second commercial slot for the years 1998 through 2003.   **b.** $y = 0.097x + 0.163$

Years since 1990	8	9	10	11	12	13
Academy Award Commercial Cost (in millions of dollars)	0.9	1.0	1.2	1.35	1.25	1.4

  a. Plot the data points.   See graphing answer section.

  b. Use a linear regression to fit a line to the data.

  c. Find the rate at which the cost of a 30-second commercial on the Academy Awards presentation is increasing per year.   Increasing at the rate of $0.097 million per year

  d. If the rate continues to increase in the same manner, predict the cost of a 30-second commercial in the year 2010. $2,103,000

26. Super Bowl Commercials. The Super Bowl is the most widely watched event on television. The table below gives the average cost for a 30-second commercial for the years 1987 through 2003. **b.** $y = 0.082x - 0.660$ **c.** The cost is increasing at the rate of 0.082 million, or, $82,000 per year.

Years since 1970	7	9	17	21	25	29	30	31	33
Super Bawl Commercial Cost (in millions of dollars)	0.1	0.25	0.5	0.75	1.2	1.5	2.2	2.0	2.2

  a. Plot the data points.   See graphing answer section.

  b. Use a linear regression to fit a line to the data.

  c. Find the rate at which the cost of a 30-second commercial during the Super Bowl is increasing per year.

  d. If the rate continues to increase in the same manner, predict the cost of a 30-second commercial during the Super Bowl in the year 2009.   2.553 million or $2,553,000

REVIEW AND PREVIEW

Solve each equation algebraically. See Section 1.5.

27. $7x + 2 = 9x - 14$ 8

28. $5(x - 2) = 4(x + 7)$ 38

29. $\dfrac{y}{7} + \dfrac{y}{3} = \dfrac{1}{21}$ $\dfrac{1}{10}$

30. $y + 0.8 = 0.3(y - 2)$ -2

Concept Extensions

31. In the United States, eating at a restaurant, or "dining in," is decreasing while "take-out," that is picking up and eating restaurant food at a different location, is increasing. The data below represents the annual restaurant meals per person in each category. Write a linear regression equation for "dining in" and one for "take-out." Use these equations and their graphs to approximate the year that the two types of dining occurred the same number of times. (*Source: Wall Street Journal Almanac*)   1997 Let $x - 0$ represent the number of years since 1980.

Year	1984	1988	1990	1992	1997	2000	2003
Dining In	69	68	64	63	64	69	70
Take-Out	43	53	55	57	66	73	76

Source: Wall Street Journal Almanac

32. GROUP ACTIVITY.   Divide into groups of 5–6 students to see if a relationship exists between the length of a person's forearm (in cm) and their height (in cm).

  a. Complete a table recording the data of each student in the group by measuring the forearm length in cm in L1 and the height in cm in L2.

  b. Make a scatter plot of the data. What type of graph does the data seem to best fit?

  c. Find the regression equation that best fits the data.

  d. Interpret the slope and the y-intercept of the regression equation.   answers may vary

33. GROUP ACTIVITY.   Complete a similar table comparing the circumference of a student's neck to the circumference of his or her wrist.   answers may vary

CHAPTER 2 PROJECT

Modeling Real Data

The number of children who live with only one parent has been steadily increasing in the United States since the 1960s. According to the U.S. Bureau of the Census, the percent of children living with both parents is declining. The following table shows the percent of children (under age 18) living with *both* parents during selected years from 1980 to 2000. In this project, you will

have the opportunity to use the data in the table to find a linear function $f(x)$ that represents the data, reflecting the change in living arrangements for children. This project may be completed by working in groups or individually. For Questions 1 through 6, use paper and pencil. For Questions 7 and 8, use your graphing calculator.

PERCENT OF U.S. CHILDREN WHO LIVE WITH BOTH PARENTS

Year	1980	1985	1990	1995	2000
x	0	5	10	15	20
Percent, y	77	74	73	69	67

Source: U.S. Bureau of the Census

1. Using paper and pencil, plot the data given in the table as ordered pairs.

2. Use a straight edge to draw on your graph what appears to be the line that "best fits" the data you plotted.

3. Estimate the coordinates of two points that fall on your best-fitting line. Use these points to find a linear function $f(x)$ for the line.

4. What is the slope of your line? Interpret its meaning. Does it make sense in the context of this situation?

5. Find the value of $f(50)$. Write a sentence interpreting its meaning in context.

6. Compare your linear function with that of another student or group. Are they different? If so, explain why.

7. Enter the data from the table into a graphing calculator. Use the linear regression feature of the calculator to find a linear function for the data. Compare this function to the one you found in Question 3. How are they alike or different? Find the value of $f(50)$ using the model you found with the graphing calculator. Compare it to the value of $f(50)$ you found in Question 5.

8. If the trend continues, use your function from Question 7 to predict when the percent of U.S. children who live with both parents falls below 50%. In the year 2035

STUDY SKILLS REMINDER

Are You Preparing for a Test on Chapter 2?

Below I have listed some common trouble areas for students in Chapter 2. After studying for your test— but before taking your test—read these.

▶ Don't forget that the graph of an ordered pair is a *single* point in the rectangular coordinate plane.

▶ Remember that the slope of a horizontal line is 0 while a vertical line has undefined slope or no slope.

▶ For a linear equation such as $2y = 3x - 6$, the slope is not the coefficient of x unless the equation is solved for y. Solving this equation for y, we have $y = \frac{3}{2}x - 3$. The slope is $\frac{3}{2}$ and the y-intercept is $(0, -3)$.

▶ Parallel lines have the same slope while perpendicular lines have negative reciprocal slopes.

Slope	Parallel line	Perpendicular line
$m = 6$	$m = 6$	$m = -\dfrac{1}{6}$
$m = -\dfrac{2}{3}$	$m = -\dfrac{2}{3}$	$m = \dfrac{3}{2}$

▶ Don't forget that the statement $f(2) = 3$ corresponds to the ordered pair $(2, 3)$.

Remember: this is simply a checklist of common trouble areas. For a review of Chapter 2, see the Highlights and Chapter Review at the end of this chapter.

2 CHAPTER VOCABULARY CHECK

Fill in each blank with one of the words or phrases listed below.

relation	standard	slope–intercept	range	point–slope
line	slope	x	parallel	perpendicular
function	domain	y	linear function	

1. A <u>relation</u> is a set of ordered pairs.
2. The graph of every linear equation in two variables is a <u>line</u>.
3. The equation $y - 8 = -5(x + 1)$ is written in <u>point-slope</u> form.
4. <u>Standard</u> form of linear equation in two variables is $Ax + By = C$.
5. The <u>range</u> of a relation is the set of all second components of the ordered pairs of the relation.
6. <u>Parallel</u> lines have the same slope and different y-intercepts.
7. <u>Slope-intercept</u> form of a linear equation in two variables is $y = mx + b$.
8. A <u>function</u> is a relation in which each first component in the ordered pairs corresponds to exactly one second component.
9. In the equation $y = 4x - 2$, the coefficient of x is the <u>slope</u> of its corresponding graph.
10. Two lines are <u>perpendicular</u> if the product of their slopes is -1.
11. To find the x-intercept of a linear equation, let <u>y</u> $= 0$ and solve for the other variable.
12. The <u>domain</u> of a relation is the set of all first components of the ordered pairs of the relation.
13. A <u>linear function</u> is a function that can be written in the form $f(x) = mx + b$.
14. To find the y-intercept of a linear equation, let <u>x</u> $= 0$ and solve for the other variable.

CHAPTER 2 HIGHLIGHTS

Definitions and Concepts	**Examples**
Section 2.1 Introduction to Graphing	

The **rectangular coordinate system,** or **Cartesian coordinate system,** consists of a vertical and a horizontal number line intersecting at their 0 coordinate. The vertical number line is called the **y-axis**, and the horizontal number line is called the **x-axis**. The point of intersection of the axes is called the **origin**. The axes divide the plane into four regions called **quadrants**.

(continued)

Definitions and Concepts	**Examples**

Section 2.1 Introduction to Graphing

To **plot** or **graph** an ordered pair means to find its corresponding point on a rectangular coordinate system.

To plot or graph the ordered pair $(-2, 5)$, start at the origin. Move 2 units to the left along the x-axis, then 5 units upward parallel to the y-axis.

A graphing utility can be used to display a portion of the rectangular coordinate system. The portion being viewed is called a **viewing window** or simply **window**.

Ordered pair $(-2, 5)$ plotted

To graph an equation using a graphing utility, solve the equation for y and enter it into the $Y=$ editor.

Graph $y = x^3 + 2$.

An ordered pair is a **solution** of an equation in two variables if replacing the variables by the corresponding coordinates results in a true statement.

Determine whether $(-2, 3)$ is a solution of

$$3x + 2y = 0$$
$$3(-2) + 2(3) = 0$$
$$-6 + 6 = 0$$
$$0 = 0 \quad \text{True}$$

$(-2, 3)$ is a solution.

A **linear equation in two variables** is an equation that can be written in the form $Ax + By = C$, where A, B, and C are real numbers and A and B are not both 0. The form $Ax + By = C$ is called **standard form**.

Linear Equations in Two Variables

$$y = -2x + 5, \quad x = 7$$
$$y - 3 = 0, \quad 6x - 4y = 10$$

$6x - 4y = 10$ is in standard form.

(continued)

Definitions and Concepts	Examples

The graph of a linear equation in two variables is a line. To graph a linear equation in two variables, find three ordered pair solutions. Plot the solution points, and draw the line connecting the points.

Graph $3x + y = -6$.

x	y
0	−6
−2	0
−3	3

To graph an equation that is not linear, find a sufficient number of ordered pair solutions so that a pattern may be discovered.

Graph $y = x^3 + 2$.

x	y
−2	−6
−1	1
0	2
1	3
2	10

A **relation** is a set of ordered pairs. The **domain** of the relation is the set of all first components of the ordered pairs. The **range** of the relation is the set of all second components of the ordered pairs.

Relation

Input: Words
Output: Number of Vowels

Domain: {cat, dog, too, give}
Range: {1, 2}

A **function** is a relation in which each element of the first set corresponds to exactly one element of the second set.

The previous relation is a function. Each word contains exactly one number of vowels.

(continued)

Definitions and Concepts	**Examples**

Vertical Line Test

If no vertical line can be drawn so that it intersects a graph more than once, the graph is the graph of a function.

Find the domain and the range of the relation. Also determine whether the relation is a function.

By the vertical line test, this graph is the graph of a function.

The symbol $f(x)$ means **function of x** and is called **function notation**.

If $f(x) = 2x^2 - 5$, find $f(-3)$.

$$f(-3) = 2(-3)^2 - 5 = 2(9) - 5 = 13$$

A **linear function** is a function that can be written in the form $f(x) = mx + b$.

To graph a linear function, find three ordered pair solutions. Graph the solutions and draw a line through the plotted points.

Linear Functions

$$f(x) = -3, g(x) = 5x, h(x) = -\frac{1}{3}x - 7$$

Graph $f(x) = -2x$.

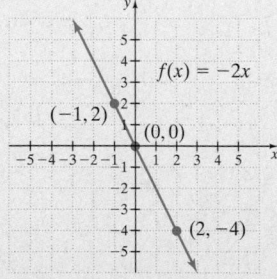

x	y or $f(x)$
-1	2
0	0
2	-4

The graph of $y = mx + b$ is the same as the graph of $y = mx$, but shifted b units up if b is positive and b units down if b is negative.

Graph $g(x) = -2x + 3$.

This is the same as the graph of $f(x) = -2x$ shifted 3 units up.

(continued)

Definitions and Concepts	**Examples**

Section 2.3 Graphing Linear Functions

The x-coordinate of a point where a graph crosses the x-axis is called an **x-intercept**. The y-coordinate of a point where a graph crosses the y-axis is called a **y-intercept**.

The x-intercepts of the graph are -2 and 2.
The y-intercept is -2.

To find an x-intercept, let $y = 0$ or $f(x) = 0$ and solve for x.

To find a y-intercept, let $x = 0$ and solve for y.

Graph $5x - y = -5$ by finding intercepts.

$$
\begin{array}{ll}
\text{If } x = 0, \text{then} & \text{If } y = 0, \text{then} \\
5x - y = -5 & 5x - y = -5 \\
5 \cdot 0 - y = -5 & 5x - 0 = -5 \\
-y = -5 & 5x = -5 \\
y = 5 & x = -1 \\
(0, 5) & (-1, 0)
\end{array}
$$

Ordered pairs are $(0, 5)$ and $(-1, 0)$.

The graph of $x = c$ is a vertical line with x-intercept $(c, 0)$.

The graph of $y = c$ is a horizontal line with y-intercept $(0, c)$.

(continued)

Definitions and Concepts	Examples

The **slope** m of the line through (x_1, y_1) and (x_2, y_2) is given by

$$m = \frac{y_2 - y_1}{x_2 - x_1} \text{ as long } x_2 \neq x_1$$

Find the slope of the line through $(-1, 7)$ and $(-2, -3)$.

$$m = \frac{y_2 - y_1}{x_2 - x_1} = \frac{-3 - 7}{-2 - (-1)} = \frac{-10}{-1} = 10$$

The **slope-intercept form** of a linear equation is $y = mx + b$, where m is the slope of the line and b is the y-intercept.

Find the slope and y-intercept of $-3x + 2y = -8$.

$$2y = 3x - 8$$
$$\frac{2y}{2} = \frac{3x}{2} - \frac{8}{2}$$
$$y = \frac{3}{2}x - 4$$

The slope the line is $\frac{3}{2}$, and the y-intercept is $(0, -4)$.

Nonvertical parallel lines have the same slope.

If the product of the slopes of two lines is -1, then the lines are perpendicular.

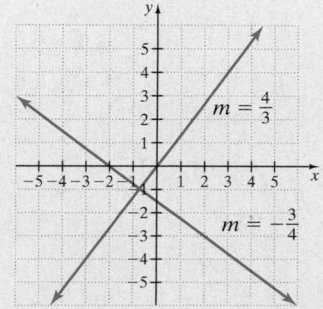

The slope of a horizontal line is 0.

The slope of a vertical line is undefined.

The slope of $y = -2$ is 0.

The slope of $x = 5$ is undefined.

(continued)

Definitions and Concepts	Examples
Section 2.5 Equations of Lines	

We can use the slope–intercept form to write an equation of a line given its slope and y-intercept.

Write an equation of the line with y-intercept $(0, -1)$ and slope $\frac{2}{3}$.

$$y = mx + b$$
$$y = \frac{2}{3}x - 1$$

The point–slope form of the equation of a line is $y - y_1 = m(x - x_1)$, where m is the slope of the line and (x_1, y_1) is a point on the line.

Find an equation of the line with slope 2 containing the point $(1, -4)$. Write the equation in standard form: $Ax + By = C$.

$$y - y_1 = m(x - x_1)$$
$$y - (-4) = 2(x - 1)$$
$$y + 4 = 2x - 2$$
$$-2x + y = -6 \qquad \text{Standard form.}$$

Section 2.6 Interpreting Data: Linear Models

Regression analysis is the process of fitting a line or a curve to a set of data points.

CHAPTER REVIEW

(2.1) *Plot the points and name the quadrant or axis in which each point lies.* **1.–2.** See graphing answer section.

1. $A(2, -1), B(-2, 1), C(0, 3), D(-3, -5)$
2. $A(-3, 4), B(4, -3), C(-2, 0), D(-4, 1)$

Determine whether each ordered pair is a solution to the given equation.

3. $7x - 8y = 56; (0, 56), (8, 0)$ no, yes
4. $-2x + 5y = 10; (-5, 0), (1, 1)$ yes, no

5. $x = 13; (13, 5), (13, 13)$ yes, yes
6. $y = 2; (7, 2), (2, 7)$ yes, no

Determine whether each equation is linear or not. Then graph the equation already written below. **7.–18.** See graphing answer section.

7. $y = 3x$ linear
8. $y = 5x$ linear
9. $3x - y = 4$ linear
10. $x - 3y = 2$ linear
11. $y = |x| + 4$ nonlinear
12. $y = x^2 + 4$ nonlinear
13. $y = -\frac{1}{2}x + 2$ linear
14. $y = -x + 5$ linear

Match each graph with its equation.

15.

D

16.

A

17.

C

18.

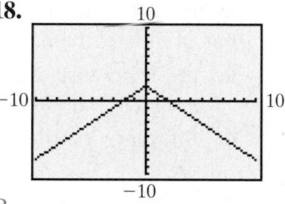

B

a. $y = x^2 - 4$ **b.** $y = -|x| + 2$

c. $y = |x| + 2$ **d.** $y = x^2 + 2$

(2.2) Find the domain and range of each relation. Also determine whether the relation is a function.

19. $\left\{\left(-\dfrac{1}{2}, \dfrac{3}{4}\right), (6, 0.75), (0, -12), (25, 25)\right\}$

20. $\left\{\left(\dfrac{3}{4}, -\dfrac{1}{2}\right), (0.75, 6), (-12, 0), (25, 25)\right\}$

21.

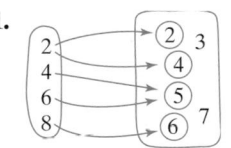

domain: $\{2, 4, 6, 8\}$; range: $\{2, 4, 5, 6\}$; not a function

22.

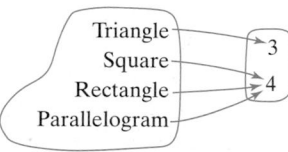

domain:{triangle, square, rectangle, parallelogram}; range: $\{3, 4\}$; function

23.

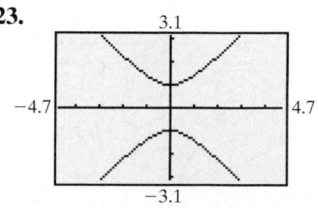

domain: $(-\infty, \infty)$; range: $(-\infty, -1] \cup [1, \infty)$; not a function

24.

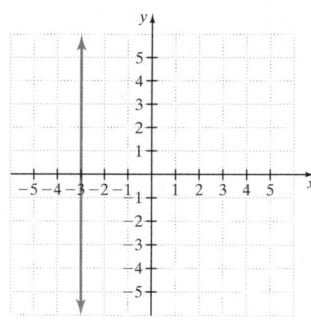

domain: $\{-3\}$; range: $(-\infty, \infty)$; not a function

25.

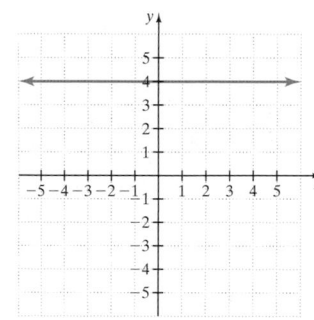

domain: $(-\infty, \infty)$; range: $\{4\}$; function

26.

domain: $[-1, 1]$; range: $[-1, 1]$; not a function

If $f(x) = x - 5$, $g(x) = -3x$, and $h(x) = 2x^2 - 6x + 1$, find the following.

27. $f(2)$ -3 **28.** $g(0)$ 0

29. $g(-6)$ 18 **30.** $h(-1)$ 9

31. $h(1)$ -3 **32.** $f(5)$ 0

The function $J(x) = 2.54x$ may be used to calculate the weight of an object on Jupiter J given its weight on Earth x.

33. If a person weighs 150 pounds on Earth, find the equivalent weight on Jupiter.   381 lb

34. A 2000-pound probe on Earth weighs how many pounds on Jupiter?   5080 lb

Use the graph of the function below to answer Exercises 35 through 38.

35. Find $f(-1)$. 0 **36.** Find $f(1)$. -2

37. Find all values of x such that $f(x) = 1$. $-2, 4$

19. domain: $\left\{-\dfrac{1}{2}, 6, 0, 25\right\}$; range: $\left\{\dfrac{3}{4}, \text{ or } 0.75, -12, 25\right\}$; function

20. domain: $\left\{\dfrac{3}{4}, \text{ or } 0.75, -12, 25\right\}$; range: $\left\{-\dfrac{1}{2}, 6, 0, 25\right\}$; not a function

38. Find all values of x such that $f(x) = -1$. 0, 2

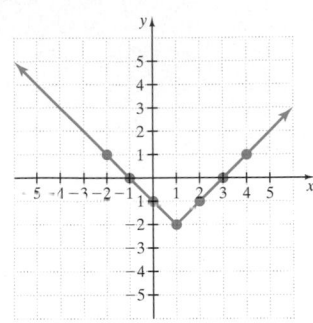

(2.3) *Graph each linear function.*

39. $f(x) = x$

40. $f(x) = -\dfrac{1}{3}x$

41. $g(x) = 4x - 1$ **39.–41.** See graphing answer section.

The graph of $f(x) = 3x$ is sketched below. Use this graph to match each linear function with its graph.

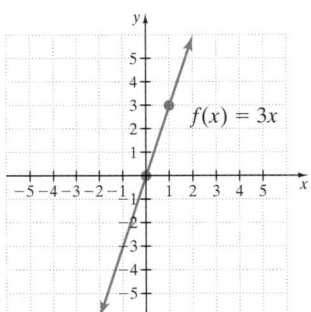

A	B
	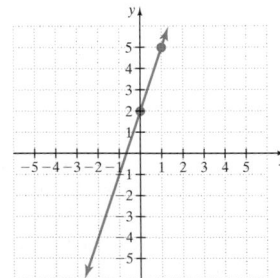

C	D
	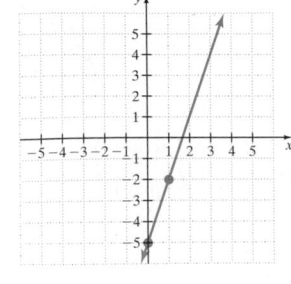

42. $f(x) = 3x + 1$ C

43. $f(x) = 3x - 2$ A

44. $f(x) = 3x + 2$ B

45. $f(x) = 3x - 5$ D

Graph each linear equation by finding intercepts if possible.
46.–51. See graphing answer section.

46. $4x + 5y = 20$

47. $3x - 2y = -9$

48. $4x - y = 3$

49. $2x + 6y = 9$

50. $y = 5$

51. $x = -2$

Graph each linear equation. **52.–54.b.** See graphing answer section.

52. $x - 2 = 0$

53. $y + 3 = 0$

54. The cost C, in dollars, of renting a minivan for a day is given by the linear function $C(x) = 0.3x + 42$, where x is number of miles driven.

  a. Find the cost of renting the minivan for a day and driving it 150 miles. $87

  b. Graph $C(x) = 0.3x + 42$.

(2.4) *Find the slope of the line through each pair of points.*

55. $(2, 8)$ and $(6, -4)$ -3

56. $(-3, 9)$ and $(5, 13)$ $\frac{1}{2}$

57. $(-7, -4)$ and $(-3, 6)$ $\frac{5}{2}$

58. $(7, -2)$ and $(-5, 7)$ $-\frac{3}{4}$

Find the slope and y-intercept of each line.

59. $6x - 15y = 20$
  $m = \frac{2}{5}, b = -\frac{4}{3}$

60. $4x + 14y = 21$
  $m = -\frac{2}{7}, b = \frac{3}{2}$

Find the slope of each line.

61. $y - 3 = 0$ 0

62. $x = -5$ undefined

Two lines are graphed on each set of axes. Decide whether l_1 or l_2 has the greater slope.

63. l_2

64. l_2

65. l_2

66. l_1

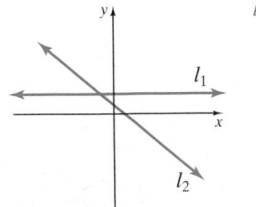

67. Recall from Exercise 54, that the cost C, in dollars, of renting a minivan for a day is given by the linear equation $y = 0.3x + 42$, where x is number of miles driven.

  a. Find and interpret the slope of this equation.

  b. Find and interpret the y-intercept of this equation.

Decide whether the lines are parallel, perpendicular, or neither.

△ **68.** $f(x) = -2x + 6$ neither
   $g(x) = 2x - 1$

△ **69.** $-x + 3y = 2$ parallel
   $6x - 18y = 3$

(2.5) Graph each linear equation using the slope and y-intercept.   **70.–73.** See graphing answer section.

70. $y = -x + 1$

71. $y = 4x - 3$

72. $3x - y = 6$

73. $y = -5x$

Find an equation of the line satisfying the conditions given.

74.   Horizontal, through $(3, -1)$
   $y = -1$

75.   Vertical, through $(-2, 1)$
   $x = -2$

△ **76.** Parallel to the line $x = 6$; through $(-4, -3)$ $x = -4$

77. Slope 0; through $(2, 5)$ $y = 5$

Find the standard form equation of each line satisfying the conditions given.

78. Through $(-3, 5)$; slope 3 $3x - y = -14$

79. Slope 2; through $(5, -2)$ $2x - y = 12$

80. Through $(-6, -1)$ and $(-4, -2)$ $x + 2y = -8$

81. Through $(-5, 3)$ and $(-4, -8)$ $11x + y = -52$

△ **82.** Through $(-2, 3)$; perpendicular to $x = 4$ $y = 3$

△ **83.** Through $(-2, -5)$; parallel to $y = 8$ $y = -5$

Find the equation of each line satisfying the given conditions. Write each equation using function notation.

84. Slope $-\dfrac{2}{3}$; y-intercept $(0, 4)$ $f(x) = -\dfrac{2}{3}x + 4$

85. Slope -1; y-intercept $(0, -2)$ $f(x) = -x - 2$

△ **86.** Through $(2, -6)$; parallel to $6x + 3y = 5$

△ **87.** Through $(-4, -2)$; parallel to $3x + 2y = 8$

△ **88.** Through $(-6, -1)$; perpendicular to $4x + 3y = 5$

△ **89.** Through $(-4, 5)$; perpendicular to $2x - 3y = 6$

67.a. $m = 0.3$; The cost increases by \$0.30 for each additional mile driven.   **67.b.** $b = 42$; The cost for 0 miles driven is \$42.

86. $f(x) = -2x - 2$ **87.** $f(x) = -\dfrac{3}{2}x - 8$ **88.** $f(x) = \dfrac{3}{4}x + \dfrac{7}{2}$ **89.** $f(x) = -\dfrac{3}{2}x - 1$ **90.a.** $y = \dfrac{4}{3}x + 65$

90. In 2002, the percent of U.S. drivers wearing seat belts was 81%. The number of drivers wearing seat belts in 1990 was 65%. Let y be the number of drivers wearing seat belts in the year x, where $x = 0$ represents 1990. (*Source:* Strategis Group for Personal Communications Asso.)

a. Write a linear equation that models the percent of U.S. drivers wearing seat belts in terms of the year x. [*Hint:* Write 2 ordered pairs of the form (years past 1990, percent of drivers).]

b. Use this equation to predict the number of U.S. drivers wearing seat belts in the year 2009. (Round to the nearest percent.)   90%

91. In 1998, the number of people (in millions) reporting arthritis was 43. The number of people (in millions) predicted to be reporting arthritis in 2020 is 60. Let y be the number of people (in millions) reporting arthritis in the year x, where $x = 0$ represents 1998. (*Source:* Arthritis Foundation)

a. Write a linear equation that models the number of people (in millions) reporting arthritis in terms of the year x (See the hint for Exercise 90.) $y = \dfrac{17}{22}x + 43$

b. Use this equation to predict the number of people reporting arthritis in 2010. (Round to the nearest million.)   52 million

(2.6)

92. According to the U.S. Bureau of the Census, the U.S. population in thousands is as given in the table below.

Year	1970	1980	1990	2000
Population (in Millions)	203, 302	226, 548	248, 710	275, 306

(Let x represent the number of years since 1970.)

a. Plot the data using the statistical plotting feature of your graphing utility.   See graphing answer section.

b. Use the data for the years given to find a linear regression equation that models the data. Graph this line with the data. $y = 2381.74x + 202,740.4$

c. Use this equation to predict the population of the United States in the year 2010.   298.010 thousand

CHAPTER 2 TEST

Remember to use your Chapter Test Prep Video CD to help you study and view solutions to the test questions you need help with.

1., 3–6. See graphing answer section.   1. A: quadrant IV; B: no quadrant; C: quadrant II

1. Plot the points, and name the quadrant in which each is located: $A(6, -2)$, $B(4, 0)$, $C(-1, 6)$.

2. Complete the ordered pair solution $(-6, \;\;)$ of the equation $2y - 3x = 12$. $(-6, -3)$

Graph each line.

3. $2x - 3y = -6$

4. $4x + 6y = 7$

5. $f(x) = \dfrac{2}{3}x$

6. $y = -3$

7. Find the slope of the line that passes through $(5, -8)$ and $(-7, 10)$. $-\dfrac{3}{2}$

8. Find the slope and the y-intercept of the line $3x + 12y = 8$. $m = -\dfrac{1}{4}, b = \dfrac{2}{3}$

Find an equation of each line satisfying the conditions given. Write Exercises 9 through 13 in standard form. Write Exercises 14 through 16 using function notation.

9. Horizontal; through $(2, -8)$ $y = -8$

10. Vertical; through $(-4, -3)$ $x = -4$

△ 11. Perpendicular to $x = 5$; through $(3, -2)$ $y = -2$

12. Through $(4, -1)$; slope -3 $3x + y = 11$

13. Through $(0, -2)$; slope 5 $5x - y = 2$

14. Through $(4, -2)$ and $(6, -3)$ $f(x) = -\dfrac{1}{2}x$

△ 15. Through $(-1, 2)$; perpendicular to $3x - y = 4$

△ 16. Parallel to $2y + x = 3$; through $(3, -2)$

△ 17. Line L_1 has the equation $2x - 5y = 8$. Line L_2 passes through the points $(1, 4)$ and $(-1, -1)$. Determine whether these lines are parallel lines, perpendicular lines, of neither.   neither

15. $f(x) = -\dfrac{1}{3}x + \dfrac{5}{3}$ 16. $f(x) = -\dfrac{1}{2}x - \dfrac{1}{2}$

Match each graph with its equation.

18.
B

19.
A

20.
D

21.
C

A. $y = 2|x - 1| + 3$
B. $y = x^2 + 2x + 3$
C. $y = 2(x - 1)^3 + 3$
D. $y = 2x + 3$

Find the domain and range of each relation. Also determine whether the relation is a function.

22.

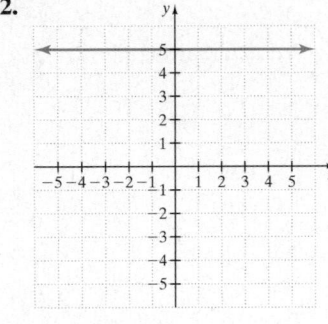

domain: $(-\infty, \infty)$;
range: $\{5\}$; function

23.

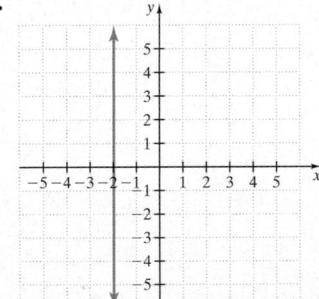

domain: $\{-2\}$;
range: $(-\infty, \infty)$;
not a function

24.

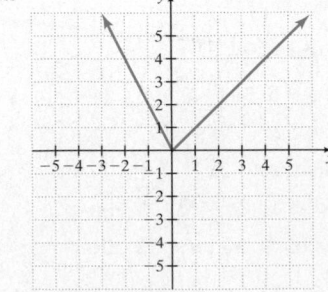

domain: $(-\infty, \infty)$;
range: $[0, \infty)$; function

25.

domain: $(-\infty, \infty)$;
range: $(-\infty, \infty)$;
function

26. The average yearly earnings for high school graduates age 18 and older is given by the linear function

$$f(x) = 732x + 21{,}428$$

where x is the number of years since 1996 that a person graduated. (*Source:* U.S. Census Bureau)

a. Find the average earnings in 1998 for high school graduates. $22,892

b. Predict the average earnings for high school graduates in the year 2005. $28,016   **c.** 2008

c. Predict the first year that the average earnings for high school graduates will be greater than $30,000.

d. Find and interpret the slope of this equation.

e. Find and interpret the y-intercept of this equation.
The average yearly earnings for a high school graduate in 1996 was $21,428.

27. According to the U.S. Bureau of the Census, Statistical Abstract of the United States, 1994, the population of the state of Florida in thousands is as given in the table below.

Year	1970	1980	1990	2000
Population (in thousands)	6791	9746	12,938	15,233

(Let x represent the number of years since 1970.)

a. Plot the data using the statistical plotting feature of your graphing utility.   See graphing answer section.

b. Use the data given to find a linear regression equation that models the data. Graph this line with the data.
b. $y = 285.18x + 6899.3$

c. Use this equation to predict the population of Florida in the year 2010.   18,306.5 thousand

26.d. The average yearly earnings for high school graduates increases $732 per year.

2 CHAPTER CUMULATIVE REVIEW

1. Evaluate: $3x - y$ when $x = 15$ and $y = 4$.   41 (Sec. 1.2, Ex. 2)

2. Add.

a. $-4 + (-3)$ -7

b. $\dfrac{1}{2} - \left(-\dfrac{1}{3}\right)$ $\dfrac{5}{6}$

c. $7 - 20$ -13 (Sec. 1.2)

3. Determine whether the following statements are true or false.

a. 3 is a real number.   true

b. $\dfrac{1}{5}$ is an irrational number.   false

c. Every rational number is an integer.   false

d. $\{1, 5\} \subseteq \{2, 3, 4, 5\}$ false (Sec. 1.2, Ex. 5)

4. Write the opposite of each.

a. -7 7

b. 0   0

c. $\dfrac{1}{4}$ $-\dfrac{1}{4}$ (Sec. 1.2)

5. Subtract.

a. $2 - 8$ -6
b. $-8 - (-1)$ -7

c. $-11 - 5$ -16
d. $10.7 - (-9.8)$ 20.5

e. $\dfrac{2}{3} - \dfrac{1}{2}$ $\dfrac{1}{6}$
f. $1 - 0.06$ 0.94

g. Subtract 7 from 4. -3 (Sec. 1.3, Ex. 2)

6. Multiply or divide.

a. $\dfrac{-42}{-6}$ 7
b. $\dfrac{0}{14}$ 0

c. $-1(-5)(-2)$ -10 (Sec. 1.3)

7. Simplify each expression.

a. 3^2 9
b. $\left(\dfrac{1}{2}\right)^4$ $\dfrac{1}{16}$

c. -5^2 -25
d. $(-5)^2$ 25

e. -5^3 -125

f. $(-5)^3$ -125 (Sec. 1.3, Ex. 6)

8. Which property is illustrated?

a. $5(x + 7) = 5 \cdot x + 5 \cdot 7$ distributive property

b. $5(x + 7) = 5(7 + x)$ commutative property for addition (Sec. 1.3)

9. Insert $<$, $>$, or $=$ between each pair of numbers to form a true statement.

   a. $-1 \quad -2 \quad >$

   b. $\dfrac{12}{4} \quad 3 \quad =$

   c. $-5 \quad 0 \quad <$

   d. $-3.5 \quad -3.05 \quad <$
   (Sec. 1.4, Ex. 2)

10. Evaluate $2x^2$ for

    a. $x = 7 \quad 98$

    b. $x = -7 \quad 98$ (Sec. 1.2)

11. Write the multiplicative inverse, or reciprocal, of each.

    a. $11 \quad \frac{1}{11}$

    b. $-9 \quad -\frac{1}{9}$

    c. $\dfrac{7}{4} \quad \dfrac{4}{7}$ (Sec. 1.4, Ex. 5)

12. Simplify $-2 + 3[5 - (7 - 10)]$.   22 (Sec. 1.3)

13. Solve: $0.6 = 2 - 3.5c$ 0.4 (Sec. 1.5, Ex. 2)

14. Solve: $2(x - 3) = -40$. -17 (Sec. 1.5)

15. Solve for x: $3x + 5 = 3(x + 2)$ $\{\ \}$ or $\varnothing$ (Sec. 1.5, Ex. 8)

16. Solve: $5(x - 7) = 4x - 35 + x$.   all real numbers (Sec. 1.5)

17. Find 16% of 25.   4 (Sec. 1.6, Ex. 4)

18. Find 25% of 16.   4 (Sec. 1.6)

19. Kelsey Ohleger was helping her friend Benji Burnstine study for an algebra exam. Kelsey told Benji that her three latest art history quiz scores are three consecutive even integers whose sum is 264. Help Benji find the scores.
    86, 88, and 90 (Sec. 1.6, Ex. 7)

20. Find 3 consecutive odd integers whose sum is 213.
    69,71,73 (Sec. 1.6)

21. Solve $V = lwh$ for h. $\dfrac{V}{lw} = h$ (Sec. 1.8, Ex. 1)

22. Solve $7x + 3y = 21$ for y

23. Find the median and the mode of the following set of numbers. These numbers were high temperatures for fourteen consecutive days in a city in Montana.

    76, 80, 85, 86, 89, 87, 82, 77, 76, 79, 82, 89, 89, 92
    median: 83.5; mode: 89 (Sec 1.9, Ex. 2)

24. Find the mean, median, and mode of:

    106, 118, 98, 56, 112, 58, 73, 118
    median: 102; mode:118

25. Name the quadrant in which the point is located.

    a. $(2, -1)$ quadrant IV

    b. $(0, 5)$ no quadrant

    c. $(-3, 5)$ quadrant II

    d. $(-2, 0)$ no quadrant

    e. $\left(-\dfrac{1}{2}, -4\right)$ quadrant III

    f. $(1.5, 1.5)$ quadrant I (Sec. 2.1. Ex.1)

26. Name the quadrant or axis in which each point is located.

    a. $(0, -2)$ y-axis

    b. $(-1, -2.5)$ quadrant III

    c. $\left(\frac{1}{2}, 0\right)$ x-axis

    d. $(4, -0.5)$ quadrant IV

27. Determine whether $(0, -12)$, $(1, 9)$, and $(2, -6)$ are solutions of the equation $3x - y = 12$.

28. Find the slope and y-intercept of $7x + 2y = 10$.

29. Is the relation $y = 2x + 1$ also a function?   yes (Sec. 2.2, Ex. 3)

30. Determine whether the graph below is the graph of a function.   no (Sec. 2.2)

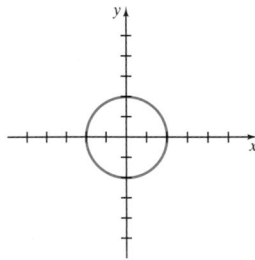

31. Find the y-intercept of the graph of each equation.

    a. $f(x) = \dfrac{1}{2}x + \dfrac{3}{7}$ $\left(0, \dfrac{3}{7}\right)$

    b. $y = -2.5x - 3.2$ $(0, -3.2)$ (Sec. 2.3, Ex. 4)

32. Find the slope of the line through $(-1, 6)$ and $(0, 9)$.

33. Write an equation of the line with y-intercept $(0, -3)$ and slope of $\dfrac{1}{4}$. $y = \dfrac{1}{4}x - 3$ (Sec. 2.5, Ex. 1)

34. Find an equation of the vertical line through $\left(-2, -\frac{3}{4}\right)$.

35. Find an equation of the horizontal line through $\left(-2, -\frac{3}{4}\right)$.

36. Write an equation of the line through $(-2, 5)$ and $(-4, 7)$.
    $x + y = 3$ (Sec. 2.5)

22. $y = \dfrac{-7x + 21}{3}$ or $y = -\dfrac{7}{3}x + 7$ (Sec. 1.8)   27. solutions: $(2, -6)$, $(0, -12)$;

not a solution: $(1, 9)$ (Sec. 2.1, Ex. 3)   28. $m = -\dfrac{7}{2}$, y-intercept $(0, 5)$ (Sec. 2.5)   32. $m = 3$ (Sec. 2.4)   34. $x = -2$ (Sec. 2.3)

35. $y = -\dfrac{3}{4}$ (Sec. 2.3)

Equations and Inequalities

Mathematics is a tool for solving problems in such diverse fields as transportation, engineering, economics, medicine, business, and biology. We solve problems using mathematics by modeling real-world phenomena with mathematical equations or inequalities. Our ability to solve problems using mathematics, then, depends in part on our ability to solve equations and inequalities. In this chapter, we solve linear equations and inequalities and graph their solutions.

S ince the beginning of civilization, education has been necessary for humanity to survive. The first public schools in the United States were established in Massachusetts in 1647. These schools were open to all children and were supported by public funds. After the Revolutionary War, attempts to unify public education resulted in the establishment of state public school systems.

Today, a large team of educators, principals, counselors, librarians, school board members, and a school superintendent administer today's complex system of public schools. Administrators establish policy and procedure as well as manage the budget and other business aspects involved in the maintenance of a large organization. Most jobs in school administration require a master's or doctoral degree in an education-related field of study.

In the Spotlight on Decision Making feature on page 236, you will have the opportunity to make a decision about long-term planning related to school enrollment as a school superintendent.

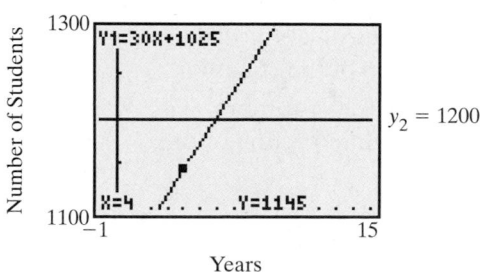

Link: www.aasa.org *(American Association of School Administrators)*
Sources of text: www.bls.gov/oco/ocos007.htm *(Bureau of Labor Statistics), World Book Millennium 2000 (encyclopedia on CD) "Education"*

3.1 SOLVING LINEAR EQUATIONS GRAPHICALLY

Objectives

1 Solve linear equations using the intersection-of-graphs method.

2 Solve problems using graphing methods.

3 Solve an identity or a contradiction.

4 Solve linear equations using the x-intercept method.

In Chapter 1, we solved linear equations algebraically and checked the solution numerically with a calculator. Recall that a solution of an equation is a value for the variable that makes the equation a true statement. In this section we solve equations algebraically and graphically. Then we use various methods to check the solution. We need to stress the importance of the check. Technology allows us to reduce the time spent on solving and aids us in verifying solutions. Consequently, by using technology we can increase the accuracy of our work.

TECHNOLOGY NOTE

Most graphing utilities have the ability to select and deselect graphs, graph equations simultaneously, and define different graph styles. See your graphing utility manual to check its capabilities. Also, find the instructions for using the intersection and root, or zero, features.

1 There are two methods we use to graphically obtain a solution of an equation. The first method we refer to as the **intersection-of-graphs method**. For this method, we graph

$$y_1 = \text{left side of equation and}$$

$$y_2 = \text{right side of equation}.$$

Recall that when a solution is substituted for a variable in an equation, the left side of the equation is equal to the right side of the equation. This means that graphically a solution occurs when $y_1 = y_2$ or where the graphs of y_1 and y_2 intersect.

Discover the Concept

Consider the equation $2x - 5 = 27$ and graph $y_1 = 2x - 5$ and $y_2 = 27$ in an integer window.

a. Use the trace feature to estimate the point of intersection of the two graphs.

b. Solve the equation algebraically and compare the x-coordinate of the point of intersection found in part (a) to the algebraic solution of the equation.

c. Locate the intersect feature on your graphing utility, sometimes found on the calculate menu. Use this feature to find the point of intersection of the two graphs.

The above discovery indicates that the solution of the equation $2x - 5 = 27$ is 16, the x-coordinate of the point of intersection, as shown above.

To check this solution, we may replace x with 16 in the original equation.

$$2x - 5 = 27$$

$$2 \cdot 16 - 5 = 27 \qquad \text{Let } x = 16.$$

$$32 - 5 = 27$$

$$27 = 27 \qquad \text{True}$$

Recall that 27 is the y-coordinate of the point of intersection. Why? Because it is the value of both the left side and the right side of the equation when x is replaced with the solution.

The steps below may be used to solve an equation by the intersection-of-graphs method.

Intersection-of-Graphs Method for Solving an Equation

Step 1: Graph $y_1 = $ left side of the equation and $y_2 = $ right side of the equation.

Step 2: Find the point(s) of intersection of the two graphs.

Step 3: The x-coordinate of a point of intersection is a solution to the equation.

Step 4: The y-coordinate of the point of intersection is the value of both the left side and the right side of the original equation when x is replaced with the solution.

CLASSROOM EXAMPLE

Solve the equation $9(x + 5) - 30 = -3$

answer: -2

EXAMPLE 1

Solve the equation $5(x - 2) + 15 = 20$.

Algebraic Solution:

$$5(x - 2) + 15 = 20$$

$$5x - 10 + 15 = 20 \quad \text{Use the distributive property.}$$

$$5x + 5 = 20 \quad \text{Combine like terms.}$$

$$5x = 15 \quad \text{Subtract 5 from both sides.}$$

$$x = 3 \quad \text{Divide both sides by 5.}$$

To check, we replace x with 3 and see that a true statement results.

$$5(x - 2) + 15 = 20$$

$$5(3 - 2) + 15 = 20$$

$$5(1) + 15 = 20$$

$$20 = 20 \quad \text{True}$$

The solution is 3.

Graphical Solution:

Graph $y_1 = 5(x - 2) + 15 \quad$ left side of equation

$\qquad y_2 = 20 \qquad$ right side of equation

Since the graph of the equation $y = 20$ is a horizontal line with y-intercept 20, use the window $[-25, 25, 5]$ by $[-25, 25, 5]$.

The x-coordinate of the point of intersection is 3. The solution is 3.

We can also use a table feature to check the solution. If $y_1 = 5(x - 2) + 15$ and $y_2 = 20$, scroll to $x = 3$ and see that y_1 and y_2 are both 20.

When solving equations graphically, we often have no indication where the intersection lies before looking at the graphs of the two equations. The zoom feature of your graphing utility may be used to quickly look for an appropriate window. If the graphs are present but the intersection point is missing, you can assess the situation and decide which component to adjust on the window setting. You will quickly get used to this assessment process as you solve more equations.

EXAMPLE 2

Graphically solve the equation $2(x - 30) + 6(x - 10) - 70 = 35 - x$.

Solution   Graph $y_1 = 2(x - 30) + 6(x - 10) - 70$ and $y_2 = 35 - x$ in an integer window.

The point of intersection $(25, 10)$ indicates that the solution is 25. To check numerically with a calculator, see the screen below.

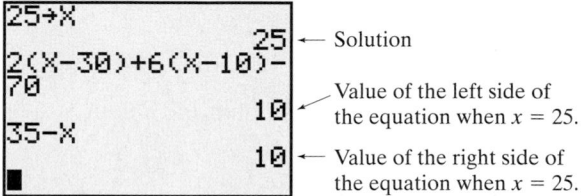

In the above example, recall that the intersection point $(25, 10)$ indicates that if $x = 25$, then the value of the expression on the left side of the equation $2(x - 30) + 6(x - 10) - 70$ is 10, and the value of the expression on the right side of the equation, $35 - x$, is also 10.

> **Helpful Hint**
>
> The integer window allows the cursor to move along a graph with integer x-values only. Using this window does *not* mean that the calculator will give only integer values when calculating a point of intersection.

> **Helpful Hint**
>
> In general, when using the intersection method, the x-coordinate of the point of intersection is a solution of the equation, and the y-coordinate of the point of intersection is the value of each side of the original equation when the variable is replaced with the solution.

EXAMPLE 3

Solution

Solve the equation using the intersection-of-graphs method.

$$5.1x + 3.78 = x + 4.7$$

Define $y_1 = 5.1x + 3.78$ and $y_2 = x + 4.7$ and graph in a standard window. The screen below shows the intersection point.

Notice the coordinates of the intersection point. When this happens, we will approximate the solution. For this example, the solution rounded to four decimal places is 0.2244.

To check when the solution is an approximation, note that the value of the left side of the equation and the value of the right side may differ slightly because of rounding. In the screen below we see a check of the equation for $x \approx 0.2244$.

The solution is approximately 0.2244.

2   Next, we use the intersection-of-graphs method to solve a problem.

EXAMPLE 4

Model 1   Model 2

$575
Special offer!

$825
Energy efficient

$0.07 per hour $0.04 per hour

PURCHASING A REFRIGERATOR

The McDonalds are purchasing a new refrigerator. They find two models that fit their needs. Model 1 sells for $575 and costs $0.07 per hour to run. Model 2 is the energy efficient model that sells for $825, but only costs $0.04 per hour to run. If x represents number of hours, then the costs to purchase and run the refrigerators are modeled by the equations

$$C_1(x) = 575 + 0.07x \qquad \textit{Cost to run model 1}$$

$$C_2(x) = 825 + 0.04x \qquad \textit{Cost to run model 2}$$

a. If the McDonalds buy Model 2, how many hours must it run before they save money?

b. How many days must it run before they save money?

Solution Graph $y_1 = 575 + 0.07x$ and $y_2 = 825 + 0.04x$. Here, the ordered pairs represent (x, y).

hours   cost

a. Find the point of intersection of the graphs.

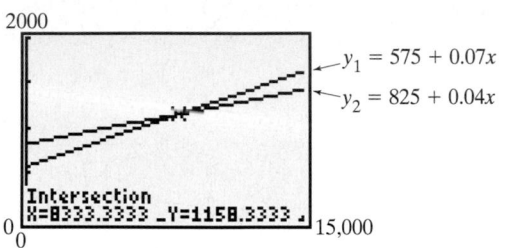

The point of intersection is $(8333.\overline{3}, 1158.\overline{3})$. This means that in approximately 8333.33 hours, both machines cost a total of approximately $1158.33 to buy and run. Before that point of intersection, the graph of y_1 is below the graph of y_2. This means that the cost of Model 1 is less than the cost of Model 2. Likewise, after the point of intersection, the graph of y_2 is below the graph of y_1. This means that the cost of Model 2 is less than the cost of Model 1. Thus, in approximately 8333.33 hours, the McDonalds start saving money.

b. To find the approximate number of days, find

$$\frac{8333.33}{24} \approx 347.22 \text{ days}.$$

---

3   Recall in Section 1.5 that we solved two special types of equations: contradictions and identities. An equation in one variable that has no solution is called a contradiction and an equation in one variable that has every number (for which the equation is defined) as a solution is called an identity. We now look at these two special types of equations graphically.

EXAMPLE 5

Solve: $4(x + 6) - 2(x - 3) = 2x + 30$

Algebraic Solution:

$$4(x + 6) - 2(x - 3) = 2x + 30$$

$$4x + 24 - 2x + 6 = 2x + 30 \qquad \text{Multiply.}$$

$$2x + 30 = 2x + 30 \qquad \text{Simplify.}$$

Since both sides are the same, we see that replacing x with any real number will result in a true statement.

Graphical Solution:

Graph $y_1 = 4(x + 6) - 2(x - 3)$ and
   $y_2 = 2x + 30$

The graphs shown below appear to be the same since there appears to be a single line only. When we trace along y_1, we have the same ordered pairs as when we trace along y_2. (If 2 points of y_1 are the same as 2 points of y_2, we know the lines are identical because 2 points uniquely determine a line.)

Both solutions show that the equation is an identity and the solution set is the set of real numbers or $\{x | x \text{ is a real number}\}$.

EXAMPLE 6

Solve $3x - 8 = 5(x - 1) - 2(x + 6)$.

Algebraic Solution:

$3x - 8 = 5(x - 1) - 2(x + 6)$

$3x - 8 = 5x - 5 - 2x - 12$ Multiply.

$3x - 8 = 3x - 17$ Simplify.

$3x - 8 - 3x = 3x - 17 - 3x$ Subtract $3x$.

$-8 = -17$

This equation is a false statement no matter what value the variable x might have.

Graphical Solution:

Graph $y_1 = 3x - 8$ and

$y_2 = 5(x - 1) - 2(x + 6)$

The graph below shows that the lines appear to be parallel and, therefore, will never intersect.

To be sure, check the slope of each graph.

$y_1 = 3x - 8 \qquad y_2 = 5(x - 1) - 2(x + 6)$

$y_2 = 3x - 17$

The lines have the same slope, 3, and their graphs are distinct, so they are indeed parallel.

The equation has no solution and is a contradiction. The solution set is $\{ \}$ or $\emptyset$.

4   We now look at another method for solving equations graphically, called the **x-intercept method**. It is actually a form of the intersection-of-graphs method, but for this method we write an equivalent equation with one side of the equation 0. Since $y = 0$ is the x-axis, we look for points where the graph intersects the x-axis or x-intercepts. Recall that an x-intercept is of the form $(x, 0)$. (An x-intercept of $(a, 0)$ is sometimes simply called an x-intercept of a.) Thus, the solutions that lie on the x-axis are called the **zeros** of the equation since this is where $y = 0$. They are also referred to as **roots** of the equation. The built-in feature on graphing utilities to find these solutions is referred to as root or zero on different graphing utilities.

EXAMPLE 7

Solve the following equation using the x-intercept method.

$$-3.1(x + 1) + 8.3 = -x + 12.4$$

Solution We begin by writing the equation so that one side equals 0 by adding x to both sides and subtracting 12.4 from both sides.

$$-3.1(x + 1) + 8.3 = -x + 12.4$$

$$-3.1(x + 1) + 8.3 + x - 12.4 = 0$$

Define $y_1 = -3.1(x + 1) + 8.3 + x - 12.4$ and graph in a standard window. Choose the root, or zero, feature to solve. You will be prompted to indicate a left bound—an x-value less than the x-intercept and a right bound—an x-value greater than the x-intercept. The guess portion of the prompting asks you to move the cursor close to the x-intercept.

The x-intercept is approximately $(-3.428571, 0)$, which means that the solution to the equation is approximately -3.428571.

To check, replace x with -3.428571 and see that the left side of the equation is approximately the right side. Rounded to two decimal places, $x \approx -3.43$ or the solution is approximately -3.43.

CLASSROOM EXAMPLE

Solve using the x-intercept method. Round the solution to 2 decimal places.
$-2x + 14.3 = 2.1(x - 5) + 3.6x$

answer: 3.22

TECHNOLOGY NOTE

The real solutions of an equation in the form $f(x) = 0$ occur at the x-intercepts of the graph $y = f(x)$. Because of this, most graphing utilities have a root or zero feature to find the x-intercepts.

Suppose you are a reporter for the *Marston Gazette*, a daily newspaper serving the medium-sized industrial city of Marston. Your editor has assigned you to a feature story about food pantries, soup kitchens, and other efforts to alleviate hunger among the city's homeless, poor, and working poor.

While researching your story assignment, you find that, according to the U.S. Department of Agriculture, 10.2% of all households in the United States do not have access to enough food to meet their basic needs. A survey conducted by Marston Social Services reveals that approximately 4950 of the estimated 39,400 Marston households must cope with hunger.

Armed with these basic facts, you now need to decide what angle to take with your story. Which of the following approaches would you choose? Why? What other information would you want to consider?

a. Marston lags behind nation in fight against hunger.
b. Marston mirrors national hunger picture.
c. Marston makes progress against hunger.

MENTAL MATH

Mentally solve the following equations.

1. $4x = 24$ 6

2. $6x = -12$ -2

3. $2x + 10 = 20$ 5

4. $5x + 25 = 30$ 1

5. $-3x = 0$ 0

6. $-2x = -14$ 7

11. $\varnothing$ **12.** $\varnothing$

EXERCISE SET 3.1

STUDY GUIDE/SSM   CD/VIDEO   PH MATH TUTOR CENTER   MathXL®Tutorials ON CD   MathXL®   MyMathLab®

Solve each equation algebraically and graphically. See Examples 1 through 3, and 7.

1. $5x + 2 = 3x + 6$ 2

2. $2x + 9 = 3x + 7$ 2

3. $9 - x = 2x + 12$ -1

4. $3 - 4x = 2 - 3x$ 1

5. $8 - (2x - 1) = 13$ -2

6. $3(2 - x) + 4 = 2x + 3$ $\frac{7}{5}$

For each given screen, write an equation in x and its solution.

7.

$y_2 = 3x - 13$ $3x - 13 = 2(x - 5) + 3$: 6

$y_1 = 2(x - 5) + 3$

8.

$y_1 = 3x - 4$ $3x - 4 = 4(x - 2)$: 4

$y_2 = 4(x - 2)$

9.

$y_1 = -(2x + 3) + 2$ $-(2x + 3) + 2 = 5x - 1 - 7x$;

all real numbers

Same Line

$y_2 = 5x - 1 - 7x$

10.

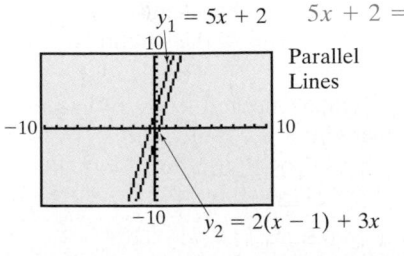

$y_1 = 5x + 2$ $5x + 2 = 2(x - 1) + 3x$; $\varnothing$

Parallel Lines

$y_2 = 2(x - 1) + 3x$

Solve each equation algebraically and graphically. See Examples 5 and 6.

11. $7(x - 6) = 5(x + 2) + 2x$ **12.** $6x - 9 = 6(x - 3)$

13. $3x - (6x + 2) = -(3x + 2)$ all real numbers

14. $8x - (2x + 3) = 6(x + 5)$ $\varnothing$

15. $5(x + 1) - 3(x - 7) = 2(x + 4) - 3$ $\varnothing$

16. $(5x + 8) - 2(x + 3) = (7x - 4) - (4x + 6)$ $\varnothing$

17. $3(x + 2) - 6(x - 5) = 36 - 3x$ all real numbers

18. $2(x + 3) - 7 = 4 + 2x$ $\varnothing$

MIXED PRACTICE

Solve each equation. If necessary, round solutions to two decimal places.

19. $(x + 2.1) - (0.5x + 3) = 12$ 25.8

20. $2(x + 1.3) - 4(5 - x) = 15$ 5.4

21. $5(a - 12) + 2(a + 15) = a - 9$ 3.5

22. $8(b + 2) - 4(b - 5) = b + 12$ -8

23. $8(p - 4) - 5(2p + 3) = 3.5(2p - 5)$ -3.28

24. $3(x + 2) - 5(x - 7) = x + 3$ $\frac{38}{3}$

25. $5(x - 2) + 2x = 7(x + 4)$ $\varnothing$

26. $3x + 2(x + 4) = 5(x + 1) + 3$ all real numbers

27. $y + 0.2 = 0.6(y + 3)$ 4

28. $-(w + 0.2) = 0.3(4 - w)$ -2

29. $2y + 5(y - 4) = 4y - 2(y - 10)$ 8

30. $9c - 3(6 - 5c) = c - 2(3c + 9)$ 0

31. $2(x - 8) + x = 3(x - 6) + 2$ all real numbers

32. $4(x + 5) = 3(x - 4) + x$ $\varnothing$

33. $\dfrac{5x - 1}{6} - 3x = \dfrac{1}{3} + \dfrac{4x + 3}{9}$ -0.32

34. $\dfrac{2r - 5}{3} - \dfrac{r}{5} = 4 - \dfrac{r + 8}{10}$ 8.59

35. $-2(b - 4) - (3b - 1) = 5b + 3$ 0.6

36. $4(t - 3) - 3(t - 2) = 2t + 8$ -14

37. $1.5(4 - x) = 1.3(2 - x)$ 17

38. $2.4(2x + 3) = -0.1(2x + 3)$ -1.5

39. $\dfrac{1}{4}(a + 2) = \dfrac{1}{6}(5 - a)$ **40.** $\dfrac{1}{3}(8 + 2c) = \dfrac{1}{5}(3c - 5)$

Solve. See Example 4.   **39.** 0.8   **40.** -55

41. Acme Mortgage Company wants to hire a student computer consultant. A first consultant charges an initial fee of $30 plus $20 per hour. A second consultant charges a flat fee of $25 per hour. If x represents number of hours, then the costs of the consultants are modeled by

$C_1(x) = 30 + 20x$ First consultant's total cost

$C_2(x) = 25x$ Second consultant's total cost

  a. Find which consultant's cost is lower if the job takes 2 hours.   second consultant

  b. Find which consultant's cost is lower if the job takes 8 hours.   first consultant   **c.** 6-hr job

  c. When is the cost of hiring each consultant the same?

42. Rod Pasch needs to hire a graphic artist. A first graphic artist charges $50 plus $35 per hour. A second one charges $50 per hour. If x represents number of hours, then the costs of the graphic artists can be modeled by

$G_1(x) = 50 + 35x$ First graphic artist's total cost

$G_2(x) = 50x$ Second graphic artist's total cost

  a. Which graphic artist costs less if the job takes 3 hours?   second artist

  b. Which graphic artist costs less if the job takes 5 hours?   first artist

  c. When is the cost of hiring each artist the same?   3 hr 20 min

43. One car rental agency charges $25 a day plus $0.30 a mile. A second car rental agency charges $28 a day plus $0.25 a mile. If x represents the number of miles driven, then the costs of the car rental agencies for a 1-day rental can be modeled by

$R_1(x) = 25 + 0.30x$ First agency's cost

$R_2(x) = 28 + 0.25x$ Second agency's cost

  a. Which agency's cost is lower if 50 miles are driven?   first agency

  b. Which agency's cost is lower if 100 miles are driven?   second agency

  c. When is the cost of using each agency the same?   60 ml

44. Copycat Printing charges $18 plus $0.03 per page for making black and white copies. Duplicate, Inc. charges $0.05 per black and white page copied. If x represents the number of

pages copied, then the charges for the companies can be modeled by   **c.** 900 pages

$C_1(x) = 18 + 0.03x$ Copycat

$C_2(x) = 0.05x$ Duplicate, Inc.

  a. For 500 copies, which company charges less?   Duplicate Inc.

  b. For 1000 copies, which company charges less?   Copycat

  c. When is the cost of using each company the same?

REVIEW AND PREVIEW

Determine which numbers in the set $\{-3, -2, -1, 0, 1, 2, 3\}$ are solutions of each inequality.

45. $x < 0$ $\{-3, -2, -1\}$ **46.** $x > 1$ $\{2, 3\}$

47. $x + 5 \le 6$ $\{-3, -2, -1, 0, 1\}$

48. $x - 3 \ge -7$ $\{-3, -2, -1, 0, 1, 2, 3\}$

49. In your own words, explain what real numbers are solutions of $x < 0$.   answers may vary

50. In your own words, explain what real numbers are solutions of $x > 1$.   answers may vary

Concept Extensions

The given screen shows the graphs of y_1 and y_2 and their intersection. Use this screen to answer the questions below.

51. Complete the ordered pair for the graphs of both y_1 and y_2: $(12, 22)$.   **57.** 1.51   **58.** -2.5 **59.** 0.91   **60.** 2.48

52. If x is less than 12, is y_1 less than, greater than, or equal to y_2?   less than   **61.** 22.87   **62.** 0.78

53. If x is greater than 12, is y_1 less than, greater than, or equal to y_2?   greater than

54. True or false? If x is 45, $y_1 < y_2$.   false

Solve each equation graphically and check by a method of your choice. Round solutions to the nearest hundredth.

55. $1.75x - 2.5 = 0$ 1.43   **56.** $3.1x + 5.6 = 0$ -1.81

57. $2.5x + 3 = 7.8x - 5$ **58.** $4.8x - 2.3 = 6.8x + 2.7$

59. $3x + \sqrt{5} = 7x - \sqrt{2}$ **60.** $0.9x + \sqrt{3} = 2.5x - \sqrt{5}$

61. $2\pi x - 5.6 = 7(x - \pi)$ **62.** $-\pi x + 1.2 = 0.3(x - 5)$

63. If the intersection-of-graphs method leads to parallel lines, explain what this means in terms of the solution of the original equation.   The equation has no solution.

64. If the intersection-of-graphs method leads to the same line, explain what this means in terms of the solution of the original equation.   All real numbers for which the equation is defined are solutions of the equation.

3.2   LINEAR INEQUALITIES AND PROBLEM SOLVING

Objectives

1 Use interval notation.

2 Solve linear inequalities using the addition property of inequality.

3 Solve linear inequalities using the multiplication property of inequality.

4 Solve problems that can be modeled by linear inequalities.

1 Relationships among measureable quantities are not always described by equations. For example, suppose that a salesperson earns a base of $600 per month plus a commission of 20% of sales. Find the minimum amount of sales needed to receive a total income of *at least* $1500 per month. Here, the phrase "at least" implies that an income of $1500 *or more* is acceptable. In symbols, we can write

$$\text{income} \geq 1500$$

This is an example of an inequality, and we will solve this problem in Example 8.

A **linear inequality** is similar to a linear equation except that the equality symbol is replaced with an inequality symbol, such as $<, >, \leq,$ or $\geq$.

Linear Inequalities in One Variable

$$3x + 5 \geq 4 \qquad 2y < 0 \qquad 3(x - 4) > 5x \qquad \frac{x}{3} \leq 5$$

↑ is greater than or equal to

↑ is less than

↑ is greater than

↑ is less than or equal to

Linear Inequality in One Variable

A linear inequality in one variable is an inequality that can be written in the form

$$ax + b < c$$

where $a, b,$ and c are real numbers and $a \neq 0$.

In this section, when we make definitions, state properties, or list steps about an inequality containing the symbol $<$, we mean that the definition, property, or steps apply to inequalities containing the symbols $>, \leq$ and $\geq$ also.

A **solution** of an inequality is a value of the variable that makes the inequality a true statement. The **solution set** of an inequality is the set of all solutions. Notice that the solution set of the inequality $x > 2$, for example, contains all numbers greater than 2. Its graph is an interval on the number line since an infinite number of values satisfy the variable. If we use open/closed-circle notation, the graph of $\{x \mid x > 2\}$ looks like the following.

In this text **interval notation** will be used to write solution sets of inequalities. To help us understand this notation, a different graphing notation will be used. Instead of an open circle, we use a parenthesis. With this new notation, the graph of $\{x|x > 2\}$ now looks like

and can be represented in interval notation as $(2, \infty)$. The symbol ∞ is read "infinity" and indicates that the interval includes *all* numbers greater than 2. The left parenthesis indicates that 2 *is not* included in the interval.

In the case 2 *is* included in the interval, we use a bracket. The graph of $\{x|x \geq 2\}$ is below

and can be represented as $[2, \infty)$. The following table shows three equivalent ways to describe an interval: in set notation, as a graph, and in interval notation.

Set Notation	Graph	Interval Notation
$\{x|x < a\}$		$(-\infty, a)$
$\{x|x > a\}$		(a, ∞)
$\{x|x \leq a\}$		$(-\infty, a]$
$\{x|x \geq a\}$		$[a, \infty)$
$\{x|a < x < b\}$		(a, b)
$\{x|a \leq x \leq b\}$		$[a, b]$
$\{x|a < x \leq b\}$		$(a, b]$
$\{x|a \leq x < b\}$		$[a, b)$

TEACHING TIP
Before doing the examples, have students write the set and interval notation for each of the following number lines.

Helpful Hint
Notice that a parenthesis is always used to enclose ∞ and $-\infty$.

Concept Check Answer:
should be $(5, \infty)$ since a parenthesis is always used to enclose

✔ **CONCEPT CHECK**

Explain what is wrong with writing the interval $(5, \infty]$.

EXAMPLE 1

Graph each set on a number line and then write in interval notation.

Solution

a. $\{x | x \geq 2\}$ **b.** $\{x | x < -1\}$ **c.** $\{x | 0.5 < x \leq 3\}$

a. 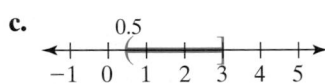 $[2, \infty)$

b. $(-\infty, -1)$

c. $(0.5, 3]$

CLASSROOM EXAMPLE

Graph on a number line and then write in interval notation.

a. $\{x | x > -3\}$

b. $\{x | x \leq 0\}$

c. $\{x | -0.5 \leq x < 2\}$

answer:

a. $(-3, \infty)$

b. $(-\infty, 0]$

c. $[-0.5, 2)$

2 Interval notation can be used to write solutions of linear inequalities. To solve a linear inequality, we use a process similar to the one used to solve a linear equation. We use properties of inequalities to write equivalent inequalities until the variable is isolated.

TEACHING TIP

Before discussing the Addition Property of Inequality, have students consider the inequality $-7 < 3$. Add 5 to both sides. Is $-7 + 5 < 3 + 5$? Subtract 5 from both sides. Is $-7 - 5 < 3 - 5$?

Addition Property of Inequality

If a, b, and c are real numbers, then

$$a < b \quad \text{and} \quad a + c < b + c$$

are equivalent inequalities.

In other words, we may add the same real number to both sides of an inequality and the resulting inequality will have the same solution set. This property also allows us to subtract the same real number from both sides.

EXAMPLE 2

Solve $x - 2 < 5$. Graph the solution set.

Solution

$$x - 2 < 5$$
$$x - 2 + 2 < 5 + 2 \qquad \text{Add 2 to both sides.}$$
$$x < 7 \qquad \text{Simplify.}$$

CLASSROOM EXAMPLE

Solve: $x + 3 < 1$. Graph the solution set.

answer: $\{x | x < -2\}, (-\infty, -2)$

The solution set is $\{x | x < 7\}$, which in interval notation is $(-\infty, 7)$. The graph of the solution set is

TEACHING TIP

If students have trouble writing interval notations, have them write interval notations from the graph of an inequality. Tell them to write the interval by looking at the shading of the graph from left to right.

Helpful Hint

In Example 2, the solution set is $\{x | x < 7\}$. This means that *all* numbers less than 7 are solutions. For example, $6.9, 0, -\pi, 1$, and -56.7 are solutions, just to name a few. To see this, replace x in $x - 2 < 5$ with each of these numbers and see that the result is a true inequality.

EXAMPLE 3

Solve $3x + 4 \geq 2x - 6$. Graph the solution set.

Solution

$$3x + 4 \geq 2x - 6$$
$$3x + 4 - 2x \geq 2x - 6 - 2x \qquad \text{Subtract } 2x \text{ from both sides.}$$
$$x + 4 \geq -6 \qquad \text{Combine like terms.}$$
$$x + 4 - 4 \geq -6 - 4 \qquad \text{Subtract 4 from both sides.}$$
$$x \geq -10 \qquad \text{Simplify.}$$

CLASSROOM EXAMPLE

Solve: $5x - 1 \geq 4x + 4$. Graph the solution set.
answer: $\{x | x \geq 5\}, [5, \infty)$

The solution set is $\{x | x \geq -10\}$, which in interval notation is $[-10, \infty)$. The graph of the solution set is

$$\xleftarrow{\quad} \overset{}{\underset{-11\,-10\,-9\,-8\,-7\,-6}{|\quad|\quad|\quad|\quad|}} \xrightarrow{\quad}$$

Discover the Concept

Let's use a graphing utility to check the solution of $3x + 4 \geq 2x - 6$ from Example 3. Let $y_1 = 3x + 4$ and $y_2 = 2x - 6$ and we can now think of the inequality as $y_1 \geq y_2$.

a. Graph y_1 and y_2 using the window $[-15, 20, 5]$ by $[-40, 30, 10]$.

b. Find the point of intersection of the graphs. What does this point represent?

c. Determine where the graph of y_1 is above the graph of y_2.

d. Now find the x-values for which $y_1 \geq y_2$.

e. Write the solution set of the inequality $y_1 \geq y_2$.

The graphs of y_1 and y_2 are shown to the left. From the above discovery, we find that

b. The point of intersection is $(-10, -26)$. This means that the solution of the equation $y_1 = y_2$ is -10.

c. The graph of y_1 is above the graph of y_2 to the right of the point of intersection, $(-10, -26)$.

d. The x-values for which $y_1 \geq y_2$ are -10 and those x-values to the right of -10 on the x-axis.

e. The solution set is $[-10, \infty)$.

3 Next, we introduce and use the multiplication property of inequality to solve linear inequalities. To understand this property, let's start with the true statement $-3 < 7$ and multiply both sides by 2.

$$-3 < 7$$
$$-3(2) < 7(2) \qquad \text{Multiply by 2.}$$
$$-6 < 14 \qquad \text{True}$$

The statement remains true.

Notice what happens if both sides of $-3 < 7$ are multiplied by -2.

$$-3 < 7$$
$$-3(-2) < 7(-2) \qquad \text{Multiply by } -2.$$
$$6 < -14 \qquad \text{False}$$

> **Helpful Hint**
>
> Whenever both sides of an inequality are multiplied or divided by a negative number, the direction of the inequality symbol **must be** reversed to form an equivalent inequality.

The inequality $6 < -14$ is a false statement. However, **if the direction of the inequality sign is reversed**, the result is true.

$$6 > -14 \qquad \text{True}$$

These examples suggest the following property.

> ## Multiplication Property of Inequality
>
> If a, b, and c are real numbers and c is **positive**, then
> $$a < b \text{ and } ac < bc$$
> are equivalent inequalities.
> If a, b, and c are real numbers and c is **negative**, then
> $$a < b \text{ and } ac > bc$$
> are equivalent inequalities.

In other words, we may multiply both sides of an inequality by the same positive real number and the result is an equivalent inequality.

We may also multiply both sides of an inequality by the same **negative number** and **reverse the direction of the inequality symbol**, and the result is an equivalent inequality. The multiplication property holds for division also, since division is defined in terms of multiplication.

EXAMPLE 4

Solve and graph the solution set.

a. $\dfrac{1}{4}x \le \dfrac{3}{8}$ **b.** $-2.3x < 6.9$

Solution   **a.**

$$\frac{1}{4}x \le \frac{3}{8}$$

$$4 \cdot \frac{1}{4}x \le 4 \cdot \frac{3}{8} \qquad \text{Multiply both sides by 4.}$$

$$x \le \frac{3}{2} \qquad \text{Simplify.}$$

> **Helpful Hint**
> The inequality symbol is the same since we are multiplying by a *positive* number.

The solution set is $\left\{ x \mid x \le \dfrac{3}{2} \right\}$, which in interval notation is $\left(-\infty, \dfrac{3}{2} \right]$. The graph of the solution set is

-3 -2 -1 0 1 $\frac{3}{2}$ 2 3

b.

$$-2.3x < 6.9$$

$$\frac{-2.3x}{-2.3} > \frac{6.9}{-2.3} \qquad \begin{array}{l}\text{Divide both sides by } -2.3 \text{ and} \\ \text{reverse the inequality symbol.}\end{array}$$

$$x > -3 \qquad \text{Simplify.}$$

> **Helpful Hint**
> The inequality symbol is *reversed* since we divided by a *negative* number.

The solution set is $\{x \mid x > -3\}$, which is $(-3, \infty)$ in interval notation. The graph of the solution set is

-5 -4 -3 -2 -1 0 1 2 3

✔ **CONCEPT CHECK**

In which of the following inequalities must the inequality symbol be reversed during the solution process?

a. $-2x > 7$

b. $2x - 3 > 10$

c. $-x + 4 + 3x < 7$

d. $-x + 4 < 5$

To solve linear inequalities in general, we follow steps similar to those for solving linear equations.

Solving A Linear Inequality in One Variable

Step 1: Clear the equation of fractions by multiplying both sides of the inequality by the least common denominator (LCD) of all fractions in the inequality.

Step 2: Use the distributive property to remove grouping symbols such as parentheses.

Step 3: Combine like terms on each side of the inequality.

Step 4: Use the addition property of inequality to write the inequality as an equivalent inequality with variable terms on one side and numbers on the other side.

Step 5: Use the multiplication property of inequality to isolate the variable.

CLASSROOM EXAMPLE

Solve: $-(2x - 6) \leq 4(2x - 4) + 2$.

answer: $[2, \infty)$

EXAMPLE 5

Solve algebraically for x: $-(x - 3) + 2 \leq 3(2x - 5) + x$. Then use a graphical approach to check the solution.

Solution

$$-(x - 3) + 2 \leq 3(2x - 5) + x$$

$-x + 3 + 2 \leq 6x - 15 + x$	Apply the distributive property.
$5 - x \leq 7x - 15$	Combine like terms.
$5 - x + x \leq 7x - 15 + x$	Add x to both sides.
$5 \leq 8x - 15$	Combine like terms.
$5 + 15 \leq 8x - 15 + 15$	Add 15 to both sides.
$20 \leq 8x$	Combine like terms.
$\dfrac{20}{8} \leq \dfrac{8x}{8}$	Divide both sides by 8.
$\dfrac{5}{2} \leq x, \quad \text{or } x \geq \dfrac{5}{2}$	Simplify.

► Helpful Hint

Don't forget that
$$\frac{5}{2} \leq x$$
means the same as
$$x \geq \frac{5}{2}.$$

To check, graph $y_1 = -(x - 3) + 2$ and $y_2 = 3(2x - 5) + x$ and find x values such that $y_1 \leq y_2$.

The point of intersection is $(2.5, 2.5)$. The graph of y_1 is equal to or below the graph of y_2 for all x-values equal to or greater than 2.5, or for the interval $\left[\frac{5}{2}, \infty\right)$.

Concept Check Answer:
a, d

EXAMPLE 6

Solve algebraically for x: $\dfrac{2}{5}(x - 6) \geq x - 1$.

Solution

$$\frac{2}{5}(x - 6) \geq x - 1$$

$$5\left[\frac{2}{5}(x - 6)\right] \geq 5(x - 1) \qquad \text{Multiply both sides by 5 to eliminate fractions.}$$

$$2x - 12 \geq 5x - 5 \qquad \text{Apply the distributive property.}$$

$$-3x - 12 \geq -5 \qquad \text{Subtract } 5x \text{ from both sides.}$$

$$-3x \geq 7 \qquad \text{Add 12 to both sides.}$$

$$\frac{-3x}{-3} \leq \frac{7}{-3} \qquad \begin{array}{l}\text{Divide both sides by } -3 \text{ and reverse} \\ \text{the inequality symbol.}\end{array}$$

$$x \leq -\frac{7}{3} \qquad \text{Simplify.}$$

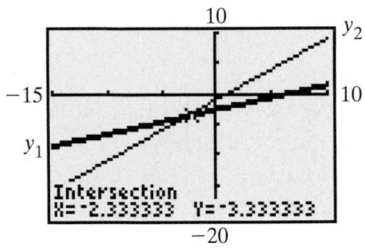

The intersection of y_1 and y_2 is the point $\left(-\frac{7}{3}, -\frac{10}{3}\right)$. $y_1 \geq y_2$ for all x values in the interval $\left(-\infty, -\frac{7}{3}\right]$ since this is where y_1 is equal to or above y_2.

The solution set written in interval notation is $\left(-\infty, -\dfrac{7}{3}\right]$. A graphing utility check is shown in the margin.

Recall that when the graph of the left side and the right side of an equation do not intersect, the equation has no solution. Let's see what a similar situation means for inequalities.

Discover the Concept

a. Use a graphing utility to solve $2(x + 3) > 2x + 1$. To do so, graph $y_1 = 2(x + 3)$ and $y_2 = 2x + 1$ and find the x-values for which the graph of y_1 is above the graph of y_2.

b. Next, solve $2(x + 3) < 2x + 1$. Notice that the graphs of y_1 and y_2 found in part (a) can be used. Find the x-values for which the graph of y_1 is below the graph of y_2.

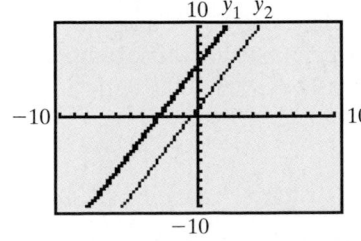

The graphs of parallel lines y_1 and y_2 from the discovery above are shown at the left. Since the graph of y_1 is *always* above the graph of y_2, all x-values satisfy the inequality $y_1 > y_2$ or $2(x + 3) > 2x + 1$. The solution set is all real numbers.

Also, notice that the graph of y_2 is *never* above the graph of y_1. This means that no x-values satisfy $y_1 < y_2$ or $2(x + 3) < 2x + 1$. The solution set is the empty set.

These results can also be obtained algebraically as we see in Example 7.

EXAMPLE 7

Solve algebraically for x:

a. $2(x + 3) > 2x + 1$ **b.** $2(x + 3) < 2x + 1$

Solution   **a.** $2(x + 3) > 2x + 1$

$2x + 6 > 2x + 1$ Distribute on the left side.

$2x + 6 - 2x > 2x + 1 - 2x$ Subtract $2x$ from both sides.

$6 > 1$ Simplify.

Since $6 > 1$ is a true statement for all values of x, this inequality and the original inequality are true for all numbers. The solution set is $\{x \mid x \text{ is a real number}\}$, or $(-\infty, \infty)$ in interval notation.

b. Solving $2(x + 3) < 2x + 1$ in a similar fashion leads us to the statement $6 < 1$. This statement, as well as the original inequality, is false for all values of x. This means that the solution set is the empty set, or $\varnothing$.

TEACHING TIP
Check students' understanding by having them write symbols for each of the following phrases:
at least
at most
between
is no more than
is more than
is less than
no less than

4   Application problems containing words such as "at least," "at most," "between," "no more than," and "no less than" usually indicate that an inequality be solved instead of an equation. In solving applications involving linear inequalities, we use the same procedure as when we solved applications involving linear equations.

EXAMPLE 8

CALCULATING INCOME WITH COMMISSION

A salesperson earns \$600 per month plus a commission of 20% of sales. Find the minimum amount of sales needed to receive a total income of at least \$1500 per month.

Solution   **1.** UNDERSTAND. Read and reread the problem. Let $x =$ amount of sales.

2. TRANSLATE. As stated in the beginning of this section, we want the income to be greater than or equal to \$1500. To write an inequality, notice that the salesperson's income consists of \$600 plus a commission (20% of sales).

In words: 600   +   commission (20% of sales) $\geq$ 1500

Translate: 600   + $0.20x$ $\geq$ 1500

3. SOLVE the inequality for x.

First, let's decide on a window.

For sales of \$2000, gross pay $= \$600 + 20\%(\$2000) = \$1000$. Similarly, for sales of \$5000, gross pay is \$1600. This means that a sales amount between \$2000 and \$5000 will give a gross pay of \$1500. Although many windows are acceptable, for this example, we graph $y_1 = 600 + 0.20x$ and $y_2 = 1500$ on a $[2000, 5000, 1000]$ by $[1000, 2000, 100]$ window.

From the graph we see that $y_1 \geq y_2$ for x-values greater than or equal to 4500, or for x-values in the interval $[4500, \infty)$.

4. INTERPRET.

Check: The income for sales of $4500 is

$$600 + 0.20(4500), \text{ or } 1500.$$

Thus, if sales are greater than or equal to $4500, income is greater than or equal to $1500.

State: The minimum amount of sales needed for the salesperson to earn at least $1500 per month is $4500 per month.

EXAMPLE 9

Solution

$y_1 = -14.25x + 598.69$

A calculator solution for Example 9.

FINDING THE ANNUAL CONSUMPTION

In the United States, the annual consumption of cigarettes is declining. The consumption c in billions of cigarettes per year since the year 1985 can be approximated by the formula

$$c = -14.25t + 598.69$$

where t is the number of years after 1985. Use this formula to predict the years that the consumption of cigarettes will be less than 200 billion per year.

1. UNDERSTAND. Read and reread the problem. To become familiar with the given formula, let's find the cigarette consumption after 25 years, which would be the year $1985 + 25$, or 2010. To do so, we substitute 25 for t in the given formula.

$$c = -14.25(25) + 598.69 = 242.44$$

Thus, in 2010, we predict cigarette consumption to be about 242.44 billion.

Variables have already been assigned in the given formula. For review, they are

c = the annual consumption of cigarettes in the United States in billions of cigarettes

t = the number of years after 1985

2. TRANSLATE. We are looking for the years that the consumption of cigarettes c is less than 200. Since we are finding years t, we substitute the expression in the formula given for c, or

$$-14.25t + 598.69 < 200$$

3. SOLVE the inequality.

$$-14.25t + 598.69 < 200 \qquad \text{Subtract 598.69 from both sides.}$$
$$-14.25t < -398.69 \qquad \text{Divide both sides by } -14.25 \text{ and round the result.}$$
$$t > 27.98$$

4. INTERPRET.

Check: We substitute a number greater than 27.98 and see that c is less than 200.

State: The annual consumption of cigarettes will be less than 200 billion for the years more than 27.98 years after 1985, or after approximately $28 + 1985 = 2013$. Thus, for the year 2013 and later, we predict cigarette consumption in the U.S. to be less than 200 billion per year.

1. $\longrightarrow$ -3 , $(-\infty, -3)$ **2.** $\longleftarrow$ -7 , $[-7, \infty)$ **3.** $\longleftarrow$ 0.3 , $[0.3, \infty)$

4. $\longleftarrow$ -0.2 , $(-\infty, -0.2)$ **5.** $\longrightarrow$ $\frac{5}{9}$, $\left(\frac{5}{9}, \infty\right)$ **6.** $\longleftarrow$ $-\frac{7}{8}$, $\left(-\infty, -\frac{7}{8}\right]$ **7.** $\longrightarrow$ $-2 \quad 5$, $(-2, 5)$

Spotlight on DECISION MAKING

Suppose you are the superintendent of Copley Public Schools. You are aware that the general population of Copley is increasing and that enrollment at the schools is steadily rising. The high school can house a maximum of 1200 students. Once this maximum has been exceeded, temporary classrooms must be erected to handle the overflow.

You have been studying the changes in population and conclude that the equation $y = 30x + 1025$ models the high school enrollment x years from now. As you prepare a long-term planning report, you must decide whether temporary classrooms will be needed in the next 10 years. If so, when is the latest that funding for temporary classrooms could be added to the annual budget?

8. $\longleftarrow$ $-5 \quad -1$, $[-5, -1]$

9. $\longleftarrow$ $-1 \quad 5$, $(-1, 5)$

10. $\longleftarrow$ $-7 \quad -3$, $[-7, -3]$

11.–12. answers may vary

MENTAL MATH

Solve each inequality.

1. $x - 2 < 4$ $\{x|x < 6\}$

2. $x - 1 > 6$ $\{x|x > 7\}$

3. $x + 5 \geq 15$ $\{x|x \geq 10\}$

4. $x + 1 \leq 8$ $\{x|x \leq 7\}$

5. $3x > 12$ $\{x|x > 4\}$

6. $5x < 20$ $\{x|x < 4\}$

7. $\frac{x}{2} \leq 1$ $\{x|x \leq 2\}$

8. $\frac{x}{4} \geq 2$ $\{x|x \geq 8\}$

EXERCISE SET 3.2

STUDY GUIDE/SSM   CD/ VIDEO   PH MATH TUTOR CENTER   MathXL®Tutorials ON CD   MathXL®   MyMathLab®

Graph the solution set of each inequality and write it in interval notation. See Example 1.

1. $\{x|x < -3\}$

2. $\{x|x \geq -7\}$

3. $\{x|x \geq 0.3\}$

4. $\{x|x < -0.2\}$

5. $\{x|\frac{5}{9} < x\}$

6. $\{x|-\frac{7}{8} \geq x\}$

7. $\{x|-2 < x < 5\}$

8. $\{x|-5 \leq x \leq -1\}$

9. $\{x|5 > x > -1\}$

10. $\{x|-3 \geq x \geq -7\}$

11. When graphing the solution set of an inequality, explain how you know whether to use a parenthesis or a bracket.

12. Explain what is wrong with the interval notation $(-6, -\infty)$.

23. ⟶ , $(-\infty, 1)$ **24.** ⟶ , $(-\infty, 5)$

Match each graph with the interval notation that describes it.

13. ⟵━━━━━━━━➤ d   **25.** ⟵━━ , $(-\infty, 2]$
  $-7 -6 -5 -4 -3 -2 -1\ \ 0$

  a. $(-5, \infty)$ **b.** $(-5, -\infty)$
  c. $(\infty, -5)$ **d.** $(-\infty, -5)$

14. ⟵━━━[━━━━➤ c   **26.** [━➤ , $[0, \infty)$
  $-13 -12 -11 -10\ -9\ -8$

  a. $(-\infty, -11]$ **b.** $(-11, \infty)$
  c. $[-11, \infty)$ **d.** $(-\infty, -11)$

15. $\frac{7}{4}$ b
  -2.5
  ⟵━(━━━━━]━━━➤   **27.** $\frac{8}{3}$ [━➤ , $\left[\frac{8}{3}, \infty\right)$
  $-3 -2 -1\ \ 0\ \ 1\ \ 2\ \ 3\ \ 4$ $2\ \ 3$

  a. $\left[\frac{7}{4}, -2.5\right)$ **b.** $\left(-2.5, \frac{7}{4}\right]$
  c. $\left[-2.5, \frac{7}{4}\right)$ **d.** $\left(\frac{7}{4}, -2.5\right)$

16. $-\frac{10}{3}$ a
  $\ \ \ \ \ \ 0.2$
  ⟵━[━━━━)━━━➤   **28.** $-\frac{48}{5}$ [━➤ , $\left[-\frac{48}{5}, \infty\right)$
  $-4 -3 -2 -1\ \ 0\ \ 1\ \ 2\ \ 3$ $-10\ -8$

  a. $\left[-\frac{10}{3}, 0.2\right)$ **b.** $\left(0.2, -\frac{10}{3}\right]$
  c. $\left(-\frac{10}{3}, 0.2\right]$ **d.** $\left[0.2, -\frac{10}{3}\right)$

Use the given screen to solve each inequality. Write the solution in interval notation.

17. $y_1 < y_2$ $(-\infty, 4)$

29. -4.7 ⟵(━━➤ , $(-\infty, -4.7)$ **30.** -2.8 ⟵(━➤ , $(-2.8, \infty)$
  $-5\ \ -4$ $-3\ \ -2$

18. $y_1 \geq 0$ $[-8, \infty)$

31. ⟵━] , $(-\infty, -3]$ **32.** $-\frac{15}{4}$ ⟵━] , $\left(-\infty, -\frac{15}{4}\right]$
  -3 $-4\ \ -3$

19. $y_1 \geq y_2$ $[-3, \infty)$

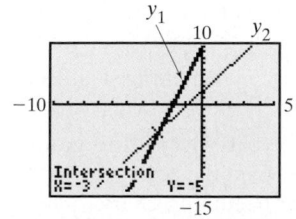

20. $y_1 > y_2$ $(-\infty, 4)$

33. ⟵(━━➤ , $(4, \infty)$ **34.** ⟶ , $(-\infty, 2)$
  4 2

Use the graph of parallel lines y_1 and y_2 to solve each inequality. Write the solution in interval notation.

37. $(-\infty, 11]$ **42.** $\left(-\frac{1}{2}, \infty\right)$
38. $(4, \infty)$ **43.** $(-\infty, \infty)$
39. $(-13, \infty)$ **46.** $(-\infty, \infty)$
40. $\left[\frac{8}{3}, \infty\right)$

21. $y_1 > y_2$ $\emptyset$ **22.** $y_1 < y_2$ $(-\infty, \infty)$

Solve. Graph the solution set and write it in interval notation. See Examples 2 through 4.

23. $7x < 6x + 1$ **24.** $11x < 10x + 5$
25. $8x - 7 \leq 7x - 5$ **26.** $7x - 1 \geq 6x - 1$
27. $\frac{3}{4}x \geq 2$ **28.** $\frac{5}{6}x \geq -8$
29. $5x < -23.5$ **30.** $4x > -11.2$
31. $-3x \geq 9$ **32.** $-4x \geq 15$
33. $-x < -4$ **34.** $-x > -2$

Solve. Write the solution set using interval notation. See Examples 5 through 7.

35. $-2x + 7 \geq 9$ $(-\infty, -1]$ **36.** $8 - 5x \leq 23$ $[-3, \infty)$
37. $15 + 2x \geq 4x - 7$ **38.** $10 + x < 6x - 10$
39. $3(x - 5) < 2(2x - 1)$ **40.** $5(x + 4) \leq 4(2x + 3)$
41. $\frac{1}{2} + \frac{2}{3} \geq \frac{x}{6}$ $(-\infty, 7]$ **42.** $\frac{3}{4} - \frac{2}{3} > -\frac{x}{6}$
43. $4(x - 1) \geq 4x - 8$ **44.** $3x + 1 < 3(x - 2)$ $\emptyset$
45. $7x < 7(x - 2)$ $\emptyset$ **46.** $8(x + 3) \leq 7(x + 5) + x$
47. $4(2x + 1) > 4$ $(0, \infty)$ **48.** $6(2 - x) \geq 12$ $(-\infty, 0]$
49. $\frac{-5x + 11}{2} \leq 7$ $\left[-\frac{3}{5}, \infty\right)$ **50.** $\frac{4x - 8}{7} < 0$ $(-\infty, 2)$
51. $8x - 16.4 \leq 10x + 2.8$ **52.** $18x - 25.6 < 10x + 60.8$

53. Explain how solving a linear inequality is similar to solving a linear equation.   answers may vary

54. Explain how solving a linear inequality is different from solving a linear equation.   answers may vary

MIXED PRACTICE

Solve. Write the solution set using interval notation.

55. $-5x + 4 \le -4(x - 1)$ $[0, \infty)$

56. $-6x + 2 < -3(x + 4)$ $\left(\frac{14}{3}, \infty\right)$

57. $\frac{1}{4}(x - 7) \ge x + 2$ $(-\infty, -5]$

58. $\frac{3}{5}(x + 1) \le x + 1$ $[-1, \infty)$

59. $\frac{2}{3}(x + 2) < \frac{1}{5}(2x + 7)$ $\left(-\infty, \frac{1}{4}\right)$

60. $\frac{1}{6}(3x + 10) > \frac{5}{12}(x - 1)$ $(-25, \infty)$

61. $4(x - 6) + 2x - 4 \ge 3(x - 7) + 10x$ $(-\infty, -1]$

62. $7(2x + 3) + 4x \le 7 + 5(3x - 4)$ $\left(-\infty, -\frac{34}{3}\right]$

63. $\frac{5x + 1}{7} - \frac{2x - 6}{4} \ge -4$ $\left[-\frac{79}{3}, \infty\right)$

64. $\frac{1 - 2x}{3} + \frac{3x + 7}{7} > 1$ $\left(-\infty, \frac{7}{5}\right)$

65. $\frac{-x + 2}{2} - \frac{1 - 5x}{8} < -1$ $(-\infty, -15)$

66. $\frac{3 - 4x}{6} - \frac{1 - 2x}{12} \le -2$ $\left[\frac{29}{6}, \infty\right)$

67. $0.8x + 0.6x \ge 4.2$ $[3, \infty)$

68. $0.7x - x > 0.45$ $(-\infty, -1.5)$

69. $\frac{x + 5}{5} - \frac{3 + x}{8} \ge -\frac{3}{10}$ $\left[-\frac{37}{3}, \infty\right)$

70. $\frac{x - 4}{2} - \frac{x - 2}{3} > \frac{5}{6}$ $(13, \infty)$

71. $\frac{x + 3}{12} + \frac{x - 5}{15} < \frac{2}{3}$ $(-\infty, 5)$

72. $\frac{3x + 2}{18} - \frac{1 + 2x}{6} \le -\frac{1}{2}$ $\left[\frac{8}{3}, \infty\right)$

Solve. See Examples 8 and 9.

73. Shureka has scores of 72, 67, 82, and 79 on her algebra tests. Use an inequality to find the minimum score she can make on the final exam to pass the course with an average of 60 or higher, given that the final exam counts as two tests.   30

74. In a Winter Olympics speed-skating event, Hans scored times of 3.52, 4.04, and 3.87 minutes on his first three trials. Use an inequality to find the maximum time he can score on his last trial so that his average time is under 4.0 minutes.   4.57 min

75. A small plane's maximum takeoff weight for passengers, luggage, and cargo is 2000 pounds. Six passengers weigh an average of 160 pounds each. Use an inequality to find the maximum weight of luggage and cargo the plane can carry. 1040 lb

76. A clerk must use the elevator to move boxes of paper. The elevator's weight limit is 1500 pounds. If each box of paper weighs 66 pounds and the clerk weighs 147 pounds, use an inequality to find the maximum number of boxes she can move on the elevator at one time.   20 boxes

77. To mail an envelope first class, the U.S. Post Office charges 37 cents for the first ounce and 23 cents per ounce for each additional ounce. Use an inequality to find the maximum number of whole ounces that can be mailed for $4.00   16 oz

78. A shopping mall parking garage charges $2 for the first half hour and $1.20 for each additional half hour or a portion of a half hour. Use an inequality to find how long you can park if you have $8.00 in cash.   3 hours

79. Northeast Telephone Company offers two billing plans for local calls. Plan 1 charges $25 per month for unlimited calls, and plan 2 charges $13 per month plus 6 cents per call. Use an inequality to find the number of monthly calls for which plan 1 is more economical than plan 2.   more than 200 calls

80. A car rental company offers two subcompact rental plans. Plan A charges $32 per day for unlimited mileage, and plan B charges $24 per day plus 15 cents per mile. Use an inequality to find the number of daily miles for which plan A is more economical than plan B.   54 miles or more

81. At room temperature, glass used in windows actually has some properties of a liquid. It has a very slow, viscous flow. (Viscosity is the property of a fluid that resists internal flow. For example, lemonade flows more easily than fudge syrup. Fudge syrup has a higher viscosity than lemonade.) Glass does not become a true liquid until temperatures are greater than or equal to 500°C. Find the Fahrenheit temperatures for which glass is a liquid. (Use the formula $F = \frac{9}{5}C + 32$.) $F \ge 932°$

82. Stibnite is a silvery white mineral with a metallic luster. It is one of the few minerals that melts easily in match flame or at temperatures of approximately 977°F or greater. Find the Celsius temperatures for which stibnite melts. (Use the formula $C = \frac{5}{9}[F - 32]$.) $C \ge 525°$

83. Although beginning salaries vary greatly according to your field of study, the equation $s = 2806.6t + 32,558$ can be used to approximate and to predict average beginning salaries for candidates with bachelor's degrees. The variable s is the starting salary and t is the number of years after 1995.

  a. Approximate when beginning salaries for candidates will be greater than $60,000.   the end of 2004

  b. Determine the year you plan to graduate from college. Use this year to find the corresponding value of t and approximate your beginning salary.   answers may vary

84. Use the formula in Example 9 to estimate the years that the consumption of cigarettes will be less than 50 billion per year.   2024 and after

The average consumption per person per year of whole milk w in gallons can be approximated by the equation

$$w = -0.13t + 8$$

where t is the number of years after 1997. The average consumption of skim milk s per person per year can be approximated by the equation

$$s = -0.16t + 4$$

where t is the number of years after 1997. The consumption of whole milk is shown on the graph in blue and the consumption of skim milk is shown on the graph in red. Use this information to answer Exercises 85 through 92.

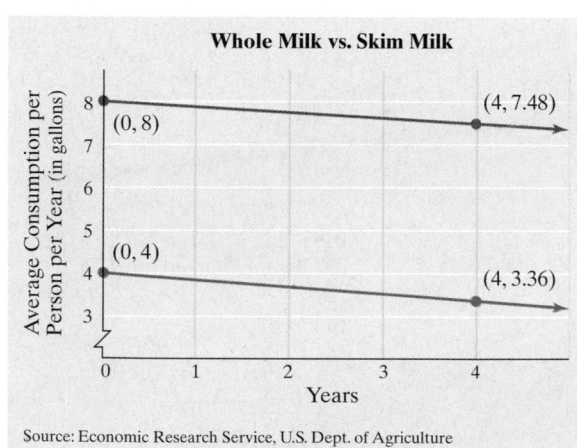

Whole Milk vs. Skim Milk

Source: Economic Research Service, U.S. Dept. of Agriculture

85. Is the consumption of whole milk increasing or decreasing over time? Explain how you arrived at your answer.

86. Is the consumption of skim milk increasing or decreasing over time? Explain how you arrived at your answer.   decreasing

87. Predict the consumption of whole milk in the year 2008 (*Hint:* Find the value of *t* that corresponds to the year 2008.)

88. Predict the consumption of skim milk in the year 2008 (*Hint:* Find the value of *t* that corresponds to the year 2008.)

89. Determine when the consumption of whole milk will be less than 7 gallons per person per year.   2004

90. Determine when the consumption of skim milk will be less than 2 gallons per person per year.   2009

91. For 1997 through 2001 the consumption of whole milk was greater than the consumption of skim milk. Explain how this can be determined from the graph.   answers may vary

92. How will the two lines in the graph appear if the consumption of whole milk is the same as the consumption of skim milk?   answers may vary

REVIEW AND PREVIEW

List or describe the integers that make both inequalities true.

93. $x < 5$ and $x > 1$ 2, 3, 4

94. $x \geq 0$ and $x \leq 7$ 0, 1, 2, 3, 4, 5, 6, 7

95. $x \geq -2$ and $x \geq 2$ 2, 3, 4 …

96. $x < 6$ and $x < -5$ -6, -7, -8, …

Graph each set on a number line and write it in interval notation. See Section 3.2.

97. $\{x \mid 0 \leq x \leq 5\}$, $[0, 5]$

98. $\{x \mid -7 < x \leq 1\}$, $(-7, 1]$

99. $\left\{x \mid -\frac{1}{2} < x < \frac{3}{2}\right\}$, $\left(-\frac{1}{2}, \frac{3}{2}\right)$

100. $\{x \mid -2.5 \leq x < 5.3\}$, $[-2.5, 5.3)$

Concept Extensions

Solve each inequality.

101. $4(x - 1) \geq 4x - 8$ $(-\infty, \infty)$

102. $3x + 1 < 3(x - 2)$ $\varnothing$

103. $7x < 7(x - 2)$ $\varnothing$

104. $8(x + 3) \leq 7(x + 5) + x$ $(-\infty, \infty)$

105. Medicare costs are projected to grow at a far faster rate than the U.S. economy in the next decade. The table below represents projected Medicare spending as the percent of the gross domestic product.

85. decreasing   87. 6.57 gal   88. 2.24 gal

Year	2000	2010	2020	2030	2040	2050	2060
Projected Percent	2.2	2.5	3.8	4.9	6.0	6.8	7.7

  a. Using the data in the table, find the regression equation that best fits the data. Let $x = 0$ represent the number of years since the year 2000. $y = 0.098x + 1.918$

  b. According to the predictions, find the annual projected rate of increase over the 60 years.   The percent is increasing at the rate of 0.098% per year.

  c. Estimate the years that the percent of Medicare spending is greater than or equal to 4%.   the years 2021 and later

INTEGRATED REVIEW   LINEAR EQUATIONS AND INEQUALITIES

Solve each equation or inequality. For inequalities, write the solution set in interval notation.

1. $-4x = 20$
-5

2. $-4x < 20$
$(-5, \infty)$

3. $\dfrac{3x}{4} \geq 2$ $\left[\dfrac{8}{3}, \infty\right)$

4. $5x + 3 \geq 2 + 4x$
$[-1, \infty)$

5. $6(y - 4) = 3(y - 8)$
0

6. $-4x \leq \dfrac{2}{5}$ $\left[-\dfrac{1}{10}, \infty\right)$

7. $-3x \geq \dfrac{1}{2}$ $\left(-\infty, -\dfrac{1}{6}\right]$

8. $5(y + 4) = 4(y + 5)$
0

9. $7x < 7(x - 2)$
$\varnothing$

10. $\dfrac{-5x + 11}{2} \leq 7$ $\left[-\dfrac{3}{5}, \infty\right)$

11. $-5x + 1.5 = -19.5$
4.2

12. $-5x + 4 = -26$
6

13. $5 + 2x - x = -x + 3 - 14$
-8

14. $12x + 14 < 11x - 2$
$(-\infty, -16)$

15. $\dfrac{x}{5} - \dfrac{x}{4} = \dfrac{x - 2}{2}$ $\dfrac{20}{11}$

16. $12x - 12 = 8(x - 1)$
1

17. $2(x - 3) > 70$
$(38, \infty)$

18. $-3x - 4.7 = 11.8$
-5.5

19. $-2(b - 4) - (3b - 1) = 5b + 3$ $\dfrac{3}{5}$

20. $8(x + 3) < 7(x + 5) + x$
$(-\infty, \infty)$

21. $\dfrac{3t + 1}{8} = \dfrac{5 + 2t}{7} + 2$ 29

22. $4(x - 6) - x = 8(x - 3) - 5x$
all real numbers

23. $\dfrac{x + 3}{12} + \dfrac{x - 5}{15} < \dfrac{2}{3}$ $(-\infty, 5)$

24. $\dfrac{y}{3} + \dfrac{y}{5} = \dfrac{y + 3}{10}$ $\dfrac{9}{13}$

25. $5(x - 6) + 2x > 3(2x - 1) - 4$
$(23, \infty)$

26. $14(x - 1) - 7x \leq 2(3x - 6) + 4$
$(-\infty, 6]$

27. $\dfrac{1}{4}(3x + 2) - x \geq \dfrac{3}{8}(x - 5) + 2$

$\left(-\infty, \dfrac{3}{5}\right]$

28. $\dfrac{1}{3}(x - 10) - 4x > \dfrac{5}{6}(2x + 1) - 1$

$\left(-\infty, -\dfrac{19}{32}\right)$

STUDY SKILLS REMINDER

Have You Decided to Successfully Complete This Course?

Ask yourself if one of your current goals is to successfully complete this course.

If it is not a goal of yours, ask yourself why? One common reason is fear of failure. Amazingly enough, fear of failure alone can be strong enough to keep many of us from doing our best in any endeavor. Another common reason is that you simply haven't taken the time to make successfully completing this course one of your goals.

If you are taking this mathematics course, then successfully completing this course probably should be one of your goals. To make it a goal, start by writing this goal in your mathematics notebook. Then read or reread Section 1.1 and make a commitment to try the suggestions in this section.

If successfully completing this course is already a goal of yours, also read or reread Section 1.1 and try some suggestions in this section so that you are actively working toward your goal.

Good luck and don't forget that a positive attitude will make a big difference!

3.3 COMPOUND INEQUALITIES

Objectives

1 Find the intersection of two sets.

2 Solve compound inequalities containing **and**.

3 Find the union of two sets.

4 Solve compound inequalities containing **or**.

Two inequalities joined by the words **and** or **or** are called **compound inequalities.**

Compound Inequalities

$$x + 3 < 8 \text{ and } x > 2$$

$$\frac{2x}{3} \geq 5 \text{ or } -x + 10 < 7$$

1 The solution set of a compound inequality formed by the word **and** is the **intersection** of the solution sets of the two inequalities. We use the symbol ∩ to represent "intersection."

> **Intersection of Two Sets**
>
> The intersection of two sets, A and B, is the set of all elements common to both sets. A intersect B is denoted by
>
> $A \cap B$

EXAMPLE 1

If $A = \{x \mid x \text{ is an even number greater than 0 and less than 10}\}$ and $B = \{3, 4, 5, 6\}$, find $A \cap B$.

Solution Let's list the elements in set A.

$$A = \{2, 4, 6, 8\}$$

The numbers 4 and 6 are in both sets. The intersection is $\{4, 6\}$.

2 A value is a solution of a compound inequality formed by the word **and** if it is a solution of *both* inequalities. For example, the solution set of the compound inequality $x \le 5$ and $x \ge 3$ contains all values of x that make the inequality $x \le 5$ a true statement **and** the inequality $x \ge 3$ a true statement. The first graph shown below is the graph of $x \le 5$, the second graph is the graph of $x \ge 3$, and the third graph shows the intersection of the two graphs. The third graph is the graph of $x \le 5$ **and** $x \ge 3$.

$\{x \mid x \le 5\}$ $(-\infty, 5]$

$\{x \mid x \ge 3\}$ $[3, \infty)$

$\{x \mid x \le 5 \text{ and } x \ge 3\}$ $[3, 5]$

Since $x \ge 3$ is the same as $3 \le x$, the compound inequality $3 \le x$ and $x \le 5$ can be written in a more compact form as $3 \le x \le 5$. The solution set $\{x \mid 3 \le x \le 5\}$ includes all numbers that are greater than or equal to 3 and at the same time less than or equal to 5. In interval notation, the solution set is $[3, 5]$.

EXAMPLE 2

Solve $x - 7 < 2$ and $2x + 1 < 9$.

Solution First we solve each inequality separately.

$$\begin{aligned} x - 7 &< 2 &\text{and}&& 2x + 1 &< 9 \\ x &< 9 &\text{and}&& 2x &< 8 \\ x &< 9 &\text{and}&& x &< 4 \end{aligned}$$

Now we can graph the two intervals on two number lines and find their intersection. Their intersection is shown on the third number line on the following page.

$\{x \mid x < 9\}$ $(-\infty, 9)$

$\{x \mid x < 4\}$ $(-\infty, 4)$

$\{x \mid x < 9 \text{ and } x < 4\} = \{x \mid x < 4\}$ $(-\infty, 4)$

The solution set is $(-\infty, 4)$.

EXAMPLE 3

Solve $2x \geq 0$ and $4x - 1 \leq -9$.

Solution   First we solve each inequality separately.

CLASSROOM EXAMPLE
Solve $4x \geq 0$ and $2x + 4 \geq 2$.
answer: $[0, \infty)$

$$2x \geq 0 \quad \text{and} \quad 4x - 1 \leq -9$$
$$x \geq 0 \quad \text{and} \quad 4x \leq -8$$
$$x \geq 0 \quad \text{and} \quad x \leq -2$$

Now we can graph the two intervals and find their intersection.

$\{x \mid x \geq 0\}$ $[0, \infty)$

$\{x \mid x \leq -2\}$ $(-\infty, -2]$

$\{x \mid x \geq 0 \text{ and } x \leq -2\} = \emptyset$ $\emptyset$

There is no number that is greater than or equal to 0 *and* less than or equal to -2. The solution set is $\emptyset$.

> **Helpful Hint**
> Example 3 shows that some compound inequalities have no solution. Also, some have all real numbers as solutions.

To solve a compound inequality written in a compact form, such as $2 < 4 - x < 7$, we get x alone in the "middle part." Since a compound inequality is really two inequalities in one statement, we must perform the same operations on all three parts of the inequality.

EXAMPLE 4

Solve $2 < 4 - x < 7$.

Solution   To get x alone, we first subtract 4 from all three parts.

CLASSROOM EXAMPLE
Solve $5 < 1 - x < 9$.
answer: $(-8, -4)$

> **Helpful Hint**
> Don't forget to reverse both inequality symbols.

$$2 < 4 - x < 7$$
$$2 - 4 < 4 - x - 4 < 7 - 4 \qquad \text{Subtract 4 from all three parts.}$$
$$-2 < -x < 3 \qquad \text{Simplify.}$$
$$\frac{-2}{-1} > \frac{-x}{-1} > \frac{3}{-1} \qquad \text{Divide all three parts by } -1 \text{ and reverse the inequality symbols.}$$
$$2 > x > -3$$

This is equivalent to $-3 < x < 2$.

The solution set in interval notation is $(-3, 2)$, and its graph is shown.

Discover the Concept

For the compound inequality $2 < 4 - x < 7$, in Example 4, graph $y_1 = 2$, $y_2 = 4 - x$, and $y_3 = 7$. With this notation, we can think of our inequality as $y_1 < y_2 < y_3$.

a. Find the point of intersection of y_1 and y_2.
b. Find the point of intersection of y_2 and y_3.
c. Determine where the graph of y_2 is between the graphs of y_1 and y_3.
d. Find the x-values for which $y_1 < y_2 < y_3$.
e. Write the solution set of $y_1 < y_2 < y_3$.

In the discovery above, we find the points of intersection as shown.

The solution of $y_1 < y_2 < y_3$, or $2 < 4 - x < 7$ contains the x-values where the graph of $y_2 = 4 - x$ is between $y_1 = 2$ and $y_3 = 7$. These x-values in interval notation are $(-3, 2)$. (Parentheses are used because of the inequality symbols $<$.)

EXAMPLE 5

Solve $-1 \le \dfrac{2x}{3} + 5 \le 2$. Then use a graphical approach to check the solution.

Solution First, clear the inequality of fractions by multiplying all three parts by the LCD of 3.

$$-1 \le \frac{2x}{3} + 5 \le 2$$

$$3(-1) \le 3\left(\frac{2x}{3} + 5\right) \le 3(2) \qquad \text{Multiply all three parts by the LCD of 3.}$$

$$-3 \le 2x + 15 \le 6 \qquad \text{Use the distributive property and multiply.}$$

$$-3 - 15 \le 2x + 15 - 15 \le 6 - 15 \qquad \text{Subtract 15 from all three parts.}$$

$$-18 \le 2x \le -9 \qquad \text{Simplify.}$$

$$\frac{-18}{2} \le \frac{2x}{2} \le \frac{-9}{2} \qquad \text{Divide all three parts by 2.}$$

$$-9 \le x \le -\frac{9}{2} \qquad \text{Simplify.}$$

The graph of the solution is shown.

The solution set in interval notation is $\left[-9, -\frac{9}{2}\right]$.

To check, graph $y_1 = -1$, $y_2 = \frac{2x}{3} + 5$, and $y_3 = 2$ in a $[-15, 5, 3]$ by $[-5, 5, 1]$ window and solve $y_1 \le y_2 \le y_3$.

The solution of $y_1 \le y_2 \le y_3$ consists of the x-values where the graph of y_2 is between or equal to the graphs of y_1 and y_3. The solution set in interval notation is $[-9, -4.5]$ or $\left[-9, -\frac{9}{2}\right]$.

3 The solution set of a compound inequality formed by the word **or** is the **union** of the solution sets of the two inequalities. We use the symbol $\cup$ to denote "union."

> ### Union of Two Sets
>
> The union of two sets, A and B, is the set of elements that belong to *either* of the sets. A union B is denoted by
>
> $$A \cup B$$

EXAMPLE 6

If $A = \{x \mid x \text{ is an even number greater than 0 and less than 10}\}$ and $B = \{3, 4, 5, 6\}$, find $A \cup B$.

Solution   Recall from Example 1 that $A = \{2, 4, 6, 8\}$. The numbers that are in either set or both sets are $\{2, 3, 4, 5, 6, 8\}$. This set is the union.

4   A value is a solution of a compound inequality formed by the word **or** if it is a solution of **either** inequality. For example, the solution set of the compound inequality $x \le 1$ **or** $x \ge 3$ contains all numbers that make the inequality $x \le 1$ a true statement **or** the inequality $x \ge 3$ a true statement.

$\{x | x \le 1\}$ $(-\infty, 1]$

$\{x | x \ge 3\}$ $[3, \infty)$

$\{x | x \le 1 \text{ or } x \ge 3\}$ $(-\infty, 1] \cup [3, \infty)$

In interval notation, the set $\{x | x \le 1 \text{ or } x \ge 3\}$ is written as $(-\infty, 1] \cup [3, \infty)$.

EXAMPLE 7

Solve $5x - 3 \le 10$ or $x + 1 \ge 5$.

Solution   First we solve each inequality separately.

$$
\begin{aligned}
5x - 3 &\le 10 & \text{or} & & x + 1 &\ge 5 \\
5x &\le 13 & \text{or} & & x &\ge 4 \\
x &\le \frac{13}{5} & \text{or} & & x &\ge 4
\end{aligned}
$$

Now we can graph each interval and find their union.

$\left\{x \middle| x \le \dfrac{13}{5}\right\}$ $\left(-\infty, \dfrac{13}{5}\right]$

$\{x | x \ge 4\}$ $[4, \infty)$

$\left\{x \middle| x \le \dfrac{13}{5} \text{ or } x \ge 4\right\}$ $\left(-\infty, \dfrac{13}{5}\right] \cup [4, \infty)$

The solution set is $\left(-\infty, \dfrac{13}{5}\right] \cup [4, \infty)$.

EXAMPLE 8

Solve $-2x - 5 < -3$ or $6x < 0$.

Solution   First we solve each inequality separately.

$$
\begin{aligned}
-2x - 5 &< -3 & \text{or} & & 6x &< 0 \\
-2x &< 2 & \text{or} & & x &< 0 \\
x &> -1 & \text{or} & & x &< 0
\end{aligned}
$$

Now we can graph each interval and find their union.

$\{x | x > -1\}$ $(-1, \infty)$

$\{x | x < 0\}$ $(-\infty, 0)$

$\{x | x > -1 \text{ or } x < 0\}$
= all real numbers $(-\infty, \infty)$

The solution sct is $(-\infty, \infty)$.

STUDY SKILLS REMINDER

Continue your outline started in Section 1.5. Write how to recognize and how to solve compound inequalities in your own words. For example:

Solving Equations and Inequalities

I. Equations
   A. Linear equations—(Section 1.5)

II. Inequalities
   A. Linear inequalities—(Section 3.2)
   B. **Compound inequalities**—Recognize: *2 inequality signs or 2 inequalities separated by "and" or "or"*—Solve: if 2 inequality signs, then solve for *x* in the middle, remember that "and" means intersection and "or" means union.

See Appendix A for summary exercises.

---

4. $\{x | x \text{ is an even integeror } x = 5 \text{ or } x = 7\}$ **8.** $\{x | x \text{ is an odd integer or } x = 4 \text{ or } x = 6\}$

9. $\{x | x \text{ is an odd integer or } x = 2 \text{ or } x = 4\}$ **12.** $\{x | x \text{ is an even integer or } x = 3 \text{ or } x = 5\}$

EXERCISE SET 3.3

STUDY   CD/   PH MATH   MathXL®Tutorials   MathXL®   MyMathLab®
GUIDE/SSM   VIDEO   TUTOR CENTER   ON CD

If $A = \{x | x \text{ is an even integer}\}$, $B = \{x | x \text{ is an odd integer}\}$, $C = \{2, 3, 4, 5\}$, and $D = \{4, 5, 6, 7\}$, list the elements of each set. See Examples 1 and 6.

1. $C \cup D$ $\{2, 3, 4, 5, 6, 7\}$

2. $C \cap D$ $\{4, 5\}$

3. $A \cap D$ $\{4, 6\}$

4. $A \cup D$

5. $A \cup B$ $\{\ldots, -2, -1, 0, 1, \ldots\}$

6. $A \cap B$ $\varnothing$

7. $B \cap D$ $\{5, 7\}$

8. $B \cup D$

9. $B \cup C$

10. $B \cap C$ $\{3, 5\}$

11. $A \cap C$ $\{2, 4\}$

12. $A \cup C$

Solve each compound inequality. Graph the solution set and write it in interval notation. See Examples 2 and 3.

13. $x < 5 \text{ and } x > -2$, $(-2, 5)$

14. $x \le 7 \text{ and } x \le 1$, $(-\infty, 1]$

15. $x + 1 \ge 7 \text{ and } 3x - 1 \ge 5$, $[6, \infty)$

16. $-2x < -8 \text{ and } x - 5 < 5$, $(4, 10)$

17. $4x + 2 \le -10 \text{ and } 2x \le 0$, $(-\infty, -3]$

18. $x + 4 > 0 \text{ and } 4x > 0$, $(0, \infty)$

Solve each compound inequality. Graph the solution set and write it in interval notation. See Examples 4 and 5.

19. $5 < x - 6 < 11$ graph, $(11, 17)$

20. $-2 \le x + 3 \le 0$ graph, $[-5, -3]$

21. $-2 \le 3x - 5 \le 7$ graph, $[1, 4]$

22. $1 < 4 + 2x < 7$ graph, $\left(-\frac{3}{2}, \frac{3}{2}\right)$

🔒 **23.** $1 \le \frac{2}{3}x + 3 \le 4$ graph, $\left[-3, \frac{3}{2}\right]$

24. $-2 < \frac{1}{2}x - 5 < 1$ graph, $(6, 12)$

25. $-5 \le \frac{x + 1}{4} \le -2$ graph, $[-21, -9]$

26. $-4 \le \frac{2x + 5}{3} \le 1$ graph, $\left[-\frac{17}{2}, -1\right]$

Solve each compound inequality. Graph the solution set and write it in interval notation. See Examples 7 and 8.

27. $x < -1$ or $x > 0$ graph, $(\infty, -1) \cap (0, \infty)$

28. $x < 1$ or $x < -3$ graph, $(-\infty, 1)$

29. $-2x \le -4$ or $5x - 20 \ge 5$ graph, $[2, \infty)$

30. $x + 4 < 0$ or $6x > -12$ graph, $(-\infty, -4) \cup (-2, \infty)$

31. $3(x - 1) < 12$ or $x + 7 > 10$ graph, $(-\infty, \infty)$

32. $5(x - 1) \ge -5$ or $5 - x \le 11$ graph, $[-6, \infty)$

33. Explain how solving an and–compound inequality is similar to finding the intersection of two sets.   answers may vary

34. Explain how solving an or–compound inequality is similar to finding the union of two sets.   answers may vary

MIXED PRACTICE

Solve each compound inequality. Graph the solution set and write it in interval notation.

35. $x < 2$ and $x > -1$ graph, $(-1, 2)$

36. $x < 5$ and $x < 1$ graph, $(-\infty, 1)$

37. $x < 2$ or $x > -1$ graph, $(-\infty, \infty)$

38. $x < 5$ or $x < 1$ graph, $(-\infty, 5)$

39. $x \ge -5$ and $x \ge -1$ graph, $[-1, \infty)$

40. $x \le 0$ or $x \ge -3$ graph, $(-\infty, \infty)$

41. $x \ge -5$ or $x \ge -1$ graph, $[-5, \infty)$

42. $x \le 0$ and $x \ge -3$ graph, $[-3, 0]$

43. $0 \le 2x - 3 \le 9$ graph, $\left[\frac{3}{2}, 6\right]$

44. $3 < 5x + 1 < 11$ graph, $\left(\frac{2}{5}, 2\right)$

45. $\frac{1}{2} < x - \frac{3}{4} < 2$ graph, $\left(\frac{5}{4}, \frac{11}{4}\right)$

46. $\frac{2}{3} < x + \frac{1}{2} < 4$ graph, $\left(\frac{1}{6}, \frac{7}{2}\right)$

47. $x + 3 \ge 3$ and $x + 3 \le 2$ graph, $\varnothing$

48. $2x - 1 \ge 3$ and $-x > 2$ graph, $\varnothing$

49. $3x \ge 5$ or $-x - 6 < 1$ graph, $(-7, \infty)$

50. $\frac{3}{8}x + 1 \le 0$ or $-2x < -4$ graph, $\left(-\infty, -\frac{8}{3}\right] \cup (2, \infty)$

51. $0 < \frac{5 - 2x}{3} < 5$ graph, $\left(-5, \frac{5}{2}\right)$

52. $-2 < \frac{-2x - 1}{3} < 2$ graph, $\left(-\frac{7}{2}, \frac{5}{2}\right)$

53. $-6 < 3(x - 2) \le 8$ graph, $\left(0, \frac{14}{3}\right]$

54. $-5 < 2(x + 4) < 8$ graph, $\left(-\frac{13}{2}, 0\right)$

55. $-x + 5 > 6$ and $1 + 2x \le -5$ graph, $(-\infty, -3]$

56. $5x \le 0$ and $-x + 5 < 8$ graph, $(-3, 0]$

🔒 **57.** $3x + 2 \le 5$ or $7x > 29$ graph, $\left(-\infty, 1\right] \cup \left(\frac{29}{7}, \infty\right)$

58. $-x < 7$ or $3x + 1 < -20$ graph, $(-\infty, -7) \cup (-7, \infty)$

59. $5 - x > 7$ and $2x + 3 \ge 13$ graph, $\varnothing$

60. $-2x < -6$ or $1 - x > -2$ graph, $(-\infty, 3) \cup (3, \infty)$

61. $-\frac{1}{2} \le \frac{4x - 1}{6} < \frac{5}{6}$ graph, $\left[-\frac{1}{2}, \frac{3}{2}\right)$

62. $-\frac{1}{2} \le \frac{3x - 1}{10} < \frac{1}{2}$ graph, $\left[-\frac{4}{3}, 2\right)$

63. $\frac{1}{15} < \frac{8 - 3x}{15} < \frac{4}{5}$ graph, $\left(-\frac{4}{3}, \frac{7}{3}\right)$

67. a. $(-5, 2.5)$ **b.** $(-\infty, -5) \cup (2.5, \infty)$ **68. a.** $(-3.5, 2.5)$

64. $-\dfrac{1}{4} < \dfrac{6 - x}{12} < -\dfrac{1}{6}$, $\left(8, 9\right)$

65. $0.3 < 0.2x - 0.9 < 1.5$, $(6, 12)$

66. $-0.7 \le 0.4x + 0.8 < 0.5$, $[-3.75, -0.75)$

Solve each compound inequality using the graphing utility screens.

67. a. $y_1 < y_2 < y_3$
  b. $y_2 < y_1$ or $y_2 > y_3$

68. a. $y_1 < y_2 < y_3$
  b. $y_2 < y_1$ or $y_2 > y_3$

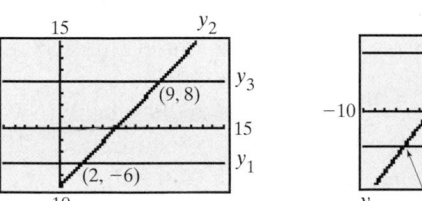

69. a. $y_1 \le y_2 \le y_3$
  b. $y_2 \le y_1$ or $y_2 \ge y_3$

70. a. $y_1 \le y_2 \le y_3$
  b. $y_2 \le y_1$ or $y_2 \ge y_3$

REVIEW AND PREVIEW   **68. b.** $(-\infty, -3.5) \cup (2.5, \infty)$

Evaluate the following. See Sections 1.2 and 1.3.

71. $|-7| - |19|$ -12

72. $|-7 - 19|$ 26

73. $-(-6) - |-10|$ -4

74. $|-4| - (-4) + |-20|$ 28

Find by inspection all values for x that make each equation true.

75. $|x| = 7$ $-7, 7$

76. $|x| = 5$ $-5, 5$

77. $|x| = 0$ 0

78. $|x| = -2$ no solution

Concept Extensions   **69. a.** $[2, 9]$ **b.** $(-\infty, 2] \cup [9, \infty)$

The formula for converting Fahrenheit temperatures to Celsius temperatures is $C = \frac{5}{9}(F - 32)$. Use this formula for Exercises 79 and 80.   **70. a.** $[-6.5, 0]$ **70. b.** $(-\infty, -6.5] \cup [0, \infty)$

79. During a recent year, the temperatures in Chicago ranged from $-29°$ to $35°C$. Use a compound inequality to convert these temperatures to Fahrenheit temperatures.

80. In Oslo, the average temperature ranges from $-10°$ to $18°$ Celsius. Use a compound inequality to convert these temperatures to the Fahrenheit scale. $14° \le F \le 64.4°$

Solve.   **79.** $-20.2° \le F \le 95°$ **86.** , $(2, \infty)$

81. Christian D'Angelo has scores of 68, 65, 75, and 78 on his algebra tests. Use a compound inequality to find the scores he can make on his final exam to receive a C in the course. The final exam counts as two tests, and a C is received if the final course average is from 70 to 79. $67 \le$ final score ≤ 94

82. Wendy Wood has scores of 80, 90, 82, and 75 on her chemistry tests. Use a compound inequality to find the range of scores she can make on her final exam to receive a B in the course. The final exam counts as two tests, and a B is received if the final course average is from 80 to 89. $76.5 \le$ final score ≤ 100

Use the graph to answer Exercises 83 and 84.

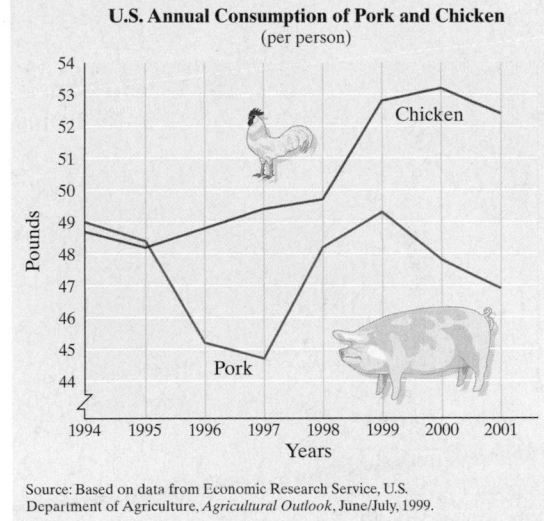

U.S. Annual Consumption of Pork and Chicken
(per person)

Source: Based on data from Economic Research Service, U.S. Department of Agriculture, *Agricultural Outlook*, June/July, 1999.

83. For what years was the consumption of pork greater than 48 pounds per person *and* the consumption of chicken greater than 48 pounds per person?   1994, 1995, 1998, 1999

84. For what years was the consumption of pork less than 48 pounds per person *or* the consumption of chicken greater than 49 pounds per person?   1997, 2000, 2001

Solve each compound inequality for x. See the example below.

To solve $x - 6 < 3x < 2x + 5$, notice that this inequality contains a variable not only in the middle, but also on the left and the right. When this occurs, we solve by rewriting the inequality using the word **and**.

$$x - 6 < 3x \quad \text{and} \quad 3x < 2x + 5$$
$$-6 < 2x \quad \text{and} \quad x < 5$$
$$-3 < x$$
$$x > -3 \quad \text{and} \quad x < 5$$

88. , $\{4\}$

89. , $(-\infty, -1)$

$x > -3$

$x < 5$

$-3 < x < 5$, or $(-3, 5)$

85. $2x - 3 < 3x + 1 < 4x - 5$, $(6, \infty)$

86. $x + 3 < 2x + 1 < 4x + 6$

87. $-3(x - 2) \le 3 - 2x \le 10 - 3x$, $[3, 7]$

88. $7x - 1 \le 7 + 5x \le 3(1 + 2x)$

89. $5x - 8 < 2(2 + x) < -2(1 + 2x)$

90. $1 + 2x < 3(2 + x) < 1 + 4x$, $(5, \infty)$

TEACHING TIP
A Group Activity for this section is available in the Instructor's Resource Manual.

3.4 ABSOLUTE VALUE EQUATIONS

Objective

1 Solve absolute value equations.

1 In Chapter 1, we defined the absolute value of a number as its distance from 0 on a number line.

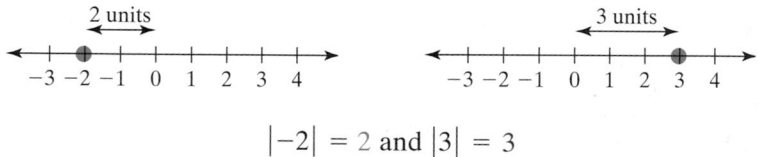

$$|-2| = 2 \text{ and } |3| = 3$$

In this section, we concentrate on solving equations containing the absolute value of a variable or a variable expression. Examples of absolute value equations are

$$|x| = 3 \qquad -5 = |2y + 7| \qquad |z - 6.7| = |3z + 1.2|$$

Since distance and absolute value are so closely related, absolute value equations and inequalities (see Section 2.7) are extremely useful in solving distance-type problems, such as calculating the possible error in a measurement. For the absolute value equation $|x| = 3$, its solution set will contain all numbers whose distance from 0 is 3 units. Two numbers are 3 units away from 0 on the number line: 3 and -3.

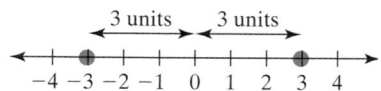

Thus, the solution set of the equation $|x| = 3$ is $\{3, -3\}$. This suggests the following:

> **Solving Equations of the Form $|X| = a$**
>
> If a is a positive number, then $|X| = a$ is equivalent to $X = a$ or $X = -a$.

> **TECHNOLOGY NOTE**
>
> Absolute value can be found under the Math Num menu or under Catolog. Check your graphing calculator manual to find the location on your calculator.

> **Helpful Hint**
>
> For the equation $|X| = a$ in the box to the right, X can be a single variable or a variable expression.

EXAMPLE 1

Solve $|p| = 2$.

Solution Since 2 is positive, $|p| = 2$ is equivalent to $p = 2$ or $p = -2$.

To check, let $p = 2$ and then $p = -2$ in the original equation.

$|p| = 2$	Original equation.		$|p| = 2$	Original equation.
$|2| = 2$	Let $p = 2$.		$|-2| = 2$	Let $p = -2$.
$2 = 2$	True		$2 = 2$	True

The solutions are 2 and -2 or the solution set is $\{2, -2\}$.

CLASSROOM EXAMPLE
Solve $|y| = 5$.
answer: $-5, 5$

TEACHING TIP
Point out the use of the word *or* in the definition of the absolute value property and then ask why "or" was used rather than "and."

To visualize the solution of Example 1, we solve $|x| = 2$ by the intersection-of-graphs method. Graph $y_1 = |x|$ and $y_2 = 2$.

The graphs of $y_1 = |x|$ and $y_2 = 2$ intersect at $(-2, 2)$ and $(2, 2)$. The solutions are thus, the x-values, -2 and 2.

If the expression inside the absolute value bars is more complicated than a single variable x, we can still apply the absolute value property.

EXAMPLE 2

Solve $|5w + 3| = 7$ algebraically. Then use a graphical approach to check.

Solution

Here the expression inside the absolute value bars is $5w + 3$. If we think of the expression $5w + 3$ as x in the absolute value property, we see that $|x| = 7$ is equivalent to

$$x = 7 \quad \text{or} \quad x = -7$$

Then substitute $5w + 3$ for x, and we have

$$5w + 3 = 7 \quad \text{or} \quad 5w + 3 = -7$$

Solve these two equations for w.

$$
\begin{aligned}
5w + 3 &= 7 &\text{or} \quad 5w + 3 &= -7 \\
5w &= 4 &\text{or} \quad 5w &= -10 \\
w &= \frac{4}{5} &\text{or} \quad w &= -2
\end{aligned}
$$

To check, graph $y_1 = |5x + 3|$ and $y_2 = 7$ in a standard window as shown.

The intersections of the two graphs are the points $(-2, 7)$ and $(0.8, 7)$. Therefore, the solutions to the equation $|5x + 3| = 7$ are $x = 0.8$, or $\frac{4}{5}$ and $x = -2$.

Both solutions check, and the solutions are -2 and $\frac{4}{5}$ or the solution set is $\left\{-2, \frac{4}{5}\right\}$.

CLASSROOM EXAMPLE

Solve $|4x + 2| = 6$.
answer: $1, -2$

TEACHING TIP

It might be a good time to point out that the check by substitution is a good way to determine if a given solution is correct but it does not check that all solutions were found.

CLASSROOM EXAMPLE

Solve $\left|\dfrac{x}{3} + 4\right| = 1$.
answer: $-9, -15$

EXAMPLE 3

Solve $\left|\dfrac{x}{2} - 1\right| = 11$.

Solution　$\left| \dfrac{x}{2} - 1 \right| = 11$ is equivalent to

$$\dfrac{x}{2} - 1 = 11 \qquad \text{or} \qquad \dfrac{x}{2} - 1 = -11$$

$$2\left(\dfrac{x}{2} - 1 \right) = 2(11) \quad \text{or} \quad 2\left(\dfrac{x}{2} - 1 \right) = 2(-11) \qquad \text{Clear fractions.}$$

$$x - 2 = 22 \qquad \text{or} \qquad x - 2 = -22 \qquad \text{Apply the distributive property.}$$

$$x = 24 \qquad \text{or} \qquad x = -20$$

The solutions are 24 and -20.

> **Helpful Hint**
>
> If the equation has a single absolute value expression containing variables, isolate the absolute value expression first.

To apply the absolute value rule, first make sure that the absolute value expression is isolated.

EXAMPLE 4

Solve $|2x| + 5 = 7$.

Solution　We want the absolute value expression alone on one side of the equation, so begin by subtracting 5 from both sides. Then apply the absolute value property.

$$|2x| + 5 = 7$$

$$|2x| = 2 \qquad \text{Subtract 5 from both sides.}$$

$$2x = 2 \quad \text{or} \quad 2x = -2$$

$$x = 1 \quad \text{or} \quad x = -1$$

The solutions are -1 and 1.

> **CLASSROOM EXAMPLE**
> Solve $|5x| + 9 = 19$.
> **answer:**　$-2, 2$

EXAMPLE 5

Solve $|y| = 0$.

Solution　We are looking for all numbers whose distance from 0 is zero units. The only number is 0. The solution is 0.

> **CLASSROOM EXAMPLE**
> Solve $|x| = 0$.
> **answer:**　0

The next two examples illustrate a special case for absolute value equations. This special case occurs when an isolated absolute value is equal to a negative number.

EXAMPLE 6

Solve $2|x| + 25 = 23$ algebraically and graphically.

Algebraic Solution:

First, isolate the absolute value.

$$2|x| + 25 = 23$$

$$2|x| = -2 \qquad \text{Subtract 25 from both sides.}$$

$$|x| = -1 \qquad \text{Divide both sides by 2.}$$

The absolute value of a number is never negative, so this equation has no solution.

> **CLASSROOM EXAMPLE**
> Solve $3|y| - 4 = 17$.
> **answer:**　$-7, 7$

Graphical Solution:

Graph $y_1 = 2|x| + 25$ and $y_2 = 23$ in a $[-47, 47, 10]$ by $[-30, 50, 10]$ window.

Since there is no point of intersection, there is no solution.

The solution set is $\{ \ \}$ or $\varnothing$.

EXAMPLE 7

Solve $\left|\dfrac{3x + 1}{2}\right| = -2$.

Solution   Again, the absolute value of any expression is never negative, so no solution exists. The solution set is $\{\ \}$ or $\varnothing$.

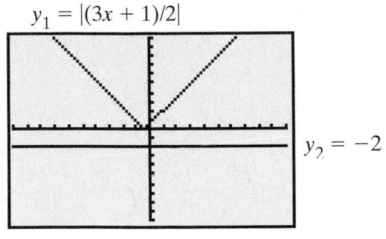

$y_1 = |(3x + 1)/2|$

$y_2 = -2$

A calculator check for Example 7.

Given two absolute value expressions, we might ask, when are the absolute values of two expressions equal? To see the answer, notice that

$$|2| = |2|, \quad |-2| = |-2|, \quad |-2| = |2|, \quad \text{and} \quad |2| = |-2|$$

$$\text{same} \qquad\quad \text{same} \qquad\quad \text{opposites} \qquad\quad \text{opposites}$$

Two absolute value expressions are equal when the expressions inside the absolute value bars are equal to or are opposites of each other.

EXAMPLE 8

Solve $|3x + 2| = |5x - 8|$ algebraically and check graphically.

Solution   This equation is true if the expressions inside the absolute value bars are equal to or are opposites of each other.

$$3x + 2 = 5x - 8 \qquad \text{or} \qquad 3x + 2 = -(5x - 8)$$

Next, solve each equation.

$$3x + 2 = 5x - 8 \quad \text{or} \quad 3x + 2 = -5x + 8$$
$$-2x + 2 = -8 \qquad \text{or} \quad 8x + 2 = 8$$
$$-2x = -10 \qquad \text{or} \qquad 8x = 6$$
$$x = 5 \qquad \text{or} \qquad x = \dfrac{3}{4}$$

Using the x-intercept method to check, rewrite the equation as

$$|3x + 2| - |5x - 8| = 0$$

and graph $y_1 = |3x + 2| - |5x - 8|$. The x-intercepts are $x = 0.75$ and $x = 5$.

The solutions are $\dfrac{3}{4}$ and 5.

EXAMPLE 9

Solve $|x - 3| = |5 - x|$.

Solution

$$x - 3 = 5 - x \quad \text{or} \qquad x - 3 = -(5 - x)$$
$$2x - 3 = 5 \qquad \text{or} \qquad x - 3 = -5 + x$$
$$2x = 8 \qquad \text{or} \quad x - 3 - x = -5 + x - x$$
$$x = 4 \qquad \text{or} \qquad -3 = -5 \qquad\qquad \text{False}$$

Recall from Section 1.5 that when an equation simplifies to a false statement, the equation has no solution. Thus, the only solution for the original absolute value equation is 4.

✔ **CONCEPT CHECK**

True or false? Absolute value equations always have two solutions. Explain your answer.

The following box summarizes the methods shown for solving absolute value equations.

Absolute Value Equations

$|X| = a$ $\begin{cases} \text{If } a \text{ is positive, then solve } X = a \text{ or } X = -a. \\ \text{If } a \text{ is } 0, \text{ solve } X = 0. \\ \text{If } a \text{ is negative, the equation } |X| = a \text{ has no solution.} \end{cases}$

$|X| = |Y|$ $\qquad$ Solve $X = Y$ or $X = -Y$.

STUDY SKILLS REMINDER

Continue your outline started in Section 1.5. Write how to recognize and how to solve absolute value equations in your own words. For example:

Solving Equations and Inequalities

I. Equations

    A. Linear equations—(Section 1.5)

    B. **Absolute value equations**—Recognize: *linear equation with absolute value bars*—Solve: if possible, first isolate absolute value: if $|x| = a$ (positive), then solve $x = a$ or $x = -a$, if $|x| = 0$, then solve $x = 0$, if $|x| = -a$, then there is no solution, if $|x| = |y|$, then solve $x = y$ or $x = -y$.

II. Inequalities

    A. Linear inequalities—(Section 3.2)

    B. Compound inequalities—(Section 3.3)

See Appendix A for summary exercises.

Concept Check Answer:
false; answers may vary

MENTAL MATH

Simplify each expression.

1. $|-7|$ 7

2. $|-8|$ 8

3. $-|5|$ -5

4. $-|10|$ -10

5. $-|-6|$ -6

6. $-|-3|$ -3

7. $|-3| + |-2| + |-7|$ 12

8. $|-1| + |-6| + |-8|$ 15

EXERCISE SET 3.4

STUDY GUIDE/SSM   CD/VIDEO   PH MATH TUTOR CENTER   MathXL®Tutorials ON CD   MathXL®   MyMathLab®

Solve each absolute value equation. See Examples 1 through 7.

 1. $|x| = 7$ $7, -7$

2. $|y| = 15$ $-15, 15$

3. $|3x| = 12.6$ $4.2, -4.2$

4. $|6n| = 12.6$ $2.1, -2.1$

5. $|2x - 5| = 9$ $7, -2$

6. $|6 + 2n| = 4$ $-5, -1$

 7. $\left|\dfrac{x}{2} - 3\right| = 1$ $8, 4$

8. $\left|\dfrac{n}{3} + 2\right| = 4$ $-18, 6$

9. $|z| + 4 = 9$ $5, -5$

10. $|x| + 1 = 3$ $2, -2$

11. $|3x| + 5 = 14$ $3, -3$

12. $|2x| - 6 = 4$ $5, -5$

13. $|2x| = 0$ 0

14. $|7z| = 0$ 0

15. $|4n + 1| + 10 = 4$ $\varnothing$

16. $|3z - 2| + 8 = 1$ $\varnothing$

17. $|5x - 1| = 0$ $\frac{1}{5}$

18. $|3y + 2| = 0$ $-\frac{2}{3}$

19. Write an absolute value equation representing all numbers x whose distance from 0 is 5 units. $|x| = 5$

20. Write an absolute value equation representing all numbers x whose distance from 0 is 2 units. $|x| = 2$

Solve. See Examples 8 and 9.

21. $|5x - 7| = |3x + 11|$ $9, -\dfrac{1}{2}$

22. $|9y + 1| = |6y + 4|$ $-\dfrac{1}{3}, 1$

23. $|z + 8| = |z - 3|$ $-\dfrac{5}{2}$

24. $|2x - 5| = |2x + 5|$ 0

25. Describe how solving an absolute value equation such as $|2x - 1| = 3$ is similar to solving an absolute value equation such as $|2x - 1| = |x - 5|$. answers may vary

26. Describe how solving an absolute value equation such as $|2x - 1| = 3$ is different from solving an absolute value equation such as $|2x - 1| = |x - 5|$. answers may vary

Use the given graphing utility screens to solve the equations shown. Write the solution set of the equation.

27. $|2x - 3| = 5$, where $y_1 = |2x - 3|$ and $y_2 = 5$ $\{-1, 4\}$

28. $|3x - 4| = 14$, where $y_1 = |3x - 4|$ and $y_2 = 14$ $\left\{-\dfrac{10}{3}, 6\right\}$

29. $|x - 4| = |1 - x|$, where $y_1 = |x - 4|$ and $y_2 = |1 - x|$ $\{2.5\}$

30. $|x + 2| = |3 - x|$, where $y_1 = |x + 2|$ and $y_2 = |3 - x|$ $\left\{\dfrac{1}{2}\right\}$

MIXED PRACTICE

Solve each absolute value equation.

31. $|x| = 4$ $4, -4$
32. $|x| = 1$ $-1, 1$
33. $|z| = -2$ $\varnothing$
34. $|y| = -9$ $\varnothing$
35. $|7 - 3x| = 7$ $0, \frac{14}{3}$
36. $|4m + 5| = 5$ $-\frac{5}{2}, 0$
37. $|6x| - 1 = 11$ $2, -2$
38. $|7z| + 1 = 22$ $-3, 3$
39. $|x - 3| + 3 = 7$ $7, -1$
40. $|x + 4| - 4 = 1$ $-9, 1$
41. $\left|\frac{z}{4} + 5\right| = -7$ $\varnothing$
42. $\left|\frac{c}{5} - 1\right| = -2$ $\varnothing$
43. $|9v - 3| = -8$ $\varnothing$
44. $|1 - 3b| = -7$ $\varnothing$
45. $|8n + 1| = 0$ $-\frac{1}{8}$
46. $|5x - 2| = 0$ $\frac{2}{5}$
47. $|1 + 6c| - 7 = -3$
48. $|2 + 3m| - 9 = -7$
49. $|5x + 1| = 11$ $2, -\frac{12}{5}$
50. $|8 - 6c| = 1$ $\frac{7}{6}, \frac{3}{2}$
51. $|4x - 2| = |-10|$ $3, -2$
52. $|3x + 5| = |-4|$
53. $|5x + 1| = |4x - 7|$
54. $|3 + 6n| = |4n + 11|$
55. $|6 + 2x| = -|-7|$ $\varnothing$
56. $|4 - 5y| = -|-3|$ $\varnothing$
57. $|2x - 6| = |10 - 2x|$ 4
58. $|4n + 5| = |4n + 3|$ -1
59. $\left|\frac{2x - 5}{3}\right| = 7$ $13, -8$
60. $\left|\frac{1 + 3n}{4}\right| = 4$ $-\frac{17}{3}, 5$
61. $2 + |5n| = 17$ $3, -3$
62. $8 + |4m| = 24$ $-4, 4$
63. $\left|\frac{2x - 1}{3}\right| = |-5|$ $8, -7$
64. $\left|\frac{5x + 2}{2}\right| = |-6|$ $-\frac{14}{5}, 2$
65. $|2y - 3| = |9 - 4y|$ $2, 3$
66. $|5z - 1| = |7 - z|$ $-\frac{3}{2}, \frac{4}{3}$
67. $\left|\frac{3n + 2}{8}\right| = |-1|$ $2, -\frac{10}{3}$
68. $\left|\frac{2r - 6}{5}\right| = |-2|$ $-2, 8$
69. $|x + 4| = |7 - x|$ $\frac{3}{2}$
70. $|8 - y| = |y + 2|$ 3
71. $\left|\frac{8c - 7}{3}\right| = -|-5|$ $\varnothing$
72. $\left|\frac{5d + 1}{6}\right| = -|-9|$ $\varnothing$

73. Explain why some absolute value equations have two solutions.   answers may vary

74. Explain why some absolute value equations have one solution.   answers may vary

REVIEW AND PREVIEW

The circle graph shows the U.S. Cheese consumption for 2001. Use this graph to answer Exercises 75–77. See Section 1.9.

U.S. Cheese Consumption

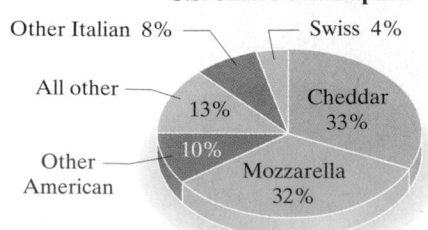

Other Italian 8% — Swiss 4%
All other 13%
Cheddar 33%
Other American 10%
Mozzarella 32%

75. What percent of cheese consumption came from cheddar cheese?   33%

76. A circle contains 360°. Find the number of degrees in the 4% sector for swiss cheese.   14.4°

77. If a family consumed 120 pounds of cheese in 2001, find the amount of mozzarella we might expect they consumed.

List five integer solutions of each inequality.

78. $|x| \leq 3$ answers may vary
79. $|x| \geq -2$ answers may vary
80. $|y| > -10$ answers may vary
81. $|y| < 0$ no solution

Concept Extensions

82. Write an absolute value equation representing all numbers x whose distance from 0 is 5 units. $|x| = 5$

83. Write an absolute value equation representing all numbers x whose distance from 0 is 2 units. $|x| = 2$

Write each as an equivalent absolute value.

84. $x = 6$ or $x = -6$ $|x| = 6$
85. $2x - 1 = 4$ or $2x - 1 = -4$ $|2x - 1| = 4$
86. $x - 2 = 3x - 4$ or $x - 2 = -(3x - 4)$

87. For what value(s) of c will an absolute value equation of the form $|ax + b| = c$ have

   a. one solution?   if $c = 0$
   b. no solution?   if c is a negative number
   c. two solutions?   if c is a positive number

Use a graphical approach to approximate the solutions of each equation. Round the solutions to two decimal places.

88. $|2.3x - 1.5| = 5$ $-1.52, 2.83$

89. $|-7.6x + 2.6| = 1.9$ $0.09. 0.59$

90. $3.6 - |4.1x - 2.6| = |x - 1.4|$ $0.08, 1.49$

91. $-1.2 + |5x + 12.1| = -|x + 7.3| + 10$ $-4, -1.37$

47. $\frac{1}{2}, -\frac{5}{6}$ **48.** $-\frac{4}{3}, 0$ **52.** $-3, -\frac{1}{3}$ **53.** $-8, \frac{2}{3}$

54. $-\frac{7}{5}, 4$ **77.** 38.4 pounds   **86.** $|x - 2| = |3x - 4|$

3.5   ABSOLUTE VALUE INEQUALITIES

Objectives

1 Solve absolute value inequalities of the form $|x| < a$.

2 Solve absolute value inequalities of the form $|x| > a$.

1 The solution set of an absolute value inequality such as $|x| < 2$ contains all numbers whose distance from 0 is less than 2 units, as shown below.

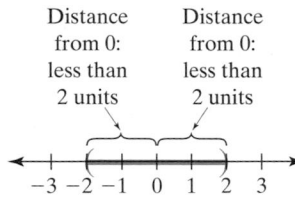

Distance from 0: less than 2 units

Distance from 0: less than 2 units

The solution set is $\{x|-2 < x < 2\}$, or $(-2, 2)$ in interval notation.

EXAMPLE 1

Solve $|x| \leq 3$.

Solution The solution set of this inequality contains all numbers whose distance from 0 is less than or equal to 3. Thus 3, -3, and all numbers between 3 and -3 are in the solution set.

The solution set is $[-3, 3]$.

In general, we have the following.

CLASSROOM EXAMPLE
Solve $|x| < 4$.
answer:

> ### Solving Absolute Value Inequalities of the Form $|X| < a$
>
> If a is a positive number, then $|X| < a$ is equivalent to $-a < X < a$.

This property also holds true for the inequality symbol $\leq$.

Discover the Concept

a. Solve the equation $|x| = 7$ by graphing $y_1 = |x|$, $y_2 = 7$, and finding the point(s) of intersection.

b. Move the trace cursor along y_1 from the left-hand point of intersection of the two graphs to the right-hand point of intersection. Observe the y-values of the points as you move the cursor.

c. Find the x-values for which $|x| < 7$ (or $y_1 < y_2$).

d. Write the solution set of $|x| < 7$ (or $y_1 < y_2$).

In the discovery above, we see that the solutions of $|x| = 7$ are 7 and -7, the x-values of the points of intersection of y_1 and y_2. The solutions of $|x| < 7$ consist of all x-values for which $y_1 < y_2$ or for which the graph of y_1 is below the graph of y_2. These x-values are between -7 and 7, or $(-7, 7)$.

Also notice that the solutions of $|x| > 7$ consist of all x-values for which $y_1 > y_2$ or for which the graph of y_1 is above the graph of y_2. These x-values are less than -7 or greater than 7, or $(-\infty, -7) \cup (7, \infty)$.

The solution set of the inequality $|x| < 7$ consists of the x-values where $y_1 < y_2$, or in interval notation $(-7, 7)$.

The solution set of the inequality $|x| > 7$ consists of the x-values where $y_1 > y_2$, or in interval notation $(-\infty, -7) \cup (7, \infty)$.

EXAMPLE 2

Solve algebraically and graphically for m: $|m - 6| < 2$.

Algebraic Solution:

From the preceding property, we see that

$$|m - 6| < 2 \text{ is equivalent to } -2 < m - 6 < 2$$

Solve this compound inequality for m by adding 6 to all three sides.

$$-2 < m - 6 < 2$$

$-2 + 6 < m - 6 + 6 < 2 + 6$ Add 6 to all three
                                   sides.

$4 < m < 8$ Simplify.

The solution set is $(4, 8)$.

Graphical Solution:

In the inequality $|m - 6| < 2$, replace m with x and proceed as usual. Graph $y_1 = |x - 6|$, $y_2 = 2$, and find x-values for which $y_1 < y_2$.

y_1 is below y_2 for x-values between 4 and 8. Thus, the solution set of $|x - 6| < 2$ in interval notation is $(4, 8)$.

The solution set is $(4, 8)$, and its graph is shown.

Helpful Hint
Before using an absolute value inequality property, isolate the absolute value expression on one side of the inequality.

EXAMPLE 3

Solve algebraically for x: $|5x + 1| + 1 \le 10$.

Solution First, isolate the absolute value expression by subtracting 1 from both sides.

$$|5x + 1| + 1 \le 10$$
$$|5x + 1| \le 10 - 1 \qquad \text{Subtract 1 from both sides.}$$
$$|5x + 1| \le 9 \qquad \text{Simplify.}$$

Since 9 is positive, we apply the absolute value property for $|X| \le a$.

$$-9 \le 5x + 1 \le 9$$
$$-9 - 1 \le 5x + 1 - 1 \le 9 - 1 \qquad \text{Subtract 1 from all three parts.}$$
$$-10 \le 5x \le 8 \qquad \text{Simplify.}$$
$$-2 \le x \le \frac{8}{5} \qquad \text{Divide all three parts by 5.}$$

The solution set is $\left[-2, \dfrac{8}{5}\right]$, and the graph is shown above.

EXAMPLE 4

Solve for x: $\left|2x - \dfrac{1}{10}\right| < -13$.

Solution The absolute value of a number is always nonnegative and can never be less than -13. Thus this absolute value inequality has no solution. The solution set is $\{\ \}$ or $\varnothing$.

2   Let us now solve an absolute value inequality of the form $|X| > a$, such as $|x| \ge 3$. The solution set contains all numbers whose distance from 0 is 3 or more units. Thus the graph of the solution set contains 3 and all points to the right of 3 on the number line or -3 and all points to the left of -3 on the number line.

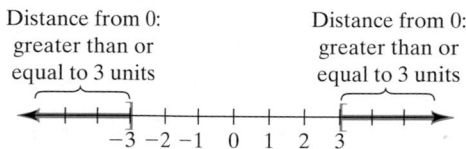

This solution set is written as $\{x \mid x \le -3 \text{ or } x \ge 3\}$. In interval notation, the solution is $(-\infty, -3] \cup [3, \infty)$, since "or" means "union." In general, we have the following.

Solving Absolute Value Inequalities of the Form $|X| > a$

If a is a positive number, then $|X| > a$ is equivalent to $X < -a$ or $X > a$.

This property also holds true for the inequality symbol $\ge$.

EXAMPLE 5

Solve algebraically and graphically for x: $|x - 3| \geq 7$.

Algebraic Solution:

Since 7 is positive,

$|x - 3| \geq 7$ is equivalent to

$x - 3 \leq -7$ or $x - 3 \geq 7$

Next, solve the compound inequality.

$$
\begin{array}{ccc}
x - 3 \leq -7 & \text{or} & x - 3 \geq 7 \\
x - 3 + 3 \leq -7 + 3 & \text{or} & x - 3 + 3 \geq 7 + 3 \\
x \leq -4 & \text{or} & x \geq 10
\end{array}
$$

The solution set is $(-\infty, -4] \cup [10, \infty)$, and its graph is shown.

Graphical Solution:

Graph $y_1 = |x - 3|$ and $y_2 = 7$. Find the x-values for which $y_1 \geq y_2$.

y_1 intersects or is above y_2 for x-values less than or equal to -4 and also x-values greater than or equal to 10, or $(-\infty, -4] \cup [10, \infty)$.

Examples 6 and 8 illustrate special cases of absolute value inequalities. These special cases occur when an isolated absolute value expression is less than, less than or equal to, greater than, or greater than or equal to a negative number or 0.

EXAMPLE 6

Solve $|2x + 9| + 5 > 3$ algebraically and check graphically.

Solution First isolate the absolute value expression by subtracting 5 from both sides.

$$
\begin{array}{ll}
|2x + 9| + 5 > 3 & \\
|2x + 9| + 5 - 5 > 3 - 5 & \text{Subtract 5 from both sides.} \\
|2x + 9| > -2 & \text{Simplify.}
\end{array}
$$

The absolute value of any number is always nonnegative and thus is always greater than -2. This inequality and the original inequality are true for all values of x. The solution set is $\{x | x \text{ is a real number}\}$ or $(-\infty, \infty)$.

$y_1 = |2x + 9| + 5$

To check graphically, see the screen to the left. Notice that $y_1 > y_2$ or $|2x + 9| + 5 > 3$ for all real numbers.

✔ CONCEPT CHECK

Without taking any solution steps, how do you know that the absolute value inequality $|3x - 2| > -9$ has a solution? What is its solution?

EXAMPLE 7

Use a graphical approach to solve $\left| \dfrac{x}{3} - 1 \right| - 2 \geq 0$.

Solution

Graph $y_1 = \left| \dfrac{x}{3} - 1 \right| - 2$ and use the graph of y_1 to solve $y_1 \geq 0$.

Find x-values where the graph of y_1 is above the x-axis.

The graph of $y = \left| \dfrac{x}{3} - 1 \right| - 2$ is above or on the x-axis for x-values less than or equal to -3 or greater than or equal to 9.

The solution set is $(-\infty, -3] \cup [9, \infty)$.

EXAMPLE 8

Solve for x: $\left| \dfrac{2(x+1)}{3} \right| \leq 0$.

Solution

Recall that "$\leq$" means "less than or equal to." The absolute value of any expression will never be less than 0, but it may be equal to 0. Thus, to solve $\left| \dfrac{2(x+1)}{3} \right| \leq 0$ we solve $\left| \dfrac{2(x+1)}{3} \right| = 0$

$$\dfrac{2(x+1)}{3} = 0$$

$$3 \left[\dfrac{2(x+1)}{3} \right] = 3(0) \qquad \text{Clear the equation of fractions.}$$

$$2x + 2 = 0 \qquad \text{Apply the distributive property.}$$

$$2x = -2 \qquad \text{Subtract 2 from both sides.}$$

$$x = -1 \qquad \text{Divide both sides by 2.}$$

The solution set is $\{-1\}$. See the screen to the left to check this solution set graphically.

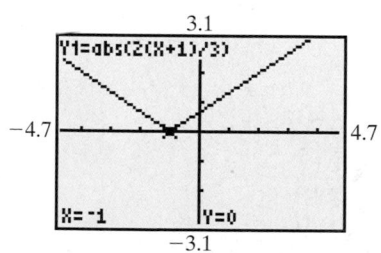

The graph touches the x-axis at $x = -1$ and is never below the x-axis. Therefore, the only point that satisfies the inequality has x-value -1.

The following box summarizes the types of absolute value equations and inequalities.

Solving Absolute Value Equations and Inequalities With $a > 0$

Algebraic Solution	Solution Graph
$\lvert X \rvert = a$ is equivalent to $X = a$ or $X = -a$.	
$\lvert X \rvert < a$ is equivalent to $-a < X < a$.	
$\lvert X \rvert > a$ is equivalent to $X < -a$ or $X > a$.	

23. ———, $\left(-\infty, \frac{2}{3}\right) \cup (2, \infty)$ **24.** ———, $\left(-\infty, -\frac{11}{3}\right) \cup (-3, \infty)$

1. ———, $[-4, 4]$
  −4 4

2. ———, $(-6, 6)$
  −6 6

3. ———, $(1, 5)$
  1 5

4. ———, $[-5, 5]$
  −5 5

5. ———, $(-5, -1)$
  −5 −1

6. ———, $(-10, 2)$
  −10 2

7. ———, $[-10, 3]$
  −10 3

8. ———, $\left[-3, \frac{21}{5}\right]$
  −3 $\frac{21}{5}$

9. ———, $[-5, 5]$
  −5 5

10. ———, $[-1, 1]$
  −1 1

11. ———, $\varnothing$

12. ———, $\varnothing$

13. ———, $[0, 12]$
  0 12

14. ———, $(-6, 2)$
  −6 2

15. ———, $(-\infty, -3) \cup (3, \infty)$
  −3 3

16. ———, $(-\infty, -4] \cup [4, \infty)$
  −4 4

17. ———, $(-\infty, -24] \cup [4, \infty)$
  −24 4

18. ———, $(-\infty, 7] \cup [11, \infty)$
  7 11

19. ———, $(-\infty, -4) \cup (4, \infty)$
  −4 4

20. ———, $(-\infty, -4) \cup (4, \infty)$
  −4 4

21. ———, $(-\infty, \infty)$ **22.** ———, $(-\infty, \infty)$

26. ———, $(-\infty, \infty)$

See Appendix A for summary exercises.

STUDY SKILLS REMINDER

Continue your outline started in Section 1.5. Write how to recognize and how to solve absolute value inequalities in your own words. For example:

Solving Equations and Inequalities

I. Equations
  A. Linear equations (Section 1.5)
  B. Absolute value equations (Section 3.4)

II. Inequalities
  A. Linear inequalities (Section 3.2)
  B. Compound inequalities (Section 3.3)
  C. **Absolute value inequalities**—Recognize: *inequality with absolute value bars*—Solve: if possible, first isolate absolute value: if $|x| > a$ (positive), solve $x < -a$ or $x > a$, if $|x| < a$ (positive) then solve $-a < x < a$.

See Appendix A for summary exercises.

MENTAL MATH

Match each absolute value inequality with an equivalent statement.

1. $|2x + 1| = 3$ D   A. $2x + 1 > 3$ or $2x + 1 < -3$

2. $|2x + 1| \le 3$ E   B. $2x + 1 \ge 3$ or $2x + 1 \le -3$

3. $|2x + 1| < 3$ C   C. $-3 < 2x + 1 < 3$

4. $|2x + 1| \ge 3$ B   D. $2x + 1 = 3$ or $2x + 1 = -3$

5. $|2x + 1| > 3$ A   E. $-3 \le 2x + 1 \le 3$

EXERCISE SET 3.5

 STUDY GUIDE/SSM   CD/VIDEO   PH MATH TUTOR CENTER   MathXL®Tutorials ON CD   MathXL®   MyMathLab®

Solve each inequality. Then graph the solution set. See Examples 1 through 4.

1. $|x| \le 4$ **2.** $|x| < 6$

3. $|x - 3| < 2$ **4.** $|y| \le 5$

5. $|x + 3| < 2$ **6.** $|x + 4| < 6$

7. $|2x + 7| \le 13$ **8.** $|5x - 3| \le 18$

9. $|x| + 7 \le 12$ **10.** $|x| + 6 \le 7$

11. $|3x - 1| < -5$ **12.** $|8x - 3| < -2$

13. $|x - 6| - 7 \le -1$ **14.** $|z + 2| - 7 < -3$

Solve each inequality. Graph the solution set. See Examples 5 through 7.

15. $|x| > 3$ **16.** $|y| \ge 4$

17. $|x + 10| \ge 14$ **18.** $|x - 9| \ge 2$

19. $|x| + 2 > 6$ **20.** $|x| - 1 > 3$

21. $|5x| > -4$ **22.** $|4x - 11| > -1$

23. $|6x - 8| + 3 > 7$ **24.** $|10 + 3x| + 1 > 2$

Solve each inequality. Graph the solution set. See Example 8.

25. $|x| \le 0$ ———, $\{0\}$ **26.** $|x| \ge 0$

27. $|8x + 3| > 0$

28. $|5x - 6| < 0$

🔒 **61.** $-15 + |2x - 7| \le -6$ ←+——+→ , $[-1, 8]$
-1 8

MIXED PRACTICE

29. ←+——+→ , $[-2, 2]$
-2 2

62. $-9 + |3 + 4x| < -4$ ←(——)→ , $\left(-2, \frac{1}{2}\right)$
-2 $\frac{1}{2}$

Solve each inequality. Graph the solution set.

🔒 **29.** $|x| \le 2$ **30.** $|z| < 6$

63. $\left|2x + \frac{3}{4}\right| - 7 \le -2$ ←+——+→ , $\left[-\frac{23}{8}, \frac{17}{8}\right]$
$-\frac{23}{8}$ $\frac{17}{8}$

31. $|y| > 1$ **32.** $|x| \ge 10$

33. $|x - 3| < 8$ **34.** $|-3 + x| \le 10$

64. $\left|\frac{3}{5} + 4x\right| - 6 < -1$ ←(——)→ , $\left(-\frac{7}{5}, \frac{11}{10}\right)$
$-\frac{7}{5}$ $\frac{11}{10}$

35. $|0.6x - 3| > 0.6$ **36.** $|1 + 0.3x| \ge 0.1$

37. $5 + |x| \le 2$ ←————→ , $\varnothing$

MIXED PRACTICE

81. $\left(-\infty, -\frac{25}{3}\right) \cup \left(\frac{35}{3}, \infty\right)$

38. $8 + |x| < 1$ ←————→ , $\varnothing$

Solve each equation or inequality for x.

39. $|x| > -4$ ←————→ , $(-\infty, \infty)$

65. $|2x - 3| < 7$ $(-2, 5)$ **66.** $|2x - 3| > 7$

40. $|x| \le -7$ ←————→ , $\varnothing$

67. $|2x - 3| = 7$ $5, -2$ **68.** $|5 - 6x| = 29$

41. $|2x - 7| \le 11$ ←+——+→ , $[-2, 9]$
-2 9

69. $|x - 5| \ge 12$ **70.** $|x + 4| \ge 20$

42. $|5x + 2| < 8$ ←(——)→ , $\left(-2, \frac{6}{5}\right)$
-2 $\frac{6}{5}$

71. $|9 + 4x| = 0$ $-\frac{9}{4}$ **72.** $|9 + 4x| \ge 0$ $(-\infty, \infty)$

43. $|x + 5| + 2 \ge 8$ ←+——+→ , $(-\infty, -11] \cup [1, \infty)$
-11 1

73. $|2x + 1| + 4 < 7$ **74.** $8 + |5x - 3| \ge 11$

44. $|-1 + x| - 6 > 2$ ←)——(→ , $(-\infty, -7) \cup (9, \infty)$
-7 9

75. $|3x - 5| + 4 = 5$ $2, \frac{4}{3}$ **76.** $|8x| = -5$ $\varnothing$

45. $|x| > 0$ ←—✗—→ , $(-\infty, 0) \cup (0, \infty)$
0

77. $|x + 11| = -1$ $\varnothing$ **78.** $|4x - 4| = -3$ $\varnothing$

46. $|x| < 0$ ←————→ , $\varnothing$

79. $\left|\frac{2x - 1}{3}\right| = 6$ $\frac{19}{2}, -\frac{17}{2}$ **80.** $\left|\frac{6 - x}{4}\right| = 5$ $-14, 26$

47. $9 + |x| > 7$ ←————→ , $(-\infty, \infty)$

48. $5 + |x| \ge 4$ ←————→ , $(-\infty, \infty)$

81. $\left|\frac{3x - 5}{6}\right| > 5$ **82.** $\left|\frac{4x - 7}{5}\right| < 2$

49. $6 + |4x - 1| \le 9$ ←+——+→ , $\left[-\frac{1}{2}, 1\right]$
$-\frac{1}{2}$ 1

Use the given graphing utility screen to solve each equation or inequality.

50. $-3 + |5x - 2| \le 4$ ←+——+→ , $\left[-1, \frac{9}{5}\right]$
-1 $\frac{9}{5}$

83. a. $|x - 3| - 2 = 6$
b. $|x - 3| - 2 < 6$
c. $|x - 3| - 2 \ge 6$
a. $\{-5, 11\}$
b. $(-5, 11)$
c. $(\infty, -5] \cup [11, \infty)$

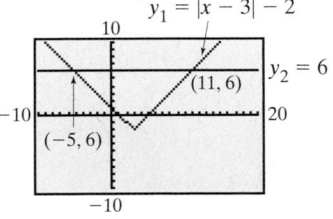

51. $\left|\frac{2}{3}x + 1\right| > 1$ ←)——(→ , $(-\infty, -3) \cup (0, \infty)$
-3 0

52. $|5x - 1| \ge 2$ ←+——+→ , $\left(-\infty, -\frac{1}{5}\right] \cup \left[\frac{3}{5}, \infty\right)$
$-\frac{1}{5}$ $\frac{3}{5}$

53. $|5x + 3| < -6$ ←————→ , $\varnothing$

54. $|4 + 9x| \ge -6$ ←————→ , $(-\infty, \infty)$

84. a. $|x + 5| - 4 = 3$
b. $|x + 5| - 4 \le 3$
c. $|x + 5| - 4 > 3$
a. $\{-12, 2\}$
b. $[-12, 2]$
c. $(-\infty, -12) \cup (2, \infty)$

55. $|8x + 3| \ge 0$ ←————→ , $(-\infty, \infty)$

56. $|5x - 6| \le 0$ ←—●—→ , $\frac{6}{5}$
$\frac{6}{5}$

57. $|1 + 3x| + 4 < 5$ ←(——)→ , $\left(-\frac{2}{3}, 0\right)$
$-\frac{2}{3}$ 0

85. a. $|x + 2| - 10 = -4$
b. $|x + 2| - 10 \le -4$
c. $|x + 2| - 10 > -4$
a. $\{-8, 4\}$
b. $[-8, 4]$
c. $(-\infty, -8) \cup (4, \infty)$

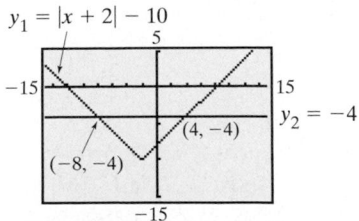

58. $|7x - 3| - 1 \le 10$ ←+——+→ , $\left[-\frac{8}{7}, 2\right]$
$-\frac{8}{7}$ 2

🔒 **59.** $\left|\frac{x + 6}{3}\right| > 2$ ←)——(→ , $(-\infty, -12) \cup (0, \infty)$
-12 0

60. $\left|\frac{7 + x}{2}\right| \ge 4$ ←+——+→ , $(-\infty, -15] \cup [1, \infty)$
-15 1

86. a. $|x + 2| + 1 = -5$
   b. $|x + 2| + 1 < -5$
   c. $|x + 2| + 1 > -5$
   a. $\emptyset$
   b. $\emptyset$
   c. $(-\infty, \infty)$

$y_1 = |x + 2| + 1$

$y_2 = -5$

95. If $y = -3$, find x 0   **96.** If $x = 4$, find y 0

Concept Extensions

97. Write an absolute value inequality representing all numbers x whose distance from 0 is less than 7 units. $|x| < 7$

98. Write an absolute value inequality representing all numbers x whose distance from 0 is greater than 4 units. $|x| > 4$

99. Write $-5 \leq x \leq 5$ as an equivalent inequality containing an absolute value. $|x| \leq 5$

100. Write $x > 1$ or $x < -1$ as an equivalent inequality containing an absolute value. $|x| > 1$

101. Describe how solving $|x - 3| = 5$ is different from solving $|x - 3| < 5$.   answers may vary

102. Describe how solving $|x + 4| = 0$ is similar to solving $|x + 4| \leq 0$.   answers may vary

The expression $|x_T - x|$ is defined to be the absolute error in x, where x_T is the true value of a quantity and x is the measured value or value as stored in a computer.

103. If the true value of a quantity is 3.5 and the absolute error must be less than 0.05, find the acceptable measured values.

104. If the true value of a quantity is 0.2 and the approximate value stored in a computer is $\dfrac{51}{256}$, find the absolute error. $\dfrac{1}{1280}$, or 0.00078125

REVIEW AND PREVIEW

Recall the formula:

$$Probability \; of \; an \; event = \frac{number \; of \; ways \; that \; the \; event \; can \; occur}{number \; of \; possible \; outcomes}$$

Find the probability of rolling each number on a single toss of a die. (Recall that a die is a cube with each of its six sides containing 1, 2, 3, 4, 5, and 6 black dots, respectively.) See Section 1.8.   **92.** 1

87. P(rolling a 2) $\frac{1}{6}$ **88.** P(rolling a 5) $\frac{1}{6}$

89. P(rolling a 7)   0   **90.** P(rolling a 0)   0

91. P(rolling a 1 or 3) $\frac{1}{3}$ **92.** P(rolling a 1, 2, 3, 4, 5, or 6)

Consider the equation $3x - 4y = 12$. For each value of x or y given, find the corresponding value of the other variable that makes the statement true. See Section 1.8.

93. If $x = 2$, find y -1.5 **94.** If $y = -1$, find x $\dfrac{8}{3}$

82. $\left(-\dfrac{3}{4}, \dfrac{17}{4} \right)$

31. $(-\infty, -1) \cup (1, \infty)$ **32.** $(-\infty, -10] \cup [10, \infty)$ **33.** $(-5, 11)$ **34.** $[-7, 13]$

35. $(-\infty, 4) \cup (6, \infty)$ **36.** $\left(-\infty, -\dfrac{11}{3} \right) \cup (-3, \infty)$ **66.** $(-\infty, -2) \cup (5, \infty)$ **68.** $-4, \dfrac{17}{3}$

30. $(-6, 6)$

69. $(-\infty, -7] \cup [17, \infty)$ **70.** $(-\infty, -24] \cup [16, \infty)$ **73.** $(-2, 1)$ **74.** $(-\infty, 0] \cup \left[\dfrac{6}{5}, \infty \right)$ **103.** $3.45 < x < 3.55$

3.6   GRAPHING LINEAR INEQUALITIES IN TWO VARIABLES

Objectives

1   Graph linear inequalities.

2   Graph the intersection or union of two linear inequalities.

1   Recall that the graph of a linear equation in two variables is the graph of all ordered pairs that satisfy the equation, and we determined that the graph is a line. Here we graph **linear inequalities** in two variables; that is, we graph all the ordered pairs that satisfy the inequality.

If the equal sign in a linear equation in two variables is replaced with an inequality symbol, the result is a linear inequality in two variables.

Examples of Linear Inequalities in Two Variables

$$3x + 5y \geq 6 \qquad 2x - 4y < -3$$
$$4x > 2 \qquad\qquad y \leq 5$$

Discover the Concept

a. Graph $x + y = 3$ by graphing $y_1 = 3 - x$ in an integer window.

b. Access the cursor that results from using the arrow key pad (not the trace cursor). Move the cursor to points above the line and notice the sum of the x- and y-coordinates.

c. Move the cursor to points below the line and notice the sum of the x- and y-coordinates.

d. Describe the set of points that satisfies $x + y < 3$ and the set of points that satisfies $x + y > 3$.

The graph of $y = 3 - x$ and a point (7, 6) above the line.

To graph the linear inequality $x + y < 3$, for example, we first graph the related **boundary** equation $x + y = 3$. The resulting boundary line contains all ordered pairs the sum of whose coordinates is 3. This line separates the plane into two **half-planes**. All points "above" the boundary line $x + y = 3$ have coordinates that satisfy the inequality $x + y > 3$, and all points "below" the line have coordinates that satisfy the inequality $x + y < 3$.

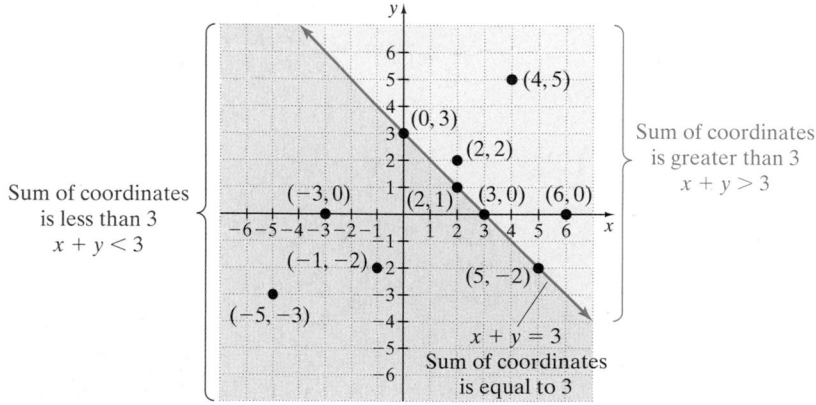

The graph, or **solution region**, for $x + y < 3$, then, is the half-plane below the boundary line and is shown shaded in the graph on the next page. The boundary line is shown dashed since it is not a part of the solution region. These ordered pairs on this line satisfy $x + y = 3$ and not $x + y < 3$.

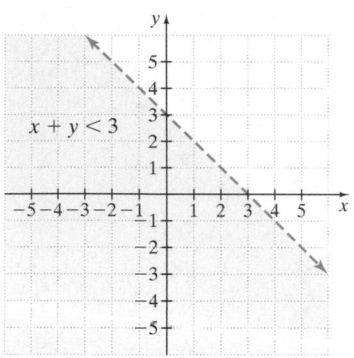

The following steps may be used to graph linear inequalities in two variables.

> ## Graphing a Linear Inequality in Two Variables
>
> **Step 1:** Graph the boundary line found by replacing the inequality sign with an equal sign. If the inequality sign is $<$ or $>$, graph a dashed line indicating that points on the line are not solutions of the inequality. If the inequality sign is $\leq$ or $\geq$, graph a solid line indicating that points on the line are solutions of the inequality.
>
> **Step 2:** Choose a **test point not on the boundary line** and substitute the coordinates of this test point into the **original inequality.**
>
> **Step 3:** If a true statement is obtained in Step 2, shade the half-plane that contains the test point. If a false statement is obtained, shade the half-plane that does not contain the test point.

EXAMPLE 1

Graph $2x - y < 6$.

Solution First, the boundary line for this inequality is the graph of $2x - y = 6$. Graph a dashed boundary line because the inequality symbol is $<$. Next, choose a test point on either side of the boundary line. The point $(0, 0)$ is not on the boundary line, so we use this point. Replacing x with 0 and y with 0 in the *original inequality* $2x - y < 6$ leads to the following:

$$2x - y < 6$$
$$2(0) - 0 < 6 \quad \text{Let } x = 0 \text{ and } y = 0.$$
$$0 < 6 \quad \text{True}$$

Because $(0, 0)$ satisfies the inequality, so does every point on the same side of the boundary line as $(0, 0)$. Shade the half-plane that contains $(0, 0)$. The half-plane graph of the inequality is shown at the top of the following page.

$2x - y < 6$
or
$y > 2x - 6$

On most graphing utilities, the boundary line will always appear solid. Thus, simply note that the boundary line should be dashed.

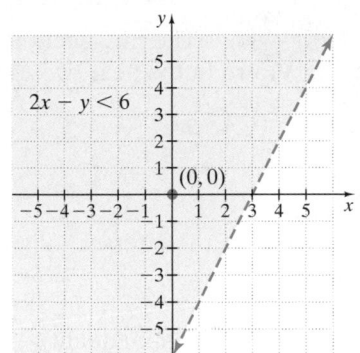

$2x - y < 6$

$(0, 0)$

TEACHING TIP

After completing Example 1, have students use the graph of the solution to predict if various points will give a true or false statement when substituted into the original inequality. Then have them substitute the values into the inequality to check their predictions.

Every point in the shaded half-plane satisfies the original inequality. Notice that the inequality $2x - y < 6$ does not describe a function since its graph does not pass the vertical line test.

In general, linear inequalities of the form $Ax + By \leq C$, when A and B are not both 0, do not describe functions.

EXAMPLE 2

Graph $3x \geq y$.

Solution

First, graph the boundary line $3x = y$. Graph a solid boundary line because the inequality symbol is $\geq$. Test a point not on the boundary line to determine which half-plane contains points that satisfy the inequality. We choose $(0, 1)$ as our test point.

$$3x \geq y$$

$$3(0) \geq 1 \quad \text{Let } x = 0 \text{ and } y = 1.$$

$$0 \geq 1 \quad \text{False}$$

This point does not satisfy the inequality, so the correct half-plane is on the opposite side of the boundary line from $(0, 1)$. The graph of $3x \geq y$ is the boundary line together with the shaded region shown.

CLASSROOM EXAMPLE

Graph $x \leq 2y$.

answer:

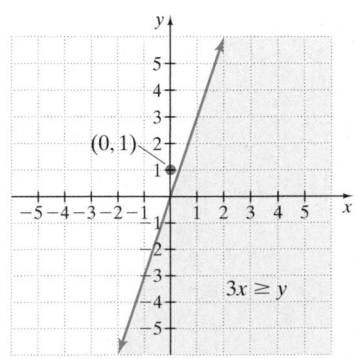

$(0, 1)$

$3x \geq y$

✔ **CONCEPT CHECK**

Concept Check Answer:
solid

If a point on the boundary line is included in the solution of an inequality in two variables, should the graph of the boundary line be solid or dashed?

TEACHING TIP
For Example 3, it may be helpful
for students to use a different color
to graph each equation.

2 The intersection and the union of linear inequalities can also be graphed, as shown in the next two examples.

EXAMPLE 3

Graph the intersection of $x \geq 1$ and $y \geq 2x - 1$.

Solution Graph each inequality. The intersection of the two graphs is all points common to both regions, as shown by the dark pink shading in the third graph.

CLASSROOM EXAMPLE

Graph the intersection of $x \leq 2$ and $y \geq x + 1$.

answer:

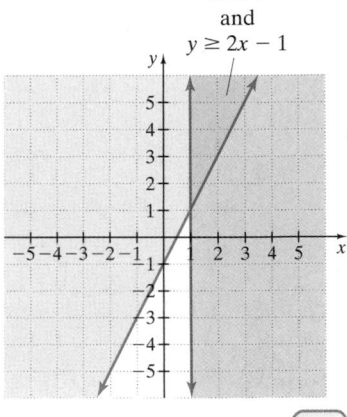

EXAMPLE 4

Graph the union of $x + \dfrac{1}{2}y \geq -4$ or $y \leq -2$.

Solution Graph each inequality. The union of the two inequalities is both shaded regions, including the solid boundary lines shown in the third graph.

CLASSROOM EXAMPLE

Graph the union of $x + 2y \leq 4$ or $y \geq -1$.

answer:

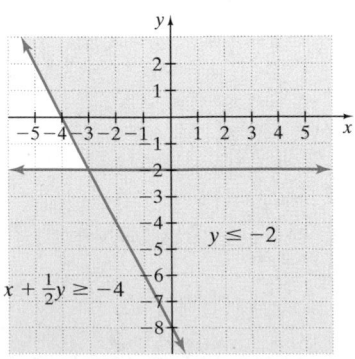

Spotlight on DECISION MAKING

Suppose you are a customer service representative for a mail-order medical supply company that sells support stockings. A customer, whose weight is 160 pounds and whose height is 5 feet 9 inches, places an order for support stockings and has asked your assistance in selecting the correct size. What size would you recommend that this customer order? Explain.

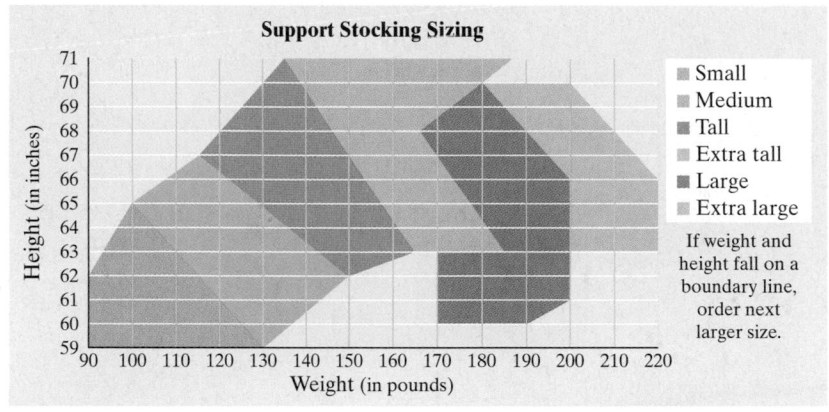

Support Stocking Sizing

- Small
- Medium
- Tall
- Extra tall
- Large
- Extra large

If weight and height fall on a boundary line, order next larger size.

Height (in inches) / Weight (in pounds)

EXERCISE SET 3.6

STUDY GUIDE/SSM   CD/ VIDEO   PH MATH TUTOR CENTER   MathXL®Tutorials ON CD   MathXL®   MyMathLab®

1.–12. See graphing answer section.

Graph each inequality. See Examples 1 and 2.

1. $x < 2$

2. $x > -3$

3. $x - y \geq 7$

4. $3x + y \leq 1$

5. $3x + y > 6$

6. $2x + y > 2$

7. $y \leq -2x$

8. $y \leq 3x$

9. $2x + 4y \geq 8$

10. $2x + 6y \leq 12$

11. $5x + 3y > -15$

12. $2x + 5y < -20$

13. Explain when a dashed boundary line should be used in the graph of an inequality.   answers may vary

14. Explain why, after the boundary line is sketched, we test a point on either side of this boundary in the original inequality.   answers may vary

15.–41. See graphing answer section.

Graph each union or intersection. See Examples 3 and 4.

15. The intersection of $x \geq 3$ and $y \leq -2$

16. The union of $x \geq 3$ or $y \leq -2$

17. The union of $x \leq -2$ or $y \geq 4$

18. The intersection of $x \leq -2$ and $y \geq 4$

19. The intersection of $x - y < 3$ and $x > 4$

20. The intersection of $2x > y$ and $y > x + 2$

21. The union of $x + y \leq 3$ or $x - y \geq 5$

22. The union of $x - y \leq 3$ or $x + y > -1$

MIXED PRACTICE

Graph each inequality.

23. $y \geq -2$

24. $y \leq 4$

25. $x - 6y < 12$

26. $x - 4y < 8$

27. $x > 5$

28. $y \geq -2$

29. $-2x + y \leq 4$

30. $-3x + y \leq 9$

31. $x - 3y < 0$

32. $x + 2y > 0$

33. $3x - 2y \leq 12$

34. $2x - 3y \leq 9$

35. The union of $x - y > 2$ or $y < 5$

36. The union of $x - y < 3$ or $x > 4$

37. The intersection of $x + y \leq 1$ and $y \leq -1$

38. The intersection of $y \geq x$ and $2x - 4y \geq 6$

39. The union of $2x + y > 4$ or $x \geq 1$

40. The union of $3x + y < 9$ or $y \leq 2$

41. The intersection of $x \geq -2$ and $x \leq 1$

42. The intersection of $x \geq -4$ and $x \leq 3$

43. The union of $x + y \leq 0$ or $3x - 6y \geq 12$

44. The intersection of $x + y \leq 0$ and $3x - 6y \geq 12$

45. The intersection of $2x - y > 3$ and $x > 0$

46. The union of $2x - y > 3$ or $x > 0$

Match each inequality with its graph.

A

B

C

D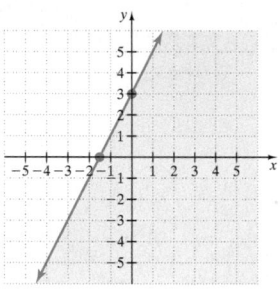

47. $y \leq 2x + 3$ D   **48.** $y < 2x + 3$ C

49. $y > 2x + 3$ A   **50.** $y \geq 2x + 3$ B

Write the inequality whose graph is given.

51.

$x \geq 2$

52.

$y \geq 1$

53.

$y \leq -3$

54.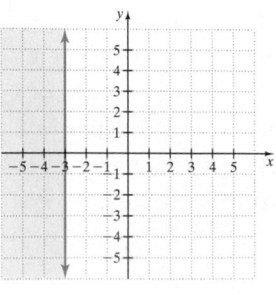

$x \leq -3$

42.–46. See graphing answer section.

55.

$y > 4$

56.

$y > -1$

57.

$x < 1$

58.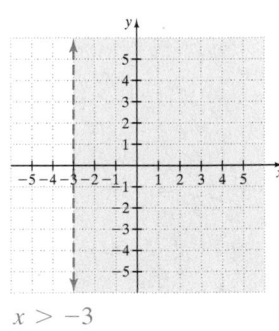

$x > -3$

REVIEW AND PREVIEW

Evaluate each expression. See Sections 1.3 and 1.4.

59. 2^3 8   **60.** 3^2 9

61. -5^2 -25 **62.** $(-5)^2$ 25

63. $(-2)^4$ 16   **64.** -2^4 -16

65. $\left(\dfrac{3}{5}\right)^3$ $\dfrac{27}{125}$ **66.** $\left(\dfrac{2}{7}\right)^2$ $\dfrac{4}{49}$

Find the domain and the range of each relation. Determine whether the relation is also a function. See Section 2.2.

67.

68.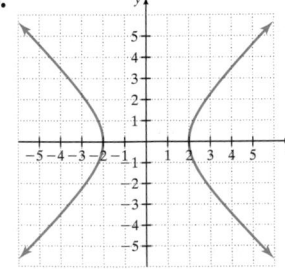

domain: $[1, 5]$; range: $[1, 3]$; no   domain: $(-\infty, -2] \cup [2, \infty)$; range: $(-\infty, \infty)$; no

Concept Extensions

Solve. **69.–71.** See graphing answer section.

69. Rheem Abo-Zahrah decides that she will study at most 20 hours every week and that she must work at least 10 hours every week. Let x represent the hours studying and y represent the hours working. Write two inequalities that model this situation and graph their intersection. $x \leq 20$ and $y \geq 10$

70. The movie and TV critic for the *New York Times* spends between 2 and 6 hours daily reviewing movies and fewer than 5 hours reviewing TV shows. Let x represent the hours watching movies and y represent the time spent watching TV. Write two inequalities that model this situation and graph their intersection. $2 < x < 6, 0 \leq y < 5$

71. Chris-Craft manufactures boats out of Fiberglas and wood. Fiberglas hulls require 2 hours work, whereas wood hulls require 4 hours work. Employees work at most 40 hours a week. The following inequalities model these restrictions, where x represents the number of Fiberglas hulls produced and y represents the number of wood hulls produced.

$$\begin{cases} x \geq 0 \\ y \geq 0 \\ 2x + 4y \leq 40 \end{cases}$$

Graph the intersection of these inequalities.

CHAPTER 3 PROJECT

Analyzing Municipal Budgets

Nearly all cities, towns, and villages operate with an annual budget. Budget items might include expenses for fire and police protection as well as for street maintenance and parks. No matter how big or small the budget, city officials need to know if municipal spending is over or under budget. In this project, you will have the opportunity to analyze a municipal budget and make budgetary recommendations. This project may be completed by working in groups or individually.

Suppose that each year your town creates a municipal budget. The next year's annual municipal budget is submitted for approval by the town's citizens at the annual town meeting. This year's budget was printed in the town newspaper earlier in the year.

You have joined a group of citizens who are concerned about your town's budgeting and spending processes. Your group plans to analyze this year's budget along with what was actually spent by the town this year. You hope to present your findings at the annual town meeting and make some budgetary recommendations for next year's budget. The municipal budget contains many different areas of spending. To help focus your group's analysis, you have decided to research spending habits only for categories in which the actual expenses differ from the budgeted amount by more than 12% of the budgeted amount.

1. For each category in the budget, write a specific absolute value inequality that describes the condition that must be met

before your group will research spending habits for that category. In each case, let the variable x represent the actual expense for a budget category.

2. For each category in the budget, write an equivalent compound inequality for the condition described in Question 1. Again, let the variable x represent the actual expense for a budget category.

3. On the next page is a listing of the actual expenditures made this year for each budget category. Use the inequalities from either Question 1 or Question 2 to complete the Budget Worksheet given at the end of this project. (The first category has been filled in.) From the Budget Worksheet, decide which categories must be researched.

4. Can you think of possible reasons why spending in the categories that must be researched were over or under budget?

5. Based on this year's municipal budget and actual expenses, what recommendations would you make for next year's budget? Explain your reasoning.

6. (Optional) Research the annual budget used by your own town or your college or university. Conduct a similar analysis of the budget with respect to actual expenses. What can you conclude?

	Department/Program	Actual Expenditure
I.	**Board of Health**	
	Immunization Programs	$14,800
	Inspections	$41,900
II.	**Fire Department**	
	Equipment	$375,000
	Salaries	$268,500
III.	**Libraries**	
	Book/Periodical Purchases	$107,300
	Equipment	$29,000
	Salaries	$118,400
IV.	**Parks and Recreation**	
	Maintenance	$82,500
	Playground Equipment	$45,000
	Salaries	$118,000
	Summer Programs	$96,200
V.	**Police Department**	
	Equipment	$328,000
	Salaries	$405,000
VI.	**Public Works**	
	Recycling	$48,100
	Sewage	$92,500
	Snow Removal & Road Salt	$268,300
	Street Maintenance	$284,000
	Water Treatment	$94,100
	TOTAL	$2,816,600

THE TOWN CRIER

Annual Budget Set at Town Meeting

ANYTOWN, USA (MG)—This year's annual budget is as follows:

	Amount Budgeted
BOARD OF HEALTH	
Immunization Programs	$15,000
Inspections	$50,000
FIRE DEPARTMENT	
Equipment	$450,000
Salaries	$275,000
LIBRARIES	
Book/Periodical Purchases	$90,000
Equipment	$30,000
Salaries	$120,000
PARKS AND RECREATION	
Maintenance	$70,000
Playground Equipment	$50,000
Salaries	$140,000
Summer Programs	$80,000
POLICE DEPARTMENT	
Equipment	$300,000
Salaries	$400,000
PUBLIC WORKS	
Recycling	$50,000
Sewage	$100,000
Snow Removal & Road Salt	$200,000
Street Maintenance	$250,000
Water Treatment	$100,000
TOTAL	**$2,770,000**

BUDGET WORKSHEET

Budget category	Budgeted amount	Minimum allowed	Actual expense	Maximum allowed	Within budget?	Amt over/ under budget
Immunization Programs	$15,000	$13,200	$14,800	$16,800	Yes	Under $200

STUDY SKILLS REMINDER

Are You Preparing for a Test on Chapter 3?

Below are listed some common trouble areas for students in Chapter 3. After studying for your test—but before taking your test—read these.

▶ Remember to solve equations both algebraically and graphically to ensure accuracy.

▶ Remember to reverse the direction of the inequality symbol when multiplying or dividing both sides of an inequality by a negative number.

$$-11x < 33 \qquad \textit{Direction of arrow is reversed.}$$
$$\frac{-11x}{-11} > \frac{33}{-11}$$
$$x > -3$$

▶ Remember the differences when solving absolute value equations and inequalities.

$$|x + 1| = 3$$
$$x + 1 = 3 \quad \text{or} \quad x + 1 = -3$$
$$x = 2 \quad \text{or} \quad x = -4$$
$$\{2, -4\}$$

$$|x + 1| < 3$$
$$-3 < x + 1 < 3$$
$$-3 - 1 < x < 3 - 1$$
$$-4 < x < 2$$
$$(-4, 2)$$

$$|x + 1| > 3$$
$$x + 1 < -3 \quad \text{or} \quad x + 1 > 3$$
$$x < -4 \quad \text{or} \quad x > 2$$
$$(-\infty, -4) \cup (2, \infty)$$

▶ Remember that an equation is not solved for a specified variable unless the variable is alone on one side of an equation *and* the other side contains *no* specified variables.

$$y = 10x + 6 - y \qquad \textit{Equation is not solved for } y.$$
$$2y = 10x + 6 \qquad \textit{Add } y \textit{ to both sides.}$$
$$y = 5x + 3 \qquad \textit{Divide both sides by } 2.$$

Remember: This is simply a checklist of common trouble areas. For a review of Chapter 3, see the Highlights and Chapter Review at the end of this chapter.

3 CHAPTER VOCABULARY CHECK

Fill in each blank with one of the words or phrases listed below.

contradiction
absolute value
linear inequality in one variable

linear equation in one variable
compound inequality
identity

intersection
solution
union

1. The statement "$x < 5$ or $x > 7$" *is called a(n)* compound inequality.

2. An equation in one variable that has no solution is called a(n) contradiction.

3. The intersection of two sets is the set of all elements common to both sets.

4. The union of two sets is the set of all elements that belong to either of the sets.

5. An equation in one variable that has every number (for which the equation is defined) as a solution is called a(n) identity.

6. A number's distance from 0 is called its absolute value.

7. When a variable in an equation is replaced by a number and the resulting equation is true, then that number is called a(n) solution of the equation.

8. The statement $5x - 0.2 < 7$ is an example of a(n) linear inequality in one variable.

9. The statement $5x - 0.2 = 7$ is an example of a(n) linear equation in one variable.

CHAPTER 3 HIGHLIGHTS

Definitions and Concepts	**Examples**

Section 3.1 Solving Linear Equations Graphically

To solve an equation graphically by the **intersection-of-graphs method:**

- Graph the left side as y_1.

- Graph the right side as y_2.

- Find any points of intersection, or where $y_1 = y_2$.

- The x-coordinate of an intersection point is a solution.

- The y-coordinate of an intersection point is the value of each side of the original equation when the variable is replaced with the solution.

Solve $5(x - 2) + 1 = 2(x - 1) + 2$

Solution $x = 3$

To solve an equation using the **x-intercept method:**

- Write the equation so that one side is 0.

- For the equation $y_1 = 0$, graph y_1.

- The x-intercepts of the graph are solutions of $y_1 = 0$.

Solve $5(x - 2) + 1 = 2(x - 1) + 2$

$5(x - 2) + 1 - 2(x - 1) - 2 = 0$

Section 3.2 Linear Inequalities and Problem Solving

A **linear inequality in one variable** is an inequality that can be written in the form $ax + b < c$, where a, b, and c are real numbers and $a \neq 0$. (The inequality symbols $\leq$, $>$, and $\geq$ also apply here.)

The **addition property of inequality** guarantees that the same number may be added to (or subtracted from) both sides of an inequality, and the resulting inequality will have the same solution set.

The **multiplication property of inequality** guarantees that both sides of an inequality may be multiplied by (or divided by) the same **positive** number, and the resulting inequality will have the same solution set. We may also multiply (or divide) both sides of an inequality by the same **negative** number and **reverse the direction of the inequality symbol**, and the result is an inequality with the same solution set.

Linear inequalities:

$$5x - 2 \leq -7 \qquad 3y > 1 \qquad \frac{z}{7} < -9(z - 3)$$

$$x - 9 \leq -16$$
$$x - 9 + 9 \leq -16 + 9 \qquad \text{Add 9.}$$
$$x \leq -7$$

$$6x < -66$$
$$\frac{6x}{6} < \frac{-66}{6} \qquad \text{Divide by 6. Do not reverse direction of inequality symbol.}$$
$$x < -11$$
$$-6x < -66 \qquad \text{Divide by } -6. \text{ Reverse direction of inequality symbol.}$$
$$\frac{-6x}{-6} > \frac{-66}{-6}$$
$$x > 11$$

(continued)

Definitions and Concepts	**Examples**

Section 3.2   Linear Inequalities and Problem Solving

To solve a linear inequality in one variable:	Solve for x: $$\frac{3}{7}(x - 4) \geq x + 2$$
1. Clear the equation of fractions.	**1.** $7\left[\frac{3}{7}(x - 4)\right] \geq 7(x + 2)$ Multiply by 7. $3(x - 4) \geq 7(x + 2)$
2. Remove grouping symbols such as parentheses. **3.** Simplify by combining like terms.	**2.** $3x - 12 \geq 7x + 14$ Apply the distributive property.
4. Write variable terms on one side and numbers on the other side using the addition property of inequality.	**4.** $-4x - 12 \geq 14$ Subtract $7x$. $-4x \geq 26$ Add 12. $\dfrac{-4x}{-4} \leq \dfrac{26}{-4}$ Divide by -4. Reverse direction of inequality symbol.
5. Isolate the variable using the multiplication property of inequality.	$x \leq -\dfrac{13}{2}$

Section 3.3   Compound Inequalities

Two inequalities joined by the words **and** or **or** are called **compound inequalities.**	Compound inequalities: $x - 7 \leq 4$ and $x \geq -21$ $2x + 7 > x - 3$ or $5x + 2 > -3$
The solution set of a compound inequality formed by the word **and** is the **intersection** $\cap$ of the solution sets of the two inequalities.	Solve for x: $x < 5$ and $x < 3$ $\{x \mid x < 5\}$ $(-\infty, 5)$ $\{x \mid x < 3\}$ $(-\infty, 3)$ $\{x \mid x < 3$ and $x < 5\}$ $(-\infty, 3)$
The solution set of a compound inequality formed by the word **or** is the **union**, $\cup$, of the solution sets of the two inequalities.	Solve for x: $x - 2 \geq -3$ or $2x \leq -4$ $x \geq -1$ or $x \leq -2$ $\{x \mid x \geq -1\}$ $[-1, \infty)$ $\{x \mid x \leq -2\}$ $(-\infty, -2]$ $\{x \mid x \leq -2$ or $x \geq -1\}$ $(-\infty, -2]$ $\cup [-1, \infty)$

(continued)

Definitions and Concepts	**Examples**

Section 3.3 Compound Inequalities

To solve a compound inequality $y_1 < y_2 < y_3$ graphically,

1. Graph separately each of the three parts y_1, y_2, and y_3, respectively, in an appropriate window.

2. Observe where the graph of y_2 is between the graphs of y_1 and y_3.

3. Find the x-coordinates of the points of intersection and determine the appropriate interval of the solution.

Solve for x: $-13 < 3x - 4 \le 8$

The solution in interval notation is $(-3, 4]$.

Section 3.4 Absolute Value Equations

If a is a positive number, then $|x| = a$ is equivalent to $x = a$ or $x = -a$.

Solve for y:

$$|5y - 1| - 7 = 4$$

$$|5y - 1| = 11$$

$5y - 1 = 11$ or $5y - 1 = -11$ Add 7.

$5y = 12$ or $\qquad 5y = -10$ Add 1.

$y = \dfrac{12}{5}$ or $y = -2$ Divide by 5.

The solutions are -2 and $\frac{12}{5}$.

If a is negative, then $|x| = a$ has no solution.

Solve for x:

$$\left| \dfrac{x}{2} - 7 \right| = -1$$

The solution set is { } or $\varnothing$.

If an absolute value equation is of the form $|x| = |y|$, solve $x = y$ or $x = -y$.

Solve for x:

$$|x - 7| = |2x + 1|$$

$x - 7 = 2x + 1$ or $x - 7 = -(2x + 1)$

$x = 2x + 8$ $\qquad\qquad$ $x - 7 = -2x - 1$

$-x = 8$ $\qquad\qquad\qquad$ $x = -2x + 6$

$x = -8$ or $\qquad\qquad$ $3x = 6$

$\qquad\qquad\qquad\qquad\qquad$ $x = 2$

The solutions are -8 and 2.

Section 3.5 Absolute Value Inequalities

If a is a positive number, then $|x| < a$ is equivalent to $-a < x < a$.

Solve for y:

$$|y - 5| \le 3$$

$-3 \le y - 5 \le 3$

$-3 + 5 \le y - 5 + 5 \le 3 + 5$ Add 5.

$2 \le y \le 8$

The solution set is $[2, 8]$.   *(continued)*

Definitions and Concepts	Examples

Section 3.5 Absolute Value Inequalities

If a is a positive number, then $|x| > a$ is equivalent to $x < -a$ or $x > a$.

Solve for x:

$$\left| \frac{x}{2} - 3 \right| > 7$$

$$\frac{x}{2} - 3 < -7 \quad \text{or} \quad \frac{x}{2} - 3 > 7$$

$$x - 6 < -14 \quad \text{or} \quad x - 6 > 14 \qquad \text{Multiply by 2.}$$

$$x < -8 \quad \text{or} \quad x > 20 \qquad \text{Add 6.}$$

The solution set is $(-\infty, -8) \cup (20, \infty)$.

Section 3.6 Graphing Linear Inequalities in Two Variables

If the equal sign in a linear equation in two variables is replaced with an inequality symbol, the result is a **linear inequality in two variables**.

Linear Inequalities in Two Variables

$$x \le -5 \quad y \ge 2$$
$$3x - 2y > 7 \quad x < -5$$

To graph a linear inequality

1. Graph the boundary line by graphing the related equation. Draw the line solid if the inequality symbol is $\le$ or $\ge$. Draw the line dashed if the inequality symbol is $<$ or $>$.

2. Choose a test point not on the line. Substitute its coordinates into the original inequality.

3. If the resulting inequality is true, shade the **half-plane** that contains the test point. If the inequality is not true, shade the half-plane that does not contain the test point.

Graph $2x - 4y \ge 4$.

1. Graph $2x - 4y = 4$. Draw a solid line because the inequality symbol is $\ge$.

2. Check the test point $(0, 0)$ in the inequality $2x - 4y > 4$.

$$2 \cdot 0 - 4 \cdot 0 > 4 \quad \text{Let } x = 0 \text{ and } y = 0.$$

$$0 > 4 \quad \text{False}$$

3. The inequality is false, so we shade the half-plane that does not contain $(0, 0)$.

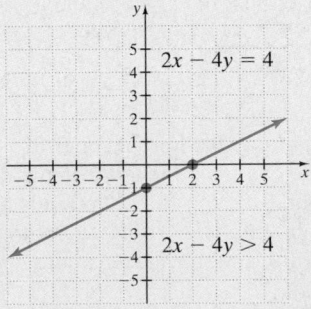

3 CHAPTER REVIEW

(3.1) Solve each equation algebraically and graphically.

1. $4(x - 6) + 3 = 27$ 12

2. $15(x + 2) - 6 = 18$ $-\frac{2}{5}$

3. $5x + 15 = 3(x + 2) + 2(x - 3)$ $\varnothing$

4. $2x - 5 + 3(x - 4) = 5(x + 2) - 27$ all real numbers

5. $14 - 2(x + 3) = 3(x - 9) + 18$ $\frac{17}{5}$

6. $16 + 2(5 - x) = 19 - 3(x + 2)$ -13

Solve each equation graphically. Round solutions to the nearest hundredth.

7. $0.4(x - 6) = \pi x + \sqrt{3}$ -1.51

8. $1.7x + \sqrt{7} = -0.4x - \sqrt{6}$ -2.43

(3.2) Solve each linear inequality.

9. $3(x - 5) > -(x + 3)$ $(3, \infty)$

10. $-2(x + 7) \geq 3(x + 2)$ $(-\infty, -4]$

11. $4x - (5 + 2x) < 3x - 1$ $(-4, \infty)$

12. $3(x - 8) < 7x + 2(5 - x)$ $(-17, \infty)$

13. $24 \geq 6x - 2(3x - 5) + 2x$ $(-\infty, 7]$

14. $48 + x \geq 5(2x + 4) - 2x$ $(-\infty, 4]$

15. $\frac{x}{3} + \frac{1}{2} > \frac{2}{3}$ $\left(\frac{1}{2}, \infty\right)$

16. $x + \frac{3}{4} < -\frac{x}{2} + \frac{9}{4}$ $(-\infty, 1)$

17. $\frac{x - 5}{2} \leq \frac{3}{8}(2x + 6)$ $[-19, \infty)$

18. $\frac{3(x - 2)}{5} > \frac{-5(x - 2)}{3}$ $(2, \infty)$

Solve.

19. George Boros can pay his housekeeper $25 per week to do his laundry, or he can have the laundromat do it at a cost of 90 cents per pound for the first 10 pounds and 80 cents for each additional pound. Use an inequality to find the weight at which it is more economical to use the housekeeper than the laundromat.   more than 30 pounds per week

20. Ceramic firing temperatures usually range from 500° to 1000° Fahrenheit. Use a compound inequality to convert this range to the Celsius scale. Round to the nearest degree. $260° \leq C \leq 538°C$

21. In the Olympic gymnastics competition, Nana must average a score of 9.65 to win the silver medal. Seven of the eight judges have reported scores of 9.5, 9.7, 9.9, 9.7, 9.7, 9.6, and 9.5. Use an inequality to find the minimum score that the last judge can give so that Nana wins the silver medal.   at least 9.6

22. Carol would like to pay cash for a car when she graduates from college and estimates that she can afford a used car that costs between $4000 and $8000. She has saved $500 so far and plans to earn the rest of the money by working the next two summers. If Carol plans to save the same amount each summer, use a compound inequality to find the range of money she must save each summer to buy the car. $1750 to $3750

(3.3) Solve each inequality.

23. $1 \leq 4x - 7 \leq 3$ $\left[2, \frac{5}{2}\right]$

24. $-2 \leq 8 + 5x < -1$ $\left[-2, -\frac{9}{5}\right)$

25. $-3 < 4(2x - 1) < 12$ $\left(\frac{1}{8}, 2\right)$

26. $-6 < x - (3 - 4x) < -3$ $\left(-\frac{3}{5}, 0\right)$

27. $\frac{1}{6} < \frac{4x - 3}{3} \leq \frac{4}{5}$ $\left(\frac{7}{8}, \frac{27}{20}\right]$

28. $0 \leq \frac{2(3x + 4)}{5} \leq 3$ $\left[-\frac{4}{3}, \frac{7}{6}\right]$

29. $x \leq 2$ and $x > -5$ $(-5, 2]$

30. $x \leq 2$ or $x > -5$ $(-\infty, \infty)$

31. $3x - 5 > 6$ or $-x < -5$ $\left(\frac{11}{3}, \infty\right)$

32. $-2x \leq 6$ and $-2x + 3 < -7$ $(5, \infty)$

(3.4) Solve each absolute value equation.

33. $|x - 7| = 9$ $16, -2$ **34.** $|8 - x| = 3$ $5, 11$

35. $|2x + 9| = 9$ $0, -9$ **36.** $|-3x + 4| = 7$ $-1, \frac{11}{3}$

37. $|3x - 2| + 6 = 10$ $2, -\frac{2}{3}$

56. ←——————→ , $\varnothing$

38. $5 + |6x + 1| = 5$ $-\dfrac{1}{6}$

39. $-5 = |4x - 3|$ $\varnothing$ 40. $|5 - 6x| + 8 = 3$ $\varnothing$

41. $|7x| - 26 = -5$ $3, -3$ 42. $-8 = |x - 3| - 10$ $1, 5$

43. $\left|\dfrac{3x - 7}{4}\right| = 2$ $5, -\dfrac{1}{3}$ 44. $\left|\dfrac{9 - 2x}{5}\right| = -3$ $\varnothing$

45. $|6x + 1| = |15 + 4x|$ $7, -\dfrac{8}{5}$

46. $|x - 3| = |7 + 2x|$ $-10, -\dfrac{4}{3}$

(3.5) Solve each absolute value inequality. Graph the solution set and write in interval notation.

47. $|5x - 1| < 9$ 48. $|6 + 4x| \geq 10$

49. $|3x| - 8 > 1$ 50. $9 + |5x| < 24$

51. $|6x - 5| \leq -1$ 52. $|6x - 5| \geq -1$

53. $\left|3x + \dfrac{2}{5}\right| \geq 4$ 54. $\left|\dfrac{4x - 3}{5}\right| < 1$

55. $\left|\dfrac{x}{3} + 6\right| - 8 > -5$ 56. $\left|\dfrac{4(x - 1)}{7}\right| + 10 < 2$

(3.6) Graph each linear inequality.   See graphing answer section.

57. $3x + y > 4$ 58. $\dfrac{1}{2}x - y < 2$

59 $5x - 2y \leq 9$ 60. $3y \geq x$

61. $y < 1$ 62. $x > -2$

63. Graph the union of $y > 2x + 3$ or $x \leq -3$.

64. Graph the intersection of $2x < 3y + 8$ and $y \geq -2$.

47. ←——( )——→ , $\left(-\dfrac{8}{5}, 2\right)$
   $-\dfrac{8}{5}$ 2

50. ←——( )——→ , $(-3, 3)$
   -3 3

53. ←——] [——→ , $\left(-\infty, -\dfrac{22}{15}\right] \cup \left[\dfrac{6}{5}, \infty\right)$
   $-\dfrac{22}{15}$ $\dfrac{6}{5}$

48. ←——] [——→ , $(-\infty, -4] \cup [1, \infty)$
   -4 1

51. ←——————→ , $\varnothing$

54. ←——( )——→ , $\left(-\dfrac{1}{2}, 2\right)$
   $-\dfrac{1}{2}$ 2

49. ←——) (——→ , $(-\infty, -3) \cup (3, \infty)$
   -3 3

52. ←——————→ , $(-\infty, \infty)$

55. ←——) (——→ , $(-\infty, -27) \cup (-9, \infty)$
   -27 -9

CHAPTER 3 TEST

Remember to use your Chapter Test Prep Video CD to help you study and view solutions to the test questions you need help with.

1. Solve $15x + 26 = -2(x + 1) - 1$ algebraically and graphically. $-\dfrac{29}{17}$

2. Solve $-3x - \sqrt{5} = \pi(x - 1)$ graphically, Round the solution to the nearest hundredth. 0.15

Solve. Write inequality solutions in interval notation.

3. $|6x - 5| - 3 = -2$ $1, \dfrac{2}{3}$

4. $|8 - 2t| = -6$ $\varnothing$

5. $3(2x - 7) - 4x > -(x + 6)$ $(5, \infty)$

6. $8 - \dfrac{x}{2} \geq 7$ $(-\infty, 2]$

7. $-3 < 2(x - 3) \leq 4$ $\left(\dfrac{3}{2}, 5\right]$

8. $|3x + 1| > 5$ $(-\infty, -2) \cup \left(\dfrac{4}{3}, \infty\right)$

9. $x \leq -2$ and $x \leq -5$ $(-\infty, -5]$

10. $x \leq -2$ or $x \leq -5$ $(-\infty, -2]$

11. $-x > 1$ and $3x + 3 \geq x - 3$ $[-3, -1)$

12. $6x + 1 > 5x + 4$ or $1 - x > -4$ $(-\infty, \infty)$

13. $|x - 5| - 4 < -2$ $(3, 7)$

14. $\left|\dfrac{5x - 7}{2}\right| = 4$ $3, -\dfrac{1}{5}$

15. $\left|17x - \dfrac{1}{5}\right| > -2$ $(-\infty, \infty)$

16. $|x - 5| = |x + 2|$ $\dfrac{3}{2}$

Graph each inequality.   17.–19.   See graphing answer section.

17. $x \leq -4$ 18. $2x - y > 5$

19. The intersection of $2x + 4y < 6$ and $y \leq -4$

Use the given screen to solve each inequality. Write the solution in interval notation.

20. $y_1 < y_2$ $(-\infty, -3)$ 21. $y_1 > y_2$ $(-3, \infty)$

3 CHAPTER CUMULATIVE REVIEW

List the elements in each set.

1. **a.** $\{x \mid x \text{ is a whole number between 1 and 6}\}$ $\{2, 3, 4, 5\}$

    b. $\{x \mid x \text{ is a natural number greater than 100}\}$

2. **a.** $\{x \mid x \text{ is an integer between } -3 \text{ and 5}\}$

    b. $\{x \mid x \text{ is a whole number between 3 and 5}\}$

3. Find each value.

    a. $|3|$ 3

    b. $|-5|$ 5

    c. $-|2|$ -2

    d. $-|-8|$ -8

    e. $|0|$ 0 (Sec. 1.2, Ex. 6)

4. Find the opposite of each number.

    a. $\frac{2}{3}$ $-\frac{2}{3}$

    b. -9 9

    c. 1.5 -1.5 (Sec. 1.2)

5. Add.

    a. $-3 + (-11)$ -14

    b. $3 + (-7)$ -4

    c. $-10 + 15$ 5

    d. $-8.3 + (-1.9)$ -10.2

    e. $-\frac{1}{4} + \frac{1}{2}$ $\frac{1}{4}$

    f. $-\frac{2}{3} + \frac{3}{7} - \frac{5}{21}$ (Sec. 1.3, Ex. 1)

6. Subtract.

    a. $-2 - (-10)$ 8

    b. $1.7 - 8.9$ -7.2

    c. $-\frac{1}{2} - \frac{1}{4}$ $-\frac{3}{4}$ (Sec. 1.3)

7. Find the square roots.

    a. $\sqrt{9}$ 3

    b. $\sqrt{25}$ 5

    c. $\sqrt{\frac{1}{4}}$ $\frac{1}{2}$ (Sec. 1.3, Ex. 7)

    d. $-\sqrt{36}$ -6

    e. $\sqrt{-36}$ not a real number

8. Multiply or divide.

    a. $-3(-2)$ 6

    b. $-\frac{3}{4}\left(-\frac{4}{7}\right)$ $\frac{3}{7}$

    c. $\frac{0}{-2}$ 0

    d. $\frac{-20}{-2}$ 10 (Sec. 1.3)

9. Evaluate each algebraic expression when $x = 2$, $y = -1$, and $z = -3$.

    a. $z - y$ -2

    b. z^2 9

    c. $\frac{2x + y}{z}$ -1 (Sec. 1.3, Ex. 11)

    d. $-x^2 - 4x$ -12

10. Find the roots.

    a. $\sqrt[4]{1}$ 1

    b. $\sqrt[3]{8}$ 2

    c. $\sqrt[4]{81}$ 3 (Sec. 1.3)

11. Write each sentence using mathematical symbols.

    a. The sum of x and 5 is 20. $x + 5 = 20$

    b. Two times the sum of 3 and y amounts to 4. $2(3 + y) = 4$

    c. Subtract 8 from x, and the difference is the same as the product of 2 and x. $x - 8 = 2x$

---

1. b. $\{101, 102, 103, \dots\}$ (Sec. 1.2, Ex. 3)   **2. a.** $\{-2, -1, 0, 1, 2, 3, 4\}$ **2. b.** $\{4\}$ (Sec. 1.2)

20. b. $12x + 4$ (Sec. 1.6)   **23.** $y = \dfrac{2x + 7}{3}$ or $y = \dfrac{2x}{3} + \dfrac{7}{3}$ (Sec. 1.8, Ex. 2)   **25.** $b = \dfrac{2A - Bh}{h}$ (Sec. 1.8, Ex. 3)

d. The quotient of z and 9 is 3 times the difference of z and 5. $\frac{z}{9} = 3(z - 5)$ (Sec. 1.4, Ex. 1)

12. Insert $<$, $>$, or $=$ between each pair of numbers to form a true statement.

   a. $-3 \quad -5 \quad >$

   b. $\frac{-12}{-4} \quad 3 \quad =$

   c. $0 \quad -2 \quad > $ (Sec. 1.4)

13. Use the commutative property of addition to write an expression equivalent to $7x + 5$. $5 + 7x$ (Sec. 1.4, Ex. 6)

14. Use the associative property of multiplication to write an expression equivalent to $5 \cdot (7x)$. Then simplify the expression. $5 \cdot (7x) = (5 \cdot 7)x = 35x$ (Sec. 1.4)

Solve for x.

15. $2x + 5 = 9$ 2 (Sec. 1.5, Ex. 1)

16. $11.2 = 1.2 - 5x$ -2 (Sec. 1.5)

17. $6x - 4 = 2 + 6(x - 1)$ all real numbers (Sec. 1.5, Ex. 9)

18. $2x + 1.5 = -0.2 + 1.6x$ -4.25 (Sec. 1.5)

19. Write the following as algebraic expressions. Then simplify.

   a. The sum of three consecutive integers, if x is the first consecutive integer. $3x + 3$

   b. The perimeter of the triangle with sides of length x, $5x$, and $6x - 3$. $12x - 3$ (Sec. 1.6, Ex. 1)

20. Write the following as algebraic expressions. Then simplify.

   a. The sum of three consecutive integers if x is the first consecutive integers. $3x + 3$

   b. The perimeter of a square with side length $3x + 1$.

21. Find two numbers such that the second number is 3 more than twice the first number and the sum of the two numbers is 72. $23, 49$ (Sec. 1.6, Ex. 3)

22. Find two numbers such that the second number is 2 more than three times the first number and the difference of the two numbers is 24. $11, 35$ (Sec. 1.6)

23. Solve $3y - 2x = 7$ for y.

24. Solve $7x - 4y = 10$ for x. $x = \dfrac{10 + 4y}{7}$ (Sec. 1.8)

25. Solve $A = \dfrac{1}{2}(B + b)h$ for b.

26. Solve $P = 2l + 2w$ for l. $l = \dfrac{P - 2w}{2}$ (Sec. 1.8)

27. Write each in interval notation.

   a. $\{x \mid x \geq 2\}$ $[2, \infty)$

   b. $\{x \mid x < -1\}$ $(-\infty, -1)$

   c. $\{x \mid 0.5 < x \leq 3\}$

28. Write each in interval notation.

   a. $\{x \mid x \leq -3\}$ $(-\infty, -3]$

   b. $\{x \mid -2 \leq x < 0.1\}$ $[-2, 0.1)$ (Sec. 3.2)

Solve.

29. $-(x - 3) + 2 \leq 3(2x - 5) + x$ $\left[\frac{5}{2}, \infty\right)$ (Sec. 3.2, Ex. 5)

30. $2(7x - 1) - 5x > -(-7x) + 4$ $(3, \infty)$ (Sec. 3.2)

31. a. $2(x + 3) > 2x + 1$ $(-\infty, \infty)$ (Sec. 3.2, Ex. 7)

   b. $2(x + 3) < 2x + 1$ $\varnothing$

32. $4(x + 1) - 3 < 4x + 1$ $\varnothing$ (Sec. 3.2)

33. If $A = \{x \mid x$ is an even number great than 0 and less than 10$\}$ and $B = \{3, 4, 5, 6\}$, find $A \cap B$.

34. Find the union: $\{-2, 0, 2, 4\} \cup \{-1, 1, 3, 5\}$

35. Solve: $x - 7 < 2$ and $2x + 1 < 9$

36. Solve: $x + 3 \leq 1$ or $3x - 1 < 8$ $(-\infty, 3)$ (Sec. 3.3)

37. If $A = \{x \mid x$ is an even number great than 0 and less than 10$\}$ and $B = \{3, 4, 5, 6\}$, find $A \cup B$.

38. Find the intersection: $\{-2, 0, 2, 4\} \cap \{-1, 1, 3, 5\}$

27. c. $(0.5, 3]$ (Sec. 3.2, Ex. 1)   **33.** $\{4, 6\}$ (Sec. 3.3, Ex. 1)   **34.** $\{-2, -1, 0, 1, 2, 3, 4, 5\}$ (Sec. 3.3)

35. $(-\infty, 4)$ (Sec. 3.3, Ex. 2)   **37.** $\{2, 3, 4, 5, 6, 8\}$ (Sec. 3.3, Ex. 6)   **38.** $\varnothing$ (Sec. 3.3)

39. Solve: $-2x - 5 < -3$ or $6x < 0$

40. Solve: $-2x - 5 < -3$ and $6x < 0$ $(-1, 0)$ (Sec. 3.3)

Solve.

41. $|p| = 2$ $2, -2$ (Sec. 3.4, Ex. 1)

42. $|x| = 5$ $\{-5, 5\}$ (Sec. 3.4)

43. $\left| \dfrac{x}{2} - 1 \right| = 11$ $24, -20$ (Sec. 3.4, Ex. 3)

44. $\left| \dfrac{y}{3} + 2 \right| = 10$ $\{24, -36\}$ (Sec. 3.4)

45. $|x - 3| = |5 - x|$ 4 (Sec. 3.4, Ex. 9)

46. $|x + 3| = |7 - x|$ 2 (Sec. 3.4)

47. $|x| \le 3$ $[-3, 3]$ (Sec. 3.5, Ex. 1)

48. $|x| > 1$ $(-\infty, -1) \cup (1, \infty)$ (Sec. 3.5)

49. $|2x + 9| + 5 > 3$ $(-\infty, \infty)$ (Sec. 3.5, Ex. 6)

50. $|3x + 1| + 9 < 1$ $\varnothing$ (Sec. 3.5)

39. $(-\infty, \infty)$ (Sec. 3.3, Ex. 8)

Systems of Equations

In this chapter, two or more equations in two or more variables are solved simultaneously. Such a collection of equations is called a **system of equations**. Systems of equations are good mathematical models for many real-world problems because these problems may involve several related patterns.

PLANNING FOR GROWTH

Development affects our lives and our environment in many ways. For example, increased highway congestion lowers the average vehicle speed in many large cities, 50 acres of farmland are lost to development each hour, and at least half of the parking spaces in most American shopping malls are vacant at least 40% of the time.

Urban or regional planners tackle problems like these. Planners analyze issues like population growth, housing needs, land use, urban or suburban sprawl, parks and recreation space, public transportation, and highways. They decide if current resources are adequate and, if not, propose ways to improve them. Planners must be problem solvers, understand how governments work, listen and communicate well, work well as part of a team, understand the principles of city design, and analyze and interpret data.

In the Spotlight on Decision Making feature on page 323, you will have the opportunity to make a decision about city bus routes as an urban planner.

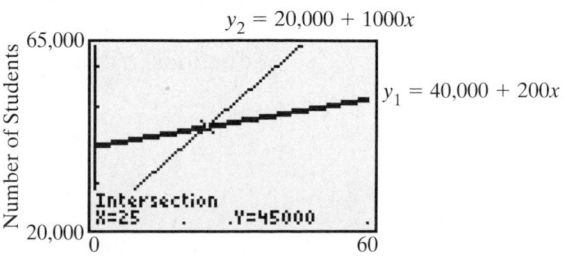

Number of Commuters.

Source: American Planning Association website

4.1 | *SOLVING SYSTEMS OF LINEAR EQUATIONS IN TWO VARIABLES*

Objectives

1 Solve a system by graphing.

2 Solve a system by substitution.

3 Solve a system by elimination.

1   An important problem that often occurs in the fields of business and economics concerns the concepts of revenue and cost. For example, suppose that a small manufacturing company begins to manufacture and sell compact disc storage units. The revenue of a company is the company's income from selling these units, and the cost is the amount of money that a company spends to manufacture these units. The following coordinate system shows the graphs of revenue and cost for the storage units.

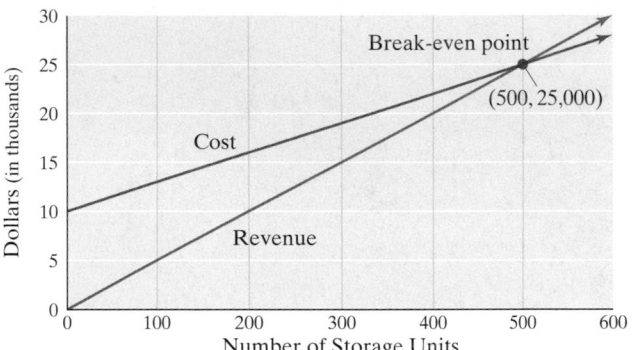

These lines intersect at the point (500, 25,000). This means that when 500 storage units are manufactured and sold, both cost and revenue are $25,000. In business, this point of intersection is called the **break-even point**. Notice that for *x*-values (units sold) less than 500, the cost graph is above the revenue graph, meaning that cost of manufacturing is greater than revenue, and so the company is losing money. For *x*-values (units sold) greater than 500, the revenue graph is above the cost graph, meaning that revenue is greater than cost, and so the company is making money.

Recall from Chapter 2 that each line is a graph of some linear equation in two variables. Both equations together form a **system of equations**. The common point of intersection is called the **solution of the system**. Some examples of systems of linear equations in two variables are

Systems of Linear Equations in Two Variables

$$\begin{cases} x - 2y = -7 \\ 3x + y = 0 \end{cases} \qquad \begin{cases} x = 5 \\ x + \dfrac{y}{2} = 9 \end{cases} \qquad \begin{cases} x - 3 = 2y + 6 \\ y = 1 \end{cases}$$

Recall that a solution of an equation in two variables is an ordered pair (x, y) that makes the equation true. A **solution of a system** of two equations in two variables is an ordered pair (x, y) that makes both equations true.

EXAMPLE 1

Determine whether the given ordered pair is a solution of the system.

a. $\begin{cases} -x + y = 2 \\ 2x - y = -3 \end{cases}$ **b.** $\begin{cases} 5x + 3y = -1 \\ x - y = 1 \end{cases}$

$(-1, 1)$ $(-2, 3)$

Solution   **a.** We replace x with -1 and y with 1 in each equation.

$-x + y = 2$ *First equation* $2x - y = -3$ *Second equation*

$-(-1) + (1) \stackrel{?}{=} 2$ *Let x = −1 and y = 1.* $2(-1) - (1) \stackrel{?}{=} -3$ *Let x = −1 and y = 1.*

$1 + 1 \stackrel{?}{=} 2$ $-2 - 1 \stackrel{?}{=} -3$

$2 = 2$ *True* $-3 = -3$ *True*

Since $(-1, 1)$ makes *both* equations true, it is a solution. Using set notation, the solution set is $\{(-1, 1)\}$.

b. We replace x with -2 and y with 3 in each equation.

$5x + 3y = -1$ *First equation* $x - y = 1$ *Second equation*

$5(-2) + 3(3) \stackrel{?}{=} -1$ *Let x = −2 and y = 3.* $(-2) - (3) \stackrel{?}{=} 1$ *Let x = −2 and y = 3.*

$-10 + 9 \stackrel{?}{=} -1$ $-5 = 1$ *False*

$-1 = -1$ *True*

Since the ordered pair $(-2, 3)$ does not make *both* equations true, it is not a solution of the system.

Example 1 above shows how to determine that an ordered pair is a solution of a system of equations, but how do we find such a solution? Actually, there are various methods to find the solution. We will investigate several in this chapter: graphing, substitution, elimination, matrices, and determinants.

To solve by graphing, we graph each equation in an appropriate window and find the coordinates of any points of intersection.

EXAMPLE 2

Solve the system by graphing.

$$\begin{cases} x + y = 2 \\ 3x - y = -2 \end{cases}$$

Solution   Since the graph of a linear equation in two variables is a line, graphing two such equations yields two lines in a plane. To use a graphing utility, solve each equation for y.

$$\begin{cases} y = -x + 2 & \text{First equation} \\ y = 3x + 2 & \text{Second equation} \end{cases}$$

Graph $y_1 = -x + 2$ and $y_2 = 3x + 2$ and find the point of intersection.

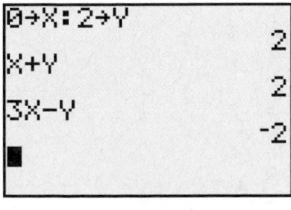

TEACHING TIP

After discussing Example 1, ask, "If you had a system of three equations, how would you determine whether an ordered pair is a solution of the system?"

TECHNOLOGY NOTE

When solving a system of equations by graphing, try the standard or integer window first and see if the intersection appears. If it does not appear in one of these windows, you may be able to see enough of the graph to estimate where the intersection will occur and adjust the window setting accordingly.

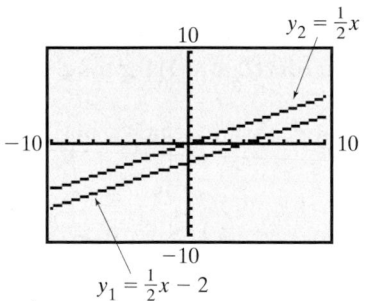

$y_2 = \frac{1}{2}x$

$y_1 = \frac{1}{2}x - 2$

> **Helpful Hint**
> - If a system of equations has *at least one solution*, the system is *consistent*.
> - If a system of equations has *no solution*, the system is *inconsistent*.

Verify the ordered pair solution $(0, 2)$ by replacing x with 0 and y with 2 in both original equations and seeing that true statements result each time. The screen on the right on the previous page shows that the ordered pair $(0, 2)$ does satisfy both equations. We conclude therefore that $(0, 2)$ is the solution of the system. A system that has at least one solution, such as this one, is said to be **consistent**.

Discover the Concept

Use your graphing utility to solve the system

$$\begin{cases} x - 2y = 4 \\ x = 2y \end{cases}$$

In the discovery above, we see that solving each equation for y produces the following:

$x - 2y = 4$ *First equation* $x = 2y$ *Second equation*

$-2y = -x + 4$ *Subtract x from both sides.* $\frac{1}{2}x = y$ *Divide both sides by 2.*

$y = \frac{1}{2}x - 2$ *Divide both sides by -2.* $y = \frac{1}{2}x$

Notice that each equation is in the form $y = mx + b$. From this form, we see that both lines have the same slope, $\frac{1}{2}$, but different y-intercepts, so they are parallel as shown to the left. Therefore, the system has no solution since the equations have no common solution (there are no intersection points). A system that has no solution is said to be **inconsistent**.

Discover the Concept

Use your graphing utility to solve the system

$$\begin{cases} 2x + 4y = 10 \\ x + 2y = 5 \end{cases}$$

In the discovery above, we see that solving each equation for y produces the following:

$2x + 4y = 10$ *First equation* $x + 2y = 5$ *Second equation*

$y = -\frac{1}{2}x + \frac{5}{2}$ $y = -\frac{1}{2}x + \frac{5}{2}$

Notice that both lines have the same slope, $-\frac{1}{2}$, and the same y-intercept, $\frac{5}{2}$. This means that the graph of each equation is the same line.

To confirm this, notice that the entries for y_1 and y_2 are the same in the table shown on the previous page. The equations have identical solutions and any ordered pair solution of one equation satisfies the other equation also. Thus, these equations are said to be **dependent equations**. The solution set of the system is $\{(x, y) \mid x + 2y = 5\}$ or, equivalently, $\{(x, y) \mid 2x + 4y = 10\}$ since the equations describe identical ordered pairs. Written this way, the solution set is read "the set of all ordered pairs (x, y), such that $2x + 4y = 10$." There are therefore an infinite number of solutions to the system.

CONCEPT CHECK

The equations in the system are dependent and the system has an infinite number of solutions. Which ordered pairs below are solutions?

$$\begin{cases} -x + 3y = 4 \\ 2x + 8 = 6y \end{cases}$$

a. $(4, 0)$ **b.** $(-4, 0)$ **c.** $(-1, 1)$

We can summarize the information thus far as follows.

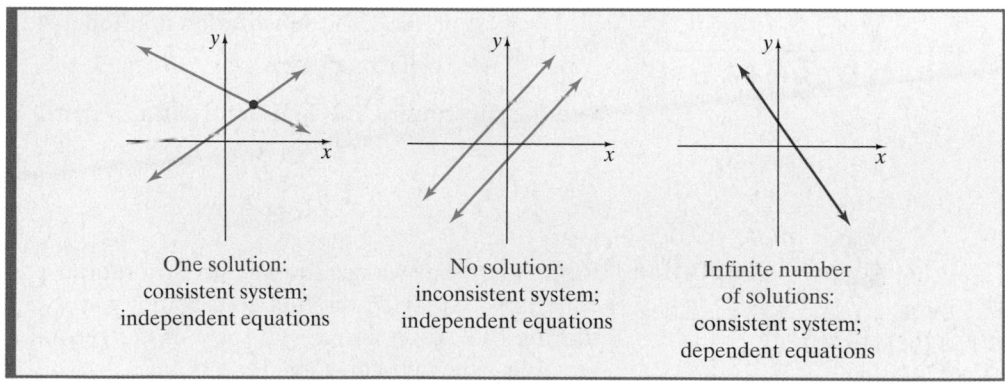

One solution: consistent system; independent equations

No solution: inconsistent system; independent equations

Infinite number of solutions: consistent system; dependent equations

A graphing calculator is a very useful tool for approximating solutions to a system of equations in two variables. See the next example.

TEACHING TIP

Remind students to set the mode to Simul (Simultaneous) if they want both equations to be graphed at the same time. Use sequential mode if they want y_1 to be graphed first.

EXAMPLE 3

Solve the system by graphing. Approximate the solution to two decimal places.

$$\begin{cases} y + 2.6x = 5.6 \\ y - 4.3x = -4.9 \end{cases}$$

Solution First use a standard window and graph both equations on a single screen. The screen on the left shows that the two lines intersect. To approximate the point of intersection, trace to the point of intersection and use an Intersect feature of the graphing calculator. We find that the approximate point of intersection is $(1.52, 1.64)$.

CLASSROOM EXAMPLE

Solve the system by graphing. Approximate the solution to two decimal places.

$$y - 5.7x = 9.8$$
$$y + 2.1x = 4.3$$

answer: $(-0.71, 5.78)$

$y_2 = 4.3x - 4.9$

$y_1 = -2.6x + 5.6$

Solving graphically. x and y values automatically stored to 14 decimal places

The numerical check with decimal approximations

Concept Check Answer:
b, c

Because the solution is an approximation, notice that the numerical check with these approximations does not show equivalent expressions. For example, instead of $y + 2.6x = 5.6$, we have $y + 2.6x = 5.592$. The number 5.592 is close to 5.6, but not equal to 5.6. Keep this in mind when checking approximations.

✔ CONCEPT CHECK

How can you tell just by looking at the following system that it has no solution?

$$\begin{cases} y = 3x + 5 \\ y = 3x - 7 \end{cases}$$

How can you tell just by looking at the following system that it has infinitely many solutions?

$$\begin{cases} x + y = 5 \\ 2x + 2y = 10 \end{cases}$$

Concept Check Answer:
Answers may vary

2   Graphing the equations of a system by hand is often a good method of finding approximate solutions of a system, but it is not a reliable method of finding exact solutions of a system. We turn instead to two algebraic methods of solving systems. We use the first method, the **substitution method**, next.

EXAMPLE 4

Use the substitution method to solve the system

$$\begin{cases} y = x + 5 & \text{First equation} \\ 3x = 2y - 9 & \text{Second equation} \end{cases}$$

Solution   Remember that we are looking for an ordered pair, if there is one, that satisfies both equations. Satisfying the first equation, $y = x + 5$, means that y must be $x + 5$. **Substituting** the expression $x + 5$ for y in the second equation yields an equation in one variable, which we can solve for x.

$$3x = 2y - 9 \qquad \text{Second equation}$$

$$3x = 2(x + 5) - 9 \qquad \text{Replace } y \text{ with } x + 5 \text{ in the second equation.}$$

$$3x = 2x + 10 - 9$$

$$x = 1$$

The x-coordinate of the solution of the system is 1. The y-coordinate is the y-value corresponding to the x-value 1. Choose either equation of the system and solve for y when x is 1.

$$y = x + 5 \qquad \text{First equation}$$

$$= 1 + 5 \qquad \text{Let } x = 1.$$

$$= 6$$

The y-coordinate is 6, so the solution of the system is $(1, 6)$. This means that, when both equations are graphed, the one point of intersection occurs at the point with coordinates $(1, 6)$. We can check this solution by substituting 1 for x and 6 for y in both equations of the system, as shown below.

CLASSROOM EXAMPLE
Use the substitution method to solve the system:
$$\begin{cases} 6x - 4y = 10 \\ y = 3x - 3 \end{cases}$$
answer: $\left(\frac{1}{3}, -2 \right)$

Satisfies $y = x + 5$

Satisfies $3x = 2y - 9$

The steps below summarize the substitution method.

> ## Solving A System of Two Equations Using the Substitution Method
>
> **Step 1:** Solve one of the equations for one of its variables.
> **Step 2:** Substitute the expression for the variable found in Step 1 into the other equation.
> **Step 3:** Find the value of one variable by solving the equation from Step 2.
> **Step 4:** Find the value of the other variable by substituting the value found in Step 3 into the equation from Step 1.
> **Step 5:** Check the ordered pair solution in *both* original equations.

EXAMPLE 5

Use the substitution method to solve the system.

$$\begin{cases} -\dfrac{x}{6} + \dfrac{y}{2} = \dfrac{1}{2} \\ \dfrac{x}{3} - \dfrac{y}{6} = -\dfrac{3}{4} \end{cases}$$

Solution First we multiply each equation by its least common denominator to clear the system of fractions. We multiply the first equation by 6 and the second equation by 12.

$$\begin{cases} 6\left(-\dfrac{x}{6} + \dfrac{y}{2}\right) = 6\left(\dfrac{1}{2}\right) \\ 12\left(\dfrac{x}{3} - \dfrac{y}{6}\right) = 12\left(-\dfrac{3}{4}\right) \end{cases} \text{ simplifies to } \begin{cases} -x + 3y = 3 & \text{First equation} \\ 4x - 2y = -9 & \text{Second equation} \end{cases}$$

To use the substitution method, we now solve the first equation for x.

$$-x + 3y = 3 \qquad \text{First equation}$$
$$3y - 3 = x \qquad \text{Solve for } x.$$

Next we replace x with $3y - 3$ in the second equation.

$$4x - 2y = -9 \qquad \text{Second equation}$$
$$4(3y - 3) - 2y = -9$$
$$12y - 12 - 2y = -9$$
$$10y = 3$$
$$y = \frac{3}{10} \qquad \text{Solve for } y.$$

To find the corresponding x-coordinate, we replace y with $\dfrac{3}{10}$ in the equation $x = 3y - 3$. Then

$$x = 3\left(\frac{3}{10}\right) - 3 = \frac{9}{10} - 3 = \frac{9}{10} - \frac{30}{10} = -\frac{21}{10}$$

The ordered pair solution is $\left(-\dfrac{21}{10}, \dfrac{3}{10}\right)$. We check this solution graphically to the left.

> ▶ **Helpful Hint**
>
> If a system of equations contains equations with fractions, first clear the equations of fractions.

3 The **elimination method**, or **addition method**, is a second algebraic technique for solving systems of equations. For this method, we rely on a version of the addition property of equality, which states that "equals added to equals are equal."

If $A = B$ and $C = D$ then $A + C = B + D$.

EXAMPLE 6

Use the elimination method to solve the system.

$$\begin{cases} x - 5y = -12 & \text{First equation} \\ -x + y = 4 & \text{Second equation} \end{cases}$$

Solution

Since the left side of each equation is equal to the right side, we add equal quantities by adding the left sides of the equations and the right sides of the equations. This sum gives us an equation in one variable, y, which we can solve for y.

$$\begin{array}{ll} x - 5y = -12 & \text{First equation} \\ \underline{-x + y = 4} & \text{Second equation} \\ -4y = -8 & \text{Add.} \\ y = 2 & \text{Solve for } y. \end{array}$$

The y-coordinate of the solution is 2. To find the corresponding x-coordinate, we replace y with 2 in either original equation of the system. Let's use the second equation.

$$\begin{array}{ll} -x + y = 4 & \text{Second equation} \\ -x + 2 = 4 & \text{Let } y = 2. \\ -x = 2 & \\ x = -2 & \end{array}$$

The ordered pair solution is $(-2, 2)$. We check numerically (to the left) to see that $(-2, 2)$ satisfies both equations of the system.

The steps below summarize the elimination method.

Satisfies
$x - 5y = -12$

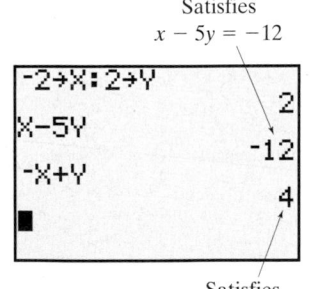

Satisfies
$-x + y = 4$

Solving A System of Two Linear Equations Using the Elimination Method

Step 1: Rewrite each equation in standard form, $Ax + By = C$.

Step 2: If necessary, multiply one or both equations by some nonzero number so that the coefficient of one variable in one equation is the opposite of its coefficient in the other equation.

Step 3: Add the equations.

Step 4: Find the value of one variable by solving the equation from Step 3.

Step 5: Find the value of the second variable by substituting the value found in Step 4 into either original equation.

Step 6: Check the proposed ordered pair solution in *both* original equations.

EXAMPLE 7

Use the elimination method to solve the system.

$$\begin{cases} 3x - 2y = 10 \\ 4x - 3y = 15 \end{cases}$$

Solution   If we add the two equations, the sum will still be an equation in two variables. Notice, however, that we can eliminate y when the equations are added if we multiply both sides of the first equation by 3 and both sides of the second equation by -2. Then

$$\begin{cases} 3(3x - 2y) = 3(10) \\ -2(4x - 3y) = -2(15) \end{cases} \quad \text{simplifies to} \quad \begin{cases} 9x - 6y = 30 \\ -8x + 6y = -30 \end{cases}$$

Next we add the left sides and add the right sides.

$$\begin{array}{r} 9x - 6y = 30 \\ -8x + 6y = -30 \\ \hline x \qquad\quad = 0 \end{array}$$

To find y, we let $x = 0$ in either equation of the system.

$$3x - 2y = 10 \qquad \text{First equation}$$
$$3(0) - 2y = 10 \qquad \text{Let } x = 0.$$
$$-2y = 10$$
$$y = -5$$

The ordered pair solution is $(0, -5)$. Check to see that $(0, -5)$ satisfies both equations.

EXAMPLE 8

Use the elimination method to solve the system.

$$\begin{cases} 3x + \dfrac{y}{2} = 2 \\ 6x + y = 5 \end{cases}$$

Solution   If we multiply both sides of the first equation by -2, the coefficients of x in the two equations will be opposites. Then

$$\begin{cases} -2\left(3x + \dfrac{y}{2}\right) = -2(2) \\ 6x + y = 5 \end{cases} \quad \text{simplifies to} \quad \begin{cases} -6x - y = -4 \\ 6x + y = 5 \end{cases}$$

Now we can add the left sides and add the right sides.

$$\begin{array}{r} -6x - y = -4 \\ 6x + y = 5 \\ \hline 0 = 1 \qquad \text{False} \end{array}$$

The resulting equation, $0 = 1$, is false for all values of y or x. Thus, the system has no solution. The solution set is $\{\ \}$ or $\varnothing$. This system is inconsistent, and the graphs of the equations are parallel lines.

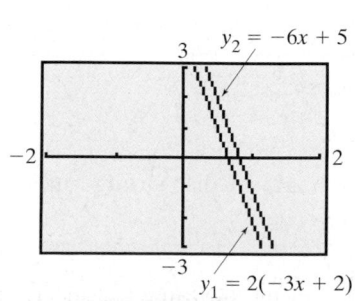

The two graphs appear to be parallel lines supporting no solution to the system of Example 8.

EXAMPLE 9

Use the elimination method to solve the system.

$$\begin{cases} -5x - 3y = 9 \\ 10x + 6y = -18 \end{cases}$$

Solution   To eliminate x when the equations are added, we multiply both sides of the first equation by 2. Then

$$\begin{cases} 2(-5x - 3y) = 2(9) \\ 10x + 6y = -18 \end{cases} \quad \text{simplifies to} \quad \begin{cases} -10x - 6y = 18 \\ 10x + 6y = -18 \end{cases}$$

Next we add the equations.

$$\begin{array}{r} -10x - 6y = 18 \\ \underline{10x + 6y = -18} \\ 0 = 0 \end{array}$$

The resulting equation, $0 = 0$, is true for all possible values of y or x. Notice in the original system that if both sides of the first equation are multiplied by -2, the result is the second equation. This means that the two equations are equivalent. They have the same solution set and there are an infinite number of solutions. Thus, the equations of this system are dependent, and the solution set of the system is

$$\{(x, y) \mid -5x - 3y = 9\} \quad \text{or, equivalently,} \quad \{(x, y) \mid 10x + 6y = -18\}.$$

The graph (shown on a standard window) and table indicate that the graph of both equations is the same line. This supports the solution above for Example 9.

> **Helpful Hint**
>
> Remember that not all ordered pairs are solutions of the system in Example 9. Only the infinite number of ordered pairs that satisfy $-5x - 3y = 9$ or equivalently $10x + 6y = -18$.

EXAMPLE 10

FINDING THE BREAK-EVEN POINT

A small manufacturing company manufactures and sells compact disc storage units. The revenue equation for these units is

$$y = 50x$$

where x is the number of units sold and y is the revenue, or income, in dollars for selling x units. The cost equation for these units is

$$y = 30x + 10,000$$

where x is the number of units manufactured and y is the total cost in dollars for manufacturing x units. Use these equations to find the number of units to be sold for the company to break even.

Solution   The break-even point is found by solving the system

$$\begin{cases} y = 50x & \text{First equation} \\ y = 30x + 10{,}000 & \text{Second equation} \end{cases}$$

To solve the system, graph $y_1 = 50x$ and $y_2 = 30x + 10{,}000$ and find the point of intersection, the break-even point.

The ordered pair solution is (500, 25,000). This means that the business must sell 500 compact disc storage units to break even. A hand-drawn graph of the equations in this system can be found at the beginning of this section.

Spotlight on

DECISION
of MAKING

Suppose you have just signed a 12-month lease for an apartment. After moving in, you find that the tap water tastes terrible. You have two options: (a) buy bottled water or (b) buy a reusable water filter pitcher and filters to filter the tap water. You estimate that you use 10 gallons of drinking water each week. Buying bottled water costs $0.50 per gallon. Using a water filter pitcher involves the following costs:

- A reusable water filter pitcher costs $50.
- A water filter costs $10 each. Each water filter lasts for 40 gallons, or 4 weeks in your situation. The cost of a water filter is $2.50 per week.
- Tap water costs $0.005 per gallon

   An equation for the cumulative cost y of the bottled water option after x weeks is
   $y = $ (cost per gallon of bottled water)(number of gallons used per week) x

   An equation for the cumulative cost y of filtered water option after x weeks is
   $y = $ (cost of reusable pitcher) $+$ (cost of water filter per week) $x +$
   (cost per gallon of tap water)(number of gallons used per week)x

Which option would you choose? Why?

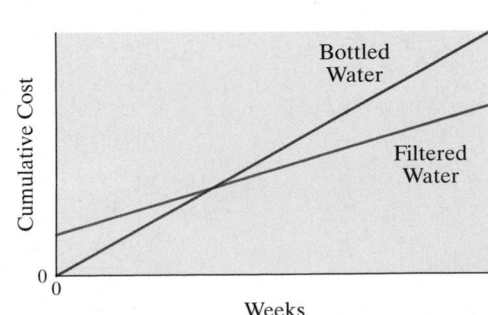

MENTAL MATH

Match each graph with the solution of the corresponding system.

TEACHING TIP
A Group Activity for this section is available in the Instructor's Resource Manual.

   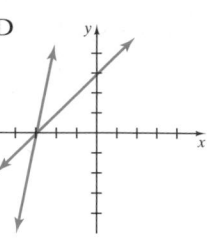

1. no solution   B   **2.** Infinite number of solutions   C   **3.** $(1, -2)$ A   **4.** $(-3, 0)$ D

7. $(2, -1)$ see graphing answer section.   **8.** $(5, 2)$ see graphing answer section.   **9.** $(1, 2)$ see graphing answer section.

EXERCISE SET 4.1

STUDY GUIDE/SSM   CD/VIDEO   PH MATH TUTOR CENTER   MathXL®Tutorials ON CD   MathXL®   MyMathLab®

10. $(2, 2)$ see graphing answer section.   **11.** ∅ see graphing answer section.   **12.** ∅ see graphing answer section.

Determine whether each given ordered pair is a solution of each system. See Example 1.

1. $\begin{cases} x - y = 3 \\ 2x - 4y = 8 \end{cases}$ $(2, -1)$ yes   **4.** no   **5.** yes

2. $\begin{cases} x - y = -4 \\ 2x + 10y = 4 \end{cases}$ $(-3, 1)$ yes

A system of equations and the graph of each equation of the system is given below. Find the solution of the system and verify that it is the solution. See Example 1.

3. $\begin{cases} 2x + 5y = 8 \\ 6x + y = 10 \end{cases}$ $(1.5, 1)$

4. $\begin{cases} x + y = 1 \\ x - 2y = 4 \end{cases}$ $(2, -1)$

5. $\begin{cases} x - 4y = -5 \\ -3x - 8y = 0 \end{cases}$ $(-2, 0.75)$

6. $\begin{cases} 2x - y = 8 \\ x - 3y = 11 \end{cases}$ $(2.6, -2.8)$

Solve each system by graphing. See Examples 2 and 3.

7. $\begin{cases} x + y = 1 \\ x - 2y = 4 \end{cases}$ **8.** $\begin{cases} 2x - y = 8 \\ x + 3y = 11 \end{cases}$

9. $\begin{cases} 2y - 4 = 0 \\ x + 2y = 5 \end{cases}$ **10.** $\begin{cases} 4x - y = 6 \\ x - y = 0 \end{cases}$

11. $\begin{cases} 3x - y = 4 \\ 6x - 2y = 4 \end{cases}$ **12.** $\begin{cases} -x + 3y = 6 \\ 3x - 9y = 9 \end{cases}$

13. Can a system consisting of two linear equations have exactly two solutions? Explain why or why not.   no; answers may vary

14. Suppose the graph of the equations in a system of two equations in two variables consists of a circle and a line. Discuss the possible number of solutions for this system.
   14. 0, 1, or 2, answers may vary

Solve each system of equations by the substitution method. See Examples 4 and 5.   **16.** $(-3, -1)$

15. $\begin{cases} x + y = 10 \\ y = 4x \end{cases}$ $(2, 8)$ **16.** $\begin{cases} 5x + 2y = -17 \\ x = 3y \end{cases}$

17. $\begin{cases} 4x - y = 9 \\ 2x + 3y = -27 \end{cases}$ $(0, -9)$ **18.** $\begin{cases} 3x - y = 6 \\ -4x + 2y = -8 \end{cases}$ $(2, 0)$

19. $\begin{cases} \dfrac{1}{2}x + \dfrac{3}{4}y = -\dfrac{1}{4} \\ \dfrac{3}{4}x - \dfrac{1}{4}y = 1 \end{cases}$ $(1, -1)$ **20.** $\begin{cases} \dfrac{2}{5}x + \dfrac{1}{5}y = -1 \\ x + \dfrac{2}{5}y = -\dfrac{8}{5} \end{cases}$ $(2, -9)$

21. $\begin{cases} \dfrac{x}{3} + y = \dfrac{4}{3} \\ -x + 2y = 11 \end{cases}$ $(-5, 3)$ **22.** $\begin{cases} \dfrac{x}{8} - \dfrac{y}{2} = 1 \\ \dfrac{x}{3} - y = 2 \end{cases}$ $(0, -2)$

Solve each system of equations by the elimination method. See Examples 6 through 9.

23. $\begin{cases} 2x - 4y = 0 \\ x + 2y = 5 \end{cases}$ $\left(\dfrac{5}{2}, \dfrac{5}{4}\right)$ **24.** $\begin{cases} 2x - 3y = 0 \\ 2x + 6y = 3 \end{cases}$ $\left(\dfrac{1}{2}, \dfrac{1}{3}\right)$

33. $\{(x, y)|3x + y = 1\}$ 34. $\{(x, y)|y = 2x - 5\}$

25. $\begin{cases} 5x + 2y = 1 \\ x - 3y = 7 \end{cases}$ $(1, -2)$ **26.** $\begin{cases} 6x - y = -5 \\ 4x - 2y = 6 \end{cases}$

27. $\begin{cases} 5x - 2y = 27 \\ -3x + 5y = 18 \end{cases}$ $(9, 9)$ **28.** $\begin{cases} 3x + 4y = 2 \\ 2x + 5y = -1 \end{cases}$ $(2, -1)$

29. $\begin{cases} 3x - 5y = 11 \\ 2x - 6y = 2 \end{cases}$ $(7, 2)$ **30.** $\begin{cases} 6x - 3y = -3 \\ 4x + 5y = -9 \end{cases}$

31. $\begin{cases} x - 2y = 4 \\ 2x - 4y = 4 \end{cases}$ $\varnothing$ **32.** $\begin{cases} -x + 3y = 6 \\ 3x - 9y = 9 \end{cases}$ $\varnothing$

33. $\begin{cases} 3x + y = 1 \\ \quad 2y = 2 - 6x \end{cases}$ **34.** $\begin{cases} \quad y = 2x - 5 \\ 8x - 4y = 20 \end{cases}$

36. $\left(-2, \dfrac{3}{4}\right)$ 39. $(-5, 3)$ 42. $\{(x, y)|x = 3y - 1\}$

MIXED PRACTICE   40. $\left(-1, -\dfrac{1}{2}\right)$ 41. $\{(x, y)|3x + 9y = 12\}$

Solve each system of equations.

35. $\begin{cases} 2x + 5y = 8 \\ 6x + y = 10 \end{cases}$ $\left(\dfrac{3}{2}, 1\right)$ **36.** $\begin{cases} x - 4y = -5 \\ -3x - 8y = 0 \end{cases}$

37. $\begin{cases} x + y = 1 \\ x - 2y = 4 \end{cases}$ $(2, -1)$ **38.** $\begin{cases} 2x - y = 8 \\ x + 3y = 11 \end{cases}$ $(5, 2)$

39. $\begin{cases} \dfrac{1}{3}x + y = \dfrac{4}{3} \\[2mm] -\dfrac{1}{4}x - \dfrac{1}{2}y = -\dfrac{1}{4} \end{cases}$ **40.** $\begin{cases} \dfrac{3}{4}x - \dfrac{1}{2}y = -\dfrac{1}{2} \\[2mm] x + y = -\dfrac{3}{2} \end{cases}$

41. $\begin{cases} 2x + 6y = 8 \\ 3x + 9y = 12 \end{cases}$ **42.** $\begin{cases} x = 3y - 1 \\ 2x - 6y = -2 \end{cases}$

43. $\begin{cases} 4x + 2y = 5 \\ 2x + y = -1 \end{cases}$ $\varnothing$ **44.** $\begin{cases} 3x + 6y = 15 \\ 2x + 4y = 3 \end{cases}$ $\varnothing$

45. $\begin{cases} 10y - 2x = 1 \\ \quad 5y = 4 - 6x \end{cases}$ **46.** $\begin{cases} 3x + 4y = 0 \\ \quad 7x = 3y \end{cases}$ $(0, 0)$

47. $\begin{cases} \dfrac{3}{4}x + \dfrac{5}{2}y = 11 \\[2mm] \dfrac{1}{16}x - \dfrac{3}{4}y = -1 \end{cases}$ 45. $\left(\dfrac{1}{2}, \dfrac{1}{5}\right)$ $(8, 2)$ **48.** $\begin{cases} \dfrac{2}{3}x + \dfrac{1}{4}y = -\dfrac{3}{2} \\[2mm] \dfrac{1}{2}x - \dfrac{1}{4}y = -2 \end{cases}$

50. $\left\{(x, y)|y = \dfrac{1}{7}x + 3\right\}$

49. $\begin{cases} x = 3y + 2 \\ 5x - 15y = 10 \end{cases}$ **50.** $\begin{cases} y = \dfrac{1}{7}x + 3 \\ x - 7y = -21 \end{cases}$

49. $\{(x, y)|x = 3y + 2\}$

51. $\begin{cases} 2x - y = -1 \\ \quad y = -2x \end{cases}$ $\left(-\dfrac{1}{4}, \dfrac{1}{2}\right)$ **52.** $\begin{cases} x = \dfrac{1}{5}y \\ x - y = -4 \end{cases}$ $(1, 5)$

53. $\begin{cases} 2x = 6 \\ y = 5 - x \end{cases}$ $(3, 2)$ **54.** $\begin{cases} x = 3y + 4 \\ -y = 5 \end{cases}$ 54. $(-11, -5)$

55. $\begin{cases} \dfrac{x + 5}{2} = \dfrac{6 - 4y}{3} \\[2mm] \dfrac{3x}{5} = \dfrac{21 - 7y}{10} \end{cases}$ $(7, -3)$ **56.** $\begin{cases} \dfrac{y}{5} = \dfrac{8 - x}{2} \\[2mm] x = \dfrac{2y - 8}{3} \end{cases}$ $(4, 10)$

57. $\begin{cases} 4x - 7y = 7 \\ 12x - 21y = 24 \end{cases}$ $\varnothing$ **58.** $\begin{cases} 2x - 5y = 12 \\ -4x + 10y = 20 \end{cases}$ $\varnothing$

59. $\begin{cases} \dfrac{2}{3}x - \dfrac{3}{4}y = -1 \\[2mm] -\dfrac{1}{6}x + \dfrac{3}{8}y = 1 \end{cases}$ $(3, 4)$ **60.** $\begin{cases} \dfrac{1}{2}x - \dfrac{1}{3}y = -3 \\[2mm] \dfrac{1}{8}x + \dfrac{1}{6}y = 0 \end{cases}$ $(-4, 3)$

61. $\begin{cases} 0.7x - 0.2y = -1.6 \\ 0.2x - y = -1.4 \end{cases}$ **62.** $\begin{cases} -0.7x + 0.6y = 1.3 \\ 0.5x - 0.3y = -0.8 \end{cases}$

63. $\begin{cases} 4x - 1.5y = 10.2 \\ 2x + 7.8y = -25.68 \end{cases}$ **64.** $\begin{cases} x - 3y = -5.3 \\ 6.3x + 6y = 3.96 \end{cases}$

61. $(-2, 1)$ 62. $(-1, 1)$ 64. $(-0.8, 1.5)$

Solve the system by graphing. Approximate the solutions to two decimal places.

65. $\begin{cases} y = -1.65x + 3.65 \\ y = 4.56x - 9.44 \end{cases}$ **66.** $\begin{cases} y = 7.61x + 3.48 \\ y = -1.26x - 6.43 \end{cases}$

67. $\begin{cases} 2.33x - 4.72y = 10.61 \\ 5.86x + 6.22y = -8.89 \end{cases}$ **68.** $\begin{cases} -7.89x - 5.68y = 3.26 \\ -3.65x + 4.98y = 11.77 \end{cases}$

65. $(2.11, 0.17)$ 66. $(-1.12, -5.02)$

REVIEW AND PREVIEW   67. $(0.57, -1.97)$ 68. $(-1.38, 1.35)$

Determine whether the given replacement values make each equation true or false. See Section 1.3.   48. $(-3, 2)$ 74. $3x - 7z = 3$

69. $3x - 4y + 2z = 5$; $x = 1$, $y = 2$, and $z = 5$ true
70. $x + 2y - z = 7$; $x = 2$, $y = -3$, and $z = 3$ false
71. $-x - 5y + 3z = 15$; $x = 0$, $y = -1$, and $z = 5$ false
72. $-4x + y - 8z = 4$; $x = 1$, $y = 0$, and $z = -1$ true

Add the equations. See Section 4.1.   75. $x + 10y = 2$

73. $3x + 2y - 5z = 10$ **74.** $x + 4y - 5z = 20$
  $-3x + 4y + z = 15$ $2x - 4y - 2z = -17$
75. $10x + 5y + 6z = 14$ **76.** $-9x - 8y - z = 31$
  $-9x + 5y - 6z = -12$ $9x + 4y - z = 12$

Concept Extensions   63. $(1.2, -3.6)$ 73. $6y - 4z = 25$
                        76. $-4y - 2z = 43$

The concept of supply and demand is used often in business. In general, as the unit price of a commodity increases, the demand for that commodity decreases. Also, as a commodity's unit price increases, the manufacturer normally increases the supply. The point where supply is equal to demand is called the equilibrium point. The following shows the graph of a demand equation and the graph of a supply equation for previously rented DVDs. The x-axis represents the number of DVDs in thousands, and the y-axis represents the cost of a DVD. Use this graph to answer Exercises 77 through 80. See Example 10.

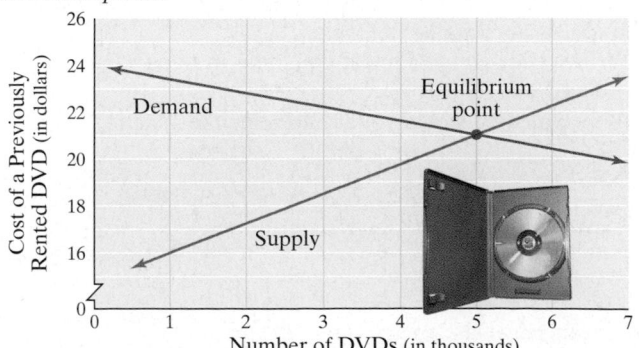

77. Find the number of DVDs and the price per DVD when supply equals demand.   5000 DVDs; $21

78. When x is between 3 and 4, is supply greater than demand or is demand greater than supply?   demand greater than supply

79. When x is greater than 7, is supply greater than demand or is demand greater than supply?   supply greater than demand

80. For what x-values are the y-values corresponding to the supply equation greater than the y-values corresponding to the demand equation?   for x values greater than 5

The revenue equation for a certain brand of toothpaste is $y = 2.5x$, where x is the number of tubes of toothpaste sold and y is the total income for selling x tubes. The cost equation is $y = 0.9x + 3000$, where x is the number of tubes of toothpaste manufactured and y is the cost of producing x tubes. The following set of axes shows the graph of the cost and revenue equations. Use this graph for Exercises 81 through 86. See Example 10.

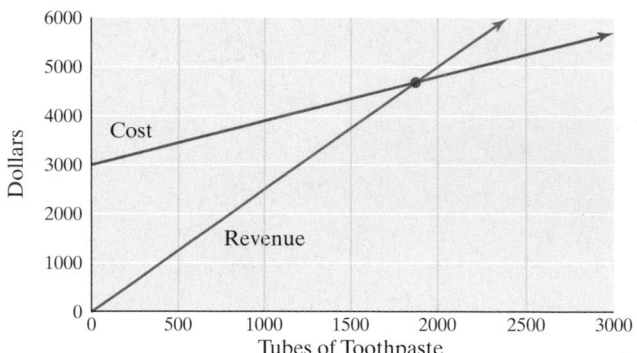

81. Find the coordinates of the point of intersection, or break-even point, by solving the system

$$\begin{cases} y = 2.5x \\ y = 0.9x + 3000 \end{cases} \quad (1875, 4687.5)$$

82. Explain the meaning of the ordered pair point of intersection.

83. If the company sells 2000 tubes of toothpaste, does the company make money or lose money?   makes money

84. If the company sells 1000 tubes of toothpaste, does the company make money or lose money?   loses money

85. For what x-values will the company make a profit? (*Hint:* For what x-values is the revenue graph "higher" than the cost graph?)   for x-values greater than 1875

86. For what x-values will the company lose money? (*Hint:* For what x-values is the revenue graph "lower" than the cost graph?)   for x-values less than 1875

87. Write a system of two linear equations in x and y that has the ordered pair solution $(2, 5)$.

88. Which method would you use to solve the system?

$$\begin{cases} 5x - 2y = 6 \\ 2x + 3y = 5 \end{cases}$$

Explain your choice.   answers may vary

82. 1875 tubes must be sold to break even.

87. answers may vary; one possibility: $\begin{cases} -2x + y = 1 \\ x - 2y = -8 \end{cases}$

89. The amount y of red meat consumed per person in the United States (in pounds) in the year x can be modeled by the linear equation $y = -x + 124.6$. The amount y of all poultry consumed per person in the United States (in pounds) in the year x can be modeled by the linear equation $y = 0.9x + 93$. In both models, $x = 0$ represents the year 1998. (*Source: Based on data and forecasts from the Economic Research Service, U.S. Department of Agriculture, 1998–2002*)

   a. What does the slope of each equation tell you about the patterns of red meat and poultry consumption in the United States?

   b. Solve this system of equations. (Round your final results to the nearest whole numbers.) $(17, 108)$

   c. Explain the meaning of your answer to part (b).

90. The amount of U.S. federal government income y (in billions of dollars) for the fiscal year x, from 1995 through 2000 ($x = 0$ represents 1995), can be modeled by the linear equation $y = 132.4x + 1328.3$. The amount of U.S. federal government expenditures y (in billions of dollars) for the same period can be modeled by the linear equation $y = 52.9x + 1504.6$. Did expenses ever equal income during this period? If so, in what year? (*Source:* Based on data from Financial Management Service, U.S. Department of the Treasury, 1995–2000)   yes; 1997

Solve each system. To do so you may want to let $a = \frac{1}{x}$ (if x is in the denominator) and let $b = \frac{1}{y}$ (if y is in the denominator).

91. $\begin{cases} \dfrac{1}{x} + y = 12 \\ \dfrac{3}{x} - y = 4 \end{cases}$ $\left(\dfrac{1}{4}, 8\right)$

92. $\begin{cases} x + \dfrac{2}{y} = 7 \\ 3x + \dfrac{3}{y} = 6 \end{cases}$ $\left(-3, \dfrac{1}{5}\right)$

93. $\begin{cases} \dfrac{1}{x} + \dfrac{1}{y} = 5 \\ \dfrac{1}{x} - \dfrac{1}{y} = 1 \end{cases}$ $\left(\dfrac{1}{3}, \dfrac{1}{2}\right)$

94. $\begin{cases} \dfrac{2}{x} + \dfrac{3}{y} = 5 \\ \dfrac{5}{x} - \dfrac{3}{y} = 2 \end{cases}$ $(1, 1)$

95. $\begin{cases} \dfrac{2}{x} + \dfrac{3}{y} = -1 \\ \dfrac{3}{x} - \dfrac{2}{y} = 18 \end{cases}$ $\left(\dfrac{1}{4}, -\dfrac{1}{3}\right)$

96. $\begin{cases} \dfrac{3}{x} - \dfrac{2}{y} = -18 \\ \dfrac{2}{x} + \dfrac{3}{y} = 1 \end{cases}$ $\left(-\dfrac{1}{4}, \dfrac{1}{3}\right)$

97. $\begin{cases} \dfrac{2}{x} - \dfrac{4}{y} = 5 \\ \dfrac{1}{x} - \dfrac{2}{y} = \dfrac{3}{2} \end{cases}$ $\varnothing$

98. $\begin{cases} \dfrac{5}{x} + \dfrac{7}{y} = 1 \\ -\dfrac{10}{x} - \dfrac{14}{y} = 0 \end{cases}$ $\varnothing$

89. a. consumption of red meat decreasing while consumption of poultry increasing

89. c. In the year 2015, red meat and poultry consumption will each be about 108 pounds per person.

4.2   SOLVING SYSTEMS OF LINEAR EQUATIONS IN THREE VARIABLES

Objective

 Solve a system of three linear equations in three variables.

In this section, the algebraic methods of solving systems of two linear equations in two variables are extended to systems of three linear equations in three variables. We call the equation $3x - y + z = -15$, for example, a **linear equation in three variables** since there are three variables and each variable is raised only to the power 1. A solution of this equation is an **ordered triple (x, y, z)** that makes the equation a true statement. For example, the ordered triple $(2, 0, -21)$ is a solution of $3x - y + z = -15$ since replacing x with 2, y with 0, and z with -21 yields the true statement $3(2) - 0 + (-21) = -15$. The graph of this equation is a plane in three-dimensional space, just as the graph of a linear equation in two variables is a line in two-dimensional space.

Although we will not discuss the techniques for graphing equations in three variables, visualizing the possible patterns of intersecting planes gives us insight into the possible patterns of solutions of a system of three three-variable linear equations. There are four possible patterns.

1. Three planes have a single point in common. This point represents the single solution of the system. This system is **consistent**.

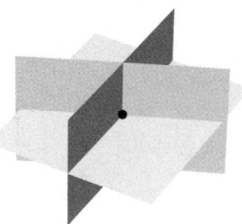

2. Three planes intersect at no point common to all three. This system has no solution. A few ways that this can occur are shown. This system is **inconsistent**.

  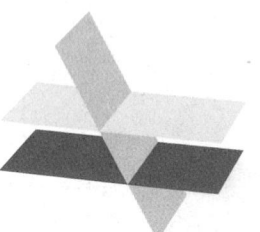

3. Three planes intersect at all the points of a single line. The system has infinitely many solutions. This system is **consistent**.

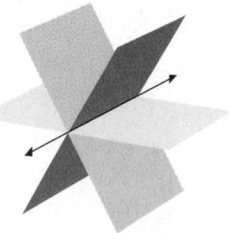

4. Three planes coincide at all points on the plane. The system is consistent, and the equations are **dependent**.

1   Just as with systems of two equations in two variables, we can use the elimination or substitution method to solve a system of three equations in three variables. To use the elimination method, we eliminate a variable and obtain a system of two equations in two variables. Then we use the methods we learned in the previous section to solve the system of two equations.

EXAMPLE 1

Solve the system.

$$\begin{cases} 3x - y + z = -15 & \text{Equation (1)} \\ x + 2y - z = 1 & \text{Equation (2)} \\ 2x + 3y - 2z = 0 & \text{Equation (3)} \end{cases}$$

Solution   Add equations (1) and (2) to eliminate z.

$$\begin{array}{r} 3x - y + z = -15 \\ x + 2y - z = 1 \\ \hline 4x + y = -14 \end{array} \quad \text{Equation (4)}$$

> **Helpful Hint**
>
> Don't forget to add two other equations besides equations (1) and (2) *and* to **eliminate the same variable**.

Next, add two *other* equations and *eliminate z again*. To do so, multiply both sides of equation (1) by 2 and add this resulting equation to equation (3). Then

$$\begin{cases} 2(3x - y + z) = 2(-15) \\ 2x + 3y - 2z = 0 \end{cases} \quad \begin{array}{c} \text{simplifies} \\ \text{to} \end{array} \quad \begin{cases} 6x - 2y + 2z = -30 \\ 2x + 3y - 2z = 0 \\ \hline 8x + y = -30 \end{cases} \quad \text{Equation (5)}$$

Now solve equations (4) and (5) for x and y. To solve by elimination, multiply both sides of equation (4) by -1 and add this resulting equation to equation (5). Then

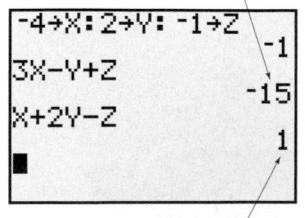

Satisfies
equation (1)

Satisfies
equation (2)

$$\begin{cases} -1(4x + y) = -1(-14) \\ 8x + y = -30 \end{cases} \quad \begin{array}{c} \text{simplifies} \\ \text{to} \end{array} \quad \begin{cases} -4x - y = 14 \\ 8x + y = -30 \\ \hline 4x = -16 \quad \text{Add the equations.} \\ x = -4 \quad \text{Solve for } x. \end{cases}$$

Replace x with -4 in equation (4) or (5).

$$\begin{array}{rl} 4x + y = -14 & \text{Equation (4)} \\ 4(-4) + y = -14 & \text{Let } x = -4. \\ y = 2 & \text{Solve for } y. \end{array}$$

Satisfies
equation (3)

Finally, replace x with -4 and y with 2 in equation (1), (2), or (3).

$$\begin{array}{rl} x + 2y - z = 1 & \text{Equation (2)} \\ -4 + 2(2) - z = 1 & \text{Let } x = -4 \text{ and } y = 2. \\ -4 + 4 - z = 1 & \\ -z = 1 & \\ z = -1 & \end{array}$$

A calculator check of Example 1.

The solution is $(-4, 2, -1)$. To check, let $x = -4$, $y = 2$, and $z = -1$ in all three original equations of the system. A paper and pencil check follows, and calculator check is shown in the margin on the previous page.

Equation (1)	*Equation (2)*	*Equation (3)*
$3x - y + z = -15$	$x + 2y - z = 1$	$2x + 3y - 2z = 0$
$3(-4) - 2 + (-1) \overset{?}{=} -15$	$-4 + 2(2) - (-1) \overset{?}{=} 1$	$2(-4) + 3(2) - 2(-1) \overset{?}{=} 0$
$-12 - 2 - 1 \overset{?}{=} -15$	$-4 + 4 + 1 \overset{?}{=} 1$	$-8 + 6 + 2 \overset{?}{=} 0$
$-15 = -15$	$1 = 1$	$0 = 0$
True	True	True

All three statements are true, so the solution is $(-4, 2, -1)$.

EXAMPLE 2

Solve the system.

$$\begin{cases} 2x - 4y + 8z = 2 & (1) \\ -x - 3y + z = 11 & (2) \\ x - 2y + 4z = 0 & (3) \end{cases}$$

Solution Add equations (2) and (3) to eliminate x, and the new equation is

$$-5y + 5z = 11 \qquad (4)$$

To eliminate x again, multiply both sides of equation (2) by 2, and add the resulting equation to equation (1). Then

$$\begin{cases} 2x - 4y + 8z = 2 \\ 2(-x - 3y + z) = 2(11) \end{cases} \begin{array}{c} \text{simplifies} \\ \text{to} \end{array} \begin{cases} 2x - 4y + 8z = 2 \\ \underline{-2x - 6y + 2z = 22} \\ -10y + 10z = 24 \quad (5) \end{cases}$$

Next, solve for y and z using equations (4) and (5). Multiply both sides of equation (4) by -2, and add the resulting equation to equation (5).

$$\begin{cases} -2(-5y + 5z) = -2(11) \\ -10y + 10z = 24 \end{cases} \begin{array}{c} \text{simplifies} \\ \text{to} \end{array} \begin{cases} 10y - 10z = -22 \\ \underline{-10y + 10z = 24} \\ 0 = 2 \quad \text{False} \end{cases}$$

Since the statement is false, this system is inconsistent and has no solution. The solution set is the empty set $\{\ \}$ or $\emptyset$.

The elimination method is summarized next.

TEACHING TIP

The most common student mistake made is not following Step 3. Emphasize this step and its importance.

▶ **Helpful Hint**

Make sure you read closely and follow Step 3.

Solving a System of Three Linear Equations by the Elimination Method

Step 1: Write each equation in standard form $Ax + By + Cz = D$.

Step 2: Choose a pair of equations and use the equations to eliminate a variable.

Step 3: Choose any other pair of equations and eliminate the **same variable** as in Step 2.

Step 4: Two equations in two variables should be obtained from Step 2 and Step 3. Use methods from Section 4.1 to solve this system for both variables.

Step 5: To solve for the third variable, substitute the values of the variables found in Step 4 into any of the original equations containing the third variable.

Step 6: Check the ordered triple solution in *all three* original equations.

✔ **CONCEPT CHECK**

In the system

$$\begin{cases} x + y + z = 6 & \text{Equation (1)} \\ 2x - y + z = 3 & \text{Equation (2)} \\ x + 2y + 3z = 14 & \text{Equation (3)} \end{cases}$$

equations (1) and (2) are used to eliminate y. Which action could be used to best finish solving? Why?

a. Use (1) and (2) to eliminate z **b.** Use (2) and (3) to eliminate y.
c. Use (1) and (3) to eliminate x

EXAMPLE 3

Solve the system.

$$\begin{cases} 2x + 4y \quad\quad = 1 & (1) \\ 4x \quad\quad - 4z = -1 & (2) \\ y - 4z = -3 & (3) \end{cases}$$

Solution

Notice that equation (2) has no term containing the variable y. Let us eliminate y using equations (1) and (3). Multiply both sides of equation (3) by -4, and add the resulting equation to equation (1). Then

$$\begin{cases} 2x + 4y \quad\quad\quad = 1 \\ -4(y - 4z) = -4(-3) \end{cases} \quad \begin{matrix} \text{simplifies} \\ \text{to} \end{matrix} \quad \begin{cases} 2x + 4y \quad\quad\quad = 1 \\ \underline{\quad\quad -4y + 16z = 12} \\ 2x \quad\quad + 16z = 13 \quad (4) \end{cases}$$

Next, solve for z using equations (4) and (2). Multiply both sides of equation (4) by -2 and add the resulting equation to equation (2).

$$\begin{cases} -2(2x + 16z) = -2(13) \\ 4x - 4z = -1 \end{cases} \quad \begin{matrix} \text{simplifies} \\ \text{to} \end{matrix} \quad \begin{cases} -4x - 32z = -26 \\ \underline{\quad 4x - 4z = -1} \\ \quad\quad -36z = -27 \\ \quad\quad\quad z = \dfrac{3}{4} \end{cases}$$

Replace z with $\dfrac{3}{4}$ in equation (3) and solve for y.

$$y - 4\left(\frac{3}{4}\right) = -3 \qquad \text{Let } z = \frac{3}{4} \text{ in equation (3).}$$
$$y - 3 = -3$$
$$y = 0$$

Replace y with 0 in equation (1) and solve for x.

$$2x + 4(0) = 1$$
$$2x = 1$$
$$x = \frac{1}{2}$$

The solution is $\left(\dfrac{1}{2}, 0, \dfrac{3}{4}\right)$. Check to see that this solution satisfies all three equations of the system A calculator check is shown in the margin.

CLASSROOM EXAMPLE

Solve the system:

$$\begin{cases} 3x + 2y \quad\quad = -1 \\ 6x \quad\quad - 2z = 4 \\ y - 3z = 2 \end{cases}$$

answer: $\left(\dfrac{1}{3}, -1, -1\right)$

A calculator check of Example 3.

EXAMPLE 4

Solve the system.

$$\begin{cases} x - 5y - 2z = 6 & (1) \\ -2x + 10y + 4z = -12 & (2) \\ \dfrac{1}{2}x - \dfrac{5}{2}y - z = 3 & (3) \end{cases}$$

Solution

Multiply both sides of equation (3) by 2 to eliminate fractions, and multiply both sides of equation (2) by $-\dfrac{1}{2}$ so that the coefficient of x is 1. The resulting system is then

$$\begin{cases} x - 5y - 2z = 6 & (1) \\ x - 5y - 2z = 6 & \text{Multiply (2) by } -\dfrac{1}{2}. \\ x - 5y - 2z = 6 & \text{Multiply (3) by 2.} \end{cases}$$

All three equations are identical, and therefore equations (1), (2), and (3) are all equivalent. There are infinitely many solutions of this system. The equations are dependent. The solution set can be written as $\{(x, y, z)\,|\,x - 5y - 2z = 6\}$.

As mentioned earlier, we can also use the substitution method to solve a system of linear equations in three variables.

EXAMPLE 5

Solve the system:

$$\begin{cases} x - 4y - 5z = 35 & (1) \\ x - 3y = 0 & (2) \\ -y + z = -55 & (3) \end{cases}$$

Solution

Notice in equations (2) and (3) that a variable is missing. Also notice that both equations contain the variable y. Let's use the substitution method by solving equation (2) for x and equation (3) for z and substituting the results in equation (1).

$$\begin{array}{ll} x - 3y = 0 & (2) \\ \quad x = 3y & \text{Solve equation (2) for } x. \\ -y + z = -55 & (3) \\ \quad z = y - 55 & \text{Solve equation (3) for } z. \end{array}$$

Now substitute $3y$ for x and $y - 55$ for z in equation (1).

$$\begin{array}{ll} x - 4y - 5z = 35 & (1) \\ 3y - 4y - 5(y - 55) = 35 & \text{Let } x = 3y \text{ and } z = y - 55. \\ 3y - 4y - 5y + 275 = 35 & \text{Use the distributive law and multiply.} \\ -6y + 275 = 35 & \text{Combine like terms.} \\ -6y = -240 & \text{Subtract 275 from both sides.} \\ y = 40 & \text{Solve.} \end{array}$$

To find x, recall that $x = 3y$ and substitute 40 for y. Then $x = 3y$ becomes $x = 3 \cdot 40 = 120$. To find z, recall that $z = y - 55$ and substitute 40 for y, also. Then $z = y - 55$ becomes $z = 40 - 55 = -15$. The solution is $(120, 40, -15)$.

STUDY SKILLS REMINDER

How Well Do You Know Your Textbook?

See if you can answer the questions below.

1. What does the 🔒 icon mean?
2. What does the ＼ icon mean?
3. What does the △ icon mean?
4. Where can you find a review for each chapter? What answers to this review can be found in the back of your text?
5. Each chapter contains an overview of the chapter along with examples. What is this feature called?
6. Does this text contain any solutions to exercises? If so, where?

11. $\{(x, y, z)|x - 2y + z = -5\}$ **12.** ∅ **14.** ∅ **16.** $(0,0,0)$ **17.** $(-3, -35, -7)$ **18.** $(-4, 5, -20)$
19. $(6, 22, -20)$ **20.** $(1, 2, -2)$ **22.** ∅ **23.** $(3, 2, 2)$ **24.** $(2, 1, 1)$

EXERCISE SET 4.2

STUDY GUIDE/SSM | CD/VIDEO | PH MATH TUTOR CENTER | MathXL®Tutorials ON CD | MathXL® | MyMathLab®

Solve. **1.** a, b, d **2.** a, c, d **10.** $(-3, -2, 5)$

1. Choose the equation(s) that has $(-1, 3, 1)$ as a solution.
   a. $x + y + z = 3$ **b.** $-x + y + z = 5$
   c. $-x + y + 2z = 0$ **d.** $x + 2y - 3z = 2$

2. Choose the equation(s) that has $(2, 1, -4)$ as a solution.
   a. $x + y + z = -1$ **b.** $x - y - z = -3$
   c. $2x - y + z = -1$ **d.** $-x - 3y - z = -1$

3. Use the result of Exercise 1 to determine whether $(-1, 3, 1)$ is a solution of the system below. Explain your answer.   yes; answers may vary **8.** $(1, 2, 5)$ **15.** $(0, 0, 0)$
$$\begin{cases} x + y + z = 3 \\ -x + y + z = 5 \\ x + 2y - 3z = 2 \end{cases}$$

4. Use the result of Exercise 2 to determine whether $(2, 1, -4)$ is a solution of the system below. Explain your answer.   no; answers may vary **9.** $(-2, 3, -1)$
$$\begin{cases} x + y + z = -1 \\ x - y - z = -3 \\ 2x - y + z = -1 \end{cases}$$

MIXED PRACTICE **5.** $(-1, 5, 2)$ **6.** $(3, -3, 1)$

Solve each system. See Examples 1 through 5. **7.** $(-2, 5, 1)$

5. $\begin{cases} x - y + z = -4 \\ 3x + 2y - z = 5 \\ -2x + 3y - z = 15 \end{cases}$ **6.** $\begin{cases} x + y - z = -1 \\ -4x - y + 2z = -7 \\ 2x - 2y - 5z = 7 \end{cases}$

7. $\begin{cases} x + y = 3 \\ 2y = 10 \\ 3x + 2y - 3z = 1 \end{cases}$ **8.** $\begin{cases} 5x = 5 \\ 2x + y = 4 \\ 3x + y - 4z = -15 \end{cases}$

9. $\begin{cases} 2x + 2y + z = 1 \\ -x + y + 2z = 3 \\ x + 2y + 4z = 0 \end{cases}$ **10.** $\begin{cases} 2x - 3y + z = 5 \\ x + y + z = 0 \\ 4x + 2y + 4z = 4 \end{cases}$

11. $\begin{cases} x - 2y + z = -5 \\ -3x + 6y - 3z = 15 \\ 2x - 4y + 2z = -10 \end{cases}$ **12.** $\begin{cases} 3x + y - 2z = 2 \\ -6x - 2y + 4z = -2 \\ 9x + 3y - 6z = 6 \end{cases}$

13. $\begin{cases} 4x - y + 2z = 5 \\ 2y + z = 4 \\ 4x + y + 3z = 10 \end{cases}$ ∅ **14.** $\begin{cases} 5y - 7z = 14 \\ 2x + y + 4z = 10 \\ 2x + 6y - 3z = 30 \end{cases}$

15. $\begin{cases} x + 5z = 0 \\ 5x + y = 0 \\ y - 3z = 0 \end{cases}$ **16.** $\begin{cases} x - 5y = 0 \\ x - z = 0 \\ -x + 5z = 0 \end{cases}$

17. $\begin{cases} 6x - 5z = 17 \\ 5x - y + 3z = -1 \\ 2x + y = -41 \end{cases}$ **18.** $\begin{cases} x + 2y = 6 \\ 7x + 3y + z = -33 \\ x - z = 16 \end{cases}$

19. $\begin{cases} x + y + z = 8 \\ 2x - y - z = 10 \\ x - 2y - 3z = 22 \end{cases}$ **20.** $\begin{cases} 5x + y + 3z = 1 \\ x - y + 3z = -7 \\ -x + y = 1 \end{cases}$

21. $\begin{cases} x + 2y - z = 5 \\ 6x + y + z = 7 \\ 2x + 4y - 2z = 5 \end{cases}$ ∅ **22.** $\begin{cases} 4x - y + 3z = 10 \\ x + y - z = 5 \\ 8x - 2y + 6z = 10 \end{cases}$

🔒 **23.** $\begin{cases} 2x - 3y + z = 2 \\ x - 5y + 5z = 3 \\ 3x + y - 3z = 5 \end{cases}$ **24.** $\begin{cases} 4x + y - z = 8 \\ x - y + 2z = 3 \\ 3x - y + z = 6 \end{cases}$

25. $\begin{cases} -2x - 4y + 6z = -8 \\ x + 2y - 3z = 4 \\ 4x + 8y - 12z = 16 \end{cases}$ $\{(x, y, z)|x + 2y - 3z = 4\}$

26. $\begin{cases} -6x + 12y + 3z = -6 \\ 2x - 4y \quad\;\; z = 2 \\ -x + 2y + \dfrac{z}{2} = -1 \end{cases}$

27. $\begin{cases} 2x + 2y - 3z = 1 \\ \quad\;\; y + 2z = -14 \\ 3x - 2y \quad\;\;\; = -1 \end{cases}$

28. $\begin{cases} 7x + 4y \quad\;\;\;\;\; = 10 \\ x - 4y + 2z = 6 \\ \quad\;\; y - 2z = -1 \end{cases}$

29. $\begin{cases} x + 2y - z = 5 \\ -3x - 2y - 3z = 11 \\ 4x + 4y + 5z = -18 \end{cases}$

30. $\begin{cases} 3x - 3y + z = -1 \\ 3x - y - z = 3 \\ -6x + 2y + 2z = -6 \end{cases}$

31. $\begin{cases} \dfrac{3}{4}x - \dfrac{1}{3}y + \dfrac{1}{2}z = 9 \\ \dfrac{1}{6}x + \dfrac{1}{3}y - \dfrac{1}{2}z = 2 \\ \dfrac{1}{2}x - \;\; y + \dfrac{1}{2}z = 2 \end{cases}$

32. $\begin{cases} \dfrac{1}{3}x - \dfrac{1}{4}y + \;\; z = -9 \\ \dfrac{1}{2}x - \dfrac{1}{3}y - \dfrac{1}{4}z = -6 \\ x - \dfrac{1}{2}y - \;\; z = -8 \end{cases}$

REVIEW AND PREVIEW

Solve. See Section 1.6.

33. The sum of two numbers is 45 and one number is twice the other. Find the numbers.   15 and 30

34. The difference between two numbers is 5. Twice the smaller number added to five times the larger number is 53. Find the numbers.   9 and 4

Solve. See Section 1.5.

35. $2(x - 1) - 3x = x - 12$ 5

36. $7(2x - 1) + 4 = 11(3x - 2)$ 1

37. $-y - 5(y + 5) = 3y - 10$

38. $z - 3(z + 7) = 6(2z + 1)$

39. Write a single linear equation in three variables that has $(-1, 2, -4)$ as a solution. (There are many possibilities.) Explain the process you used to write an equation.

40. Write a system of three linear equations in three variables that has $(2, 1, 5)$ as a solution. (There are many possibilities.) Explain the process you used to write an equation.

41. Write a system of linear equations in three variables that has the solution $(-1, 2, -4)$. Explain the process you used to write your system.   39.–42. answers may vary

42. When solving a system of three equation in three unknowns, explain how to determine that a system has no solution.

43. The fraction $\dfrac{1}{24}$ can be written as the following sum:

$$\frac{1}{24} = \frac{x}{8} + \frac{y}{4} + \frac{z}{3}$$

where the numbers x, y, and z are solutions of

$$\begin{cases} x + y + z = 1 \\ 2x - y + z = 0 \\ -x + 2y + 2z = -1 \end{cases}$$

Solve the system and see that the sum of the fractions is $\dfrac{1}{24}$. $(1, 1, -1)$

44. The fraction $\dfrac{1}{18}$ can be written as the following sum:

$$\frac{1}{18} = \frac{x}{2} + \frac{y}{3} + \frac{z}{9}$$

where the numbers x, y, and z are solutions of

$$\begin{cases} x + 3y + z = -3 \\ -x + y + 2z = -14 \\ 3x + 2y - z = 12 \end{cases}$$

Solve the system and see that the sum of the fractions is $\dfrac{1}{18}$. $(1, 1, -7)$

Solving systems involving more than three variables can be accomplished with methods similar to those encountered in this section. Apply what you already know to solve each system of equations in four variables.

45. $\begin{cases} x + y \quad\;\;\; - w = 0 \\ \quad\; y + 2z + w = 3 \\ x \quad\;\;\; - z \quad\;\;\; = 1 \\ 2x - y \quad\;\;\; - w = -1 \end{cases}$ $(1, 1, 0, 2)$

46. $\begin{cases} 5x + 4y \quad\;\;\;\;\;\; = 29 \\ \quad\; y + z - w = -2 \\ 5x \quad\;\;\; + z \quad\;\;\; = 23 \\ \quad\; y - z + w = 4 \end{cases}$ $(5, 1, -2, 1)$

47. $\begin{cases} x + y + z + w = 5 \\ 2x + y + z + w = 6 \\ x + y + z \quad\;\;\; = 2 \\ x + y \quad\;\;\;\;\;\;\; = 0 \end{cases}$ $(1, -1, 2, 3)$

48. $\begin{cases} 2x \quad\;\; - z \quad\;\;\;\;\; = -1 \\ \quad\; y + z + w = 9 \\ \quad\; y \quad\;\; -2w = -6 \\ x + y \quad\;\;\;\;\;\;\; = 3 \end{cases}$ $(1, 2, 3, 4)$

49. Write a system of three linear equations in three variables that are dependent equations.   answers may vary

50. What is the solution to the system in Exercise 49?   infinite number of solutions

---

26. $\{(x, y, z)|\; -6x + 12y + 3z = -6\}$ **27.** $(-3, -4, -5)$ **28.** $(2, -1, 0)$ **29.** $(0, \frac{1}{2}, -4)$

30. $(\frac{1}{3}, 0, -2)$ **31.** $(12, 6, 4)$ **32.** $(-6, 12, -4)$ **37.** $-\dfrac{5}{3}$ **38.** $-\dfrac{27}{14}$

4.3 SYSTEMS OF LINEAR EQUATIONS AND PROBLEM SOLVING

Objectives

1. Solve problems that can be modeled by a system of two linear equations.
2. Solve problems with cost and revenue functions.
3. Solve problems that can be modeled by a system of three linear equations.

1 Thus far, we have solved problems by writing one-variable equations and solving for the variable. Some of these problems can be solved, perhaps more easily, by writing a system of equations, as illustrated in this section.

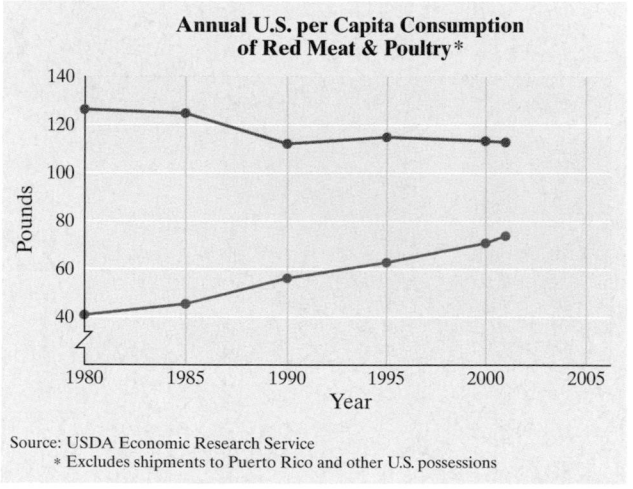

Annual U.S. per Capita Consumption of Red Meat & Poultry*

Source: USDA Economic Research Service
* Excludes shipments to Puerto Rico and other U.S. possessions

EXAMPLE 1

PREDICTING EQUAL CONSUMPTION OF RED MEAT AND POULTRY

America's consumption of red meat has decreased most years since 1980 while consumption of poultry has increased. The function $y = -0.71x + 125.6$ approximates the annual pounds of red meat consumed per capita, where x is the number of years since 1980. The function $y = 1.56x + 39.7$ approximates the annual pounds of poultry consumed per capita, where x is also the number of years since 1980. If this trend continues, determine the year when the annual consumption of red meat and poultry are equal.

Solution

1. UNDERSTAND. Read and reread the problem and guess a year. Let's guess the year 2010. This year is 30 years since 1980, so $x = 30$. Now let $x = 30$ in each given function.

Red meat: $y = -0.71x + 125.6 = -0.71(30) + 125.6 = 104.3$ pounds

Poultry: $y = 1.56x + 39.7 \quad = 1.56(30) + 39.7 \quad = 86.5$ pounds

Since the projected pounds in 2010 for red meat and poultry are not the same, we guessed incorrectly, but we do have a better understanding of the problem. We also know that the year will be later than 2010 since projected consumption of red meat is still greater than poultry that year.

2. TRANSLATE. We are already given the system of equations.

3. SOLVE. We want to know the year x in which pounds y are the same, so we solve the system:

$$\begin{cases} y = -0.71x + 125.6 \\ y = 1.56x + 39.7 \end{cases}$$

Since both equations are solved for y, one way to solve is to use the substitution method.

$$y = -0.71x + 125.6 \qquad \text{First equation}$$

$$\underbrace{1.56x + 39.7} = -0.71x + 125.6 \qquad \text{Let } y = 1.56x + 39.7.$$

$$2.27x = 85.9$$

$$x = \frac{85.9}{2.27} \approx 37.84$$

4. INTERPRET. Since we are only asked to find the year, we need only solve for x.
Check: To check, see whether $x \approx 37.84$ gives approximately the same number of pounds of red meat and poultry.

Red meat: $\quad y = -0.71x + 125.6 = -0.71(37.84) + 125.6 \approx 98.7336$ pounds

Poultry: $\qquad y = 1.56x + 39.7 \quad = 1.56(37.84) + 39.7 \quad \approx 98.7304$ pounds

Since we rounded the number of years, the number of pounds do differ slightly. They differ only by 0.0032, so we can assume that we solved correctly.

State: The consumption of red meat and poultry will be the same about 37.84 years after 1980, or 2017.84. Thus, in the year 2017, we predict the consumption will be the same.*

*Note: A similar exercise is found in Section 4.1, Exercise 85. In the example above, the data years used to generate the equations are 1980–2002. In Section 4.1, the data years used are 1998–2002. Notice the differing equations and answers.

EXAMPLE 2

FINDING UNKNOWN NUMBERS

A first number is 4 less than a second number. Four times the first number is 6 more than twice the second. Find the numbers.

Solution

1. UNDERSTAND. Read and reread the problem and guess a solution. If a first number is 10 and this is 4 less than a second number, the second number is 14. Four times the first number is 4(10), or 40. This is not equal to 6 more than twice the second number, which is 2(14) + 6 or 34. Although we guessed incorrectly, we now have a better understanding of the problem.

Since we are looking for two numbers, we will let

$$x = \text{first number}$$
$$y = \text{second number}$$

2. TRANSLATE. Since we have assigned two variables to this problem, we will translate the given facts into two equations. For the first statement we have

In words: | the first number | is | 4 less than the second number |
|---|---|---|
| ↓ | ↓ | ↓ |
Translate: | x | $=$ | $y - 4$ |

TEACHING TIP
Classroom Activity
After reviewing Example 2, have students work in pairs to create a number problem. Instruct them to begin by selecting two numbers. Then have them describe one number in terms of the other in two ways using the example as a model. Finally, have them give their word problem to another pair to solve.

Next we translate the second statement into an equation.

In words:

Translate: $4x$ $=$ $2y + 6$

3. SOLVE. Here we solve the system

$$\begin{cases} x = y - 4 \\ 4x = 2y + 6 \end{cases}$$

Since the first equation expresses x in terms of y, we will use substitution. We substitute $y - 4$ for x in the second equation and solve for y.

$$4x = 2y + 6 \qquad \textit{Second equation}$$
$$4(y - 4) = 2y + 6$$
$$4y - 16 = 2y + 6 \qquad \textsf{Let } x = y - 4.$$
$$2y = 22$$
$$y = 11$$

Now we replace y with 11 in the equation $x = y - 4$ and solve for x. Then $x = y - 4$ becomes $x = 11 - 4 = 7$. The ordered pair solution of the system is $(7, 11)$.

4. INTERPRET. Since the solution of the system is $(7, 11)$, then the first number we are looking for is 7 and the second number is 11.
Check: Notice that 7 *is* 4 less than 11, and 4 times 7 *is* 6 more than twice 11. The proposed numbers, 7 and 11, are correct.
State: The numbers are 7 and 11.

EXAMPLE 3

FINDING THE RATE OF SPEED

Two cars leave Indianapolis, one traveling east and the other west. After 3 hours they are 297 miles apart. If one car is traveling 5 mph faster than the other, what is the speed of each?

Solution

CLASSROOM EXAMPLE

Two trains leave Tulsa, one traveling north and the other south. After 4 hours, they are 376 miles apart. If one train is traveling 10 mph faster than the other, what is the speed of each?
answer: 42 mph; 52 mph

1. UNDERSTAND. Read and reread the problem. Let's guess a solution and use the formula $d = rt$ to check. Suppose that one car is traveling at a rate of 55 miles per hour. This means that the other car is traveling at a rate of 50 miles per hour since we are told that one car is traveling 5 mph faster than the other. To find the distance apart after 3 hours, we will first find the distance traveled by each car. One car's distance is rate · time $= 55(3) = 165$ miles. The other car's distance is rate · time $= 50(3) = 150$ miles. Since one car is traveling east and the other west, their distance apart is the sum of their distances, or 165 miles $+$ 150 miles $= 315$ miles. Although this distance apart is not the required distance of 297 miles, we now have a better understanding of the problem.

$50(3) = 150$ mi $55(3) = 165$ mi

$150 + 165 = 315$ mi

Let's model the problem with a system of equations. We will let

$$x = \text{speed of one car}$$
$$y = \text{speed of the other car}$$

We summarize the information on the following chart. Both cars have traveled 3 hours. Since distance = rate · time, their distances are $3x$ and $3y$ miles, respectively.

	Rate	·	*Time*	=	*Distance*
One Car	x		3		$3x$
Other Car	y		3		$3y$

2. TRANSLATE. We can now translate the stated conditions into two equations.

In words:

one car's distance	added to	the other car's distance	is	297
↓	↓	↓	↓	↓

Translate: $3x$ $+$ $3y$ $=$ 297

In words:

one car's speed	is	5 mph faster than the other
↓	↓	↓

Translate: x $=$ $y + 5$

3. SOLVE. Here we solve the system

$$\begin{cases} 3x + 3y = 297 \\ x \qquad = y + 5 \end{cases}$$

Again, the substitution method is appropriate. We replace x with $y + 5$ in the first equation and solve for y.

$$3x + 3y = 297 \qquad \text{First equation}$$
$$3\,(y + 5) + 3y = 297 \qquad \text{Let } x = y = 5.$$
$$3y + 15 + 3y = 297$$
$$6y = 282$$
$$y = 47$$

To find x, we replace y with 47 in the equation $x = y + 5$. Then $x = 47 + 5 = 52$. The ordered pair solution of the system is $(52, 47)$.

4. INTERPRET. The solution $(52, 47)$ means that the cars are traveling at 52 mph and 47 mph, respectively.

Check: Notice that one car is traveling 5 mph faster than the other. Also, if one car travels 52 mph for 3 hours, the distance is $3(52) = 156$ miles. The other car traveling for 3 hours at 47 mph travels a distance of $3(47) = 141$ miles. The sum of the distances $156 + 141$ is 297 miles, the required distance.

State: The cars are traveling at 52 mph and 47 mph.

Helpful Hint

Don't forget to attach units, if appropriate.

> ### EXAMPLE 4

MIXING SOLUTIONS

Lynn Pike, a pharmacist, needs 70 liters of a 50% alcohol solution. She has available a 30% alcohol solution and an 80% alcohol solution. How many liters of each solution should she mix to obtain 70 liters of a 50% alcohol solution?

Solution

1. UNDERSTAND. Read and reread the problem. Next, guess the solution. Suppose that we need 20 liters of the 30% solution. Then we need $70 - 20 = 50$ liters of the 80% solution. To see if this gives us 70 liters of a 50% alcohol solution, let's find the amount of pure alcohol in each solution.

$$\underset{\text{of liters}}{\begin{array}{c}\text{number}\end{array}} \times \underset{\text{strength}}{\begin{array}{c}\text{alcohol}\end{array}} = \underset{\text{pure alcohol}}{\begin{array}{c}\text{amount of}\end{array}}$$

20 liters	×	0.30	=	6 liters
50 liters	×	0.80	=	40 liters
70 liters	×	0.50	=	35 liters

Since 6 liters + 40 liters = 46 liters and not 35 liters, our guess is incorrect, but we have gained some insight as to how to model and check this problem. We will let

$$x = \text{amount of 30\% solution, in liters}$$
$$y = \text{amount of 80\% solution, in liters}$$

and use a table to organize the given data.

	Number of Liters	Alcohol Strength	Amount of Pure Alcohol
30% Solution	x	30%	$0.30x$
80% Solution	y	80%	$0.80y$
50% Solution Needed	70	50%	$(0.50)(70)$

A calculator check for Example 4. (We already see by inspection that $42 + 28 = 70$, as needed.)

2. TRANSLATE. We translate the stated conditions into two equations.

$$\text{In words:} \quad \underset{\text{30\% solution}}{\begin{array}{c}\text{amount of}\end{array}} + \underset{\text{80\% solution}}{\begin{array}{c}\text{amount of}\end{array}} = 70$$

$$\text{Translate:} \quad x \quad + \quad y \quad = \quad 70$$

In words:	amount of pure alcohol in 30% solution	+	amount of pure alcohol in 80% solution	=	amount of pure alcohol in 50% solution
	↓		↓		↓
Translate:	$0.30x$	+	$0.80y$	=	$(0.50)(70)$

3. SOLVE. Here we solve the system

$$\begin{cases} x + y = 70 \\ 0.30x + 0.80y = (0.50)(70) \end{cases}$$

To solve this system, we use the elimination method. We multiply both sides of the first equation by -3 and both sides of the second equation by 10. Then

$$\begin{cases} -3(x + y) = -3(70) \\ 10(0.30x + 0.80y) = 10(0.50)(70) \end{cases} \quad \begin{matrix} \text{simplifies} \\ \text{to} \end{matrix} \quad \begin{cases} -3x - 3y = -210 \\ \underline{3x + 8y = 350} \\ \quad\quad 5y = 140 \\ \quad\quad\, y = 28 \end{cases}$$

Now we replace y with 28 in the equation $x + y = 70$ and find that $x + 28 = 70$, or $x = 42$.

The ordered pair solution of the system is $(42, 28)$.

4. INTERPRET.

Check: Check the solution in the same way that we checked our guess.

State: The pharmacist needs to mix 42 liters of 30% solution and 28 liters of 80% solution to obtain 70 liters of 50% solution.

✔ **CONCEPT CHECK**

Suppose you mix an amount of 25% acid solution with an amount of 60% acid solution. You then calculate the acid strength of the resulting acid mixture. For which of the following results should you suspect an error in your calculation? Why?

a. 14%   **b.** 32%   **c.** 55%

2   Recall that businesses are often computing cost and revenue functions or equations to predict sales, to determine whether prices need to be adjusted, and to see whether the company is making or losing money. Recall also that the value at which revenue equals cost is called the break-even point. When revenue is less than cost, the company is losing money; when revenue is greater than cost, the company is making money.

EXAMPLE 5

FINDING A BREAK-EVEN POINT

A manufacturing company recently purchased $3000 worth of new equipment to offer new personalized stationery to its customers. The cost of producing a package of personalized stationery is $3.00, and it is sold for $5.50. Find the number of packages that must be sold for the company to break even.

Solution

1. UNDERSTAND. Read and reread the problem. Notice that the cost to the company will include a one-time cost of $3000 for the equipment and then $3.00 per package produced. The revenue will be $5.50 per package sold.
   To model this problem, we will let

   x = number of packages of personalized stationery
   $C(x)$ = total cost for producing x packages of stationery
   $R(x)$ = total revenue for selling x packages of stationery

2. TRANSLATE. The revenue equation is

   In words:

   The cost equation is

   In words:

   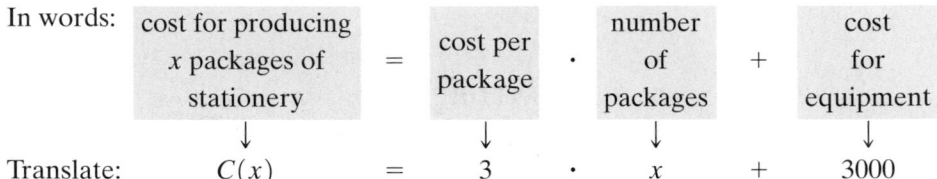

   Since the break-even point is when $R(x) = C(x)$, we solve the equation

   $$5.5x = 3x + 3000$$

3. SOLVE.

   $$5.5x = 3x + 3000$$
   $$2.5x = 3000 \qquad \text{Subtract } 3x \text{ from both sides.}$$
   $$x = 1200 \qquad \text{Divide both sides by } 2.5.$$

4. INTERPRET.
   Check: To see whether the break-even point occurs when 1200 packages are produced and sold, see if revenue equals cost when $x = 1200$. When $x = 1200$, $R(x) = 5.5x = 5.5(1200) = 6600$ and $C(x) = 3x + 3000 = 3(1200) + 3000 = 6600$. Since $R(1200) = C(1200) = 6600$, the break-even point is 1200.
   State: The company must sell 1200 packages of stationery to break even. The graph of this system is shown.

$y_2 = 3x + 3000$

$y_1 = 5.5x$

A calculator graph of Example 5.

3   To introduce problem solving by writing a system of three linear equations in three variables, we solve a problem about triangles.

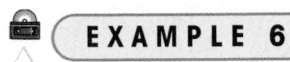 **EXAMPLE 6**

FINDING ANGLE MEASURES

The measure of the largest angle of a triangle is 80° more than the measure of the smallest angle, and the measure of the remaining angle is 10° more than the measure of the smallest angle. Find the measure of each angle.

Solution

TEACHING TIP
Classroom Activity
After reviewing Example 6 have students work in pairs to write their own triangle word problem. Have them begin by deciding the measure of each angle in a triangle. Then have them write a word problem based on a mathematical relationship between one pair of angles and between another pair of angles. Finally, have them give their word problem to another pair to solve.

1. UNDERSTAND. Read and reread the problem. Recall that the sum of the measures of the angles of a triangle is 180°. Then guess a solution. If the smallest angle measures 20°, the measure of the largest angle is 80° more, or $20° + 80° = 100°$. The measure of the remaining angle is 10° more than the measure of the smallest angle, or $20° + 10° = 30°$. The sum of these three angles is $20° + 100° + 30° = 150°$, not the required 180°. We now know that the measure of the smallest angle is greater than 20°.

To model this problem we will let

$$x = \text{degree measure of the smallest angle}$$
$$y = \text{degree measure of the largest angle}$$
$$z = \text{degree measure of the remaining angle}$$

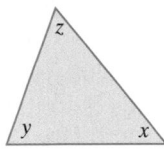

CLASSROOM EXAMPLE
The measure of the largest angle of a triangle is 90° more than the measure of the smallest angle, and the measure of the remaining angle is 30° more than the measure of the smallest angle. Find the measure of each angle.

answer: 20°, 50°, 110°

2. TRANSLATE. We translate the given information into three equations.

In words: the sum of the measures   =   180

Translate: $x + y + z$ =   180

In words: the largest angle   is   80 more than the smallest angle

Translate y = $x + 80$

In words: the remaining angle   is   10 more than the smallest angle

Translate: z = $x + 10$

3. SOLVE. We solve the system

$$\begin{cases} x + y + z = 180 \\ y = x + 80 \\ z = x + 10 \end{cases}$$

Since y and z are both expressed in terms of x, we will solve using the substitution method. We substitute $y = x + 80$ and $z = x + 10$ in the first equation. Then

$$x + y + z = 180 \qquad \text{First equation}$$

$$x + \overbrace{(x + 80)} + \overbrace{(x + 10)} = 180 \qquad \text{Let } y = x + 80 \text{ and } z = x + 10.$$

$$3x + 90 = 180$$
$$3x = 90$$
$$x = 30$$

Then $y = x + 80 = 30 + 80 = 110$, and $z = x + 10 = 30 + 10 = 40$. The ordered triple solution is $(30, 110, 40)$.

4. INTERPRET.

Check: Notice that $30° + 40° + 110° = 180°$. Also, the measure of the largest angle, $110°$, is $80°$ more than the measure of the smallest angle, $30°$. The measure of the remaining angle, $40°$, is $10°$ more than the measure of the smallest angle, $30°$.

State: The angles measure $30°, 110°,$ and $40°$.

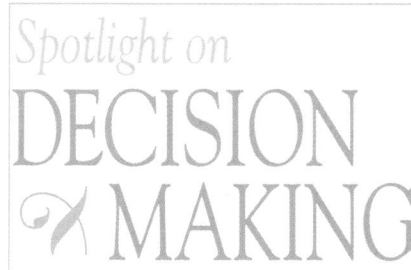

Spotlight on
DECISION
ℐ MAKING

Suppose you are choosing a long-distance telephone plan. You have narrowed your choices to the One Rate® 7¢ Plan offered by AT&T and the Qwest Communicator plan offered by Qwest Communications International, Inc. Under the AT&T One Rate® 7¢ Plan, calls made any time cost $0.07 per minute and users are charged a $5.95 monthly fee. Under the Qwest Communicator plan, calls made any time cost $0.05 per minute and users are charged a $9.95 monthly fee. Which long-distance plan would you choose? Why? What other factors would you want to consider? Would you change your choice if you knew that your long-distance calls averaged 4 hours per month? Explain your reasoning.

3. a. Enterprise class: 1101 ft; Nimitz class: 1092 ft

EXERCISE SET 4.3

STUDY GUIDE/SSM | CD/ VIDEO | PH MATH TUTOR CENTER | MathXL®Tutorials ON CD | MathXL® | MyMathLab®

Solve. See Examples 1 through 4.

1. One number is two more than a second number. Twice the first is 4 less than 3 times the second. Find the numbers.   10 and 8

2. Three times one number minus a second is 8, and the sum of the numbers is 12. Find the numbers.   5 and 7

3. The United States has the world's only "large deck" aircraft carriers which can hold up to 72 aircraft. The Enterprise class carrier is longest in length while the Nimitz class carrier is the second longest. The total length of these two carriers is 2193 feet while the difference of their lengths is only 9 feet. (*Source:* U.S.A. Today, May, 2001)
   a. Find the length of each class carrier.

b. If a football field has a length of 100 yards, determine the length of the Enterprise class carrier in terms of number of football fields.   3.67 football fields

4. The rate of growth of participation (age 6 and older) in sports featured in the X-Games is surpassing that for older sports such as football and baseball. The most popular X-Game sport is in-line roller skating, followed by skateboarding. In 2000, the total number of participants in both sports was 40.6 million. If the number of participants in roller skating was 5.8 million more than twice the number of participants in skateboarding, find the number of participants in each sport. (*Source:* January 2001 Sporting Goods Manufacturers Association Sports Participation Topline Report)   in-line roller skating: 29 million; skateboarding: 11.6 million

13. −10 and −8
14. 12 and 18

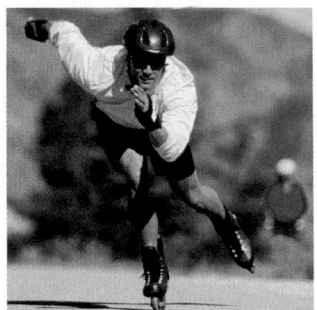

6. still water; 8.7 km/hr; current: 1.9 km/hr
9. United Kingdom: 27,720 students; Spain: 12,292 students
5. A Delta 727 traveled 560 mph with the wind and 480 mph against the wind. Find the speed of the plane in still air and the speed of the wind.   plane: 520 mph; wind: 40 mph

6. Terry Watkins can row about 10.6 kilometers in 1 hour downstream and 6.8 kilometers upstream in 1 hour. Find how fast he can row in still water, and find the speed of the current.

7. Find how many quarts of 4% butterfat milk and 1% butterfat milk should be mixed to yield 60 quarts of 2% butterfat milk.   20 qt of 4%; 40 qt of 1%

8. A pharmacist needs 500 milliliters of a 20% phenobarbital solution but has only 5% and 25% phenobarbital solutions available. Find how many milliliters of each he should mix to get the desired solution.   125 ml of 5%; 375 ml of 25%

9. In a recent year, the United Kingdom was the most popular host country for U.S. students traveling abroad to study. Spain was the second most popular destination. A total of 40,012 students visited one of the two countries. If 15,428 more U.S. students studied in the United Kingdom than in Spain, find how many students studied abroad in each country. (*Source:* Institute of International Education, Open Doors 2000)

10. In 2000, Washington D.C. had the most conventions while Orlando was the second most popular convention location. A total of 353 conventions were held in one of the two destinations. If Washington D.C. had 27 more conventions than Orlando, find the number of conventions held in each location. (*Source: Successful Meetings* magazine)

11. Karen Karlin bought some large frames for $15 each and some small frames for $8 each at a closeout sale. If she bought 22 frames for $239, find how many of each type she bought.   9 large frames; 13 small frames

12. Hilton University Drama Club sold 311 tickets for a play. Student tickets cost 50 cents each; nonstudent tickets cost $1.50.

21. **a.** answers may vary but notice the slope of each function

If total receipts were $385.50, find how many tickets of each type were sold.   81 student tickets; 230 nonstudent tickets

13. One number is two less than a second number. Twice the first is 4 more than 3 times the second. Find the numbers.

14. Twice one number plus a second number is 42, and the first number minus the second number is −6. Find the numbers.

15. In the United States, the percent of women using the Internet is increasing faster than the percent of men. For the years 1996–2001, the function $y = 7x + 18.7$ can be used to estimate the percent of females using the Internet while the function $y = 6x + 27.7$ can be used to estimate the percent of males. For both functions, x is the number of years since 1996. If this trend continues, predict the year in which the percent of females using the Internet is equal to the percent of males. (*Source:* Pew Internet & American Life Project)   2005

16. The percent of car-vehicle sales is decreasing while the percent of light-truck (pickups, sport-utility vans and mini-vans) vehicle sales is increasing. For the years 1997–2000, the function $y = -x + 54.5$ can be used to estimate the percent of vehicle sales being cars while the function $y = x + 45.5$ can be used to estimate the percent of vehicle sales being light trucks. For both functions, x is the number of years since 1997.   **10.** Washington D.C.: 190; Orlando: 163

   a. If this trend continues, predict the year in which the percent of car sales equals the percent of light-truck sales.   2001

   b. Before the actual 2001 vehicle sales data was published, *USA Today* predicted that light-truck sales would likely be greater than car sales in the year 2001. Does your prediction from part a agree with this statement? (*Source: USA Today* and Autodata)   yes   **20.** 12 mph and 16 mph

17. An office supply store in San Diego sells 7 writing tablets and 4 pens for $6.40. Also, 2 tablets and 19 pens cost $5.40. Find the price of each.   tablets: $0.80; pens: $0.20

18. A Candy Barrel shop manager mixes M&M's worth $2.00 per pound with trail mix worth $1.50 per pound. Find how many pounds of each she should use to get 50 pounds of a party mix worth $1.80 per pound.   30 lb of M&M's; 20 lb of trail mix

19. An airplane takes 3 hours to travel a distance of 2160 miles with the wind. The return trip takes 4 hours against the wind. Find the speed of the plane in still air and the speed of the wind.   speed of plane: 630 mph; wind: 90 mph

20. Two cyclists start at the same point and travel in opposite directions. One travels 4 mph faster than the other. In 4 hours they are 112 miles apart. Find how fast each is traveling.

21. The percent of men 65 years of age or older in the labor force has decreased most years since 1900 while the percent of women 65 years of age or older has slightly increased. For the years 1900–2000, the function $y = -0.52x + 64.5$ approximates the percent of men 65 years of age or older in the labor force while the function $y = 0.02x + 7.2$ approximates the percent of women 65 years of age or older in the labor force. For both functions, x is the number of years after 1900. (*Source: The World Almanac*, 2002)

   a. Explain how the decrease of men described can be verified by their given function while the slight increase of women described can be verified by their given function.

b. If this trend continues, determine the year when the percent of men and woman 65 years of age or older in the labor force are the same.   2006

22. The annual U.S. per capita consumption of butter has remained about the same since 1980 while consumption of margarine has decreased. For the years 1995–1999, the function $y = 0.05x + 4.3$ approximates the annual U.S. per capita consumption of butter in pounds and the function $y = -0.31x + 9.3$ approximates the annual U.S. per capita consumption of margarine in pounds. For both functions, x is the number of years after 1995. If this trend continues, determine the year when the pounds of butter consumed equals the pounds of margarine consumed.   2008

△ **23.** The perimeter of a triangle is 93 centimeters. If two sides are equally long and the third side is 9 centimeters longer than the others, find the lengths of the three sides.   28 cm; 28 cm; 37 cm

24. Jack Reinholt, a car salesman, has a choice of two pay arrangements: a weekly salary of $200 plus 5% commission on sales, or a straight 15% commission. Find the amount of weekly sales for which Jack's earnings are the same regardless of the pay arrangement. $2000

25. Hertz car rental agency charges $25 daily plus 10 cents per mile. Budget charges $20 daily plus 25 cents per mile. Find the daily mileage for which the Budget charge for the day is twice that of the Hertz charge for the day.   600 mi

26. Carroll Blakemore, a drafting student, bought three templates and a pencil one day for $6.45. Another day he bought two pads of paper and four pencils for $7.50. If the price of a pad of paper is three times the price of a pencil, find the price of each type of item.   template: $1.90; pencil: $0.75; pad: $2.25

△ **27.** In the figure, line l and line m are parallel lines cut by transversal t. Find the values of x and y. $x = 75$; $y = 105$

△ **28.** Find the values of x and y in the following isosceles triangle. $x = 40$; $y = 70$

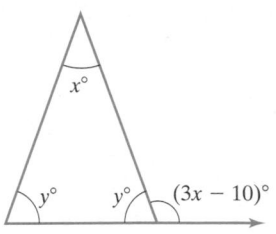

Given the cost function $C(x)$ and the revenue function $R(x)$, find the number of units x that must be sold to break even. See Example 5.

29. $C(x) = 30x + 10,000$ $R(x) = 46x$ 625 units

30. $C(x) = 12x + 15,000$ $R(x) = 32x$ 750 units

31. $C(x) = 1.2x + 1500$ $R(x) = 1.7x$ 3000 units

32. $C(x) = 0.8x + 900$ $R(x) = 2x$ 750 units

33. $C(x) = 75x + 160,000$ $R(x) = 200x$ 1280 units

34. $C(x) = 105x + 70,000$ $R(x) = 245x$ 500 units

35. The planning department of Abstract Office Supplies has been asked to determine whether the company should introduce a new computer desk next year. The department estimates that $6000 of new manufacturing equipment will need to be purchased and that the cost of constructing each desk will be $200. The department also estimates that the revenue from each desk will be $450.

   a. Determine the revenue function $R(x)$ from the sale of x desks. $R(x) = 450x$

   b. Determine the cost function $C(x)$ for manufacturing x desks. $C(x) = 200x + 6000$

   c. Find the break-even point.   24 desks

36. Baskets, Inc., is planning to introduce a new woven basket. The company estimates that $500 worth of new equipment will be needed to manufacture this new type of basket and that it will cost $15 per basket to manufacture. The company also estimates that the revenue from each basket will be $31.

   a. Determine the revenue function $R(x)$ from the sale of x baskets. $R(x) = 31x$

   b. Determine the cost function $C(x)$ for manufacturing x baskets. $C(x) = 15x + 500$

   c. Find the break-even point.   31.25, or 32 baskets

Solve. See Example 6.   **37.** 2 units of Mix A; 3 units of Mix B; 1 unit of Mix C

37. Rabbits in a lab are to be kept on a strict daily diet that includes 30 grams of protein, 16 grams of fat, and 24 grams of carbohydrates. The scientist has only three food mixes available with the following grams of nutrients per unit.

	Protein	*Fat*	*Carbohydrate*
Mix A	4	6	3
Mix B	6	1	2
Mix C	4	1	12

Find how many units of each mix are needed daily to meet each rabbit's dietary need.

38. Gerry Gunderson mixes different solutions with concentrations of 25%, 40%, and 50% to get 200 liters of a 32% solution. If he uses twice as much of the 25% solution as of the 40% solution, find how many liters of each kind he uses.   120 L of 25%; 60 L of 40%; 20 L of 50%

△ **39.** The perimeter of a quadrilateral (four-sided polygon) is 29 inches. The longest side is twice as long as the shortest side. The other two sides are equally long and are 2 inches longer than the shortest side. Find the length of all four sides.   5 in.; 7 in.; 10 in.

△ **40.** The measure of the largest angle of a triangle is 90° more than the measure of the smallest angle, and the measure of the remaining angle is 30° more than the measure of the smallest angle. Find the measure of each angle.   20°, 50°, 110°

41. The sum of three numbers is 40. One number is five more than a second number. It is also twice the third. Find the numbers.   18, 13, and 9

42. The sum of the digits of a three-digit number is 15. The tens-place digit is twice the hundreds-place digit, and the ones-place digit is 1 less than the hundreds-place digit. Find the three-digit number.   483

43. In 2003, the WNBA's top scorer was Lauren Jackson of the Seattle Storm. She scored a total of 698 points during the regular season. The number of two-point field goals Jackson made was 19 less than six times the number of three-point field goals she made. The number of free throws (each worth one point) she made was 64 fewer than the number of two-point field goals she made. Find how many free throws, two-point field goals, and three-point field goals Lauren Jackson made during the 2003 regular season. (*Source:* Women's National Basketball Association)   151 free throws; 215 two-point field goals; 39 three-point field goals

44. During the 2003 NBA playoffs, the top-scoring player was Kobe Bryant of the Los Angeles Lakers . Bryant scored a total of 385 points during the playoffs. The number of free throws (each worth one point) he made was 11 more than three times the number of three-point field goals he made. Bryant also made 26 fewer free throws than two-point field goals. How many free throws, two-point field goals, and three-point field goals did Kobe Bryant make during the 2003 playoffs? (*Source:* National Basketball Association)   86 free throws; 112 two-point field goals; 25 three-point field goals

45. Find the values of x, y, and z in the following triangle. $x = 60$; $y = 55$; $z = 65$

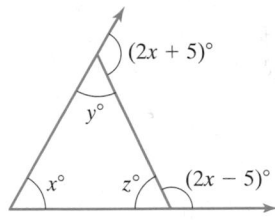

△ **46.** The sum of the measures of the angles of a quadrilateral is 360°. Find the value of x, y, and z in the following quadrilateral. $x = 95$; $y = 123$; $z = 70$

REVIEW AND PREVIEW   **47.** $5x + 5z = 10$

Multiply both sides of equation (1) by 2, and add the resulting equation to equation (2). See Section 4.2.   **48.** $3y + 8z = 18$

47. $3x - y + z = 2$ (1)
    $-x + 2y + 3z = 6$ (2)

48. $2x + y + 3z = 7$ (1)
    $-4x + y + 2z = 4$ (2)

Multiply both sides of equation (1) by -3, and add the resulting equation to equation (2). See Section 4.2.

49. $x + 2y - z = 0$ (1)
    $3x + y - z = 2$ (2)
    $-5y + 2z = 2$

50. $2x - 3y + 2z = 5$ (1)
    $x - 9y + z = -1$ (2)
    $-5x - 5z = -16$

Concept Extensions

51. The number of personal bankruptcy petitions filed in the United States has been on the rise since the early 1980s. In 2001, the number of petitions filed was 200,000 more than four times the number of petitions filed in 1980. This is equivalent to an increase of 1,100,000 petitions filed from 1980 to 2001. Find how many personal bankruptcy petitions were filed in each year. (*Source:* Based on data from the Administrative Office of the United States Courts)

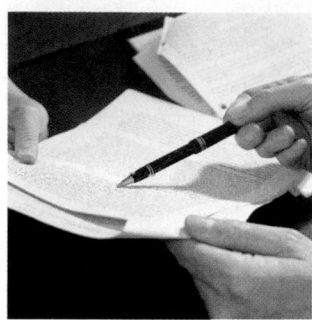

52. In 2000, the median weekly earnings for male bus drivers in the United States was $100 more than the median weekly earnings for female bus drivers. The median weekly earnings for female bus drivers was 0.75 times that of their male counterparts. Also in 2000, the median weekly earnings for female financial managers in the United States was $400 less than the median weekly earnings for male financial managers. The median weekly earnings for male financial managers was 1.5 times that of their female counterparts. (*Source:* Based on data from the U.S. Bureau of Labor Statistics)

a. Find the median weekly earnings for female bus drivers in the United States in 2000. $300

b. Find the median weekly earnings for female financial managers in the United States in 2000. $800

51. 1980: 300,000; 2001: 1,400,000

c. Of the four groups of workers described in the problem, which group makes the greatest weekly earnings? Which group makes the least weekly earnings?

53. Find the values of a, b, and c such that the equation $y = ax^2 + bx + c$ has ordered pair solutions $(1, 6)$, $(-1, -2)$, and $(0, -1)$. To do so, substitute each ordered pair solution into the equation. Each time, the result is an equation in three unknowns: a, b, and c. Then solve the resulting system of three linear equations in three unknowns, a, b, and c. $a = 3, b = 4, c = -1$

54. Find the values of a, b, and c such that the equation $y = ax^2 + bx + c$ has ordered pair solutions $(1, 2), (2, 3)$ and $(-1, 6)$. (*Hint:* See Exercise 53.) $a = 1, b = -2, c = 3$

55. Data (x, y) for the total number y (in thousands) of college-bound students who took the ACT assessment in the year x are $(0, 1065)$, $(1, 1070)$ and $(3, 1175)$, where $x = 0$ represents 2000 and $x = 1$ represents 2001. Find the values of a, b, and c such that the equation $y = ax^2 + bx + c$ models this data. According to your model, how many students will take the ACT in 2009? (*Source:* ACT, Inc.)

56. Monthly normal rainfall data (x, y) for Portland, Oregon, are $(4, 2.47)$, $(7, 0.6)$, $(8, 1.1)$, where x represents time in months (with $x = 1$ representing January) and y represents rainfall in inches. Find the values of a, b, and c rounded to 2 decimal places such that the equation $y = ax^2 + bx + c$ models this data. According to your model, how much rain should Portland expect during September? (*Source:* National Climatic Data Center) $a = 0.28, b = -3.71, c = 12.83$; 2.12 in. in Sept.

52. c. male financial managers; female bus drivers

55. $a = 15\frac{5}{6}, b = -10\frac{5}{6}, c = 1065$; 2250 students in 2009

INTEGRATED REVIEW   SYSTEMS OF LINEAR EQUATIONS

The graphs of various systems of equations are shown. Match each graph with the solution of its corresponding system.

A   B   C   D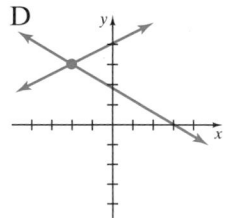

1. Solution: $(1, 2)$ C   **2.** Solution: $(-2, 3)$ D
3. No solution   A   **4.** Infinite number of solutions   B

Solve each system by elimination or substitution.

5. $\begin{cases} x + y = 4 \\ \quad y = 3x \end{cases}$ $(1, 3)$ **6.** $\begin{cases} x - y = -4 \\ \quad y = 4x \end{cases}$ $\left(\dfrac{4}{3}, \dfrac{16}{3}\right)$ **7.** $\begin{cases} x + y = 1 \\ x - 2y = 4 \end{cases}$ $(2, -1)$

8. $\begin{cases} 2x - y = 8 \\ x + 3y = 11 \end{cases}$ (5, 2)   **9.** $\begin{cases} 2x + 5y = 8 \\ 6x + y = 10 \end{cases}$ $\left(\dfrac{3}{2}, 1\right)$ **10.** $\begin{cases} x - 4y = -5 \\ -3x - 8y = 0 \end{cases}$ $\left(-2, \dfrac{3}{4}\right)$

11. $\begin{cases} 4x - 7y = 7 \\ 12x - 21y = 24 \end{cases}$ $\varnothing$ **12.** $\begin{cases} 2x - 5y = 3 \\ -4x + 10y = -6 \end{cases}$ **13.** $\begin{cases} x + y = 2 \\ -3y + z = -7 \\ 2x + y - z = -1 \end{cases}$ (-1, 3, 2)

14. $\begin{cases} y + 2z = -3 \\ x - 2y = 7 \\ 2x - y + z = 5 \end{cases}$ **15.** $\begin{cases} 2x + 4y - 6z = 3 \\ -x + y - z = 6 \\ x + 2y - 3z = 1 \end{cases}$ $\varnothing$ **16.** $\begin{cases} x - y + 3z = 2 \\ -2x + 2y - 6z = -4 \\ 3x - 3y + 9z = 6 \end{cases}$

17. $\begin{cases} x + y - 4z = 5 \\ x - y + 2z = -2 \\ 3x + 2y + 4z = 18 \end{cases}$ **18.** $\begin{cases} 2x - y + 3z = 2 \\ x + y - 6z = 0 \\ 3x + 4y - 3z = 6 \end{cases}$ $\left(1, 1, \dfrac{1}{3}\right)$

12. $\{(x, y) \mid 2x - 5y = 3\}$
14. $(1, -3, 0)$
16. $\{(x, y, z) \mid x - y + 3z = 2\}$

19. A first number is 8 less than a second number. Twice the first number is 11 more than the second number. Find the numbers.

△**20.** The sum of the measures of the angles of a quadrilateral is 360°. The two smallest angles of the quadrilateral have the same measure. The third angle measures 30° more than the measure of one of the smallest

angles and the fourth angle measures 50° more than the measure of one of the smallest angles. Find the measure of each angle.
70°; 70°; 100°; 120°

17. $\left(2, 5, \dfrac{1}{2}\right)$ **19.** 19 and 27

4.4   SOLVING SYSTEMS OF EQUATIONS BY MATRICES

Objectives

1 Use matrices to solve a system of two equations.

2 Use matrices to solve a system of three equations.

By now, you may have noticed that the solution of a system of equations depends on the coefficients of the equations in the system and not on the variables. In this section, we introduce solving a system of equations by a **matrix**.

1 A matrix (plural: **matrices**) is a rectangular array of numbers. The following are examples of matrices.

$$\begin{bmatrix} 1 & 0 \\ 0 & 1 \end{bmatrix} \quad \begin{bmatrix} 2 & 1 & 3 & -1 \\ 0 & -1 & 4 & 5 \\ -6 & 2 & 1 & 0 \end{bmatrix} \quad \begin{bmatrix} a & b & c \\ d & e & f \end{bmatrix}$$

The numbers aligned horizontally in a matrix are in the same **row**. The numbers aligned vertically are in the same **column**.

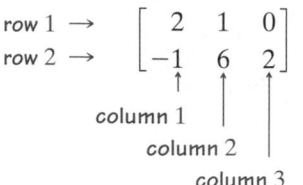

This matrix has 2 rows and 3 columns. It is called a 2×3 (read "two by three") matrix.

To see the relationship between systems of equations and matrices, study the example below.

> ### Helpful Hint
>
> Before writing the corresponding matrix associated with a system of equations, make sure that the equations are written in standard form.

System of Equations (in standard form)

$$\begin{cases} 2x - 3y = 6 & \text{Equation 1} \\ x + y = 0 & \text{Equation 2} \end{cases}$$

Corresponding Matrix

$$\begin{bmatrix} 2 & -3 & \vdots & 6 \\ 1 & 1 & \vdots & 0 \end{bmatrix} \begin{matrix} \text{Row 1} \\ \text{Row 2} \end{matrix}$$

Notice that the rows of the matrix correspond to the equations in the system. The coefficients of each variable are placed to the left of a vertical dashed line. The constants are placed to the right. Each of these numbers in the matrix is called an **element**.

The method of solving systems by matrices is to write this matrix as an equivalent matrix from which we easily identify the solution. Two matrices are equivalent if they represent systems that have the same solution set. The following **row operations** can be performed on matrices, and the result is an equivalent matrix.

TEACHING TIP

Point out that matrices provide a shorthand notation for representing systems of equations.

Elementary Row Operations

1. Any two rows in a matrix may be interchanged.
2. The elements of any row may be multiplied (or divided) by the same nonzero number.
3. The elements of any row may be multiplied (or divided) by a nonzero number and added to their corresponding elements in any other row.

> ### Helpful Hint
>
> Notice that these *row* operations are the same operations that we can perform on *equations* in a system.

To solve a system of two equations in x and y by matrices, write the corresponding matrix associated with the system. Then use elementary row operations to write equivalent matrices until you have a matrix of the form

$$\begin{bmatrix} 1 & a & \vdots & b \\ 0 & 1 & \vdots & c \end{bmatrix},$$

where a, b, and c are constants. Why? If a matrix associated with a system of equations is in this form, we can easily solve for x and y. For example,

Matrix

$$\begin{bmatrix} 1 & 2 & \vdots & -3 \\ 0 & 1 & \vdots & 5 \end{bmatrix}$$

corresponds to

System of Equations

$$\begin{cases} 1x + 2y = -3 \\ 0x + 1y = 5 \end{cases} \quad \text{or} \quad \begin{cases} x + 2y = -3 \\ y = 5 \end{cases}$$

In the second equation, we have $y = 5$. Substituting this in the first equation, we have $x + 2(5) = -3$ or $x = -13$. The solution of the system is the ordered pair $(-13, 5)$.

EXAMPLE 1

Use matrices to solve the system.

$$\begin{cases} x + 3y = 5 \\ 2x - y = -4 \end{cases}$$

Solution   The corresponding matrix is $\begin{bmatrix} 1 & 3 & | & 5 \\ 2 & -1 & | & -4 \end{bmatrix}$. We use elementary row operations to

CLASSROOM EXAMPLE

Use matrices to solve:

$$\begin{cases} x + 2y = -4 \\ 2x - 3y = 13 \end{cases}$$

answer: $(2, -3)$

write an equivalent matrix that looks like $\begin{bmatrix} 1 & a & | & b \\ 0 & 1 & | & c \end{bmatrix}$.

For the matrix given, the element in the first row, first column is already 1, as desired. Next we write an equivalent matrix with a 0 below the 1. To do this, we multiply row 1 by -2 and add to row 2. *We will change only row 2.*

$$\begin{bmatrix} 1 & 3 & | & 5 \\ -2(1) + 2 & -2(3) + (-1) & | & -2(5) + (-4) \end{bmatrix} \text{ simplifies to } \begin{bmatrix} 1 & 3 & | & 5 \\ 0 & -7 & | & -14 \end{bmatrix}$$

$$\begin{matrix} \uparrow & \uparrow & \uparrow & \uparrow & \uparrow & \uparrow \\ \text{row 1} & \text{row 2} & \text{row 1} & \text{row 2} & \text{row 1} & \text{row 2} \\ \text{element} & \text{element} & \text{element} & \text{element} & \text{element} & \text{element} \end{matrix}$$

Now we change the -7 to a 1 by use of an elementary row operation. We divide row 2 by -7, then

$$\begin{bmatrix} 1 & 3 & | & 5 \\ \dfrac{0}{-7} & \dfrac{-7}{-7} & | & \dfrac{-14}{-7} \end{bmatrix} \quad \text{simplifies to} \quad \begin{bmatrix} 1 & 3 & | & 5 \\ 0 & 1 & | & 2 \end{bmatrix}$$

This last matrix corresponds to the system

$$\begin{cases} x + 3y = 5 \\ y = 2 \end{cases}$$

To find x, we let $y = 2$ in the first equation, $x + 3y = 5$.

$$\begin{array}{ll} x + 3y = 5 & \text{First equation} \\ x + 3(2) = 5 & \text{Let } y = 2. \\ x = -1 & \end{array}$$

The ordered pair solution is $(-1, 2)$. Check to see that this ordered pair satisfies both equations.

EXAMPLE 2

Use matrices to solve the system.

$$\begin{cases} 2x - y = 3 \\ 4x - 2y = 5 \end{cases}$$

Solution The corresponding matrix is $\begin{bmatrix} 2 & -1 & | & 3 \\ 4 & -2 & | & 5 \end{bmatrix}$. To get 1 in the row 1, column 1 position, we divide the elements of row 1 by 2.

$$\begin{bmatrix} \dfrac{2}{2} & -\dfrac{1}{2} & | & \dfrac{3}{2} \\ 4 & -2 & | & 5 \end{bmatrix} \quad \text{simplifies to} \quad \begin{bmatrix} 1 & -\dfrac{1}{2} & | & \dfrac{3}{2} \\ 4 & -2 & | & 5 \end{bmatrix}$$

To get 0 under the 1, we multiply the elements of row 1 by -4 and add the new elements to the elements of row 2.

$$\begin{bmatrix} 1 & -\dfrac{1}{2} & | & \dfrac{3}{2} \\ -4(1) + 4 & -4\left(-\dfrac{1}{2}\right) - 2 & | & -4\left(\dfrac{3}{2}\right) + 5 \end{bmatrix} \quad \text{simplifies to} \quad \begin{bmatrix} 1 & -\dfrac{1}{2} & | & \dfrac{3}{2} \\ 0 & 0 & | & -1 \end{bmatrix}$$

The corresponding system is $\begin{cases} x - \dfrac{1}{2}y = \dfrac{3}{2} \\ 0 = -1 \end{cases}$. The equation $0 = -1$ is false for all y or x values; hence the system is inconsistent and has no solution.

✔ **CONCEPT CHECK**

Consider the system
$$\begin{cases} 2x - 3y = 8 \\ x + 5y = -3 \end{cases}$$
What is wrong with its corresponding matrix shown below?

$$\begin{bmatrix} 2 & -3 & | & 8 \\ 0 & 5 & | & -3 \end{bmatrix}$$

2 To solve a system of three equations in three variables using matrices, we will write the corresponding matrix in the form

$$\begin{bmatrix} 1 & a & b & | & d \\ 0 & 1 & c & | & e \\ 0 & 0 & 1 & | & f \end{bmatrix}$$

Thus far, we have solved systems of equations by matrices by writing the augmented matrix of the system as an equivalent matrix with 1's along the diagonal we'll call the main diagonal, and 0's below the 1's. Another way to solve a system of equations by matrices is to use the elementary row operations and write the augmented matrix of the system as an equivalent matrix with 1's along this main diagonal and 0's *above* and below these 1's. This form of a matrix is given a special name: the **reduced row echelon form**.

Although we will not practice writing matrices in reduced row echelon form in this text, we can easily see an advantage of this form.

Below is a system of equations, its corresponding augmented matrix and an equivalent matrix in reduced row echelon form found by performing elementary row operations on the augmented matrix.

System of Equations	*Augmented Matrix of the System*	*Matrix in Reduced Row Echelon Form*
$3x + y + 2z = 5$ $x - y + z = 0$ $-2x + 2y - z = -1$	$\begin{bmatrix} 3 & 1 & 2 & \vert & 5 \\ 1 & -1 & 1 & \vert & 0 \\ -2 & 2 & -1 & \vert & -1 \end{bmatrix}$	$\begin{bmatrix} 1 & 0 & 0 & \vert & 2 \\ 0 & 1 & 0 & \vert & 1 \\ 0 & 0 & 1 & \vert & -1 \end{bmatrix}$

The corresponding system equivalent to the original system is

$$\begin{cases} 1x + 0y + 0z = 2 \\ 0x + 1y + 0z = 1 \\ 0x + 0y + 1z = -1 \end{cases} \quad \text{or} \quad \begin{cases} x = 2 \\ y = 1 \\ z = -1 \end{cases}$$

Notice that the solution of the system is $(2, 1, -1)$, which is found by reading the last column of the associated matrix written in reduced row echelon form.

Although using elementary row operations to write an augmented matrix in equivalent reduced row echelon form may be tedious for us, a calculator can do this quickly and accurately.

EXAMPLE 3

Solve the system of equations in Example 1 using matrices and a calculator.

$$\begin{cases} x + 3y = 5 \\ 2x - y = -4 \end{cases}$$

Solution   This is the same system of equations solved in Example 1. Recall that the augmented matrix associated with this system is

$$\begin{bmatrix} 1 & 3 & \vdots & 5 \\ 2 & -1 & \vdots & -4 \end{bmatrix}$$

Using the matrix edit feature, put the corresponding entries in matrix A. Then use a feature of your calculator to find the equivalent reduced row echelon form.

Row 2 column 3 entry, for example

Augmented matrix in reduced row echelon form

The augmented matrix in reduced row echelon form is $\begin{bmatrix} 1 & 0 & \vdots & -1 \\ 0 & 1 & \vdots & 2 \end{bmatrix}$, which corresponds to the system

$$\begin{cases} 1x + 0y = -1 \\ 0x + 1y = 2 \end{cases} \quad \text{or} \quad \begin{cases} x = -1 \\ y = 2 \end{cases}$$

The solution of the system is $(-1, 2)$, as confirmed in Example 1.

EXAMPLE 4

Use matrices to solve the system.

$$\begin{cases} x + 2y + z = 2 \\ -2x - y + 2z = 5 \\ x + 3y - 2z = -8 \end{cases}$$

Solution   The corresponding matrix is $\begin{bmatrix} 1 & 2 & 1 & \vdots & 2 \\ -2 & -1 & 2 & \vdots & 5 \\ 1 & 3 & -2 & \vdots & -8 \end{bmatrix}$. Our goal is to write an

equivalent matrix with 1's along the diagonal (see the numbers in red) and 0's below the 1's. The element in row 1, column 1 is already 1. Next we get 0's for each element in the rest of column 1. To do this, first we multiply the elements of row 1 by 2 and add the new elements to row 2. Also, we multiply the elements of row 1 by -1 and add the new elements to the elements of row 3. We *do not change row 1*. Then

CLASSROOM EXAMPLE

Use matrices to solve:

$$\begin{cases} x + 3y + z = 5 \\ -3x + y - 3z = 5 \\ x + 2y - 2z = 9 \end{cases}$$

answer: $(1, 2, -2)$

$$\begin{bmatrix} 1 & 2 & 1 & \vdots & 2 \\ 2(1) - 2 & 2(2) - 1 & 2(1) + 2 & \vdots & 2(2) + 5 \\ -1(1) + 1 & -1(2) + 3 & -1(1) - 2 & \vdots & -1(2) - 8 \end{bmatrix}$$ simplifies to $\begin{bmatrix} 1 & 2 & 1 & \vdots & 2 \\ 0 & 3 & 4 & \vdots & 9 \\ 0 & 1 & -3 & \vdots & -10 \end{bmatrix}$

We continue down the diagonal and use elementary row operations to get 1 where the element 3 is now. To do this, we interchange rows 2 and 3.

$$\begin{bmatrix} 1 & 2 & 1 & \vdots & 2 \\ 0 & 3 & 4 & \vdots & 9 \\ 0 & 1 & -3 & \vdots & -10 \end{bmatrix}$$ is equivalent to $\begin{bmatrix} 1 & 2 & 1 & \vdots & 2 \\ 0 & 1 & -3 & \vdots & -10 \\ 0 & 3 & 4 & \vdots & 9 \end{bmatrix}$

Next we want the new row 3, column 2 element to be 0. We multiply the elements of row 2 by -3 and add the result to the elements of row 3.

$$\begin{bmatrix} 1 & 2 & 1 & \vdots & 2 \\ 0 & 1 & -3 & \vdots & -10 \\ -3(0) + 0 & -3(1) + 3 & -3(-3) + 4 & \vdots & -3(-10) + 9 \end{bmatrix}$$ simplifies to

$$\begin{bmatrix} 1 & 2 & 1 & \vdots & 2 \\ 0 & 1 & -3 & \vdots & -10 \\ 0 & 0 & 13 & \vdots & 39 \end{bmatrix}$$

Finally, we divide the elements of row 3 by 13 so that the final diagonal element is 1.

$$\begin{bmatrix} 1 & 2 & 1 & \vdots & 2 \\ 0 & 1 & -3 & \vdots & -10 \\ \frac{0}{13} & \frac{0}{13} & \frac{13}{13} & \vdots & \frac{39}{13} \end{bmatrix}$$ simplifies to $\begin{bmatrix} 1 & 2 & 1 & \vdots & 2 \\ 0 & 1 & -3 & \vdots & -10 \\ 0 & 0 & 1 & \vdots & 3 \end{bmatrix}$

This matrix corresponds to the system

$$\begin{cases} x + 2y + z = 2 \\ y - 3z = -10 \\ z = 3 \end{cases}$$

We identify the z-coordinate of the solution as 3. Next we replace z with 3 in the second equation and solve for y.

$$y - 3z = -10 \quad \textit{Second equation}$$
$$y - 3(3) = -10 \quad \textit{Let } z = 3.$$
$$y = -1$$

To find x, we let $z = 3$ and $y = -1$ in the first equation.

$$x + 2y + z = 2 \quad \textit{First equation}$$
$$x + 2(-1) + 3 = 2 \quad \textit{Let } z = 3 \text{ and } y = -1.$$
$$x = 1$$

The ordered triple solution is $(1, -1, 3)$. Check to see that it satisfies all three equations in the original system.

EXAMPLE 5

Solve the system in Example 4 using matrices and a calculator.

$$\begin{cases} x + 2y + z = 2 \\ -2x - y + 2z = 5 \\ x + 3y - 2z = -8 \end{cases}$$

Solution   Enter the augmented matrix in matrix A and then use your calculator to find the reduced row echelon form of matrix A.

Dimensions of augmented matrix

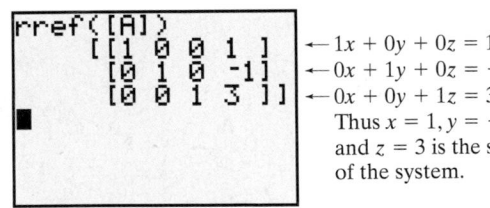

$\leftarrow 1x + 0y + 0z = 1$
$\leftarrow 0x + 1y + 0z = -1$
$\leftarrow 0x + 0y + 1z = 3$
Thus $x = 1, y = -1,$ and $z = 3$ is the solution of the system.

Augmented matrix in reduced row echelon form.

This confirms that the solution of the system of equations is $(1, -1, 3)$.

Spotlight on
DECISION
✈ MAKING

Suppose you are an urban planner working for the public transportation authority of a large city. Currently, more commuters travel into the downtown area during the morning rush hour than travel out of it. However, a strong economy is creating new jobs in the suburbs. As suburban job growth continues, more city dwellers will travel out of the downtown area during the morning rush hour. You have been assigned to study the situation, focusing on how the trend will impact existing bus routes and schedules and how soon.

After a detailed study of public transportation utilization, you find that the number of commuters into downtown each morning can be described by the equation $y = 40,000 + 200x$, where x is the number of months from now. The number of commuters out of downtown each morning can be described by the equation $y = 20,000 + 1000x$, where x is the number of months from now. In the short term, as the numbers of commuters in each direction increase, additional buses can be put on the existing routes to handle the load. But once the number of commuters leaving downtown exceeds the number of commuters traveling into downtown, the bus routes must be totally revamped.

It usually takes the public transportation authority $1\frac{1}{2}$ years to plan major changes to bus routes. If changes are needed more quickly than that, a consulting firm can be hired to speed up the process. Decide whether a consulting firm will be needed to help revamp the bus routes.

EXERCISE SET 4.4

STUDY GUIDE/SSM   CD/ VIDEO   PH MATH TUTOR CENTER   MathXL®Tutorials ON CD   MathXL®   MyMathLab®

Solve each system of linear equations using matrices. See Examples 1 and 3.

1. $\begin{cases} x + y = 1 \\ x - 2y = 4 \end{cases}$ $(2, -1)$ **2.** $\begin{cases} 2x - y = 8 \\ x + 3y = 11 \end{cases}$ $(5, 2)$

3. $\begin{cases} x + 3y = 2 \\ x + 2y = 0 \end{cases}$ $(-4, 2)$ **4.** $\begin{cases} 4x - y = 5 \\ 3x - 3y = 0 \end{cases}$ $(1, -1)$

7. $\{(x, y) \mid 3x - 3y = 9\}$ **8.** $\{(x, y) \mid 9x - 3y = 6\}$

Solve each system of linear equations using matrices. See Example 2.

5. $\begin{cases} x - 2y = 4 \\ 2x - 4y = 4 \end{cases}$ $\varnothing$ **6.** $\begin{cases} -x + 3y = 6 \\ 3x - 9y = 9 \end{cases}$ $\varnothing$

7. $\begin{cases} 3x - 3y = 9 \\ 2x - 2y = 6 \end{cases}$ **8.** $\begin{cases} 9x - 3y = 6 \\ -18x + 6y = -12 \end{cases}$

9. $(-2, 5, -2)$ **10.** $(1, 2, 4)$ **11.** $(1, -2, 3)$ **12.** $(1, 1, -2)$

Solve each system of linear equations using matrices. See Examples 4 and 5.

9. $\begin{cases} x + y = 3 \\ 2y = 10 \\ 3x + 2y - 4z = 12 \end{cases}$ **10.** $\begin{cases} 5x = 5 \\ 2x + y = 4 \\ 3x + y - 5z = -15 \end{cases}$

11. $\begin{cases} 2y - z = -7 \\ x + 4y + z = -4 \\ 5x - y + 2z = 13 \end{cases}$ **12.** $\begin{cases} 4y + 3z = -2 \\ 5x - 4y = 1 \\ -5x + 4y + z = -3 \end{cases}$

15. $(2, 1, -1)$ **16.** $(5, -2, 4)$ **18.** $(0, -9)$

MIXED PRACTICE

Solve each system of linear equations using matrices.

13. $\begin{cases} x - 4 = 0 \\ x + y = 1 \end{cases}$ $(4, -3)$ **14.** $\begin{cases} 3y = 6 \\ x + y = 7 \end{cases}$ $(5, 2)$

15. $\begin{cases} x + y + z = 2 \\ 2x - z = 5 \\ 3y + z = 2 \end{cases}$ **16.** $\begin{cases} x + 2y + z = 5 \\ x - y - z = 3 \\ y + z = 2 \end{cases}$

17. $\begin{cases} 5x - 2y = 27 \\ -3x + 5y = 18 \end{cases}$ $(9, 9)$ **18.** $\begin{cases} 4x - y = 9 \\ 2x + 3y = -27 \end{cases}$

19. $\begin{cases} 4x - 7y = 7 \\ 12x - 21y = 24 \end{cases}$ $\varnothing$ **20.** $\begin{cases} 2x - 5y = 12 \\ -4x + 10y = 20 \end{cases}$ $\varnothing$

21. $\begin{cases} 4x - y + 2z = 5 \\ 2y + z = 4 \\ 4x + y + 3z = 10 \end{cases}$ $\varnothing$ **22.** $\begin{cases} 5y - 7z = 14 \\ 2x + y + 4z = 10 \\ 2x + 6y - 3z = 30 \end{cases}$ $\varnothing$

23. $\begin{cases} 4x + y + z = 3 \\ -x + y - 2z = -11 \\ x + 2y + 2z = -1 \end{cases}$ **24.** $\begin{cases} x + y + z = 9 \\ 3x - y + z = -1 \\ -2x + 2y - 3z = -2 \end{cases}$

REVIEW AND PREVIEW

23. $(1, -4, 3)$ **24.** $(0, 5, 4)$

Determine whether each graph is the graph of a function. See Section 2.2.

25. function

26. function

27.

not a function

28.

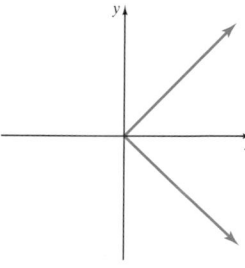

not a function

Evaluate. See Section 1.3.

29. $(-1)(-5) - (6)(3)$ -13 **30.** $(2)(-8) - (-4)(1)$ -12

31. $(4)(-10) - (2)(-2)$ -36 **32.** $(-7)(3) - (-2)(-6)$ -33

33. $(-3)(-3) - (-1)(-9)$ 0 **34.** $(5)(6) - (10)(10)$ -70

Combining Concepts

35. The percent y of U.S. households that owned a black-and-white television set between the years 1980 and 1993 can be modeled by the linear equation $2.3x + y = 52$, where x represents the number of years after 1980. Similarly, the percent y of U.S. households that owned a microwave oven during this same period can be modeled by the linear equation $-5.4x + y = 14$. (*Source:* Based on data from the Energy Information Administration, U.S. Department of Energy)

   a. The data used to form these two models was incomplete. It is impossible to tell from the data the year in which the percent of households owning black-and-white television sets was the same as the percent of households owning microwave ovens. Use matrix methods to estimate the year in which this occurred.   end of 1984

   b. Did more households own black-and-white television sets or microwave ovens in 1980? In 1993? What trends do these models show? Does this seem to make sense? Why or why not?   **c.** in 2002

   c. According to the models, when will the percent of households owning black-and-white television sets reach 0%?

d. Do you think your answer to part **c** is accurate? Why or why not?   answers may vary

36. The most popular amusement park in the world (according to annual attendance) is Tokyo Disneyland, whose yearly attendance in thousands can be approximated by the equation $y = 1201x + 16{,}507$ where x is the number of years after 2000. In second place is Walt Disney World's Magic Kingdom, whose yearly attendance, in thousands, can be approximated by

35. b. black-and-white sets; microwave ovens; the percent of households owning black-and-white television sets is decreasing and the percent of households owning microwave ovens is increasing; answers may vary.

$y - 616x + 15{,}400$. Find the last year when attendance in Magic Kingdom was greater than attendance in Tokyo Disneyland. (*Source*: Amusement Business)   1999

37. For the system $\begin{cases} 2x - 3y = 8 \\ x + 5y = -3 \end{cases}$, explain what is wrong with

writing the corresponding matrix as $\begin{bmatrix} 2 & 3 & \vdots & 8 \\ 0 & 5 & \vdots & -3 \end{bmatrix}$.   answers may vary

---

4.5   SOLVING SYSTEMS OF EQUATIONS BY DETERMINANTS

Objectives

1   Define and evaluate a 2×2 determinant.

2   Use Cramer's rule to solve a system of two linear equations in two variables.

3   Define and evaluate a 3×3 determinant.

4   Use Cramer's rule to solve a linear system of three equations in three variables.

1   We have solved systems of two linear equations in two variables in four different ways: graphically, by substitution, by elimination, and by matrices. Now we analyze another method called **Cramer's rule**.

Recall that a matrix is a rectangular array of numbers. If a matrix has the same number of rows and columns, it is called a **square matrix**. Examples of square matrices are

$$\begin{bmatrix} 1 & 6 \\ 5 & 2 \end{bmatrix} \qquad \begin{bmatrix} 2 & 4 & 1 \\ 0 & 5 & 2 \\ 3 & 6 & 9 \end{bmatrix}$$

A **determinant** is a real number associated with a square matrix. The determinant of a square matrix is denoted by placing vertical bars about the array of numbers. Thus,

The determinant of the square matrix $\begin{bmatrix} 1 & 6 \\ 5 & 2 \end{bmatrix}$ is $\begin{vmatrix} 1 & 6 \\ 5 & 2 \end{vmatrix}$.

The determinant of the square matrix $\begin{bmatrix} 2 & 4 & 1 \\ 0 & 5 & 2 \\ 3 & 6 & 9 \end{bmatrix}$ is $\begin{vmatrix} 2 & 4 & 1 \\ 0 & 5 & 2 \\ 3 & 6 & 9 \end{vmatrix}$.

We define the determinant of a 2×2 matrix first. (Recall that 2×2 is read "two by two." It means that the matrix has 2 rows and 2 columns.)

Determinant of a 2 × 2 Matrix

$$\begin{vmatrix} a & b \\ c & d \end{vmatrix} = ad - bc$$

EXAMPLE 1

Evaluate each determinant.

a $\begin{vmatrix} -1 & 2 \\ 3 & -4 \end{vmatrix}$

b. $\begin{vmatrix} 2 & 0 \\ 7 & -5 \end{vmatrix}$

Solution   First we identify the values of *a, b, c,* and *d.* Then we perform the evaluation.

a. Here $a = -1$, $b = 2$, $c = 3$, and $d = -4$.

$$\begin{vmatrix} -1 & 2 \\ 3 & -4 \end{vmatrix} = ad - bc = (-1)(-4) - (2)(3) = -2$$

b. In this example, $a = 2$, $b = 0$, $c = 7$, and $d = -5$.

$$\begin{vmatrix} 2 & 0 \\ 7 & -5 \end{vmatrix} = ad - bc = 2(-5) - (0)(7) = -10$$

A calculator may be used to find determinants. In Example 1a, we found that $\begin{vmatrix} -1 & 2 \\ 3 & -4 \end{vmatrix} = -2$.

To check with a calculator, enter the elements of the determinant $\begin{vmatrix} -1 & 2 \\ 3 & -4 \end{vmatrix}$ in matrix *A* and then find the determinant of matrix *A*.

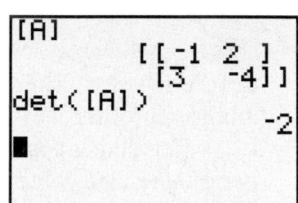

Calculators may be used to find determinants, and therefore, may also be used to solve systems of equations by Cramer's rule.

2   To develop Cramer's rule, we solve the system $\begin{cases} ax + by = h \\ cx + dy = k \end{cases}$ using elimination.

First, we eliminate *y* by multiplying both sides of the first equation by *d* and both sides of the second equation by $-b$ so that the coefficients of *y* are opposites. The result is that

$$\begin{cases} d(ax + by) = d \cdot h \\ -b(cx + dy) = -b \cdot k \end{cases} \quad \text{simplifies to} \quad \begin{cases} adx + bdy = hd \\ -bcx - bdy = -kb \end{cases}$$

We now add the two equations and solve for *x.*

$$\begin{array}{rll} adx + bdy &= hd & \\ \underline{-bcx - bdy} &= \underline{-kb} & \\ adx - bcx &= hd - kb & \text{Add the equations.} \\ (ad - bc)x &= hd - kb & \\ x &= \dfrac{hd - kb}{ad - bc} & \text{Solve for } x. \end{array}$$

When we replace x with $\dfrac{hd - kb}{ad - bc}$ in the equation $ax + by = h$ and solve for y, we find that $y = \dfrac{ak - ch}{ad - bc}$.

Notice that the numerator of the value of x is the determinant of

$$\begin{vmatrix} h & b \\ k & d \end{vmatrix} = hd - kb$$

Also, the numerator of the value of y is the determinant of

$$\begin{vmatrix} a & h \\ c & k \end{vmatrix} = ak - hc$$

Finally, the denominators of the values of x and y are the same and are the determinant of

$$\begin{vmatrix} a & b \\ c & d \end{vmatrix} = ad - bc$$

This means that the values of x and y can be written in determinant notation.

$$x = \frac{\begin{vmatrix} h & b \\ k & d \end{vmatrix}}{\begin{vmatrix} a & b \\ c & d \end{vmatrix}} \quad \text{and} \quad y = \frac{\begin{vmatrix} a & h \\ c & k \end{vmatrix}}{\begin{vmatrix} a & b \\ c & d \end{vmatrix}}$$

For convenience, we label the determinants D, D_x, and D_y.

x-coefficients

y-coefficients

$$\begin{vmatrix} a & b \\ c & d \end{vmatrix} = D \qquad \begin{vmatrix} h & b \\ k & d \end{vmatrix} = D_x \qquad \begin{vmatrix} a & h \\ c & k \end{vmatrix} = D_y$$

x-column replaced by constants

y-column replaced by constants

These determinant formulas for the coordinates of the solution of a system are known as **Cramer's rule.**

Cramer's Rule for Two Linear Equations in Two Variables

The solution of the system $\begin{cases} ax + by = h \\ cx + dy = k \end{cases}$ is given by

$$x = \frac{\begin{vmatrix} h & b \\ k & d \end{vmatrix}}{\begin{vmatrix} a & b \\ c & d \end{vmatrix}} = \frac{D_x}{D} \qquad y = \frac{\begin{vmatrix} a & h \\ c & k \end{vmatrix}}{\begin{vmatrix} a & b \\ c & d \end{vmatrix}} = \frac{D_y}{D}$$

as long as $D = ad - bc$ is not 0.

When $D = 0$, the system is either inconsistent or the equations are dependent. When this happens, we need to use another method to see which is the case.

EXAMPLE 2

Use Cramer's rule to solve each system.

a. $\begin{cases} 3x + 4y = -7 \\ x - 2y = -9 \end{cases}$ **b.** $\begin{cases} 5x + y = 5 \\ -7x - 2y = -7 \end{cases}$

Solution **a.** First we find $D, D_x,$ and D_y.

$$\begin{array}{ccc} a & b & h \\ \downarrow & \downarrow & \downarrow \end{array}$$
$$\begin{cases} 3x + 4y = -7 \\ x - 2y = -9 \end{cases}$$
$$\begin{array}{ccc} \uparrow & \uparrow & \uparrow \\ c & d & k \end{array}$$

$$D = \begin{vmatrix} a & b \\ c & d \end{vmatrix} = \begin{vmatrix} 3 & 4 \\ 1 & -2 \end{vmatrix} = 3(-2) - 4(1) = -10$$

$$D_x = \begin{vmatrix} h & b \\ k & d \end{vmatrix} = \begin{vmatrix} -7 & 4 \\ -9 & -2 \end{vmatrix} = (-7)(-2) - 4(-9) = 50$$

$$D_y = \begin{vmatrix} a & h \\ c & k \end{vmatrix} = \begin{vmatrix} 3 & -7 \\ 1 & -9 \end{vmatrix} = 3(-9) - (-7)(1) = -20$$

Then $x = \dfrac{D_x}{D} = \dfrac{50}{-10} = -5$ and $y = \dfrac{D_y}{D} = \dfrac{-20}{-10} = 2$. The ordered pair solution is $(-5, 2)$.

As always, check the solution in both original equations.

b. Find $D, D_x,$ and D_y for $\begin{cases} 5x + y = 5 \\ -7x - 2y = -7 \end{cases}$.

$$D = \begin{vmatrix} 5 & 1 \\ -7 & -2 \end{vmatrix} = 5(-2) - (1)(-7) = -3$$

$$D_x = \begin{vmatrix} 5 & 1 \\ -7 & -2 \end{vmatrix} = 5(-2) - (1)(-7) = -3$$

$$D_y = \begin{vmatrix} 5 & 5 \\ -7 & -7 \end{vmatrix} = 5(-7) - 5(-7) = 0$$

Then $x = \dfrac{D_x}{D} = \dfrac{-3}{-3} = 1$ and $y = \dfrac{D_y}{D} = \dfrac{0}{-3} = 0$.

The ordered pair solution is $(1, 0)$. Check this solution in both original equations.

Let's solve the system in Example 2a using Cramer's rule and a calculator.

$$\begin{cases} 3x + 4y = -7 \\ x - 2y = -9 \end{cases}$$

Recall that,

$$D = \begin{vmatrix} 3 & 4 \\ 1 & -2 \end{vmatrix}, \quad D_x = \begin{vmatrix} -7 & 4 \\ -9 & -2 \end{vmatrix}, \quad D_y = \begin{vmatrix} 3 & -7 \\ 1 & -9 \end{vmatrix}$$

To evaluate D, D_x, and D_y, enter the elements of D in matrix A, D_x in matrix B, and D_y in matrix C and evaluate the determinant of each matrix. Then

$$x = \frac{D_x}{D} = \frac{\det[B]}{\det[A]} = -5 \quad \text{and} \quad y = \frac{D_y}{D} = \frac{\det[C]}{\det[A]} = 2$$

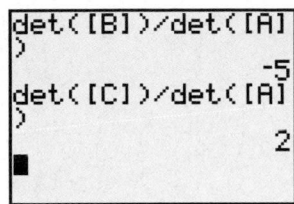

The solution is $(-5, 2)$, as expected.

3   A 3×3 determinant can be used to solve a system of three equations in three variables. The determinant of a 3×3 matrix, however, is considerably more complex than a 2×2 one.

Determinant of a 3 × 3 Matrix

$$\begin{vmatrix} a_1 & b_1 & c_1 \\ a_2 & b_2 & c_2 \\ a_3 & b_3 & c_3 \end{vmatrix} = a_1 \cdot \begin{vmatrix} b_2 & c_2 \\ b_3 & c_3 \end{vmatrix} - a_2 \cdot \begin{vmatrix} b_1 & c_1 \\ b_3 & c_3 \end{vmatrix} + a_3 \cdot \begin{vmatrix} b_1 & c_1 \\ b_2 & c_2 \end{vmatrix}$$

Notice that the determinant of a 3×3 matrix is related to the determinants of three 2×2 matrices. Each determinant of these 2×2 matrices is called a **minor**, and every element of a 3×3 matrix has a minor associated with it. For example, the minor of c_2 is the determinant of the 2×2 matrix found by deleting the row and column containing c_2.

$$\begin{matrix} a_1 & b_1 & c_1 \\ a_2 & b_2 & c_2 \\ a_3 & b_3 & c_3 \end{matrix} \qquad \text{The minor of } c_2 \text{ is} \qquad \begin{vmatrix} a_1 & b_1 \\ a_3 & b_3 \end{vmatrix}$$

Also, the minor of element a_1 is the determinant of the 2×2 matrix that has no row or column containing a_1.

$$\begin{matrix} a_1 & b_1 & c_1 \\ a_2 & b_2 & c_2 \\ a_3 & b_3 & c_3 \end{matrix} \qquad \text{The minor of } a_1 \text{ is} \qquad \begin{vmatrix} b_2 & c_2 \\ b_3 & c_3 \end{vmatrix}$$

So the determinant of a 3×3 matrix can be written as

$$a_1 \cdot (\text{minor of } a_1) - a_2 \cdot (\text{minor of } a_2) + a_3 \cdot (\text{minor of } a_3)$$

Finding the determinant by using minors of elements in the first column is called **expanding** by the minors of the first column. *The value of a determinant can be found by expanding by the minors of any row or column.* The following **array of signs** is helpful in determining whether to add or subtract the product of an element and its minor.

$$\begin{matrix} + & - & + \\ - & + & - \\ + & - & + \end{matrix}$$

If an element is in a position marked $+$, we add. If marked $-$, we subtract.

Suppose you are interested in finding the determinant of a 4 × 4 matrix. Study the pattern shown in the array of signs for a 3 × 3 matrix. Use the pattern to expand the array of signs for use with a 4 × 4 matrix.

EXAMPLE 3

Evaluate by expanding by the minors of the given row or column.

$$\begin{vmatrix} 0 & 5 & 1 \\ 1 & 3 & -1 \\ -2 & 2 & 4 \end{vmatrix}$$

a. First column **b.** Second row

Solution **a.** The elements of the first column are $0, 1$, and -2. The first column of the array of signs is $+, -, +$.

$$\begin{vmatrix} 0 & 5 & 1 \\ 1 & 3 & -1 \\ -2 & 2 & 4 \end{vmatrix} = 0 \cdot \begin{vmatrix} 3 & -1 \\ 2 & 4 \end{vmatrix} - 1 \cdot \begin{vmatrix} 5 & 1 \\ 2 & 4 \end{vmatrix} + (-2) \cdot \begin{vmatrix} 5 & 1 \\ 3 & -1 \end{vmatrix}$$

$$= 0(12 - (-2)) - 1(20 - 2) + (-2)(-5 - 3)$$

$$= 0 - 18 + 16 = -2$$

b. The elements of the second row are $1, 3$, and -1. This time, the signs begin with $-$ and again alternate.

$$\begin{vmatrix} 0 & 5 & 1 \\ 1 & 3 & -1 \\ -2 & 2 & 4 \end{vmatrix} = -1 \cdot \begin{vmatrix} 5 & 1 \\ 2 & 4 \end{vmatrix} + 3 \cdot \begin{vmatrix} 0 & 1 \\ -2 & 4 \end{vmatrix} - (-1) \cdot \begin{vmatrix} 0 & 5 \\ -2 & 2 \end{vmatrix}$$

$$= -1(20 - 2) + 3(0 - (-2)) - (-1)(0 - (-10))$$

$$= -18 + 6 + 10 = -2$$

Notice that the determinant of the 3 × 3 matrix is the same regardless of the row or column you select to expand by.

CLASSROOM EXAMPLE

Evaluate by expanding by the minors of the given row or column.

a. First column **b.** Third row

$$\begin{vmatrix} 2 & 0 & 1 \\ -1 & 3 & 2 \\ 5 & 1 & 4 \end{vmatrix}$$

answer: **a.** 4 **b.** 4

Why would expanding by minors of the second row be a good choice for the determinant

$$\begin{vmatrix} 3 & 4 & -2 \\ 5 & 0 & 0 \\ 6 & -3 & 7 \end{vmatrix}?$$

Concept Check Answer:

$$\begin{array}{cccc} + & - & + & - \\ - & + & - & + \\ + & - & + & - \\ - & + & - & + \end{array}$$

Concept Check Answer:
Two elements of the second row are 0, which makes calculations easier.

4   A system of three equations in three variables may be solved with Cramer's rule also. Using the elimination process to solve a system with unknown constants as coefficients leads to the following.

Cramer's Rule for Three Equations in Three Variables

The solution of the system $\begin{cases} a_1x + b_1y + c_1z = k_1 \\ a_2x + b_2y + c_2z = k_2 \\ a_3x + b_3y + c_3z = k_3 \end{cases}$ is given by

$$x = \frac{D_x}{D} \qquad y = \frac{D_y}{D} \qquad \text{and} \qquad z = \frac{D_z}{D}$$

where

$$D = \begin{vmatrix} a_1 & b_1 & c_1 \\ a_2 & b_2 & c_2 \\ a_3 & b_3 & c_3 \end{vmatrix} \qquad D_x = \begin{vmatrix} k_1 & b_1 & c_1 \\ k_2 & b_2 & c_2 \\ k_3 & b_3 & c_3 \end{vmatrix}$$

$$D_y = \begin{vmatrix} a_1 & k_1 & c_1 \\ a_2 & k_2 & c_2 \\ a_3 & k_3 & c_3 \end{vmatrix} \qquad D_z = \begin{vmatrix} a_1 & b_1 & k_1 \\ a_2 & b_2 & k_2 \\ a_3 & b_3 & k_3 \end{vmatrix}$$

as long as D is not 0.

EXAMPLE 4

Use Cramer's rule to solve the system.

$$\begin{cases} x - 2y + z = 4 \\ 3x + y - 2z = 3 \\ 5x + 5y + 3z = -8 \end{cases}$$

Solution   First we find $D, D_x, D_y,$ and D_z. Beginning with D, we expand by the minors of the first column.

$$D = \begin{vmatrix} 1 & -2 & 1 \\ 3 & 1 & -2 \\ 5 & 5 & 3 \end{vmatrix} = 1 \cdot \begin{vmatrix} 1 & -2 \\ 5 & 3 \end{vmatrix} - 3 \cdot \begin{vmatrix} -2 & 1 \\ 5 & 3 \end{vmatrix} + 5 \cdot \begin{vmatrix} -2 & 1 \\ 1 & -2 \end{vmatrix}$$

$$= 1(3 - (-10)) - 3(-6 - 5) + 5(4 - 1)$$

$$= 13 + 33 + 15 = 61$$

$$D_x = \begin{vmatrix} 4 & -2 & 1 \\ 3 & 1 & -2 \\ -8 & 5 & 3 \end{vmatrix} = 4 \cdot \begin{vmatrix} 1 & -2 \\ 5 & 3 \end{vmatrix} - 3 \cdot \begin{vmatrix} -2 & 1 \\ 5 & 3 \end{vmatrix} + (-8) \cdot \begin{vmatrix} -2 & 1 \\ 1 & -2 \end{vmatrix}$$

$$= 4(3 - (-10)) - 3(-6 - 5) + (-8)(4 - 1)$$

$$= 52 + 33 - 24 = 61$$

$$D_y = \begin{vmatrix} 1 & 4 & 1 \\ 3 & 3 & -2 \\ 5 & -8 & 3 \end{vmatrix} = 1 \cdot \begin{vmatrix} 3 & -2 \\ -8 & 3 \end{vmatrix} - 3 \cdot \begin{vmatrix} 4 & 1 \\ -8 & 3 \end{vmatrix} + 5 \cdot \begin{vmatrix} 4 & 1 \\ 3 & -2 \end{vmatrix}$$

$$= 1(9 - 16) - 3(12 - (-8)) + 5(-8 - 3)$$

$$= -7 - 60 - 55 = -122$$

$$D_z = \begin{vmatrix} 1 & -2 & 4 \\ 3 & 1 & 3 \\ 5 & 5 & -8 \end{vmatrix} = 1 \cdot \begin{vmatrix} 1 & 3 \\ 5 & -8 \end{vmatrix} - 3 \cdot \begin{vmatrix} -2 & 4 \\ 5 & -8 \end{vmatrix} + 5 \cdot \begin{vmatrix} -2 & 4 \\ 1 & 3 \end{vmatrix}$$

$$= 1(-8 - 15) - 3(16 - 20) + 5(-6 - 4)$$

$$= -23 + 12 - 50 = -61$$

From these determinants, we calculate the solution.

$$x = \frac{D_x}{D} = \frac{61}{61} = 1 \qquad y = \frac{D_y}{D} = \frac{-122}{61} = -2 \qquad z = \frac{D_z}{D} = \frac{-61}{61} = -1$$

The ordered triple solution is $(1, -2, -1)$. Check this solution by verifying that it satisfies each equation of the system.

8. $(1, -1)$ **9.** $\{(x, y) \mid 3x + y = 1\}$ **10.** $\{(x, y) \mid y = 2x - 5\}$

59.
$$\begin{array}{cccc} + & - & + & - \\ - & + & - & + \\ + & - & + & - \\ - & + & - & + \end{array}$$

EXERCISE SET 4.5

STUDY GUIDE/SSM | CD/VIDEO | PH MATH TUTOR CENTER | MathXL®Tutorials ON CD | MathXL® | MyMathLab®

Evaluate. See Example 1.

 1. $\begin{vmatrix} 3 & 5 \\ -1 & 7 \end{vmatrix}$ 26

2. $\begin{vmatrix} -5 & 1 \\ 0 & -4 \end{vmatrix}$ 20

3. $\begin{vmatrix} 9 & -2 \\ 4 & -3 \end{vmatrix}$ -19

4. $\begin{vmatrix} 4 & 0 \\ 9 & 8 \end{vmatrix}$ 32

5. $\begin{vmatrix} -2 & 9 \\ 4 & -18 \end{vmatrix}$ 0

6. $\begin{vmatrix} -40 & 8 \\ 70 & -14 \end{vmatrix}$ 0

Use Cramer's rule, if possible, to solve each system of linear equations. See Example 2.

7. $\begin{cases} 2y - 4 = 0 \\ x + 2y = 5 \end{cases}$ $(1, 2)$

8. $\begin{cases} 4x - y = 5 \\ 3x - 3 = 0 \end{cases}$

9. $\begin{cases} 3x + y = 1 \\ 2y = 2 - 6x \end{cases}$

10. $\begin{cases} y = 2x - 5 \\ 8x - 4y = 20 \end{cases}$

11. $\begin{cases} 5x - 2y = 27 \\ -3x + 5y = 18 \end{cases}$ $(9, 9)$

12. $\begin{cases} 4x - y = 9 \\ 2x + 3y = -27 \end{cases}$ $(0, -9)$

Evaluate. See Example 3. **19.** $(-2, 0, 5)$ **20.** $(1, 1, 2)$

13. $\begin{vmatrix} 2 & 1 & 0 \\ 0 & 5 & -3 \\ 4 & 0 & 2 \end{vmatrix}$ 8

14. $\begin{vmatrix} -6 & 4 & 2 \\ 1 & 0 & 5 \\ 0 & 3 & 1 \end{vmatrix}$ 92

15. $\begin{vmatrix} 4 & -6 & 0 \\ -2 & 3 & 0 \\ 4 & -6 & 1 \end{vmatrix}$ 0

16. $\begin{vmatrix} 5 & 2 & 1 \\ 3 & -6 & 0 \\ -2 & 8 & 0 \end{vmatrix}$ 12

17. $\begin{vmatrix} 3 & 6 & -3 \\ -1 & -2 & 3 \\ 4 & -1 & 6 \end{vmatrix}$ 54

18. $\begin{vmatrix} 2 & -2 & 1 \\ 4 & 1 & 3 \\ 3 & 1 & 2 \end{vmatrix}$ -3

Use Cramer's rule, if possible, to solve each system of linear equations. See Example 4. **21.** $(6, -2, 4)$ **22.** $(-1, 0, 2)$

19. $\begin{cases} 3x + z = -1 \\ -x - 3y + z = 7 \\ 3y + z = 5 \end{cases}$

20. $\begin{cases} 4y - 3z = -2 \\ 8x - 4y = 4 \\ -8x + 4y + z = -2 \end{cases}$

21. $\begin{cases} x + y + z = 8 \\ 2x - y - z = 10 \\ x - 2y + 3z = 22 \end{cases}$

22. $\begin{cases} 5x + y + 3z = 1 \\ x - y - 3z = -7 \\ -x + y = 1 \end{cases}$

Evaluate. **25.** 15 **31.** 56 **32.** -56 **37.** $(-2, 3, -1)$

23. $\begin{vmatrix} 10 & -1 \\ -4 & 2 \end{vmatrix}$ 16

24. $\begin{vmatrix} -6 & 2 \\ 5 & -1 \end{vmatrix}$ -4

25. $\begin{vmatrix} 1 & 0 & 4 \\ 1 & -1 & 2 \\ 3 & 2 & 1 \end{vmatrix}$

26. $\begin{vmatrix} 0 & 1 & 2 \\ 3 & -1 & 2 \\ 3 & 2 & -2 \end{vmatrix}$ 30

27. $\begin{vmatrix} \frac{3}{4} & \frac{5}{2} \\ -\frac{1}{6} & \frac{7}{3} \end{vmatrix}$ $\frac{13}{6}$

28. $\begin{vmatrix} \frac{5}{7} & \frac{1}{3} \\ \frac{6}{7} & \frac{2}{3} \end{vmatrix}$ $\frac{4}{21}$

29. $\begin{vmatrix} 4 & -2 & 2 \\ 6 & -1 & 3 \\ 2 & 1 & 1 \end{vmatrix}$ 0

30. $\begin{vmatrix} 1 & 5 & 0 \\ 7 & 9 & -4 \\ 3 & 2 & -2 \end{vmatrix}$ 0

31. $\begin{vmatrix} -2 & 5 & 4 \\ 5 & -1 & 3 \\ 4 & 1 & 2 \end{vmatrix}$

32. $\begin{vmatrix} 5 & -2 & 4 \\ -1 & 5 & 3 \\ 1 & 4 & 2 \end{vmatrix}$

MIXED PRACTICE **38.** $(-3, -2, 5)$ **40.** $(-4, 3)$

Use Cramer's rule, if possible, to solve each system of linear equations. **41.** $(-2, 1)$ **42.** $(-1, 1)$ **45.** $(0, 2, -1)$

33. $\begin{cases} 2x - 5y = 4 \\ x + 2y = -7 \end{cases}$ $(-3, -2)$

34. $\begin{cases} 3x - y = 2 \\ -5x + 2y = 0 \end{cases}$ $(4, 10)$

35. $\begin{cases} 4x + 2y = 5 \\ 2x + y = -1 \end{cases}$ $\varnothing$

36. $\begin{cases} 3x + 6y = 15 \\ 2x + 4y = 3 \end{cases}$ $\varnothing$

37. $\begin{cases} 2x + 2y + z = 1 \\ -x + y + 2z = 3 \\ x + 2y + 4z = 0 \end{cases}$

38. $\begin{cases} 2x - 3y + z = 5 \\ x + y + z = 0 \\ 4x + 2y + 4z = 4 \end{cases}$

39. $\begin{cases} \frac{2}{3}x - \frac{3}{4}y = -1 \\ -\frac{1}{6}x + \frac{3}{4}y = \frac{5}{2} \end{cases}$ $(3, 4)$

40. $\begin{cases} \frac{1}{2}x - \frac{1}{3}y = -3 \\ \frac{1}{8}x + \frac{1}{6}y = 0 \end{cases}$

41. $\begin{cases} 0.7x - 0.2y = -1.6 \\ 0.2x \quad - y = -1.4 \end{cases}$ **42.** $\begin{cases} -0.7x + 0.6y = 1.3 \\ 0.5x - 0.3y = -0.8 \end{cases}$

43. $\begin{cases} -2x + 4y - 2z = 6 \\ x - 2y + z = -3 \\ 3x - 6y + 3z = -9 \end{cases}$ **44.** $\begin{cases} -x - y + 3z = 2 \\ 4x + 4y - 12z = -8 \\ -3x - 3y + 9z = 6 \end{cases}$

45. $\begin{cases} x - 2y + z = -5 \\ 3y + 2z = 4 \\ 3x - y \quad = -2 \end{cases}$ **46.** $\begin{cases} 4x + 5y \quad = 10 \\ 3y + 2z = -6 \\ x + y + z = 3 \end{cases}$

REVIEW AND PREVIEW **43.** $\{(x, y, z) \mid x - 2y + z = -3\}$

Simplify each expression. See Section 1.4. **46.** $(5, -2, 0)$

47. $5x - 6 + x - 12$ **48.** $4y + 3 - 15y - 1$

49. $2(3x - 6) + 3(x - 1)$ $9x - 15$ **47.** $6x - 18$

50. $-3(2y - 7) - 1(11 + 12y)$ $-18y + 10$ **48.** $-11y + 2$

Graph each function. See Section 2.2.

51. $f(x) = 5x - 6$ **52.** $g(x) = -x + 1$

53. $h(x) = 3$ **54.** $f(x) = -3$

Concept Extensions **44.** $\{(x, y, z) \mid -x - y + 3z = 2\}$

Find the value of x such that each is a true statement.

55. $\begin{vmatrix} 1 & x \\ 2 & 7 \end{vmatrix} = -3$ 5 **56.** $\begin{vmatrix} 6 & 1 \\ -2 & x \end{vmatrix} = 26$ 4

51.–54. See graphing answer section.

57. If all the elements in a single row of a determinant are zero, to what does the determinant evaluate? Explain your answer. 0

58. If all the elements in a single column of a determinant are 0, to what does the determinant evaluate? Explain your answer. 0

59. Suppose you are interested in finding the determinant of a 4×4 matrix. Study the pattern shown in the array of signs for a 3×3 matrix. Use the pattern to expand the array of signs for use with a 4×4 matrix. **60.** answers may vary

60. Why would expanding by minors of the second row be a good choice for the determinant $\begin{vmatrix} 3 & 4 & -2 \\ 5 & 0 & 0 \\ 6 & -3 & 7 \end{vmatrix}$?

Find the value of each determinant. To evaluate a 4×4 determinant, select any row or column and expand by the minors. The array of signs for a 4×4 determinant is the same as for a 3×3 determinant except expanded.

61. $\begin{vmatrix} 5 & 0 & 0 & 0 \\ 0 & 4 & 2 & -1 \\ 1 & 3 & -2 & 0 \\ 0 & -3 & 1 & 2 \end{vmatrix}$ -125 **62.** $\begin{vmatrix} 1 & 7 & 0 & -1 \\ 1 & 3 & -2 & 0 \\ 1 & 0 & -1 & 2 \\ 0 & -6 & 2 & 4 \end{vmatrix}$ 0

63. $\begin{vmatrix} 4 & 0 & 2 & 5 \\ 0 & 3 & -1 & 1 \\ 0 & 0 & 2 & 0 \\ 0 & 0 & 0 & 1 \end{vmatrix}$ 24 **64.** $\begin{vmatrix} 2 & 0 & -1 & 4 \\ 6 & 0 & 4 & 1 \\ 2 & 4 & 3 & -1 \\ 4 & 0 & 5 & -4 \end{vmatrix}$ 56

CHAPTER 4 PROJECT

Locating Lightning Strikes

Lightning, most often produced during thunderstorms, is a rapid discharge of high-current electricity into the atmosphere. Around the world, lightning occurs at a rate of approximately 100 flashes per second. Because of lightning's potentially destructive nature, meteorologists track lightning activity by recording and plotting the positions of lightning strikes. In this project, you will have the opportunity to pinpoint the location of a lightning strike. This project may be completed by working in groups or individually.

Weather recording stations use a directional antenna to detect and measure the electromagnetic field emitted by a lightning bolt. The antenna can determine the angle between a fixed point and the position of the lightning strike but cannot determine the distance to the lightning strike. However, the angle measured by the antenna can be used to find the slope of the line connecting the positions of the weather station and the lightning strike. From there, the equation of the line connecting these points may be found.

If two such lines may be found—that is, if another weather station's antenna detects the same lightning flash—the coordinates of the lightning strike's position may be pinpointed.

Weather-recording station *A*

Lightning strike

Weather-recording station *B*

1. A weather recording station A is located at the coordinates $(35, 28)$. A second weather recording station B is located at the coordinates $(52, 12)$. Plot the positions of the two weather recording stations.

2. A lightning strike is detected by both stations. Station A uses a measured angle to find the slope of the line from the station to the lightning strike as $m = -1.732$. Station B computes a slope of $m = 0.577$ from the angle it measured. Use this information to find the equations of the lines connecting each station to the position of the lightning strike.

3. Solve the resulting system of equations in each of the following ways (or work with other students in your group so that each student solves the system in one of the following ways).

   (a) Using a graph. Graph the two equations on your plot of the positions of the two weather recording stations. Estimate the coordinates of their point of intersection.

   (b) Using either the method of substitution or of elimination (whichever you prefer)

   (c) Using matrices

   (d) Using Cramer's rule

   (e) (Optional) Using a graphing calculator to graph the lines and an intersect feature to estimate the coordinates of their point of intersection

4. Compare the results from each method. What are the coordinates of the lightning strike? Which method do you prefer? Why?

CHAPTER VOCABULARY CHECK

Fill in each blank with one of the words or phrases listed below.

matrix	determinant	consistent	system of equations
solution	inconsistent	square	

1. Two or more linear equations in two variables form a system of equations.

2. A solution of a system of two equations in two variables is an ordered pair that makes both equations true.

3. A(n) consistent system of equations has at least one solution.

4. If a matrix has the same number of rows and columns, it is called a square matrix.

5. A real number associated with a square matrix is called its determinant.

6. A(n) inconsistent system of equations has no solution.

7. A matrix is a rectangular array of numbers.

STUDY SKILLS REMINDER

What Should You Do The Day of an Exam?

On the day of an exam, try the following:

- ► Allow yourself plenty of time to arrive.
- ► Read the directions on the test carefully.
- ► Read each problem carefully as you take your test. Make sure that you answer the question asked.
- ► Watch your time and pace yourself so that you may attempt each problem on your test.
- ► If you have time, check your work and answers.
- ► Do not turn your test in early. If you have extra time, spend it double-checking your work.

Good luck!

CHAPTER 4 HIGHLIGHTS

Definitions and Concepts	**Examples**

Section 4.1 Solving Systems of Linear Equations in Two Variables

A **system of linear equations** consists of two or more linear equations.

A **solution** of a system of two equations in two variables is an ordered pair (x, y) that makes both equations true.

Geometrically, a solution of a system in two variables is a point common to the graphs of the equations.

A system of equations with at least one solution is a **consistent system**. A system that has no solution is an **inconsistent system**.

If the graphs of two linear equations are identical, the equations are **dependent**.

If their graphs are different, the equations are **independent**.

One solution: consistent and independent

No solution: inconsistent and independent

Infinite number of solutions; consistent and dependent

Solve by graphing $\begin{cases} y = 2x - 1 \\ x + 2y = 13 \end{cases}$

or

the solution is $(3, 5)$.

To solve a system of linear equations by the **substitution method**:

Step 1: Solve one equation for a variable.

Step 2: Substitute the expression for the variable into the other equation.

Step 3: Solve the equation from *Step 2* to find the value of one variable.

Step 4: Substitute the value from *Step 3* in either original equation to find the value of the other variable.

Step 5: Check the solution in both equations.

Solve by substitution:

$$\begin{cases} y = x + 2 \\ 3x - 2y = -5 \end{cases}$$

Substitute $x + 2$ for y in the second equation.

$$3x - 2y = -5$$
$$3x - 2(x + 2) = -5$$

$$3x - 2x - 4 = -5$$
$$x - 4 = -5 \qquad \text{Simplify.}$$
$$x = -1 \qquad \text{Add 4.}$$

To find y, let $x = -1$ in $y = x + 2$, so $y = -1 + 2 = 1$. The solution $(-1, 1)$ checks.

To solve a system of linear equations by the **elimination method:**

Step 1: Rewrite each equation in standard form $Ax + By = C$.

Step 2: Multiply one or both equations by a nonzero number so that the coefficients of a variable are opposites.

Step 3: Add the equations.

Step 4: Find the value of one variable by solving the resulting equation.

Solve by elimination:

$$\begin{cases} x - 3y = -3 \\ -2x + y = 6 \end{cases}$$

Multiply both sides of the first equation by 2.

$$\begin{array}{r} 2x - 6y = -6 \\ -2x + y = 6 \\ \hline -5y = 0 \\ y = 0 \end{array}$$

Add.

Divide by -5.

(continued)

Definitions and Concepts	Examples

Section 4.1 Solving Systems of Linear Equations in Two Variables

Step 5: Substitute the value from *Step 4* into either original equation to find the value of the other variable.

Step 6: Check the solution in both equations.

To find x, let $y = 0$ in an original equation.
$$x - 3y = -3$$
$$x - 3 \cdot 0 = -3$$
$$x = -3$$

The solution $(-3, 0)$ checks.

Section 4.2 Solving Systems of Linear Equations in Three Variables

A **solution** of an equation in three variables x, y, and z is an **ordered triple** (x, y, z) that makes the equation a true statement.

Verify that $(-2, 1, 3)$ is a solution of
$2x + 3y - 2z = -7$.

Replace x with -2, y with 1, and z with 3.
$$2(-2) + 3(1) - 2(3) = -7$$
$$-4 + 3 - 6 = -7$$
$$-7 = -7 \qquad \text{True}$$

$(-2, 1, 3)$ is a solution.

To solve a system of three linear equations by the elimination method:

Step 1: Write each equation in standard form, $Ax + By + Cz = D$.

Step 2: Choose a pair of equations and use the equations to eliminate a variable.

Step 3: Choose any other pair of equations and eliminate the same variable.

Step 4: Solve the system of two equations in two variables from *Steps 1* and *2*.

Step 5: Solve for the third variable by substituting the values of the variables from *Step 4* into any of the original equations.

Solve

$$\begin{cases} 2x + y - z = 0 \ (1) \\ x - y - 2z = -6 \ (2) \\ -3x - 2y + 3z = -22 \ (3) \end{cases}$$

1. Each equation is written in standard form.

2.
$$\begin{array}{rl} 2x + y - z = & 0 \ (1) \\ x - y - 2z = & -6 \ (2) \qquad \text{Add.} \\ \hline 3x \quad\quad - 3z = & -6 \ (4) \end{array}$$

3. Eliminate y from equations (1) and (3) also.

$$\begin{array}{rl} 4x + 2y - 2z = & 0 \\ -3x - 2y + 3z = & -22 \ (3) \qquad \text{Multiply equation} \\ \hline x \quad\quad + z = & -22 \ (5) \qquad \text{(1) by 2.} \end{array}$$

Add.

4. Solve

$$\begin{cases} 3x - 3z = -6 \ (4) \\ x + z = -22 \ (5) \end{cases}$$

$$\begin{array}{rl} x - z = & -2 \\ x + z = & -22 \ (5) \\ \hline 2x \quad = & -24 \qquad \text{Divide equation} \\ x \quad = & -12 \qquad \text{(4) by 3.} \end{array}$$

To find z, use equation (5).

$$x + z = -22$$
$$-12 + z = -22$$
$$z = -10$$

(continued)

Definitions and Concepts	**Examples**

A numerical check of the solution.

5. To find y, use equation (1).

$$2x + y - z = 0$$
$$2(-12) + y - (-10) = 0$$
$$-24 + y + 10 = 0$$
$$y = 14$$

The solution is $(-12, 14, -10)$.

Section 4.3 Systems of Linear Equations and Problem Solving

1. UNDERSTAND the problem.

2. TRANSLATE.

3. SOLVE.

4. INTERPRET.

Two numbers have a sum of 11. Twice one number is 3 less than 3 times the other. Find the numbers.

1. Read and reread.
$$x = \text{one number}$$
$$y = \text{other number}$$

2. In words:

sum of numbers	is	11
↓	↓	↓

Translate: $x + y$ $=$ 11

twice one number	is	3 less than 3 times the other number
↓	↓	↓

Translate: $2x$ $=$ $3y - 3$

3. Solve the system: $\begin{cases} x + y = 11 \\ 2x = 3y - 3 \end{cases}$

In the first equation $x = 11 - y$. Substitute into the other equation.

$$2x = 3y - 3$$
$$2(11 - y) = 3y - 3$$
$$22 - 2y = 3y - 3$$
$$-5y = -25$$
$$y = 5$$

Replace y with 5 in the equation $x = 11 - y$. Then $x = 11 - 5 = 6$. The solution is $(6, 5)$.

4. *Check:* See that $6 + 5 = 11$ is the required sum and that twice 6 is 3 times 5 less 3. *State:* The numbers are 6 and 5.

(continued)

Definitions and Concepts	**Examples**

Section 4.4 Solving Systems of Equations by Matrices

A **matrix** is a rectangular array of numbers.

$$\begin{bmatrix} -7 & 0 & 3 \\ 1 & 2 & 4 \end{bmatrix} \quad \begin{bmatrix} a & b & c \\ d & e & f \\ g & h & i \end{bmatrix}$$

The **corresponding matrix of the system** is obtained by writing a matrix composed of the coefficients of the variables and the constants of the system.

The corresponding matrix of the system

$$\begin{cases} x - y = 1 \\ 2x + y = 11 \end{cases} \quad \text{is} \quad \begin{bmatrix} 1 & -1 & | & 1 \\ 2 & 1 & | & 11 \end{bmatrix}$$

The following **row operations** can be performed on matrices, and the result is an equivalent matrix.

Use matrices to solve: $\begin{cases} x - y = 1 \\ 2x + y = 11 \end{cases}$

Elementary row operations

1. Interchange any two rows.
2. Multiply (or divide) the elements of one row by the same nonzero number.
3. Multiply (or divide) the elements of one row by the same nonzero number and add to its corresponding elements in any other row.

The corresponding matrix is

$$\begin{bmatrix} 1 & -1 & | & 1 \\ 2 & 1 & | & 11 \end{bmatrix}$$

Use row operations to write an equivalent matrix with 1's along the diagonal and 0's below each 1 in the diagonal. Multiply row 1 by -2 and add to row 2. Change row 2 only.

$$\begin{bmatrix} 1 & -1 & | & 1 \\ -2(1) + 2 & -2(-1) + 1 & | & -2(1) + 11 \end{bmatrix}$$

simplifies to $\begin{bmatrix} 1 & -1 & | & 1 \\ 0 & 3 & | & 9 \end{bmatrix}$

Divide row 2 by 3.

$$\begin{bmatrix} 1 & -1 & | & 1 \\ \dfrac{0}{3} & \dfrac{3}{3} & | & \dfrac{9}{3} \end{bmatrix} \quad \text{simplifies to} \quad \begin{bmatrix} 1 & -1 & | & 1 \\ 0 & 1 & | & 3 \end{bmatrix}$$

This matrix corresponds to the system

$$\begin{cases} x - y = 1 \\ y = 3 \end{cases}$$

Let $y = 3$ in the first equation.

$$x - 3 = 1$$
$$x = 4$$

The ordered pair solution is $(4, 3)$.

Alternatively, you can use a calculator to enter and write the matrix in reduced row echelon form.

Enter matrix values.

Under matrix math select reduced row echelon form (rref).

(continued)

Definitions and Concepts	Examples

Section 4.4 Solving Systems of Equations by Matrices

Thus, we have

Reduced row echelon form   **Equivalent System of Equations**

$$\begin{bmatrix} 1 & 0 & \vdots & 4 \\ 0 & 1 & \vdots & 3 \end{bmatrix} \qquad \begin{matrix} 1x + 0y = 4 \\ 0x + 1y = 3 \end{matrix} \text{ or } \begin{matrix} x = 4 \\ y = 3. \end{matrix}$$

The ordered pair solution is $(4, 3)$.

Section 4.5 Solving Systems of Equations by Determinants

A **square matrix** is a matrix with the same number of rows and columns.

$$\begin{bmatrix} -2 & 1 \\ 6 & 8 \end{bmatrix} \qquad \begin{bmatrix} 4 & -1 & 6 \\ 0 & 2 & 5 \\ 1 & 1 & 2 \end{bmatrix}$$

A **determinant** is a real number associated with a square matrix. To denote the determinant, place vertical bars about the array of numbers.

The determinant of $\begin{bmatrix} -2 & 1 \\ 6 & 8 \end{bmatrix}$ is $\begin{vmatrix} -2 & 1 \\ 6 & 8 \end{vmatrix}$.

The determinant of a 2×2 matrix is

$$\begin{vmatrix} a & b \\ c & d \end{vmatrix} = ad - bc$$

$$\begin{vmatrix} -2 & 1 \\ 6 & 8 \end{vmatrix} = -2 \cdot 8 - 1 \cdot 6 = -22$$

Cramer's Rule for Two Linear Equations in Two Variables

The solution of the system $\begin{cases} ax + by = h \\ cx + dy = k \end{cases}$ is given by

$$x = \frac{\begin{vmatrix} h & b \\ k & d \end{vmatrix}}{\begin{vmatrix} a & b \\ c & d \end{vmatrix}} = \frac{D_x}{D} \qquad y = \frac{\begin{vmatrix} a & h \\ c & k \end{vmatrix}}{\begin{vmatrix} a & b \\ c & d \end{vmatrix}} = \frac{D_y}{D}$$

as long as $D = ad - bc$ is not 0.

Use Cramer's rule to solve

$$\begin{cases} 3x + 2y = 8 \\ 2x - y = -11 \end{cases}$$

$$D = \begin{vmatrix} 3 & 2 \\ 2 & -1 \end{vmatrix} = 3(-1) - 2(2) = -7$$

$$D_x = \begin{vmatrix} 8 & 2 \\ -11 & -1 \end{vmatrix} = 8(-1) - 2(-11) = 14$$

$$D_y = \begin{vmatrix} 3 & 8 \\ 2 & -11 \end{vmatrix} = 3(-11) - 8(2) = -49$$

$$x = \frac{D_x}{D} = \frac{14}{-7} = -2 \qquad y = \frac{D_y}{D} = \frac{-49}{-7} = 7$$

The ordered pair solution is $(-2, 7)$.

Determinant of a 3×3 Matrix

$$\begin{vmatrix} a_1 & b_1 & c_1 \\ a_2 & b_2 & c_2 \\ a_3 & b_3 & c_3 \end{vmatrix} = a_1 \cdot \begin{vmatrix} b_2 & c_2 \\ b_3 & c_3 \end{vmatrix} - a_2 \cdot \begin{vmatrix} b_1 & c_1 \\ b_3 & c_3 \end{vmatrix} + a_3 \cdot \begin{vmatrix} b_1 & c_1 \\ b_2 & c_2 \end{vmatrix}$$

Each 2×2 matrix above is called a **minor**.

$$\begin{vmatrix} 0 & 2 & -1 \\ 5 & 3 & 0 \\ 2 & -2 & 4 \end{vmatrix} = 0 \begin{vmatrix} 3 & 0 \\ -2 & 4 \end{vmatrix} - 2 \begin{vmatrix} 5 & 0 \\ 2 & 4 \end{vmatrix} + (-1) \begin{vmatrix} 5 & 3 \\ 2 & -2 \end{vmatrix}$$

$$= 0(12 - 0) - 2(20 - 0) - 1(-10 - 6)$$

$$= 0 - 40 + 16 = -24$$

(continued)

Definitions and Concepts	Examples

Section 4.5 Solving Systems of Equations by Determinants

Cramer's Rule for Three Equations in Three Variables

The solution of the system $\begin{cases} a_1x + b_1y + c_1z = k_1 \\ a_2x + b_2y + c_2z = k_2 \\ a_3x + b_3y + c_3z = k_3 \end{cases}$

is given by

$$x = \frac{D_x}{D}, \quad y = \frac{D_y}{D}, \quad \text{and} \quad z = \frac{D_z}{D}$$

where

$$D = \begin{vmatrix} a_1 & b_1 & c_1 \\ a_2 & b_2 & c_2 \\ a_3 & b_3 & c_3 \end{vmatrix} \quad D_x = \begin{vmatrix} k_1 & b_1 & c_1 \\ k_2 & b_2 & c_2 \\ k_3 & b_3 & c_3 \end{vmatrix}$$

$$D_y = \begin{vmatrix} a_1 & k_1 & c_1 \\ a_2 & k_2 & c_2 \\ a_3 & k_3 & c_3 \end{vmatrix} \quad D_z = \begin{vmatrix} a_1 & b_1 & k_1 \\ a_2 & b_2 & k_2 \\ a_3 & b_3 & k_3 \end{vmatrix}$$

as long as D is not 0.

Use Cramer's rule to solve

$$\begin{cases} 3y + 2z = 8 \\ x + y + z = 3 \\ 2x - y + z = 2 \end{cases}$$

$$D = \begin{vmatrix} 0 & 3 & 2 \\ 1 & 1 & 1 \\ 2 & -1 & 1 \end{vmatrix} = -3$$

$$D_x = \begin{vmatrix} 8 & 3 & 2 \\ 3 & 1 & 1 \\ 2 & -1 & 1 \end{vmatrix} = 3$$

$$D_y = \begin{vmatrix} 0 & 8 & 2 \\ 1 & 3 & 1 \\ 2 & 2 & 1 \end{vmatrix} = 0$$

$$D_z = \begin{vmatrix} 0 & 3 & 8 \\ 1 & 1 & 3 \\ 2 & -1 & 2 \end{vmatrix} = -12$$

$$x = \frac{D_x}{D} = \frac{3}{-3} = -1 \quad y = \frac{D_y}{D} = \frac{0}{-3} = 0$$

$$z = \frac{D_z}{D} = \frac{-12}{-3} = 4$$

The ordered triple solution is $(-1, 0, 4)$.

CHAPTER REVIEW

1.–5. See graphing answer section.

(4.1) *Solve each system of equations in two variables by each of three methods: (1) graphing, (2) substitution, and (3) elimination.*

1. $\begin{cases} 3x + 10y = 1 \\ x + 2y = -1 \end{cases}$ $(-3, 1)$

2. $\begin{cases} y = \frac{1}{2}x + \frac{2}{3} \\ 4x + 6y = 4 \end{cases}$ $\left(0, \frac{2}{3}\right)$

3. $\begin{cases} 2x - 4y = 22 \\ 5x - 10y = 16 \end{cases}$ $\varnothing$

4. $\begin{cases} 3x - 6y = 12 \\ 2y = x - 4 \end{cases}$

5. $\begin{cases} \frac{1}{2}x - \frac{3}{4}y = -\frac{1}{2} \\ \frac{1}{8}x + \frac{3}{4}y = \frac{19}{8} \end{cases}$ $\left(3, \frac{8}{3}\right)$

4. $\{(x, y) \mid 3x - 6y = 12\}$

6. The revenue equation for a certain style of backpack is

$$y = 32x$$

where x is the number of backpacks sold and y is the income in dollars for selling x backpacks. The cost equation for these units is   1500 backpacks   **8.** $(2, 0, -3)$ **10.** $(-1, 2, 0)$

$$y = 15x + 25,500$$

where x is the number of backpacks manufactured and y is the cost in dollars for manufacturing x backpacks. Find the number of units to be sold for the company to break even.

(4.2) *Solve each system of equations in three variables.*

7. $\begin{cases} x + z = 4 \\ 2x - y = 4 \\ x + y - z = 0 \end{cases}$ $(2, 0, 2)$

8. $\begin{cases} 2x + 5y = 4 \\ x - 5y + z = -1 \\ 4x - z = 11 \end{cases}$

9. $\begin{cases} 4y + 2z = 5 \\ 2x + 8y = 5 \\ 6x + 4z = 1 \end{cases}$ $\left(-\frac{1}{2}, \frac{3}{4}, 1\right)$

10. $\begin{cases} 5x + 7y = 9 \\ 14y - z = 28 \\ 4x + 2z = -4 \end{cases}$

11. $\begin{cases} 3x - 2y + 2z = 5 \\ -x + 6y + z = 4 \\ 3x + 14y + 7z = 20 \end{cases}$ $\emptyset$ **12.** $\begin{cases} x + 2y + 3z = 11 \\ y + 2z = 3 \\ 2x + 2z = 10 \end{cases}$

13. $\begin{cases} 7x - 3y + 2z = 0 \\ 4x - 4y - z = 2 \\ 5x + 2y + 3z = 1 \end{cases}$ **14.** $\begin{cases} x - 3y - 5z = -5 \\ 4x - 2y + 3z = 13 \\ 5x + 3y + 4z = 22 \end{cases}$

(4.3) *Use systems of equations to solve the following applications.*

15. The sum of three numbers is 98. The sum of the first and second is two more than the third number, and the second is four times the first. Find the numbers.   10, 40, and 48

16. One number is 3 times a second number, and twice the sum of the numbers is 168. Find the numbers.   63 and 21

17. Two cars leave Chicago, one traveling east and the other west. After 4 hours they are 492 miles apart. If one car is traveling 7 mph faster than the other, find the speed of each.

△ **18.** The foundation for a rectangular Hardware Warehouse has a length three times the width and is 296 feet around. Find the dimensions of the building.   width, 37 ft; length, 111 ft

19. James Callahan has available a 10% alcohol solution and a 60% alcohol solution. Find how many liters of each solution he should mix to make 50 liters of a 40% alcohol solution.

20. An employee at a See's Candy Store needs a special mixture of candy. She has creme-filled chocolates that sell for $3.00 per pound, chocolate-covered nuts that sell for $2.70 per pound, and chocolate-covered raisins that sell for $2.25 per pound. She wants to have twice as many raisins as nuts in the mixture. Find how many pounds of each she should use to make 45 pounds worth $2.80 per pound.   **48.** $\{(x, y) \mid x - 2y = 4\}$

21. Chris Kringler has $2.77 in his coin jar—all in pennies, nickels, and dimes. If he has 53 coins in all and four more nickels than dimes, find how many of each type of coin he has.

22. If $10,000 and $4000 are invested such that $1250 is earned in one year, and if the rate of interest on the larger investment is 2% more than that of the smaller investment, find the rates of interest.   larger investment, 9.5%: smaller investment, 7.5%

△ **23.** The perimeter of an isosceles (two sides equal) triangle is 73 centimeters. If two sides are of equal length and the third side is 7 centimeters longer than the others, find the lengths of the three sides.   Two sides, 22 cm each: third side, 29 cm

24. The sum of three numbers is 295. One number is five more than a second and twice the third. Find the numbers.

(4.4) *Use matrices to solve each system.*

25. $\begin{cases} 3x + 10y = 1 \\ x + 2y = -1 \end{cases}$ $(-3, 1)$ **26.** $\begin{cases} 3x - 6y = 12 \\ 2y = x - 4 \end{cases}$

27. $\begin{cases} 3x - 2y = -8 \\ 6x + 5y = 11 \end{cases}$ $\left(-\frac{2}{3}, 3\right)$ **28.** $\begin{cases} 6x - 6y = -5 \\ 10x - 2y = 1 \end{cases}$ $\left(\frac{1}{3}, \frac{7}{6}\right)$

29. $\begin{cases} 3x - 6y = 0 \\ 2x + 4y = 5 \end{cases}$ $\left(\frac{5}{4}, \frac{5}{8}\right)$ **30.** $\begin{cases} 5x - 3y = 10 \\ -2x + y = -1 \end{cases}$

31. $\begin{cases} 0.2x - 0.3y = -0.7 \\ 0.5x + 0.3y = 1.4 \end{cases}$ $(1, 3)$ **32.** $\begin{cases} 3x + 2y = 8 \\ 3x - y = 5 \end{cases}$ $(2, 1)$

33. $\begin{cases} x + z = 4 \\ 2x - y = 0 \\ x + y - z = 0 \end{cases}$ $(1, 2, 3)$ **34.** $\begin{cases} 2x + 5y = 4 \\ x - 5y + z = -1 \\ 4x - z = 11 \end{cases}$

35. $\begin{cases} 3x - y = 11 \\ x + 2z = 13 \\ y - z = -7 \end{cases}$ **36.** $\begin{cases} 5x + 7y + 3z = 9 \\ 14y - z = 28 \\ 4x + 2z = -4 \end{cases}$

37. $\begin{cases} 7x - 3y + 2z = 0 \\ 4x - 4y - z = 2 \\ 5x + 2y + 3z = 1 \end{cases}$ **38.** $\begin{cases} x + 2y + 3z = 14 \\ y + 2z = 3 \\ 2x - 2z = 10 \end{cases}$ $\emptyset$

(4.5) *Evaluate.*   **34.** $(2, 0, -3)$ **35.** $(3, -2, 5)$

39. $\begin{vmatrix} -1 & 3 \\ 5 & 2 \end{vmatrix}$ -17 **40.** $\begin{vmatrix} 3 & -1 \\ 2 & 5 \end{vmatrix}$ 17

41. $\begin{vmatrix} 2 & -1 & -3 \\ 1 & 2 & 0 \\ 3 & -2 & 2 \end{vmatrix}$ 34 **42.** $\begin{vmatrix} -2 & 3 & 1 \\ 4 & 4 & 0 \\ 1 & -2 & 3 \end{vmatrix}$ -72

Use Cramer's rule, if possible, to solve each system of equations.

43. $\begin{cases} 3x - 2y = -8 \\ 6x + 5y = 11 \end{cases}$ $\left(-\frac{2}{3}, 3\right)$ **44.** $\begin{cases} 6x - 6y = -5 \\ 10x - 2y = 1 \end{cases}$

45. $\begin{cases} 3x + 10y = 1 \\ x + 2y = -1 \end{cases}$ $(-3, 1)$ **46.** $\begin{cases} y = \frac{1}{2}x + \frac{2}{3} \\ 4x + 6y = 4 \end{cases}$

47. $\begin{cases} 2x - 4y = 22 \\ 5x - 10y = 16 \end{cases}$ $\emptyset$ **48.** $\begin{cases} 3x - 6y = 12 \\ 2y = x - 4 \end{cases}$

49. $\begin{cases} x + z = 4 \\ 2x - y = 0 \\ x + y - z = 0 \end{cases}$ $(1, 2, 3)$ **50.** $\begin{cases} 2x + 5y = 4 \\ x - 5y + z = -1 \\ 4x - z = 11 \end{cases}$

51. $\begin{cases} x + 3y - z = 5 \\ 2x - y - 2z = 3 \\ x + 2y + 3z = 4 \end{cases}$ $(2, 1, 0)$ **52.** $\begin{cases} 2x - z = 1 \\ 3x - y + 2z = 3 \\ x + y + 3z = -2 \end{cases}$

53. $\begin{cases} x + 2y + 3z = 14 \\ y + 2z = 3 \\ 2x - 2z = 10 \end{cases}$ $\emptyset$ **54.** $\begin{cases} 5x + 7y = 9 \\ 14y - z = 28 \\ 4x + 2z = -4 \end{cases}$

12. $(5, 3, 0)$ **13.** $(1, 1, -2)$ **14.** $(3, 1, 1)$ **17.** 58 mph, 65 mph   **19.** 20 liters of 10% solution; 30 liters of 60% solution

20. 30 lb of creme-filled: 5 lb of chocolate-covered nuts: 10 lb of chocolate-covered raisins   **21.** 17 pennies, 20 nickels, and 16 dimes

24. 120, 115, and 60   **26.** $\{(x, y) \mid x - 2y = 4\}$ **30.** $(-7, -15)$ **36.** $(-1, 2, 0)$ **37.** $(1, 1, -2)$ **44.** $\left(\frac{1}{3}, \frac{7}{6}\right)$ **46.** $\left(0, \frac{2}{3}\right)$

50. $(2, 0, -3)$ **52.** $\left(\frac{3}{7}, -2, -\frac{1}{7}\right)$ **54.** $(-1, 2, 0)$

CHAPTER 4 TEST

Remember to use your Chapter Test Prep Video CD to help you study and view solutions to the test questions you need help with.

Evaluate each determinant. **3.–4.** See graphing answer section.

1. $\begin{vmatrix} 4 & -7 \\ 2 & 5 \end{vmatrix}$ 34

2. $\begin{vmatrix} 4 & 0 & 2 \\ 1 & -3 & 5 \\ 0 & -1 & 2 \end{vmatrix}$ −6

Solve each system of equations graphically and then solve by the elimination method or the substitution method.

3. $\begin{cases} 2x - y = -1 \\ 5x + 4y = 17 \end{cases}$

4. $\begin{cases} 7x - 14y = 5 \\ x = 2y \end{cases}$ ∅

6. $\{(x, y) \mid 10x + 4y = 10\}$ **7.** $(-1, -2, 4)$ **8.** ∅

Solve each system. **3.** $(1, 3)$ see graphing answer section.

5. $\begin{cases} 4x - 7y = 29 \\ 2x + 5y = -11 \end{cases}$ $(2, -3)$

6. $\begin{cases} 15x + 6y = 15 \\ 10x + 4y = 10 \end{cases}$

7. $\begin{cases} 2x - 3y = 4 \\ 3y + 2z = 2 \\ x - z = -5 \end{cases}$

8. $\begin{cases} 3x - 2y - z = -1 \\ 2x - 2y = 4 \\ 2x - 2z = -12 \end{cases}$

9. $\begin{cases} \dfrac{x}{2} + \dfrac{y}{4} = -\dfrac{3}{4} \\ x + \dfrac{3}{4}y = -4 \end{cases}$ $\left(\dfrac{7}{2}, -10\right)$

11. $(3, -1, 2)$

12. $\{(x, y) \mid x - y = -2\}$

13. $(5, -3)$

Use Cramer's rule, if possible, to solve each system.

10. $\begin{cases} 3x - y = 7 \\ 2x + 5y = -1 \end{cases}$ $(2, -1)$

11. $\begin{cases} x + y + z = 4 \\ 2x + 5y = 1 \\ x - y - 2z = 0 \end{cases}$

Use matrices to solve each system. For Exercise 13, use matrices and a graphing utility.

12. $\begin{cases} x - y = -2 \\ 3x - 3y = -6 \end{cases}$

13. $\begin{cases} x + 2y = -1 \\ 2x + 5y = -5 \end{cases}$

14. $\begin{cases} x - y - z = 0 \\ 3x - y - 5z = -2 \\ 2x + 3y = -5 \end{cases}$ $(-1, -1, 0)$

15. A motel in New Orleans charges $90 per day for double occupancy and $80 per day for single occupancy. If 80 rooms are occupied for a total of $6930, how many rooms of each kind are there?   53 double rooms; 27 single rooms

16. The research department of a company that manufactures children's fruit drinks is experimenting with a new flavor. A 17.5% fructose solution is needed, but only 10% and 20% solutions are available. How many gallons of a 10% fructose solution should be mixed with a 20% fructose solution in order to obtain 20 gallons of a 17.5% fructose solution?   5 gal of 10%, 15 gal of 20%

17. A company that manufactures boxes recently purchased $2000 worth of new equipment to offer gift boxes to its customers. The cost of producing a package of gift boxes is $1.50 and it is sold for $4.00. Find the number of packages that must be sold for the company to break even.   800 packages

18. The measure of the largest angle of a triangle is three less than 5 times the measure of the smallest angle. The measure of the remaining angle is 1 less than twice the measure of the smallest angle. Find the measure of each angle.   23°, 45°, 112°

CHAPTER CUMULATIVE REVIEW

1. Determine whether each statement is true or false.

  a. $3 \in \{x \mid x \text{ is a natural number}\}$ true

  b. $7 \notin \{1, 2, 3\}$ true (Sec. 1.2, Ex. 4)

2. Determine whether each statement is true or false.

  a. $\{0, 7\} \subseteq \{0, 2, 4, 6, 8\}$ false

  b. $\{1, 3, 5\} \subseteq \{1, 3, 5, 7\}$ true; (Sec. 1.2)

3. Simplify the following expressions.

  a. $11 + 2 - 7$ 6   **b.** $-5 - 4 + 2$ −7 (Sec. 1.3, Ex. 3)

4. Subtract.

  a. $-7 - (-2)$ −5   **b.** $14 - 38$ −24; (Sec. 1.3)

5. Write the additive inverse, or opposite, of each.

  a. 8 −8   **b.** $\dfrac{1}{5}$ $-\dfrac{1}{5}$ **c.** -9.6 9.6 (Sec. 1.4, Ex. 4)

6. Write the reciprocal of each.

  a. 5 $\dfrac{1}{5}$ **b.** $-\dfrac{2}{3}$ $-\dfrac{3}{2}$; (Sec. 1.4)

7. Use the distributive property to multiply.

  a. $3(2x + y)$ $6x + 3y$ **b.** $-(3x - 1)$ **c.** $0.7a(b - 2)$

8. Multiply. **8. a.** $21x - 14y + 28$ **b.** $2s + 3t$ (Sec. 1.4)

  a. $7(3x - 2y + 4)$ **b.** $-(-2s - 3t)$

9. Use the distributive property to simplify each expression.

  a. $3x - 5x + 4$ **b.** $7yz + yz$ $8yz$ **c.** $4z + 6.1$

10. Simplify.   **7. b.** $-3x + 1$ **c.** $0.7ab - 1.4a$ (Sec. 1.4, Ex. 8)

  a. $5y^2 - 1 + 2(y^2 + 2)$ $7y^2 + 3$

  b. $(7.8x - 1.2) - (5.6x - 2.4)$ $2.2x + 1.2$ (Sec. 1.4)

Solve.   **9. a.** $-2x + 4$ **9. c.** $4z + 6.1$ (Sec. 1.4, Ex. 12)

11. $-6x - 1 + 5x = 3$ −4 (Sec. 1.5, Ex. 3)

12. $8y - 14 = 6y - 14$ 0, (Sec. 1.5)

13. $0.3x + 0.1 = 0.27x - 0.02$ −4 (Sec. 1.5, Ex. 7)

14. $2(m - 6) - m = 4(m - 3) - 3m$ all real numbers, (Sec. 1.5)

15. A pennant in the shape of an isosceles triangle is to be constructed for the Slidell High School Athletic Club and sold at a fund-raiser. The company manufacturing the pennant charges according to perimeter, and the athletic club has determined that a perimeter of 149 centimeters should make a nice profit. If each equal side of the triangle is twice the length of the third side, increased by 12 centimeters, find the lengths of the sides of the triangular pennant.   25 cm, 62 cm, 62 cm (Sec. 1.6, Ex. 6)

16. A quadrilateral has 4 angles whose sum is 360°. In a particular quadrilateral, two angles have the same measure. A third angle is 10° more than the measure of one of the equal angles, and the fourth angle is half the measure of one of the equal angles. Find the measures of the angles.   100°, 100°, 110°, 50° (Sec. 1.6)

17. Solve: $3x + 4 \geq 2x - 6$. Graph the solution set.

18. Solve: $5(2x - 1) > -5$ $(0, \infty)$ (Sec. 3.2)

19. Solve: $2 < 4 - x < 7$ $(-3, 2)$ (Sec. 3.3, Ex. 4)

20. Solve: $-1 < \dfrac{-2x - 1}{3} < 1$ $(-2, 1)$ (Sec. 3.3)

21. Solve: $|2x| + 5 = 7$ $-1, 1$ (Sec. 3.4, Ex. 4)

22. Solve: $|x - 5| = 4$ $1, 9$ (Sec. 3.4)

23. Solve for m: $|m - 6| < 2$ $(4, 8)$ (Sec. 3.5, Ex. 2)

24. $|2x + 1| > 5$ $(-\infty, -3) \cup (2, \infty)$ (Sec. 3.5)

25. Plot each ordered pair on a Cartesian coordinate system and name the quadrant in which the point is located.
a. $(2, -1)$ **b.** $(0, 5)$ **c.** $(-3, 5)$
d. $(-2, 0)$ **e.** $\left(-\dfrac{1}{2}, -4\right)$ **f.** $(1.5, 1.5)$

26. Name the quadrant or axis each point is located.
a. $(-1, -5)$ Q III   **b.** $(4, -2)$ Q IV   **c.** $(0, 2)$

27. Is the relation $y = 2x + 1$ also a function?   yes (Sec. 2.2, Ex. 3)

28. Graph: $-2x + \dfrac{1}{2}y = -2$ **26. c.** y-axis (Sec. 2.1)

29. If $f(x) = 7x^2 - 3x + 1, g(x) = 3x - 2$, and $h(x) = x^2$, find the following.
a. $f(1)$ 5   **b.** $g(3)$ 7
c. $h(-2)$ 4 (Sec. 2.2, Ex. 7)

30. If $f(x) = 3x^2$, find the following.
a. $f(5)$ 75   **b.** $f(-2)$ 12 (Sec. 2.2)

31. Find the y-intercept of the graph of each equation.
a. $f(x) + \dfrac{1}{2}x + \dfrac{3}{7}$ $\left(0, \dfrac{3}{7}\right)$ **b.** $y = -2.5x - 3.2$

32. Find the slope of the line containing $(-2, 6)$ and $(0, 9)$.

33. Find the slope and the y-intercept of the line $3x - 4y = 4$.

34. Find the slope and y-intercept of the line defined by $y = 2$.

35. Are the following pairs of lines parallel, perpendicular, or neither?   **35. b.** neither (Sec. 2.4, Ex. 7)
a. $3x + 7y = 4$ $6x + 14y = 7$ parallel
b. $-x + 3y = 2$ $2x + 6y = 5$

36. Find an equation of the line through $(0, -9)$ with slope $\dfrac{1}{5}$.

37. Find an equation of the line through points $(4, 0)$ and $(-4, -5)$. Write the equation using function notation.

38. Find an equation of the line through $(-2, 6)$ perpendicular to $f(x) = \dfrac{1}{2}x - \dfrac{1}{3}$. $y = -2x + 2$ (Sec. 2.5)

39. Graph $3x \geq y$.   See graphing answer section; (Sec. 3.6, Ex. 2)

40. Graph: $x \geq 1$.

41. Determine whether the given ordered pair is a solution of the system.   **34.** $m = 0$, y-intercept: $(0, 2)$ (Sec. 2.4)
a. $\begin{cases} -x + y = 2 \\ 2x - y = -3 \end{cases}$ $(-1, 1)$ yes
b. $\begin{cases} 5x + 3y = -1 \\ x - y = 1 \end{cases}$ $(-2, 3)$ no (Sec. 4.1, Ex. 1)

42. Solve the system: **37.** $f(x) = \dfrac{5}{8}x - \dfrac{5}{2}$ (Sec. 2.5, Ex. 5)
$\begin{cases} 5x + y = -2 \\ 4x - 2y = -10 \end{cases}$ $(-1, 3)$, (Sec. 4.2)

43. Solve the system.
$\begin{cases} 3x - y + z = -15 \\ x + 2y - z = 1 \\ 2x + 3y - 2z = 0 \end{cases}$ $(-4, 2, -1)$ (Sec. 4.2, Ex. 1)

44. Solve the system:
$\begin{cases} x - 2y + z = 0 \\ 3x - y - 2z = -15 \\ 2x - 3y + 3z = 7 \end{cases}$ $(-1, 2, 5)$ (Sec. 4.3)

45. Use matrices to solve the system.
$\begin{cases} x + 3y = 5 \\ 2x - y = -4 \end{cases}$ $(-1, 2)$ (Sec. 4.4, Ex. 1)

46. Solve the system:
$\begin{cases} -6x + 8y = 0 \\ 9x - 12y = 2 \end{cases}$ $\varnothing$, (Sec. 4.2)

17. $\{x \mid x \geq -10\};$ ⟵———⟶ (Sec. 3.2, Ex. 3) -10

25. See graphing answer section; (Sec. 2.1, Ex. 1)

33. slope: $\dfrac{3}{4}$; y-intercept: $(0, -1)$ (Sec. 2.4, Ex. 3)

28. See graphing answer section.

31. b. $(0, -3.2)$ (Sec. 2.3, Ex. 4)   **32.** $m = \dfrac{3}{2}$ (Sec. 2.4)   **36.** $y = \dfrac{1}{5}x - 9$ (Sec. 2.5)   **40.** See graphing answer section.

Exponents, Polynomials, and Polynomial Functions

Linear equations are important for solving problems. They are not sufficient, however, to solve all problems. Many real-world phenomena are modeled by polynomials. We begin this chapter by reviewing exponents. We will then study operations on polynomials and how polynomials can be used in problem solving.

TOO SMALL TO BE SEEN WITH THE NAKED EYE

A microbe is a tiny living organism that is too small to be seen with the naked eye. Microbes are everywhere, both indoors and outdoors. A single gram of ordinary soil may contain up to one billion microbes belonging to over 10,000 different species of microbes. We even find microbes in our food: Bread, chocolate, yogurt, and cheese are produced with the help of certain microbes.

A microbiologist is a person who studies microbes. Microbiologists identify harmful microbes, develop vaccines or treatments for disease, protect the environment, and conduct research. They need an understanding of the sciences, especially biology, chemistry, and physics, as well as mathematics.

In the Spotlight on Decision Making feature on page 354, you will have the opportunity to make a decision about microscope magnification as a microbiologist.

Source of text : American Society for Microbiology website

5.1   EXPONENTS AND SCIENTIFIC NOTATION

Objectives

1   Use the product rule for exponents.

2   Evaluate expressions raised to the *0* power.

3   Use the quotient rule for exponents.

4   Evaluate expressions raised to the negative *n*th power.

5   Convert between scientific notation and standard notation.

1   Recall that exponents may be used to write repeated factors in a more compact form. As we have seen in the previous chapters, exponents can be used when the repeated factor is a number or a variable. For example,

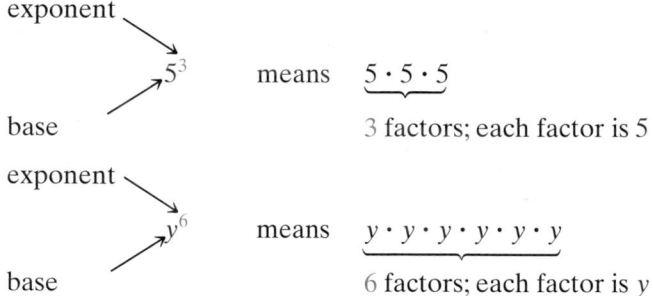

exponent

5^3 means $\underline{5 \cdot 5 \cdot 5}$

base   3 factors; each factor is 5

exponent

y^6 means $\underbrace{y \cdot y \cdot y \cdot y \cdot y \cdot y}$

base   6 factors; each factor is y

Expressions such as 5^3 and y^6 that contain exponents are called **exponential expressions.** Exponential expressions can be multiplied, divided, added, subtracted, and themselves raised to powers. In this section, we review operations on exponential expressions.

Discover the Concept

a. Use a calculator to evaluate each pair of expressions. Then compare the results.

$$2^5 \cdot 2^4 \text{ and } 2^9$$
$$(3.7)^2 \cdot (3.7)^3 \text{ and } (3.7)^5$$
$$(-5)^4 \cdot (-5)^2 \text{ and } (-5)^6$$

b. From the results of part a, write a pattern you observed for multiplying exponential expressions with the same base.

c. Use your findings to simplify the product

$$x^2 \cdot x^3$$

In part a in the box above, we see that each pair of expressions is equal since they simplify to the same number as shown in the screens on the top of the next page. It appears that the product of exponential expressions with the same base is the base raised to the sum of the exponents. From this, we reason that

$$x^2 \cdot x^3 = x^{2+3} = x^5$$

To check, use the definition of a^n:

$$x^2 \cdot x^3 = \underbrace{(x \cdot x)(x \cdot x \cdot x)}_{x \text{ is a factor } 5 \text{ times}} = x^5$$

This suggests the following.

Product Rule for Exponents

If m and n are positive integers and a is a real number, then

$$a^m \cdot a^n = a^{m+n}$$

In other words, the *product* of exponential expressions with a common base is the common base raised to a power equal to the *sum* of the exponents of the factors.

EXAMPLE 1

Use the product rule to simplify.

a. $2^2 \cdot 2^5$ b. $x^7 x^3$ c. $y \cdot y^2 \cdot y^4$

Solution

a. $2^2 \cdot 2^5 = 2^{2+5} = 2^7$

b. $x^7 x^3 = x^{7+3} = x^{10}$

c. $y \cdot y^2 \cdot y^4 = (y^1 \cdot y^2) \cdot y^4$
$= y^3 \cdot y^4$
$= y^7$

EXAMPLE 2

Use the product rule to simplify.

a. $(3x^6)(5x)$ b. $(-2x^3p^2)(4xp^{10})$

Solution

Here, we use properties of multiplication to group together like bases.

a. $(3x^6)(5x) = 3(5)x^6x^1 = 15x^7$

b. $(-2x^3p^2)(4xp^{10}) = -2(4)x^3x^1p^2p^{10} = -8x^4p^{12}$

2   The definition of a^n does not include the possibility that n might be 0. But if it did, then, by the product rule,

$$\underline{a^0 \cdot a^n} = a^{0+n} = a^n = \underline{1 \cdot a^n}$$

From this, we reasonably define that $a^0 = 1$, as long as a does not equal 0.

Zero Exponent

If a does not equal 0, then $a^0 = 1$.

EXAMPLE 3

Evaluate the following.

a. 7^0 **b.** -7^0 **c.** $(2x + 5)^0$ **d.** $2x^0$

Solution   **a.** $7^0 = 1$

b. Without parentheses, only 7 is raised to the 0 power.

$$-7^0 = -(7^0) = -(1) = -1$$

c. $(2x + 5)^0 = 1$

d. $2x^0 = 2(1) = 2$

3   Next, we discover a pattern that occurs when dividing exponential expressions with a common base.

Discover the Concept

a. Use a calculator to evaluate each pair of expressions. Then compare the results.

$$\frac{7^5}{7^2} \text{ and } 7^3$$

$$\frac{(-3)^9}{(-3)^4} \text{ and } (-3)^5$$

b. From the results of part a, write a pattern you observed for dividing exponential expressions with the same base.

c. Use your findings to simplify the quotient $\dfrac{x^9}{x^2}$.

```
7^5/7^2
              343
7^3
              343
■
```

In part a in the box above, we see that each pair of expressions is equal since they simplify to the same number as shown in the screens to the left and on the next page. It appears that the quotient of exponential expressions with the same base is the base raised to the difference of the exponents. From this, we reason that

$$\frac{x^9}{x^2} = x^{9-2} = x^7$$

```
(-3)^9/(-3)^4
            -243
(-3)^5
            -243
■
```

To check, we begin with the definition of a^n to simplify $\dfrac{x^9}{x^2}$.

$$\frac{x^9}{x^2} = \frac{x \cdot x \cdot x \cdot x \cdot x \cdot x \cdot x \cdot x \cdot x}{x \cdot x} = x^7$$

(Assume for the next two sections that denominators containing variables are not 0.) Notice again, that the result is exactly the same if we subtract the exponents.

$$\frac{x^9}{x^2} = x^{9-2} = x^7$$

This suggests the following.

Quotient Rule for Exponents

If a is a nonzero real number and n and m are integers, then

$$\frac{a^m}{a^n} = a^{m-n}$$

In other words, the *quotient* of exponential expressions with a common base is the common base raised to a power equal to the *difference* of the exponents.

EXAMPLE 4

Use the quotient rule to simplify.

a. $\dfrac{x^7}{x^4}$ **b.** $\dfrac{5^8}{5^2}$ **c.** $\dfrac{20x^6}{4x^5}$ **d.** $\dfrac{12y^{10}z^7}{14y^8z^7}$

Solution

a. $\dfrac{x^7}{x^4} = x^{7-4} = x^3$

b. $\dfrac{5^8}{5^2} = 5^{8-2} = 5^6$

c. $\dfrac{20x^6}{4x^5} = 5x^{6-5} = 5x^1$, or $5x$

d. $\dfrac{12y^{10}z^7}{14y^8z^7} = \dfrac{6}{7}y^{10-8} \cdot z^{7-7} = \dfrac{6}{7}y^2z^0 = \dfrac{6}{7}y^2$,   or $\dfrac{6y^2}{7}$

4   When the exponent of the denominator is larger than the exponent of the numerator, applying the quotient rule yields a negative exponent. For example,

$$\frac{x^3}{x^5} = x^{3-5} = x^{-2}$$

Using the definition of a^n, though, gives us

TEACHING TIP
Consider also using patterns
to show that for $a \neq 0$ and
$n > 0$, $a^{-n} = \dfrac{1}{a^n}$.

$2^3 = 8$	$3^3 = 27$
$2^2 = 4$	$3^2 = 9$
$2^1 = 2$	$3^1 = 3$
$2^0 = 1$	$3^0 = 1$
$2^{-1} = \dfrac{1}{2}$	$3^{-1} = \dfrac{1}{3}$
$2^{-2} = \dfrac{1}{4}$	$3^{-2} = \dfrac{1}{9}$
$2^{-3} = \dfrac{1}{8}$	$3^{-3} = \dfrac{1}{27}$

$$\frac{x^3}{x^5} = \frac{x \cdot x \cdot x}{x \cdot x \cdot x \cdot x \cdot x} = \frac{1}{x^2}$$

From this, we reasonably define $x^{-2} = \dfrac{1}{x^2}$ or, in general, $a^{-n} = \dfrac{1}{a^n}$.

Negative Exponents

If a is a real number other than 0 and n is a positive integer, then

$$a^{-n} = \frac{1}{a^n}$$

EXAMPLE 5

Simplify and write with positive exponents only.

a. 5^{-2} **b.** $(-4)^{-4}$ **c.** $2x^{-3}$ **d.** $(3x)^{-1}$

e. $\dfrac{m^5}{m^{15}}$ **f.** $\dfrac{3^3}{3^6}$ **g.** $2^{-1} + 3^{-2}$ **h.** $\dfrac{1}{t^{-5}}$

CLASSROOM EXAMPLE
Simplify and write with positive exponents
only.

a. 7^{-2} **b.** $(-3)^{-4}$

c. $5x^{-4}$ **d.** $(2x)^{-1}$

e. $\dfrac{x^3}{x^{10}}$ **f.** $\dfrac{4^2}{4^5}$

g. $3^{-1} + 2^{-2}$ **h.** $\dfrac{1}{y^{-4}}$

answer:

a. $\dfrac{1}{49}$ **b.** $\dfrac{1}{81}$ **c.** $\dfrac{5}{x^4}$ **d.** $\dfrac{1}{2x}$

e. $\dfrac{1}{x^7}$ **f.** $\dfrac{1}{64}$ **g.** $\dfrac{7}{12}$ **h.** y^4

Solution

a. $5^{-2} = \dfrac{1}{5^2} = \dfrac{1}{25}$

b. $(-4)^{-4} = \dfrac{1}{(-4)^4} = \dfrac{1}{256}$

c. $2x^{-3} = 2 \cdot \dfrac{1}{x^3} = \dfrac{2}{x^3}$ Without parentheses, only x is raised to the -3 power.

d. $(3x)^{-1} = \dfrac{1}{(3x)^1} = \dfrac{1}{3x}$ With parentheses, both 3 and x are raised to the -1 power.

e. $\dfrac{m^5}{m^{15}} = m^{5-15} = m^{-10} = \dfrac{1}{m^{10}}$

f. $\dfrac{3^3}{3^6} = 3^{3-6} = 3^{-3} = \dfrac{1}{3^3} = \dfrac{1}{27}$

g. $2^{-1} + 3^{-2} = \dfrac{1}{2^1} + \dfrac{1}{3^2} = \dfrac{1}{2} + \dfrac{1}{9} = \dfrac{9}{18} + \dfrac{2}{18} = \dfrac{11}{18}$

h. $\dfrac{1}{t^{-5}} = \dfrac{1}{\dfrac{1}{t^5}} = 1 \div \dfrac{1}{t^5} = 1 \cdot \dfrac{t^5}{1} = t^5$

```
5^-2▸Frac
            1/25
2-1+3^-2▸Frac
            11/18
■
```

A calculator check for Example 5a and g.

> ### Helpful Hint
>
> Notice that when a factor containing an exponent is moved from the numerator to the denominator or from the denominator to the numerator, the sign of its exponent changes.
>
> $$x^{-3} = \frac{1}{x^3}, \qquad 5^{-2} = \frac{1}{5^2} = \frac{1}{25}$$
>
> $$\frac{1}{y^{-4}} = y^4, \qquad \frac{1}{2^{-3}} = 2^3 = 8$$

EXAMPLE 6

Simplify and write with positive exponents only.

a. $\dfrac{x^{-9}}{x^2}$ **b.** $\dfrac{p^4}{p^{-3}}$ **c.** $\dfrac{2^{-3}}{2^{-1}}$ **d.** $\dfrac{2x^{-7}y^2}{10xy^{-5}}$ **e.** $\dfrac{(3x^{-3})(x^2)}{x^6}$

Solution

a. $\dfrac{x^{-9}}{x^2} = x^{-9-2} = x^{-11} = \dfrac{1}{x^{11}}$

b. $\dfrac{p^4}{p^{-3}} = p^{4-(-3)} = p^7$

c. $\dfrac{2^{-3}}{2^{-1}} = 2^{-3-(-1)} = 2^{-2} = \dfrac{1}{2^2} = \dfrac{1}{4}$

d. $\dfrac{2x^{-7}y^2}{10xy^{-5}} = \dfrac{x^{-7-1} \cdot y^{2-(-5)}}{5} = \dfrac{x^{-8}y^7}{5} = \dfrac{y^7}{5x^8}$

e. Simplify the numerator first.

$$\dfrac{(3x^{-3})(x^2)}{x^6} = \dfrac{3x^{-3+2}}{x^6} = \dfrac{3x^{-1}}{x^6} = 3x^{-1-6} = 3x^{-7} = \dfrac{3}{x^7}$$

CLASSROOM EXAMPLE

Simplify and write with positive exponents only.

a. $\dfrac{y^{-10}}{y^3}$ **b.** $\dfrac{q^5}{q^{-4}}$ **c.** $\dfrac{5^{-4}}{5^{-2}}$

d. $\dfrac{10x^{-8}y^5}{20xy^{-5}}$ **e.** $\dfrac{(4x^{-1})(x^5)}{x^7}$

answer:

a. $\dfrac{1}{y^{13}}$ **b.** q^9 **c.** $\dfrac{1}{25}$

d. $\dfrac{y^{10}}{2x^9}$ **e.** $\dfrac{4}{x^3}$

✔ **CONCEPT CHECK**

Find and correct the error in the following:

$$\dfrac{y^{-6}}{y^{-2}} = y^{-6-2} = y^{-8} = \dfrac{1}{y^8}$$

EXAMPLE 7

Simplify. Assume that a and t are nonzero integers and that x is not 0.

a. $x^{2a} \cdot x^3$ **b.** $\dfrac{x^{2t-1}}{x^{t-5}}$

Concept Check Answer:

$\dfrac{y^{-6}}{y^{-2}} = y^{-6-(-2)} = y^{-4} = \dfrac{1}{y^4}$

Solution

a. $x^{2a} \cdot x^3 = x^{2a+3}$ Use the product rule.

b. $\dfrac{x^{2t-1}}{x^{t-5}} = x^{(2t-1)-(t-5)}$ Use the quotient rule.

$$= x^{2t-1-t+5} = x^{t+4}$$

5   Very large and very small numbers occur frequently in nature. For example, the distance between the Earth and the Sun is approximately 150,000,000 kilometers. A helium atom has a diameter of 0.000 000 022 centimeters. It can be tedious to write these very large and very small numbers in standard notation like this. **Scientific notation** is a convenient shorthand notation for writing very large and very small numbers.

Helium atom

0.000 000 022
centimeters

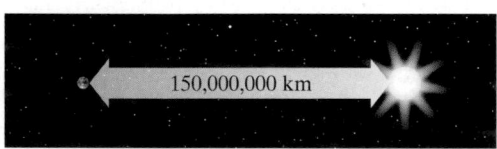

150,000,000 km

Scientific Notation

A positive number is written in **scientific notation** if it is written as the product of a number a, where $1 \le a < 10$ and an integer power r of 10:

$$a \times 10^r$$

The following are examples of numbers written in scientific notation.

diameter of helium atom $\rightarrow 2.2 \times 10^{-8}$ cm; 1.5×10^8 Km $\leftarrow$ approximate distance between Earth and Sun

---

Writing a Positive Number in Scientific Notation

Step 1: Move the decimal point in the original number until the new number has a value between 1 and 10.

Step 2: Count the number of decimal places the decimal point was moved in Step 1. If the original number is 10 or greater, the count is positive. If the original number is less than 1, the count is negative.

Step 3: Write the product of the new number in Step 1 by 10 raised to an exponent equal to the count found in Step 2.

EXAMPLE 8

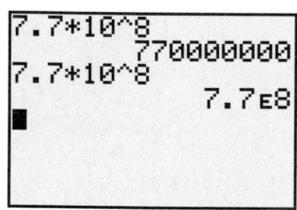

Write each number in scientific notation.

a. 730,000   **b.** 0.00000104

Solution

a. Step 1: Move the decimal point until the number is between 1 and 10.

$$730,000.$$

Step 2: The decimal point is moved 5 places and the original number is 10 or greater, so the count is positive 5.

Step 3: $730,000 = 7.3 \times 10^5$.

b. Step 1: Move the decimal point until the number is between 1 and 10.

$$0.00000104$$

Step 2: The decimal point is moved 6 places and the original number is less then 1, so the count is −6.

Step 3: $0.00000104 = 1.04 \times 10^{-6}$.

The top screen to the left verifies the results of Example 8. This screen was generated with the calculator in scientific notation mode.

To write a scientific notation number in standard form, we reverse the preceding steps.

Writing a Scientific Notation Number in Standard Notation

Move the decimal point in the number the same number of places as the exponent on 10. If the exponent is positive, move the decimal point to the right. If the exponent is negative, move the decimal point to the left.

EXAMPLE 9

Write each number in standard notation.

a. 7.7×10^8 **b.** 1.025×10^{-3}

Solution

a. $7.7 \times 10^8 = 770,000,000$ *Since the exponent is positive, move the decimal point 8 places to the right. Add zeros as needed.*

b. $1.025 \times 10^{-3} = 0.001025$ *Since the exponent is negative, move the decimal point 3 places to the left. Add zeros as needed.*

✔ CONCEPT CHECK

Which of the following numbers have values that are less than 1?

Concept Check Answer:
a, c, d

a. 3.5×10^{-5}　　　　　　　**b.** 3.5×10^5

c. -3.5×10^5　　　　　　　**d.** -3.5×10^{-5}

Spotlight on
DECISION MAKING

Suppose you are a microbiologist. You know that when an image is viewed through a microscope, its magnification is the number of times the image is enlarged. For example, if a 4-millimeter-long object is viewed at five times magnification (denoted 5 × magnification), it appears as an object that is $5 \times 4 = 20$ millimeters long.

Suppose you are studying the *Ebola Zaire* virus, which has an average length of 9.2×10^{-5} centimeters. You would like to view an Ebola virus with a microscope so that it appears to be 4 centimeters long. Decide what magnification setting (rounded to the nearest thousand) you will need to use on the microscope.

MENTAL MATH

Use positive exponents to state each expression.

TEACHING TIP
A Group Activity for this section is available in the Instructor's Resource Manual.

1. $5x^{-1}y^{-2}$ 　 $\dfrac{5}{xy^2}$

2. $7xy^{-4}$ 　 $\dfrac{7x}{y^4}$

3. $a^2b^{-1}c^{-5}$ 　 $\dfrac{a^2}{bc^5}$

4. $a^{-4}b^2c^{-6}$ 　 $\dfrac{b^2}{a^4c^6}$

5. $\dfrac{y^{-2}}{x^{-4}}$ 　 $\dfrac{x^4}{y^2}$

6. $\dfrac{x^{-7}}{z^{-3}}$ 　 $\dfrac{z^3}{x^7}$

EXERCISE SET 5.1

    Math XP

STUDY GUIDE/SSM　　CD/VIDEO　　PH MATH TUTOR CENTER　　MathXL®Tutorials ON CD　　MathXL®　　MyMathLab®

Use the product rule to simplify each expression. See Examples 1 and 2. 　**9.** $-16x^6y^3p^2$ 　**10.** $18a^3b^6$

1. $4^2 \cdot 4^3$ 　 4^5

2. $3^3 \cdot 3^5$ 　 3^8

3. $x^5 \cdot x^3$ 　 x^8

4. $a^2 \cdot a^9$ 　 a^{11}

5. $-7x^3 \cdot 20x^9$ 　 $-140x^{12}$

6. $-3y \cdot -9y^4$ 　 $27y^5$

7. $(4xy)(-5x)$ 　 $-20x^2y$

8. $(7xy)(7aby)$ 　 $49abxy^2$

9. $(-4x^3p^2)(4y^3x^3)$

10. $(-6a^2b^3)(-3ab^3)$

Evaluate the following. See Example 3. **17.–18.** answers may vary

11. -8^0 -1

12. $(-9)^0$ 1

13. $(4x + 5)^0$ 1

14. $8x^0 + 1$ 9

15. $(5x)^0 + 5x^0$ 6

16. $4y^0 - (4y)^0$ 3

17. Explain why $(-5)^0$ simplifies to 1 but -5^0 simplifies to -1.

18. Explain why both $4x^0 - 3y^0$ and $(4x - 3y)^0$ simplify to 1.

Find each quotient. See Example 4.

19. $\dfrac{a^5}{a^2}$ a^3

20. $\dfrac{x^9}{x^4}$ x^5

21. $\dfrac{x^9 y^6}{x^8 y^6}$ x

22. $\dfrac{a^{12} b^2}{a^9 b}$ $a^3 b$

23. $-\dfrac{26z^{11}}{2z^7}$ $-13z^4$

24. $\dfrac{16x^5}{8x}$ $2x^4$

25. $\dfrac{-36a^5 b^7 c^{10}}{6ab^3 c^4}$ $-6a^4 b^4 c^6$

26. $\dfrac{49a^3 bc^{14}}{-7abc^8}$ $-7a^2 c^6$

Simplify each expression. Write answers with positive exponents. See Examples 5 and 6.

27. 4^{-2} $\dfrac{1}{16}$

28. 2^{-3} $\dfrac{1}{8}$

29. $\dfrac{x^7}{x^{15}}$ $\dfrac{1}{x^8}$

30. $\dfrac{z}{z^3}$ $\dfrac{1}{z^2}$

31. $5a^{-4}$ $\dfrac{5}{a^4}$

32. $10b^{-1}$ $\dfrac{10}{b}$

33. $\dfrac{x^{-2}}{x^5}$ $\dfrac{1}{x^7}$

34. $\dfrac{y^{-6}}{y^{-9}}$ y^3

35. $\dfrac{8r^4}{2r^{-4}}$ $4r^8$

36. $\dfrac{3s^3}{15s^{-3}}$ $\dfrac{s^6}{5}$

37. $\dfrac{x^{-9} x^4}{x^{-5}}$ 1

38. $\dfrac{y^{-7} y}{y^8}$ $\dfrac{1}{y^{14}}$

MIXED PRACTICE

Simplify the following. Write answers with positive exponents.

39. $4^{-1} + 3^{-2}$ $\dfrac{13}{36}$

40. $1^{-3} - 4^{-2}$ $\dfrac{15}{16}$

41. $4x^0 + 5$ 9

42. $-5x^0$ -5

43. $x^7 \cdot x^8 \cdot x$ x^{16}

44. $y^6 \cdot y \cdot y^4$ y^{11}

45. $2x^3 \cdot 5x^7$ $10x^{10}$

46. $-3z^4 \cdot 10z^7$ $-30z^{11}$

47. $\dfrac{z^{12}}{z^{15}}$ $\dfrac{1}{z^3}$

48. $\dfrac{x^{11}}{x^{20}}$ $\dfrac{1}{x^9}$

49. $\dfrac{y^{-3}}{y^{-7}}$ y^4

50. $\dfrac{z^{-12}}{z^{10}}$ $\dfrac{1}{z^{22}}$

51. $3x^{-1}$ $\dfrac{3}{x}$

52. $(4x)^{-1}$ $\dfrac{1}{4x}$

53. $3^0 - 3t^0$ -2

54. $4^0 + 4x^0$ 5

55. $\dfrac{r^4}{r^{-4}}$ r^8

56. $\dfrac{x^{-5}}{x^3}$ $\dfrac{1}{x^8}$

57. $\dfrac{x^{-7} y^{-2}}{x^2 y^2}$ $\dfrac{1}{x^9 y^4}$

58. $\dfrac{a^{-5} b^7}{a^{-2} b^{-3}}$ $\dfrac{b^{10}}{a^3}$

59. $\dfrac{2a^{-6} b^2}{18ab^{-5}}$ $\dfrac{b^7}{9a^7}$

60. $\dfrac{18ab^{-6}}{3a^{-3} b^6}$ $\dfrac{6a^4}{b^{12}}$

61. $\dfrac{(24x^8)(x)}{20x^{-7}}$ $\dfrac{6x^{16}}{5}$

62. $\dfrac{(30z^2)(z^5)}{55z^{-4}}$ $\dfrac{6z^{11}}{11}$

Write each number in scientific notation. See Example 8.

63. 31,250,000 3.125×10^7

64. 678,000 6.78×10^5

65. 0.016 1.6×10^{-2}

66. 0.007613 7.613×10^{-3}

67. 67,413 6.7413×10^4

68. 36,800,000 3.68×10^7

69. 0.0125 1.25×10^{-2}

70. 0.00084 8.4×10^{-4}

71. 0.000053 5.3×10^{-5}

72. 98,700,000,000 9.87×10^{10}

Write each number in scientific notation.

73. The approximate distance between Jupiter and the sun is 778,300,000 kilometers. (*Source:* National Space Data Center) 7.783×10^8

74. Total revenues for Wal-Mart in fiscal year 2003 were $245,525,000,000. (*Source:* Wal-Mart Stores, Inc.) 2.45525×10^{11}

75. At the various hotels at the Walt Disney World Resort in Florida, there are 737,000 square feet of meeting facilities. (*Source:* The Walt Disney Company) 7.37×10^5

76. In 2002, the American toy industry had retail sales of $30,606,000,000. (*Source:* Toy Industry Asso.) 3.0606×10^{10}

77. In 2002, the New York City subway system carried a total of 1,410,000,000 passengers. (*Source:* Metropolitan Transit Authority) 1.41×10^9

78. The center of the sun is about 27,000,000° F. 2.7×10^7

79. A pulsar is a rotating neutron star that gives off sharp, regular pulses of radio waves. For one particular pulsar, the rate of pulses is every 0.001 second. 1×10^{-3}

80. To convert from cubic inches to cubic meters, multiply by 0.0000164. 1.64×10^{-5}

Write each number in standard notation, without exponents. See Example 9. **87.** 7,350,000,000,000

81. 3.6×10^{-9} 0.0000000036

82. 2.7×10^{-5} 0.000027

83. 9.3×10^{7} 93,000,000

84. 6.378×10^{8} 637,800,000

85. 1.278×10^{6} 1,278,000

86. 7.6×10^{4} 76,000

87. 7.35×10^{12}

88. 1.66×10^{-5} 0.0000166

89. 4.03×10^{-7} 0.000000403

90. 8.007×10^{8} 800,700,000

Write each number in standard notation.

91. The estimated world population in 1 A.D. was 2.0×10^{8}. (*Source: World Almanac and Book of Facts*)   200,000,000

92. There are 3.949×10^{6} miles of highways, roads, and streets in the United States. (*Source:* Bureau of Transportation Statistics)   3,949,000

93. In 2005, teenagers and children are expected to spend 4.9×10^{9} dollars on purchases and transactions made online. (*Source:* Jupiter Research)   4,900,000,000

94. Each day, an estimated 2.0×10^{7} adults in America drink gourmet coffee beverages. (*Source:* National Coffee Association)   20,000,000

REVIEW AND PREVIEW

Evaluate. See Sections 1.3 and 5.1.

95. $(5 \cdot 2)^{2}$ 100

96. $5^{2} \cdot 2^{2}$ 100

97. $\left(\dfrac{3}{4}\right)^{3}$ $\dfrac{27}{64}$

98. $\dfrac{3^{3}}{4^{3}}$ $\dfrac{27}{64}$

99. $(2^{3})^{2}$ 64

100. $(2^{2})^{3}$ 64

101. $(2^{-1})^{4}$ $\dfrac{1}{16}$

102. $(2^{4})^{-1}$ $\dfrac{1}{16}$

Concept Extensions

103. Explain how to convert a number from standard notation to scientific notation.   answers may vary

104. Explain how to convert a number from scientific notation to standard notation.   answers may vary

105. Simplify where possible.

  a. $x^{a} \cdot x^{a}$ x^{2a}

  b. $x^{a} + x^{a}$ $2x^{a}$

  c. $\dfrac{x^{a}}{x^{b}}$ x^{a-b}

  d. $x^{a} \cdot x^{b}$ x^{a+b}

  e. $x^{a} + x^{b}$ $x^{a} + x^{b}$

106. Which numbers are equal to 36,000? Of these, which is written in scientific notation?   **a, b, c, d; d**

  a. 36×10^{3}

  b. 360×10^{2}

  c. 0.36×10^{5}

  d. 3.6×10^{4}

Without calculating, determine which number is larger.

107. 7^{11} or 7^{13} 7^{13}

108. 5^{10} or 5^{9} 5^{10}

109. 7^{-11} or 7^{-13} 7^{-11}

110. 5^{-10} or 5^{-9} 5^{-9}

Simplify. Assume that variables in the exponent represent nonzero integers and that x, y, and z are not 0. See Example 7.

111. $x^{5} \cdot x^{7a}$ x^{7a+5}

112. $y^{2p} \cdot y^{9p}$ y^{11p}

113. $\dfrac{x^{3t-1}}{x^{t}}$ x^{2t-1}

114. $\dfrac{y^{4p-2}}{y^{3p}}$ y^{p-2}

115. $x^{4a} \cdot x^{7}$ x^{4a+7}

116. $x^{9y} \cdot x^{-7y}$ x^{2y}

117. $\dfrac{z^{6x}}{z^{7}}$ z^{6x-7}

118. $\dfrac{y^{6}}{y^{4z}}$ y^{6-4z}

119. $\dfrac{x^{3t} \cdot x^{4t-1}}{x^{t}}$ x^{6t-1}

120. $\dfrac{z^{5x} \cdot z^{x-7}}{z^{x}}$ z^{5x-7}

121. $x^{9+b} \cdot x^{3a-b}$ x^{3a+9}

122. $z^{2a-b} \cdot z^{5a-b}$ z^{7a-2b}

5.2   *MORE WORK WITH EXPONENTS AND SCIENTIFIC NOTATION*

Objectives

1   Use the power rules for exponents.

2   Use exponent rules and definitions to simplify exponential expressions.

3   Compute, using scientific notation.

1   The volume of the cube shown whose side measures x^2 units is $(x^2)^3$ cubic units. To simplify an expression such as $(x^2)^3$, we use the definition of a^n. Then

$$(x^2)^3 = \underbrace{(x^2)(x^2)(x^2)}_{x^2 \text{ is a factor 3 times}} = x^{2+2+2} = x^6$$

x^2
units

Notice that the result is exactly the same if the exponents are multiplied.

$$(x^2)^3 = x^{2 \cdot 3} = x^6$$

This suggests that the power of an exponential expression raised to a power is the product of the exponents. To discover additional rules for exponents, see below.

Discover the Concept

a. Use a calculator to evaluate each pair of expressions. Then compare the results.

$$(-2 \cdot 3)^7 \quad \text{and} \quad (-2)^7 \cdot 3^7$$
$$(3 \cdot 5)^{-2} \quad \text{and} \quad 3^{-2} \cdot 5^{-2}$$

$$\left(\frac{6}{7}\right)^4 \quad \text{and} \quad \frac{6^4}{7^4}$$
$$\left(\frac{3}{4}\right)^{-5} \quad \text{and} \quad \frac{3^{-5}}{4^{-5}}$$

b. From the results of part a, write a pattern you observed for raising a product to a power and for raising a quotient to a power.

c. Use your findings to simplify $(xy)^3$ and $\left(\dfrac{r}{s}\right)^7$.

The results of part a are below.

It appears that a product raised to a power is equal to the product of each factor raised to the power. Also, a quotient raised to a power is equal to the quotient of the numerator raised to the power and the denominator raised to the power. From this, we reason that

$$(xy)^3 = x^3y^3 \quad \text{and} \quad \left(\frac{r}{s}\right)^7 = \frac{r^7}{s^7}.$$

These power rules suggested above are given in the following box.

The Power Rule and Power of a Product or Quotient Rules for Exponents

If a and b are real numbers and m and n are integers, then

$$(a^m)^n = a^{m \cdot n} \qquad \text{Power rule}$$
$$(ab)^m = a^m b^m \qquad \text{Power of a product}$$
$$\left(\frac{a}{b}\right)^n = \frac{a^n}{b^n} (b \neq 0) \qquad \text{Power of a quotient}$$

EXAMPLE 1

Use the power rule to simplify the following expressions. Use positive exponents to write all results.

a. $(x^5)^7$ **b.** $(2^2)^3$ **c.** $(5^{-1})^2$ **d.** $(y^{-3})^{-4}$

Solution

a. $(x^5)^7 = x^{5 \cdot 7} = x^{35}$

b. $(2^2)^3 = 2^{2 \cdot 3} = 2^6 = 64$

c. $(5^{-1})^2 = 5^{-1 \cdot 2} = 5^{-2} = \frac{1}{5^2} = \frac{1}{25}$

d. $(y^{-3})^{-4} = y^{-3(-4)} = y^{12}$

EXAMPLE 2

Use the power rules to simplify the following. Use positive exponents to write all results.

a. $(5x^2)^3$ **b.** $\left(\frac{2}{3}\right)^3$ **c.** $\left(\frac{3p^4}{q^5}\right)^2$ **d.** $\left(\frac{2^{-3}}{y}\right)^{-2}$ **e.** $(x^{-5}y^2z^{-1})^7$

Solution

a. $(5x^2)^3 = 5^3 \cdot (x^2)^3 = 5^3 \cdot x^{2 \cdot 3} = 125x^6$

b. $\left(\frac{2}{3}\right)^3 = \frac{2^3}{3^3} = \frac{8}{27}$

c. $\left(\dfrac{3p^4}{q^5}\right)^2 = \dfrac{(3p^4)^2}{(q^5)^2} = \dfrac{3^2 \cdot (p^4)^2}{(q^5)^2} = \dfrac{9p^8}{q^{10}}$

d. $\left(\dfrac{2^{-3}}{y}\right)^{-2} = \dfrac{(2^{-3})^{-2}}{y^{-2}}$

$$= \dfrac{2^6}{y^{-2}} = 64y^2 \qquad \text{Use the negative exponent rule.}$$

e. $(x^{-5}y^2z^{-1})^7 = (x^{-5})^7 \cdot (y^2)^7 \cdot (z^{-1})^7$

$$= x^{-35}y^{14}z^{-7} = \dfrac{y^{14}}{x^{35}z^7}$$

2   In the next few examples, we practice the use of several of the rules and definitions for exponents. The following is a summary of these rules and definitions.

Summary of Rules for Exponents

If a and b are real numbers and m and n are integers, then

Product rule	$a^m \cdot a^n = a^{m+n}$	
Zero exponent	$a^0 = 1$	$(a \neq 0)$
Negative exponent	$a^{-n} = \dfrac{1}{a^n}$	$(a \neq 0)$
Quotient rule	$\dfrac{a^m}{a^n} = a^{m-n}$	$(a \neq 0)$
Power rule	$(a^m)^n = a^{m \cdot n}$	
Power of a product	$(ab)^m = a^m \cdot b^m$	
Power of a quotient	$\left(\dfrac{a}{b}\right)^m = \dfrac{a^m}{b^m}$	$(b \neq 0)$

EXAMPLE 3

Simplify each expression. Use positive exponents to write the answers.

a. $(2x^0y^{-3})^{-2}$ **b.** $\left(\dfrac{x^{-5}}{x^{-2}}\right)^{-3}$ **c.** $\left(\dfrac{2}{7}\right)^{-2}$ **d.** $\dfrac{5^{-2}x^{-3}y^{11}}{x^2y^{-5}}$

Solution   **a.** $(2x^0y^{-3})^{-2} = 2^{-2}(x^0)^{-2}(y^{-3})^{-2}$

$$= 2^{-2}x^0y^6$$

$$= \dfrac{1(y^6)}{2^2} \qquad \text{Write } x^0 \text{ as } 1.$$

$$= \dfrac{y^6}{4}$$

b. $\left(\dfrac{x^{-5}}{x^{-2}}\right)^{-3} = \dfrac{(x^{-5})^{-3}}{(x^{-2})^{-3}} = \dfrac{x^{15}}{x^6} = x^{15-6} = x^9$

c. $\left(\dfrac{2}{7}\right)^{-2} = \dfrac{2^{-2}}{7^{-2}} = \dfrac{7^2}{2^2} = \dfrac{49}{4}$

d. $\dfrac{5^{-2}x^{-3}y^{11}}{x^2y^{-5}} = (5^{-2})\left(\dfrac{x^{-3}}{x^2}\right)\left(\dfrac{y^{11}}{y^{-5}}\right) = 5^{-2}x^{-3-2}y^{11-(-5)} = 5^{-2}x^{-5}y^{16}$

$$= \dfrac{y^{16}}{5^2x^5} = \dfrac{y^{16}}{25x^5}$$

EXAMPLE 4

Simplify each expression. Use positive exponents to write the answers.

a. $\left(\dfrac{3x^2y}{y^{-9}z}\right)^{-2}$ **b.** $\left(\dfrac{3a^2}{2x^{-1}}\right)^3\left(\dfrac{x^{-3}}{4a^{-2}}\right)^{-1}$

Solution There is often more than one way to simplify exponential expressions. Here, we will simplify inside the parentheses if possible before we apply the power rules for exponents.

a. $\left(\dfrac{3x^2y}{y^{-9}z}\right)^{-2} = \left(\dfrac{3x^2y^{10}}{z}\right)^{-2} = \dfrac{3^{-2}x^{-4}y^{-20}}{z^{-2}} = \dfrac{z^2}{3^2x^4y^{20}} = \dfrac{z^2}{9x^4y^{20}}$

b. $\left(\dfrac{3a^2}{2x^{-1}}\right)^3\left(\dfrac{x^{-3}}{4a^{-2}}\right)^{-1} = \dfrac{27a^6}{8x^{-3}}\cdot\dfrac{x^3}{4^{-1}a^2}$

$$= \dfrac{27\cdot 4\cdot a^6x^3x^3}{8\cdot a^2} = \dfrac{27a^4x^6}{2}$$

EXAMPLE 5

Simplify each expression. Assume that a and b are integers and that x and y are not 0.

a. $x^{-b}(2x^b)^2$ **b.** $\dfrac{(y^{3a})^2}{y^{a-6}}$

Solution **a.** $x^{-b}(2x^b)^2 = x^{-b}2^2x^{2b} = 4x^{-b+2b} = 4x^b$

b. $\dfrac{(y^{3a})^2}{y^{a-6}} = \dfrac{y^{6a}}{y^{a-6}} = y^{6a-(a-6)} = y^{6a-a+6} = y^{5a+6}$

3   To perform operations on numbers written in scientific notation, we use properties of exponents.

EXAMPLE 6

Perform the indicated operations. Write each result in scientific notation. Use a calculator to check the results.

a. $(8.1 \times 10^5)(5 \times 10^{-7})$ **b.** $\dfrac{1.2 \times 10^4}{3 \times 10^{-2}}$

Solution   **a.** $(8.1 \times 10^5)(5 \times 10^{-7}) = 8.1 \times 5 \times 10^5 \times 10^{-7}$

$$= 40.5 \times 10^{-2}$$ Not in scientific notation because 40.5 is not between 1 and 10.

$$= (4.05 \times 10^1) \times 10^{-2}$$

$$= 4.05 \times 10^{-1}$$

b. $\dfrac{1.2 \times 10^4}{3 \times 10^{-2}} = \left(\dfrac{1.2}{3}\right)\left(\dfrac{10^4}{10^{-2}}\right) = 0.4 \times 10^{4-(-2)}$

$$= 0.4 \times 10^6 = (4 \times 10^{-1}) \times 10^6 = 4 \times 10^5$$

A calculator check for Example 6.

EXAMPLE 7

Use scientific notation to simplify $\dfrac{2000 \times 0.000021}{700}$. Write the result in scientific notation. Use a calculator to check the results.

Solution

$$\dfrac{2000 \times 0.000021}{700} = \dfrac{(2 \times 10^3)(2.1 \times 10^{-5})}{7 \times 10^2} = \dfrac{2(2.1)}{7} \cdot \dfrac{10^3 \cdot 10^{-5}}{10^2}$$

$$= 0.6 \times 10^{-4}$$
$$= (6 \times 10^{-1}) \times 10^{-4}$$
$$= 6 \times 10^{-5}$$

A calculator check for Example 7.

STUDY SKILLS REMINDER

Are You Satisfied with Your Performance on a Particular Quiz or Exam?

If not, analyze your quiz or exam like you would a good mystery novel. Look for common themes in your errors.

Were most of your errors a result of

▶ *Carelessness*? If your errors were careless, did you turn in your work before the allotted time expired? If so, resolve next time to use the entire time allotted. Any extra time can be spent checking your work.

▶ *Running out of time*? If so, make a point to better manage your time on your next exam. A few suggestions are to work any questions that you are unsure of last and to check your work after all of the questions have been answered.

▶ *Not understanding a concept*? If so, review that concept and correct your work. Remember next time to make sure that all concepts on a quiz or exam are understood before the exam.

MENTAL MATH

Simplify. See Examples 1 through 4.

1. $(x^4)^5$ x^{20} **2.** $(5^6)^2$ 5^{12} **3.** $x^4 \cdot x^5$ x^9 **4.** $x^7 \cdot x^8$ x^{15} **5.** $(y^6)^7$ y^{42}

6. $(x^3)^4$ x^{12} **7.** $(z^4)^9$ z^{36} **8.** $(z^3)^7$ z^{21} **9.** $(z^{-6})^{-3}$ z^{18} **10.** $(y^{-4})^{-2}$ y^8

EXERCISE SET 5.2

STUDY   CD/   PH MATH   MathXL®Tutorials   MathXL®   MyMathLab®
GUIDE/SSM   VIDEO   TUTOR CENTER   ON CD

Simplify. Write each answer using positive exponents only. See Examples 1 and 2.

1. $(3^{-1})^2$ $\dfrac{1}{9}$

2. $(2^{-2})^2$ $\dfrac{1}{16}$

3. $(x^4)^{-9}$ $\dfrac{1}{x^{36}}$

4. $(y^7)^{-3}$ $\dfrac{1}{y^{21}}$

5. $(y)^{-5}$ $\dfrac{1}{y^5}$

6. $(z^{-1})^{10}$ $\dfrac{1}{z^{10}}$

7. $(3x^2y^3)^2$ $9x^4y^6$

8. $(4x^3yz)^2$ $16x^6y^2z^2$

9. $\left(\dfrac{2x^5}{y^{-3}}\right)^4$ $16x^{20}y^{12}$

10. $\left(\dfrac{3a^{-4}}{b^7}\right)^3$ $\dfrac{27}{a^{12}b^{21}}$

11. $(a^2bc^{-3})^{-6}$ $\dfrac{c^{18}}{a^{12}b^6}$

12. $(6x^{-6}y^7z^0)^{-2}$ $\dfrac{x^{12}}{36y^{14}}$

13. $\left(\dfrac{x^7y^{-3}}{z^{-4}}\right)^{-5}$ $\dfrac{y^{15}}{x^{35}z^{20}}$

14. $\left(\dfrac{a^{-2}b^{-5}}{c^{-11}}\right)^{-6}$ $\dfrac{a^{12}b^{30}}{c^{66}}$

15. $(5^{-1})^3$ $\dfrac{1}{125}$

Simplify. Write each answer using positive exponents only. See Examples 3 and 4.

16. $\left(\dfrac{a^{-4}}{a^{-5}}\right)^{-2}$ $\dfrac{1}{a^2}$

17. $\left(\dfrac{x^{-9}}{x^{-4}}\right)^{-3}$ x^{15}

 18. $\left(\dfrac{2a^{-2}b^5}{4a^2b^7}\right)^{-2}$ $4a^8b^4$

19. $\left(\dfrac{5x^7y^4}{10x^3y^{-2}}\right)^{-3}$ $\dfrac{8}{x^{12}y^{18}}$

20. $\dfrac{4^{-1}x^2yz}{x^{-2}yz^3}$ $\dfrac{x^4}{4z^2}$

21. $\dfrac{8^{-2}x^{-3}y^{11}}{x^2y^{-5}}$ $\dfrac{y^{16}}{64x^5}$

22. $\left(\dfrac{6p^6}{p^{12}}\right)^2$ $\dfrac{36}{p^{12}}$

23. $\left(\dfrac{4p^6}{p^9}\right)^3$ $\dfrac{64}{p^9}$

24. $(-8y^3xa^{-2})^{-3}$ $-\dfrac{a^6}{512x^3y^9}$

25. $(-xy^0x^2a^3)^{-3}$ $-\dfrac{1}{x^9a^9}$

26. $\left(\dfrac{x^{-2}y^{-2}}{a^{-3}}\right)^{-7}$ $\dfrac{x^{14}y^{14}}{a^{21}}$

27. $\left(\dfrac{x^{-1}y^{-2}}{5^{-3}}\right)^{-5}$ $\dfrac{x^5y^{10}}{5^{15}}$

MIXED PRACTICE

Simplify. Write each answer using positive exponents.

28. $(8^2)^{-1}$ $\dfrac{1}{64}$

29. $(x^7)^{-9}$ $\dfrac{1}{x^{63}}$

30. $(y^{-4})^5$ $\dfrac{1}{y^{20}}$

31. $\left(\dfrac{7}{8}\right)^3$ $\dfrac{343}{512}$

32. $\left(\dfrac{4}{3}\right)^2$ $\dfrac{16}{9}$

 33. $(4x^2)^2$ $16x^4$

34. $(-8x^3)^2$ $64x^6$

35. $(-2^{-2}y)^3$ $-\dfrac{y^3}{64}$

36. $(-4^{-6}y^{-6})^{-4}$ $4^{24}y^{24}$

37. $\left(\dfrac{4^{-4}}{y^3x}\right)^{-2}$ $4^8x^2y^6$

38. $\left(\dfrac{7^{-3}}{ab^2}\right)^{-2}$ $7^6a^2b^4$

39. $\left(\dfrac{1}{4}\right)^{-3}$ 64

40. $\left(\dfrac{1}{8}\right)^{-2}$ 64

41. $\left(\dfrac{3x^5}{6x^4}\right)^4$ $\dfrac{x^4}{16}$

42. $\left(\dfrac{8^{-3}}{y^2}\right)^{-2}$ 8^6y^4

 43. $\dfrac{(y^3)^{-4}}{y^3}$ $\dfrac{1}{y^{15}}$

44. $\dfrac{2(y^3)^{-3}}{y^{-3}}$ $\dfrac{2}{y^6}$

45. $\left(\dfrac{2x^{-3}}{y^{-1}}\right)^{-3}$ $\dfrac{x^9}{8y^3}$

46. $\left(\dfrac{n^5}{2m^{-2}}\right)^{-4}$ $\dfrac{16}{n^{20}m^8}$

47. $\dfrac{3^{-2}a^{-5}b^6}{4^{-2}a^{-7}b^{-3}}$ $\dfrac{16a^2b^9}{9}$

48. $\dfrac{2^{-3}m^{-4}n^{-5}}{5^{-2}m^{-5}n}$ $\dfrac{25m}{8n^6}$

49. $(4x^6y^5)^{-2}(6x^4y^3)$ $\dfrac{3}{8x^8y^7}$

50. $(5xy)^3(z^{-2})^{-3}$ $125x^3y^3z^6$

51. $x^6(x^6bc)^{-6}$ $\dfrac{1}{x^{30}b^6c^6}$

52. $2(y^2b)^{-4}$ $\dfrac{2}{y^8b^4}$

53. $\dfrac{2^{-3}x^2y^{-5}}{5^{-2}x^7y^{-1}}$ $\dfrac{25}{8x^5y^4}$

54. $\dfrac{7^{-1}a^{-3}b^5}{a^2b^{-2}}$ $\dfrac{b^7}{7a^5}$

55. $\left(\dfrac{2x^2}{y^4}\right)^3\left(\dfrac{2x^5}{y}\right)^{-2}$ $\dfrac{2}{x^4y^{10}}$

56. $\left(\dfrac{3z^{-2}}{y}\right)^2\left(\dfrac{9y^{-4}}{z^{-3}}\right)^{-1}$ $\dfrac{y^2}{z^7}$

Perform each indicated operation. Write each answer in scientific notation. See Examples 6 and 7.   **57.** 1.45×10^9 **58.** 2.16×10^{-2}

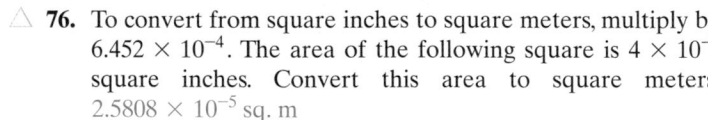 **57.** $(5 \times 10^{11})(2.9 \times 10^{-3})$

58. $(3.6 \times 10^{-12})(6 \times 10^9)$

59. $(2 \times 10^5)^3$ 8×10^{15}

60. $(3 \times 10^{-7})^3$ 2.7×10^{-20}

61. $\dfrac{3.6 \times 10^{-4}}{9 \times 10^2}$ 4×10^{-7}

62. $\dfrac{1.2 \times 10^9}{2 \times 10^{-5}}$ 6×10^{13}

63. $\dfrac{0.0069}{0.023}$ 3×10^{-1}

64. $\dfrac{0.00048}{0.0016}$ 3×10^{-1}

65. $\dfrac{18,200 \times 100}{91,000}$ 2×10^1

66. $\dfrac{0.0003 \times 0.0024}{0.0006 \times 20}$ 6×10^{-5}

67. $\dfrac{6000 \times 0.006}{0.009 \times 400}$ 1×10^1

68. $\dfrac{0.00016 \times 300}{0.064 \times 100}$ 7.5×10^{-3}

69. $\dfrac{0.00064 \times 2000}{16,000}$ 8×10^{-5} **70.** $\dfrac{0.00072 \times 0.003}{0.00024}$ 9×10^{-3}

71. $\dfrac{66,000 \times 0.001}{0.002 \times 0.003}$ 1.1×10^7 **72.** $\dfrac{0.0007 \times 11,000}{0.001 \times 0.0001}$ 7.7×10^7

73. $\dfrac{1.25 \times 10^{15}}{(2.2 \times 10^{-2})(6.4 \times 10^{-5})}$ $8.877840909 \times 10^{20}$

74. $\dfrac{(2.6 \times 10^{-3})(4.8 \times 10^{-4})}{1.3 \times 10^{-12}}$ 9.6×10^5

Solve.

75. A computer can add two numbers in about 10^{-8} second. Express in scientific notation how long it would take this computer to do this task 200,000 times. 2×10^{-3} sec

△ **76.** To convert from square inches to square meters, multiply by 6.452×10^{-4}. The area of the following square is 4×10^{-2} square inches. Convert this area to square meters. 2.5808×10^{-5} sq. m

4×10^{-2} sq. in.

△ **77.** To convert from cubic inches to cubic meters, multiply by 1.64×10^{-5}. A grain of salt is in the shape of a cube. If an average size of a grain of salt is 3.8×10^{-6} cubic inches, convert this volume to cubic meters. 6.232×10^{-11} cu. m

REVIEW AND PREVIEW

Simplify each expression. See Section 1.4.

78. $-5y + 4y - 18 - y$

79. $12m - 14 - 15m - 1$

80. $-3x - (4x - 2)$

81. $-9y - (5 - 6y)$

82. $3(z - 4) - 2(3z + 1)$

83. $5(x - 3) - 4(2x - 5)$

Concept Extensions

Simplify the following. Assume that variables in the exponents represent integers and that all other variables are not 0. See Example 5.

84. $(x^{3a+6})^3$ x^{9a+18}

85. $(x^{2b+7})^2$ x^{4b+14}

86. $\dfrac{x^{4a}(x^{4a})^3}{x^{4a-2}}$ x^{12a+2}

87. $\dfrac{x^{-5y+2}x^{2y}}{x}$ x^{-3y+1}

88. $(b^{5x-2})^{2x}$ b^{10x^2-4x}

89. $(c^{2a+3})^3$ c^{6a+9}

90. $\dfrac{(y^{2a})^8}{y^{a-3}}$ y^{15a+3}

91. $\dfrac{(y^{4a})^7}{y^{2a-1}}$ y^{26a+1}

92. $\left(\dfrac{2x^{3t}}{x^{2t-1}}\right)^4$ $16x^{4t+4}$

93. $\left(\dfrac{3y^{5a}}{y^{-a+1}}\right)^2$ $9y^{12a-2}$

94. $\dfrac{(z^{a+2})^b}{(z^{b-1})^a}$ z^{2b+a}

95. $\dfrac{(y^{3-a})^b}{(y^{1-b})^a}$ y^{3b-a}

96. $\dfrac{x^{2a+1}y^{a-1}}{x^{3a+1}y^{2a-3}}$ $x^{-a}y^{-a+2}$

97. $\dfrac{x^{-5-3a}y^{-2a-b}}{x^{-5+3b}y^{-2b-a}}$ $x^{-3a-3b}y^{b-a}$

△ **98.** Each side of the cube shown is $\dfrac{2x^{-2}}{y}$ meters. Find its volume. $\dfrac{8}{x^6y^3}$ cu. m

$\dfrac{2x^{-2}}{y}$ m

△ **99.** The lot shown is in the shape of a parallelogram with base $\dfrac{3x^{-1}}{y^{-3}}$ feet and height $5x^{-7}$ feet. Find its area. $\dfrac{15y^3}{x^8}$ sq. ft

$5x^{-7}$ ft

$\dfrac{3x^{-1}}{y^{-3}}$ ft

100. The density D of an object is equivalent to the quotient of its mass M and volume V. Thus $D = \dfrac{M}{V}$. Express in scientific notation the density of an object whose mass is 500,000 pounds and whose volume is 250 cubic feet. 2×10^3

101. The density of ordinary water is 3.12×10^{-2} tons per cubic foot. The volume of water in the largest of the Great Lakes, Lake Superior, is 4.269×10^{14} cubic feet. Use the formula $D = \dfrac{M}{V}$ (see Exercise 100) to find the mass (in tons) of the water in Lake Superior. Express your answer in scientific notation. (*Source:* National Ocean Service) 1.331928×10^{13}

102. Is there a number a such that $a^{-1} = a^1$? If so, give the value of a. yes; $a = \pm 1$

103. Is there a number a such that a^{-2} is a negative number? If so, give the value of a. no

104. Explain whether 0.4×10^{-5} is written in scientific notation. answers may vary

105. The estimated population of the United States in 2003 was 2.927×10^8 people. The land area of the United States is 3.536×10^6 square miles. Find the population density (number of people per square mile) for the United States in 2003. Round to the nearest whole number. (*Source:* U.S. Census Bureau) 83 people per sq. mi

106. In 2002, the value of goods and services imported into the United States was $\$1.141 \times 10^{12}$. The estimated population of the United States in 2002 was 2.821×10^8 people. Find the average value of imports per person in the United States for 2002. Round to the nearest dollar. (*Sources:* U.S. Census Bureau, Bureau of Economic Analysis) $\$4045$ per person

107. In 2002, the population of Japan was 1.27×10^8 people. At the same time, the population of Oceania (including the countries of Australia, Fiji, New Zealand, etc.) was 3.22×10^7 people. How many times greater was the population of Japan than the population of Oceania? Round to the nearest tenth. (*Source:* Population Reference Bureau) 3.9

108. Explain whether 0.4×10^{-5} is written in scientific notation.

109. The Moscow subway system has a passenger volume of 3.16×10^9 passengers. The São Paulo subway system has a passenger volume of 7.01×10^8 passengers. How many times greater is the Moscow subway volume than the São Paulo volume? Round to the nearest tenth. (*Source:* New York City Transit Authority) 4.5 times

110. China's fighting force numbers 2.93×10^6 soldiers. Taiwan's fighting force numbers 4.25×10^5. How many times greater is China's armed forces than Taiwan's? Round to the nearest whole number. (*Source:* Russell Ash, *The Top 10 of Everything*) 7 times

108. answers may vary

78. $-2y - 18$ **79.** $-3m - 15$ **80.** $-7x + 2$ **81.** $-3y - 5$
82. $-3z - 14$ **83.** $-3x + 5$

5.3   POLYNOMIALS AND POLYNOMIAL FUNCTIONS

Objectives

1 Identify term, constant, polynomial, monomial, binomial, trinomial, and the degree of a term and of a polynomial.

2 Define polynomial functions.

3 Review combining like terms.

4 Add polynomials.

5 Subtract polynomials.

6 Recognize the graph of a polynomial function from the degree of the polynomial.

1 A **term** is a number or the product of a number and one or more variables raised to powers. The **numerical coefficient**, or simply the **coefficient**, is the numerical factor of a term.

Term	Numerical Coefficient
$-12x^5$	-12
x^3y	1
$-z$	-1
2	2

If a term contains only a number, it is called a **constant term**, or simply a **constant**.

A **polynomial** is a finite sum of terms in which all variables are raised to nonnegative integer powers and no variables appear in any denominator.

TEACHING TIP

Point out that polynomials are expressions not equations.

Polynomials	Not Polynomials	
$4x^5y + 7xz$	$5x^{-3} + 2x$	Negative integer exponent
$-5x^3 + 2x + \dfrac{2}{3}$	$\dfrac{6}{x^2} - 5x + 1$	Variable in denominator

TEACHING TIP

After defining polynomial, ask students to name the prefix for three. Then have them suggest a name for polynomials with exactly three terms such as $4x^2 - 3x + 9$. Finally, ask them to guess what binomials and monomials are by writing an example of each.

A polynomial that contains only one variable is called a **polynomial in one variable**. For example, $3x^2 - 2x + 7$ is a **polynomial in x**. This polynomial in x is written in *descending order* since the terms are listed in descending order of the variable's exponents. (The term 7 can be thought of as $7x^0$.) The following examples are polynomials in one variable written in **descending order**.

$$4x^3 - 7x^2 + 5 \qquad y^2 - 4 \qquad 8a^4 - 7a^2 + 4a$$

A **monomial** is a polynomial consisting of one term. A **binomial** is a polynomial consisting of two terms. A **trinomial** is a polynomial consisting of three terms.

Monomials	Binomials	Trinomials
ax^2	$x + y$	$x^2 + 4xy + y^2$
$-3x$	$6y^2 - 2$	$-x^4 + 3x^3 + 1$
4	$\dfrac{5}{7}z^3 - 2z$	$8y^2 - 2y - 10$

By definition, all monomials, binomials, and trinomials are also polynomials.
Each term of a polynomial has a **degree**.

> ### Degree of a Term
>
> The **degree of a term** is the sum of the exponents on the *variables* contained in the term.

EXAMPLE 1

Find the degree of each term.

a. $3x^2$　　**b.** -2^3x^5　　**c.** y　　**d.** $12x^2yz^3$　　**e.** 5

Solution

a. The exponent on x is 2, so the degree of the term is 2.

b. The exponent on x is 5, so the degree of the term is 5. (Recall that the degree is the sum of the exponents on only the *variables*.)

c. The degree of y, or y^1, is 1.

d. The degree is the sum of the exponents on the variables, or $2 + 1 + 3 = 6$.

e. The degree of 5, which can be written as $5x^0$, is 0.

　　From the preceding example, we can say that the degree of a constant is 0. Also, the term 0 has no degree.

Each polynomial also has a degree.

> ### Degree of a Polynomial
>
> The **degree of a polynomial** is the largest degree of all its terms.

EXAMPLE 2

Find the degree of each polynomial and indicate whether the polynomial is also a monomial, binomial, or trinomial.

	Polynomial	Degree	Classification
a.	$7x^3 - 3x + 2$	3	Trinomial
b.	$-xyz$	$1 + 1 + 1 = 3$	Monomial
c.	$x^4 - 16$	4	Binomial

EXAMPLE 3

Find the degree of the polynomial

$$3xy + x^2y^2 - 5x^2 - 6.$$

Solution   The degree of each term is

$$3xy + x^2y^2 - 5x^2 - 6$$

Degree: 2   4   2   0

The largest degree of any term is 4, so the degree of this polynomial is 4.

2   At times, it is convenient to use function notation to represent polynomials. For example, we may write $P(x)$ to represent the polynomial $3x^2 - 2x - 5$. In symbols, this is

$$P(x) = 3x^2 - 2x - 5$$

This function is called a **polynomial function** because the expression $3x^2 - 2x - 5$ is a polynomial.

> **Helpful Hint**
>
> Recall that the symbol $P(x)$ **does not mean** P times x. It is a special symbol used to denote a function.

If $P(x) = 3x^2 - 2x - 5$, let's evaluate $P(2)$. Below is a review of methods that we have used to evaluate functions at given values.

$P(x) = 3(x)^2 - 2x - 5$
$P(2) = 3 \cdot 2^2 - 2 \cdot 2 - 5$
$P(2) = 3$

$P(2) = 3$

$P(2) = 3$

$P(2) = 3$

EXAMPLE 4

If $P(x) = 3x^2 - 2x - 5$, find each function value. Use a graph to check part a and a table to check part b.

a. $P(1)$ **b.** $P(-2)$

Solution   **a.** Substitute 1 for x in $P(x) = 3x^2 - 2x - 5$ and simplify.

$$P(x) = 3x^2 - 2x - 5$$
$$P(1) = 3(1)^2 - 2(1) - 5 = -4$$

b. Substitute -2 for x in $P(x) = 3x^2 - 2x - 5$ and simplify.

$$P(x) = 3x^2 - 2x - 5$$
$$P(-2) = 3(-2)^2 - 2(-2) - 5 = 11$$

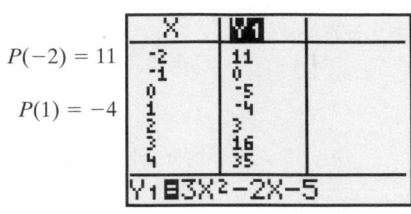

$P(-2) = 11$

$P(1) = -4$

$P(1) = -4$

Calculator checks for Example 4a and b.

Many real-world phenomena are modeled by polynomial functions. If the polynomial function model is given, we can often find the solution of a problem by evaluating the function at a certain value.

EXAMPLE 5

CLASSROOM EXAMPLE

Use Example 5 and find the height of the object when $t = 0$ seconds and $t = 2$ seconds.

answer: $P(0) = 1053$ ft, $P(2) = 989$ ft

EXAMPLE 5

FINDING THE HEIGHT OF AN OBJECT

The world's highest bridge, Royal Gorge suspension bridge in Colorado, is 1053 feet above the Arkansas River. An object is dropped from the top of this bridge. Neglecting air resistance, the height of the object at time t seconds is given by the polynomial function $P(t) = -16t^2 + 1053$.

a. Use a table to find the height of the object when $x = 3$ seconds and when $x = 7$ seconds.

b. Approximate, to the nearest second, the time when the object hits the ground.

c. Approximate, to the nearest tenth of a second, the time when the object hits the ground.

Solution Enter the polynomial function $y_1 = -16x^2 + 1053$ in your graphing utility, where x is time in seconds and y_1 is height in feet.

a. Since we are asked to find height at 3 seconds and 7 seconds, we may set table start at 3 and table increment at 1.

$P(3) = 909$

$P(7) = 269$

Because $P(3) = 909$, the height of the object at 3 seconds is 909 feet. Because $P(7) = 269$, the height of the object at 7 seconds is 269 feet.

b. The object will hit the ground when $P(x)$ or $y = 0$. Since

$$P(8) = 29 \text{ and } P(9) = -243,$$

the object hits the ground between 8 and 9 seconds. Since 29 is closer to 0 than -243, the time to the nearest second is 8 seconds.

TEACHING TIP

After discussing Example 5, ask, "What does the constant term represent in a height problem?" Follow up with "How high is a bridge if the height at time t of an object dropped from the bridge is given by $P(t) = -16t^2 + 934$?"

c. Given the table on the previous page, graph y_1 in an appropriate window. We graph y_1 in a $[-1, 10, 1]$ by $[-200, 1500, 100]$ window. Recall that the object hits the ground when $P(x)$ or $y = 0$. Thus, use a zero or root function of your graphing utility to find the x-intercept of the graph in the first quadrant. To the nearest tenth, the object hits the ground at 8.1 seconds.

Notice from the table or the first quadrant portion of the graph that as time x increases the height of the object decreases.

3   Before we add polynomials, recall that terms are considered to be **like terms** if they contain exactly the same variables raised to exactly the same powers.

Like Terms	Unlike Terms
$-5x^2, -x^2$	$4x^2, 3x$
$7xy^3z, -2xzy^3$	$12x^2y^3, -2xy^3$

To simplify a polynomial, **combine like terms** by using the distributive property. For example, by the distributive property,

$$5x + 7x = (5 + 7)x = 12x$$

EXAMPLE 6

Simplify by combining like terms.

a. $-12x^2 + 7x^2 - 6x$ **b.** $3xy - 2x + 5xy - x$

Solution   By the distributive property,

a. $-12x^2 + 7x^2 - 6x = (-12 + 7)x^2 - 6x = -5x^2 - 6x$

b. Use the associative and commutative properties to group together like terms; then combine.

$$3xy - 2x + 5xy - x = 3xy + 5xy - 2x - x$$
$$= (3 + 5)xy + (-2 - 1)x$$
$$= 8xy - 3x$$

4   Now we have reviewed the necessary skills to add polynomials.

Adding Polynomials

Combine all like terms.

EXAMPLE 7

Add.

a. $(7x^3y - xy^3 + 11) + (6x^3y - 4)$ **b.** $(3a^3 - b + 2a - 5) + (a + b + 5)$

Solution   **a.** To add, remove the parentheses and group like terms.

$$(7x^3y - xy^3 + 11) + (6x^3y - 4)$$
$$= 7x^3y - xy^3 + 11 + 6x^3y - 4$$
$$= 7x^3y + 6x^3y - xy^3 + 11 - 4 \qquad \text{Group like terms.}$$
$$= 13x^3y - xy^3 + 7 \qquad \text{Combine like terms.}$$

b.
$$(3a^3 - b + 2a - 5) + (a + b + 5)$$
$$= 3a^3 - b + 2a - 5 + a + b + 5$$
$$= 3a^3 - b + b + 2a + a - 5 + 5 \qquad \text{Group like terms.}$$
$$= 3a^3 + 3a \qquad \text{Combine like terms.}$$

EXAMPLE 8

Add $11x^3 - 12x^2 + x - 3$ and $x^3 - 10x + 5$.

Solution
$$(11x^3 - 12x^2 + x - 3) + (x^3 - 10x + 5)$$
$$= 11x^3 + x^3 - 12x^2 + x - 10x - 3 + 5 \qquad \text{Group like terms.}$$
$$= 12x^3 - 12x^2 - 9x + 2 \qquad \text{Combine like terms.}$$

Sometimes it is more convenient to add polynomials vertically. To do this, line up like terms beneath one another and add like terms.

5 The definition of subtraction of real numbers can be extended to apply to polynomials. To subtract a number, we add its opposite.

$$a - b = a + (-b)$$

Likewise, to subtract a polynomial, we add its opposite. In other words, if P and Q are polynomials, then

$$P - Q = P + (-Q)$$

The polynomial $-Q$ is the **opposite**, or **additive inverse**, of the polynomial Q. We can find $-Q$ by writing the opposite of each term of Q.

Subtracting Polynomials

To subtract a polynomial, add its opposite.

For example,

To subtract, add its opposite (found by writing the opposite of each term).

$$(3x^2 + 4x - 7) - (3x^2 - 2x - 5) = (3x^2 + 4x - 7) + (-3x^2 + 2x + 5)$$
$$= 3x^2 + 4x - 7 - 3x^2 + 2x + 5$$
$$= 6x - 2 \qquad \text{Combine like terms.}$$

✔ **CONCEPT CHECK**

Which polynomial is the opposite of $16x^3 - 5x + 7$?

a. $-16x^3 - 5x + 7$ **b.** $-16x^3 + 5x - 7$

c. $16x^3 + 5x + 7$ **d.** $-16x^3 + 5x + 7$

EXAMPLE 9

Subtract $(12z^5 - 12z^3 + z) - (-3z^4 + z^3 + 12z)$.

Solution To subtract, add the opposite of the second polynomial to the first polynomial.

$$(12z^5 - 12z^3 + z) - (-3z^4 + z^3 + 12z)$$
$$= 12z^5 - 12z^3 + z + 3z^4 - z^3 - 12z) \qquad \text{Add the opposite of the polynomial being subtracted.}$$
$$= 12z^5 + 3z^4 - 12z^3 - z^3 + z - 12z \qquad \text{Group like terms.}$$
$$= 12z^5 + 3z^4 - 13z^3 - 11z \qquad \text{Combine like terms.}$$

✔ **CONCEPT CHECK**

Why is the following subtraction incorrect?

$$(7z - 5) - (3z - 4)$$
$$= 7z - 5 - 3z - 4$$
$$= 4z - 9$$

EXAMPLE 10

Subtract $4x^3y^2 - 3x^2y^2 + 2y^2$ from $10x^3y^2 - 7x^2y^2$.

Solution If we subtract 2 from 8, the difference is $8 - 2 = 6$. Notice the order of the numbers, and then write "Subtract $4x^3y^2 - 3x^2y^2 + 2y^2$ from $10x^3y^2 - 7x^2y^2$" as a mathematical expression.

Concept Check Answer:

b

Concept Check Answer:

With parentheses removed, the expression should be

$7z - 5 - 3z + 4 = 4z - 1$

TEACHING TIP

Before going over vertical addition or subtraction, ask students to rewrite the sum $364.82 + 19,472.1$ vertically. Then have a student show the vertical sum on the board. Ask why they aligned the 3 in the first number with the 4 in the second number.

TEACHING TIP

After discussing Example 10, have students subtract $10x^3y^2 - 7x^2y^2$ from $4x^3y^2 - 3x^2y^2 + 2y^2$. Is the result the same?

$$(10x^3y^2 - 7x^2y^2) - (4x^3y^2 - 3x^2y^2 + 2y^2)$$

$$= 10x^3y^2 - 7x^2y^2 - 4x^3y^2 + 3x^2y^2 - 2y^2 \quad \text{Remove parentheses.}$$

$$= 6x^3y^2 - 4x^2y^2 - 2y^2 \quad \text{Combine like terms.}$$

To add or subtract polynomials vertically, just remember to line up like terms. For example, perform the subtraction $(10x^3y^2 - 7x^2y^2) - (4x^3y^2 - 3x^2y^2 + 2y^2)$ vertically. Add the opposite of the second polynomial.

$$\begin{array}{r} 10x^3y^2 - 7x^2y^2 \\ -(4x^3y^2 - 3x^2y^2 + 2y^2) \end{array} \quad \text{is equivalent to} \quad \begin{array}{r} 10x^3y^2 - 7x^2y^2 \\ -4x^3y^2 + 3x^2y^2 - 2y^2 \\ \hline 6x^3y^2 - 4x^2y^2 - 2y^2 \end{array}$$

Polynomial functions, like polynomials, can be added, subtracted, multiplied, and divided. For example, if

$$P(x) = x^2 + x + 1$$

then

$$2P(x) = 2(x^2 + x + 1) = 2x^2 + 2x + 2 \quad \text{Use the distributive property.}$$

Also, if $Q(x) = 5x^2 - 1$, then $P(x) + Q(x) = (x^2 + x + 1) + (5x^2 - 1)$
$$= 6x^2 + x.$$

A useful business and economics application of subtracting polynomial functions is finding the profit function $P(x)$ when given a revenue function $R(x)$ and a cost function $C(x)$. In business, it is true that

$$\text{profit} = \text{revenue} - \text{cost}, \text{ or}$$
$$P(x) = R(x) - C(x)$$

For example, if the revenue function is $R(x) = 7x$ and the cost function is $C(x) = 2x + 5000$, then the profit function is

$$P(x) = R(x) - C(x)$$

or

$$P(x) = 7x - (2x + 5000) \quad \text{Substitute } R(x) = 7x$$
$$P(x) = 5x - 5000 \quad \text{and } C(x) = 2x + 5000.$$

Problem-solving exercises involving profit are in the exercise set.

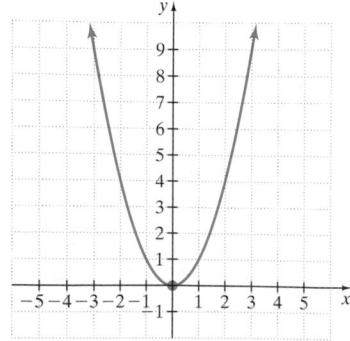

6 In this section, we reviewed how to find the degree of a polynomial. Knowing the degree of a polynomial can help us recognize the graph of the related polynomial function. For example, we know from Section 2.1 that the graph of the polynomial function $f(x) = x^2$ is a parabola as shown to the left.

The polynomial x^2 has degree 2. The graphs of all polynomial functions of degree 2 will have this same general shape—opening upward, as shown, or downward. Graphs of polynomial functions of degree 2 or 3 will, in general, resemble one of the graphs shown next.

General Shapes of Graphs of Polynomial Functions

Degree 2

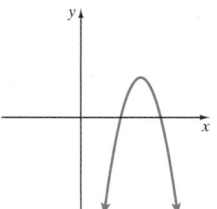

Coefficient of x^2
is a positive number.

Coefficient of x^2
is a negative number.

Degree 3

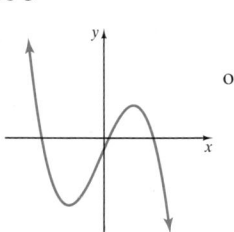

Coefficient of x^3
is a positive number.

Coefficient of x^3
is a negative number.

EXAMPLE 11

Determine which of the following graphs most closely resembles the graph of
$f(x) = 5x^3 - 6x^2 + 2x + 3$

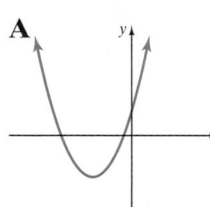

A   **B**   **C**   **D**

Solution The degree of $f(x)$ is 3, which means that its graph has the shape of B or D. The coefficient of x^3 is 5, a positive number, so the graph has the shape of B.

In addition to determining the general shape of the graph of a function from the degree of the related polynomial, we also can find the y-intercept from the constant term. That is, if $p(x) = x^2 + 3x - 2$, then $p(0) = -2$, so the point $(0, -2)$ is the y-intercept.

EXAMPLE 12

Match each graph with its equation and give the coordinates of the y-intercept of each graph.

a. $f(x) = x^2 + x - 5$ **b.** $g(x) = x^3 - x^2 + x + 2$

c. $h(x) = x^3 - 4x^2 + 2x - 3$ **d.** $F(x) = -x^3 + x - 6$

e. $H(x) = -x^2 + 5x + 1$

1.

2.

3.

4.

5.

Solution   **a.** The polynomial function $f(x) = x^2 + x - 5$ has degree 2, so the shape of its graph is a parabola. Since the coefficient of x^2 is the positive number 1, the parabola opens upward and has the shape of option 4. Its y-intercept is $(0, -5)$.

      b. The polynomial function $g(x) = x^3 - x^2 + x + 2$ has degree 3, so the shape of its graph is option 1, 3, or 5. Since the coefficient of x^3 is the positive number 1, it has the shape of option 3 or 5. The y-intercept of $g(x)$ is $(0, 2)$, so the graph resembles option 5.

      c. The graph of $h(x)$ resembles option 3. The y-intercept of $h(x)$ is $(0, -3)$.

      d. Since the coefficient of x^3 is -1, the graph of $F(x)$ resembles option 1. The y-intercept is $(0, -6)$.

      e. The graph of $H(x)$ is a parabola opening downward, since the coefficient of x^2 is -1. Its graph is option 2. The y-intercept is $(0, 1)$.

EXERCISE SET 5.3

| STUDY GUIDE/SSM | CD/ VIDEO | PH MATH TUTOR CENTER | MathXL®Tutorials ON CD | MathXL® | MyMathLab® |

Find the degree of each term. See Example 1.

1. 4   0   **2.** 7   0

 3. $5x^2$ 2   **4.** $-z^3$ 3

 5. $-3xy^2$ 3   **6.** $12x^3z$ 4

Find the degree of each polynomial and indicate whether the polynomial is a monomial, binomial, trinomial, or none of these. See Examples 2 and 3.   **7.** degree 1; binomial   **8.** degree 1; binomial

7. $6x + 3$ **8.** $7x - 8$

9. $3x^2 - 2x + 5$ **10.** $5x^2 - 3x^2y - 2x^3$

11. $-xyz$ **12.** -9

 13. $x^2y - 4xy^2 + 5x + y$ degree 3; none of these

14. $-2x^2y - 3y^2 + 4x + y^5$ degree 5; none of these

15. In your own words, describe how to find the degree of a term.   answers may vary

16. In your own words, describe how to find the degree of a polynomial.   answers may vary

If $P(x) = x^2 + x + 1$ and $Q(x) = 5x^2 - 1$, find the following. See Example 4.

 17. $P(7)$ 57   **18.** $Q(4)$ 79

19. $Q(-10)$ 499   **20.** $P(-4)$ 13

21. $Q\left(\frac{1}{4}\right)$ $-\frac{11}{16}$ **22.** $P\left(\frac{1}{2}\right)$ $\frac{7}{4}$

Solve. See Example 5.

23. Enter $y_1 = -16x^2 + 75$ on a graphing utility. Look at a table of values with the table starting at 0 and incrementing by 1.
  a. Between what two x-values do the y-values change from positive to negative?   2 and 3
  b. Graph y_1 in a $[0, 5, 1]$ by $[-10, 100, 10]$ window. Between what two integers does the graph cross the x-axis?   2 and 3

24. Enter $y_1 = -16x^2 - 50x + 400$ on a graphing utility. Look at a table of values with the table starting at 0 and incrementing by 1.
  a. Between what two x-values do the y-values change from positive to negative?   3 and 4
  b. Graph y_1 in a $[0, 5, 1]$ by $[-100, 500, 100]$ window. Between what two integers does the graph cross the x-axis?   3 and 4

25. An object is dropped off the top of a building that is 525 feet high. The distance, in feet, above the ground at x seconds is given by $P(x) = -16x^2 + 525$. Approximate to the nearest tenth of a second when the object hits the ground.   5.7 sec

26. An object is dropped off the top of a building that is 350 feet high. The distance, in feet, above the ground at x seconds is given by $P(x) = -16x^2 + 350$. Approximate to the nearest tenth of a second when the object will pass a window that is 100 feet above ground level.   4.0 sec

27. Two intermediate algebra students perform the following experiment. An egg is dropped from the top of a 100-foot building at the same instant that another egg is thrown down with a speed of 85 feet per second from a nearby building 200 feet high. The distance, in feet, above the ground at x seconds for the dropped egg is given by $P_1(x) = -16x^2 + 100$, and for the egg thrown with a speed of 85 feet per second, it is $P_2(x) = -16x^2 - 85x + 200$.
  a. Which egg hits the ground first?   the second egg
  b. Are the eggs ever the same distance above the ground at the same time? If yes, give the time this happens and approximate your answer to the nearest second.   1 sec

28. A stone is dropped off a bridge that is 150 feet above a river. The distance, in feet, above the ground at x seconds is given by the equation $P_1(x) = -16x^2 + 150$. One second later, another stone is thrown down with a speed of 80 feet per second. The distance, in feet, above the ground at x seconds is given by $P_2(x) = -16(x - 1)^2 - 80(x - 1) + 150$.
  a. Which stone hits the water first?   the second stone
  b. Does the second stone ever catch up with the first stone so that they are the same distance above the water at the same time? If yes, give the time this happens and approximate your answer to the nearest tenth of a second.   1.3 sec

Simplify by combining like terms. See Example 6.

29. $5y + y$ $6y$ **30.** $-x + 3x$ $2x$

31. $4x + 7x - 3$ $11x - 3$ **32.** $-8y + 9y + 4y^2$ $4y^2 + y$

33. $4xy + 2x - 3xy - 1$ $xy + 2x - 1$

34. $-8xy^2 + 4x - x + 2xy^2$ $-6xy^2 + 3x$

35. $7x^2 - 2xy + 5y^2 - x^2 + xy + 11y^2$ $6x^2 - xy + 16y^2$

36. $-a^2 + 18ab - 2b^2 + 14a^2 - 12ab - b^2$ $13a^2 + 6ab - 3b^2$

MIXED PRACTICE

Perform the indicated operations. See Examples 7 through 10.

37. $(9y^2 - 8) + (9y^2 - 9)$ $18y^2 - 17$

38. $(x^2 + 4x - 7) + (8x^2 + 9x - 7)$ $9x^2 + 13x - 14$

 39. Add $(x^2 + xy - y^2)$ and $(2x^2 - 4xy + 7y^2)$.

40. Add $(4x^3 - 6x^2 + 5x + 7)$ and $(2x^2 + 6x - 3)$.

41. $x^2 - 6x + 3$ **42.** $-2x^2 + 3x - 9$
  $\underline{+\ \ (2x + 5)}$ $\underline{+\ \ \ \ \ (2x - 3)}$

9. degree 2; trinomial   **10.** degree 3; trinomial   **11.** degree 3; monomial   **12.** degree 0; monomial   **39.** $3x^2 - 3xy + 6y^2$
40. $4x^3 - 4x^2 + 11x + 4$ **41.** $x^2 - 4x + 8$ **42.** $-2x^2 + 5x - 12$

43. $(9y^2 - 7y + 5) - (8y^2 - 7y + 2)$ $y^2 + 3$

44. $(2x^2 + 3x + 12) - (5x - 7)$ $2x^2 - 2x + 19$

45. Subtract $(6x^2 - 3x)$ from $(4x^2 + 2x)$. $-2x^2 + 5x$

46. Subtract $(xy + x - y)$ from $(xy + x - 3)$. $y - 3$

47. $3x^2 - 4x + 8$ **48.** $-3x^2 - 4x + 8$
  $-\quad (5x^2 - 7)$ $-\quad (5x + 12)$

49. $(5x - 11) + (-x - 2)$ $4x - 13$

50. $(3x^2 - 2x) + (5x^2 - 9x)$ $8x^2 - 11x$

51. $(7x^2 + x + 1) - (6x^2 + x - 1)$ $x^2 + 2$

52. $(4x - 4) - (-x - 4)$ $5x$

53. $(7x^3 - 4x + 8) + (5x^3 + 4x + 8x)$ $12x^3 + 8x + 8$

54. $(9xyz + 4x - y) + (-9xyz - 3x + y + 2)$ $x + 2$

55. $(9x^3 - 2x^2 + 4x - 7) - (2x^3 - 6x^2 - 4x + 3)$

56. $(3x^2 + 6xy + 3y^2) - (8x^2 - 6xy - y^2)$

57. Add $(y^2 + 4yx + 7)$ and $(-19y^2 + 7yx + 7)$.

58. Subtract $(x - 4)$ from $(3x^2 - 4x + 5)$. $3x^2 - 5x + 9$

59. $(3x^3 - b + 2a - 6) + (-4x^3 + b + 6a - 6)$

60. $(5x^2 - 6) + (2x^2 - 4x + 8)$ $7x^2 - 4x + 2$

61. $(4x^2 - 6x + 2) - (-x^2 + 3x + 5)$ $5x^2 - 9x - 3$

62. $(5x^2 + x + 9) - (2x^2 - 9)$ $3x^2 + x + 18$

63. $(-3x + 8) + (-3x^2 + 3x - 5)$ $-3x^2 + 3$

64. $(5y^2 - 2y + 4) + (3y + 7)$ $5y^2 + y + 11$

65. $(-3 + 4x^2 + 7xy^2) + (2x^3 - x^2 + xy^2)$

66. $(-3x^2y + 4) - (-7x^2y - 8y)$ $4x^2y + 8y + 4$

67. $6y^2 - 6y + 4$ **68.** $-4x^3 + 4x^2 - 4x$
  $-(-y^2 - 6y + 7)$ $-(2x^3 - 2x^2 + 3x)$

69. $3x^2 + 15x + 8$ **70.** $9x^2 + 9x - 4$
  $+(2x^2 + 7x + 8)$ $+(7x^2 - 3x - 4)$

71. $\left(\dfrac{1}{2}x^2 - \dfrac{1}{3}x^2y + 2y^3\right) + \left(\dfrac{1}{4}x^2 - \dfrac{8}{3}x^2y^2 - \dfrac{1}{2}y^3\right)$

72. $\left(\dfrac{2}{5}a^2 - ab + \dfrac{4}{3}b^2\right) + \left(\dfrac{1}{5}a^2b - ab + \dfrac{5}{6}b^2\right)$

73. Find the sum of $(5q^4 - 2q^2 - 3q)$ and $(-6q^4 + 3q^2 + 5)$.

74. Find the sum of $(5y^4 - 7y^2 + x^2 - 3)$ and $(-3y^4 + 2y^2 + 4)$. $2y^4 - 5y^2 + x^2 + 1$

75. Subtract $(3x + 7)$ from the sum of $(7x^2 + 4x + 9)$ and $(8x^2 + 7x - 8)$. $15x^2 + 8x - 6$

76. Subtract $(9x + 8)$ from the sum of $(3x^2 - 2x - x^3 + 2)$ and $(5x^2 - 8x - x^3 + 4)$. $-2x^3 + 8x^2 - 19x - 2$

77. Find the sum of $(4x^4 - 7x^2 + 3)$ and $(2 - 3x^4)$.

78. Find the sum of $(8x^4 - 14x^2 + 6)$ and $(-12x^6 - 21x^4 - 9x^2)$. $-12x^6 - 13x^4 - 23x^2 + 6$

79. $\left(\dfrac{2}{3}x^2 - \dfrac{1}{6}x + \dfrac{5}{6}\right) - \left(\dfrac{1}{3}x^2 + \dfrac{5}{6}x - \dfrac{1}{6}\right)$ $\dfrac{1}{3}x^2 - x + 1$

80. $\left(\dfrac{3}{16}x^2 + \dfrac{5}{8}x - \dfrac{1}{4}\right) - \left(\dfrac{5}{16}x^2 - \dfrac{3}{8}x + \dfrac{3}{4}\right)$ $-\dfrac{1}{8}x^2 + x - 1$

Solve. See Example 5.

The surface area of a rectangular box is given by the polynomial
$$2HL + 2LW + 2HW$$
and is measured in square units. In business, surface area is often calculated to help determine cost of materials.

47. $-2x^2 - 4x + 15$
48. $-3x^2 - 9x - 4$
55. $7x^3 + 4x^2 + 8x - 10$
56. $-5x^2 + 12xy + 4y^2$
57. $-18y^2 + 11yx + 14$
59. $-x^3 + 8a - 12$
65. $8xy^2 + 2x^3 + 3x^2 - 3$
67. $7y^2 - 3$

△ **81.** A rectangular box is to be constructed to hold a new camcorder. The box is to have dimensions 5 inches by 4 inches by 9 inches. Find the surface area of the box.   202 sq. in.

△ **82.** Suppose it has been determined that a box of dimensions 4 inches by 4 inches by 8.5 inches can be used to contain the camcorder in Exercise 81. Find the surface area of this box and calculate the square inches of material saved by using this box instead of the box in Exercise 81.   168 sq. in.; 34 sq. in. saved

A projectile is fired upward from the ground with an initial velocity of 300 feet per second. Neglecting air resistance, the height of the projectile at any time t can be described by the polynomial function
$$P(t) = -16t^2 + 300t$$

Use the following table of values to answer Exercise 83.

X	Y1
1	284
2	536
3	756
4	944
5	1100
6	1224
7	1316

Y1◻-16X²+300X

83. Find the height of the projectile at the given times.

   a. $t = 1$ second   284 ft   **b.** $t = 2$ seconds   536 ft

   c. $t = 3$ seconds   756 ft   **d.** $t = 4$ seconds   944 ft

   e. Generate the table above for positive integer x-values and explain why the height increases and then decreases as time passes.   answers may vary

   f. Use the table from part (e) and approximate (to the nearest second) how long before the object hits the ground.   19 sec

68. $-6x^3 + 6x^2 - 7x$ **69.** $5x^2 + 22x + 16$ **70.** $16x^2 + 6x - 8$ **71.** $\dfrac{3}{4}x^2 - \dfrac{1}{3}x^2y - \dfrac{8}{3}x^2y^2 + \dfrac{3}{2}y^3$ **72.** $\dfrac{2}{5}a^2 + \dfrac{1}{5}a^2b - 2ab + \dfrac{13}{6}b^2$

73. $-q^4 + q^2 - 3q + 5$ **77.** $x^4 - 7x^2 + 5$

84. An object is thrown upward with an initial velocity of 25 feet per second from the top of the 984-foot-high Eiffel Tower in Paris, France. The height of the object at any time t can be described by the polynomial function $P(t) = -16t^2 + 25t + 984$. Find the height of the projectile at each given time.(*Source:* Council on Tall Buildings and Urban Habitat, Lehigh University)

  a. $t = 1$ second 993 ft

  b. $t = 3$ seconds 915 ft

  c. $t = 5$ seconds 709 ft

  d. Approximate (to the nearest second) how long before the object hits the ground. 9 sec

85. The polynomial function $P(x) = 45x - 100,000$ models the relationship between the number of computer briefcases x that a company sells and the profit the company makes, $P(x)$. Find $P(4000)$, the profit from selling 4000 computer briefcases. $80,000

86. The total cost (in dollars) for MCD, Inc., Manufacturing Company to produce x blank audiocassette tapes per week is given by the polynomial function $C(x) = 0.8x + 10,000$. Find the total cost of producing 20,000 tapes per week. $26,000

87. The total revenues (in dollars) for MCD, Inc., Manufacturing Company to sell x blank audiocasette tapes per week is given by the polynomial function $R(x) = 2x$. Find the total revenue from selling 20,000 tapes per week. $40,000

88. In business, profit equals revenue minus cost, or $P(x) = R(x) - C(x)$. Find the profit function for MCD, Inc. by subtracting the given functions in Exercises 86 and 87.

Match each equation with its graph. See Examples 11 and 12.

89. $f(x) = 3x^2 - 2$ **90.** $h(x) = 5x^3 - 6x + 2$

91. $g(x) = -2x^3 - 3x^2 + 3x - 2$ **92.** $g(x) = -2x^2 - 6x + 2$

89. A 90. B 91. D 92. C

A

B

C

D
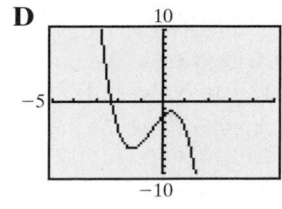

REVIEW AND PREVIEW

Multiply. See Section 1.4.

93. $5(3x - 2)$ **94.** $-7(2z - 6y)$

95. $-2(x^2 - 5x + 6)$ **96.** $5(-3y^2 - 2y + 7)$

Concept Extensions

Perform indicated operations.

97. $(4x^{2a} - 3x^a + 0.5) - (x^{2a} - 5x^a - 0.2)$

98. $(9y^{5a} - 4y^{3a} + 1.5y) - (6y^{5a} - y^{3a} + 4.7y)$

99. $(8x^{2y} - 7x^y + 3) + (-4x^{2y} + 9x^y - 14)$

100. $(14z^{5x} + 3z^{2x} + z) - (2z^{5x} - 10z^{2x} + 3z)$

Find each perimeter.

△ **101.** $(6x^2 + 14y)$ units

$(x + 5y)$ units $(3x^2 - x + 2y)$ units

△ **102.** $(z^3 + 2z^2 - 2z + 3)$ units

$(z + 2)$ units $(2z^2 + z)$ units $(z^3 - 4z + 1)$ units

If $P(x) = 3x + 3, Q(x) = 4x^2 - 6x + 3$, and $R(x) = 5x^2 - 7$, find the following.

103. $P(x) + Q(x)$ **104.** $R(x) + P(x)$

105. $Q(x) - R(x)$ **106.** $P(x) - Q(x)$

107. $2[Q(x)] - R(x)$ **108.** $-5[P(x)] - Q(x)$

109. $3[R(x)] + 4[P(x)]$ **110.** $2[Q(x)] + 7[R(x)]$

*If $P(x)$ is the polynomial given, find **a.** $P(a)$, **b.** $P(-x)$, and **c.** $P(x + h)$.*

111. $P(x) = 2x - 3$ **112.** $P(x) = 8x + 3$

113. $P(x) = 4x$ **114.** $P(x) = -4x$

115. $P(x) = 4x - 1$ **116.** $P(x) = 3x - 2$

88. $P(x) = 1.2x - 10,000$ **93.** $15x - 10$ **94.** $-14z + 42y$ **95.** $-2x^2 + 10x - 12$ **96.** $-15y^2 - 10y + 35$ **97.** $3x^{2a} + 2x^a + 0.7$ **98.** $3y^{5a} - 3y^{3a} - 3.2y$ **99.** $4x^{2y} + 2x^y - 11$ **100.** $12z^{5x} + 13z^{2x} - 2z$ **103.** $4x^2 - 3x + 6$ **104.** $5x^2 + 3x - 4$ **105.** $-x^2 - 6x + 10$ **106.** $-4x^2 + 9x$ **107.** $3x^2 - 12x + 13$ **108.** $-4x^2 - 9x - 18$ **109.** $15x^2 + 12x - 9$ **110.** $43x^2 - 12x - 43$ **111.a.** $2a - 3$ **b.** $-2x - 3$ **c.** $2x + 2h - 3$ **112.a.** $8a + 3$ **b.** $-8x + 3$ **c.** $8x + 8h + 3$ **113.a.** $4a$ **b.** $-4x$ **c.** $4x + 4h$ **114.a.** $-4a$ **b.** $4x$ **c.** $-4x - 4h$ **115.a.** $4a - 1$ **b.** $-4x - 1$ **c.** $4x + 4h - 1$ **116. a.** $3a - 2$ **b.** $-3x - 2$ **c.** $3x + 3h - 2$

117. The function $f(x) = -246.7x^2 + 1887.9x + 1016.9$ can be used to approximate the increasing number of radio stations on the Internet during the years 1998–2003 where x is the number of years after 1998 and $f(x)$ is the number of stations. Round answers to the nearest whole. (*Source:* BRS Media, Inc.)

   a. Approximate the number of radio stations on the Internet in 1998.   1017 stations

   b. Approximate the number of radio stations on the Internet in 2000.   3806 stations

   c. Use the function to predict the number of radio stations on the Internet in 2006.   331 stations

   d. From parts (a), (b), and (c), determine whether the number of radio stations on the Internet is increasing at a steady rate. Explain why or why not.   **d.** answers may vary

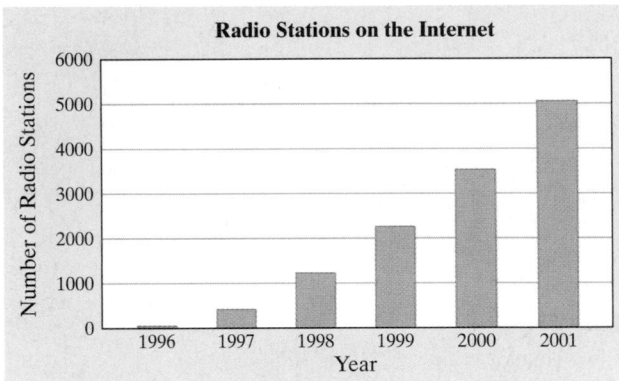

118. The function $f(x) = 0.21x^2 - 0.55x + 8.57$ can be used to approximate the number of Americans enrolled in health maintenance organizations (HMOs) during the period 1980–2001 where x is the number of years after 1980 and $f(x)$ is the number of millions of Americans. Round answers to the nearest tenth of a million. (*Source:* Based on data from *Health, United States, 2001*, National Center for Health Statistics)

   a. Approximate the number of Americans enrolled in HMOs in 1995.   47.6 million

   b. Approximate the number of Americans enrolled in HMOs in 2000.   81.6 million

   c. Use the function to predict the number of Americans enrolled in HMOs in 2005.   126.1 million

   d. From parts (a), (b), and (c), determine whether the number of Americans enrolled in HMOs is changing at a steady rate. Explain why or why not.   answers may vary

119. Sport utility vehicle (SUV) sales in the Uninted States. have increased since 1992. The function $f(x) = 0.014x^2 + 0.12x + 0.85$ can be used to approximate the number of SUV sales during the years 1990–2000 where x is the number of years after 1990 and $f(x)$ is the SUV sales (in millions). Round answers to the nearest tenth of a million. (*Source:* Wards Communications)   **a.** 3.1 million SUVs

   a. Approximate the number of SUVs sold in 1999.

   b. Use the function to predict the number of SUVs sold in 2005.   5.8 million SUVs

120. Digital camera sales have increased since 1996. The function $f(x) = 34.7x^2 + 68.2x + 377.3$ can be used to approximate the revenue from selling digital cameras for the years 1996–2001 where x is the number of years after 1996 and $f(x)$ is the sales revenue (in million of dollars). Round answers to the nearest whole million. (*Source:* International Data Corporation)

   a. Approximate the revenue from digital camera sales in 2000. $1205 million

   b. Use the function to predict the revenue from digital camera sales in 2004. $3144 million

121. The function $f(x) = 1.4x^2 + 129.6x + 939$ can be used to approximate spending for health care in the United States, where x is the number of years since 1980 and $f(x)$ is the amount of money spent per capita. (*Source:* U.S. Health Care Financing Administration)

   a. Approximate the amount of money spent on health care per capita in the year 1985. $1622

   b. Approximate the amount of money spent on health care per capita in the year 1995. $3198

   c. Use the given function to predict the amount of money that will be spent on health care per capita in the year 2010. $6087

   d. From parts **a, b,** and **c,** is the amount of money spent rising at a steady rate? Why or why not? No, $f(x)$ is not linear.

122. The function $f(x) = 904x^2 - 10{,}311x + 70{,}479$ can be used to approximate the number of AIDS cases reported in the United States from 1995 to 2002, where x is the number of years since 1995. (*Source:* Based on data from the U.S. Centers for Disease Control and Prevention)

   a. Approximate the number of AIDS cases reported in the United States in 1995.   70,479 cases

   b. Approximate the number of AIDS cases reported in the United States in 2002.   42,598 cases

   c. Approximate the number of AIDS cases reported in the United States in 2000.   41,524 cases

   d. Describe the trend in the number of AIDS cases reported during the period covered by the model.   the number of cases decreased, and increased

5.4   *MULTIPLYING POLYNOMIALS*

Objectives

1. Multiply two polynomials.
2. Multiply binomials.
3. Square binomials.
4. Multiply the sum and difference of two terms.
5. Multiply three or more polynomials.
6. Evaluate polynomial functions.

1 Properties of real numbers and exponents are used continually in the process of multiplying polynomials. To multiply monomials, for example, we apply the commutative and associative properties of real numbers and the product rule for *exponents*.

EXAMPLE 1

Multiply.

a. $(2x^3)(5x^6)$ **b.** $(7y^4z^4)(-xy^{11}z^5)$

Solution   Group like bases and apply the product rule for exponents.

a. $(2x^3)(5x^6) = 2(5)(x^3)(x^6) = 10x^9$

b. $(7y^4z^4)(-xy^{11}z^5) = 7(-1)x(y^4y^{11})(z^4z^5) = -7xy^{15}z^9$

> **Helpful Hint**
> See Sections 5.1 and 5.2 to review exponential expressions further.

To multiply a monomial by a polynomial other than a monomial, we use an expanded form of the distributive property.

$$a(b + c + d + \cdots + z) = ab + ac + ad + \cdots + az$$

Notice that the monomial a is multiplied by each term of the polynomial.

EXAMPLE 2

Multiply.

a. $2x(5x - 4)$ **b.** $-3x^2(4x^2 - 6x + 1)$ **c.** $-xy(7x^2y + 3xy - 11)$

Solution   Apply the distributive property.

a. $2x(5x - 4) = 2x(5x) + 2x(-4)$ Use the distributive property.

$ = 10x^2 - 8x$ Multiply.

b. $-3x^2(4x^2 - 6x + 1) = -3x^2(4x^2) + (-3x^2)(-6x) + (-3x^2)(1)$
$$= -12x^4 + 18x^3 - 3x^2$$

c. $-xy(7x^2y + 3xy - 11) = -xy(7x^2y) + (-xy)(3xy) + (-xy)(-11)$
$$= -7x^3y^2 - 3x^2y^2 + 11xy$$

To multiply any two polynomials, we can use the following.

Multiplying Two Polynomials

To multiply any two polynomials, use the distributive property and multiply each term of one polynomial by each term of the other polynomial. Then combine any like terms.

✔ **CONCEPT CHECK**

Find the error:
$$4x(x - 5) + 2x$$
$$= 4x(x) + 4x(-5) + 4x(2x)$$
$$= 4x^2 - 20x + 8x^2$$
$$= 12x^2 - 20x$$

EXAMPLE 3

Multiply and simplify the product if possible.

a. $(x + 3)(2x + 5)$ **b.** $(2x - 3)(5x^2 - 6x + 7)$

Solution **a.** Multiply each term of $(x + 3)$ by $(2x + 5)$.

$(x + 3)(2x + 5) = x(2x + 5) + 3(2x + 5)$ Apply the distributive property.
$$= 2x^2 + 5x + 6x + 15 \qquad \text{Apply the distributive property again.}$$
$$= 2x^2 + 11x + 15 \qquad \text{Combine like terms.}$$

b. Multiply each term of $(2x - 3)$ by each term of $(5x^2 - 6x + 7)$.

$(2x - 3)(5x^2 - 6x + 7) = 2x(5x^2 - 6x + 7) + (-3)(5x^2 - 6x + 7)$

$$= 10x^3 - 12x^2 + 14x - 15x^2 + 18x - 21$$

$$= 10x^3 - 27x^2 + 32x - 21 \qquad \text{Combine like terms.}$$

Sometimes polynomials are easier to multiply vertically, in the same way we multiply real numbers. When multiplying vertically, we line up like terms in the **partial products** vertically. This makes combining like terms easier.

CLASSROOM EXAMPLE

Multiply.

a. $(x + 2)(3x + 1)$

b. $(7x - 2)(2x^2 - x + 1)$

answer:

a. $3x^2 + 7x + 2$

b. $14x^3 - 11x^2 + 9x - 2$

Concept Check Answer:
$4x(x - 5) + 2x$
$$= 4x(x) + 4x(-5) + 2x$$
$$= 4x^2 - 20x + 2x$$
$$= 4x^2 - 18x$$

EXAMPLE 4

Multiply vertically $(4x^2 + 7)(x^2 + 2x + 8)$.

Solution

$$
\begin{array}{r}
x^2 +\ 2x\ +\ 8 \\
4x^2\ +\ 7 \\
\hline
7x^2 +\ 14x\ +\ 56 \\
4x^4 + 8x^3 + 32x^2 \\
\hline
4x^4 + 8x^3 + 39x^2 + 14x + 56
\end{array}
$$

$7(x^2 + 2x + 8)$

$4x^2(x^2 + 2x + 8)$

Combine like terms.

2   When multiplying a binomial by a binomial, we can use a special order of multiplying terms, called the **FOIL** order. The letters of FOIL stand for "**F**irst-**O**uter-**I**nner-**L**ast." To illustrate this method, let's multiply $(2x - 3)$ by $(3x + 1)$.

F

Multiply the **F**irst terms of each binomial. $(2x - 3)(3x + 1)$ $2x(3x) = 6x^2$

O

Multiply the **O**uter terms of each binomial. $(2x - 3)(3x + 1)$ $2x(1) = 2x$

I

Multiply the **I**nner terms of each binomial. $(2x - 3)(3x + 1)$ $-3(3x) = -9x$

L

Multiply the **L**ast terms of each binomial. $(2x - 3)(3x + 1)$ $-3(1) = -3$
        Combine like terms.

$$6x^2 + 2x - 9x - 3 = 6x^2 - 7x - 3$$

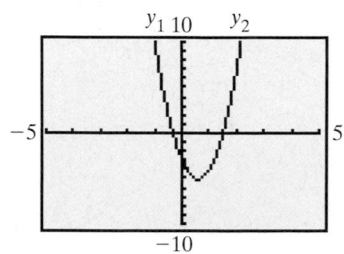

Technology Note

A graphing utility may be used to visualize operations on polynomials. To visualize a multiplication statement such as

$$(2x - 3)(3x + 1) = 6x^2 - 7x - 3,$$

graph $y_1 = (2x - 3)(3x + 1)$ and $y_2 = 6x^2 - 7x - 3$ on the same set of axes and see that their graphs coincide.

Note: If the graphs do not coincide, we can be sure that a mistake has been made in multiplying the polynomials or in entering keystrokes. If the graphs appear to coincide, we cannot be sure that our work is correct. This is because it is possible for the graphs to differ so slightly that we do not notice it.

   The graphs of y_1 and y_2 are shown to the left. The graphs *appear* to coincide so the multiplication statement

$$(2x - 3)(3x + 1) = 6x^2 - 7x - 3$$

appears to be correct.

   A table of values, such as the one to the left showing y_1 and y_2, can also help confirm operations on polynomials.

EXAMPLE 5

Use the FOIL order to multiply $(x - 1)(x + 2)$.

Solution

$$\underset{\text{First}}{\downarrow} \quad \underset{\text{Outer}}{\downarrow} \quad \underset{\text{Inner}}{\downarrow} \quad \underset{\text{Last}}{\downarrow}$$

$$(x - 1)(x + 2) = x \cdot x + 2 \cdot x + (-1)x + (-1)(2)$$

$$- x^2 + 2x - x - 2$$

$$= x^2 + x - 2 \quad \text{Combine like terms.}$$

EXAMPLE 6

Multiply.

a. $(2x - 7)(3x - 4)$ **b.** $(3x + y)(5x - 2y)$

Solution

$$\underset{\text{First}}{\downarrow} \quad \underset{\text{Outer}}{\downarrow} \quad \underset{\text{Inner}}{\downarrow} \quad \underset{\text{Last}}{\downarrow}$$

a. $(2x - 7)(3x - 4) = 2x(3x) + 2x(-4) + (-7)(3x) + (-7)(-4)$

$$= 6x^2 - 8x - 21x + 28$$

$$= 6x^2 - 29x + 28$$

$$\underset{\text{F}}{\downarrow} \quad \underset{\text{O}}{\downarrow} \quad \underset{\text{I}}{\downarrow} \quad \underset{\text{L}}{\downarrow}$$

b. $(3x + y)(5x - 2y) = 15x^2 - 6xy + 5xy - 2y^2$

$$= 15x^2 - xy - 2y^2$$

3 The **square of a binomial** is a special case of the product of two binomials. By the FOIL order for multiplying two binomials, we have

$$(a + b)^2 = (a + b)(a + b)$$

$$\underset{\text{F}}{\downarrow} \quad \underset{\text{O}}{\downarrow} \quad \underset{\text{I}}{\downarrow} \quad \underset{\text{L}}{\downarrow}$$

$$= a^2 + ab + ba + b^2$$

$$= a^2 + 2ab + b^2$$

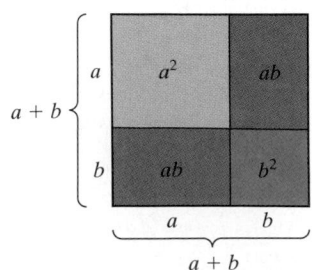

This product can be visualized geometrically by analyzing areas.

Area of larger square: $(a + b)^2$

Sum of areas of smaller rectangles: $a^2 + 2ab + b^2$

Thus, $(a + b)^2 = a^2 + 2ab + b^2$

The same pattern occurs for the square of a difference. In general,

---

Square of a Binomial

$$(a + b)^2 = a^2 + 2ab + b^2 \qquad (a - b)^2 = a^2 - 2ab + b^2$$

In other words, a binomial squared is the sum of the first term squared, twice the product of both terms, and the second term squared.

EXAMPLE 7

Multiply.

a. $(x + 5)^2$ **b.** $(x - 9)^2$ **c.** $(3x + 2z)^2$ **d.** $(4m^2 - 3n)^2$

Solution

$$(a + b)^2 = a^2 + 2 \cdot a \cdot b + b^2$$

a.
$$(x + 5)^2 = x^2 + 2 \cdot x \cdot 5 + 5^2 = x^2 + 10x + 25$$

b. $(x - 9)^2 = x^2 - 2 \cdot x \cdot 9 + 9^2 = x^2 - 18x + 81$

c. $(3x + 2z)^2 = (3x)^2 + 2(3x)(2z) + (2z)^2 = 9x^2 + 12xz + 4z^2$

d. $(4m^2 - 3n)^2 = (4m^2)^2 - 2(4m^2)(3n) + (3n)^2 = 16m^4 - 24m^2n + 9n^2$

> ▶ **Helpful Hint**
>
> Note that $(a + b)^2 = a^2 + 2ab + b^2$, **not** $a^2 + b^2$. Also,
> $(a - b)^2 = a^2 - 2ab + b^2$, **not** $a^2 - b^2$.

---

4   Another special product applies to the sum and difference of the same two terms. Multiply $(a + b)(a - b)$ to see a pattern.

$$(a + b)(a - b) = a^2 - ab + ba - b^2$$
$$= a^2 - b^2$$

Product of the Sum and Difference of Two Terms

$$(a + b)(a - b) = a^2 - b^2$$

The product of the sum and difference of the same two terms is the difference of the first term squared and the second term squared.

EXAMPLE 8

Multiply.

a. $(x - 3)(x + 3)$

b. $(4y + 1)(4y - 1)$

c. $(x^2 + 2y)(x^2 - 2y)$

d. $\left(3m^2 - \dfrac{1}{2}\right)\left(3m^2 + \dfrac{1}{2}\right)$

Solution

a.
$$(a + b)(a - b) = a^2 - b^2$$
$$(x + 3)(x - 3) = x^2 - 3^2 = x^2 - 9$$

b. $(4y + 1)(4y - 1) = (4y)^2 - 1^2 = 16y^2 - 1$

c. $(x^2 + 2y)(x^2 - 2y) = (x^2)^2 - (2y)^2 = x^4 - 4y^2$

d. $\left(3m^2 - \dfrac{1}{2}\right)\left(3m^2 + \dfrac{1}{2}\right) = (3m^2)^2 - \left(\dfrac{1}{2}\right)^2 = 9m^4 - \dfrac{1}{4}$

EXAMPLE 9

Multiply $[3 + (2a + b)]^2$.

Solution

Think of 3 as the first term and $(2a + b)$ as the second term, and apply the method for squaring a binomial.

$$[a \quad + \quad b]^2 = a^2 + 2(a)\cdot \quad b \quad + \quad b^2$$
$$[3 + (2a + b)]^2 = 3^2 + 2(3)(2a + b) + (2a + b)^2$$
$$= 9 + 6(2a + b) + (2a + b)^2$$
$$= 9 + 12a + 6b + (2a)^2 + 2(2a)(b) + b^2 \qquad \text{Square } (2a + b).$$
$$= 9 + 12a + 6b + 4a^2 + 4ab + b^2$$

EXAMPLE 10

Multiply $[(5x - 2y) - 1][(5x - 2y) + 1]$.

Solution

Think of $(5x - 2y)$ as the first term and 1 as the second term, and apply the method for the product of the sum and difference of two terms.

$$\overbrace{(a \quad - b)}\ \overbrace{(a \quad + b)} = \overbrace{a^2 \quad - b^2}$$
$$[\overbrace{(5x - 2y)} - 1][\overbrace{(5x - 2y)} + 1] = \overbrace{(5x - 2y)^2} - 1^2$$
$$= (5x)^2 - 2(5x)(2y) + (2y)^2 - 1 \quad \text{Square } (5x - 2y).$$
$$= 25x^2 - 20xy + 4y^2 - 1$$

5   To multiply three or more polynomials, more than one method may be needed.

EXAMPLE 11

Multiply $(x - 3)(x + 3)(x^2 - 9)$.

Solution   We multiply the first two binomials, the sum and difference of two terms. Then we multiply the resulting two binomials, the square of a binomial.

$$(x - 3)(x + 3)(x^2 - 9) = (x^2 - 9)(x^2 - 9) \quad \text{Multiply } (x - 3)(x + 3).$$
$$= (x^2 - 9)^2$$
$$= x^4 - 18x^2 + 81 \quad \text{Square } (x^2 - 9).$$

6   Our work in multiplying polynomials is often useful in evaluating polynomial functions.

EXAMPLE 12

If $f(x) = x^2 + 5x - 2$, find $f(a + 1)$.

Solution   To find $f(a + 1)$, replace x with the expression $a + 1$ in the polynomial function $f(x)$.

$$f(x) = x^2 + 5x - 2$$
$$f(a + 1) = (a + 1)^2 + 5(a + 1) - 2$$
$$= a^2 + 2a + 1 + 5a + 5 - 2$$
$$= a^2 + 7a + 4$$

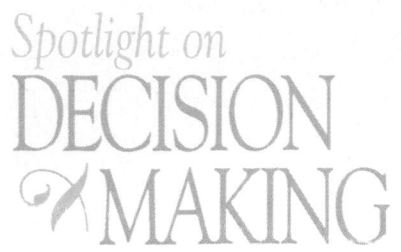

Spotlight on
DECISION
MAKING

Suppose you would like to sign up for an online service and have received the following advertisements from internet service providers in the mail.

US Online

Try our Internet service FREE for 30 days! See why everyone is talking about us:
• Unlimited Internet access and e-mail for a low monthly fee of $21.95
• No set-up fee
• Your own 6 MB Web site
• Thousands of local access numbers across the nation
• Around-the-clock help with our toll-free 888 number
Give us a call to set up your service today!

Interconnect

When you sign up with Interconnect as your Internet service provider, you get:
• Unlimited access to the Internet and e-mail
• Local access numbers in major metropolitan areas
• Online expert help
all for just $11.95 per month*! And, for a limited time, you can try Interconnect for a full month for FREE!
*A one-time $20 set-up fee applies, however.

e-Link

The sky's the limit with e-Link! If you act now, you can get your first 60 days for FREE, as well as a waiver of the $25 set-up fee. For just $19.95 per month, you get:
• 150 hours of Internet access and e-mail*
• a personal 3 MB Web page
• toll-free help, 24 hours a day, 7 days a week
• local access numbers around the country
*Each additional hour costs $2.95.

You construct a decision grid to help make your choice. In the decision grid, give each of the decision criteria a rank reflecting its importance to you, with 1 being not important to 10 being very important. Then for each online service, decide how well the criteria are supported, assigning a 1 in the rating column for poor support to a rating of 10 for excellent support. For US Online, fill in the Score column by multiplying rank by rating for each criteria. Repeat for each online service. Finally, total the scores in each column for each online service. The service with the highest score is likely to be the best choice for you.

Based on your decision-grid analysis, which online service would you choose? Explain.

Criteria	Rank	US Online Rating	US Online Score	Interconnect Rating	Interconnect Score	e-Link Rating	e-Link Score
Free Trial Period							
Monthly Fee							
Unlimited Access							
Includes E-mail							
Includes Personal Web Page							
Set-up Fee							
Toll-free Help							
Local Access Numbers							
TOTAL							

EXERCISE SET 5.4

STUDY GUIDE/SSM   CD/VIDEO   PH MATH TUTOR CENTER   MathXL®Tutorials ON CD   MathXL®   MyMathLab®

Multiply. See Examples 1 through 4.

1. $(-4x^3)(3x^2)$ $-12x^5$

2. $(-6a)(4a)$ $-24a^2$

3. $3x(4x + 7)$ $12x^2 + 21x$

4. $5x(6x - 4)$ $30x^2 - 20x$

5. $-6xy(4x + y)$

6. $-8y(6xy + 4x)$

7. $-4ab(xa^2 + ya^2 - 3)$

8. $-6b^2z(z^2a + baz - 3b)$

9. $(x - 3)(2x + 4)$

10. $(y + 5)(3y - 2)$

11. $(2x + 3)(x^3 - x + 2)$

12. $(a + 2)(3a^2 - a + 5)$

13.
$$\begin{array}{r} 3x - 2 \\ \times\ 5x + 1 \end{array}$$

14.
$$\begin{array}{r} 2z - 4 \\ \times\ 6z - 2 \end{array}$$

15.
$$\begin{array}{r} 3m^2 + 2m - 1 \\ \times\qquad 5m + 2 \end{array}$$

16.
$$\begin{array}{r} 2x^2 - 3x - 4 \\ \times\qquad x + 5 \end{array}$$

17. Explain how to multiply a polynomial by a polynomial.

18. Explain why $(3x + 2)^2$ does not equal $9x^2 + 4$.

17.–18. answers may vary

Multiply the binomials. See Examples 5 and 6.

19. $(x - 3)(x + 4)$

20. $(c - 3)(c + 1)$

21. $(5x + 8y)(2x - y)$

22. $(2n - 9m)(n - 7m)$

23. $(3x - 1)(x + 3)$

24. $(5d - 3)(d + 6)$

25. $\left(3x + \dfrac{1}{2}\right)\left(3x - \dfrac{1}{2}\right)$

26. $\left(2x - \dfrac{1}{3}\right)\left(2x + \dfrac{1}{3}\right)$

Multiply, using special product methods. See Examples 7 and 8.

27. $(x + 4)^2$ $x^2 + 8x + 16$

28. $(x - 5)^2$ $x^2 - 10x + 25$

29. $(6y - 1)(6y + 1)$

30. $(x - 9)(x + 9)$

31. $(3x - y)^2$

32. $(4x - z)^2$

33. $(3b - 6y)(3b + 6y)$

34. $(2x - 4y)(2x + 4y)$

Multiply, using special product methods. See Examples 9 and 10.

35. $[3 + (4b + 1)]^2$ $16b^2 + 32b + 16$

36. $[5 - (3b - 3)]^2$ $9b^2 - 48b + 64$

37. $[(2s - 3) - 1][(2s - 3) + 1]$ $4s^2 - 12s + 8$

38. $[(2y + 5) + 6][(2y + 5) - 6]$ $4y^2 + 20y - 11$

39. $[(xy + 4) - 6]^2$ $x^2y^2 - 4xy + 4$

40. $[(2a^2 + 4a) + 1]^2$ $4a^2 + 16a^3 + 20a^2 + 8a + 1$

41. Explain when the FOIL method can be used to multiply polynomials.   answers may vary

42. Explain why the product of $(a + b)$ and $(a - b)$ is not a trinomial.   answers may vary

Multiply. See Example 11.

43. $(x + y)(2x - 1)(x + 1)$ $2x^3 + 2x^2y + x^2 + xy - x - y$

44. $(z + 2)(z - 3)(2z + 1)$ $2z^3 - z^2 - 13z - 6$

45. $(x - 2)^4$ $x^4 - 8x^3 + 24x^2 - 32x + 16$

46. $(x - 1)^4$ $x^4 - 4x^3 + 6x^2 - 4x + 1$

47. $(x - 5)(x + 5)(x^2 + 25)$ $x^4 - 625$

48. $(x + 3)(x - 3)(x^2 + 9)$ $x^4 - 81$

MIXED PRACTICE

Multiply.

49. $(3x + 1)(3x + 5)$ $9x^2 + 18x + 5$

50. $(4x - 5)(5x + 6)$ $20x^2 - x - 30$

51. $(2x^3 + 5)(5x^2 + 4x + 1)$

52. $(3y^3 - 1)(3y^3 - 6y + 1)$ $9y^6 - 18y^4 + 6y - 1$

53. $(7x - 3)(7x + 3)$ $49x^2 - 9$

54. $(4x + 1)(4x - 1)$ $16x^2 - 1$

55.
$$\begin{array}{r} 3x^2 + 4x - 4 \\ \times\qquad 3x + 6 \end{array}$$ $9x^3 + 30x^2 + 12x - 24$

56.
$$\begin{array}{r} 6x^2 + 2x - 1 \\ \times\qquad 3x - 6 \end{array}$$ $18x^3 - 30x^2 - 15x + 6$

57. $\left(4x + \dfrac{1}{3}\right)\left(4x - \dfrac{1}{2}\right)$ $16x^2 - \dfrac{2}{3}x - \dfrac{1}{6}$

58. $\left(4y - \dfrac{1}{3}\right)\left(3y - \dfrac{1}{8}\right)$ $12y^2 - \dfrac{3}{2}y + \dfrac{1}{24}$

59. $(6x + 1)^2$ $36x^2 + 12x + 1$

60. $(4x + 7)^2$ $16x^2 + 56x + 49$

61. $(x^2 + 2y)(x^2 - 2y)$ $x^4 - 4y^2$

62. $(3x + 2y)(3x - 2y)$ $9x^2 - 4y^2$

63. $-6a^2b^2[5a^2b^2 - 6a - 6b]$ $-30a^4b^4 + 36a^3b^2 + 36a^2b^3$

64. $7x^2y^3(-3ax - 4xy + z)$ $-21ax^3y^3 - 28x^3y^4 + 7x^2y^3z$

65. $(a - 4)(2a - 4)$ $2a^2 - 12a + 16$

66. $(2x - 3)(x + 1)$ $2x^2 - x - 3$

67. $(7ab + 3c)(7ab - 3c)$ $49a^2b^2 - 9c^2$

68. $(3xy - 2b)(3xy + 2b)$ $9x^2y^2 - 4b^2$

69. $(m - 4)^2$ $m^2 - 8m + 16$

70. $(x + 2)^2$ $x^2 + 4x + 4$

71. $(3x + 1)^2$ $9x^2 + 6x + 1$

5. $-24x^2y - 6xy^2$ **6.** $-48xy^2 - 32xy$ **7.** $-4a^3bx - 4a^3by + 12ab$ **8.** $-6b^2z^3a - 6b^3az^2 + 18b^3z$ **9.** $2x^2 - 2x - 12$
10. $3y^2 + 13y - 10$ **11.** $2x^4 + 3x^3 - 2x^2 + x + 6$ **12.** $3a^3 + 5a^2 + 3a + 10$ **13.** $15x^2 - 7x - 2$ **14.** $12z^2 - 28z + 8$
15. $15m^3 + 16m^2 - m - 2$ **16.** $2x^3 + 7x^2 - 19x - 20$ **19.** $x^2 + x - 12$ **20.** $c^2 - 2c - 3$ **21.** $10x^2 + 11xy - 8y^2$
22. $2n^2 - 23mn + 63m^2$ **23.** $3x^2 + 8x - 3$ **24.** $5d^2 + 27d - 18$ **25.** $9x^2 - \dfrac{1}{4}$ **26.** $4x^2 - \dfrac{1}{9}$ **29.** $36y^2 - 1$ **30.** $x^2 - 81$
31. $9x^2 - 6xy + y^2$ **32.** $16x^2 - 8xz + z^2$ **33.** $9b^2 - 36y^2$ **34.** $4x^2 - 16y^2$ **51.** $10x^5 + 8x^4 + 2x^3 + 25x^2 + 20x + 5$

72. $(4x + 6)^2$

73. $(y - 4)(y - 3)$

74. $(c - 8)(c + 2)$

75. $(x + y)(2x - 1)(x + 1)$

76. $(z + 2)(z - 3)(2z + 1)$

77. $(3x^2 + 2x - 1)^2$

78. $(4x^2 + 4x - 4)^2$

79. $(3x + 1)(4x^2 - 2x + 5)$

80. $(2x - 1)(5x^2 - x - 2)$ $10x^3 - 7x^2 - 3x + 2$

If $f(x) = x^2 - 3x$, find the following. See Example 12.

81. $f(a)$ $a^2 - 3a$

82. $f(c)$ $c^2 - 3c$

83. $f(a + h)$

84. $f(a + 5)$ $a^2 + 7a + 10$

85. $f(b - 2)$ $b^2 - 7b + 10$

86. $f(a - b)$

REVIEW AND PREVIEW

Use the slope-intercept form of a line, $y = mx + b$, to find the slope of each line. See Section 2.4.

87. $y = -2x + 7$ -2

88. $y = \dfrac{3}{2}x - 1$ $\dfrac{3}{2}$

89. $3x - 5y = 14$ $\dfrac{3}{5}$

90. $x + 7y = 2$ $-\dfrac{1}{7}$

Use the vertical line test to determine which of the following are graphs of functions. See Section 2.2.

91.

function

92.

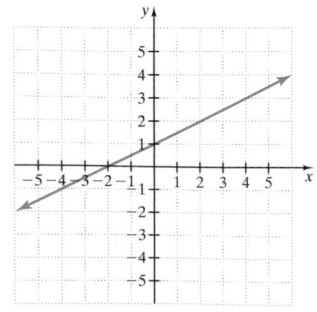

function

Concept Extensions

93. If $F(x) = x^2 + 3x + 2$, find

  a. $F(a + h)$ $a^2 + 2ah + h^2 + 3a + 3h + 2$

  b. $F(a)$ $a^2 + 3a + 2$

  c. $F(a + h) - F(a)$ $2ah + h^2 + 3h$

94. If $g(x) = x^2 + 2x + 1$, find

  a. $g(a + h)$ $a^2 + 2ah + h^2 + 2a + 2h + 1$

  b. $g(a)$ $a^2 + 2a + 1$

  c. $g(a + h) - g(a)$ $2ah + h^2 + 2h$

Multiply. Assume that variables represent positive integers.

95. $5x^2y^n(6y^{n+1} - 2)$

96. $-3yz^n(2y^3z^{2n} - 1)$

97. $(x^a + 5)(x^{2a} - 3)$

98. $(x^a + y^{2b})(x^a - y^{2b})$

For Exercises 99 through 102, write the result as a simplified polynomial.

△ **99.** Find the area of the circle. Do not approximate π.
  $\pi(25x^2 - 20x + 4)$ sq. km

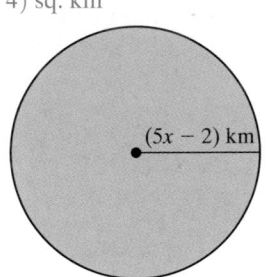

$(5x - 2)$ km

△ **100.** Find the volume of the cylinder. Do not approximate π.
  $\pi(7y^3 - 42y^2 + 63y)$ cu. cm

$(y - 3)$ cm

$7y$ cm

Find the area of each shaded region.

△ **101.**

$(3x - 2)$ in.

$(3x - 2)$ in.

x

x

$(8x^2 - 12x + 4)$ sq. in.

---

72. $16x^2 + 48x + 36$ **73.** $y^2 - 7y + 12$ **74.** $c^2 - 6c - 16$ **75.** $2x^3 + 2x^2y + x^2 + xy - x - y$ **76.** $2z^3 - z^2 - 13z - 6$
77. $9x^4 + 12x^3 - 2x^2 - 4x + 1$ **78.** $16x^4 + 32x^3 - 16x^2 - 32x + 16$ **79.** $12x^3 - 2x^2 + 13x + 5$ **83.** $a^2 + 2ah + h^2 - 3a - 3h$
86. $a^2 - 2ab + b^2 - 3a + 3b$ **95.** $30x^2y^{2n+1} - 10x^2y^n$ **96.** $-6y^4z^{3n} + 3yz^n$ **97.** $x^{3a} + 5x^{2a} - 3x^a - 15$ **98.** $x^{2a} - y^{4b}$

△ **102.**

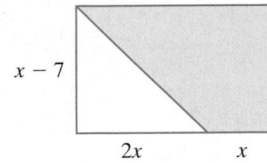

$(2x^2 - 14x)$ sq. units

$x - 7$

$2x$ x

103. Perform each indicated operation. Explain the difference between the two problems.

a. $(3x + 5) + (3x + 7)$ $6x + 12; 9x^2 + 36x + 35$

b. $(3x + 5)(3x + 7)$ one operation is addition, the other is multiplication

104. Explain when the FOIL method can be used to multiply polynomials.   answers may vary

If $R(x) = x + 5, Q(x) = x^2 - 2$, *and* $P(x) = 5x$, *find the following.*

105. $P(x) \cdot R(x)$ $5x^2 + 25x$

106. $P(x) \cdot Q(x)$ $5x^3 - 10x$

107. $[Q(x)]^2$ $x^4 - 4x^2 + 4$

108. $[R(x)]^2$ $x^2 + 10x + 25$

109. $R(x) \cdot Q(x)$ $x^3 + 5x^2 - 2x - 10$

110. $P(x) \cdot R(x) \cdot Q(x)$ $5x^4 + 25x^3 - 10x^2 - 50x$

STUDY SKILLS REMINDER

Are You Getting all the Mathematics Help that You Need?

Remember that, in addition to your instructor, there are many places to get help with your mathematics course. For example, see which of the list below are available.

▶ This text has an accompanying video lesson for every section in this text.

▶ The back of this book contains answers to odd-numbered exercises.

▶ A tutorial software program with lessons corresponding to each section in the text is available.

▶ A student solutions manual is available that contains worked-out solutions to odd-numbered exercises as well as solutions to every exercise in the Integrated Reviews, Chapter Reviews, Chapter Tests, and Cumulative Reviews.

▶ Don't forget to check with your instructor for other local resources available to you, such as a tutor center.

5.5   *THE GREATEST COMMON FACTOR AND FACTORING BY GROUPING*

Objectives

1 Identify the GCF.

2 Factor out the GCF of a polynomial's terms.

3 Factor polynomials by grouping.

1   **Factoring** is the reverse process of multiplying. It is the process of writing a polynomial as a product.

$$6x^2 + 13x - 5 = (3x - 1)(2x + 5)$$

factoring / multiplying

In the next few sections, we review techniques for factoring polynomials. These techniques are used at the end of this chapter to solve polynomial equations.

To factor a polynomial, we first factor out the greatest common factor (GCF) of its terms, using the distributive property. The GCF of a list of terms or monomials is the product of the GCF of the numerical coefficients and each GCF of the powers of a common variable.

Finding the GCF of a List of Monomials

Step 1: Find the GCF of the numerical coefficients.

Step 2: Find the GCF of the variable factors.

Step 3: The product of the factors found in Steps 1 and 2 is the GCF of the monomials.

EXAMPLE 1

Find the GCF of $20x^3y$, $10x^2y^2$, and $35x^3$.

Solution   The GCF of the numerical coefficients 20, 10, and 35 is 5, the largest integer that is a factor of each integer. The GCF of the variable factors x^3, x^2, and x^3 is x^2 because x^2 is the largest factor common to all three powers of x. The variable y is not a common factor because it does not appear in all three monomials. The GCF is thus

$$5 \cdot x^2, \quad \text{or} \quad 5x^2$$

CLASSROOM EXAMPLE
Find the GCF of $8ab^4$, $20a^2b^4$ and $28ab^3$.
answer: $4ab^3$

To see this in factored form,

$$20x^3y = 2 \cdot 2 \cdot 5 \cdot x^2 \cdot x \cdot y$$
$$10x^2y^2 = 2 \cdot 5 \cdot x^2 \cdot y$$
$$35x^3 = 3 \cdot 5 \cdot x^2 \cdot x$$
$$\text{GCF} = 5 \cdot x^2$$

2   A first step in factoring polynomials is to use the distributive property and write the polynomial as a product of the GCF of its monomial terms and a simpler polynomial. This is called **factoring out** the GCF.

EXAMPLE 2

Factor.

a. $8x^2 + 4$ **b.** $5y - 2z^4$ **c.** $6x^2 - 3x^3$

Solution

a. The GCF of terms $8x^2$ and 4 is 4.

$$8x^2 + 4 = 4 \cdot 2x^2 + 4 \cdot 1 \qquad \text{Factor out 4 from each term.}$$
$$= 4(2x^2 + 1) \qquad \text{Apply the distributive property.}$$

The factored form of $8x^2 + 4$ is $4(2x^2 + 1)$. To check, multiply $4(2x^2 + 1)$ to see that the product is $8x^2 + 4$.

b. There is no common factor of the terms $5y$ and $-2z^4$ other than 1 (or -1).

c. The greatest common factor of $6x^2$ and $-3x^3$ is $3x^2$. Thus,

$$6x^2 - 3x^3 = 3x^2 \cdot 2 - 3x^2 \cdot x$$
$$= 3x^2(2 - x)$$

> **Helpful Hint**
>
> To verity that the GCF has been factored out correctly, multiply the factors together and see that their product is the original polynomial.

EXAMPLE 3

Factor $17x^3y^2 - 34x^4y^2$.

Solution

The GCF of the two terms is $17x^3y^2$, which we factor out of each term.

$$17x^3y^2 - 34x^4y^2 = 17x^3y^2 \cdot 1 - 17x^3y^2 \cdot 2x$$
$$= 17x^3y^2(1 - 2x)$$

> **Helpful Hint**
>
> If the GCF happens to be one of the terms in the polynomial, a factor of 1 will remain for this term when the GCF is factored out. For example, in the polynomial $21x^2 + 7x$, the GCF of $21x^2$ and $7x$ is $7x$, so
>
> $$21x^2 + 7x = 7x \cdot 3x + 7x \cdot 1 = 7x(3x + 1)$$

✔ **CONCEPT CHECK**

Which factorization of $12x^2 + 9x - 3$ is correct?

a. $3(4x^2 + 3x + 1)$ **b.** $3(4x^2 + 3x - 1)$ **c.** $3(4x^2 + 3x - 3)$ **d.** $3(4x^2 + 3x)$

EXAMPLE 4

Factor $-3x^3y + 2x^2y - 5xy$.

Solution Two possibilities are shown for factoring this polynomial. First, the common factor xy is factored out.

$$-3x^3y + 2x^2y - 5xy = xy(-3x^2 + 2x - 5)$$

Also, the common factor $-xy$ can be factored out as shown.

$$-3x^3y + 2x^2y - 5xy = -xy(3x^2) + (-xy)(-2x) + (-xy)(5)$$
$$= -xy(3x^2 - 2x + 5)$$

Both of these alternatives are correct.

EXAMPLE 5

Factor $2(x - 5) + 3a(x - 5)$.

Solution The greatest common factor is the binomial factor $(x - 5)$.

$$2(x - 5) + 3a(x - 5) = (x - 5)(2 + 3a)$$

EXAMPLE 6

Factor $7x(x^2 + 5y) - (x^2 + 5y)$.

Solution

$$7x(x^2 + 5y) - (x^2 + 5y) = 7x(x^2 + 5y) - 1(x^2 + 5y)$$
$$= (x^2 + 5y)(7x - 1)$$

▶ **Helpful Hint**

Notice that we wrote $-(x^2 + 5y)$ as $-1(x^2 + 5y)$ to aid in factoring.

3   Sometimes it is possible to factor a polynomial by grouping the terms of the polynomial and looking for common factors in each group. This method of factoring is called **factoring by grouping**.

EXAMPLE 7

Factor $ab - 6a + 2b - 12$.

Solution   First look for the GCF of all four terms. The GCF of all four terms is 1. Next group the first two terms and the last two terms and factor out common factors from each group.

$$ab - 6a + 2b - 12 = (ab - 6a) + (2b - 12)$$

Factor a from the first group and 2 from the second group.

$$= a(b - 6) + 2(b - 6)$$

Now we see a GCF of $(b - 6)$. Factor out $(b - 6)$ to get

$$a(b - 6) + 2(b - 6) = (b - 6)(a + 2)$$

Check: To check, multiply $(b - 6)$ and $(a + 2)$ to see that the product is $ab - 6a + 2b - 12$.

> ### Helpful Hint
> Notice that the polynomial $a(b - 6) + 2(b - 6)$ is *not* in factored form. It is a *sum*, not a *product*. The factored form is $(b - 6)(a + 2)$.

CLASSROOM EXAMPLE

Factor: $xy - 5y + 3x - 15$
answer: $(x - 5)(y + 3)$

TEACHING TIP

After reviewing Example 7, have students try to factor the problem by grouping different pairs of terms. Is there more than one way to group the terms? Is there more than one way to factor the expression? Can all pairings be factored?

TEACHING TIP

Point out that factoring by grouping should be considered when the polynomial has 4 terms.

EXAMPLE 8

Factor $x^3 + 5x^2 + 3x + 15$.

Solution

$$\begin{aligned}
x^3 + 5x^2 + 3x + 15 &= (x^3 + 5x^2) + (3x + 15) &&\text{Group pairs of terms.} \\
&= x^2(x + 5) + 3(x + 5) &&\text{Factor each binomial.} \\
&= (x + 5)(x^2 + 3) &&\text{Factor out the common factor,} \\
& &&(x + 5).
\end{aligned}$$

CLASSROOM EXAMPLE

Factor: $y^3 + 6y^2 + 4y + 24$
answer: $(y + 6)(y^2 + 4)$

EXAMPLE 9

Factor $m^2n^2 + m^2 - 2n^2 - 2$.

Solution

$$\begin{aligned}
m^2n^2 + m^2 - 2n^2 - 2 &= (m^2n^2 + m^2) + (-2n^2 - 2) &&\text{Group pairs of terms.} \\
&= m^2(n^2 + 1) - 2(n^2 + 1) &&\text{Factor each binomial.} \\
&= (n^2 + 1)(m^2 - 2) &&\text{Factor out the common factor,} \\
& &&(n^2 + 1).
\end{aligned}$$

CLASSROOM EXAMPLE

Factor: $a^2b^2 + a^2 - 3b^2 - 3$
answer: $(b^2 + 1)(a^2 - 3)$

EXAMPLE 10

Factor $xy + 2x - y - 2$.

Solution

$$xy + 2x - y - 2 = (xy + 2x) + (-y - 2) \quad \text{Group pairs of terms.}$$
$$= x(y + 2) - 1(y + 2) \quad \text{Factor each binomial.}$$
$$= (y + 2)(x - 1) \quad \text{Factor out the common factor, } (y + 2).$$

CLASSROOM EXAMPLE

Factor: $ab + 5a - b - 5$

answer: $(b + 5)(a - 1)$

MENTAL MATH

Find the GCF of each list of monomials.

1. $6, 12$ 6
2. $9, 27$ 9
3. $15x, 10$ 5
4. $9x, 12$ 3
5. $13x, 2x$ x
6. $4y, 5y$ y
7. $7x, 14x$ $7x$
8. $8z, 4z$ $4z$

EXERCISE SET 5.5

Find the GCF of each list of monomials. See Example 1.

1. a^8, a^5, a^3 a^3
2. b^9, b^2, b^5 b^2
3. $x^2y^3z^3, y^2z^3, xy^2z^2$ y^2z^2
4. $xy^2z^3, x^2y^2z^2, x^2y^3$ xy^2
5. $6x^3y, 9x^2y^2, 12x^2y$ $3x^2y$
6. $4xy^2, 16xy^3, 8x^2y^2$ $4xy^2$
7. $10x^3yz^3, 20x^2z^5, 45xz^3$
8. $12y^2z^4, 9xy^3z^4, 15x^2y^2z^3$

Factor out the GCF in each polynomial. See Examples 2 through 6.

9. $18x - 12$ $6(3x - 2)$
10. $21x + 14$ $7(3x + 2)$
11. $4y^2 - 16xy^3$
12. $3z - 21xz^4$ $3z(1 - 7xz^3)$
13. $6x^5 - 8x^4 + 2x^3$
14. $9x + 3x^2 - 6x^3$
15. $8a^3b^3 - 4a^2b^2 + 4ab + 16ab^2$ $4ab(2a^2b^2 - ab + 1 + 4b)$
16. $12a^3b - 6ab + 18ab^2 - 18a^2b$ $6ab(2a^2 - 1 + 3b - 3a)$
17. $6(x + 3) + 5a(x + 3)$
18. $2(x - 4) + 3y(x - 4)$
19. $2x(z + 7) + (z + 7)$
20. $x(y - 2) + (y - 2)$
21. $3x(x^2 + 5) - 2(x^2 + 5)$
22. $4x(2y + 3) - 5(2y + 3)$
23. When $3x^2 - 9x + 3$ is factored, the result is $3(x^2 - 3x + 1)$. Explain why it is necessary to include the term 1 in this factored form.   answers may vary

24. Construct a trinomial whose GCF is $5x^2y^3$.   answers may vary

Factor each polynomial by grouping. See Examples 7 through 10.

25. $ab + 3a + 2b + 6$
26. $ab + 2a + 5b + 10$
27. $ac + 4a - 2c - 8$
28. $bc + 8b - 3c - 24$
29. $2xy - 3x - 4y + 6$
30. $12xy - 18x - 10y + 15$
31. $12xy - 8x - 3y + 2$
32. $20xy - 15x - 4y + 3$

MIXED PRACTICE

Factor each polynomial.

33. $6x^3 + 9$ $3(2x^3 + 3)$
34. $6x^2 - 8$ $2(3x^2 - 4)$
35. $x^3 + 3x^2$ $x^2(x + 3)$
36. $x^4 - 4x^3$ $x^3(x - 4)$
37. $8a^3 - 4a$ $4a(2a^2 - 1)$
38. $12b^4 + 3b^2$ $3b^2(4b^2 + 1)$
39. $-20x^2y + 16xy^3$ $4xy(-5x + 4y^2)$ or $-4xy(5x - 4y^2)$
40. $-18xy^3 + 27x^4y$ $9xy(-2y^2 + 3x^3)$ or $-9xy(2y^2 - 3x^3)$
41. $10a^2b^3 + 5ab^2 - 15ab^3$ $5ab^2(2ab + 1 - 3b)$
42. $10ef - 20e^2f^3 + 30e^3f$ $10ef(1 - 2ef^2 + 3e^2)$
43. $9abc^2 + 6a^2bc - 6ab + 3bc$ $3b(3ac^2 + 2a^2c - 2a + c)$
44. $4a^2b^2c - 6ab^2c - 4ac + 8a$ $2a(2ab^2c - 3b^2c - 2c + 4)$

---

7. $5xz^3$ 8. $3y^2z^3$ 11. $4y^2(1 - 4xy)$ 13. $2x^3(3x^2 - 4x + 1)$ 14. $3x(3 + x - 2x^2)$ 17. $(x + 3)(6 + 5a)$ 18. $(x - 4)(2 + 3y)$
19. $(z + 7)(2x + 1)$ 20. $(y - 2)(x + 1)$ 21. $(x^2 + 5)(3x - 2)$ 22. $(2y + 3)(4x - 5)$ 25. $(a + 2)(b + 3)$ 26. $(a + 5)(b + 2)$
27. $(a - 2)(c + 4)$ 28. $(b - 3)(c + 8)$ 29. $(x - 2)(2y - 3)$ 30. $(6x - 5)(2y - 3)$ 31. $(4x - 1)(3y - 2)$
32. $-(4y - 3)(5x - 1)$

45. $4x(y - 2) - 3(y - 2)$ **46.** $8y(z + 8) - 3(z + 8)$

47. $6xy + 10x + 9y + 15$ **48.** $15xy + 20x + 6y + 8$

49. $xy + 3y - 5x - 15$ **50.** $xy + 4y - 3x - 12$

51. $6ab - 2a - 9b + 3$ **52.** $16ab - 8a - 6b + 3$

53. $12xy + 18x + 2y + 3$ **54.** $20xy + 8x + 5y + 2$

55. $2m(n - 8) - (n - 8)$ **56.** $3a(b - 4) - (b - 4)$

57. $15x^3y^2 - 18x^2y^2$ **58.** $12x^4y^2 - 16x^3y^3$

59. $2x^2 + 3xy + 4x + 6y$ **60.** $3x^2 + 12x + 4xy + 16y$

61. $5x^2 + 5xy - 3x - 3y$ **62.** $4x^2 + 2xy - 10x - 5y$

63. $x^3 + 3x^2 + 4x + 12$ **64.** $x^3 + 4x^2 + 3x + 12$

65. $x^3 - x^2 - 2x + 2$ **66.** $x^3 - 2x^2 - 3x + 6$

REVIEW EXERCISES

Simplify the following. See Section 5.1.

67. $(5x^2)(11x^5)$ $55x^7$ **68.** $(7y)(-2y^3)$ $-14y^4$

69. $(5x^2)^3$ $125x^6$ **70.** $(-2y^3)^4$ $16y^{12}$

Find each product by using the FOIL order of multiplying binomials. See Section 5.4.

71. $(x + 2)(x - 5)$ $x^2 - 3x - 10$

72. $(x - 7)(x - 1)$ $x^2 - 8x + 7$

73. $(x + 3)(x + 2)$ $x^2 + 5x + 6$

74. $(x - 4)(x + 2)$ $x^2 - 2x - 8$

75. $(y - 3)(y - 1)$ $y^2 - 4y + 3$

76. $(s + 8)(s + 10)$ $s^2 + 18s + 80$

Concept Extensions

77. A factored polynomial can be in many forms. For example, a factored form of $xy - 3x - 2y + 6$ is $(x - 2)(y - 3)$. Which of the following is not a factored form of $xy - 3x - 2y + 6$?   none

   a. $(2 - x)(3 - y)$ **b.** $(-2 + x)(-3 + y)$

   c. $(x - 2)(y - 3)$ **d.** $(-x + 2)(-y + 3)$

78. Consider the following sequence of algebraic steps:

$$x^3 - 6x^2 + 2x - 10 = (x^3 - 6x^2) + (2x - 10)$$
$$= x^2(x - 6) + 2(x - 5)$$

Explain whether the final result is the factored form of the original polynomial.   answers may vary

79. Which factorization of $12x^2 + 9x + 3$ is correct?   a

   a. $3(4x^2 + 3x + 1)$ **b.** $3(4x^2 + 3x - 1)$

   c. $3(4x^2 + 3x - 3)$ **d.** $3(4x^2 + 3x)$

Solve.

△ **80.** The material needed to manufacture a tin can is given by the polynomial $2\pi r^2 + 2\pi rh$ where the radius is r and height is h. Factor this expression.

$2\pi r(r + h)$

81. The amount E of voltage in an electrical circuit is given by the formula

$$IR_1 + IR_2 = E$$

Write an equivalent equation by factoring the expression $IR_1 + IR_2$. $I(R_1 + R_2) = E$

82. At the end of T years, the amount of money A in a savings account earning simple interest from an initial investment of P dollars at rate R is given by the formula

$$A = P + PRT$$

Write an equivalent equation by factoring the expression $P + PRT$. $A = P(1 + RT)$

△ **83.** An open-topped box has a square base and a height of 10 inches. If each of the bottom edges of the box has length x inches, find the amount of material needed to construct the box. Write the answer in factored form. $x(x + 40)$ sq. in.

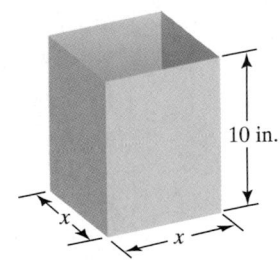

10 in.

84. An object is thrown upward from the ground with an initial velocity of 64 feet per second. The height $h(t)$ of the object after t seconds is given by the polynomial function

$$h(t) = -16t^2 + 64t$$

   a. Write an equivalent factored expression for the function $h(t)$ by factoring $-16t^2 + 64t$. $h(t) = -16t(t - 4)$

   b. Find $h(1)$ by using $h(t) = -16t^2 + 64t$ and then by using the factored form of $h(t)$.   48 ft

   c. Explain why the values found in part **b** are the same.   answers may vary

45. $(y - 2)(4x - 3)$ **46.** $(z + 8)(8y - 3)$ **47.** $(2x + 3)(3y + 5)$ **48.** $(3y + 4)(5x + 2)$ **49.** $(x + 3)(y - 5)$
50. $(x + 4)(y - 3)$ **51.** $(2a - 3)(3b - 1)$ **52.** $(2b - 1)(8a - 3)$ **53.** $(6x + 1)(2y + 3)$ **54.** $(5y + 2)(4x + 1)$
55. $(n - 8)(2m - 1)$ **56.** $(b - 4)(3a - 1)$ **57.** $3x^2y^2(5x - 6)$ **58.** $4x^3y^2(3x - 4y)$ **59.** $(2x + 3y)(x + 2)$
60. $(x + 4)(3x + 4y)$ **61.** $(5x - 3)(x + y)$ **62.** $(2x + y)(2x - 5)$ **63.** $(x^2 + 4)(x + 3)$ **64.** $(x + 4)(x^2 + 3)$
65. $(x^2 - 2)(x - 1)$ **66.** $(x - 2)(x^2 - 3)$

85. An object is dropped from the gondola of a hot-air balloon at a height of 224 feet. The height $h(t)$ of the object after t seconds is given by the polynomial function

$$h(t) = -16t^2 + 224$$

224 ft

a. Write an equivalent factored expression for the function $h(t)$ by factoring $-16t^2 + 224$. $h(t) = -16(t^2 - 14)$

b. Find $h(2)$ by using $h(t) = -16t^2 + 224$ and then by using the factored form of the function.   160 ft

c. Explain why the values found in part **b** are the same. answers may vary

Factor. Assume that variables used as exponents represent positive integers.

86. $x^{3n} - 2x^{2n} + 5x^n$ $x^n(x^{2n} - 2x^n + 5)$

87. $3y^n + 3y^{2n} + 5y^{8n}$ $y^n(3 + 3y^n + 5y^{7n})$

88. $6x^{8a} - 2x^{5a} - 4x^{3a}$ $2x^{3a}(3x^{5a} - x^{2a} - 2)$

89. $3x^{5a} - 6x^{3a} + 9x^{2a}$ $3x^{2a}(x^{3a} - 2x^a + 3)$

5.6   FACTORING TRINOMIALS

Objectives

1 Factor trinomials of the form $x^2 + bx + c$.

2 Factor trinomials of the form $ax^2 + bx + c$.

a. Method 1 - Trial and Check

b. Method 2 - Grouping

3 Factor by substitution.

1   In the previous section, we used factoring by grouping to factor four-term polynomials. In this section, we present techniques for factoring trinomials. Since $(x - 2)(x + 5) = x^2 + 3x - 10$, we say that $(x - 2)(x + 5)$ is a factored form of $x^2 + 3x - 10$. Taking a close look at how $(x - 2)$ and $(x + 5)$ are multiplied suggests a pattern for factoring trinomials of the form

$$x^2 + bx + c$$

$$(x - 2)(x + 5) = x^2 + 3x - 10$$

$$-2 + 5$$
$$-2 \cdot 5$$

The pattern for factoring is summarized next.

> ### Factoring a Trinomial of the Form $x^2 + bx + c$
>
> Find two numbers whose product is c and whose sum is b. The factored form of $x^2 + bx + c$ is
>
> $$(x + \text{one number})(x + \text{other number})$$

EXAMPLE 1

Factor $x^2 + 10x + 16$.

Solution   We look for two integers whose product is 16 and whose sum is 10. Since our integers must have a positive product and a positive sum, we look at only positive factors of 16.

Positive Factors of 16	*Sum of Factors*	
1, 16	$1 + 16 = 17$	
4, 4	$4 + 4 = 8$	
2, 8	$2 + 8 = 10$	*Correct pair*

The correct pair of numbers is 2 and 8 because their product is 16 and their sum is 10. Thus,

$$x^2 + 10x + 16 = (x + 2)(x + 8)$$

Check: To check, see that $(x + 2)(x + 8) = x^2 + 10x + 16$.

> **Technology Note**
>
> Just as a graphing utility can be used to visualize the multiplication of polynomials, a graphing utility can also be used to visualize the factorization of a polynomial. For example, to visualize the factorization from Example 1, graph $y_1 = x^2 + 10x + 16$ and $y_2 = (x + 2)(x + 8)$ and see that the graphs coincide.

EXAMPLE 2

Factor $x^2 - 12x + 35$.

Solution   We need to find two integers whose product is 35 and whose sum is -12. Since our integers must have a positive product and a negative sum, we consider only negative factors of 35.

Negative Factors of 35	*Sum of Factors*	
$-1, -35$	$-1 + (-35) = -36$	
$-5, -7$	$-5 + (-7) = -12$	*Correct pair*

The numbers are -5 and -7.

$$x^2 - 12x + 35 = [x + (-5)][x + (-7)]$$
$$= (x - 5)(x - 7)$$

Check: To check, see that $(x - 5)(x - 7) = x^2 - 12x + 35$.

EXAMPLE 3

Factor $5x^3 - 30x^2 - 35x$.

Solution

First we factor out the greatest common factor, $5x$.

$$5x^3 - 30x^2 - 35x = 5x(x^2 - 6x - 7)$$

Next we try to factor $x^2 - 6x - 7$ by finding two numbers whose product is -7 and whose sum is -6. The numbers are 1 and -7.

$$5x^3 - 30x^2 - 35x = 5x(x^2 - 6x - 7)$$
$$= 5x(x + 1)(x - 7)$$

> **Helpful Hint**
>
> If the polynomial to be factored contains a common factor that is factored out, don't forget to include that common factor in the final factored form of the original polynomial.

EXAMPLE 4

Factor $2n^2 - 38n + 80$.

Solution

The terms of this polynomial have a greatest common factor of 2, which we factor out first.

$$2n^2 - 38n + 80 = 2(n^2 - 19n + 40)$$

Next we factor $n^2 - 19n + 40$ by finding two numbers whose product is 40 and whose sum is -19. Both numbers must be negative since their sum is -19. Possibilities are

$$-1 \text{ and } -40, \qquad -2 \text{ and } -20, \qquad -4 \text{ and } -10, \qquad -5 \text{ and } -8$$

None of the pairs has a sum of -19, so no further factoring with integers is possible. The factored form of $2n^2 - 38n + 80$ is

$$2n^2 - 38n + 80 = 2(n^2 - 19n + 40)$$

We call a polynomial such as $n^2 - 19n + 40$ that cannot be factored further, a **prime polynomial**.

2 Next, we factor trinomials of the form $ax^2 + bx + c$, where the coefficient a of x^2 is not 1. Don't forget that the first step in factoring any polynomial is to factor out the greatest common factor of its terms. We will review two methods here. The first method we'll call trial and check.

EXAMPLE 5

METHOD 1—TRIAL AND CHECK

Factor $2x^2 + 11x + 15$.

Solution Factors of $2x^2$ are $2x$ and x. Let's try these factors as first terms of the binomials.

$$2x^2 + 11x + 15 = (2x +)(x +)$$

Next we try combinations of factors of 15 until the correct middle term, $11x$, is obtained. We will try only positive factors of 15 since the coefficient of the middle term, 11, is positive. Positive factors of 15 are 1 and 15 and 3 and 5.

$(2x + 1)(x + 15)$
$1x$
$30x$
$31x$ *Incorrect middle term*

$(2x + 3)(x + 5)$
$3x$
$10x$
$13x$ *Incorrect middle term*

$(2x + 15)(x + 1)$
$15x$
$2x$
$17x$ *Incorrect middle term*

$(2x + 5)(x + 3)$
$5x$
$6x$
$11x$ *Correct middle term*

Thus, the factored form of $2x^2 + 11x + 15$ is $(2x + 5)(x + 3)$.

Factoring a Trinomial of the Form $ax^2 + bx + c$

Step 1: Write all pairs of factors of ax^2.

Step 2: Write all pairs of factors of c, the constant term.

Step 3: Try various combinations of these factors until the correct middle term bx is found.

Step 4: If no combination exists, the polynomial is **prime**.

EXAMPLE 6

Factor $3x^2 - x - 4$.

Solution Factors of $3x^2$: $3x \cdot x$
Factors of -4: $-1 \cdot 4$, $1 \cdot -4$, $-2 \cdot 2$, $2 \cdot -2$

Let's try possible combinations of these factors.

$(3x - 1)(x + 4)$
$-1x$
$12x$
$11x$ *Incorrect middle term*

$(3x + 4)(x - 1)$
$4x$
$-3x$
$1x$ *Incorrect middle term*

$(3x - 4)(x + 1)$
$-4x$
$3x$
$-1x$ *Correct middle term*

Thus, $3x^2 - x - 4 = (3x - 4)(x + 1)$.

> ### Helpful Hint—Sign Patterns
>
> A positive constant in a trinomial tells us to look for two numbers with the same sign. The sign of the coefficient of the middle term tells us whether the signs are both positive or both negative.
>
> both · same
> positive · sign
> ↓ · ↓
>
> both · same
> negative · sign
> ↓ · ↓
>
> $$2x^2 + 7x + 3 = (2x + 1)(x + 3) \qquad 2x^2 - 7x + 3 = (2x - 1)(x - 3)$$
>
> A negative constant in a trinomial tells us to look for two numbers with opposite signs.
>
> opposite
> signs
> ↓
>
> opposite
> signs
> ↓
>
> $$2x^2 - 5x - 3 = (2x + 1)(x - 3) \qquad 2x^2 + 5x - 3 = (2x - 1)(x + 3)$$

EXAMPLE 7

Factor $12x^3y - 22x^2y + 8xy$.

Solution

First we factor out the greatest common factor of the terms of this trinomial, $2xy$.

$$12x^3y - 22x^2y + 8xy = 2xy(6x^2 - 11x + 4)$$

Now we try to factor the trinomial $6x^2 - 11x + 4$.
Factors of $6x^2$: $2x \cdot 3x$, $\quad 6x \cdot x$

Let's try $2x$ and $3x$.

$$2xy(6x^2 - 11x + 4) = 2xy(2x + \quad)(3x + \quad)$$

The constant term, 4, is positive and the coefficient of the middle term, -11, is negative, so we factor 4 into negative factors only.
Negative factors of 4: $-4(-1)$, $\quad -2(-2)$

Let's try -4 and -1.

$$2xy(2x - \underset{-12x}{\underline{4)(3x}} - 1)$$

$$\underbrace{}$$

$$\begin{array}{r} -2x \\ \hline -14x \end{array} \quad \text{Incorrect middle term}$$

This combination cannot be correct, because one of the factors, $(2x - 4)$, has a common factor of 2. This cannot happen if the polynomial $6x^2 - 11x + 4$ has no common factors.

Now let's try -1 and -4.

$$2xy(2x - \underset{-3x}{\underline{1)(3x}} - 4)$$

$$\begin{array}{r} -8x \\ \hline -11x \end{array} \quad \text{Correct middle term}$$

CLASSROOM EXAMPLE

Factor: $24x^2y^2 - 42xy^2 + 9y^2$
answer: $3y^2(2x - 3)(4x - 1)$

Thus,

$$12x^3y - 22x^2y + 8xy = 2xy(2x - 1)(3x - 4)$$

If this combination had not worked, we would have tried -2 and -2 as factors of 4 and then $6x$ and x as factors of $6x^2$.

> ### Helpful Hint
>
> If a trinomial has no common factor (other than 1), then none of its binomial factors will contain a common factor (other than 1).

EXAMPLE 8

Factor $16x^2 + 24xy + 9y^2$.

Solution   No greatest common factor can be factored out of this trinomial.

Factors of $16x^2$: $16x \cdot x$, $8x \cdot 2x$, $4x \cdot 4x$

Factors of $9y^2$: $y \cdot 9y$, $3y \cdot 3y$

We try possible combinations until the correct factorization is found.

$$16x^2 + 24xy + 9y^2 = (4x + 3y)(4x + 3y) \quad \text{or} \quad (4x + 3y)^2$$

The trinomial $16x^2 + 24xy + 9y^2$ in Example 8 is an example of a **perfect square trinomial** since its factors are two identical binomials. In the next section, we examine a special method for factoring perfect square trinomials.

METHOD 2— GROUPING

There is another method we can use when factoring trinomials of the form $ax^2 + bx + c$: Write the trinomial as a four-term polynomial, and then factor by grouping.

> ### Factoring a Trinomial of the Form $ax^2 + bx + c$ by Grouping
>
> **Step 1:** Find two numbers whose product is $a \cdot c$ and whose sum is b.
> **Step 2:** Write the term bx as a sum by using the factors found in Step 1.
> **Step 3:** Factor by grouping.

EXAMPLE 9

Factor $6x^2 + 13x + 6$.

In this trinomial, $a = 6$, $b = 13$, and $c = 6$.

Solution   **Step 1:** Find two numbers whose product is $a \cdot c$, or $6 \cdot 6 = 36$, and whose sum is b, 13. The two numbers are 4 and 9.

Step 2: Write the middle term, $13x$, as the sum $4x + 9x$.

$$6x^2 + 13x + 6 = 6x^2 + 4x + 9x + 6$$

Step 3: Factor $6x^2 + 4x + 9x + 6$ by grouping.

$$(6x^2 + 4x) + (9x + 6) = 2x(3x + 2) + 3(3x + 2)$$
$$= (3x + 2)(2x + 3)$$

✔ **CONCEPT CHECK**

Name one way that a factorization can be checked.

EXAMPLE 10

Factor $18x^2 - 9x - 2$.

Solution   In this trinomial, $a = 18$, $b = -9$, and $c = -2$.

Step 1: Find two numbers whose product is $a \cdot c$ or $18(-2) = -36$ and whose sum is b, -9. The two numbers are -12 and 3.

Step 2: Write the middle term, $-9x$, as the sum $-12x + 3x$.

$$18x^2 - 9x - 2 = 18x^2 - 12x + 3x - 2$$

Step 3: Factor by grouping.

$$(18x^2 - 12x) + (3x - 2) = 6x(3x - 2) + 1(3x - 2)$$
$$= (3x - 2)(6x + 1).$$

3   A complicated looking polynomial may be a simpler trinomial "in disguise." Revealing the simpler trinomial is possible by substitution.

EXAMPLE 11

Factor $2(a + 3)^2 - 5(a + 3) - 7$.

Solution   The quantity $(a + 3)$ is in two of the terms of this polynomial. **Substitute** x for $(a + 3)$, and the result is the following simpler trinomial.

$$2(a + 3)^2 - 5(a + 3) - 7 \quad \text{Original trinomial.}$$
$$\downarrow \qquad\qquad \downarrow$$
$$= 2(x)^2 \quad - \quad 5(x) - 7 \quad \text{Substitute } x \text{ for } (a + 3).$$

Now factor $2x^2 - 5x - 7$.

$$2x^2 - 5x - 7 = (2x - 7)(x + 1)$$

But the quantity in the original polynomial was $(a + 3)$, not x. Thus, we need to reverse the substitution and replace x with $(a + 3)$.

$$(2x - 7)(x + 1) \qquad \text{Factored expression.}$$
$$= [2(a + 3) - 7][(a + 3) + 1] \quad \text{Substitute } (a + 3) \text{ for } x.$$
$$= (2a + 6 - 7)(a + 3 + 1) \quad \text{Remove inside parentheses.}$$
$$= (2a - 1)(a + 4) \quad \text{Simplify.}$$

Thus, $2(a + 3)^2 - 5(a + 3) - 7 = (2a - 1)(a + 4)$.

Concept Check Answer:
Answers may vary. A sample is: By multiplying the factors to see that the product is the original polynomial.

EXAMPLE 12

Factor $5x^4 + 29x^2 - 42$.

Solution

Again, substitution may help us factor this polynomial more easily. Since this polynomial contains the variable x, we will choose a different substitution variable. Let $y = x^2$, so $y^2 = (x^2)^2$, or x^4. Then

$$5x^4 + 29x^2 - 42$$

becomes

$$5y^2 + 29y - 42$$

which factors as

$$5y^2 + 29y - 42 = (5y - 6)(y + 7)$$

Next, replace y with x^2 to get

$$(5x^2 - 6)(x^2 + 7)$$

Discover the Concept

a. The following polynomials are factored. Graph each related polynomial function shown in a standard window. Compare the x-intercepts of the graph and the factors of the polynomial.

Polynomial	Factors	Related Polynomial Function	x-Intercepts
$x^2 - x - 12$	$(x - 4)(x + 3)$	$y = x^2 - x - 12$	
$x^2 + 10x + 16$	$(x + 2)(x + 8)$	$y = x^2 + 10x + 16$	
$x^2 - 8x + 15$	$(x - 3)(x - 5)$	$y = x^2 - 8x + 15$	
$2x^2 - 17x + 30$	$(2x - 5)(x - 6)$	$y = 2x^2 - 17x + 30$	

b. Find a connection between the factors and the x-intercepts above. If you know the x-intercepts of the related graph, can it help you factor the polynomial?

c. If a function has x-intercepts of 2 and -8, what are two factors of the related polynomial? Graph a related function in factored form and see if the x-intercepts are 2 and -8.

In the above discovery, we find that if the graph of a polynomial function has an x-intercept at a, then the related polynomial has a factor of $(x - a)$. Also, if a polynomial has a factor of $(x - a)$, the graph of the related polynomial function has an x-intercept at a.

MENTAL MATH

1. Find two numbers whose product is 10 and whose sum is 7.   5 and 2
2. Find two numbers whose product is 12 and whose sum is 8.   2 and 6
3. Find two numbers whose product is 24 and whose sum is 11.   8 and 3
4. Find two numbers whose product is 30 and whose sum is 13.   10 and 3

EXERCISE SET 5.6

STUDY GUIDE/SSM   CD/VIDEO   PH MATH TUTOR CENTER   MathXL®Tutorials ON CD   MathXL®   MyMathLab®

Factor each trinomial. See Examples 1 through 4.

1. $x^2 + 9x + 18$　**2.** $x^2 + 9x + 20$

3. $x^2 - 12x + 32$　**4.** $x^2 - 12x + 27$

5. $x^2 + 10x - 24$　**6.** $x^2 + 3x - 54$

7. $x^2 - 2x - 24$　**8.** $x^2 - 9x - 36$

9. $3x^2 - 18x + 24$　**10.** $x^2y^2 + 4xy^2 + 3y^2$

11. $4x^2z + 28xz + 40z$　**12.** $5x^2 - 45x + 70$

13. $2x^2 + 30x - 108$　**14.** $3x^2 + 12x - 96$

15. Find all positive and negative integers b such that $x^2 + bx + 6$ factors.　$\pm 5, \pm 7$

16. Find all positive and negative integers b such that $x^2 + bx - 10$ factors.　$\pm 3, \pm 9$

A polynomial function is graphed on each screen. Use the graph to write factors of the related polynomial. Each tick mark on the x-axis is 1 unit, and it can be assumed that both x-intercepts are integer values.

17.

$(x + 5), (x - 3)$

1. $(x + 3)(x + 6)$
2. $(x + 4)(x + 5)$
3. $(x - 8)(x - 4)$
4. $(x - 3)(x - 9)$
5. $(x + 12)(x - 2)$
6. $(x - 6)(x + 9)$
7. $(x - 6)(x + 4)$

18.

$(x + 2), (x - 1)$

8. $(x + 3)(x - 12)$
9. $3(x - 2)(x - 4)$
10. $y^2(x + 1)(x + 3)$
11. $4z(x + 2)(x + 5)$
12. $5(x - 2)(x - 7)$
13. $2(x + 18)(x - 3)$
14. $3(x - 4)(x + 8)$

19.

$(x - 2), (x - 6)$

21. $(2x - 3)(x - 4)$
22. $(3x - 4)(x - 5)$
23. prime polynomial
24. $(3x - 8)(2x + 1)$

20.

$(x + 1), (x + 5)$

25. $2(3x - 5)(2x + 5)$
26. $3(2y - 3)(2y - 5)$

Factor each trinomial. See Examples 5 through 10.

21. $2x^2 - 11x + 12$　**22.** $3x^2 - 19x + 20$

23. $2x^2 + 25x - 20$　**24.** $6x^2 - 13x - 8$

25. $12x^2 + 10x - 50$　**26.** $12y^2 - 48y + 45$

27. $3y^4 - y^3 - 10y^2$　**28.** $2x^2z + 5xz - 12z$

29. $6x^3 + 8x^2 + 24x$　**30.** $18y^3 + 12y^2 + 2y$

31. $x^2 + 8xz + 7z^2$　**32.** $a^2 - 2ab - 15b^2$

33. $2x^2 - 5xy - 3y^2$　**34.** $6x^2 + 11xy + 4y^2$

35. $x^2 - x - 12$　**36.** $x^2 + 4x - 5$

37. $28y^2 + 22y + 4$　**38.** $24y^3 - 2y^2 - y$

39. $2x^2 + 15x - 27$　**40.** $3x^2 + 14x + 15$

41. Find all positive and negative integers b such that $3x^2 + bx + 5$ factors.　$\pm 8, \pm 16$

42. Find all positive and negative integers b such that $2x^2 + bx + 7$ factors.　$\pm 9, \pm 15$

27. $y^2(3y + 5)(y - 2)$　**28.** $z(2x - 3)(x + 4)$

29. $2x(3x^2 + 4x + 12)$　**30.** $2y(3y + 1)^2$

Use substitution to factor each polynomial completely. See Examples 11 and 12.

43. $x^4 + x^2 - 6$　**44.** $x^4 - x^2 - 20$

45. $(5x + 1)^2 + 8(5x + 1) + 7$　$(5x + 8)(5x + 2)$

46. $(3x - 1)^2 + 5(3x - 1) + 6$　$(3x + 1)(3x + 2)$

47. $x^6 - 7x^3 + 12$　**48.** $x^6 - 4x^3 - 12$

49. $(a + 5)^2 - 5(a + 5) - 24$　$(a - 3)(a + 8)$

50. $(3c + 6)^2 + 12(3c + 6) - 28$　$(3c + 4)(3c + 20)$

31. $(x + 7z)(x + z)$　**32.** $(a + 3b)(a - 5b)$

33. $(2x + y)(x - 3y)$　**34.** $(3x + 4y)(2x + y)$

Solve.

51. The volume $V(x)$ of a box in terms of its height x is given by the function $V(x) = 3x^3 - 2x^2 - 8x$. Factor this expression for $V(x)$.　$x(3x + 4)(x - 2)$

35. $(x - 4)(x + 3)$
36. $(x - 1)(x + 5)$
37. $2(7y + 2)(2y + 1)$
38. $y(6y + 1)(4y - 1)$
39. $(2x - 3)(x + 9)$
40. $(3x + 5)(x + 3)$
43. $(x^2 + 3)(x^2 - 2)$
44. $(x^2 - 5)(x^2 + 4)$
47. $(x^3 - 4)(x^3 - 3)$
48. $(x^3 + 2)(x^3 - 6)$

52. Based on your results from Exercise 51, find the length and width of the box if the height is 5 inches and the dimensions of the box are whole numbers.　width, 3 in., height, 19 in.

MIXED PRACTICE

Factor each polynomial completely.

53. $x^2 - 24x - 81$ **54.** $x^2 - 48x - 100$

55. $x^2 - 15x - 54$ **56.** $x^2 - 15x + 54$

57. $3x^2 - 6x + 3$ $3(x - 1)^2$ **58.** $8x^2 - 8x + 2$ $2(2x - 1)^2$

🔒 **59.** $3x^2 - 5x - 2$ **60.** $5x^2 - 14x - 3$

61. $8x^2 - 26x + 15$ **62.** $12x^2 - 17x + 6$

🔒 **63.** $18x^4 + 21x^3 + 6x^2$ **64.** $20x^5 + 54x^4 + 10x^3$

65. $3a^2 + 12ab + 12b^2$ **66.** $2x^2 + 16xy + 32y^2$

67. $x^2 + 4x + 5$ **68.** $x^2 + 6x + 8$

69. $2(x + 4)^2 + 3(x + 4) - 5$ $(2x + 13)(x + 3)$

70. $3(x + 3)^2 + 2(x + 3) - 5$ $(3x + 14)(x + 2)$

71. $6x^2 - 49x + 30$ **72.** $4x^2 - 39x + 27$

73. $x^4 - 5x^2 - 6$ **74.** $x^4 - 5x^2 + 6$

75. $6x^3 - x^2 - x$ **76.** $12x^3 + x^2 - x$

77. $12a^2 - 29ab + 15b^2$ **78.** $16y^2 + 6yx - 27x^2$

79. $9x^2 + 30x + 25$ **80.** $4x^2 + 6x + 9$

81. $3x^2y - 11xy + 8y$ **82.** $5xy^2 - 9xy + 4x$

83. $2x^2 + 2x - 12$ **84.** $3x^2 + 6x - 45$

85. $(x - 4)^2 + 3(x - 4) - 18$ $(x + 2)(x - 7)$

86. $(x - 3)^2 - 2(x - 3) - 8$ $(x - 1)(x - 7)$

87. $2x^6 + 3x^3 - 9$ **88.** $3x^6 - 14x^3 + 8$

89. $72xy^4 - 24xy^2z + 2xz^2$ **90.** $36xy^2 - 48xyz^2 + 16xz^4$

REVIEW AND PREVIEW

Multiply. See Section 5.4.

91. $(x - 3)(x + 3)$ **92.** $(x - 4)(x + 4)$

93. $(2x + 1)^2$ **94.** $(3x + 5)^2$

95. $(x - 2)(x^2 + 2x + 4)$ **96.** $(y + 1)(y^2 - y + 1)$

Concept Extensions

97. Suppose that a movie is being filmed in New York City. An action shot requires an object to be thrown upward with an initial velocity of 80 feet per second off the top of 1 Madison Square Plaza, a height of 576 feet. The height $h(t)$ in feet of the object after t seconds is given by the function $h(t) = -16t^2 + 80t + 576$. (*Source:* The World Almanac)

    a. 576 ft; 672 ft; 640 ft; 480 ft   **c.** $-16(t + 4)(t - 9)$

    a. Find the height of the object at $t = 0$ seconds, $t = 2$ seconds, $t = 4$ seconds, and $t = 6$ seconds.

    b. Explain why the height of the object increases and then decreases as time passes.   answers may vary

---

53. $(x - 27)(x + 3)$

54. $(x + 2)(x - 50)$

55. $(x - 18)(x + 3)$

61. $(4x - 3)(2x - 5)$
62. $(4x - 3)(3x - 2)$
63. $3x^2(2x + 1)(3x + 2)$
64. $2x^3(2x + 5)(5x + 1)$
65. $3(a + 2b)^2$
66. $2(x + 4y)^2$
67. prime polynomial
68. $(x + 2)(x + 4)$

71. $(3x - 2)(2x - 15)$
72. $(4x - 3)(x - 9)$

---

    c. Factor the polynomial $-16t^2 + 80t + 576$.

576 ft

98. Suppose that an object is thrown upward with an initial velocity of 64 feet per second off the edge of a 960-foot-cliff. The height $h(t)$ in feet of the object after t seconds is given by the function   **a.** 960 ft; 1008 ft; 768 ft; 240 ft

$$h(t) = -16t^2 + 64t + 960$$

    a. Find the height of the object at $t = 0$ seconds, $t = 3$ seconds, $t = 6$ seconds, and $t = 9$ seconds.

    b. Explain why the height of the object increases and then decreases as time passes.   answers may vary

    c. Factor the polynomial $-16t^2 + 64t + 960$.
     $-16(t - 10)(t + 6)$

Factor. Assume that variables used as exponents represent positive integers.

99. $x^{2n} + 10x^n + 16$ **100.** $x^{2n} - 7x^n + 12$

101. $x^{2n} - 3x^n - 18$ **102.** $x^{2n} + 7x^n - 18$

103. $2x^{2n} + 11x^n + 5$ **104.** $3x^{2n} - 8x^n + 4$

105. $4x^{2n} - 12x^n + 9$ **106.** $9x^{2n} + 24x^n + 16$

Recall that a graphing calculator may be used to check addition, subtraction, and multiplication of polynomials. In the same manner, a graphing calculator may be used to check factoring of polynomials in one variable. For example, to see that

$$2x^3 - 9x^2 - 5x = x(2x + 1)(x - 5)$$

graph $Y_1 = 2x^3 - 9x^2 - 5x$ *and* $Y_2 = x(2x + 1)(x - 5)$. *Then trace along both graphs to see that they coincide. Factor the following and use this method to check your results.*

107. $x^4 + 6x^3 + 5x^2$ $x^2(x + 5)(x + 1)$

108. $x^3 + 6x^2 + 8x$ $x(x + 2)(x + 4)$

109. $30x^3 + 9x^2 - 3x$ $3x(5x - 1)(2x + 1)$

110. $-6x^4 + 10x^3 - 4x^2$ $-2x^2(3x - 2)(x - 1)$

---

73. $(x^2 - 6)(x^2 + 1)$ **74.** $(x^2 - 2)(x^2 - 3)$ **75.** $x(3x + 1)(2x - 1)$

76. $x(4x - 1)(3x + 1)$ **77.** $(4a - 3b)(3a - 5b)$

78. $(8y - 9x)(2y + 3x)$ **79.** $(3x + 5)^2$ **80.** prime polynomial   **81.** $y(3x - 8)(x - 1)$ **82.** $x(5y - 4)(y - 1)$

83. $2(x + 3)(x - 2)$ **84.** $3(x - 3)(x + 5)$ **87.** $(2x^3 - 3)(x^3 + 3)$ **88.** $(3x^3 - 2)(x^3 - 4)$ **89.** $2x(6y^2 - z)^2$

90. $4x(3y - 2z^2)^2$ **91.** $x^2 - 9$ **92.** $x^2 - 16$ **93.** $4x^2 + 4x + 1$ **94.** $9x^2 + 30x + 25$ **95.** $x^3 - 8$ **96.** $y^3 + 1$

99. $(x^n + 2)(x^n + 8)$ **100.** $(x^n - 3)(x^n - 4)$ **101.** $(x^n - 6)(x^n + 3)$ **102.** $(x^n - 2)(x^n + 9)$ **103.** $(2x^n + 1)(x^n + 5)$

104. $(3x^n - 2)(x^n - 2)$ **105.** $(2x^n - 3)^2$ **106.** $(3x^n + 4)$

5.7   *FACTORING BY SPECIAL PRODUCTS*

Objectives

1.   Factor a perfect square trinomial.
2.   Factor the difference of two squares.
3.   Factor the sum or difference of two cubes.

1   In the previous section, we considered a variety of ways to factor trinomials of the form $ax^2 + bx + c$. In one particular example, we factored $16x^2 + 24xy + 9y^2$ as

$$16x^2 + 24xy + 9y^2 = (4x + 3y)^2$$

Recall that $16x^2 + 24xy + 9y^2$ is a perfect square trinomial because its factors are two identical binomials. A perfect square trinomial can be factored quickly if you recognize the trinomial as a perfect square.

A trinomial is a perfect square trinomial if it can be written so that its first term is the square of some quantity a, its last term is the square of some quantity b, and its middle term is twice the product of the quantities a and b. The following special formulas can be used to factor perfect square trinomials.

TEACHING TIP

Show that the rectangular model of a perfect square trinomial forms a square because the length equals the width.

	$4x$	$+$	3
$4x$	$16x^2$		$12x$
$+$			
3	$12x$		9

> **Perfect Square Trinomials**
>
> $$a^2 + 2ab + b^2 = (a + b)^2$$
> $$a^2 - 2ab + b^2 = (a - b)^2$$

Notice that these formulas above are the same special products from Section 5.4 for the square of a binomial.

From

$$a^2 + 2ab + b^2 = (a + b)^2,$$

we see that

$$16x^2 + 24xy + 9y^2 = (4x)^2 + 2(4x)(3y) + (3y)^2 = (4x + 3y)^2$$

EXAMPLE 1

Factor $m^2 + 10m + 25$.

Solution   Notice that the first term is a square: $m^2 = (m)^2$, the last term is a square: $25 = 5^2$; and $10m = 2 \cdot 5 \cdot m$.

Thus,

$$m^2 + 10m + 25 = m^2 + 2(m)(5) + 5^2 = (m + 5)^2$$

EXAMPLE 2

Factor $3a^2x - 12abx + 12b^2x$.

Solution   The terms of this trinomial have a GCF of $3x$, which we factor out first.

$$3a^2x - 12abx + 12b^2x = 3x(a^2 - 4ab + 4b^2)$$

Now, the polynomial $a^2 - 4ab + 4b^2$ is a perfect square trinomial. Notice that the first term is a square: $a^2 = (a)^2$; the last term is a square: $4b^2 = (2b)^2$; and $4ab = 2(a)(2b)$. The factoring can now be completed as

$$3x(a^2 - 4ab + 4b^2) = 3x(a - 2b)^2$$

> **Helpful Hint**
>
> If you recognize a trinomial as a perfect square trinomial, use the special formulas to factor. However, methods for factoring trinomials in general from Section 5.6 will also result in the correct factored form.

2   We now factor special types of binomials, beginning with the **difference of two squares.** The special product pattern presented in Section 5.4 for the product of a sum and a difference of two terms is used again here. However, the emphasis is now on factoring rather than on multiplying.

> **Difference of Two Squares**
>
> $$a^2 - b^2 = (a + b)(a - b)$$

Notice that a binomial is a difference of two squares when it is the difference of the square of some quantity a and the square of some quantity b.

EXAMPLE 3

Factor the following.

a. $x^2 - 9$ **b.** $16y^2 - 9$ **c.** $50 - 8y^2$ **d.** $x^2 - \dfrac{1}{4}$

Solution   **a.** $x^2 - 9 = x^2 - 3^2$ **b.** $16y^2 - 9 = (4y)^2 - 3^2$
         $= (x + 3)(x - 3)$ $= (4y + 3)(4y - 3)$

c. First factor out the common factor of 2.
   $50 - 8y^2 = 2(25 - 4y^2)$
       $= 2(5 + 2y)(5 - 2y)$

d. $x^2 - \dfrac{1}{4} = x^2 - \left(\dfrac{1}{2}\right)^2 = \left(x + \dfrac{1}{2}\right)\left(x - \dfrac{1}{2}\right)$

✔ **CONCEPT CHECK**

Is $(x - 4)(y^2 - 9)$ completely factored? Why or why not?

The binomial $x^2 + 9$ is a **sum of two squares** and cannot be factored by using real numbers. **In general, except for factoring out a GCF, the sum of two squares usually cannot be factored by using real numbers.**

> **Helpful Hint**
> The sum of two squares whose GCF is 1 usually cannot be factored by using real numbers. For example, $x^2 + 9$ is called a prime polynomial.

EXAMPLE 4

Factor the following.

a. $p^4 - 16$

b. $(x + 3)^2 - 36$

Solution

a. $p^4 - 16 = (p^2)^2 - 4^2$

$$= (p^2 + 4)(p^2 - 4)$$

The binomial factor $p^2 + 4$ cannot be factored by using real numbers, but the binomial factor $p^2 - 4$ is a difference of squares.

$$(p^2 + 4)(p^2 - 4) = (p^2 + 4)(p + 2)(p - 2)$$

b. Factor $(x + 3)^2 - 36$ as the difference of squares.

$$(x + 3)^2 - 36 = (x + 3)^2 - 6^2$$

$$= [(x + 3) + 6][(x + 3) - 6] \quad \text{Factor.}$$

$$= [x + 3 + 6][x + 3 - 6] \quad \text{Remove parentheses.}$$

$$= (x + 9)(x - 3) \quad \text{Simplify.}$$

EXAMPLE 5

Factor $x^2 + 4x + 4 - y^2$.

Solution

Factoring by grouping comes to mind since the sum of the first three terms of this polynomial is a perfect square trinomial.

$$x^2 + 4x + 4 - y^2 = (x^2 + 4x + 4) - y^2 \quad \text{Group the first three terms.}$$

$$= (x + 2)^2 - y^2 \quad \text{Factor the perfect square trinomial.}$$

This is not factored yet since we have a *difference*, not a *product*. Since $(x + 2)^2 - y^2$ is a difference of squares, we have

$$(x + 2)^2 - y^2 = [(x + 2) + y][(x + 2) - y]$$

$$= (x + 2 + y)(x + 2 - y)$$

3   Although the sum of two squares usually cannot be factored, the sum of two cubes, as well as the difference of two cubes, can be factored as follows.

> **Sum and Difference of Two Cubes**
>
> $$a^3 + b^3 = (a + b)(a^2 - ab + b^2)$$
> $$a^3 - b^3 = (a - b)(a^2 + ab + b^2)$$

To check the first pattern, let's find the product of $(a + b)$ and $(a^2 - ab + b^2)$.

$$(a + b)(a^2 - ab + b^2) = a(a^2 - ab + b^2) + b(a^2 - ab + b^2)$$
$$= a^3 - a^2b + ab^2 + a^2b - ab^2 + b^3$$
$$= a^3 + b^3$$

EXAMPLE 6

Factor $x^3 + 8$.

Solution   First we write the binomial in the form $a^3 + b^3$. Then we use the formula

$$a^3 + b^3 = (a + b)(a^2 - a \cdot b + b^2), \quad \text{where } a \text{ is } x \text{ and } b \text{ is } 2.$$
$$x^3 + 8 = x^3 + 2^3 = (x + 2)(x^2 - x \cdot 2 + 2^2)$$

Thus, $x^3 + 8 = (x + 2)(x^2 - 2x + 4)$

EXAMPLE 7

Factor $p^3 + 27q^3$.

Solution

$$p^3 + 27q^3 = p^3 + (3q)^3$$
$$= (p + 3q)[p^2 - (p)(3q) + (3q)^2]$$
$$= (p + 3q)(p^2 - 3pq + 9q^2)$$

EXAMPLE 8

Factor $y^3 - 64$.

Solution   This is a difference of cubes since $y^3 - 64 = y^3 - 4^3$.

From

$$a^3 - b^3 = (a - b)(a^2 + a \cdot b + b^2)$$
$$y^3 - 4^3 = (y - 4)(y^2 + y \cdot 4 + 4^2)$$
$$= (y - 4)(y^2 + 4y + 16)$$

> **Helpful Hint**
>
> When factoring sums or differences of cubes, be sure to notice the sign patterns.
>
> same sign
>
> $$x^3 + y^3 = (x + y)(x^2 - xy + y^2)$$
>
> opposite sign   always positive
>
> same sign
>
> $$x^3 - y^3 = (x - y)(x^2 + xy + y^2)$$
>
> opposite sign   always positive

EXAMPLE 9

Factor $125q^2 - n^3q^2$.

Solution First we factor out a common factor of q^2.

$$125q^2 - n^3q^2 = q^2(125 - n^3)$$
$$= q^2(5^3 - n^3)$$

opposite sign   positive

$$= q^2(5 - n)[5^2 + (5)(n) + n^2]$$
$$= q^2(5 - n)(25 + 5n + n^2)$$

Thus, $125q^2 - n^3q^2 = q^2(5 - n)(25 + 5n + n^2)$. The trinomial $25 + 5n + n^2$ cannot be factored further.

EXERCISE SET 5.7

STUDY GUIDE/SSM   CD/VIDEO   PH MATH TUTOR CENTER   MathXL®Tutorials ON CD   MathXL®   MyMathLab®

Factor the following. See Examples 1 and 2.   6. $x(x + 7)^2$

1. $x^2 + 6x + 9$ $(x + 3)^2$ **2.** $x^2 - 10x + 25$ $(x - 5)^2$

3. $4x^2 - 12x + 9$ $(2x - 3)^2$ **4.** $25x^2 + 10x + 1$ $(5x + 1)^2$

5. $3x^2 - 24x + 48$ $3(x - 4)^2$ **6.** $x^3 + 14x^2 + 49x$

7. $9y^2x^2 + 12yx^2 + 4x^2$ **8.** $32x^2 - 16xy + 2y^2$
$x^2(3y + 2)^2$ $2(4x - y)^2$

Factor the following. See Examples 3 and 4.

9. $x^2 - 25$ $(x + 5)(x - 5)$ **10.** $y^2 - 100$

11. $9 - 4z^2$ $(3 + 2z)(3 - 2z)$ **12.** $16x^2 - y^2$

13. $(y + 2)^2 - 49$ **14.** $(x - 1)^2 - z^2$

15. $64x^2 - 100$ **16.** $4x^2 - 36$
$4(4x + 5)(4x - 5)$ $4(x + 3)(x - 3)$

Factor the following. See Examples 6 through 9.

17. $x^3 + 27$ **18.** $y^3 + 1$

19. $z^3 - 1$ **20.** $x^3 - 8$

21. $m^3 + n^3$ **22.** $r^3 + 125$

23. $x^3y^2 - 27y^2$ **24.** $64 - p^3$

25. $a^3b + 8b^4$ **26.** $8ab^3 + 27a^4$

27. $125y^3 - 8x^3$ **28.** $54y^3 - 128$

Factor the following. See Example 5.

29. $x^2 + 6x + 9 - y^2$ **30.** $x^2 + 12x + 36 - y^2$

31. $x^2 - 10x + 25 - y^2$ **32.** $x^2 - 18x + 81 - y^2$

33. $4x^2 + 4x + 1 - z^2$ **34.** $9y^2 + 12y + 4 - x^2$

MIXED PRACTICE

Factor each polynomial completely.

35. $9x^2 - 49$ **36.** $25x^2 - 4$

37. $x^2 - 12x + 36$ $(x - 6)^2$ **38.** $x^2 - 18x + 81$ $(x - 9)^2$

39. $x^4 - 81$ **40.** $x^4 - 256$

41. $x^2 + 8x + 16 - 4y^2$ **42.** $x^2 + 14x + 49 - 9y^2$

43. $(x + 2y)^2 - 9$ **44.** $(3x + y)^2 - 25$

45. $x^3 - 216$ **46.** $8 - a^3$

47. $x^3 + 125$ **48.** $x^3 + 216$

49. $4x^2 + 25$ **50.** $16x^2 + 25$

51. $4a^2 + 12a + 9$ **52.** $9a^2 - 30a + 25$

53. $18x^2y - 2y$ **54.** $12xy^2 - 108x$

55. $8x^3 + y^3$ **56.** $27x^3 - y^3$

57. $x^6 - y^3$ **58.** $x^3 - y^6$

59. $x^2 + 16x + 64 - x^4$ **60.** $x^2 + 20x + 100 - x^4$

61. $3x^6y^2 + 81y^2$ **62.** $x^2y^9 + x^2y^3$

63. $(x + y)^3 + 125$ **64.** $(x + y)^3 + 27$

65. $(2x + 3)^3 - 64$ **66.** $(4x + 2)^3 - 125$

REVIEW AND PREVIEW

Solve the following equations. See Section 1.5.

67. $x - 5 = 0$ 5 **68.** $x + 7 = 0$ -7

69. $3x + 1 = 0$ $-\frac{1}{3}$ **70.** $5x - 15 = 0$ 3

71. $-2x = 0$ 0 **72.** $3x = 0$ 0

73. $-5x + 25 = 0$ 5 **74.** $-4x - 16 = 0$ -4

Concept Extensions

△ **75.** A manufacturer of metal washers needs to determine the cross-sectional area of each washer. If the outer radius of the washer is R and the radius of the hole is r, express the area of the washer as a polynomial. Factor this polynomial completely. $\pi R^2 - \pi r^2 = \pi(R + r)(R - r)$

△ **76.** Express the area of the shaded region as a polynomial. Factor the polynomial completely.
$A = x^2 - y^2 = (x + y)(x - y)$

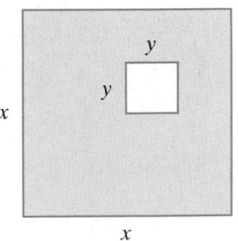

10. $(y + 10)(y - 10)$ **12.** $(4x + y)(4x - y)$ **13.** $(y + 9)(y - 5)$ **14.** $(x - 1 + z)(x - 1 - z)$ **17.** $(x + 3)(x^2 - 3x + 9)$
18. $(y + 1)(y^2 - y + 1)$ **19.** $(z - 1)(z^2 + z + 1)$ **20.** $(x - 2)(x^2 + 2x + 4)$ **21.** $(m + n)(m^2 - mn + n^2)$
22. $(r + 5)(r^2 - 5r + 25)$ **23.** $y^2(x - 3)(x^2 + 3x + 9)$ **24.** $(4 - p)(16 + 4p + p^2)$ **25.** $b(a + 2b)(a^2 - 2ab + 4b^2)$
26. $a(2b + 3a)(4b^2 - 6ab + 9a^2)$ **27.** $(5y - 2x)(25y^2 + 10yx + 4x^2)$ **28.** $2(3y - 4)(9y^2 + 12y + 16)$ **29.** $(x + 3 + y)(x + 3 - y)$
30. $(x + 6 + y)(x + 6 - y)$ **31.** $(x - 5 + y)(x - 5 - y)$ **32.** $(x - 9 + y)(x - 9 - y)$ **33.** $(2x + 1 + z)(2x + 1 - z)$
34. $(3y + 2 + x)(3y + 2 - x)$ **35.** $(3x + 7)(3x - 7)$ **36.** $(5x + 2)(5x - 2)$ **39.** $(x^2 + 9)(x + 3)(x - 3)$
40. $(x^2 + 16)(x + 4)(x - 4)$ **41.** $(x + 4 + 2y)(x + 4 - 2y)$ **42.** $(x + 7 + 3y)(x + 7 - 3y)$ **43.** $(x + 2y + 3)(x + 2y - 3)$
44. $(3x + y + 5)(3x + y - 5)$ **45.** $(x - 6)(x^2 + 6x + 36)$ **46.** $(2 - a)(4 + 2a + a^2)$ **47.** $(x + 5)(x^2 - 5x + 25)$
48. $(x + 6)(x^2 - 6x + 36)$ **49.** prime polynomial **50.** prime polynomial

Express the volume of each solid as a polynomial. To do so, subtract the volume of the "hole" from the volume of the larger solid. Then factor the resulting polynomial.

△ **77.**

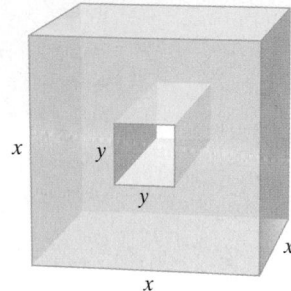

$x^3 - y^2x; \; x(x+y)(x-y)$

△ **78.**

$\pi x^2h - \pi y^2h; \; \pi h(x+y)(x-y)$

Find the value of c that makes each trinomial a perfect square trinomial.

79. $x^2 + 6x + c$ $c = 9$ **80.** $y^2 + 10y + c$ $c = 25$

81. $m^2 - 14m + c$ $c = 49$ **82.** $n^2 - 2n + c$ $c = 1$

83. $x^2 + cx + 16$ $c = \pm 8$ **84.** $x^2 + cx + 36$ $c = \pm 12$

85. Factor $x^6 - 1$ completely, using the following methods from this chapter.   **a.** $(x+1)(x^2-x+1)(x-1)(x^2+x+1)$

 a. Factor the expression by treating it as the difference of two squares, $(x^3)^2 - 1^2$.

 b. Factor the expression treating it as the difference of two cubes, $(x^2)^3 - 1^3$. $(x+1)(x-1)(x^4 + x^2 + 1)$

 ✎ **c.** Are the answers to parts **a** and **b** the same? Why or why not?   answers may vary

Factor. Assume that variables used as exponents represent positive integers.

86. $x^{2n} - 25$ $(x^n + 5)(x^n - 5)$

87. $x^{2n} - 36$ $(x^n + 6)(x^n - 6)$

88. $36x^{2n} - 49$ $(6x^n + 7)(6x^n - 7)$

89. $25x^{2n} - 81$ $(5x^n + 9)(5x^n - 9)$

90. $x^{4n} - 16$ $(x^{2n} + 4)(x^n + 2)(x^n - 2)$

91. $x^{4n} - 625$ $(x^{2n} + 25)(x^n + 5)(x^n - 5)$

51. $(2a+3)^2$ **52.** $(3a-5)^2$ **53.** $2y(3x+1)(3x-1)$ **54.** $12x(y+3)(y-3)$ **55.** $(2x+y)(4x^2-2xy+y^2)$
56. $(3x-y)(9x^2+3xy+y^2)$ **57.** $(x^2-y)(x^4+x^2y+y^2)$ **58.** $(x-y^2)(x^2+xy^2+y^4)$ **59.** $(x+8+x^2)(x+8-x^2)$
60. $(x+10+x^2)(x+10-x^2)$ **61.** $3y^2(x^2+3)(x^4-3x^2+9)$ **62.** $x^2y^3(y^2+1)(y^4-y^2+1)$
63. $(x+y+5)(x^2+2xy+y^2-5x-5y+25)$ **64.** $(x+y+3)(x^2+2xy+y^2-3x-3y+9)$
65. $(2x-1)(4x^2+20x+37)$ **66.** $(4x-3)(16x^2+36x+39)$

INTEGRATED REVIEW   OPERATIONS ON POLYNOMIALS AND FACTORING STRATEGIES

OPERATIONS ON POLYNOMIALS

Perform each indicated operation.

1. $(-y^2 + 6y - 1) + (3y^2 - 4y - 10)$
$2y^2 + 2y - 11$

2. $(5z^4 - 6z^2 + z + 1) - (7z^4 - 2z + 1)$
$-2z^4 - 6z^2 + 3z$

3. Subtract $(x - 5)$ from $(x^2 - 6x + 2)$.
$x^2 - 7x + 7$

4. $(2x^2 + 6x - 5) + (5x^2 - 10x)$
$7x^2 - 4x - 5$

5. $(5x - 3)^2$
$25x^2 - 30x + 9$

6. $(5x^2 - 14x - 3) \div (5x + 1)$
$x - 3$

7. $(2x^4 - 3x^2 + 5x - 2) \div (x + 2)$

$2x^3 - 4x^2 + 5x - 5 + \dfrac{8}{x+2}$

8. $(4x - 1)(x^2 - 3x - 2)$
$4x^3 - 13x^2 - 5x + 2$

FACTORING STRATEGIES

The key to proficiency in factoring polynomials is to practice until you are comfortable with each technique. A strategy for factoring polynomials completely is given next.

Factoring a Polynomial

Step 1: Are there any common factors? If so, factor out the greatest common factor.

Step 2: How many terms are in the polynomial?

  a. If there arc *two* terms, decide if one of the following formulas may be applied:

  i. Difference of two squares: $a^2 - b^2 = (a - b)(a + b)$

  ii. Difference of two cubes: $a^3 - b^3 = (a - b)(a^2 + ab + b^2)$

  iii. Sum of two cubes: $a^3 + b^3 = (a + b)(a^2 - ab + b^2)$

  b. If there are *three* terms, try one of the following:

  i. Perfect square trinomial: $a^2 + 2ab + b^2 = (a + b)^2$
  $$a^2 - 2ab + b^2 = (a - b)^2$$

  ii. If not a perfect square trinomial, factor by using the methods presented in Section 5.5.

  c. If there are *four* or more terms, try factoring by grouping.

Step 3: See whether any factors in the factored polynomial can be factored further.

A few examples are worked for you below.

EXAMPLE 1

Factor each polynomial complctely.

a. $8a^2b - 4ab$ **b.** $36x^2 - 9$ **c.** $2x^2 - 5x - 7$

d. $5p^2 + 5 + qp^2 + q$ **e.** $9x^2 + 24x + 16$ **f.** $y^2 + 25$

Solution **a. Step 1:** The terms have a common factor of *4ab*, which we factor out.

$$8a^2b - 4ab = 4ab(2a - 1)$$

Step 2: There are two terms, but the binomial $2a - 1$ is not the difference of two squares or the sum or difference of two cubes.

Step 3: The factor $2a - 1$ cannot be factored further.

b. Step 1: Factor out a common factor of 9.

$$36x^2 - 9 = 9(4x^2 - 1)$$

Step 2: The factor $4x^2 - 1$ has two terms, and it is the difference of two squares.

$$9(4x^2 - 1) = 9(2x + 1)(2x - 1)$$

Step 3: No factor with more than one term can be factored further.

c. Step 1: The terms of $2x^2 - 5x - 7$ contain no common factor other than 1 or -1.

Step 2: There are three terms. The trinomial is not a perfect square, so we factor by methods from Section 5.6.
$$2x^2 - 5x - 7 = (2x - 7)(x + 1)$$

Step 3: No factor with more than one term can be factored further.

d. Step 1: There is no common factor of all terms of $5p^2 + 5 + qp^2 + q$.

Step 2: The polynomial has four terms, so try factoring by grouping.
$$5p^2 + 5 + qp^2 + q = (5p^2 + 5) + (qp^2 + q) \qquad \text{Group the terms.}$$
$$= 5(p^2 + 1) + q(p^2 + 1)$$
$$= (p^2 + 1)(5 + q)$$

Step 3: No factor can be factored further.

e. Step 1: The terms of $9x^2 + 24x + 16$ contain no common factor other than 1 or -1.

Step 2: The trinomial $9x^2 + 24x + 16$ is a perfect square trinomial, and $9x^2 + 24x + 16 = (3x + 4)^2$.

Step 3: No factor can be factored further.

f. Step 1: There is no common factor of $y^2 + 25$ other than 1.

Step 2: This binomial is the sum of two squares and is prime.

Step 3: The binomial $y^2 + 25$ cannot be factored further.

EXAMPLE 2

Factor each completely.

a. $27a^3 - b^3$ **b.** $3n^2m^4 - 48m^6$ **c.** $2x^2 - 12x + 18 - 2z^2$
d. $8x^4y^2 + 125xy^2$ **e.** $(x - 5)^2 - 49y^2$

Solution **a.** This binomial is the difference of two cubes.
$$27a^3 - b^3 = (3a)^3 - b^3$$
$$= (3a - b)[(3a)^2 + (3a)(b) + b^2]$$
$$= (3a - b)(9a^2 + 3ab + b^2)$$

b. $3n^2m^4 - 48m^6 = 3m^4(n^2 - 16m^2)$ Factor out the GCF, $3m^4$.
$$= 3m^4(n + 4m)(n - 4m) \quad \text{Factor the difference of squares.}$$

c. $2x^2 - 12x + 18 - 2z^2 = 2(x^2 - 6x + 9 - z^2)$ The GCF is 2.
$$= 2[(x^2 - 6x + 9) - z^2] \quad \begin{array}{l}\text{Group the first three}\\\text{terms together.}\end{array}$$
$$= 2[(x - 3)^2 - z^2] \quad \begin{array}{l}\text{Factor the perfect}\\\text{square trinomial.}\end{array}$$
$$= 2[(x - 3) + z][(x - 3) - z] \quad \begin{array}{l}\text{Factor the difference}\\\text{of squares.}\end{array}$$
$$= 2(x - 3 + z)(x - 3 - z)$$

d. $8x^4y^2 + 125xy^2 = xy^2(8x^3 + 125)$ The GCF is xy^2.

$= xy^2[(2x)^3 + 5^3]$

$= xy^2(2x + 5)[(2x)^2 - (2x)(5) + 5^2]$ Factor the sum of cubes.

$= xy^2(2x + 5)(4x^2 - 10x + 25)$

e. This binomial is the difference of squares.

$(x - 5)^2 - 49y^2 = (x - 5)^2 - (7y)^2$

$= [(x - 5) + 7y][(x - 5) - 7y]$

$= (x - 5 + 7y)(x - 5 - 7y)$

Factor completely.

9. $x^2 - 8x + 16 - y^2$

$(x - 4 + y)$

$(x - 4 - y)$

10. $12x^2 - 22x - 20$

$2(3x + 2)(2x - 5)$

11. $x^4 - x$

$x(x - 1)$

$(x^2 + x + 1)$

12. $(2x + 1)^2 - 3(2x + 1) + 2$

$2x(2x - 1)$

13. $14x^2y - 2xy$

$2xy(7x - 1)$

14. $24ab^2 - 6ab$

$6ab(4b - 1)$

15. $4x^2 - 16$

$4(x + 2)(x - 2)$

16. $9x^2 - 81$

$9(x + 3)(x - 3)$

17. $3x^2 - 8x - 11$

$(3x - 11)(x + 1)$

18. $5x^2 - 2x - 3$

$(5x + 3)(x - 1)$

19. $4x^2 + 8x - 12$

$4(x + 3)(x - 1)$

20. $6x^2 - 6x - 12$

$6(x + 1)(x - 2)$

21. $4x^2 + 36x + 81$

$(2x + 9)^2$

22. $25x^2 + 40x + 16$

$(5x + 4)^2$

23. $8x^3 + 125y^3$

$(2x + 5y)$

$(4x^2 - 10xy + 25y^2)$

24. $27x^3 - 64y^3$

$(3x - 4y)$

$(9x^2 + 12xy + 16y^2)$

25. $64x^2y^3 - 8x^2$

$8x^2(2y - 1)$

$(4y^2 + 2y + 1)$

26. $27x^5y^4 - 216x^2y$

$27x^2y(xy - 2)$

$(x^2y^2 + 2xy + 4)$

27. $(x + 5)^3 + y^3$

$(x + 5 + y)$

$(x^2 + 10x - xy$

$-5y + y^2 + 25)$

28. $(y - 1)^3 + 27x^3$

$(y - 1 + 3x)$

$(y^2 - 2y + 1$

$-3xy + 3x + 9x^2)$

29. $(5a - 3)^2 - 6(5a - 3) + 9$

$(5a - 6)^2$

30. $(4r + 1)^2 + 8(4r + 1) + 16$

$(4r + 5)^2$

31. $7x^2 - 63x$

$7x(x - 9)$

32. $20x^2 + 23x + 6$

$(4x + 3)(5x + 2)$

33. $ab - 6a + 7b - 42$
$(a + 7)(b - 6)$

34. $20x^2 - 220x + 600$
$20(x - 6)(x - 5)$

35. $x^4 - 1$
$(x^2 + 1)(x - 1)$
$(x + 1)$

36. $15x^2 - 20x$
$5x(3x - 4)$

37. $10x^2 - 7x - 33$
$(5x - 11)(2x + 3)$

38. $45m^3n^3 - 27m^2n^2$
$9m^2n^2(5mn - 3)$

39. $5a^3b^3 - 50a^3b$
$5a^3b(b^2 - 10)$

40. $x^4 + x$
$x(x + 1)$
$(x^2 - x + 1)$

41. $16x^2 + 25$
prime

42. $20x^3 + 20y^3$
$20(x + y)$
$(x^2 - xy + y^2)$

43. $10x^3 - 210x^2 + 1100x$
$10x(x - 10)$
$(x - 11)$

44. $9y^2 - 42y + 49$
$(3y - 7)^2$

45. $64a^3b^4 - 27a^3b$
$a^3b(4b - 3)$
$(16b^2 + 12b + 9)$

46. $y^4 - 16$
$(y^2 + 4)(y + 2)$
$(y - 2)$

47. $2x^3 - 54$
$2(x - 3)$
$(x^2 + 3x + 9)$

48. $2sr + 10s - r - 5$
$(2s - 1)(r + 5)$

49. $3y^5 - 5y^4 + 6y - 10$
$(y^4 + 2)(3y - 5)$

50. $64a^2 + b^2$
prime

51. $100z^3 + 100$
$100(z + 1)$
$(z^2 - z + 1)$

52. $250x^4 - 16x$
$2x(5x - 2)$
$(25x^2 + 10x + 4)$

53. $4b^2 - 36b + 81$
$(2b - 9)^2$

54. $2a^5 - a^4 + 6a - 3$
$(a^4 + 3)(2a - 1)$

55. $(y - 6)^2 + 3(y - 6) + 2$
$(y - 4)(y - 5)$

56. $(c + 2)^2 - 6(c + 2) + 5$
$(c - 3)(c + 1)$

△**57.** Express the area of the shaded region as a polynomial. Factor the polynomial completely.

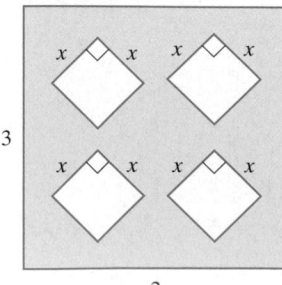

$A = 9 - 4x^2$
$= (3 + 2x)(3 - 2x)$

5.8   SOLVING EQUATIONS BY FACTORING AND PROBLEM SOLVING

Objectives

1. Solve polynomial equations by factoring.
2. Solve problems that can be modeled by polynomial equations.
3. Find the x-intercepts of a polynomial function.

1   In this section, your efforts to learn factoring start to pay off. We use factoring to solve polynomial equations, which in turn helps us solve problems that can be modeled by polynomial equations and also helps us sketch the graph of polynomial functions.

A **polynomial equation** is the result of setting two polynomials equal to each other. Examples of polynomial equations are

$$3x^3 - 2x^2 = x^2 + 2x - 1 \qquad 2.6x + 7 = -1.3 \qquad -5x^2 - 5 = -9x^2 - 2x + 1$$

A polynomial equation is in **standard form** if one side of the equation is 0. In standard form the polynomial equations above are

$$3x^3 - 3x^2 - 2x + 1 = 0 \qquad 2.6x + 8.3 = 0 \qquad 4x^2 + 2x - 6 = 0$$

The degree of a simplified polynomial equation in standard form is the same as the highest degree of any of its terms. A polynomial equation of degree 2 is also called a **quadratic equation**.

A solution of a polynomial equation in one variable is a value of the variable that makes the equation true. The method presented in this section for solving polynomial equations is called the **factoring method**. This method is based on the **zero-factor property**.

Zero-Factor Property

If a and b are real numbers and $a \cdot b = 0$, then $a = 0$ or $b = 0$. This property is true for three or more factors also.

In other words, if the product of two or more real numbers is zero, then at least one number must be zero.

EXAMPLE 1

Solve $(x + 2)(x - 6) = 0$.

Solution   By the zero-factor property, $(x + 2)(x - 6) = 0$ only if $x + 2 = 0$ or $x - 6 = 0$.

$$x + 2 = 0 \quad \text{or} \quad x - 6 = 0 \quad \text{Apply the zero-factor property.}$$
$$x = -2 \quad \text{or} \quad x = 6 \quad \text{Solve each linear equation.}$$

```
-2→X: (X+2)(X-6)
                    0
6→X: (X+2)(X-6)
                    0
■
```

To check, let $x = -2$ and then let $x = 6$ in the original equation as shown in the screen to the left.

Both -2 and 6 check, so they are both solutions. The solution set is $\{-2, 6\}$.

EXAMPLE 2

Solve $2x^2 + 9x - 5 = 0$.

Solution To use the zero-factor property, one side of the equation must be 0, and the other side must be in factored form.

CLASSROOM EXAMPLE

Solve: $3x^2 + 5x - 2 = 0$

answer: $-2, \dfrac{1}{3}$

$$2x^2 + 9x - 5 = 0$$
$$(2x - 1)(x + 5) = 0 \qquad \text{Factor.}$$
$$2x - 1 = 0 \quad \text{or} \quad x + 5 = 0 \quad \text{Set each factor equal to zero.}$$
$$2x = 1$$
$$x = \frac{1}{2} \quad \text{or} \qquad x = -5 \quad \text{Solve each linear equation.}$$

```
1/2→X: 2X²+9X-5
                    0
-5→X: 2X²+9X-5
                    0
■
```

The solutions are -5 and $\dfrac{1}{2}$. To check, let $x = \dfrac{1}{2}$ in the original equation; then let $x = -5$ in the original equation as shown in the screen to the left.

TECHNOLOGY NOTE

We now have a variety of ways to check a solution to an equation. We can check numerically as above or graphically. We will only check each equation one way, but remember that you can select the method that you prefer.

Solving Polynomial Equations by Factoring

Step 1: Write the equation in standard form so that one side of the equation is 0.

Step 2: Factor the polynomial completely.

Step 3: Set each factor containing a variable equal to 0.

Step 4: Solve the resulting equations.

Step 5: Check each solution in the original equation.

Since it is not always possible to factor a polynomial, not all polynomial equations can be solved by factoring. Other methods of solving polynomial equations are presented in Chapter 8.

EXAMPLE 3

Solve $x(2x - 7) = 4$.

CLASSROOM EXAMPLE

Solve: $x(5x - 7) = -2$

answer: $\dfrac{2}{5}, 1$

Solution First, write the equation in standard form; then, factor.

$$x(2x - 7) = 4$$

$$2x^2 - 7x = 4 \qquad \text{Multiply.}$$

$$2x^2 - 7x - 4 = 0 \qquad \text{Write in standard form.}$$

$$(2x + 1)(x - 4) = 0 \qquad \text{Factor.}$$

$$2x + 1 = 0 \quad \text{or} \quad x - 4 = 0 \qquad \text{Set each factor equal to zero.}$$

$$2x = -1 \qquad\qquad\qquad \text{Solve.}$$

$$x = -\frac{1}{2} \quad \text{or} \quad x = 4$$

The solutions are $-\dfrac{1}{2}$ and 4. Check both solutions in the original equation as shown to the left.

> ### Helpful Hint
>
> To apply the zero-factor property, one side of the equation must be 0, and the other side of the equation must be factored. To solve the equation $x(2x - 7) = 4$, for example, you may **not** set each factor equal to 4.

EXAMPLE 4

Solve $3(x^2 + 4) + 5 = -6(x^2 + 2x) + 13$.

Solution   Rewrite the equation so that one side is 0.

$$3(x^2 + 4) + 5 = -6(x^2 + 2x) + 13.$$

$$3x^2 + 12 + 5 = -6x^2 - 12x + 13 \qquad \text{Apply the distributive property.}$$

$$9x^2 + 12x + 4 = 0 \qquad \text{Rewrite the equation so that one side is } 0.$$

$$(3x + 2)(3x + 2) = 0 \qquad \text{Factor.}$$

$$3x + 2 = 0 \quad \text{or} \quad 3x + 2 = 0 \qquad \text{Set each factor equal to } 0.$$

$$3x = -2 \quad \text{or} \qquad 3x = -2$$

$$x = -\frac{2}{3} \quad \text{or} \qquad x = -\frac{2}{3} \qquad \text{Solve each equation.}$$

Notice that the identical factors $(3x + 2)$, led to a single solution, $-\frac{2}{3}$. The solution is $-\frac{2}{3}$.

We choose to check this time by graphing and using the intersection-of-graphs method as shown in the screen to the left. The intersection is at $x \approx -0.66667$, which is an approximation for $-\frac{2}{3}$.

$y_1 = 3(x^2 + 4) + 5$

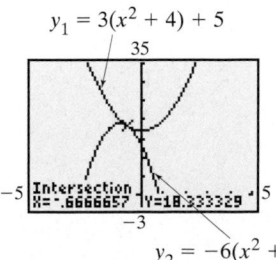

$y_2 = -6(x^2 + 2x) + 13$

If the equation contains fractions, we clear the equation of fractions as a first step.

EXAMPLE 5

Solve $2x^2 = \dfrac{17}{3}x + 1$.

Solution

$$2x^2 = \frac{17}{3}x + 1$$

$$3(2x^2) = 3\left(\frac{17}{3}x + 1\right) \qquad \text{Clear the equation of fractions.}$$

$$6x^2 = 17x + 3 \qquad \text{Apply the distributive property.}$$

$$6x^2 - 17x - 3 = 0 \qquad \text{Rewrite the equation in standard form.}$$

$$(6x + 1)(x - 3) = 0 \qquad \text{Factor.}$$

$$6x + 1 = 0 \quad \text{or} \quad x - 3 = 0 \quad \text{Set each factor equal to zero.}$$

$$6x = -1$$

$$x = -\frac{1}{6} \quad \text{or} \quad x = 3 \quad \text{Solve each equation.}$$

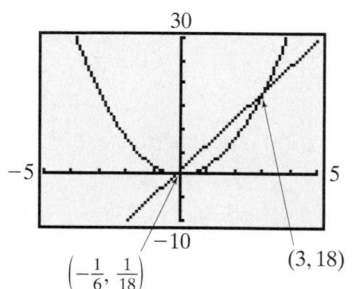

$\left(-\dfrac{1}{6}, \dfrac{1}{18}\right)$ $(3, 18)$

The two graphs intersect twice. Verify the solutions of $x = -\frac{1}{6}$ and $x = 3$.

The solutions are $-\dfrac{1}{6}$ and 3. To check, graph $y_1 = 2x^2$ and $y_2 = \frac{17}{3}x + 1$. Notice that the graph of y_1 is a parabola and the graph of y_2 is a line.

EXAMPLE 6

Solve $x^3 = 4x$.

Solution

$$x^3 = 4x$$

$$x^3 - 4x = 0 \qquad \text{Rewrite the equation so that one side is 0.}$$

$$x(x^2 - 4) = 0 \qquad \text{Factor out the GCF, } x.$$

$$x(x + 2)(x - 2) = 0 \qquad \text{Factor the difference of squares.}$$

$$x = 0 \quad \text{or} \quad x + 2 = 0 \quad \text{or} \quad x - 2 = 0 \quad \text{Set each factor equal to 0.}$$

$$x = 0 \quad \text{or} \quad x = -2 \quad \text{or} \quad x = 2 \quad \text{Solve each equation.}$$

TEACHING TIP

After reviewing Example 6, challenge students to write the factored form of an equation that has the solutions $-4, -1$, and 8.

The solutions are $-2, 0$, and 2. Check by substituting into the original equation.

Notice that the *third*-degree equation of Example 6 yielded *three* solutions. To see why, graph $y_1 = x^3$ and $y_2 = 4x$.

The graphs intersect in three points representing the three solutions of the equation. The solutions of the equation $x^3 = 4x$ are $x = -2, x = 0$, and $x = 2$.

EXAMPLE 7

Solve $x^3 + 5x^2 = x + 5$.

Solution   First, write the equation so that one side is 0.

$$x^3 + 5x^2 - x - 5 = 0$$
$$(x^3 - x) + (5x^2 - 5) = 0 \qquad \text{Factor by grouping.}$$
$$x(x^2 - 1) + 5(x^2 - 1) = 0$$
$$(x^2 - 1)(x + 5) = 0$$
$$(x + 1)(x - 1)(x + 5) = 0 \qquad \text{Factor the difference of squares.}$$
$$x + 1 = 0 \quad \text{or} \quad x - 1 = 0 \quad \text{or} \quad x + 5 = 0 \qquad \text{Set each factor equal to 0.}$$
$$x = -1 \quad \text{or} \qquad x = 1 \quad \text{or} \qquad x = -5 \qquad \text{Solve each equation.}$$

The solutions are $-5, -1$, and 1. The screen to the left verifies the solutions. Check in the original equation.

✔ **CONCEPT CHECK**

Which solution strategies are incorrect? Why?

a. Solve $(y - 2)(y + 2) = 4$ by setting each factor equal to 4.
b. Solve $(x + 1)(x + 3) = 0$ by setting each factor equal to 0.
c. Solve $z^2 + 5z + 6 = 0$ by factoring $z^2 + 5z + 6$ and setting each factor equal to 0.
d. Solve $x^2 + 6x + 8 = 10$ by factoring $x^2 + 6x + 8$ and setting each factor equal to 0.

②   Some problems may be modeled by polynomial equations. To solve these problems, we use the same problem-solving steps that were introduced in Section 1.6. When solving these problems, keep in mind that a solution of an equation that models a problem is not always a solution to the problem. For example, a person's weight or the length of a side of a geometric figure is always a positive number. Discard solutions that do not make sense as solutions of the problem.

EXAMPLE 8

FINDING THE RETURN TIME OF A ROCKET

An Alpha III model rocket is launched from the ground with an A8–3 engine. Without a parachute, the height of the rocket h at time t seconds is approximated by the equation,

$$h = -16t^2 + 144t$$

Find how long it takes the rocket to return to the ground.

Solution   1. UNDERSTAND.   Read and reread the problem. The equation $h = -16t^2 + 144t$ models the height of the rocket. Familiarize yourself with this equation by finding a few values.

When $t = 1$ second, the height of the rocket is

$$h = -16(1)^2 + 144(1) = 128 \text{ feet}$$

When $t = 2$ seconds, the height of the rocket is

$$h = -16(2)^2 + 144(2) = 224 \text{ feet}$$

2. TRANSLATE. To find how long it takes the rocket to return to the ground, we want to know what value of t makes the height h equal to 0. That is, we want to solve $h = 0$.

$$-16t^2 + 144t = 0$$

3. SOLVE the quadratic equation by factoring.

$$-16t^2 + 144t = 0$$
$$-16t(t - 9) = 0$$
$$-16t = 0 \quad \text{or} \quad t - 9 = 0$$
$$t = 0 \qquad\qquad t = 9$$

4. INTERPRET. The height h is 0 feet at time 0 seconds (when the rocket is launched) and at time 9 seconds.

> **Check:** Graph $y_1 = -16x^2 + 144x$ and see that $(9, 0)$ is an x-intercept as shown to the left.
>
> **State:** The rocket returns to the ground 9 seconds after it is launched.

The rocket is at ground level at 0 seconds and 9 seconds.

Some of the exercises at the end of this section make use of the **Pythagorean theorem.** Before we review this theorem, recall that a **right triangle** is a triangle that contains a 90° angle, or right angle. The **hypotenuse** of a right triangle is the side opposite the right angle and is the longest side of the triangle. The **legs** of a right triangle are the other sides of the triangle.

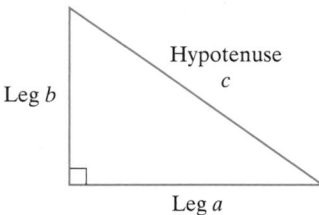

Pythagorean Theorem

In a right triangle, the sum of the squares of the lengths of the two legs is equal to the square of the length of the hypotenuse.

$$(\text{leg})^2 + (\text{leg})^2 = (\text{hypotenuse})^2 \quad \text{or} \quad a^2 + b^2 = c^2$$

EXAMPLE 9

USING THE PYTHAGOREAN THEOREM

While framing an addition to an existing home, Kim Menzies, a carpenter, used the Pythagorean theorem to determine whether a wall was "square"—that is, whether the wall formed a right angle with the floor. He used a triangle whose sides are three consecutive integers. Find a right triangle whose sides are three consecutive integers.

Solution   1. UNDERSTAND.   Read and reread the problem.

Let x, $x + 1$, and $x + 2$ be three consecutive integers. Since these integers represent lengths of the sides of a right triangle, we have

$$x = \text{one leg}$$
$$x + 1 = \text{other leg}$$
$$x + 2 = \text{hypotenuse (longest side)}$$

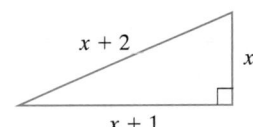

2. TRANSLATE.   By the Pythagorean theorem, we have

In words: | $(\text{leg})^2$ | $+$ | $(\text{leg})^2$ | $=$ | $(\text{hypotenuse})^2$ |
|---|---|---|---|---|
| $\downarrow$ | | $\downarrow$ | | $\downarrow$ |

Translate: $(x)^2$ $+$ $(x + 1)^2$ $=$ $(x + 2)^2$

3. SOLVE   the equation.

$$x^2 + (x + 1)^2 = (x + 2)^2$$
$$x^2 + x^2 + 2x + 1 = x^2 + 4x + 4 \qquad \text{Multiply.}$$
$$2x^2 + 2x + 1 = x^2 + 4x + 4$$
$$x^2 - 2x - 3 = 0 \qquad \text{Write in standard form.}$$
$$(x - 3)(x + 1) = 0$$
$$x - 3 = 0 \quad \text{or} \quad x + 1 = 0$$
$$x = 3 \qquad\qquad x = -1$$

4. INTERPRET.   Discard $x = -1$ since length cannot be negative. If $x = 3$, then $x + 1 = 4$ and $x + 2 = 5$.

Check: To check, see that $(\text{leg})^2 + (\text{leg})^2 = (\text{hypotenuse})^2$

$$3^2 + 4^2 = 5^2$$

$$9 + 16 = 25 \quad \text{True}$$

State: The lengths of the sides of the right triangle are 3, 4, and 5 units. Kim used this information, for example, by marking off lengths of 3 and 4 feet on the floor and framing respectively. If the diagonal length between these marks was 5 feet, the wall was "square." If not, adjustments were made.

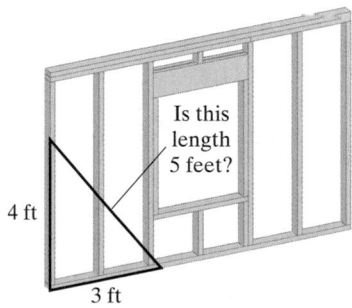

3   Recall that to find the *x*-intercepts of the graph of a function, let $f(x) = 0$, or $y = 0$, and solve for *x*. This fact gives us a visual interpretation of the results of this section.

From Example 1, we know that the solutions of the equation $(x + 2)(x - 6) = 0$ are -2 and 6. These solutions give us important information about the related polynomial function $p(x) = (x + 2)(x - 6)$. We know that when *x* is -2 or when *x* is 6, the value of $p(x)$ is 0.

$$p(x) = (x + 2)(x - 6)$$

$$p(-2) = (-2 + 2)(-2 - 6) = (0)(-8) = 0$$

$$p(6) = (6 + 2)(6 - 6) = (8)(0) = 0$$

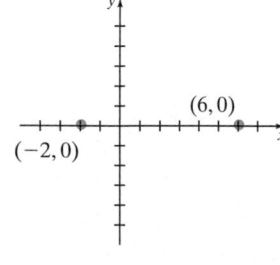

Thus, we know that $(-2, 0)$ and $(6, 0)$ are the *x*-intercepts of the graph of $p(x)$.

We also know that the graph of $p(x)$ does not cross the *x*-axis at any other point. For this reason, and the fact that $p(x) = (x + 2)(x - 6) = x^2 - 4x - 12$ has degree 2, we conclude that the graph of *p* must look something like one of these two graphs:

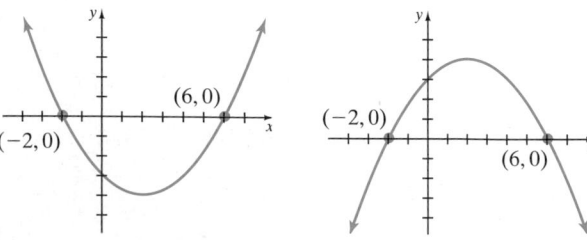

In a later chapter, we explore these graphs more fully. For the moment, know that the solutions of a polynomial equation are the *x*-intercepts of the graph of the related function and that the *x*-intercepts of the graph of a polynomial function are the solutions of the related polynomial equation. These values are also called **roots**, or **zeros**, of a polynomial function.

EXAMPLE 10

MATCH EACH FUNCTION WITH ITS GRAPH

$$f(x) = (x - 3)(x + 2) \qquad g(x) = x(x + 2)(x - 2) \qquad h(x) = (x - 2)(x + 2)(x - 1)$$

CLASSROOM EXAMPLE

Match each function with its graph.

a.

b.

c.

answer:
a. $g(x) = (x - 5)(x - 1)(x + 1)$
b. $f(x) = (x + 1)(x - 3)$
c. $h(x) = x(x - 1)(x + 1)$

A

B

C
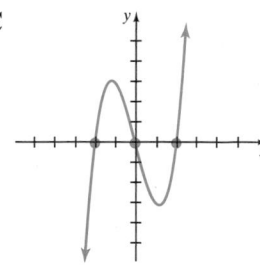

Solution

The graph of the function $f(x) = (x - 3)(x + 2)$ has two x-intercepts, $(3, 0)$ and $(-2, 0)$, because the equation $0 = (x - 3)(x + 2)$ has two solutions, 3 and -2.

The graph of $f(x)$ is graph B.

The graph of the function $g(x) = x(x + 2)(x - 2)$ has three x-intercepts $(0, 0)$, $(-2, 0)$, and $(2, 0)$, because the equation $0 = x(x + 2)(x - 2)$ has three solutions, 0, -2, and 2.

The graph of $g(x)$ is graph C.

The graph of the function $h(x) = (x - 2)(x + 2)(x - 1)$ has three x-intercepts, $(-2, 0)$, $(1, 0)$, and $(2, 0)$, because the equation $0 = (x - 2)(x + 2)(x - 1)$ has three solutions, -2, 1, and 2.

The graph of $h(x)$ is graph A.

STUDY SKILLS REMINDER

Continue your outline started in Section 1.5. Write how to recognize and how to solve quadratic and higher degree equations by factoring in your own words. For example:

Solving Equations and Inequalities

I. Equations

    A. Linear equations (Sec. 1.5)

    B. Absolute value equations (Sec. 3.4)

    C. Quadratic and Higher Degree Equations–Recognize: *highest power on variable in equation is at least 2 when written in standard form (one side 0)*–Solve: write in standard form and try to factor the polynomial on one side of the equation. If it factors, set each factor equal to 0.

II. Inequalities

    A. Linear inequalities (Sec. 3.2)

    B. Compound inequalities (Sec. 3.3)

    C. Absolute value inequalities (Sec. 3.5)

See Appendix A for summary exercises.

MENTAL MATH

Solve each equation for the variable. See Example 1.

1. $(x - 3)(x + 5) = 0$ $3, -5$

2. $(y + 5)(y + 3) = 0$ $-5, -3$

3. $(z - 3)(z + 7) = 0$ $3, -7$

4. $(c - 2)(c - 4) = 0$ $2, 4$

5. $x(x - 9) = 0$ $0, 9$

6. $w(w + 7) = 0$ $0, -7$

EXERCISE SET 5.8

    Math XL

STUDY GUIDE/SSM   CD/ VIDEO   PH MATH TUTOR CENTER   MathXL®Tutorials ON CD   MathXL®   MyMathLab®

Solve each equation. See Example 1.

5. $-3, -8$

7. $\frac{1}{4}, -\frac{2}{3}$

25. $2, 1, -1$

26. $-3, 0, 4$

35. $-\frac{1}{2}, \frac{1}{3}$

36. $-1, -\frac{5}{8}$

1. $(x + 3)(3x - 4) = 0$ $-3, \frac{4}{3}$

2. $(5x + 1)(x - 2) = 0$ $-\frac{1}{5}, 2$

3. $3(2x - 5)(4x + 3) = 0$ $\frac{5}{2}, -\frac{3}{4}$

4. $8(3x - 4)(2x - 7) = 0$ $\frac{4}{3}, \frac{7}{2}$

Solve each equation. See Examples 2 through 5.

5. $x^2 + 11x + 24 = 0$

6. $y^2 - 10y + 24 = 0$ $4, 6$

7. $12x^2 + 5x - 2 = 0$

8. $3y^2 - y - 14 = 0$ $-2, \frac{7}{3}$

9. $z^2 + 9 = 10z$ $1, 9$

10. $n^2 + n = 72$ $-9, 8$

11. $x(5x + 2) = 3$ $\frac{3}{5}, -1$

12. $n(2n - 3) = 2$ $-\frac{1}{2}, 2$

13. $x^2 - 6x = x(8 + x)$ 0

14. $n(3 + n) = n^2 + 4n$ 0

15. $\frac{z^2}{6} - \frac{z}{2} - 3 = 0$ $6, -3$

16. $\frac{c^2}{20} - \frac{c}{4} + \frac{1}{5} = 0$ $1, 4$

17. $\frac{x^2}{2} + \frac{x}{20} = \frac{1}{10}$ $\frac{2}{5}, -\frac{1}{2}$

18. $\frac{y^2}{30} = \frac{y}{15} + \frac{1}{2}$ $-3, 5$

19. $\frac{4t^2}{5} = \frac{t}{5} + \frac{3}{10}$ $\frac{3}{4}, -\frac{1}{2}$

20. $\frac{5x^2}{6} - \frac{7x}{2} + \frac{2}{3} = 0$ $\frac{1}{5}, 4$

Solve each equation. See Examples 6 and 7.

21. $(x + 2)(x - 7)(3x - 8) = 0$ $-2, 7, \frac{8}{3}$

22. $(4x + 9)(x - 4)(x + 1) = 0$ $-\frac{9}{4}, -1, 4$

23. $y^3 = 9y$ $0, 3, -3$

24. $n^3 = 16n$ $-4, 0, 4$

25. $x^3 - x = 2x^2 - 2$

26. $m^3 = m^2 + 12m$

27. Explain how solving $2(x - 3)(x - 1) = 0$ differs from solving $2x(x - 3)(x - 1) = 0$.   answers may vary

28. Explain why the zero-factor property works for more than two numbers whose product is 0.   answers may vary

MIXED PRACTICE

Solve each equation.   **29.** $-\frac{7}{2}, 10$ **30.** $-4, \frac{1}{5}$

29. $(2x + 7)(x - 10) = 0$

30. $(x + 4)(5x - 1) = 0$

31. $3x(x - 5) = 0$ $0, 5$

32. $4x(2x + 3) = 0$ $-\frac{3}{2}, 0$

33. $x^2 - 2x - 15 = 0$ $-3, 5$ **34.** $x^2 + 6x - 7 = 0$ $-7, 1$

35. $12x^2 + 2x - 2 = 0$

36. $8x^2 + 13x + 5 = 0$

37. $w^2 - 5w = 36$ $-4, 9$

38. $x^2 + 32 = 12x$ $4, 8$

39. $25x^2 - 40x + 16 = 0$ $\frac{4}{5}$

40. $9n^2 + 30n + 25 = 0$ $-\frac{5}{3}$

41. $2r^3 + 6r^2 = 20r$ $-5, 0, 2$

42. $-2t^3 = 108t - 30t^2$ $0, 6, 9$

43. $z(5z - 4)(z + 3) = 0$ $-3, 0, \frac{4}{5}$

44. $2r(r + 3)(5r - 4) = 0$ $-3, 0, \frac{4}{5}$

45. $2z(z + 6) = 2z^2 + 12z - 8$ $\varnothing$

46. $3c^2 - 8c + 2 = c(3c - 8)$ $\varnothing$

47. $(x - 1)(x + 4) = 24$ $-7, 4$

48. $(2x - 1)(x + 2) = -3$ $-1, -\frac{1}{2}$

49. $\frac{x^2}{4} - \frac{5}{2}x + 6 = 0$ $4, 6$

50. $\frac{x^2}{18} + \frac{x}{2} + 1 = 0$ $-6, -3$

51. $y^2 + \frac{1}{4} = -y$ $-\frac{1}{2}$

52. $\dfrac{x^2}{10} + \dfrac{5}{2} = x$ 5

53. $y^3 + 4y^2 = 9y + 36$ **54.** $x^3 + 5x^2 = x + 5$

55. $2x^3 = 50x$ $-5, 0, 5$ **56.** $m^5 = 36m^3$ $-6, 0, 6$

57. $x^2 + (x + 1)^2 = 61$ **58.** $y^2 + (y + 2)^2 = 34$

59. $m^2(3m - 2) = m$ **60.** $x^2(5x + 3) = 26x$

61. $3x^2 = -x$ $-\frac{1}{3}, 0$ **62.** $y^2 = -5y$ $-5, 0$

63. $x(x - 3) = x^2 + 5x + 7$ **64.** $z^2 - 4z + 10 = z(z - 5)$

65. $3(t - 8) + 2t = 7 + t$ $\frac{31}{4}$

66. $7c - 2(3c + 1) = 5(4 - 2c)$ 2

67. $-3(x - 4) + x = 5(3 - x)$ 1

68. $-4(a + 1) - 3a = -7(2a - 3)$ $\frac{25}{7}$

69. Which solution strategies are incorrect? Why?
   a. Solve $(y - 2)(y + 2) = 4$ by setting each factor equal to 4.   incorrect
   b. Solve $(x + 1)(x + 3) = 0$ by setting each factor equal to 0.   correct
   c. Solve $z^2 + 5z + 6 = 0$ by factoring $z^2 + 5z + 6$ and setting each factor equal to 0.   correct
   d. Solve $x^2 + 6x + 8 = 10$ by factoring $x^2 + 6x + 8$ and setting each factor equal to 0.   incorrect

70. Describe two ways a linear equation differs from a quadratic equation.   answers may vary

Solve. See Examples 8 and 9.

71. One number exceeds another by five, and their product is 66. Find the numbers. -11 and -6 or 6 and 11

72. If the sum of two numbers is 4 and their product is $\dfrac{15}{4}$, find the numbers. $\dfrac{3}{2}$ and $\dfrac{5}{2}$

 73. An electrician needs to run a cable from the top of a 60-foot tower to a transmitter box located 45 feet away from the base of the tower. Find how long he should make the cable.   75 ft

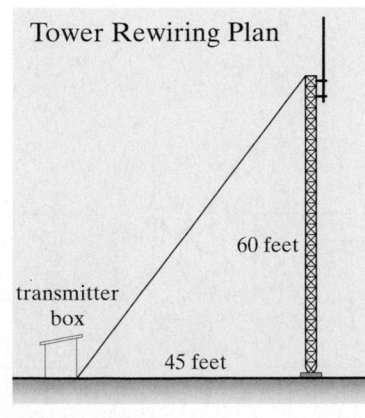

Tower Rewiring Plan

60 feet

transmitter box

45 feet

53. $-4, -3, 3$
54. $-5, -1, 1$
57. $-6, 5$
58. $-5, 3$
59. $-\dfrac{1}{3}, 0, 1$
60. $-\dfrac{13}{5}, 0, 2$
63. $-\dfrac{7}{8}$
64. -10

△ **74.** A stereo system installer needs to run speaker wire along the two diagonals of a rectangular room whose dimensions are 40 feet by 75 feet. Find how much speaker wire she needs.   170 ft

40 ft   75 ft

75. If the cost, $C(x)$, for manufacturing x units of a certain product is given by $C(x) = x^2 - 15x + 50$, find the number of units manufactured at a cost of $9500.   105 units

△ **76.** Determine whether any three consecutive integers represent the lengths of the sides of a right triangle.   yes; 3, 4, 5

△ **77.** The shorter leg of a right triangle is 3 centimeters less than the other leg. Find the length of the two legs if the hypotenuse is 15 centimeters.   12 cm and 9 cm

△ **78.** The longer leg of a right triangle is 4 feet longer than the other leg. Find the length of the two legs if the hypotenuse is 20 feet.   12 ft and 16 ft

△ **79.** Marie Mulroney has a rectangular board 12 inches by 16 inches around which she wants to put a uniform border of shells. If she has enough shells for a border whose area is 128 square inches, determine the width of the border.   2 in.

12 in.

16 in.

x

x

△ **80.** A gardener has a rose garden that measures 30 feet by 20 feet. He wants to put a uniform border of pine bark around the outside of the garden. Find how wide the border should be if he has enough pine bark to cover 336 square feet.   3 ft

x

20 ft

30 ft

x

81. While hovering near the top of Ribbon Falls in Yosemite National Park at 1600 feet, a helicopter pilot accidentally drops his sunglasses. The height $h(t)$ of the sunglasses after t seconds is given by the polynomial function

$$h(t) = -16t^2 + 1600$$

When will the sunglasses hit the ground?   10 sec

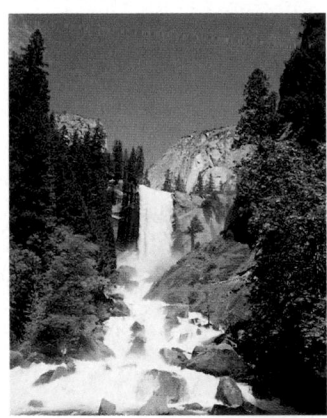

82. After t seconds, the height $h(t)$ of a model rocket launched from the ground into the air is given by the function

$$h(t) = -16t^2 + 80t$$

Find how long it takes the rocket to reach a height of 96 feet.   2 sec and 3 sec (ascending and descending)

△ **83.** The floor of a shed has an area of 90 square feet. The floor is in the shape of a rectangle whose length is 3 feet less than twice the width. Find the length and the width of the floor of the shed.   width: $7\frac{1}{2}$ ft length: 12 ft

△ **84.** A vegetable garden with an area of 200 square feet is to be fertilized. If the length of the garden is 1 foot less than three times the width, find the dimensions of the garden. width: $8\frac{1}{3}$ ft length: 24 ft

85. The function $W(x) = 0.5x^2$ gives the number of servings of wedding cake that can be obtained from a two-layer x-inch square wedding cake tier. What size square wedding cake tier is needed to serve 50 people? (*Source:* Based on data from the *Wilton 2000 Yearbook of Cake Decorating*)   10-in. square tier

86. Use the function in Exercise 85 to determine what size wedding cake tier is needed to serve 200 people.   20-in. square tier

87. Suppose that a movie is being filmed in New York City. An action shot requires an object to be thrown upward with an initial velocity of 80 feet per second off the top of 1 Madison Square Plaza, a height of 576 feet. The height $h(t)$ in feet of the object after t seconds is given by the function

$$h(t) = -16t^2 + 80t + 576.$$

Determine how long before the object strikes the ground. (See Exercise 97, Section 5.6) (*Source: The World Almanac*)   9 sec

576 ft

88. Suppose that an object is thrown upward with an initial velocity of 64 feet per second off the edge of a 960-foot-cliff. The height $h(t)$ in feet of the object after t seconds is given by the function

$$h(t) = -16t^2 + 64t + 960$$

Determine how long before the object strikes the ground. (See Exercise 98, Section 5.6)   10 sec

Match each polynomial function with its graph (A–F). See Example 10.

89. $f(x) = (x - 2)(x + 5)$ E

90. $g(x) = (x + 1)(x - 6)$ D

91. $h(x) = x(x + 3)(x - 3)$ F

92. $F(x) = (x + 1)(x - 2)(x + 5)$ A

93. $G(x) = 2x^2 + 9x + 4$ B

94. $H(x) = 2x^2 - 7x - 4$ C

A

B

C

D

E

F

REVIEW AND PREVIEW

Write the x- and y-intercepts for each graph and determine whether the graph is the graph of a function. See Sections 2.1 and 2.2.

95.

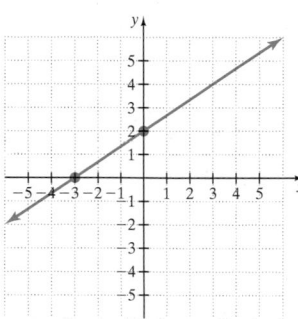

$(-3, 0), (0, 2)$; function

96.

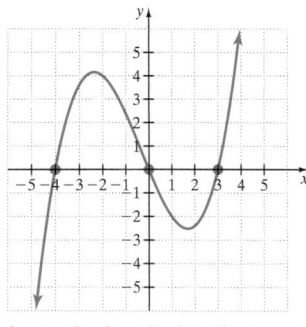

$(-4, 0), (0, 0), (3, 0)$; function

97.

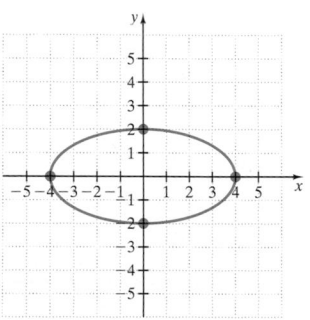

$(-4, 0), (0, 2),$
$(4, 0), (0, -2);$
not a function

98.

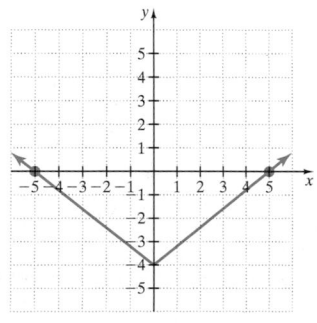

$(-5, 0), (5, 0),$
$(0, -4)$; function

99. Draw a function with intercepts $(-3, 0), (5, 0),$ and $(0, 4)$. answers may vary

100. Draw a function with intercepts $(-7, 0), \left(-\dfrac{1}{2}, 0\right), (4, 0),$ and $(0, -1)$.   answers may vary

Concept Extensions

Solve.

101. $(x^2 + x - 6)(3x^2 - 14x - 5) = 0$ $\left\{-3, -\dfrac{1}{3}, 2, 5\right\}$

102. $(x^2 - 9)(x^2 + 8x + 16) = 0$ $\{-4, -3, 3\}$

103. Is the following step correct? Why or why not?

$$x(x - 3) = 5$$
$$x = 5 \text{ or } x - 3 = 5$$

no; answers may vary

Write a quadratic equation that has the given numbers as solutions.

104. $5, 3$ **105.** $6, 7$

106. $-1, 2$ **107.** $4, -3$

104. answers may vary Ex.: $f(x) = x^2 - 8x + 15$
105. answers may vary Ex.: $f(x) = x^2 - 13x + 42$
106. answers may vary Ex.: $f(x) = x^2 - x - 2$
107. answers may vary Ex.: $f(x) = x^2 - x - 12$

CHAPTER 5 PROJECT

Investigating Earth's Water

Earth is covered by water. In fact, oceans cover nearly three-fourths of the surface of Earth. However, oceans aren't the only source of Earth's water. The melting of one of the other main sources of Earth's water, icecaps and glaciers, is expected to contribute to a global rise in ocean level due to global warming over the next 100 years. In this project, you will have the opportunity to investigate where Earth's water exists and how the ocean level will change. This project may be completed by working in groups or individually.

1. Refer to Table 1. Which accounts for more of Earth's water: groundwater or icecaps and glaciers?

2. Find the total volume of water that exists on planet Earth. Add this figure to the table.

3. Using the total you computed in Question 2, complete the percent column of the table. Discuss your findings.

Widespread industrialization during the nineteenth and twentieth centuries has led to an increase in the presence of carbon dioxide, methane, nitrous oxide, and chlorofluorocarbons in Earth's atmosphere. These so-called greenhouse gases are believed by some scientists to be responsible for an increase in the average global temperature of 0.2°C to 0.3°C in the last half of the twentieth century. If this global warming trend continues, one of its consequences may be a global increase in the level of the oceans. An overall increase in ocean level will be due to changes in icecaps and glaciers, as well as thermal expansion. Higher global temperatures lead to warming in the top layers of the ocean, causing the water to expand and elevate the sea level.

Table 2 lists each contributor to overall changes in ocean level along with a polynomial model describing the projected rise y (in centimeters) each is expected to contribute x years after 2000.

4. By 2020, how much will the ocean level have risen due to thermal expansion?

5. Using the polynomial models given in Table 2, find a single polynomial model that gives the overall rise in ocean level from 1990 to 2100.

6. Using your model from Question 5, find the projected overall increase in ocean level for
   a. 2025   b. 2050
   c. 2075   d. 2100

7. Discuss the impact of your findings in Question 6.

Table 1 Where Earth's Water Exists

	Water Volume (cubic kilometers)	Percent
Atmosphere	1.3×10^4	
Average in stream channels	1.0×10^3	
Freshwater lakes	1.2×10^5	
Groundwater	8.3×10^6	
Icecaps and glaciers	2.9×10^7	
Oceans	1.32×10^9	
Saline lakes and inland seas	1.0×10^5	
Water in soil above groundwater	6.7×10^5	
Total		

(*Source:* Data from B.J. Skinner, *Earth Resources*, 2nd Ed., Prentice Hall, 1976)

Table 2 Projections of Global Ocean Level Rise by Contributor, 1990–2100

Alpine glaciers:
$$y = 0.0006x^2 + 0.0936x + 0.8788$$

Greenland ice sheet:
$$y = 0.0004x^2 + 0.0164x + 0.1212$$

Antarctica ice sheet:
$$y = -0.0001x^2 - 0.0002x + 0.0076$$

Thermal expansion:
$$y = 0.0011x^2 + 0.1564x + 1.4545$$

(*Source:* Based on data from Frederick K. Lutgens, Edward J. Tarbuck, *The Atmosphere: An Introduction to Meteorology*, 7th Ed., Prentice Hall, 1998)

5 CHAPTER VOCABULARY CHECK

Fill in each blank with one of the words or phrases listed below.

quadratic equation	scientific notation	polynomial	exponents	1 0	monomial
binomial	trinomial	degree of a polynomial	degree of a term		factoring

1. A <u>polynomial</u> is a finite sum of terms in which all variables are raised to nonnegative integer powers and no variables appear in any denominator.

2. <u>Factoring</u> is the process of writing a polynomial as a product.

3. <u>Exponents</u> are used to write repeated factors in a more compact form.

4. The <u>degree of a term</u> is the sum of the exponents on the variables contained in the term.

5. A <u>monomial</u> is a polynomial with one term.

6. If a is not 0, $a^0 = $ <u>1</u>.

7. A <u>trinomial</u> is a polynomial with three terms.

8. A polynomial equation of degree 2 is also called a <u>quadratic equation</u>.

9. A positive number is written in <u>scientific notation</u> if it is written as the product of a number a, such that $1 \leq a < 10$ and a power of 10.

10. The <u>degree of a polynomial</u> is the largest degree of all of its terms.

11. A <u>binomial</u> is a polynomial with two terms.

12. If a and b are real numbers and $a \cdot b = $ <u>0</u>, then $a = 0$ or $b = 0$.

CHAPTER 5 HIGHLIGHTS

Definitions and Concepts	Examples
Section 5.1 Exponents and Scientific Notation	
Product rule: $a^m \cdot a^n = a^{m+n}$	$x^2 \cdot x^3 = x^5$
Zero exponent: $a^0 = 1$, $a \neq 0$	$7^0 = 1, (-10)^0 = 1$
Quotient rule: $\dfrac{a^m}{a^n} = a^{m-n}$	$\dfrac{y^{10}}{y^4} = y^{10-4} = y^6$
Negative exponent: $a^{-n} = \dfrac{1}{a^n}$	$3^{-2} = \dfrac{1}{3^2} = \dfrac{1}{9}, \dfrac{x^{-5}}{x^{-7}} = x^{-5-(-7)} = x^2$
A positive number is written in **scientific notation** if it is written as the product of a number a, where $1 \leq a < 10$, and an integer power of 10: $a \times 10^r$.	**Numbers written in scientific notation** $$568,000 = 5.68 \times 10^5$$ $$0.0002117 = 2.117 \times 10^{-4}$$

Definitions and Concepts	**Examples**

Section 5.2 More Work with Exponents and Scientific Notation

Power rules:

$$(a^m)^n = a^{m \cdot n}$$
$$(ab)^m = a^m b^m$$
$$\left(\frac{a}{b}\right)^n = \frac{a^n}{b^n}$$

$$(7^8)^2 = 7^{16}$$
$$(2y)^3 = 2^3 y^3 = 8y^3$$
$$\left(\frac{5x^{-3}}{x^2}\right)^{-2} = \frac{5^{-2}x^6}{x^{-4}}$$
$$= 5^{-2} \cdot x^{6-(-4)}$$
$$= \frac{x^{10}}{5^2}, \text{ or } \frac{x^{10}}{25}$$

Section 5.3 Polynomials and Polynomial Functions

A **polynomial** is a finite sum of terms in which all variables have exponents raised to nonnegative integer powers and no variables appear in the denominator.

Polynomials

$$1.3x^2 \quad \text{(monomial)}$$
$$-\frac{1}{3}y + 5 \quad \text{(binomial)}$$
$$6z^2 - 5z + 7 \quad \text{(trinomial)}$$

A function P is a **polynomial function** if $P(x)$ is a polynomial.

For the polynomial function

$$P(x) = -x^2 + 6x - 12, \text{find } P(-2)$$
$$P(-2) = -(-2)^2 + 6(-2) - 12 = -28.$$

To add polynomials, combine all like terms.

Add

$$(3y^2x - 2yx + 11) + (-5y^2x - 7)$$
$$= -2y^2x - 2yx + 4$$

To subtract polynomials, change the signs of the terms of the polynomial being subtracted, then add.

Subtract

$$(-2z^3 - z + 1) - (3z^3 + z - 6)$$
$$= -2z^3 - z + 1 - 3z^3 - z + 6$$
$$= -5z^3 - 2z + 7$$

Section 5.4 Multiplying Polynomials

To multiply two polynomials, use the distributive property and multiply each term of one polynomial by each term of the other polynomial; then combine like terms.

Special products

$$(a + b)^2 = a^2 + 2ab + b^2$$
$$(a - b)^2 = a^2 - 2ab + b^2$$
$$(a + b)(a - b) = a^2 - b^2$$

Multiply

$$(x^2 - 2x)(3x^2 - 5x + 1)$$
$$= 3x^4 - 5x^3 + x^2 - 6x^3 + 10x^2 - 2x$$
$$= 3x^4 - 11x^3 + 11x^2 - 2x$$
$$(3m + 2n)^2 = 9m^2 + 12mn + 4n^2$$
$$(z^2 - 5)^2 = z^4 - 10z^2 + 25$$
$$(7y + 1)(7y - 1) = 49y^2 - 1$$

(continued)

Definitions and Concepts	**Examples**

Section 5.4   Multiplying Polynomials

The FOIL method may be used when multiplying two binomials.	*Multiply*

$$(x^2 + 5)(2x^2 - 9)$$

$$\begin{array}{cccc} \text{F} & \text{O} & \text{I} & \text{L} \end{array}$$
$$= x^2(2x^2) + x^2(-9) + 5(2x^2) + 5(-9)$$
$$= 2x^4 - 9x^2 + 10x^2 - 45$$
$$= 2x^4 + x^2 - 45$$

Section 5.5   The Greatest Common Factor and Factoring by Grouping

The greatest common factor (GCF) of the terms of a polynomial is the product of the GCF of the numerical coefficients and the GCF of the variable factors.	Factor $14xy^3 - 2xy^2 = 2 \cdot 7 \cdot x \cdot y^3 - 2 \cdot x \cdot y^2$. The GCF is $2 \cdot x \cdot y^2$, or $2xy^2$. $$14xy^3 - 2xy^2 = 2xy^2(7y - 1)$$
To factor a polynomial by grouping, group the terms so that each group has a common factor. Factor out these common factors. Then see if the new groups have a common factor.	Factor $x^4y - 5x^3 + 2xy - 10$. $$\begin{aligned} x^4y - 5x^3 + 2xy - 10 &= x^3(xy - 5) + 2(xy - 5) \\ &= (xy - 5)(x^3 + 2) \end{aligned}$$

Section 5.6   Factoring Trinomials

To factor $ax^2 + bx + c$, by trial and check,	Factor $28x^2 - 27x - 10$.
Step 1: Write all pairs of factors of ax^2.	Factors of $28x^2$: $28x$ and x, $2x$ and $14x$, $4x$ and $7x$.
Step 2: Write all pairs of factors of c.	Factors of -10: -2 and 5, 2 and -5, -10 and 1, 10 and -1.
Step 3: Try combinations of these factors until the middle term bx is found.	$$28x^2 - 27x - 10 = (7x + 2)(4x - 5)$$
To factor $ax^2 + bx + c$ by grouping, Find two numbers whose product is $a \cdot c$ and whose sum is b, write the term bx using the two numbers found, and then factor by grouping.	$$\begin{aligned} 2x^2 - 3x - 5 &= 2x^2 - 5x + 2x - 5 \\ &= x(2x - 5) + 1(2x - 5) \\ &= (2x - 5)(x + 1) \end{aligned}$$

Section 5.7   Factoring by Special Products

Perfect square trinomial $$a^2 + 2ab + b^2 = (a + b)^2$$ $$a^2 - 2ab + b^2 = (a - b)^2$$ Difference of two squares $$a^2 - b^2 = (a + b)(a - b)$$	*Factor* $$25x^2 + 30x + 9 = (5x + 3)^2$$ $$49z^2 - 28z + 4 = (7z - 2)^2$$ $$36x^2 - y^2 = (6x + y)(6x - y)$$ *(continued)*

Definitions and Concepts	Examples

Section 5.7   Factoring by Special Products

Sum and difference of two cubes

$$a^3 + b^3 = (a + b)(a^2 - ab + b^2)$$

$$a^3 - b^3 = (a - b)(a^2 + ab + b^2)$$

To factor a polynomial

Step 1: Factor out the GCF.

Step 2: If the polynomial is a binomial, see if it is a difference of two squares or a sum or difference of two cubes. If it is a trinomial, see if it is a perfect square trinomial. If not, try factoring by methods of Section 5.6. If it is a polynomial with 4 or more terms, try factoring by grouping.

Step 3: See if any factors can be factored further.

$$8y^3 + 1 = (2y + 1)(4y^2 - 2y + 1)$$

$$27p^3 - 64q^3 = (3p - 4q)(9p^2 + 12pq + 16q^2)$$

Factor $10x^4y + 5x^2y - 15y$.

$$10x^4y + 5x^2y - 15y = 5y(2x^4 + x^2 - 3)$$
$$= 5y(2x^2 + 3)(x^2 - 1)$$
$$= 5y(2x^2 + 3)(x + 1)(x - 1)$$

Section 5.8   Solving Equations by Factoring and Problem Solving

To solve polynomial equations by factoring:

Step 1: Write the equation so that one side is 0.

Step 2: Factor the polynomial completely.

Step 3: Set each factor equal to 0.

Step 4: Solve the resulting equations.

Step 5: Check each solution.

Solve

$$2x^3 - 5x^2 = 3x$$
$$2x^3 - 5x^2 - 3x = 0$$
$$x(2x + 1)(x - 3) = 0$$

$x = 0$ or $2x + 1 = 0$ or $x - 3 = 0$

$x = 0$ or $x = -\dfrac{1}{2}$ or $x = 3$

The solutions are 0, $-\dfrac{1}{2}$, and 3.

CHAPTER REVIEW

(5.1) *Evaluate.*

1. $(-2)^2$ 4
2. $(-3)^4$ 81
3. -2^2 -4
4. -3^4 -81
5. 8^0 1
6. -9^0 -1
7. -4^{-2} $-\frac{1}{16}$
8. $(-4)^{-2}$ $\frac{1}{16}$

Simplify each expression. Use only positive exponents.

9. $-xy^2 \cdot y^3 \cdot xy^2z$ $-x^2y^7z$
10. $(-4xy)(-3xy^2b)$ $12x^2y^3b$
11. $a^{-14} \cdot a^5$ $\frac{1}{a^9}$
12. $\frac{a^{16}}{a^{17}}$ $\frac{1}{a}$
13. $\frac{x^{-7}}{x^4}$ $\frac{1}{x^{11}}$
14. $\frac{9a(a^{-3})}{18a^{15}}$ $\frac{1}{2a^{17}}$
15. $\frac{y^{6p-3}}{y^{6p+2}}$ $\frac{1}{y^5}$

Write in scientific notation.

16. 36,890,000 3.689×10^7
17. -0.000362 -3.62×10^{-4}

Write each number without exponents.

18. 1.678×10^{-6} 0.000001678
19. 4.1×10^5 410,000

(5.2) *Simplify. Use only positive exponents.*

20. $(8^5)^3$ 8^{15}
21. $\left(\dfrac{a}{4}\right)^2$ $\dfrac{a^2}{16}$
22. $(3x)^3$ $27x^3$
23. $(-4x)^{-2}$ $\dfrac{1}{16x^2}$
24. $\left(\dfrac{6x}{5}\right)^2$ $\dfrac{36x^2}{25}$
25. $(8^6)^{-3}$ $\dfrac{1}{8^{18}}$
26. $\left(\dfrac{4}{3}\right)^{-2}$ $\dfrac{9}{16}$
27. $(-2x^3)^{-3}$ $-\dfrac{1}{8x^9}$

28. $\left(\dfrac{8p^6}{4p^4}\right)^{-2}$ $\dfrac{1}{4p^4}$

29. $(-3x^{-2}y^2)^3$ $\dfrac{-27y^6}{x^6}$

30. $\left(\dfrac{x^{-5}y^{-3}}{z^3}\right)^{-5}$ $x^{25}y^{15}z^{15}$

31. $\dfrac{4^{-1}x^3yz}{x^{-2}yx^4}$ $\dfrac{xz}{4}$

32. $(5xyz)^{-4}(x^{-2})^{-3}$ $\dfrac{x^2}{625y^4z^4}$

33. $\dfrac{2(3yz)^{-3}}{y^{-3}}$ $\dfrac{2}{27z^3}$

Simplify each expression.

34. $x^{4a}(3x^{5a})^3$ $27x^{19a}$

35. $\dfrac{4y^{3x-3}}{2y^{2x+4}}$ $2y^{x-7}$

Use scientific notation to find the quotient. Express each quotient in scientific notation.

36. $\dfrac{(0.00012)(144{,}000)}{0.0003}$ 5.76×10^4

37. $\dfrac{(-0.00017)(0.00039)}{3000}$ -2.21×10^{-11}

Simplify. Use only positive exponents.

38. $\dfrac{27x^{-5}y^5}{18x^{-6}y^2} \cdot \dfrac{x^4y^{-2}}{x^{-2}y^3}$ $\dfrac{3x^7}{2y^2}$

39. $\dfrac{3x^5}{y^{-4}} \cdot \dfrac{(3xy^{-3})^{-2}}{(z^{-3})^{-4}}$ $\dfrac{x^3y^{10}}{3z^{12}}$

40. $\dfrac{(x^w)^2}{(x^{w-4})^{-2}}$ x^{4w-8}

(5.3) Find the degree of each polynomial.

41. $x^2y - 3xy^3z + 5x + 7y$ 5 **42.** $3x + 2$ 1

Simplify by combining like terms.

43. $4x + 8x - 6x^2 - 6x^2y$ $12x - 6x^2 - 6x^2y$

44. $-8xy^3 + 4xy^3 - 3x^3y$ $-4xy^3 - 3x^3y$

48. $-4x^3 + 4x^2 + 16xy - 9x + 18$

Add or subtract as indicated. **47.** $8x^2 + 2b - 22$

45. $(3x + 7y) + (4x^2 - 3x + 7) + (y - 1)$ $4x^2 + 8y + 6$

46. $(4x^2 - 6xy + 9y^2) - (8x^2 - 6xy - y^2)$ $-4x^2 + 10y^2$

47. $(3x^2 - 4b + 28) + (9x^2 - 30) - (4x^2 - 6b + 20)$

48. Add $(9xy + 4x^2 + 18)$ and $(7xy - 4x^3 - 9x)$.

49. Subtract $(x - 7)$ from the sum of $(3x^2y - 7xy - 4)$ and $(9x^2y + x)$. $12x^2y - 7xy + 3$

50. $\begin{array}{l} x^2 - 5x + 7 \\ \underline{-(x + 4)} \end{array}$ $x^2 - 6x + 3$

51. $\begin{array}{l} x^3 + 2xy^2 - y \\ + (x - 4xy^2 -7) \end{array}$ $x^3 + x - 2xy^2 - y - 7$

If $P(x) = 9x^2 - 7x + 8$, find the following.

52. $P(6)$ 290 **53.** $P(-2)$ 58

54. $P(-3)$ 110

If $P(x) = 2x - 1$ and $Q(x) = x^2 + 2x - 5$, find the following.

55. $P(x) + Q(x)$ $x^2 + 4x - 6$

56. $2[P(x)] - Q(x)$ $-x^2 + 2x + 3$

△ **57.** Find the perimeter of the rectangle. $(6x^2y - 12x + 12)$ cm

$x^2y + 5$ cm

$2x^2y - 6x + 1$ cm

58. $-24x^3 + 36x^2 - 6x$ **59.** $-12a^2b^5 - 28a^2b^3 - 4ab^2$
60. $2x^2 + x - 36$ **61.** $9x^2a^2 - 24xab + 16b^2$
62. $36x^3 - 11x^2 - 8x - 3$ **63.** $15x^2 + 18xy - 81y^2$

(5.4) Multiply.

58. $-6x(4x^2 - 6x + 1)$ **59.** $-4ab^2(3ab^3 + 7ab + 1)$

60. $(x - 4)(2x + 9)$ **61.** $(-3xa + 4b)^2$

62. $(9x^2 + 4x + 1)(4x - 3)$ **63.** $(5x - 9y)(3x + 9y)$

64. $\left(x - \dfrac{1}{3}\right)\left(x + \dfrac{2}{3}\right)$ **65.** $(x^2 + 9x + 1)^2$

64. $x^2 + \dfrac{1}{3}x - \dfrac{2}{9}$ **65.** $x^4 + 18x^3 + 83x^2 + 18x + 1$

Multiply, using special products.

66. $(3x - y)^2$ **67.** $(4x + 9)^2$

68. $(x + 3y)(x - 3y)$ $x^2 - 9y^2$

69. $[4 + (3a - b)][4 - (3a - b)]$ $16 - 9a^2 + 6ab - b^2$

70. If $P(x) = 2x - 1$ and $Q(x) = x^2 + 2x - 5$, find $P(x) \cdot Q(x)$. $2x^3 + 3x^2 - 12x + 5$

△ **71.** Find the area of the rectangle. $(9y^2 - 49z^2)$ sq. units

$3y - 7z$ units

$3y + 7z$ units

66. $9x^2 - 6xy + y^2$ **67.** $16x^2 + 72x + 81$

Multiply. Assume that all variable exponents represent integers.

72. $4a^b(3a^{b+2} - 7)$ $12a^{2b+2} - 28a^b$

73. $(4xy^z - b)^2$ $16x^2y^{2z} - 8xy^zb + b^2$

74. $(3x^a - 4)(3x^a + 4)$ $9x^{2a} - 16$

(5.5) Factor out the greatest common factor.

75. $16x^3 - 24x^2$ $8x^2(2x - 3)$ **76.** $36y - 24y^2$ $12y(3 - 2y)$

77. $6ab^2 + 8ab - 4a^2b^2$ **78.** $14a^2b^2 - 21ab^2 + 7ab$

79. $6a(a + 3b) - 5(a + 3b)$ **80.** $4x(x - 2y) - 5(x - 2y)$

81. $xy - 6y + 3x - 18$ **82.** $ab - 8b + 4a - 32$

83. $pq - 3p - 5q + 15$ **84.** $x^3 - x^2 - 2x + 2$

△ **85.** A smaller square is cut from a larger rectangle. Write the area of the shaded region as a factored polynomial. $x(2y - x)$

77. $2ab(3b + 4 - 2ab)$
78. $7ab(2ab - 3b + 1)$
79. $(a + 3b)(6a - 5)$
80. $(x - 2y)(4x - 5)$
81. $(x - 6)(y + 3)$
82. $(a - 8)(b + 4)$
83. $(p - 5)(q - 3)$
84. $(x^2 - 2)(x - 1)$

(5.6) Completely factor each polynomial.

86. $x^2 - 14x - 72$ **87.** $x^2 + 16x - 80$

88. $2x^2 - 18x + 28$ **89.** $3x^2 + 33x + 54$

90. $2x^3 - 7x^2 - 9x$ **91.** $3x^2 + 2x - 16$

92. $6x^2 + 17x + 10$ **93.** $15x^2 - 91x + 6$

94. $4x^2 + 2x - 12$ **95.** $9x^2 - 12x - 12$

96. $y^2(x + 6)^2 - 2y(x + 6)^2 - 3(x + 6)^2$

97. $(x + 5)^2 + 6(x + 5) + 8$

98. $x^4 - 6x^2 - 16$ **99.** $x^4 + 8x^2 - 20$

(5.7) Factor each polynomial completely.

100. $x^2 - 100$ **101.** $x^2 - 81$

102. $2x^2 - 32$ **103.** $6x^2 - 54$

104. $81 - x^4$ **105.** $16 - y^4$

106. $(y + 2)^2 - 25$ **107.** $(x - 3)^2 - 16$

108. $x^3 + 216$ **109.** $y^3 + 512$

110. $8 - 27y^3$ **111.** $1 - 64y^3$

112. $6x^4y + 48xy$ **113.** $2x^5 + 16x^2y^3$

114. $x^2 - 2x + 1 - y^2$ **115.** $x^2 - 6x + 9 - 4y^2$

116. $4x^2 + 12x + 9$ **117.** $16a^2 - 40ab + 25b^2$

△ **118.** The volume of the cylindrical shell is $\pi R^2 h - \pi r^2 h$ cubic units. Write this volume as a factored expression.

86. $(x - 18)(x + 4)$
87. $(x - 4)(x + 20)$
88. $2(x - 2)(x - 7)$
89. $3(x + 2)(x + 9)$
90. $x(2x - 9)(x + 1)$
91. $(3x + 8)(x - 2)$
92. $(6x + 5)(x + 2)$
93. $(15x - 1)(x - 6)$
94. $2(2x - 3)(x + 2)$
95. $3(x - 2)(3x + 2)$
96. $(x + 6)^2(y - 3)(y + 1)$

97. $(x + 7)(x + 9)$
98. $(x^2 - 8)(x^2 + 2)$
99. $(x^2 - 2)(x^2 + 10)$
100. $(x + 10)(x - 10)$
101. $(x + 9)(x - 9)$
102. $2(x + 4)(x - 4)$
103. $6(x + 3)(x - 3)$
104. $(9 + x^2)(3 + x)(3 - x)$
105. $(4 + y^2)(2 + y)(2 - y)$
106. $(y + 7)(y - 3)$
107. $(x - 7)(x + 1)$

108. $(x + 6)(x^2 - 6x + 36)$
109. $(y + 8)(y^2 - 8y + 64)$
110. $(2 - 3y)(4 + 6y + 9y^2)$
111. $(1 - 4y)(1 + 4y + 16y^2)$
112. $6xy(x + 2)(x^2 - 2x + 4)$
113. $2x^2(x + 2y)(x^2 - 2xy + 4y^2)$
114. $(x - 1 - y)(x - 1 + y)$
115. $(x - 3 - 2y)(x - 3 + 2y)$
116. $(2x + 3)^2$
117. $(4a - 5b)^2$
118. $\pi h(R - r)(R + r)$ cu.units

(5.8) Solve each polynomial equation for the variable.

119. $(3x - 1)(x + 7) = 0$ $\frac{1}{3}, -7$

120. $3(x + 5)(8x - 3) = 0$ $-5, \frac{3}{8}$

121. $5x(x - 4)(2x - 9) = 0$ $0, 4, \frac{9}{2}$

122. $6(x + 3)(x - 4)(5x + 1) = 0$ $-3, -\frac{1}{5}, 4$

123. $2x^2 = 12x$ $0, 6$

124. $4x^3 - 36x = 0$ $-3, 0, 3$

125. $(1 - x)(3x + 2) = -4x$ $-\frac{1}{3}, 2$

126. $2x(x - 12) = -40$ $2, 10$

127. $3x^2 + 2x = 12 - 7x$ $-4, 1$

128. $2x^2 + 3x = 35$ $\frac{7}{2}, -5$

129. $x^3 - 18x = 3x^2$ $0, 6, -3$

130. $19x^2 - 42x = -x^3$ $-21, 0, 2$

131. $12x = 6x^3 + 6x^2$ $0, -2, 1$

132. $8x^3 + 10x^2 = 3x$ $-\frac{3}{2}, 0, \frac{1}{4}$

133. The sum of a number and twice its square is 105. Find the number. $-\frac{15}{2}, 7$

△ **134.** The length of a rectangular piece of carpet is 5 meters less than twice its width. Find the dimensions of the carpet if its area is 33 square meters. $5\frac{1}{2}$ meters, 6 meters

135. A scene from an adventure film calls for a stunt dummy to be dropped from above the second-story platform of the Eiffel Tower, a distance of 400 feet. Its height $h(t)$ at time t seconds is given by

$$h(t) = -16t^2 + 400$$

Determine when the stunt dummy will reach the ground. 5 sec

400 ft

CHAPTER 5 TEST

Remember to use your Chapter Test Prep Video CD to help you study and view solutions to the test questions you need help with.

Simplify. Use positive exponents to write the answers.

1. $(-9x)^{-2}$ $\dfrac{1}{81x^2}$

2. $-3xy^{-2}(4xy^2)z$ $-12x^2z$

3. $\dfrac{6^{-1}a^2b^{-3}}{3^{-2}a^{-5}b^2}$ $\dfrac{3a^7}{2b^5}$

4. $\left(\dfrac{-xy^{-5}z}{xy^3}\right)^{-5}$ $-\dfrac{y^{40}}{z^5}$

Write in scientific notation.

5. 630,000,000 6.3×10^8

6. 0.01200 1.2×10^{-2}

7. Write 5×10^{-6} without exponents.   0.000005

8. Use scientific notation to find the quotient.

$$\dfrac{(0.0024)(0.00012)}{0.00032} \quad 0.0009$$

Perform the indicated operations.

9. $(4x^3y - 3x - 4) - (9x^3y + 8x + 5)$ $-5x^3y - 11x - 9$

10. $-3xy(4x + y)$

11. $(3x + 4)(4x - 7)$

12. $(5a - 2b)(5a + 2b)$

13. $(6m + n)^2$

14. $(2x - 1)(x^2 - 6x + 4)$

Factor each polynomial completely.

15. $16x^3y - 12x^2y^4$

16. $x^2 - 13x - 30$

17. $4y^2 + 20y + 25$

18. $6x^2 - 15x - 9$

19. $4x^2 - 25$

20. $x^3 + 64$

21. $3x^2y - 27y^3$

22. $6x^2 + 24$

23. $16y^3 - 2$

24. $x^2y - 9y - 3x^2 + 27$

Solve the equation for the variable.

25. $3n(7n - 20) = 96$ $4, -\frac{8}{7}$

26. $(x + 2)(x - 2) = 5(x + 4)$ $-3, 8$

27. $2x^3 + 5x^2 = 8x + 20$ $-\frac{5}{2}, -2, 2$

28. Write the area of the shaded region as a factored polynomial. $(x + 2y)(x - 2y)$

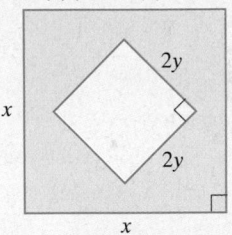

29. A pebble is hurled upward from the top of the Canada Trust Tower, which is 880 feet tall, with an initial velocity of 96 feet per second. Neglecting air resistance, the height $h(t)$ of the pebble after t seconds is given by the polynomial function

$$h(t) = -16t^2 + 96t + 880$$

a. Find the height of the pebble when $t = 1$.   960 ft

b. Find the height of the pebble when $t = 5.1$.   953.44 ft

c. When will the pebble hit the ground?   11 sec

10. $-12x^2y - 3xy^2$ **11.** $12x^2 - 5x - 28$ **12.** $25a^2 - 4b^2$ **13.** $36m^2 + 12mn + n^2$ **14.** $2x^3 - 13x^2 + 14x - 4$
15. $4x^2y(4x - 3y^3)$ **16.** $(x - 15)(x + 2)$ **17.** $(2y + 5)^2$ **18.** $3(2x + 1)(x - 3)$ **19.** $(2x + 5)(2x - 5)$
20. $(x + 4)(x^2 - 4x + 16)$ **21.** $3y(x + 3y)(x - 3y)$ **22.** $6(x^2 + 4)$ **23.** $2(2y - 1)(4y^2 + 2y + 1)$ **24.** $(x + 3)(x - 3)(y - 3)$

CHAPTER CUMULATIVE REVIEW

5. \$11,607.55 (Sec. 1.8, Ex. 5)   **7.** See graphing answer section.

1. Find the roots.

a. $\sqrt[3]{27}$ 3

b. $\sqrt[5]{1}$ 1

c. $\sqrt[4]{16}$ 2 (Sec. 1.3, Ex. 8)

2. Find the roots.

a. $\sqrt[3]{64}$ 4

b. $\sqrt[4]{81}$ 3 (Sec. 1.3)

c. $\sqrt[5]{32}$ 2

3. Solve: $2(x - 3) = 5x - 9$.   1 (Sec. 1.5, Ex. 4)

4. Solve. $0.3y + 2.4 = 0.1y + 4$ 8 (Sec. 1.5)

5. Karen Estes just received an inheritance of \$10,000 and plans to place all the money in a savings account that pays 5%

compounded quarterly to help her son go to college in 3 years. How much money will be in the account in 3 years?

6. A gallon of latex paint can cover 400 square feet. How many gallon containers of paint should be bought to paint two coats on each wall of a rectangular room whose dimensions are 14 feet by 18 feet? (Assume 8-foot ceilings).   3 gals (Sec. 1.8)

7. Solve and graph the solution set.

a. $\dfrac{1}{4}x \le \dfrac{3}{8}$ $\left\{x \middle| x \le \dfrac{3}{2}\right\}$; (Sec. 3.2, Ex. 4)

b. $-2.3x < 6.9$ $\{x | x > -3\}$; (Sec. 3.2, Ex. 4)

8. Solve. Graph the solution set and write it in interval

notation. $x + 2 \le \dfrac{1}{4}(x - 7)$ (Sec. 3.2) ⟵──┤── , $(-\infty, -5]$
─5

Solve.   **15.–16., 19.–20., 25.** See graphing answer section.

9. $-1 \le \dfrac{2x}{3} + 5 \le 2$ $\left[-9, -\dfrac{9}{2}\right]$ (Sec. 3.3, Ex. 5)

10. $-\dfrac{1}{3} < \dfrac{3x + 1}{6} \le \dfrac{1}{3}$ $\left(-1, \dfrac{1}{3}\right]$ (Sec. 3.3)

11. $|y| = 0$ 0 (Sec. 3.4, Ex. 5)

12. $8 + |4c| = 24$ $-4, 4$ (Sec. 3.4)

13. $\left|2x - \dfrac{1}{10}\right| < -13$ Ø (Sec. 3.5, Ex. 4)

14. $|5x - 1| + 9 > 5$ $(-\infty, \infty)$ (Sec. 3.5)

15. Graph the linear equation $y = -2x + 3$.   (Sec. 2.1, Ex. 5)

16. Graph the linear equation $y = 3x$.   (Sec. 2.1)

17. Is the relation $x = y^2$ also a function?   no (Sec. 2.2, Ex. 4)

18. If $f(x) = 3x^2 + 2x + 3$, find $f(-3)$.   24   (Sec. 2.2)

19. Graph $x = 2$.   (Sec. 2.3, Ex. 5)

20. Graph $y - 5 = 0$ (Sec. 2.3)

21. Find the slope of the line $y = 2$.   0 (Sec. 2.4, Ex. 6)

22. Find the slope of the line, $f(x) = -2x - 3$. -2 (Sec. 2.4)

23. Find an equation of the horizontal line containing the point (2, 3). $y = 3$ (Sec. 2.5, Ex. 7)

24. Find the equation of the vertical line containing the point $(-3, 2)$. $x = -3$ (Sec. 2.5)

25. Graph the union of $x + \dfrac{1}{2}y \ge -4$ or $y \le -2$.
(Sec. 3.6, Ex. 4)

26. Find the equation of the line containing the point $(-2, 3)$ and slope of 0. $y = 3$ (Sec. 3.6)

27. Use the substitution method to solve the system.
$\begin{cases} y = x + 5 \\ 3x = 2y - 9 \end{cases}$ (1, 6)   (Sec. 4.1, Ex. 4)

28. Use the substitution method to solve the system.
$\begin{cases} 4x - 2y = 8 \\ y = 3x - 6 \end{cases}$ (2, 0)   (Sec. 4.1)

29. Solve the system.
$\begin{cases} 2x + 4y \quad\;\; = 1 \\ 4x \quad\quad - 4z = -1 \\ \quad\;\; y - 4z = -3 \end{cases}$ $\left(\dfrac{1}{2}, 0, \dfrac{3}{4}\right)$ (Sec. 4.2, Ex. 3)

30. Solve the system.
$\begin{cases} x + y - \dfrac{3}{2}z = \dfrac{1}{2} \\ \quad\;\; -y - 2z = 14 \\ x - \dfrac{2}{3}y \quad\quad = -\dfrac{1}{3} \end{cases}$ $(-3, -4, -5)$ (Sec. 4.2)

31. A first number is 4 less than a second number. Four times the first number is 6 more than twice the second. Find the numbers.   7, 11 (Sec. 4.3, Ex. 2)

32. One solution contains 20% acid and a second solution con-

tains 60% acid. How many ounces of each solution should be mixed in order to have 50 ounces of a 30% acid solution?
37.5 oz. of 20% solution; 12.5 oz. of 60% solution (Sec. 4.3)

33. Use matrices to solve the system.
$\begin{cases} 2x - y = 3 \\ 4x - 2y = 5 \end{cases}$ no solution (Sec. 4.4, Ex. 2)

34. Use matrices to solve the system.
$\begin{cases} 4y = 8 \\ x + y = 7 \end{cases}$ (5, 2)   (Sec. 4.4)

35. Use Cramer's rule to solve the system.
$\begin{cases} x - 2y + z = 4 \\ 3x + y - 2z = 3 \\ 5x + 5y + 3z = -8 \end{cases}$ $(1, -2, -1)$ (Sec. 4.5, Ex. 4)

36. Use Cramer's Rule to solve the system.
$\begin{cases} x + y + z = 0 \\ 2x - 3y + z = 5 \\ 2x + y + 2z = 2 \end{cases}$ $(-3, -2, 5)$ (Sec. 4.5)

37. Write each number in scientific notation.
  a. 730,000 7.3×10^5 **b.** 0.00000104 1.04×10^{-6}
(Sec. 5.1, Ex. 8)

38. Write each number in scientific notation.
  a. 8,250,000 8.25×10^6 **b.** 0.0000346 3.46×10^{-5}

39. Simplify each expression. Use positive exponents to write (Sec. 5.1) the answers.
  a. $(2x^0y^{-3})^{-2}$ $\dfrac{y^6}{4}$ **b.** $\left(\dfrac{x^{-5}}{x^{-2}}\right)^{-3}$ x^9
  c. $\left(\dfrac{2}{7}\right)^{-2}$ $\dfrac{49}{4}$ **d.** $\dfrac{5^{-2}x^{-3}y^{11}}{x^2y^{-5}}$ $\dfrac{y^{16}}{25x^5}$ (Sec. 5.2, Ex. 3)

40. Simplify each expression. Use positive exponents to write the answers.
  a. $(4a^{-1}b^o)^{-3}$ $\dfrac{a^3}{64}$ **b.** $\left(\dfrac{a^{-6}}{a^{-8}}\right)^{-2}$ $\dfrac{1}{a^4}$
  c. $\left(\dfrac{2}{3}\right)^{-3}$ $\dfrac{27}{8}$ **d.** $\dfrac{3^{-2}a^{-2}b^{12}}{a^4b^{-5}}$ $\dfrac{b^{17}}{9a^6}$ (Sec. 5.2)

41. Find the degree of the polynomial $3xy + x^2y^2 - 5x^2 - 6$.

42. Subtract $(5x^2 + 3x)$ from $(3x^2 - 2x)$. $-2x^2 - 5x$ (Sec. 5.3)

43. Multiply.
  a. $(2x^3)(5x^6)$ $10x^9$ **b.** $(7y^4z^4)(-xy^{11}z^5)$
  $-7xy^{15}z^9$ (Sec. 5.4, Ex. 1)

44. Multiply.   **41.** 4 (Sec. 5.3, Ex. 3)
  a. $(3y^6)(4y^2)$ $12y^8$ **b.** $(6a^3b^2)(-a^2bc^4)$ $-6a^5b^3c^4$

Factor.   (Sec. 5.4)

45. $17x^3y^2 - 34x^4y^2$ $17x^3y^2(1 - 2x)$ (Sec. 5.5, Ex. 3)

46. $12x^3y - 3xy^3$ $3xy(2x + y)(2x - y)$ (Sec. 5.7)

47. $x^2 + 10x + 16$ $(x + 2)(x + 8)$ (Sec. 5.6, Ex. 1)

48. $5a^2 + 14a - 3$ $(5a - 1)(a + 3)$ (Sec. 5.6)

49. Solve $2x^2 + 9x - 5 = 0$. $-5, \frac{1}{2}$ (Sec. 5.8, Ex. 2)

50. Solve $3x^2 - 10x - 8 = 0$ $4, -\dfrac{2}{3}$ (Sec. 5.8)

Rational Expressions

Polynomials are to algebra what integers are to arithmetic. We have added, subtracted, multiplied, and raised polynomials to powers, each operation yielding another polynomial, just as these operations on integers yield another integer. But when we divide one integer by another, the result may or may not be another integer. Likewise, when we divide one polynomial by another, we may or may not get a polynomial in return. The quotient $x \div (x + 1)$ is not a polynomial; it is a *rational expression* that can be written as $\dfrac{x}{x + 1}$.

In this chapter, we study these new algebraic forms known as rational expressions and the *rational functions* they generate.

Have you ever thought about how many feet, or even miles, of wiring are needed in your house, dormitory, or apartment building to make all of your lights and electrical appliances work? Without electricians to wire our homes and buildings, we all would probably be in the dark right now.

In addition to installing wiring and coaxial or fiber-optic cable, electricians also may repair or maintain electrical components. Most electricians learn their trade through a four-or-five year apprenticeship program that includes both on-the-job training and classes such as electrical theory and mathematics. Electricians use math and problem-solving skills in tasks such as estimating job costs, testing circuits, and reading blueprints.

In the Spotlight on Decision Making feature on page 448, you will have the opportunity to make a decision as an electrician about which resistor to use to repair a power supply.

The table and the graph representing the power dissipated by a resistor in the Spotlight on Decision Making feature on page 448. According to the graph, the power decreases as the resistance increases.

Source: National Electrical Contractors Association Website

6.1 RATIONAL FUNCTIONS AND MULTIPLYING AND DIVIDING RATIONAL EXPRESSIONS

Objectives

1   Find the domain of a rational expression.

2   Simplify rational expressions.

3   Multiply rational expressions.

4   Divide rational expressions.

5   Use rational functions in applications.

Recall that a *rational number*, or *fraction*, is a number that can be written as the quotient $\frac{p}{q}$ of two integers p and q as long as q is not 0. A **rational expression** is an expression that can be written as the quotient $\frac{P}{Q}$ of two polynomials P and Q as long as Q is not 0.

Examples of Rational Expressions

$$\frac{3x + 7}{2} \qquad \frac{7x + 2}{x - 3} \qquad \frac{7x - 2}{2x^2 + 7x + 6}$$

Rational expressions are sometimes used to describe functions. For example, we call the function $f(x) = \frac{x^2 + 2}{x - 3}$ a **rational function** since $\frac{x^2 + 2}{x - 3}$ is a rational expression.

1   As with fractions, a rational expression is **undefined** if the denominator is 0. If a variable in a rational expression is replaced with a number that makes the denominator 0, we say that the rational expression is **undefined** for this value of the variable. For example, the rational expression $\frac{x^2 + 2}{x - 3}$ is undefined when x is 3, because replacing x with 3 results in a denominator of 0. For this reason, we must exclude 3 from the domain of the function $f(x) = \frac{x^2 + 2}{x - 3}$.

The domain of f is then

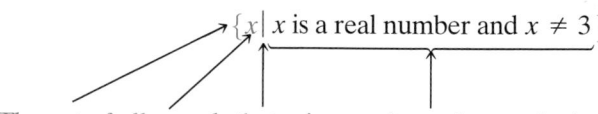

$$\{x \mid x \text{ is a real number and } x \neq 3\}$$

"The set of all x such that x is a real number and x is not equal to 3."
In this section, we will use this set builder notations to write domains.

Unless told otherwise, we assume that the domain of a function described by an equation is the set of all real numbers for which the equation is defined.

EXAMPLE 1

Find the domain of each rational function.

a. $f(x) = \dfrac{8x^3 + 7x^2 + 20}{2}$ **b.** $f(x) = \dfrac{7x + 2}{x - 3}$ **c.** $f(x) = \dfrac{7x - 2}{x^2 - 2x - 15}$

Solution   The domain of each function will contain all real numbers except those values that make the denominator 0.

a. No matter what the value of x, the denominator of $f(x) = \dfrac{8x^3 + 7x^2 + 20}{2}$ is never 0, so the domain of f is $\{x | x \text{ is a real number}\}$.

b. To find the values of x that make the denominator of $f(x)$ equal to 0, we solve the equation "denominator = 0":

$$x - 3 = 0, \quad \text{or} \quad x = 3$$

The domain must exclude 3 since the rational expression is undefined when x is 3. The domain of g is $\{x | x \text{ is a real number and } x \neq 3\}$.

c. We find the domain by setting the denominator equal to 0.

$$x^2 - 2x - 15 = 0 \quad \text{\small Set the denominator equal to 0 and solve.}$$
$$(x - 5)(x + 3) = 0$$
$$x - 5 = 0 \quad \text{or} \quad x + 3 = 0$$
$$x = 5 \quad \text{or} \quad x = -3$$

If x is replaced with 5 or with -3, the rational expression is undefined.

The domain of f is $\{x | x \text{ is a real number and } x \neq 5, x \neq -3\}$.

CLASSROOM EXAMPLE

Find the domain of each rational function.

a. $\dfrac{x^2 + 1}{x - 6}$

b. $\dfrac{5x + 4}{x^2 - 3x - 10}$

c. $\dfrac{x^2 - 9}{4}$

answer:

a. $\{x | x \text{ is a real number}, x \neq 6\}$

b. $\{x | x \text{ is a real number}, x \neq 5, x \neq -2\}$

c. $\{x | x \text{ is a real number}\}$

Let's use a graphing utility to confirm the domain of the function in Example 1**b**, $f(x) = \frac{7x + 2}{x - 3}$.

To confirm this domain, graph $y_1 = \frac{7x + 2}{x - 3}$. The domain of $f(x)$ does not include 3, so the graph of $f(x)$ does not exist at $x = 3$. If we graph $f(x)$ in dot mode, the graph shows that the function is undefined at $x = 3$ and no ordered pair solutions exist with an x-value of 3. See the graph in the margin.

The graph of $f(x) = \frac{7x + 2}{x - 3}$ confirms that the domain is $\{x | x \text{ is a real number and } x \neq 3\}$.

TECHNOLOGY NOTE

Most graphing utilities have a dot mode. When graphing a function in dot mode only the calculated points of the function are plotted. Consult your owner's manual for specific instructions on using this feature.

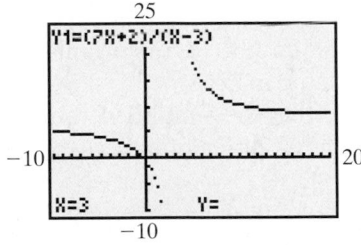

In dot mode the graphing utility plots only the ordered pairs that make the sentence true. Notice y_1 is undefined when $x = 3$ and thus a point does not exist on the graph of the function.

Technology Note

Use care when interpreting the graph of a rational expression on a graphing utility. For example, if we graph $f(x) = \frac{7x + 2}{x - 3}$ in *connected* mode, the graphing utility *connects* the last point plotted before $x = 3$ with the next point plotted after $x = 3$. The result of this is a vertical line that appears on the graph.

We know that this vertical line is not part of the graph because the function is undefined at $x = 3$. We also know that this vertical line is not part of the graph since the graph of this function would then not pass the vertical line test.

The graph of $y = \frac{7x + 2}{x - 3}$ is undefined at $x = 3$.

In connected mode the graphing utility shows a vertical line where it connects the point plotted to the left of $x = 3$ with the next point plotted to the right of $x = 3$.

✔ **CONCEPT CHECK**

For which of these values (if any) is the rational expression $\dfrac{x - 3}{x^2 + 2}$ undefined?

Concept Check Answer: e

a. 2   **b.** 3   **c.** -2 **d.** 0   **e.** None of these

2   Recall that a fraction is in lowest terms or simplest form if the numerator and denominator have no common factors other than 1 (or -1). For example, $\dfrac{3}{13}$ is in lowest terms since 3 and 13 have no common factors other than 1 (or -1).

To **simplify** a rational expression, or to write it in lowest terms, we use the fundamental principle of rational expressions.

Fundamental Principle of Rational Expressions

For any rational expression $\dfrac{P}{Q}$ and any polynomial R, where $R \neq 0$,

$$\frac{PR}{QR} = \frac{P}{Q}$$

In this table, $y_1 = \dfrac{(x+2)^2}{x^2 - 4}$

and $y_2 = \dfrac{x+2}{x-2}$.

Notice that both expressions are undefined at $x = 2$ (and an error message is given), and when $x = -2$, y_1 is not defined.

Thus, the fundamental principle says that multiplying or dividing the numerator and denominator of a rational expression by the same nonzero polynomial yields an equivalent rational expression.

To simplify a rational expression such as $\dfrac{(x+2)^2}{x^2-4}$, factor the numerator and the denominator and then use the fundamental principle of rational expressions to divide out common factors.

$$\frac{(x+2)^2}{x^2-4} = \frac{(x+2)(x+2)}{(x+2)(x-2)} = \frac{x+2}{x-2}$$

This means that the rational expression $\dfrac{(x+2)^2}{x^2-4}$ has the same value as the rational expression $\dfrac{x+2}{x-2}$ for all values of x except 2 and -2. (Remember that when x is 2, the denominators of both rational expressions are 0 and that when x is -2, the original rational expression has a denominator of 0.) See the table in the margin.

As we simplify rational expressions, we will assume that the simplified rational expression is equivalent to the original rational expression for all real numbers except those for which either denominator is 0.

In general, the following steps may be used to simplify rational expressions or to write a rational expression in lowest terms.

Simplifying or Writing a Rational Expression in Lowest Terms

Step 1: Completely factor the numerator and denominator of the rational expression.

Step 2: Apply the fundamental principle of rational expressions to divide out factors common to both the numerator and denominator.

For now, we assume that variables in a rational expression do not represent values that make the denominator 0.

EXAMPLE 2

Simplify $\dfrac{2x^2}{10x^3 - 2x^2}$.

Solution Factor out the GCF of $2x^2$ from the denominator. Then divide numerator and denominator by their GCF, $2x^2$.

$$\frac{2x^2}{10x^3 - 2x^2} = \frac{2x^2 \cdot 1}{2x^2(5x - 1)} = \frac{1}{5x - 1}$$

When the terms in the numerator of a rational expression differ by sign from the terms of the denominator, the polynomials are opposites of each other and the expression simplifies to -1. To see this, study Example 3b below.

EXAMPLE 3

Simplify each rational expression.

a. $\dfrac{2 + x}{x + 2}$ **b.** $\dfrac{2 - x}{x - 2}$

TEACHING TIP

Before discussing Example 3b, have students rewrite the problem so the like terms in the numerator and denominator are aligned. Then ask them if they notice any similarities in the two expressions.

Solution **a.** $\dfrac{2 + x}{x + 2} = \dfrac{x + 2}{x + 2} = 1$ By the commutative property of addition, $2 + x = x = 2$.

b. $\dfrac{2 - x}{x - 2}$

The terms in the numerator of $\dfrac{2 - x}{x - 2}$ differ by sign from the terms of the denominator, so the polynomials are opposites of each other and the expression simplifies to -1. To see this, we factor out -1 from the numerator or the denominator. If -1 is factored from the numerator, then

$$\frac{2 - x}{x - 2} = \frac{-1(-2 + x)}{x - 2} = \frac{-1(x - 2)}{x - 2} = \frac{-1}{1} = -1$$

If -1 is factored from the denominator, the result is the same.

$$\frac{2 - x}{x - 2} = \frac{2 - x}{-1(-x + 2)} = \frac{2 - x}{-1(2 - x)} = \frac{1}{-1} = -1$$

Helpful Hint

When the numerator and the denominator of a rational expression are opposites of each other, the expression simplifies to -1.

Discover the Concept

a. Consider the expression $\dfrac{2x^2 - 18}{x^2 - 2x - 3}$. Graph the numerator as $y_1 = 2x^2 - 18$ and the denominator as $y_2 = x^2 - 2x - 3$ in the same window.

b. Completely factor the numerator and the denominator of the original expression.

c. What do you notice about the common factor in the expression and the point of intersection of the graphs of the numerator and denominator?

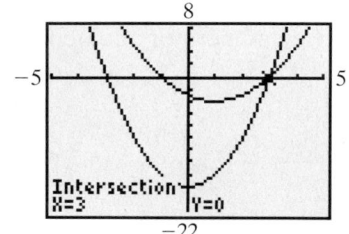

Graphs of $y_1 = 2x^2 - 18$ and $y_2 = x^2 - 2x - 3$ both intersect the x-axis at $x = 3$.

In the above discovery, we notice that both graphs intersect the x-axis at $x = 3$. In part b, we find that $(x - 3)$ is a common factor of the numerator and denominator.

$$\frac{2x^2 - 18}{x^2 - 2x - 3} = \frac{2(x^2 - 9)}{x^2 - 2x - 3} = \frac{2(x + 3)(x - 3)}{(x + 1)(x - 3)}$$

In general, if the graphs of the numerator and denominator of a rational expression share an x-intercept c, then $(x - c)$ is a common factor of the numerator and denominator of the rational expression.

EXAMPLE 4

Simplify $\dfrac{18 - 2x^2}{x^2 - 2x - 3}$.

Solution

$$\frac{18 - 2x^2}{x^2 - 2x - 3} = \frac{2(9 - x^2)}{(x + 1)(x - 3)} \qquad \text{Factor.}$$

$$= \frac{2(3 + x)(3 - x)}{(x + 1)(x - 3)} \qquad \text{Factor completely.}$$

$$= \frac{2(3 + x) \cdot -1(x - 3)}{(x + 1)(x - 3)} \qquad \begin{array}{l}\text{Notice the opposites } 3 - x \\ \text{and } x - 3. \text{ Write } 3 - x \\ \text{as } -1(x - 3) \text{ and simplify.}\end{array}$$

$$= -\frac{2(3 + x)}{x + 1}$$

Helpful Hint

Recall that for a fraction $\dfrac{a}{b}$,

$$\frac{a}{-b} = \frac{-a}{b} = -\frac{a}{b}$$

For example

$$\frac{-(x + 1)}{(x + 2)} = \frac{(x + 1)}{-(x + 2)} = -\frac{x + 1}{x + 2}$$

✔ CONCEPT CHECK

Which of the following expressions are equivalent to $\dfrac{x}{8 - x}$?

a. $\dfrac{-x}{x - 8}$ **b.** $\dfrac{-x}{8 - x}$ **c.** $\dfrac{x}{x - 8}$ **d.** $\dfrac{-x}{-8 + x}$

EXAMPLE 5

Simplify each rational expression.

a. $\dfrac{x^3 + 8}{2 + x}$ **b.** $\dfrac{2y^2 + 2}{y^3 - 5y^2 + y - 5}$

Solution

a. $\dfrac{x^3 + 8}{2 + x} = \dfrac{(x + 2)(x^2 - 2x + 4)}{x + 2}$ Factor the sum of the two cubes.

$\quad\quad = x^2 - 2x + 4$ Divide out common factors.

b. $\dfrac{2y^2 + 2}{y^3 - 5y^2 + y - 5} = \dfrac{2(y^2 + 1)}{(y^3 - 5y^2) + (y - 5)}$ Factor the numerator.

$\quad\quad = \dfrac{2(y^2 + 1)}{y^2(y - 5) + 1(y - 5)}$ Factor the denominator by grouping.

$\quad\quad = \dfrac{2(y^2 + 1)}{(y - 5)(y^2 + 1)}$

$\quad\quad = \dfrac{2}{y - 5}$ Divide out common factors.

✔ **CONCEPT CHECK**

Does $\dfrac{n}{n + 2}$ simplify to $\dfrac{1}{2}$? Why or why not?

3 Arithmetic operations on rational expressions are performed in the same way as they are on rational numbers.

Multiplying Rational Expressions

The rule for multiplying rational expressions is

$$\frac{P}{Q} \cdot \frac{R}{S} = \frac{PR}{QS} \quad \text{as long as } Q \neq 0 \text{ and } S \neq 0.$$

To multiply rational expressions, you may use these steps:

Step 1: Completely factor each numerator and denominator.

Step 2: Use the rule above and multiply the numerators and the denominators.

Step 3: Simplify the product by dividing the numerator and denominator by their common factors.

When we multiply rational expressions, notice that we factor each numerator and denominator first. This helps when we apply the fundamental principle to write the product in simplest form.

EXAMPLE 6

Multiply.

a. $\dfrac{1 + 3n}{2n} \cdot \dfrac{2n - 4}{3n^2 - 2n - 1}$ b. $\dfrac{x^3 - 1}{-3x + 3} \cdot \dfrac{15x^2}{x^2 + x + 1}$

Solution

a. $\dfrac{1 + 3n}{2n} \cdot \dfrac{2n - 4}{3n^2 - 2n - 1} = \dfrac{1 + 3n}{2n} \cdot \dfrac{2(n - 2)}{(3n + 1)(n - 1)}$ Factor.

$= \dfrac{(1 + 3n) \cdot 2(n - 2)}{2n(3n + 1)(n - 1)}$ Multiply.

$= \dfrac{n - 2}{n(n - 1)}$ Divide out common factors.

b. $\dfrac{x^3 - 1}{-3x + 3} \cdot \dfrac{15x^2}{x^2 + x + 1} = \dfrac{(x - 1)(x^2 + x + 1)}{-3(x - 1)} \cdot \dfrac{15x^2}{x^2 + x + 1}$ Factor.

$= \dfrac{(x - 1)(x^2 + x + 1) \cdot 3 \cdot 5x^2}{-1 \cdot 3(x - 1)(x^2 + x + 1)}$ Factor.

$= \dfrac{5x^2}{-1}$ Divide out common factors.

$= -5x^2$

4   Recall that two numbers are reciprocals of each other if their product is 1. Similarly, if $\dfrac{P}{Q}$ is a rational expression, then $\dfrac{Q}{P}$ is its **reciprocal**, since

$$\frac{P}{Q} \cdot \frac{Q}{P} = \frac{P \cdot Q}{Q \cdot P} = 1$$

The following are examples of expressions and their reciprocals.

Expression	*Reciprocal*
$\dfrac{3}{x}$	$\dfrac{x}{3}$
$\dfrac{2 + x^2}{4x - 3}$	$\dfrac{4x - 3}{2 + x^2}$
x^3	$\dfrac{1}{x^3}$
0	no reciprocal

Dividing Rational Expressions

The rule for dividing rational expressions is

$$\frac{P}{Q} \div \frac{R}{S} = \frac{P}{Q} \cdot \frac{S}{R} = \frac{PS}{QR} \quad \text{as long as } Q \neq 0, S \neq 0, \text{ and } R \neq 0.$$

To divide by a rational expression, use the rule above and multiply by its reciprocal. Then simplify if possible.

Notice that division of rational expressions is the same as for rational numbers.

EXAMPLE 7

Divide.

a. $\dfrac{3x}{5y} \div \dfrac{9y}{x^5}$ **b.** $\dfrac{8m^2}{3m^2 - 12} \div \dfrac{40}{2 - m}$

Solution   **a.** $\dfrac{3x}{5y} \div \dfrac{9y}{x^5} = \dfrac{3x}{5y} \cdot \dfrac{x^5}{9y}$ Multiply by the reciprocal of the divisor.

$= \dfrac{x^6}{15y^2}$ Simplify.

b. $\dfrac{8m^2}{3m^2 - 12} \div \dfrac{40}{2 - m} = \dfrac{8m^2}{3m^2 - 12} \cdot \dfrac{2 - m}{40}$ Multiply by the reciprocal of the divisor.

$= \dfrac{8m^2(2 - m)}{3(m + 2)(m - 2) \cdot 40}$ Factor and multiply.

$= \dfrac{8m^2 \cdot -1(m - 2)}{3(m + 2)(m - 2) \cdot 8 \cdot 5}$ Write $(2 - m)$ as $-1(m - 2)$.

$= -\dfrac{m^2}{15(m + 2)}$ Simplify.

CLASSROOM EXAMPLE

Divide.

a. $\dfrac{7x}{2y^2} \div \dfrac{8y}{3x^4}$ **b.** $\dfrac{12y^3}{5y^2 - 5} \div \dfrac{6}{1 - y}$

answer:

a. $\dfrac{21x^5}{16y^3}$ **b.** $-\dfrac{2y^3}{5(y + 1)}$

> ▶ **Helpful Hint**
>
> When dividing rational expressions, do not divide out common factors until the division problem is rewritten as a multiplication problem.

EXAMPLE 8

Perform each indicated operation.

$$\dfrac{x^2 - 25}{(x + 5)^2} \cdot \dfrac{3x + 15}{4x} \div \dfrac{x^2 - 3x - 10}{x}$$

Solution $\dfrac{x^2 - 25}{(x + 5)^2} \cdot \dfrac{3x + 15}{4x} \div \dfrac{x^2 - 3x - 10}{x}$

$= \dfrac{x^2 - 25}{(x + 5)^2} \cdot \dfrac{3x + 15}{4x} \cdot \dfrac{x}{x^2 - 3x - 10}$ To divide, multiply by the reciprocal.

$= \dfrac{(x + 5)(x - 5)}{(x + 5)(x + 5)} \cdot \dfrac{3(x + 5)}{4x} \cdot \dfrac{x}{(x - 5)(x + 2)}$

$= \dfrac{3}{4(x + 2)}$

CLASSROOM EXAMPLE

Perform each indicated operation.

$\dfrac{(x + 3)^2}{x^2 - 9} \cdot \dfrac{2x - 6}{5x} \div \dfrac{x^2 + 7x + 12}{x}$

answer:

$\dfrac{2}{5(x + 4)}$

5   Rational functions occur often in real-life situations.

EXAMPLE 9

COST FOR PRESSING COMPACT DISCS

For the ICL Production Company, the rational function $C(x) = \dfrac{2.6x + 10,000}{x}$ describes the company's cost per disc of pressing x compact discs. Find the cost per disc for pressing

a. 100 compact discs

b. 1000 compact discs

Solution

a. $C(100) = \dfrac{2.6(100) + 10,000}{100} = \dfrac{10,260}{100} = 102.6$

The cost per disc for pressing 100 compact discs is $102.60.

b. $C(1000) = \dfrac{2.6(1000) + 10,000}{1000} = \dfrac{12,600}{1000} = 12.6$

The cost per disc for pressing 1000 compact discs is $12.60. Notice that as more compact discs are produced, the cost per disc decreases as indicated by the graph to the left.

CLASSROOM EXAMPLE

A company's cost per book for printing x particular books is given by the rational function $C(x) = \dfrac{0.8x + 5000}{x}$.

Find the cost per book for printing:

a. 100 books   b. 1000 books

answer:

a. $50.80   b. $5.80

Spotlight on
DECISION MAKING

Suppose you are an electrician at a small packaging plant. You are repairing machinery that heats the hot glue gun used for sealing boxes. You have determined that a resistor in the machinery's 50-volt direct current power supply must be replaced. To keep the glue warm, the power supply must dissipate about 2000 watts of power.

You know that the power P (in watts) dissipated by a resistor in a direct current circuit is given by the formula $P = \dfrac{V^2}{R}$, where V is the voltage (in volts) and R is the resistance (in ohms). Which of the three resistors shown in the parts list would you use to replace the faulty resistor? Why?

Parts List Resistors

Part Number	Material	Resistance (in ohms)
1298	Aluminum	0.95
3169	Nickel	1.81
4203	Tungsten	1.22

EXERCISE SET 6.1

STUDY GUIDE/SSM   CD/ VIDEO   PH MATH TUTOR CENTER   MathXL®Tutorials ON CD   MathXL®   MyMathLab®

Find each function value. See Example 9.

 1. $f(x) = \dfrac{x + 8}{2x - 1}; f(2), f(0), f(-1)$ $\quad \dfrac{10}{3}, -8, -\dfrac{7}{3}$

2. $f(y) = \dfrac{y - 2}{-5 + y}; f(-5), f(0), f(10)$ $\quad \dfrac{7}{10}, \dfrac{2}{5}, \dfrac{8}{5}$

3. $g(x) = \dfrac{x^2 + 8}{x^3 - 25x}; g(3), g(-2), g(1)$ $\quad -\dfrac{17}{48}, \dfrac{2}{7}, -\dfrac{3}{8}$

4. $s(t) = \dfrac{t^3 + 1}{t^2 + 1}; s(-1), s(1), s(2)$ $\quad 0, 1, \dfrac{9}{5}$

Find the domain of each rational function. See Example 1.

5. $f(x) = \dfrac{5x - 7}{4}$

6. $g(x) = \dfrac{4 - 3x}{2}$

7. $s(t) = \dfrac{t^2 + 1}{2t}$

8. $v(t) = -\dfrac{5t + t^2}{3t}$

9. $f(x) = \dfrac{3x}{7 - x}$

10. $f(x) = \dfrac{-4x}{-2 + x}$

11. $f(x) = \dfrac{x}{3x - 1}$

12. $g(x) = \dfrac{-2}{2x + 5}$

13. $R(x) = \dfrac{3 + 2x}{x^3 + x^2 - 2x}$

14. $h(x) = \dfrac{5 - 3x}{2x^2 - 14x + 20}$

15. $C(x) = \dfrac{x + 3}{x^2 - 4}$

16. $R(x) = \dfrac{5}{x^2 - 7x}$

17. In your own words, explain how to find the domain of a rational function.   answers may vary

18. In your own words, explain how to simplify a rational expression or to write it in lowest terms.   answers may vary

Simplify each rational expression. See Examples 2 through 5.

19. $\dfrac{4x - 8}{3x - 6}$ $\dfrac{4}{3}$

20. $\dfrac{12 - 6x}{30 - 15x}$ $\dfrac{2}{5}$

21. $\dfrac{2x - 14}{7 - x}$ -2

22. $\dfrac{9 - x}{5x - 45}$ $-\dfrac{1}{5}$

23. $\dfrac{x^2 - 2x - 3}{x^2 - 6x + 9}$ $\dfrac{x + 1}{x - 3}$

24. $\dfrac{x^2 + 10x + 25}{x^2 + 8x + 15}$ $\dfrac{x + 5}{x + 3}$

25. $\dfrac{2x^2 + 12x + 18}{x^2 - 9}$ $\dfrac{2(x + 3)}{x - 3}$

26. $\dfrac{x^2 - 4}{2x^2 + 8x + 8}$ $\dfrac{x - 2}{2(x + 2)}$

27. $\dfrac{3x + 6}{x^2 + 2x}$ $\dfrac{3}{x}$

28. $\dfrac{3x + 4}{9x^2 + 4}$ $\dfrac{3x + 4}{9x^2 + 4}$

29. $\dfrac{2x^2 - x - 3}{2x^3 - 3x^2 + 2x - 3}$

30. $\dfrac{3x^2 - 5x - 2}{6x^3 + 2x^2 + 3x + 1}$

31. $\dfrac{8q^2}{16q^3 - 16q^2}$ $\dfrac{1}{2(q - 1)}$

32. $\dfrac{3y}{6y^2 - 30y}$ $\dfrac{1}{2(y - 5)}$

33. $\dfrac{x^2 + 6x - 40}{10 + x}$ $x - 4$

34. $\dfrac{x^2 - 8x + 16}{4 - x}$ $4 - x$

35. $\dfrac{x^3 - 125}{5 - x}$ $-x^2 - 5x - 25$

36. $\dfrac{4x + 4}{2x^3 + 2}$ $\dfrac{2}{x^2 - x + 1}$

37. $\dfrac{8x^3 - 27}{4x - 6}$ $\dfrac{4x^2 + 6x + 9}{2}$

38. $\dfrac{9x^2 - 15x + 25}{27x^3 + 125}$ $\dfrac{1}{3x + 5}$

39. Which expression below does not simplify to 1?

a. $\dfrac{2 + x}{x + 2}$ **b.** $\dfrac{5 - x}{-x + 5}$ **c.** $\dfrac{-x - y}{-y - x}$ **d.** $\dfrac{x - 3}{3 - x}$ d

40. Which expression below does not simplify to -1?

a. $\dfrac{2 - x}{x - 2}$ **b.** $\dfrac{y + 5}{y - 5}$ **c.** $\dfrac{x + y}{-x - y}$ **d.** $\dfrac{-5 + z}{5 - z}$ b

MIXED PRACTICE

Multiply or divide as indicated. Simplify all answers. See Examples 6 through 8.

41. $\dfrac{3xy^3}{4x^3y^2} \cdot \dfrac{-8x^3y^4}{9x^4y^7}$ $-\dfrac{2}{3x^3y^2}$

42. $-\dfrac{2xyz^3}{5x^2z^2} \cdot \dfrac{10xy}{x^3}$ $-\dfrac{4y^2z}{x^3}$

43. $\dfrac{8a}{3a^4b^2} \div \dfrac{4b^5}{6a^2b}$ $\dfrac{4}{ab^6}$

44. $\dfrac{3y^3}{14x^4} \div \dfrac{8y^3}{7x}$ $\dfrac{3}{16x^3}$

45. $\dfrac{a^2b}{a^2 - b^2} \cdot \dfrac{a + b}{4a^3b}$ $\dfrac{1}{4a(a - b)}$

46. $\dfrac{3ab^2}{a^2 - 4} \cdot \dfrac{a - 2}{6a^2b^2}$ $\dfrac{1}{2a(a + 2)}$

47. $\dfrac{x^2 - 9}{4} \div \dfrac{x^2 - 6x + 9}{x^2 - x - 6}$ $\dfrac{(x + 2)(x + 3)}{4}$

48. $\dfrac{a - 5b}{a^2 + ab} \div \dfrac{15b - 3a}{b^2 - a^2}$ $\dfrac{a - b}{3a}$

49. $\dfrac{9x + 9}{4x + 8} \cdot \dfrac{2x + 4}{3x^2 - 3}$ $\dfrac{3}{2(x - 1)}$

50. $\dfrac{x^2 - 1}{10x + 30} \cdot \dfrac{12x + 36}{3x - 3}$ $\dfrac{2(x + 1)}{5}$

51. $\dfrac{a + b}{ab} \div \dfrac{a^2 - b^2}{4a^3b}$ $\dfrac{4a^2}{a - b}$

52. $\dfrac{6a^2b^2}{a^2 - 4} \div \dfrac{3ab^2}{a - 2}$ $\dfrac{2a}{a + 2}$

53. $\dfrac{2x^2 - 4x - 30}{5x^2 - 40x - 75} \div \dfrac{x^2 - 8x + 15}{x^2 - 6x + 9}$ $\dfrac{2(x + 3)(x - 3)}{5(x^2 - 8x - 15)}$

54. $\dfrac{4a + 36}{a^2 - 7a - 18} \div \dfrac{a^2 - a - 6}{a^2 - 81}$ $\dfrac{4(a + 9)^2}{(a + 2)^2(a - 3)}$

55. $\dfrac{2x^3 - 16}{6x^2 + 6x - 36} \cdot \dfrac{9x + 18}{3x^2 + 6x + 12}$ $\dfrac{x + 2}{x + 3}$

56. $\dfrac{x^2 - 3x + 9}{5x^2 - 20x - 105} \cdot \dfrac{x^2 - 49}{x^3 + 27}$ $\dfrac{x + 7}{5(x + 3)^2}$

57. $\dfrac{15b - 3a}{b^2 - a^2} \div \dfrac{a - 5b}{ab + b^2}$ $\dfrac{3b}{a - b}$

58. $\dfrac{4x + 4}{x - 1} \div \dfrac{x^2 - 4x - 5}{x^2 - 1}$ $\dfrac{4(x + 1)}{x - 5}$

59. $\dfrac{a^3 + a^2b + a + b}{a^3 + a} \cdot \dfrac{6a^2}{2a^2 - 2b^2}$ $\dfrac{3a}{a - b}$

60. $\dfrac{a^2 - 2a}{ab - 2b + 3a - 6} \cdot \dfrac{8b + 24}{3a + 6}$ $\dfrac{8a}{3(a + 2)}$

61. $\dfrac{5a}{12} \cdot \dfrac{2}{25a^2} \cdot \dfrac{15a}{2}$ $\dfrac{1}{4}$

62. $\dfrac{4a}{7} \div \dfrac{a^2}{14} \cdot \dfrac{3}{a}$ $\dfrac{24}{a^2}$

63. $\dfrac{3x - x^2}{x^3 - 27} \div \dfrac{x}{x^2 + 3x + 9}$ -1

64. $\dfrac{x^2 - 3x}{x^3 - 27} \div \dfrac{2x}{2x^2 + 6x + 18}$ 1

65. $\dfrac{4a}{7} \div \left(\dfrac{a^2}{14} \cdot \dfrac{3}{a}\right)$ $\dfrac{8}{3}$

66. $\dfrac{a^2}{14} \cdot \dfrac{3}{a} \div \dfrac{4a}{7}$ $\dfrac{3}{8}$

67. $\dfrac{8b + 24}{3a + 6} \div \dfrac{ab - 2b + 3a - 6}{a^2 - 4a + 4}$ $\dfrac{8(a - 2)}{3(a + 2)}$

68. $\dfrac{2a^2 - 2b^2}{a^3 + a^2b + a + b} \div \dfrac{6a^2}{a^3 + a}$ $\dfrac{a - b}{3a}$

29. $\dfrac{x + 1}{x^2 + 1}$

30. $\dfrac{x - 2}{2x^2 + 1}$

57. $\dfrac{3b}{a - b}$

58. $\dfrac{4(x + 1)}{x - 5}$

61. $\dfrac{1}{4}$

62. $\dfrac{24}{a^2}$

5. $\{x \mid x$ is a real number$\}$
6. $\{x \mid x$ is a real number$\}$
7. $\{t \mid t$ is a real number and $t \neq 0\}$
8. $\{t \mid t$ is a real number and $t \neq 0\}$
9. $\{x \mid x$ is a real number and $x \neq 7\}$
10. $\{x \mid x$ is a real number and $x \neq 2\}$
11. $\left\{x \mid x$ is a real number and $x \neq \dfrac{1}{3}\right\}$
12. $\left\{x \mid x$ is a real number and $x \neq -\dfrac{5}{2}\right\}$
13. $\{x \mid x$ is a real number and $x \neq -2, x \neq 0, x \neq 1\}$
14. $\{x \mid x$ is a real number and $x \neq 5, x \neq 2\}$
15. $\{x \mid x$ is a real number and $x \neq 2, x \neq -2\}$
16. $\{x \mid x$ is a real number and $x \neq 0, x \neq 7\}$

69. $\dfrac{4}{x} \div \dfrac{3xy}{x^2} \cdot \dfrac{6x^2}{x^4}$ $\dfrac{8}{x^2 y}$

70. $\dfrac{4}{x} \cdot \dfrac{3xy}{x^2} \div \dfrac{6x^2}{x^4}$ $2y$

71. $\dfrac{3x^2 - 5x - 2}{y^2 + y - 2} \cdot \dfrac{y^2 + 4y - 5}{12x^2 + 7x + 1} \div \dfrac{5x^2 - 9x - 2}{8x^2 - 2x - 1}$

72. $\dfrac{x^2 + x - 2}{3y^2 - 5y - 2} \cdot \dfrac{12y^2 + y - 1}{x^2 + 4x - 5} \div \dfrac{8y^2 - 6y + 1}{5y^2 - 9y - 2}$

73. $\dfrac{5a^2 - 20}{3a^2 - 12a} \div \dfrac{a^3 + 2a^2}{2a^2 - 8a} \cdot \dfrac{9a^3 + 6a^2}{2a^2 - 4a}$ $\dfrac{5(3a + 2)}{a}$

74. $\dfrac{5a^2 - 20}{3a^2 - 12a} \div \left(\dfrac{a^3 + 2a^2}{2a^2 - 8a} \cdot \dfrac{9a^3 + 6a^2}{2a^2 - 4a} \right)$ $\dfrac{20(a - 2)^2}{9a^3(3a + 2)}$

75. $\dfrac{5x^4 + 3x^2 - 2}{x - 1} \cdot \dfrac{x + 1}{x^4 - 1} \cdot \dfrac{5x^2 - 2}{(x - 1)^2}$ **71.** $\dfrac{(y + 5)(2x - 1)}{(y + 2)(5x + 1)}$

76. $\dfrac{3x^4 - 10x^2 - 8}{x - 2} \cdot \dfrac{3x + 6}{15x^2 + 10}$ $\dfrac{3(x + 2)^2}{5}$

REVIEW AND PREVIEW

72. $\dfrac{(x + 2)(5y + 1)}{(x + 5)(2y - 1)}$

Perform the indicated operations. See Section 1.3.

77. $\dfrac{4}{5} + \dfrac{3}{5}$ $\dfrac{7}{5}$

78. $\dfrac{4}{10} - \dfrac{7}{10}$ $-\dfrac{3}{10}$

79. $\dfrac{5}{28} - \dfrac{2}{21}$ $\dfrac{1}{12}$

80. $\dfrac{5}{13} + \dfrac{2}{7}$ $\dfrac{61}{91}$

81. $\dfrac{3}{8} + \dfrac{1}{2} - \dfrac{3}{16}$ $\dfrac{11}{16}$

82. $\dfrac{2}{9} - \dfrac{1}{6} + \dfrac{2}{3}$ $\dfrac{13}{18}$

Concept Extensions

△ **83.** Find the area of the rectangle. $\dfrac{5}{x - 2}$ sq. m

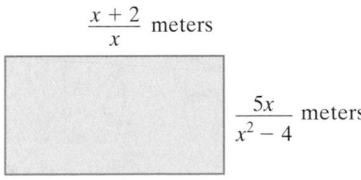

$\dfrac{x + 2}{x}$ meters

$\dfrac{5x}{x^2 - 4}$ meters

△ **84.** Find the area of the triangle. $\dfrac{1}{6y}$ sq. in.

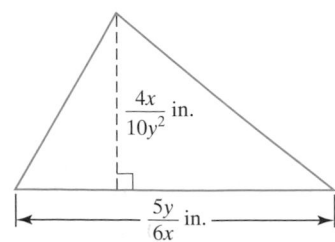

$\dfrac{4x}{10y^2}$ in.

$\dfrac{5y}{6x}$ in.

85. The function $f(x) = \dfrac{100{,}000x}{100 - x}$ models the cost in dollars for removing x percent of the pollutants from a bayou in which a nearby company dumped creosote.

a. What is the domain of $f(x)$? $\{x | 0 \le x < 100\}$

b. Find the cost of removing 30% of the pollutants from the bayou. (*Hint:* Find $f(30)$.) $\$42{,}857.14$

c. Find the cost of removing 60% of the pollutants and then 80% of the pollutants. $\$150{,}000$; $\$400{,}000$

↘ **d.** Find $f(90)$, then $f(95)$, and then $f(99)$. What happens to the cost as x approaches 100%? $\$900{,}000$; $\$1{,}900{,}000$; $\$9{,}900{,}000$; answers may vary

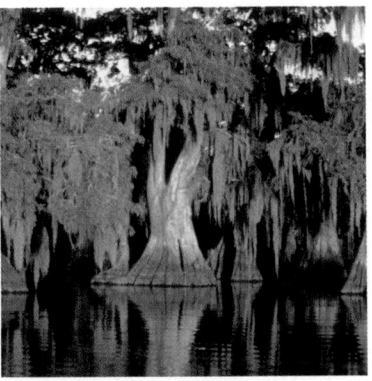

86. The total revenue from the sale of a popular book is approximated by the rational function $R(x) = \dfrac{1000x^2}{x^2 + 4}$ where x is the number of years since publication and $R(x)$ is the total revenue in millions of dollars.

a. Find the total revenue at the end of the first year.

b. Find the total revenue at the end of the second year.

c. Find the revenue during the second year only.

↘ **87.** In our definition of division for

86. a. $\$200$ million
    b. $\$500$ million
    c. $\$300$ million

$$\dfrac{P}{Q} \div \dfrac{R}{S}$$

we stated that $Q \ne 0$, $S \ne 0$, and $R \ne 0$. Explain why R cannot equal 0.   answers may vary

88. Find the polynomial in the second numerator such that the following statement is true. $(x - 5)(2x + 7)$

$$\dfrac{x^2 - 4}{x^2 - 7x + 10} \cdot \dfrac{?}{2x^2 + 11x + 14} = 1$$

△ **89.** A parallelogram has area $\dfrac{x^2 + x - 2}{x^3}$ square feet and height $\dfrac{x^2}{x - 1}$ feet. Express the length of its base as a rational expression in x. (*Hint:* Since $A = b \cdot h$, then $b = \dfrac{A}{h}$ or $b = A \div h$.) $\dfrac{(x + 2)(x - 1)^2}{x^5}$ ft

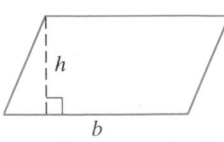

90. A lottery prize of $\dfrac{15x^3}{y^2}$ dollars is to be divided among $5x$ people. Express the amount of money each person is to receive as a rational expression in x and y.

91. Graph a portion of the function $f(x) = \dfrac{20x}{100 - x}$. To do so, complete the given table, plot the points, and then connect the plotted points with a smooth curve.

90. $\dfrac{3x^2}{y^2}$ dollars per person

x	0	10	30	50	70	90	95	99
y or $f(x)$	0	$\dfrac{20}{9}$	$\dfrac{60}{7}$	20	$\dfrac{140}{3}$	180	380	1980

92. The domain of the function $f(x) = \dfrac{1}{x}$ is all real numbers except 0. This means that the graph of this function will be in two pieces: one piece corresponding to x values less than 0 and one piece corresponding to x values greater than 0. Graph the function by completing the following tables, separately plotting the points, and connecting each set of plotted points with a smooth curve.

x	$\dfrac{1}{4}$	$\dfrac{1}{2}$	1	2	4
y or $f(x)$	4	2	1	$\dfrac{1}{2}$	$\dfrac{1}{4}$

x	-4	-2	-1	$-\dfrac{1}{2}$	$-\dfrac{1}{4}$
y or $f(x)$	$-\dfrac{1}{4}$	$-\dfrac{1}{2}$	-1	-2	-4

Perform the indicated operation. Write all answers in lowest terms.

93. $\dfrac{x^{2n} - 4}{7x} \cdot \dfrac{14x^3}{x^n - 2}$ $2x^2(x^n + 2)$

94. $\dfrac{x^{2n} + 4x^n + 4}{4x - 3} \cdot \dfrac{8x^2 - 6x}{x^n + 2}$ $2x(x^n + 2)$

95. $\dfrac{y^{2n} + 9}{10y} \cdot \dfrac{y^n - 3}{y^{4n} - 81}$ $\dfrac{1}{10y(y^n + 3)}$

96. $\dfrac{y^{4n} - 16}{y^{2n} + 4} \cdot \dfrac{6y}{y^n + 2}$ $6y(y^n - 2)$

97. $\dfrac{y^{2n} - y^n - 2}{2y^n - 4} \div \dfrac{y^{2n} - 1}{1 + y^n}$ $\dfrac{y^n + 1}{2(y^n - 1)}$

98. $\dfrac{y^{2n} + 7y^n + 10}{10} \div \dfrac{y^{2n} + 4y^n + 4}{5y^n + 25}$ $\dfrac{(y^n + 5)^2}{2(y^n + 2)}$

91.–92. See graphing answer section.

6.2   ADDING AND SUBTRACTING RATIONAL EXPRESSIONS

Objectives

1 Add or subtract rational expressions with common denominators.

2 Identify the least common denominator of two or more rational expressions.

3 Add or subtract rational expressions with unlike denominators.

TEACHING TIP

Ask students to write a summary of how to add or subtract fractions. Then review this process with them. As you add or subtract rational expressions, remind students of the similarity.

1 Rational expressions, like rational numbers, can be added or subtracted. We define the sum or difference of rational expressions in the same way that we defined the sum or difference of rational numbers (fractions).

TEACHING TIP

For Example 1, remind students that the result is true for all values for which the rational expression is defined. For example, verify that each result is true when $x = 3$.

Adding or Subtracting Rational Expressions with Common Denominators

If $\dfrac{P}{Q}$ and $\dfrac{R}{Q}$ are rational expressions, then

$$\frac{P}{Q} + \frac{R}{Q} = \frac{P + R}{Q} \quad \text{and} \quad \frac{P}{Q} - \frac{R}{Q} = \frac{P - R}{Q}$$

To add or subtract rational expressions with common denominators, add or subtract the numerators and write the sum or difference over the common denominator.

EXAMPLE 1

Add or subtract.

a. $\dfrac{x}{4} + \dfrac{5x}{4}$ **b.** $\dfrac{x^2}{x + 7} - \dfrac{49}{x + 7}$ **c.** $\dfrac{x}{3y^2} - \dfrac{x + 1}{3y^2}$

Solution   The rational expressions have common denominators, so add or subtract their numerators and place the sum or difference over their common denominator.

a. $\dfrac{x}{4} + \dfrac{5x}{4} = \dfrac{x + 5x}{4} = \dfrac{6x}{4} = \dfrac{3x}{2}$ Add the numerators and write the result over the common denominator.

b. $\dfrac{x^2}{x + 7} - \dfrac{49}{x + 7} = \dfrac{x^2 - 49}{x + 7}$ Subtract the numerators and write the result over the common denominator.

$\qquad = \dfrac{(x + 7)(x - 7)}{x + 7}$ Factor the numerator.

$\qquad = x - 7$ Simplify.

c. $\dfrac{x}{3y^2} - \dfrac{x + 1}{3y^2} = \dfrac{x - (x + 1)}{3y^2}$ Subtract the numerators.

$\qquad = \dfrac{x - x - 1}{3y^2}$ Use the distributive property.

$\qquad = -\dfrac{1}{3y^2}$ Simplify.

CLASSROOM EXAMPLE

Add or subtract.

a. $\dfrac{x}{6} + \dfrac{7x}{6}$

b. $\dfrac{x^2}{x + 3} - \dfrac{9}{x + 3}$ **c.** $\dfrac{a}{5b^3} - \dfrac{a + 2}{5b^3}$

answer:

a. $\dfrac{4x}{3}$ **b.** $x - 3$ **c.** $-\dfrac{2}{5b^3}$

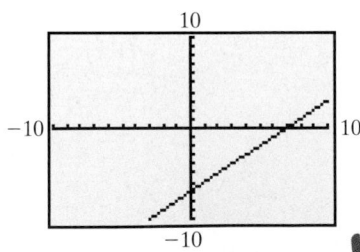

A graphing utility can be used to provide partial support of the result of operations on rational expressions. For example, the graphs of

$y_1 = \dfrac{x^2}{x + 7} - \dfrac{49}{x + 7}$ and

$y_2 = x - 7$ appear to coincide. This provides partial support for the algebraic solution in part b of Example 1. The trace feature can be used to verify that the two graphs coincide (except when x is -7).

Concept Check Answer:

$\dfrac{3 + 2y}{y^2 - 1} - \dfrac{y + 3}{y^2 - 1} =$

$\qquad \dfrac{3 + 2y - y - 3}{y^2 - 1} = \dfrac{y}{y^2 - 1}$

Helpful Hint

Be sure to insert parentheses here so that the entire numerator is subtracted.

✔ CONCEPT CHECK

Find and correct the **error**.

$\dfrac{3 + 2y}{y^2 - 1} - \dfrac{y + 3}{y^2 - 1}$

$= \dfrac{3 + 2y - y + 3}{y^2 - 1}$

$= \dfrac{y + 6}{y^2 - 1}$

2   To add or subtract rational expressions with unlike denominators, first write the rational expressions as equivalent rational expressions with common denominators.

CLASSROOM EXAMPLE

Find the LCD of each rational expression.

a. $\dfrac{7}{20a^2b^3}, \dfrac{9}{15ab^4}$ b. $\dfrac{6x}{x-2}, \dfrac{5}{x+2}$

c. $\dfrac{x+4}{x^2-36}, \dfrac{x}{x^2+12x+36}, \dfrac{x^3}{3x^2+19x+6}$

d. $\dfrac{6}{x^2-1}, \dfrac{7}{2-2x}$

answer:

a. $60a^2b^4$ b. $(x-2)(x+2)$

c. $(x-6)(3x+1)(x+6)^2$

d. $-2(x+1)(x-1)$

The **least common denominator (LCD)** is usually the easiest common denominator to work with. The LCD of a list of rational expressions is a polynomial of least degree whose factors include the denominator factors in the list.

Use the following steps to find the LCD.

> **Finding the Least Common Denominator (LCD)**
>
> **Step 1:** Factor each denominator completely.
>
> **Step 2:** The LCD is the product of all unique factors each raised to the greatest power that appears in any factored denominator.

EXAMPLE 2

Find the LCD of the rational expressions in each list.

a. $\dfrac{2}{3x^5y^2}, \dfrac{3z}{5xy^3}$

b. $\dfrac{7}{z+1}, \dfrac{z}{z-1}$

c. $\dfrac{m-1}{m^2-25}, \dfrac{2m}{2m^2-9m-5}, \dfrac{7}{m^2-10m+25}$

d. $\dfrac{x}{x^2-4}, \dfrac{11}{6-3x}$

Solution

a. First we factor each denominator.

$$3x^5y^2 = 3\cdot x^5 \cdot y^2$$
$$5xy^3 = 5\cdot x \cdot y^3$$
$$\text{LCD} = 3\cdot5\cdot x^5\cdot y^3 = 15x^5y^3$$

> **Helpful Hint**
> The greatest power of x is 5, so we have a factor of x^5. The greatest power of y is 3, so we have a factor of y^3.

b. The denominators $z+1$ and $z-1$ do not factor further. Thus,

$$\text{LCD} = (z+1)(z-1)$$

c. We first factor each denominator.

$$m^2-25 = (m+5)(m-5)$$
$$2m^2-9m-5 = (2m+1)(m-5)$$
$$m^2-10m+25 = (m-5)(m-5)$$
$$\text{LCD} = (m+5)(2m+1)(m-5)^2$$

d. Factor each denominator.

$$x^2-4 = (x+2)(x-2)$$
$$6-3x = 3(2-x) = 3(-1)(x-2)$$
$$\text{LCD} = 3(-1)(x+2)(x-2)$$
$$= -3(x+2)(x-2)$$

> **Helpful Hint**
> $(x-2)$ and $(2-x)$ are opposite factors. Notice that -1 was factored from $(2-x)$ so that the factors are identical.

> **Helpful Hint**
> If opposite factors occur, do not use both in the LCD. Instead, factor -1 from one of the opposite factors so that the factors are then identical.

3   To add or subtract rational expressions with unlike denominators, we write each rational expression as an equivalent rational expression so that their denominators are alike.

TEACHING TIP

Before discussing Example 3c, ask students to write the fraction $\frac{-7}{8}$ as a fraction with a negative denominator and $\frac{5}{-6}$ as a fraction with a positive denominator. Then have them rewrite $\frac{3}{8} + \frac{2}{-8}$ so both fractions have a denominator of 8.

Adding or Subtracting Rational Expressions with Unlike Denominators

Step 1: Find the LCD of the rational expressions.

Step 2: Write each rational expression as an equivalent rational expression whose denominator is the LCD found in Step 1.

Step 3: Add or subtract numerators, and write the result over the common denominator.

Step 4: Simplify the resulting rational expression.

EXAMPLE 3

Perform the indicated operation.

a. $\dfrac{2}{x^2 y} + \dfrac{5}{3x^3 y}$ **b.** $\dfrac{3x}{x + 2} + \dfrac{2x}{x - 2}$ **c.** $\dfrac{x}{x - 1} - \dfrac{4}{1 - x}$

Solution   **a.** The LCD is $3x^3 y$. Write each fraction as an equivalent fraction with denominator $3x^3 y$. To do this, we multiply both the numerator and denominator of each fraction by the factors needed to obtain the LCD as denominator.

The first fraction is multiplied by $\dfrac{3x}{3x}$ so that the new denominator is the LCD.

$$\dfrac{2}{x^2 y} + \dfrac{5}{3x^3 y} = \dfrac{2 \cdot 3x}{x^2 y \cdot 3x} + \dfrac{5}{3x^3 y} \quad \text{The second expression already has a denominator of } 3x^3 y.$$

$$= \dfrac{6x}{3x^3 y} + \dfrac{5}{3x^3 y}$$

$$= \dfrac{6x + 5}{3x^3 y} \quad \text{Add the numerators.}$$

CLASSROOM EXAMPLE

Perform the indicated operation.

a. $\dfrac{7}{a^3 b} + \dfrac{9}{2a^4 b}$ **b.** $\dfrac{x}{x + 5} + \dfrac{6x}{x - 5}$

c. $\dfrac{m}{m - 6} - \dfrac{8}{6 - m}$

answer:

a. $\dfrac{14a + 9}{2a^4 b}$ **b.** $\dfrac{7x^2 + 25x}{(x + 5)(x - 5)}$

c. $\dfrac{m + 8}{m - 6}$

b. The LCD is the product of the two denominators: $(x + 2)(x - 2)$.

$$\dfrac{3x}{x + 2} + \dfrac{2x}{x - 2} = \dfrac{3x \cdot (x - 2)}{(x + 2) \cdot (x - 2)} + \dfrac{2x \cdot (x + 2)}{(x - 2) \cdot (x + 2)} \quad \text{Write equivalent rational expressions.}$$

$$= \dfrac{3x(x - 2) + 2x(x + 2)}{(x + 2)(x - 2)} \quad \text{Add the numerators.}$$

$$= \dfrac{3x^2 - 6x + 2x^2 + 4x}{(x + 2)(x - 2)} \quad \text{Apply the distributive property.}$$

$$= \dfrac{5x^2 - 2x}{(x + 2)(x - 2)} \quad \text{Simplify the numerator.}$$

X	Y1	Y2
-3	10.2	10.2
-2	ERROR	ERROR
-1	-2.333	-2.333
0	0	0
1	-1	-1
2	ERROR	ERROR
3	7.8	7.8

Y1■3X/(X+2)+2X/...

This table of $y_1 = \dfrac{3x}{x + 2} + \dfrac{2x}{x - 2}$ and $y_2 = \dfrac{5x^2 - 2x}{(x + 2)(x - 2)}$ provides partial support for the result of part b in Example 3. Notice that both −2 and 2 give error messages indicating each would make the denominator zero.

c. The LCD is either $x - 1$ or $1 - x$. To get a common denominator of $x - 1$, we factor -1 from the denominator of the second rational expression.

$$\frac{x}{x - 1} - \frac{4}{1 - x} = \frac{x}{x - 1} - \frac{4}{-1(x - 1)} \qquad \text{Write } 1 - x \text{ as } -1(x - 1).$$

$$= \frac{x}{x - 1} - \frac{-1 \cdot 4}{x - 1} \qquad \text{Write } \frac{4}{-1(x - 1)} \text{ as } \frac{-1 \cdot 4}{x - 1}.$$

$$= \frac{x - (-4)}{x - 1}$$

$$= \frac{x + 4}{x - 1} \qquad \text{Simplify.}$$

EXAMPLE 4

Subtract $\dfrac{5k}{k^2 - 4} - \dfrac{2}{k^2 + k - 2}$.

Solution $\dfrac{5k}{k^2 - 4} - \dfrac{2}{k^2 + k - 2} = \dfrac{5k}{(k + 2)(k - 2)} - \dfrac{2}{(k + 2)(k - 1)}$ Factor each denominator to find the LCD.

The LCD is $(k + 2)(k - 2)(k - 1)$. We write equivalent rational expressions with the LCD as denominators.

$$\frac{5k}{(k + 2)(k - 2)} - \frac{2}{(k + 2)(k - 1)} = \frac{5k \cdot (k - 1)}{(k + 2)(k - 2) \cdot (k - 1)} - \frac{2 \cdot (k - 2)}{(k + 2)(k - 1) \cdot (k - 2)} \qquad \text{Subtract the numerators.}$$

$$= \frac{5k(k - 1) - 2(k - 2)}{(k + 2)(k - 2)(k - 1)}$$

$$= \frac{5k^2 - 5k - 2k + 4}{(k + 2)(k - 2)(k - 1)} \qquad \text{Multiply in the numerator.}$$

$$= \frac{5k^2 - 7k + 4}{(k + 2)(k - 2)(k - 1)} \qquad \text{Simplify.}$$

CLASSROOM EXAMPLE

Subtract:

$$\frac{2x}{x^2 - 9} - \frac{3}{x^2 - 4x + 3}$$

answer: $\dfrac{2x^2 - 5x - 9}{(x + 3)(x - 3)(x - 1)}$

EXAMPLE 5

Add $\dfrac{2x - 1}{2x^2 - 9x - 5} + \dfrac{x + 3}{6x^2 - x - 2}$.

Solution $\dfrac{2x - 1}{2x^2 - 9x - 5} + \dfrac{x + 3}{6x^2 - x - 2} = \dfrac{2x - 1}{(2x + 1)(x - 5)} + \dfrac{x + 3}{(2x + 1)(3x - 2)}$ Factor the denominators.

The LCD is $(2x + 1)(x - 5)(3x - 2)$.

$$= \frac{(2x - 1) \cdot (3x - 2)}{(2x + 1)(x - 5) \cdot (3x - 2)} + \frac{(x + 3) \cdot (x - 5)}{(2x + 1)(3x - 2) \cdot (x - 5)}$$

$$= \frac{(2x - 1)(3x - 2) + (x + 3)(x - 5)}{(2x + 1)(x - 5)(3x - 2)} \qquad \text{Add the numerators.}$$

$$= \frac{6x^2 - 7x + 2 + x^2 - 2x - 15}{(2x + 1)(x - 5)(3x - 2)} \qquad \text{Multiply in the numerator.}$$

$$= \frac{7x^2 - 9x - 13}{(2x + 1)(x - 5)(3x - 2)} \qquad \text{Simplify.}$$

CLASSROOM EXAMPLE

Add:

$$\frac{x + 1}{x^2 + x - 12} + \frac{2x - 1}{x^2 + 6x + 8}$$

answer: $\dfrac{3x^2 - 4x + 5}{(x + 2)(x - 3)(x + 4)}$

EXAMPLE 6

Perform each indicated operation.

$$\frac{7}{x-1} + \frac{10x}{x^2-1} - \frac{5}{x+1}$$

Solution

$$\frac{7}{x-1} + \frac{10x}{x^2-1} - \frac{5}{x+1} = \frac{7}{x-1} + \frac{10x}{(x-1)(x+1)} - \frac{5}{x+1} \qquad \text{Factor the denominators.}$$

The LCD is $(x-1)(x+1)$.

$$= \frac{7 \cdot (x+1)}{(x-1) \cdot (x+1)} + \frac{10x}{(x-1)(x+1)} - \frac{5 \cdot (x-1)}{(x+1) \cdot (x-1)}$$

$$= \frac{7(x+1) + 10x - 5(x-1)}{(x-1)(x+1)} \qquad \text{Add and subtract the numerators.}$$

$$= \frac{7x + 7 + 10x - 5x + 5}{(x-1)(x+1)} \qquad \text{Multiply in the numerator.}$$

$$= \frac{12x + 12}{(x-1)(x+1)} \qquad \text{Simplify.}$$

$$= \frac{12(x+1)}{(x-1)(x+1)} \qquad \text{Factor the numerator.}$$

$$= \frac{12}{x-1} \qquad \text{Divide out common factors.}$$

STUDY SKILLS REMINDER

Is Your Notebook Still Organized?

Is your notebook still organized? It it's not, it's not too late to start organizing it. Start writing your notes and completing your homework assignment in a notebook with pockets (spiral or ring binder). Take class notes in this notebook, and then follow the notes with your completed homework assignment. When you receive graded papers or handouts, place them in the notebook pocket so that you will not lose them.

Remember to mark (possibly with an exclamation point) any note(s) that seem extra important to you. Also remember to mark (possibly with a question mark) any notes or homework that you are having trouble with. Don't forget to see your instructor or a math tutor to help you with the concepts or exercises that you are having trouble understanding.

Also don't forget to write neatly and keep a positive attitude.

MENTAL MATH

Name the operation(s) below that make each statement true.

a. Addition   **b.** Subtraction   **c.** Multiplication   **d.** Division

1. The denominators must be the same before performing the operation.   a, b

2. To perform this operation, you multiply the first rational expression by the reciprocal of the second rational expression.   d

3. Numerator times numerator all over denominator times denominator.   c

4. These operations are commutative (order doesn't matter.)   a, c

For the rational expressions $\dfrac{5}{y}$ and $\dfrac{7}{y}$, perform each operation mentally.

5. Addition $\dfrac{12}{y}$ **6.** Subtraction $-\dfrac{2}{y}$ **7.** Multiplication $\dfrac{35}{y^2}$ **8.** Division $\dfrac{5}{7}$

EXERCISE SET 6.2

STUDY GUIDE/SSM   CD/VIDEO   PH MATH TUTOR CENTER   MathXL®Tutorials ON CD   MathXL®   MyMathLab®

Perform the indicated operation. If possible, simplify your answer. See Example 1.

1. $\dfrac{2}{x} - \dfrac{5}{x}$ $-\dfrac{3}{x}$

2. $\dfrac{4}{x^2} + \dfrac{2}{x^2}$ $\dfrac{6}{x^2}$

 3. $\dfrac{2}{x-2} + \dfrac{x}{x-2}$ $\dfrac{x+2}{x-2}$

4. $\dfrac{x}{5-x} + \dfrac{2}{5-x}$ $\dfrac{x+2}{5-x}$

5. $\dfrac{x^2}{x+2} - \dfrac{4}{x+2}$ $x-2$

6. $\dfrac{4}{x-2} - \dfrac{x^2}{x-2}$ $-2-x$

7. $\dfrac{2x-6}{x^2+x-6} + \dfrac{3-3x}{x^2+x-6}$ $\dfrac{1}{2-x}$

8. $\dfrac{5x+2}{x^2+2x-8} + \dfrac{2-4x}{x^2+2x-8}$ $\dfrac{1}{x-2}$

△ **9.** Find the perimeter and the area of the square.

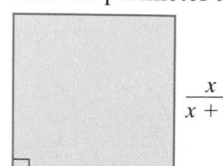

$\dfrac{4x}{x+5}$ ft; $\dfrac{x^2}{x^2+10x+25}$ sq. ft

15. $(x+7)(x-7)$
16. $(2x-1)(2x+1)$
17. $6(x+2)(x-2)$

△ **10.** Find the perimeter of the quadrilateral. $(2x+3)$ cm

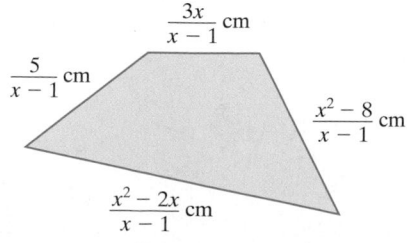

Find the LCD of the rational expressions in each list. See Example 2.

11. $\dfrac{2}{7}, \dfrac{3}{5x}$ $35x$

12. $\dfrac{4}{5y}, \dfrac{3}{4y^2}$ $20y^2$

13. $\dfrac{3}{x}, \dfrac{2}{x+1}$ $x(x+1)$

14. $\dfrac{5}{2x}, \dfrac{7}{2+x}$ $2x(2+x)$

15. $\dfrac{12}{x+7}, \dfrac{8}{x-7}$

16. $\dfrac{1}{2x-1}, \dfrac{x}{2x+1}$

17. $\dfrac{5}{3x+6}, \dfrac{2x}{2x-4}$

18. $\dfrac{2}{3a+9}, \dfrac{5}{5a-15}$

19. $\dfrac{5+x}{(3x-1)(x+2)}, \dfrac{2}{3x-1}$ $(3x-1)(x+2)$

20. $\dfrac{6-x}{(x+3)(x-3)}, \dfrac{9}{x+3}$ $(x+3)(x-3)$

21. $\dfrac{2a}{a^2-b^2}, \dfrac{1}{a^2-2ab+b^2}$ $(a+b)(a-b)^2$

22. $\dfrac{2a}{a^2+8a+16}, \dfrac{7a}{a^2+a-12}$ $(a+4)^2(a-3)$

23. $\dfrac{x}{x^2-9}, \dfrac{5x}{x}, \dfrac{7}{12-4x}$

24. $\dfrac{9}{x^2-25}, \dfrac{1}{50-10x}, \dfrac{6}{x}$

↘ **25.** When is the LCD of two rational expressions equal to the product of their denominators? (*Hint:* What is the LCD of $\dfrac{1}{x}$ and $\dfrac{7}{x+5}$?)   answers may vary

18. $15(a+3)(a-3)$ **23.** $-4x(x+3)(x-3)$
24. $-10x(x+5)(x-5)$

26. When is the LCD of two rational expressions with different denominators equal to one of the denominators?(*Hint:* What is the LCD of $\dfrac{3x}{x+2}$ and $\dfrac{7x+1}{(x+2)^3}$?)   answers may vary

Perform the indicated operation. If possible, simplify your answer. See Example 3a and 3b.

27. $\dfrac{4}{3x} + \dfrac{3}{2x}$ $\dfrac{17}{6x}$

28. $\dfrac{10}{7x} - \dfrac{5}{2x} - \dfrac{15}{14x}$

29. $\dfrac{5}{2y^2} - \dfrac{2}{7y}$ $\dfrac{35-4y}{14y^2}$

30. $\dfrac{4}{11x^4y} - \dfrac{1}{4x^2y^3}$ $\dfrac{16y^2 \cdot 11x^2}{44x^4y^3}$

31. $\dfrac{x-3}{x+4} - \dfrac{x+2}{x-4}$

32. $\dfrac{x-1}{x-5} - \dfrac{x+2}{x+5}$

33. $\dfrac{1}{x-5} + \dfrac{x}{x^2-x-20}$ $\dfrac{2x+4}{(x-5)(x+4)}$ **31.** $\dfrac{-13x+4}{(x+4)(x-4)}$

34. $\dfrac{x+1}{x^2-x-20} - \dfrac{2}{x+4}$ $\dfrac{-x+11}{(x+4)(x-5)}$ **32.** $\dfrac{7x+5}{(x-5)(x+5)}$

Perform the indicated operation. If possible, simplify your answer. See Example 3c.

35. $\dfrac{1}{a-b} + \dfrac{1}{b-a}$ 0

36. $\dfrac{1}{a-3} - \dfrac{1}{3-a}$ $\dfrac{2}{a-3}$

37. $\dfrac{x+1}{1-x} + \dfrac{1}{x-1}$ $-\dfrac{x}{x-1}$

38. $\dfrac{5}{1-x} - \dfrac{1}{x-1}$ $-\dfrac{6}{x-1}$

39. $\dfrac{5}{x-2} + \dfrac{x+4}{2-x}$ $\dfrac{-x+1}{x-2}$

40. $\dfrac{3}{5-x} + \dfrac{x+2}{x-5}$ $\dfrac{x-1}{x-5}$

Perform each indicated operation. If possible, simplify your answer. See Examples 4 through 6.

41. $\dfrac{y+1}{y^2-6y+8} - \dfrac{3}{y^2-16}$ $\dfrac{y^2+2y+10}{(y+4)(y-4)(y-2)}$

42. $\dfrac{x+2}{x^2-36} - \dfrac{x}{x^2+9x+18}$ $\dfrac{6+11x}{(x-6)(x+6)(x+3)}$

43. $\dfrac{x+4}{3x^2+11x+6} + \dfrac{x}{2x^2+x-15}$ $\dfrac{5(x^2+x-4)}{(3x+2)(x+3)(2x-5)}$

44. $\dfrac{x+3}{5x^2+12x+4} + \dfrac{6}{x^2-x-6}$ $\dfrac{x^2+30x+3}{(5x+2)(x-3)(x+2)}$

45. $\dfrac{7}{x^2-x-2} + \dfrac{x}{x^2+4x+3}$ $\dfrac{x^2+5x+21}{(x-2)(x+1)(x+3)}$

46. $\dfrac{a}{a^2+10a+25} + \dfrac{4}{a^2+6a+5}$ $\dfrac{a^2+5a+20}{(a+5)^2(a+1)}$

47. $\dfrac{2}{x+1} - \dfrac{3x}{3x+3} + \dfrac{1}{2x+2}$ $\dfrac{-2x+5}{2(x+1)}$

48. $\dfrac{5}{3x-6} - \dfrac{x}{x-2} + \dfrac{3+2x}{5x-10}$ $\dfrac{34-9x}{15(x-2)}$

49. $\dfrac{3}{x+3} + \dfrac{5}{x^2+6x+9} - \dfrac{x}{x^2-9}$ $\dfrac{2(x^2+x-21)}{(x+3)^2(x-3)}$

50. $\dfrac{x+2}{x^2-2x-3} + \dfrac{x}{x-3} - \dfrac{4}{x+1}$ $\dfrac{x^2-2x+14}{(x-3)(x+1)}$

MIXED PRACTICE

55. $\dfrac{25}{6(x+5)}$ **56.** $\dfrac{73}{21(x-1)}$

Add or subtract as indicated. If possible, simplify your answer.

51. $\dfrac{4}{3x^2y^3} + \dfrac{5}{3x^2y^3}$ $\dfrac{3}{x^2y^3}$

52. $\dfrac{7}{2xy^4} + \dfrac{1}{2xy^4}$ $\dfrac{4}{xy^4}$

53. $\dfrac{x-5}{2x} - \dfrac{x+5}{2x}$ $-\dfrac{5}{x}$

54. $\dfrac{x+4}{4x} - \dfrac{x-4}{4x}$ $\dfrac{2}{x}$

55. $\dfrac{3}{2x+10} + \dfrac{8}{3x+15}$

56. $\dfrac{10}{3x-3} + \dfrac{1}{7x-7}$

57. $\dfrac{-2}{x^2-3x} - \dfrac{1}{x^3-3x^2}$

58. $\dfrac{-3}{2a+8} - \dfrac{8}{a^2+4a}$

59. $\dfrac{ab}{a^2-b^2} + \dfrac{b}{a+b}$

60. $\dfrac{x}{25-x^2} + \dfrac{2}{3x-15}$

61. $\dfrac{5}{x^2-4} - \dfrac{3}{x^2+4x+4}$

62. $\dfrac{3z}{z^2-9} - \dfrac{2}{3-z}$

63. $\dfrac{2}{a^2+2a+1} + \dfrac{3}{a^2-1}$ $\dfrac{5a+1}{(a+1)^2(a-1)}$

64. $\dfrac{9x+2}{3x^2-2x-8} + \dfrac{7}{3x^2+x-4}$ $\dfrac{3x-4}{(x-2)(x-1)}$

65. In your own words, explain how to add rational expressions with different denominators.   answers may vary

66. In your own words, explain how to multiply rational expressions.   answers may vary

67. In your own words, explain how to divide rational expressions.   answers may vary

68. In your own words, explain how to subtract rational expressions with different denominators.   answers may vary

57. $\dfrac{-2x-1}{x^2(x-3)}$ **58.** $\dfrac{-3a-16}{2a(a+4)}$

MIXED PRACTICE

Perform the indicated operation. If possible, simplify your answer.

69. $\left(\dfrac{2}{3} - \dfrac{1}{x}\right) \cdot \left(\dfrac{3}{x} + \dfrac{1}{2}\right)$ $\dfrac{2x^2+9x-18}{6x^2}$

70. $\left(\dfrac{2}{3} - \dfrac{1}{x}\right) \div \left(\dfrac{3}{x} + \dfrac{1}{2}\right)$ $\dfrac{2(2x-3)}{3(6+x)}$

59. $\dfrac{2ab-b^2}{(a+b)(a-b)}$

71. $\left(\dfrac{1}{x} + \dfrac{2}{3}\right) - \left(\dfrac{1}{x} - \dfrac{2}{3}\right)$ $\dfrac{4}{3}$ **72.** $\left(\dfrac{1}{2} + \dfrac{2}{x}\right) - \left(\dfrac{1}{2} - \dfrac{1}{x}\right)$ $\dfrac{3}{x}$

73. $\left(\dfrac{2a}{3}\right)^2 \div \left(\dfrac{a^2}{a+1} - \dfrac{1}{a+1}\right)$ $\dfrac{4a^2}{9(a-1)}$

60. $\dfrac{-x+10}{3(x-5)(x+5)}$ **61.** $\dfrac{2x+16}{(x+2)^2(x-2)}$ **62.** $\dfrac{5z+6}{(z-3)(z+3)}$

74. $\left(\dfrac{x+2}{2x} - \dfrac{x-2}{2x}\right) \cdot \left(\dfrac{5x}{4}\right)^2$ $\dfrac{25x}{8}$

75. $\left(\dfrac{2x}{3}\right)^2 \div \left(\dfrac{x}{3}\right)^2$ 4   **76.** $\left(\dfrac{2x}{3}\right)^2 \cdot \left(\dfrac{3}{x}\right)^2$ 4

77. $\dfrac{x}{x^2-9} + \dfrac{3}{x^2-6x+9} - \dfrac{1}{x+3}$ $\dfrac{6x}{(x+3)(x-3)^2}$

78. $\dfrac{3}{x^2-9} - \dfrac{x}{x^2-6x+9} + \dfrac{1}{x+3}$ $-\dfrac{6x}{(x-3)^2(x+3)}$

79. $\left(\dfrac{x}{x+1} - \dfrac{x}{x-1}\right) \div \dfrac{x}{2x+2}$ $-\dfrac{4}{x-1}$

80. $\dfrac{x}{2x+2} \div \left(\dfrac{x}{x+1} + \dfrac{x}{x-1}\right)$ $\dfrac{x-1}{4x}$

81. $\dfrac{4}{x} \cdot \left(\dfrac{2}{x+2} - \dfrac{2}{x-2}\right)$ $-\dfrac{32}{x(x+2)(x-2)}$

82. $\dfrac{1}{x+1} \cdot \left(\dfrac{5}{x} + \dfrac{2}{x-3}\right)$ $\dfrac{7x-15}{x(x-3)(x+1)}$

REVIEW AND PREVIEW

Use the distributive property to multiply the following. See Section 1.4.

83. $12\left(\dfrac{2}{3} + \dfrac{1}{6}\right)$ 10   **84.** $14\left(\dfrac{1}{7} + \dfrac{3}{14}\right)$ 5

85. $x^2\left(\dfrac{4}{x^2} + 1\right)$ $4 + x^2$ **86.** $5y^2\left(\dfrac{1}{y^2} - \dfrac{1}{5}\right)$ $5 - y^2$

Find each root. See Section 1.3.

87. $\sqrt{100}$ 10   **88.** $\sqrt{25}$ 5

89. $\sqrt[3]{8}$ 2   **90.** $\sqrt[3]{27}$ 3

91. $\sqrt[4]{81}$ 3   **92.** $\sqrt[4]{16}$ 2

Use the Pythagorean theorem to find each unknown length of a right triangle. See Section 5.8.

△ **93.**

5 m
3 meters
4 meters

△ **94.**

25 ft
7 feet
24 feet

Concept Extensions

Perform the indicated operation. Begin by writing each term with positive exponents only.

95. $x^{-1} + (2x)^{-1}$ $\dfrac{3}{2x}$ **96.** $3y^{-1} + (4y)^{-1}$ $\dfrac{13}{4y}$

97. $4x^{-2} - 3x^{-1}$ $\dfrac{4-3x}{x^2}$ **98.** $(4x)^{-2} - (3x)^{-1}$ $\dfrac{3-16x}{48x^2}$

99. $x^{-3}(2x+1) - 5x^{-2}$ $\dfrac{1-3x}{x^3}$ **100.** $4x^{-3} + x^{-4}(5x+7)$ $\dfrac{9x+7}{x^4}$

Use a graphing calculator to support the results of each exercise.

101. Exercise 3   **102.** Exercise 4

103. Exercise 31   **104.** Exercise 32

101.–104. See graphing answer section.

6.3   SIMPLIFYING COMPLEX FRACTIONS

Objectives

1 Simplify complex fractions by simplifying the numerator and denominator and then dividing.

2 Simplify complex fractions by multiplying by a common denominator.

3 Simplify expressions with negative exponents.

1   A rational expression whose numerator, denominator, or both contain one or more rational expressions is called a **complex rational expression** or a **complex fraction**.

Complex Fractions

$$\dfrac{\dfrac{1}{a}}{\dfrac{b}{2}} \qquad \dfrac{\dfrac{x}{2y^2}}{\dfrac{6x-2}{9y}} \qquad \dfrac{x+\dfrac{1}{y}}{y+1}$$

The parts of a complex fraction are

$$\dfrac{\left.\dfrac{x}{y+2}\right\}}{7+\dfrac{1}{y}\Big\}} \qquad \begin{array}{l} \leftarrow \text{ Numerator of complex fraction.} \\[4pt] \leftarrow \text{ Main fraction bar.} \\[8pt] \leftarrow \text{ Denominator of complex fraction.} \end{array}$$

Our goal in this section is to simplify complex fractions. A complex fraction is simplified when it is in the form $\dfrac{P}{Q}$, where P and Q are polynomials that have no common factors. Two methods of simplifying complex fractions are introduced. The first method evolves from the definition of a fraction as a quotient.

Simplifying A Complex Fraction: Method I

Step 1: Simplify the numerator and the denominator of the complex fraction so that each is a single fraction.

Step 2: Perform the indicated division by multiplying the numerator of the complex fraction by the reciprocal of the denominator of the complex fraction.

Step 3: Simplify if possible.

EXAMPLE 1

Simplify each complex fraction.

a. $\dfrac{\dfrac{2x}{27y^2}}{\dfrac{6x^2}{9}}$ **b.** $\dfrac{\dfrac{5x}{x+2}}{\dfrac{10}{x-2}}$ **c.** $\dfrac{\dfrac{x}{y^2}+\dfrac{1}{y}}{\dfrac{y}{x^2}+\dfrac{1}{x}}$

Solution   **a.** The numerator of the complex fraction is already a single fraction, and so is the denominator. Perform the indicated division by multiplying the numerator, $\dfrac{2x}{27y^2}$, by the reciprocal of the denominator, $\dfrac{6x^2}{9}$. Then simplify.

$$\dfrac{\dfrac{2x}{27y^2}}{\dfrac{6x^2}{9}} = \dfrac{2x}{27y^2} \div \dfrac{6x^2}{9}$$

$$= \dfrac{2x}{27y^2} \cdot \dfrac{9}{6x^2} \qquad \text{Multiply by the reciprocal of } \dfrac{6x^2}{9}.$$

$$= \dfrac{2x \cdot 9}{27y^2 \cdot 6x^2}$$

$$= \dfrac{1}{9xy^2}$$

> ▶ **Helpful Hint**
>
> Both the numerator and denominator are single fractions, so we perform the indicated division.

b. $\dfrac{\left\{\dfrac{5x}{x+2}\right.}{\left\{\dfrac{10}{x-2}\right.} = \dfrac{5x}{x+2} \div \dfrac{10}{x-2} = \dfrac{5x}{x+2} \cdot \dfrac{x-2}{10}$ Multiply by the reciprocal of $\dfrac{10}{x-2}$.

$$= \dfrac{5x(x-2)}{2 \cdot 5(x+2)}$$

$$= \dfrac{x(x-2)}{2(x+2)} \qquad \text{Simplify.}$$

c. First simplify the numerator and the denominator of the complex fraction separately so that each is a single fraction. Then perform the indicated division.

$$\dfrac{\dfrac{x}{y^2}+\dfrac{1}{y}}{\dfrac{y}{x^2}+\dfrac{1}{x}} = \dfrac{\dfrac{x}{y^2}+\dfrac{1 \cdot y}{y \cdot y}}{\dfrac{y}{x^2}+\dfrac{1 \cdot x}{x \cdot x}} \qquad \begin{array}{l}\text{Simplify the numerator. The LCD is } y^2.\\ \text{Simplify the denominator. The LCD is } x^2.\end{array}$$

$$= \dfrac{\dfrac{x+y}{y^2}}{\dfrac{y+x}{x^2}} \qquad \text{Add.}$$

$$= \dfrac{x+y}{y^2} \div \dfrac{y+x}{x^2} \qquad \text{Add.}$$

$$= \dfrac{x+y}{y^2} \cdot \dfrac{x^2}{y+x} \qquad \text{Multiply by the reciprocal of } \dfrac{y+x}{x^2}.$$

$$= \dfrac{x^2(x+y)}{y^2(y+x)}$$

$$= \dfrac{x^2}{y^2} \qquad \text{Simplify.}$$

CONCEPT CHECK

Which of the following are equivalent to $\dfrac{\dfrac{1}{x}}{\dfrac{3}{y}}$?

a. $\dfrac{1}{x} \div \dfrac{3}{y}$ **b.** $\dfrac{1}{x} \cdot \dfrac{y}{3}$ **c.** $\dfrac{1}{x} \div \dfrac{y}{3}$

2 Next we look at another method of simplifying complex fractions. With this method we multiply the numerator and the denominator of the complex fraction by the LCD of all fractions in the complex fraction.

TEACHING TIP

Ask students to explain why the value of an expression is not changed when you multiply the numerator and denominator by the LCD.

Simplifying A Complex Fraction: Method II

Step 1: Multiply the numerator and the denominator of the complex fraction by the LCD of the fractions in both the numerator and the denominator.

Step 2: Simplify.

EXAMPLE 2

Simplify each complex fraction.

a. $\dfrac{\dfrac{5x}{x+2}}{\dfrac{10}{x-2}}$ **b.** $\dfrac{\dfrac{x}{y^2}+\dfrac{1}{y}}{\dfrac{y}{x^2}+\dfrac{1}{x}}$

Solution **a.** The least common denominator of $\dfrac{5x}{x+2}$ and $\dfrac{10}{x-2}$ is $(x+2)(x-2)$. Multiply both the numerator, $\dfrac{5x}{x+2}$, and the denominator, $\dfrac{10}{x-2}$, by the LCD.

CLASSROOM EXAMPLE

Use Method 2 to simplify:

a. $\dfrac{\dfrac{6x}{x-5}}{\dfrac{12}{x+5}}$ **b.** $\dfrac{\dfrac{x}{y^2}-\dfrac{1}{y}}{\dfrac{y}{x^2}-\dfrac{1}{x}}$

answer:

a. $\dfrac{x(x+5)}{2(x-5)}$ **b.** $-\dfrac{x^2}{y^2}$

$$\dfrac{\dfrac{5x}{x+2}}{\dfrac{10}{x-2}} = \dfrac{\left(\dfrac{5x}{x+2}\right) \cdot (x+2)(x-2)}{\left(\dfrac{10}{x-2}\right) \cdot (x+2)(x-2)}$$ Multiply numerator and denominator by the LCD.

$$= \dfrac{5x \cdot (x-2)}{2 \cdot 5 \cdot (x+2)}$$ Simplify.

$$= \dfrac{x(x-2)}{2(x+2)}$$ Simplify.

Concept Check Answer:
a and b

b. The least common denominator of $\dfrac{x}{y^2}, \dfrac{1}{y}, \dfrac{y}{x^2},$ and $\dfrac{1}{x}$ is x^2y^2.

$$\dfrac{\dfrac{x}{y^2} + \dfrac{1}{y}}{\dfrac{y}{x^2} + \dfrac{1}{x}} = \dfrac{\left(\dfrac{x}{y^2} + \dfrac{1}{y}\right) \cdot x^2y^2}{\left(\dfrac{y}{x^2} + \dfrac{1}{x}\right) \cdot x^2y^2}$$ *Multiply the numerator and denominator by the LCD.*

$$= \dfrac{\dfrac{x}{y^2} \cdot x^2y^2 + \dfrac{1}{y} \cdot x^2y^2}{\dfrac{y}{x^2} \cdot x^2y^2 + \dfrac{1}{x} \cdot x^2y^2}$$ *Use the distributive property.*

$$= \dfrac{x^3 + x^2y}{y^3 + xy^2}$$ *Simplify.*

$$= \dfrac{x^2(x + y)}{y^2(y + x)}$$ *Factor.*

$$= \dfrac{x^2}{y^2}$$ *Simplify.* $\ominus$

3   If an expression contains negative exponents, write the expression as an equivalent expression with positive exponents.

EXAMPLE 3

Simplify.

$$\dfrac{x^{-1} + 2xy^{-1}}{x^{-2} - x^{-2}y^{-1}}$$

Solution   This fraction does not appear to be a complex fraction. If we write it by using only positive exponents, however, we see that it is a complex fraction.

$$\dfrac{x^{-1} + 2xy^{-1}}{x^{-2} - x^{-2}y^{-1}} = \dfrac{\dfrac{1}{x} + \dfrac{2x}{y}}{\dfrac{1}{x^2} - \dfrac{1}{x^2y}}$$

The LCD of $\dfrac{1}{x}, \dfrac{2x}{y}, \dfrac{1}{x^2},$ and $\dfrac{1}{x^2y}$ is x^2y. Multiply both the numerator and denominator by x^2y.

$$= \dfrac{\left(\dfrac{1}{x} + \dfrac{2x}{y}\right) \cdot x^2y}{\left(\dfrac{1}{x^2} - \dfrac{1}{x^2y}\right) \cdot x^2y}$$

$$= \dfrac{\dfrac{1}{x} \cdot x^2y + \dfrac{2x}{y} \cdot x^2y}{\dfrac{1}{x^2} \cdot x^2y - \dfrac{1}{x^2y} \cdot x^2y}$$ *Apply the distributive property.*

$$= \dfrac{xy + 2x^3}{y - 1} \quad \text{or} \quad \dfrac{x(y + 2x^2)}{y - 1}$$ *Simplify.* $\ominus$

EXAMPLE 4

Simplify: $\dfrac{(2x)^{-1} + 1}{2x^{-1} - 1}$

Solution

$\dfrac{(2x)^{-1} + 1}{2x^{-1} - 1} = \dfrac{\dfrac{1}{2x} + 1}{\dfrac{2}{x} - 1}$

Write using positive exponents.

> **Helpful Hint**
> Don't forget that
> $(2x)^{-1} = \dfrac{1}{2x}$, but
> $2x^{-1} = 2 \cdot \dfrac{1}{x} = \dfrac{2}{x}$.

$= \dfrac{\left(\dfrac{1}{2x} + 1\right) \cdot 2x}{\left(\dfrac{2}{x} - 1\right) \cdot 2x}$

The LDC of $\dfrac{1}{2x}$ and $\dfrac{2}{x}$ is 2x.

$= \dfrac{\dfrac{1}{2x} \cdot 2x + 1 \cdot 2x}{\dfrac{2}{x} \cdot 2x - 1 \cdot 2x}$

Use distributive property.

$= \dfrac{1 + 2x}{4 - 2x} \quad \text{or} \quad \dfrac{1 + 2x}{2(2 - x)}$

Simplify.

CLASSROOM EXAMPLE

Simplify: $\dfrac{5 - 3x^{-1}}{2 + (3x)^{-1}}$

answer: $\dfrac{15x - 9}{6x + 1}$

STUDY SKILLS REMINDER

Are You Satisfied With Your Performance in this Course thus Far?

If not, ask yourself the following questions:

▶ Am I attending all class periods and arriving on time?

▶ Am I working and checking my homework assignments?

▶ Am I getting help when I need it?

▶ In addition to my instructor, am I using the supplements to this text that could help me? For example, the tutorial video lessons? The tutorial software?

▶ Am I satisfied with my performance on quizzes and tests?

If you answered no to *any* of these questions, read or reread Section 1.1 for suggestions in these areas. Also, you may want to contact your instructor for additional feedback.

EXERCISE SET 6.3

STUDY GUIDE/SSM CD/VIDEO PH MATH TUTOR CENTER MathXL®Tutorials ON CD MathXL® MyMathLab®

Simplify each complex fraction. See Examples 1 and 2.

1. $\dfrac{\frac{10}{3x}}{\frac{5}{6x}}$ 4

2. $\dfrac{\frac{15}{2x}}{\frac{5}{6x}}$ 9

3. $\dfrac{1 + \frac{2}{5}}{2 + \frac{3}{5}}$ $\dfrac{7}{13}$

4. $\dfrac{2 + \frac{1}{7}}{3 - \frac{4}{7}}$ $\dfrac{15}{17}$

5. $\dfrac{\frac{4}{x-1}}{\frac{x}{x-1}}$ $\dfrac{4}{x}$

6. $\dfrac{\frac{x}{x+2}}{\frac{2}{x+2}}$ $\dfrac{x}{2}$

7. $\dfrac{1 - \frac{2}{x}}{x + \frac{4}{9x}}$ $\dfrac{9(x-2)}{9x^2 + 4}$

8. $\dfrac{5 - \frac{3}{x}}{x + \frac{2}{3x}}$ $\dfrac{3(5x-3)}{3x^2 + 2}$

9. $\dfrac{\frac{4x^2 - y^2}{xy}}{\frac{2}{y} - \frac{1}{x}}$ $2x + y$

10. $\dfrac{\frac{x^2 - 9y^2}{xy}}{\frac{1}{y} - \frac{3}{x}}$ $x + 3y$

11. $\dfrac{\frac{x+1}{3}}{\frac{2x-1}{6}}$ $\dfrac{2(x+1)}{2x-1}$

12. $\dfrac{\frac{x+3}{12}}{\frac{4x-5}{15}}$ $\dfrac{5(x+3)}{4(4x-5)}$

13. $\dfrac{\frac{2}{x} + \frac{3}{x^2}}{\frac{4}{x^2} - \frac{9}{x}}$ $\dfrac{2x+3}{4-9x}$

14. $\dfrac{\frac{2}{x^2} + \frac{1}{x}}{\frac{4}{x^2} - \frac{1}{x}}$ $\dfrac{2+x}{4-x}$

15. $\dfrac{\frac{1}{x} + \frac{2}{x^2}}{x + \frac{8}{x^2}}$ $\dfrac{1}{x^2 - 2x + 4}$

16. $\dfrac{\frac{1}{y} + \frac{3}{y^2}}{y + \frac{27}{y^2}}$ $\dfrac{1}{y^2 - 3y + 9}$

17. $\dfrac{\frac{4}{5-x} + \frac{5}{x-5}}{\frac{2}{x} + \frac{3}{x-5}}$ $\dfrac{x}{5x-10}$

18. $\dfrac{\frac{3}{x-4} - \frac{2}{4-x}}{\frac{2}{x-4} - \frac{2}{x}}$ $\dfrac{5x}{8}$

19. $\dfrac{\frac{x+2}{x} - \frac{2}{x-1}}{\frac{x+1}{x} + \frac{x+1}{x-1}}$ $\dfrac{x-2}{2x-1}$

20. $\dfrac{\frac{5}{a+2} - \frac{1}{a-2}}{\frac{3}{2+a} + \frac{6}{2-a}}$ $-\dfrac{4(a-3)}{3(a+6)}$

21. $\dfrac{\frac{2}{x} + 3}{\frac{4}{x^2} - 9}$ $\dfrac{x}{2 - 3x}$

22. $\dfrac{2 + \frac{1}{x}}{4x - \frac{1}{x}}$ $\dfrac{1}{2x-1}$

23. $\dfrac{1 - \frac{x}{y}}{\frac{x^2}{y^2} - 1} - \dfrac{y}{x+y}$

24. $\dfrac{1 - \frac{2}{x}}{x - \frac{4}{x}}$ $\dfrac{1}{x+2}$

25. $\dfrac{\frac{-2x}{x-y}}{\frac{y}{x^2}} - \dfrac{2x^3}{y(x-y)}$

26. $\dfrac{\frac{7y}{x^2 + xy}}{\frac{y^2}{x^2}}$ $\dfrac{7x}{y(x+y)}$

27. $\dfrac{\frac{2}{x} + \frac{1}{x^2}}{\frac{y}{x^2}}$ $\dfrac{2x+1}{y}$

28. $\dfrac{\frac{5}{x^2} - \frac{2}{x}}{\frac{1}{x} + 2}$ $\dfrac{5 - 2x}{x + 2x^2}$

29. $\dfrac{\frac{x}{9} - \frac{1}{x}}{1 + \frac{3}{x}}$ $\dfrac{x-3}{9}$

30. $\dfrac{\frac{x}{4} - \frac{4}{x}}{1 - \frac{4}{x}}$ $\dfrac{x+4}{4}$

31. $\dfrac{\frac{x-1}{x^2-4}}{1 + \frac{1}{x-2}}$ $\dfrac{1}{x+2}$

32. $\dfrac{\frac{2}{x+5} + \frac{4}{x+3}}{\frac{3x+13}{x^2+8x+15}}$ 2

Simplify. See Examples 3 and 4.

33. $\dfrac{x^{-1}}{x^{-2} + y^{-2}}$ $\dfrac{xy^2}{x^2 + y^2}$

34. $\dfrac{a^{-3} + b^{-1}}{a^{-2}}$ $\dfrac{b + a^3}{ab}$

35. $\dfrac{2a^{-1} + 3b^{-2}}{a^{-1} - b^{-1}}$ $\dfrac{2b^2 + 3a}{b(b-a)}$

36. $\dfrac{x^{-1} + y^{-1}}{3x^{-2} + 5y^{-2}}$ $\dfrac{xy(x+y)}{3y^2 + 5x^2}$

37. $\dfrac{1}{x - x^{-1}}$ $\dfrac{x}{(x+1)(x-1)}$

38. $\dfrac{x^{-2}}{x + 3x^{-1}}$ $\dfrac{1}{x(x^2 + 3)}$

39. $\dfrac{a^{-1} + 1}{a^{-1} - 1}$ $\dfrac{1+a}{1-a}$

40. $\dfrac{a^{-1} - 4}{4 + a^{-1}}$ $\dfrac{1 - 4a}{4a + 1}$

41. $\dfrac{3x^{-1} + (2y)^{-1}}{x^{-2}}$ $\dfrac{x(x+6y)}{2y}$

42. $\dfrac{5x^{-2} - 3y^{-1}}{x^{-1} + y^{-1}}$ $\dfrac{5y - 3x^2}{xy + x^2}$

43. $\dfrac{2a^{-1} + (2a)^{-1}}{a^{-1} + 2a^{-2}}$ $\dfrac{5a}{2a+4}$

44. $\dfrac{a^{-1} + 2a^{-2}}{2a^{-1} + (2a)^{-1}}$ $\dfrac{2a+4}{5a}$

45. $\dfrac{5x^{-1} + 2y^{-1}}{x^{-2}y^{-2}}$ $5xy^2 + 2x^2y$

46. $\dfrac{x^{-2}y^{-2}}{5x^{-1} + 2y^{-1}}$ $\dfrac{1}{xy(5y + 2x)}$

47. $\dfrac{5x^{-1} - 2y^{-1}}{25x^{-2} - 4y^{-2}}$ $\dfrac{xy}{2x + 5y}$

48. $\dfrac{3x^{-1} + 3y^{-1}}{4x^{-2} - 9y^{-2}}$
$\dfrac{3xy(y + x)}{(2y - 3x)(2y + 3x)}$

REVIEW AND PREVIEW

Simplify. See Sections 5.1 and 5.2.

49. $\dfrac{3x^3y^2}{12x} \quad \dfrac{x^2y^2}{4}$

50. $\dfrac{-36xb^3}{9xb^2} \quad -4b$

51. $\dfrac{144x^5y^5}{-16x^2y} \quad -9x^3y^4$

52. $\dfrac{48x^3y^2}{-4xy} \quad -12x^2y$

Solve the following. See Section 2.6.

53. $|x - 5| = 9 \quad -4, 14$

54. $|2y + 1| = 1 \quad -1, 0$

Concept Extensions

55. When the source of a sound is traveling toward a listener, the pitch that the listener hears due to the Doppler effect is given by the complex rational compression $\dfrac{a}{1 - \dfrac{s}{770}}$, where a is the actual pitch of the sound and s is the speed of the sound source. Simplify this expression. $\dfrac{770a}{770 - s}$

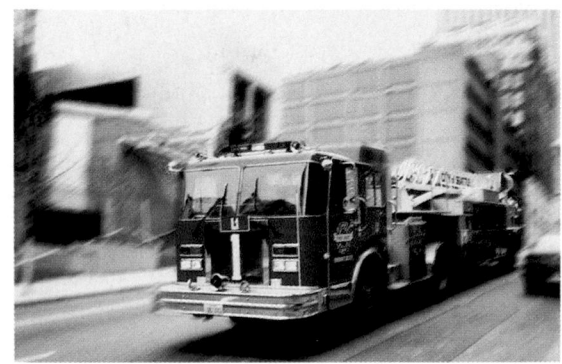

56. In baseball, the earned run average (ERA) statistic gives the average number of earned runs scored on a pitcher per game. It is computed with the following expression: $\dfrac{E}{\dfrac{I}{9}}$, where E is the number of earned runs scored on a pitcher and I is the total number of innings pitched by the pitcher. Simplify this expression. $\dfrac{9E}{I}$

57. Which of the following are equivalent to $\dfrac{\dfrac{1}{x}}{\dfrac{3}{y}}$?

a. $\dfrac{1}{x} \div \dfrac{3}{y}$ **b.** $\dfrac{1}{x} \cdot \dfrac{y}{3}$ **c.** $\dfrac{1}{x} \div \dfrac{y}{3}$ a, b

58. In your own words, explain one method for simplifying a complex fraction. answers may vary

Simplify.

59. $\dfrac{1}{1 + (1 + x)^{-1}} \quad \dfrac{1 + x}{2 + x}$

60. $\dfrac{(x + 2)^{-1} + (x - 2)^{-1}}{(x^2 - 4)^{-1}} \quad 2x$

61. $\dfrac{x}{1 - \dfrac{1}{1 + \dfrac{1}{x}}} \quad x^2 + x$

62. $\dfrac{x}{1 - \dfrac{1}{1 - \dfrac{1}{x}}} \quad -x(x - 1)$

63. $\dfrac{\dfrac{2}{y^2} - \dfrac{5}{xy} - \dfrac{3}{x^2}}{\dfrac{2}{y^2} + \dfrac{7}{xy} + \dfrac{3}{x^2}} \quad \dfrac{x - 3y}{x + 3y}$

64. $\dfrac{\dfrac{2}{x^2} - \dfrac{1}{xy} - \dfrac{1}{y^2}}{\dfrac{1}{x^2} - \dfrac{3}{xy} + \dfrac{2}{y^2}} \quad \dfrac{2y + x}{y - 2x}$

65. $\dfrac{3(a + 1)^{-1} + 4a^{-2}}{(a^3 + a^2)^{-1}}$

$3a^2 + 4a + 4$

66. $\dfrac{9x^{-1} - 5(x - y)^{-1}}{4(x - y)^{-1}}$

$\dfrac{4x - 9y}{4x}$

In the study of calculus, the difference quotient $\dfrac{f(a + h) - f(a)}{h}$ *is often found and simplified. Find and simplify this quotient for each function f(x) by following steps **a** through **d**.*

a. Find $(a + h)$.

b. Find $f(a)$.

c. Use steps **a** and **b** to find $\dfrac{f(a + h) - f(a)}{h}$

d. Simplify the result of step **c**.

67. $f(x) = \dfrac{1}{x}$

68. $f(x) = \dfrac{5}{x}$

69. $\dfrac{3}{x + 1}$

70. $\dfrac{2}{x^2}$

67. a. $\dfrac{1}{a + h}$ **b.** $\dfrac{1}{a}$

c. $\dfrac{\dfrac{1}{a + h} - \dfrac{1}{a}}{h}$ **d.** $\dfrac{-1}{a(a + h)}$

68. a. $\dfrac{5}{a + h}$ **b.** $\dfrac{5}{a}$

c. $\dfrac{\dfrac{5}{a + h} - \dfrac{5}{a}}{h}$ **d.** $\dfrac{-5}{a(a + h)}$

69. a. $\dfrac{3}{a + h + 1}$ **b.** $\dfrac{3}{a + 1}$

c. $\dfrac{\dfrac{3}{a + h + 1} - \dfrac{3}{a + 1}}{h}$

d. $\dfrac{-3}{(a + h + 1)(a + 1)}$

70. a. $\dfrac{2}{(a + h)^2}$ **b.** $\dfrac{2}{a^2}$

c. $\dfrac{\dfrac{2}{(a + h)^2} - \dfrac{2}{a^2}}{h}$

d. $\dfrac{-2(2a + h)}{a^2(a + h)^2}$

6.4   DIVIDING POLYNOMIALS

Objectives

1. Divide a polynomial by a monomial.
2. Divide by a polynomial.

1   Recall that a rational expression is a quotient of polynomials. An equivalent form of a rational expression can be obtained by performing the indicated division. For example, the rational expression $\dfrac{10x^3 - 5x^2 + 20x}{5x}$ can be thought of as the polynomial $10x^3 - 5x^2 + 20x$ divided by the monomial $5x$. To perform this division of a polynomial by a monomial (which we do below) recall the following addition fact for fractions with a common denominator.

$$\frac{a}{c} + \frac{b}{c} = \frac{a + b}{c}$$

If a, b, and c are monomials, we might read this equation from right to left and gain insight into dividing a polynomial by a monomial.

> **Dividing A Polynomial by a Monomial**
>
> Divide each term in the polynomial by the monomial.
> $$\frac{a + b}{c} = \frac{a}{c} + \frac{b}{c}, \text{ where } c \neq 0$$

EXAMPLE 1

Divide $10x^3 - 5x^2 + 20x$ by $5x$.

Solution   We divide each term of $10x^3 - 5x^2 + 20x$ by $5x$ and simplify.

$$\frac{10x^3 - 5x^2 + 20x}{5x} = \frac{10x^3}{5x} - \frac{5x^2}{5x} + \frac{20x}{5x} = 2x^2 - x + 4$$

Check: To check, see that (quotient) (divisor) = dividend, or

$$(2x^2 - x + 4)(5x) = 10x^3 - 5x^2 + 20x.$$

CLASSROOM EXAMPLE

Divide
$$16y^3 - 8y^2 + 6y \text{ by } 2y.$$

answer: $8y^2 - 4y + 3$

TEACHING TIP

Remind students that the goal in a division problem is to find the expression which when multiplied by the divisor results in the dividend.

EXAMPLE 2

Divide $\dfrac{3x^5y^2 - 15x^3y - x^2y - 6x}{x^2y}$

Solution We divide each term in the numerator by x^2y.

$$\frac{3x^5y^2 - 15x^3y - x^2y - 6x}{x^2y} = \frac{3x^5y^2}{x^2y} - \frac{15x^3y}{x^2y} - \frac{x^2y}{x^2y} - \frac{6x}{x^2y}$$

$$= 3x^3y - 15x - 1 - \frac{6}{xy}$$

2 To divide a polynomial by a polynomial other than a monomial, we use **long division.** Polynomial long division is similar to long division of real numbers. We review long division of real numbers by dividing 7 into 296.

$$\text{Divisor:} \quad 7\overline{)296} \quad \begin{array}{r} 42 \\ \end{array}$$

$$\begin{array}{r} -28 \\ \hline 16 \\ -14 \\ \hline 2 \end{array}$$

$4(7) = 28.$

Subtract and bring down the next digit in the dividend.

$2(7) = 14.$

Subtract. The remainder is 2.

The quotient is $42\dfrac{2 \text{ (remainder)}}{7 \text{ (divisor)}}$.

Check: To check, notice that

$$42(7) + 2 = 296, \text{ the dividend.}$$

This same division process can be applied to polynomials, as shown next.

EXAMPLE 3

Divide $2x^2 - x - 10$ by $x + 2$.

Solution $2x^2 - x - 10$ is the dividend, and $x + 2$ is the divisor.

Step 1: Divide $2x^2$ by x.

$$x + 2\overline{)2x^2 - x - 10} \quad \begin{array}{r} 2x \\ \end{array}$$

$\dfrac{2x^2}{x} = 2x$, so $2x$ is the first term of the quotient.

Step 2: Multiply $2x(x + 2)$.

$$x + 2\overline{)2x^2 - x - 10} \quad \begin{array}{r} 2x \\ \end{array}$$
$$2x^2 + 4x \qquad 2x(x + 2)$$

Like terms are lined up vertically.

By graphing $y_1 = 2x^2 - x - 10$ and $y_2 = x + 2$ in the same window, we see that the graphs share the x-intercept $x = -2$. We know this means that $x + 2$ is a common factor of $2x^2 - x - 10$ and $x + 2$. In other words, $x + 2$ divides into $2x^2 - x - 10$ evenly without a remainder.

Step 3: Subtract $(2x^2 + 4x)$ from $(2x^2 - x - 10)$ by changing the signs of $(2x^2 + 4x)$ and adding.

$$
\begin{array}{r}
2x \\
x + 2{\overline{\smash{\big)}\,2x^2 -\ x - 10}} \\
\underline{-2x^2 - 4x} \\
-5x
\end{array}
$$

Step 4: Bring down the next term, -10, and start the process over.

$$
\begin{array}{r}
2x \\
x + 2{\overline{\smash{\big)}\,2x^2 -\ x - 10}} \\
\underline{-2x^2 - 4x} \\
-5x - 10
\end{array}
$$

Step 5: Divide $-5x$ by x.

$$
\begin{array}{r}
2x -\ 5 \\
x + 2{\overline{\smash{\big)}\,2x^2 -\ x - 10}} \\
\underline{-2x^2 - 4x} \\
-5x - 10
\end{array}
$$

$\dfrac{-5x}{x} = -5$, so -5 is the second term of the quotient.

Step 6: Multiply $-5(x + 2)$.

$$
\begin{array}{r}
2x -\ 5 \\
x + 2{\overline{\smash{\big)}\,2x^2 -\ x - 10}} \\
\underline{-2x^2 - 4x} \\
-5x - 10 \\
-5x - 10
\end{array}
$$

$-5(x + 2)$

Like terms are lined up vertically.

Step 7: Subtract $(-5x - 10)$ from $(-5x - 10)$.

$$
\begin{array}{r}
2x -\ 5 \\
x + 2{\overline{\smash{\big)}\,2x^2 -\ x - 10}} \\
\underline{-2x^2 - 4x} \\
-5x - 10 \\
\underline{+5x + 10} \\
0
\end{array}
$$

Then $\dfrac{2x^2 - x - 10}{x + 2} = 2x - 5$. There is no remainder.

Check: Check this result by multiplying $2x - 5$ by $x + 2$. Their product is $(2x - 5)(x + 2) = 2x^2 - x - 10$, the dividend.

TEACHING TIP

After reviewing Example 3, have students complete the table below. Ask them what pattern they notice in the table.

x	$x + 2$	$2x^2 - x - 10$	$\dfrac{2x^2 - x - 10}{x + 2}$	$2x - 5$
1				
2				
3				

EXAMPLE 4

Divide: $(6x^2 - 19x + 12) \div (3x - 5)$

Solution

$$
\begin{array}{r}
2x \\
3x - 5 \overline{)6x^2 - 19x + 12} \\
\underline{6x^2 - 10x} \\
-9x + 12
\end{array}
$$

Divide $\dfrac{6x^2}{3x} = 2x$.

Multiply $2x(3x - 5)$.
Subtract by adding the opposite.
Bring down the next term, $+12$.

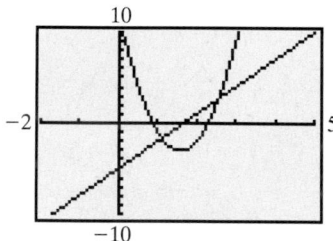

$$
\begin{array}{r}
2x - 3 \\
3x - 5 \overline{)6x^2 - 19x + 12} \\
\underline{6x^2 - 10x} \\
-9x + 12 \\
\underline{-9x + 15} \\
-3
\end{array}
$$

Divide $\dfrac{-9x}{3x} = -3$.

Multiply $-3(3x - 5)$.
Subtract by adding the opposite.

Check: divisor   ·   quotient   +   remainder

$$(3x - 5) \quad (2x - 3) \quad + 1(-3) = 6x^2 - 19x + 15 - 3$$

$$= 6x^2 - 19x + 12 \qquad \text{The dividend}$$

The division checks, so

$$\frac{6x^2 - 19x + 12}{3x - 5} = 2x - 3 - \frac{3}{3x - 5}$$

> **Helpful Hint**
> This fraction is the remainder over the divisor.

EXAMPLE 5

Divide: $(7x^3 + 16x^2 + 2x - 1) \div (x + 4)$.

Solution

$$
\begin{array}{r}
7x^2 - 12x + 50 \\
x + 4 \overline{)7x^3 + 16x^2 + 2x - 1} \\
\underline{7x^3 + 28x^2} \\
-12x^2 + 2x \\
\underline{-12x^2 - 48x} \\
50x - 1 \\
\underline{50x + 200} \\
-201
\end{array}
$$

Divide $\dfrac{7x^3}{x} = 7x^2$.

$7x^2(x + 4)$
Subtract. Bring down $2x$.
$\dfrac{-12x^2}{x} = -12x$, a term of the quotient.
$-12x(x + 4)$
Subtract. Bring down -1.
$\dfrac{50x}{x} = 50$, a term of the quotient.
$50(x + 4)$.
Subtract.

Thus, $\dfrac{7x^3 + 16x^2 + 2x - 1}{x + 4} = 7x^2 - 12x + 50 + \dfrac{-201}{x + 4}$ or

$$7x^2 - 12x + 50 + \frac{201}{x + 4}.$$

EXAMPLE 6

Divide $3x^4 + 2x^3 - 8x + 6$ by $x^2 - 1$.

Solution   Before dividing, we represent any "missing powers" by the product of 0 and the variable raised to the missing power. There is no x^2 term in the dividend, so we include $0x^2$ to represent the missing term. Also, there is no x term in the divisor, so we include $0x$ in the divisor.

$$
\begin{array}{r}
3x^2 + 2x + 3 \\
x^2 + 0x - 1\overline{)3x^4 + 2x^3 + 0x^2 - 8x + 6} \\
3x^4 + 0x^3 - 3x^2 \\
\hline
2x^3 + 3x^2 - 8x \\
2x^3 - 0x^2 - 2x \\
\hline
3x^2 - 6x + 6 \\
3x^2 + 0x - 3 \\
\hline
-6x + 9
\end{array}
$$

$\dfrac{3x^4}{x^2} = 3x^2$

$3x^2(x^2 + 0x - 1)$

Subtract. Bring down $-8x$.

$\dfrac{2x^3}{x^2} = 2x$, a term of the quotient.

$2x(x^2 + 0x - 1)$

Subtract. Bring down 6.

$\dfrac{3x^2}{x^2} = 3$, a term of the quotient.

$3(x^2 + 0x - 1)$

Subtract.

The division process is finished when the degree of the remainder polynomial is less than the degree of the divisor. Thus,

$$\frac{3x^4 + 2x^3 - 8x + 6}{x^2 - 1} = 3x^2 + 2x + 3 + \frac{-6x + 9}{x^2 - 1}$$

EXAMPLE 7

Divide $27x^3 + 8$ by $3x + 2$.

Solution   We replace the missing terms in the dividend with $0x^2$ and $0x$.

$$
\begin{array}{r}
9x^2 - 6x + 4 \\
3x + 2\overline{)27x^3 + 0x^2 + 0x + 8} \\
27x^3 + 18x^2 \\
\hline
-18x^2 + 0x \\
-18x^2 - 12x \\
\hline
12x + 8 \\
12x + 8 \\
\hline
0
\end{array}
$$

$9x^2(3x + 2)$

Subtract. Bring down $0x$.

$-6x(3x + 2)$

Subtract. Bring down 8.

$4(3x + 2)$

Thus, $\dfrac{27x^3 + 8}{3x + 2} = 9x^2 - 6x + 4$.

✔ **CONCEPT CHECK**

In a division problem, the divisor is $4x^3 - 5$. The division process can be stopped when which of these possible remainder polynomials is reached?

   a. $2x^4 + x^2 - 3$

   b. $x^3 - 5^2$

   c. $4x^2 + 25$

Concept Check Answer: c

EXERCISE SET 6.4

STUDY GUIDE/SSM · CD/VIDEO · PH MATH TUTOR CENTER · MathXL®Tutorials ON CD · MathXL® · MyMathLab®

Divide. See Examples 1 and 2.

1. Divide $4a^2 + 8a$ by $2a$. $2a + 4$

2. Divide $6x^4 - 3x^3$ by $3x^2$. $2x^2 - x$

3. $\dfrac{12a^5b^2 + 16a^4b}{4a^4b}$ $3ab + 4$

5. $\dfrac{4x^2y^2 + 6xy^2 - 4y^2}{2x^2y}$

7. $\dfrac{4x^2 + 8x + 4}{4}$

4. $x^2 + 3xy - y^2$

5. $2y + \dfrac{3y}{x} - \dfrac{2y}{x^2}$

6. $3x^2 + 37x + 12$

4. $\dfrac{4x^3y + 12x^2y^2 - 4xy^3}{4xy}$

6. $\dfrac{6x^5 + 74x^4 + 24x^3}{2x^3}$

8. $\dfrac{15x^3 - 5x^2 + 10x}{5x^2}$

9. A board of length A $(3x^4 + 6x^2 - 18)$ meters is to be cut into three pieces of the same length. Find the length of each piece. $(x^4 + 2x^2 - 6)$ m

$(3x^4 + 6x^2 - 18)$ m

△ **10.** The perimeter of a regular hexagon is given to be $12x^5 - 48x^3 + 3$ miles. Find the length of each side.

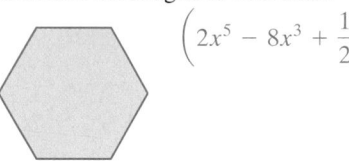

$\left(2x^5 - 8x^3 + \dfrac{1}{2}\right)$ mi

7. $x^2 + 2x + 1$ **8.** $3x - 1 + \dfrac{2}{x}$

Divide. See Examples 3 through 7.

11. $(x^2 + 3x + 2) \div (x + 2)$ $x + 1$

12. $(y^2 + 7y + 10) \div (y + 5)$ $y + 2$

13. $(2x^2 - 6x - 7) \div (x + 1)$ $2x - 8 + \dfrac{1}{x + 1}$

14. $(3x^2 + 19x + 18) \div (x + 5)$ $3x + 4 + \dfrac{2}{x + 5}$

15. $2x^2 + 3x - 2$ by $2x + 4$ $x - \dfrac{1}{2}$

16. $6x^2 - 17x - 3$ by $3x - 9$ $2x + \dfrac{1}{3}$

17. $(4x^3 + 7x^2 + 8x + 20) \div (2x + 4)$ $2x^2 - \dfrac{1}{2}x + 5$

18. $(18x^3 + x^2 - 90x - 5) \div (9x^2 - 45)$ $2x + \dfrac{1}{9}$

△ **19.** If the area of the rectangle is $(15x^2 - 29x - 14)$ square inches and its length is $(5x + 2)$ inches, find its width. $(3x - 7)$ in.

?

$(5x + 2)$ in.

△ **20.** If the area of a parallelogram is $(2x^2 - 17x + 35)$ square centimeters and its base is $(2x - 7)$ centimeters, find its height. $(x - 5)$ cm

?

$(2x - 7)$ cm

MIXED PRACTICE

Divide.

21. $25a^2b^{12}$ by $10a^5b^7$ $\dfrac{5b^5}{2a^3}$

22. $12a^2b^3$ by $8a^7b$ $\dfrac{3b^2}{2a^5}$

23. $(x^6y^6 - x^3y^3) \div x^3y^3$ $x^3y^3 - 1$

24. $(25xy^2 + 75xyz + 125x^2yz) \div -5x^2y$ $-\dfrac{5y}{x} - \dfrac{15z}{x} - 25z$

25. $(a^2 + 4a + 3) \div (a + 1)$ $a + 3$

26. $(3x^2 - 14x + 16) \div (x - 2)$ $3x - 8$

27. $(2x^2 + x - 10) \div (x - 2)$ $2x + 5$

28. $(x^2 - 7x + 12) \div (x - 5)$ $x - 2 + \dfrac{2}{x - 5}$

29. $-16y^3 + 24y^4$ by $-4y^2$ $4y - 6y^2$

30. $-20a^2b + 12ab^2$ by $-4ab$ $5a - 3b$

31. $(2x^2 + 13x + 15) \div (x - 5)$ $2x + 23 + \dfrac{130}{x - 5}$

32. $(2x^2 + 13x + 5) \div (2x + 3)$ $x + 5 - \dfrac{10}{2x + 3}$

33. $(20x^2y^3 + 6xy^4 - 12x^3y^5) \div 2xy^3$ $10x + 3y - 6x^2y^2$

34. $(3x^2y + 6x^2y^2 + 3xy) \div 3xy$ $x + 2xy + 1$

35. $(6x^2 + 16x + 8) \div (3x + 2)$ $2x + 4$

36. $(x^2 - 25) \div (x + 5)$ $x - 5$

37. $(2y^2 + 7y - 15) \div (2y - 3)$ $y + 5$

38. $(3x^2 - 4x + 6) \div (x - 2)$ $3x + 2 + \dfrac{10}{x - 2}$

39. $4x^2 - 9$ by $2x - 3$

40. $8x^2 + 6x - 27$ by $4x + 9$

41. $2x^3 + 6x - 4$ by $x + 4$

42. $4x^3 - 5x$ by $2x - 1$

43. $3x^2 - 4$ by $x - 1$

44. $x^2 - 9$ by $x + 4$

45. $(-13x^3 + 2x^4 + 16x^2 - 9x + 20) \div (5 - x)$

46. $(5x^2 - 5x + 2x^3 + 20) \div (4 + x)$ $2x^2 - 3x + 7 - \dfrac{8}{x + 4}$

🔒 **47.** $3x^5 - x^3 + 4x^2 - 12x - 8$ by $x^2 - 2$

48. $-8x^3 + 2x^4 + 19x^2 - 33x + 15$ by $x^2 - x + 5$

49. $(3x^3 - 5) \div 3x^2$ $x - \dfrac{5}{3x^2}$

50. $(14x^3 - 2) \div (7x - 1)$ $2x^2 + \dfrac{2}{7}x + \dfrac{2}{49} - \dfrac{96}{49(7x - 1)}$

REVIEW AND PREVIEW

Insert $<$, $>$, or $=$ to make each statement true. See Section 1.3.

51. 3^2 ___ $= (-3)^2$

52. $(-5)^2$ ___ $= 5^2$

53. -2^3 ___ $= (-2)^3$

54. 3^4 ___ $= (-3)^4$

Solve each inequality. See Section 3.5.

55. $|x + 5| < 4$ $(-9, -1)$

56. $|x - 1| \le 8$ $[-7, 9]$

57. $|2x + 7| \ge 9$
$(-\infty, -8] \cup [1, \infty)$

58. $|4x + 2| > 10$
$(-\infty, -3) \cup (2, \infty)$

Concept Extensions

59. Find $P(1)$ for the polynomial function
$P(x) = 3x^3 + 2x^2 - 4x + 3$. Next, divide
$3x^3 + 2x^2 - 4x + 3$ by $x - 1$. Compare the remainder
with $P(1)$.   4

60. Find $P(-2)$ for the polynomial function
$P(x) = x^3 - 4x^2 - 3x + 5$. Next, divide
$x^3 - 4x^2 - 3x + 5$ by $x + 2$. Compare the remainder with
$P(-2)$. -13

61. Find $P(-3)$ for the polynomial
$P(x) = 5x^4 - 2x^2 + 3x - 6$. Next, divide
$5x^4 - 2x^2 + 3x - 6$ by $x + 3$. Compare the remainder
with $P(-3)$.   372

62. Find $P(2)$ for the polynomial function
$P(x) = -4x^4 + 2x^3 - 6x + 3$. Next, divide
$-4x^4 + 2x^3 - 6x + 3$ by $x - 2$. Compare the remainder
with $P(2)$. -57

39. $2x + 3$

40. $2x - 3$

✎ **63.** Write down any patterns you noticed from Exercises 59–62.
answers may vary

✎ **64.** Explain how to check polynomial long division.   answers may vary

Divide.

65. $x^3 + \dfrac{5}{3}x^2 + \dfrac{5}{3}x + \dfrac{8}{3} + \dfrac{8}{3(x - 1)}$

65. $\left(x^4 + \dfrac{2}{3}x^3 + x\right) \div (x - 1)$

66. $\left(2x^3 + \dfrac{9}{2}x^2 - 4x - 10\right) \div (x + 2)$ $2x^2 + \dfrac{1}{2}x - 5$

67. $\left(3x^4 - x - x^3 + \dfrac{1}{2}\right) \div (2x - 1)$

68. $\left(2x^4 + \dfrac{1}{2}x^3 + x^2 + x\right) \div (x - 2)$

69. $(5x^4 - 2x^2 + 10x^3 - 4x) \div (5x + 10)$ $x^3 - \dfrac{2}{5}x$

70. $(9x^5 + 6x^4 - 6x^2 - 4x) \div (3x + 2)$ $3x^4 - 2x$

67. $\dfrac{3}{2}x^3 + \dfrac{1}{4}x^2 + \dfrac{1}{8}x - \dfrac{7}{16} + \dfrac{1}{16(2x - 1)}$

For each given $f(x)$ and $g(x)$, find $\dfrac{f(x)}{g(x)}$. Also find any x-values that are not in the domain of $\dfrac{f(x)}{g(x)}$. (Note: Since $g(x)$ is in the denominator, $g(x)$ cannot be 0).

48. $2x^2 - 6x + 3$

68. $2x^3 + \dfrac{9}{2}x^2 + 10x + 21 + \dfrac{42}{x - 2}$

71. $f(x) = 25x^2 - 5x + 30$; $g(x) = 5x$ $5x - 1 + \dfrac{6}{x}$; $x \ne 0$

72. $f(x) = 12x^4 - 9x^3 + 3x - 1$; $g(x) = 3x$
$4x^3 - 3x^2 + 1 - \dfrac{1}{3x}$; $x \ne 0$

73. $f(x) = 7x^4 - 3x^2 + 2$; $g(x) = x - 2$
$7x^3 + 14x^2 + 25x + 50 + \dfrac{102}{x - 2}$; $x \ne 2$

74. $f(x) = 2x^3 - 4x^2 + 1$; $g(x) = x + 3$
$2x^2 - 10x + 30 - \dfrac{89}{x + 3}$; $x \ne -3$

✎ **75.** Try performing the following division without changing the order of the terms. Describe why this makes the process more complicated. Then perform the division again after putting the terms in the dividend in descending order of exponents.
$$\frac{4x^2 - 12x - 12 + 3x^3}{x - 2}$$
answers may vary

76. Gateway is a leading direct marketer of personal computers. Gateway's annual net profit can be modeled by the polynomial function
$$P(x) = -54x^3 + 305x^2 - 363x + 245$$
where $P(x)$ is net profit in millions of dollars in the year x. Gateway's annual revenue can be modeled by the function

41. $2x^2 - 8x + 38 - \dfrac{156}{x + 4}$ **42.** $2x^2 + x - 2 - \dfrac{2}{2x - 1}$ **43.** $3x + 3 - \dfrac{1}{x - 1}$ **44.** $x - 4 + \dfrac{7}{x + 4}$ **45.** $-2x^3 + 3x^2 - x + 4$

$R(x) = 1140x + 5295$, where $R(x)$ is revenue in millions of dollars in the year x. In both models, $x = 0$ represents the year 1996. (*Source:* Gateway, Inc., 1996–2000)

a. Suppose that a market analyst has found the model $P(x)$, and another analyst at the same firm has found the model $R(x)$. The analysts have been asked by their manager to work together to find a model for Gateway's net profit margin. The analysts know that a company's net profit margin is the ratio of its net profit to its revenue. Describe how these two analysts could collaborate to find a function $m(x)$ that models Gateway's net profit margin based on the work they have done independently.

answers may vary

b. Without actually finding $m(x)$, give a general description of what you would expect the form of the result to be.

answers may vary

6.5 SYNTHETIC DIVISION AND THE REMAINDER THEOREM

Objectives

1 Use synthetic division to divide a polynomial by a binomial.

2 Use the remainder theorem to evaluate polynomials.

1 When a polynomial is to be divided by a binomial of the form $x - c$, a shortcut process called **synthetic division** may be used. On the left is an example of long division, and on the right, the same example showing the coefficients of the variables only.

$$
\begin{array}{r}
2x^2 + 5x + 2 \\
x - 3 \overline{)2x^3 - x^2 - 13x + 1} \\
\underline{2x^3 - 6x^2} \\
5x^2 - 13x \\
\underline{5x^2 - 15x} \\
2x + 1 \\
\underline{2x - 6} \\
7
\end{array}
\qquad
\begin{array}{r}
2 \quad 5 \quad 2 \\
1 - 3 \overline{)2 - 1 - 13 + 1} \\
\underline{2 - 6} \\
5 - 13 \\
\underline{5 - 15} \\
2 + 1 \\
\underline{2 - 6} \\
7
\end{array}
$$

Notice that as long as we keep coefficients of powers of x in the same column, we can perform division of polynomials by performing algebraic operations on the coefficients only. This shortcut process of dividing with coefficients only in a special format is called synthetic division. To find $(2x^3 - x^2 - 13x + 1) \div (x - 3)$ by synthetic division, follow the next example.

EXAMPLE 1

Use synthetic division to divide $2x^3 - x^2 - 13x + 1$ by $x - 3$.

Solution To use synthetic division, the divisor must be in the form $x - c$. Since we are dividing by $x - 3$, c is 3. Write down 3 and the coefficients of the dividend.

$$c \quad\quad \begin{array}{r|rrrr} 3 & 2 & -1 & -13 & 1 \\ & \downarrow & & & \\ \hline & 2 & & & \end{array}$$

Next, draw a line and bring down the first coefficient of the dividend.

$$\begin{array}{r|rrrr} 3 & 2 & -1 & -13 & 1 \\ & & 6 & & \\ \hline & 2 & & & \end{array}$$

Multiply $3 \cdot 2$ and write down the product, 6.

$$\begin{array}{r|rrrr} 3 & 2 & -1 & -13 & 1 \\ & & 6 & & \\ \hline & 2 & 5 & & \end{array}$$

Add $-1 + 6$. Write down the sum, 5.

$$\begin{array}{r|rrrr} 3 & 2 & -1 & -13 & 1 \\ & & 6 & 15 & \\ \hline & 2 & 5 & 2 & \end{array}$$

$3 \cdot 5 = 15$.
$-13 + 15 = 2$.

$$\begin{array}{r|rrrr} 3 & 2 & -1 & -13 & 1 \\ & & 6 & 15 & 6 \\ \hline & 2 & 5 & 2 & 7 \end{array}$$

$3 \cdot 2 = 6$.
$1 + 6 = 7$.

The quotient is found in the bottom row. The numbers 2, 5, and 2 are the coefficients of the quotient polynomial, and the number 7 is the remainder. The degree of the quotient polynomial is one less than the degree of the dividend. In our example, the degree of the dividend is 3, so the degree of the quotient polynomial is 2. As we found when we performed the long division, the quotient is

$$2x^2 + 5x + 2, \quad \text{remainder 7}$$

or

$$2x^2 + 5x + 2 + \frac{7}{x - 3}$$

EXAMPLE 2

Use synthetic division to divide $x^4 - 2x^3 - 11x^2 + 5x + 34$ by $x + 2$.

Solution The divisor is $x + 2$, which we write in the form $x - c$ as $x - (-2)$. Thus, c is -2. The dividend coefficients are 1, -2, -11, 5, and 34.

$$c \quad\quad \begin{array}{r|rrrrr} -2 & 1 & -2 & -11 & 5 & 34 \\ & & -2 & 8 & 6 & -22 \\ \hline & 1 & -4 & -3 & 11 & 12 \end{array}$$

The dividend is a fourth-degree polynomial, so the quotient polynomial is a third-degree polynomial. The quotient is $x^3 - 4x^2 - 3x + 11$ with a remainder of 12. Thus,

$$\frac{x^4 - 2x^3 - 11x^2 + 5x + 34}{x + 2} = x^3 - 4x^2 - 3x + 11 + \frac{12}{x + 2}$$

✔ **CONCEPT CHECK**

Which division problems are candidates for the synthetic division process?

a. $(3x^2 + 5) \div (x + 4)$ **b.** $(x^3 - x^2 + 2) \div (3x^3 - 2)$

c. $(y^4 + y - 3) \div (x^2 + 1)$ **d.** $x^5 \div (x - 5)$

> **Helpful Hint**
>
> Before dividing by synthetic division, write the dividend in descending order of variable exponents. Any "missing powers" of the variable should be represented by 0 times the variable raised to the missing power.

Discover the Concept

Let $P(x) = 2x^3 - 4x^2 + 5$.

a. Find $P(2)$ by substitution.

b. Use synthetic division to find the remainder when $P(x)$ is divided by $x - 2$.

c. Compare the results of parts a and b.

d. Now read Example 3 below.

EXAMPLE 3

If $P(x) = 2x^3 - 4x^2 + 5$

a. Find $P(2)$ by substitution.

b. Use synthetic division to find the remainder when $P(x)$ is divided by $x - 2$.

Solution   **a.** $P(x) = 2x^3 - 4x^2 + 5$

$P(2) = 2(2)^3 - 4(2)^2 + 5$

$= 2(8) - 4(4) + 5 = 16 - 16 + 5 = 5$

Thus, $P(2) = 5$.

b. The coefficients of $P(x)$ are 2, -4, 0, and 5. The number 0 is a coefficient of the missing power of x^1. The divisor is $x - 2$, so c is 2.

$$
\begin{array}{c}
c \longrightarrow \quad 2 \,\big|\;\; 2 \quad -4 \quad 0 \quad 5 \\
\underline{ \quad 4 \quad 0 \quad 0} \\
2 \quad 0 \quad 0 \quad 5 \quad \text{remainder}
\end{array}
$$

The remainder when $P(x)$ is divided by $x - 2$ is 5.

> **2** Notice in the preceding example that $P(2) = 5$ and that the remainder when $P(x)$ is divided by $x - 2$ is 5. This is no accident. This illustrates the **remainder theorem.**

CLASSROOM EXAMPLE

If $P(x) = 3x^3 - 7x + 1$,

a. Find $P(3)$ by substitution

b. Use synthetic division to find the remainder when $P(x)$ is divided by $x - 3$

answer:

a. 61   **b.** 61

$P(2) = 5$

Remainder Theorem

If a polynomial $P(x)$ is divided by $x - c$, then the remainder is $P(c)$.

EXAMPLE 4

Use the remainder theorem and synthetic division to find $P(4)$ if

$$P(x) = 4x^6 - 25x^5 + 35x^4 + 17x^2.$$

Concept Check Answer:

a and d

Solution   To find $P(4)$ by the remainder theorem, we divide $P(x)$ by $x - 4$. The coefficients of $P(x)$ are $4, -25, 35, 0, 17, 0$, and 0. Also, c is 4.

CLASSROOM EXAMPLE

Use the remainder theorem and synthetic division to find $P(2)$ if
$P(x) = 6x^5 - 4x^4 - 40x - 10$

answer: 38

$$
c \longrightarrow \underline{4} \begin{array}{|rrrrrrr} 4 & -25 & 35 & 0 & 17 & 0 & 0 \\ & 16 & -36 & -4 & -16 & 4 & 16 \\ \hline 4 & -9 & -1 & -4 & 1 & 4 & 16 \end{array} \text{ remainder}
$$

Thus, $P(4) = 16$, the remainder.

EXERCISE SET 6.5

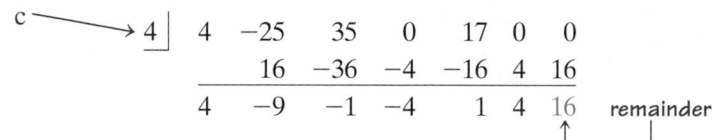

STUDY GUIDE/SSM   CD/VIDEO   PH MATH TUTOR CENTER   MathXL®Tutorials ON CD   MathXL®   MyMathLab®

Use synthetic division to divide. See Examples 1 and 2.

1. $(x^2 + 3x - 40) \div (x - 5)$ $x + 8$

2. $(x^2 - 14x + 24) \div (x - 2)$ $x - 12$

3. $(x^2 + 5x - 6) \div (x + 6)$ $x - 1$

4. $(x^2 + 12x + 32) \div (x + 4)$ $x + 8$

5. $(x^3 - 7x^2 - 13x + 5) \div (x - 2)$ $x^2 - 5x - 23 - \frac{41}{x - 2}$

6. $(x^3 + 6x^2 + 4x - 7) \div (x + 5)$ $x^2 + x - 1 - \frac{2}{x + 5}$

7. $(4x^2 - 9) \div (x - 2)$ **8.** $(3x^2 - 4) \div (x - 1)$

7. $4x + 8 + \frac{7}{x - 2}$

8. $3x + 3 - \frac{1}{x - 1}$

For the given polynomial P(x) and the given c, find P(c) by (a) direct substitution and (b) the remainder theorem. See Examples 3 and 4.

9. $P(x) = 3x^2 - 4x - 1; P(2)$ 3

10. $P(x) = x^2 - x + 3; P(5)$ 23

11. $P(x) = 4x^4 + 7x^2 + 9x - 1; P(-2)$ 73

12. $P(x) = 8x^5 + 7x + 4; P(-3)$ -1961

13. $P(x) = x^5 + 3x^4 + 3x - 7; P(-1)$ -8

14. $P(x) = 5x^4 - 4x^3 + 2x - 1; P(-1)$ 6

MIXED PRACTICE

19. $2x^3 - 3x^2 + x - 4$

Use synthetic division to divide.

15. $(x^3 - 3x^2 + 2) \div (x - 3)$ $x^2 + \frac{2}{x - 3}$

16. $(x^2 + 12) \div (x + 2)$ $x - 2 + \frac{16}{x + 2}$

17. $(6x^2 + 13x + 8) \div (x + 1)$ $6x + 7 + \frac{1}{x + 1}$

18. $(x^3 - 5x^2 + 7x - 4) \div (x - 3)$ $x^2 - 2x + 1 - \frac{1}{x - 3}$

19. $(2x^4 - 13x^3 + 16x^2 - 9x + 20) \div (x - 5)$

20. $(3x^4 + 5x^3 - x^2 + x - 2) \div (x + 2)$ $3x^3 - x^2 + x - 1$

23. $3x^2 - \frac{9}{2}x + \frac{7}{4} + \frac{47}{8\left(x - \frac{1}{2}\right)}$

21. $(3x^2 - 15) \div (x + 3)$ $3x - 9 + \frac{12}{x + 3}$

22. $(3x^2 + 7x - 6) \div (x + 4)$ $3x - 5 + \frac{14}{x + 4}$

23. $(3x^3 - 6x^2 + 4x + 5) \div \left(x - \frac{1}{2}\right)$

24. $(8x^3 - 6x^2 - 5x + 3) \div \left(x + \frac{3}{4}\right)$ $8x^2 - 12x + 4$

25. $(3x^3 + 2x^2 - 4x + 1) \div \left(x - \frac{1}{3}\right)$ $3x^2 + 3x - 3$

26. $(9y^3 + 9y^2 - y + 2) \div \left(y + \frac{2}{3}\right)$

27. $(7x^2 - 4x + 12 + 3x^3) \div (x + 1)$ $3x^2 + 4x - 8 + \frac{20}{x + 1}$

28. $(x^4 + 4x^3 - x^2 - 16x - 4) \div (x - 2)$

29. $(x^3 - 1) \div (x - 1)$ $x^2 + x + 1$

30. $(y^3 - 8) \div (y - 2)$ $y^2 + 2y + 4$

31. $(x^2 - 36) \div (x + 6)$ $x - 6$

32. $(4x^3 + 12x^2 + x - 12) \div (x + 3)$ $4x^2 + 1 - \frac{15}{x + 3}$

For the given polynomial P(x) and the given c, use the remainder theorem to find P(c).

33. $P(x) = x^3 + 3x^2 - 7x + 4; 1$ 1

34. $P(x) = x^3 + 5x^2 - 4x - 6; 2$ 14

35. $P(x) = 3x^3 - 7x^2 - 2x + 5; -3$ -133

36. $P(x) = 4x^3 + 5x^2 - 6x - 4; -2$ -4

37. $P(x) = 4x^4 + x^2 - 2; -1$ 3

38. $P(x) = x^4 - 3x^2 - 2x + 5; -2$ 13

39. $P(x) = 2x^4 - 3x^2 - 2; \frac{1}{3}$ $-\frac{187}{81}$

26. $9y^2 + 3y - 3 + \dfrac{4}{y + \frac{2}{3}}$ **28.** $x^3 + 6x^2 + 11x + 6 + \dfrac{8}{x - 2}$

40. $P(x) = 4x^4 - 2x^3 + x^2 - x - 4; \dfrac{1}{2} \quad -\dfrac{17}{4}$

41. $P(x) = x^5 + x^4 - x^3 + 3; \dfrac{1}{2} \quad \dfrac{95}{32}$

42. $P(x) = x^5 - 2x^3 + 4x^2 - 5x + 6; \dfrac{2}{3} \quad \dfrac{968}{243}$

43. Explain an advantage of using the remainder theorem instead of direct substitution.   answers may vary

44. Explain an advantage of using synthetic division instead of long division.   answers may vary

REVIEW AND PREVIEW

Solve each equation for x. See Sections 1.5 and 5.8.

45. $7x + 2 = x - 3 \quad -\dfrac{5}{6}$

46. $4 - 2x = 17 - 5x \quad \dfrac{13}{3}$

47. $x^2 = 4x - 4 \quad 2$

48. $5x^2 + 10x = 15 \quad -3, 1$

49. $\dfrac{x}{3} - 5 = 13 \quad 54$

50. $\dfrac{2x}{9} + 1 = \dfrac{7}{9} \quad -1$

Factor the following. See Sections 5.5 and 5.7.

51. $x^3 - 1$

52. $8y^3 + 1$

53. $125z^3 + 8$

54. $a^3 - 27$

55. $xy + 2x + 3y + 6$

56. $x^2 - x + xy - y$

57. $x^3 - 9x \quad x(x + 3)(x - 3)$

58. $2x^3 - 32x$

Concept Extensions

We say that 2 is a factor of 8 because 2 divides 8 evenly, or with a remainder of 0. In the same manner, the polynomial $x - 2$ is a factor of the polynomial $x^3 - 14x^2 + 24x$ because the remainder is 0 when $x^3 - 14x^2 + 24x$ is divided by $x - 2$. Use this information for Exercises 59 through 61.

59. Use synthetic division to show that $x + 3$ is a factor of $x^3 + 3x^2 + 4x + 12$.
$(x + 3)(x^2 + 4) = x^3 + 3x^2 + 4x + 12$

60. Use synthetic division to show that $x - 2$ is a factor of $x^3 - 2x^2 - 3x + 6$.
$(x - 2)(x^2 - 3) = x^3 - 2x^2 - 3x + 6$

61. From the remainder theorem, the polynomial $x - c$ is a factor of a polynomial function $P(x)$ if $P(c)$ is what value?   0

62. If a polynomial is divided by $x - 5$, the quotient is $2x^2 + 5x - 6$ and the remainder is 3. Find the original polynomial. $2x^3 - 5x^2 - 31x + 33$

63. If a polynomial is divided by $x + 3$, the quotient is $x^2 - x + 10$ and the remainder is -2. Find the original polynomial. $x^3 + 2x^2 + 7x + 28$

△ **64.** If the area of a parallelogram is $(x^4 - 23x^2 + 9x - 5)$ square centimeters and its base is $(x + 5)$ centimeters, find its height. $(x^3 - 5x^2 + 2x - 1)$ cm

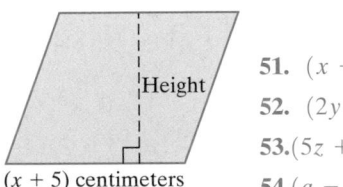

Height

$(x + 5)$ centimeters

51. $(x - 1)(x^2 + x + 1)$
52. $(2y + 1)(4y^2 - 2y + 1)$
53. $(5z + 2)(25z^2 - 10z + 4)$
54. $(a - 3)(a^2 + 3a + 9)$

△ **65.** If the volume of a box is $(x^4 + 6x^3 - 7x^2)$ cubic meters, its height is x^2 meters, and its length is $(x + 7)$ meters, find its width. $(x - 1)$ m

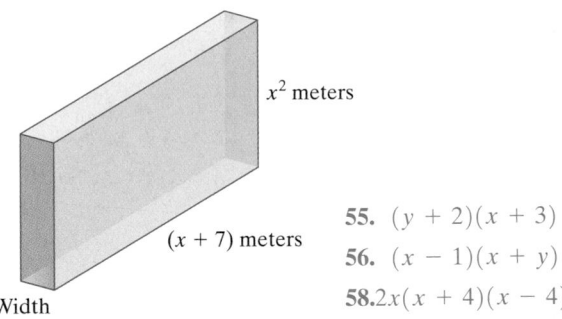

x^2 meters

$(x + 7)$ meters

Width

55. $(y + 2)(x + 3)$
56. $(x - 1)(x + y)$
58. $2x(x + 4)(x - 4)$

6.6   SOLVING EQUATIONS CONTAINING RATIONAL EXPRESSIONS

Objective

1   Solve equations containing rational expressions.

1   In this section, we solve equations containing rational expressions. Before beginning this section, make sure that you understand the difference between an *equation* and an *expression*. An **equation** contains an equal sign and an **expression** does not.

Equation	*Expression*
$\dfrac{x}{2} + \dfrac{x}{6} = \dfrac{2}{3}$	$\dfrac{x}{2} + \dfrac{x}{6}$

Solving Equations Containing Rational Expressions

To solve *equations* containing rational expressions, first clear the equation of fractions by multiplying both sides of the equation by the LCD of all rational expressions. Then solve as usual.

✔ **CONCEPT CHECK**

True or false? Clearing fractions is valid when solving an equation and when simplifying rational expressions. Explain.

EXAMPLE 1

Solve : $\dfrac{4x}{5} + \dfrac{3}{2} = \dfrac{3x}{10}$ algebraically and graphically.

Solution   The LCD of $\dfrac{4x}{5}, \dfrac{3}{2}$, and $\dfrac{3x}{10}$ is 10. We multiply both sides of the equation by 10.

$$\frac{4x}{5} + \frac{3}{2} = \frac{3x}{10}$$

$$10\left(\frac{4x}{5} + \frac{3}{2}\right) = 10\left(\frac{3x}{10}\right) \quad \text{Multiply both sides by the LCD.}$$

$$10 \cdot \frac{4x}{5} + 10 \cdot \frac{3}{2} = 10 \cdot \frac{3x}{10} \quad \text{Use the distributive property.}$$

$$8x + 15 = 3x \quad \text{Simplify.}$$

$$15 = -5x \quad \text{Subtract } 8x \text{ from both sides.}$$

$$-3 = x \quad \text{Solve.}$$

Check: To check this solution graphically, we use the intersection-of-graphs method shown below.

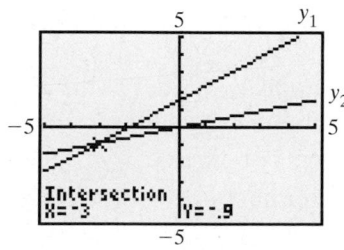

The point of intersection of $y_1 = \dfrac{4x}{5} + \dfrac{3}{2}$ and $y_2 = \dfrac{3x}{10}$ is $(-3, -0.9)$. Thus, the solution of the equation $\dfrac{4x}{5} + \dfrac{3}{2} = \dfrac{3x}{10}$ is $x = -3$.

The solution is -3 or the solution set is $\{-3\}$.

The important difference of the equations in this section is that the denominator of a rational expression may contain a variable. Recall that a rational expression is undefined for values of the variable that make the denominator 0. If a proposed solution makes the denominator 0, then it must be rejected as a solution of the original equation. Such proposed solutions are called **extraneous solutions**.

▱ **EXAMPLE 2** ▱

Solve: $\dfrac{3}{x} - \dfrac{x+21}{3x} = \dfrac{5}{3}$ algebraically and check graphically.

Solution The LCD of the denominators x, $3x$, and 3 is $3x$. We multiply both sides by $3x$.

$$\frac{3}{x} - \frac{x+21}{3x} = \frac{5}{3}$$

$$3x\left(\frac{3}{x} - \frac{x+21}{3x}\right) = 3x\left(\frac{5}{3}\right) \quad \text{Multiply both sides by the LCD.}$$

$$3x\cdot\frac{3}{x} - 3x\cdot\frac{x+21}{3x} = 3x\cdot\frac{5}{3} \quad \text{Use the distributive property.}$$

$$9 - (x+21) = 5x \quad \text{Simplify.}$$

$$9 - x - 21 = 5x$$

$$-12 = 6x$$

$$-2 = x \quad \text{Solve.}$$

The proposed solution is -2.

Check: A check using a graphing utility is shown below.

The intersection of $y_1 = \dfrac{3}{x} - \dfrac{x+21}{3x}$ and $y_2 = \dfrac{5}{3}$ verifies that the solution of the equation $\dfrac{3}{x} - \dfrac{x+21}{3x} = \dfrac{5}{3}$ is $x = -2$.

The solution is -2 or the solution set is $\{-2\}$.

The following steps may be used to algebraically solve equations containing rational expressions.

Solving an Equation Containing Rational Expressions

Step 1: Multiply both sides of the equation by the LCD of all rational expressions in the equation.

Step 2: Simplify both sides.

Step 3: Determine whether the equation is linear, quadratic, or higher degree and solve accordingly.

Step 4: Check the solution in the original equation.

▱ **EXAMPLE 3** ▱

Solve: $\dfrac{x+6}{x-2} = \dfrac{2(x+2)}{x-2}$

Solution First we multiply both sides of the equation by the LCD, $x - 2$.

$$\frac{x + 6}{x - 2} = \frac{2(x + 2)}{x - 2}$$

$$(x - 2) \cdot \frac{x + 6}{x - 2} = (x - 2) \cdot \frac{2(x + 2)}{x - 2} \quad \text{Multiply both sides by } x - 2.$$

$$x + 6 = 2(x + 2) \qquad\qquad \text{Simplify.}$$

$$x + 6 = 2x + 4 \qquad\qquad \text{Use the distributive property.}$$

$$2 = x \qquad\qquad\qquad \text{Solve.}$$

Check: The proposed solution is 2. Notice that 2 makes a denominator 0 in the original equation. This can also be seen in a check. Check the proposed solution 2 in the original equation.

$$\frac{x + 6}{x - 2} = \frac{2(x + 2)}{x - 2}$$

$$\frac{2 + 6}{2 - 2} = \frac{2(2 + 2)}{2 - 2}$$

$$\frac{8}{0} = \frac{2(4)}{0}$$

The denominators are 0, so 2 is not a solution of the original equation. The solution is $\{\ \ \}$ or $\varnothing$.

EXAMPLE 4

Solve: $\dfrac{2x}{2x - 1} + \dfrac{1}{x} = \dfrac{1}{2x - 1}$

Solution The LCD is $x(2x - 1)$. Multiply both sides by $x(2x - 1)$. By the distributive property, this is the same as multiplying each term by $x(2x - 1)$.

$$x(2x - 1) \cdot \frac{2x}{2x - 1} + x(2x - 1) \cdot \frac{1}{x} = x(2x - 1) \cdot \frac{1}{2x - 1}$$

$$x(2x) + (2x - 1) = x \quad \text{Simplify.}$$

$$2x^2 + 2x - 1 - x = 0$$

$$2x^2 + x - 1 = 0$$

$$(x + 1)(2x - 1) = 0$$

$$x + 1 = 0 \quad \text{or} \quad 2x - 1 = 0$$

$$x = -1 \qquad\qquad x = \frac{1}{2}$$

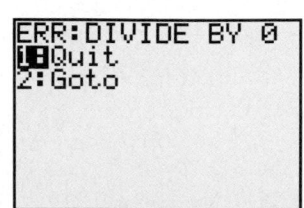

A numerical check that -1 is the solution to the equation in Example 4.

The error message that results when $\frac{1}{2}$ is checked as a proposed solution in Example 4.

The number $\dfrac{1}{2}$ makes the denominator $2x - 1$ equal 0, so it is not a solution. The solution is -1.

EXAMPLE 5

Solve: $\dfrac{2x}{x-3} + \dfrac{6-2x}{x^2-9} = \dfrac{x}{x+3}$

Solution   We factor the second denominator to find that the LCD is $(x+3)(x-3)$. We multiply both sides of the equation by $(x+3)(x-3)$. By the distributive property, this is the same as multiplying each term by $(x+3)(x-3)$.

$$\frac{2x}{x-3} + \frac{6-2x}{x^2-9} = \frac{x}{x+3}$$

$$(x+3)(x-3) \cdot \frac{2x}{x-3} + (x+3)(x-3) \cdot \frac{6-2x}{(x+3)(x-3)}$$

$$= (x+3)(x-3)\left(\frac{x}{x+3}\right)$$

$$2x(x+3) + (6-2x) = x(x-3) \qquad \text{Simplify.}$$

$$2x^2 + 6x + 6 - 2x = x^2 - 3x \qquad \text{Use the distributive property.}$$

Next we solve this quadratic equation by the factoring method. To do so, we first write the equation so that one side is 0.

$$x^2 + 7x + 6 = 0$$

$$(x+6)(x+1) = 0 \qquad \text{Factor.}$$

$$x = -6 \text{ or } x = -1 \qquad \text{Set each factor equal to } 0.$$

Neither -6 nor -1 makes any denominator 0 so they are both solutions. The solutions are -6 and -1.

CLASSROOM EXAMPLE

Solve: $\dfrac{2x}{x-4} + \dfrac{10-5x}{x^2-16} = \dfrac{x}{x+4}$

answer: $-5, -2$

STUDY SKILLS REMINDER

Continue your outline started in Section 1.5. Write how to recognize and how to solve equations containing rational expressions in your own words. For example:

Solving Equations and Inequalities

 I. Equations

     A. Linear equations (Sec. 1.5)

     B. Absolute value equations (Sec. 3.4)

     C. Quadratic Equations (Sec. 5.8)

     D. Equations with Rational Expressions—Recognize: *equation with rational expressions* — Solve: multiply both sides of the equation by the LCD—then simplify to decide whether the equation is linear or quadratic

 II. Inequalities

     A. Linear inequalities (Sec. 3.2)

     B. Compound inequalities (Sec. 3.3)

     C. Absolute value inequalities (Sec. 3.5)

EXAMPLE 6

Solve: $\dfrac{z}{2z^2 + 3z - 2} - \dfrac{1}{2z} = \dfrac{3}{z^2 + 2z}$

Solution   Factor the denominators to find that the LCD is $2z(z + 2)(2z - 1)$. Multiply both sides by the LCD. Remember, by using the distributive property, this is the same as multiplying each term by $2z(z + 2)(2z - 1)$.

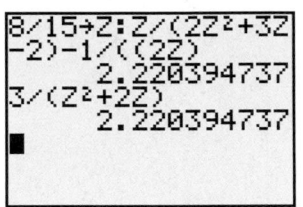

The solution is $z = \dfrac{8}{15}$.

$$\dfrac{z}{2z^2 + 3z - 2} - \dfrac{1}{2z} = \dfrac{3}{z^2 + 2z}$$

$$\dfrac{z}{(2z - 1)(z + 2)} - \dfrac{1}{2z} = \dfrac{3}{z(z + 2)}$$

$$2z(z + 2)(2z - 1) \cdot \dfrac{z}{(2z - 1)(z + 2)} - 2z(z + 2)(2z - 1) \cdot \dfrac{1}{2z}$$

$$= 2z(z + 2)(2z - 1) \cdot \dfrac{3}{z(z + 2)} \quad \text{Apply the distributive property.}$$

$$2z(z) - (z + 2)(2z - 1) = 3 \cdot 2(2z - 1) \qquad \text{Simplify.}$$

$$2z^2 - (2z^2 + 3z - 2) = 12z - 6$$

$$2z^2 - 2z^2 - 3z + 2 = 12z - 6$$

$$-3z + 2 = 12z - 6$$

$$-15z = -8$$

$$z = \dfrac{8}{15} \qquad \text{Solve.}$$

The proposed solution $\dfrac{8}{15}$ does not make any denominator 0; the solution is $\dfrac{8}{15}$.

A graph can be helpful in visualizing solutions of equations. For example, to visualize the solution of the equation $\dfrac{3}{x} - \dfrac{x + 21}{3x} = \dfrac{5}{3}$ in Example 2, the graph of the related rational function $f(x) = \dfrac{3}{x} - \dfrac{x + 21}{3x}$ is shown. A solution of the equation is an x-value that corresponds to a y-value of $\dfrac{5}{3}$.

Notice that an x-value of -2 corresponds to a y-value of $\dfrac{5}{3}$. The solution of the equation is indeed -2 as shown in Example 2.

EXERCISE SET 6.6

STUDY GUIDE/SSM   CD/VIDEO   PH MATH TUTOR CENTER   MathXL®Tutorials ON CD   MathXL®   MyMathLab®

Solve each equation. See Examples 1 and 2.

1. $\dfrac{x}{2} - \dfrac{x}{3} = 12$ 72

2. $x = \dfrac{x}{2} - 4$ -8

3. $\dfrac{x}{3} = \dfrac{1}{6} + \dfrac{x}{4}$ 2

4. $\dfrac{x}{2} = \dfrac{21}{10} - \dfrac{x}{5}$ 3

5. $\dfrac{2}{x} + \dfrac{1}{2} = \dfrac{5}{x}$ 6

6. $\dfrac{5}{3x} + 1 = \dfrac{7}{6}$ 10

7. $\dfrac{x^2 + 1}{x} = \dfrac{5}{x}$ $2, -2$

8. $\dfrac{x^2 - 14}{2x} = -\dfrac{5}{2x}$ $-3, 3$

Solve each equation. See Examples 3 through 6.

9. $\dfrac{x + 5}{x + 3} = \dfrac{2}{x + 3}$ $\varnothing$

10. $\dfrac{x - 7}{x - 1} = \dfrac{11}{x - 1}$ 18

11. $\dfrac{5}{x - 2} - \dfrac{2}{x + 4} = -\dfrac{4}{x^2 + 2x - 8}$ $-\dfrac{28}{3}$

12. $\dfrac{1}{x - 1} + \dfrac{1}{x + 1} = \dfrac{2}{x^2 - 1}$ $\varnothing$

13. $\dfrac{1}{x - 1} = \dfrac{2}{x + 1}$ 3

14. $\dfrac{6}{x + 3} = \dfrac{4}{x - 3}$ 15

15. $\dfrac{x^2 - 23}{2x^2 - 5x - 3} + \dfrac{2}{x - 3} = \dfrac{-1}{2x + 1}$ -8

16. $\dfrac{4x^2 - 24x}{3x^2 - x - 2} + \dfrac{3}{3x + 2} = \dfrac{-4}{x - 1}$ $\dfrac{5}{4}$

17. $\dfrac{1}{x - 4} - \dfrac{3x}{x^2 - 16} = \dfrac{2}{x + 4}$ 3

18. $\dfrac{3}{2x + 3} - \dfrac{1}{2x - 3} = \dfrac{4}{4x^2 - 9}$ 4

19. $\dfrac{1}{x - 4} = \dfrac{8}{x^2 - 16}$ $\varnothing$

20. $\dfrac{2}{x^2 - 4} = \dfrac{1}{2x - 4}$ $\varnothing$

21. $\dfrac{1}{x - 2} - \dfrac{2}{x^2 - 2x} = 1$ 1

22. $\dfrac{12}{3x^2 + 12x} = 1 - \dfrac{1}{x + 4}$ 1

MIXED PRACTICE

Solve each equation.

23. $\dfrac{5}{x} = \dfrac{20}{12}$ 3

24. $\dfrac{2}{x} = \dfrac{10}{5}$ 1

25. $1 - \dfrac{4}{a} = 5$ -1

26. $7 + \dfrac{6}{a} = 5$ -3

27. $\dfrac{x^2 + 5}{x} - 1 = \dfrac{5(x + 1)}{x}$ 6

28. $\dfrac{x^2 + 6}{x} + 5 = \dfrac{2(x + 3)}{x}$ -3

29. $\dfrac{1}{2x} - \dfrac{1}{x + 1} = \dfrac{1}{3x^2 + 3x}$ $\dfrac{1}{3}$

30. $\dfrac{2}{x - 5} + \dfrac{1}{2x} = \dfrac{5}{3x^2 - 15x}$ $\dfrac{5}{3}$

31. $\dfrac{1}{x} - \dfrac{x}{25} = 0$ $-5, 5$

32. $\dfrac{x}{4} + \dfrac{5}{x} = 3$ $10, 2$

33. $5 - \dfrac{2}{2y - 5} = \dfrac{3}{2y - 5}$ 3

34. $1 - \dfrac{5}{y + 7} = \dfrac{4}{y + 7}$ 2

35. $\dfrac{x - 1}{x + 2} = \dfrac{2}{3}$ 7

36. $\dfrac{6x + 7}{2x + 9} = \dfrac{5}{3}$ 3

37. $\dfrac{x + 3}{x + 2} = \dfrac{1}{x + 2}$ $\varnothing$

38. $\dfrac{2x + 1}{4 - x} = \dfrac{9}{4 - x}$ $\varnothing$

39. $\dfrac{1}{a - 3} + \dfrac{2}{a + 3} = \dfrac{1}{a^2 - 9}$ $\dfrac{4}{3}$

40. $\dfrac{12}{9 - a^2} + \dfrac{3}{3 + a} = \dfrac{2}{3 - a}$ $\varnothing$

41. $\dfrac{64}{x^2 - 16} + 1 = \dfrac{2x}{x - 4}$ -12

42. $2 + \dfrac{3}{x} = \dfrac{2x}{x + 3}$ -1

43. $\dfrac{-15}{4y + 1} + 4 = y$ $1, \dfrac{11}{4}$

44. $\dfrac{36}{x^2 - 9} + 1 = \dfrac{2x}{x + 3}$ 9

45. $\dfrac{28}{x^2 - 9} + \dfrac{2x}{x - 3} + \dfrac{6}{x + 3} = 0$ $-5, -1$

46. $\dfrac{x^2 - 20}{x^2 - 7x + 12} = \dfrac{3}{x - 3} + \dfrac{5}{x - 4}$ $1, 7$

47. $\dfrac{x + 2}{x^2 + 7x + 10} = \dfrac{1}{3x + 6} - \dfrac{1}{x + 5}$ $-\dfrac{7}{5}$

48. $\dfrac{3}{2x - 5} + \dfrac{2}{2x + 3} = 0$ $\dfrac{1}{10}$

51. length, 15 in.; width, 10 in.

REVIEW AND PREVIEW

Write each sentence as an equation and solve. See Section 1.6.

49. Four more than 3 times a number is 19.   5

50. The sum of two consecutive integers is 147.   73 and 74

51. The length of a rectangle is 5 inches more than the width. Its perimeter is 50 inches. Find the length and width.

52. The sum of a number and its reciprocal is $\dfrac{5}{2}$. $\dfrac{1}{2}$ and 2

The following graph is from a survey of state and federal prisons. Use this histogram to answer Exercises 53–56.

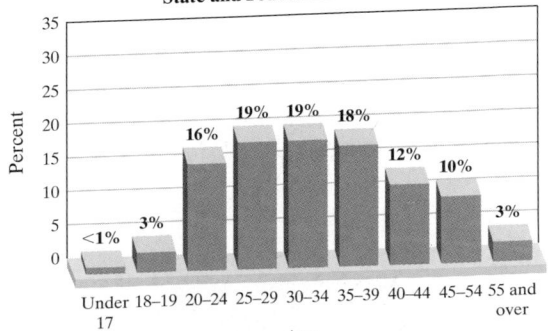

State and Federal Prison Inmates

Source: Bureau of Justice Statistics, U.S. Department of Justice

53. What percent of state and federal prison inmates are age 45 to 54?   10%

54. What percent of state and federal prison inmates are 55 years old or older?   3%

55. What age category shows the highest percent of prison inmates?   25–29 and 30–34

56. What percent of state and federal prison inmates are 20 to 34 years old?   54%

57. At the end of 2001, there were 35,710 inmates under the jurisdiction of state and federal correctional authorities in the state of Louisiana. Approximately how many 25- to 29-year-old inmates would you expect to have been held in Louisiana at the end of 2001? Round to the nearest whole. (*Source:* Bureau of Justice Statistics)   6785 inmates

Concept Extensions

58. The average cost of producing x game disks for a computer is given by the function $f(x) = 3.3 + \dfrac{5400}{x}$. Find the number of game disks that must be produced for the average cost to be $5.10.   3000 game disks

59. The average cost of producing x electric pencil sharpeners is given by the function $f(x) = 20 + \dfrac{4000}{x}$. Find the number of electric pencil sharpeners that must be produced for the average cost to be $25.

800 pencil sharpeners

Solve each equation. Begin by writing each equation with positive exponents only.

60. $x^{-2} - 19x^{-1} + 48 = 0$ $\frac{1}{16}, \frac{1}{3}$ **61.** $x^{-2} - 5x^{-1} - 36 = 0$ $\frac{1}{9}, -\frac{1}{4}$

62. $p^{-2} + 4p^{-1} - 5 = 0$ $-\frac{1}{5}, 1$ **63.** $6p^{-2} - 5p^{-1} + 1 = 0$ $3, 2$

Solve each equation. Round solutions to two decimal places.

64. $\dfrac{1.4}{x - 2.6} = \dfrac{-3.5}{x + 7.1}$ **65.** $\dfrac{-8.5}{x + 1.9} = \dfrac{5.7}{x - 3.6}$

66. $\dfrac{10.6}{y} - 14.7 = \dfrac{9.92}{3.2} + 7.6$ **67.** $\dfrac{12.2}{x} + 17.3 = \dfrac{9.6}{x} - 14.7$

64. -0.17 **65.** 1.39 **66.** 0.42 **67.** -0.08

Solve each equation by substitution.

For example, to solve Exercise 68, first let $u = x - 1$. After substituting, we have $u^2 + 3u + 2 = 0$. Solve for u and then substitute back to solve for x.

68. $(x - 1)^2 + 3(x - 1) + 2 = 0$ $-1, 0$

69. $(4 - x)^2 - 5(4 - x) + 6 = 0$ $1, 2$

70. $\left(\dfrac{3}{x - 1}\right)^2 + 2\left(\dfrac{3}{x - 1}\right) + 1 = 0$ -2

71. $\left(\dfrac{5}{2 + x}\right)^2 + \left(\dfrac{5}{2 + x}\right) - 20 = 0$ $-3, -\dfrac{3}{4}$

Use a graphing calculator to verify the solution of each given exercise.   **72.–75.** See graphing answer section.

72. Exercise 23   **73.** Exercise 24

74. Exercise 35   **75.** Exercise 36

EXPRESSIONS AND EQUATIONS CONTAINING RATIONAL EXPRESSIONS

INTEGRATED REVIEW

It is very important that you understand the difference between an expression and an equation containing rational expressions. An equation contains an equal sign; an expression does not.

Expression to be Simplified	**Equation to be Solved**
$\dfrac{x}{2} + \dfrac{x}{6}$	$\dfrac{x}{2} + \dfrac{x}{6} = \dfrac{2}{3}$

Write both rational expressions with the LCD, 6, as the denominator.

Multiply both sides by the LCD, 6.

$$\dfrac{x}{2} + \dfrac{x}{6} = \dfrac{x \cdot 3}{2 \cdot 3} + \dfrac{x}{6}$$

$$6\left(\dfrac{1}{2} + \dfrac{x}{6}\right) = 6\left(\dfrac{2}{3}\right)$$

$$= \dfrac{3x}{6} + \dfrac{x}{6}$$

$$3 + x = 4$$

$$= \dfrac{4x}{6} = \dfrac{2x}{3}$$

$$x = 1$$

Check to see that the solution is 1.

> ► **Helpful Hint**
> Remember: Equations can be cleared of fractions; expressions cannot.

Perform each indicated operation and simplify, or solve the equation for the variable.

1. $\dfrac{x}{2} = \dfrac{1}{8} + \dfrac{x}{4}$

$\dfrac{1}{2}$

2. $\dfrac{x}{4} = \dfrac{3}{2} + \dfrac{x}{10}$

10

3. $\dfrac{1}{8} + \dfrac{x}{4}$

$\dfrac{1 + 2x}{8}$

4. $\dfrac{3}{2} + \dfrac{x}{10}$

$\dfrac{15 + x}{10}$

5. $\dfrac{4}{x + 2} - \dfrac{2}{x - 1}$

$\dfrac{2(x - 4)}{(x + 2)(x - 1)}$

6. $\dfrac{5}{x - 2} - \dfrac{10}{x + 4}$

$-\dfrac{5(x - 8)}{(x - 2)(x + 4)}$

7. $\dfrac{4}{x + 2} = \dfrac{2}{x - 1}$

4

8. $\dfrac{5}{x - 2} = \dfrac{10}{x + 4}$

8

9. $\dfrac{2}{x^2 - 4} = \dfrac{1}{x + 2} - \dfrac{3}{x - 2}$

-5

10. $\dfrac{3}{x^2 - 25} = \dfrac{1}{x + 5} + \dfrac{2}{x - 5}$

$-\dfrac{2}{3}$

11. $\dfrac{5}{x^2 - 3x} + \dfrac{4}{2x - 6}$

$\dfrac{2x + 5}{x(x - 3)}$

12. $\dfrac{5}{x^2 - 3x} \div \dfrac{4}{2x - 6}$

$\dfrac{5}{2x}$

13. $\dfrac{x - 1}{x + 1} + \dfrac{x + 7}{x - 1} = \dfrac{4}{x^2 - 1}$

-2

14. $\left(1 - \dfrac{y}{x}\right) \div \left(1 - \dfrac{x}{y}\right)$

$-\dfrac{y}{x}$

15. $\dfrac{a^2 - 9}{a - 6} \cdot \dfrac{a^2 - 5a - 6}{a^2 - a - 6}$

$\dfrac{(a + 3)(a + 1)}{a + 2}$

16. $\dfrac{2}{a - 6} + \dfrac{3a}{a^2 - 5a - 6} - \dfrac{a}{5a + 5}$

$\dfrac{-a^2 + 31a + 10}{5(a - 6)(a + 1)}$

17. $\dfrac{2x + 3}{3x - 2} = \dfrac{4x + 1}{6x + 1}$

$-\dfrac{1}{5}$

18. $\dfrac{5x - 3}{2x} = \dfrac{10x + 3}{4x + 1}$

$-\dfrac{3}{13}$

19. $\dfrac{a}{9a^2 - 1} + \dfrac{2}{6a - 2}$

$\dfrac{4a + 1}{(3a + 1)(3a - 1)}$

20. $\dfrac{3}{4a - 8} - \dfrac{a + 2}{a^2 - 2a}$

$\dfrac{-a - 8}{4a(a - 2)}$ or $-\dfrac{a + 8}{4a(a - 2)}$

21. $-\dfrac{3}{x^2} - \dfrac{1}{x} + 2 = 0$

$-1, \dfrac{3}{2}$

22. $\dfrac{x}{2x + 6} + \dfrac{5}{x^2 - 9}$

$\dfrac{x^2 - 3x + 10}{2(x + 3)(x - 3)}$

23. $\dfrac{x - 8}{x^2 - x - 2} + \dfrac{2}{x - 2}$

$\dfrac{3}{x + 1}$

24. $\dfrac{x - 8}{x^2 - x - 2} + \dfrac{2}{x - 2} = \dfrac{3}{x + 1}$

$\{x \mid x \text{ is a real number and } x \neq 2, x \neq -1\}$

25. $\dfrac{3}{a} - 5 = \dfrac{7}{a} - 1$ **26.** $\dfrac{7}{3z - 9} + \dfrac{5}{z}$

-1

$\dfrac{22z - 45}{3z(z - 3)}$

Use $\dfrac{x}{5} - \dfrac{x}{4} = \dfrac{1}{10}$ *and* $\dfrac{x}{5} - \dfrac{x}{4} + \dfrac{1}{10}$ *for Exercises 27 and 28.*

27. a. Which one above is an expression?

$\dfrac{x}{5} - \dfrac{x}{4} + \dfrac{1}{10}$

b. Describe the first step to simplify this expression.

Write each rational expression term so that the denominator is the LCD, 20.

c. Simplify the expression.

$\dfrac{-x + 2}{20}$

28. a. Which one above is an equation?

$\dfrac{x}{5} - \dfrac{x}{4} = \dfrac{1}{10}$

b. Describe the first step to solve this equation.

Clear the equation of fractions by multiplying each term by the LCD, 20.

c. Solve the equation.

-2

For each exercise, choose the correct statement.* Each figure represents a real number and no denominators are 0.

29. a. $\dfrac{\triangle + \square}{\triangle} = \square$

b. $\dfrac{\triangle + \square}{\triangle} = 1 + \dfrac{\square}{\triangle}$ b

c. $\dfrac{\triangle + \square}{\triangle} = \dfrac{\square}{\triangle}$

d. $\dfrac{\triangle + \square}{\triangle} = 1 + \square$

e. $\dfrac{\triangle + \square}{\triangle - \square} = -1$

30. a. $\dfrac{\triangle}{\square} + \dfrac{\square}{\triangle} = \dfrac{\triangle + \square}{\square + \triangle} = 1$

b. $\dfrac{\triangle}{\square} + \dfrac{\square}{\triangle} = \dfrac{\triangle + \square}{\triangle\square}$ d

c. $\dfrac{\triangle}{\square} + \dfrac{\square}{\triangle} = \triangle\triangle + \square\square$

d. $\dfrac{\triangle}{\square} + \dfrac{\square}{\triangle} = \dfrac{\triangle\triangle + \square\square}{\square\triangle}$

e. $\dfrac{\triangle}{\square} + \dfrac{\square}{\triangle} = \dfrac{\triangle\square}{\square\triangle} = 1$

31. a. $\dfrac{\triangle}{\square} \cdot \dfrac{\bigcirc}{\square} = \dfrac{\triangle\bigcirc}{\square}$

b. $\dfrac{\triangle}{\square} \cdot \dfrac{\bigcirc}{\square} = \triangle\bigcirc$ d

c. $\dfrac{\triangle}{\square} \cdot \dfrac{\bigcirc}{\square} = \dfrac{\triangle + \bigcirc}{\square + \square}$

d. $\dfrac{\triangle}{\square} \cdot \dfrac{\bigcirc}{\square} = \dfrac{\triangle\bigcirc}{\square\square}$

32. a. $\dfrac{\triangle}{\square} \div \dfrac{\bigcirc}{\triangle} = \dfrac{\triangle\triangle}{\square\bigcirc}$

b. $\dfrac{\triangle}{\square} \div \dfrac{\bigcirc}{\triangle} = \dfrac{\bigcirc\square}{\triangle\triangle}$ a

c. $\dfrac{\triangle}{\square} \div \dfrac{\bigcirc}{\triangle} = \dfrac{\bigcirc}{\square}$

d. $\dfrac{\triangle}{\square} \div \dfrac{\bigcirc}{\triangle} = \dfrac{\triangle + \triangle}{\square + \bigcirc}$

33. a. $\dfrac{\dfrac{\triangle + \square}{\bigcirc}}{\dfrac{\triangle}{\bigcirc}} = \square$

b. $\dfrac{\dfrac{\triangle + \square}{\bigcirc}}{\dfrac{\triangle}{\bigcirc}} = \dfrac{\triangle\triangle + \triangle\square}{\bigcirc\bigcirc}$

c. $\dfrac{\dfrac{\triangle + \square}{\bigcirc}}{\dfrac{\triangle}{\bigcirc}} = 1 + \square$ d

d. $\dfrac{\dfrac{\triangle + \square}{\bigcirc}}{\dfrac{\triangle}{\bigcirc}} = \dfrac{\triangle + \square}{\triangle}$

*My thanks to Kelly Champagne for permission to use her exercises for 29 through 33.

6.7   RATIONAL EQUATIONS AND PROBLEM SOLVING

Objectives

1   Solve an equation containing rational expressions for a specified variable.

2   Solve problems by writing equations containing rational expressions.

1   In Section 1.8 we solved equations for a specified variable. In this section, we continue practicing this skill by solving equations containing rational expressions for a specified variable. The steps given in Section 1.8 for solving equations for a specified variable are repeated here.

Solving Equations for a Specified Variable

Step 1: Clear the equation of fractions or rational expressions by multiplying each side of the equation by the least common denominator (LCD) of all denominators in the equation.

Step 2: Use the distributive property to remove grouping symbols such as parentheses.

Step 3: Combine like terms on each side of the equation.

Step 4: Use the addition property of equality to rewrite the equation as an equivalent equation with terms containing the specified variable on one side and all other terms on the other side.

Step 5: Use the distributive property and the multiplication property of equality to get the specified variable alone.

EXAMPLE 1

Solve $\dfrac{1}{x} + \dfrac{1}{y} = \dfrac{1}{z}$ for x.

Solution   To clear this equation of fractions, we multiply both sides of the equation by xyz, the LCD of $\dfrac{1}{x}, \dfrac{1}{y}$, and $\dfrac{1}{z}$.

$$\frac{1}{x} + \frac{1}{y} = \frac{1}{z}$$

$$xyz\left(\frac{1}{x} + \frac{1}{y}\right) = xyz\left(\frac{1}{z}\right) \quad \text{Multiply both sides by } xyz.$$

$$xyz\left(\frac{1}{x}\right) + xyz\left(\frac{1}{y}\right) = xyz\left(\frac{1}{z}\right) \quad \text{Use the distributive property.}$$

$$yz + xz = xy \quad \text{Simplify.}$$

Notice the two terms that contain the specified variable x.

CLASSROOM EXAMPLE

Solve $\dfrac{1}{x} + \dfrac{1}{y} = \dfrac{1}{z}$ for y.

answer: $y = \dfrac{xz}{x - z}$

Next, we subtract xz from both sides so that all terms containing the specified variable x are on one side of the equation and all other terms are on the other side.

$$yz = xy - xz$$

Now we use the distributive property to factor x from $xy - xz$ and then the multiplication property of equality to solve for x.

$$yz = x(y - z)$$

$$\frac{yz}{y - z} = x \quad \text{or} \quad x = \frac{yz}{y - z} \qquad \text{Divide both sides by } y - z.$$

2   Problem solving sometimes involves modeling a described situation with an equation containing rational expressions. In Examples 2 through 5, we practice solving such problems and use the problem-solving steps first introduced in Section 2.2.

EXAMPLE 2

FINDING AN UNKNOWN NUMBER

If a certain number is subtracted from the numerator and added to the denominator of $\frac{9}{19}$, the new fraction is equivalent to $\frac{1}{3}$. Find the number.

Solution   **1.** UNDERSTAND the problem.   Read and reread the problem and try guessing the solution. For example, if the unknown number is 3, we have

$$\frac{9 - 3}{19 + 3} = \frac{1}{3}$$

To see if this is a true statement, we simplify the fraction on the left side.

$$\frac{6}{22} = \frac{1}{3} \quad \text{or} \quad \frac{3}{11} = \frac{1}{3} \quad \text{False}$$

Since this is not a true statement, 3 is not the correct number. Remember that the purpose of this step is not to guess the correct solution but to gain an understanding of the problem posed.

We will let n = the number to be subtracted from the numerator and added to the denominator.

2. TRANSLATE the problem.

In words:

when the number is subtracted from the numerator and added to the denominator of the fraction $\frac{9}{19}$	this is equivalent to	$\frac{1}{3}$
↓	↓	↓

Translate:

$$\frac{9 - n}{19 + n} \qquad = \qquad \frac{1}{3}$$

3. SOLVE the equation for n.

$$\frac{9 - n}{19 + n} = \frac{1}{3}$$

To solve for n, we begin by multiplying both sides by the LCD of $3(19 + n)$.

$$3(19 + n) \cdot \frac{9 - n}{19 + n} = 3(19 + n) \cdot \frac{1}{3} \qquad \text{Multiply both sides by the LCD.}$$
$$3(9 - n) = 19 + n \qquad \text{Simplify.}$$
$$27 - 3n = 19 + n$$
$$8 = 4n$$
$$2 = n \qquad \text{Solve.}$$

4. INTERPRET the results.

Check: If we subtract 2 from the numerator and add 2 to the denominator of $\frac{9}{19}$, we

have $\frac{9 - 2}{19 + 2} = \frac{7}{21} = \frac{1}{3}$, and the problem checks.

State: The unknown number is 2.

A **ratio** is the quotient of two number or two quantities. Since rational expressions are quotients of quantities, rational expressions are ratios, also. A **proportion** is a mathematical statement that two ratios are equal.

EXAMPLE 3

CALCULATING HOMES HEAT BY ELECTRICITY

In the United States, 7 out of every 25 homes are heated by electricity. At this rate, how many homes in a community of 36,000 homes would you predict are heated by electricity? (*Source:* 2000 Census Survey)

Solution

1. UNDERSTAND. Read and reread the problem. Try to estimate a reasonable solution.

For example, since 7 is less than $\frac{1}{3}$ of 25, we might reason that the solution

would be less than $\frac{1}{3}$ of 36,000 or 12,000.

Let's let x = number of homes in the community heated by electricity.

2. TRANSLATE.

homes heated by electricity → $\dfrac{7}{25} = \dfrac{x}{36,000}$ ← homes heated by electricity

total homes → $\phantom{\dfrac{7}{25} = \dfrac{x}{36,000}}$ ← total homes

3. SOLVE. To solve this proportion we can multiply both sides by the LCD, 36,000, or we can set cross products equal. We will set cross products equal.

$$\frac{7}{25} = \frac{x}{36,000}$$

$$25x = 7 \cdot 36,000$$
$$x = \frac{252,000}{25}$$
$$x = 10,080$$

4. INTERPRET.

Check: To check, replace x with 10,080 in the proportion and see that a true statement results. Notice that our answer is reasonable since it is less than 12,000 as we stated above.

State: We predict that 10,080 homes are heated by electricity.

The following work example leads to an equation containing rational expressions.

EXAMPLE 4

CALCULATING WORK HOURS

Melissa Scarlatti can clean the house in 4 hours, whereas her husband, Zack, can do the same job in 5 hours. They have agreed to clean together so that they can finish in time to watch a movie on TV that starts in 2 hours. How long will it take them to clean the house together? Can they finish before the movie starts?

Solution

1. UNDERSTAND. Read and reread the problem. The key idea here is the relationship between the *time* (in hours) it takes to complete the job and the *part of the job* completed in 1 unit of time (1 hour). For example, if the *time* it takes Melissa to complete the job is 4 hours, the part of the job she can complete in 1 hour is $\frac{1}{4}$. Similarly, Zack can complete $\frac{1}{5}$ of the job in 1 hour.

We will let $t = $ *the time* in hours it takes Melissa and Zack to clean the house together. Then $\frac{1}{t}$ represents the *part of the job* they complete in 1 hour. We summarize the given information in a chart.

	Hours to Complete the Job	*Part of Job Completed in 1 Hour*
MELISSA ALONE	4	$\frac{1}{4}$
ZACK ALONE	5	$\frac{1}{5}$
TOGETHER	t	$\frac{1}{t}$

2. TRANSLATE.

In words:	part of job Melissa can complete in 1 hour	added to	part of job Zack can complete in 1 hour	is equal to	part of job they can complete together in 1 hour
	↓	↓	↓	↓	↓
Translate:	$\frac{1}{4}$	$+$	$\frac{1}{5}$	$=$	$\frac{1}{t}$

$y_2 = \dfrac{1}{x}$

$y_1 = \dfrac{1}{4} + \dfrac{1}{5}$

The solution of $\dfrac{1}{4} + \dfrac{1}{5} = \dfrac{1}{t}$ is $2.\overline{2}$ or $2\frac{2}{9}$.

3. SOLVE.

$$\frac{1}{4} + \frac{1}{5} = \frac{1}{t}$$

$$20t\left(\frac{1}{4} + \frac{1}{5}\right) = 20t\left(\frac{1}{t}\right) \qquad \text{Multiply both sides by the LCD, } 20t.$$

$$5t + 4t = 20$$

$$9t = 20$$

$$t = \frac{20}{9} \quad \text{or} \quad 2\frac{2}{9} \quad \text{Solve.}$$

4. INTERPRET.

Check: The proposed solution is $2\frac{2}{9}$. That is, Melissa and Zack would take $2\frac{2}{9}$ hours to clean the house together. This proposed solution is reasonable since $2\frac{2}{9}$ hours is more than half of Melissa's time and less than half of Zack's time. Check this solution in the originally stated problem.

State: Melissa and Zack can clean the house together in $2\frac{2}{9}$ hours. They cannot complete the job before the movie starts.

⊂⊃

EXAMPLE 5

FINDING THE SPEED OF A CURRENT

Steve Deitmer takes $1\frac{1}{2}$ times as long to go 72 miles upstream in his boat as he does to return. If the boat cruises at 30 mph in still water, what is the speed of the current?

Solution

1. UNDERSTAND. Read and reread the problem. Guess a solution. Suppose that the current is 4 mph. The speed of the boat upstream is slowed down by the current: $30 - 4$, or 26 mph, and the speed of the boat downstream is speeded up by the current: $30 + 4$, or 34 mph. Next let's find out how long it takes to travel 72 miles upstream and 72 miles downstream. To do so, we use the formula $d = rt$, or $\dfrac{d}{r} = t$.

Upstream	*Downstream*
$\dfrac{d}{r} = t$	$\dfrac{d}{r} = t$
$\dfrac{72}{26} = t$	$\dfrac{72}{34} = t$
$2\dfrac{10}{13} = t$	$2\dfrac{2}{17} = t$

Since the time upstream $\left(2\frac{10}{13}\text{hours}\right)$ is not $1\frac{1}{2}$ times the time downstream $\left(2\frac{2}{17}\text{hours}\right)$, our guess is not correct. We do, however, have a better understanding of the problem.

We will let

$$x = \text{the speed of the current}$$
$$30 + x = \text{the speed of the boat downstream}$$
$$30 - x = \text{the speed of the boat upstream}$$

This information is summarized in the following chart, where we use the formula $\frac{d}{r} = t$.

	Distance	*Rate*	*Time* $\left(\frac{d}{r}\right)$
UPSTREAM	72	$30 - x$	$\dfrac{72}{30 - x}$
DOWNSTREAM	72	$30 + x$	$\dfrac{72}{30 + x}$

2. TRANSLATE.   Since the time spent traveling upstream is $1\frac{1}{2}$ times the time spent traveling downstream, we have

In words:

time upstream	is	$1\frac{1}{2}$	times	time downstream
↓	↓	↓	↓	↓

Translate: $\dfrac{72}{30 - x}$ $=$ $\dfrac{3}{2}$ $\cdot$ $\dfrac{72}{30 + x}$

3. SOLVE. $\dfrac{72}{30 - x} = \dfrac{3}{2} \cdot \dfrac{72}{30 + x}$

First we multiply both sides by the LCD, $2(30 + x)(30 - x)$.

$$2(30 + x)(30 - x) \cdot \frac{72}{30 - x} = 2(30 + x)(30 - x)\left(\frac{3}{2} \cdot \frac{72}{30 + x}\right)$$

$$72 \cdot 2(30 + x) = 3 \cdot 72 \cdot (30 - x) \quad \text{Simplify.}$$
$$2(30 + x) = 3(30 - x) \quad \text{Divide both sides by 72.}$$
$$60 + 2x = 90 - 3x \quad \text{Use the distributive property.}$$
$$5x = 30$$
$$x = 6 \quad \text{Solve.}$$

4. INTERPRET.

Check: Check the proposed solution of 6 mph in the originally stated problem.

State: The current's speed is 6 mph.

Spotlight on

DECISION ⤵ MAKING

Suppose you are an aviation safety inspector. You are testing the accuracy of the radioaltimeter on an airplane. To be acceptable, the altitude reading given by the altimeter must be within 3% of the actual altitude. You know that you can check the altimeter's altitude reading with the equation $t = \dfrac{2a}{c}$, where t is the time it takes a radar pulse aimed downward from the airplane to bounce off Earth's surface and return to the radioaltimeter, a is the altitude, and c is the speed of light, 3×10^8 meters per second.

During your test, you find that it takes 5×10^{-5} second for the radar pulse emitted by the altimeter to be returned. The altimeter reads an altitude of 7420 meters. Is the altimeter reading acceptable? Explain.

EXERCISE SET 6.7

| STUDY GUIDE/SSM | CD/ VIDEO | PH MATH TUTOR CENTER | MathXL®Tutorials ON CD | MathXL® | MyMathLab® |

Solve each equation for the specified variable. See Example 1.

1. $F = \dfrac{9}{5}C + 32$ for C (Meteorology) $C = \dfrac{5}{9}(F - 32)$

2. $V = \dfrac{1}{3}\pi r^2 h$ for h (Volume) $h = \dfrac{3V}{\pi r^2}$

3. $Q = \dfrac{A - I}{L}$ for I (Finance) $I = A - QL$

4. $P = 1 - \dfrac{C}{S}$ for S (Finance) $S = \dfrac{C}{1 - P}$

5. $\dfrac{1}{R} = \dfrac{1}{R_1} + \dfrac{1}{R_2}$ for R (Electronics) $R = \dfrac{R_1 R_2}{R_1 + R_2}$

6. $\dfrac{1}{R} = \dfrac{1}{R_1} + \dfrac{1}{R_2}$ for R_1 (Electronics) $R_1 = \dfrac{RR_2}{R_2 - R}$

7. $S = \dfrac{n(a + L)}{2}$ for n (Sequences) $n = \dfrac{2S}{a + L}$

8. $S = \dfrac{n(a + L)}{2}$ for a (Sequences) $a = \dfrac{2S - nL}{n}$

9. $A = \dfrac{h(a + b)}{2}$ for b (Geometry) $b = \dfrac{2A - ah}{h}$

10. $A = \dfrac{h(a + b)}{2}$ for h (Geometry) $h = \dfrac{2A}{a + b}$

11. $\dfrac{P_1 V_1}{T_1} = \dfrac{P_2 V_2}{T_2}$ for T_2 (Chemistry) $T_2 = \dfrac{P_2 V_2 T_1}{P_1 V_1}$

12. $H = \dfrac{kA(T_1 - T_2)}{L}$ for T_2 (Physics) $T_2 = \dfrac{kAT_1 - LH}{kA}$

13. $f = \dfrac{f_1 f_2}{f_1 + f_2}$ for f_2 (Optics) $f_2 = \dfrac{f_1 f}{f_1 - f}$

14. $I = \dfrac{E}{R + r}$ for r (Electronics) $r = \dfrac{E - IR}{I}$

15. $\lambda = \dfrac{2L}{n}$ for L (Physics) $L = \dfrac{n\lambda}{2}$

16. $S = \dfrac{a_1 - a_n r}{1 - r}$ for a_1 (Sequences) $a_1 = S(1 - r) + a_n r$

17. $\dfrac{\theta}{\omega} = \dfrac{2L}{c}$ for c $c = \dfrac{2L\omega}{\theta}$

18. $F = \dfrac{-GMm}{r^2}$ for M (Physics) $M = -\dfrac{Fr^2}{Gm}$

Solve. See Example 2.

19. The sum of a number and 5 times its reciprocal is 6. Find the number(s).   1 and 5

20. The quotient of a number and 9 times its reciprocal is 1. Find the number(s).   −3 and 3

21. If a number is added to the numerator of $\frac{12}{41}$ and twice the number is added to the denominator of $\frac{12}{41}$, the resulting fraction is equivalent to $\frac{1}{3}$. Find the number.   5

22. If a number is subtracted from the numerator of $\frac{13}{8}$ and added to the denominator of $\frac{13}{8}$, the resulting fraction is equivalent to $\frac{2}{5}$. Find the number.   7

Solve. See Example 3.

23. An Arabian camel can drink 15 gallons of water in 10 minutes. At this rate, how much water can the camel drink in 3 minutes? (*Source: Grolier, Inc.*)   4.5 gal

24. An Arabian camel can travel 20 miles in 8 hours, carrying a 300-pound load on its back. At this rate, how far can the camel travel in 10 hours? (*Source: Grolier, Inc.*)   25 mi

25. In 2000, 10.2 out of every 100 Coast Guard personnel were women. If there are 35,712 total Coast Guard personnel on active duty, estimate the number of women. Round to the nearest whole. (*Source: The World Almanac*, 2003)   3643 women

26. In 2000, 43 out of every 50 Navy personnel were men. If there are 375,618 total Navy personnel on active duty, estimate the number of men. Round to the nearest whole. (*Source: The World Almanac*, 2003)   323,031 men

Solve. See Example 4.

27. An experienced roofer can roof a house in 26 hours. A beginning roofer needs 39 hours to complete the same job. Find how long it takes for the two to do the job together.   15.6 hr

28. Alan Cantrell can word process a research paper in 6 hours. With Steve Isaac's help, the paper can be processed in 4 hours. Find how long it takes Steve to word process the paper alone.   12 hr

29. Three postal workers can sort a stack of mail in 20 minutes, 30 minutes, and 60 minutes, respectively. Find how long it takes them to sort the mail if all three work together.   10 min

30. A new printing press can print newspapers twice as fast as the old one can. The old one can print the afternoon edition in 4 hours. Find how long it takes to print the afternoon edition if both printers are operating. $1\frac{1}{3}$ hr

Solve. See Example 5.

31. Mattie Evans drove 150 miles in the same amount of time that it took a turbopropeller plane to travel 600 miles. The speed of the plane was 150 mph faster than the speed of the car. Find the speed of the plane.   200 mph

32. An F-100 plane and a Toyota truck leave the same town at sunrise and head for a town 450 miles away. The speed of the plane is three times the speed of the truck, and the plane arrives 6 hours ahead of the truck. Find the speed of the truck.   50 mph

33. The speed of Lazy River's current is 5 mph. If a boat travels 20 miles downstream in the same time that it takes to travel 10 miles upstream, find the speed of the boat in still water.   15 mph

34. The speed of a boat in still water is 24 mph. If the boat travels 54 miles upstream in the same time that it takes to travel 90 miles downstream, find the speed of the current.   6 mph

MIXED PRACTICE

Solve.

35. The sum of the reciprocals of two consecutive integers is $-\frac{15}{56}$. Find the two integers.   −8 and −7

36. The sum of the reciprocals of two consecutive odd integers is $\frac{20}{99}$. Find the two integers.   9 and 11

37. One hose can fill a goldfish pond in 45 minutes, and two hoses can fill the same pond in 20 minutes. Find how long it takes the second hose alone to fill the pond.   36 min

38. $3\frac{1}{3}$ hr

38. If Sarah Clark can do a job in 5 hours and Dick Belli and Sarah working together can do the same job in 2 hours, find how long it takes Dick to do the job alone.

39. Two trains going in opposite directions leave at the same time. One train travels 15 mph faster than the other. In 6 hours the trains are 630 miles apart. Find the speed of each.   45 mph; 60 mph

40. The speed of a bicyclist is 10 mph faster than the speed of a walker. If the bicyclist travels 26 miles in the same amount of

time that the walker travels 6 miles, find the speed of the bicyclist.   13 mph

41. A giant tortoise can travel 0.17 miles in 1 hour. At this rate, how long would it take the tortoise to travel 1 mile? Round to the nearest tenth of an hour. (*Source: The World Almanac*, 2003)   5.9 hr

42. A black mamba snake can travel 88 feet in 3 seconds. At this rate, how long does it take to travel 300 feet (the length of a football field)? Round to the nearest tenth of a second. (*Source: The World Almanac*, 2003)   10.2 sec

43. Moo Dairy has three machines to fill half-gallon milk cartons. The machines can fill the daily quota in 5 hours, 6 hours, and 7.5 hours, respectively. Find how long it takes to fill the daily quota if all three machines are running.   2 hr

44. The inlet pipe of an oil tank can fill the tank in 1 hour, 30 minutes. The outlet pipe can empty the tank in 1 hour. Find how long it takes to empty a full tank if both pipes are open.   3 hr

45. A plane flies 465 miles with the wind and 345 miles against the wind in the same length of time. If the speed of the wind is 20 mph, find the speed of the plane in still air.   135 mph

46. Two rockets are launched. The first travels at 9000 mph. Fifteen minutes later the second is launched at 10,000 mph. Find the distance at which both rockets are an equal distance from Earth.   22,500 mi

47. Two joggers, one averaging 8 mph and one averaging 6 mph, start from a designated initial point. The slower jogger arrives at the end of the run a half-hour after the other jogger. Find the distance of the run.   12 mi

48. A semi truck travels 300 miles through the flatland in the same amount of time that it travels 180 miles through the Great Smoky Mountains. The rate of the truck is 20 miles per hour slower in the mountains than in the flatland. Find both the flatland rate and mountain rate.   flatland rate: 50 mph; mountain rate: 30 mph

49. Smith Engineering is in the process of reviewing the salaries of their surveyors. During this review, the company found that an experienced surveyor can survey a roadbed in 4 hours. An apprentice surveyor needs 5 hours to survey the same stretch of road. If the two work together, find how long it takes them to complete the job.

50. The numerator of a fraction is 4 less than the denominator. If both the numerator and the denominator are increased by 2, the resulting fraction is equivalent to $\frac{2}{3}$. Find the fraction.

51. The denominator of a fraction is 1 more than the numerator. If both the numerator and the denominator are decreased by 3, the resulting fraction is equivalent to $\frac{4}{5}$. Find the fraction.

52. Cyclist Lance Armstrong of the United States won the 2003 Tour de France. This inspired an amateur cyclist to train for a local road race. He rode the first 20-mile portion of his workout at a constant rate. For the 16-mile cooldown portion of his workout, he reduced his speed by 2 miles per hour. Each portion of the workout took equal time. Find the cyclist's rate during the first portion and his rate during the cooldown portion.   10 mph; 8 mph

53. In 2 minutes, a conveyor belt can move 300 pounds of recyclable aluminum from the delivery truck to a storage area. A smaller belt can move the same quantity of cans the same distance in 6 minutes. If both belts are used, find how long it takes to move the cans to the storage area.

54. Gary Marcus and Tony Alva work at Lombardo's Pipe and Concrete. Mr. Lombardo is preparing an estimate for a customer. He knows that Gary can lay a slab of concrete in 6 hours. Tony can lay the same size slab in 4 hours. If both work on the job and the cost of labor is $45.00 per hour, determine what the labor estimate should be. $108

55. While road testing a new make of car, the editor of a consumer magazine finds that she can go 10 miles into a 3-mile-per-hour wind in the same amount of time that she can go 11 miles with a 3-mile-per-hour wind behind her. Find the speed of the car in still air.   63 mph

56. Mr. Dodson can paint his house by himself in four days. His son will need an additional day to complete the job if he works by himself. If they work together, find how long it takes to paint the house.

57. The world record for the largest white bass caught is held by Ronald Sprouse of Virginia. The bass weighed 6 pounds 13 ounces. If Ronald rows to his favorite fishing spot 9 miles downstream in the same amount of time that he rows 3 miles upstream and if the current is 6 mph, find how long it takes him to cover the 12 miles.   1 hr

58. A marketing manager travels 1080 miles in a corporate jet and then an additional 240 miles by car. If the car ride takes 1 hour longer, and if the rate of the jet is 6 times the rate of the car, find the time the manager travels by jet and find the time she travels by car.   by jet: 3 hr; by car: 4 hr

59. An experienced bricklayer can construct a small wall in 3 hours. An apprentice can complete the job in 6 hours. Find how long it takes if they work together.   2 hr

49. $2\frac{2}{9}$ hr   **50.** $\frac{6}{10}$ **51.** $\frac{7}{8}$ **53.** $1\frac{1}{2}$ min   **56.** $2\frac{2}{9}$ days

REVIEW AND PREVIEW

Solve each equation for x. See Section 1.5.

60. $\dfrac{x}{5} = \dfrac{x+2}{3}$ -5

61. $\dfrac{x}{4} = \dfrac{x+3}{6}$ 6

62. $\dfrac{x-3}{2} = \dfrac{x-5}{6}$ 2

63. $\dfrac{x-6}{4} = \dfrac{x-2}{5}$ 22

Concept Extensions

Calculating body-mass index (BMI) is a way to gauge whether a person should lose weight. Doctors recommend that body-mass index values fall between 19 and 25. The formula for body-mass index B is $B = \dfrac{705w}{h^2}$, where w is weight in pounds and h is height in inches. Use this formula to answer Exercises 64 and 65.

64. A patient is 5 ft 8 in. tall. What should his or her weight be to have a body-mass index of 25? Round to the nearest whole pound.   164 lb

65. A doctor recorded a body-mass index of 47 on a patient's chart. Later, a nurse notices that the doctor recorded the patient's weight as 240 pounds but neglected to record the patient's height. Explain how the nurse can use the information from the chart to find the patient's height. Then find the height.   answers may vary; 60 in. or 5 ft

In physics, when the source of a sound is traveling toward an observer, the relationship between the actual pitch a of the sound and the pitch h that the observer hears due to the Doppler effect is described by the formula $h = \dfrac{a}{1 - \dfrac{s}{770}}$, where s is the speed of the sound source in miles per hour. Use this formula to answer Exercise 66.

66. An emergency vehicle has a single-tone siren with the pitch of the musical note E. As it approaches an observer standing by the road, the vehicle is traveling 50 mph. Is the pitch that the observer hears due to the Doppler effect lower or higher than the actual pitch? To which musical note is the pitch that the observer hears closest?   higher; F

Pitch of an Octave of Musical Notes in Hertz (HZ)

Note	Pitch
Middle C	261.63
D	293.66
E	329.63
F	349.23
G	392.00
A	440.00
B	493.88

Note: Greater numbers indicate higher pitches (acoustically).
(*Source*: American Standards Association)

In electronics, the relationship among the resistances R_1 and R_2 of two resistors wired in a parallel circuit and their combined resistance R is described by the formula $\dfrac{1}{R} = \dfrac{1}{R_1} + \dfrac{1}{R_2}$. Use this formula to solve Exercises 67 through 69.

67. If the combined resistance is 2 ohms and one of the two resistances is 3 ohms, find the other resistance.   6 ohms

68. Find the combined resistance of two resistors of 12 ohms each when they are wired in a parallel circuit.   6 ohms

69. The relationship among resistance of two resistors wired in a parallel circuit and their combined resistance may be extended to three resistors of resistances R_1, R_2, and R_3. Write an equation you believe may describe the relationship, and use it to find the combined resistance if R_1 is 5, R_2 is 6, and R_3 is 2.

$\dfrac{1}{R} = \dfrac{1}{R_1} + \dfrac{1}{R_2} + \dfrac{1}{R_3}$; $R = \dfrac{15}{13}$ ohms

6.8   VARIATION AND PROBLEM SOLVING

Objectives

1   Solve problems involving direct variation.

2   Solve problems involving inverse variation.

3   Solve problems involving joint variation.

4   Solve problems involving combined variation.

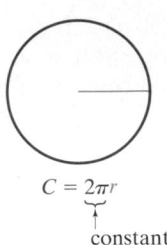

$C = 2\pi r$

constant

1 A very familiar example of direct variation is the relationship of the circumference C of a circle to its radius r. The formula $C = 2\pi r$ expresses that the circumference is always 2π times the radius. In other words, C is always a constant multiple (2π) of r. Because it is, we say that **C varies directly as r**, that **C varies directly with r**, or that **C is directly proportional to r**.

Direct Variation

y varies directly as x, or **y is directly proportional to x**, if there is a nonzero constant k such that

$$y = kx$$

The number k is called the **constant of variation** or the **constant of proportionality**.

In the above definition, the relationship described between x and y is a linear one. In other words, the graph of $y = kx$ is a line. The slope of the line is k, and the line passes through the origin.

For example, the graph of the direct variation equation $C = 2\pi r$ is shown. The horizontal axis represents the radius r, and the vertical axis is the circumference C. From the graph we can read that when the radius is 6 units, the circumference is approximately 38 units. Also, when the circumference is 45 units, the radius is between 7 and 8 units. Notice that as the radius increases, the circumference increases.

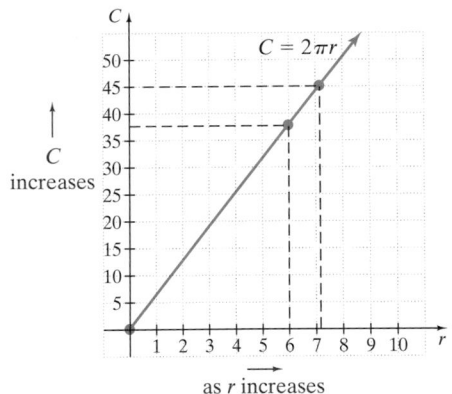

EXAMPLE 1

Suppose that y varies directly as x. If y is 5 when x is 30, find the constant of variation and the direct variation equation.

Solution Since y varies directly as x, we write $y = kx$. If $y = 5$ when $x = 30$, we have that

$$y = kx$$
$$5 = k(30) \qquad \text{Replace } y \text{ with 5 and } x \text{ with 30.}$$
$$\frac{1}{6} = k \qquad \text{Solve for } k.$$

The constant of variation is $\frac{1}{6}$.

After finding the constant of variation k, the direct variation equation can be written as

$$y = \frac{1}{6}x.$$

EXAMPLE 2

USING DIRECT VARIATION AND HOOKE'S LAW

Hooke's law states that the distance a spring stretches is directly proportional to the weight attached to the spring. If a 40-pound weight attached to the spring stretches the spring 5 inches, find the distance that a 65-pound weight attached to the spring stretches the spring.

Solution   1. UNDERSTAND.   Read and reread the problem. Notice that we are given that the distance a spring stretches is **directly proportional** to the weight attached. We let

$$d = \text{the distance stretched}$$
$$w = \text{the weight attached}$$

The constant of variation is represented by k.

TEACHING TIP

After discussing Example 1, have students make a table of ordered pairs that satisfy the direct variation equation. For each ordered pair, ask them to find the ratio, $\dfrac{y}{x}$, and the product, xy. What patterns, if any, are present? Is the ratio defined for all possible ordered pairs?

65 lbs

CLASSROOM EXAMPLE

Use Hooke's law as stated in Example 2. If a 56-pound weight attached to a spring stretches the spring 8 inches, find the distance that an 85-pound weight attached to the spring stretches the spring.

answer: $12\frac{1}{7}$

2. TRANSLATE.   Because d is directly proportional to w, we write

$$d = kw$$

3. SOLVE.   When a weight of 40 pounds is attached, the spring stretches 5 inches. That is, when $w = 40$, $d = 5$.

$$d = kw$$
$$5 = k(40) \quad \text{Replace } d \text{ with 5 and } w \text{ with 40.}$$
$$\frac{1}{8} = k \qquad \text{Solve for } k.$$

Now when we replace k with $\dfrac{1}{8}$ in the equation

$$d = kw, \text{ we have}$$

$$d = \frac{1}{8}w$$

To find the stretch when a weight of 65 pounds is attached, we replace w with 65 to find d.

$$d = \frac{1}{8}(65)$$

$$= \frac{65}{8} = 8\frac{1}{8} \quad \text{or} \quad 8.125$$

4. INTERPRET.

Check: Check the proposed solution of 8.125 inches in the original problem.
State: The spring stetches 8.125 inches when a 65-pound weight is attached.

2 When y is proportional to the **reciprocal** of another variable x, we say that **y varies inversely as x**, or that **y is inversely proportional to x**. An example of the inverse variation relationship is the relationship between the pressure that a gas exerts and the volume of its container. As the volume of a container decreases, the pressure of the gas it contains increases.

Inverse Variation

y **varies inversely as** x, or y **is inversely proportional to** x, if there is a nonzero constant k such that

$$y = \frac{k}{x}$$

The number k is called the **constant of variation** or the **constant of proportionality.**

Notice that $y = \dfrac{k}{x}$ is a rational equation. Its graph for $k > 0$ and $x > 0$ is shown. From the graph, we can see that as x increases, y decreases.

$$y = \frac{k}{x}, k > 0, x > 0$$

EXAMPLE 3

Suppose that u varies inversely as w. If u is 3 when w is 5, find the constant of variation and the inverse variation equation.

Solution Since u varies inversely as w, we have $u = \dfrac{k}{w}$. We let $u = 3$ and $w = 5$, and we solve for k.

$$u = \frac{k}{w}$$

$$3 = \frac{k}{5} \quad \text{Let } u = 3 \text{ and } w = 5.$$

$$15 = k \quad \text{Multiply both sides by 5.}$$

The constant of variation k is 15. This gives the inverse variation equation

$$u = \frac{15}{w}$$

EXAMPLE 4

USING INVERSE VARIATION AND BOYLE'S LAW

Boyle's law says that if the temperature stays the same, the pressure P of a gas is inversely proportional to the volume V. If a cylinder in a steam engine has a pressure of 960 kilopascals when the volume is 1.4 cubic meters, find the pressure when the volume increases to 2.5 cubic meters.

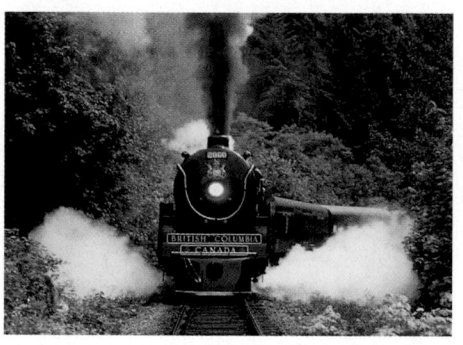

Solution   1. UNDERSTAND.   Read and reread the problem. Notice that we are given that the pressure of a gas is *inversely proportional* to the volume. We will let P = the pressure and V = the volume. The constant of variation is represented by k.

2. TRANSLATE.   Because P is inversely proportional to V, we write

$$P = \frac{k}{V}$$

When P = 960 kilopascals, the volume V = 1.4 cubic meters. We use this information to find k.

$$960 = \frac{k}{1.4} \qquad \text{Let } P = 960 \text{ and } V = 1.4.$$
$$1344 = k \qquad \text{Multiply both sides by 1.4.}$$

Thus, the value of k is 1344. Replacing k with 1344 in the variation equation, we have

$$P = \frac{1344}{V}$$

Next we find P when V is 2.5 cubic meters.

3. SOLVE.

$$P = \frac{1344}{2.5} \qquad \text{Let } V = 2.5.$$
$$= 537.6$$

4. INTERPRET.

Check: Check the proposed solution in the original problem.
State: When the volume is 2.5 cubic meters, the pressure is 537.6 kilopascals.

CLASSROOM EXAMPLE

The speed r at which one needs to drive in order to travel a constant distance is inversely proportional to the time t. A fixed distance can be driven in 5 hours at a rate of 24 mph. Find the rate needed to drive the same distance in 4 hours.

answer: 30 mph

3 Sometimes the ratio of a variable to the product of many other variables is constant. For example, the ratio of distance traveled to the product of speed and time traveled is always 1.

$$\frac{d}{rt} = 1 \quad \text{or} \quad d = rt$$

Such a relationship is called **joint variation.**

Joint Variation

If the ratio of a variable y to the product of two or more variables is constant, then **y varies jointly as,** or **is jointly proportional to,** the other variables. If

$$y = kxz$$

then the number k is the **constant of variation** or the **constant of proportionality.**

✔ **CONCEPT CHECK**

Which type of variation is represented by the equation $xy = 8$? Explain.

   a. Direct variation

   b. Inverse variation

   c. Joint variation

△ **EXAMPLE 5**

EXPRESSING SURFACE AREA

The lateral surface area of a cylinder varies jointly as its radius and height. Express this surface area S in terms of radius r and height h.

Solution Because the surface area varies jointly as the radius r and the height h, we equate S to a constant multiple of r and h.

$$S = krh$$

In the equation, $S = krh$, it can be determined that the constant k is 2π, and we then have the formula $S = 2\pi rh$. (The lateral surface area formula does not include the areas of the two circular bases.)

4 Some examples of variation involve combinations of direct, inverse, and joint variation. We will call these variations **combined variation.**

EXAMPLE 6

FINDING COLUMN WEIGHT

The maximum weight that a circular column can support is directly proportional to the fourth power of its diameter and is inversely proportional to the square of its height. A 2-meter-diameter column that is 8 meters in height can support 1 ton. Find the weight that a 1-meter-diameter column that is 4 meters in height can support.

1 ton

8 m   ?

4 m

2 m   1 m

Solution

1. UNDERSTAND.   Read and reread the problem. Let w = weight, d = diameter, h = height, and k = the constant of variation.

2. TRANSLATE.   Since w is directly proportional to d^4 and inversely proportional to h^2, we have

$$w = \frac{kd^4}{h^2}$$

3. SOLVE.   To find k, we are given that a 2-meter-diameter column that is 8 meters in height can support 1 ton. That is, $w = 1$ when $d = 2$ and $h = 8$, or

$$1 = \frac{k \cdot 2^4}{8^2} \qquad \text{Let } w = 1, d = 2, \text{ and } h = 8.$$

$$1 = \frac{k \cdot 16}{64}$$

$$4 = k \qquad \text{Solve for } k.$$

Now replace k with 4 in the equation $w = \dfrac{kd^4}{h^2}$ and we have

$$w = \frac{4d^4}{h^2}$$

To find weight w for a 1-meter-diameter column that is 4 meters in height, let $d = 1$ and $h = 4$.

$$w = \frac{4 \cdot 1^4}{4^2}$$

$$w = \frac{4}{16} = \frac{1}{4}$$

4. INTERPRET.   **Check:** Check the proposed solution in the original problem.

   State: The 1-meter-diameter column that is 4 meters in height can hold $\dfrac{1}{4}$ ton of weight.

Spotlight on
DECISION MAKING

Suppose you are a painting contractor. You have been hired to paint the ceilings of a one-story home, whose layout is shown in the figure. The amount of paint you need is directly proportional to the area that is to be painted. You know that 450 square feet can be painted with 4 quarts of paint. Quarts of paint cost $5.95 each and gallons of paint cost $21.50 each. You have estimated the cost of the paint to be about $50. Is your estimate correct? If so, explain why. If not, how much more should the estimate be?

MENTAL MATH

State whether each equation represents direct, inverse, or joint variation.

1. $y = 5x$ direct

2. $y = \dfrac{700}{x}$ inverse

3. $y = 5xz$ joint

4. $y = \dfrac{1}{2}abc$ joint

5. $y = \dfrac{9.1}{x}$ inverse

6. $y = 2.3x$ direct

7. $y = \dfrac{2}{3}x$ direct

8. $y = 3.1\,st$ joint

TEACHING TIP
A Group Activity for this section is available in the Instructor's Resource Manual.

EXERCISE SET 6.8

STUDY GUIDE/SSM   CD/VIDEO   PH MATH TUTOR CENTER   MathXL®Tutorials ON CD   MathXL®   MyMathLab®

MIXED PRACTICE

Write an equation to describe each variation. Use k for the constant of proportionality. See Examples 1 through 6.

1. y varies directly as x $y = kx$

2. p varies directly as q $p = kq$

3. a varies inversely as b $a = \dfrac{k}{b}$

4. y varies inversely as x $y = \dfrac{k}{x}$

5. y varies jointly as x and z $y = kxz$

6. y varies jointly as $q, r,$ and t $y = kqrt$

7. y varies inversely as x^3 $y = \dfrac{k}{x^3}$

8. y varies inversely as a^4 $y = \dfrac{k}{a^4}$

9. y varies directly as x and inversely as p^2 $y = \dfrac{kx}{p^2}$

10. y varies directly as a^5 and inversely as b $y = \dfrac{ka^5}{b}$

If y varies directly as x, find the constant of variation k and the direct variation equation for each situation. See Example 1.

11. $y = 4$ when $x = 20$ $k = \dfrac{1}{5}; y = \dfrac{1}{5}x$

12. $y = 5$ when $x = 30$ $k = \dfrac{1}{6}; y = \dfrac{1}{6}x$

13. $y = 6$ when $x = 4$ $k = \dfrac{3}{2}; y = \dfrac{3}{2}x$

14. $y = 12$ when $x = 8$ $k = \dfrac{3}{2}; y = \dfrac{3}{2}x$

15. $y = 7$ when $x = \dfrac{1}{2}$ $k = 14; y = 14x$

16. $y = 11$ when $x = \dfrac{1}{3}$ $k = 33; y = 33x$

17. $y = 0.2$ when $x = 0.8$ $k = 0.25; y = 0.25x$

18. $y = 0.4$ when $x = 2.5$ $k = 0.16; y = 0.16x$

Solve. See Example 2.

 19. The weight of a synthetic ball varies directly with the cube of its radius. A ball with a radius of 2 inches weighs 1.20 pounds. Find the weight of a ball of the same material with a 3-inch radius.   4.05 lb

20. At sea, the distance to the horizon is directly proportional to the square root of the elevation of the observer. If a person who is 36 feet above the water can see 7.4 miles, find how far a person 64 feet above the water can see. Round answer to one decimal place.   9.9 mi

21. The amount P of pollution varies directly with the population N of people. Kansas City has a population of 450,000 and produces 260,000 tons of pollutants. Find how many tons of pollution we should expect St. Louis to produce, if we know that its population is 980,000. Round answer to the nearest whole ton. $P = 566{,}222$ tons

22. Charles' law states that if the pressure P stays the same, the volume V of a gas is directly proportional to its temperature T. If a balloon is filled with 20 cubic meters of a gas at a temperature of 300 K, find the new volume if the temperature rises to 360 K while the pressure stays the same.   24 cu. m

If y varies inversely as x, find the constant of variation k and the inverse variation equation for each situation. See Example 3.

 23. $y = 6$ when $x = 5$ $k = 30; y = \dfrac{30}{x}$

24. $y = 20$ when $x = 9$ $k = 180; y = \dfrac{180}{x}$

25. $y = 100$ when $x = 7$ $k = 700; y = \dfrac{700}{x}$

26. $y = 63$ when $x = 3$ $k = 189; y = \dfrac{189}{x}$

27. $y = \dfrac{1}{8}$ when $x = 16$ $k = 2; y = \dfrac{2}{x}$

28. $y = \dfrac{1}{10}$ when $x = 40$ $k = 4; y = \dfrac{4}{x}$

29. $y = 0.2$ when $x = 0.7$ $k = 0.14; y = \dfrac{0.14}{x}$

30. $y = 0.6$ when $x = 0.3$ $k = 0.18; y = \dfrac{0.18}{x}$

Solve. See Example 4.

31. Pairs of markings a set distance apart are made on highways so that police can detect drivers exceeding the speed limit. Over a fixed distance, the speed R varies inversely with the time T. In one particular pair of markings, R is 45 mph when T is 6 seconds. Find the speed of a car that travels the given distance in 5 seconds.   54 mph

32. The weight of an object on or above the surface of Earth varies inversely as the square of the distance between the object and Earth's center. If a person weighs 160 pounds on Earth's surface, find the individual's weight if he moves 200 miles above Earth. Round answer to the nearest pound. (Assume that Earth's radius is 4000 miles.)   145 lb

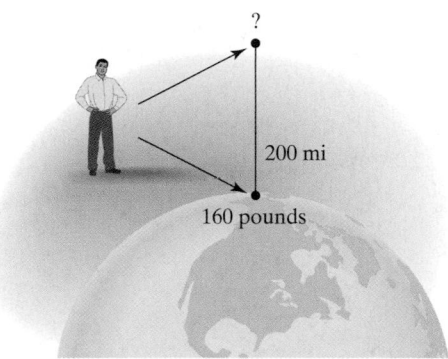

33. If the voltage V in an electric circuit is held constant, the current I is inversely proportional to the resistance R. If the current is 40 amperes when the resistance is 270 ohms, find the current when the resistance is 150 ohms.   72 amps

34. Because it is more efficient to produce larger numbers of items, the cost of producing recordable disks is inversely proportional to the number produced. If 4000 can be produced at a cost of $0.45 each, find the cost per disk when 6000 are produced. $0.30

35. The intensity I of light varies inversely as the square of the distance d from the light source. If the distance from the light source is doubled (see the figure at the top of the next page

and the figure directly below it), determine what happens to the intensity of light at the new location.   divided by 4

12 in.

24 in.

△ **36.** The maximum weight that a circular column can hold is inversely proportional to the square of its height. If an 8-foot column can hold 2 tons, find how much weight a 10-foot column can hold.   1.28 tons

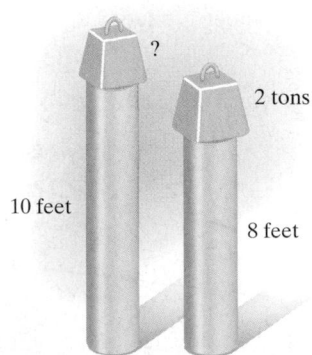

Write each statement as an equation. See Example 5.

37. x varies jointly as y and z. $x = kyz$

38. P varies jointly as R and the square of S. $P = kRS^2$

39. r varies jointly as s and the cube of t. $r = kst^3$

40. a varies jointly as b and c. $a = kbc$

For each statement, find the constant of variation and the variation equation. See Examples 5 and 6.

41. y varies directly as the cube of x; $y = 9$ when $x = 3$
$k = \frac{1}{3}; y = \frac{1}{3}x^3$

42. y varies directly as the cube of x; $y = 32$ when $x = 4$
$k = \frac{1}{2}; y = \frac{1}{2}x^3$

43. y varies directly as the square root of x; $y = 0.4$ when $x = 4$
$k = 0.2; y = 0.2\sqrt{x}$

44. y varies directly as the square root of x; $y = 2.1$ when $x = 9$
$k = 0.7; y = 0.7\sqrt{x}$

45. y varies inversely as the square of x; $y = 0.052$ when $x = 5$
$k = 1.3; y = \frac{1.3}{x^2}$

46. y varies inversely as the square of x; $y = 0.011$ when $x = 10$
$k = 1.1; y = \frac{1.1}{x^2}$

47. y varies jointly as x and the cube of z; $y = 120$ when $x = 5$ and $z = 2$ $k = 3; y = 3xz^3$

48. y varies jointly as x and the square of z; $y = 360$ when $x = 4$ and $z = 3$ $k = 10; y = 10xz^2$

Solve. See Examples 5 and 6.

🔒 **49.** The maximum weight that a rectangular beam can support varies jointly as its width and the square of its height and inversely as its length. If a beam $\frac{1}{2}$ foot wide, $\frac{1}{3}$ foot high, and 10 feet long can support 12 tons, find how much a similar beam can support if the beam is $\frac{2}{3}$ foot wide, $\frac{1}{2}$ foot high, and 16 feet long.   22.5 tons

50. The number of cars manufactured on an assembly line at a General Motors plant varies jointly as the number of workers and the time they work. If 200 workers can produce 60 cars in 2 hours, find how many cars 240 workers should be able to make in 3 hours.   108 cars

△ **51.** The volume of a cone varies jointly as the square of its radius and its height. If the volume of a cone is 32π cubic inches when the radius is 4 inches and the height is 6 inches, find the volume of a cone when the radius is 3 inches and the height is 5 inches. 15π cu. in.

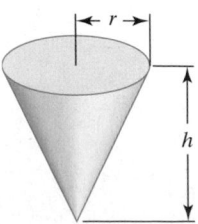

52. When a wind blows perpendicularly against a flat surface, its force is jointly proportional to the surface area and the speed of the wind. A sail whose surface area is 12 square feet experiences a 20-pound force when the wind speed is 10 miles per hour. Find the force on an 8-square-foot sail if the wind speed is 12 miles per hour.   16 lb

53. The horsepower that can be safely transmitted to a shaft varies jointly as the shaft's angular speed of rotation (in revolutions per minute) and the cube of its diameter. A 2-inch shaft making 120 revolutions per minute safely transmits 40 horsepower. Find how much horsepower can be safely transmitted by a 3-inch shaft making 80 revolutions per minute.   90 hp

△ **54.** The maximum weight that a rectangular beam can support varies jointly as its width and the square of its height and inversely as its length. If a beam $\frac{1}{3}$ foot wide, 1 foot high, and 10 feet long can support 3 tons, find how much weight a similar beam can support if it is 1 foot wide, $\frac{1}{3}$ foot high, and 9 feet long. $1\frac{1}{9}$ tons

55. The atmospheric pressure y (in millibars) is inversely proportional to the altitude x (in kilometers). If the atmospheric pressure is 400 millibars at an altitude of 8 kilometers, find the atmospheric pressure at an altitude of 4 kilometers.   800 millibars

REVIEW AND PREVIEW

Find the exact circumference and area of each circle. See the inside cover for a list of geometric formulas.

△ **56.** $C = 8\pi$ in.; $A = 16\pi$ sq. in.

△ **57.** 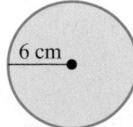 $C = 12\pi$ cm; $A = 36\pi$ sq. cm

△ **58.** $C = 18\pi$ cm; $A = 81\pi$ sq. cm

△ **59.** 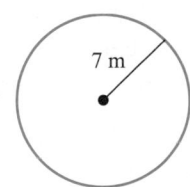 $C = 14\pi$ m; $A = 49\pi$ sq. m

Find each square root. See Section 1.3.

60. $\sqrt{81}$ 9

61. $\sqrt{36}$ 6

62. $\sqrt{1}$ 1

63. $\sqrt{4}$ 2

64. $\sqrt{\frac{1}{4}}$ $\frac{1}{2}$

65. $\sqrt{\frac{1}{25}}$ $\frac{1}{5}$

66. $\sqrt{\frac{4}{9}}$ $\frac{2}{3}$

67. $\sqrt{\frac{25}{121}}$ $\frac{5}{11}$

Concept Extensions

68. The horsepower to drive a boat varies directly as the cube of the speed of the boat. If the speed of the boat is to double, determine the corresponding increase in horsepower required.   multiplied by 8

69. The volume of a cylinder varies jointly as the height and the square of the radius. If the height is halved and the radius is doubled, determine what happens to the volume.   multiplied by 2

70. Suppose that y varies directly as x. If x is doubled, what is the effect on y?   multiplied by 2

71. Suppose that y varies directly as x^2. If x is doubled, what is the effect on y?   multiplied by 4

Complete the following table for the inverse variation $y = \dfrac{k}{x}$ over each given value of k. Plot the points on a rectangular coordinate system.

x	$\frac{1}{4}$	$\frac{1}{2}$	1	2	4
$y = \dfrac{k}{x}$					

72. $k = 1$ See graphing answer section.

73. $k = 3$ See graphing answer section.

74. $k = 5$ See graphing answer section.

75. $k = \dfrac{1}{2}$ See graphing answer section.

CHAPTER 6 PROJECT

Modeling Electricity Production

According to the U.S. Department of Energy, energy produced by renewable sources (including hydroelectric, geothermal, wind, and solar powers) accounted for nearly 8% of the United States' total production of energy in 2004. Wind energy can be harnessed by windmills to produce electricity, but it is the least utilized of these renewable energy sources. However, progressive communities are experimenting with fields of windmills for communal electricity needs, examining exactly how wind speed affects the amount of electricity produced.

A community in California is experimenting with electricity generated by windmills. City engineers are analyzing data they have gathered about their field of windmills. The engineers are familiar with other research demonstrating that the amount of electricity that a windmill generates hourly (in watt-hours) is directly proportional to the cube of the wind speed (in miles per hour). In this project, you will use this fact to help the engineers analyze the amount of electricity generated by the windmill field. This project may be completed by working in groups or individually.

1. The city engineers have documented that when the wind speed is exactly 10 miles per hour, a windmill generates electricity at a rate of 15 watt-hours. Find a formula that models the relationship between the wind speed and the amount of electricity generated hourly by a windmill.

2. Complete the following table for the given wind speeds. Then use the table to estimate the wind speed required to obtain: **a.** 100 watt-hours, and **b.** 400 watt-hours.

3. Plot the ordered pairs from the table. Describe the trend shown by the graph.

Wind Speed (miles per hour)	Electricity Generated (watt-hours)
15	
17	
19	
21	
23	
25	
27	
29	
31	
33	
35	

4. The engineers' data show that for several days the wind speed was more or less steady at 20 miles per hour, and the windmills generated the expected 120 watt-hours. According to the weather forecast for the coming few days, wind speed will fluctuate wildly, but will still average 20 miles per hour. Should the engineers still expect the windmills to generate 120 watt-hours? Demonstrate your reasoning with a numerical example.

5. During one three-day period, each windmill generated 150 watt-hours. The forecast predicts that the wind speed for the coming few days will drop by half. How many watt-hours should the engineers now expect each windmill to generate? In general, if the wind speed yields c watt-hours, how many watt-hours does half the wind speed yield?

STUDY SKILLS REMINDER

Are You Preparing for a Test on Chapter 6?

Below I have listed some common trouble areas for students in Chapter 6. After studying for your test—but before taking your test—read these.

▶ Make sure you know the difference in the following:

Simplify: $\dfrac{\dfrac{3}{x}}{\dfrac{1}{x} - \dfrac{5}{y}}$

Solve: $\dfrac{5x}{6} - \dfrac{1}{2} = \dfrac{5x}{12}$

Subtract: $\dfrac{1}{2x} - \dfrac{7}{x-3}$

Multiply numerator and denominator by the LCD.

$\dfrac{\dfrac{3}{x} \cdot xy}{\dfrac{1}{x} \cdot xy - \dfrac{5}{y} \cdot xy}$

$= \dfrac{3y}{y - 5x}$

Multiply both sides by the LCD.

$12 \cdot \dfrac{5x}{6} - 12 \cdot \dfrac{1}{2} = 12 \cdot \dfrac{5x}{12}$

$2 \cdot 5x - 6 = 5x$

$10x - 6 = 5x$

$5x = 6$

$x = \dfrac{6}{5}$

Write each expression as an equivalent expression with the LCD.

$\dfrac{1 \cdot (x-3)}{2x \cdot (x-3)} - \dfrac{7 \cdot 2x}{(x-3) \cdot 2x}$

$= \dfrac{x-3}{2x(x-3)} - \dfrac{14x}{2x(x-3)}$

$= \dfrac{-13x - 3}{2x(x-3)}$

Remember: This is simply a checklist of common trouble areas. For a review of Chapter 6, see the Highlights and Chapter Review at the end of this chapter.

CHAPTER VOCABULARY CHECK

Fill in each blank with one of the words or phrases listed below.

rational expression   equation   complex fraction   opposites   synthetic division
least common denominator expression   long division   jointly   directly   inversely

1. A rational expression whose numerator, denominator, or both contain one or more rational expressions is called a <u>complex fraction</u>.
2. To divide a polynomial by a polynomial other than a monomial, we use <u>long division</u>.
3. In the equation $y = kx$, y varies <u>directly</u> as x.
4. In the equation $y = \dfrac{k}{x}$, y varies <u>inversely</u> as x.
5. The <u>least common denominator</u> of a list of rational expressions is a polynomial of least degree whose factors include the denominator factors in the list.
6. When a polynomial is to be divided by a binomial of the form $x - c$, a shortcut process called <u>synthetic division</u> may be used.
7. In the equation $y = kxz$, y varies <u>jointly</u> as x and z.
8. The expressions $(x - 5)$ and $(5 - x)$ are called <u>opposites</u>.
9. A <u>rational expression</u> is an expression that can be written as the quotient $\dfrac{P}{Q}$ of two polynomials P and Q as long as Q is not 0.
10. Which is an expression and which is an equation? An example of an <u>equation</u> is $\dfrac{2}{x} + \dfrac{2}{x^2} = 7$ and an example of an <u>expression</u> is $\dfrac{2}{x} + \dfrac{5}{x^2}$.

CHAPTER 6 HIGHLIGHTS

Definitions and Concepts	Examples

Section 6.1 Rational Functions and Multiplying and Dividing Rational Expressions

A **rational expression** is the quotient $\frac{P}{Q}$ of two polynomials P and Q, as long as Q is not 0.

$$\frac{2x - 6}{7}, \qquad \frac{t^2 - 3t + 5}{t - 1}$$

To Simplify a Rational Expression

Step 1: Completely factor the numerator and the denominator.

Step 2: Apply the fundamental principle of rational expressions.

Simplify.

$$\frac{2x^2 + 9x - 5}{x^2 - 25} = \frac{(2x - 1)(x + 5)}{(x - 5)(x + 5)}$$
$$= \frac{2x - 1}{x - 5}$$

To Multiply Rational Expressions

Step 1: Completely factor numerators and denominators.

Step 2: Multiply the numerators and multiply the denominators.

Step 3: Apply the fundamental principle of rational expressions.

Multiply $\dfrac{x^3 + 8}{12x - 18} \cdot \dfrac{14x^2 - 21x}{x^2 + 2x}$.

$$= \frac{(x + 2)(x^2 - 2x + 4)}{6(2x - 3)} \cdot \frac{7x(2x - 3)}{x(x + 2)}$$
$$= \frac{7(x^2 - 2x + 4)}{6}$$

To Divide Rational Expressions

Multiply the first rational expression by the reciprocal of the second rational expression.

Divide $\dfrac{x^2 + 6x + 9}{5xy - 5y} \div \dfrac{x + 3}{10y}$.

$$= \frac{(x + 3)(x + 3)}{5y(x - 1)} \cdot \frac{2 \cdot 5y}{x + 3}$$
$$= \frac{2(x + 3)}{x - 1}$$

A **rational function** is a function described by a rational expression.

$$f(x) = \frac{2x - 6}{7}, \qquad h(t) = \frac{t^2 - 3t + 5}{t - 1}$$

Section 6.2 Adding and Subtracting Rational Expressions

To Add or Subtract Rational Expressions

Step 1: Find the LCD.

Step 2: Write each rational expression as an equivalent rational expression whose denominator is the LCD.

Step 3: Add or subtract numerators and write the result over the common denominator.

Step 4: Simplify the resulting rational expression.

Subtract $\dfrac{3}{x + 2} - \dfrac{x + 1}{x - 3}$.

$$= \frac{3 \cdot (x - 3)}{(x + 2) \cdot (x - 3)} - \frac{(x + 1) \cdot (x + 2)}{(x - 3) \cdot (x + 2)}$$
$$= \frac{3(x - 3) - (x + 1)(x + 2)}{(x + 2)(x - 3)}$$
$$= \frac{3x - 9 - (x^2 + 3x + 2)}{(x + 2)(x - 3)}$$
$$= \frac{3x - 9 - x^2 - 3x - 2}{(x + 2)(x - 3)}$$
$$= \frac{-x^2 - 11}{(x + 2)(x - 3)}$$

Definitions and Concepts	Examples

Section 6.3 Simplifying Complex Fractions

Method 1: Simplify the numerator and the denominator so that each is a single fraction. Then perform the indicated division and simplify if possible.

Simplify $\dfrac{\dfrac{x+2}{x}}{x-\dfrac{4}{x}}$.

Method 1: $\dfrac{\dfrac{x+2}{x}}{\dfrac{x\cdot x}{1\cdot x}-\dfrac{4}{x}}=\dfrac{\dfrac{x+2}{x}}{\dfrac{x^2-4}{x}}$

$=\dfrac{x+2}{x}\cdot\dfrac{x}{(x+2)(x-2)}=\dfrac{1}{x-2}$

Method 2: Multiply the numerator and the denominator of the complex fraction by the LCD of the fractions in both the numerator and the denominator. Then simplify if possible.

Method 2: $\dfrac{\left(\dfrac{x+2}{x}\right)\cdot x}{\left(x-\dfrac{4}{x}\right)\cdot x}=\dfrac{x+2}{x\cdot x-\dfrac{4}{x}\cdot x}$

$=\dfrac{x+2}{x^2-4}=\dfrac{x+2}{(x+2)(x-2)}=\dfrac{1}{x-2}$

Section 6.4 Dividing Polynomials

To divide a polynomial by a monomial: Divide each term in the polynomial by the monomial.

Divide $\dfrac{12a^5b^3-6a^2b^2+ab}{6a^2b^2}$

$=\dfrac{12a^5b^3}{6a^2b^2}-\dfrac{6a^2b^2}{6a^2b^2}+\dfrac{ab}{6a^2b^2}$

$=2a^3b-1+\dfrac{1}{6ab}$

To divide a polynomial by a polynomial, other than a monomial:

Use long division.

Divide $2x^3-x^2-8x-1$ by $x-2$.

$$
\begin{array}{r}
2x^2+3x-2 \\
x-2\overline{)2x^3-\ x^2-8x-1} \\
\underline{2x^3-4x^2} \\
3x^2-8x \\
\underline{3x^2-6x} \\
-2x-1 \\
\underline{-2x+4} \\
-5
\end{array}
$$

The quotient is $2x^2+3x-2-\dfrac{5}{x-2}$.

Section 6.5 Synthetic Division and the Remainder Theorem

A shortcut method called **synthetic division** may be used to divide a polynomial by a binomial of the form $x-c$.

Use synthetic division to divide $2x^3-x^2-8x-1$ by $x-2$.

$$
\begin{array}{r|rrrr}
2 & 2 & -1 & -8 & -1 \\
  & \downarrow & 4 & 6 & -4 \\
\hline
  & 2 & 3 & -2 & -5
\end{array}
$$

The quotient is $2x^2+3x-2-\dfrac{5}{x-2}$.

Definitions and Concepts	**Examples**

Section 6.6 Solving Equations Containing Rational Expressions

To solve an equation containing rational expressions: Multiply both sides of the equation by the LCD of all rational expressions. Then apply the distributive property and simplify. Solve the resulting equation and then check each proposed solution to see whether it makes the denominator 0. If so, it is an **extraneous solution.**

Solve $x - \dfrac{3}{x} = \dfrac{1}{2}$.

$$2x\left(x - \frac{3}{x}\right) = 2x\left(\frac{1}{2}\right)$$ The LCD is $2x$.

$$2x \cdot x - 2x\left(\frac{3}{x}\right) = 2x\left(\frac{1}{2}\right)$$ Distribute.

$$2x^2 - 6 = x$$

$$2x^2 - x - 6 = 0$$ Subtract x.

$$(2x + 3)(x - 2) = 0$$ Factor.

$$x = -\frac{3}{2} \quad \text{or} \quad x = 2$$ Solve.

Both $-\dfrac{3}{2}$ and 2 check. The solutions are 2 and $-\dfrac{3}{2}$.

Section 6.7 Rational Equations and Problem Solving

Solving an Equation for a Specified Variable

Treat the specified variable as the only variable of the equation and solve as usual.

Solve for x.

$$A = \frac{2x + 3y}{5}$$

$$5A = 2x + 3y$$ Multiply both sides by 5.

$$5A - 3y = 2x$$ Subtract $3y$ from both sides.

$$\frac{5A - 3y}{2} = x$$ Divide both sides by 2.

Jeanee and David Dillon volunteer every year to clean a strip of Lake Ponchartrain Beach. Jeanee can clean all the trash in this area of beach in 6 hours; David takes 5 hours. Find how long it will take them to clean the area of beach together.

Problem-Solving Steps to Follow

1. UNDERSTAND.

1. Read and reread the problem.
Let x = time in hours that it takes Jeanee and David to clean the beach together.

	Hours to Complete	*Part Completed in 1 Hour*
Jeanee Alone	6	$\dfrac{1}{6}$
David Alone	5	$\dfrac{1}{5}$
Together	x	$\dfrac{1}{x}$

(continued)

Definitions and Concepts	**Examples**

Section 6.7 Rational Equations and Problem Solving

2. TRANSLATE.

2. In words:

part Jeanee can complete in 1 hour	+	part David can complete in 1 hour	=	part they can complete together in 1 hour
↓		↓		↓

Translate:

$$\frac{1}{6} + \frac{1}{5} = \frac{1}{x}$$

3. SOLVE.

3. $\dfrac{1}{6} + \dfrac{1}{5} = \dfrac{1}{x}$ *Multiply both sides by* $30x$.

$$5x + 6x = 30$$

$$11x = 30$$

$$x = \frac{30}{11} \quad \text{or} \quad 2\frac{8}{11}$$

4. INTERPRET.

4. *Check* and then *state*. Together, they can clean the beach in $2\dfrac{8}{11}$ hours.

Section 6.8 Variation and Problem Solving

y **varies directly as** *x*, or *y* is **directly proportional to** *x*, if there is a nonzero constant *k* such that

$$y = kx$$

y **varies inversely as** *x*, or *y* is **inversely proportional to** x, if there is a nonzero constant *k* such that

$$y = \frac{k}{x}$$

y **varies jointly as** *x* and *z* or *y* is **jointly proportional to** *x* and *z* if there is a nonzero constant *k* such that

$$y = kxz$$

The circumference of a circle *C* varies directly as its radius *r*.

$$C = \underbrace{2\pi}_{k} r$$

Pressure *P* varies inversely with volume *V*.

$$P = \frac{k}{V}$$

The lateral surface area *S* of a cylinder varies jointly as its radius *r* and height *h*.

$$S = \underbrace{2\pi}_{k} rh$$

6 CHAPTER REVIEW

Find the domain for each rational function.

1. $f(x) = \dfrac{3 - 5x}{7}$ $\{x \mid x \text{ is a real number}\}$

2. $g(x) = \dfrac{2x + 4}{11}$ $\{x \mid x \text{ is a real number}\}$

3. $F(x) = \dfrac{-3x^2}{x - 5}$ $\{x \mid x \text{ is a real number and } x \neq 5\}$

4. $h(x) = \dfrac{4x}{3x - 12}$ $\{x \mid x \text{ is a real number and } x \neq 4\}$

5. $f(x) = \dfrac{x^3 + 2}{x^2 + 8x}$ $\{x \mid x \text{ is real number and } x \neq 0, x \neq -8\}$

6. $G(x) = \dfrac{20}{3x^2 - 48}$ $\{x \mid x \text{ is a real number and } x \neq -4, x \neq 4\}$

Write each rational expression in lowest terms.

7. $\dfrac{15x^4}{45x^2}$ $\dfrac{x^2}{3}$

8. $\dfrac{x + 2}{2 + x}$ 1

9. $\dfrac{18m^6p^2}{10m^4p}$ $\dfrac{9m^2p}{5}$

10. $\dfrac{x - 12}{12 - x}$ -1

11. $\dfrac{5x - 15}{25x - 75}$ $\dfrac{1}{5}$

12. $\dfrac{22x + 8}{11x + 4}$ 2

13. $\dfrac{2x}{2x^2 - 2x}$ $\dfrac{1}{x - 1}$

14. $\dfrac{x + 7}{x^2 - 49}$ $\dfrac{1}{x - 7}$

15. $\dfrac{2x^2 + 4x - 30}{x^2 + x - 20}$ $\dfrac{2(x - 3)}{x - 4}$

16. $\dfrac{xy - 3x + 2y - 6}{x^2 + 4x + 4}$ $\dfrac{y - 3}{x + 2}$

17. The average cost of manufacturing x bookcases is given by the rational function

$$C(x) = \dfrac{35x + 4200}{x}$$

a. Find the average cost per bookcase of manufacturing 50 bookcases. $\$119$

b. Find the average cost per bookcase of manufacturing 100 bookcases. $\$77$

c. As the number of bookcases increases, does the average cost per bookcase increase or decrease? (See parts **a** and **b**.) decrease

Perform the indicated operation. If possible, simplify your answer.

18. $\dfrac{5}{x^3} \cdot \dfrac{x^2}{15}$ $\dfrac{1}{3x}$

19. $\dfrac{3x^4yz^3}{15x^2y^2} \cdot \dfrac{10xy}{z^6}$ $\dfrac{2x^3}{z^3}$

20. $\dfrac{4 - x}{5} \cdot \dfrac{15}{2x - 8}$ $-\dfrac{3}{2}$

21. $\dfrac{x^2 - 6x + 9}{2x^2 - 18} \cdot \dfrac{4x + 12}{5x - 15}$ $\dfrac{2}{5}$

22. $\dfrac{a - 4b}{a^2 + ab} \cdot \dfrac{b^2 - a^2}{8b - 2a}$ $\dfrac{a - b}{2a}$

23. $\dfrac{x^2 - x - 12}{2x^2 - 32} \cdot \dfrac{x^2 + 8x + 16}{3x^2 + 21x + 36}$ $\dfrac{1}{6}$

24. $\dfrac{2x^3 + 54}{5x^2 + 5x - 30} \cdot \dfrac{6x + 12}{3x^2 - 9x + 27}$ $\dfrac{4(x + 2)}{5(x - 2)}$

25. $\dfrac{3}{4x} \div \dfrac{8}{2x^2}$ $\dfrac{3x}{16}$

26. $\dfrac{4x + 8y}{3} \div \dfrac{5x + 10y}{9}$ $\dfrac{12}{5}$

27. $\dfrac{5ab}{14c^3} \div \dfrac{10a^4b^2}{6ac^5}$ $\dfrac{3c^2}{14a^2b}$

28. $\dfrac{2}{5x} \div \dfrac{4 - 18x}{6 - 27x}$ $\dfrac{3}{5x}$

29. $\dfrac{x^2 - 25}{3} \div \dfrac{x^2 - 10x + 25}{x^2 - x - 20}$

30. $\dfrac{a - 4b}{a^2 + ab} \div \dfrac{20b - 5a}{b^2 - a^2}$

31. $\dfrac{7x + 28}{2x + 4} \div \dfrac{x^2 + 2x - 8}{x^2 - 2x - 8}$

32. $\dfrac{3x + 3}{x - 1} \div \dfrac{x^2 - 6x - 7}{x^2 - 1}$

33. $\dfrac{2x - x^2}{x^3 - 8} \div \dfrac{x^2}{x^2 + 2x + 4}$ $-\dfrac{1}{x}$

34. $\dfrac{5a^2 - 20}{a^3 + 2a^2 + a + 2} \div \dfrac{7a}{a^3 + a}$ $\dfrac{5(a - 2)}{7}$

35. $\dfrac{2a}{21} \div \dfrac{3a^2}{7} \cdot \dfrac{4}{a}$ $\dfrac{8}{9a^2}$

36. $\dfrac{5x - 15}{3 - x} \cdot \dfrac{x + 2}{10x + 20} \cdot \dfrac{x^2 - 9}{x^2 - x - 6}$ $-\dfrac{x + 3}{2(x + 2)}$

37. $\dfrac{4a + 8}{5a^2 - 20} \cdot \dfrac{3a^2 - 6a}{a + 3} \div \dfrac{2a^2}{5a + 15}$ $\dfrac{6}{a}$

29. $\dfrac{(x + 4)(x + 5)}{3}$

30. $\dfrac{a - b}{5a}$

31. $\dfrac{7(x - 4)}{2(x - 2)}$

32. $\dfrac{3(x + 1)}{x - 7}$

***(6.2)** Find the LCD of the rational expressions in the list.*

38. $\dfrac{4}{9}, \dfrac{5}{2}$ 18

39. $\dfrac{5}{4x^2y^5}, \dfrac{3}{10x^2y^4}, \dfrac{x}{6y^4}$ $60x^2y^5$

40. $\dfrac{5}{2x}, \dfrac{7}{x - 2}$ $2x(x - 2)$

41. $\dfrac{3}{5x}, \dfrac{2}{x - 5}$ $5x(x - 5)$

42. $\dfrac{1}{5x^3}, \dfrac{4}{x^2 + 3x - 28}, \dfrac{11}{10x^2 - 30x}$ $10x^3(x - 4)(x + 7)(x - 3)$

Perform the indicated operation. If possible, simplify your answer.

43. $\dfrac{2}{15} + \dfrac{4}{15}$ $\dfrac{2}{5}$

44. $\dfrac{4}{x - 4} + \dfrac{x}{x - 4}$ $\dfrac{4 + x}{x - 4}$

45. $\dfrac{4}{3x^2} + \dfrac{2}{3x^2}$ $\dfrac{2}{x^2}$

46. $\dfrac{1}{x-2} - \dfrac{1}{4-2x}$ $\dfrac{3}{2(x-2)}$

47. $\dfrac{2x+1}{x^2+x-6} + \dfrac{2-x}{x^2+x-6}$ $\dfrac{1}{x-2}$

48. $\dfrac{7}{2x} + \dfrac{5}{6x}$ $\dfrac{13}{3x}$

49. $\dfrac{1}{3x^2y^3} - \dfrac{1}{5x^4y}$ $\dfrac{5x^2-3y^2}{15x^4y^3}$

50. $\dfrac{1}{10-x} + \dfrac{x-1}{x-10}$ $\dfrac{x-2}{x-10}$

51. $\dfrac{x-2}{x+1} - \dfrac{x-3}{x-1}$ $\dfrac{-x+5}{(x+1)(x-1)}$

52. $\dfrac{x}{9-x^2} - \dfrac{2}{5x-15}$ $\dfrac{-7x-6}{5(x+3)(x-3)}$

53. $2x+1 - \dfrac{1}{x-3}$ $\dfrac{2x^2-5x-4}{x-3}$

54. $\dfrac{2}{a^2-2a+1} + \dfrac{3}{a^2-1}$ $\dfrac{5a-1}{(a-1)^2(a+1)}$

55. $\dfrac{x}{9x^2+12x+16} - \dfrac{3x+4}{27x^3-64}$ $\dfrac{3x^2-7x-4}{(3x-4)(9x^2+12x+16)}$

Perform the indicated operation. If possible, simplify your answer.

56. $\dfrac{2}{x-1} - \dfrac{3x}{3x-3} + \dfrac{1}{2x-2}$ $\dfrac{5-2x}{2(x-1)}$

57. $\dfrac{3}{2x} \cdot \left(\dfrac{2}{x+1} - \dfrac{2}{x-3} \right)$ $-\dfrac{12}{x(x+1)(x-3)}$

58. $\left(\dfrac{2}{x} - \dfrac{1}{5} \right) \cdot \left(\dfrac{2}{x} + \dfrac{1}{3} \right)$ $\dfrac{(10-x)(6+x)}{15x^2}$

59. $\dfrac{2}{x^2-16} - \dfrac{3x}{x^2+8x+16} + \dfrac{3}{x+4}$ $\dfrac{14x-40}{(x+4)^2(x-4)}$

△ **60.** Find the perimeter of the heptagon (polygon with 7 sides). $\dfrac{11}{x}$

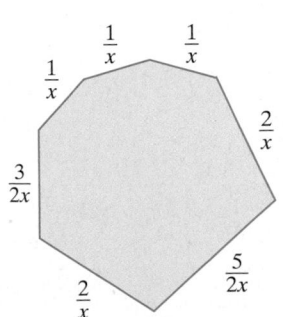

(6.3) Simplify each complex fraction.

61. $\dfrac{\frac{2}{5}}{\frac{3}{5}}$ $\dfrac{2}{3}$

62. $\dfrac{1 - \frac{3}{4}}{2 + \frac{1}{4}}$ $\dfrac{1}{9}$

63. $\dfrac{\frac{1}{x} - \frac{2}{3x}}{\frac{5}{2x} - \frac{1}{3}}$ $\dfrac{2}{15-2x}$

64. $\dfrac{\frac{x^2}{15}}{\frac{x+1}{5x}}$ $\dfrac{x^3}{3(x+1)}$

65. $\dfrac{\frac{3}{y^2}}{\frac{6}{y^3}}$ $\dfrac{y}{2}$

66. $\dfrac{\frac{x+2}{3}}{\frac{5}{x-2}}$ $\dfrac{(x+2)(x-2)}{15}$

67. $\dfrac{2 - \frac{3}{2x}}{x - \frac{2}{5x}}$ $\dfrac{20x-15}{10x^2-4}$

68. $\dfrac{1 + \frac{x}{y}}{\frac{x^2}{y^2} - 1}$ $\dfrac{y}{x-y}$

69. $\dfrac{\frac{5}{x} + \frac{1}{xy}}{\frac{3}{x^2}}$ $\dfrac{5xy+x}{3y}$

70. $\dfrac{\frac{x}{3} - \frac{3}{x}}{1 + \frac{3}{x}}$ $\dfrac{x-3}{3}$

71. $\dfrac{\frac{1}{x-1} + 1}{\frac{1}{x+1} - 1}$ $\dfrac{1+x}{1-x}$

72. $\dfrac{2}{1 - \frac{2}{x}}$ $\dfrac{2x}{x-2}$

73. $\dfrac{1}{1 + \dfrac{2}{1 - \frac{1}{x}}}$ $\dfrac{x-1}{3x-1}$

74. $\dfrac{\frac{x^2+5x-6}{4x+3}}{\frac{(x+6)^2}{8x+6}}$ $\dfrac{2(x-1)}{x+6}$

75. $\dfrac{\frac{x-3}{x+3} + \frac{x+3}{x-3}}{\frac{x-3}{x+3} - \frac{x+3}{x-3}}$ $-\dfrac{x^2+9}{6x}$

76. $\dfrac{\frac{3}{x-1} - \frac{2}{1-x}}{\frac{2}{x-1} - \frac{2}{x}}$ $\dfrac{5x}{2}$

77. If $f(x) = \dfrac{3}{x}$, find each of the following:

a. $f(a+h)$ $\dfrac{3}{a+h}$

b. $f(a)$ $\dfrac{3}{a}$

c. Use parts **a** and **b** to find $\dfrac{f(a+h) - f(a)}{h}$.

77. c. $\dfrac{\frac{3}{a+h} - \frac{3}{a}}{h}$

d. Simplify the results of part **c.** $\dfrac{-3}{a(a+h)}$

(6.4) *Divide.*

78. Divide $3x^5yb^9$ by $9xy^7$. $\dfrac{x^4b^9}{3y^6}$

79. Divide $-9xb^4z^3$ by $-4axb^2$. $\dfrac{9b^2z^3}{4a}$

80. $(4xy + 2x^2 - 9) \div 4xy$ $1 + \dfrac{x}{2y} - \dfrac{9}{4xy}$

81. Divide $12xb^2 + 16xb^4$ by $4xb^3$. $\dfrac{3}{b} + 4b$

82. $(3x^4 - 25x^2 - 20) \div (x - 3)$

83. $(-x^2 + 2x^4 + 5x - 12) \div (x + 2)$

84. $(2x^4 - x^3 + 2x^2 - 3x + 1) \div (2x - 1)$ $x^3 + x - 1$

85. $(2x^3 + 3x^2 - 2x + 2) \div (2x + 3)$ $x^2 - 1 + \dfrac{5}{2x + 3}$

86. $(3x^4 + 5x^3 + 7x^2 + 3x - 2) \div (x^2 + x + 2)$

87. $(9x^4 - 6x^3 + 3x^2 - 12x - 30) \div (3x^2 - 2x - 5)$ $3x^2 + 6$

(6.5) *Use synthetic division to find each quotient.*

88. $(3x^3 + 12x - 4) \div (x - 2)$ $3x^2 + 6x + 24 + \dfrac{44}{x - 2}$

89. $(3x^3 + 2x^2 - 4x - 1) \div \left(x + \dfrac{3}{2}\right)$

90. $(x^5 - 1) \div (x + 1)$ $x^4 - x^3 + x^2 - x + 1 - \dfrac{2}{x + 1}$

91. $(x^3 - 81) \div (x - 3)$ $x^2 + 3x + 9 - \dfrac{54}{x - 3}$

92. $(x^3 - x^2 + 3x^4 - 2) \div (x - 4)$

93. $(3x^4 - 2x^2 + 10) \div (x + 2)$

If $P(x) = 3x^5 - 9x + 7$, use the remainder theorem to find the following.

94. $P(4)$ 3043

95. $P(-5)$ -9323

96. $P\left(\dfrac{2}{3}\right)$ $\dfrac{113}{81}$

△ **97.** $P\left(-\dfrac{1}{2}\right)$ $\dfrac{365}{32}$

98. If the area of the rectangle is $(x^4 - x^3 - 6x^2 - 6x + 18)$ square miles and its width is $(x - 3)$ miles, find the length. $(x^3 + 2x^2 - 6)$ mi

$$x^4 - x^3 - 6x^2 - 6x + 18 \text{ square miles} \qquad x - 3 \text{ miles}$$

(6.6) *Solve each equation for x.*

99. $\dfrac{2}{5} = \dfrac{x}{15}$ 6

100. $\dfrac{3}{x} + \dfrac{1}{3} = \dfrac{5}{x}$ 6

101. $4 + \dfrac{8}{x} = 8$ 2

102. $\dfrac{2x + 3}{5x - 9} = \dfrac{3}{2}$ 3

103. $\dfrac{1}{x - 2} - \dfrac{3x}{x^2 - 4} = \dfrac{2}{x + 2}$ $\dfrac{3}{2}$

104. $\dfrac{7}{x} - \dfrac{x}{7} = 0$ $-7, 7$

105. $\dfrac{x - 2}{x^2 - 7x + 10} = \dfrac{1}{5x - 10} - \dfrac{1}{x - 5}$ $\dfrac{5}{3}$

Solve the equations for x or perform the indicated operation. Simplify.

106. $\dfrac{5}{x^2 - 7x} + \dfrac{4}{2x - 14}$ $\dfrac{2x + 5}{x(x - 7)}$

107. $3 - \dfrac{5}{x} - \dfrac{2}{x^2} = 0$ $-\dfrac{1}{3}, 2$

108. $\dfrac{4}{3 - x} - \dfrac{7}{2x - 6} + \dfrac{5}{x}$ $\dfrac{-5x - 30}{2x(x - 3)}$

82. $3x^3 + 9x^2 + 2x + 6 - \dfrac{2}{x - 3}$ **83.** $2x^3 - 4x^2 + 7x - 9 + \dfrac{6}{x + 2}$ **86.** $3x^2 + 2x - 1$

92. $3x^3 + 13x^2 + 51x + 204 + \dfrac{814}{x - 4}$ **93.** $3x^3 - 6x^2 + 10x - 20 + \dfrac{50}{x + 2}$

89. $3x^2 - \dfrac{5}{2}x - \dfrac{1}{4} - \dfrac{5}{8\left(x + \dfrac{3}{2}\right)}$

(6.7) Solve the equation for the specified variable.

△ **109.** $A = \dfrac{h(a + b)}{2}, a$ $a = \dfrac{2A - hb}{h}$

110. $\dfrac{1}{R} = \dfrac{1}{R_1} + \dfrac{1}{R_2}, R_2$ $R_2 = \dfrac{RR_1}{R_1 - R}$

111. $I = \dfrac{E}{R + r}, R$ $R = \dfrac{E - Ir}{I}$

112. $A = P + Prt, r$ $r = \dfrac{A - P}{Pt}$

113. $H = \dfrac{kA(T_1 - T_2)}{L}, A$ $A = \dfrac{HL}{k(T_1 - T_2)}$

Solve.

114. The sum of a number and twice its reciprocal is 3. Find the number(s). $1, 2$

115. If a number is added to the numerator of $\dfrac{3}{7}$, and twice that number is added to the denominator of $\dfrac{3}{7}$, the result is equivalent to $\dfrac{10}{21}$. Find the number.   7

116. The denominator of a fraction is 2 more than the numerator. If the numerator is decreased by 3 and the denominator is increased by 5, the resulting fraction is equivalent to $\dfrac{2}{3}$. Find the fraction. $\dfrac{23}{25}$

117. The sum of the reciprocals of two consecutive even integers is $-\dfrac{9}{40}$. Find the two integers. -10 and -8

118. Three boys can paint a fence in 4 hours, 5 hours, and 6 hours, respectively. Find how long it will take all three boys to paint the fence. $1\frac{23}{37}$ hr

119. If Sue Katz can type a certain number of mailing labels in 6 hours and Tom Neilson and Sue working together can type the same number of mailing labels in 4 hours, find how long it takes Tom alone to type the mailing labels.   12 hr

120. The inlet pipe of a water tank can fill the tank in 2 hours and 30 minutes. The outlet pipe can empty the tank in 2 hours. Find how long it takes to empty a full tank if both pipes are open.   10 hr

121. Timmy Garnica drove 210 miles in the same amount of time that it took a DC-10 jet to travel 1715 miles. The speed of the jet was 430 mph faster than the speed of the car. Find the speed of the jet.   490 mph

122. The combined resistance R of two resistors in parallel with resistances R_1 and R_2 is given by the formula $\dfrac{1}{R} = \dfrac{1}{R_1} + \dfrac{1}{R_2}$. If the combined resistance is $\dfrac{30}{11}$ ohms and the resistance of one of the two resistors is 5 ohms, find the resistance of the other resistor.   6 ohms

123. The speed of a Ranger boat in still water is 32 mph. If the boat travels 72 miles upstream in the same time that it takes to travel 120 miles downstream, find the speed of the current.   8 mph

124. A B737 jet flies 445 miles with the wind and 355 miles against the wind in the same length of time. If the speed of the jet in still air is 400 mph, find the speed of the wind.   45 mph

125. The speed of a jogger is 3 mph faster than the speed of a walker. If the jogger travels 14 miles in the same amount of time that the walker travels 8 miles, find the speed of the walker.   4 mph

126. Two Amtrak trains traveling on parallel tracks leave Tucson at the same time. The speed of one train is 18 mph faster than the other. If the faster train travels 378 miles in the same time that the other train travels 270 miles, find the speed of each train.   63 mph and 45 mph

(6.8) Solve each variation problem.

127. A is directly proportional to B. If $A = 6$ when $B = 14$, find A when $B = 21$.   9

128. C is inversely proportional to D. If $C = 12$ when $D = 8$, find C when $D = 24$.   4

129. According to Boyle's law, the pressure exerted by a gas is inversely proportional to the volume, as long as the temperature stays the same. If a gas exerts a pressure of 1250 pounds per square inch when the volume is 2 cubic feet, find the volume when the pressure is 800 pounds per square inch.   3.125 cu. ft

△ **130.** The surface area of a sphere varies directly as the square of its radius. If the surface area is 36π square inches when the radius is 3 inches, find the surface area when the radius is 4 inches. 64π sq. in.

CHAPTER 6 TEST

Remember to use your Chapter Test Prep Video CD to help you study and view solutions to the test questions you need help with.

Find the domain of each rational function.

1. $f(x) = \dfrac{5x^2}{1 - x}$

2. $g(x) = \dfrac{9x^2 - 9}{x^2 + 4x + 3}$

Write each rational expression in lowest terms.

3. $\dfrac{7x - 21}{24 - 8x}$ $-\dfrac{7}{8}$

4. $\dfrac{x^2 - 4x}{x^2 + 5x - 36}$ $\dfrac{x}{x + 9}$

Perform the indicated operation. If possible, simplify your answer.

5. $\dfrac{2x^3 + 16}{6x^2 + 12x} \cdot \dfrac{5}{x^2 - 2x + 4}$ $\dfrac{5}{3x}$

6. $\dfrac{3x^2 - 12}{x^2 + 2x - 8} \div \dfrac{6x + 18}{x + 4}$ $\dfrac{x + 2}{2(x + 3)}$

7. $\dfrac{4x - 12}{2x - 9} \div \dfrac{3 - x}{4x^2 - 81} \cdot \dfrac{x + 3}{5x + 15}$ $\dfrac{-4(2x + 9)}{5}$

8. $\dfrac{3 + 2x}{10 - x} + \dfrac{13 + x}{x - 10}$ -1

9. $\dfrac{3}{x^2 - x - 6} + \dfrac{2}{x^2 - 5x + 6}$ $\dfrac{5x - 2}{(x - 3)(x + 2)(x - 2)}$

10. $\dfrac{5}{x - 7} - \dfrac{2x}{3x - 21} + \dfrac{x}{2x - 14}$ $\dfrac{-x + 30}{6(x - 7)}$

11. $\dfrac{3x}{5} \cdot \left(\dfrac{5}{x} - \dfrac{5}{2x} \right)$ $\dfrac{3}{2}$

Simplify each complex fraction.

12. $\dfrac{\dfrac{4x}{13}}{\dfrac{20x}{13}}$ $\dfrac{1}{5}$

13. $\dfrac{\dfrac{5}{x} - \dfrac{7}{3x}}{\dfrac{9}{8x} - \dfrac{1}{x}}$ $\dfrac{64}{3}$

Divide.

14. $(4x^2y + 9x + 3xz) \div 3xz$ $\dfrac{4xy}{3z} + \dfrac{3}{z} + 1$

15. $(4x^3 - 5x) \div (2x + 1)$ $2x^2 - x - 2 + \dfrac{2}{2x + 1}$

16. Use synthetic division to divide $(4x^4 - 3x^3 - x - 1)$ by $(x + 3)$. $4x^3 - 15x^2 + 45x - 136 + \dfrac{407}{x + 3}$

17. If $P(x) = 4x^4 + 7x^2 - 2x - 5$, use the remainder theorem to find $P(-2)$.   91

Solve each equation for x.

18. $\dfrac{3}{x + 2} - \dfrac{1}{5x} = \dfrac{2}{5x^2 + 10x}$ $\dfrac{2}{7}$

19. $\dfrac{x^2 + 8}{x} - 1 = \dfrac{2(x + 4)}{x}$ 3

20. Solve for x: $\dfrac{x + b}{a} = \dfrac{4x - 7a}{b}$ $x = \dfrac{7a^2 + b^2}{4a - b}$

21. The product of one more than a number and twice the reciprocal of the number is $\dfrac{12}{5}$. Find the number.   5

22. If Jan can weed the garden in 2 hours and her husband can weed it in 1 hour and 30 minutes, find how long, it takes them to weed the garden together. $\frac{6}{7}$ hr

23. Suppose that W is inversely proportional to V. If $W = 20$ when $V = 12$, find W when $V = 15$.   16

24. Suppose that Q is jointly proportional to R and the square of S. If $Q = 24$ when $R = 3$ and $S = 4$, find Q when $R = 2$ and $S = 3$.   9

25. When an anvil is dropped into a gorge, the speed with which it strikes the ground is directly proportional to the square root of the distance it falls. An anvil that falls 400 feet hits the ground at a speed of 160 feet per second. Find the height of a cliff over the gorge if a dropped anvil hits the ground at a speed of 128 feet per second.   256 ft

1. $\{x \mid x \text{ is a real number and } x \neq 1\}$ **2.** $\{x \mid x \text{ is a real number and } x \neq -3, x \neq -1\}$

6 CHAPTER CUMULATIVE REVIEW

1. Translate each phrase to an algebraic expression. Use the variable x to represent each unknown number.
   a. Eight times a number $8x$
   b. Three more than eight times a number $8x + 3$
   c. The quotient of a number and -7 $x \div -7$ or $\dfrac{x}{-7}$
   d. One and six-tenths subtracted from twice a number
   $2x - 1.6$ (Sec. 1.2, Ex. 8)

2. Translate each phrase to an algebraic expression. Use the variable x to represent each unknown number.
   a. One third subtracted from a number $x - \frac{1}{3}$
   b. Six less than five times a number $5x - 6$
   c. Three more than eight times a number $8x + 3$
   d. The quotient of seven and the difference of two and a number. $\dfrac{7}{2 - x}$ (Sec. 1.2)

3. Solve for y: $\dfrac{y}{3} - \dfrac{y}{4} = \dfrac{1}{6}$ 2 (Sec. 1.5, Ex. 5)

4. Solve $\dfrac{x}{7} + \dfrac{x}{5} = \dfrac{12}{5}$ 7 (Sec. 1.5)

5. In the United States, the annual consumption of cigarettes is declining. The consumption c in billions of cigarettes per year since the year 1985 can be approximated by the formula $c = -14.25t + 598.69$ where t is the number of years after 1985. Use this formula to predict the years that the consumption of cigarettes will be less than 200 billion per year. 2013 and after (Sec. 3.2, Ex. 9)

6. Olivia has scores of 78, 65, 82, and 79 on her algebra tests. Use an inequality to find the minimum score she can make or her final exam to pass the course with a 78 average or higher, given that the final exam counts as two tests.   82 (Sec. 3.2)

7. Solve: $\left| \dfrac{3x + 1}{2} \right| = -2$ Ø (Sec. 3.4, Ex. 7)

8. Solve: $\left| \dfrac{2x - 1}{3} \right| + 6 = 3$ Ø (Sec. 3.4)

9. Solve for x: $\left| \dfrac{2(x + 1)}{3} \right| \le 0$ -1 (Sec. 3.5, Ex. 8)

10. Solve for x: $\left| \dfrac{3(x - 1)}{4} \right| \ge 2$ $\left(\infty, -\frac{5}{3}\right] \cup \left[\frac{11}{3}, \infty\right)$ (Sec. 3.5)

11. Graph the equation $y = -2x + 3$.   See graphing answer section. (Sec. 2.1, Ex. 5)

12. Graph the equation $y = -x + 3$.   See graphing answer section. (Sec. 2.1)

13. Which of the following relations are also functions?
    a. $\{(-2, 5), (2, 7), (-3, 5), (9, 9)\}$ function
    b.   not a function
    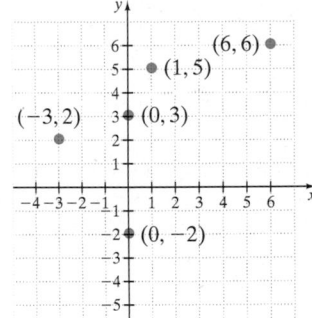
    c.   Input   Correspondence   Output
       People in a   Each person's   The set of
       certain city   age   nonegative integers
       function (Sec. 2.2, Ex. 2)

14. If $f(x) = -x^2 + 3x - 2$, find
    a. f(0) -2 b. $f(-3)$ -20 c. $f\left(\frac{1}{3}\right)$ $-\dfrac{10}{9}$ (Sec. 2.2)

15. Graph $x - 3y = 6$ by plotting intercept points.   See graphing answer section. (Sec. 2.3, Ex. 3)

16. Graph $3x - y = 6$ by plotting x- and y-intercepts.   See graphing answer section. (Sec. 2.3)

17. Find an equation of the line with slope -3 containing the point $(1, -5)$. Write the equation in slope–intercept form $y = mx + b$. $y = -3x - 2$ (Sec. 2.5, Ex. 4)

18. Find an equation of the line with slope $\frac{1}{2}$ containing the point $(-1, 3)$. Use function notation to write the equation.
    $f(x) = \dfrac{1}{2}x + \dfrac{7}{2}$ (Sec. 2.5)

19. Graph the intersection of $x \ge 1$ and $y \ge 2x - 1$.   See graphing anwer section (Sec. 3.6, Ex. 3)

20. Graph the union of $2x + y \le 4$ or $y > 2$.   See graphing answer section. (Sec. 3.6)

21. Use the elimination method to solve the system.
    $\begin{cases} 3x - 2y = 10 \\ 4x - 3y = 15 \end{cases}$ $(0, -5)$ (Sec. 4.1, Ex. 7)

22. Use the substitution method to solve the system.
    $\begin{cases} -2x + 3y = 6 \\ 3x - y = 5 \end{cases}$ $(3, 4)$ (Sec. 4.1)

23. Solve the system. $\begin{cases} 2x - 4y + 8z = 2 \\ -x - 3y + z = 11 \\ x - 2y + 4z = 0 \end{cases}$ $\varnothing$ (Sec. 4.2, Ex. 2)

24. Solve the system. $\begin{cases} 2x - 2y + 4z = 6 \\ -4x - y + z = -8 \\ 3x - y + z = 6 \end{cases}$ (2, 1, 1) (Sec. 4.2)

25. The measure of the largest angle of a triangle is 80° more than the measure of the smallest angle, and the measure of the remaining angle is 10° more than the measure of the smallest angle. Find the measure of each angle. 30°, 110°, 40° (Sec. 4.3, Ex. 6)

26. Kernersville office supply sold three reams of paper and two boxes of manila folders for $21.90. Also, five reams of paper and one box of manila folders cost $24.25. Find the price of a ream of paper and a box of manila folders. Paper, $3.80; folders, $5.25 (Sec. 4.3)

27. Use matrices to solve the system. $\begin{cases} x + 2y + z = 2 \\ -2x - y + 2z = 5 \\ x + 3y - 2z = -8 \end{cases}$ (1, −1, 3) (Sec. 4.4, Ex. 4)

28. Use matrices to solve the system. $\begin{cases} x + y + z = 9 \\ 2x - 2y + 3z = 2 \\ -3x + y - z = 1 \end{cases}$ (0, 5, 4) (Sec. 4.4)

29. Evaluate the following.
   a. 7^0 1
   b. -7^0 −1
   c. $(2x + 5)^0$ 1
   d. $2x^0$ 2 (Sec. 5.1, Ex. 3)

30. Simplify the following. Write answers with positive exponents.
   a. $2^{-2} + 3^{-1}$ $\dfrac{7}{12}$
   b. $-6a^0$ −6
   c. $\dfrac{x^{-5}}{x^{-2}}$ $\dfrac{1}{x^3}$ (Sec. 5.1)

31. Simplify each. Assume that a and b are integers and that x and y are not 0.
   a. $x^{-b}(2x^b)^2$ $4x^b$
   b. $\dfrac{(y^{3a})^2}{y^{a-6}}$ y^{5a+6} (Sec. 5.2, Ex. 5)

32. Simplify each. Assume that a and b are integers and that x and y are not 0.
   a. $3x^{4a}(4x^{-a})^2$ $48a^{2a}$
   b. $\dfrac{(y^{4b})^3}{y^{2b-3}}$ y^{10b+3} (Sec. 5.2)

33. Find the degree of each term.
   a. $3x^2$ 2
   b. -2^3x^5 5
   c. y 1
   d. $12x^2yz^3$ 6
   e. 5 0 (Sec. 5.3, Ex. 1)

34. Subtract $(2x - 7)$ from $2x^2 + 8x - 3$ (Sec. 5.3)

35. Multiply $[3 + (2a + b)]^2$

36. Multiply $[4 + (3x - y)]^2$

37. Factor $ab - 6a + 2b - 12$

38. Factor $xy + 2x - 5y - 10$ $(x - 5)(y + 2)$ (Sec. 5.5)

34. $2x^2 + 6x + 4$ (Sec. 5.3,)

35. $9 + 12a + 6b + 4a^2 + 4ab + b^2$ (Sec. 5.4, Ex. 9) **36.** $16 + 24x - 8y + 9x^2 - 6xy + y^2$ (Sec.5.4)

39. $2n^2 - 38n + 80$ $2(n^2 - 19n + 40)$ (Sec. 5.6, Ex. 4)

40. Factor $6x^2 - x - 35$ $(2x - 5)(3x + 7)$ (Sec. 5.6)

41. Factor $x^2 + 4x + 4 - y^2$
   $(x + 2 + y)(x + 2 - y)$ (Sec. 5.7, Ex. 5)

42. Factor $4x^2 - 4x + 1 - 9y^2$ $(2x + 3y - 1)(2x - 3y - 1)$ (Sec. 5.7)

43. Solve $(x + 2)(x - 6) = 0$ −2, 6 (Sec. 5.8, Ex. 1)

44. Solve $2x(3x + 1)(x - 3) = 0$ $0, -\frac{1}{3}, 3$ (Sec. 5.8)

45. Simplify $\dfrac{2x^2}{10x^3 - 2x^2} \cdot \dfrac{1}{5x - 1}$ (Sec. 6.1, Ex. 2)

46. For the graph of $f(x)$, answer the following:

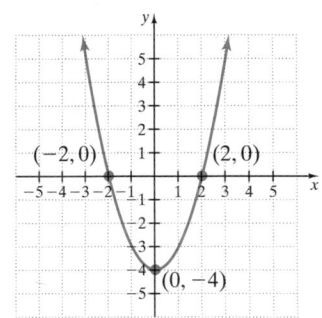

   a. Find the domain and range. domain: $(-\infty, \infty)$ range: $[-4, \infty)$

   b. List the x- and y-intercepts. x-intercepts: $(-2, 0)$, $(2, 0)$ y-intercept: $(0, -4)$

   c. Find the coordinates of the point with the greatest y-value. There is no such point

   d. Find the coordinates of the point with the least y-value. $(0, -4)$

   e. List the x-values whose y-values are equal to 0. −2, 2

   f. List the x-values whose y-values are less than 0. between $x = -2$ and $x = 2$

   g. Find the solutions of $f(x) = 0$. −2, 2 (Sec. 2.2)

47. Subtract $\dfrac{5k}{k^2 - 4} - \dfrac{2}{k^2 + k - 2}$.

48. Subtract $\dfrac{5a}{a^2 - 4} - \dfrac{3}{2 - a} \cdot \dfrac{8a + 6}{(a + 2)(a - 2)}$ (Sec. 6.2)

49. Solve: $\dfrac{3}{x} - \dfrac{x + 21}{3x} = \dfrac{5}{3}$. −2 (Sec. 6.6, Ex. 2)

50. Solve $\dfrac{3x - 4}{2x} = -\dfrac{8}{x}$. −4 (Sec. 6.6)

37. $(b - 6)(a + 2)$ (Sec. 5.5, Ex. 7)

47. $\dfrac{5k^2 - 7k + 4}{(k + 2)(k - 2)(k - 1)}$ (Sec. 6.2, Ex. 4)

Rational Exponents, Radicals, and Complex Numbers

In this chapter, radical notation is reviewed, and then rational exponents are introduced. As the name implies, rational exponents are exponents that are rational numbers. We present an interpretation of rational exponents that is consistent with the meaning and rules already established for integer exponents, and we present two forms of notation for roots: radical and exponent. We conclude this chapter with complex numbers, a natural extension of the real number system.

The National Aeronautics and Space Administration (NASA) has been making advances in science and technology since it was established in 1958. NASA employees proudly strive to accomplish their mission to explore space, learn about our own planet, and encourage future exploration of the universe. Future projects include the Kepler mission, which will search for habitable planets in deep space, and the Space Interferometry Mission, which will make continuous observations of space over 95 million kilometers from Earth.

In addition to astronauts, NASA employs many scientists of different disciplines, among them engineers, mathematicians, computer programmers, microbiologists, geologists, and medical doctors. NASA also offers summer internship and student employment programs, both paid and unpaid.

In the Spotlight on Decision Making feature on page 528, you will have the opportunity to explore a new moon orbiting the planet Uranus as a NASA scientist.

```
6.67E-11→G
          6.67E-11
9.07E20→M
          9.07E20
6.2E5→R
           620000
√(2GM/R)
```

The numbers and formula to the left are from the Spotlight on Decision Making feature on page 528. The use of scientific notation makes it easier to calculate with large numbers.

Source: www.nasa.gov *(National Aeronautics and Space Administration)*

7.1 RADICALS AND RADICAL FUNCTIONS

Objectives

1. Find square roots.
2. Approximate roots using a calculator.
3. Find cube roots.
4. Find nth roots.
5. Find $\sqrt[n]{a^n}$ where a is a real number.
6. Graph square and cube root functions.

1 Recall from Section 1.3 that to find a **square root** of a number a, we find a number that was squared to get a.

Thus, because

$$5^2 = 25 \quad \text{and} \quad (-5)^2 = 25, \text{ then}$$

both 5 and -5 are square roots of 25.

Recall that we denote the **nonnegative**, or **principal, square root** with the **radical sign**.

$$\sqrt{25} = 5$$

We denote the **negative square root** with the **negative radical sign**.

$$-\sqrt{25} = -5$$

An expression containing a radical sign is called a **radical expression**. An expression within, or "under," a radical sign is called a **radicand**.

radical expression: $\sqrt{a}$ ← radical sign

↖ radicand

Principal and Negative Square Roots

If a is a nonnegative number, then

$\sqrt{a}$ is the **principal**, or **nonnegative square root** of a

$-\sqrt{a}$ is the **negative square root** of a

EXAMPLE 1

Simplify. Assume that all variables represent positive numbers.

a. $\sqrt{36}$ **b.** $\sqrt{0}$ **c.** $\sqrt{\dfrac{4}{49}}$ **d.** $\sqrt{0.25}$

e. $\sqrt{x^6}$ **f.** $\sqrt{9x^{12}}$ **g.** $-\sqrt{81}$

Solution   **a.** $\sqrt{36} = 6$ because $6^2 = 36$ and 6 is not negative.

b. $\sqrt{0} = 0$ because $0^2 = 0$ and 0 is not negative.

c. $\sqrt{\dfrac{4}{49}} = \dfrac{2}{7}$ because $\left(\dfrac{2}{7}\right)^2 = \dfrac{4}{49}$ and $\dfrac{2}{7}$ is not negative.

d. $\sqrt{0.25} = 0.5$ because $(0.5)^2 = 0.25$.

e. $\sqrt{x^6} = x^3$ because $(x^3)^2 = x^6$.

f. $\sqrt{9x^{12}} = 3x^6$ because $(3x^6)^2 = 9x^{12}$.

g. $-\sqrt{81} = -9$. The negative in front of the radical indicates the negative square root of 81.

Can we find the square root of a negative number, say $\sqrt{-4}$? That is, can we find a real number whose square is -4? No, there is no real number whose square is -4, and we say that $\sqrt{-4}$ is not a real number. In general:

The square root of a negative number is not a real number.

> **Helpful Hint**
> * Remember: $\sqrt{0} = 0$
> * Don't forget, the square root of a negative number, such as $\sqrt{-9}$, is not a real number. In Section 7.7, we will see what kind of a number $\sqrt{-9}$ is.

2   Recall that numbers such as 1, 4, 9, and 25 are called **perfect squares**, since $1 = 1^2, 4 = 2^2, 9 = 3^2$, and $25 = 5^2$. Square roots of perfect square radicands simplify to rational numbers. What happens when we try to simplify a root such as $\sqrt{3}$? Since there is no rational number whose square is 3, then $\sqrt{3}$ is not a rational number. It is called an **irrational number**, and we can find a decimal **approximation** of it. To find decimal approximations, use a calculator. For example, an approximation for $\sqrt{3}$ is

$$\sqrt{3} \approx 1.732$$
$$\uparrow$$
approximation symbol

To see if the approximation is reasonable, notice that since

$$1 < 3 < 4, \text{ then}$$
$$\sqrt{1} < \sqrt{3} < \sqrt{4}, \text{ or}$$
$$1 < \sqrt{3} < 2.$$

We found $\sqrt{3} \approx 1.732$, a number between 1 and 2, so our result is reasonable.

EXAMPLE 2

Use a calculator to approximate $\sqrt{20}$. Round the approximation to 3 decimal places and check to see that your approximation is reasonable.

$$\sqrt{20} \approx 4.472$$

Solution   Is this reasonable? Since $16 < 20 < 25$, then $\sqrt{16} < \sqrt{20} < \sqrt{25}$, or $4 < \sqrt{20} < 5$. The approximation is between 4 and 5 and thus is reasonable.

3 Finding roots can be extended to other roots such as cube roots. For example, since $2^3 = 8$, we call 2 the **cube root** of 8. In symbols, we write

$$\sqrt[3]{8} = 2$$

Cube Root

The **cube root** of a real number a is written as $\sqrt[3]{a}$, and

$$\sqrt[3]{a} = b \text{ only if } b^3 = a$$

From this definition, we have

$$\sqrt[3]{64} = 4 \text{ since } 4^3 = 64$$
$$\sqrt[3]{-27} = -3 \text{ since } (-3)^3 = -27$$
$$\sqrt[3]{x^3} = x \text{ since } x^3 = x^3$$

Notice that, unlike with square roots, *it is possible to have a negative radicand when finding a cube root.* This is so because the *cube* of a negative number is a negative number. Therefore, the *cube root* of a negative number is a negative number.

EXAMPLE 3

Find the cube roots.

a. $\sqrt[3]{1}$ **b.** $\sqrt[3]{-64}$ **c.** $\sqrt[3]{\dfrac{8}{125}}$

d. $\sqrt[3]{x^6}$ **e.** $\sqrt[3]{-27x^9}$

Solution

a. $\sqrt[3]{1} = 1$ because $1^3 = 1$.

b. $\sqrt[3]{-64} = -4$ because $(-4)^3 = -64$.

c. $\sqrt[3]{\dfrac{8}{125}} = \dfrac{2}{5}$ because $\left(\dfrac{2}{5}\right)^3 = \dfrac{8}{125}$.

d. $\sqrt[3]{x^6} = x^2$ because $(x^2)^3 = x^6$.

e. $\sqrt[3]{-27x^9} = -3x^3$ because $(-3x^3)^3 = -27x^9$.

CLASSROOM EXAMPLE

Find each cube root.

a. $\sqrt[3]{0}$ b. $\sqrt[3]{-8}$ c. $\sqrt[3]{\dfrac{1}{64}}$

d. $\sqrt[3]{x^9}$ e. $\sqrt[3]{-64x^6}$

answer: a. 0 b. -2 c. $\dfrac{1}{4}$

d. x^3 e. $-4x^2$

TEACHING TIP

For Examples 3d and 3e, have students verify that the answer is correct for positive and negative values of x.

4 Just as we can raise a real number to powers other than 2 or 3, we can find roots other than square roots and cube roots. In fact, we can find the **nth root** of a number, where n is any natural number. In symbols, the *n*th root of a is written as $\sqrt[n]{a}$, where n is called the **index**. The index 2 is usually omitted for square roots.

> **Helpful Hint**
>
> If the index is even, such as $\sqrt{}$, $\sqrt[4]{}$, $\sqrt[6]{}$, and so on, the radicand must be non-negative for the root to be a real number. For example,
>
> $$\sqrt[4]{16} = 2, \text{ but } \sqrt[4]{-16} \text{ is not a real number.}$$
> $$\sqrt[6]{64} = 2, \text{ but } \sqrt[6]{-64} \text{ is not a real number.}$$
>
> If the index is odd, such as $\sqrt[3]{}$, $\sqrt[5]{}$, and so on, the radicand may be any real number. For example,
>
> $$\sqrt[3]{64} = 4 \text{ and } \sqrt[3]{-64} = -4$$
> $$\sqrt[5]{32} = 2 \text{ and } \sqrt[5]{-32} = -2$$

✔ **CONCEPT CHECK**

Which one is not a real number?

a. $\sqrt[3]{-15}$ **b.** $\sqrt[4]{-15}$ **c.** $\sqrt[5]{-15}$ **d.** $\sqrt{(-15)^2}$

EXAMPLE 4

Simplify the following expressions.

a. $\sqrt[4]{81}$ **b.** $\sqrt[5]{-243}$ **c.** $-\sqrt{25}$

d. $\sqrt[4]{-81}$ **e.** $\sqrt[3]{64x^3}$

Solution   **a.** $\sqrt[4]{81} = 3$ because $3^4 = 81$ and 3 is positive.
b. $\sqrt[5]{-243} = -3$ because $(-3)^5 = -243$.
c. $-\sqrt{25} = -5$ because -5 is the opposite of $\sqrt{25}$.
d. $\sqrt[4]{-81}$ is not a real number. There is no real number that, when raised to the fourth power, is -81.
e. $\sqrt[3]{64x^3} = 4x$ because $(4x)^3 = 64x^3$.

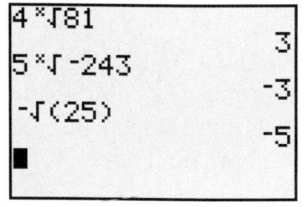

A calculator check for Example 4a, b, c.

5 Recall that the notation $\sqrt{a^2}$ indicates the positive square root of a^2 only. For example,

$$\sqrt{(-5)^2} = \sqrt{25} = 5$$

When variables are present in the radicand and it is unclear whether the variable represents a positive number or a negative number, absolute value bars are sometimes needed to ensure that the result is a positive number. For example,

$$\sqrt{x^2} = |x|$$

This ensures that the result is positive. This same situation may occur when the index is any *even* positive integer. When the index is any *odd* positive integer, absolute value bars are not necessary.

> **Finding $\sqrt[n]{a^n}$**
>
> If n is an *even* positive integer, then $\sqrt[n]{a^n} = |a|$.
>
> If n is an *odd* positive integer, then $\sqrt[n]{a^n} = a$.

EXAMPLE 5

Simplify.

a. $\sqrt{(-3)^2}$ **b.** $\sqrt{x^2}$ **c.** $\sqrt[4]{(x-2)^4}$ **d.** $\sqrt[3]{(-5)^3}$

e. $\sqrt[5]{(2x-7)^5}$ **f.** $\sqrt{25x^2}$ **g.** $\sqrt{x^2 + 2x + 1}$

Solution   **a.** $\sqrt{(-3)^2} = |-3| = 3$ When the index is even, the absolute value bars ensure us that our result is not negative.

b. $\sqrt{x^2} = |x|$

c. $\sqrt[4]{(x-2)^4} = |x-2|$

d. $\sqrt[3]{(-5)^3} = -5$

e. $\sqrt[5]{(2x-7)^5} = 2x - 7$ Absolute value bars are not needed when the index is odd.

f. $\sqrt{25x^2} = 5|x|$

g. $\sqrt{x^2 + 2x + 1} = \sqrt{(x+1)^2} = |x+1|$

CLASSROOM EXAMPLE

Simplify. Assume that the variables represent any real number.

a. $\sqrt{(-5)^2}$ **b.** $\sqrt{x^6}$

c. $\sqrt[4]{(x+6)^4}$ **d.** $\sqrt[3]{(-3)^3}$

e. $\sqrt[5]{(7x-1)^5}$ **f.** $\sqrt{36x^2}$

g. $\sqrt{x^2 + 6x + 9}$

answer: **a.** 5   **b.** $|x^3|$ **c.** $|x+6|$
d. -3 **e.** $7x - 1$ **f.** $6|x|$ **g.** $|x+3|$

6   Recall that an equation in x and y describes a function if each x-value is paired with exactly one y-value. With this in mind, does the equation

$$y = \sqrt{x}$$

describe a function? First, notice that replacement values for x must be nonnegative real numbers, since $\sqrt{x}$ is not a real number if $x < 0$. The notation $\sqrt{x}$ denotes the principal square root of x, so for every nonnegative number x, there is exactly one number, $\sqrt{x}$. Therefore, $y = \sqrt{x}$ describes a function, and we may write it as

$$f(x) = \sqrt{x}$$

In general, radical functions are functions of the form

$$f(x) = \sqrt[n]{x}.$$

TEACHING TIP

Have students work in pairs to determine the domain of the functions in the Examples. Then have students share results with the class. If results vary, have students try to convince others of their solution until agreement is reached.

Recall that the domain of a function in x is the set of all possible replacement values of x. This means that if n is even, the domain is the set of all nonnegative numbers, or $\{x \mid x \geq 0\}$. If n is odd, the domain is the set of all real numbers. Keep this in mind as we find function values.

EXAMPLE 6

If $f(x) = \sqrt{x - 4}$ and $g(x) = \sqrt[3]{x + 2}$, find each function value.

a. $f(8)$ **b.** $f(6)$

c. $g(-1)$ **d.** $g(1)$

Solution   **a.** $f(8) = \sqrt{8 - 4} = \sqrt{4} = 2$

b. $f(6) = \sqrt{6 - 4} = \sqrt{2}$

c. $g(-1) = \sqrt[3]{-1 + 2} = \sqrt[3]{1} = 1$

d. $g(1) = \sqrt[3]{1 + 2} = \sqrt[3]{3}$

> **Helpful Hint**
>
> Notice that for the function $f(x) = \sqrt{x - 4}$, the domain includes all real numbers that make the radicand ≥ 0. To see what numbers these are, solve $x - 4 \geq 0$ and find that $x \geq 4$. The domain is $\{x \mid x \geq 4\}$.
>
> The domain of the cube root function $g(x) = \sqrt[3]{x + 2}$ is the set of real numbers.

EXAMPLE 7

Identify the domain of the square root function $f(x) = \sqrt{x}$ and then graph the function.

Solution   To become familiar with this function, we find function values for $f(x)$. For example,

$$f(0) = \sqrt{0} = 0$$
$$f(1) = \sqrt{1} = 1$$
$$f(4) = \sqrt{4} = 2$$
$$f(9) = \sqrt{9} = 3$$

Choosing perfect squares for x ensures us that $f(x)$ is a rational number, but it is important to stress that $f(x) = \sqrt{x}$ is defined for all nonnegative real numbers. For example,

$$f(3) = \sqrt{3} \approx 1.732$$

The domain of this function is the set of all nonnegative numbers or $\{x \mid x \geq 0\}$. Below, we graph $y_1 = \sqrt{x}$ in a $[-3, 10, 1]$ by $[-2, 10, 1]$ window. The table below shows that a negative x-value gives an error message since $\sqrt{x}$ is not a real number when x is negative.

Notice that the graph of this function passes the vertical line test, as expected.

The equation $f(x) = \sqrt[3]{x}$ also describes a function. Here x may be any real number, so the domain of this function is the set of all real numbers, or $(-\infty, \infty)$. A few function values are given next.

$$f(0) = \sqrt[3]{0} = 0$$
$$f(1) = \sqrt[3]{1} = 1$$

$$f(-1) = \sqrt[3]{-1} = -1$$

$$f(6) = \sqrt[3]{6}$$
$$f(-6) = \sqrt[3]{-6}$$
$\left.\begin{array}{c}\\\\\end{array}\right\}$ Here, there is no rational number whose cube is 6. Thus, the radicals do not simplify to rational numbers.

$$f(8) = \sqrt[3]{8} = 2$$
$$f(-8) = \sqrt[3]{-8} = -2$$

EXAMPLE 8

Identify the domain of the function $f(x) = \sqrt[3]{x}$ and then graph the function.

Solution   The domain of this function is the set of all real numbers. Below, we graph $y_1 = \sqrt[3]{x}$ in a $[-10, 10, 1]$ by $[-5, 5, 1]$ window.

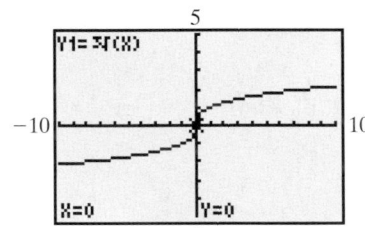

The graph of this function passes the vertical line test, as expected.

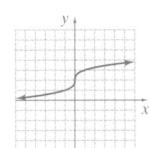

Spotlight on
DECISION
＆ MAKING

Suppose you are a scientist working for NASA. A new moon, S/2001U1, has been discovered orbiting the planet Uranus in our outer solar system. You have been asked to check whether it is possible for this moon to have an oxygen atmosphere. You can do so by comparing the average speed of an oxygen molecule (480 meters per second) to the moon's **escape velocity**, the speed an object must travel to permanently leave the moon's gravitational pull. If the moon's escape velocity is greater than the average speed of oxygen molecules, then it is possible for the moon to retain oxygen in its atmosphere—that is, if oxygen exists on the moon at all.

Data about the new moon are listed in the table. Use that along with the escape velocity formula given to the right to decide whether it is possible for S/2001U1 to have an oxygen atmosphere.

$$v = \sqrt{\frac{2GM}{r}}, \text{ where}$$

v is the escape velocity (in meters per second, m/s),

M is the mass of the moon (in kilograms, kg),

r is the radius of the moon (in meters, m), and

G is the universal constant of gravitation where

$$\left(G = 6.67 \times 10^{-11} \frac{\text{m}^3}{\text{kg} \cdot \text{s}^2}\right).$$

S/2001U1 Parameters	
Mass	9.07×10^{20} kg
Radius	620,000 m
Visual geometric albedo	0.07
Orbital period	7.2 days

MENTAL MATH

Choose the correct letter. No pencil is needed, just think your way through these.

1. Which radical is not a real number?

   a. $\sqrt{3}$ **b.** $-\sqrt{11}$ **c.** $\sqrt[3]{-10}$ **d.** $\sqrt{-10}$ d

2. Which radical(s) simplify to 3?

   a. $\sqrt{9}$ **b.** $\sqrt{-9}$ **c.** $\sqrt[3]{27}$ **d.** $\sqrt[3]{-27}$ a, c

3. Which radical(s) simplify to −3?

   a. $\sqrt{9}$ **b.** $\sqrt{-9}$ **c.** $\sqrt[3]{27}$ **d.** $\sqrt[3]{-27}$ d

4. Which radical docs not simplify to a whole number?

   a. $\sqrt{64}$ **b.** $\sqrt[3]{64}$ **c.** $\sqrt{8}$ **d.** $\sqrt[3]{8}$ c

EXERCISE SET 7.1

STUDY GUIDE/SSM   CD/ VIDEO   PH MATH TUTOR CENTER   MathXL®Tutorials ON CD   MathXL®   MyMathLab®

31. not a real number   **32.** not a real number

Simplify. Assume that variables represent positive real numbers. See Example 1.

1. $\sqrt{100}$ 10   **2.** $\sqrt{400}$ 20

3. $\sqrt{\dfrac{1}{4}}$ $\dfrac{1}{2}$ **4.** $\sqrt{\dfrac{9}{25}}$ $\dfrac{3}{5}$

5. $\sqrt{0.0001}$ 0.01   **6.** $\sqrt{0.04}$ 0.2

7. $-\sqrt{36}$ −6   **8.** $-\sqrt{9}$ −3

9. $\sqrt{x^{10}}$ x^5 **10.** $\sqrt{x^{16}}$ x^8

11. $\sqrt{16y^6}$ $4y^3$ **12.** $\sqrt{64y^{20}}$ $8y^{10}$

Use a calculator to approximate each square root to 3 decimal places. Check to see that each approximation is reasonable. See Example 2.

13. $\sqrt{7}$ 2.646   **14.** $\sqrt{11}$ 3.317

15. $\sqrt{38}$ 6.164   **16.** $\sqrt{56}$ 7.483

17. $\sqrt{200}$ 14.142   **18.** $\sqrt{300}$ 17.321

Find each cube root. See Example 3.

19. $\sqrt[3]{64}$ 4   **20.** $\sqrt[3]{27}$ 3

21. $\sqrt[3]{\dfrac{1}{8}}$ $\dfrac{1}{2}$ **22.** $\sqrt[3]{\dfrac{27}{64}}$ $\dfrac{3}{4}$

23. $\sqrt[3]{-1}$ −1   **24.** $\sqrt[3]{-125}$ −5

25. $\sqrt[3]{x^{12}}$ x^4 **26.** $\sqrt[3]{x^{15}}$ x^5

27. $\sqrt[3]{-27x^9}$ $-3x^3$ **28.** $\sqrt[3]{-64x^6}$ $-4x^2$

Find each root. Assume that all variables represent nonnegative real numbers. See Example 4.

29. $-\sqrt[4]{16}$ −2   **30.** $\sqrt[5]{-243}$ −3

31. $\sqrt[4]{-16}$ **32.** $\sqrt{-16}$

33. $\sqrt[5]{-32}$ −2   **34.** $\sqrt[5]{-1}$ −1

35. $\sqrt[5]{x^{20}}$ x^4 **36.** $\sqrt[4]{x^{20}}$ x^5

37. $\sqrt[6]{64x^{12}}$ $2x^2$ **38.** $\sqrt[5]{-32x^{15}}$ $-2x^3$

39. $\sqrt{81x^4}$ $9x^2$ **40.** $\sqrt[4]{81x^4}$ $3x$

41. $\sqrt[4]{256x^8}$ $4x^2$ **42.** $\sqrt{256x^8}$ $16x^4$

Simplify. Assume that the variables represent any real number. See Example 5.

43. $\sqrt{(-8)^2}$ 8   **44.** $\sqrt{(-7)^2}$ 7

45. $\sqrt[3]{(-8)^3}$ −8   **46.** $\sqrt[5]{(-7)^5}$ −7

47. $\sqrt{4x^2}$ $2|x|$ **48.** $\sqrt[4]{16x^4}$ $4|x|$

49. $\sqrt[3]{x^3}$ x **50.** $\sqrt[5]{x^5}$ x

51. $\sqrt{(x-5)^2}$ $|x-5|$ **52.** $\sqrt{(y-6)^2}$ $|y-6|$

53. $\sqrt{x^2+4x+4}$

   (*Hint:* Factor the polynomial first.) $|x+2|$

54. $\sqrt{x^2-8x+16}$

   (*Hint:* Factor the polynomial first.) $|x-4|$

MIXED PRACTICE   **61.** $5ab^{10}$ **63.** $-3x^4y^3$ **64.** $-2a^7b^2$

Simplify each radical. Assume that all variables represent positive real numbers.

55. $-\sqrt{121}$ -11 **56.** $-\sqrt[3]{125}$ -5 **57.** $\sqrt[3]{8x^3}$ $2x$

58. $\sqrt{16x^8}$ $4x^4$ **59.** $\sqrt{y^{12}}$ y^6 **60.** $\sqrt[3]{y^{12}}$ y^4

61. $\sqrt{25a^2b^{20}}$ **62.** $\sqrt{9x^4y^6}$ $3x^2y^3$ **63.** $\sqrt[3]{-27x^{12}y^9}$

64. $\sqrt[3]{-8a^{21}b^6}$ **65.** $\sqrt[4]{a^{16}b^4}$ a^4b **66.** $\sqrt[4]{x^8y^{12}}$ x^2y^3

67. $\sqrt[5]{-32x^{10}y^5}$ $-2x^2y$ **68.** $\sqrt[5]{-243z^{15}}$ $-3z^3$

69. $\sqrt{\dfrac{25}{49}}$ $\dfrac{5}{7}$ **70.** $\sqrt{\dfrac{4}{81}}$ $\dfrac{2}{9}$ **71.** $\sqrt{\dfrac{x^2}{4y^2}}$ $\dfrac{x}{2y}$

72. $\sqrt{\dfrac{y^{10}}{9x^6}}$ $\dfrac{y^5}{3x^3}$ **73.** $-\sqrt[3]{\dfrac{z^{21}}{27x^3}}$ $-\dfrac{z^7}{3x}$ **74.** $-\sqrt[3]{\dfrac{64a^3}{b^9}}$ $-\dfrac{4a}{b^3}$

 75. $\sqrt[4]{\dfrac{x^4}{16}}$ $\dfrac{x}{2}$ **76.** $\sqrt[4]{\dfrac{y^4}{81x^4}}$ $\dfrac{y}{3x}$

If $f(x) = \sqrt{2x+3}$ *and* $g(x) = \sqrt[3]{x-8}$, *find the following function values. See Example 6.*

77. $f(0)$ $\sqrt{3}$ **78.** $g(0)$ -2 **79.** $g(7)$ -1 **80.** $f(-1)$ 1

81. $g(-19)$ -3 **82.** $f(3)$ 3 **83.** $f(2)$ $\sqrt{7}$ **84.** $g(1)$ $\sqrt[3]{-7}$

Identify the domain and then graph each function. See Example 7.

85. $f(x) = \sqrt{x}+2$ $[0,\infty)$.   **86.** $f(x) = \sqrt{x}-2$ $[0,\infty)$

87. $f(x) = \sqrt{x-3}$; use the following table.
$[3,\infty)$

85.–90. See graphing answer section.

x	$f(x)$
3	0
4	1
7	2
12	3

88. $f(x) = \sqrt{x+1}$; use the following table.
$[-1,\infty)$

x	$f(x)$
-1	0
0	1
3	2
8	3

Match the graph with its equation. All graphs are in a $[-10,10,1]$ *by* $[-5,5,5]$ *window.*

A.   **B.**

C.   **D.**

89. $f(x) = \sqrt{x}+2$ C   **90.** $f(x) = \sqrt{x}-2$ A

91. $f(x) = \sqrt{x}-3$ D   **92.** $f(x) = \sqrt{x}+1$ B

Match the graph with its equation. All graphs are in a $[-4.7,4.7,1]$ *by* $[-5,5,1]$ *window.*

A.   **B.**

C.   **D.**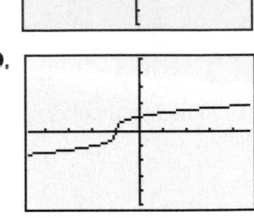

93. $f(x) = \sqrt[3]{x}+1$ A   **94.** $f(x) = \sqrt[3]{x}-2$ C

95. $g(x) = \sqrt[3]{x-1}$ B   **96.** $g(x) = \sqrt[3]{x+1}$ D

Identify the domain and then graph each function. See Example 8.

97. $f(x) = \sqrt[3]{x}+1$ $(-\infty,\infty)$

98. $f(x) = \sqrt[3]{x}-2$ $(-\infty,\infty)$

99. $g(x) = \sqrt[3]{x-1}$; use the following table. $(-\infty,\infty)$

x	$g(x)$
-1	0
0	1
-2	-1
7	2
-9	-2

99.–100. See graphing answer section.

100. $g(x) = \sqrt[3]{x+1}$; use the following table. $(-\infty,\infty)$

x	$g(x)$
1	0
2	1
0	-1
9	2
-7	-2

REVIEW AND PREVIEW   **103.** $-60x^7y^{10}z^5$ **104.** $-28a^6b^2c^6$

Simplify each exponential expression. See Sections 5.1 and 5.2.

101. $(-2x^3y^2)^5$ $-32x^{15}y^{10}$ **102.** $(4y^6z^7)^3$ $64y^{18}z^{21}$

103. $(-3x^2y^3z^5)(20x^5y^7)$ **104.** $(-14a^5bc^2)(2abc^4)$

105. $\dfrac{7x^{-1}y}{14(x^5y^2)^{-2}}$ $\dfrac{x^9y^5}{2}$ **106.** $\dfrac{(2a^{-1}b^2)^3}{(8a^2b)^{-2}}$ $512ab^8$

Concept Extensions

107. Explain why $\sqrt{-64}$ is not a real number.   answers may vary

108. Explain why $\sqrt[3]{-64}$ is a real number.   answers may vary

For Exercises 109 through 112, do not use a calculator.

109. $\sqrt{160}$ is closest to

    a. 10   **b.** 13   **c.** 20   **d.** 40 13

110. $\sqrt{1000}$ is closest to

    a. 10   **b.** 30   **c.** 100   **d.** 500 30

111. The perimeter of the triangle is closest to

    a. 12   **b.** 18   **c.** 66   **d.** 132 18

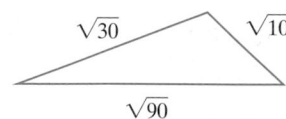

112. The length of the bent wire is closest to

    a. 5   **b.** $\sqrt{28}$ **c.** 7   **d.** 14 7

The Mosteller formula for calculating adult body surface area is $B = \sqrt{\dfrac{hw}{3131}}$, *where B is an individual's body surface area in square meters, h is the individual's height in inches, and w is the individual's weight in pounds. Use this information to answer Exercises 113 and 114. Round answers to 2 decimal places.*

113. Find the body surface area of an individual who is 66 inches tall and who weighs 135 pounds.   1.69 sq. m

114. Find the body surface area of an individual who is 74 inches tall and who weighs 225 pounds.   2.31 sq. m

115. Suppose that a friend tells you that $\sqrt{13} \approx 5.7$. Without a calculator, how can you convince your friend that he or she must have made an error?   answers may vary

116. Escape velocity is the minimum speed that an object must reach to escape a planet's pull of gravity. Escape velocity v is given by the equation $v = \sqrt{\dfrac{2Gm}{r}}$, where m is the mass of the planet, r is its radius, and G is the universal gravitational constant, which has a value of $G = 6.67 \times 10^{-11}$ m^3/kg·s^2. The mass of Earth is 5.97×10^{24} kg and its radius is 6.37×10^6 m. Use this information to find the escape velocity for Earth. Round to the nearest whole number. (Source: National Space Science Data Center)   11,181 m per sec

7.2   RATIONAL EXPONENTS

Objectives

  1 Understand the meaning of $a^{1/n}$.

  2 Understand the meaning of $a^{m/n}$.

  3 Understand the meaning of $a^{-m/n}$.

  4 Use rules for exponents to simplify expressions that contain rational exponents.

  5 Use rational exponents to simplify radical expressions.

1   So far in this text, we have not defined expressions with rational exponents such as $3^{1/2}$, $x^{2/3}$, and $-9^{-1/4}$. We will define these expressions so that the rules for exponents will apply to these rational exponents as well.

Suppose that $x = 5^{1/3}$. Then

using rules for
⌈exponents⌉
↓

$$x^3 = \left(5^{1/3}\right)^3 = 5^{1/3 \cdot 3} = 5^1 \text{ or } 5$$

Since $x^3 = 5$, then x is the number whose cube is 5, or $x = \sqrt[3]{5}$. Notice that we also know that $x = 5^{1/3}$. This means

$$5^{1/3} = \sqrt[3]{5}$$

> **Definition of $a^{1/n}$**
>
> If n is a positive integer greater than 1 and $\sqrt[n]{a}$ is a real number, then
>
> $$a^{1/n} = \sqrt[n]{a}$$

Notice that the denominator of the rational exponent corresponds to the index of the radical.

EXAMPLE 1

Use radical notation to write the following. Simplify if possible.

a. $4^{1/2}$ **b.** $64^{1/3}$

c. $x^{1/4}$ **d.** $0^{1/6}$

e. $-9^{1/2}$ **f.** $(81x^8)^{1/4}$

g. $(5y)^{1/3}$

Solution
a. $4^{1/2} = \sqrt{4} = 2$ **b.** $64^{1/3} = \sqrt[3]{64} = 4$

c. $x^{1/4} = \sqrt[4]{x}$ **d.** $0^{1/6} = \sqrt[6]{0} = 0$

e. $-9^{1/2} = -\sqrt{9} = -3$ **f.** $(81x^8)^{1/4} = \sqrt[4]{81x^8} = 3x^2$

g. $(5y)^{1/3} = \sqrt[3]{5y}$

2   As we expand our use of exponents to include $\dfrac{m}{n}$, we define their meaning so that rules for exponents still hold true. For example, by properties of exponents,

$$8^{2/3} = (8^{1/3})^2 = \left(\sqrt[3]{8}\right)^2 \quad \text{or}$$
$$8^{2/3} = (8^2)^{1/3} = \sqrt[3]{8^2}$$

> **Definition of $a^{m/n}$**
>
> If m and n are positive integers greater than 1 with $\dfrac{m}{n}$ in lowest terms, then
>
> $$a^{m/n} = \sqrt[n]{a^m} = \left(\sqrt[n]{a}\right)^m$$
>
> as long as $\sqrt[n]{a}$ is a real number.

Notice that the denominator n of the rational exponent corresponds to the index of the radical. The numerator m of the rational exponent indicates that the base is to be raised to the mth power. This means

$$8^{2/3} = \sqrt[3]{8^2} = \sqrt[3]{64} = 4 \quad \text{or}$$
$$8^{2/3} = \left(\sqrt[3]{8}\right)^2 = 2^2 = 4$$

From simplifying $8^{2/3}$, can you see that it doesn't matter whether you raise to a power first and then take the nth root or you take the nth root first and then raise to a power?

> **Helpful Hint**
>
> Most of the time, $\left(\sqrt[n]{a} \right)^m$ will be easier to calculate than $\sqrt[n]{a^m}$.

EXAMPLE 2

Use radical notation to write the following. Then simplify if possible.

a. $4^{3/2}$ **b.** $-16^{3/4}$ **c.** $(-27)^{2/3}$

d. $\left(\dfrac{1}{9} \right)^{3/2}$ **e.** $(4x - 1)^{3/5}$

Solution

a. $4^{3/2} = \left(\sqrt{4} \right)^3 = 2^3 = 8$

b. $-16^{3/4} = -\left(\sqrt[4]{16} \right)^3 = -(2)^3 = -8$

c. $(-27)^{2/3} = \left(\sqrt[3]{-27} \right)^2 = (-3)^2 = 9$

d. $\left(\dfrac{1}{9} \right)^{3/2} = \left(\sqrt{\dfrac{1}{9}} \right)^3 = \left(\dfrac{1}{3} \right)^3 = \dfrac{1}{27}$

e. $(4x - 1)^{3/5} = \sqrt[5]{(4x - 1)^3}$

CLASSROOM EXAMPLE

Use radical notation to rewrite each expression. Simplify if possible.

a. $9^{3/2}$ **b.** $-256^{3/4}$ **c.** $(-32)^{2/5}$

d. $\left(\dfrac{1}{4} \right)^{3/2}$ **e.** $(2x + 1)^{2/7}$

answer: **a.** 27 **b.** -64 **c.** 4

d. $\dfrac{1}{8}$ **e.** $\sqrt[7]{(2x + 1)^2}$

> **Helpful Hint**
>
> The *denominator* of a rational exponent is the index of the corresponding radical. For example, $x^{1/5} = \sqrt[5]{x}$ and $z^{2/3} = \sqrt[3]{z^2}$, or $z^{2/3} = \left(\sqrt[3]{z} \right)^2$.

3   The rational exponents we have given meaning to exclude negative rational numbers. To complete the set of definitions, we define $a^{-m/n}$.

TEACHING TIP

Ask students, "What is 3^{-2}? What is a^{-2}? What is a^{-n} when n is an integer? Using this information, what is $a^{-m/n}$ where m and n are integers?"

> **Definition of $a^{-m/n}$**
>
> $$a^{-m/n} = \dfrac{1}{a^{m/n}}$$
>
> as long as $a^{m/n}$ is a nonzero real number.

EXAMPLE 3

Write each expression with a positive exponent, and then simplify.

a. $16^{-3/4}$ **b.** $(-27)^{-2/3}$

Solution

a. $16^{-3/4} = \dfrac{1}{16^{3/4}} = \dfrac{1}{\left(\sqrt[4]{16}\right)^3} = \dfrac{1}{2^3} = \dfrac{1}{8}$

b. $(-27)^{-2/3} = \dfrac{1}{(-27)^{2/3}} = \dfrac{1}{\left(\sqrt[3]{-27}\right)^2} = \dfrac{1}{(-3)^2} = \dfrac{1}{9}$

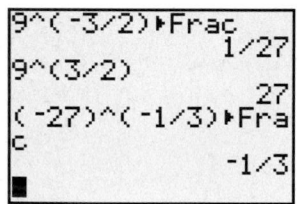

A calculator check for
Example 3.

A calculator check for the
Helpful Hint.

> **Helpful Hint**
>
> If an expression contains a negative rational exponent, such as $9^{-3/2}$, you may want to first write the expression with a positive exponent and then interpret the rational exponent. Notice that the sign of the base is not affected by the sign of its exponent. For example,
>
> $$9^{-3/2} = \frac{1}{9^{3/2}} = \frac{1}{\left(\sqrt{9}\right)^3} = \frac{1}{27}$$
>
> Also,
>
> $$(-27)^{-1/3} = \frac{1}{(-27)^{1/3}} = -\frac{1}{3}$$

✔ **CONCEPT CHECK**

Which one is correct?

a. $-8^{2/3} = \dfrac{1}{4}$ **b.** $8^{-2/3} = -\dfrac{1}{4}$ **c.** $8^{-2/3} = -4$ **d.** $-8^{-2/3} = -\dfrac{1}{4}$

④   It can be shown that the properties of integer exponents hold for rational exponents. By using these properties and definitions, we can now simplify expressions that contain rational exponents.

These rules are repeated here for review.

Note: For the remainder of this chapter, we will assume that variables represent positive real numbers. Since this is so, we need not insert absolute value bars when we simplify even roots.

TEACHING TIP
Please note the comment about variables representing positive real numbers only.

Summary of Exponent Rules

If m and n are rational numbers, and a, b, and c are numbers for which the expressions below exist, then

Product rule for exponents: $\qquad\qquad\qquad a^m \cdot a^n = a^{m+n}$

Power rule for exponents: $\qquad\qquad\qquad (a^m)^n = a^{m \cdot n}$

Power rules for products and quotients: $\qquad (ab)^n = a^n b^n \qquad\qquad$ and

$$\left(\frac{a}{c}\right)^n = \frac{a^n}{c^n}, c \neq 0$$

Quotient rule for exponents: $\qquad\qquad \dfrac{a^m}{a^n} = a^{m-n}, a \neq 0$

Zero exponent: $\qquad\qquad\qquad\qquad\quad a^0 = 1, a \neq 0$

Negative exponent: $\qquad\qquad\qquad\quad a^{-n} = \dfrac{1}{a^n}, a \neq 0$

EXAMPLE 4

A calculator check for Example 4c.

Use properties of exponents to simplify. Write results with only positive exponents.

a. $b^{1/3} \cdot b^{5/3}$ $\qquad$ **b.** $x^{1/2} x^{1/3}$ $\qquad$ **c.** $\dfrac{7^{1/3}}{7^{4/3}}$

d. $y^{-4/7} \cdot y^{6/7}$ $\qquad$ **e.** $\dfrac{(2x^{2/5} y^{-1/3})^5}{x^2 y}$

Solution

a. $b^{1/3} \cdot b^{5/3} = b^{(1/3+5/3)} = b^{6/3} = b^2$

b. $x^{1/2} x^{1/3} = x^{(1/2+1/3)} = x^{3/6+2/6} = x^{5/6}$ $\qquad$ *Use the product rule.*

c. $\dfrac{7^{1/3}}{7^{4/3}} = 7^{1/3-4/3} = 7^{-3/3} = 7^{-1} = \dfrac{1}{7}$ $\qquad$ *Use the quotient rule.*

d. $y^{-4/7} \cdot y^{6/7} = y^{-4/7+6/7} = y^{2/7}$ $\qquad$ *Use the product rule.*

e. We begin by using the power rule $(ab)^m = a^m b^m$ to simplify the numerator.

$$\frac{(2x^{2/5} y^{-1/3})^5}{x^2 y} = \frac{2^5 (x^{2/5})^5 (y^{-1/3})^5}{x^2 y} = \frac{32 x^2 y^{-5/3}}{x^2 y} \qquad \text{Use the power rule and simplify}$$

$$= 32 x^{2-2} y^{-5/3-3/3} \qquad \text{Apply the quotient rule.}$$

$$= 32 x^0 y^{-8/3}$$

$$= \frac{32}{y^{8/3}}$$

CLASSROOM EXAMPLE
Use the properties of exponents to simplify.

a. $x^{1/3} x^{1/4}$ $\quad$ **b.** $\dfrac{9^{2/5}}{9^{12/5}}$ $\quad$ **c.** $y^{-3/10} \cdot y^{6/10}$

d. $(11^{2/9})^3$ $\quad$ **e.** $\dfrac{(3x^{-2/3})^3}{x^2}$

answer: **a.** $x^{7/12}$ $\quad$ **b.** $\dfrac{1}{81}$ $\quad$ **c.** $y^{3/10}$

d. $11^{2/3}$ $\quad$ **e.** $\dfrac{27}{x^4}$

EXAMPLE 5

Multiply.

a. $z^{2/3}(z^{1/3} - z^5)$ **b.** $(x^{1/3} - 5)(x^{1/3} + 2)$

Solution

a. $z^{2/3}(z^{1/3} - z^5) = z^{2/3}z^{1/3} - z^{2/3}z^5$ Apply the distributive property.

$= z^{(2/3+1/3)} - z^{(2/3+5)}$ Use the product rule.

$= z^{3/3} - z^{(2/3+15/3)}$

$= z - z^{17/3}$

b. $(x^{1/3} - 5)(x^{1/3} + 2) = x^{2/3} + 2x^{1/3} - 5x^{1/3} - 10$ Think of $(x^{1/3} - 5)$ and $(x^{1/3} + 2)$ as 2 binomials, and FOIL.

$= x^{2/3} - 3x^{1/3} - 10$

EXAMPLE 6

Factor $x^{-1/2}$ from the expression $3x^{-1/2} - 7x^{5/2}$. Assume that all variables represent positive numbers.

Solution

$$3x^{-1/2} - 7x^{5/2} = (x^{-1/2})(3) - (x^{-1/2})(7x^{6/2})$$
$$= x^{-1/2}(3 - 7x^3)$$

To check, multiply $x^{-1/2}(3 - 7x^3)$ to see that the product is $3x^{-1/2} - 7x^{5/2}$.

5   Some radical expressions are easier to simplify when we first write them with rational exponents. We can simplify some radical expressions by first writing the expression with rational exponents. Use properties of exponents to simplify, and then convert back to radical notation.

EXAMPLE 7

Use rational exponents to simplify. Assume that variables represent positive numbers.

a. $\sqrt[8]{x^4}$ **b.** $\sqrt[6]{25}$ **c.** $\sqrt[4]{r^2 s^6}$

Solution

a. $\sqrt[8]{x^4} = x^{4/8} = x^{1/2} = \sqrt{x}$

b. $\sqrt[6]{25} = 25^{1/6} = (5^2)^{1/6} = 5^{2/6} = 5^{1/3} = \sqrt[3]{5}$

c. $\sqrt[4]{r^2 s^6} = (r^2 s^6)^{1/4} = r^{2/4} s^{6/4} = r^{1/2} s^{3/2} = (rs^3)^{1/2} = \sqrt{rs^3}$

```
6×√25
        1.709975947
³√(5)
        1.709975947
■
```

A calculator check for
Example 7b.

EXAMPLE 8

Use rational exponents to write as a single radical.

a. $\sqrt{x} \cdot \sqrt[4]{x}$ **b.** $\dfrac{\sqrt{x}}{\sqrt[3]{x}}$ **c.** $\sqrt[3]{3} \cdot \sqrt{2}$

Solution

a. $\sqrt{x} \cdot \sqrt[4]{x} = x^{1/2} \cdot x^{1/4} = x^{1/2+1/4}$
$$= x^{3/4} = \sqrt[4]{x^3}$$

b. $\dfrac{\sqrt{x}}{\sqrt[3]{x}} = \dfrac{x^{1/2}}{x^{1/3}} = x^{1/2-1/3} = x^{3/6-2/6}$
$$= x^{1/6} = \sqrt[6]{x}$$

c. $\sqrt[3]{3} \cdot \sqrt{2} = 3^{1/3} \cdot 2^{1/2}$ Write with rational exponents.

$\phantom{\sqrt[3]{3} \cdot \sqrt{2}} = 3^{2/6} \cdot 2^{3/6}$ Write the exponents so that they have the same denominator.

$\phantom{\sqrt[3]{3} \cdot \sqrt{2}} = (3^2 \cdot 2^3)^{1/6}$ Use $a^n b^n = (ab)^n$

$\phantom{\sqrt[3]{3} \cdot \sqrt{2}} = \sqrt[6]{3^2 \cdot 2^3}$ Write with radical notation.

$\phantom{\sqrt[3]{3} \cdot \sqrt{2}} = \sqrt[6]{72}$ Multiply $3^2 \cdot 2^3$.

```
³√(3)*√(2)
         2.039648903
6 ˣ√72
         2.039648903
■
```

A calculator check for
Example 8c.

Spotlight on DECISION & MAKING

Suppose you are a telecommunications industry analyst. A colleague has just formulated a mathematical model for the number of cellular telephone subscriptions in the United States from 1985 to 2002. The model is $y = 0.162x^{22/5}$ where y is the number of cellular telephone subscriptions x years after 1980. The actual data from 1985 to 2002 are listed in the table.

Your colleague has asked your help in evaluating whether this model represents the actual data well. By comparing the numbers of subscriptions given by the model to the actual data given in the table, decide whether this mathematical model is acceptable. Explain your reasoning.

U.S. Cellular Telephone Subscriptions, 1985–2002

Year	Subscriptions (in thousands)	Year	Subscriptions (in thousands)
1985	204	1994	19,283
1986	500	1995	28,154
1987	884	1996	38,195
1988	1607	1997	48,706
1989	2692	1998	60,831
1990	4367	1999	76,285
1991	6380	2000	97,036
1992	8893	2001	118,398
1993	13,067	2002	134,561

(*Source:* The CTIA Semi-Annual Wireless Survey)

MENTAL MATH

Choose the correct letter for each exercise. Letters will be used more than once. No pencil is needed. Just think about the meaning of each expression.

A = 2, B = −2, C = not a real number

1. $4^{1/2}$ A
2. $-4^{1/2}$ B
3. $(-4)^{1/2}$ C
4. $8^{1/3}$ A
5. $-8^{1/3}$ B

6. $(-8)^{1/3}$ B
7. $(-32)^{1/5}$ B
8. $(-16)^{1/4}$ C
9. $-16^{1/4}$ B
10. $-32^{1/5}$ B

58. $x + x^2$ **60.** $3x^{3/2} + 3x^{1/2}y$ **62.** $y + 10y^{1/2} + 25$ **65.** $x^{1/5}(x^{1/5} - 3)$
66. $x^{2/7}(x^{1/7} - 2)$ **67.** $x^{-1/3}(5 + x)$ **68.** $x^{-3/4}(1 + 3x)$

EXERCISE SET 7.2

STUDY GUIDE/SSM   CD/VIDEO   PH MATH TUTOR CENTER   MathXL®Tutorials ON CD   MathXL®   MyMathLab®

Use radical notation to write each expression. Simplify if possible. See Example 1.

 1. $49^{1/2}$ 7
2. $64^{1/3}$ 4
3. $27^{1/3}$ 3
4. $8^{1/3}$ 2

5. $\left(\dfrac{1}{16}\right)^{1/4}$ $\dfrac{1}{2}$
6. $\left(\dfrac{1}{64}\right)^{1/2}$ $\dfrac{1}{8}$

7. $169^{1/2}$ 13
8. $81^{1/4}$ 3

 9. $2m^{1/3}$ $2\sqrt[3]{m}$
10. $(2m)^{1/3}$ $\sqrt[3]{2m}$

11. $(9x^4)^{1/2}$ $3x^2$
12. $(16x^8)^{1/2}$ $4x^4$

13. $(-27)^{1/3}$ −3
14. $-64^{1/2}$ −8

15. $-16^{1/4}$ −2
16. $(-32)^{1/5}$ −2

Use radical notation to write each expression. Simplify if possible. See Example 2.   **22.** not a real number

 17. $16^{3/4}$ 8
18. $4^{5/2}$ 32

19. $(-64)^{2/3}$ 16
20. $(-8)^{4/3}$ 16

21. $(-16)^{3/4}$ not a real number
22. $(-9)^{3/2}$

23. $(2x)^{3/5}$ $\sqrt[5]{(2x)^3}$
24. $2x^{3/5}$ $2\sqrt[5]{x^3}$

25. $(7x + 2)^{2/3}$ $\sqrt[3]{(7x + 2)^2}$
26. $(x - 4)^{3/4}$ $\sqrt[4]{(x - 4)^3}$

27. $\left(\dfrac{16}{9}\right)^{3/2}$ $\dfrac{64}{27}$
28. $\left(\dfrac{49}{25}\right)^{3/2}$ $\dfrac{343}{125}$

Write with positive exponents. Simplify if possible. See Example 3.   **34.** not a real number

29. $8^{-4/3}$ $\dfrac{1}{16}$
30. $64^{-2/3}$ $\dfrac{1}{16}$

31. $(-64)^{-2/3}$ $\dfrac{1}{16}$
32. $(-8)^{-4/3}$ $\dfrac{1}{16}$

33. $(-4)^{-3/2}$ not a real number
34. $(-16)^{-5/4}$

35. $x^{-1/4}$ $\dfrac{1}{x^{1/4}}$
36. $y^{-1/6}$ $\dfrac{1}{y^{1/6}}$

37. $\dfrac{1}{a^{-2/3}}$ $a^{2/3}$
38. $\dfrac{1}{n^{-8/9}}$ $n^{8/9}$

39. $\dfrac{5}{7x^{-3/4}}$ $\dfrac{5x^{3/4}}{7}$
40. $\dfrac{2}{3y^{-5/7}}$ $\dfrac{2y^{5/7}}{3}$

41. Explain how writing x^{-7} with positive exponents is similar to writing $x^{-1/4}$ with positive exponents.   answers may vary

42. Explain how writing $2x^{-5}$ with positive exponents is similar to writing $2x^{-3/4}$ with positive exponents.   answers may vary

Use the properties of exponents to simplify each expression. Write with positive exponents. See Example 4.

43. $a^{2/3}a^{5/3}$ $a^{7/3}$
44. $b^{9/5}b^{8/5}$ $b^{17/5}$

45. $x^{-2/5} \cdot x^{7/5}$ x
46. $y^{4/3} \cdot y^{-1/3}$ y

47. $3^{1/4} \cdot 3^{3/8}$ $3^{5/8}$
48. $5^{1/2} \cdot 5^{1/6}$ $5^{2/3}$

49. $\dfrac{y^{1/3}}{y^{1/6}}$ $y^{1/6}$
50. $\dfrac{x^{3/4}}{x^{1/8}}$ $x^{5/8}$

51. $(4u^2)^{3/2}$ $8u^3$
52. $(32^{1/5}x^{2/3})^3$ $8x^2$

53. $\dfrac{b^{1/2}b^{3/4}}{-b^{1/4}}$ $-b$
54. $\dfrac{a^{1/4}a^{-1/2}}{a^{2/3}}$ $\dfrac{1}{a^{11/12}}$

55. $\dfrac{(3x^{1/4})^3}{x^{1/12}}$ $27x^{2/3}$
56. $\dfrac{(2x^{1/5})^4}{x^{3/10}}$ $16x^{1/2}$

Multiply. See Example 5.   **61.** $4x^{2/3} - 9$

57. $y^{1/2}(y^{1/2} - y^{2/3})$ $y - y^{7/6}$
58. $x^{1/2}(x^{1/2} + x^{3/2})$

59. $x^{2/3}(2x - 2)$ $2x^{5/3} - 2x^{2/3}$
60. $3x^{1/2}(x + y)$

61. $(2x^{1/3} + 3)(2x^{1/3} - 3)$
62. $(y^{1/2} + 5)(y^{1/2} + 5)$

Factor the common factor from the given expression. See Example 6.   **63.** $x^{8/3}(1 + x^{2/3})$ **64.** $x^{3/2}(x - 1)$

63. $x^{8/3}$; $x^{8/3} + x^{10/3}$
64. $x^{3/2}$; $x^{5/2} - x^{3/2}$

65. $x^{1/5}$; $x^{2/5} - 3x^{1/5}$
66. $x^{2/7}$; $x^{3/7} - 2x^{2/7}$

67. $x^{-1/3}$; $5x^{-1/3} + x^{2/3}$
68. $x^{-3/4}$; $x^{-3/4} + 3x^{1/4}$

Use rational exponents to simplify each radical. Assume that all variables represent positive numbers. See Example 7.

 69. $\sqrt[6]{x^3}$ $\sqrt{x}$

70. $\sqrt[9]{a^3}$ $\sqrt[3]{a}$

71. $\sqrt[6]{4}$ $\sqrt[3]{2}$

72. $\sqrt[4]{36}$ $\sqrt{6}$

 73. $\sqrt[4]{16x^2}$ $2\sqrt{x}$

74. $\sqrt[8]{4y^2}$ $\sqrt[4]{2y}$

75. $\sqrt[8]{x^4y^4}$ $\sqrt{xy}$

76. $\sqrt[9]{y^6z^3}$ $\sqrt[3]{y^2z}$

Use rational expressions to write as a single radical expression. See Example 8.

77. $\sqrt[3]{y} \cdot \sqrt[5]{y^2}$ $\sqrt[15]{y^{11}}$

78. $\sqrt[3]{y^2} \cdot \sqrt[6]{y}$ $\sqrt[6]{y^5}$

79. $\dfrac{\sqrt[3]{b^2}}{\sqrt[4]{b}}$ $\sqrt[12]{b^5}$

80. $\dfrac{\sqrt[4]{a}}{\sqrt[5]{a}}$ $\sqrt[20]{a}$

81. $\dfrac{\sqrt[3]{a^2}}{\sqrt[6]{a}}$ $\sqrt{a}$

82. $\dfrac{\sqrt[5]{b^2}}{\sqrt[10]{b^3}}$ $\sqrt[10]{b}$

83. $\sqrt{3} \cdot \sqrt[3]{4}$ $\sqrt[6]{432}$

84. $\sqrt[3]{5} \cdot \sqrt{2}$ $\sqrt[6]{200}$

85. $\sqrt[5]{7} \cdot \sqrt[3]{y}$ $\sqrt[15]{343y^5}$

86. $\sqrt[4]{5} \cdot \sqrt[3]{x}$ $\sqrt[12]{125x^4}$

REVIEW AND PREVIEW

Write each integer as a product of two integers such that one of the factors is a perfect square. For example, write 18 as $9 \cdot 2$, because 9 is a perfect square.

87. 75 $25 \cdot 3$

88. 20 $4 \cdot 5$

89. 48 $16 \cdot 3$ or $4 \cdot 12$

90. 45 $9 \cdot 5$

Write each integer as a product of two integers such that one of the factors is a perfect cube. For example, write 24 as $8 \cdot 3$, because 8 is a perfect cube.

91. 16 $8 \cdot 2$

92. 56 $8 \cdot 7$

93. 54 $27 \cdot 2$

94. 80 $8 \cdot 10$

Concept Extensions

Basal metabolic rate (BMR) is the number of calories per day a person needs to maintain life. A person's basal metabolic rate $B(w)$ in calories per day can be estimated with the function $B(w) = 70w^{3/4}$, where w is the person's weight in kilograms. Use this information to answer Exercises 95 and 96.

95. Estimate the BMR for a person who weighs 60 kilograms. Round to the nearest calorie. (*Note:* 60 kilograms is approximately 132 pounds.) 1509 calories

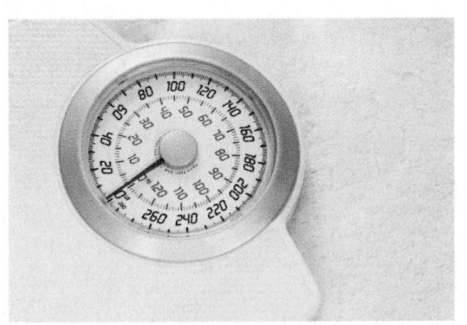

96. Estimate the BMR for a person who weighs 90 kilograms. Round to the nearest calorie. (*Note:* 90 kilograms is approximately 198 pounds.) 2045 calories

The number of cellular telephone subscriptions in the United States from 1994 through 2002 can be modeled by the function $f(x) = 1.54x^{9/5}$, where y is the number of cellular telephone subscriptions in millions, x years after 1990. (Source: Based on data from the Cellular Telecommunications & Internet Association, 1994–2000) Use this information to answer Exercises 97 and 98.

97. Use this model to estimate the number of cellular telephone subscriptions in the United States in 2000. Round to the nearest tenth of a million. 97.2 million

98. Predict the number of cellular telephone subscriptions in the United States in 2007. Round to the nearest tenth of a million. 252.5 million

Fill in each box with the correct expression.   **100.** $x^{3/8}$ **102.** $y^{1/4}$

99. $\square \cdot a^{2/3} = a^{3/3}$, or a $a^{1/3}$

100. $\square \cdot x^{1/8} = x^{4/8}$, or $x^{1/2}$

101. $\dfrac{\square}{x^{-2/5}} = x^{3/5}$ $x^{1/5}$

102. $\dfrac{\square}{y^{-3/4}} = y^{4/4}$, or y

Use a calculator to write a four-decimal-place approximation of each number.

103. $8^{1/4}$ 1.6818

104. $20^{1/5}$ 1.8206

105. $18^{3/5}$ 5.6645

106. $76^{5/7}$ 22.0515

107. In physics, the speed of a wave traveling over a stretched string with tension t and density u is given by the expression $\dfrac{\sqrt{t}}{\sqrt{u}}$. Write this expression with rational exponents. $\dfrac{t^{1/2}}{u^{1/2}}$

108. In electronics, the angular frequency of oscillations in a certain type of circuit is given by the expression $(LC)^{-1/2}$. Use radical notation to write this expression. $\dfrac{1}{\sqrt{LC}}$

7.3 SIMPLIFYING RADICAL EXPRESSIONS

Objectives

1 Use the product rule for radicals.

2 Use the quotient rule for radicals.

3 Simplify radicals.

1 It is possible to simplify some radicals that do not evaluate to rational numbers. To do so, we use a product rule and a quotient rule for radicals. To discover the product rule, notice the following pattern.

$$\sqrt{9} \cdot \sqrt{4} = 3 \cdot 2 = 6$$
$$\sqrt{9 \cdot 4} = \sqrt{36} = 6$$

Since both expressions simplify to 6, it is true that

$$\sqrt{9} \cdot \sqrt{4} = \sqrt{9 \cdot 4}$$

This pattern suggests the following product rule for radicals.

Product Rule for Radicals

If $\sqrt[n]{a}$ and $\sqrt[n]{b}$ are real numbers, then

$$\sqrt[n]{a} \cdot \sqrt[n]{b} = \sqrt[n]{ab}$$

Notice that the product rule is the relationship $a^{1/n} \cdot b^{1/n} = (ab)^{1/n}$ stated in radical notation.

EXAMPLE 1

CLASSROOM EXAMPLE

Use the product rule to multiply.

a. $\sqrt{2} \cdot \sqrt{7}$ b. $\sqrt{17} \cdot \sqrt{y}$

c. $\sqrt[3]{2} \cdot \sqrt[3]{32}$ d. $\sqrt[4]{6} \cdot \sqrt[4]{3x^2}$

e. $\sqrt{\dfrac{3}{x}} \cdot \sqrt{\dfrac{y}{2}}$

answer: a. $\sqrt{14}$ b. $\sqrt{17y}$ c. 4

d. $\sqrt[4]{18x^2}$ e. $\sqrt{\dfrac{3y}{2x}}$

Multiply.

a. $\sqrt{3} \cdot \sqrt{5}$ b. $\sqrt{21} \cdot \sqrt{x}$ c. $\sqrt[3]{4} \cdot \sqrt[3]{2}$

d. $\sqrt[4]{5y^2} \cdot \sqrt[4]{2x^3}$ e. $\sqrt{\dfrac{2}{a}} \cdot \sqrt{\dfrac{b}{3}}$

Solution a. $\sqrt{3} \cdot \sqrt{5} = \sqrt{3 \cdot 5} = \sqrt{15}$

b. $\sqrt{21} \cdot \sqrt{x} = \sqrt{21x}$

c. $\sqrt[3]{4} \cdot \sqrt[3]{2} = \sqrt[3]{4 \cdot 2} = \sqrt[3]{8} = 2$

d. $\sqrt[4]{5y^2} \cdot \sqrt[4]{2x^3} = \sqrt[4]{5y^2 \cdot 2x^3} = \sqrt[4]{10y^2x^3}$

e. $\sqrt{\dfrac{2}{a}} \cdot \sqrt{\dfrac{b}{3}} = \sqrt{\dfrac{2}{a} \cdot \dfrac{b}{3}} = \sqrt{\dfrac{2b}{3a}}$

2   To discover a quotient rule for radicals, notice the following pattern.

$$\sqrt{\frac{4}{9}} = \frac{2}{3}$$

$$\frac{\sqrt{4}}{\sqrt{9}} = \frac{2}{3}$$

Since both expressions simplify to $\frac{2}{3}$, it is true that

$$\sqrt{\frac{4}{9}} = \frac{\sqrt{4}}{\sqrt{9}}$$

This pattern suggests the following quotient rule for radicals.

Quotient Rule for Radicals

If $\sqrt[n]{a}$ and $\sqrt[n]{b}$ are real numbers and $\sqrt[n]{b}$ is not zero, then

$$\sqrt[n]{\frac{a}{b}} = \frac{\sqrt[n]{a}}{\sqrt[n]{b}}$$

Notice that the quotient rule is the relationship $\left(\frac{a}{b}\right)^{1/n} = \frac{a^{1/n}}{b^{1/n}}$ stated in radical notation. We can use the quotient rule to simplify radical expressions by reading the rule from left to right, or to divide radicals by reading the rule from right to left.

For example,

$$\sqrt{\frac{x}{16}} = \frac{\sqrt{x}}{\sqrt{16}} = \frac{\sqrt{x}}{4} \qquad \text{Using } \sqrt[n]{\frac{a}{b}} = \frac{\sqrt[n]{a}}{\sqrt[n]{b}}$$

$$\frac{\sqrt{75}}{\sqrt{3}} = \sqrt{\frac{75}{3}} = \sqrt{25} = 5 \qquad \text{Using } \frac{\sqrt[n]{a}}{\sqrt[n]{b}} = \sqrt[n]{\frac{a}{b}}$$

Note: *Recall that from Section 7.2 on, we assume that variables represent positive real numbers. Since this is so, we need not insert absolute value bars when we simplify even roots.*

EXAMPLE 2

Use the quotient rule to simplify.

a. $\sqrt{\frac{25}{49}}$ **b.** $\sqrt{\frac{x}{9}}$

c. $\sqrt[3]{\frac{8}{27}}$ **d.** $\sqrt[4]{\frac{3}{16y^4}}$

Solution

CLASSROOM EXAMPLE

Use the quotient rule to simplify. Assume that all variables represent positive real numbers.

a. $\sqrt{\dfrac{9}{25}}$ b. $\sqrt{\dfrac{y}{36}}$

c. $\sqrt[3]{\dfrac{27}{64}}$ d. $\sqrt[5]{\dfrac{7}{32x^5}}$

answer:

a. $\dfrac{3}{5}$ b. $\dfrac{\sqrt{y}}{6}$ c. $\dfrac{3}{4}$ d. $\dfrac{\sqrt[5]{7}}{2x}$

a. $\sqrt{\dfrac{25}{49}} = \dfrac{\sqrt{25}}{\sqrt{49}} = \dfrac{5}{7}$

b. $\sqrt{\dfrac{x}{9}} = \dfrac{\sqrt{x}}{\sqrt{9}} = \dfrac{\sqrt{x}}{3}$

c. $\sqrt[3]{\dfrac{8}{27}} = \dfrac{\sqrt[3]{8}}{\sqrt[3]{27}} = \dfrac{2}{3}$

d. $\sqrt[4]{\dfrac{3}{16y^4}} = \dfrac{\sqrt[4]{3}}{\sqrt[4]{16y^4}} = \dfrac{\sqrt[4]{3}}{2y}$

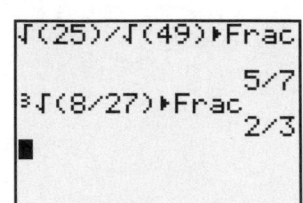

A calculator check for Examples 2a and c.

3   Both the product and quotient rules can be used to simplify a radical. If the product rule is read from right to left, we have that $\sqrt[n]{ab} = \sqrt[n]{a} \cdot \sqrt[n]{b}$. This is used to simplify the following radicals.

EXAMPLE 3

CLASSROOM EXAMPLE

Simplify.

a. $\sqrt{18}$ b. $\sqrt[3]{40}$ c. $\sqrt{14}$ d. $\sqrt[4]{162}$

answer:

a. $3\sqrt{2}$ b. $2\sqrt[3]{5}$ c. $\sqrt{14}$ d. $3\sqrt[4]{2}$

Simplify the following.

a. $\sqrt{50}$ b. $\sqrt[3]{24}$ c. $\sqrt{26}$ d. $\sqrt[4]{32}$

Solution   a. Factor 50 such that one factor is the largest perfect square that divides 50. The largest perfect square factor of 50 is 25, so we write 50 as $25 \cdot 2$ and use the product rule for radicals to simplify.

Helpful Hint

Don't forget that, for example, $5\sqrt{2}$ means $5 \cdot \sqrt{2}$.

$$\sqrt{50} = \sqrt{25 \cdot 2} = \sqrt{25} \cdot \sqrt{2} = 5\sqrt{2}$$

The largest perfect square factor of 50.

b. $\sqrt[3]{24} = \sqrt[3]{8 \cdot 3} = \sqrt[3]{8} \cdot \sqrt[3]{3} = 2\sqrt[3]{3}$

The largest perfect cube factor of 24.

c. $\sqrt{26}$ The largest perfect square factor of 26 is 1, so $\sqrt{26}$ cannot be simplified further.

d. $\sqrt[4]{32} = \sqrt[4]{16 \cdot 2} = \sqrt[4]{16} \cdot \sqrt[4]{2} = 2\sqrt[4]{2}$

The largest fourth power factor of 32.

```
√(50)
        7.071067812
5√(2)
        7.071067812
■
```

A calculator check for Example 3a.

After simplifying a radical such as a square root, always check the radicand to see that it contains no other perfect square factors. It may, if the largest perfect square factor of the radicand was not originally recognized. For example,

$$\sqrt{200} = \sqrt{4 \cdot 50} = \sqrt{4} \cdot \sqrt{50} = 2\sqrt{50}$$

Notice that the radicand 50 still contains the perfect square factor 25. This is because 4 is not the largest perfect square factor of 200. We continue as follows.

$$2\sqrt{50} = 2\sqrt{25 \cdot 2} = 2 \cdot \sqrt{25} \cdot \sqrt{2} = 2 \cdot 5 \cdot \sqrt{2} = 10\sqrt{2}$$

The radical is now simplified since 2 contains no perfect square factors (other than 1).

> **Helpful Hint**
>
> To help you recognize largest perfect power factors of a radicand, it will help if you are familiar with some perfect powers. A few are listed below.
>
> | Perfect Squares | 1, | 4, | 9, | 16, | 25, | 36, | 49, | 64, | 81, | 100, | 121, | 144 |
> |---|---|---|---|---|---|---|---|---|---|---|---|---|
> | | 1^2 | 2^2 | 3^2 | 4^2 | 5^2 | 6^2 | 7^2 | 8^2 | 9^2 | 10^2 | 11^2 | 12^2 |
>
> | Perfect Cubes | 1, | 8, | 27, | 64, | 125 |
> |---|---|---|---|---|---|
> | | 1^3 | 2^3 | 3^3 | 4^3 | 5^3 |
>
> | Perfect Fourth Powers | 1, | 16, | 81, | 256 |
> |---|---|---|---|---|
> | | 1^4 | 2^4 | 3^4 | 4^4 |

In general, we say that a radicand of the form $\sqrt[n]{a}$ is simplified when the radicand a contains no factors that are perfect nth powers (other than 1 or -1).

EXAMPLE 4

Use the product rule to simplify.

a. $\sqrt{25x^3}$ **b.** $\sqrt[3]{54x^6y^8}$ **c.** $\sqrt[4]{81z^{11}}$

Solution

a. $\sqrt{25x^3} = \sqrt{25x^2 \cdot x}$ *Find the largest perfect square factor.*

$\qquad\quad = \sqrt{25x^2} \cdot \sqrt{x}$ *Apply the product rule.*

$\qquad\quad = 5x\sqrt{x}$ *Simplify.*

b. $\sqrt[3]{54x^6y^8} = \sqrt[3]{27 \cdot 2 \cdot x^6 \cdot y^6 \cdot y^2}$ *Factor the radicand and identify perfect cube factors.*

$\qquad\qquad = \sqrt[3]{27x^6y^6 \cdot 2y^2}$

$\qquad\qquad = \sqrt[3]{27x^6y^6} \cdot \sqrt[3]{2y^2}$ *Apply the product rule.*

$\qquad\qquad = 3x^2y^2\sqrt[3]{2y^2}$ *Simplify.*

c. $\sqrt[4]{81z^{11}} = \sqrt[4]{81 \cdot z^8 \cdot z^3}$ *Factor the radicand and identify perfect fourth power factors.*

$\qquad\qquad = \sqrt[4]{81z^8} \cdot \sqrt[4]{z^3}$ *Apply the product rule.*

$\qquad\qquad = 3z^2\sqrt[4]{z^3}$ *Simplify.*

EXAMPLE 5

Use the quotient rule to divide, and simplify if possible.

a. $\dfrac{\sqrt{20}}{\sqrt{5}}$ **b.** $\dfrac{\sqrt{50x}}{2\sqrt{2}}$

c. $\dfrac{7\sqrt[3]{48x^4y^8}}{\sqrt[3]{6y^2}}$ **d.** $\dfrac{2\sqrt[4]{32a^8b^6}}{\sqrt[4]{a^{-1}b^2}}$

Solution

a. $\dfrac{\sqrt{20}}{\sqrt{5}} = \sqrt{\dfrac{20}{5}}$ Apply the quotient rule.

$= \sqrt{4}$ Simplify.

$= 2$ Simplify.

b. $\dfrac{\sqrt{50x}}{2\sqrt{2}} = \dfrac{1}{2} \cdot \sqrt{\dfrac{50x}{2}}$ Apply the quotient rule.

$= \dfrac{1}{2} \cdot \sqrt{25x}$ Simplify.

$= \dfrac{1}{2} \cdot \sqrt{25} \cdot \sqrt{x}$ Factor $25x$.

$= \dfrac{1}{2} \cdot 5 \cdot \sqrt{x}$ Simplify.

$= \dfrac{5}{2}\sqrt{x}$

c. $\dfrac{7\sqrt[3]{48x^4y^8}}{\sqrt[3]{6y^2}} = 7 \cdot \sqrt[3]{\dfrac{48x^4y^8}{6y^2}}$ Apply the quotient rule.

$= 7 \cdot \sqrt[3]{8x^4y^6}$ Simplify.

$= 7\sqrt[3]{8x^3y^6 \cdot x}$ Factor.

$= 7 \cdot \sqrt[3]{8x^3y^6} \cdot \sqrt[3]{x}$ Apply the product rule.

$= 7 \cdot 2xy^2 \cdot \sqrt[3]{x}$ Simplify.

$= 14xy^2\sqrt[3]{x}$

d. $\dfrac{2\sqrt[4]{32a^8b^6}}{\sqrt[4]{a^{-1}b^2}} = 2\sqrt[4]{\dfrac{32a^8b^6}{a^{-1}b^2}} = 2\sqrt[4]{32a^9b^4} = 2\sqrt[4]{16 \cdot a^8 \cdot b^4 \cdot 2 \cdot a}$

$= 2\sqrt[4]{16a^8b^4} \cdot \sqrt[4]{2a} = 2 \cdot 2a^2b \cdot \sqrt[4]{2a} = 4a^2b\sqrt[4]{2a}$

CLASSROOM EXAMPLE

Use the quotient rule to divide. Then simplify if possible.

a. $\dfrac{\sqrt{75}}{\sqrt{3}}$ **b.** $\dfrac{\sqrt{80y}}{3\sqrt{5}}$

c. $\dfrac{5\sqrt[3]{162x^8}}{\sqrt[3]{3x^2}}$ **d.** $\dfrac{3\sqrt[4]{243x^9y^6}}{\sqrt[4]{x^{-3}y}}$

answer:

a. 5 **b.** $\dfrac{4}{3}\sqrt{y}$

c. $15x^2\sqrt[3]{2}$ **d.** $9x^3y\sqrt[4]{3y}$

✔ **CONCEPT CHECK**

Find and correct the error:

$\dfrac{\sqrt[3]{27}}{\sqrt{9}} = \sqrt[3]{\dfrac{27}{9}} = \sqrt[3]{3}$

Concept Check Answer:

$\dfrac{\sqrt[3]{27}}{\sqrt{9}} = \dfrac{3}{3} = 1$

EXERCISE SET 7.3

STUDY GUIDE/SSM   CD/VIDEO   PH MATH TUTOR CENTER   MathXL®Tutorials ON CD   MathXL®   MyMathLab®

12. $\sqrt[4]{27a^2b^3}$

Use the product rule to multiply. See Example 1.

1. $\sqrt{7}\cdot\sqrt{2}$ $\sqrt{14}$

2. $\sqrt{11}\cdot\sqrt{10}$ $\sqrt{110}$

3. $\sqrt[4]{8}\cdot\sqrt[4]{2}$ 2

4. $\sqrt[4]{27}\cdot\sqrt[4]{3}$ 3

5. $\sqrt[3]{4}\cdot\sqrt[3]{9}$ $\sqrt[3]{36}$

6. $\sqrt[3]{10}\cdot\sqrt[3]{5}$ $\sqrt[3]{50}$

7. $\sqrt{2}\cdot\sqrt{3x}$ $\sqrt{6x}$

8. $\sqrt{3y}\cdot\sqrt{5x}$ $\sqrt{15xy}$

9. $\sqrt{\dfrac{7}{x}}\cdot\sqrt{\dfrac{2}{y}}$ $\sqrt{\dfrac{14}{xy}}$

10. $\sqrt{\dfrac{6}{m}}\cdot\sqrt{\dfrac{n}{5}}$ $\sqrt{\dfrac{6n}{5m}}$

11. $\sqrt[4]{4x^3}\cdot\sqrt[4]{5}$ $\sqrt[4]{20x^3}$

12. $\sqrt[4]{ab^2}\cdot\sqrt[4]{27ab}$

Use the quotient rule to simplify. See Examples 2 and 3.

13. $\sqrt{\dfrac{6}{49}}$ $\dfrac{\sqrt{6}}{7}$

14. $\sqrt{\dfrac{8}{81}}$ $\dfrac{2\sqrt{2}}{9}$

15. $\sqrt{\dfrac{2}{49}}$ $\dfrac{\sqrt{2}}{7}$

16. $\sqrt{\dfrac{5}{121}}$ $\dfrac{\sqrt{5}}{11}$

17. $\sqrt[4]{\dfrac{x^3}{16}}$ $\dfrac{\sqrt[4]{x^3}}{2}$

18. $\sqrt[4]{\dfrac{y}{81x^4}}$ $\dfrac{\sqrt[4]{y}}{3x}$

19. $\sqrt[3]{\dfrac{4}{27}}$ $\dfrac{\sqrt[3]{4}}{3}$

20. $\sqrt[3]{\dfrac{3}{64}}$ $\dfrac{\sqrt[3]{3}}{4}$

21. $\sqrt[4]{\dfrac{8}{x^8}}$ $\dfrac{\sqrt[4]{8}}{x^2}$

22. $\sqrt[4]{\dfrac{a^3}{81}}$ $\dfrac{\sqrt[4]{a^3}}{3}$

23. $\sqrt[3]{\dfrac{2x}{81y^{12}}}$ $\dfrac{\sqrt[3]{2x}}{3y^4\sqrt[3]{3}}$

24. $\sqrt[3]{\dfrac{3}{8x^6}}$ $\dfrac{\sqrt[3]{3}}{2x^2}$

25. $\sqrt{\dfrac{x^2y}{100}}$ $\dfrac{x\sqrt{y}}{10}$

26. $\sqrt{\dfrac{y^2z}{36}}$ $\dfrac{y\sqrt{z}}{6}$

27. $\sqrt{\dfrac{5x^2}{4y^2}}$ $\dfrac{\sqrt{5}x}{2y}$

28. $\sqrt{\dfrac{y^{10}}{9x^6}}$ $\dfrac{y^5}{3x^3}$

29. $-\sqrt[3]{\dfrac{z^7}{27x^3}}$ $-\dfrac{z^2\sqrt[3]{z}}{3x}$

30. $-\sqrt[3]{\dfrac{64a}{b^9}}$ $-\dfrac{4\sqrt[3]{a}}{b^3}$

Simplify. See Examples 3 and 4.

31. $\sqrt{32}$ $4\sqrt{2}$

32. $\sqrt{27}$ $3\sqrt{3}$

33. $\sqrt[3]{192}$ $4\sqrt[3]{3}$

34. $\sqrt[3]{108}$ $3\sqrt[3]{4}$

35. $5\sqrt{75}$ $25\sqrt{3}$

36. $3\sqrt{8}$ $6\sqrt{2}$

37. $\sqrt{24}$ $2\sqrt{6}$

38. $\sqrt{20}$ $2\sqrt{5}$

39. $\sqrt{100x^5}$ $10x^2\sqrt{x}$

40. $\sqrt{64y^9}$ $8y^4\sqrt{y}$

41. $\sqrt[3]{16y^7}$ $2y^2\sqrt[3]{2y}$

42. $\sqrt[3]{64y^9}$ $4y^3$

43. $\sqrt[4]{a^8b^7}$ $a^2b\sqrt[4]{b^3}$

44. $\sqrt[5]{32z^{12}}$ $2z^2\sqrt[5]{z^2}$

45. $\sqrt{y^5}$ $y^2\sqrt{y}$

46. $\sqrt[3]{y^5}$ $y\sqrt[3]{y^2}$

47. $\sqrt{25a^2b^3}$ $5ab\sqrt{b}$

48. $\sqrt{9x^5y^7}$ $3x^2y^3\sqrt{xy}$

49. $\sqrt[5]{-32x^{10}y}$ $-2x^2\sqrt[5]{y}$

50. $\sqrt[5]{-243z^9}$ $-3z\sqrt[5]{z^4}$

51. $\sqrt[3]{50x^{14}}$ $x^4\sqrt[3]{50x^2}$

52. $\sqrt[3]{40y^{10}}$ $2y^3\sqrt[3]{5y}$

53. $-\sqrt{32a^8b^7}$ $-4a^4b^3\sqrt{2b}$

54. $-\sqrt{20ab^6}$ $-2b^3\sqrt{5a}$

55. $\sqrt{9x^7y^9}$ $3x^3y^4\sqrt{xy}$

56. $\sqrt{12r^9s^{12}}$ $2r^4s^6\sqrt{3r}$

57. $\sqrt[3]{125r^9s^{12}}$ $5r^3s^4$

58. $\sqrt[3]{8a^6b^9}$ $2a^2b^3$

Use the quotient rule to divide. Then simplify if possible. See Example 5.

59. $\dfrac{\sqrt{14}}{\sqrt{7}}$ $\sqrt{2}$

60. $\dfrac{\sqrt{45}}{\sqrt{9}}$ $\sqrt{5}$

61. $\dfrac{\sqrt[3]{24}}{\sqrt[3]{3}}$ 2

62. $\dfrac{\sqrt[3]{10}}{\sqrt[3]{2}}$ $\sqrt[3]{5}$

63. $\dfrac{5\sqrt[4]{48}}{\sqrt[4]{3}}$ 10

64. $\dfrac{7\sqrt[4]{162}}{\sqrt[4]{2}}$ 21

65. $\dfrac{\sqrt{x^5y^3}}{\sqrt{xy}}$ x^2y

66. $\dfrac{\sqrt{a^7b^6}}{\sqrt{a^3b^2}}$ a^2b^2

67. $\dfrac{8\sqrt[3]{54m^7}}{\sqrt[3]{2m}}$ $24m^2$

68. $\dfrac{\sqrt[3]{128x^3}}{-3\sqrt[3]{2x}}$ $-\dfrac{4\sqrt[3]{x^2}}{3}$

69. $\dfrac{3\sqrt{100x^2}}{2\sqrt{2x^{-1}}}$

70. $\dfrac{\sqrt{270y^2}}{5\sqrt{3y^{-4}}}$

71. $\dfrac{\sqrt[4]{96a^{10}b^3}}{\sqrt[4]{3a^2b^3}}$ $2a^2\sqrt[4]{2}$

72. $\dfrac{\sqrt[5]{64x^{10}y^3}}{\sqrt[5]{2x^3y^{-7}}}$ $2xy^2\sqrt[5]{x^2}$

REVIEW AND PREVIEW

Perform each indicated operation. See Sections 1.4 and 5.4.

73. $6x + 8x$ $14x$

74. $(6x)(8x)$ $48x^2$

75. $(2x + 3)(x - 5)$

76. $(2x + 3) + (x - 5)$

77. $9y^2 - 8y^2$ y^2

78. $(9y^2)(-8y^2)$ $-72y^4$

79. $-3(x + 5)$ $-3x - 15$

80. $-3 + x + 5$ $x + 2$

81. $(x - 4)^2$ $x^2 - 8x + 16$

82. $(2x + 1)^2$

Concept Extensions

△ **83.** The formula for the surface area A of a cone with height h and radius r is given by

$$A = \pi r\sqrt{r^2 + h^2}$$

a. Find the surface area of a cone whose height is 3 centimeters and whose radius is 4 centimeters. 20π sq. cm

69. $\dfrac{15x\sqrt{2x}}{2}$ or $\dfrac{15x}{2}\sqrt{2x}$ **70.** $\dfrac{3y^3\sqrt{10}}{5}$ or $\dfrac{3y^3}{5}\sqrt{10}$ **75.** $2x^2 - 7x - 15$ **76.** $3x - 2$ **82.** $4x^2 + 4x + 1$

b. Approximate to two decimal places the surface area of a cone whose height is 7.2 feet and whose radius is 6.8 feet.
211.57 sq. ft

△ **84.** Before Mount Vesuvius, a volcano in Italy, erupted violently in 79 A.D., its height was 4190 feet. Vesuvius was roughly cone-shaped, and its base had a radius of approximately 25,200 feet. Use the formula for the surface area of a cone, given in Exercise 83, to approximate the surface area this volcano had before it erupted. (*Source:* Global Volcanism Network)   2,022,426,050 sq. ft

4190 ft

25,200 ft

85. The owner of Knightime Video has determined that the demand equation for renting older releases is given by the equation $F(x) = 0.6\sqrt{49 - x^2}$, where x is the price in dollars per two-day rental and $F(x)$ is the number of times the video is demanded per week.

a. Approximate to one decimal place the demand per week of an older release if the rental price is $3 per two-day rental.   3.8 times

b. Approximate to one decimal place the demand per week of an older release if the rental price is $5 per two-day rental.   2.9 times

c. Explain how the owner of the video store can use this equation to predict the number of copies of each tape that should be in stock.   answers may vary

---

7.4   ADDING, SUBTRACTING, AND MULTIPLYING RADICAL EXPRESSIONS

Objectives

1 Add or subtract radical expressions.

2 Multiply radical expressions.

1   We have learned that sums or differences of like terms can be simplified. To simplify these sums or differences, we use the distributive property. For example,

$$2x + 3x = (2 + 3)x = 5x \quad \text{and} \quad 7x^2y - 4x^2y = (7 - 4)x^2y = 3x^2y$$

The distributive property can also be used to add **like radicals**.

TEACHING TIP

Before discussing like radicals, have students evaluate $2x + 3x$ for $x = \sqrt{7}$.

> ### Like Radicals
> Radicals with the same index and the same radicand are like radicals.

For example, $2\sqrt{7} + 3\sqrt{7} = (2 + 3)\sqrt{7} = 5\sqrt{7}$. Also,

$$5\sqrt{3x} - 7\sqrt{3x} = (5 - 7)\sqrt{3x} = -2\sqrt{3x}$$

The expression $2\sqrt{7} + 2\sqrt[3]{7}$ cannot be simplified further since $2\sqrt{7}$ and $2\sqrt[3]{7}$ are not like radicals.

EXAMPLE 1

Add or subtract as indicated.

a. $4\sqrt{11} + 8\sqrt{11}$ **b.** $5\sqrt[3]{3x} - 7\sqrt[3]{3x}$ **c.** $2\sqrt{7} + 2\sqrt[3]{7}$

Solution   **a.** $4\sqrt{11} + 8\sqrt{11} = (4 + 8)\sqrt{11} = 12\sqrt{11}$

b. $5\sqrt[3]{3x} - 7\sqrt[3]{3x} = (5 - 7)\sqrt[3]{3x} = -2\sqrt[3]{3x}$

c. $2\sqrt{7} + 2\sqrt[3]{7}$

This expression cannot be simplified since $2\sqrt{7}$ and $2\sqrt[3]{7}$ do not contain like radicals.

When adding or subtracting radicals, always check first to see whether any radicals can be simplified.

CLASSROOM EXAMPLE

Add or subtract as indicated.
a. $5\sqrt{15} + 2\sqrt{15}$
b. $9\sqrt[3]{2y} - 15\sqrt[3]{2y}$
c. $6\sqrt{10} - 3\sqrt[3]{10}$
answer: **a.** $7\sqrt{15}$ **b.** $-6\sqrt[3]{2y}$
c. $6\sqrt{10} - 3\sqrt[3]{10}$

✔ **CONCEPT CHECK**

True or false:

$$\sqrt{a} + \sqrt{b} = \sqrt{a + b}?$$

Explain.

EXAMPLE 2

Add or subtract. Assume that variables represent positive real numbers.

a. $\sqrt{20} + 2\sqrt{45}$ **b.** $\sqrt[3]{54} - 5\sqrt[3]{16} + \sqrt[3]{2}$ **c.** $\sqrt{27x} - 2\sqrt{9x} + \sqrt{72x}$

d. $\sqrt[3]{98} + \sqrt{98}$ **e.** $\sqrt[3]{48y^4} + \sqrt[3]{6y^4}$

Solution   First, simplify each radical. Then add or subtract any like radicals.

a. $\sqrt{20} + 2\sqrt{45} = \sqrt{4 \cdot 5} + 2\sqrt{9 \cdot 5}$ Factor 20 and 45.

$= \sqrt{4} \cdot \sqrt{5} + 2 \cdot \sqrt{9} \cdot \sqrt{5}$ Use the product rule.

$= 2 \cdot \sqrt{5} + 2 \cdot 3 \cdot \sqrt{5}$ Simplify $\sqrt{4}$ and $\sqrt{9}$.

$= 2\sqrt{5} + 6\sqrt{5}$ Add like radicals.

$= 8\sqrt{5}$

b. $\sqrt[3]{54} - 5\sqrt[3]{16} + \sqrt[3]{2}$

$= \sqrt[3]{27} \cdot \sqrt[3]{2} - 5 \cdot \sqrt[3]{8} \cdot \sqrt[3]{2} + \sqrt[3]{2}$ Factor and use the product rule.

$= 3 \cdot \sqrt[3]{2} - 5 \cdot 2 \cdot \sqrt[3]{2} + \sqrt[3]{2}$ Simplify $\sqrt[3]{27}$ and $\sqrt[3]{8}$.

$= 3\sqrt[3]{2} - 10\sqrt[3]{2} + \sqrt[3]{2}$ Write $5 \cdot 2$ as 10.

$= -6\sqrt[3]{2}$ Combine like radicals.

```
³√(54)−5³√(16)+³
√(2)
            -7.559526299
-6³√(2)
            -7.559526299
■
```

A calculator check for
Example 2b.

c. $\sqrt{27x} - 2\sqrt{9x} + \sqrt{72x}$

$= \sqrt{9} \cdot \sqrt{3x} - 2 \cdot \sqrt{9} \cdot \sqrt{x} + \sqrt{36} \cdot \sqrt{2x}$ Factor and use the product rule.

$= 3 \cdot \sqrt{3x} - 2 \cdot 3 \cdot \sqrt{x} + 6 \cdot \sqrt{2x}$ Simplify $\sqrt{9}$ and $\sqrt{36}$.

$= 3\sqrt{3x} - 6\sqrt{x} + 6\sqrt{2x}$ Write $2 \cdot 3$ as 6.

> ▶ **Helpful Hint**
> None of these terms contain like radicals. We can simplify no further.

d. $\sqrt[3]{98} + \sqrt{98} = \sqrt[3]{98} + \sqrt{49} \cdot \sqrt{2}$ Factor and use the product rule.

$= \sqrt[3]{98} + 7\sqrt{2}$ No further simplification is possible.

e. $\sqrt[3]{48y^4} + \sqrt[3]{6y^4} = \sqrt[3]{8y^3} \cdot \sqrt[3]{6y} + \sqrt[3]{y^3} \cdot \sqrt[3]{6y}$ Factor and use the product rule.

$= 2y\sqrt[3]{6y} + y\sqrt[3]{6y}$ Simplify $\sqrt[3]{8y^3}$ and $\sqrt[3]{y^3}$.

$= 3y\sqrt[3]{6y}$ Combine like radicals.

EXAMPLE 3

Add or subtract as indicated.

a. $\dfrac{\sqrt{45}}{4} - \dfrac{\sqrt{5}}{3}$ **b.** $\sqrt[3]{\dfrac{7x}{8}} + 2\sqrt[3]{7x}$

Solution **a.** $\dfrac{\sqrt{45}}{4} - \dfrac{\sqrt{5}}{3} = \dfrac{3\sqrt{5}}{4} - \dfrac{\sqrt{5}}{3}$ To subtract, notice that the LCD is 12.

$= \dfrac{3\sqrt{5} \cdot 3}{4 \cdot 3} - \dfrac{\sqrt{5} \cdot 4}{3 \cdot 4}$ Write each expression as an equivalent expression with a denominator of 12.

$= \dfrac{9\sqrt{5}}{12} - \dfrac{4\sqrt{5}}{12}$ Multiply factors in the numerator and the denominator.

$= \dfrac{5\sqrt{5}}{12}$ Subtract.

b. $\sqrt[3]{\dfrac{7x}{8}} + 2\sqrt[3]{7x} = \dfrac{\sqrt[3]{7x}}{\sqrt[3]{8}} + 2\sqrt[3]{7x}$ Apply the quotient rule for radicals.

$= \dfrac{\sqrt[3]{7x}}{2} + 2\sqrt[3]{7x}$ Simplify.

$= \dfrac{\sqrt[3]{7x}}{2} + \dfrac{2\sqrt[3]{7x} \cdot 2}{2}$ Write each expression as an equivalent expression with a denominator of 2.

$= \dfrac{\sqrt[3]{7x}}{2} + \dfrac{4\sqrt[3]{7x}}{2}$

$= \dfrac{5\sqrt[3]{7x}}{2}$ Add.

2   We can multiply radical expressions by using many of the same properties used to multiply polynomial expressions. For instance, to multiply $\sqrt{2}(\sqrt{6} - 3\sqrt{2})$, we use the distributive property and multiply $\sqrt{2}$ by each term inside the parentheses.

$$\sqrt{2}(\sqrt{6} - 3\sqrt{2}) = \sqrt{2}(\sqrt{6}) - \sqrt{2}(3\sqrt{2}) \qquad \textit{Use the distributive property.}$$
$$= \sqrt{2\cdot 6} - 3\sqrt{2\cdot 2}$$
$$= \sqrt{2\cdot 2\cdot 3} - 3\cdot 2 \qquad \textit{Use the product rule for radicals.}$$
$$= 2\sqrt{3} - 6$$

EXAMPLE 4

Multiply.

a. $\sqrt{3}(5 + \sqrt{30})$ **b.** $(\sqrt{5} - \sqrt{6})(\sqrt{7} + 1)$ **c.** $(7\sqrt{x} + 5)(3\sqrt{x} - \sqrt{5})$

d. $(4\sqrt{3} - 1)^2$ **e.** $(\sqrt{2x} - 5)(\sqrt{2x} + 5)$ **f.** $(\sqrt{x-3} + 5)^2$

Solution

a. $\sqrt{3}(5 + \sqrt{30}) = \sqrt{3}(5) + \sqrt{3}(\sqrt{30})$
$$= 5\sqrt{3} + \sqrt{3\cdot 30}$$
$$= 5\sqrt{3} + \sqrt{3\cdot 3\cdot 10}$$
$$= 5\sqrt{3} + 3\sqrt{10}$$

b. To multiply, we can use the FOIL method.

$$(\sqrt{5} - \sqrt{6})(\sqrt{7} + 1) = \overset{\text{First}}{\sqrt{5}\cdot\sqrt{7}} + \overset{\text{Outer}}{\sqrt{5}\cdot 1} - \overset{\text{Inner}}{\sqrt{6}\cdot\sqrt{7}} - \overset{\text{Last}}{\sqrt{6}\cdot 1}$$
$$= \sqrt{35} + \sqrt{5} - \sqrt{42} - \sqrt{6}$$

c. $(7\sqrt{x} + 5)(3\sqrt{x} - \sqrt{5}) = 7\sqrt{x}(3\sqrt{x}) - 7\sqrt{x}(\sqrt{5}) + 5(3\sqrt{x}) - 5(\sqrt{5})$
$$= 21x - 7\sqrt{5x} + 15\sqrt{x} - 5\sqrt{5}$$

d. $(4\sqrt{3} - 1)^2 = (4\sqrt{3} - 1)(4\sqrt{3} - 1)$
$$= 4\sqrt{3}(4\sqrt{3}) - 4\sqrt{3}(1) - 1(4\sqrt{3}) - 1(-1)$$
$$= 16\cdot 3 - 4\sqrt{3} - 4\sqrt{3} + 1$$
$$= 48 - 8\sqrt{3} + 1$$
$$= 49 - 8\sqrt{3}$$

e. $(\sqrt{2x} - 5)(\sqrt{2x} + 5) = \sqrt{2x}\cdot\sqrt{2x} + 5\sqrt{2x} - 5\sqrt{2x} - 5\cdot 5$
$$= 2x - 25$$

f. $\underbrace{(\sqrt{x-3}}_{a} + \underbrace{5)^2}_{b} = \underbrace{(\sqrt{x-3})^2}_{a^2} + \underbrace{2\cdot}_{+\,2\cdot}\underbrace{\sqrt{x-3}}_{a}\cdot\underbrace{5}_{\cdot\,b} + \underbrace{5^2}_{b^2}$

$$= x - 3 + 10\sqrt{x-3} + 25 \qquad \textit{Simplify.}$$
$$= x + 22 + 10\sqrt{x-3} \qquad \textit{Combine like terms.}$$

MENTAL MATH

Simplify. Assume that all variables represent positive real numbers.

1. $2\sqrt{3} + 4\sqrt{3}$ $6\sqrt{3}$

2. $5\sqrt{7} + 3\sqrt{7}$ $8\sqrt{7}$

3. $8\sqrt{x} - 5\sqrt{x}$ $3\sqrt{x}$

4. $3\sqrt{y} + 10\sqrt{y}$ $13\sqrt{y}$

5. $7\sqrt[3]{x} + 5\sqrt[3]{x}$ $12\sqrt[3]{x}$

6. $8\sqrt[3]{z} - 2\sqrt[3]{z}$ $6\sqrt[3]{z}$

Add or subtract if possible.

7. $\sqrt{11} + \sqrt{11}$ $2\sqrt{11}$

8. $\sqrt{11} + \sqrt[3]{11}$ $\sqrt{11} + \sqrt[3]{11}$

9. $9\sqrt{13} - \sqrt{13}$ $8\sqrt{13}$

10. $9\sqrt{13} - \sqrt[4]{13}$ $9\sqrt{13} - \sqrt[4]{13}$

11. $8\sqrt[3]{2x} + 3\sqrt[3]{2x} - \sqrt[3]{2x}$ $10\sqrt[3]{2x}$

12. $8\sqrt[3]{2x} + 3\sqrt[3]{2x^2} - \sqrt[3]{2x}$ $7\sqrt[3]{2x} + 3\sqrt[3]{2x^2}$

3. $10x\sqrt{2x}$ **4.** $10x\sqrt{5x}$ **8.** $-7a\sqrt[3]{3a}$ **16.** $\dfrac{7x\sqrt{7}}{10}$ **17.** $14 + \sqrt{3}$ **18.** $11 - 5\sqrt{10}$ **24.** $-8\sqrt[3]{4}$ **27.** $20y\sqrt{2y}$ **28.** $-2xy\sqrt{2y}$

31. $6\sqrt[3]{11} - 4\sqrt{11}$ **32.** $3\sqrt[3]{5} + 4\sqrt{5}$ **40.** $-\dfrac{7\sqrt{11}}{5x}$ **43.** $\dfrac{14x\sqrt[3]{2x}}{9}$ **44.** $\dfrac{11y\sqrt[3]{y^2}}{8}$

EXERCISE SET 7.4

STUDY GUIDE/SSM · CD/VIDEO · PH MATH TUTOR CENTER · MathXL®Tutorials ON CD · MathXL® · MyMathLab®

Add or subtract. See Examples 1 through 3.

1. $\sqrt{8} - \sqrt{32}$ $-2\sqrt{2}$

2. $\sqrt{27} - \sqrt{75}$ $-2\sqrt{3}$

3. $2\sqrt{2x^3} + 4x\sqrt{8x}$

4. $3\sqrt{45x^3} + x\sqrt{5x}$

 5. $2\sqrt{50} - 3\sqrt{125} + \sqrt{98}$ $17\sqrt{2} - 15\sqrt{5}$

6. $4\sqrt{32} - \sqrt{18} + 2\sqrt{128}$ $29\sqrt{2}$

7. $\sqrt[3]{16x} - \sqrt[3]{54x}$ $-\sqrt[3]{2x}$ **8.** $2\sqrt[3]{3a^4} - 3a\sqrt[3]{81a}$

9. $\sqrt{9b^3} - \sqrt{25b^3} + \sqrt{49b^3}$ $5b\sqrt{b}$

10. $\sqrt{4x^7} + 9x^2\sqrt{x^3} - 5x\sqrt{x^5}$ $6x^3\sqrt{x}$

11. $\dfrac{5\sqrt{2}}{3} + \dfrac{2\sqrt{2}}{5}$ $\dfrac{31\sqrt{2}}{15}$

12. $\dfrac{\sqrt{3}}{2} + \dfrac{4\sqrt{3}}{3}$ $\dfrac{11\sqrt{3}}{6}$

 13. $\sqrt[3]{\dfrac{11}{8}} - \dfrac{\sqrt[3]{11}}{6}$ $\dfrac{\sqrt[3]{11}}{3}$

14. $\dfrac{2\sqrt[4]{4}}{7} - \dfrac{\sqrt[4]{4}}{14}$ $\dfrac{3\sqrt[4]{4}}{14}$

15. $\dfrac{\sqrt{20x}}{9} + \sqrt{\dfrac{5x}{9}}$ $\dfrac{5\sqrt{5x}}{9}$

16. $\dfrac{3x\sqrt{7}}{5} + \sqrt{\dfrac{7x^2}{100}}$

17. $7\sqrt{9} - 7 + \sqrt{3}$

18. $\sqrt{16} - 5\sqrt{10} + 7$

19. $2 + 3\sqrt{y^2} - 6\sqrt{y^2} + 5$ $7 - 3y$

20. $3\sqrt{7} - \sqrt[3]{x} + 4\sqrt{7} - 3\sqrt[3]{x}$ $7\sqrt{7} - 4\sqrt[3]{x}$

21. $3\sqrt{108} - 2\sqrt{18} - 3\sqrt{48}$ $6\sqrt{3} - 6\sqrt{2}$

22. $-\sqrt{75} + \sqrt{12} - 3\sqrt{3}$ $-6\sqrt{3}$

23. $-5\sqrt[3]{625} + \sqrt[3]{40}$ $-23\sqrt[3]{5}$ **24.** $-2\sqrt[3]{108} - \sqrt[3]{32}$

25. $\sqrt{9b^3} - \sqrt{25b^3} + \sqrt{16b^3}$ $2b\sqrt{b}$

26. $\sqrt{4x^7y^5} + 9x^2\sqrt{x^3y^5} - 5xy\sqrt{x^5y^3}$ $6x^3y^2\sqrt{xy}$

27. $5y\sqrt{8y} + 2\sqrt{50y^3}$ **28.** $3\sqrt{8x^2y^3} - 2x\sqrt{32y^3}$

29. $\sqrt[3]{54xy^3} - 5\sqrt[3]{2xy^3} + y\sqrt[3]{128x}$ $2y\sqrt[3]{2x}$

30. $2\sqrt[3]{24x^3y^4} + 4x\sqrt[3]{81y^4}$ $16xy\sqrt[3]{3y}$

31. $6\sqrt[3]{11} + 8\sqrt{11} - 12\sqrt{11}$

32. $3\sqrt[3]{5} + 4\sqrt{5}$

33. $-2\sqrt[4]{x^7} + 3\sqrt[4]{16x^7}$ $4x\sqrt[4]{x^3}$

34. $6\sqrt[3]{24x^3} - 2\sqrt[3]{81x^3} - x\sqrt[3]{3}$ $5x\sqrt[3]{3}$

35. $\dfrac{4\sqrt{3}}{3} - \dfrac{\sqrt{12}}{3}$ $\dfrac{2\sqrt{3}}{3}$

36. $\dfrac{\sqrt{45}}{10} + \dfrac{7\sqrt{5}}{10}$ $\sqrt{5}$

37. $\dfrac{\sqrt[3]{8x^4}}{7} + \dfrac{3x\sqrt[3]{x}}{7}$ $\dfrac{5x\sqrt[3]{x}}{7}$

38. $\dfrac{\sqrt[4]{48}}{5x} - \dfrac{2\sqrt[4]{3}}{10x}$ $\dfrac{\sqrt[4]{3}}{5x}$

39. $\sqrt{\dfrac{28}{x^2}} + \sqrt{\dfrac{7}{4x^2}}$ $\dfrac{5\sqrt{7}}{2x}$

40. $\dfrac{\sqrt{99}}{5x} - \sqrt{\dfrac{44}{x^2}}$

41. $\sqrt[3]{\dfrac{16}{27}} - \dfrac{\sqrt[3]{54}}{6}$ $\dfrac{\sqrt[3]{2}}{6}$

42. $\dfrac{\sqrt[3]{3}}{10} + \sqrt[3]{\dfrac{24}{125}}$ $\dfrac{\sqrt[3]{3}}{2}$

43. $-\dfrac{\sqrt[3]{2x^4}}{9} + \sqrt[3]{\dfrac{250x^4}{27}}$

44. $\dfrac{\sqrt[3]{y^5}}{8} + \dfrac{5y\sqrt[3]{y^2}}{4}$

△ **45.** Find the perimeter of the trapezoid. $15\sqrt{3}$ in.

$2\sqrt{12}$ in.

$3\sqrt{3}$ in. $\sqrt{12}$ in.

$2\sqrt{27}$ in.

△ **46.** Find the perimeter of the triangle. $\left(6\sqrt{2} + 3\sqrt{5}\right)$ meters

$\sqrt{8}$ m $\sqrt{32}$ m

$\sqrt{45}$ m

Multiply, and then simplify if possible. See Example 4.

47. $\sqrt{7}(\sqrt{5} + \sqrt{3})$ $\sqrt{35} + \sqrt{21}$ **48.** $\sqrt{5}(\sqrt{15} - \sqrt{35})$

49. $(\sqrt{5} - \sqrt{2})^2$ $7 - 2\sqrt{10}$ **50.** $(3x - \sqrt{2})(3x - \sqrt{2})$

51. $\sqrt{3x}(\sqrt{3} - \sqrt{x})$ **52.** $\sqrt{5y}(\sqrt{y} + \sqrt{5})$

53. $(2\sqrt{x} - 5)(3\sqrt{x} + 1)$ **54.** $(8\sqrt{y} + z)(4\sqrt{y} - 1)$

55. $(\sqrt[3]{a} - 4)(\sqrt[3]{a} + 5)$ **56.** $(\sqrt[3]{a} + 2)(\sqrt[3]{a} + 7)$

57. $6(\sqrt{2} - 2)$ $6\sqrt{2} - 12$ **58.** $\sqrt{5}(6 - \sqrt{5})$

59. $\sqrt{2}(\sqrt{2} + x\sqrt{6})$ **60.** $\sqrt{3}(\sqrt{3} - 2\sqrt{5x})$

61. $(2\sqrt{7} + 3\sqrt{5})(\sqrt{7} - 2\sqrt{5})$ $-16 - \sqrt{35}$

62. $(\sqrt{6} - 4\sqrt{2})(3\sqrt{6} + 1)$ $18 + \sqrt{6} - 24\sqrt{3} - 4\sqrt{2}$

63. $(\sqrt{x} - y)(\sqrt{x} + y)$ $x - y^2$

64. $(3\sqrt{x} + 2)(\sqrt{3x} - 2)$

65. $(\sqrt{3} + x)^2$ **66.** $(\sqrt{y} - 3x)^2$

67. $(\sqrt{5x} - 3\sqrt{2})(\sqrt{5x} - 3\sqrt{3})$

68. $(5\sqrt{3x} - \sqrt{y})(4\sqrt{x} + 1)$

69. $(\sqrt[3]{4} + 2)(\sqrt[3]{2} - 1)$ $2\sqrt[3]{2} - \sqrt[3]{4}$

70. $(\sqrt[3]{3} + \sqrt[3]{2})(\sqrt[3]{9} - \sqrt[3]{4})$ $1 - \sqrt[3]{12} + \sqrt[3]{18}$

71. $(\sqrt[3]{x} + 1)(\sqrt[3]{x} - 4\sqrt{x} + 7)$

72. $(\sqrt[3]{3x} + 3)(\sqrt[3]{2x} - 3x - 1)$

73. $(\sqrt{x - 1} + 5)^2$ $x + 24 + 10\sqrt{x - 1}$

74. $(\sqrt{3x + 1} + 2)^2$ $3x + 5 + 4\sqrt{3x + 1}$

75. $(\sqrt{2x + 5} - 1)^2$ $2x + 6 - 2\sqrt{2x + 5}$

76. $(\sqrt{x - 6} - 7)^2$ $x + 43 - 14\sqrt{x - 6}$

REVIEW AND PREVIEW

Factor each numerator and denominator. Then simplify if possible. See Section 6.1.

77. $\dfrac{2x - 14}{2}$ $x - 7$ **78.** $\dfrac{8x - 24y}{4}$ $2(x - 3y)$

79. $\dfrac{7x - 7y}{x^2 - y^2}$ $\dfrac{7}{x + y}$ **80.** $\dfrac{x^3 - 8}{4x - 8}$ $\dfrac{x^2 + 2x + 4}{4}$

81. $\dfrac{6a^2b - 9ab}{3ab}$ $2a - 3$ **82.** $\dfrac{14r - 28r^2s^2}{7rs}$ $\dfrac{2(1 - 2rs^2)}{s}$

83. $\dfrac{-4 + 2\sqrt{3}}{6}$ $\dfrac{-2 + \sqrt{3}}{3}$ **84.** $\dfrac{-5 + 10\sqrt{7}}{5}$ $-1 + 2\sqrt{7}$

Concept Extensions

△ **85.** Find the perimeter and area of the rectangle. $22\sqrt{5}$ ft; 150 sq. ft

53. $6x - 13\sqrt{x} - 5$

48. $5\sqrt{3} - 5\sqrt{7}$

50. $9x^2 - 6x\sqrt{2} + 2$

51. $3\sqrt{x} - x\sqrt{3}$

52. $y\sqrt{5} + 5\sqrt{y}$

$\sqrt{125}$ ft

$3\sqrt{20}$ ft

△ **86.** Find the area and perimeter of the trapezoid. (*Hint:* The area of a trapezoid is the product of half the height $6\sqrt{3}$ meters and the sum of the bases $2\sqrt{63}$ and $7\sqrt{7}$ meters.)

$39\sqrt{21}$ sq. m; $(12\sqrt{3} + 13\sqrt{7})$ m

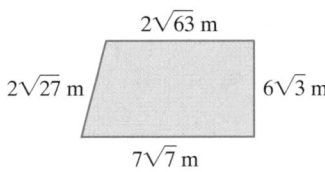

$2\sqrt{63}$ m

$2\sqrt{27}$ m

$6\sqrt{3}$ m

$7\sqrt{7}$ m

87. **a.**   Add: $\sqrt{3} + \sqrt{3}$ $2\sqrt{3}$

 b. Multiply: $\sqrt{3} \cdot \sqrt{3}$ 3

 c. Describe the differences in parts **a** and **b**.   answers may vary

88. Multiply: $(\sqrt{2} + \sqrt{3} - 1)^2$ $2\sqrt{6} - 2\sqrt{2} - 2\sqrt{3} + 6$

89. Explain how simplifying $2x + 3x$ is similar to simplifying $2\sqrt{x} + 3\sqrt{x}$.   answers may vary

90. Explain how multiplying $(x - 2)(x + 3)$ is similar to multiplying $(\sqrt{x} - \sqrt{2})(\sqrt{x} + 3)$.   answers may vary

---

| 7.5 | **RATIONALIZING DENOMINATORS AND NUMERATORS OF RADICAL EXPRESSIONS** |

54. $32y - 8\sqrt{y} + 4z\sqrt{y} - z$ **55.** $\sqrt[3]{a^2} + \sqrt[3]{a} - 20$ **56.** $\sqrt[3]{a^2} + 9\sqrt[3]{a} + 14$ **58.** $6\sqrt{5} - 5$ **59.** $2 + 2x\sqrt{3}$ **60.** $3 - 2\sqrt{15x}$

64. $3x\sqrt{3} - 6\sqrt{x} + 2\sqrt{3x} - 4$

65. $3 + 2x\sqrt{3} + x^2$

66. $y - 6x\sqrt{y} + 9x^2$

67. $5x - 3\sqrt{15x} - 3\sqrt{10x} + 9\sqrt{6}$

68. $20x\sqrt{3} + 5\sqrt{3x} - 4\sqrt{xy} - \sqrt{y}$

Objectives

1 Rationalize denominators.

2 Rationalize denominators having two terms.

3 Rationalize numerators.

71 $-4\sqrt[6]{x^5} + \sqrt[3]{x^2} + 8\sqrt[3]{x} - 4\sqrt{x} + 7$

72. $\sqrt[3]{6x^2} - 3x\sqrt[3]{3x} - \sqrt[3]{3x} + 3\sqrt[3]{2x} - 9x - 3$

1   Often in mathematics, it is helpful to write a radical expression such as $\dfrac{\sqrt{3}}{\sqrt{2}}$ either without a radical in the denominator or without a radical in the numerator. The process of writing this expression as an equivalent expression but without a radical in the denominator is called **rationalizing the denominator**. To rationalize the denominator of $\dfrac{\sqrt{3}}{\sqrt{2}}$, we use the fundamental principle of fractions and multiply the numerator and the denominator by $\sqrt{2}$. Recall that this is the same as multiplying by $\dfrac{\sqrt{2}}{\sqrt{2}}$, which simplifies to 1.

$$\frac{\sqrt{3}}{\sqrt{2}} = \frac{\sqrt{3} \cdot \sqrt{2}}{\sqrt{2} \cdot \sqrt{2}} = \frac{\sqrt{6}}{\sqrt{4}} = \frac{\sqrt{6}}{2}$$

EXAMPLE 1

Rationalize the denominator of each expression.

a. $\dfrac{2}{\sqrt{5}}$ **b.** $\dfrac{2\sqrt{16}}{\sqrt{9x}}$ **c.** $\sqrt[3]{\dfrac{1}{2}}$

Solution

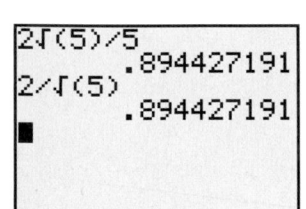

A calculator check for Example 1a.

a. To rationalize the denominator, we multiply the numerator and denominator by a factor that makes the radicand in the denominator a perfect square.

$$\frac{2}{\sqrt{5}} = \frac{2 \cdot \sqrt{5}}{\sqrt{5} \cdot \sqrt{5}} = \frac{2\sqrt{5}}{5}$$ The denominator is now rationalized

b. First, we simplify the radicals and then rationalize the denominator.

$$\frac{2\sqrt{16}}{\sqrt{9x}} = \frac{2(4)}{3\sqrt{x}} = \frac{8}{3\sqrt{x}}$$

To rationalize the denominator, multiply the numerator and denominator by $\sqrt{x}$. Then

$$\frac{8}{3\sqrt{x}} = \frac{8 \cdot \sqrt{x}}{3\sqrt{x} \cdot \sqrt{x}} = \frac{8\sqrt{x}}{3x}$$

CLASSROOM EXAMPLE

Rationalize each denominator.

a. $\dfrac{7}{\sqrt{2}}$ **b.** $\dfrac{2\sqrt{9}}{\sqrt{16y}}$ **c.** $\sqrt[3]{\dfrac{2}{25}}$

answer: **a.** $\dfrac{7\sqrt{2}}{2}$ **b.** $\dfrac{3\sqrt{y}}{2y}$ **c.** $\dfrac{\sqrt[3]{10}}{5}$

c. $\sqrt[3]{\dfrac{1}{2}} = \dfrac{\sqrt[3]{1}}{\sqrt[3]{2}} = \dfrac{1}{\sqrt[3]{2}}$. Now we rationalize the denominator. Since $\sqrt[3]{2}$ is a cube root, we want to multiply by a value that will make the radicand 2 a perfect cube. If we multiply $\sqrt[3]{2}$ by $\sqrt[3]{2^2}$, we get $\sqrt[3]{2^3} = \sqrt[3]{8} = 2$.

$$\frac{1 \cdot \sqrt[3]{2^2}}{\sqrt[3]{2} \cdot \sqrt[3]{2^2}} = \frac{\sqrt[3]{4}}{\sqrt[3]{2^3}} = \frac{\sqrt[3]{4}}{2}$$ Multiply the numerator and denominator by $\sqrt[3]{2^2}$ and then simplify.

CONCEPT CHECK

Determine by which number both the numerator and denominator can be multiplied to rationalize the denominator of the radical expression.

a. $\dfrac{1}{\sqrt[3]{7}}$ **b.** $\dfrac{1}{\sqrt[4]{8}}$

EXAMPLE 2

Rationalize the denominator of $\sqrt{\dfrac{7x}{3y}}$.

Solution

$\sqrt{\dfrac{7x}{3y}} = \dfrac{\sqrt{7x}}{\sqrt{3y}}$ Use the quotient rule. No radical may be simplified further.

$= \dfrac{\sqrt{7x} \cdot \sqrt{3y}}{\sqrt{3y} \cdot \sqrt{3y}}$ Multiply numerator and denominator by $\sqrt{3y}$ so that the radicand in the denominator is a perfect square.

$= \dfrac{\sqrt{21xy}}{3y}$ Use the product rule in the numerator and denominator. Remember that $\sqrt{3y} \cdot \sqrt{3y} = 3y$.

CLASSROOM EXAMPLE

Rationalize the denominator of $\sqrt{\dfrac{5m}{11n}}$.

answer: $\dfrac{\sqrt{55mn}}{11n}$

EXAMPLE 3

Rationalize the denominator of $\dfrac{\sqrt[4]{x}}{\sqrt[4]{81y^5}}$.

Solution First, simplify each radical if possible.

$\dfrac{\sqrt[4]{x}}{\sqrt[4]{81y^5}} = \dfrac{\sqrt[4]{x}}{\sqrt[4]{81y^4} \cdot \sqrt[4]{y}}$ Use the product rule in the denominator.

$= \dfrac{\sqrt[4]{x}}{3y\sqrt[4]{y}}$ Write $\sqrt[4]{81y^4}$ as $3y$.

$= \dfrac{\sqrt[4]{x} \cdot \sqrt[4]{y^3}}{3y\sqrt[4]{y} \cdot \sqrt[4]{y^3}}$ Multiply numerator and denominator by $\sqrt[4]{y^3}$ so that the radicand in the denominator is a perfect fourth power.

$= \dfrac{\sqrt[4]{xy^3}}{3y\sqrt[4]{y^4}}$ Use the product rule in the numerator and denominator.

$= \dfrac{\sqrt[4]{xy^3}}{3y^2}$ In the denominator, $\sqrt[4]{y^4} = y$ and $3y \cdot y = 3y^2$.

CLASSROOM EXAMPLE

Rationalize the denominator of $\dfrac{\sqrt[5]{a^2}}{\sqrt[5]{32b^{12}}}$.

answer: $\dfrac{\sqrt[5]{a^2b^3}}{2b^3}$

---

2 Remember the product of the sum and difference of two terms?

$$(a + b)(a - b) = a^2 - b^2$$

These two expressions are called conjugates of each other.

Concept Check Answer:

a. $\sqrt[3]{7^2}$ or $\sqrt[3]{49}$ **b.** $\sqrt[4]{2}$

To rationalize a numerator or denominator that is a sum or difference of two terms, we use conjugates. To see how and why this works, let's rationalize the denominator of the expression $\dfrac{5}{\sqrt{3} - 2}$. To do so, we multiply both the numerator and the denominator by $\sqrt{3} + 2$, the **conjugate** of the denominator $\sqrt{3} - 2$, and see what happens.

$$\frac{5}{\sqrt{3} - 2} = \frac{5(\sqrt{3} + 2)}{(\sqrt{3} - 2)(\sqrt{3} + 2)}$$

$$= \frac{5(\sqrt{3} + 2)}{(\sqrt{3})^2 - 2^2} \qquad \text{Multiply the sum and difference of two terms: } (a + b)(a - b) = a^2 - b^2.$$

$$= \frac{5(\sqrt{3} + 2)}{3 - 4}$$

$$= \frac{5(\sqrt{3} + 2)}{-1}$$

$$= -5(\sqrt{3} + 2) \quad \text{or} \quad -5\sqrt{3} - 10$$

Notice in the denominator that the product of $(\sqrt{3} - 2)$ and its conjugate, $(\sqrt{3} + 2)$, is -1. In general, the product of an expression and its conjugate will contain no radical terms. This is why, when rationalizing a denominator or a numerator containing two terms, we multiply by its conjugate. Examples of conjugates are

$$\sqrt{a} - \sqrt{b} \quad \text{and} \quad \sqrt{a} + \sqrt{b}$$
$$x + \sqrt{y} \quad \text{and} \quad x - \sqrt{y}$$

EXAMPLE 4

Rationalize each denominator.

a. $\dfrac{2}{3\sqrt{2} + 4}$ b. $\dfrac{\sqrt{6} + 2}{\sqrt{5} - \sqrt{3}}$ c. $\dfrac{2\sqrt{m}}{3\sqrt{x} + \sqrt{m}}$

Solution   a. Multiply the numerator and denominator by the conjugate of the denominator, $3\sqrt{2} + 4$.

$$\frac{2}{3\sqrt{2} + 4} = \frac{2(3\sqrt{2} - 4)}{(3\sqrt{2} + 4)(3\sqrt{2} - 4)}$$

$$= \frac{2(3\sqrt{2} - 4)}{(3\sqrt{2})^2 - 4^2}$$

$$= \frac{2(3\sqrt{2} - 4)}{18 - 16}$$

$$= \frac{2(3\sqrt{2} - 4)}{2}, \quad \text{or} \quad 3\sqrt{2} - 4$$

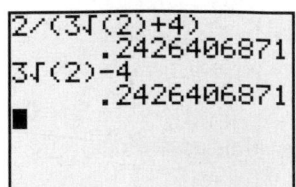

A calculator check for Example 4a. Notice the use of parentheses in the first expression.

It is often helpful to leave a numerator in factored form to help determine whether the expression can be simplified.

A calculator check for Example 4b. Notice the use of parentheses in the numerators and the denominator containing more than one term.

b. Multiply the numerator and denominator by the conjugate of $\sqrt{5} - \sqrt{3}$.

$$\frac{\sqrt{6} + 2}{\sqrt{5} - \sqrt{3}} = \frac{(\sqrt{6} + 2)(\sqrt{5} + \sqrt{3})}{(\sqrt{5} - \sqrt{3})(\sqrt{5} + \sqrt{3})}$$

$$= \frac{\sqrt{6}\sqrt{5} + \sqrt{6}\sqrt{3} + 2\sqrt{5} + 2\sqrt{3}}{(\sqrt{5})^2 - (\sqrt{3})^2}$$

$$= \frac{\sqrt{30} + \sqrt{18} + 2\sqrt{5} + 2\sqrt{3}}{5 - 3}$$

$$= \frac{\sqrt{30} + 3\sqrt{2} + 2\sqrt{5} + 2\sqrt{3}}{2}$$

c. Multiply by the conjugate of $3\sqrt{x} + \sqrt{m}$ to eliminate the radicals from the denominator.

$$\frac{2\sqrt{m}}{3\sqrt{x} + \sqrt{m}} = \frac{2\sqrt{m}(3\sqrt{x} - \sqrt{m})}{(3\sqrt{x} + \sqrt{m})(3\sqrt{x} - \sqrt{m})} = \frac{6\sqrt{mx} - 2m}{(3\sqrt{x})^2 - (\sqrt{m})^2}$$

$$= \frac{6\sqrt{mx} - 2m}{9x - m}$$

TEACHING TIP

After discussing rationalizing the numerator, ask students, "Is it possible to rationalize both the denominator and numerator? Explain."

3   As mentioned earlier, it is also often helpful to write an expression such as $\dfrac{\sqrt{3}}{\sqrt{2}}$ as an equivalent expression without a radical in the numerator. This process is called **rationalizing the numerator**. To rationalize the numerator of $\dfrac{\sqrt{3}}{\sqrt{2}}$, we multiply the numerator and the denominator by $\sqrt{3}$.

$$\frac{\sqrt{3}}{\sqrt{2}} = \frac{\sqrt{3} \cdot \sqrt{3}}{\sqrt{2} \cdot \sqrt{3}} = \frac{\sqrt{9}}{\sqrt{6}} = \frac{3}{\sqrt{6}}$$

EXAMPLE 5

Rationalize the numerator of $\dfrac{\sqrt{7}}{\sqrt{45}}$.

Solution   First we simplify $\sqrt{45}$.

$$\frac{\sqrt{7}}{\sqrt{45}} = \frac{\sqrt{7}}{\sqrt{9 \cdot 5}} = \frac{\sqrt{7}}{3\sqrt{5}}$$

CLASSROOM EXAMPLE

Rationalize the numerator of $\dfrac{\sqrt{18}}{\sqrt{75}}$.

answer: $\dfrac{6}{5\sqrt{6}}$

Next we rationalize the numerator by multiplying the numerator and the denominator by $\sqrt{7}$.

$$\frac{\sqrt{7}}{3\sqrt{5}} = \frac{\sqrt{7} \cdot \sqrt{7}}{3\sqrt{5} \cdot \sqrt{7}} = \frac{7}{3\sqrt{5 \cdot 7}} = \frac{7}{3\sqrt{35}}$$

EXAMPLE 6

Rationalize the numerator of $\dfrac{\sqrt[3]{2x^2}}{\sqrt[3]{5y}}$.

Solution

The numerator and the denominator of this expression are already simplified. To rationalize the numerator, $\sqrt[3]{2x^2}$, we multiply the numerator and denominator by a factor that will make the radicand a perfect cube. If we multiply $\sqrt[3]{2x^2}$ by $\sqrt[3]{4x}$, we get $\sqrt[3]{8x^3} = 2x$.

$$\frac{\sqrt[3]{2x^2}}{\sqrt[3]{5y}} = \frac{\sqrt[3]{2x^2} \cdot \sqrt[3]{4x}}{\sqrt[3]{5y} \cdot \sqrt[3]{4x}} = \frac{\sqrt[3]{8x^3}}{\sqrt[3]{20xy}} = \frac{2x}{\sqrt[3]{20xy}}$$

CLASSROOM EXAMPLE

Rationalize the numerator of $\dfrac{\sqrt[3]{3a}}{\sqrt[3]{7b}}$.

answer: $\dfrac{3a}{\sqrt[3]{63a^2b}}$

EXAMPLE 7

Rationalize the numerator of $\dfrac{\sqrt{x} + 2}{5}$.

Solution

We multiply the numerator and the denominator by the conjugate of the numerator, $\sqrt{x} + 2$.

$$\frac{\sqrt{x} + 2}{5} = \frac{(\sqrt{x} + 2)(\sqrt{x} - 2)}{5(\sqrt{x} - 2)} \quad \text{Multiply by } \sqrt{x} - 2, \text{ the conjugate of } \sqrt{x} + 2.$$

$$= \frac{(\sqrt{x})^2 - 2^2}{5(\sqrt{x} - 2)} \quad (a+b)(a-b) = a^2 - b^2.$$

$$= \frac{x - 4}{5(\sqrt{x} - 2)}$$

CLASSROOM EXAMPLE

Rationalize the numerator of $\dfrac{\sqrt{x} + 5}{3}$.

answer: $\dfrac{x - 25}{3(\sqrt{x} - 5)}$

STUDY SKILLS REMINDER

How Are Your Homework Assignments Going?

By now, you should have good homework habits. If not, it's never too late to begin. Why is it so important in mathematics to keep up with homework? You probably now know the answer to that question. You have probably realized by now that many concepts in mathematics build on each other. Your understanding of one chapter in mathematics usually depends on your understanding of the previous chapter's material.

Don't forget that completing your homework assignment involves a lot more than attempting a few of the problems assigned.

To complete a homework assignment, remember these four things:

1. Attempt all of it.

2. Check it.

3. Correct it.

4. If needed, ask questions about it.

MENTAL MATH

Find the conjugate of each expression.

1. $\sqrt{2} + x$ $\sqrt{2} - x$ **2.** $\sqrt{3} + y$ $\sqrt{3} - y$ **3.** $5 - \sqrt{a}$ $5 + \sqrt{a}$

4. $6 - \sqrt{b}$ $6 + \sqrt{b}$ **5.** $7\sqrt{5} + 8\sqrt{x}$ $7\sqrt{5} - 8\sqrt{x}$ **6.** $9\sqrt{2} - 6\sqrt{y}$ $9\sqrt{2} + 6\sqrt{y}$

EXERCISE SET 7.5

STUDY GUIDE/SSM   CD/VIDEO   PH MATH TUTOR CENTER   MathXL®Tutorials ON CD   MathXL®   MyMathLab®

Rationalize each denominator. See Examples 1 through 3.

1. $\dfrac{\sqrt{2}}{\sqrt{7}}$ $\dfrac{\sqrt{14}}{7}$ **2.** $\dfrac{\sqrt{3}}{\sqrt{2}}$ $\dfrac{\sqrt{6}}{2}$

3. $\sqrt{\dfrac{1}{5}}$ $\dfrac{\sqrt{5}}{5}$ **4.** $\sqrt{\dfrac{1}{2}}$ $\dfrac{\sqrt{2}}{2}$

5. $\sqrt[3]{\dfrac{3}{4}}$ $\dfrac{\sqrt[3]{6}}{2}$ **6.** $\sqrt[3]{\dfrac{2}{9}}$ $\dfrac{\sqrt[3]{6}}{3}$

7. $\dfrac{4}{\sqrt[3]{3}}$ $\dfrac{4\sqrt[3]{9}}{3}$ **8.** $\dfrac{6}{\sqrt[3]{9}}$ $2\sqrt[3]{3}$

9. 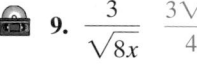 $\dfrac{3}{\sqrt{8x}}$ $\dfrac{3\sqrt{2x}}{4x}$ **10.** $\dfrac{5}{\sqrt{27a}}$ $\dfrac{5\sqrt{3a}}{9a}$

11. $\dfrac{3}{\sqrt[3]{4x^2}}$ $\dfrac{3\sqrt[3]{2x}}{2x}$ **12.** $\dfrac{5}{\sqrt[3]{3y}}$ $\dfrac{5\sqrt[3]{9y^2}}{3y}$

13. $\sqrt{\dfrac{4}{x}}$ $\dfrac{2\sqrt{x}}{x}$ **14.** $\sqrt{\dfrac{25}{y}}$ $\dfrac{5\sqrt{y}}{y}$

15. $\dfrac{9}{\sqrt{3a}}$ $\dfrac{3\sqrt{3a}}{a}$ **16.** $\dfrac{x}{\sqrt{5}}$ $\dfrac{x\sqrt{5}}{5}$

17. $\dfrac{3}{\sqrt[3]{2}}$ $\dfrac{3\sqrt[3]{4}}{2}$ **18.** $\dfrac{5}{\sqrt[3]{9}}$ $\dfrac{5\sqrt[3]{3}}{3}$

19. $\dfrac{2\sqrt{3}}{\sqrt{7}}$ $\dfrac{2\sqrt{21}}{7}$ **20.** $\dfrac{-5\sqrt{2}}{\sqrt{11}}$ $\dfrac{-5\sqrt{22}}{11}$

21. $\sqrt{\dfrac{2x}{5y}}$ $\dfrac{\sqrt{10xy}}{5y}$ **22.** $\sqrt{\dfrac{13a}{2b}}$ $\dfrac{\sqrt{26ab}}{2b}$

23. $\sqrt[4]{\dfrac{81}{8}}$ $\dfrac{3\sqrt[4]{2}}{2}$ **24.** $\sqrt[4]{\dfrac{1}{9}}$ $\dfrac{\sqrt[4]{9}}{3}$

25. $\sqrt[4]{\dfrac{16}{9x^7}}$ $\dfrac{2\sqrt[4]{9x}}{3x^2}$ **26.** $\sqrt[5]{\dfrac{32}{m^6n^{13}}}$ $\dfrac{2\sqrt[5]{m^4n^2}}{m^2n^3}$

27. $\dfrac{5a}{\sqrt[5]{8a^9b^{11}}}$ $\dfrac{5a\sqrt[5]{4ab^4}}{2a^2b^3}$ **28.** $\dfrac{9y}{\sqrt[4]{4y^9}}$ $\dfrac{9\sqrt[4]{4y^3}}{2y^2}$

Rationalize each denominator. See Example 4.

29. $\dfrac{6}{2 - \sqrt{7}}$ $-2(2 + \sqrt{7})$ **30.** $\dfrac{3}{\sqrt{7} - 4}$ $-\dfrac{\sqrt{7} + 4}{3}$

31. $\dfrac{-7}{\sqrt{x} - 3}$ $\dfrac{7(3 + \sqrt{x})}{9 - x}$ **32.** $\dfrac{-8}{\sqrt{y} + 4}$ $\dfrac{32 - 8\sqrt{y}}{y - 16}$

33. $\dfrac{\sqrt{2} - \sqrt{3}}{\sqrt{2} + \sqrt{3}}$ $-5 + 2\sqrt{6}$ **34.** $\dfrac{\sqrt{3} + \sqrt{4}}{\sqrt{2} + \sqrt{3}}$

35. $\dfrac{\sqrt{a} + 1}{2\sqrt{a} - \sqrt{b}}$ **36.** $\dfrac{2\sqrt{a} - 3}{2\sqrt{a} - \sqrt{b}}$

37. $\dfrac{8}{1 + \sqrt{10}}$ $-\dfrac{8(1 - \sqrt{10})}{9}$ **38.** $\dfrac{-3}{\sqrt{6} - 2}$

39. $\dfrac{\sqrt{x}}{\sqrt{x} + \sqrt{y}}$ $\dfrac{x - \sqrt{xy}}{x - y}$ **40.** $\dfrac{2\sqrt{a}}{2\sqrt{x} - \sqrt{y}}$

41. $\dfrac{2\sqrt{3} + \sqrt{6}}{4\sqrt{3} - \sqrt{6}}$ $\dfrac{5 + 3\sqrt{2}}{7}$ **42.** $\dfrac{4\sqrt{5} + \sqrt{2}}{2\sqrt{5} - \sqrt{2}}$

Rationalize each numerator. See Examples 5 and 6.

43. $\sqrt{\dfrac{5}{3}}$ $\dfrac{5}{\sqrt{15}}$ **44.** $\sqrt{\dfrac{3}{2}}$ $\dfrac{3}{\sqrt{6}}$

45. $\sqrt{\dfrac{18}{5}}$ $\dfrac{6}{\sqrt{10}}$ **46.** $\sqrt{\dfrac{12}{7}}$ $\dfrac{6}{\sqrt{21}}$

34. $3 - \sqrt{6} - 2\sqrt{2} + 2\sqrt{3}$ **35.** $\dfrac{2a + 2\sqrt{a} + \sqrt{ab} + \sqrt{b}}{4a - b}$ **36.** $\dfrac{4a + 2\sqrt{ab} - 6\sqrt{a} - 3\sqrt{b}}{4a - b}$ **38.** $\dfrac{-3(\sqrt{6} + 2)}{2}$

40. $\dfrac{4\sqrt{ax} + 2\sqrt{ay}}{4x - y}$ **42.** $\dfrac{7 + \sqrt{10}}{3}$

47. $\dfrac{\sqrt{4x}}{7}$ $\dfrac{2x}{7\sqrt{x}}$

48. $\dfrac{\sqrt{3x^5}}{6}$ $\dfrac{x^3}{2\sqrt{3x}}$

49. $\dfrac{\sqrt[3]{5y^2}}{\sqrt[3]{4x}}$ $\dfrac{5y}{\sqrt[3]{100xy}}$

50. $\dfrac{\sqrt[3]{4x}}{\sqrt[3]{z^4}}$ $\dfrac{2x}{z\sqrt[3]{2x^2z}}$

51. $\sqrt{\dfrac{2}{5}}$ $\dfrac{2}{\sqrt{10}}$

52. $\sqrt{\dfrac{3}{7}}$ $\dfrac{3}{\sqrt{21}}$

53. $\dfrac{\sqrt{2x}}{11}$ $\dfrac{2x}{11\sqrt{2x}}$

54. $\dfrac{\sqrt{y}}{7}$ $\dfrac{y}{7\sqrt{y}}$

55. $\sqrt[3]{\dfrac{7}{8}}$ $\dfrac{7}{2\sqrt[3]{49}}$

56. $\sqrt[3]{\dfrac{25}{2}}$ $\dfrac{5}{\sqrt[3]{10}}$

57. $\dfrac{\sqrt[3]{3x^5}}{10}$ $\dfrac{3x^2}{10\sqrt[3]{9x}}$

58. $\sqrt[3]{\dfrac{9y}{7}}$ $\dfrac{3y}{\sqrt[3]{21y^2}}$

59. $\sqrt{\dfrac{18x^4y^6}{3z}}$ $\dfrac{6x^2y^3}{\sqrt{6z}}$

60. $\sqrt{\dfrac{8x^5y}{2z}}$ $\dfrac{2x^3y}{\sqrt{xyz}}$

61. When rationalizing the denominator of $\dfrac{\sqrt{5}}{\sqrt{7}}$, explain why both the numerator and the denominator must be multiplied by $\sqrt{7}$.   answers may vary

62. When rationalizing the numerator of $\dfrac{\sqrt{5}}{\sqrt{7}}$, explain why both the numerator and the denominator must be multiplied by $\sqrt{5}$.   answers may vary

Rationalize each numerator. See Example 7.

63. $\dfrac{2 - \sqrt{11}}{6}$ $\dfrac{-7}{12 + 6\sqrt{11}}$

64. $\dfrac{\sqrt{15} + 1}{2}$ $\dfrac{7}{\sqrt{15} - 1}$

65. $\dfrac{2 - \sqrt{7}}{-5}$ $\dfrac{3}{10 + 5\sqrt{7}}$

66. $\dfrac{\sqrt{5} + 2}{\sqrt{2}}$ $\dfrac{1}{\sqrt{10} - 2\sqrt{2}}$

67. $\dfrac{\sqrt{x} + 3}{\sqrt{x}}$ $\dfrac{x - 9}{x - 3\sqrt{x}}$

68. $\dfrac{5 + \sqrt{2}}{\sqrt{2x}}$

69. $\dfrac{\sqrt{2} - 1}{\sqrt{2} + 1}$ $\dfrac{1}{3 + 2\sqrt{2}}$

70. $\dfrac{\sqrt{8} - \sqrt{3}}{\sqrt{2} + \sqrt{3}}$ $\dfrac{5}{7 + 3\sqrt{6}}$

71. $\dfrac{\sqrt{x} + 1}{\sqrt{x} - 1}$ $\dfrac{x - 1}{x - 2\sqrt{x} + 1}$

72. $\dfrac{\sqrt{x} + \sqrt{y}}{\sqrt{x} - \sqrt{y}}$

REVIEW AND PREVIEW

Solve each equation. See Sections 1.5 and 5.8.

73. $2x - 7 = 3(x - 4)$ 5

74. $9x - 4 = 7(x - 2)$ -5

75. $(x - 6)(2x + 1) = 0$ $-\dfrac{1}{2}, 6$

76. $(y + 2)(5y + 4) = 0$ $-2, -\dfrac{4}{5}$

77. $x^2 - 8x = -12$ $2, 6$

78. $x^3 = x$ $0, 1, -1$

Concept Extensions

△ **79.** The formula of the radius r of a sphere with surface area A is

$$r = \sqrt{\dfrac{A}{4\pi}}$$

Rationalize the denominator of the radical expression in this formula. $r = \dfrac{\sqrt{A\pi}}{2\pi}$

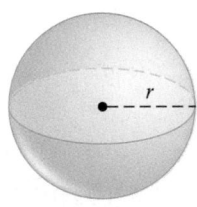

△ **80.** The formula for the radius r of a cone with height 7 centimeters and volume V is

$$r = \sqrt{\dfrac{3V}{7\pi}}$$

Rationalize the numerator of the radical expression in this formula. $r = \dfrac{3V}{\sqrt{21\pi V}}$

7 cm

81. Explain why rationalizing the denominator does not change the value of the original expression.   answers may vary

82. Explain why rationalizing the numerator does not change the value of the original expression.   answers may vary

68. $\dfrac{23}{5\sqrt{2x} - 2\sqrt{x}}$

72. $\dfrac{x - y}{x - 2\sqrt{xy} + y}$

RADICALS AND RATIONAL EXPONENTS

INTEGRATED REVIEW

Throughout this review, assume that all variables represent positive real numbers. Find each root.

1. $\sqrt{81}$
9

2. $\sqrt[3]{-8}$
-2

3. $\sqrt[4]{\dfrac{1}{16}}$
$\dfrac{1}{2}$

4. $\sqrt{x^6}$
x^3

5. $\sqrt[3]{y^9}$
y^3

6. $\sqrt{4y^{10}}$
$2y^5$

7. $\sqrt[5]{-32y^5}$
$-2y$

8. $\sqrt[4]{81b^{12}}$
$3b^3$

Use radical notation to rewrite each expression. Simplify if possible.

9. $36^{1/2}$
6

10. $(3y)^{1/4}$
$\sqrt[4]{3y}$

11. $64^{-2/3}$
$\dfrac{1}{16}$

12. $(x + 1)^{3/5}$
$\sqrt[5]{(x + 1)^3}$

Use the properties of exponents to simplify each expression. Write with positive exponents.

13. $y^{-1/6} \cdot y^{7/6}$
y

14. $\dfrac{(2x^{1/3})^4}{x^{5/6}}$
$16x^{1/2}$

15. $\dfrac{x^{1/4} x^{3/4}}{x^{-1/4}}$
$x^{5/4}$

16. $4^{1/3} \cdot 4^{2/5}$
$4^{11/15}$

Use rational exponents to simplify each radical.

17. $\sqrt[3]{8x^6}$
$2x^2$

18. $\sqrt[12]{a^9 b^6}$
$\sqrt[4]{a^3 b^2}$

Use rational exponents to write each as a single radical expression.

19. $\sqrt[4]{x} \cdot \sqrt{x}$
$\sqrt[4]{x^3}$

20. $\sqrt{5} \cdot \sqrt[3]{2}$
$\sqrt[6]{500}$

Simplify.

21. $\sqrt{40}$
$2\sqrt{10}$

22. $\sqrt[4]{16x^7 y^{10}}$
$2xy^2 \sqrt[4]{x^3 y^2}$

23. $\sqrt[3]{54x^4}$
$3x\sqrt[3]{2x}$

24. $\sqrt[5]{-64b^{10}}$
$-2b^2 \sqrt[5]{2}$

Multiply or divide. Then simplify if possible.

25. $\sqrt{5} \cdot \sqrt{x}$
$\sqrt{5x}$

26. $\sqrt[3]{8x} \cdot \sqrt[3]{8x^2}$
$4x$

27. $\dfrac{\sqrt{98y^6}}{\sqrt{2y}}$
$7y^2\sqrt{y}$

28. $\dfrac{\sqrt[4]{48a^9 b^3}}{\sqrt[4]{ab^3}}$
$2a^2 \sqrt[4]{3}$

Perform each indicated operation.

29. $\sqrt{20} - \sqrt{75} + 5\sqrt{7}$
$2\sqrt{5} - 5\sqrt{3} + 5\sqrt{7}$

30. $\sqrt[3]{54y^4} - y\sqrt[3]{16y}$
$y\sqrt[3]{2y}$

31. $\sqrt{3}(\sqrt{5} - \sqrt{2})$
$\sqrt{15} - \sqrt{6}$

32. $(\sqrt{7} + \sqrt{3})^2$
$10 + 2\sqrt{21}$

33. $(2x - \sqrt{5})(2x + \sqrt{5})$
$4x^2 - 5$

34. $(\sqrt{x+1} - 1)^2$
$x + 2 - 2\sqrt{x+1}$

Rationalize each denominator.

35. $\sqrt{\dfrac{7}{3}}$
$\dfrac{\sqrt{21}}{3}$

36. $\dfrac{5}{\sqrt[3]{2x^2}}$
$\dfrac{5\sqrt[3]{4x}}{2x}$

37. $\dfrac{\sqrt{3} - \sqrt{7}}{2\sqrt{3} + \sqrt{7}}$
$\dfrac{13 - 3\sqrt{21}}{5}$

Rationalize each numerator.

38. $\sqrt{\dfrac{7}{3}}$
$\dfrac{7}{\sqrt{21}}$

39. $\sqrt[3]{\dfrac{9y}{11}}$
$\dfrac{3y}{\sqrt[3]{33y^2}}$

40. $\dfrac{\sqrt{x} - 2}{\sqrt{x}}$
$\dfrac{x - 4}{x + 2\sqrt{x}}$

7.6 RADICAL EQUATIONS AND PROBLEM SOLVING

Objectives

1 Solve equations that contain radical expressions.

2 Use the Pythagorean theorem to model problems.

1 In this section, we present techniques to solve equations containing radical expressions such as

$$\sqrt{2x - 3} = 9$$

We use the power rule to help us solve these radical equations.

> **Power Rule**
>
> If both sides of an equation are raised to the same power, **all** solutions of the original equation are **among** the solutions of the new equation.

This property *does not* say that raising both sides of an equation to a power yields an equivalent equation. A solution of the new equation *may or may not* be a solution of the original equation. For example, $(-2)^2 = 2^2$, but $-2 \neq 2$. Thus, *each solution of the new equation must be checked* to make sure it is a solution of the original equation. Recall that a proposed solution that is not a solution of the original equation is called an **extraneous solution**.

EXAMPLE 1

Solve $\sqrt{2x - 3} = 9$. Use a graphing utility to check.

Solution   We use the power rule to square both sides of the equation to eliminate the radical.

$$\sqrt{2x - 3} = 9$$
$$\left(\sqrt{2x - 3}\right)^2 = 9^2$$
$$2x - 3 = 81$$
$$2x = 84$$
$$x = 42$$

Now check the solution using a graphing utility. To check using the intersection-of-graphs method, graph $y_1 = \sqrt{2x - 3}$ and $y_2 = 9$ in a $[0, 95, 10]$ by $[-5, 15, 1]$ window. The intersection has x-value 42, the solution, as shown.

The solution checks, so we conclude that the solution is 42 or the solution set is $\{42\}$.

In the next example, we choose to solve the equation graphically, using the x-intercept method since the equation is written so that one side is 0.

CLASSROOM EXAMPLE

Solve $\sqrt{3x - 2} = 5$.

answer: 9

EXAMPLE 2

Solve $\sqrt{-10x - 1} + 3x = 0$ for x.

Solution   Graph $y_1 = \sqrt{-10x - 1} + 3x$ in a decimal window.

The equation has two solutions since there are two x-intercepts. The x-intercepts are $x = -1$ and $x = -0.11111\ldots$, a repeating decimal. Use the fraction command to write $-0.\overline{1}$ as the equivalent fraction $-\frac{1}{9}$.

Check the solutions numerically as in the screen to the left.

The solutions are -1 and $-\frac{1}{9}$.

CLASSROOM EXAMPLE

Solve $\sqrt{9x - 2} - 2x = 0$.

answer: $\frac{1}{4}, 2$

A numerical check showing -1 and $-\frac{1}{9}$ are both solutions of the equation in Example 2.

Solving a Radical Equation Algebraically

Step 1: Isolate one radical on one side of the equation.

Step 2: Raise each side of the equation to a power equal to the index of the radical and simplify.

Step 3: If the equation still contains a radical term, repeat Steps 1 and 2. If not, solve the equation.

Step 4: Check all proposed solutions in the original equation.

▶ **Helpful Hint**

To solve a radical equation graphically, use the intersection-of-graphs method or the x-intercept method.

When do we solve a radical equation algebraically and when do we solve it graphically? If given a choice, it depends on the complexity of the equation itself. An equation that contains only one radical may be a good candidate for the algebraic process. Recall that by raising each side to a power, you are introducing the possibility of extraneous roots or solutions. For this reason, a check is mandatory. However, when solving graphically, you can visualize the numbers of solutions immediately. For the intersection-of-graphs method, the number of real solutions is equal to the number of intersections of the two graphs. For the x-intercept method, the number of real solutions is equal to the number of x-intercepts. We check after solving by a graphing method to confirm that the solutions are exact.

In Example 3, we choose to solve graphically by the intersection-of-graphs method and we confirm the solution algebraically.

EXAMPLE 3

Solve for x: $\sqrt[3]{x + 1} + 5 = 3$.

Solution

Graph $y_1 = \sqrt[3]{x + 1} + 5$ and $y_2 = 3$. The intersection of the two graphs has an x-value of -9, the solution of the equation.

To check algebraically, first, isolate the radical by subtracting 5 from both sides of the equation.

$$\sqrt[3]{x + 1} + 5 = 3$$
$$\sqrt[3]{x + 1} = -2$$

Next, raise both sides of the equation to the third power to eliminate the radical.

$$\left(\sqrt[3]{x + 1}\right)^3 = (-2)^3$$
$$x + 1 = -8$$
$$x = -9$$

The solution checks, so the solution is -9.

In Example 4, we solve algebraically and check graphically. Notice that in solving algebraically, an extraneous solution is introduced. However, graphically we see that there is only one solution because there is only one point of intersection for the two graphs.

EXAMPLE 4

Solve $\sqrt{4 - x} = x - 2$ for x.

Solution

$$\sqrt{4 - x} = x - 2$$
$$\left(\sqrt{4 - x}\right)^2 = (x - 2)^2$$
$$4 - x = x^2 - 4x + 4 \qquad \text{Write the quadratic equation in standard form.}$$
$$x^2 - 3x = 0$$
$$x(x - 3) = 0 \qquad \text{Factor.}$$
$$x = 0 \quad \text{or} \quad x - 3 = 0$$
$$x = 3$$

Check the possible solutions.

$$\sqrt{4 - x} = x - 2 \qquad\qquad\qquad \sqrt{4 - x} = x - 2$$
$$\sqrt{4 - 0} = 0 - 2 \quad \text{Let } x = 0. \qquad \sqrt{4 - x} = 3 - 2 \quad \text{Let } x = 3.$$
$$2 = -2 \quad \text{False} \qquad\qquad\qquad 1 = 1 \quad \text{True}$$

The proposed solution 3 checks, but 0 does not. When a proposed solution does not check, it is an **extraneous root or solution**. Since 0 is an extraneous solution, the solution is 3.

> **Helpful Hint**
> In Example 4, notice that $(x - 2)^2 = x^2 - 4x + 4$. Make sure binomials are squared correctly.

CLASSROOM EXAMPLE

Solve $\sqrt{9 + x} = x + 3$.

answer: 0

TEACHING TIP

Warn students about a common mistake in equations like Example 4. They must remember to square binomials like $(x - 2)^2$ correctly.

To check the results of Example 4, graph $y_1 = \sqrt{4 - x}$ and $y_2 = x - 2$. The x-value of the point of intersection is 3, as expected.

✔ CONCEPT CHECK

How can you immediately tell that the equation $\sqrt{2y + 3} = -4$ has no real solution?

Concept Check Answer:
answers may vary

In Example 5 we solve graphically. Use the intersection-of-graphs method and confirm the solution algebraically.

EXAMPLE 5

Solve $\sqrt{2x + 5} + \sqrt{2x} = 3$.

Solution
Graph $y_1 = \sqrt{2x + 5} + \sqrt{2x}$ and $y_2 = 3$ in a $[-10, 10, 1]$ by $[-1, 10, 1]$ window. The intersection of the two graphs is $x = 0.22222\ldots$, a repeating decimal. Convert the value of x to fraction form using the fraction command. The solution is $\frac{2}{9}$.

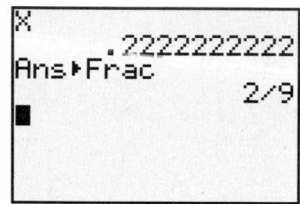

To check, we confirm the solution algebraically.

Isolate a radical by subtracting $\sqrt{2x}$ from both sides.

$$\sqrt{2x + 5} + \sqrt{2x} = 3$$
$$\sqrt{2x + 5} = 3 - \sqrt{2x}$$

Use the power rule to begin eliminating the radicals. Square both sides.

$$\left(\sqrt{2x + 5}\right)^2 = \left(3 - \sqrt{2x}\right)^2$$
$$2x + 5 = 9 - 6\sqrt{2x} + 2x \qquad \text{Multiply } \left(3 - \sqrt{2x}\right)\left(3 - \sqrt{2x}\right).$$

There is still a radical in the equation, so isolate the radical again. Then square both sides.

$$2x + 5 = 9 - 6\sqrt{2x} + 2x$$

$$6\sqrt{2x} = 4 \qquad\qquad \text{Isolate the radical.}$$

$$\left(6\sqrt{2x}\right)^2 = 4^2 \qquad\qquad \text{Square both sides of the equation to eliminate the radical.}$$

$$36(2x) = 16$$

$$72x = 16 \qquad\qquad \text{Multiply.}$$

$$x = \frac{16}{72} \qquad\qquad \text{Solve.}$$

$$x = \frac{2}{9} \qquad\qquad \text{Simplify.}$$

The proposed solution, $\frac{2}{9}$, checks in the original equation. The solution is $\frac{2}{9}$.

> **Helpful Hint**
>
> Make sure expressions are squared correctly. In Example 5, we squared $\left(3 - \sqrt{2x}\right)$ as
>
> $$\left(3 - \sqrt{2x}\right)^2 = \left(3 - \sqrt{2x}\right)\left(3 - \sqrt{2x}\right)$$
> $$= 3 \cdot 3 - 3\sqrt{2x} - 3\sqrt{2x} + \sqrt{2x} \cdot \sqrt{2x}$$
> $$= 9 - 6\sqrt{2x} + 2x$$

✔ **CONCEPT CHECK**

What is wrong with the following solution?

$$\sqrt{2x + 5} + \sqrt{4 - x} = 8$$
$$\left(\sqrt{2x + 5} + \sqrt{4 - x}\right)^2 = 8^2$$
$$(2x + 5) + (4 - x) = 64$$
$$x + 9 = 64$$
$$x = 55$$

2   Recall that the Pythagorean theorem states that in a right triangle, the length of the hypotenuse squared equals the sum of the lengths of each of the legs squared.

TEACHING TIP

When discussing the Pythagorean theorem with students, remind them that in $a^2 + b^2 = c^2$, c *must be* the hypotenuse.

Pythagorean Theorem

If a and b are the lengths of the legs of a right triangle and c is the length of the hypotenuse, then $a^2 + b^2 = c^2$.

⚠ **EXAMPLE 6**

Find the length of the unknown leg of the right triangle.

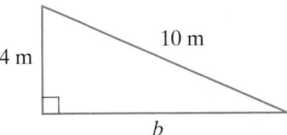

CLASSROOM EXAMPLE

Find the length of the unknown leg of the right triangle.

answer: $3\sqrt{5}$ cm

Solution   In the formula $a^2 + b^2 = c^2$, c is the hypotenuse. Here, $c = 10$, the length of the hypotenuse, and $a = 4$. We solve for b. Then $a^2 + b^2 = c^2$ becomes

$$4^2 + b^2 = 10^2$$
$$16 + b^2 = 100$$
$$b^2 = 84 \qquad \text{Subtract 16 from both sides.}$$
$$b = \pm\sqrt{84} = \pm\sqrt{4 \cdot 21} = \pm 2\sqrt{21}$$

Concept Check Answer:

$\left(\sqrt{2x + 5} + \sqrt{4 - x}\right)^2$ is not $(2x + 5) + (4 - x)$.

Since b is a length and thus is positive, we will use the positive value only.

The unknown leg of the triangle is $2\sqrt{21}$ meters long.

EXAMPLE 7

CALCULATING PLACEMENT OF A WIRE

A 50-foot supporting wire is to be attached to a 75-foot antenna. Because of surrounding buildings, sidewalks, and roadways, the wire must be anchored exactly 20 feet from the base of the antenna.

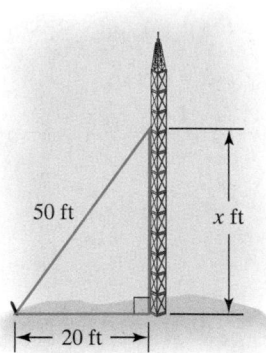

a. How high from the base of the antenna is the wire attached?

b. Local regulations require that a supporting wire be attached at a height no less than $\frac{3}{5}$ of the total height of the antenna. From part **a**, have local regulations been met?

Solution

1. UNDERSTAND. Read and reread the problem. From the diagram we notice that a right triangle is formed with hypotenuse 50 feet and one leg 20 feet. Let x be the height from the base of the antenna to the attached wire.

2. TRANSLATE. Use the Pythagorean theorem.

$$a^2 + b^2 = c^2$$
$$20^2 + x^2 = 50^2 \qquad a = 20, c = 50$$

3. SOLVE.
$$20^2 + x^2 = 50^2$$
$$400 + x^2 = 2500$$
$$x^2 = 2100 \qquad \text{Subtract 400 from both sides.}$$
$$x = \pm\sqrt{2100}$$
$$= \pm 10\sqrt{21}$$

4. INTERPRET. *Check* the work and *state* the solution.

Check: We will use only the positive value, $x = 10\sqrt{21}$ because x represents length. The wire is attached exactly $10\sqrt{21}$ feet from the base of the pole, or approximately 45.8 feet.

State: The supporting wire must be attached at a height no less than $\frac{3}{5}$ of the total height of the antenna. This height is $\frac{3}{5}$ (75 feet), or 45 feet. Since we know from part **a** that the wire is to be attached at a height of approximately 45.8 feet, local regulations have been met.

STUDY SKILLS REMINDER

Continue your outline started in Section 1.5. Write how to recognize and how to solve equations containing radical terms in your own words. For example:

Solving Equations and Inequalities

1. I. Equations
   - **A.** Linear equations (Sec. 1.5)
   - **B.** Absolute value equations (Sec. 3.4)
   - **C.** Quadratic Equations (Sec. 5.8)
   - **D.** Equations with Rational Expressions (Sec. 6.6)
   - **E. Equations with Radicals**—Recognize: *equation with a radical term*—Solve: isolate a radical and then raise each side of the equation to a power equal to the index of the radical. You may have to repeat this. Remember to check possible solutions in the original equation.

2. II. Inequalities
   - **A.** Linear inequalities (Sec. 3.2)
   - **B.** Compound inequalities (Sec. 3.3)
   - **C.** Absolute value inequalities (Sec. 3.5)

Spotlight on DECISION ℱ MAKING

Suppose you are a psychologist studying a person's ability to recognize patterns. You theorize that IQ is linked to pattern-recognition ability and design a research study using human subjects to test your theory. As part of your study, you have decided to form three groups of test subjects based roughly on IQ. Group A will consist of subjects having IQ's under 90, Group B will consist of subjects having IQ's from 90 to 105, and Group C will consist of subjects having IQ's over 105.

While preparing for your research study, you came across the findings of another psychologist suggesting that the number S of nonsense syllables that a person can repeat consecutively depends on his or her IQ score I according to the equation $S = 2\sqrt{I} - 9$. Because administering IQ tests can be time-consuming and because your groupings by IQ need only be approximate, you decide to use this equation as a quick way to assign test subjects to Groups A, B, and C. Each subject is individually screened by listening to a string of 20 random nonsense syllables and then repeating as many as possible. The results for the first 5 test subjects are listed in the table. For each subject, decide to which group—A, B, or C—the subject should be assigned.

Test Subject Screening

Subject	S—the number of Nonsense Syllables Successfully Repeated
1	11
2	13
3	9
4	12
5	10

EXERCISE SET 7.6

STUDY GUIDE/SSM   CD/VIDEO   PH MATH TUTOR CENTER   MathXL®Tutorials ON CD   MathXL®   MyMathLab®

Solve. See Examples 1 and 2.

1. $\sqrt{2x} = 4$ 8

2. $\sqrt{3x} = 3$ 3

3. $\sqrt{x-3} = 2$ 7

4. $\sqrt{x+1} = 5$ 24

5. $\sqrt{2x} = -4$ ∅

6. $\sqrt{5x} = -5$ ∅

7. $\sqrt{4x-3} - 5 = 0$ 7

8. $\sqrt{x-3} - 1 = 0$ 4

9. $\sqrt{2x-3} - 2 = 1$ 6

10. $\sqrt{3x+3} - 4 = 8$ 47

Solve. See Example 3.

11. $\sqrt[3]{6x} = -3$ $-\dfrac{9}{2}$

12. $\sqrt[3]{4x} = -2$ −2

13. $\sqrt[3]{x-2} - 3 = 0$ 29

14. $\sqrt[3]{2x-6} - 4 = 0$ 35

Solve. See Examples 4 and 5.

15. $\sqrt{13-x} = x-1$ 4

16. $\sqrt{2x-3} = 3-x$ 2

17. $x - \sqrt{4-3x} = -8$ −4

18. $2x + \sqrt{x+1} = 8$ 3

19. $\sqrt{y+5} = 2 - \sqrt{y-4}$ ∅

20. $\sqrt{x+3} + \sqrt{x-5} = 3$ $\dfrac{181}{36}$

21. $\sqrt{x-3} + \sqrt{x+2} = 5$ 7

22. $\sqrt{2x-4} - \sqrt{3x+4} = -2$ 4, 20

MIXED PRACTICE

Solve. See Examples 1 through 5. **38.** 3 **41.** 1 **46.** $\dfrac{4}{7}$

23. $\sqrt{3x-2} = 5$ 9

24. $\sqrt{5x-4} = 9$ 17

25. $-\sqrt{2x} + 4 = -6$ 50

26. $-\sqrt{3x+9} = -12$ 45

27. $\sqrt{3x+1} + 2 = 0$ ∅

28. $\sqrt{3x+1} - 2 = 0$ 1

29. $\sqrt[4]{4x+1} - 2 = 0$ $\dfrac{15}{4}$

30. $\sqrt[4]{2x-9} - 3 = 0$ 45

31. $\sqrt{4x-3} = 7$ 13

32. $\sqrt{3x+9} = 6$ 9

33. $\sqrt[3]{6x-3} - 3 = 0$ 5

34. $\sqrt[3]{3x} + 4 = 7$ 9

35. $\sqrt[3]{2x-3} - 2 = -5$ −12

36. $\sqrt[3]{x-4} - 5 = -7$ −4

37. $\sqrt{x+4} = \sqrt{2x-5}$ 9

38. $\sqrt{3y+6} = \sqrt{7y-6}$

39. $x - \sqrt{1-x} = -5$ −3

40. $x - \sqrt{x-2} = 4$ 6

41. $\sqrt[3]{-6x-1} = \sqrt[3]{-2x-5}$

42. $x + \sqrt{x+5} = 7$ 4

43. $\sqrt{5x-1} - \sqrt{x+2} = 3$ 1

44. $\sqrt{2x-1} - 4 = -\sqrt{x-4}$ 5

45. $\sqrt{2x-1} = \sqrt{1-2x}$ $\dfrac{1}{2}$

46. $\sqrt{7x-4} = \sqrt{4-7x}$

47. $\sqrt{3x+4} - 1 = \sqrt{2x+1}$ 0, 4

48. $\sqrt{x-2} + 3 = \sqrt{4x+1}$ 2, 6

49. $\sqrt{y+3} - \sqrt{y-3} = 1$ $\dfrac{37}{4}$

50. $\sqrt{x+1} - \sqrt{x-1} = 2$ ∅

Find the length of the unknown side of each triangle. See Example 6.

△ **51.** $3\sqrt{5}$ ft   △ **52.** $\sqrt{113}$ in.

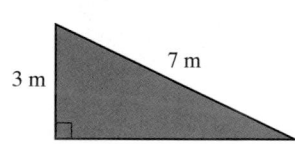

△ **53.** $2\sqrt{10}$ m

△ **54.** $\sqrt{33}$ cm

Find the length of the unknown side of each triangle. Give the exact length and a one-decimal-place approximation. See Example 6.

△ **55.** $2\sqrt{131}$ m ≈ 22.9 m

△ **56.** $5\sqrt{7}$ cm ≈ 13.2 cm

TEACHING TIP

A Group Activity for this section is available in the Instructor's Resource Manual.

57.

7 mm   7.2 mm

$\sqrt{100.84}$ mm ≈ 10.0 mm

58.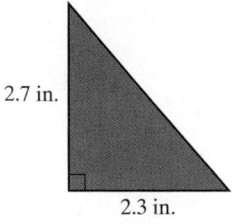

2.7 in.

2.3 in.

$\sqrt{12.58}$ in. ≈ 3.5 in.

Solve. See Example 7. Give exact answers and two-decimal-place approximations where appropriate.

59. A wire is needed to support a vertical pole 15 feet high. The cable will be anchored to a stake 8 feet from the base of the pole. How much cable is needed?   17 ft

15 ft

8 ft

60. The tallest structure in the United States is a TV tower in Blanchard, North Dakota. Its height is 2063 feet. A 2382-foot length of wire is to be used as a guy wire attached to the top of the tower. Approximate to the nearest foot how far from the base of the tower the guy wire must be anchored. (*Source:* U.S. Geological Survey)   1191 ft

2382 ft   2063 ft

?

61. A spotlight is mounted on the eaves of a house 12 feet above the ground. A flower bed runs between the house and the sidewalk, so the closest the ladder can be placed to the house is 5 feet. How long a ladder is needed so that an electrician can reach the place where the light is mounted?   13 ft

12 ft

5 ft

62. A wire is to be attached to support a telephone pole. Because of surrounding buildings, sidewalks, and roadways, the wire must be anchored exactly 15 feet from the base of the pole. Telephone company workers have only 30 feet of cable, and 2 feet of that must be used to attach the cable to the pole and to the stake on the ground. How high from the base of the pole can the wire be attached? $\sqrt{559}$ ft ≈ 23.64 ft

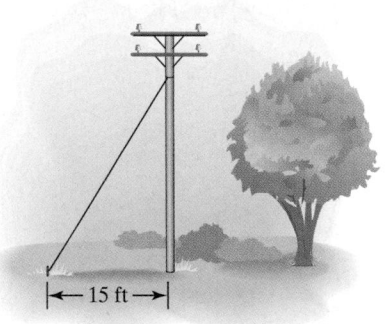

15 ft

63. The radius of the Moon is 1080 miles. Use the formula for the radius r of a sphere given its surface area A,

$$r = \sqrt{\frac{A}{4\pi}}$$

to find the surface area of the Moon. Round to the nearest square mile. (*Source:* National Space Science Data Center)

14,657,415 sq. mi

64. Police departments find it very useful to be able to approximate the speed of a car when they are given the distance that the car skidded before it came to a stop. If the road surface is wet concrete, the function $S(x) = \sqrt{10.5x}$ is used, where $S(x)$ is the speed of the car in miles per hour and x is the distance skidded in feet. Find how fast a car was moving if it skidded 280 feet on wet concrete. $14\sqrt{15}$ mph ≈ 54.22 mph

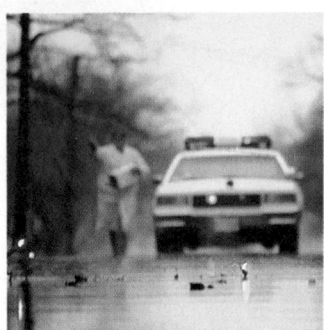

65. The formula $v = \sqrt{2gh}$ gives the velocity v, in feet per second, of an object when it falls h feet accelerated by gravity g, in feet per second squared. If g is approximately 32 feet per second squared, find how far an object has fallen if its velocity is 80 feet per second.   100 ft

66. Two tractors are pulling a tree stump from a field. If two forces A and B pull at right angles (90°) to each other, the size of the resulting force R is given by the formula $R = \sqrt{A^2 + B^2}$. If tractor A is exerting 600 pounds of force and the resulting force is 850 pounds, find how much force tractor B is exerting. $50\sqrt{145}$ lb. ≈ 602.08 lb

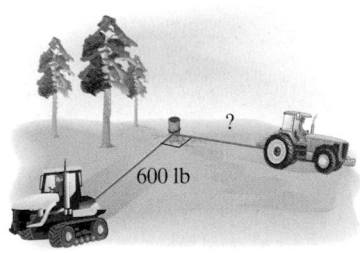

In psychology, it has been suggested that the number S of nonsense syllables that a person can repeat consecutively depends on his or her IQ score I according to the equation $S = 2\sqrt{I} - 9$.

67. Use this relationship to estimate the IQ of a person who can repeat 11 nonsense syllables consecutively.   100

68. Use this relationship to estimate the IQ of a person who can repeat 15 nonsense syllables consecutively.   144

*The **period** of a pendulum is the time it takes for the pendulum to make one full back-and-forth swing. The period of a pendulum depends on the length of the pendulum. The formula for the period P, in seconds, is $P = 2\pi\sqrt{\dfrac{l}{32}}$ where l is the length of the pendulum in feet. Use this formula for Exercises 69 through 74.*

69. Find the period of a pendulum whose length is 2 feet. Give an exact answer and a two-decimal-place approximation. $\dfrac{\pi}{2}$ sec ≈ 1.57 sec

70. Klockit sells a 43-inch lyre pendulum. Find the period of this pendulum. Round your answer to 2 decimal places. (Hint: First convert inches to feet.)   2.10 sec

71. Find the length of a pendulum whose period is 4 seconds. Round your answer to 2 decimal places.   12.97 ft

72. Find the length of a pendulum whose period is 3 seconds. Round your answer to 2 decimal places.   7.30 ft

73. Study the relationship between period and pendulum length in Exercises 69 through 72 and make a conjecture about this relationship.   answers may vary

74. Galileo experimented with pendulums. He supposedly made conjectures about pendulums of equal length with different bob weights. Try this experiment. Make two pendulums 3 feet long. Attach a heavy weight (lead) to one and a light weight (a cork) to the other. Pull both pendulums back the same angle measure and release. Make a conjecture from your observations. (There is more about pendulums in the Chapter 7 Project.)   answers may vary

If the three lengths of the sides of a triangle are known, Heron's formula can be used to find its area. If a, b, and c are the three lengths of the sides, Heron's formula for area is:

$$A = \sqrt{s(s - a)(s - b)(s - c)}$$

where s is half the perimeter of the triangle, or $s = \dfrac{1}{2}(a + b + c)$.

Use this formula to find the area of each triangle. Give an exact answer and then a two-decimal place approximation.

 75. $15\sqrt{3}$ sq. mi $\approx$ 25.98 sq. mi

 76. $2\sqrt{2}$ sq. cm $\approx$ 2.83 sq. cm

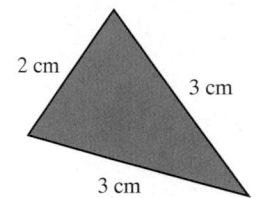

77. Describe when Heron's formula might be useful.   answers may vary

78. In your own words, explain why you think *s* in *Heron's formula* is called the *semiperimeter*.   answers may vary

The maximum distance D(h) in kilometers that a person can see from a height h kilometers above the ground is given by the function $D(h) = 111.7\sqrt{h}$. Use this function for Exercises 79 and 80. Round your answers to two decimal places.

79. Find the height that would allow a person to see 80 kilometers.   0.51 km

80. Find the height that would allow a person to see 40 kilometers.   0.13 km

REVIEW AND PREVIEW

Use the vertical line test to determine whether each graph represents the graph of a function. See Section 3.2.

81.

function

82.

not a function

83.

function

84.

not a function

85.

not a function

86.

not a function

Simplify. See Section 6.3.

87. $\dfrac{\dfrac{x}{6}}{\dfrac{2x}{3} + \dfrac{1}{2}}$ $\dfrac{x}{4x + 3}$

88. $\dfrac{\dfrac{1}{y} + \dfrac{4}{5}}{\dfrac{-3}{20}}$ $-\dfrac{20 + 16y}{3y}$

89. $\dfrac{\dfrac{z}{5} + \dfrac{1}{10}}{\dfrac{z}{20} - \dfrac{z}{5}}$ $-\dfrac{4z + 2}{3z}$

90. $\dfrac{\dfrac{1}{y} + \dfrac{1}{x}}{\dfrac{1}{y} - \dfrac{1}{x}}$ $\dfrac{x + y}{x - y}$

Concept Extensions

91. Solve: $\sqrt{\sqrt{x + 3} + \sqrt{x}} = \sqrt{3}$ 1

92. Explain why proposed solutions of radical equations must be checked.   answers may vary

93. The cost $C(x)$ in dollars per day to operate a small delivery service is given by $C(x) = 80\sqrt[3]{x} + 500$, where x is the number of deliveries per day. In July, the manager decides that it is necessary to keep delivery costs below $1620.00. Find the greatest number of deliveries this company can make per day and still keep overhead below $1620.00. 2743 deliveries

94. Consider the equations $\sqrt{2x} = 4$ and $\sqrt[3]{2x} = 4$.

   a. Explain the difference in solving these equations.

   b. Explain the similarity in solving these equations.

      a.–b. answers may vary

Example

For Exercises 95 through 98, see the example below.

Solve $(t^2 - 3t) - 2\sqrt{t^2 - 3t} = 0$.

Solution

Substitution can be used to make this problem somewhat simpler. Since $t^2 - 3t$ occurs more than once, let $x = t^2 - 3t$.

$$(t^2 - 3t) - 2\sqrt{t^2 - 3t} = 0$$
$$x - 2\sqrt{x} = 0 \qquad \text{Let } x = t^2 - 3t.$$
$$x = 2\sqrt{x}$$
$$x^2 = (2\sqrt{x})^2$$
$$x^2 = 4x$$
$$x^2 - 4x = 0$$
$$x(x - 4) = 0$$
$$x = 0 \quad \text{or} \quad x - 4 = 0$$
$$x = 4$$

Now we "undo" the substitution.

$x = 0$ Replace x with $t^2 - 3t$.

$$t^2 - 3t = 0$$
$$t(t - 3) = 0$$
$$t = 0 \quad \text{or} \quad t - 3 = 0$$
$$t = 3$$

$x = 4$ Replace x with $t^2 - 3t$.

$$t^2 - 3t = 4$$
$$t^2 - 3t - 4 = 0$$
$$(t - 4)(t + 1) = 0$$
$$t - 4 = 0 \quad \text{or} \quad t + 1 = 0$$
$$t = 4 \qquad t = -1$$

In this problem, we have four possible solutions: 0, 3, 4, and -1. All four solutions check in the original equation, so the solutions are $-1, 0, 3, 4$.

Solve. See the preceding example.

95. $3\sqrt{x^2 - 8x} = x^2 - 8x$ $-1, 0, 8, 9$

96. $\sqrt{(x^2 - x) + 7} = 2(x^2 - x) - 1$ $-1, 2$

97. $7 - (x^2 - 3x) = \sqrt{(x^2 - 3x) + 5}$ $-1, 4$

98. $x^2 + 6x = 4\sqrt{x^2 + 6x}$ $-8, -6, 0, 2$

7.7 COMPLEX NUMBERS

Objectives

1 Define imaginary and complex numbers.

2 Add or subtract complex numbers.

3 Multiply complex numbers.

4 Divide complex numbers.

5 Raise i to powers.

1 Our work with radical expressions has excluded expressions such as $\sqrt{-16}$ because $\sqrt{-16}$ is not a real number; there is no real number whose square is -16. In this section, we discuss a number system that includes roots of negative numbers. This number system is the **complex number system**, and it includes the set of real numbers as a subset. The complex number system allows us to solve equations such as $x^2 + 1 = 0$ that have no real number solutions. The set of complex numbers includes the **imaginary unit**.

> **Imaginary Unit**
>
> The imaginary unit, written i, is the number whose square is -1. That is,
>
> $$i^2 = -1 \quad \text{and} \quad i = \sqrt{-1}$$

To write the square root of a negative number in terms of i, use the property that if a is a positive number, then

$$\sqrt{-a} = \sqrt{-1} \cdot \sqrt{a}$$
$$= i \cdot \sqrt{a}$$

Using i, we can write $\sqrt{-16}$ as

$$\sqrt{-16} = \sqrt{-1 \cdot 16} = \sqrt{-1} \cdot \sqrt{16} = i \cdot 4, \text{ or } 4i$$

EXAMPLE 1

Write with i notation.

a. $\sqrt{-36}$　　　b. $\sqrt{-5}$　　　c. $-\sqrt{-20}$

> **Helpful Hint**
> Since $\sqrt{5}i$ can easily be confused with $\sqrt{5i}$, we write $\sqrt{5}i$ as $i\sqrt{5}$.

Solution   a. $\sqrt{-36} = \sqrt{-1 \cdot 36} = \sqrt{-1} \cdot \sqrt{36} = i \cdot 6, \text{ or } 6i$

b. $\sqrt{-5} = \sqrt{-1(5)} = \sqrt{-1} \cdot \sqrt{5} = i\sqrt{5}$.

c. $-\sqrt{-20} = -\sqrt{-1 \cdot 20} = -\sqrt{-1} \cdot \sqrt{4 \cdot 5} = -i \cdot 2\sqrt{5} = -2i\sqrt{5}$

This is an approximate answer whereas $i\sqrt{5}$ is an exact answer.

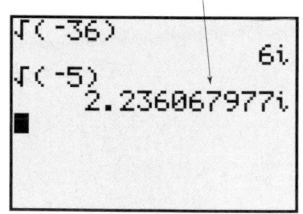

A calculator check of Example 1a and b.

The product rule for radicals does not necessarily hold true for imaginary numbers. *To multiply square roots of negative numbers, first we write each number in terms of the imaginary unit i.* For example, to multiply $\sqrt{-4}$ and $\sqrt{-9}$, we first write each number in the form bi.

$$\sqrt{-4}\sqrt{-9} = 2i(3i) = 6i^2 = 6(-1) = -6$$

We will also use this method to simplify quotients of square roots of negative numbers.

EXAMPLE 2

Multiply or divide as indicated.

a. $\sqrt{-3} \cdot \sqrt{-5}$

b. $\sqrt{-36} \cdot \sqrt{-1}$

c. $\sqrt{8} \cdot \sqrt{-2}$

d. $\dfrac{\sqrt{-125}}{\sqrt{5}}$

Solution   a. $\sqrt{-3} \cdot \sqrt{-5} = i\sqrt{3}(i\sqrt{5}) = i^2\sqrt{15} = -1\sqrt{15} = -\sqrt{15}$

b. $\sqrt{-36} \cdot \sqrt{-1} = 6i(i) = 6i^2 = 6(-1) = -6$

c. $\sqrt{8} \cdot \sqrt{-2} = 2\sqrt{2}(i\sqrt{2}) = 2i(\sqrt{2}\sqrt{2}) = 2i(2) = 4i$

d. $\dfrac{\sqrt{-125}}{\sqrt{5}} = \dfrac{i\sqrt{125}}{\sqrt{5}} = i\sqrt{25} = 5i$

Now that we have practiced working with the imaginary unit, we define complex numbers.

> ### Complex Numbers
>
> A **complex number** is a number that can be written in the form $a + bi$, where a and b are real numbers.

Notice that the set of real numbers is a subset of the complex numbers since any real number can be written in the form of a complex number. For example,

$$16 = 16 + 0i$$

In general, a complex number $a + bi$ is a real number if $b = 0$. Also, a complex number is called a **pure imaginary number** if $a = 0$. For example,

$$3i = 0 + 3i \quad \text{and} \quad i\sqrt{7} = 0 + i\sqrt{7}$$

are pure imaginary numbers.

The following diagram shows the relationship between complex numbers and their subsets.

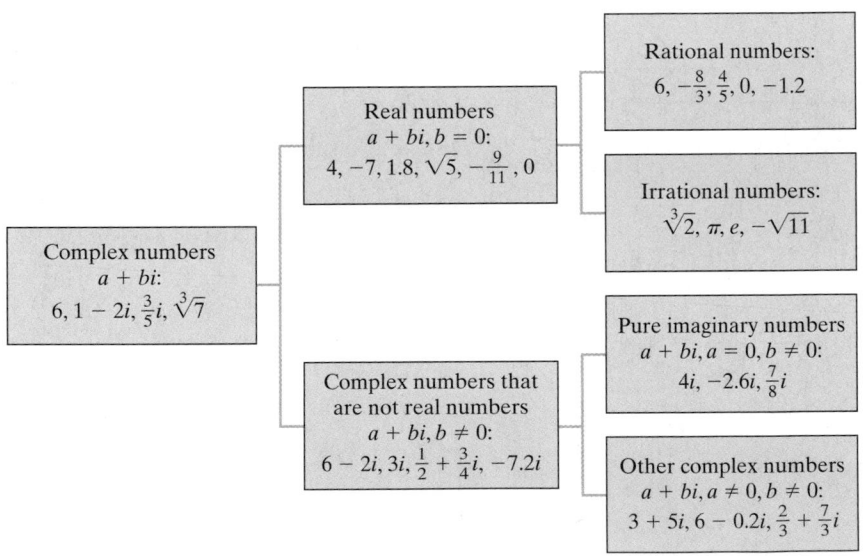

✔ **CONCEPT CHECK**

True or false? Every complex number is also a real number.

2   Two complex numbers $a + bi$ and $c + di$ are equal if and only if $a = c$ and $b = d$. Complex numbers can be added or subtracted by adding or subtracting their real parts and then adding or subtracting their imaginary parts.

Sum or Difference of Complex Numbers

If $a + bi$ and $c + di$ are complex numbers, then their sum is

$$(a + bi) + (c + di) = (a + c) + (b + d)i$$

Their difference is

$$(a + bi) - (c + di) = a + bi - c - di = (a - c) + (b - d)i$$

EXAMPLE 3

Add or subtract the complex numbers. Write the sum or difference in the form $a + bi$.

a. $(2 + 3i) + (-3 + 2i)$ **b.** $5i - (1 - i)$ **c.** $(-3 - 7i) - (-6)$

Solution   **a.** $(2 + 3i) + (-3 + 2i) = (2 - 3) + (3 + 2)i = -1 + 5i$

b. $5i - (1 - i) = 5i - 1 + i$

$$= -1 + (5 + 1)i$$

$$= -1 + 6i$$

c. $(-3 - 7i) - (-6) = -3 - 7i + 6$

$$= (-3 + 6) - 7i$$

$$= 3 - 7i$$

A calculator check for Example 3.

3   To multiply two complex numbers of the form $a + bi$, we multiply as though they are binomials. Then we use the relationship $i^2 = -1$ to simplify.

EXAMPLE 4

Multiply the complex numbers. Write the product in the form $a + bi$.

a. $-7i \cdot 3i$ **b.** $3i(2 - i)$ **c.** $(2 - 5i)(4 + i)$

d. $(2 - i)^2$ **e.** $(7 + 3i)(7 - 3i)$

Solution

a. $-7i \cdot 3i = -21i^2$

$= -21(-1)$ Replace i^2 with -1.

$= 21$

b. $3i(2 - i) = 3i \cdot 2 - 3i \cdot i$ Use the distributive property.

$= 6i - 3i^2$ Multiply.

$= 6i - 3(-1)$ Replace i^2 with -1.

$= 6i + 3$

$= 3 + 6i$

Use the FOIL order below. (First, Outer, Inner, Last)

c. $(2 - 5i)(4 + i) = 2(4) + 2(i) - 5i(4) - 5i(i)$

                    F   O   I   L

$= 8 + 2i - 20i - 5i^2$

$= 8 - 18i - 5(-1)$ $i^2 = -1$.

$= 8 - 18i + 5$

$= 13 - 18i$

d. $(2 - i)^2 = (2 - i)(2 - i)$

$= 2(2) - 2(i) - 2(i) + i^2$

$= 4 - 4i + (-1)$ $i^2 = -1$.

$= 3 - 4i$

e. $(7 + 3i)(7 - 3i) = 7(7) - 7(3i) + 3i(7) - 3i(3i)$

$= 49 - 21i + 21i - 9i^2$

$= 49 - 9(-1)$ $i^2 = -1$.

$= 49 + 9$

$= 58$

Notice that if you add, subtract, or multiply two complex numbers, just like real numbers, the result is a complex number.

4   From Example 4e, notice that the product of $7 + 3i$ and $7 - 3i$ is a real number. These two complex numbers are called **complex conjugates** of one another. In general, we have the following definition.

Complex Conjugates

The complex numbers $(a + bi)$ and $(a - bi)$ are called **complex conjugates** of each other, and $(a + bi)(a - bi) = a^2 + b^2$.

To see that the product of a complex number $a + bi$ and its conjugate $a - bi$ is the real number $a^2 + b^2$ we multiply.

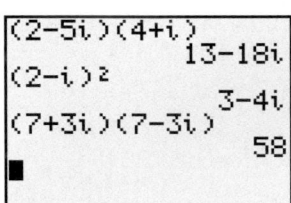

A calculator check for Example 4c, d, and e.

$$(a + bi)(a - bi) = a^2 - abi + abi - b^2i^2$$
$$= a^2 - b^2(-1)$$
$$= a^2 + b^2$$

We use complex conjugates to divide by a complex number.

EXAMPLE 5

Divide. Write in the form $a + bi$.

a. $\dfrac{2 + i}{1 - i}$ 　　　　　　　　　**b.** $\dfrac{7}{3i}$

Solution

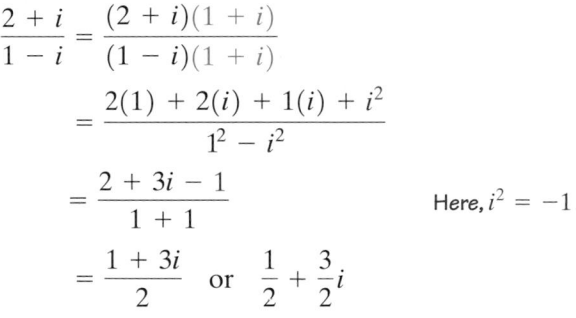

A check for Example 5. Notice the need for parentheses in the denominator when simplifying part b.

a. Multiply the numerator and denominator by the complex conjugate of $1 - i$ to eliminate the imaginary number in the denominator.

$$\frac{2 + i}{1 - i} = \frac{(2 + i)(1 + i)}{(1 - i)(1 + i)}$$
$$= \frac{2(1) + 2(i) + 1(i) + i^2}{1^2 - i^2}$$
$$= \frac{2 + 3i - 1}{1 + 1} \qquad \text{Here, } i^2 = -1.$$
$$= \frac{1 + 3i}{2} \quad \text{or} \quad \frac{1}{2} + \frac{3}{2}i$$

> **Helpful Hint**
>
> Recall that division can be checked by multiplication.
>
> To check that $\dfrac{2 + i}{1 - i} = \dfrac{1}{2} + \dfrac{3}{2}i$, in Example 5a, multiply $\left(\dfrac{1}{2} + \dfrac{3}{2}i\right)(1 - i)$ to verify that the product is $2 + i$.

b. Multiply the numerator and denominator by the conjugate of $3i$. Note that $3i = 0 + 3i$, so its conjugate is $0 - 3i$ or $-3i$.

$$\frac{7}{3i} = \frac{7(-3i)}{(3i)(-3i)} = \frac{-21i}{-9i^2} = \frac{-21i}{-9(-1)} = \frac{-21i}{9} = \frac{-7i}{3} \quad \text{or} \quad 0 - \frac{7}{3}i$$

5   We can use the fact that $i^2 = -1$ to find higher powers of i. To find i^3, we rewrite it as the product of i^2 and i.

$$i^3 = i^2 \cdot i = (-1)i = -i$$
$$i^4 = i^2 \cdot i^2 = (-1) \cdot (-1) = 1$$

We continue this process and use the fact that $i^4 = 1$ and $i^2 = -1$ to simplify i^5 and i^6.

$$i^5 = i^4 \cdot i = 1 \cdot i = i$$
$$i^6 = i^4 \cdot i^2 = 1 \cdot (-1) = -1$$

If we continue finding powers of i, we generate the following pattern. Notice that the values i, -1, $-i$, and 1 repeat as i is raised to higher and higher powers.

$i^1 = i$	$i^5 = i$	$i^9 = i$
$i^2 = -1$	$i^6 = -1$	$i^{10} = -1$
$i^3 = -i$	$i^7 = -i$	$i^{11} = -i$
$i^4 = 1$	$i^8 = 1$	$i^{12} = 1$

This pattern allows us to find other powers of i. To do so, we will use the fact that $i^4 = 1$ and rewrite a power of i in terms of i^4.

For example, $i^{22} = i^{20} \cdot i^2 = (i^4)^5 \cdot i^2 = 1^5 \cdot (-1) = 1 \cdot (-1) = -1$.

EXAMPLE 6

Find the following powers of i.

a. i^7 **b.** i^{20} **c.** i^{46} **d.** i^{-12}

Solution   **a.** $i^7 = i^4 \cdot i^3 = 1(-i) = -i$

b. $i^{20} = (i^4)^5 = 1^5 = 1$

c. $i^{46} = i^{44} \cdot i^2 = (i^4)^{11} \cdot i^2 = 1^{11}(-1) = -1$

d. $i^{-12} = \dfrac{1}{i^{12}} = \dfrac{1}{(i^4)^3} = \dfrac{1}{(1)^3} = \dfrac{1}{1} = 1$

CLASSROOM EXAMPLE

Find the powers of i.

a. i^{11} **b.** i^{40} **c.** i^{50} **d.** i^{-10}

answer: **a.** $-i$ **b.** 1 **c.** -1 **d.** -1

MENTAL MATH

Simplify. See Example 1.

 1. $\sqrt{-81}$ $9i$ **2.** $\sqrt{-49}$ $7i$ **3.** $\sqrt{-7}$ $i\sqrt{7}$ **4.** $\sqrt{-3}$ $i\sqrt{3}$

  5. $-\sqrt{16}$ -4 **6.** $-\sqrt{4}$ -2 **7.** $\sqrt{-64}$ $8i$ **8.** $\sqrt{-100}$ $10i$

EXERCISE SET 7.7

STUDY GUIDE/SSM   CD/VIDEO   PH MATH TUTOR CENTER   MathXL®Tutorials ON CD   MathXL®   MyMathLab®

Write in terms of i. See Example 1.

1. $\sqrt{-24}$ $2i\sqrt{6}$ **2.** $\sqrt{-32}$ $4i\sqrt{2}$

3. $-\sqrt{-36}$ $-6i$ **4.** $-\sqrt{-121}$ $-11i$

5. $8\sqrt{-63}$ $24i\sqrt{7}$ **6.** $4\sqrt{-20}$ $8i\sqrt{5}$

7. $-\sqrt{54}$ $-3\sqrt{6}$ **8.** $\sqrt{-63}$ $3i\sqrt{7}$

Multiply or divide. See Example 2.

9. $\sqrt{-2}\cdot\sqrt{-7}$ $-\sqrt{14}$ **10.** $\sqrt{-11}\cdot\sqrt{-3}$ $-\sqrt{33}$

 11. $\sqrt{-5}\cdot\sqrt{-10}$ $-5\sqrt{2}$ **12.** $\sqrt{-2}\cdot\sqrt{-6}$ $-2\sqrt{3}$

13. $\sqrt{16}\cdot\sqrt{-1}$ $4i$ **14.** $\sqrt{3}\cdot\sqrt{-27}$ $9i$

15. $\dfrac{\sqrt{-9}}{\sqrt{3}}$ $i\sqrt{3}$ **16.** $\dfrac{\sqrt{49}}{\sqrt{-10}}$ $-\dfrac{7i\sqrt{10}}{10}$

17. $\dfrac{\sqrt{-80}}{\sqrt{-10}}$ $2\sqrt{2}$ **18.** $\dfrac{\sqrt{-40}}{\sqrt{-8}}$ $\sqrt{5}$

Add or subtract. Write the sum or difference in the form a + bi.
See Example 3.   **19.** $6-4i$ **21.** $-2+6i$ **22.** 0

 19. $(4-7i)+(2+3i)$ **20.** $(2-4i)-(2-i)$ $-3i$

21. $(6+5i)-(8-i)$ **22.** $(8-3i)+(-8+3i)$

23. $6-(8+4i)$ $-2-4i$

24. $(9-4i)-9$ $-4i$

Multiply. Write the product in the form a + bi. See Example 4.   **28.** 30

25. $6i(2-3i)$ $18+12i$ **26.** $5i(4-7i)$ $35+20i$

27. $(\sqrt{3}+2i)(\sqrt{3}-2i)$ 7   **28.** $(\sqrt{5}-5i)(\sqrt{5}+5i)$

29. $(4-2i)^2$ $12-16i$ **30.** $(6-3i)^2$ $27-36i$

Write each quotient in the form a + bi. See Example 5.

31. $\dfrac{4}{i}$ $-4i$ **32.** $\dfrac{5}{6i}$ $-\dfrac{5}{6}i$

33. $\dfrac{7}{4+3i}$ $\dfrac{28}{25}-\dfrac{21}{25}i$ **34.** $\dfrac{9}{1-2i}$ $\dfrac{9}{5}+\dfrac{18}{5}i$

 35. $\dfrac{3+5i}{1+i}$ $4+i$ **36.** $\dfrac{6+2i}{4-3i}$ $\dfrac{18}{25}+\dfrac{26}{25}i$

37. $\dfrac{5-i}{3-2i}$ $\dfrac{17}{13}+\dfrac{7}{13}i$ **38.** $\dfrac{6-i}{2+i}$ $\dfrac{11}{5}-\dfrac{8}{5}i$

MIXED PRACTICE

Perform the indicated operation. Write the result in the form
a + bi.   **41.** $2-i$ **42.** $-8+4i$

39. $(7i)(-9i)$ 63   **40.** $(-6i)(-4i)$ -24

41. $(6-3i)-(4-2i)$ **42.** $(-2-4i)-(6-8i)$

 43. $(6-2i)(3+i)$ 20   **44.** $(2-4i)(2-i)$ $-10i$

45. $(8-3i)+(2+3i)$ 10

46. $(7+4i)+(4-4i)$ 11

47. $(1-i)(1+i)$ 2   **48.** $(6+2i)(6-2i)$ 40

49. $\dfrac{16+15i}{-3i}$ $-5+\dfrac{16}{3}i$ **50.** $\dfrac{2-3i}{-7i}$ $\dfrac{3}{7}+\dfrac{2}{7}i$

51. $(9+8i)^2$ $17+144i$ **52.** $(4-7i)^2$ $-33-56i$

53. $\dfrac{2}{3+i}$ $\dfrac{3}{5}-\dfrac{1}{5}i$ **54.** $\dfrac{5}{3-2i}$ $\dfrac{15}{13}+\dfrac{10}{13}i$

55. $(5-6i)-4i$ $5-10i$ **56.** $(6-2i)+7i$ $6+5i$

57. $\dfrac{2-3i}{2+i}$ $\dfrac{1}{5}-\dfrac{8}{5}i$ **58.** $\dfrac{6+5i}{6-5i}$ $\dfrac{11}{61}+\dfrac{60}{61}i$

59. $(2+4i)+(6-5i)$ **60.** $(5-3i)+(7-8i)$

Find each power of i. See Example 6.   **59.** $8-i$ **60.** $12-11i$

61. i^8 1   **62.** i^{10} -1

63. i^{21} i **64.** i^{15} $-i$

65. i^{11} $-i$ **66.** i^{40} 1

67. i^{-6} -1 **68.** i^{-9} $-i$

69. $(2i)^6$ -64 **70.** $(5i)^4$ 625

71. $(-3i)^5$ $-243i$ **72.** $(-2i)^7$ $128i$

REVIEW AND PREVIEW

Recall that the sum of the measures of the angles of a triangle is
180°. Find the unknown angle in each triangle. See Section 4.3.

73.   40°   **74.** 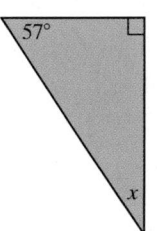   33°

Use synthetic division to divide the following. See Section 6.5.

75. $(x^3-6x^2+3x-4)\div(x-1)$ $x^2-5x-2-\dfrac{6}{x-1}$

76. $(5x^4-3x^2+2)\div(x+2)$

$5x^3-10x^2+17x-34+\dfrac{70}{x+2}$

Thirty people were recently polled about their average monthly balance in their checking accounts. The results of this poll are shown in the following histogram. Use this graph to answer Exercises 77 through 82. See Section 1.2.

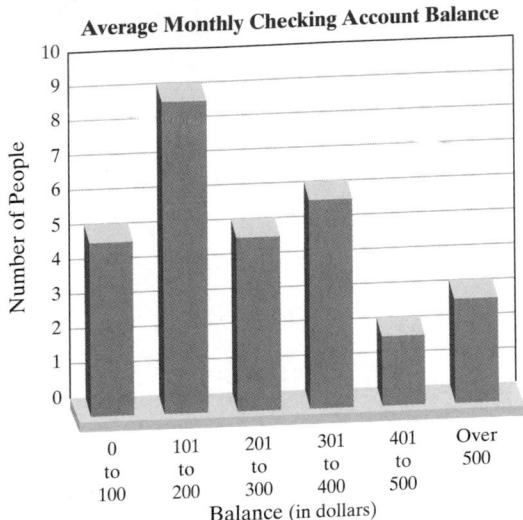

Average Monthly Checking Account Balance

Number of People

Balance (in dollars)

77. How many people polled reported an average checking balance of $201 to $300?   5 people

78. How many people polled reported an average checking balance of $0 to $100?   5 people

79. How many people polled reported an average checking balance of $200 or less?   14 people

80. How many people polled reported an average checking balance of $301 or more?   11 people

81. What percent of people polled reported an average checking balance of $201 to $300?   16.7%

82. What percent of people polled reported an average checking balance of $0 to $100?   16.7%

Concept Extensions

Write in the form $a + bi$.

83. $i^3 + i^4$ $1 - i$

84. $i^8 - i^7$ $1 + i$

85. $i^6 + i^8$ 0

86. $i^4 + i^{12}$ 2

87. $2 + \sqrt{-9}$ $2 + 3i$

88. $5 - \sqrt{-16}$ $5 - 4i$

89. $\dfrac{6 + \sqrt{-18}}{3}$ $2 + i\sqrt{2}$

90. $\dfrac{4 - \sqrt{-8}}{2}$ $2 - i\sqrt{2}$

91. $\dfrac{5 - \sqrt{-75}}{10}$ $\dfrac{1}{2} - \dfrac{\sqrt{3}}{2}i$

92. Describe how to find the conjugate of a complex number.   answers may vary

93. Explain why the product of a complex number and its complex conjugate is a real number.   answers may vary

Simplify.

94. $\left(8 - \sqrt{-3}\right) - \left(2 + \sqrt{-12}\right)$ $6 - 3i\sqrt{3}$

95. $\left(8 - \sqrt{-4}\right) - \left(2 + \sqrt{-16}\right)$ $6 - 6i$

96. Determine whether $2i$ is a solution of $x^2 + 4 = 0$.   yes

97. Determine whether $-1 + i$ is a solution of $x^2 + 2x = -2$.   yes

CHAPTER 7 PROJECT

Calculating the Length and Period of a Pendulum

Materials:

- string (at least 1 meter long)
- meter stick
- calculator
- weight
- stopwatch

This activity may be completed by working in groups or individually. Make a simple pendulum by securely tying the string to a weight.

   The formula relating a pendulum's period T (in seconds) to its length l (in centimeters) is

$$T = 2\pi\sqrt{\dfrac{l}{980}}$$

   The **period** of a pendulum is defined as the time it takes the pendulum to complete one full back-and-forth swing. In this activity, you will be measuring your simple pendulum's period with a stopwatch. Because the periods will be only a few seconds long, it will be more accurate for you to time a total of five complete swings and then find the average time of one complete swing.

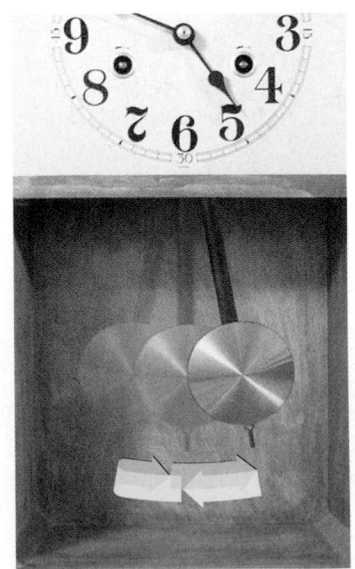

1. For each of the pendulum (string) lengths given in Table 1, measure the time required for 5 complete swings and record it in the appropriate column. Next, divide this value by 5 to find the measured period of the pendulum for the given length and record it in the Measured Period T_m column in the table. Use the given formula to calculate the theoretical period T for the same pendulum length and record it in the appropriate column. (Round to two decimal places.) Find and record in the last column the difference between the measured period and the theoretical period.

2. For each of the periods T given in Table 2, use the given formula and calculate the theoretical pendulum length l required

to yield the given period. Record l in the appropriate column; round to one decimal place. Next, use this length l and measure and record the time for 5 complete swings. Divide this value by 5 to find the measured period T_m, and record it. Then find and record in the last column the difference between the theoretical period and the measured period.

3. Use the general trends you find in the tables to describe the relationship between a pendulum's period and its length.

4. Discuss the differences you found between the values of the theoretical period and the measured period. What factors contributed to these differences?

Table 1

Length l (centimeters)	Time for 5 Swings (seconds)	Measured Period T_m (seconds)	Theoretical Period T (seconds)	Difference $|T - T_m|$
30				
55				
70				

Table 2

Period T (seconds)	Theoretical Length l (centimeters)	Time for 5 Swings (seconds)	Measured Period T_m (seconds)	Difference $|T - T_m|$
1				
1.25				
2				

STUDY SKILLS REMINDER

Are You Preparing for a Test on Chapter 7?

Below I have listed some common trouble areas for students in Chapter 7. After studying for your test, but before taking your test, read these.

▶ Remember how to convert an expression with rational expressions to one with radicals and one with radicals to one with rational expressions.

$$7^{2/3} = \sqrt[3]{7^2} \text{ or } \left(\sqrt[3]{7}\right)^2$$
$$\sqrt[5]{4^3} = 4^{3/5}$$

▶ Remember the difference between $\sqrt{x} + \sqrt{x}$ and $\sqrt{x} \cdot \sqrt{x}, x > 0$.

$$\sqrt{x} + \sqrt{x} = 2\sqrt{x}$$
$$\sqrt{x} \cdot \sqrt{x} = x$$

▶ Don't forget the difference between rationalizing the denominator of $\sqrt{\dfrac{2}{x}}$ and rationalizing the denominator of $\dfrac{\sqrt{2}}{\sqrt{x}+1}, x > 0$.

$$\sqrt{\frac{2}{x}} = \frac{\sqrt{2}}{\sqrt{x}} = \frac{\sqrt{2} \cdot \sqrt{x}}{\sqrt{x} \cdot \sqrt{x}} = \frac{\sqrt{2x}}{x}$$
$$\frac{\sqrt{2}}{\sqrt{x}+1} = \frac{\sqrt{2}\left(\sqrt{x}-1\right)}{\left(\sqrt{x}+1\right)\left(\sqrt{x}-1\right)} = \frac{\sqrt{2}\left(\sqrt{x}-1\right)}{x-1}$$

Remember: This is simply a checklist of common trouble areas. For a review of Chapter 7, see the Highlights and Chapter Review at the end of this chapter.

CHAPTER VOCABULARY CHECK

Fill in each blank with one of the words or phrases listed below.

index	rationalizing	conjugate	principal square root	cube root
complex number	like radicals	radicand	imaginary unit	

1. The <u>conjugate</u> of $\sqrt{3} + 2$ is $\sqrt{3} - 2$.

2. The <u>principal square root</u> of a nonnegative number a is written as $\sqrt{a}$.

3. The process of writing a radical expression as an equivalent expression but without a radical in the denominator is called <u>rationalizing</u> the denominator.

4. The <u>imaginary unit</u>, written i, is the number whose square is -1.

5. The <u>cube root</u> of a number is written as $\sqrt[3]{a}$.

6. In the notation $\sqrt[n]{a}$, n is called the <u>index</u> and a is called the <u>radicand</u>.

7. Radicals with the same index and the same radicand are called <u>like radicals</u>.

8. A <u>complex number</u> is a number that can be written in the form $a + bi$ where a and b are real numbers.

CHAPTER 7 HIGHLIGHTS

Definitions and Concepts	Examples

Section 7.1 Radicals and Radical Functions

The **positive**, or **principal**, **square root** of a nonnegative number a is written as $\sqrt{a}$.

$$\sqrt{a} = b \text{ only if } b^2 = a \text{ and } b \geq 0$$

The **negative square root of** a is written as $-\sqrt{a}$.

The **cube root** of a real number a is written as $\sqrt[3]{a}$.

$$\sqrt[3]{a} = b \text{ only if } b^3 = a$$

If n is an even positive integer, then $\sqrt[n]{a^n} = |a|$.

If n is an odd positive integer, then $\sqrt[n]{a^n} = a$.

A **radical function** in x is a function defined by an expression containing a root of x.

$$\sqrt{36} = 6 \qquad \sqrt{\frac{9}{100}} = \frac{3}{10}$$

$$-\sqrt{36} = -6 \qquad \sqrt{0.04} = 0.2$$

$$\sqrt[3]{27} = 3 \qquad \sqrt[9]{-\frac{1}{8}} = -\frac{1}{2}$$

$$\sqrt[3]{y^6} = y^2 \qquad \sqrt[3]{64x^9} = 4x^3$$

$$\sqrt{(-3)^2} = |-3| = 3$$

$$\sqrt[3]{(-7)^3} = -7$$

If $\qquad f(x) = \sqrt{x} + 2,$

$$f(1) = \sqrt{(1)} + 2 = 1 + 2 = 3$$

$$f(3) = \sqrt{(3)} + 2 \approx 3.73$$

Section 7.2 Rational Exponents

$a^{1/n} = \sqrt[n]{a}$ if $\sqrt[n]{a}$ is a real number.

If m and n are positive integers greater than 1 with $\dfrac{m}{n}$ in lowest terms and $\sqrt[n]{a}$ is a real number, then

$$a^{m/n} = (a^{1/n})^m = (\sqrt[n]{a})^m$$

$a^{-m/n} = \dfrac{1}{a^{m/n}}$ as long as $a^{m/n}$ is a nonzero number.

Exponent rules are true for rational exponents.

$$81^{1/2} = \sqrt{81} = 9$$

$$(-8x^3)^{1/3} = \sqrt[3]{-8x^3} = -2x$$

$$4^{5/2} = (\sqrt{4})^5 = 2^5 = 32$$

$$27^{2/3} = (\sqrt[3]{27})^2 = 3^2 = 9$$

$$16^{-3/4} = \frac{1}{16^{3/4}} = \frac{1}{(\sqrt[4]{16})^3} = \frac{1}{2^3} = \frac{1}{8}$$

$$x^{2/3} \cdot x^{-5/6} = x^{2/3 - 5/6} = x^{-1/6} = \frac{1}{x^{1/6}}$$

$$(8^4)^{1/2} = 8^2 = 64$$

$$\frac{a^{4/5}}{a^{-2/5}} = a^{4/5 - (-2/5)} = a^{6/5}$$

(continued)

Definitions and Concepts	Examples

Section 7.3 Simplifying Radical Expressions

Product and Quotient Rules

If $\sqrt[n]{a}$ and $\sqrt[n]{b}$ are real numbers,

$$\sqrt[n]{a} \cdot \sqrt[n]{b} = \sqrt[n]{a \cdot b}$$

$$\frac{\sqrt[n]{a}}{\sqrt[n]{b}} = \sqrt[n]{\frac{a}{b}}, \text{ provided } \sqrt[n]{b} \neq 0$$

A radical of the form $\sqrt[n]{a}$ is **simplified** when a contains no factors that are perfect nth powers.

Multiply or divide as indicated:

$$\sqrt{11} \cdot \sqrt{3} = \sqrt{33}$$

$$\frac{\sqrt[3]{40x}}{\sqrt[3]{5x}} = \sqrt[3]{8} = 2$$

$$\sqrt{40} = \sqrt{4 \cdot 10} = 2\sqrt{10}$$

$$\sqrt{36x^5} = \sqrt{36x^4 \cdot x} = 6x^2\sqrt{x}$$

$$\sqrt[3]{24x^7y^3} = \sqrt[3]{8x^6y^3 \cdot 3x} = 2x^2y\sqrt[3]{3x}$$

$$\sqrt{36x^4 \cdot x} = 6x^2\sqrt{x}$$

Section 7.4 Adding, Subtracting, and Multiplying Radical Expressions

Radicals with the same index and the same radicand are **like radicals**.

The distributive property can be used to add like radicals.

Radical expressions are multiplied by using many of the same properties used to multiply polynomials.

$$5\sqrt{6} + 2\sqrt{6} = (5 + 2)\sqrt{6} = 7\sqrt{6}$$
$$= \sqrt[3]{3x} - 10\sqrt[3]{3x} + 3\sqrt[3]{10x}$$
$$= (-1 - 10)\sqrt[3]{3x} + 3\sqrt[3]{10x}$$
$$= -11\sqrt[3]{3x} + 3\sqrt[3]{10x}$$

Multiply.

$$\left(\sqrt{5} - \sqrt{2x}\right)\left(\sqrt{2} + \sqrt{2x}\right)$$
$$= \sqrt{10} + \sqrt{10x} - \sqrt{4x} - 2x$$
$$= \sqrt{10} + \sqrt{10x} - 2\sqrt{x} - 2x$$
$$\left(2\sqrt{3} - \sqrt{8x}\right)\left(2\sqrt{3} + \sqrt{8x}\right)$$
$$= 4(3) - 8x = 12 - 8x$$

Section 7.5 Rationalizing Denominators and Numerators of Radical Expressions

The **conjugate** of $a + b$ is $a - b$.

The process of writing the denominator of a radical expression without a radical is called **rationalizing the denominator.**

The conjugate of $\sqrt{7} + \sqrt{3}$ is $\sqrt{7} - \sqrt{3}$

Rationalize each denominator.

$$\frac{\sqrt{5}}{\sqrt{3}} = \frac{\sqrt{5} \cdot \sqrt{3}}{\sqrt{3} \cdot \sqrt{3}} = \frac{\sqrt{15}}{3}$$

$$\frac{6}{\sqrt{7} + \sqrt{3}} = \frac{6\left(\sqrt{7} - \sqrt{3}\right)}{\left(\sqrt{7} + \sqrt{3}\right)\left(\sqrt{7} - \sqrt{3}\right)}$$

$$= \frac{6\left(\sqrt{7} - \sqrt{3}\right)}{7 - 3}$$

$$= \frac{6\left(\sqrt{7} - \sqrt{3}\right)}{4} = \frac{3\left(\sqrt{7} - \sqrt{3}\right)}{2}$$

(continued)

Definitions and Concepts	Examples

Section 7.5 Rationalizing Denominators and Numerators of Radical Expressions

The process of writing the numerator of a radical expression without a radical is called **rationalizing the numerator.**

Rationalize each numerator.

$$\frac{\sqrt[3]{9}}{\sqrt[3]{5}} = \frac{\sqrt[3]{9} \cdot \sqrt[3]{3}}{\sqrt[3]{5} \cdot \sqrt[3]{3}} = \frac{\sqrt[3]{27}}{\sqrt[3]{15}} = \frac{3}{\sqrt[3]{15}}$$

$$\frac{\sqrt{9} + \sqrt{3x}}{12} = \frac{\left(\sqrt{9} + \sqrt{3x}\right)\left(\sqrt{9} - \sqrt{3x}\right)}{12\left(\sqrt{9} - \sqrt{3x}\right)}$$

$$= \frac{9 - 3x}{12\left(\sqrt{9} - \sqrt{3x}\right)}$$

$$= \frac{3(3 - x)}{3 \cdot 4\left(3 - \sqrt{3x}\right)} = \frac{3 - x}{4\left(3 - \sqrt{3x}\right)}$$

Section 7.6 Radical Equations and Problem Solving

To Solve a Radical Equation Algebraically

Step 1: Write the equation so that one radical is by itself on one side of the equation.

Step 2: Raise each side of the equation to a power equal to the index of the radical and simplify.

Step 3: If the equation still contains a radical, repeat Steps 1 and 2. If not, solve the equation.

Step 4: Check all proposed solutions in the original equation.

Solve $x = \sqrt{4x + 9} + 3$.

1. $x - 3 = \sqrt{4x + 9}$

2. $(x - 3)^2 = \left(\sqrt{4x + 9}\right)^2$

   $x^2 - 6x + 9 = 4x + 9$

3. $x^2 - 10x = 0$

   $x(x - 10) = 0$

   $x = 0$ or $x = 10$

4. The proposed solution 10 checks, but 0 does not. The solution is 10.

To Solve a Radical Equation Graphically

Use the intersection-of-graphs method or the *x*-intercept method.

Solve $x = \sqrt{4x + 9} + 3$ using the intersection-of-graphs method.

Graph $y_1 = x$ and $y_2 = \sqrt{4x + 9} + 3$ in a $[-5, 20, 5]$ by $[-10, 15, 5]$ window.

The intersection of the two graphs is at $x = 10$. The solution of the equation is 10.

See the algebraic solution for a check.

(continued)

Definitions and Concepts	**Examples**

Section 7.7 Complex Numbers

$$i^2 = -1 \text{ and } i = \sqrt{-1}$$

A **complex number** is a number that can be written in the form $a + bi$, where a and b are real numbers.

Simplify $\sqrt{-9}$.

$$\sqrt{-9} = \sqrt{-1 \cdot 9} = \sqrt{-1} \cdot \sqrt{9} = i \cdot 3 \text{ or } 3i$$

Complex Numbers	*Written in form a + bi*
12	$12 + 0i$
$-5i$	$0 + (-5)i$
$-2 - 3i$	$-2 + (-3)i$

Multiply,

$$\sqrt{-3} \cdot \sqrt{-7} = i\sqrt{3} \cdot i\sqrt{7}$$
$$= i^2\sqrt{21}$$
$$= -\sqrt{21}$$

To add or subtract complex numbers, add or subtract their real parts and then add or subtract their imaginary parts.

Perform each indicated operation.

$$(-3 + 2i) - (7 - 4i) = -3 + 2i - 7 + 4i$$
$$= -10 + 6i$$

To multiply complex numbers, multiply as though they are binomials.

$$(-7 - 2i)(6 + i) = -42 - 7i - 12i - 2i^2$$
$$= -42 - 19i - 2(-1)$$
$$= -42 - 19i + 2$$
$$= -40 - 19i$$

The complex numbers $(a + bi)$ and $(a - bi)$ are called **complex conjugates.**

The complex conjugate of

$$(3 + 6i) \text{ is } (3 - 6i).$$

Their product is a real number.

$$(3 - 6i)(3 + 6i) = 9 - 36i^2$$
$$= 9 - 36(-1) = 9 + 36 = 45$$

To divide complex numbers, multiply the numerator and the denominator by the conjugate of the denominator.

Divide.

$$\frac{4}{2 - i} = \frac{4(2 + i)}{(2 - i)(2 + i)}$$
$$= \frac{4(2 + i)}{4 - i^2}$$
$$= \frac{4(2 + i)}{5}$$
$$= \frac{8 + 4i}{5} = \frac{8}{5} + \frac{4}{5}i$$

CHAPTER REVIEW

(7.1) *Find the root. Assume that all variables represent positive numbers.* **4.** not a real number

1. $\sqrt{81}$ 9

2. $\sqrt[4]{81}$ 3

3. $\sqrt[3]{-8}$ -2

4. $\sqrt[4]{-16}$

5. $-\sqrt{\dfrac{1}{49}}$ $-\dfrac{1}{7}$

6. $\sqrt{x^{64}}$ x^{32}

7. $-\sqrt{36}$ -6

8. $\sqrt[3]{64}$ 4

9. $\sqrt[3]{-a^6b^9}$ $-a^2b^3$

10. $\sqrt{16a^4b^{12}}$ $4a^2b^6$

11. $\sqrt[5]{32a^5b^{10}}$ $2ab^2$

12. $\sqrt[5]{-32x^{15}y^{20}}$ $-2x^3y^4$

13. $\sqrt{\dfrac{x^{12}}{36y^2}}$ $\dfrac{x^6}{6y}$

14. $\sqrt[3]{\dfrac{27y^3}{z^{12}}}$ $\dfrac{3y}{z^4}$

Simplify. Use absolute value bars when necessary.

15. $\sqrt{(-x)^2}$ $|-x|$

16. $\sqrt[4]{(x^2-4)^4}$ $|x^2-4|$

17. $\sqrt[3]{(-27)^3}$ -27

18. $\sqrt[5]{(-5)^5}$ -5

19. $-\sqrt[5]{x^5}$ $-x$

20. $\sqrt[4]{16(2y+z)^{12}}$ $2|(2y+z)^3|$

21. $\sqrt{25(x-y)^{10}}$ $5|(x-y)^5|$

22. $\sqrt[5]{-y^5}$ $-y$

23. $\sqrt[9]{-x^9}$ $-x$

Identify the domain and then graph each function.

24. $f(x) = \sqrt{x} + 3$ $[0, \infty)$; See graphing answer section.

25. $g(x) = \sqrt[3]{x} - 3$; use the accompanying table. $(-\infty, \infty)$; See graphing answer section.

x	-5	2	3	4	11
$g(x)$	-2	-1	0	1	2

(7.2) *Evaluate the following.*

26. $\left(\dfrac{1}{81}\right)^{1/4}$ $\dfrac{1}{3}$

27. $\left(-\dfrac{1}{27}\right)^{1/3}$ $-\dfrac{1}{3}$

28. $(-27)^{-1/3}$ $-\dfrac{1}{3}$

29. $(-64)^{-1/3}$ $-\dfrac{1}{4}$

30. $-9^{3/2}$ -27

31. $64^{-1/3}$ $\dfrac{1}{4}$

32. $(-25)^{5/2}$ not a real number

33. $\left(\dfrac{25}{49}\right)^{-3/2}$ $\dfrac{343}{125}$

34. $\left(\dfrac{8}{27}\right)^{-2/3}$ $\dfrac{9}{4}$

35. $\left(-\dfrac{1}{36}\right)^{-1/4}$ not a real number

Write with rational exponents.

36. $\sqrt[3]{x^2}$ $x^{2/3}$

37. $\sqrt[5]{5x^2y^3}$ $5^{1/5}x^{2/5}y^{3/5}$

Write with radical notation.

38. $y^{4/5}$ $\sqrt[5]{y^4}$

39. $5(xy^2z^5)^{1/3}$ $5\sqrt[3]{xy^2z^5}$

40. $(x+2y)^{-1/2}$ $\dfrac{1}{\sqrt{x+2y}}$

Simplify each expression. Assume that all variables represent positive numbers. Write with only positive exponents.

41. $a^{1/3}a^{4/3}a^{1/2}$ $a^{13/6}$

42. $\dfrac{b^{1/3}}{b^{4/3}}$ $\dfrac{1}{b}$

43. $(a^{1/2}a^{-2})^3$ $\dfrac{1}{a^{9/2}}$

44. $(x^{-3}y^6)^{1/3}$ $\dfrac{y^2}{x}$

45. $\left(\dfrac{b^{3/4}}{a^{-1/2}}\right)^8$ a^4b^6

46. $\dfrac{x^{1/4}x^{-1/2}}{x^{2/3}}$ $\dfrac{1}{x^{11/12}}$

47. $\left(\dfrac{49c^{5/3}}{a^{-1/4}b^{5/6}}\right)^{-1}$ $\dfrac{b^{5/6}}{49a^{1/4}c^{5/3}}$

48. $a^{-1/4}(a^{5/4}-a^{9/4})$ $a-a^2$

Use a calculator and write a three-decimal-place approximation.

49. $\sqrt{20}$ 4.472

50. $\sqrt[3]{-39}$ -3.391

51. $\sqrt[4]{726}$ 5.191

52. $56^{1/3}$ 3.826

53. $-78^{3/4}$ -26.246

54. $105^{-2/3}$ 0.045

Use rational exponents to write each radical with the same index. Then multiply.

55. $\sqrt[3]{2} \cdot \sqrt{7}$ $\sqrt[6]{1372}$

56. $\sqrt[3]{3} \cdot \sqrt[4]{x}$ $\sqrt[12]{81x^3}$

(7.3) *Perform the indicated operations and then simplify if possible. For the remainder of this review, assume that variables represent positive numbers only.*

57. $\sqrt{3} \cdot \sqrt{8}$ $2\sqrt{6}$

58. $\sqrt[3]{7y} \cdot \sqrt[3]{x^2z}$ $\sqrt[3]{7x^2yz}$

59. $\dfrac{\sqrt{44x^3}}{\sqrt{11x}}$ $2x$

60. $\dfrac{\sqrt[4]{a^6b^{13}}}{\sqrt[4]{a^2b}}$ ab^3

Simplify.

61. $\sqrt{60}$ $2\sqrt{15}$

62. $-\sqrt{75}$ $-5\sqrt{3}$

63. $\sqrt[3]{162}$ $3\sqrt[3]{6}$

64. $\sqrt[3]{-32}$ $-2\sqrt[3]{4}$

65. $\sqrt{36x^7}$ $6x^3\sqrt{x}$

66. $\sqrt[3]{24a^5b^7}$ $2ab^2\sqrt[3]{3a^2b}$

67. $\sqrt{\dfrac{p^{17}}{121}}$ $\dfrac{p^8\sqrt{p}}{11}$

68. $\sqrt[3]{\dfrac{y^5}{27x^6}}$ $\dfrac{y\sqrt[3]{y^2}}{3x^2}$

69. $\sqrt[4]{\dfrac{xy^6}{81}}$ $\dfrac{y\sqrt[4]{xy^2}}{3}$

70. $\sqrt{\dfrac{2x^3}{49y^4}}$ $\dfrac{x\sqrt{2x}}{7y^2}$

△ **71.** The formula for the radius r of a circle of area A is
$$r = \sqrt{\dfrac{A}{\pi}}$$
71.a. $\dfrac{5}{\sqrt{\pi}}$ m or $\dfrac{5\sqrt{\pi}}{\pi}$ m

a. Find the exact radius of a circle whose area is 25 square meters.

b. Approximate to two decimal places the radius of a circle whose area is 104 square inches. 5.75 in.

(7.4) Perform the indicated operation.

72. $x\sqrt{75xy} - \sqrt{27x^3y}$ **73.** $2\sqrt{32x^2y^3} - xy\sqrt{98y}$

74. $\sqrt[3]{128} + \sqrt[3]{250}$ $9\sqrt[3]{2}$ **75.** $3\sqrt[4]{32a^5} - a\sqrt[4]{162a}$

76. $\dfrac{5}{\sqrt{4}} + \dfrac{\sqrt{3}}{3}$ $\dfrac{15 + 2\sqrt{3}}{6}$ **77.** $\sqrt{\dfrac{8}{x^2}} - \sqrt{\dfrac{50}{16x^2}}$ $\dfrac{3\sqrt{2}}{4x}$

78. $2\sqrt{50} - 3\sqrt{125} + \sqrt{98}$ $17\sqrt{2} - 15\sqrt{5}$

79. $2a\sqrt[4]{32b^5} - 3b\sqrt[4]{162a^4b} + \sqrt[4]{2a^4b^5}$ $-4ab\sqrt[4]{2b}$

Multiply and then simplify if possible.

80. $\sqrt{3}\left(\sqrt{27} - \sqrt{3}\right)$ 6   **81.** $\left(\sqrt{x} - 3\right)^2$

82. $\left(\sqrt{5} - 5\right)\left(2\sqrt{5} + 2\right)$ $-8\sqrt{5}$

83. $\left(2\sqrt{x} - 3\sqrt{y}\right)\left(2\sqrt{x} + 3\sqrt{y}\right)$ $4x - 9y$

84. $\left(\sqrt{a} + 3\right)\left(\sqrt{a} - 3\right)$ **85.** $\left(\sqrt[3]{a} + 2\right)^2$

86. $\left(\sqrt[3]{5x} + 9\right)\left(\sqrt[3]{5x} - 9\right)$ $\sqrt[3]{25x^2} - 81$

87. $\left(\sqrt[3]{a} + 4\right)\left(\sqrt[3]{a^2} - 4\sqrt[3]{a} + 16\right)$ $a + 64$

(7.5) Rationalize each denominator.

88. $\dfrac{3}{\sqrt{7}}$ $\dfrac{3\sqrt{7}}{7}$ **89.** $\sqrt{\dfrac{x}{12}}$ $\dfrac{\sqrt{3x}}{6}$

90. $\dfrac{5}{\sqrt[3]{4}}$ $\dfrac{5\sqrt[3]{2}}{2}$ **91.** $\sqrt{\dfrac{24x^5}{3y^2}}$ $\dfrac{2x^2\sqrt{2x}}{y}$

92. $\sqrt[3]{\dfrac{15x^6y^7}{z^2}}$ $\dfrac{x^2y^2\sqrt[3]{15yz}}{z}$ **93.** $\dfrac{5}{2 - \sqrt{7}}$ $-\dfrac{10 + 5\sqrt{7}}{3}$

94. $\dfrac{3}{\sqrt{y} - 2}$ $\dfrac{3\sqrt{y} + 6}{y - 4}$ **95.** $\dfrac{\sqrt{2} - \sqrt{3}}{\sqrt{2} + \sqrt{3}}$ $-5 + 2\sqrt{6}$

Rationalize each numerator.

96. $\dfrac{\sqrt{11}}{3}$ $\dfrac{11}{3\sqrt{11}}$ **97.** $\sqrt{\dfrac{18}{y}}$ $\dfrac{6}{\sqrt{2y}}$ **98.** $\dfrac{\sqrt[3]{9}}{7}$ $\dfrac{3}{7\sqrt[3]{3}}$

99. $\sqrt{\dfrac{24x^5}{3y^2}}$ $\dfrac{4x^3}{y\sqrt{2x}}$ **100.** $\sqrt[3]{\dfrac{xy^2}{10z}}$ $\dfrac{xy}{\sqrt[3]{10x^2yz}}$

101. $\dfrac{\sqrt{x} + 5}{-3}$ $\dfrac{x - 25}{-3\sqrt{x} + 15}$

(7.6) Solve each equation for the variable.

102. $\sqrt{y - 7} = 5$ 32   **103.** $\sqrt{2x} + 10 = 4$ $\varnothing$

104. $\sqrt[3]{2x - 6} = 4$ 35   **105.** $\sqrt{x + 6} = \sqrt{x + 2}$ $\varnothing$

106. $2x - 5\sqrt{x} = 3$ 9   **107.** $\sqrt{x + 9} = 2 + \sqrt{x - 7}$ 16

Find each unknown length.

△ **108.**

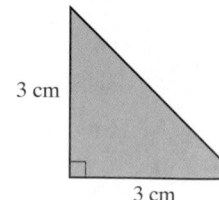

3√2

72. $2x\sqrt{3xy}$ **73.** $xy\sqrt{2y}$
75. $3a\sqrt[4]{2a}$ **81.** $x - 6\sqrt{x} + 9$
84. $a - 9$ **85.** $\sqrt[3]{a^2} + 4\sqrt[3]{a} + 4$
116. $15 - 4i$ **117.** $-13 - 3i$
123. 87

△ **109.**

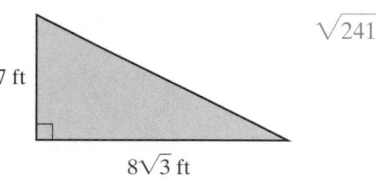

$\sqrt{241}$

110. Beverly Hillis wants to determine the distance x across a pond on her property. She is able to measure the distances shown on the following diagram. Find how wide the lake is at the crossing point, indicated by the triangle, to the nearest tenth of a foot.   51.2 ft

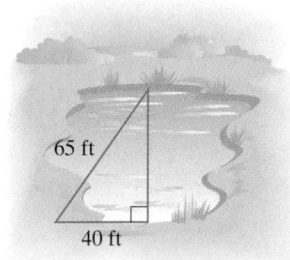

△ **111.** A pipe fitter needs to connect two underground pipelines that are offset by 3 feet, as pictured in the diagram. Neglecting the joints needed to join the pipes, find the length of the shortest possible connecting pipe rounded to the nearest hundredth of a foot.   4.24 ft

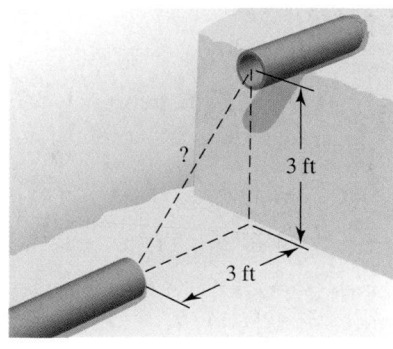

(7.7) Perform the indicated operation and simplify. Write the result in the form $a + bi$.

112. $\sqrt{-8}$ $2i\sqrt{2}$ **113.** $-\sqrt{-6}$ $-i\sqrt{6}$

114. $\sqrt{-4} + \sqrt{-16}$ $6i$ **115.** $\sqrt{-2} \cdot \sqrt{-5}$ $-\sqrt{10}$

116. $(12 - 6i) + (3 + 2i)$ **117.** $(-8 - 7i) - (5 - 4i)$

118. $\left(\sqrt{3} + \sqrt{2}\right) + \left(3\sqrt{2} - \sqrt{-8}\right)$ $\sqrt{3} + 4\sqrt{2} - 2i\sqrt{2}$

119. $2i(2 - 5i)$ $10 + 4i$ **120.** $-3i(6 - 4i)$ $-12 - 18i$

121. $(3 + 2i)(1 + i)$ $1 + 5i$ **122.** $(2 - 3i)^2$ $-5 - 12i$

123. $\left(\sqrt{6} - 9i\right)\left(\sqrt{6} + 9i\right)$ **124.** $\dfrac{2 + 3i}{2i}$ $\dfrac{3}{2} - i$

125. $\dfrac{1 + i}{-3i}$ $-\dfrac{1}{3} + \dfrac{1}{3}i$

CHAPTER 7 TEST

Remember to use your Chapter Test Prep Video CD to help you study and view solutions to the test questions you need help with.

Raise to the power or find the root. Assume that all variables represent positive numbers. Write with only positive exponents.

1. $\sqrt{216}$ $6\sqrt{6}$

2. $-\sqrt[4]{x^{64}}$ $-x^{16}$

3. $\left(\dfrac{1}{125}\right)^{1/3}$ $\dfrac{1}{5}$

4. $\left(\dfrac{1}{125}\right)^{-1/3}$ 5

5. $\left(\dfrac{8x^3}{27}\right)^{2/3}$ $\dfrac{4x^2}{9}$

6. $\sqrt[3]{-a^{18}b^9}$ $-a^6b^3$

7. $\left(\dfrac{64c^{4/3}}{a^{-2/3}b^{5/6}}\right)^{1/2}$ $\dfrac{8a^{1/3}c^{2/3}}{b^{5/12}}$

8. $a^{-2/3}(a^{5/4} - a^3)$

Find the root. Use absolute value bars when necessary.

9. $\sqrt[4]{(4xy)^4}$ $|4xy|$ or $4|xy|$ **10.** $\sqrt[3]{(-27)^3}$ -27

Rationalize the denominator. Assume that all variables represent positive numbers.

11. $\sqrt{\dfrac{9}{y}}$ $\dfrac{3\sqrt{y}}{y}$

12. $\dfrac{4 - \sqrt{x}}{4 + 2\sqrt{x}}$

13. $\dfrac{\sqrt[3]{ab}}{\sqrt[3]{ab^2}}$ $\dfrac{\sqrt[3]{b^2}}{b}$

14. Rationalize the numerator of $\dfrac{\sqrt{6} + x}{8}$ and simplify.

$\dfrac{6 - x^2}{8(\sqrt{6} - x)}$

Perform the indicated operations. Assume that all variables represent positive numbers.

15. $\sqrt{125x^3} - 3\sqrt{20x^3}$ $-x\sqrt{5x}$

16. $\sqrt{3}(\sqrt{16} - \sqrt{2})$ $4\sqrt{3} - \sqrt{6}$

17. $(\sqrt{x} + 1)^2$ $x + 2\sqrt{x} + 1$

18. $(\sqrt{2} - 4)(\sqrt{3} + 1)$ $\sqrt{6} - 4\sqrt{3} + \sqrt{2} - 4$

19. $(\sqrt{5} + 5)(\sqrt{5} - 5)$ -20

Use a calculator to approximate each to three decimal places.

20. $\sqrt{561}$ 23.685

21. $386^{-2/3}$ 0.019

8. $a^{7/12} - a^{7/3}$ **12.** $\dfrac{8 - 6\sqrt{x} + x}{8 - 2x}$

Solve.

22. $x = \sqrt{x - 2} + 2$ $2, 3$

23. $\sqrt{x^2 - 7} + 3 = 0$ $\varnothing$

24. $\sqrt[3]{x + 5} = \sqrt[3]{2x - 1}$ 6

Perform the indicated operation and simplify. Write the result in the form $a + bi$.

25. $\sqrt{-2}$ $i\sqrt{2}$

26. $-\sqrt{-8}$ $-2i\sqrt{2}$

27. $(12 - 6i) - (12 - 3i)$ $-3i$

28. $(6 - 2i)(6 + 2i)$ 40

29. $(4 + 3i)^2$ $7 + 24i$

30. $\dfrac{1 + 4i}{1 - i}$ $-\dfrac{3}{2} + \dfrac{5}{2}i$

△ **31.** Find x. $\dfrac{5\sqrt{2}}{2}$

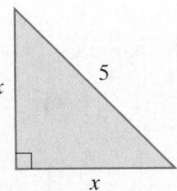

32. Identify the domain of $g(x)$. Then graph the function.

$$g(x) = \sqrt{x + 2}$$

$[-2, \infty)$; see graphing answer section.

Solve.

33. The function $V(r) = \sqrt{2.5r}$ can be used to estimate the maximum safe velocity V in miles per hour at which a car can travel if it is driven along a curved road with a *radius of curvature* r in feet. To the nearest whole number, find the maximum safe speed if a cloverleaf exit on an expressway has a radius of curvature of 300 feet.   27 mph

34. Use the formula from Exercise 33 to find the radius of curvature if the safe velocity is 30 mph.   360 ft

CHAPTER CUMULATIVE REVIEW

1. Simplify each expression.   **1. c.** $3.1x - 0.3$ (Sec. 1.4, Ex. 13)
   a. $3xy - 2xy + 5 - 7 + xy$ $2xy - 2$
   b. $7x^2 + 3 - 5(x^2 - 4)$ $2x^2 + 23$
   c. $(2.1x - 5.6) - (-x - 5.3)$
   d. $\frac{1}{2}(4a - 6b) - \frac{1}{3}(9a + 12b - 1) + \frac{1}{4}$ $-a - 7b + \frac{7}{12}$

2. Simplify each expression.
   a. $2(x - 3) + (5x + 3)$ $7x - 3$
   b. $4(3x + 2) - 3(5x - 1)$ $-3x + 11$
   c. $7x + 2(x - 7) - 3x$ $6x - 14$ (Sec. 1.4)

3. Solve for x: $\dfrac{x + 5}{2} + \dfrac{1}{2} = 2x - \dfrac{x - 3}{8}$
   3. $\dfrac{21}{11}$ (Sec. 1.5, Ex. 6)

4. Solve $\dfrac{a - 1}{2} + a = 2 - \dfrac{2a + 7}{8}$ $\dfrac{13}{14}$ (Sec. 1.5)

   5. $4500 per month (Sec. 3.2, Ex. 8)

5. A salesperson earns $600 per month plus a commission of 20% of sales. Find the minimum amount of sales needed to receive a total income of at least $1500 per month.

6. The Smith family owns a lake house 121.5 miles from home. If it takes them $4\frac{1}{2}$ hours round-trip to drive from their house to their lake house, find their average speed.   54 mph (Sec. 1.6)

7. Solve: $2|x| + 25 = 23$ $\varnothing$ (Sec. 3.4, Ex. 6)

8. Solve: $|3x - 2| + 5 = 5$ $\frac{2}{3}$ (Sec. 3.4)

9. Solve: $\left|\dfrac{x}{3} - 1\right| - 2 \geq 0$
   9. $(-\infty, -3] \cup [9, \infty)$ (Sec. 3.5, Ex. 7)

10. Solve: $\left|\dfrac{x}{2} - 1\right| \leq 0$.   2 (Sec. 3.5)

    11. See graphing answer section. (Sec. 2.1, Ex. 9)

11. Graph the equation $y = |x|$.

12. Graph $y = |x - 2|$.   See graphing answer section.

13. Determine the domain and range of each relation.
    a. $\{(2, 3), (2, 4), (0, -1), (3, -1)\}$ domain: $\{2, 0, 3\}$; range: $\{3, 4, -1\}$
    b.

    domain: $\{-4, -3, -2, -1, 0, 1, 2, 3\}$; range: $\{1\}$

    c.   Input: Output:

    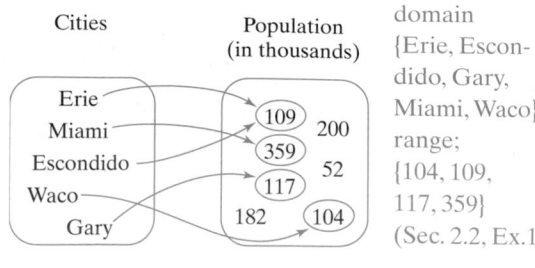

    Cities   Population (in thousands)

    Erie
    Miami   109   200
              359
    Escondido   52
              117
    Waco
              182   104
    Gary

    domain {Erie, Escondido, Gary, Miami, Waco}; range; {104, 109, 117, 359} (Sec. 2.2, Ex.1)

14. Find the domain and the range of each relation. Use the vertical line test to determine whether each graph is the graph of a function.

    a.

    domain: $(-\infty, 0]$; range: $(-\infty, \infty)$; not a function

    b.

    domain: $(-\infty, \infty)$; range: $(-\infty, \infty)$; function

    c.

    $(-2, 0)$ $(2, 0)$

    domain: $(-\infty, -2] \cup [2, \infty)$; range: $(-\infty, \infty)$; not a function (Sec. 2.2)

    15. See graphing answer section. (Sec. 2.3, Ex. 6)
    17. undefined (Sec. 2.4, Ex. 5)

15. Graph $y = -3$.

16. Graph $f(x) = -2$.   See graphing answer section. (Sec. 2.3)

17. Find the slope of the line $x = -5$.

18. Find the slope of $y = -3$.   0 (Sec. 2.4)

19. Use the substitution method to solve the system.
$$\begin{cases} -\dfrac{x}{6} + \dfrac{y}{2} = \dfrac{1}{2} \\ \dfrac{x}{3} - \dfrac{y}{6} = -\dfrac{3}{4} \end{cases}$$ $\left(-\dfrac{21}{10}, \dfrac{3}{10}\right)$ (Sec. 4.1, Ex. 5)

20. Use the substitution method to solve the system.
$$\begin{cases} \dfrac{x}{6} - \dfrac{y}{2} = 1 \\ \dfrac{x}{3} - \dfrac{y}{4} = 2 \end{cases}$$ $(6, 0)$ (Sec. 4.1)

21. Use the product rule to simplify.
    a. $2^2 \cdot 2^5$ 2^7
    b. $x^7 x^3$ x^{10}
    c. $y \cdot y^2 \cdot y^4$ y^7 (Sec. 5.1, Ex. 1)

22. At a seasonal clearance sale, Nana Long spent $33.75. She paid $3.50 for tee-shirts and $4.25 for shorts. If she bought items, how many of each item did she buy?   6 shirts; 3 shorts (Sec. 4.3)

23. Use scientific notation to simplify $\dfrac{2000 \times 0.000021}{700}$.

   23. 6×10^{-5} (Sec. 5.2, Ex. 7)

24. Use scientific notation to simplify and write the answer in scientific notation. $\dfrac{0.0000035 \times 4000}{0.28}$ 5×10^{-2} (Sec. 5.2)

25. If $P(x) = 3x^2 - 2x - 5$, find the following.

   a. $P(1)$ -4

   b. $P(-2)$ 11 (Sec. 5.3, Ex. 4)

26. Subtract $(2x - 5)$ from the sum of $(5x^2 - 3x + 6)$ and $(4x^2 + 5x - 3)$. $9x^2 + 8$ (Sec. 5.3)

27. Multiply and simplify the product if possible.

   a. $(x + 3)(2x + 5)$ $2x^2 + 11x + 15$

   b. $(2x - 3)(5x^2 - 6x + 7)$ **27. b.** $10x^3 - 27x^2 + 32x - 21$ (Sec. 5.4, Ex. 3)

28. Multiply and simplify the product if possible.

   a. $(y - 2)(3y + 4)$ $3y^2 - 2y - 8$ **29.** $5x^2$ (Sec. 5.5, Ex. 1)

   b. $(3y - 1)(2y^2 + 3y - 1)$ $6y^3 + 7y^2 - 6y + 1$ (Sec. 5.4)

29. Find the GCF of $20x^3y$, $10x^2y^2$, and $35x^3$.

30. Factor $x^3 - x^2 + 4x - 4$ $(x - 1)(x^2 + 4)$ (Sec. 5.5)

31. Simplify each rational expression.

   a. $\dfrac{x^3 + 8}{2 + x}$ $x^2 - 2x + 4$

   b. $\dfrac{2y^2 + 2}{y^3 - 5y^2 + y - 5}$ $\dfrac{2}{y - 5}$ (Sec. 6.1, Ex. 5)

32. Simplify each rational expression.

   a. $\dfrac{a^3 - 8}{2 - a}$ $-a^2 - 2a - 4$

   b. $\dfrac{3a^2 - 3}{a^3 + 5a^2 - a - 5}$ $\dfrac{3}{a + 5}$ (Sec. 6.1)

33. Perform the indicated operation.

   a. $\dfrac{2}{x^2y} + \dfrac{5}{3x^3y}$ $\dfrac{6x + 5}{3x^3y}$

   b. $\dfrac{3x}{x + 2} + \dfrac{2x}{x - 2}$ $\dfrac{5x^2 - 2x}{(x + 2)(x - 2)}$

   c. $\dfrac{x}{x - 1} - \dfrac{4}{1 - x}$ $\dfrac{x + 4}{x - 1}$ (Sec. 6.2, Ex. 3)

34. Perform the indicated operations.

   a. $\dfrac{3}{xy^2} - \dfrac{2}{3x^2y}$ $\dfrac{9x - 2y}{3x^2y^2}$

   b. $\dfrac{5x}{x + 3} - \dfrac{2x}{x - 3}$ $\dfrac{3x^2 - 21x}{(x + 3)(x - 3)}$

   c. $\dfrac{x}{x - 2} - \dfrac{5}{2 - x}$ $\dfrac{x + 5}{x - 2}$ (Sec. 6.2)

35. Simplify each complex fraction.

   a. $\dfrac{\dfrac{5x}{x + 2}}{\dfrac{10}{x - 2}}$ $\dfrac{x(x - 2)}{2(x + 2)}$

   b. $\dfrac{\dfrac{x}{y^2} + \dfrac{1}{y}}{\dfrac{y}{x^2} + \dfrac{1}{x}}$ $\dfrac{x^2}{y^2}$ (Sec. 6.3, Ex. 2)

36. Simplify each complex fraction.

   a. $\dfrac{\dfrac{y - 2}{16}}{\dfrac{2y + 3}{12}}$ $\dfrac{3(y - 2)}{4(2y + 3)}$

   b. $\dfrac{\dfrac{x}{16} - \dfrac{1}{x}}{1 - \dfrac{4}{x}}$ $\dfrac{x + 4}{16}$ (Sec. 6.3)

   37. $2x^2 - x + 4$ (Sec. 6.4, Ex. 1)

37. Divide $10x^3 - 5x^2 + 20x$ by $5x$.

38. Divide $x^3 - 2x^2 + 3x - 6$ by $x - 2$. $x^2 + 3$ (Sec. 6.4)

39. Use synthetic division to divide $2x^3 - x^2 - 13x + 1$ by $x - 3$. $2x^2 + 5x + 2 + \dfrac{7}{x - 3}$ (Sec. 6.5, Ex. 1)

40. Use synthetic division to divide $4y^3 - 12y^2 - y + 12$ by $y - 3$. $4y^2 - 1 + \dfrac{9}{y - 3}$ (Sec. 6.5)

41. Solve: $\dfrac{x + 6}{x - 2} = \dfrac{2(x + 2)}{x - 2}$ $\varnothing$ (Sec. 6.6, Ex. 3)

42. Solve: $\dfrac{28}{9 - a^2} = \dfrac{2a}{a - 3} + \dfrac{6}{a + 3}$ $-1, -5$ (Sec. 6.6)

43. Solve $\dfrac{1}{x} + \dfrac{1}{y} = \dfrac{1}{z}$ for x. $x = \dfrac{yz}{y - z}$ (Sec. 6.7, Ex. 1)

44. Solve $A = \dfrac{h(a + b)}{2}$ for a. $a = \dfrac{2A - bh}{h}$ (Sec. 6.7)

45. Suppose that u varies inversely as w. If u is 3 when w is 5, find the constant of variation and the inverse variation equation.

46. Suppose that y varies directly as x. If $y = 0.51$ when $x = 3$, find the constant of variation and the direct variation equation. $k = 0.17$; $y = 0.17x$ (Sec. 6.8)

47. Write each expression with a positive exponent, and then simplify.

   45. constant of variation: 15; $u = \dfrac{15}{w}$ (Sec. 6.8, Ex. 3)

   a. $16^{-3/4}$ $\dfrac{1}{8}$

   b. $(-27)^{-2/3}$ $\dfrac{1}{9}$ (Sec. 7.2, Ex. 3)

48. Write each expression with a positive exponent, and then simplify.

a. $(81)^{-3/4}$ $\dfrac{1}{27}$

b. $(-125)^{-2/3}$ $\dfrac{1}{25}$

49. Rationalize the numerator of $\dfrac{\sqrt{x} + 2}{5}$. $\dfrac{x - 4}{5(\sqrt{x} - 2)}$

(Sec. 7.5, Ex. 7)

50. Add or subtract.

a. $\sqrt{36a^3} - \sqrt{144a^3} + \sqrt{4a^3}$ $-4a\sqrt{a}$

b. $\sqrt[3]{128ab^3} - 3\sqrt[3]{2ab^3} + b\sqrt[3]{16a}$ $3b\sqrt[3]{2a}$

c. $\dfrac{\sqrt[3]{81}}{10} + \sqrt[3]{\dfrac{192}{125}}$ $\dfrac{11\sqrt[3]{3}}{10}$ (Sec. 7.4)

Quadratic Equations and Functions

An important part of the study of algebra is learning to model and solve problems. Often, the model of a problem is a quadratic equation or a function containing a second-degree polynomial. In this chapter, we continue the work begun in Chapter 5, when we solved polynomial equations in one variable by factoring. Two additional methods of solving quadratic equations are analyzed, as well as methods of solving nonlinear inequalities in one variable.

Community theater groups began to formulate in the United States in the early part of the twentieth century. While many major cities have large theater districts, over 5000 theater companies nationwide entertain smaller communities. Community theater groups provide a creative outlet for aspiring actors, musicians, technicians, producers, and directors.

Producers are responsible for the business end of the show: obtaining financing, setting a budget, securing the performance location, advertising, setting ticket prices, and hiring the director. Directors are in charge of the artistic decisions in the show. They choose the script, audition actors, select costumes and sets, conduct rehearsals, and oversee the cast and crew.

In the Spotlight on Decision Making feature on page 651, you will have the opportunity to decide on optimal ticket prices as a community theater member.

Link: `www.communitytheater.org` *(Community Theater Green Room)*

Source of text: World Book Millenium 2000 (encyclopedia on CD) "Theater," `http://www.bls.gov/oco/ocos093.htm` *(Bureau of Labor Statistics)*

8.1 SOLVING QUADRATIC EQUATIONS BY COMPLETING THE SQUARE

Objectives

1. Use the square root property to solve quadratic equations.
2. Solve quadratic equations by completing the square.
3. Use quadratic equations to solve problems.

1 In Chapter 5, we solved quadratic equations by factoring. Recall that a **quadratic,** or **second-degree, equation** is an equation that can be written in the form $ax^2 + bx + c = 0$, where a, b, and c are real numbers and a is not 0. To solve a quadratic equation such as $x^2 = 9$ by factoring, we use the zero-factor theorem. To use the zero-factor theorem, the equation must first be written in standard form, $ax^2 + bx + c = 0$.

$$x^2 = 9$$
$$x^2 - 9 = 0 \qquad \text{Subtract 9 from both sides.}$$
$$(x + 3)(x - 3) = 0 \qquad \text{Factor.}$$
$$x + 3 = 0 \quad \text{or} \quad x - 3 = 0 \qquad \text{Set each factor equal to } 0.$$
$$x = -3 \qquad x = 3 \qquad \text{Solve.}$$

The solution set is $\{-3, 3\}$, the positive and negative square roots of 9. Not all quadratic equations can be solved by factoring, so we need to explore other methods. Notice that the solutions of the equation $x^2 = 9$ are two numbers whose square is 9.

$$3^2 = 9 \qquad \text{and} \qquad (-3)^2 = 9$$

TEACHING TIP
Point out that $\sqrt{9}$ evaluates to positive 3 only, but when solving a problem such as $x^2 = 9$, solutions are the positive and negative square roots of 9, or $\pm\sqrt{9} = \pm3$.

Thus, we can solve the equation $x^2 = 9$ by taking the square root of both sides. Be sure to include both $\sqrt{9}$ and $-\sqrt{9}$ as solutions since both $\sqrt{9}$ and $-\sqrt{9}$ are numbers whose square is 9.

$$x^2 = 9$$
$$\sqrt{x^2} = \pm\sqrt{9} \qquad \text{The notation } \pm\sqrt{9} \text{ (read as "plus or minus } \sqrt{9}\text{")}$$
$$\qquad\qquad\qquad \text{indicates the pair of numbers } +\sqrt{9} \text{ and } -\sqrt{9}.$$
$$x = \pm3$$

This illustrates the square root property.

> **Helpful Hint**
>
> The notation ±3, for example, is read as "plus or minus 3." It is a shorthand notation for the pair of numbers $+3$ and -3.

Square Root Property

If b is a real number and if $a^2 = b$, then $a = \pm\sqrt{b}$.

Helpful Hint

Examples of Real Numbers: $-3, 7.6, \sqrt{5}, \frac{13}{20}$
Examples of Complex Numbers that are not Real Numbers: $2 - 6i, -1.2i,$
$3 + \sqrt{11}i$

EXAMPLE 1

Use the square root property to solve $x^2 = 50$.

Solution

$x^2 = 50$

$x = \pm\sqrt{50}$ Use the square root property.

$x = \pm 5\sqrt{2}$ Simplify the radical.

CLASSROOM EXAMPLE
Solve $x^2 = 20$.

answer: $\pm 2\sqrt{5}$

Check   A calculator check is in the margin while a pencil and paper check is below.

Let $x = 5\sqrt{2}$.   Let $x = -5\sqrt{2}$.

$x^2 = 50$ $x^2 = 50$

$\left(5\sqrt{2}\right)^2 \overset{?}{=} 50$ $\left(-5\sqrt{2}\right)^2 \overset{?}{=} 50$

$25 \cdot 2 \overset{?}{=} 50$ $25 \cdot 2 \overset{?}{=} 50$

$50 = 50$ True $50 = 50$ True

```
5√(2)→X:X²
             50
-5√(2)→X:X²
             50
■
```

The solutions are $5\sqrt{2}$ and $-5\sqrt{2}$, or the solution set is $\{-5\sqrt{2}, 5\sqrt{2}\}$.

EXAMPLE 2

Use the square root property to solve $2x^2 = 14$.

Solution   First we get the squared variable alone on one side of the equation.

$$2x^2 = 14$$

$$x^2 = 7 \qquad \text{Divide both sides by 2.}$$

$$x = \pm\sqrt{7} \qquad \text{Use the square root property.}$$

CLASSROOM EXAMPLE
Solve $5x^2 = 55$.

answer: $\pm\sqrt{11}$

Check to see that the solutions are $\sqrt{7}$ and $-\sqrt{7}$, or the solution set is $\{-\sqrt{7}, \sqrt{7}\}$.

EXAMPLE 3

Use the square root property to solve $(x + 1)^2 = 12$. Use a graphing utility to check.

Solution

$(x + 1)^2 = 12$

$x + 1 = \pm\sqrt{12}$ Use the square root property.

$x + 1 = \pm 2\sqrt{3}$ Simplify the radical.

$x = -1 \pm 2\sqrt{3}$ Subtract 1 from both sides.

CLASSROOM EXAMPLE

Solve $(x + 2)^2 - 18$.

answer: $-2 \pm 3\sqrt{2}$

Check

Below is a check for $-1 + 2\sqrt{3}$. The check for $-1 - 2\sqrt{3}$ is almost the same and is left for you to do on your own.

$$(x + 1)^2 = 12$$

$$\left(-1 + 2\sqrt{3} + 1\right)^2 \stackrel{?}{=} 12$$

$$\left(2\sqrt{3}\right)^2 \stackrel{?}{=} 12$$

$$4 \cdot 3 \stackrel{?}{=} 12$$

$$12 = 12 \quad \text{True}$$

The solutions are $-1 + 2\sqrt{3}$ and $-1 - 2\sqrt{3}$.

To check graphically, we use the intersection-of-graphs method and graph

$$y_1 = (x + 1)^2 \quad \text{and} \quad y_2 = 12$$

The approximate points of intersection are shown below in the first two screens.

Notice that this graphical check gives approximate solutions only since $-1 + 2\sqrt{3}$ and $-1 - 2\sqrt{3}$ are irrational. The approximate solutions are $x \approx -4.464$ and $x \approx 2.464$. The exact solutions are $x = -1 + 2\sqrt{3}$ and $x = -1 - 2\sqrt{3}$.

Next, see that the x-values of the points of intersection approximate the exact solutions. The screen above and to the right confirms this.

EXAMPLE 4

Use the square root property to solve $(2x - 5)^2 = -16$.

Solution

$(2x - 5)^2 = -16$

$2x - 5 = \pm\sqrt{-16}$ Use the square root property.

$2x - 5 = \pm 4i$ Simplify the radical.

$2x = 5 \pm 4i$ Add 5 to both sides.

$x = \dfrac{5 \pm 4i}{2}$ Divide both sides by 2.

CLASSROOM EXAMPLE

Solve $(3x - 1)^2 = -4$.

answer: $\dfrac{1 \pm 2i}{3}$

The solutions are $\dfrac{5 + 4i}{2}$ and $\dfrac{5 - 4i}{2}$.

✔ **CONCEPT CHECK**

How do you know just by looking that $(x - 2)^2 = -4$ has complex, but not real solutions?

Discover the Concept

From the example above, we know that the equation $(2x - 5)^2 = -16$ has two complex, but not real solutions. Use the intersection-of-graphs method to check.

a. Graph $y_1 = (2x - 5)^2$ and $y_2 = -16$.

b. Locate any points of intersection of the graphs.

c. Summarize the results of parts (a) and (b), and what you think occurred.

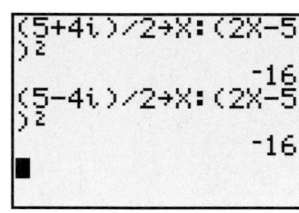

The x-axis and the y-axis of the rectangular coordinate system include real numbers only. Thus, coordinates of points of the associated plane are real numbers only. This means that the intersection-of-graphs method on the rectangular coordinate system gives real number solutions of a related equation only. Since the graphs above do not intersect, the equation has no real number solutions.

To check Example 4 numerically, see the screen to the left.

2   Notice from Examples 3 and 4 that, if we write a quadratic equation so that one side is the square of a binomial, we can solve by using the square root property. To write the square of a binomial, we write perfect square trinomials. Recall that a perfect square trinomial is a trinomial that can be factored into two identical binomial factors.

TEACHING TIP

Before beginning objective 2 have students work in pairs to determine the relationship between coefficients b and c (in $ax^2 + bx + c = 0$) for equations of the form $(x + n)^2 = 0$, where n is an integer. Have them share their conclusions with the class.

Perfect Square Trinomials	*Factored Form*
$x^2 + 8x + 16$	$(x + 4)^2$
$x^2 - 6x + 9$	$(x - 3)^2$
$x^2 + 3x + \dfrac{9}{4}$	$\left(x + \dfrac{3}{2}\right)^2$

Notice that for each perfect square trinomial, **the constant term of the trinomial is the square of half the coefficient of the x-term.** For example,

$$x^2 + 8x + 16 \qquad\qquad x^2 - 6x + 9$$

$$\frac{1}{2}(8) = 4 \text{ and } 4^2 = 16 \qquad \frac{1}{2}(-6) = -3 \text{ and } (-3)^2 = 9$$

The process of writing a quadratic equation so that one side is a perfect square trinomial is called **completing the square.**

EXAMPLE 5

Solve $p^2 + 2p = 4$ by completing the square.

Solution   First, add the square of half the coefficient of p to both sides so that the resulting trinomial will be a perfect square trinomial. The coefficient of p is 2.

Concept Check Answer:
answers may vary

$$\frac{1}{2}(2) = 1 \quad \text{and} \quad 1^2 = 1$$

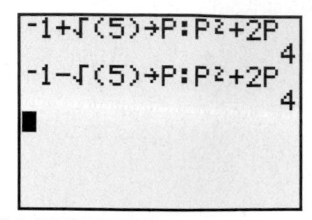

TEACHING TIP

Remind the students of the necessity of learning to solve both algebraically and graphically. Solving graphically is useful if an approximate solution is acceptable.

Add 1 to both sides of the original equation.

$$p^2 + 2p = 4$$

$$p^2 + 2p + 1 = 4 + 1 \qquad \text{Add 1 to both sides.}$$

$$(p + 1)^2 = 5 \qquad \text{Factor the trinomial; simplify the right side.}$$

We may now use the square root property and solve for p.

$$p + 1 = \pm\sqrt{5} \qquad \text{Use the square root property.}$$

$$p = -1 \pm \sqrt{5} \qquad \text{Subtract 1 from both sides.}$$

Notice that there are two solutions: $-1 + \sqrt{5}$ and $-1 - \sqrt{5}$. A numerical check is shown to the left.

EXAMPLE 6

Solve $m^2 - 7m - 1 = 0$ for m.

Algebraic Solution:

To solve by completing the square, we first add 1 to both sides of the equation so that the left side has no constant term.

$$m^2 - 7m - 1 = 0$$

$$m^2 - 7m = 1$$

Now find the constant term that makes the left side a perfect square trinomial by squaring half the coefficient of m. Add this constant to both sides of the equation.

$$\frac{1}{2}(-7) = -\frac{7}{2}$$

and

$$\left(-\frac{7}{2}\right)^2 = \frac{49}{4}$$

$$m^2 - 7m + \frac{49}{4} = 1 + \frac{49}{4} \qquad \text{Add } \frac{49}{4} \text{ to both sides of the equation.}$$

$$\left(m - \frac{7}{2}\right)^2 = \frac{53}{4} \qquad \text{Factor the perfect square trinomial and simplify the right side.}$$

$$m - \frac{7}{2} = \pm\sqrt{\frac{53}{4}} \qquad \text{Apply the square root property.}$$

$$m = \frac{7}{2} \pm \frac{\sqrt{53}}{2} \qquad \text{Add } \frac{7}{2} \text{ to both sides and simplify } \sqrt{\frac{53}{4}}.$$

$$m = \frac{7 \pm \sqrt{53}}{2} \qquad \text{Simplify.}$$

The solutions are $\dfrac{7 + \sqrt{53}}{2}$ and $\dfrac{7 - \sqrt{53}}{2}$.

Graphical Solution:

Enter $y_1 = x^2 - 7x - 1$ and find the x-intercepts of the graph, or the zeros by using the zero option on the Calc menu.

Rounded to the nearest hundredth we have x (or m) ≈ -0.14 and x (or m) ≈ 7.14. These are approximations of the exact solutions as shown below.

EXAMPLE 7

Solve $2x^2 - 8x + 3 = 0$.

Solution   Our procedure for finding the constant term to complete the square works only if the coefficient of the squared variable term is 1. Therefore, to solve this equation, the first step is to divide both sides by 2, the coefficient of x^2.

$$2x^2 - 8x + 3 = 0$$

$$x^2 - 4x + \frac{3}{2} = 0 \qquad \text{Divide both sides by 2.}$$

$$x^2 - 4x = -\frac{3}{2} \qquad \text{Subtract } \frac{3}{2} \text{ from both sides.}$$

Next find the square of half of -4.

$$\frac{1}{2}(-4) = -2 \quad \text{and} \quad (-2)^2 = 4$$

Add 4 to both sides of the equation to complete the square.

$$x^2 - 4x + 4 = -\frac{3}{2} + 4$$

$$(x - 2)^2 = \frac{5}{2} \qquad \text{Factor the perfect square and simplify the right side.}$$

$$x - 2 = \pm\sqrt{\frac{5}{2}} \qquad \text{Apply the square root property.}$$

$$x - 2 = \pm\frac{\sqrt{10}}{2} \qquad \text{Rationalize the denominator.}$$

$$x = 2 \pm \frac{\sqrt{10}}{2} \qquad \text{Add 2 to both sides.}$$

$$= \frac{4}{2} \pm \frac{\sqrt{10}}{2} \qquad \text{Find the common denominator.}$$

$$= \frac{4 \pm \sqrt{10}}{2} \qquad \text{Simplify.}$$

The solutions are $\dfrac{4 + \sqrt{10}}{2}$ and $\dfrac{4 - \sqrt{10}}{2}$.

The following steps may be used to solve a quadratic equation such as $ax^2 + bx + c = 0$ by completing the square. This method may be used whether or not the polynomial $ax^2 + bx + c$ is factorable.

Solving a Quadratic Equation in x By Completing the Square

Step 1: If the coefficient of x^2 is 1, go to Step 2. Otherwise, divide both sides of the equation by the coefficient of x^2.

Step 2: Isolate all variable terms on one side of the equation.

Step 3: Complete the square for the resulting binomial by adding the square of half of the coefficient of x to both sides of the equation.

Step 4: Factor the resulting perfect square trinomial and write it as the square of a binomial.

Step 5: Use the square root property to solve for x.

EXAMPLE 8

Solve $3x^2 - 9x + 8 = 0$ by completing the square.

Solution $3x^2 - 9x + 8 = 0$

CLASSROOM EXAMPLE

Solve $2x^2 - 2x + 7 = 0$.

answer: $\dfrac{1 \pm i\sqrt{13}}{2}$

Step 1: $x^2 - 3x + \dfrac{8}{3} = 0$ Divide both sides of the equation by 3.

Step 2: $x^2 - 3x = -\dfrac{8}{3}$ Subtract $\dfrac{8}{3}$ from both sides.

Since $\dfrac{1}{2}(-3) = -\dfrac{3}{2}$ and $\left(-\dfrac{3}{2}\right)^2 = \dfrac{9}{4}$, we add $\dfrac{9}{4}$ to both sides of the equation.

Step 3: $x^2 - 3x + \dfrac{9}{4} = -\dfrac{8}{3} + \dfrac{9}{4}$

TEACHING TIP

Remind students that the calculator must be in $a + bi$ mode to check solutions that are complex, but not real numbers.

Step 4: $\left(x - \dfrac{3}{2}\right)^2 = -\dfrac{5}{12}$ Factor the perfect square trinomial.

Step 5: $x - \dfrac{3}{2} = \pm\sqrt{-\dfrac{5}{12}}$ Apply the square root property.

$x - \dfrac{3}{2} = \pm\dfrac{i\sqrt{5}}{2\sqrt{3}}$ Simplify the radical.

$x - \dfrac{3}{2} = \pm\dfrac{i\sqrt{15}}{6}$ Rationalize the denominator.

$x = \dfrac{3}{2} \pm \dfrac{i\sqrt{15}}{6}$ Add $\dfrac{3}{2}$ to both sides.

$= \dfrac{9}{6} \pm \dfrac{i\sqrt{15}}{6}$ Find a common denominator.

$= \dfrac{9 \pm i\sqrt{15}}{6}$ Simplify.

TEACHING TIP

After discussing Example 8, have students work in pairs to describe when an equation of the form $ax^2 + bx + c = 0$ will have real solutions if a is positive. To help them get started ask, "Why will the solution always be real when $c \leq 0$?" Then ask, "Are the solutions always complex when $c > 0$? Explain."

The solutions are $\dfrac{9 + i\sqrt{15}}{6}$ and $\dfrac{9 - i\sqrt{15}}{6}$.

3   Recall the **simple interest** formula $I = Prt$, where I is the interest earned, P is the principal, r is the rate of interest, and t is time. If \$100 is invested at a simple interest rate of 5% annually, at the end of 3 years the total interest I earned is

$$I = P \cdot r \cdot t$$

or

$$I = 100 \cdot 0.05 \cdot 3 = \$15$$

and the new principal is

$$\$100 + \$15 = \$115$$

Most of the time, the interest computed on money borrowed or money deposited is **compound interest.** Compound interest, unlike simple interest, is computed on original principal *and* on interest already earned. To see the difference between simple interest and compound interest, suppose that \$100 is invested at a rate of 5% compounded annually. To find the total amount of money at the end of 3 years, we calculate as follows.

$$I = P \cdot r \cdot t$$

First year: Interest $= \$100 \cdot 0.05 \cdot 1 = \$5.00$
New principal $= \$100.00 + \$5.00 = \$105.00$

Second year: Interest $= \$105.00 \cdot 0.05 \cdot 1 = \$5.25$
New principal $= \$105.00 + \$5.25 = \$110.25$

Third year: Interest $= \$110.25 \cdot 0.05 \cdot 1 \approx \$5.51$
New principal $= \$110.25 + \$5.51 = \$115.76$

At the end of the third year, the total compound interest earned is \$15.76, whereas the total simple interest earned is \$15.

It is tedious to calculate compound interest as we did above, so we use a compound interest formula. The formula for calculating the total amount of money when interest is compounded annually is

$$A = P(1 + r)^t$$

where P is the original investment, r is the interest rate per compounding period, and t is the number of periods. For example, the amount of money A at the end of 3 years if \$100 is invested at 5% compounded annually is

$$A = \$100(1 + 0.05)^3 \approx \$100(1.1576) = \$115.76$$

as we previously calculated.

EXAMPLE 9

FINDING INTEREST RATES

Find the interest rate r if \$2000 compounded annually grows to \$2420 in 2 years.

Solution   1. UNDERSTAND the problem. Since the \$2000 is compounded annually, we use the compound interest formula. For this example, make sure that you understand the formula for compounding interest annually.

2. TRANSLATE. We substitute the given values into the formula.

$$A = P(1 + r)^t$$

$$2420 = 2000(1 + r)^2 \qquad \text{Let } A = 2420, P = 2000, \text{ and } t = 2.$$

3. SOLVE. Solve the equation for r. We will solve algebraically and graphically.

Algebraic Solution:

$$2420 = 2000(1 + r)^2$$

$$\frac{2420}{2000} = (1 + r)^2 \qquad \text{Divide both sides by 2000.}$$

$$\frac{121}{100} = (1 + r)^2 \qquad \text{Simplify the fraction.}$$

$$\pm\sqrt{\frac{121}{100}} = 1 + r \qquad \text{Use the square root property.}$$

$$\pm\frac{11}{10} = 1 + r \qquad \text{Simplify.}$$

$$-1 \pm \frac{11}{10} = r$$

$$-\frac{10}{10} \pm \frac{11}{10} = r$$

$$\frac{1}{10} = r \quad \text{or} \quad -\frac{21}{10} = r$$

Graphical Solution:

Enter $y_1 = 2000(1 + x)^2$ and $y_2 = 2420$. You may need to experiment a little before you find an appropriate window, but keep the following in mind. Since $y_2 = 2420$, use a window for the y-axis that includes this number. Since x represents the rate, use a much smaller window for the x-axis. The viewing window for the graph below is $[-2.5, 1, 0.1]$ for x and $[2000, 3000, 200]$ for y.

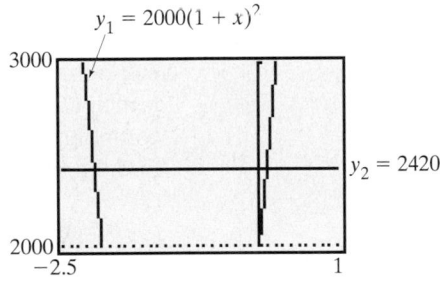

The two points of intersection have x-coordinates of -2.1 and 0.1.

4. INTERPRET. The rate cannot be negative, so we reject $-\frac{21}{10}$.

Check: $\frac{1}{10} = 0.10 = 10\%$ per year. If we invest \$2000 at 10% compounded annually, in 2 years the amount in the account would be $2000(1 + 0.10)^2 = 2420$ dollars, the desired amount.

State: The interest rate is 10% compounded annually.

13. $\dfrac{3 - 2\sqrt{2}}{2}, \dfrac{3 + 2\sqrt{2}}{2}$ **14.** $\dfrac{-9 - \sqrt{6}}{4}, \dfrac{-9 + \sqrt{6}}{4}$

EXERCISE SET 8.1

STUDY GUIDE/SSM   CD/VIDEO   PH MATH TUTOR CENTER   MathXL®Tutorials ON CD   MathXL®   MyMathLab®

20. $-2i\sqrt{3}, 2i\sqrt{3}$

Use the square root property to solve each equation. These equations have real-number solutions. Check each solution graphically. See Examples 1 through 3.   **4.** $-\sqrt{11}, \sqrt{11}$ **7.** $-\sqrt{10}, \sqrt{10}$

 1. $x^2 = 16$ $-4, 4$

2. $x^2 = 49$ $-7, 7$

3. $x^2 - 7 = 0$ $-\sqrt{7}, \sqrt{7}$

4. $x^2 - 11 = 0$

5. $x^2 = 18$ $-3\sqrt{2}, 3\sqrt{2}$

6. $y^2 = 20$ $-2\sqrt{5}, 2\sqrt{5}$

7. $3z^2 - 30 = 0$

8. $2x^2 = 4$ $-\sqrt{2}, \sqrt{2}$

9. $(x + 5)^2 = 9$ $-8, -2$

10. $(y - 3)^2 = 4$ $1, 5$

11. $(z - 6)^2 = 18$

12. $(y + 4)^2 = 27$

13. $(2x - 3)^2 = 8$

14. $(4x + 9)^2 = 6$

11. $6 - 3\sqrt{2}, 6 + 3\sqrt{2}$ **12.** $-4 - 3\sqrt{3}, -4 + 3\sqrt{3}$

Use the square root property to solve each equation. Check each solution numerically. See Examples 1 through 4.   **18.** $-\sqrt{10}, \sqrt{10}$

19. $-2i\sqrt{2}, 2i\sqrt{2}$ **21.** $1 - 4i, 1 + 4i$ **22.** $-2 - 5i, -2 + 5i$

15. $x^2 + 9 = 0$ $-3i, 3i$

16. $x^2 + 4 = 0$ $-2i, 2i$

17. $x^2 - 6 = 0$ $-\sqrt{6}, \sqrt{6}$

18. $y^2 - 10 = 0$

19. $2z^2 + 16 = 0$

20. $3p^2 + 36 = 0$

21. $(x - 1)^2 = -16$

22. $(y + 2)^2 = -25$

23. $(z + 7)^2 = 5$

24. $(x + 10)^2 = 11$

25. $(x + 3)^2 = -8$

26. $(y - 4)^2 = -18$

23. $-7 - \sqrt{5}, -7 + \sqrt{5}$ **24.** $-10 - \sqrt{11}, -10 + \sqrt{11}$

25. $-3 - 2i\sqrt{2}, -3 + 2i\sqrt{2}$ **26.** $4 - 3i\sqrt{2}, 4 + 3i\sqrt{2}$

Add the proper constant to each binomial so that the resulting tri-nomial is a perfect square trinomial. Then factor the trinomial.

27. $x^2 + 16x +$ ____

28. $y^2 + 2y +$ ____

29. $z^2 - 12z +$ ____

30. $x^2 - 8x +$ ____

31. $p^2 + 9p +$ ____

32. $n^2 + 5n +$ ____

33. $x^2 + x +$ ____

34. $y^2 - y +$ ____

27. $x^2 + 16x + 64 = (x + 8)^2$ **28.** $y^2 + 2y + 1 = (y + 1)^2$

29. $z^2 - 12z + 36 = (z - 6)^2$ **30.** $x^2 - 8x + 16 = (x - 4)^2$

MIXED PRACTICE

Solve each equation by completing the square. These equations have real number solutions. Check each solution graphically. See Examples 5 through 7.

35. $x^2 + 8x = -15$ $-5, -3$

36. $y^2 + 6y = -8$ $-4, -2$

37. $x^2 + 6x + 2 = 0$
$-3 - \sqrt{7}, -3 + \sqrt{7}$

38. $x^2 - 2x - 2 = 0$
$1 - \sqrt{3}, 1 + \sqrt{3}$

39. $x^2 + x - 1 = 0$
$\frac{-1 - \sqrt{5}}{2}, \frac{-1 + \sqrt{5}}{2}$

40. $x^2 + 3x - 2 = 0$
$\frac{-3 - \sqrt{17}}{2}, \frac{-3 + \sqrt{17}}{2}$

41. $x^2 + 2x - 5 = 0$
$-1 - \sqrt{6}, -1 + \sqrt{6}$

42. $y^2 + y - 7 = 0$
$\frac{-1 - \sqrt{29}}{2}, \frac{-1 + \sqrt{29}}{2}$

43. $3p^2 - 12p + 2 = 0$
$\frac{6 - \sqrt{30}}{3}, \frac{6 + \sqrt{30}}{3}$

44. $2x^2 + 14x - 1 = 0$
$\frac{-7 - \sqrt{51}}{2}, \frac{-7 + \sqrt{51}}{2}$

45. $4y^2 - 12y - 2 = 0$
$\frac{3 - \sqrt{11}}{2}, \frac{3 + \sqrt{11}}{2}$

46. $6x^2 - 3 = 6x$
$\frac{1 - \sqrt{3}}{2}, \frac{1 + \sqrt{3}}{2}$

47. $2x^2 + 7x = 4$
$-4, \frac{1}{2}$

48. $3x^2 - 4x = 4$
$-\frac{2}{3}, 2$

49. $x^2 - 4x - 5 = 0$
$-1, 5$

50. $y^2 + 6y - 8 = 0$
$-3 - \sqrt{17}, -3 + \sqrt{17}$

51. $x^2 + 8x + 1 = 0$
$-4 - \sqrt{15}, -4 + \sqrt{15}$

52. $x^2 - 10x + 2 = 0$
$5 - \sqrt{23}, 5 + \sqrt{23}$

53. $3y^2 + 6y - 4 = 0$
$\frac{-3 - \sqrt{21}}{3}, \frac{-3 + \sqrt{21}}{3}$

54. $2y^2 + 12y + 3 = 0$
$\frac{-6 - \sqrt{30}}{2}, \frac{-6 + \sqrt{30}}{2}$

55. $2x^2 - 3x - 5 = 0$
$-1, \frac{5}{2}$

56. $5x^2 + 3x - 2 = 0$
$-1, \frac{2}{5}$

Solve each equation by completing the square. Check numerically or graphically. See Examples 5 through 8.

57. $y^2 + 2y + 2 = 0$
$-1 - i, -1 + i$

58. $x^2 + 4x + 6 = 0$
$-2 - i\sqrt{2}, -2 + i\sqrt{2}$

59. $x^2 - 6x + 3 = 0$
$3 - \sqrt{6}, 3 + \sqrt{6}$

60. $x^2 - 7x - 1 = 0$
$\frac{7 - \sqrt{53}}{2}, \frac{7 + \sqrt{53}}{2}$

61. $2a^2 + 8a = -12$
$-2 - i\sqrt{2}, -2 + i\sqrt{2}$

62. $3x^2 + 12x = -14$
$\frac{-6 - i\sqrt{6}}{3}, \frac{-6 + i\sqrt{6}}{3}$

63. $5x^2 + 15x - 1 = 0$
$\frac{-15 - 7\sqrt{5}}{10}, \frac{-15 + 7\sqrt{5}}{10}$

64. $16y^2 + 16y - 1 = 0$
$\frac{-2 - \sqrt{5}}{4}, \frac{-2 + \sqrt{5}}{4}$

65. $2x^2 - x + 6 = 0$
$\frac{1 - i\sqrt{47}}{4}, \frac{1 + i\sqrt{47}}{4}$

66. $4x^2 - 2x + 5 = 0$
$\frac{1 - i\sqrt{19}}{4}, \frac{1 + i\sqrt{19}}{4}$

67. $x^2 + 10x + 28 = 0$
$-5 - i\sqrt{3}, -5 + i\sqrt{3}$

68. $y^2 + 8y + 18 = 0$
$-4 - i\sqrt{2}, -4 + i\sqrt{2}$

69. $z^2 + 3z - 4 = 0$
$-4, 1$

70. $y^2 + y - 2 = 0$
$-2, 1$

71. $2x^2 - 4x + 3 = 0$
$\frac{2 - i\sqrt{2}}{2}, \frac{2 + i\sqrt{2}}{2}$

72. $9x^2 - 36x = -40$
$\frac{6 - 2i}{3}, \frac{6 + 2i}{3}$

73. $3x^2 + 3x = 5$
$\frac{-3 - \sqrt{69}}{6}, \frac{-3 + \sqrt{69}}{6}$

74. $5y^2 - 15y = 1$
$\frac{15 - 7\sqrt{5}}{10}, \frac{15 + 7\sqrt{5}}{10}$

Use the graph to determine how many real-number solutions exist for each equation.

75. $2x^2 - 3x - 5 = 0$
$y = 2x^2 - 3x - 5$

76. $5x^2 + 3x - 2 = 0$
$y = 5x^2 + 3x - 2$

2 real number solutions

2 real number solutions

77. $x^2 + 2x + 2 = 0$
$y = x^2 + 2x + 2$

78. $x^2 + 4x + 6 = 0$
$y = x^2 + 4x + 6$

no real number solutions

no real number solutions

Use the formula $A = P(1 + r)^t$ to solve Exercises 79–82. See Example 9.

79. Find the rate r at which \$3000 grows to \$4320 in 2 years.   20%

80. Find the rate r at which \$800 grows to \$882 in 2 years.   5%

81. Find the rate at which \$810 grows to \$1000 in 2 years. $11\frac{1}{9}$%

82. Find the rate at which \$2000 grows to \$2880 in 2 years.   20%

83. In your own words, what is the difference between simple interest and compound interest?   answers may vary

84. If you are depositing money in an account that pays 4%, would you prefer the interest to be simple or compound? Explain why.   compound

85. If you are borrowing money at a rate of 10%, would you prefer the interest to be simple or compound? Explain why.   simple

31. $p^2 + 9p + \frac{81}{4} = \left(p + \frac{9}{2}\right)^2$ **32.** $n^2 + 5n + \frac{25}{4} = \left(n + \frac{5}{2}\right)^2$

REVIEW AND PREVIEW

33. $x^2 + x + \frac{1}{4} = \left(x + \frac{1}{2}\right)^2$

Simplify each expression. See Section 7.1.

86. $\frac{3}{4} - \sqrt{\frac{25}{16}}$ $-\frac{1}{2}$

87. $\frac{3}{5} + \sqrt{\frac{16}{25}}$ $\frac{7}{5}$

88. $\frac{1}{2} - \sqrt{\frac{9}{4}}$ -1

89. $\frac{9}{10} - \sqrt{\frac{49}{100}}$ $\frac{1}{5}$

34. $y^2 - y + \frac{1}{4} = \left(y - \frac{1}{2}\right)^2$

Simplify each expression. See Section 7.5.

90. $\dfrac{6 + 4\sqrt{5}}{2}$ $3 + 2\sqrt{5}$ **91.** $\dfrac{10 - 20\sqrt{3}}{2}$ $5 - 10\sqrt{3}$

92. $\dfrac{3 - 9\sqrt{5}}{6}$ $\frac{1\,-\,3\sqrt{2}}{2}$ **93.** $\dfrac{12 - 8\sqrt{7}}{16}$ $\frac{3\,-\,2\sqrt{7}}{4}$

Evaluate $\sqrt{b^2 - 4ac}$ for each set of values. See Section 7.3.

94. $a = 2, b = 4, c = -1$ **95.** $a = 1, b = 6, c = 2$
$2\sqrt{6}$ $2\sqrt{7}$

96. $a = 3, b = -1, c = -2$ **97.** $a = 1, b = -3, c = -1$
5 $\sqrt{13}$

Concept Extensions

Find two possible missing terms so that each is a perfect square trinomial.

98. $x^2 + \quad + 16$ $-8x, 8x$ **99.** $y^2 + \quad + 9$ $-6y, 6y$

100. $z^2 + \quad + \dfrac{25}{4}$ $-5z, 5z$ **101.** $x^2 + \quad + \dfrac{1}{4}$ $-x, x$

Neglecting air resistance, the distance $s(t)$ in feet traveled by a freely falling object is given by the function $s(t) = 16t^2$, where t is time in seconds. Use this formula to solve Exercises 102 through 105. Round answers to two decimal places.

102. The Petronas Towers in Kuala Lumpur, built in 1997, are the tallest buildings in Malaysia. Each tower is 1483 feet tall. How long would it take an object to fall to the ground from the top of one of the towers? (*Source:* Council on Tall Buildings and Urban Habitat, Lehigh University)   9.63 sec

103. The height of the Chicago Beach Tower Hotel, built in 1998 in Dubai, United Arab Emirates, is 1053 feet. How long would it take an object to fall to the ground from the top of the building? (*Source:* Council on Tall Buildings and Urban Habitat, Lehigh University)   8.11 sec

104. The height of the Nurek Dam in Tajikistan (part of the former USSR that borders Afghanistan) is 984 feet. How long would it take an object to fall from the top to the base of the dam? (*Source:* U.S. Committee on Large Dams of the International Commission on Large Dams)   7.84 sec

105. The Hoover Dam, located on the Colorado River on the border of Nevada and Arizona near Las Vegas, is 725 feet tall. How long would it take an object to fall from the top to the

base of the dam? (*Source:* U.S. Committee on Large Dams of the International Commission on Large Dams)   6.73 sec

Solve.

△ **106.** The area of a square room is 225 square feet. Find the dimensions of the room.   15 ft by 15 ft

△ **107.** The area of a circle is 36π square inches. Find the radius of the circle.   6 in.

△ **108.** An isosceles right triangle has legs of equal length. If the hypotenuse is 20 centimeters long, find the length of each leg. $10\sqrt{2}$ cm

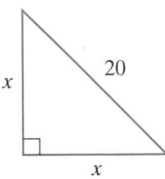

△ **109.** A 27-inch TV is advertised in the *Daily Sentry* newspaper. If 27 inches is the measure of the diagonal of the picture tube, find the measure of each side of the picture tube.   16.2 in. by 21.6 in.

A common equation used in business is a demand equation. It expresses the relationship between the unit price of some commodity and the quantity demanded. For Exercises 110 and 111, p represents the unit price and x represents the quantity demanded in thousands.

110. A manufacturing company has found that the demand equation for a certain type of scissors is given by the equation $p = -x^2 + 47$. Find the demand for the scissors if the price is $11 per pair.   6 thousand scissors

111. Acme, Inc., sells desk lamps and has found that the demand equation for a certain style of desk lamp is given by the equation $p = -x^2 + 15$. Find the demand for the desk lamp if the price is $7 per lamp.   2.828 thousand units or 2828 units

8.2   *SOLVING QUADRATIC EQUATIONS BY THE QUADRATIC FORMULA*

Objectives

1   Solve quadratic equations by using the quadratic formula.

2   Determine the number and type of solutions of a quadratic equation by using the discriminant.

3   Solve geometric problems modeled by quadratic equations.

1   Any quadratic equation can be solved by completing the square. Since the same sequence of steps is repeated each time we complete the square, let's complete the square for a general quadratic equation, $ax^2 + bx + c = 0, a \neq 0$. By doing so, we find a pattern for the solutions of a quadratic equation known as the **quadratic formula.**

Recall that to complete the square for an equation such as $ax^2 + bx + c = 0$, we first divide both sides by the coefficient of x^2.

$$ax^2 + bx + c = 0$$

$$x^2 + \frac{b}{a}x + \frac{c}{a} = 0 \qquad \text{Divide both sides by } a, \text{ the coefficient of } x^2.$$

$$x^2 + \frac{b}{a}x = -\frac{c}{a} \qquad \text{Subtract the constant } \frac{c}{a} \text{ from both sides.}$$

Next, find the square of half $\frac{b}{a}$, the coefficient of x.

$$\frac{1}{2}\left(\frac{b}{a}\right) = \frac{b}{2a} \quad \text{and} \quad \left(\frac{b}{2a}\right)^2 = \frac{b^2}{4a^2}$$

Add this result to both sides of the equation.

$$x^2 + \frac{b}{a}x + \frac{b^2}{4a^2} = -\frac{c}{a} + \frac{b^2}{4a^2} \qquad \text{Add } \frac{b^2}{4a^2} \text{ to both sides.}$$

$$x^2 + \frac{b}{a}x + \frac{b^2}{4a^2} = \frac{-c \cdot 4a}{a \cdot 4a} + \frac{b^2}{4a^2} \qquad \text{Find a common denominator on the right side.}$$

$$x^2 + \frac{b}{a}x + \frac{b^2}{4a^2} = \frac{b^2 - 4ac}{4a^2} \qquad \text{Simplify the right side.}$$

$$\left(x + \frac{b}{2a}\right)^2 = \frac{b^2 - 4ac}{4a^2} \qquad \text{Factor the perfect square trinomial on the left side.}$$

$$x + \frac{b}{2a} = \pm\sqrt{\frac{b^2 - 4ac}{4a^2}} \qquad \text{Apply the square root property.}$$

$$x + \frac{b}{2a} = \pm\frac{\sqrt{b^2 - 4ac}}{2a} \qquad \text{Simplify the radical.}$$

$$x = -\frac{b}{2a} \pm \frac{\sqrt{b^2 - 4ac}}{2a} \qquad \text{Subtract } \frac{b}{2a} \text{ from both sides.}$$

$$x = \frac{-b \pm \sqrt{b^2 - 4ac}}{2a} \qquad \text{Simplify.}$$

This equation identifies the solutions of the general quadratic equation in standard form and is called the quadratic formula. It can be used to solve any equation written in standard form $ax^2 + bx + c = 0$ as long as a is not 0.

TECHNOLOGY NOTE

When evaluating the quadratic formula using a calculator, it is sometimes more convenient to evaluate the radicand separately. This prevents incorrect placement of parentheses.

Quadratic Formula

A quadratic equation written in the form $ax^2 + bx + c = 0$ has the solutions

$$x = \frac{-b \pm \sqrt{b^2 - 4ac}}{2a}$$

EXAMPLE 1

Solve $3x^2 + 16x + 5 = 0$ for x.

CLASSROOM EXAMPLE

Solve $2x^2 + 9x + 10 = 0$.

answer: $-\dfrac{5}{2}, -2$

Solution This equation is in standard form, so $a = 3$, $b = 16$, and $c = 5$. Substitute these values into the quadratic formula.

$$x = \frac{-b \pm \sqrt{b^2 - 4ac}}{2a} \qquad \text{Quadratic formula.}$$

$$= \frac{-16 \pm \sqrt{16^2 - 4(3)(5)}}{2 \cdot 3} \qquad \text{Use } a = 3, b = 16, \text{ and } c = 5.$$

$$= \frac{-16 \pm \sqrt{256 - 60}}{6}$$

$$= \frac{-16 \pm \sqrt{196}}{6} = \frac{-16 \pm 14}{6}$$

$$x = \frac{-16 + 14}{6} = -\frac{1}{3} \quad \text{or} \quad x = \frac{-16 - 14}{6} = -\frac{30}{6} = -5$$

TEACHING TIP

Have students work in pairs to solve the equation $2x^2 + 7x + 5 = 0$ using factoring, completing the square, and the quadratic formula to illustrate that the solutions are the same no matter which method is used. Have them show each step. Then have them verify that each solution is correct.

The solutions are $-\dfrac{1}{3}$ and -5, or the solution set is $\left\{-\frac{1}{3}, -5\right\}$.

A numerical check is shown to the left. Notice the correct placement of parentheses.

As usual, another way to check the solution of an equation is to use a graphical method such as the intersection-of-graphs method. Remember that this graphical method shows real number solutions only.

Solving Example 1 numerically by evaluating the formula using the calculator.

EXAMPLE 2

Solve $2x^2 - 4x = 3$.

Solution First write the equation in standard form by subtracting 3 from both sides.

$$2x^2 - 4x - 3 = 0$$

> **Helpful Hint**
>
> To replace a, b, and c correctly in the quadratic formula, write the quadratic equation in standard form $ax^2 + bx + c = 0$.

Now $a = 2$, $b = -4$, and $c = -3$. Substitute these values into the quadratic formula.

$$x = \frac{-b \pm \sqrt{b^2 - 4ac}}{2a}$$

$$= \frac{-(-4) \pm \sqrt{(-4)^2 - 4(2)(-3)}}{2 \cdot 2}$$

$$= \frac{4 \pm \sqrt{16 + 24}}{4}$$

$$= \frac{4 \pm \sqrt{40}}{4} = \frac{4 \pm 2\sqrt{10}}{4}$$

$$= \frac{2(2 \pm \sqrt{10})}{2 \cdot 2} = \frac{2 \pm \sqrt{10}}{2}$$

The solutions are $\dfrac{2 + \sqrt{10}}{2}$ and $\dfrac{2 - \sqrt{10}}{2}$, or the solution set is $\left\{ \dfrac{2 - \sqrt{10}}{2}, \dfrac{2 + \sqrt{10}}{2} \right\}$.

Since the solutions are real numbers, let's check using the intersection-of-graphs method. To do so, graph $y_1 = 2x^2 - 4x$ and $y_2 = 3$. Find the approximate points of intersection as shown next, and see that the exact solutions have the same approximations.

The screen to the right above shows that the exact solutions using the quadratic formula have the same approximations as those found by the intersection of graphs method.

> **Helpful Hint**
>
> To simplify the expression $\dfrac{4 \pm 2\sqrt{10}}{4}$ in the preceding example, note that 2 is factored out of both terms of the numerator *before* simplifying.
>
> $$\frac{4 \pm 2\sqrt{10}}{4} = \frac{2(2 \pm \sqrt{10})}{2 \cdot 2} = \frac{2 \pm \sqrt{10}}{2}$$

✔ **CONCEPT CHECK**

For the quadratic equation $x^2 = 7$, which substitution is correct?

   a. $a = 1$, $b = 0$, and $c = -7$

   b. $a = 1$, $b = 0$, and $c = 7$

   c. $a = 0$, $b = 0$, and $c = 7$

   d. $a = 1$, $b = 1$, and $c = -7$

EXAMPLE 3

Solve $\frac{1}{4}m^2 - m + \frac{1}{2} = 0$.

Solution We could use the quadratic formula with $a = \frac{1}{4}$, $b = -1$, and $c = \frac{1}{2}$. Instead, we find a simpler, equivalent standard form equation whose coefficients are not fractions.

Multiply both sides of the equation by 4 to clear fractions.

$$4\left(\frac{1}{4}m^2 - m + \frac{1}{2}\right) = 4 \cdot 0$$

$$m^2 - 4m + 2 = 0 \qquad \text{Simplify.}$$

Substitute $a = 1$, $b = -4$, and $c = 2$ into the quadratic formula and simplify.

$$m = \frac{-(-4) \pm \sqrt{(-4)^2 - 4(1)(2)}}{2 \cdot 1} = \frac{4 \pm \sqrt{16 - 8}}{2}$$

$$= \frac{4 \pm \sqrt{8}}{2} = \frac{4 \pm 2\sqrt{2}}{2} = \frac{2\left(2 \pm \sqrt{2}\right)}{2}$$

$$= 2 \pm \sqrt{2}$$

The solutions are $2 + \sqrt{2}$ and $2 - \sqrt{2}$. Check these real number solutions numerically or graphically.

EXAMPLE 4

Solve $x = -3x^2 - 3$.

Solution The equation in standard form is $3x^2 + x + 3 = 0$. Thus, let $a = 3$, $b = 1$, and $c = 3$ in the quadratic formula.

$$x = \frac{-1 \pm \sqrt{1^2 - 4(3)(3)}}{2 \cdot 3} = \frac{-1 \pm \sqrt{1 - 36}}{6} = \frac{-1 \pm \sqrt{-35}}{6} = \frac{-1 \pm i\sqrt{35}}{6}$$

The solutions are $\frac{-1 + i\sqrt{35}}{6}$ and $\frac{-1 - i\sqrt{35}}{6}$.

Since the solutions are not real numbers, we check numerically as shown below.

✔ CONCEPT CHECK

What is the first step in solving $-3x^2 = 5x - 4$ using the quadratic formula?

In Example 1, the equation $3x^2 + 16x + 5 = 0$ had 2 real roots, $-\dfrac{1}{3}$ and -5. In Example 4, the equation $3x^2 + x + 3 = 0$ (written in standard form) had no real roots. How do their related graphs compare? Recall that the x-intercepts of $f(x) = 3x^2 + 16x + 5$ occur where $f(x) = 0$ or where $3x^2 + 16x + 5 = 0$. Since this equation has 2 real roots, the graph has 2 x-intercepts. Similarly, since the equation $3x^2 + x + 3 = 0$ has no real roots, the graph of $f(x) = 3x^2 + x + 3$ has no x-intercepts.

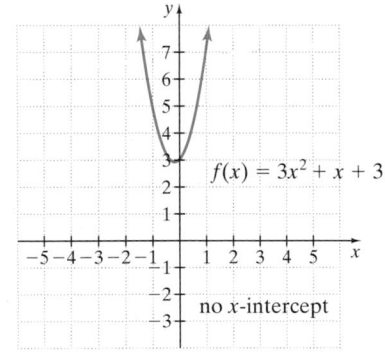

2 In the quadratic formula, $x = \dfrac{-b \pm \sqrt{b^2 - 4ac}}{2a}$, the radicand $b^2 - 4ac$ is called the **discriminant** because, by knowing its value, we can **discriminate** among the possible number and type of solutions of a quadratic equation. Possible values of the discriminant and their meanings are summarized next.

Discriminant

The following table corresponds the discriminant $b^2 - 4ac$ of a quadratic equation of the form $ax^2 + bx + c = 0$ with the number and type of solutions of the equation.

$b^2 - 4ac$	Number and Type of Solutions
Positive	Two real solutions
Zero	One real solution
Negative	Two complex but not real solutions

To see the results of the discriminant graphically, study the screens below showing examples of graphs of $y = ax^2 + bx + c$.

$b^2 - 4ac$ is positive
Two x-intercepts
Two real solutions

$b^2 - 4ac = 0$
One x-intercept
One real solution
(double root)

$b^2 - 4ac$ is negative
No x-intercept
Two complex, but not real solutions

EXAMPLE 5

Use the discriminant to determine the number and type of solutions of each quadratic equation.

a. $x^2 + 2x + 1 = 0$ **b.** $3x^2 + 2 = 0$ **c.** $2x^2 - 7x - 4 = 0$

Solution

a. In $x^2 + 2x + 1 = 0$, $a = 1$, $b = 2$, and $c = 1$. Thus,

$$b^2 - 4ac = 2^2 - 4(1)(1) = 0$$

Since $b^2 - 4ac = 0$, this quadratic equation has one real solution.

b. In this equation, $a = 3$, $b = 0$, $c = 2$. Then $b^2 - 4ac = 0 - 4(3)(2) = -24$. Since $b^2 - 4ac$ is negative, the quadratic equation has two complex but not real solutions.

c. In this equation, $a = 2$, $b = -7$, and $c = -4$. Then

$$b^2 - 4ac = (-7)^2 - 4(2)(-4) = 81$$

Since $b^2 - 4ac$ is positive, the quadratic equation has two real solutions.

To confirm graphically, see the results below.

a.

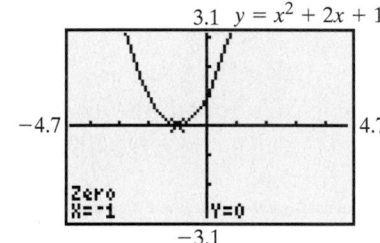

One x-intercept indicates one real solution.

b.

No x-intercepts indicate no real solutions, but two complex, not real solutions.

c.

Two x-intercepts indicate two real solutions.

3   The quadratic formula is useful in solving problems that are modeled by quadratic equations.

⚠ **EXAMPLE 6**

CALCULATING DISTANCE SAVED

At a local university, students often leave the sidewalk and cut across the lawn to save walking distance. Given the diagram below of a favorite place to cut across the lawn, approximate how many feet of walking distance a student saves by cutting across the lawn instead of walking on the sidewalk.

Solution

CLASSROOM EXAMPLE

Given the diagram below, approximate to the nearest foot how many feet of walking distance a person saves by cutting across the lawn instead of walking on the sidewalk.

answer: 8 feet

1. **UNDERSTAND.** Read and reread the problem. In the diagram, notice that a triangle is formed. Since the corner of the block forms a right angle, we use the Pythagorean theorem for right triangles. You may want to review this theorem.

2. **TRANSLATE.** By the Pythagorean theorem, we have

   In words: $(\text{leg})^2 + (\text{leg})^2 = (\text{hypotenuse})^2$

   Translate: $x^2 + (x + 20)^2 = 50^2$

3. **SOLVE.**

Algebraic Solution:

$x^2 + x^2 + 40x + 400 = 2500$ Square $(x + 20)$ and 50.

$2x^2 + 40x - 2100 = 0$ Set the equation equal to 0.

$x^2 + 20x - 1050 = 0$ Divide by 2.

Here, $a = 1, b = 20, c = -1050$. By the quadratic formula,

$$x = \frac{-20 \pm \sqrt{20^2 - 4(1)(-1050)}}{2 \cdot 1}$$

$$= \frac{-20 \pm \sqrt{400 + 4200}}{2} = \frac{-20 \pm \sqrt{4600}}{2}$$

$$= \frac{-20 \pm \sqrt{100 \cdot 46}}{2} = \frac{-20 \pm 10\sqrt{46}}{2}$$

$$= -10 \pm 5\sqrt{46}$$ Simplify.

Graphical Solution:

Let $y_1 = x^2 + (x + 20)^2$ and $y_2 = 50^2$. Since x represents feet, we are interested in positive solutions only. We also estimate the length labeled x to be less than 50, so we choose the x-window to be $[0, 50, 10]$. The y-window is chosen to be $[2000, 3000, 100]$ so it contains 2500. Find $x \approx -44$ and $x \approx 24$.

The positive solution is approximately 24.

4. **INTERPRET.**

   Check: The length of a side of a triangle can't be negative, so we reject $-10 - 5\sqrt{46}$. Since $-10 + 5\sqrt{46} \approx 24$ feet, the walking distance along the sidewalk is

   $$x + (x + 20) \approx 24 + (24 + 20) = 68 \text{ feet}.$$

   State: A student saves about $68 - 50$ or 18 feet of walking distance by cutting across the lawn.

EXAMPLE 7

CALCULATING LANDING TIME

An object is thrown upward from the top of a 200-foot cliff with a velocity of 12 feet per second. The height h in feet of the object after t seconds is

$$h = -16t^2 + 12t + 200$$

How long after the object is thrown will it strike the ground? Round to the nearest tenth of a second.

200 ft

Solution

1. UNDERSTAND. Read and reread the problem.

2. TRANSLATE. Since we want to know when the object strikes the ground, we want to know when the height $h = 0$, or

$$0 = -16t^2 + 12t + 200$$

3. SOLVE. First we divide both sides of the equation by -4.

$$0 = 4t^2 - 3t - 50 \qquad \text{Divide both sides by } -4.$$

Here, $a = 4$, $b = -3$, and $c = -50$. By the quadratic formula,

$$t = \frac{-(-3) \pm \sqrt{(-3)^2 - 4(4)(-50)}}{2 \cdot 4}$$

$$= \frac{3 \pm \sqrt{9 + 800}}{8}$$

$$= \frac{3 \pm \sqrt{809}}{8}$$

4. INTERPRET. **Check:** We check our calculations from the quadratic formula. Since the time won't be negative, we reject the proposed solution

$$\frac{3 - \sqrt{809}}{8}.$$

State: The time it takes for the object to strike the ground is exactly

$$\frac{3 + \sqrt{809}}{8} \text{ seconds} \approx 3.9 \text{ seconds}.$$

STUDY SKILLS REMINDER

Continue your outline started in Section 1.5. For this section, you only need to expand on your earlier section – **I.C.**– on solving quadratic equations. For example:

Solving Equations and Inequalities

I. Equations

    A. Linear equations (Sec. 1.5)

    B. Absolute value equations (Sec. 3.4)

    C. Quadratic and Higher Degree Equations—

    Solve: first write the equation in standard form (one side is 0.)

        1. If the polynomial on one side factors, solve by factoring.

        2. If the polynomial is of the form $(ax + b)^2$, solve by the square root property.

        3. If the polynomial does not factor, solve by the quadratic formula.

    D. Equations with rational expressions (Sec. 6.6)

    E. Equations with radicals (Sec. 7.6)

II. Inequalities

    A. Linear inequalities (Sec. 3.2)

    B. Compound inequalities (Sec. 3.3)

    C. Absolute value inequalities (Sec. 3.5)

Spotlight on DECISION MAKING

Suppose you are a registered dietician. Recently, you read an article in a nutrition journal that described a relationship between weight and the Recommended Dietary Allowance (RDA) for vitamin A in children up to age 10. The relationship is $y = 0.149x^2 - 4.475x + 406.478$, where y is the RDA for vitamin A in micrograms for a child whose weight is x pounds. (*Source:* Food and Nutrition Board, National Academy of Sciences—Institute of Medicine, 1989)

    You are working with a 4-year-old patient who weighs 40 pounds. After analyzing her diet, you are able to determine that she is currently getting an average of 400 micrograms of vitamin A daily. Decide whether her current vitamin A intake is adequate. If not, how much more is needed each day? In either case, determine how much weight she will need to gain before a daily intake of 500 micrograms of vitamin A is appropriate.

MENTAL MATH

Identify the values of a, b, and c in each quadratic equation.

1. $x^2 + 3x + 1 = 0$
$a = 1, b = 3, c = 1$

2. $2x^2 - 5x - 7 = 0$
$a = 2, b = -5, c = -7$

3. $7x^2 - 4 = 0$
$a = 7, b = 0, c = -4$

4. $x^2 + 9 = 0$
$a = 1, b = 0, c = 9$

5. $6x^2 - x = 0$
$a = 6, b = -1, c = 0$

6. $5x^2 + 3x = 0$
$a = 5, b = 3, c = 0$

14. $-3 - \sqrt{7}, -3 + \sqrt{7}$ **17.** $\dfrac{3 - \sqrt{11}}{2}, \dfrac{3 + \sqrt{11}}{2}$

EXERCISE SET 8.2

STUDY GUIDE/SSM | CD/VIDEO | PH MATH TUTOR CENTER | MathXL®Tutorials ON CD | MathXL® | MyMathLab®

2. $-12, 1$ **3.** $-\dfrac{3}{5}, 1$ **4.** $-\dfrac{1}{5}, 3$

Use the quadratic formula to solve each equation. These equations have real number solutions and can be checked graphically. See Examples 1 through 3.

1. $m^2 + 5m - 6 = 0$ $-6, 1$ **2.** $p^2 + 11p - 12 = 0$

3. $2y = 5y^2 - 3$ **4.** $5x^2 - 3 = 14x$

5. $x^2 - 6x + 9 = 0$ 3 **6.** $y^2 + 10y + 25 = 0$ -5

7. $x^2 + 7x + 4 = 0$ **8.** $y^2 + 5y + 3 = 0$

9. $8m^2 - 2m = 7$ **10.** $11n^2 - 9n = 1$

11. $3m^2 - 7m = 3$ **12.** $x^2 - 13 = 5x$

13. $\dfrac{1}{2}x^2 - x - 1 = 0$ **14.** $\dfrac{1}{6}x^2 + x + \dfrac{1}{3} = 0$

15. $\dfrac{2}{5}y^2 + \dfrac{1}{5}y = \dfrac{3}{5}$ $-\dfrac{3}{2}, 1$ **16.** $\dfrac{1}{8}x^2 + x = \dfrac{5}{2}$ $-10, 2$

17. $\dfrac{1}{3}y^2 - y - \dfrac{1}{6} = 0$ **18.** $\dfrac{1}{2}y^2 = y + \dfrac{1}{2}$

19. Solve Exercise 1 by factoring. Explain the result.

20. Solve Exercise 2 by factoring. Explain the result.

7. $\dfrac{-7 - \sqrt{33}}{2}, \dfrac{-7 + \sqrt{33}}{2}$ **8.** $\dfrac{-5 - \sqrt{13}}{2}, \dfrac{-5 + \sqrt{13}}{2}$

Use the quadratic formula to solve each equation. Check numerically. See Example 4. **13.** $1 - \sqrt{3}, 1 + \sqrt{3}$

21. $6 = -4x^2 + 3x$ **22.** $9x^2 + x + 2 = 0$

23. $(x + 5)(x - 1) = 2$ **24.** $x(x + 6) = 2$

25. $10y^2 + 10y + 3 = 0$ **26.** $3y^2 + 6y + 5 = 0$

9. $\dfrac{1 - \sqrt{57}}{8}, \dfrac{1 + \sqrt{57}}{8}$ **10.** $\dfrac{9 - 5\sqrt{5}}{22}, \dfrac{9 + 5\sqrt{5}}{22}$

The solutions of the quadratic equation $ax^2 + bx + c = 0$ are

$$\dfrac{-b + \sqrt{b^2 - 4ac}}{2a} \quad and \quad \dfrac{-b - \sqrt{b^2 - 4ac}}{2a}$$

11. $\dfrac{7 - \sqrt{85}}{6}, \dfrac{7 + \sqrt{85}}{6}$ **12.** $\dfrac{5 - \sqrt{77}}{2}, \dfrac{5 + \sqrt{77}}{2}$

Use the discriminant to determine the number and type of solutions of each equation. Confirm your result by graphing the related equation of the form $y = ax^2 + bx + c$. See Example 5.

27. $9x - 2x^2 + 5 = 0$ two real solutions

28. $5 - 4x + 12x^2 = 0$ two complex but not real solutions

29. $4x^2 + 12x = -9$ one real solution

30. $9x^2 + 1 = 6x$ one real solution

31. $3x = -2x^2 + 7$ two real solutions

32. $3x^2 = 5 - 7x$ two real solutions

33. $6 = 4x - 5x^2$ two complex but not real solutions

34. $8x = 3 - 9x^2$ two real solutions

26. $\dfrac{-3 - i\sqrt{6}}{3}, \dfrac{-3 + i\sqrt{6}}{3}$

50. $\dfrac{3 - \sqrt{5}}{4}, \dfrac{3 + \sqrt{5}}{4}$

40. $\dfrac{-9 - \sqrt{105}}{2}, \dfrac{-9 + \sqrt{105}}{2}$ **41.** $\dfrac{-1 - \sqrt{19}}{6}, \dfrac{-1 + \sqrt{19}}{6}$ **42.** $\dfrac{-1 - \sqrt{57}}{14}, \dfrac{-1 + \sqrt{57}}{14}$ **49.** $\dfrac{19 - \sqrt{345}}{2}, \dfrac{19 + \sqrt{345}}{2}$

MIXED PRACTICE **18.** $1 - \sqrt{2}, 1 + \sqrt{2}$

19.–20. answers may vary **21.** $\dfrac{3 - i\sqrt{87}}{8}, \dfrac{3 + i\sqrt{87}}{8}$

Use the quadratic formula to solve each equation. These equations have real number solutions. **22.** $\dfrac{-1 - i\sqrt{71}}{18}, \dfrac{-1 + i\sqrt{71}}{18}$

35. $x^2 + 5x = -2$ **36.** $y^2 - 8 = 4y$

37. $(m + 2)(2m - 6) = 5(m - 1) - 12$

38. $7p(p - 2) + 2(p + 4) = 3$ **37.** $\dfrac{5}{2}, 1$ **38.** $\dfrac{5}{7}, 1$

39. $\dfrac{x^2}{3} - x = \dfrac{5}{3}$ **40.** $\dfrac{x^2}{2} - 3 = -\dfrac{9}{2}x$

41. $x(6x + 2) - 3 = 0$ **42.** $x(7x + 1) = 2$

23. $-2 - \sqrt{11}, -2 + \sqrt{11}$ **24.** $-3 - \sqrt{11}, -3 + \sqrt{11}$

Use the quadratic formula to solve each equation.

43. $x^2 + 6x + 13 = 0$ $-3 - 2i, -3 + 2i$

44. $x^2 + 2x + 2 = 0$ $-1 - i, -1 + i$

45. $\dfrac{2}{5}y^2 + \dfrac{1}{5}y + \dfrac{3}{5} = 0$ $\dfrac{-1 - i\sqrt{23}}{4}, \dfrac{-1 + i\sqrt{23}}{4}$

46. $\dfrac{1}{8}x^2 + x + \dfrac{5}{2} = 0$ $-4 - 2i, -4 + 2i$

47. $\dfrac{1}{2}y^2 = y - \dfrac{1}{2}$ 1 **48.** $\dfrac{2}{3}x^2 - \dfrac{20}{3}x = -\dfrac{100}{6}$ 5

49. $(n - 2)^2 = 15n$ **50.** $\left(p - \dfrac{1}{2}\right)^2 = \dfrac{p}{2}$

25. $\dfrac{-5 - i\sqrt{5}}{10}, \dfrac{-5 + i\sqrt{5}}{10}$

Given the following graphs of quadratic equations of the form $y = f(x)$, find the number of real-number solutions of the related equation $f(x) = 0$. **36.** $2 - 2\sqrt{3}, 2 + 2\sqrt{3}$

51. a.   **b.**

2 real solutions | There is one x-intercept. 1 real solution

52. a.   **b.**

no real solutions | 2 real solutions

35. $\dfrac{-5 - \sqrt{17}}{2}, \dfrac{-5 + \sqrt{17}}{2}$

39. $\dfrac{3 - \sqrt{29}}{2}, \dfrac{3 + \sqrt{29}}{2}$

Solve. See Examples 6 and 7.

53. Nancy, Thelma, and John Varner live on a corner lot. Often, neighborhood children cut across their lot to save walking distance. Given the diagram below, approximate to the nearest foot how many feet of walking distance is saved by cutting across their property instead of walking around the lot.   14 ft

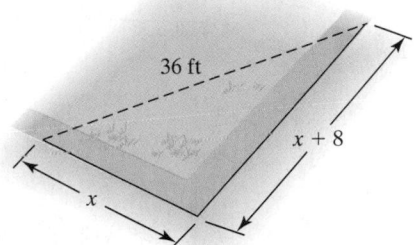

54. Given the diagram below, approximate to the nearest foot how many feet of walking distance a person saves by cutting across the lawn instead of walking on the sidewalk.   16 ft

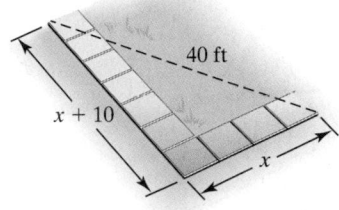

55. The hypotenuse of an isosceles right triangle is 2 centimeters longer than either of its legs. Find the exact length of each side. (*Hint:* An isosceles right triangle is a right triangle whose legs are the same length.)
$2 + 2\sqrt{2}$ cm, $2 + 2\sqrt{2}$ cm, $4 + 2\sqrt{2}$ cm

56. The hypotenuse of an isosceles right triangle is one meter longer than either of its legs. Find the length of each side.
$1 + \sqrt{2}$ m, $1 + \sqrt{2}$ m, $2 + \sqrt{2}$ m

57. Uri Chechov's rectangular dog pen for his Irish setter must have an area of 400 square feet. Also, the length must be 10 feet longer than the width. Find the dimensions of the pen.
width, $-5 + 5\sqrt{17}$ ft; length, $5 + 5\sqrt{17}$ ft

58. An entry in the Peach Festival Poster Contest must be rectangular and have an area of 1200 square inches. Furthermore, its length must be 20 inches longer than its width. Find the dimensions each entry must have.
width, $-10 + 10\sqrt{13}$ in. length, $10 + 10\sqrt{13}$ in.

59. A holding pen for cattle must be square and have a diagonal length of 100 meters.
a. Find the length of a side of the pen. $50\sqrt{2}$ m
b. Find the area of the pen.   5000 sq. m

60. A rectangle is three times longer than it is wide. It has a diagonal of length 50 centimeters.
a. Find the dimensions of the rectangle.
b. Find the perimeter of the rectangle. $40\sqrt{10}$ cm

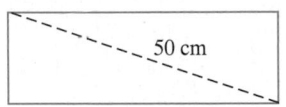

61. The heaviest reported door in the world is the 708.6 ton radiation shield door in the National Institute for Fusion Science at Toki, Japan. If the height of the door is 1.1 feet longer than its width, and its front area (neglecting depth) is 1439.9 square feet, find its width and height [Interesting note: the door is 6.6 feet thick.] (Source: Guiness World Records.)   37.4 ft by 38.5 ft

62. Christi and Robbie Wegmann are constructing a rectangular stained glass window whose length is 7.3 inches longer than its width. If the area of the window is 569.9 square inches, find its width and length.   20.5 in. by 27.8 in.

63. If a point *B* divides a line segment such that the smaller portion is to the larger portion as the larger is to the whole, the whole is the length of the *golden ratio.*

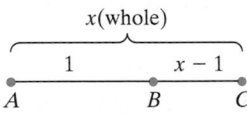

The golden ratio was thought by the Greeks to be the most pleasing to the eye, and many of their buildings contained numerous examples of the golden ratio. The value of the golden ratio is the positive solution of

(smaller) $\dfrac{x - 1}{1} = \dfrac{1}{x}$ (larger)
(larger) (whole)

Find this value.

64. The base of a triangle is four more than twice its height. If the area of the triangle is 42 square centimeters, find its base and height.   base, $2 + 2\sqrt{43}$ cm; height, $-1 + \sqrt{43}$ cm

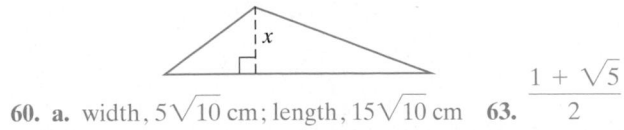

60. a. width, $5\sqrt{10}$ cm; length, $15\sqrt{10}$ cm   **63.** $\dfrac{1 + \sqrt{5}}{2}$

The Wollomombi Falls in Australia have a height of 1100 feet. A pebble is thrown upward from the top of the falls with an initial velocity of 20 feet per second. The height of the pebble h after t seconds is given by the equation $h = -16t^2 + 20t + 1100$. Use this equation for Exercises 65 and 66.

65. How long after the pebble is thrown will it hit the ground? Round to the nearest tenth of a second.   8.9 sec

66. How long after the pebble is thrown will it be 550 feet from the ground? Round to the nearest tenth of a second.   6.5 sec

A ball is thrown downward from the top of a 180-foot building with an initial velocity of 20 feet per second. The height of the ball h after t seconds is given by the equation $h = -16t^2 - 20t + 180$. Use this equation to answer Exercises 67 and 68.

180 ft

50 ft

67. How long after the ball is thrown will it strike the ground? Round the result to the nearest tenth of a second.   2.8 sec

68. How long after the ball is thrown will it be 50 feet from the ground? Round the result to the nearest tenth of a second.   2.3 sec

REVIEW AND PREVIEW   78. $-1.4, 0.9$ 81. Wednesday

Solve each equation. See Sections 6.6 and 7.6.

69. $\sqrt{5x - 2} = 3$ $\frac{11}{5}$ **70.** $\sqrt{y + 2} + 7 = 12$ 23

71. $\frac{1}{x} + \frac{2}{5} = \frac{7}{x}$ 15   **72.** $\frac{10}{z} = \frac{5}{z} - \frac{1}{3}$ -15

Factor. See Section 5.7.

73. $x^4 + x^2 - 20$ **74.** $2y^4 + 11y^2 - 6$

75. $z^4 - 13z^2 + 36$ **76.** $x^4 - 1$

Concept Extensions   82. $33°$ F

Use the quadratic formula and a calculator to approximate each solution to the nearest tenth.

77. $2x^2 - 6x + 3 = 0$ $0.6, 2.4$ **78.** $3.6x^2 + 1.8x - 4.3 = 0$

The accompanying graph shows the daily low temperatures for one week in New Orleans, Louisiana.

79. Which day of the week shows the greatest decrease in low temperature?   Sunday to Monday

80. Which day of the week shows the greatest increase in low temperature?   Friday to Saturday

81. Which day of the week had the lowest low temperature?

82. Use the graph to estimate the low temperature on Thursday.

Notice that the shape of the temperature graph is similar to the curve drawn. In fact, this graph can be modeled by the quadratic function $f(x) = 3x^2 - 18x + 56$, where f(x) is the temperature in degrees Fahrenheit and x is the number of days from Sunday. (This graph is shown in red.) Use this function to answer Exercises 83 and 84.

83. Use the quadratic function given to approximate the temperature on Thursday. Does your answer agree with the graph above? $f(4) = 32$; yes

84. Use the function given and the quadratic formula to find when the temperature was 35° F. [*Hint:* Let $f(x) = 35$ and solve for x.] Round your answer to one decimal place and interpret your result. Does your answer agree with the graph above? $x \approx 4.4$ or $x \approx 1.6$. The temperature was 35° on Monday and Thursday. Yes.

85. Wal-Mart Stores' net income can be modeled by the quadratic function $f(x) = 112.5x^2 + 498.7x + 5454$, where f(x) is net income in millions of dollars and x is the number of years after 2000. (*Source:* Based on data from Wal-Mart Stores, Inc.)

a. Find Wal-Mart's net income in 2002. $6901.4 million

b. If the trend described by the model continues, predict the year after 2000 in which Wal-Mart's net income will be $15,000 million. Round to the nearest whole year.   2007

86. The number of inmates in custody in U.S. prisons and jails can be modeled by the quadratic function $p(x) = -3892.4x^2 + 91{,}152x + 1{,}576{,}254$ where p(x) is the number of inmates and x is the number of years after 1995. (*Source:* Based on data from the Bureau of Justice Statistics, U.S. Department of Justice, 1995–2002) Round **a** and **b** to the nearest ten thousand.

73. $(x^2 + 5)(x + 2)(x - 2)$ **74.** $(2y^2 - 1)(y^2 + 6)$ **75.** $(z + 3)(z - 3)(z + 2)(z - 2)$ **76.** $(x^2 + 1)(x - 1)(x + 1)$

a. Find the number of prison inmates in the United States in 2000.   1,930,000 inmates

b. Find the number of prison inmates in the United States in 2002.   2,020,000 inmates

The solutions of the quadratic equation $ax^2 + bx + c = 0$ are $\dfrac{-b + \sqrt{b^2 - 4ac}}{2a}$ *and* $\dfrac{-b - \sqrt{b^2 - 4ac}}{2a}$.

87. Show that the sum of these solutions is $\dfrac{-b}{a}$.

88. Show that the product of these solutions is $\dfrac{c}{a}$.

Use the quadratic formula to solve each quadratic equation.

89. $3x^2 - \sqrt{12}x + 1 = 0$, $\dfrac{\sqrt{3}}{3}$

   (*Hint:* $a = 3, b = -\sqrt{12}, c = 1$)

90. $5x^2 + \sqrt{20}x + 1 = 0$ $-\dfrac{\sqrt{5}}{5}$

91. $x^2 + \sqrt{2}x + 1 = 0$ $\dfrac{-\sqrt{2} - i\sqrt{2}}{2}, \dfrac{-\sqrt{2} + i\sqrt{2}}{2}$

92. $x^2 - \sqrt{2}x + 1 = 0$ $\dfrac{\sqrt{2} - i\sqrt{2}}{2}, \dfrac{\sqrt{2} + i\sqrt{2}}{2}$

93. $2x^2 - \sqrt{3}x - 1 = 0$ $\dfrac{\sqrt{3} - \sqrt{11}}{4}, \dfrac{\sqrt{3} + \sqrt{11}}{4}$

94. $7x^2 + \sqrt{7}x - 2 = 0$ $\dfrac{-2\sqrt{7}}{7}, \dfrac{\sqrt{7}}{7}$

87.–88. answers may vary

8.3   SOLVING EQUATIONS BY USING QUADRATIC METHODS

Objectives

1   Solve various equations that are quadratic in form.

2   Solve problems that lead to quadratic equations.

1   In this section, we discuss various types of equations that can be solved in part by using the methods for solving quadratic equations.

Once each equation is simplified, you may want to use these steps when deciding what method to use to solve the quadratic equation.

TECHNOLOGY NOTE

When using the intersection-of-graphs method to solve an equation with radicals, it is necessary that the cursor is on a point where the function is defined.

> ### Solving a Quadratic Equation
>
> **Step 1:** If the equation is in the form $(ax + b)^2 = c$, use the square root property and solve. If not, go to Step 2.
>
> **Step 2:** Write the equation in standard form: $ax^2 + bx + c = 0$.
>
> **Step 3:** Try to solve the equation by the factoring method. If not possible, go to Step 4.
>
> **Step 4:** Solve the equation by the quadratic formula.

The first example is a radical equation that becomes a quadratic equation once we square both sides.

EXAMPLE 1

Solve $x - \sqrt{x} - 6 = 0$.

Solution   Recall that to solve a radical equation, first get the radical alone on one side of the equation. Then square both sides.

$$x - 6 = \sqrt{x} \qquad \text{Add } \sqrt{x} \text{ to both sides.}$$

$$(x - 6)^2 = \left(\sqrt{x}\right)^2 \qquad \text{Square both sides.}$$

$$x^2 - 12x + 36 = x$$

CLASSROOM EXAMPLE

Solve $x - \sqrt{x - 1} - 3 = 0$.

answer: 5

$$x^2 - 13x + 36 = 0 \qquad \textit{Set the equation equal to } 0.$$

$$(x - 9)(x - 4) = 0$$

$$x - 9 = 0 \quad \text{or} \quad x - 4 = 0$$

$$x = 9 \qquad\qquad x = 4$$

Check

Let $x = 9$

$$x - \sqrt{x} - 6 = 0$$

$$9 - \sqrt{9} - 6 \stackrel{?}{=} 0$$

$$9 - 3 - 6 \stackrel{?}{=} 0$$

$$0 = 0 \qquad \textit{True}$$

Let $x = 4$

$$x - \sqrt{x} - 6 = 0$$

$$4 - \sqrt{4} - 6 \stackrel{?}{=} 0$$

$$4 - 2 - 6 \stackrel{?}{=} 0$$

$$-4 = 0 \qquad \textit{False}$$

The solution is 9 or the solution set is $\{9\}$.

Intersection
X=9   **Y=3**

EXAMPLE 2

Solve $\dfrac{3x}{x - 2} - \dfrac{x + 1}{x} = \dfrac{6}{x(x - 2)}$.

Solution In this equation, x cannot be either 2 or 0, because these values cause denominators to equal zero. To solve for x, we first multiply both sides of the equation by $x(x - 2)$ to clear the fractions. By the distributive property, this means that we multiply each term by $x(x - 2)$.

$$x(x - 2)\left(\frac{3x}{x - 2}\right) - x(x - 2)\left(\frac{x + 1}{x}\right) = x(x - 2)\left[\frac{6}{x(x - 2)}\right]$$

$$3x^2 - (x - 2)(x + 1) = 6 \qquad \textit{Simplify.}$$

$$3x^2 - (x^2 - x - 2) = 6 \qquad \textit{Multiply.}$$

$$3x^2 - x^2 + x + 2 = 6$$

$$2x^2 + x - 4 = 0 \qquad \textit{Simplify.}$$

This equation cannot be factored using integers, so we solve by the quadratic formula.

$$x = \frac{-1 \pm \sqrt{1^2 - 4(2)(-4)}}{2 \cdot 2} \qquad \begin{array}{l}\textit{Use } a = 2, b = 1, \text{ and } c = -4 \text{ in} \\ \textit{the quadratic formula.}\end{array}$$

$$= \frac{-1 \pm \sqrt{1 + 32}}{4} \qquad \textit{Simplify.}$$

$$= \frac{-1 \pm \sqrt{33}}{4}$$

Neither proposed solution will make the denominators 0.

The solutions are $\dfrac{-1 + \sqrt{33}}{4}$ and $\dfrac{-1 - \sqrt{33}}{4}$ or the solution set is

$$\left\{\frac{-1 + \sqrt{33}}{4}, \frac{-1 - \sqrt{33}}{4}\right\}.$$

To check, we evaluate the left and right sides of the equation with the proposed solutions and see that a true statement results. Careful placement of parentheses is important, as shown below.

EXAMPLE 3

Solve $p^4 - 3p^2 - 4 = 0$.

Solution

First we factor the trinomial.

$$p^4 - 3p^2 - 4 = 0$$

$$(p^2 - 4)(p^2 + 1) = 0 \qquad \text{Factor.}$$

$$(p - 2)(p + 2)(p^2 + 1) = 0 \qquad \text{Factor further.}$$

$$p - 2 = 0 \quad \text{or} \quad p + 2 = 0 \quad \text{or} \quad p^2 + 1 = 0 \qquad \begin{array}{l}\text{Set each factor equal}\\\text{to } 0 \text{ and solve.}\end{array}$$

$$p = 2 \qquad\qquad p = -2 \qquad\qquad p^2 = -1$$

$$p = \pm\sqrt{-1} = \pm i$$

The solutions are $2, -2, i$ and $-i$.

Example 3 numerical check for the imaginary solutions.

TEACHING TIP

Before discussing Example 4, have students use $y = x + 2$ to rewrite $y^2 + 6y + 7 = 0$ in terms of x.

> **Helpful Hint**
>
> Example 3 can be solved using substitution also. Think of $p^4 - 3p^2 - 4 = 0$ as
>
> $(p^2)^2 - 3p^2 - 4 = 0$ Then let $x = p^2$, and solve and substitute back. The
> solutions will be the same.
>
> $x^2 - 3x - 4 = 0$

✔ **CONCEPT CHECK**

a. True or false? The maximum number of solutions that a quadratic equation can have is 2.

b. True or false? The maximum number of solutions that an equation in quadratic form can have is 2.

EXAMPLE 4

Solve $(x - 3)^2 - 3(x - 3) - 4 = 0$.

Solution

Notice that the quantity $(x - 3)$ is repeated in this equation. Sometimes it is helpful to substitute a variable (in this case other than x) for the repeated quantity. We will let $y = x - 3$. Then

$$(x - 3)^2 - 3(x - 3) - 4 = 0$$

becomes

$$y^2 - 3y - 4 = 0 \quad \text{Let } x - 3 = y.$$

$$(y - 4)(y + 1) = 0 \quad \text{Factor.}$$

Graphical check for Example 4.

TEACHING TIP

After discussing Example 4, have students solve the equation by first simplifying it. Then have them compare their results to those in the example.

To solve, we use the zero factor property.

$$y - 4 = 0 \quad \text{or} \quad y + 1 = 0 \qquad \text{Set each factor equal to } 0.$$

$$y = 4 \qquad\qquad y = -1 \qquad \text{Solve.}$$

To find values of x, we substitute back. That is, we substitute $x - 3$ for y.

> **Helpful Hint**
>
> When using substitution, don't forget to substitute back to the original variable.

$$x - 3 = 4 \quad \text{or} \quad x - 3 = -1$$

$$x = 7 \qquad\qquad x = 2$$

Both 2 and 7 check. The solutions are 2 and 7.

TEACHING TIP

Before discussing Example 5, have students use $y = x^{1/3}$ to rewrite $y^2 + 8y + 15 = 0$ in terms of x.

EXAMPLE 5

Solve $x^{2/3} - 5x^{1/3} + 6 = 0$.

Solution The key to solving this equation is recognizing that $x^{2/3} = (x^{1/3})^2$. We replace $x^{1/3}$ with m so that

$$(x^{1/3})^2 - 5x^{1/3} + 6 = 0$$

becomes

$$m^2 - 5m + 6 = 0$$

CLASSROOM EXAMPLE

Solve $x^{2/3} - 7x^{1/3} + 10 = 0$.

answer: 8, 125

Now we solve by factoring.

$$m^2 - 5m + 6 = 0$$

$$(m - 3)(m - 2) = 0 \qquad\qquad \text{Factor.}$$

$$m - 3 = 0 \quad \text{or} \quad m - 2 = 0 \qquad \text{Set each factor equal to } 0.$$

$$m = 3 \qquad\qquad m = 2$$

Since $m = x^{1/3}$, we have

$$x^{1/3} = 3 \qquad\qquad \text{or} \qquad x^{1/3} = 2$$

$$x = 3^3 = 27 \quad \text{or} \qquad x = 2^3 = 8$$

Both 8 and 27 check. The solutions are 8 and 27.

To visualize these solutions, graph $y_1 = x^{2/3} - 5x^{1/3} + 6$ and find the x-intercepts of the graph.

The x-intercepts are 8 and 27, so the solutions are 8 and 27.

2   The next example is a work problem. This problem is modeled by a rational equation that simplifies to a quadratic equation.

EXAMPLE 6

FINDING WORK TIME

Together, an experienced word processor and an apprentice word processor can create a word document in 6 hours. Alone, the experienced word processor can create the document 2 hours faster than the apprentice word processor can. Find the time in which each person can create the word document alone.

Solution

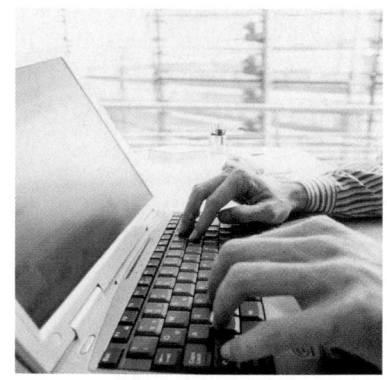

1. UNDERSTAND. Read and reread the problem. The key idea here is the relationship between the *time* (hours) it takes to complete the job and the *part of the job* completed in one unit of time (hour). For example, because they can complete the job together in 6 hours, the *part of the job* they can complete in 1 hour is $\frac{1}{6}$.

   Let

   x = the *time* in hours it takes the apprentice word processor to complete the job alone

   $x - 2$ = the *time* in hours it takes the experienced word processor to complete the job alone

   We can summarize in a chart the information discussed:

	Total Hours to Complete Job	*Part of Job Completed in 1 Hour*
Apprentice word processor	x	$\frac{1}{x}$
Experienced word processor	$x - 2$	$\frac{1}{x - 2}$
Together	6	$\frac{1}{6}$

2. TRANSLATE.

   In words:

part of job completed by apprentice word processor in 1 hour	added to	part of job completed by experienced word processor in 1 hour	is equal to	part of job completed together in 1 hour
↓	↓	↓	↓	↓

   Translate:

   $$\frac{1}{x} \quad + \quad \frac{1}{x - 2} \quad = \quad \frac{1}{6}$$

3. SOLVE.

Algebraic Solution:

$$\frac{1}{x} + \frac{1}{x-2} = \frac{1}{6}$$

$$6x(x-2)\left(\frac{1}{x} + \frac{1}{x-2}\right) = 6x(x-2)\cdot\frac{1}{6}$$

Multiply both sides by the LCD, $6x(x-2)$.

$$6x(x-2)\cdot\frac{1}{x} + 6x(x-2)\cdot\frac{1}{x-2} = 6x(x-2)\cdot\frac{1}{6}$$

Use the distributive property.

$$6(x-2) + 6x = x(x-2)$$

$$6x - 12 + 6x = x^2 - 2x$$

$$0 = x^2 - 14x + 12$$

Now we can substitute $a = 1$, $b = -14$, and $c = 12$ into the quadratic formula and simplify.

$$x = \frac{-(-14) \pm \sqrt{(-14)^2 - 4(1)(12)}}{2\cdot 1} = \frac{14 \pm \sqrt{148}}{2}$$

$$x \approx \frac{14 + 12.2}{2} = 13.1 \quad \text{or} \quad x \approx \frac{14 - 12.2}{2} = 0.9$$

Graphical Solution:

Graph $y_1 = \frac{1}{x} + \frac{1}{x-2}$ and $y_2 = \frac{1}{6}$ in a $[-4, 18.8, 2]$ by $[-1, 1, 1]$ window. The approximate intersections are $x \approx 0.9$ and $x \approx 13.1$. A hand sketch is shown below.

4. INTERPRET.

Check: If the apprentice word processor completes the job alone in 0.9 hours, the experienced word processor completes the job alone in $x - 2 = 0.9 - 2 = -1.1$ hours. Since this is not possible, we reject the solution of 0.9. The approximate solution thus is 13.1 hours.

State: The apprentice word processor can complete the job alone in approximately 13.1 hours, and the experienced word processor can complete the job alone in approximately

$$x - 2 = 13.1 - 2 = 11.1 \text{ hours}.$$

EXAMPLE 7

Beach ← x mph Fargo

400 mi

$x + 10$ mph →

400 mi

FINDING DRIVING SPEEDS

Beach and Fargo are about 400 miles apart. A salesperson travels from Fargo to Beach one day at a certain speed. She returns to Fargo the next day and drives 10 mph faster. Her total travel time was $14\frac{2}{3}$ hours. Find her speed to Beach and the return speed to Fargo.

Solution **1.** UNDERSTAND. Read and reread the problem. Let

$$x = \text{the speed to Beach, so}$$

$$x + 10 = \text{the return speed to Fargo.}$$

Then organize the given information in a table.

	distance	=	rate	·	time	
To Beach	400		x		$\dfrac{400}{x}$	← distance ← rate
Return to Fargo	400		$x + 10$		$\dfrac{400}{x + 10}$	← distance ← rate

```
WINDOW
 Xmin=0
 Xmax=65
 Xscl=10
 Ymin=0
 Ymax=20
 Yscl=2
 Xres=1■
```

Below is a graphical solution for Example 7. Since x represents mph, we choose $[0, 65]$. Since the y-window should include $\frac{44}{3}$, we choose $[0, 20]$.

Here $x = 50$ indicates the speed.

2. TRANSLATE.

In words:

$$\boxed{\text{time to Beach}} \;+\; \boxed{\text{return time to Fargo}} \;=\; \boxed{14\tfrac{2}{3} \text{ hours}}$$

Translate:

$$\dfrac{400}{x} \;+\; \dfrac{400}{x + 10} \;=\; \dfrac{44}{3}$$

3. SOLVE.

$$\dfrac{400}{x} + \dfrac{400}{x + 10} = \dfrac{44}{3}$$

$$\dfrac{100}{x} + \dfrac{100}{x + 10} = \dfrac{11}{3} \qquad \text{Divide both sides by 4.}$$

$$3x(x + 10)\left(\dfrac{100}{x} + \dfrac{100}{x + 10}\right) = 3x(x + 10) \cdot \dfrac{11}{3} \qquad \text{Multiply both sides by the LCD } 3x(x + 10).$$

$$3x(x + 10) \cdot \dfrac{100}{x} + 3x(x + 10) \cdot \dfrac{100}{x + 10} = 3x(x + 10) \cdot \dfrac{11}{3} \qquad \text{Use the distributive property.}$$

$$3(x + 10) \cdot 100 + 3x \cdot 100 = x(x + 10) \cdot 11$$

$$300x + 3000 + 300x = 11x^2 + 110x$$

$$0 = 11x^2 - 490x - 3000 \qquad \text{Set equation equal to 0.}$$

$$0 = (11x + 60)(x - 50) \qquad \text{Factor.}$$

$$11x + 60 = 0 \quad \text{or} \quad x - 50 = 0 \qquad \text{Set each factor equal to 0.}$$

$$x = -\dfrac{60}{11} \text{ or } -5\dfrac{5}{11}; \quad x = 50$$

4. INTERPRET.

Check: The speed is not negative, so it's not $-5\dfrac{5}{11}$. The number 50 does check.

State: The speed to Beach was 50 mph and her return speed to Fargo was 60 mph.

EXERCISE SET 8.3

STUDY GUIDE/SSM   CD/VIDEO   PH MATH TUTOR CENTER   MathXL®Tutorials ON CD   MathXL®   MyMathLab®

Solve. See Example 1.

1. $2x = \sqrt{10 + 3x}$ 2

2. $3x = \sqrt{8x + 1}$ 1

3. $x - 2\sqrt{x} = 8$ 16

4. $x - \sqrt{2x} = 4$ 8

5. $\sqrt{9x} = x + 2$ 1, 4

6. $\sqrt{16x} = x + 3$ 1, 9

Solve. See Example 2.

7. $\dfrac{2}{x} + \dfrac{3}{x - 1} = 1$ $3 - \sqrt{7}, 3 + \sqrt{7}$

8. $\dfrac{6}{x^2} = \dfrac{3}{x + 1}$ $1 - \sqrt{3}, 1 + \sqrt{3}$

9. $\dfrac{3}{x} + \dfrac{4}{x + 2} = 2$ $\dfrac{3 - \sqrt{57}}{4}, \dfrac{3 + \sqrt{57}}{4}$

10. $\dfrac{5}{x - 2} + \dfrac{4}{x + 2} = 1$ $\dfrac{9 - \sqrt{105}}{2}, \dfrac{9 + \sqrt{105}}{2}$

11. $\dfrac{7}{x^2 - 5x + 6} = \dfrac{2x}{x - 3} - \dfrac{x}{x - 2}$ $\dfrac{1 - \sqrt{29}}{2}, \dfrac{1 + \sqrt{29}}{2}$

12. $\dfrac{11}{2x^2 + x - 15} = \dfrac{5}{2x - 5} - \dfrac{x}{x + 3}$ $\dfrac{5 - \sqrt{33}}{2}, \dfrac{5 + \sqrt{33}}{2}$

Solve. See Example 3.

13. $p^4 - 16 = 0$

14. $x^4 + 2x^2 - 3 = 0$

15. $4x^4 + 11x^2 = 3$

16. $z^4 = 81$

17. $z^4 - 13z^2 + 36 = 0$

18. $9x^4 + 5x^2 - 4 = 0$

Solve. See Examples 4 and 5.   **19.** $125, -8$

19. $x^{2/3} - 3x^{1/3} - 10 = 0$

20. $x^{2/3} + 2x^{1/3} + 1 = 0$ -1

21. $(5n + 1)^2 + 2(5n + 1) - 3 = 0$

22. $(m - 6)^2 + 5(m - 6) + 4 = 0$ 2, 5

23. $2x^{2/3} - 5x^{1/3} = 3$

24. $3x^{2/3} + 11x^{1/3} = 4$

25. $1 + \dfrac{2}{3t - 2} = \dfrac{8}{(3t - 2)^2}$

26. $2 - \dfrac{7}{x + 6} = \dfrac{15}{(x + 6)^2}$

27. $20x^{2/3} - 6x^{1/3} - 2 = 0$

28. $4x^{2/3} + 16x^{1/3} = -15$

MIXED PRACTICE

Solve. See Examples 1 through 5.

29. $a^4 - 5a^2 + 6 = 0$ $-\sqrt{2}, \sqrt{2}, -\sqrt{3}, \sqrt{3}$

30. $x^4 - 12x^2 + 11 = 0$ $-1, 1, -\sqrt{11}, \sqrt{11}$

31. $\dfrac{2x}{x - 2} + \dfrac{x}{x + 3} = -\dfrac{5}{x + 3}$ $\dfrac{-9 - \sqrt{201}}{6}, \dfrac{-9 + \sqrt{201}}{6}$

32. $\dfrac{5}{x - 3} + \dfrac{x}{x + 3} = \dfrac{19}{x^2 - 9}$ $-1 - \sqrt{5}, -1 + \sqrt{5}$

33. $(p + 2)^2 = 9(p + 2) - 20$ 2, 3

34. $2(4m - 3)^2 - 9(4m - 3) = 5$ $\dfrac{5}{8}, 2$

35. $2x = \sqrt{11x + 3}$ 3   **36.** $4x = \sqrt{2x + 3}$ $\dfrac{1}{2}$

37. $x^{2/3} - 8x^{1/3} + 15 = 0$ 27, 125

38. $x^{2/3} - 2x^{1/3} - 8 = 0$ $-8, 64$

39. $y^3 + 9y - y^2 - 9 = 0$ $1, -3i, 3i$

40. $x^3 + x - 3x^2 - 3 = 0$ $3, -i, i$

41. $2x^{2/3} + 3x^{1/3} - 2 = 0$ $\dfrac{1}{8}, -8$

42. $6x^{2/3} - 25x^{1/3} - 25 = 0$ $-\dfrac{125}{216}, 125$

43. $x^{-2} - x^{-1} - 6 = 0$ $-\dfrac{1}{2}, \dfrac{1}{3}$

44. $y^{-2} - 8y^{-1} + 7 = 0$ $\dfrac{1}{7}, 1$

45. $x - \sqrt{x} = 2$ 4   **46.** $x - \sqrt{3x} = 6$ 12

47. $\dfrac{x}{x - 1} + \dfrac{1}{x + 1} = \dfrac{2}{x^2 - 1}$ -3

48. $\dfrac{x}{x - 5} + \dfrac{5}{x + 5} = -\dfrac{1}{x^2 - 25}$ $-12, 2$

49. $p^4 - p^2 - 20 = 0$ $-\sqrt{5}, \sqrt{5}, -2i, 2i$

50. $x^4 - 10x^2 + 9 = 0$ $-1, 1, -3, 3$

51. $2x^3 = -54$ $-3, \dfrac{3 - 3i\sqrt{3}}{2}, \dfrac{3 + 3i\sqrt{3}}{2}$

52. $y^3 - 216 = 0$ $6, -3 - 3i\sqrt{3}, -3 + 3i\sqrt{3}$

53. $1 = \dfrac{4}{x - 7} + \dfrac{5}{(x - 7)^2}$ 6, 12

54. $3 + \dfrac{1}{2p + 4} = \dfrac{10}{(2p + 4)^2}$ $-\dfrac{7}{6}, -3$

55. $27y^4 + 15y^2 = 2$ $-\dfrac{1}{3}, \dfrac{1}{3}, -\dfrac{i\sqrt{6}}{3}, \dfrac{i\sqrt{6}}{3}$

56. $8z^4 + 14z^2 = -5$ $-\dfrac{i\sqrt{2}}{2}, \dfrac{i\sqrt{2}}{2}, -\dfrac{i\sqrt{5}}{2}, \dfrac{i\sqrt{5}}{2}$

21. $-\dfrac{4}{5}, 0$

23. $-\dfrac{1}{8}, 27$

24. $-64, \dfrac{1}{27}$

25. $-\dfrac{2}{3}, \dfrac{4}{3}$

26. $-\dfrac{15}{2}, -1$

27. $-\dfrac{1}{125}, \dfrac{1}{8}$

28. $-\dfrac{125}{8}, -\dfrac{27}{8}$

13. $-2, 2, -2i, 2i$ **14.** $-1, 1, -i\sqrt{3}, i\sqrt{3}$ **15.** $-\dfrac{1}{2}, \dfrac{1}{2}, -i\sqrt{3}, i\sqrt{3}$ **16.** $-3, 3, -3i, 3i$ **17.** $-3, 3, -2, 2$ **18.** $-\dfrac{2}{3}, \dfrac{2}{3}, -i, i$

Solve. See Examples 6 and 7.

57. A jogger ran 3 miles, decreased her speed by 1 mile per hour, and then ran another 4 miles. If her total time jogging was $1\frac{3}{5}$ hours, find her speed for each part of her run.
5 mph, then 4 mph

58. Mark Keaton's workout consists of jogging for 3 miles, and then riding his bike for 5 miles at a speed 4 miles per hour faster than he jogs. If his total workout time is 1 hour, find his jogging speed and his biking speed.   jogging, 6 mph; biking, 10 mph

59. A Chinese restaurant in Mandeville, Louisiana, has a large goldfish pond around the restaurant. Suppose that an inlet pipe and a hose together can fill the pond in 8 hours. The inlet pipe alone can complete the job in one hour less time than the hose alone. Find the time that the hose can complete the job alone and the time that the inlet pipe can complete the job alone. Round each to the nearest tenth of an hour.   inlet pipe, 15.5 hr; hose, 16.5hr

60. A water tank on a farm in Flatonia, Texas, can be filled with a large inlet pipe and a small inlet pipe in 3 hours. The large inlet pipe alone can fill the tank in 2 hours less time than the small inlet pipe alone. Find the time to the nearest tenth of an hour each pipe can fill the tank alone.   small pipe, 7.2 hr; large pipe, 5.2 hr

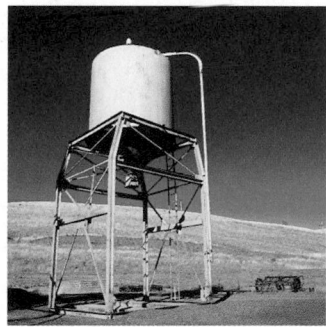

61. Roma Sherry drove 330 miles from her hometown to Tucson. During her return trip, she was able to increase her speed by 11 mph. If her return trip took 1 hour less time, find her original speed and her speed returning home.   55 mph, 66 mph

62. A salesperson drove to Portland, a distance of 300 miles. During the last 80 miles of his trip, heavy rainfall forced him to decrease his speed by 15 mph. If his total driving time was 6 hours, find his original speed and his speed during the rainfall.   55 mph, 40 mph

63. Bill Shaughnessy and his son Billy can clean the house together in 4 hours. When the son works alone, it takes him an hour longer to clean than it takes his dad alone. Find how long to the nearest tenth of an hour it takes the son to clean alone.   8.5 hr

64. Together, Noodles and Freckles eat a 50-pound bag of dog food in 30 days. Noodles by himself eats a 50-pound bag in 2 weeks less time than Freckles does by himself. How many days to the nearest whole day would a 50-pound bag of dog food last Freckles?   68 days

65. The product of a number and 4 less than the number is 96. Find the number.   12 or −8

66. A whole number increased by its square is two more than twice itself. Find the number.   2

△ **67.** Suppose that an open box is to be made from a square sheet of cardboard by cutting out squares from each corner as shown and then folding along the dotted lines. If the box is to have a volume of 300 cubic centimeters, find the original dimensions of the sheet of cardboard.

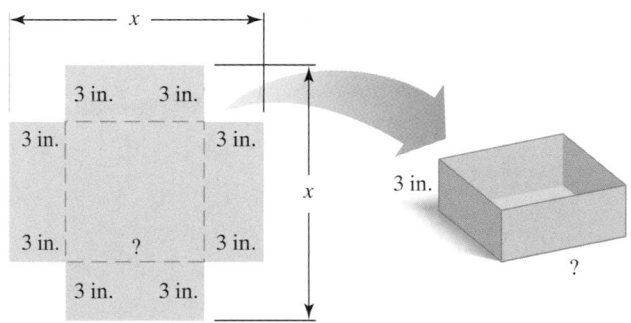

a. The ? in the drawing above will be the length (and also the width) of the box as shown in the drawing above. Represent this length in terms of x. $(x − 6)$ in.

b. Use the formula for volume of a box, $V = l \cdot w \cdot h$, to write an equation in x. $300 = (x − 6) \cdot (x − 6) \cdot 3$

c. Solve the equation for x and give the dimensions of the sheet of cardboard. Check your solution.   16 cm by 16 cm

△ **68.** Suppose that an open box is to be made from a square sheet of cardboard by cutting out squares from each corner as shown and then folding along the dotted lines. If the box is to have a volume of 128 cubic inches, find the original dimensions of the sheet of cardboard. (*Hint:* Use Exercise 67 Parts **a**, **b**, and **c** to help you.)   12 in. by 12 in.

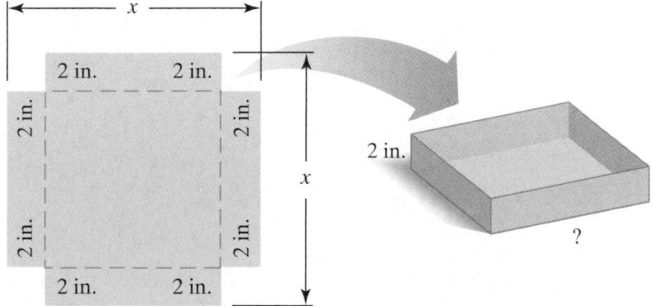

△ **69.** A sprinkler that sprays water in a circular motion is to be used to water a square garden. If the area of the garden is 920 square feet, find the smallest whole number *radius* that the sprinkler can be adjusted to so that the entire garden is watered.   22 feet

△ **70.** Suppose that a square field has an area of 6270 square feet. See Exercise 69 and find a new sprinkler radius.   56 feet

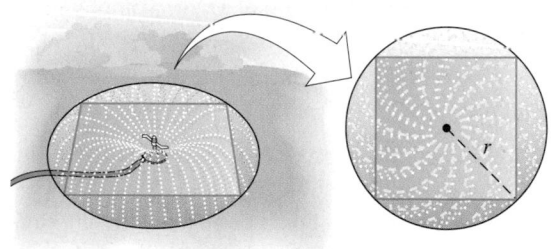

REVIEW AND PREVIEW

Solve each inequality. See Section 3.2.

71. $\dfrac{5x}{3} + 2 \le 7$ $(-\infty, 3]$ **72.** $\dfrac{2x}{3} + \dfrac{1}{6} \ge 2$ $\left[\dfrac{11}{4}, \infty\right)$

73. $\dfrac{y - 1}{15} > -\dfrac{2}{5}$ $(-5, \infty)$ **74.** $\dfrac{z - 2}{12} < \dfrac{1}{4}$ $(-\infty, 5)$

Find the domain and range of each graphed relation. Decide which relations are also functions. See Section 2.2.

75.

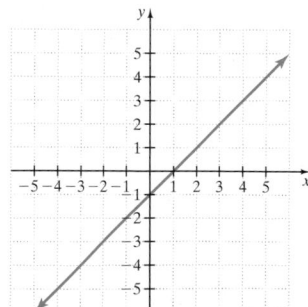

domain: $\{x \mid x$ is a real number$\}$ or $(-\infty, \infty)$
range: $\{y \mid y$ is a real number$\}$ or $(-\infty, \infty)$ function

76.

domain: $\{x \mid -3 \le x \le 3\}$ or $[-3, 3]$
range: $\{y \mid -2 \le y \le 2\}$ or $[-2, 2]$ not a function

77.

domain: $\{x \mid x$ is a real number$\}$ or $(-\infty, \infty)$
range: $\{y \mid y \ge -1\}$ or $[-1, \infty)$ function

78.

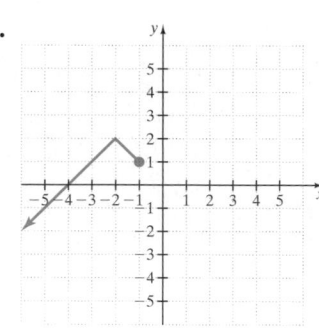

domain: $\{x \mid x \le -1\}$ or $(-\infty, -1]$
range: $\{y \mid y \le 2\}$ or $(-\infty, 2]$ function

Concept Extensions

79. Write a polynomial equation that has three solutions: 2, 5, and -7.   answers may vary

80. Write a polynomial equation that has three solutions: 0, $2i$, and $-2i$.   answers may vary

81. During the 2003 Grand Prix of Miami auto race, Adrian Fernandez posted the fastest lap speed but Mario Dominguez won the race. The track is 7920 feet (1.5 miles) long. Fernandez's fastest lap speed was 0.88 feet per second faster than Dominguez's fastest lap speed. Traveling at these fastest speeds, Mario Dominguez would have taken 0.38 seconds longer than Fernandez to complete a lap. (*Source:* Championship Auto Racing Teams, Inc.)

   a. Find Mario Dominguez's fastest lap speed during the race. Round to two decimal places.   134.05 ft per sec

   b. Find Adrian Fernandez's fastest lap speed during the race. Round to two decimal places.   134.93 ft per sec

   c. Convert each speed to miles per hour. Round to one decimal place.   Fernandez: 92.0 mph; Dominguez: 91.4 mph

82. Use a graphing calculator to solve Exercise 29. Compare the solution with the solution from Exercise 29. Explain any differences.   answers may vary

SUMMARY ON SOLVING QUADRATIC EQUATIONS   **INTEGRATED REVIEW**

Use the square root property to solve each equation.

1. $x^2 - 10 = 0$

$\{-\sqrt{10}, \sqrt{10}\}$

2. $x^2 - 14 = 0$

$\{-\sqrt{14}, \sqrt{14}\}$

3. $(x - 1)^2 = 8$

$\{1 - 2\sqrt{2}, 1 + 2\sqrt{2}\}$

4. $(x + 5)^2 = 12$

$\{-5 - 2\sqrt{3}, -5 + 2\sqrt{3}\}$

Solve each equation by completing the square.

5. $x^2 + 2x - 12 = 0$

$\{-1 - \sqrt{13}, -1 + \sqrt{13}\}$

6. $x^2 - 12x + 11 = 0$

$\{1, 11\}$

7. $3x^2 + 3x = 5$

$\left\{\dfrac{-3 - \sqrt{69}}{6}, \dfrac{-3 + \sqrt{69}}{6}\right\}$

8. $16y^2 + 16y = 1$

$\left\{\dfrac{-2 - \sqrt{5}}{4}, \dfrac{-2 + \sqrt{5}}{4}\right\}$

Use the quadratic formula to solve each equation

9. $2x^2 - 4x + 1 = 0$

$\left\{\dfrac{2 - \sqrt{2}}{2}, \dfrac{2 + \sqrt{2}}{2}\right\}$

10. $\dfrac{1}{2}x^2 + 3x + 2 = 0$

$\{-3 - \sqrt{5}, -3 + \sqrt{5}\}$

11. $x^2 + 4x = -7$

$\{-2 + i\sqrt{3}, -2 - i\sqrt{3}\}$

12. $x^2 + x = -3$

$\left\{\dfrac{-1 - i\sqrt{11}}{2}, \dfrac{-1 + i\sqrt{11}}{2}\right\}$

Solve each equation. Use a method of your choice.

13. $x^2 + 3x + 6 = 0$

$\left\{\dfrac{-3 + i\sqrt{15}}{2}, \dfrac{-3 - i\sqrt{15}}{2}\right\}$

14. $2x^2 + 18 = 0$

$\{3i, -3i\}$

15. $x^2 + 17x = 0$

$\{0, -17\}$

16. $4x^2 - 2x - 3 = 0$

$\left\{\dfrac{1 + \sqrt{13}}{4}, \dfrac{1 - \sqrt{13}}{4}\right\}$

17. $(x - 2)^2 = 27$

$\{2 + 3\sqrt{3}, 2 - 3\sqrt{3}\}$

18. $\dfrac{1}{2}x^2 - 2x + \dfrac{1}{2} = 0$

$\{2 + \sqrt{3}, 2 - \sqrt{3}\}$

19. $3x^2 + 2x = 8$

$\left\{-2, \dfrac{4}{3}\right\}$

20. $2x^2 = -5x - 1$

$\left\{\dfrac{-5 + \sqrt{17}}{4}, \dfrac{-5 - \sqrt{17}}{4}\right\}$

21. $x(x - 2) = 5$

$\{1 - \sqrt{6}, 1 + \sqrt{6}\}$

22. $x^2 - 31 = 0$

$\{-\sqrt{31}, \sqrt{31}\}$

23. $5x^2 - 55 = 0$

$\{-\sqrt{11}, \sqrt{11}\}$

24. $5x^2 + 55 = 0$

$\{-i\sqrt{11}, i\sqrt{11}\}$

25. $x(x + 5) = 66$

$\{-11, 6\}$

26. $5x^2 + 6x - 2 = 0$

$\left\{\dfrac{-3 + \sqrt{19}}{5}, \dfrac{-3 - \sqrt{19}}{5}\right\}$

27. $2x^2 + 3x = 1$

$\left\{\dfrac{-3 + \sqrt{17}}{4}, \dfrac{-3 - \sqrt{17}}{4}\right\}$

△ **28.** The diagonal of a square room measures 20 feet. Find the exact length of a side of the room. Then approximate the length to the nearest tenth of a foot. $10\sqrt{2}$ ft ≈ 14.1 ft

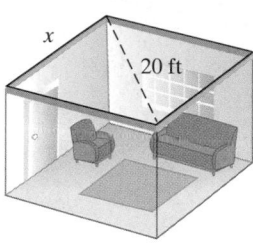

29. Jack and Lucy Hoag together can prepare a crawfish boil for a large party in 4 hours. Lucy alone can complete the job in 2 hours less time than Jack alone. Find the time that each person can prepare the crawfish boil alone. Round each time to the nearest tenth of an hour. Jack: 9.1 hr; Lucy: 7.1 hr

30. Diane Gray exercises at Total Body Gym. On the treadmill, she runs 5 miles, then increases her speed by 1 mile per hour and runs an additional 2 miles. If her total time on the treadmill is $1\frac{1}{3}$ hours, find her speed during each part of her run.   5 mph during the first part, then 6 mph

8.4 NONLINEAR INEQUALITIES IN ONE VARIABLE

Objectives

1 Solve polynomial inequalities of degree 2 or greater.

2 Solve inequalities that contain rational expressions with variables in the denominator.

1 Just as we can solve linear inequalities in one variable, so can we also solve quadratic inequalities in one variable. A **quadratic inequality** is an inequality that can be written so that one side is a quadratic expression and the other side is 0. Here are examples of quadratic inequalities in one variable. Each is written in **standard form.**

$$x^2 - 10x + 7 \le 0 \quad 3x^2 + 2x - 6 > 0$$
$$2x^2 + 9x - 2 < 0 \quad x^2 - 3x + 11 \ge 0$$

A solution of a quadratic inequality in one variable is a value of the variable that makes the inequality a true statement.

Discover the Concept

Graph the quadratic function $y = x^2 - 3x - 10$.

a. Find the values of x for which $y = 0$. (Recall that these are the x-intercepts of the graph.)

b. Find the x-values for which $y < 0$.

c. Find the x-values for which $y > 0$.

d. What is the solution of $y = 0$ or $x^2 - 3x - 10 = 0$?

e. What is the solution of $y < 0$ or $x^2 - 3x - 10 < 0$?

f. What is the solution of $y > 0$ or $x^2 - 3x - 10 > 0$?

x-values
corresponding to *negative*
y-values

TEACHING TIP

Have students use a graphing calculator to check solutions in this section.

Notice that the x-values for which y is positive are separated from the x values for which y is negative by the x-intercepts. (Recall that the x-intercepts correspond to values of x for which $y = 0$.) Thus, the solution set of $x^2 - 3x - 10 < 0$ consists of all real numbers from -2 to 5, or in interval notation, $(-2, 5)$.

It is not necessary to graph $y = x^2 - 3x - 10$ to solve the related inequality $x^2 - 3x - 10 < 0$. Instead, we can draw a number line representing the x-axis and keep the following in mind: *A region on the number line for which the value of* $x^2 - 3x - 10$ *is positive is separated from a region on the number line for which the value of* $x^2 - 3x - 10$ *is negative by a value for which the expression is* 0.

Let's find these values for which the expression is 0 by solving the related equation:

$$x^2 - 3x - 10 = 0$$
$$(x - 5)(x + 2) = 0 \qquad \text{Factor.}$$
$$x - 5 = 0 \quad \text{or} \quad x + 2 = 0 \qquad \text{Set each factor equal to } 0.$$
$$x = 5 \quad x = -2 \qquad \text{Solve.}$$

These two numbers -2 and 5, divide the number line into three regions. We will call the regions A, B, and C. These regions are important because, if the value of $x^2 - 3x - 10$ is negative when a number from a region is substituted for x, then $x^2 - 3x - 10$ is negative when any number in that region is substituted for x. The same is true if the value of $x^2 - 3x - 10$ is positive for a particular value of x in a region.

To see whether the inequality $x^2 - 3x - 10 < 0$ is true or false in each region, we choose a test point from each region and substitute its value for x in the inequality $x^2 - 3x - 10 < 0$. If the resulting inequality is true, the region containing the test point is a solution region.

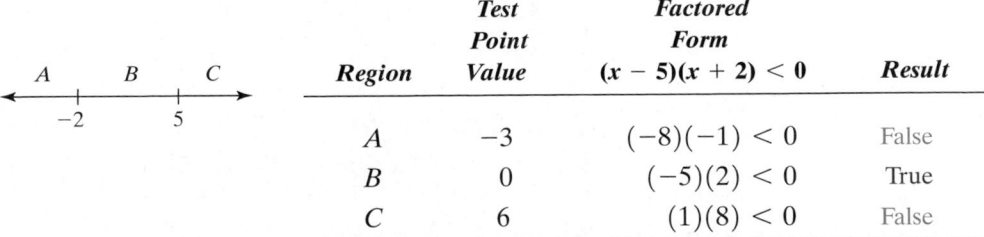

Region	Test Point Value	Factored Form $(x - 5)(x + 2) < 0$	Result
A	-3	$(-8)(-1) < 0$	False
B	0	$(-5)(2) < 0$	True
C	6	$(1)(8) < 0$	False

The values in region B satisfy the inequality. The numbers -2 and 5 are not included in the solution set since the inequality symbol is $<$. The solution set is $(-2, 5)$, and its graph is shown.

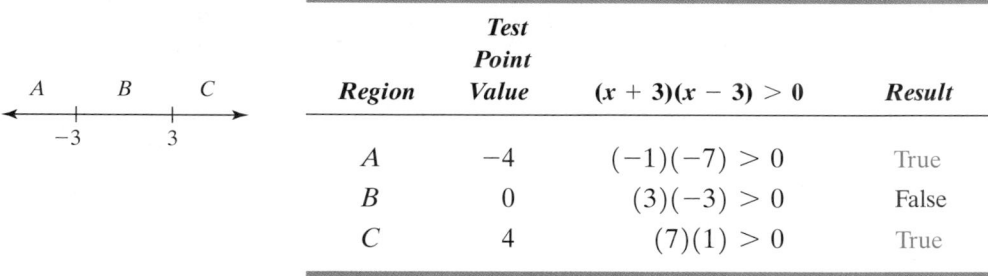

EXAMPLE 1

Solve $(x + 3)(x - 3) > 0$.

Solution   First we solve the related equation $(x + 3)(x - 3) = 0$.

$$(x + 3)(x - 3) = 0$$

$$x + 3 = 0 \quad \text{or} \quad x - 3 = 0$$

$$x = -3 \qquad\qquad x = 3$$

The two numbers -3 and 3 separate the number line into three regions, A, B, and C.

Now we substitute the value of a test point from each region. If the test value satisfies the inequality, every value in the region containing the test value is a solution.

Region	Test Point Value	$(x + 3)(x - 3) > 0$	Result
A	-4	$(-1)(-7) > 0$	True
B	0	$(3)(-3) > 0$	False
C	4	$(7)(1) > 0$	True

The points in regions A and C satisfy the inequality. The numbers -3 and 3 are not included in the solution since the inequality symbol is $>$. The solution set is $(-\infty, -3) \cup (3, \infty)$, and its graph is shown.

The following steps may be used to solve a polynomial inequality.

Solving a Polynomial Inequality

Step 1: Write the inequality in standard form and then solve the related equation.

Step 2: Separate the number line into regions with the solutions from Step 1.

Step 3: For each region, choose a test point and determine whether its value satisfies the *original inequality*.

Step 4: The solution set includes the regions whose test point value is a solution. If the inequality symbol is $\leq$ or $\geq$, the values from Step 1 are solutions; if $<$ or $>$, they are not.

✔ **CONCEPT CHECK**

When choosing a test point in Step 4, why would the solutions from Step 2 not make good choices for test points?

EXAMPLE 2

Solve $x^2 - 4x \leq 0$.

Solution

First we solve the related equation $x^2 - 4x = 0$.

$$x^2 - 4x = 0$$

$$x(x - 4) = 0$$

$$x = 0 \quad \text{or} \quad x = 4$$

The numbers 0 and 4 separate the number line into three regions, A, B, and C.

Check a test value in each region in the original inequality. Values in region B satisfy the inequality. The numbers 0 and 4 are included in the solution since the inequality symbol is $\leq$. The solution set is $[0, 4]$, and its graph is shown.

Next, we solve a quadratic inequality similar to Example 2 graphically.

EXAMPLE 3

Use a graphical approach to solve $x^2 - 4x \geq 0$.

Solution

The x-intercepts of the graph of $y = x^2 - 4x$ are found by solving $x^2 - 4x = 0$. We found these intercepts earlier to be $x = 0$ and $x = 4$. Next, graph $y = x^2 - 4x$. Solutions of the quadratic inequality are x-values where $y \geq 0$, or where the graph is on or above the x-axis.

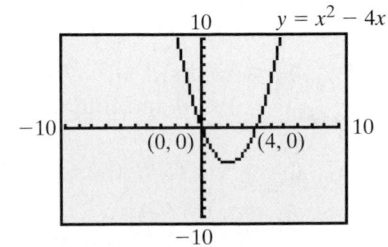

The x-values where the graph is on or above the x-axis are the x-intercepts of 0 and 4 and the x-values to the left of the x-intercept 0 and to the right of the x-intercept 4. In interval notation, the solution set is $(-\infty, 0] \cup [4, \infty)$.

EXAMPLE 4

Use a graphical approach to solve $(x + 2)(x - 1)(x - 5) < 0$.

Solution

The x-intercepts of the graph are found by solving $(x + 2)(x - 1)(x - 5) = 0$. These x-intercepts are $-2, 1,$ and 5. Graph $y = (x + 2)(x - 1)(x - 5)$. To solve $y < 0$, find x-values where the graph lies below the x-axis.

Again, the solution set of $y < 0$ contains x-values where the graph of y is below the x-axis. In interval notation, the solution set is $(-\infty, -2) \cup (1, 5)$.

$y = (x + 2)(x - 1)(x - 5)$

EXAMPLE 5

Use a graphical approach to solve $x^2 + 2x > -4$.

Solution

In standard form, the quadratic inequality is $x^2 + 2x + 4 > 0$. The graph of $y = x^2 + 2x + 4$ is shown below. Notice that the entire graph lies above the x-axis, so $y > 0$ for all values of x. The solution set in interval notation is $(-\infty, \infty)$.

TEACHING TIP

Remind students to write the related equation so that the right side = 0 when solving inequalities graphically.

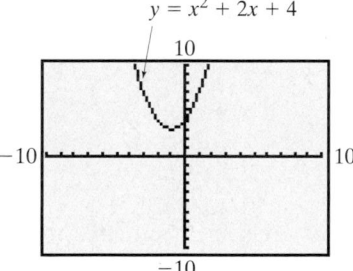

$y = x^2 + 2x + 4$

2 Inequalities containing rational expressions with variables in the denominator are solved by using a similar procedure.

EXAMPLE 6

Solve $\dfrac{x + 2}{x - 3} \leq 0$.

Solution

First we find all values that make the denominator equal to 0. To do this, we solve $x - 3 = 0$ and find that $x = 3$.

Next, we solve the related equation $\dfrac{x + 2}{x - 3} = 0$.

$$\frac{x + 2}{x - 3} = 0$$

$$x + 2 = 0 \qquad \text{Multiply both sides by the LCD, } x - 3.$$

$$x = -2$$

Now we place these numbers on a number line and proceed as before, checking test point values in the original inequality.

$$
\begin{array}{ccc}
A & B & C \\
\hline
-2 & 3 &
\end{array}
$$

Choose -3 from region A.

$$\frac{x + 2}{x - 3} \le 0$$

$$\frac{-3 + 2}{-3 - 3} \le 0$$

$$\frac{-1}{-6} \le 0$$

$$\frac{1}{6} \le 0 \quad \text{False}$$

Choose 0 from region B.

$$\frac{x + 2}{x - 3} \le 0$$

$$\frac{0 + 2}{0 - 3} \le 0$$

$$-\frac{2}{3} \le 0 \quad \text{True}$$

Choose 4 from region C.

$$\frac{x + 2}{x - 3} \le 0$$

$$\frac{4 + 2}{4 - 3} \le 0$$

$$6 \le 0 \quad \text{False}$$

The solution set is $[-2, 3)$. This interval includes -2 because -2 satisfies the original inequality. This interval does not include 3, because 3 would make the denominator 0.

$$
\begin{array}{ccccc}
& A & B & C & \\
\hline
F & -2 & T & 3 & F
\end{array}
$$

To visualize this graphically, graph

$$y = \frac{x + 2}{x - 3}$$

and find x-values where the graph is on or below the x-axis. The graph shown to the left is on the x-axis at $x = -2$. The graph is below the x-axis from -2 to the undefined value $x = 3$. Thus, we verify the solution set as $[-2, 3)$.

The following steps may be used to algebraically solve a rational inequality with variables in the denominator.

Solving a Rational Inequality Algebraically

Step 1: Solve for values that make all denominators 0.

Step 2: Solve the related equation.

Step 3: Separate the number line into regions with the solutions from Steps 1 and 2.

Step 4: For each region, choose a test point and determine whether its value satisfies the *original inequality.*

Step 5: The solution set includes the regions whose test point value is a solution. Check whether to include values from Step 2. Be sure *not* to include values that make any denominator 0.

To graphically solve a rational inequality with variables in the denominator, use the steps below.

> ## Solving a Rational Inequality Graphically
>
> **Step 1:** Write the related equation so that the right side is 0.
>
> **Step 2:** Enter the left side expression from Step 1 into the Y = editor and graph.
>
> **Step 3:** If < or ≤, determine the intervals where the graph is below the *x*-axis. If > or ≥, determine the intervals where the graph is above the *x*-axis.
>
> **Step 4:** Write down the solution intervals from Step 3. If ≤ or ≥, include the endpoints of the intervals as solutions by using brackets. Remember not to include values that make any expression undefined.

EXAMPLE 7

Solve $\dfrac{5}{x+1} < -2$.

Solution   First we find values for x that make the denominator equal to 0.

$$x + 1 = 0$$
$$x = -1$$

Next we solve $\dfrac{5}{x+1} = -2$.

$$(x+1) \cdot \frac{5}{x+1} = (x+1) \cdot -2 \qquad \text{Multiply both sides by the LCD, } x+1.$$
$$5 = -2x - 2 \qquad \text{Simplify.}$$
$$7 = -2x$$
$$-\frac{7}{2} = x$$

We use these two solutions to divide a number line into three regions and choose test points. Only a test point value from region B satisfies the *original inequality*. The solution set is $\left(-\dfrac{7}{2}, -1\right)$, and its graph is shown.

The graph of $y = \dfrac{5}{x+1} + 2$. The solution is found by the portion of the graph below the *x*-axis. That is, from the *x*-intercept $x = -3.5$ to where the expression is undefined at $x = -1$.

STUDY SKILLS REMINDER

Continue your outline started in Section 1.5. Write how to recognize and how to solve nonlinear inequalities in your own words. For example:

Solving Equations and Inequalities

I. Equations

    A. Linear equations (Sec. 1.5)

    B. Absolute value equations (Sec. 3.4)

    C. Quadratic and higher degree equations (Sec. 5.8 and Chapter 8)

    D. Equations with rational expressions (Sec. 6.6)

    E. Equations with radicals (Sec. 7.6)

II. Inequalities

    A. Linear inequalities (Sec. 3.2)

    B. Compound inequalities (Sec. 3.3)

    C. Absolute value inequalities (Sec. 3.5)

    D. Nonlinear inequalities—

        1. Polynomial inequalities — Recognize: *inequalities that are quadratic or higher degree* — Solve: write in standard form (equal to 0). Solve the related equation and use solutions to divide the number line into regions. Check regions with test values.

        2. Rational inequalities — Recognize: *inequalities that contain a variable in the denominator* — Solve: find values that make all denominators 0 and find solutions to the related equation. Use these numbers to divide the number line into regions. Check regions with test values in the original inequality.

1. ⟷ , $(-\infty, -5) \cup (-1, \infty)$
$\quad\quad$ −5 −1

2. ⟷ , $[-5, -1]$
$\quad\quad$ −5 −1

3. ⟷ , $[-4, 3]$
$\quad\quad$ −4 $\quad$ 3

4. ⟷ , $(-\infty, -4) \cup (1, \infty)$
$\quad\quad$ −4 $\quad$ 1

5. ⟷ , $[2, 5]$
$\quad\quad$ 2 $\quad$ 5

6. ⟷ , $(-\infty, -5] \cup [-3, \infty)$
$\quad\quad$ −5 −3

7. ⟷ , $\left(-5, -\frac{1}{3}\right)$
$\quad\quad$ −5 $\quad$ $-\frac{1}{3}$

8. ⟷ , $\left(-1, \frac{7}{2}\right)$
$\quad\quad$ −1 $\quad$ $\frac{7}{2}$

9. ⟷ , $(2, 4) \cup (6, \infty)$
$\quad\quad$ 2 $\quad$ 4 $\quad$ 6

10. ⟷ , $(-\infty, 2] \cup [4, 6]$
$\quad\quad$ 2 $\quad$ 4 $\quad$ 6

11. ⟷ , $(-\infty, -4] \cup [0, 1]$
$\quad\quad$ −4 $\quad$ 0 $\quad$ 1

12. ⟷ , $(-2, 0) \cup (6, \infty)$
$\quad\quad$ −2 $\quad$ 0 $\quad$ 6

EXERCISE SET 8.4

STUDY GUIDE/SSM $\quad$ CD/VIDEO $\quad$ PH MATH TUTOR CENTER $\quad$ MathXL®Tutorials ON CD $\quad$ MathXL® $\quad$ MyMathLab®

Solve each quadratic inequality. Graph the solution set and write the solution set in interval notation. See Examples 1 through 5.

1. $(x + 1)(x + 5) > 0$ $\qquad$ **2.** $(x + 1)(x + 5) \leq 0$

3. $(x - 3)(x + 4) \leq 0$ $\qquad$ **4.** $(x + 4)(x - 1) > 0$

5. $x^2 - 7x + 10 \leq 0$ $\qquad$ **6.** $x^2 + 8x + 15 \geq 0$

7. $3x^2 + 16x < -5$ $\qquad$ **8.** $2x^2 - 5x < 7$

9. $(x - 6)(x - 4)(x - 2) > 0$

10. $(x - 6)(x - 4)(x - 2) \leq 0$

11. $x(x - 1)(x + 4) \leq 0$

12. $x(x - 6)(x + 2) > 0$

Use the related graph to solve each inequality. Write the solution set in interval notation. See Examples 3 through 5.

13. $(x^2 - 9)(x^2 - 4) \leq 0$ $[-3, -2] \cup [2, 3]$

14. $(x^2 - 16)(x^2 - 1) \leq 0$ $[-4, -1] \cup [1, 4]$

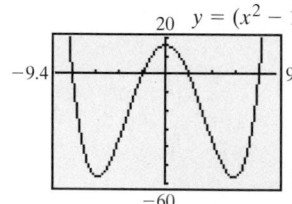

Solve each inequality. Graph the solution set and write the solution set in interval notation. See Example 6.

15. $\dfrac{x + 7}{x - 2} < 0$ **16.** $\dfrac{x - 5}{x - 6} > 0$ **17.** $\dfrac{5}{x + 1} > 0$

18. $\dfrac{3}{y - 5} < 0$

Use the related graph to solve each inequality. Write the solution set in interval notation. See Example 7.

19. $\dfrac{x + 1}{x - 4} \geq 0$ $(-\infty, -1] \cup (4, \infty)$

$y = \dfrac{x + 1}{x - 4}$

20. $\dfrac{x + 1}{x - 4} \leq 0$ $[-1, 4)$

$y = \dfrac{x + 1}{x - 4}$

Solve each inequality. Graph the solution set and write the solution set in interval notation. See Example 7.

21. $\dfrac{3}{x - 2} < 4$ **22.** $\dfrac{-2}{y + 3} > 2$ **23.** $\dfrac{x^2 + 6}{5x} \geq 1$

24. $\dfrac{y^2 + 15}{8y} \leq 1$

---

15. ⟵⟶ , $(-7, 2)$
   −7   2

16. ⟵⟶ , $(-\infty, 5) \cup (6, \infty)$
   5   6

MIXED PRACTICE

Solve each inequality. Graph the solution set and write the solution set in interval notation.

25. $(x - 8)(x + 7) > 0$ **26.** $(x - 5)(x + 1) < 0$

27. $(2x - 3)(4x + 5) \leq 0$ **28.** $(6x + 7)(7x - 12) > 0$

29. $x^2 > x$ **30.** $x^2 < 25$

31. $(2x - 8)(x + 4)(x - 6) \leq 0$

32. $(3x - 12)(x + 5)(2x - 3) \geq 0$

33. $6x^2 - 5x \geq 6$

34. $12x^2 + 11x \leq 15$

35. $4x^3 + 16x^2 - 9x - 36 > 0$

36. $x^3 + 2x^2 - 4x - 8 < 0$

37. $x^4 - 26x^2 + 25 \geq 0$

38. $16x^4 - 40x^2 + 9 \leq 0$

39. $(2x - 7)(3x + 5) > 0$

40. $(4x - 9)(2x + 5) < 0$

41. $\dfrac{x}{x - 10} < 0$

42. $\dfrac{x + 10}{x - 10} > 0$

43. $\dfrac{x - 5}{x + 4} \geq 0$

44. $\dfrac{x - 3}{x + 2} \leq 0$

45. $\dfrac{x(x + 6)}{(x - 7)(x + 1)} \geq 0$

46. $\dfrac{(x - 2)(x + 2)}{(x + 1)(x - 4)} \leq 0$

47. $\dfrac{-1}{x - 1} > -1$

48. $\dfrac{4}{y + 2} < -2$

49. $\dfrac{x}{x + 4} \leq 2$

50. $\dfrac{4x}{x - 3} \geq 5$

51. $\dfrac{z}{z - 5} \geq 2z$

52. $\dfrac{p}{p + 4} \leq 3p$

53. $\dfrac{(x + 1)^2}{5x} > 0$

54. $\dfrac{(2x - 3)^2}{x} < 0$

---

17. ⟵⟶ , $(-1, \infty)$
   −1

18. ⟵⟶ , $(-\infty, 5)$
   5

22. ⟵⟶ , $(-4, -3)$
   −4   −3

21. ⟵⟶ , $(-\infty, 2) \cup \left(\frac{11}{4}, \infty\right)$
   2   $\frac{11}{4}$

23. ⟵⟶ , $(0, 2] \cup [3, \infty)$
   0   2   3

40. ⟵⟶ , $\left(-\frac{5}{2}, \frac{9}{4}\right)$
   $-\frac{5}{2}$ $\frac{9}{4}$

41. ⟵⟶ , $(0, 10)$
   0   10

44. ⟵⟶ , $(-2, 3]$
   −2   3

24. ⟵⟶ , $(-\infty, 0) \cup [3, 5]$
   0   3   5

25. ⟵⟶ , $(-\infty, -7) \cup (8, \infty)$
   −7   8

26. ⟵⟶ , $(-1, 5)$
   −1   5

27. ⟵⟶ , $\left[-\frac{5}{4}, \frac{3}{2}\right]$
   $-\frac{5}{4}$ $\frac{3}{2}$

48. ⟵⟶ , $(-4, -2)$
   −4   −2

---

30. ⟵⟶ , $(-5, 5)$
   −5   5

28. ⟵⟶ , $\left(-\infty, -\frac{7}{6}\right) \cup \left(\frac{12}{7}, \infty\right)$
   $-\frac{7}{6}$ $\frac{12}{7}$

34. ⟵⟶ , $\left[-\frac{5}{3}, \frac{3}{4}\right]$
   $-\frac{5}{3}$ $\frac{3}{4}$

29. ⟵⟶ , $(-\infty, 0) \cup (1, \infty)$
   0   1

31. ⟵⟶ , $(-\infty, -4] \cup [4, 6]$
   −4   4   6

32. ⟵⟶ , $\left[-5, \frac{3}{2}\right] \cup [4, \infty)$
   −5 $\frac{3}{2}$ 4

33. ⟵⟶ , $\left(-\infty, -\frac{2}{3}\right] \cup \left[\frac{3}{2}, \infty\right)$
   $-\frac{2}{3}$ $\frac{3}{2}$

35. ⟵⟶ , $\left(-4, -\frac{3}{2}\right) \cup \left(\frac{3}{2}, \infty\right)$
   −4 $-\frac{3}{2}$ $\frac{3}{2}$

36. ⟵⟶ , $(-\infty, -2) \cup (-2, 2)$
   −2   2

37. ⟵⟶ , $(-\infty, -5] \cup [-1, 1] \cup [5, \infty)$
   −5 −1   1   5

38. ⟵⟶ , $\left[-\frac{3}{2}, -\frac{1}{2}\right] \cup \left[\frac{1}{2}, \frac{3}{2}\right]$
   $-\frac{3}{2}$ $-\frac{1}{2}$ $\frac{1}{2}$ $\frac{3}{2}$

39. ⟵⟶ , $\left(-\infty, -\frac{5}{3}\right) \cup \left(\frac{7}{2}, \infty\right)$
   $-\frac{5}{3}$ $\frac{7}{2}$

42. ⟵⟶ , $(-\infty, -10) \cup (10, \infty)$
   −10   10

43. ⟵⟶ , $(-\infty, -4) \cup [5, \infty)$
   −4   5

45. ⟵⟶ , $(-\infty, -6] \cup (-1, 0] \cup (7, \infty)$
   −6 −1   0   7

46. ⟵⟶ , $[-2, -1) \cup [2, 4)$
   −2 −1   2   4

47. ⟵⟶ , $(-\infty, 1) \cup (2, \infty)$
   1   2

49. ⟵⟶ , $(-\infty, -8] \cup (-4, \infty)$
   −8   −4

50. ⟵⟶ , $(3, 15]$
   3   15

51. ⟵⟶ , $(-\infty, 0] \cup \left(5, \frac{11}{2}\right]$
   0   5 $\frac{11}{2}$

52. ⟵⟶ , $\left(-4, -\frac{11}{3}\right] \cup [0, \infty)$
   −4 $-\frac{11}{3}$ 0

---

53. ⟵⟶ , $(0, \infty)$
   0

54. ⟵⟶ , $(-\infty, 0)$
   0

REVIEW AND PREVIEW

Recall that the graph of $f(x) + K$ is the same as the graph of $f(x)$ shifted K units upward if $K > 0$ and $|K|$ units downward if $K < 0$. Use the graph of $f(x) = |x|$ below to sketch the graph of each function. (See Sections 2.1 and 2.3.)   55.–58. See graphing answer section.

55. $g(x) = |x| + 2$ **56.** $H(x) = |x| - 2$

57. $F(x) = |x| - 1$ **58.** $h(x) = |x| + 5$

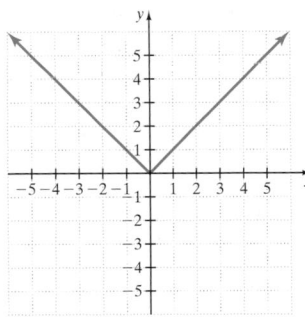

Use the graph of $f(x) = x^2$ below to sketch the graph of each function.   59.–62.   See graphing answer section.

59. $F(x) = x^2 - 3$ **60.** $h(x) = x^2 - 4$

61. $H(x) = x^2 + 1$ **62.** $g(x) = x^2 + 3$

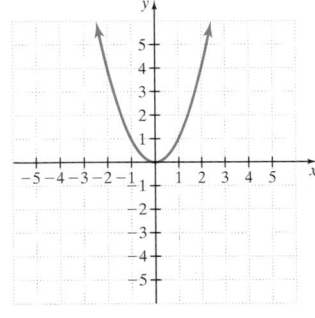

Concept Extensions

Find all numbers that satisfy each of the following.

63. Explain why $\dfrac{x + 2}{x - 3} > 0$ and $(x + 2)(x - 3) > 0$ have the same solutions.   answers may vary

64. Explain why $\dfrac{x + 2}{x - 3} \geq 0$ and $(x + 2)(x - 3) \geq 0$ do not have the same solutions.   answers may vary

65. A number minus its reciprocal is less than zero. Find the numbers.   any number less than -1 or between 0 and 1

66. Twice a number added to its reciprocal is nonnegative. Find the numbers.   any number greater than 0

67. The total profit function $P(x)$ for a company producing x thousand units is given by
$$P(x) = -2x^2 + 26x - 44$$
Find the values of x for which the company makes a profit. [*Hint:* The company makes a profit when $P(x) > 0$.]   x is between 2 and 11

68. A projectile is fired straight up from the ground with an initial velocity of 80 feet per second. Its height $s(t)$ in feet at any time t is given by the function
$$s(t) = -16t^2 + 80t$$
Find the interval of time for which the height of the projectile is greater than 96 feet.   between 2 and 3 sec

Use a graphing calculator to check each exercise.   69.–72.   See graphing answer section.

69. Exercise 25.   **70.** Exercise 26.

71. Exercise 37.   **72.** Exercise 38.

8.5   *QUADRATIC FUNCTIONS AND THEIR GRAPHS*

Objectives

1 Graph quadratic functions of the form $f(x) = x^2 + k$.

2 Graph quadratic functions of the form $f(x) = (x - h)^2$.

3 Graph quadratic functions of the form $f(x) = (x - h)^2 + k$.

4 Graph quadratic functions of the form $f(x) = ax^2$.

5 Graph quadratic functions of the form $f(x) = a(x - h)^2 + k$.

1 We first graphed the quadratic equation $y = x^2$ in Section 2.1. In Section 2.2, we learned that this graph defines a function, and we wrote $y = x^2$ as $f(x) = x^2$. In these sections, we discovered that the graph of a quadratic function is a parabola opening upward or downward. In this section, we continue our study of quadratic functions and their graphs.

First, let's recall the definition of a quadratic function.

TECHNOLOGY NOTE

An Apps for the TI-83 plus can be downloaded from http://www.education.ti.com/apps called Transform that can be used to discover changes to the parabola when a, b, or c varies in $f(x) = ax^2 + bx + c$.

Quadratic Function

A quadratic function is a function that can be written in the form $f(x) = ax^2 + bx + c$, where a, b, and c are real numbers and $a \neq 0$.

Notice that equations of the form $y = ax^2 + bx + c$, where $a \neq 0$, define quadratic functions, since y is a function of x or $y = f(x)$.

Recall that if $a > 0$, the parabola opens upward and if $a < 0$, the parabola opens downward. Also, the vertex of a parabola is the lowest point if the parabola opens upward and the highest point if the parabola opens downward. The axis of symmetry is the vertical line that passes through the vertex.

TEACHING TIP

Have students use a graphing calculator to graph several equations of the form $y = x^2 + k$ or $y = x^2 - k$, where k is any number. Suggest that they sketch the graph first and then use the graphing calculator to verify.

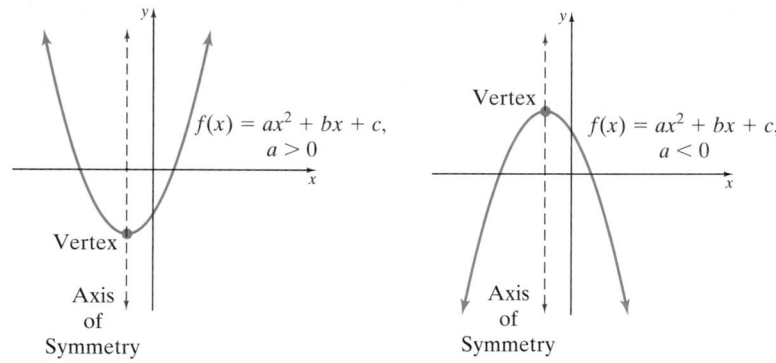

Discover the Concept

Graph $f(x) = x^2$, $g(x) = x^2 + 3$ and $h(x) = x^2 - 5$. Compare the graphs of $g(x)$ and $h(x)$ to the graph of $f(x)$.

The graphs of $y_1 = x^2$, $y_2 = x^2 + 3$, $y_3 = x^2 - 5$ and corresponding tables are shown below.

Each y-value is increased by 3

Each y-value is decreased by 5

Notice that the graph of $y_2 = x^2 + 3$ is the same as the graph of $y_1 = x^2$, but shifted *upward* 3 units. Also, the graph of $y_3 = x^2 - 5$ is the same as the graph of $y_1 = x^2$, but shifted *downward* 5 units. The axis of symmetry is the same for all graphs: $x = 0$.

In general, we have the following properties.

Graphing the Parabola Defined by $f(x) = x^2 + k$

If k is positive, the graph of $f(x) = x^2 + k$ is the graph of $y = x^2$ shifted upward k units.
If k is negative, the graph of $f(x) = x^2 + k$ is the graph of $y = x^2$ shifted downward $|k|$ units.
The vertex is $(0, k)$, and the axis of symmetry is the y-axis.

EXAMPLE 1

Graph each function.

a. $F(x) = x^2 + 2$ **b.** $g(x) = x^2 - 3$

Solution   **a.** $F(x) = x^2 + 2$

The graph of $F(x) = x^2 + 2$ is obtained by shifting the graph of $y = x^2$ upward 2 units.

b. $g(x) = x^2 - 3$

The graph of $g(x) = x^2 - 3$ is obtained by shifting the graph of $y = x^2$ downward 3 units.

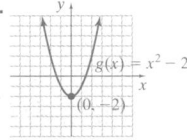

2 Now we will graph functions of the form $f(x) = (x - h)^2$.

Discover the Concept

Graph $f(x) = x^2$, $g(x) = (x - 2)^2$, and $h(x) = (x + 4)^2$. Compare the graphs of $g(x)$ and $h(x)$ to the graph of $f(x)$.

The graphs of $y_1 = x^2$, $y_2 = (x - 2)^2$, and $y_3 = (x + 4)^2$ are shown below.

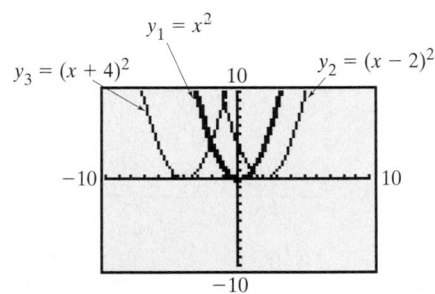

The graph of $y_2 = (x - 2)^2$ is the same as the graph of $y_1 = x^2$ shifted *to the right* 2 units. Also, the graph of $y_3 = (x + 4)^2$ is the same as the graph of $y_1 = x^2$ shifted *to the left* 4 units. Notice that the axes of symmetry have shifted also.

In general, we have the following properties.

> ### Graphing the Parabola Defined by $f(x) = (x - h)^2$
>
> If h is positive, the graph of $f(x) = (x - h)^2$ is the graph of $y = x^2$ shifted to the right h units.
> If h is negative, the graph of $f(x) = (x - h)^2$ is the graph of $y = x^2$ shifted to the left $|h|$ units.
> The vertex is $(h, 0)$, and the axis of symmetry is the vertical line $x = h$.

EXAMPLE 2

Graph each function.

a. $G(x) = (x - 3)^2$ **b.** $F(x) = (x + 1)^2$

Solution **a.** The graph of $G(x) = (x - 3)^2$ is obtained by shifting the graph of $y = x^2$ to the right 3 units. The graph of $G(x)$ is below on the left.

b. The equation $F(x) = (x + 1)^2$ can be written as $F(x) = [x - (-1)]^2$. The graph of $F(x) = [x - (-1)]^2$ is obtained by shifting the graph of $y = x^2$ to the left 1 unit. The graph of $F(x)$ is below on the right.

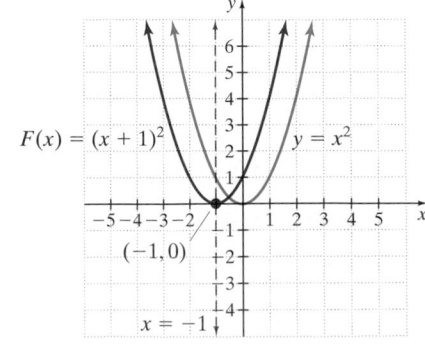

TEACHING TIP
You or your students use a graphing calculator to graph several equations of the form $y = (x - h)^2$ or $y = (x + h)^2$, where h is any number. Sketch the graph first and then use a graphing calculator to verify.

3 As we will see in graphing functions of the form $f(x) = (x - h)^2 + k$, it is possible to combine vertical and horizontal shifts.

> **Graphing the Parabola Defined by $f(x) = (x - h)^2 + k$**
>
> The parabola has the same shape as $y = x^2$.
> The vertex is (h, k), and the axis of symmetry is the vertical line $x = h$.

EXAMPLE 3

Graph $F(x) = (x - 3)^2 + 1$.

Solution   The graph of $F(x) = (x - 3)^2 + 1$ is the graph of $y = x^2$ shifted 3 units to the right and 1 unit up. The vertex is then $(3, 1)$, and the axis of symmetry is $x = 3$. A few ordered pair solutions are plotted to aid in graphing.

CLASSROOM EXAMPLE

Graph $F(x) = (x - 2)^2 + 3$.

answer:

x	$F(x) = (x - 3)^2 + 1$
1	5
2	2
4	2
5	5

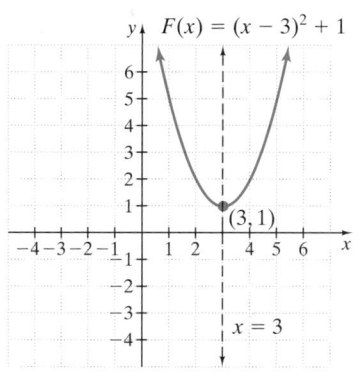

4 Next, we discover the change in the shape of the graph when the coefficient of x^2 is not 1.

Discover the Concept

Graph $f(x) = ax^2$ for $a = 1, 3$ and $\frac{1}{2}$. Compare the other graphs to the graph of $f(x) = 1x^2$ or $f(x) = x^2$.

The graphs of $y_1 = x^2$, $y_2 = 3x^2$, and $y_3 = \frac{1}{2}x^2$ and corresponding tables are shown below.

Compare the table of values. We see that for each x-value the corresponding value of y_2 is triple the corresponding value of y_1. Similarly, the value of y_3 is half the value of y_1. The result is that the graph of $y_2 = 3x^2$ is narrower than the graph of $f(x) = x^2$ and the graph of $y_3 = \frac{1}{2}x^2$ is wider. The vertex for each graph is $(0, 0)$, and the axis of symmetry is the y-axis.

> ### Graphing the Parabola Defined by $f(x) = ax^2$
>
> If a is positive, the parabola opens upward, and if a is negative, the parabola opens downward.
> If $|a| > 1$, the graph of the parabola is narrower than the graph of $y = x^2$.
> If $|a| < 1$, the graph of the parabola is wider than the graph of $y = x^2$.

EXAMPLE 4

Graph $f(x) = -2x^2$.

Solution   Because $a = -2$, a negative value, this parabola opens downward. Since $|-2| = 2$ and $2 > 1$, the parabola is narrower than the graph of $y = x^2$. The vertex is $(0, 0)$, and the axis of symmetry is the y-axis. We verify this by plotting a few points.

x	$f(x) = -2x^2$
-2	-8
-1	-2
0	0
1	-2
2	-8

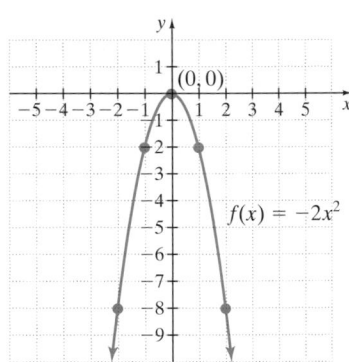

5   Now we will see the shape of the graph of a quadratic function of the form $f(x) = a(x - h)^2 + k$.

EXAMPLE 5

Find the vertex and the axis of symmetry of the graph of $g(x) = \frac{1}{2}(x + 2)^2 + 5$. Use a graphing utility to verify the results.

Solution   The function $g(x) = \frac{1}{2}(x + 2)^2 + 5$ may be written as $g(x) = \frac{1}{2}[x - (-2)]^2 + 5$. Thus, this graph is the same shape as the graph of $y = x^2$ shifted 2 units to the left and 5 units up, and it is wider because a is $\frac{1}{2}$. The vertex is $(-2, 5)$, and the axis of symmetry is $x = -2$. The screen below shows the graph with the vertex indicated.

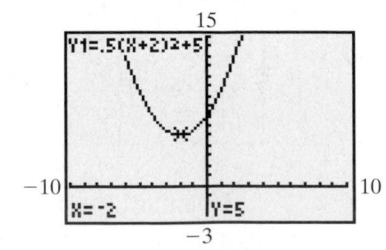

In general, the following holds.

Graph of a Quadratic Function

The graph of a quadratic function written in the form $f(x) = a(x - h)^2 + k$ is a parabola with vertex (h, k). If $a > 0$, the parabola opens upward, and if $a < 0$, the parabola opens downward. The axis of symmetry is the line whose equation is $x = h$.

✔ **CONCEPT CHECK**

Which description of the graph of $f(x) = -0.35(x + 3)^2 - 4$ is correct?

a. The graph opens downward and has its vertex at $(-3, 4)$.

b. The graph opens upward and has its vertex at $(-3, 4)$.

c. The graph opens downward and has its vertex at $(-3, -4)$.

d. The graph is narrower than the graph of $y = x^2$.

Concept Check Answer:
c

TEACHING TIP

Have students write an equation that satisfies given criteria such as "narrow parabola with maximum in third quadrant" or "wide parabola with line of symmetry of $x = -5$." Then have them give their equation to another student to check.

MENTAL MATH

State the vertex of the graph of each quadratic function.   **7.** $(-1, 5)$ **8.** $(10, -7)$

1. $f(x) = x^2$ $(0, 0)$ 　　　　**2.** $f(x) = -5x^2$ $(0, 0)$ 　　　　**3.** $g(x) = (x - 2)^2$ $(2, 0)$ 　　　　**4.** $g(x) = (x + 5)^2$ $(-5, 0)$

5. $f(x) = 2x^2 + 3$ $(0, 3)$ 　　**6.** $h(x) = x^2 - 1$ $(0, -1)$ 　　**7.** $g(x) = (x + 1)^2 + 5$ 　　　　**8.** $h(x) = (x - 10)^2 - 7$

TEACHING TIP

A Group Activity for this section is available in the Instructor's Resource Manual.

EXERCISE SET 8.5

STUDY GUIDE/SSM　　CD/VIDEO　　PH MATH TUTOR CENTER　　MathXL®Tutorials ON CD　　MathXL®　　MyMathLab®

1.–54. See graphing answer section.

Sketch the graph of each quadratic function. Label the vertex, and sketch and label the axis of symmetry. See Examples 1 through 3.

1. $f(x) = x^2 - 1$ 　　　　**2.** $g(x) = x^2 + 3$

 3. $h(x) = x^2 + 5$ 　　　**4.** $h(x) = x^2 - 4$

5. $g(x) = x^2 + 7$ 　　　　**6.** $f(x) = x^2 - 2$

7. $f(x) = (x - 5)^2$ 　　　**8.** $g(x) = (x + 5)^2$

9. $h(x) = (x + 2)^2$ 　　　**10.** $H(x) = (x - 1)^2$

11. $G(x) = (x + 3)^2$ 　　　**12.** $f(x) = (x - 6)^2$

13. $f(x) = (x - 2)^2 + 5$ 　　**14.** $g(x) = (x - 6)^2 + 1$

15. $h(x) = (x + 1)^2 + 4$ **16.** $G(x) = (x + 3)^2 + 3$

17. $g(x) = (x + 2)^2 - 5$ **18.** $h(x) = (x + 4)^2 - 6$

Sketch the graph of each quadratic function. Label the vertex, and sketch and label the axis of symmetry. See Example 4.

19. $g(x) = -x^2$ **20.** $f(x) = 5x^2$

21. $h(x) = \dfrac{1}{3}x^2$ **22.** $g(x) = -3x^2$

23. $H(x) = 2x^2$ **24.** $f(x) = -\dfrac{1}{4}x^2$

Sketch the graph of each quadratic function. Label the vertex, and sketch and label the axis of symmetry.

25. $f(x) = 2(x - 1)^2 + 3$ **26.** $g(x) = 4(x - 4)^2 + 2$

27. $h(x) = -3(x + 3)^2 + 1$ **28.** $f(x) = -(x - 2)^2 - 6$

29. $H(x) = \dfrac{1}{2}(x - 6)^2 - 3$ **30.** $G(x) = \dfrac{1}{5}(x + 4)^2 + 3$

MIXED PRACTICE   **63.** $-5 - \sqrt{26}, -5 + \sqrt{26}$

Sketch the graph of each quadratic function. Label the vertex, and sketch and label the axis of symmetry. See Example 5.

31. $f(x) = -(x - 2)^2$ **32.** $g(x) = -(x + 6)^2$

33. $F(x) = -x^2 + 4$ **34.** $H(x) = -x^2 + 10$

35. $F(x) = 2x^2 - 5$ **36.** $g(x) = \dfrac{1}{2}x^2 - 2$

37. $h(x) = (x - 6)^2 + 4$ **38.** $f(x) = (x - 5)^2 + 2$

39. $F(x) = \left(x + \dfrac{1}{2}\right)^2 - 2$ **40.** $H(x) = \left(x + \dfrac{1}{2}\right)^2 - 3$

41. $F(x) = \dfrac{3}{2}(x + 7)^2 + 1$ **42.** $g(x) = -\dfrac{3}{2}(x - 1)^2 - 5$

43. $f(x) = \dfrac{1}{4}x^2 - 9$ **44.** $H(x) = \dfrac{3}{4}x^2 - 2$

45. $G(x) = 5\left(x + \dfrac{1}{2}\right)^2$ **46.** $F(x) = 3\left(x - \dfrac{3}{2}\right)^2$

47. $h(x) = -(x - 1)^2 - 1$ **48.** $f(x) = -3(x + 2)^2 + 2$

49. $g(x) = \sqrt{3}(x + 5)^2 + \dfrac{3}{4}$

50. $G(x) = \sqrt{5}(x - 7)^2 - \dfrac{1}{2}$

51. $h(x) = 10(x + 4)^2 - 6$

52. $h(x) = 8(x + 1)^2 + 9$

53. $f(x) = -2(x - 4)^2 + 5$

54. $G(x) = -4(x + 9)^2 - 1$

REVIEW AND PREVIEW   **64.** $-7 - \sqrt{29}, -7 + \sqrt{29}$

Add the proper constant to each binomial so that the resulting trinomial is a perfect square trinomial. See Section 8.1.

55. $x^2 + 8x$ $x^2 + 8x + 16$ **56.** $y^2 + 4y$ $y^2 + 4y + 4$

57. $z^2 - 16z$ $z^2 - 16z + 64$ **58.** $x^2 - 10x$ $x^2 - 10x + 25$

59. $y^2 + y$ $y^2 + y + \dfrac{1}{4}$ **60.** $z^2 - 3z$ $z^2 - 3z + \dfrac{9}{4}$

Solve by completing the square. See Section 8.1.

61. $x^2 + 4x = 12$ $-6, 2$ **62.** $y^2 + 6y = -5$ $-5, -1$

63. $z^2 + 10z - 1 = 0$ **64.** $x^2 + 14x + 20 = 0$

65. $z^2 - 8z = 2$ **66.** $y^2 - 10y = 3$

Concept Extensions   **65.** $4 - 3\sqrt{2}, 4 + 3\sqrt{2}$

Write the equation of the parabola that has the same shape as $f(x) = 5x^2$ but with the following vertex.

67. $(2, 3)$ **68.** $(1, 6)$

69. $(-3, 6)$ **70.** $(4, -1)$ $f(x) = 5(x - 4)^2 - 1$

The shifting properties covered in this section apply to the graphs of all functions. Given the accompanying graph of $y = f(x)$, sketch the graph of each of the following.   **71.–76.** See graphing answer section.

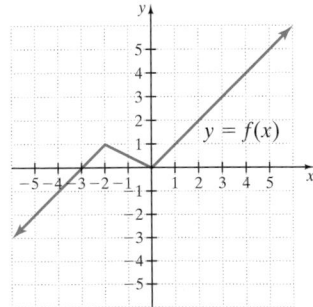

71. $y = f(x) + 1$ **72.** $y = f(x) - 2$

73. $y = f(x - 3)$ **74.** $y = f(x + 3)$

75. $y = f(x + 2) + 2$ **76.** $y = f(x - 1) + 1$

66. $5 - 2\sqrt{7}, 5 + 2\sqrt{7}$ **67.** $f(x) = 5(x - 2)^2 + 3$ **68.** $f(x) = 5(x - 1)^2 + 6$ **69.** $f(x) = 5(x + 3)^2 + 6$

8.6 FURTHER GRAPHING OF QUADRATIC FUNCTIONS

Objectives

1. Write quadratic functions in the form $y = a(x - h)^2 + k$.
2. Derive a formula for finding the vertex of a parabola.
3. Find the minimum or maximum value of a quadratic function.

1 We know that the graph of a quadratic function is a parabola. If a quadratic function is written in the form

$$f(x) = a(x - h)^2 + k$$

we can easily find the vertex (h, k) and graph the parabola. To write a quadratic function in this form, complete the square. (See Section 8.1 for a review of completing the square.)

EXAMPLE 1

Graph $f(x) = x^2 - 4x - 12$. Find the vertex and any intercepts.

Solution The graph of this quadratic function is a parabola. To find the vertex of the parabola, we will write the function in the form $y = (x - h)^2 + k$. To do this, we complete the square on the binomial $x^2 - 4x$. To simplify our work, we let $f(x) = y$.

$$y = x^2 - 4x - 12 \qquad \text{Let } f(x) = y.$$

$$y + 12 = x^2 - 4x \qquad \text{Add 12 to both sides to get the } x\text{-variable terms alone.}$$

Now we add the square of half of -4 to both sides.

$$\frac{1}{2}(-4) = -2 \quad \text{and} \quad (-2)^2 = 4$$

$$y + 12 + 4 = x^2 - 4x + 4 \qquad \text{Add 4 to both sides.}$$

$$y + 16 = (x - 2)^2 \qquad \text{Factor the trinomial.}$$

$$y = (x - 2)^2 - 16 \qquad \text{Subtract 16 from both sides.}$$

$$f(x) = (x - 2)^2 - 16 \qquad \text{Replace } y \text{ with } f(x).$$

From this equation, we can see that the vertex of the parabola is $(2, -16)$, a point in quadrant IV, and the axis of symmetry is the line $x = 2$.

Notice that $a = 1$. Since $a > 0$, the parabola opens upward. This parabola opening upward with vertex $(2, -16)$ will have two x-intercepts and one y-intercept. (See the Helpful Hint after this example.)

x-intercepts: let y or $f(x) = 0$ y-intercept: let $x = 0$

$$f(x) = x^2 - 4x - 12 \qquad\qquad f(x) = x^2 - 4x - 12$$

$$0 = x^2 - 4x - 12 \qquad\qquad f(0) = 0^2 - 4 \cdot 0 - 12$$

$$0 = (x - 6)(x + 2) \qquad\qquad\quad = -12$$

$$0 = x - 6 \quad \text{or} \quad 0 = x + 2$$

$$6 = x \qquad\qquad -2 = x$$

TEACHING TIP
For Example 1, have students use the y-intercept and the axis of symmetry to identify another point on the graph.

A check for Example 1.

The two x-intercepts are $(6, 0)$ and $(-2, 0)$. The y-intercept is $(0, -12)$. The sketch of $f(x) = x^2 - 4x - 12$ is shown.

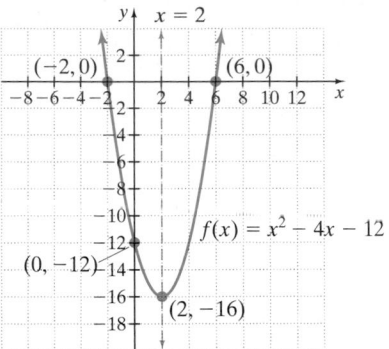

Notice that the axis of symmetry is always halfway between the x-intercepts. For the example above, halfway between -2 and 6 is $\dfrac{-2 + 6}{2} = 2$, and the axis of symmetry is $x = 2$.

CLASSROOM EXAMPLE

Graph $f(x) = 2x^2 + 2x + 5$. Find the vertex and any intercepts.

answer:

vertex: $\left(-\dfrac{1}{2}, \dfrac{9}{2}\right)$; y-intercept: $(0, 5)$

> ### Helpful Hint
>
> Parabola Opens Upward
> Vertex in I or II: no x-intercept
> Vertex in III or IV: 2 x-intercepts
>
> Parabola Opens Downward
> Vertex in I or II: 2 x-intercepts
> Vertex in III or IV: no x-intercept.
>
>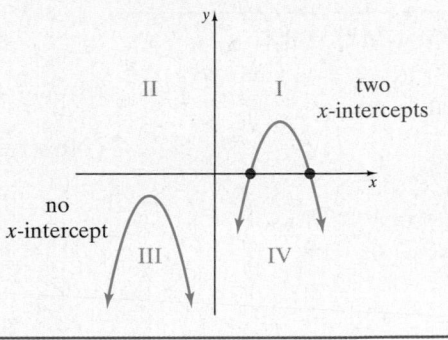

From the helpful hint above, you can see that the vertex of an upward parabola is its minimum point and the vertex of a downward parabola is its maximum point. We can find these points using the Calc menu and the maximum and minimum feature on a graphing utility.

EXAMPLE 2

TEACHING TIP (EXAMPLE 2)

You may want to mention to students why, for example, the calculator may report the maximum x-value as 1.2500005 when the exact x-value is 1.25. Tell students that the Maximum or Minimum is calculated *(continued)*

Use a graphing utility to find the vertex and the axis of symmetry of the graph of $y = -2x^2 + 5x - 1$.

Solution Graph $y_1 = -2x^2 + 5x - 1$ in a decimal window. Since the coefficent of x^2 is negative, the parabola opens downward and the function has a maximum value. Press maximum on the Calc menu. The calculator will find the maximum (or minimum) over a specified interval only, so we must first enter left and right endpoints or bounds for an interval. Next, we move the cursor as close as possible to what appears to be the vertex for a Guess.

Left Bound

Right Bound

Guess

Maximum

using a Taylor Polynomial, a concept usually covered in calculus. For this example, the accuracy of the answer given by the calculator depends on the interval chosen with the left and right bounds.

The maximum value of the function is 2.125. The vertex is then $(1.25, 2.125)$. Since it is a downward parabola, the axis of symmetry is a vertical line through the vertex with the equation $x = 1.25$.

Note: If the parabola is an upward parabola, then use the minimum feature on the Calc menu.

EXAMPLE 3

Graph $f(x) = 3x^2 + 3x + 1$. Find the vertex and any intercepts. Check using a graphing utility.

Solution   Replace $f(x)$ with y and complete the square on x to write the equation in the form $y = a(x - h)^2 + k$.

$$y = 3x^2 + 3x + 1 \qquad \text{Replace } f(x) \text{ with } y.$$

$$y - 1 = 3x^2 + 3x \qquad \text{Isolate } x\text{-variable terms.}$$

Factor 3 from the terms $3x^2 + 3x$ so that the coefficient of x^2 is 1.

$$y - 1 = 3(x^2 + x) \qquad \text{Factor out 3.}$$

The coefficient of x in the parentheses above is 1. Then $\frac{1}{2}(1) = \frac{1}{2}$ and $\left(\frac{1}{2}\right)^2 = \frac{1}{4}$.

Since we are adding $\frac{1}{4}$ inside the parentheses, we are really adding $3\left(\frac{1}{4}\right)$, so we *must* add $3\left(\frac{1}{4}\right)$ to the left side.

$$y - 1 + 3\left(\frac{1}{4}\right) = 3\left(x^2 + x + \frac{1}{4}\right)$$

$$y - \frac{1}{4} = 3\left(x + \frac{1}{2}\right)^2 \qquad \begin{array}{l}\text{Simplify the left side and}\\ \text{factor the right side.}\end{array}$$

$$y = 3\left(x + \frac{1}{2}\right)^2 + \frac{1}{4} \qquad \text{Add } \frac{1}{4} \text{ to both sides.}$$

$$f(x) = 3\left(x + \frac{1}{2}\right)^2 + \frac{1}{4} \qquad \text{Replace } y \text{ with } f(x).$$

Then $a = 3$, $h = -\frac{1}{2}$, and $k = \frac{1}{4}$. This means that the parabola opens upward with vertex $\left(-\frac{1}{2}, \frac{1}{4}\right)$ and that the axis of symmetry is the line $x = -\frac{1}{2}$.

CLASSROOM EXAMPLE

Graph $f(x) = -x^2 - 2x + 8$. Find the vertex and any intercepts.

answer:

vertex: $(-1, 9)$;

x-intercepts: $(-4, 0), (2, 0)$;

y-intercept: $(0, 8)$

A check for Example 3. This parabola opens up, so we used the minimum feature to calculate the vertex.

To find the y-intercept, let $x = 0$. Then

$$f(0) = 3(0)^2 + 3(0) + 1 = 1$$

Thus the y-intercept is $(0, 1)$.

This parabola has no x-intercepts since the vertex is in the second quadrant and opens upward. Use the vertex, axis of symmetry, and y-intercept to sketch the parabola.

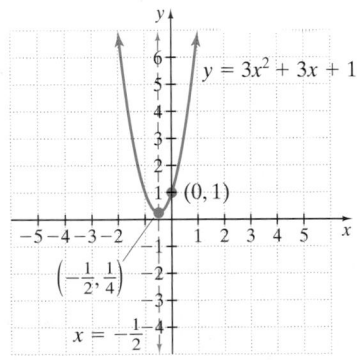

2   There is also a formula that may be used to find the vertex of a parabola. Now that we have practiced completing the square, we will show that the x-coordinate of the vertex of the graph of $f(x)$ or $y = ax^2 + bx + c$ can be found by the formula $x = \dfrac{-b}{2a}$. To do so, we complete the square on x and write the equation in the form $y = a(x - h)^2 + k$.

First, isolate the x-variable terms by subtracting c from both sides.

$$y = ax^2 + bx + c$$

$$y - c = ax^2 + bx$$

Next, factor a from the terms $ax^2 + bx$.

$$y - c = a\left(x^2 + \frac{b}{a}x\right)$$

Next, add the square of half of $\dfrac{b}{a}$, or $\left(\dfrac{b}{2a}\right)^2 = \dfrac{b^2}{4a^2}$, to the right side inside the parentheses.

Because of the factor a, what we really added was $a\left(\dfrac{b^2}{4a^2}\right)$ and this must be added to the left side.

$$y - c + a\left(\frac{b^2}{4a^2}\right) = a\left(x^2 + \frac{b}{a}x + \frac{b^2}{4a^2}\right)$$

$$y - c + \frac{b^2}{4a} = a\left(x + \frac{b}{2a}\right)^2$$

Simplify the left side and factor the right side.

$$y = a\left(x + \frac{b}{2a}\right)^2 + c - \frac{b^2}{4a}$$

Add c to both sides and subtract $\dfrac{b^2}{4a}$ from both sides.

Compare this form with $f(x)$ or $y = a(x - h)^2 + k$ and see that h is $\dfrac{-b}{2a}$, which means that the x-coordinate of the vertex of the graph of $f(x) = ax^2 + bx + c$ is $\dfrac{-b}{2a}$.

> ## Vertex Formula
>
> The graph of $f(x) = ax^2 + bx + c$, when $a \neq 0$, is a parabola with vertex
>
> $$\left(\frac{-b}{2a}, f\left(\frac{-b}{2a}\right) \right)$$

Let's use this formula to find the vertex of the parabola we graphed in Example 1.

EXAMPLE 4

Find the vertex of the graph of $f(x) = x^2 - 4x - 12$.

Solution In the quadratic function $f(x) = x^2 - 4x - 12$, notice that $a = 1, b = -4$, and $c = -12$. Then

$$\frac{-b}{2a} = \frac{-(-4)}{2(1)} = 2$$

The x-value of the vertex is 2. To find the corresponding $f(x)$ or y-value, find $f(2)$. Then

$$f(2) = 2^2 - 4(2) - 12 = 4 - 8 - 12 = -16$$

The vertex is $(2, -16)$. These results agree with our findings in Example 1.

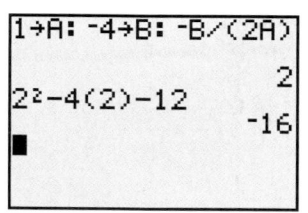

A calculator check of Example 4.

EXAMPLE 5

Use a graphing utility to graph $f(x) = 3 - 2x - x^2$. Find the vertex and any intercepts algebraically.

Solution For $f(x) = 3 - 2x - x^2$, we have $a = -1, b = -2$, and $c = 3$. Since $a < 0$, the parabola opens downward. We use the vertex formula to find the vertex.

$$x = \frac{-b}{2a} = \frac{-(-2)}{2(-1)} = -1$$

Also, $f(-1) = 4$, so the vertex has coordinates $(-1, 4)$. The vertex is in the second quadrant and the parabola opens downward, so the parabola has two x-intercepts and a y-intercept as shown to the left.

To find the x-intercepts, let y or $f(x) = 0$ and solve for x.

$$f(x) = 3 - 2x - x^2$$
$$0 = 3 - 2x - x^2$$
$$0 = (3 + x)(1 - x)$$
$$3 + x = 0 \quad \text{or} \quad 1 - x = 0$$
$$x = -3 \quad \text{or} \quad 1 = x$$

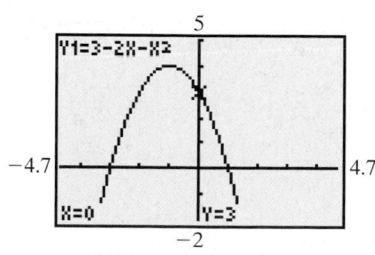

The x-intercepts are $(-3, 0)$ and $(1, 0)$. To find the y-intercept, let $x = 0$.

$$f(0) = 3 - 2 \cdot 0 - 0^2 = 3$$

The y-intercept is $(0, 3)$.

3 The vertex of a parabola gives us some important information about its corresponding quadratic function. The quadratic function whose graph is a parabola that opens upward has a minimum value, and the quadratic function whose graph is a parabola that opens downward has a maximum value. The $f(x)$ or y-value of the vertex is the minimum or maximum value of the function.

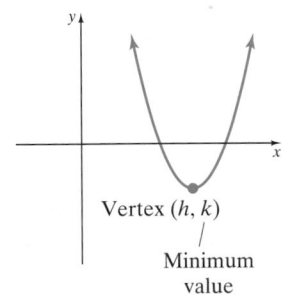

✔ **CONCEPT CHECK**

Without making any calculations, tell whether the graph of $f(x) = 7 - x - 0.3x^2$ has a maximum value or a minimum value. Explain your reasoning.

EXAMPLE 6

FINDING MAXIMUM HEIGHT

A rock is thrown upward from the ground. Its height in feet above ground after t seconds is given by the function $f(t) = -16t^2 + 20t$. Find the maximum height of the rock and the number of seconds it took for the rock to reach its maximum height.

Solution

1. UNDERSTAND. The maximum height of the rock is the largest value of $f(t)$. Since the function $f(t) = -16t^2 + 20t$ is a quadratic function, its graph is a parabola. It opens downward since $-16 < 0$. Thus, the maximum value of $f(t)$ is the $f(t)$ or y-value of the vertex of its graph.

2. TRANSLATE. To find the vertex (h, k), notice that for $f(t) = -16t^2 + 20t$, $a = -16, b = 20$, and $c = 0$. We will use these values and the vertex formula

$$\left(\frac{-b}{2a}, f\left(\frac{-b}{2a} \right) \right)$$

3. SOLVE.

$$h = \frac{-b}{2a} = \frac{-20}{-32} = \frac{5}{8}$$

$$f\left(\frac{5}{8} \right) = -16\left(\frac{5}{8} \right)^2 + 20\left(\frac{5}{8} \right)$$

CLASSROOM EXAMPLE

An object is thrown upward from the top of a 100-foot cliff. Its height in feet above ground after t seconds is given by the function $f(t) = -16t^2 + 10t + 100$. Find the maximum height of the object and the number of seconds it took for the object to reach its maximum height.

answer: maximum height:

$$101\frac{9}{16} \text{ ft in } \frac{5}{16} \text{ sec}$$

Concept Check Answer:
$f(x)$ has a maximum value since it opens downward.

TEACHING TIP
After discussing Example 6, ask students to determine when the rock will reach a height of 12 feet. Have them find their answer in two ways.

$$= -16\left(\frac{25}{64}\right) + \frac{25}{2}$$

$$= -\frac{25}{4} + \frac{50}{4} = \frac{25}{4}$$

4. INTERPRET. Check: The graph of $f(t)$ is a parabola opening downward with vertex $\left(\frac{5}{8}, \frac{25}{4}\right)$. To *check*, graph $y_1 = -16x^2 + 20x$ in a $[0, 2, 1]$ by $[0, 10, 1]$ window. The maximum height occurs when $x = 0.625$ or $\frac{5}{8}$ seconds and is $6\frac{1}{4}$ feet.

State: This means that the rock's maximum height is $\frac{25}{4}$ feet, or $6\frac{1}{4}$ feet, which was reached in $\frac{5}{8}$ second.

Spotlight on

DECISION MAKING

Suppose you are a member of a community theater group, the Slidell Players. For an upcoming performance of *Grease*, your group must decide on a ticket price. The graph shows the relationship between ticket price and box office receipts for past Slidell Players performances.
What ticket price would you suggest that the Slidell Players charge for its performance of *Grease*? Explain your reasoning.

EXERCISE SET 8.6

STUDY
GUIDE/SSM

CD/
VIDEO

PH MATH
TUTOR CENTER

MathXL®Tutorials
ON CD

MathXL®

MyMathLab®

Find the vertex of the graph of each quadratic function. See Examples 1 through 4.

1. $f(x) = x^2 + 8x + 7$ **2.** $f(x) = x^2 + 6x + 5$ **5.** $f(x) = 5x^2 - 10x + 3$ **6.** $f(x) = -3x^2 + 6x + 4$

3. $f(x) = -x^2 + 10x + 5$ **4.** $f(x) = -x^2 - 8x + 2$ **7.** $f(x) = -x^2 + x + 1$ **8.** $f(x) = x^2 - 9x + 8$

1. $(-4, -9)$ **2.** $(-3, -4)$ **3.** $(5, 30)$ **4.** $(-4, 18)$ **5.** $(1, -2)$ **6.** $(1, 7)$ **7.** $\left(\frac{1}{2}, \frac{5}{4}\right)$ **8.** $\left(\frac{9}{2}, -\frac{49}{4}\right)$

Match each function with its graph. See Examples 1 through 4.

A

$(-1, -4)$

B
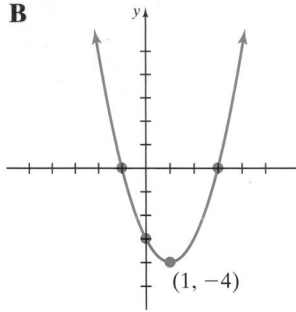
$(1, -4)$

C

$(-2, -1)$

D

$(2, -1)$

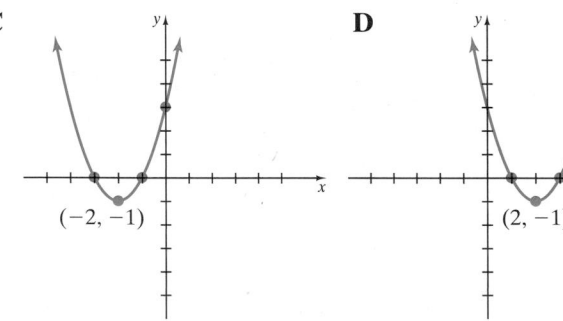

9. $f(x) = x^2 - 4x + 3$ D   **10.** $f(x) = x^2 + 2x - 3$ A

11. $f(x) = x^2 - 2x - 3$ B   **12.** $f(x) = x^2 + 4x + 3$ C

MIXED PRACTICE **13.–44.** See graphing answer section.

Find the vertex of the graph of each quadratic function. Determine whether the graph opens upward or downward, find any intercepts, and sketch the graph. See Examples 1 through 5.

13. $f(x) = x^2 + 4x - 5$ **14.** $f(x) = x^2 + 2x - 3$

15. $f(x) = -x^2 + 2x - 1$ **16.** $f(x) = -x^2 + 4x - 4$

17. $f(x) = x^2 - 4$ **18.** $f(x) = x^2 - 1$

19. $f(x) = 4x^2 + 4x - 3$ **20.** $f(x) = 2x^2 - x - 3$

21. $f(x) = x^2 + 8x + 15$ **22.** $f(x) = x^2 + 10x + 9$

23. $f(x) = x^2 - 6x + 5$ **24.** $f(x) = x^2 - 4x + 3$

25. $f(x) = x^2 - 4x + 5$ **26.** $f(x) = x^2 - 6x + 11$

27. $f(x) = 2x^2 + 4x + 5$ **28.** $f(x) = 3x^2 + 12x + 16$

29. $f(x) = -2x^2 + 12x$ **30.** $f(x) = -4x^2 + 8x$

31. $f(x) = x^2 + 1$ **32.** $f(x) = x^2 + 4$

33. $f(x) = x^2 - 2x - 15$ **34.** $f(x) = x^2 - 4x + 3$

35. $f(x) = -5x^2 + 5x$ **36.** $f(x) = 3x^2 - 12x$

37. $f(x) = -x^2 + 2x - 12$ **38.** $f(x) = -x^2 + 8x - 17$

39. $f(x) = 3x^2 - 12x + 15$ **40.** $f(x) = 2x^2 - 8x + 11$

41. $f(x) = x^2 + x - 6$ **42.** $f(x) = x^2 + 3x - 18$

43. $f(x) = -2x^2 - 3x + 35$ **44.** $f(x) = 3x^2 - 13x - 10$

Solve. See Example 6.

45. If a projectile is fired straight upward from the ground with an initial speed of 96 feet per second, then its height h in feet after t seconds is given by the equation

$$h(t) = -16t^2 + 96t$$

Find the maximum height of the projectile.   144 ft

46. The cost C in dollars of manufacturing x bicycles at Holladay's Production Plant is given by the function $C(x) = 2x^2 - 800x + 92{,}000$.

   a. Find the number of bicycles that must be manufactured to minimize the cost.   200 bicycles

   b. Find the minimum cost. $12,000

47. If Rheam Gaspar throws a ball upward with an initial speed of 32 feet per second, then its height h in feet after t seconds is given by the equation

$$h(t) = -16t^2 + 32t$$

Find the maximum height of the ball.   16 ft

48. The Utah Ski Club sells calendars to raise money. The profit P, in cents, from selling x calendars is given by the equation $P(x) = 360x - x^2$.

   a. Find how many calendars must be sold to maximize profit.   180 calendars

   b. Find the maximum profit. $32,400

49. Find two numbers whose sum is 60 and whose product is as large as possible. [*Hint:* Let x and $60 - x$ be the two positive numbers. Their product can be described by the function $f(x) = x(60 - x)$.]   30 and 30

50. Find two numbers whose sum is 11 and whose product is as large as possible. (Use the hint for Exercise 49.)   5.5, 5.5

51. Find two numbers whose difference is 10 and whose product is as small as possible. (Use the hint for Exercise 49.)   5, -5

52. Find two numbers whose difference is 8 and whose product is as small as possible.   4, -4

△ **53.** The length and width of a rectangle must have a sum of 40. Find the dimensions of the rectangle that will have the maximum area. (Use the hint for Exercise 49.)   length, 20 units; width, 20 units

△ **54.** The length and width of a rectangle must have a sum of 50. Find the dimensions of the rectangle that will have maximum area.   length, 25; width, 25

REVIEW AND PREVIEW   **55.–64.** See graphing answer section.

Sketch the graph of each function. See Section 8.5.

55. $f(x) = x^2 + 2$ **56.** $f(x) = (x - 3)^2$

57. $g(x) = x + 2$ **58.** $h(x) = x - 3$

59. $f(x) = (x + 5)^2 + 2$ **60.** $f(x) = 2(x - 3)^2 + 2$

61. $f(x) = 3(x - 4)^2 + 1$ **62.** $f(x) = (x + 1)^2 + 4$

63. $f(x) = -(x - 4)^2 + \dfrac{3}{2}$ **64.** $f(x) = -2(x + 7)^2 + \dfrac{1}{2}$

Concept Extensions

Find the vertex of the graph of each quadratic function. Determine whether the graph opens upward or downward, find the y-intercept, approximate the x-intercepts to one decimal place, and sketch the graph.   **65.–68.** See graphing answer section.

65. $f(x) = x^2 + 10x + 15$ **66.** $f(x) = x^2 - 6x + 4$

67. $f(x) = 3x^2 - 6x + 7$ **68.** $f(x) = 2x^2 + 4x - 1$

Find the maximum or minimum value of each function. Approximate to two decimal places.

69. $f(x) = 2.3x^2 - 6.1x + 3.2$ −0.84

70. $f(x) = 7.6x^2 + 9.8x - 2.1$ −5.26

71. $f(x) = -1.9x^2 + 5.6x - 2.7$ 1.43

72. $f(x) = -5.2x^2 - 3.8x + 5.1$ 5.79

73. The number of inmates in custody in U.S. prisons and jails can be modeled by the quadratic function

$$p(x) = -x^2 + 93x + 1128$$

where $p(x)$ is the number of inmates in thousands and x is the number of years after 1990. (*Source*: Based on data from the Bureau of Justice Statistics, U.S. Department of Justice, 1990–2000)

＼ **a.** Will this function have a maximum or a minimum? How can you tell?   maximum; answers may vary

b. According to this model, in what year will the number of prison and jail inmates in custody in the United States be at its maximum/minimum?   2036

c. What is the maximum/minimum number of inmates predicted?   32,902,500 or about 3290.3 thousands

74. Methane is a gas produced by landfills, natural gas systems, and coal mining that contributes to the greenhouse effect and global warming. Projected methane emissions in the United States can be modeled by the quadratic function

$$f(x) = -0.072x^2 + 1.93x + 173.9$$

where $f(x)$ is the amount of methane produced in million metric tons and x is the number of years after 2000. (*Source*: Based on data from the U.S. Environmental Protection Agency, 2000–2020) Round answers to two decimal places.

a. According to this model, what will U.S. emissions of methane be in 2009?   185.44 million metric tons

＼ **b.** Will this function have a maximum or a minimum? How can you tell?   maximum; answers may vary

c. In what year will methane emissions in the United States be at their maximum/minimum? Round to the nearest whole year.   2013

d. What is the level of methane emissions for that year? (Use your rounded answer from part **c.**)   186.82 million metric tons

---

Objective

1   Plot data points and recognize whether a linear or quadratic model best fits the data.

As we have seen thus far in our text, many situations arise that involve two related quantities. We saw that we can investigate and record certain facts about these situations, and these are called **data**. To *model* the data means to find an equation that describes the relationship between the given data quantities. Visually, the graph of the *best fit equation* should have a majority of the plotted ordered pair data points on the graph or close to it. Recall from Section 2.6 that a technique called **regression** is used to determine an equation that best fits two-quantity data. In this text, we concentrate on modeling ordered pair data points with an equation that visually appears to best fit those data points. We can also check the *r* factor. The best fit occurs when *r* is close to 1 or −1.

Recall that an equation that best fits a set of ordered pair data points is called a regression equation. Keep in mind that no model, or equation, may fit the data exactly, but a model should fit the data closely enough so that useful predictions may be made using it. Remember that when we use an equation to make predictions in the future, we are assuming that future data is also modelled by the equation.

1   In this section, we will concentrate on two models, linear and quadratic. These models are shown below for your review.

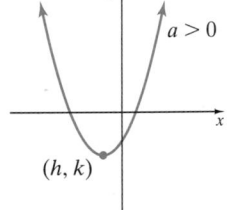

Linear Model	*Quadratic Model*
$y = mx + b$	$y = ax^2 + bx + c$
slope: m	$a > 0$, *parabola opens upward*
y-intercept: (0, b)	$a < 0$, *parabola opens downward*
	vertex: (h, k) where $h = \frac{-b}{2a}$

A quadratic model may be used when the plotted data points suggest a parabola.

EXAMPLE 1

A QUADRATIC MODEL FOR FUEL CONSUMPTION

The table below shows the average miles per gallon (MPG) for U.S. vehicles in the years shown. (*Source:* U.S. Federal Highway Administration, *Highway Statistics*, annual.)

Year	1960	1970	1980	1990	1998	2000	2003
Average MPG	13.4	12	13.2	16.2	17.1	18.5	19.8

a. Graph the data points.

b. Find and graph a quadratic regression equation that models the data.

c. Use the equation to predict the average MPG for the year 2010.

Solution   **a.** List the years in L_1 and the average MPG in L_2. (For L_1, enter the number of years past 1960.) Then draw a scatterplot as shown below.

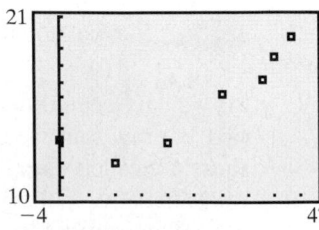

Notice that the graph resembles the graph of a parabola.

b. To find and graph the quadratic regression equation that models this data, select QuadReg from the Stat Calc menu. Enter L_1, L_2, Y_1 so that the screen reads QuadReg L_1, L_2, Y_1. This instructs the calculator to use L_1 and L_2 as x and y lists and enters the regression equation in the Y = editor as Y_1.

The quadratic regression equation is $y = 0.006x^2 - 0.115x + 13.104$ where a, b, and c values are rounded to three decimal places.

c. To predict the average MPG for the year 2010, we evaluate the quadratic regression equation we have stored in Y_1 for 50 $(2010 - 1960 = 50)$. The predicted MPG for the automobiles in the year 2010 is 23.0 MPG rounded to the nearest tenth.

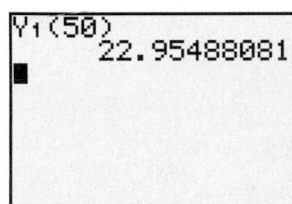

Americans are buying more sport-utility vehicles and other types of trucks, reducing passenger cars' share of the total vehicle market.

Passenger Car and Truck Retail Sales, 1997 and 2001

For the next example, we will study the increase in total truck sales.

EXAMPLE 2

TRUCK RETAIL SALES

The table below represents the retail sales in thousands of passenger trucks for selected years 1992 through 2001.

Year	Truck Sales (in thousands)
1992	4903
1993	5681
1994	6421
1995	6481
1996	6930
1997	7226
2000	8697
2001	9185

(Source: American Automobile Manufacturer's Assoc.)

a. Graph the data points and identify the type of function that best fits the data.

b. Find the corresponding regression equation for the data and then graph the results.

c. If the sales trend continues in the same manner, predict the number of cars and the number of trucks that we would expect to sell in the year 2010.

Solution   **a.** List the years in L_1 and the truck sales in L_2. (For L_1, enter the number of years past 1990.) Then draw a scatterplot as shown below.

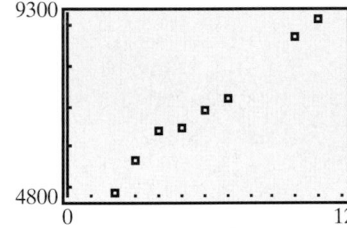

Since a linear function appears to best fit the data, we will model the data with a linear regression equation.

b. Select LinReg from the Stat Calc menu. Enter L_1, L_2, Y_1, so that the screen reads LinReg $(ax + b)$ L_1, L_2, Y_1. This instructs the calculator to use L_1 and L_2 as x and y lists and enters the regression equation in the $Y=$ editor as Y_1.

Linear regression equation.

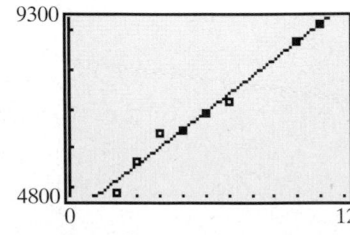

Graph of linear regression equation.

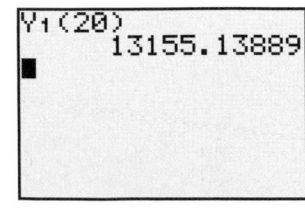

Projected truck sales in 2010.

The linear regression equation is

$$y = 443.903x + 4277.083$$

with a and b values rounded to 3 decimal places.

c. To predict truck sales for the year 2010, we evaluate the linear regression equation we have stored in Y_1 for 2010. The predicted truck sales for the year 2010 is 13,155 thousand trucks, rounded to the nearest thousand trucks.

See Exercise 12 for a table similar to the one in Example 2 except containing passenger car sales.

In Example 3, we see a need for research institutions to forecast the projected amount of funding available for use in the future.

EXAMPLE 3

ALZHEIMER'S RESEARCH

The funding for Alzheimer's disease research at the National Institute of Health has almost doubled since 1994. The table below represents the funding history.

Year	1994	1996	1998	2000	2002
Funding (millions)	298.1	310	358	462	594.7

a. Let $x =$ the number of years since 1990 and plot the data points. Identify the type of function that appears to best fit the data.

b. Find an appropriate regression equation for the data and then graph the results.

c. If the trend continues in the same pattern, find the amount of funding expected in 2009.

Solution   **a.** Enter the data in L1 (years since 1990) and L2 (funding in millions) and draw a scatter plot.

Graphed data point.

The graph resembles a part of a parabola so we will model the data with a quadratic regression equation.

b. The regression results and the graph of the equation are shown on the screens on the following page. The equation is $y = 5.314x^2 - 47.769x + 404.08$ with a and b rounded to three decimal places.

c. To forecast the amount of funding expected in the year 2009, we evaluate the quadratic regression equation we stored in Y1 for $x = 19$. The expected funding for the year 2009 is approximately $1415 million rounded to the nearest million dollars.

Quadratic regression equation.

Graph of quadratic regression equation.

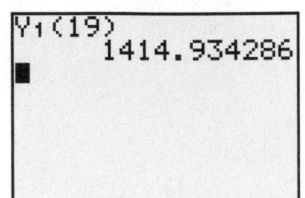

Predicted funding for 2009 in millions of dollars.

EXAMPLE 4

DVD AND VCR SALES

The graph below shows the increase and decrease in VCR unit sales and the increase in DVD unit sales.

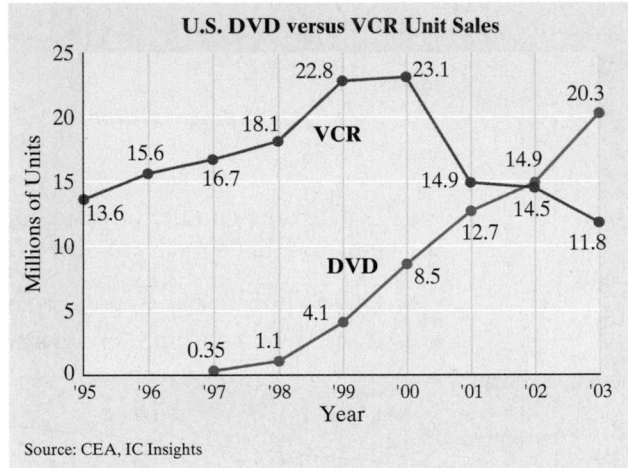

Source: CEA, IC Insights

Although the market did fluctuate, notice that from 2000 through 2003, the sales of VCR units are decreasing and appear to be linear for 3 out of 4 data points. Also, from 1997 through 2003 the DVD unit sales are increasing and appear linear.

For VCR sales, years 2000 through 2003,

a. Graph the data, letting $x = $ the number of years since 1990.
b. Find the linear regression equation that models the data.
c. Graph the linear regression equation.
d. Use the equation to predict the VCR unit sales for 2006.

For DVD sales, years 1997 through 2003,

e. Graph the data, letting $x = $ the number of years since 1990.
f. Find the linear regression equation that models the data.
g. Graph the linear regression equation.
h. Use the equation to predict the DVD unit sales for 2011.

Solution

a. Graphed data points.

b. Linear regression equation.

c. Graph of linear regression data.

d. Predicted VCR unit sales for 2011.

As you can see in part c above, the linear regression equation is

$$y = -3.43x + 55.52$$

From part d, we see that in 2006, the predicted sales for VCR units is 0.64 million, or 640,000 units.

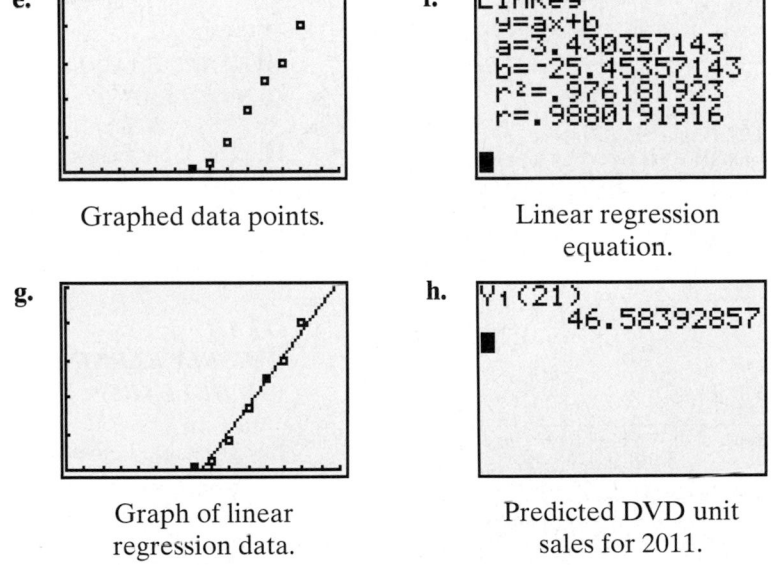

e. Graphed data points.

f. Linear regression equation.

g. Graph of linear regression data.

h. Predicted DVD unit sales for 2011.

As you can see in part f above, the linear regression equation is

$$y = 3.430x - 25.454$$

where a and b are rounded to 3 decimal places.

From part h, we see that in 2011 the predicted sales for DVD units is 46.584 million, or 46,584,000 units, rounded to the nearest thousand.

In this section, we concentrated on linear and quadratic regression equations only. There are many other types of regression equations that can be used as models and there are advantages and disadvantages of each. Proceed with regression equations carefully and know that we have only looked at the tip of the regression analysis iceberg.

EXERCISE SET 8.7

STUDY
GUIDE/SSM   CD/
VIDEO   PH MATH
TUTOR CENTER   MathXL®Tutorials
ON CD   MathXL®   MyMathLab®

Solve. Predictions using regression equations may differ slightly depending on the rounding of coefficients in the equation to different place values. Given the following graphs of data points, tell whether the graph is likely to be best represented by a linear model, a quadratic model, or neither. See Examples 1 through 4.

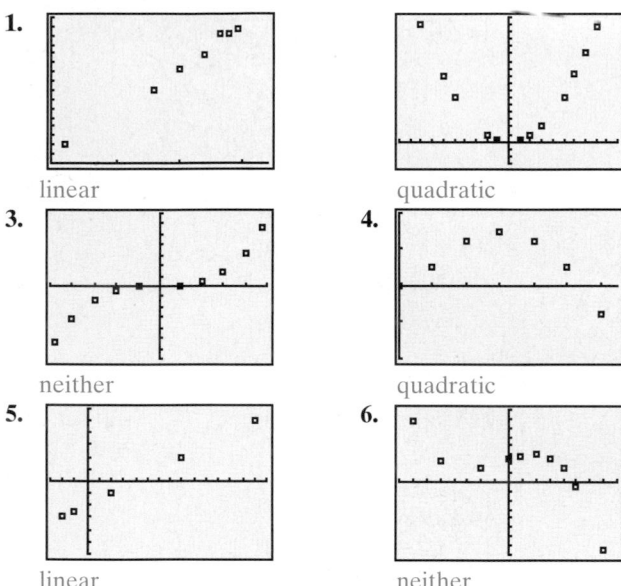

1. linear

2. quadratic

3. neither

4. quadratic

5. linear

6. neither

7.

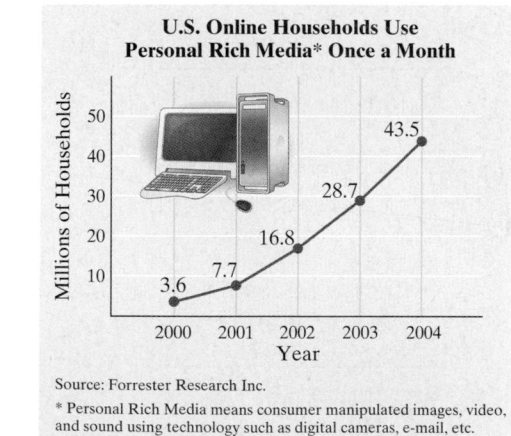

U.S. Online Households Use Personal Rich Media* Once a Month

3.6   7.7   16.8   28.7   43.5

Source: Forrester Research Inc.

* Personal Rich Media means consumer manipulated images, video, and sound using technology such as digital cameras, e-mail, etc.

quadratic   **10.** $y = 19.801x + 643.076$; $1039.09

8.

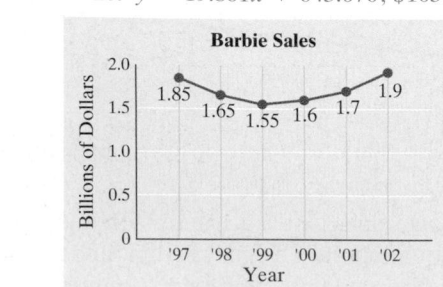

Barbie Sales

1.85   1.65   1.55   1.6   1.7   1.9

quadratic   **11.** $y = 9.547x + 467.446$; 658.39; or
$y = 0.710x^2 + 2630x + 476.247$; 812.95

For each exercise tell whether the graph is likely to be best represented by a linear model, a quadratic model, or neither. If linear or quadratic, find a linear or quadratic regression equation that models the given data. Round the coefficients to three decimal places. Then use the equation to answer the question. Unless told otherwise, let x = the number of years since 1990. If the graphed data does not appear linear or quadratic, write "not linear or quadratic" and you are through with that exercise. See Examples 1 through 4.

9. The table below shows the number of U.S. wildfires in the years 1999–2003. If there is a trend, predict the number of U.S. wildfires in 2005.   not linear or quadratic

YEAR	1999	2000	2001	2002	2003
NO. OF WILDFIRES (IN THOUSANDS)	93.7	122.8	84.1	88.4	57.6

10. In order to see the value of attaining a 4-year college degree, study the following data from the U.S. Bureau of Labor Statistics. This table shows median weekly earnings of full-time wage and salary workers 25 years and older with a 4-year degree. If there is a trend, predict the median weekly earnings in 2010.

YEAR	1990	1994	1996	1997	2000
WEEKLY EARNINGS (IN DOLLARS)	639	733	758	779	841

11. The table below shows the median weekly earnings of full-time wage and salary workers 25 years and older with 1 to 3 years of college. If this trend continues, predict the median weekly earnings in 2010. For this exercise, find a linear and a quadratic regression equation.

YEAR	1990	1994	1996	1997	2000
WEEKLY EARNINGS (IN DOLLARS)	476	499	518	528	574

12. The table below shows the retail sales, in thousands, of passenger cars for the years 1992 through 2001. If there is a trend, predict the number of cars to be sold in the year 2005. not linear or quadratic

Year	Car Sales (in thousands)
1992	8213
1993	8518
1994	8991
1995	8635
1996	8527
1997	8272
1998	8142
1999	8677
2000	8852
2001	8422

13. Use the regression equation in Example 4, part C to predict the number of VCRs to be sold in the year 2007. Explain what happened and why. The predicted number of VCR sales is negative; answers may vary.

14. Use the data in Exercise 7 to continue to project the number of U.S. online households to use personal rich media once a month. Find the projection for the year 2008. $y = 1.729x^2 - 31.406x + 144.557$; 139.3 million households

15. Use the data in Exercise 8 to project the billions of dollars from Barbie doll sales in the year 2009. $y = 0.05x^2 - 0.937x + 5.953$; $6.197 billion

16. Today more older workers are electing to stay on the job rather than retiring. The graph below represents the percent of Americans ages 60 to 64 that choose to stay in the labor force. Use the table below to find a quadratic regression equation that best fits the data. Predict the percent of workers that are staying on the job in the year 2008. Let x = number of years since 1980. $y = 0.025x^2 - 0.386x + 45.381$; 54.362% of Americans aged 60 to 64 in work force

Year	Percent of workers
1980	46
1983	43.8
1985	43.4
1990	44.8
1991	44
1992	45
1995	45
1999	46.5
2000	47.2

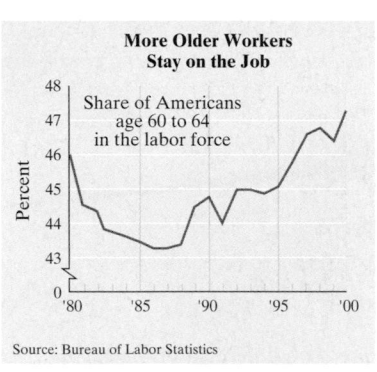

More Older Workers Stay on the Job

Share of Americans age 60 to 64 in the labor force

Source: Bureau of Labor Statistics

17. A tennis shoe manufacturer is studying the suggested selling price of a certain brand of tennis shoe. The suggested selling price cannot be too low or the cost of manufacturing the shoes will not be covered, and it cannot be too high or people will not purchase the shoes. Below are data collected for selling tennis shoes at various prices and the corresponding profit. **b.** $y = -13.010x^2 + 1727.626x - 37,489.948$
c. $17,457 profit at $80; $19,332 profit at $60

PRICE (IN DOLLARS)	39	49	65	75	85	95	105
PROFIT (IN DOLLARS)	9500	16,750	19,600	19,500	15,475	7500	1500

a. Plot the data points and use the graph to determine whether a linear or quadratic regression equation would best represent the data.   quadratic

b. Find a regression equation for this data where x is price of the tennis shoes and y is profit.

c. Using the model, find the projected profit if the shoes are sold for $80 and for $60.

d. Use the table to find the profit if the shoes are sold for $105. Is this a good selling price? Why or why not?

e. Using the model, find the selling price, to the nearest dollar, that will yield the maximum profit. $66

d. $1500; no; answers may vary

Given the following data for four houses sold in comparable neighborhoods and their corresponding number of square feet, draw a scatter plot and find a regression equation that best fits the relationship between the number of square feet and the selling price of the house.

18.

Square Feet	1519	2593	3005	3016
Selling Price	$91,238	$220,000	$254,000	$269,000

$y = 114.453x - 81,378.675$

19.

Square Feet	1754	2151	2587	2671
Selling Price	$90,900	$130,000	$155,000	$172,500

$y = 82.509x - 51,908.333$

20. Use the equation found in Exercise 18 to predict the selling price of a house with 2200 square feet. $170,418

21. Use the equation found in Exercise 19 to predict the selling price of a house with 2400 square feet. $146,113

22. A quadratic equation that models the average miles per gallon (MPG) for U.S. cars only is given by:

$$y = -0.134x^2 + 2.200x + 12.72$$

where x is the year. Use this model to fill in the predicted MPG below and compare with the actual given data. (Round predicted MPG to one decimal place.)

x (year)	Actual MPG	Predicted MPG	Difference between Actual and Predicted MPG
1996	21.2	21.1	0.1
1997	21.5	21.6	−0.1
1998	21.6	21.7	−0.1
1999	21.4	21.7	−0.3
2000	22	21.3	0.7
2001	20.4	20.7	−0.3

23. Find a linear equation that models the annual consumption of cigarettes per person where x is the year and y is the number of cigarettes. Use this model to fill in the predicted consumption of cigarettes below and compare with the actual given data.

x (year)	Actual Consumption (per person)	Predicted Consumption (per person)	Difference between Actual and Predicted
2000	1637	1648.2	$-.2$
2001	1626	1618.9	7.1
2002	1609	1589.6	19.4
2003	1545	1560.5	-15.5

Plot the following points and then find a regression equation that best models the data. Let $x = 3$ represent March, $x = 4$ represent April, and so on. For each exercise, explain the limitation of the regression equation that you found.

24. The table below shows the average temperature in Louisville, Kentucky, for the months of March through October. $y = -1.994x^2 + 28.220x - 23.339$; answers may vary

March	April	May	June	July	August	Sept	Oct
45	57	65	74	78	76	70	58

25. The table below shows the average temperature in Wellington, New Zealand, for the months of March through October. $y = 0.75x^2 - 10.774x + 86.655$; answers may vary

March	April	May	June	July	August	Sept	Oct
60	57	52	49	47	48	51	54

State whether the relationship described can best be modeled by a linear function or a quadratic function.

26. The area of a square *and* the length of one side   quadratic

27. A salary of $300 per week plus 5% commission on sales *and* weekly sales   linear

28. A number of calculators *and* their cost, plus tax   linear

29. The area of a circle *and* the radius of the circle   quadratic

30. The average temperature of Cleveland, Ohio, *and* the months of March through October   quadratic

31. An oven is turned on and set to reach 400 degrees. After it reaches the desired temperature, it is shut off. Temperatures are recorded every 2 minutes, from the time the oven is turned on until it cools to room temperature.   quadratic

32. The number of bacteria present in a person with respect to time when he is coming down with the flu, has the flu, and then recovers   quadratic

33. The distance traveled during a recent trip with respect to the amount of time elapsed   linear

REVIEW AND PREVIEW

Solve each of the following equations. See Sections 1.4 and 5.6.

34. $(x + 3)(x - 5) = 0$ $-3, 5$

35. $(x - 1)(x + 19) = 0$ $-19, 1$

36. $2x^2 - 7x - 15 = 0$ $-\frac{3}{2}, 5$

37. $6x^2 + 13x - 5 = 0$ $-\frac{5}{2}, \frac{1}{3}$

38. $3(x - 4) + 2 = 5(x - 6)$ **10**

39. $4(2x + 7) = 2(x - 4)$ -6

Find the y-intercept of the graph of each function. See Section 2.3.

40. $f(x) = x^3 + 3x^2 - 5x - 8$ $(0, -8)$

41. $f(x) = 2x^3 + x^2 - 7x + 12$ $(0, 12)$

42. $g(x) = x^2 - 3x + 5$ $(0, 5)$

43. $g(x) = 3x^2 - 5x - 10$ $(0, -10)$

CHAPTER 8 PROJECT

Fitting a Quadratic Model to Data

Throughout the twentieth century, the eating habits of Americans changed noticeably. Americans started consuming less whole milk and butter, and started consuming more skim and low-fat milk and margarine. We also started eating more poultry and fish. In this project, you will have the opportunity to investigate trends in per capita consumption of poultry during the twentieth century. This project may be completed by working in groups or individually.

We will start by finding a quadratic model, $y = ax^2 + bx + c$, that has ordered pair solutions that correspond to the data for U.S. per capita consumption of poultry given in the table. To do so, substitute each data pair into the equation. Each time, the result is an equation in three unknowns: a, b, and c. Because there are three pairs of data, we can form a system of three linear equations in three unknowns. Solving for the values of a, b, and c gives a quadratic model that represents the given data.

U.S. PER CAPITA CONSUMPTION OF POULTRY (IN POUNDS)

Year	x	Poultry Consumption, y (in pounds)
1909	9	11
1969	69	33
2001	101	66

(Source: Economic research service, U.S. Department of Agriculture)

1. Write the system of equations that must be solved to find the values of *a, b,* and *c* needed for a quadratic model of the given data.

2. Solve the system of equations for *a, b,* and *c.* Recall the various methods of solving linear systems used in Chapter 4. You might consider using matrices, Cramer's rule, or a graphing calculator to do so. Round to the nearest thousandth.

3. Write the quadratic model for the data. Note that the variable *x* represents the number of years after 1900.

4. In 1939, the actual U.S. per capita consumption of poultry was 12 pounds per person. Based on this information, how accurate do you think this model is for years other than those given in the table?

5. Use your model to estimate the per capita consumption of poultry in 1950.

6. According to the model, in what year was per capita consumption of poultry 50 pounds per person?

7. In what year was the per capita consumption of poultry at its lowest level? What was that level?

8. Who might be interested in a model like this and how would it be helpful?

STUDY SKILLS REMINDER

Are You Preparing for a Test on Chapter 8?

Below I have listed some common trouble areas for students in Chapter 8. After studying for your test—but before taking your test—read these.

▶ Don't forget that to solve a quadratic equation such as $x^2 + 6x = 1$, by completing the square, add the square of half of 6 to both sides.

$$x^2 + 6x = 1$$

$$x^2 + 6x + 9 = 1 + 9 \quad \text{Add 9 to both sides, } \left(\frac{1}{2}(6) = 3 \text{ and } 3^2 = 9\right)$$

$$(x + 3)^2 = 10$$

$$x + 3 = \pm\sqrt{10}$$

$$x = -3 \pm \sqrt{10}$$

▶ Remember to write a quadratic equation in standard form $(ax^2 + bx + c = 0)$ before using the quadratic formula to solve.

$$x(4x - 1) = 1$$

$$4x^2 - x - 1 = 0 \qquad \text{Write in standard form.}$$

$$x = \frac{-(-1) \pm \sqrt{(-1)^2 - 4(4)(-1)}}{2 \cdot 4} \qquad \begin{array}{l}\text{Use the quadratic} \\ \text{formula with} \\ a = 4, b = -1, \text{and} \\ c = -1.\end{array}$$

$$x = \frac{1 \pm \sqrt{17}}{8} \qquad \text{Simplify.}$$

▶ Review the steps for solving a quadratic equation in general on page 617.

▶ Don't forget how to graph a quadratic function in the form $f(x) = a(x - h)^2 + k$. The graph of $f(x) = -2(x - 3)^2 - 1$

opens
downward

shift 3
units right

shift 1
unit down

narrower

(continued)

Remember: This is simply a checklist of common trouble areas. For a review of Chapter 8, see the Highlights and Chapter Review at the end of this chapter.

CHAPTER VOCABULARY CHECK

Fill in each blank with one of the words or phrases listed below.

quadratic formula	quadratic	discriminant	$\pm\sqrt{b}$
completing the square	quadratic inequality	(h, k)	$(0, k)$
$(h, 0)$	$\dfrac{-b}{2a}$		

1. The <u>discriminant</u> helps us know find the number and type of solutions of a quadratic equation.

2. If $a^2 = b$, then $a = $ <u>$\pm\sqrt{b}$</u>.

3. The graph of $f(x) = ax^2 + bx + c$ where a is not 0 is a parabola whose vertex has x-value of <u>$\dfrac{-b}{2a}$</u>.

4. A(n) <u>quadratic inequality</u> is an inequality that can be written so that one side is a quadratic expression and the other side is 0.

5. The process of writing a quadratic equation so that one side is a perfect square trinomial is called <u>completing the square</u>.

6. The graph of $f(x) = x^2 + k$ has vertex <u>$(0, k)$</u>.

7. The graph of $f(x) = (x - h)^2$ has vertex <u>$(h, 0)$</u>.

8. The graph of $f(x) = (x - h)^2 + k$ has vertex <u>(h, k)</u>.

9. The formula $x = \dfrac{-b \pm \sqrt{b^2 - 4ac}}{2a}$ is called the <u>quadratic formula</u>.

10. A <u>quadratic</u> equation is one that can be written in the form $ax^2 + bx + c = 0$ where $a, b,$ and c are real numbers and a is not 0.

CHAPTER 8 HIGHLIGHTS

Definitions and Concepts	**Examples**

Section 8.1 Solving Quadratic Equations by Completing the Square

Square root property

If b is a real number and if $a^2 = b$, then $a = \pm\sqrt{b}$.

To solve a quadratic equation in x by completing the square

Step 1: If the coefficient of x^2 is not 1, divide both sides of the equation by the coefficient of x^2.

Step 2: Isolate the variable terms.

Step 3: Complete the square by adding the square of half of the coefficient of x to both sides.

Step 4: Write the resulting trinomial as the square of a binomial.

Step 5: Apply the square root property and solve for x.

Solve $(x + 3)^2 = 14$.
$$x + 3 = \pm\sqrt{14}$$
$$x = -3 \pm \sqrt{14}$$

Solve $3x^2 - 12x - 18 = 0$.

1. $x^2 - 4x - 6 = 0$

2. $x^2 - 4x = 6$

3. $\frac{1}{2}(-4) = -2$ and $(-2)^2 = 4$
$$x^2 - 4x + 4 = 6 + 4$$

4. $(x - 2)^2 = 10$

5. $x - 2 = \pm\sqrt{10}$
$$x = 2 \pm \sqrt{10}$$

Section 8.2 Solving Quadratic Equations by the Quadratic Formula

A quadratic equation written in the form $ax^2 + bx + c = 0$ has solutions

$$x = \frac{-b \pm \sqrt{b^2 - 4ac}}{2a}$$

Solve $x^2 - x - 3 = 0$.

$$a = 1, b = -1, c = -3$$
$$x = \frac{-(-1) \pm \sqrt{(-1)^2 - 4(1)(-3)}}{2 \cdot 1}$$
$$x = \frac{1 \pm \sqrt{13}}{2}$$

Section 8.3 Solving Equations by Using Quadratic Methods

Substitution is often helpful in solving an equation that contains a repeated variable expression.

Solve $(2x + 1)^2 - 5(2x + 1) + 6 = 0$.
Let $m = 2x + 1$. Then

$$m^2 - 5m + 6 = 0 \qquad \text{Let } m = 2x + 1.$$
$$(m - 3)(m - 2) = 0$$
$$m = 3 \quad \text{or} \quad m = 2$$
$$2x + 1 = 3 \quad \text{or} \quad 2x + 1 = 2 \quad \text{Substitute back.}$$
$$x = 1 \quad \text{or} \quad x = \frac{1}{2}$$

Definitions and Concepts	Examples

Section 8.4 Nonlinear Inequalities in One Variable

To solve a polynomial inequality

Step 1: Write the inequality in standard form.

Step 2: Solve the related equation.

Step 3: Use solutions from Step 2 to separate the number line into regions.

Step 4: Use test points to determine whether values in each region satisfy the original inequality.

Step 5: Write the solution set as the union of regions whose test point value is a solution.

Solve $x^2 \geq 6x$.

1. $x^2 - 6x \geq 0$

2. $x^2 - 6x = 0$

   $x(x - 6) = 0$

   $x = 0$ or $x = 6$

3.

4.

Region	Test Point Value	$x^2 \geq 6x$	Result
A	-2	$(-2)^2 \geq 6(-2)$	True
B	1	$1^2 \geq 6(1)$	False
C	7	$7^2 \geq 6(7)$	True

5.

The solution set is $(-\infty, 0] \cup [6, \infty)$.

To solve a rational inequality

Step 1: Solve for values that make all denominators 0.

Step 2: Solve the related equation.

Step 3: Use solutions from Steps 1 and 2 to separate the number line into regions.

Step 4: Use test points to determine whether values in each region satisfy the original inequality.

Step 5: Write the solution set as the union of regions whose test point value is a solution.

Solve $\dfrac{6}{x - 1} < -2$.

1. $x - 1 = 0$ Set denominator equal to 0.

   $x = 1$

2. $\dfrac{6}{x - 1} = -2$

   $6 = -2(x - 1)$ Multiply by $(x - 1)$.

   $6 = -2x + 2$

   $4 = -2x$

   $-2 = x$

3.

4. Only a test value from region *B* satisfies the original inequality.

5.

The solution set is $(-2, 1)$.

Definitions and Concepts	**Examples**

Section 8.5 Quadratic Functions and Their Graphs

Graph of a quadratic function

The graph of a quadratic function written in the form $f(x) = a(x - h)^2 + k$ is a parabola with vertex (h, k). If $a > 0$, the parabola opens upward; if $a < 0$, the parabola opens downward. The axis of symmetry is the line whose equation is $x = h$.

$$f(x) = a(x - h)^2 + k$$

Graph $g(x) = 3(x - 1)^2 + 4$.

The graph is a parabola with vertex $(1, 4)$ and axis of symmetry $x = 1$. Since $a = 3$ is positive, the graph opens upward.

Section 8.6 Further Graphing of Quadratic Functions

The graph of $f(x) = ax^2 + bx + c$, where $a \neq 0$, is a parabola with vertex

$$\left(\frac{-b}{2a}, f\left(\frac{-b}{2a} \right) \right)$$

Graph $f(x) = x^2 - 2x - 8$. Find the vertex and x- and y-intercepts.

$$\frac{-b}{2a} = \frac{-(-2)}{2 \cdot 1} = 1$$

$$f(1) = 1^2 - 2(1) - 8 = -9$$

The vertex is $(1, -9)$.

$$0 = x^2 - 2x - 8$$
$$0 = (x - 4)(x + 2)$$
$$x = 4 \quad \text{or} \quad x = -2$$

The x-intercepts are $(4, 0)$ and $(-2, 0)$.

$$f(0) = 0^2 - 2 \cdot 0 - 8 = -8$$

The y-intercept is $(0, -8)$.

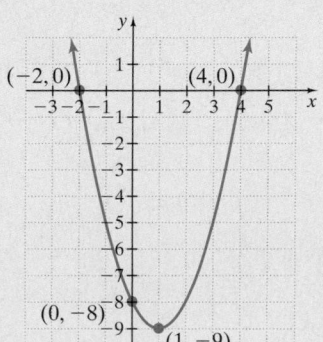

Definitions and Concepts	**Examples**

Section 8.7 Interpreting Data: Linear and Quadratic Models

Linear Model
$$y = mx + b$$
slope: m
y-intercept: (0, b)

Quadratic Model
$$y = ax^2 + bx + c$$
a > 0, parabola opens up
a < 0, parabola opens down
vertex; (h, k) where $h = \frac{-b}{2a}$

Given the following graphs of data points, tell whether the graph can best be modeled by a linear model or a quadratic model.

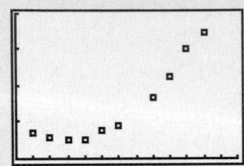

A linear model   *A quadratic model*

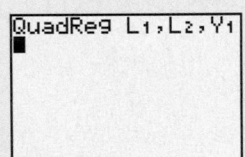

Indicates appropriate regression, x-list, y-list,
y = position to store equation.

Regression equations

Graphs of regression equations.

CHAPTER REVIEW

23. $\dfrac{5 - i\sqrt{143}}{12}, \dfrac{5 + i\sqrt{143}}{12}$ **24.** $\dfrac{1 - i\sqrt{35}}{9}, \dfrac{1 + i\sqrt{35}}{9}$ **25.** $\dfrac{21 - \sqrt{41}}{50}, \dfrac{21 + \sqrt{41}}{50}$

30. $-4, 2 - 2i\sqrt{3}, 2 + 2i\sqrt{3}$

(8.1) *Solve by factoring.* **3.** $\dfrac{4}{5}, -\dfrac{1}{2}$ **4.** $-\dfrac{6}{7}, 5$

1. $x^2 - 15x + 14 = 0$ $14, 1$

2. $x^2 - x - 30 = 0$ $-5, 6$

3. $10x^2 = 3x + 4$ **4.** $7a^2 = 29a + 30$

Solve by using the square root property. **7.** $-\dfrac{4}{9}, \dfrac{2}{9}$

5. $4m^2 = 196$ $-7, 7$ **6.** $9y^2 = 36$ $-2, 2$

9. $\dfrac{-3 - \sqrt{5}}{2}, \dfrac{-3 + \sqrt{5}}{2}$

7. $(9n + 1)^2 = 9$

8. $(5x - 2)^2 = 2$ **8.** $\dfrac{2 - \sqrt{2}}{5}, \dfrac{2 + \sqrt{2}}{5}$

10. $\dfrac{-1 - 3i\sqrt{3}}{2}, \dfrac{-1 + 3i\sqrt{3}}{2}$

11. $\dfrac{-3 - i\sqrt{7}}{8}, \dfrac{-3 + i\sqrt{7}}{8}$

Solve by completing the square.

9. $z^2 + 3z + 1 = 0$ **10.** $x^2 + x + 7 = 0$ **12.** $\dfrac{17 - \sqrt{145}}{9}, \dfrac{17 + \sqrt{145}}{9}$

11. $(2x + 1)^2 = x$ **12.** $(3x - 4)^2 = 10x$

13. If P dollars are originally invested, the formula $A = P(1 + r)^2$ gives the amount A in an account paying interest rate r compounded annually after 2 years. Find the interest rate r such that $2500 increases to $2717 in 2 years. Round the result to the nearest hundredth of a percent. 4.25%

△ **14.** Two ships leave a port at the same time and travel at the same speed. One ship is traveling due north and the other due east. In a few hours, the ships are 150 miles apart. How many miles has each ship traveled? Give an exact answer and a one-decimal-place approximation. $75\sqrt{2}$ mi; 106.1 mi

(8.2) *If the discriminant of a quadratic equation has the given value, determine the number and type of solutions of the equation.*
15. two complex but not real solutions
15. -8

16. 48 two real solutions

17. 100 two real solutions **18.** 0 one real solution

Solve by using the quadratic formula. **21.** $-i\sqrt{11}, i\sqrt{11}$

19. $x^2 - 16x + 64 = 0$ 8 **20.** $x^2 + 5x = 0$ $-5, 0$

21. $x^2 + 11 = 0$ **22.** $2x^2 + 3x = 5$ $-\dfrac{5}{2}, 1$

23. $6x^2 + 7 = 5x$ **24.** $9a^2 + 4 = 2a$

25. $(5a - 2)^2 - a = 0$ **26.** $(2x - 3)^2 = x$ $1, \dfrac{9}{4}$

27. Cadets graduating from military school usually toss their hats high into the air at the end of the ceremony. One cadet threw his hat so that its distance $d(t)$ in feet above the ground t seconds after it was thrown was $d(t) = -16t^2 + 30t + 6$.

a. Find the distance above the ground of the hat 1 second after it was thrown. 20 ft

b. Find the time it takes the hat to hit the ground. Give an exact time and a one-decimal-place approximation.

△ **28.** The hypotenuse of an isosceles right triangle is 6 centimeters longer than either of the legs. Find the length of the legs. $\left(6 + 6\sqrt{2}\right)$ cm

27. b. $\dfrac{15 + \sqrt{321}}{16}$ sec; 2.1 sec

29. $3, \dfrac{-3 + 3i\sqrt{3}}{2}, \dfrac{-3 - 3i\sqrt{3}}{2}$

(8.3) *Solve each equation for the variable.* **36.** $8, 64$

29. $x^3 = 27$ **30.** $y^3 = -64$

31. $\dfrac{5}{x} + \dfrac{6}{x - 2} = 3$ $\dfrac{2}{3}, 5$ **32.** $\dfrac{7}{8} = \dfrac{8}{x^2}$ $-\dfrac{8\sqrt{7}}{7}, \dfrac{8\sqrt{7}}{7}$

33. $x^4 - 21x^2 - 100 = 0$ $-5, 5, -2i, 2i$

34. $5(x + 3)^2 - 19(x + 3) = 4$ $-\dfrac{16}{5}, 1$

35. $x^{2/3} - 6x^{1/3} + 5 = 0$ $1, 125$ **36.** $x^{2/3} - 6x^{1/3} = -8$

37. $a^6 - a^2 = a^4 - 1$ $-1, 1, -i, i$ **38.** $y^{-2} + y^{-1} = 20$ $-\dfrac{1}{5}, \dfrac{1}{4}$

39. Two postal workers, Jerome Grant and Tim Bozik, can sort a stack of mail in 5 hours. Working alone, Tim can sort the mail in 1 hour less time than Jerome can. Find the time that each postal worker can sort the mail alone. Round the result to one decimal place. Jerome, 10.5 hr; Tim, 9.5 hr

40. A negative number decreased by its reciprocal is $-\dfrac{24}{5}$. Find the number.   The number is -5.

(8.4) Solve each inequality for x. Graph the solution set and write each solution set in interval notation.

41. $2x^2 - 50 \le 0$ $[-5, 5]$

42. $\dfrac{1}{4}x^2 < \dfrac{1}{16}$ $\left(-\dfrac{1}{2}, \dfrac{1}{2}\right)$

43. $(2x - 3)(4x + 5) \ge 0$ $\left(-\infty, -\dfrac{5}{4}\right] \cup \left[\dfrac{3}{2}, \infty\right)$

44. $(x^2 - 16)(x^2 - 1) > 0$

45. $\dfrac{x - 5}{x - 6} < 0$ $(5, 6)$

46. $\dfrac{x(x + 5)}{4x - 3} \ge 0$ $[-5, 0] \cup \left(\dfrac{3}{4}, \infty\right)$

47. $\dfrac{(4x + 3)(x - 5)}{x(x + 6)} > 0$

48. $(x + 5)(x - 6)(x + 2) \le 0$ $(-\infty, -5] \cup [-2, 6]$

49. $x^3 + 3x^2 - 25x - 75 > 0$ $(-5, -3) \cup (5, \infty)$

50. $\dfrac{x^2 + 4}{3x} \le 1$ $(-\infty, 0)$

51. $\dfrac{(5x + 6)(x - 3)}{x(6x - 5)} < 0$ $\left(-\dfrac{6}{5}, 0\right) \cup \left(\dfrac{5}{6}, 3\right)$

52. $\dfrac{3}{x - 2} > 2$ $\left(2, \dfrac{7}{2}\right)$

(8.5) Sketch the graph of each function. Label the vertex and the axis of symmetry.   **53.–60.** See graphing answer section.

53. $f(x) = x^2 - 4$ **54.** $g(x) = x^2 + 7$

55. $H(x) = 2x^2$ **56.** $h(x) = -\dfrac{1}{3}x^2$

57. $F(x) = (x - 1)^2$ **58.** $G(x) = (x + 5)^2$

59. $f(x) = (x - 4)^2 - 2$ **60.** $f(x) = -3(x - 1)^2 + 1$

(8.6) Sketch the graph of each function. Find the vertex and the intercepts.   **61.–64.** See graphing answer section.

61. $f(x) = x^2 + 10x + 25$ **62.** $f(x) = -x^2 + 6x - 9$

63. $f(x) = 4x^2 - 1$ **64.** $f(x) = -5x^2 + 5$

65. Find the vertex of the graph of $f(x) = -3x^2 - 5x + 4$. Determine whether the graph opens upward or downward, find the *y*-intercept, approximate the *x*-intercepts to one decimal place, and sketch the graph.   See graphing answer section.

66. The function $h(t) = -16t^2 + 120t + 300$ gives the height in feet of a projectile fired from the top of a building in *t* seconds.
   a. When will the object reach a height of 350 feet? Round your answer to one decimal place.   0.4 sec and 7.1 sec
   b. Explain why part **a** has two answers.   answers may vary

67. Find two numbers whose product is as large as possible, given that their sum is 420.   210 and 210

68. Write an equation of a quadratic function whose graph is a parabola that has vertex $(-3, 7)$ and that passes through the origin. $y = -\dfrac{7}{9}(x + 3)^2 + 7$

(8.7) For each exercise, find a linear or quadratic regression equation that models the given data. Round the coefficients to three decimal places. Then use the equation to answer the question.

69. According to the U.S. Bureau of the Census, Statistical Abstract of the United States, the U.S. population in thousands is as given in the table below. If this trend continues, predict the population in the year 2005. Let $x =$ be the number of years since 1970. $y = 2361.58x + 202,874.8$; 285,530.1 thousand

Year	1970	1980	1990	2000
Population (in thousands)	203,302	226,548	248,710	274,634

70. The sales of downloaded digital music online (in millions of dollars) is shown in the graph below. If the trend is to continue to increase in this manner, predict the amount of sales for 2006. (Use quadratic regression. Let $x =$ be the number of years since 1990.) $y = 35.893x^2 - 598.393 + 2465$; \$2079.3 million

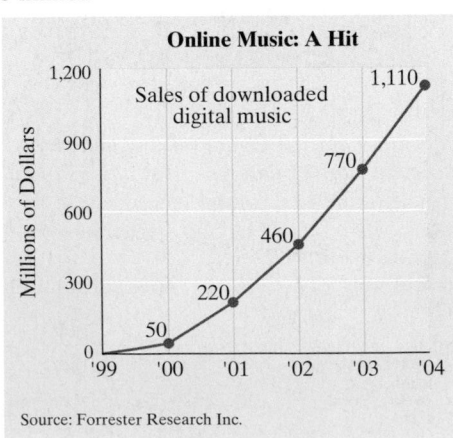

Source: Forrester Research Inc.

44. $(-\infty, -4) \cup (-1, 1) \cup (4, \infty)$

47. $(-\infty, -6) \cup \left(-\dfrac{3}{4}, 0\right) \cup (5, \infty)$

CHAPTER 8 TEST   Remember to use your Chapter Test Prep Video CD to help you study and view solutions to the test questions you need help with.

13. , $(-\infty, -3) \cup [2, 3)$

Solve each equation for the variable.   **1.** $\frac{7}{5}, -1$

1. $5x^2 - 2x = 7$ **2.** $(x + 1)^2 = 10$

3. $m^2 - m + 8 = 0$ **4.** $y^2 - 3y = 5$

5. $\frac{4}{x + 2} + \frac{2x}{x - 2} = \frac{6}{x^2 - 4}$ **6.** $x^5 + 3x^4 = x + 3$

7. $(x + 1)^2 - 15(x + 1) + 56 = 0$ **6,7**   **4.** $\frac{3 + \sqrt{29}}{2}, \frac{3 - \sqrt{29}}{2}$

Solve the equation for the variable by completing the square.

8. $x^2 - 6x = -2$ **9.** $2a^2 + 5 = 4a$

Solve each inequality for x. Graph the solution set and then write the solution set in interval notation.

10. $2x^2 - 7x > 15$ **11.** $(x^2 - 16)(x^2 - 25) \geq 0$

12. $\frac{5}{x + 3} < 1$ **13.** $\frac{7x - 14}{x^2 - 9} \leq 0$

2. $-1 - \sqrt{10}, -1 + \sqrt{10}$ **3.** $\frac{1 + i\sqrt{31}}{2}, \frac{1 - i\sqrt{31}}{2}$

Graph each function. Label the vertex.

14. $f(x) = 3x^2$ **15.** $G(x) = -2(x - 1)^2 + 5$

5. $-2 - \sqrt{11}, -2 + \sqrt{11}$ **6.** $-1, 1, -i, i, -3$ **8.** $3 - \sqrt{7}, 3 + \sqrt{7}$

Graph each function. Find and label the vertex, y-intercept, and x-intercepts (if any).   **14.–17.** See graphing answer section.

9. $\frac{2 - i\sqrt{6}}{2}, \frac{2 + i\sqrt{6}}{2}$

16. $h(x) = x^2 - 4x + 4$ **17.** $F(x) = 2x^2 - 8x + 9$

18. Dave and Sandy Hartranft can paint a room together in 4 hours. Working alone, Dave can paint the room in 2 hours less time than Sandy can. Find how long it takes Sandy to paint the room alone. $\left(5 + \sqrt{17}\right)$ hr ≈ 9.12 hr

19. A stone is thrown upward from a bridge. The stone's height in feet, $s(t)$, above the water t seconds after the stone is thrown is a function given by the equation $s(t) = -16t^2 + 32t + 256$.

a. Find the maximum height of the stone.   272 ft

b. Find the time it takes the stone to hit the water. Round the answer to two decimal places.   5.12 sec

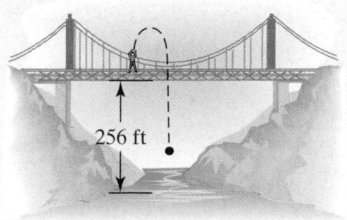

256 ft

20. Given the diagram shown, approximate to the nearest foot how many feet of walking distance a person saves by cutting across the lawn instead of walking on the sidewalk.   7 ft

20 ft

$x + 8$

x

21. After steadily falling since 1990, the nation's birth rate started rising according to the chart below. (Birth rate per 1000 population)   **a.** quadratic

Years	Birth rate
1990	16.8
1992	15.8
1994	15.1
1996	14.4
1998	14.6
2000	14.7
2001	14.5

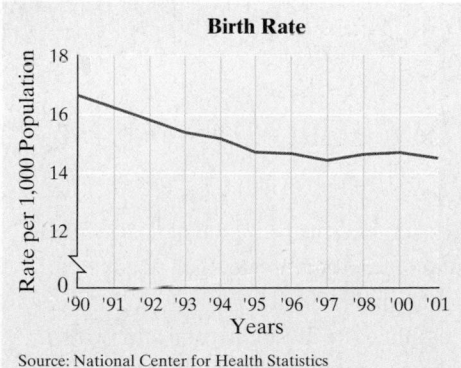

Birth Rate

Rate per 1,000 Population

'90 '91 '92 '93 '94 '95 '96 '97 '98 '00 '01

Years

Source: National Center for Health Statistics

a. Use the graph of the data points to identify the type of function (linear or quadratic) that best fits the data.

b. Find the corresponding regression equation for the data and use it to predict the birth rate for 2009. (Round regression coefficients to three decimal places. Round your answer to the nearest hundredth.) $y = 0.034x^2 - 0.563x + 16.784$; 18.26 births per 1000 people

10. , $\left(-\infty, -\frac{3}{2}\right) \cup (5, \infty)$ **11.** , $(-\infty, -5] \cup [-4, 4] \cup [5, \infty)$ **12.** , $(-\infty, -3) \cup (2, \infty)$

STUDY SKILLS REMINDER

Are You Prepared for Your Final Exam?

To prepare for your final exam, try the following study techniques.

▶ Review the material that you will be responsible for on your exam. Also check your notebook for any lecture notes that you highlighted.

▶ Review any formulas that you may need to memorize.

▶ Check to see if your instructor or math department will be conducting a final exam review.

▶ Check with your instructor to see whether there are final exams from previous semesters/quarters that are available to students for study.

▶ Use your previously taken tests as a practice final exam. To do so, rewrite the test questions in mixed order on blank sheets of paper. This will help you prepare for exam conditions.

▶ If you are unsure of a few topics, see your instructor or visit a learning lab for further assistance. Also, viewing the video segment of a troublesome section will help.

▶ If you need further exercises to work, try the chapter tests at the end of appropriate chapters.

Good luck! I hope you have enjoyed this textbook and your intermediate algebra course.

CHAPTER CUMULATIVE REVIEW

1.a. $5 + y \geq 7$ **6.a.** $\dfrac{a^6}{b^3 c^9}$

1. Write each sentence using mathematical symbols.
   a. The sum of 5 and y is greater than or equal to 7.
   b. 11 is not equal to z. $11 \neq z$
   c. 20 is less than the difference of 5 and twice x.
      $20 < 5 - 2x$ (Sec. 1.4, Ex. 3)

2. Solve $|3x - 2| = -5$. $\varnothing$ (Sec. 3.4)

3. Find the slope of the line containing the points $(0, 3)$ and $(2, 5)$. Graph the line. See graphing answer section; (Sec. 2.4, Ex. 1)

4. Use the elimination method to solve the system.
   $\begin{cases} -6x + y = 5 \\ 4x - 2y = 6 \end{cases}$ $(-2, -7)$ (Sec. 4.1)

5. Use the elimination method to solve the system:
   $\begin{cases} x - 5y = -12 \\ -x + y = 4 \end{cases}$ $(-2, 2)$ (Sec. 4.1, Ex. 6)

6. Simplify. Use positive exponents to write each answer.
   a. $(a^{-2}bc^3)^{-3}$
   b. $\left(\dfrac{a^{-4}b^2}{c^3}\right)^{-2}$ $\dfrac{a^8 c^6}{b^4}$
   c. $\left(\dfrac{3a^8 b^2}{12a^5 b^5}\right)^{-2}$ $\dfrac{16b^6}{a^6}$ (Sec. 5.2)

7. Multiply. (Sec. 5.4, Ex. 6)
   a. $(2x - 7)(3x - 4)$ $6x^2 - 29x + 28$
   b. $(3x + y)(5x - 2y)$ $15x^2 - xy - 2y^2$

8. Multiply.
   a. $(4a - 3)(7a - 2)$ $28a^2 - 29a + 6$
   b. $(2a + b)(3a - 5b)$ $6a^2 - 7ab - 5b^2$ (Sec. 5.4)

9. Factor.

   a. $8x^2 + 4$ $4(2x^2 + 1)$

   b. $5y - 2z^4$ prime polynomial

   c. $6x^2 - 3x^3$ $3x^2(2 - x)$ (Sec. 5.5, Ex. 2)

10. Factor.

   a. $9x^3 + 27x^2 - 15x$ $3x(3x^2 + 9x - 5)$

   b. $2x(3y - 2) - 5(3y - 2)$ $(3y - 2)(2x - 5)$

   c. $2xy + 6x - y - 3$ $(2x - 1)(y + 3)$ (Sec. 5.5)

Factor the polynomials in Exercises 11 through 14.

11. $x^2 - 12x + 35$ $(x - 5)(x - 7)$ (Sec. 5.6, Ex. 2)

12. $x^2 - 2x - 48$. $(x + 6)(x - 8)$ (Sec. 5.6)

13. $3a^2x - 12abx + 12b^2x$ $3x(a - 2b)^2$ (Sec. 5.7, Ex. 2)

14. $2ax^2 - 12axy + 18ay^2$ $2a(x - 3y)^2$ (Sec. 5.7)

15. Solve $3(x^2 + 4) + 5 = -6(x^2 + 2x) + 13$. $-\dfrac{2}{3}$ (Sec. 5.8, Ex. 4)

16. Solve $2(a^2 + 2) - 8 = -2a(a - 2) - 5$. $\dfrac{1}{2}$ (Sec. 5.8)

17. Solve $x^3 = 4x$. $-2, 0, 2$ (Sec. 5.8., Ex. 6)

18. Find the vertex and any intercepts of $f(x) = x^2 + x - 12$.

19. Simplify. $\dfrac{2x^2}{10x^3 - 2x^2}$ $\dfrac{1}{5x - 1}$ (Sec. 6.1, Ex. 2)

20. Simplify $\dfrac{x^2 - 4x + 4}{2 - x}$. $2 - x$ (Sec. 6.1)

21. Add $\dfrac{2x - 1}{2x^2 - 9x - 5} + \dfrac{x + 3}{6x^2 - x - 2}$.

22. Subtract. $\dfrac{a + 1}{a^2 - 6a + 8} - \dfrac{3}{16 - a^2}$ $\dfrac{a^2 + 8a - 2}{(a - 2)(a - 4)(a + 4)}$ (Sec. 6.2)

23. Simplify. $\dfrac{x^{-1} + 2xy^{-1}}{x^{-2} - x^{-2}y^{-1}}$ $\dfrac{xy + 2x^3}{y - 1}$ (Sec. 6.3, Ex. 3)

24. Simplify $\dfrac{(2a)^{-1} + b^{-1}}{a^{-1} + (2b)^{-1}}$. $\dfrac{2a + b}{a + 2b}$ (Sec. 6.3)

25. Divide: $\dfrac{3x^5y^2 - 15x^3y - x^2y - 6x}{x^2y}$

26. Divide $x^3 - 3x^2 - 10x + 24$ by $x + 3$. $x^2 - 6x + 8$ (Sec. 6.4)

27. If $P(x) = 2x^3 - 4x^2 + 5$

   a. Find $P(2)$ by substitution.   5

   b. Use synthetic division to find the remainder when $P(x)$ is divided by $x - 2$.   5 (Sec. 6.5, Ex. 3)

28. If $P(x) = 4x^3 - 2x^2 + 3$,

   a. Find $P(-2)$ by substitution. -37

   b. Use synthetic division to find the remainder when $P(x)$ is divided by $x + 2$. -37 (Sec. 6.5)

29. Solve: $\dfrac{4x}{5} + \dfrac{3}{2} = \dfrac{3x}{10}$ -3 (Sec. 6.6, Ex. 1)

30. Solve $\dfrac{x + 3}{x^2 + 5x + 6} = \dfrac{3}{2x + 4} - \dfrac{1}{x + 3}$. -1 (Sec. 6.6)

31. If a certain number is subtracted from the numerator and added to the denominator of $\dfrac{9}{19}$, the new fraction is equivalent to $\dfrac{1}{3}$. Find the number.   2 (Sec. 6.7, Ex. 2)

32. Mr. Briley can roof his house in 24 hours. His son can roof the same house in 40 hours. If they work together, how long will it take to roof the house?   15 hours (Sec. 6.7)

33. Suppose that y varies directly as x. If y is 5 when x is 30, find the constant of variation and the direct variation equation.

34. Suppose that y varies inversely as x. If y is 8 when x is 24, find the constant of variation and the inverse variation equation.

35. Simplify.   (Sec. 7.1, Ex. 5)

   a. $\sqrt{(-3)^2}$ 3   **b.** $\sqrt{x^2}$ $|x|$

   c. $\sqrt[4]{(x - 2)^4}$ $|x - 2|$ **d.** $\sqrt[3]{(-5)^3}$ -5

   e. $\sqrt[5]{(2x - 7)^5}$ $2x - 7$ **f.** $\sqrt{25x^2}$ $5|x|$

   g. $\sqrt{x^2 + 2x + 1}$ $|x + 1|$

36. Simplify. Assume that the variables represent any real number.

   a. $\sqrt{(-2)^2}$ 2   **b.** $\sqrt{y^2}$ $|y|$

   c. $\sqrt[4]{(a - 3)^4}$ $|a - 3|$ **d.** $\sqrt[3]{(-6)^3}$ -6

   e. $\sqrt[5]{(3x - 1)^5}$ $3x - 1$ (Sec. 7.1)

37. Use rational exponents to simplify. Assume that variables represent positive numbers.   (Sec. 7.2, Ex. 7)

   a. $\sqrt[8]{x^4}$ $\sqrt{x}$

   b. $\sqrt[6]{25}$ $\sqrt[3]{5}$

   c. $\sqrt[4]{r^2s^6}$ $\sqrt{rs^3}$

---

18. vertex: $\left(-\dfrac{1}{2}, -\dfrac{49}{4}\right)$; y-intercept: $(0, -12)$; x-intercepts: $(3, 0), (-4, 0)$ (Sec. 8.6)   **21.** $\dfrac{7x^2 - 9x - 13}{(2x + 1)(x - 5)(3x - 2)}$ (Sec. 6.2, Ex. 5)

25. $3x^3y - 15x - 1 - \dfrac{6}{xy}$ (Sec. 6.4, Ex. 2)   **33.** $\dfrac{1}{6}$; $y = \dfrac{1}{6}x$ (Sec. 6.8, Ex. 1)   **34.** 112; $y = \dfrac{112}{x}$ (Sec. 6.8)

38. Use rational exponents to simplify. Assume that variables represent positive numbers.

  a. $\sqrt[4]{5^2}$ $\sqrt{5}$

  b. $\sqrt[12]{x^3}$ $\sqrt[4]{x}$

  c. $\sqrt[6]{x^2y^4}$ $\sqrt[3]{xy^2}$ (Sec. 7.2)

39. Use the product rule to simplify. (Sec. 7.3, Ex. 4)

  a. $\sqrt{25x^3}$ $5x\sqrt{x}$

  b. $\sqrt[3]{54x^6y^8}$ $3x^2y^2\sqrt[3]{2y^2}$

  c. $\sqrt[4]{81z^{11}}$ $3z^2\sqrt[4]{z^3}$

40. Use the product rule to simplify. Assume that variables represent positive numbers.

  a. $\sqrt{64a^5}$ $8a^2\sqrt{a}$

  b. $\sqrt[3]{24a^7b^9}$ $2a^2b^3\sqrt[3]{3a}$

  c. $\sqrt[4]{48x^9}$ $2x^2\sqrt[4]{3x}$ (Sec. 7.3)

41. Rationalize the denominator of each expression.

  a. $\dfrac{2}{\sqrt{5}}$ $\dfrac{2\sqrt{5}}{5}$

  b. $\dfrac{2\sqrt{16}}{\sqrt{9x}}$ $\dfrac{8\sqrt{x}}{3x}$

  c. $\sqrt[3]{\dfrac{1}{2}}$ $\dfrac{\sqrt[3]{4}}{2}$ (Sec. 7.5, Ex. 1)

42. Multiply. Simplify if possible.

  a. $\left(\sqrt{3} - 4\right)\left(2\sqrt{3} + 2\right)$ $-2 - 6\sqrt{3}$

  b. $\left(\sqrt{5} - x\right)^2$ $5 - 2x\sqrt{5} + x^2$

  c. $\left(\sqrt{a} + b\right)\left(\sqrt{a} - b\right)$ $a - b^2$ (Sec. 7.4)

43. Solve $\sqrt{2x + 5} + \sqrt{2x} = 3$. $\frac{2}{9}$ (Sec. 7.6, Ex. 5)

44. Solve $\sqrt{x - 2} = \sqrt{4x + 1} - 3$. $2, 6$ (Sec. 7.6)

45. Divide. Write in the form $a + bi$.

  a. $\dfrac{2 + i}{1 - i}$ $\dfrac{1}{2} + \dfrac{3}{2}i$

  b. $\dfrac{7}{3i}$ $-\dfrac{7}{3}i$ (Sec. 7.7, Ex. 5)

46. Write each product in the form of $a + bi$.

  a. $3i(5 - 2i)$ $6 + 15i$

  b. $(6 - 5i)^2$ $11 - 60i$

  c. $\left(\sqrt{3} + 2i\right)\left(\sqrt{3} - 2i\right)$ 7 (Sec. 7.7)

47. Use the square root property to solve $(x + 1)^2 = 12$.
  $-1 + 2\sqrt{3}, -1 - 2\sqrt{3}$ (Sec. 8.1, Ex. 3)

48. Use the square root property to solve $(y - 1)^2 = 24$.
  $1 + 2\sqrt{6}; 1 - 2\sqrt{6}$ (Sec. 8.1)

49. Solve $x - \sqrt{x} - 6 = 0$. 9 (Sec. 8.3, Ex. 1)

50. Use the quadratic formula to solve. $m^2 = 4m + 8$
  $2 - 2\sqrt{3}, 2 + 2\sqrt{3}$ (Sec. 8.2)

Exponential and Logarithmic Functions

In this chapter, we discuss two closely related functions: exponential and logarithmic functions. These functions are vital to applications in economics, finance, engineering, the sciences, education, and other fields. Models of tumor growth and learning curves are two examples of the uses of exponential and logarithmic functions.

High quality of products or goods is essential to the success of any manufacturer. A quality assurance inspector is a person whose job is to determine whether a product meets industry standards for quality and safety. Inspectors test and retest samples at every step in the production process.

A quality assurance inspector is a person with a good eye for detail. Inspectors should be familiar with all aspects of the production process so that they can spot problems quickly. They should be able to work as members of a team, and they should possess the writing skills necessary to communicate problems in reports. Quality assurance inspectors also use technical tools such as problem solving, statistical quality control, and computer skills.

In the Spotlight on Decision Making feature on page 714, you will have the opportunity to make a decision about the reliability of a DVD player as a quality assurance inspector for an electronics manufacturer.

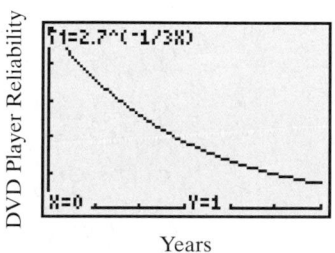

Link: www.asq.org (American Society for Quality)

Source of text: www.workforcedevelopment.com/engineering/qualitycontrol.html

9.1 THE ALGEBRA OF FUNCTIONS; COMPOSITE FUNCTIONS

Objectives

1 Add, subtract, multiply, and divide functions.

2 Construct composite functions.

1 As we have seen in earlier chapters, it is possible to add, subtract, multiply, and divide functions. Although we have not stated it as such, the sums, differences, products, and quotients of functions are themselves functions. For example, if $f(x) = 3x$ and $g(x) = x + 1$, their product, $f(x) \cdot g(x) = 3x(x + 1) = 3x^2 + 3x$, is a new function. We can use the notation $(f \cdot g)(x)$ to denote this new function. Finding the sum, difference, product, and quotient of functions to generate new functions is called the **algebra of functions.**

Algebra of Functions

Let f and g be functions. New functions from f and g are defined as follows.

Sum	$(f + g)(x) = f(x) + g(x)$
Difference	$(f - g)(x) = f(x) - g(x)$
Product	$(f \cdot g)(x) = f(x) \cdot g(x)$
Quotient	$\left(\dfrac{f}{g}\right)(x) = \dfrac{f(x)}{g(x)}, g(x) \neq 0$

EXAMPLE 1

If $f(x) = x - 1$ and $g(x) = 2x - 3$, find

a. $(f + g)(x)$

b. $(f - g)(x)$

c. $(f \cdot g)(x)$

d. $\left(\dfrac{f}{g}\right)(x)$

Solution Use the algebra of functions and replace $f(x)$ by $x - 1$ and $g(x)$ by $2x - 3$. Then we simplify.

a. $(f + g)(x) = f(x) + g(x)$

$\qquad\qquad = (x - 1) + (2x - 3)$

$\qquad\qquad = 3x - 4$

TEACHING TIP

For Example 1, have students confirm that the sum (difference, product, quotient) of $f(7)$ and $g(7)$ gives the same result as the functions summed (subtracted, multiplied, divided) and then evaluated at $x = 7$.

b. $(f - g)(x) = f(x) - g(x)$

$$= (x - 1) - (2x - 3)$$

$$= x - 1 - 2x + 3$$

$$= -x + 2$$

c. $(f \cdot g)(x) = f(x) \cdot g(x)$

$$= (x - 1)(2x - 3)$$

$$= 2x^2 - 5x + 3$$

d. $\left(\dfrac{f}{g}\right)(x) = \dfrac{f(x)}{g(x)} = \dfrac{x - 1}{2x - 3}$, where $x \neq \dfrac{3}{2}$

There is an interesting but not surprising relationship between the graphs of functions and the graphs of their sum, difference, product, and quotient. For example, the graph of $(f + g)(x)$ can be found by adding the graph of $f(x)$ to the graph of $g(x)$. We add two graphs by adding y-values of corresponding x-values. To see this, look at the following.

If $f(x) = \dfrac{1}{2}x + 2$ and $g(x) = \dfrac{1}{3}x^2 + 4$, then

$$(f + g)(x) = f(x) + g(x)$$

$$= \left(\dfrac{1}{2}x + 2\right) + \left(\dfrac{1}{3}x^2 + 4\right)$$

$$= \dfrac{1}{3}x^2 + \dfrac{1}{2}x + 6.$$

TEACHING TIP

You may want to show students another way to check addition of functions. Enter one function in Y_1, the second in Y_2, and the sum of these functions that the student has calculated in Y_3. To check this sum, use the Y Variables menu and enter $Y_4 = Y_1 + Y_2$. Y_4 is then the sum of the functions as calculated by the student's calculator. If the tables for Y_3 and Y_4 are the same, as shown on the screen to the right, the sum that was calculated by the student checks. (The Y Variables menu is found under Y-VARS. Function. Then select and enter the desired function name.)

To visualize this addition of functions with a graphing calculator, graph

$$Y_1 = \dfrac{1}{2}x + 2, \qquad Y_2 = \dfrac{1}{3}x^2 + 4, \qquad Y_3 = \dfrac{1}{3}x^2 + \dfrac{1}{2}x + 6$$

Use a TABLE feature to verify that for a given x value, $Y_1 + Y_2 = Y_3$. For example, verify that when $x = 0$, $Y_1 = 2$, $Y_2 = 4$, and $Y_3 = 2 + 4 = 6$.

2 Another way to combine functions is called **function composition.** To understand this new way of combining functions, study the diagrams below. The left diagram shows degrees Celsius $f(x)$ as a function of degrees Fahrenheit x. The right diagram shows Kelvins $g(x)$ as a function of degrees Celsius x. (The Kelvin scale is a temperature scale devised by Lord Kelvin in 1848.) The function represented by the first diagram we will call f, and the second function we will call g.

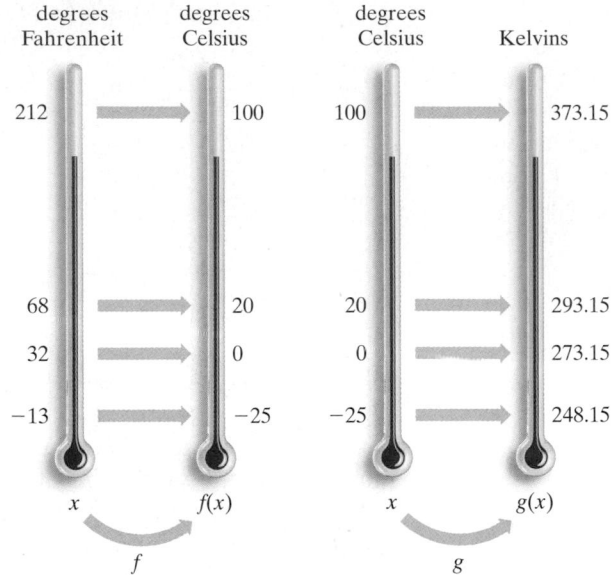

Suppose that we want a function that shows a direct conversion from degrees Fahrenheit to Kelvins. In other words, suppose that a function is needed that shows Kelvins as a function of degrees Fahrenheit. This can easily be done because the output of the first function $f(x)$ is the same as the input of the second function. If we use $f(x)$ to represent this, then we get the following diagram.

For example $g(f(-13)) = 248.15$, and so on.

Since the output of the first function is used as the input of the second function, we write the new function as $g(f(x))$. The new function is formed from the composition of the other two functions. The mathematical symbol for this composition is $(g \circ f)(x)$. Thus, $(g \circ f)(x) = g(f(x))$.

It is possible to find an equation for the composition of the two functions f and g. In other words, we can find a function that converts degrees Fahrenheit directly to Kelvins. The function $f(x) = \dfrac{5}{9}(x - 32)$ converts degrees Fahrenheit to degrees Celsius, and the function $g(x) = x + 273.15$ converts degrees Celsius to Kelvins. Thus,

$$(g \circ f)(x) = g(f(x)) = g\left(\frac{5}{9}(x - 32)\right) = \frac{5}{9}(x - 32) + 273.15$$

In general, the notation **$g(f(x))$** means "g composed with f" and can be written as $(g \circ f)(x)$. Also $f(g(x))$, or $(f \circ g)(x)$, means "f composed with g."

Composition of Functions

The composition of functions f and g is
$$(f \circ g)(x) = f(g(x))$$

Helpful Hint

$(f \circ g)(x)$ does not mean the same as $(f \cdot g)(x)$.
$$(f \circ g)(x) = f(g(x)) \text{ while } (f \cdot g)(x) = f(x) \cdot g(x)$$
$\qquad\qquad\uparrow\qquad\qquad\qquad\qquad\qquad\uparrow$
Composition of functions   Multiplication of functions

EXAMPLE 2

If $f(x) = x^2$ and $g(x) = x + 3$, find each composition.

a. $(f \circ g)(2)$ and $(g \circ f)(2)$

b. $(f \circ g)(x)$ and $(g \circ f)(x)$

Solution   **a.** $(f \circ g)(2) = f(g(2))$

$\qquad\qquad = f(5)$ Replace $g(2)$ with 5. [Since $g(x) = x + 3$, then $g(2) = 2 + 3 = 5$.]

$\qquad\qquad = 5^2 = 25$

$(g \circ f)(2) = g(f(2))$

$\qquad\qquad = g(4)$ Since $f(x) = x^2$, then $f(2) = 2^2 = 4$.

$\qquad\qquad = 4 + 3 = 7$

b. $(f \circ g)(x) = f(g(x))$

$\qquad\qquad = f(x + 3)$ Replace $g(x)$ with $x + 3$.

$\qquad\qquad = (x + 3)^2$ $f(x + 3) = (x + 3)^2$

$\qquad\qquad = x^2 + 6x + 9$ Square $(x + 3)$.

$(g \circ f)(x) = g(f(x))$

$\qquad\qquad = g(x^2)$ Replace $f(x)$ with x^2.

$\qquad\qquad = x^2 + 3$ $g(x^2) = x^2 + 3$

CLASSROOM EXAMPLE

If $f(x) = x^2$ and $g(x) = 2x + 1$, find each composition.

a. $(f \circ g)(3)$ and $(g \circ f)(3)$

b. $(f \circ g)(x)$ and $(g \circ f)(x)$

answer:
a. 49; 19   **b.** $4x^2 + 4x + 1; 2x^2 + 1$

Here are two different ways that composite functions can be evaluated using a graphing calculator. Below, we find $(f \circ g)(2)$ from Example 2a.

Enter $y_1 = x^2$ and $y_2 = x + 3$. Evaluate $(f \circ g)(2)$ by first evaluating $Y_2(2)$ and then, since $Y_2(2) = 5$, evaluating $Y_1(5)$.

Alternatively, first replace the variable in $g(x)$ with 2 and evaluate the expression. We then take that result and evaluate the $f(x)$ function there.

EXAMPLE 3

If $f(x) = |x|$ and $g(x) = x - 2$, find each composition.

a. $(f \circ g)(x)$

b. $(g \circ f)(x)$

Solution   **a.** $(f \circ g)(x) = f(g(x)) = f(x - 2) = |x - 2|$

b. $(g \circ f)(x) = g(f(x)) = g(|x|) = |x| - 2$

> **Helpful Hint**
>
> In Examples 2 and 3, notice that $(g \circ f)(x) \neq (f \circ g)(x)$. In general, $(g \circ f)(x)$ *may* or *may not* equal $(f \circ g)(x)$.

EXAMPLE 4

If $f(x) = 5x$, $g(x) = x - 2$, and $h(x) = \sqrt{x}$, write each function as a composition using two of the given functions.

a. $F(x) = \sqrt{x - 2}$

b. $G(x) = 5x - 2$

Solution   **a.** Notice the order in which the function F operates on an input value x. First, 2 is subtracted from x. This is the function $g(x) = x - 2$. Then the square root *of that result* is taken. The square root function is $h(x) = \sqrt{x}$. This means that $F = h \circ g$. To check, we find $h \circ g$.

$$F(x) = (h \circ g)(x) = h(g(x)) = h(x - 2) = \sqrt{x - 2}$$

b. Notice the order in which the function G operates on an input value x. First, x is multiplied by 5, and then 2 is subtracted from the result. This means that $G = g \circ f$. To check, we find $g \circ f$.

$$G(x) = (g \circ f)(x) = g(f(x)) = g(5x) = 5x - 2$$

MENTAL MATH

Match each function with its definition.

1. $(f \circ g)(x)$ C
2. $(f \cdot g)(x)$ E
3. $(f - g)(x)$ F
4. $(g \circ f)(x)$ A
5. $\left(\dfrac{f}{g}\right)(x)$ D
6. $(f + g)(x)$ B

A. $g[f(x)]$
B. $f(x) + g(x)$
C. $f[g(x)]$
D. $\dfrac{f(x)}{g(x)}$
E. $f(x) \cdot g(x)$
F. $f(x) - g(x)$

1. **a.** $3x - 6$ **b.** $-x - 8$
   c. $2x^2 - 13x - 7$
   d. $\dfrac{x - 7}{2x + 1}$, where $x \neq -\dfrac{1}{2}$

2. **a.** $6x + 2$ **b.** $-4x + 6$
   c. $5x^2 + 18x - 8$
   d. $\dfrac{x + 4}{5x - 2}$, where $x \neq \dfrac{2}{5}$

3. **a.** $x^2 + 5x + 1$ **b.** $x^2 - 5x + 1$
   c. $5x^3 + 5x$
   d. $\dfrac{x^2 + 1}{5x}$, where $x \neq 0$

4. **a.** $x^2 + 3x - 2$ **b.** $x^2 - 3x - 2$
   c. $3x^3 - 6x$
   d. $\dfrac{x^2 - 2}{3x}$, where $x \neq 0$

5. **a.** $\sqrt{x} + x + 5$ **b.** $\sqrt{x} - x - 5$ **c.** $x\sqrt{x} + 5\sqrt{x}$ **d.** $\dfrac{\sqrt{x}}{x + 5}$, where $x \neq -5$

EXERCISE SET 9.1

STUDY GUIDE/SSM   CD/ VIDEO   PH MATH TUTOR CENTER   MathXL®Tutorials ON CD   MathXL®   MyMathLab®

For the functions f and g, find **a.** $(f + g)(x)$, **b.** $(f - g)(x)$, **c.** $(f \cdot g)(x)$, *and* **d.** $\left(\dfrac{f}{g}\right)(x)$. *See Example 1.*

1. $f(x) = x - 7, g(x) = 2x + 1$
2. $f(x) = x + 4, g(x) = 5x - 2$
3. $f(x) = x^2 + 1, g(x) = 5x$
4. $f(x) = x^2 - 2, g(x) = 3x$
5. $f(x) = \sqrt{x}, g(x) = x + 5$
6. $f(x) = \sqrt[3]{x}, g(x) = x - 3$
7. $f(x) = -3x, g(x) = 5x^2$
8. $f(x) = 4x^3, g(x) = -6x$

6. **a.** $\sqrt[3]{x} + x - 3$
   b. $\sqrt[3]{x} - x + 3$
   c. $x\sqrt[3]{x} - 3\sqrt[3]{x}$
   d. $\dfrac{\sqrt[3]{x}}{x - 3}$, where $x \neq 3$
7. **a.** $5x^2 - 3x$
   b. $-5x^2 - 3x$
   c. $-15x^3$
   d. $-\dfrac{3}{5x}$, where $x \neq 0$

If $f(x) = x^2 - 6x + 2, g(x) = -2x$, *and* $h(x) = \sqrt{x}$, *find each composition. See Example 2.*

9. $(f \circ g)(2)$ 42
10. $(h \circ f)(-2)$ $3\sqrt{2}$
11. $(g \circ f)(-1)$ -18
12. $(f \circ h)(1)$ -3
13. $(g \circ h)(0)$ 0
14. $(h \circ g)(0)$ 0

Find $(f \circ g)(x)$ *and* $(g \circ f)(x)$. *See Examples 2 and 3.*

15. $f(x) = x^2 + 1, g(x) = 5x$
    $(f \circ g)(x) = 25x^2 + 1; (g \circ f)(x) = 5x^2 + 5$
16. $f(x) = x - 3, g(x) = x^2$
    $(f \circ g)(x) = x^2 - 3; (g \circ f)(x) = x^2 - 6x + 9$
17. $f(x) = 2x - 3, g(x) = x + 7$
    $(f \circ g)(x) = 2x + 11; (g \circ f)(x) = 2x + 4$

18. $f(x) = x + 10, g(x) = 3x + 1$
    $(f \circ g)(x) = 3x + 11; (g \circ f)(x) = 3x + 31$
19. $f(x) = x^3 + x - 2, g(x) = -2x$
20. $f(x) = -4x, g(x) = x^3 + x^2 - 6$
21. $f(x) = \sqrt{x}, g(x) = -5x + 2$
    $(f \circ g)(x) = \sqrt{-5x + 2}; (g \circ f)(x) = -5\sqrt{x} + 2$
22. $f(x) = 7x - 1, g(x) = \sqrt[3]{x}$
    $(f \circ g)(x) = 7\sqrt[3]{x} - 1; (g \circ f)(x) = \sqrt[3]{7x - 1}$

If $f(x) = 3x, g(x) = \sqrt{x}$, *and* $h(x) = x^2 + 2$, *write each function as a composition using two of the given functions. See Example 4.*

23. $H(x) = \sqrt{x^2 + 2}$ $H(x) = (g \circ h)(x)$
24. $G(x) = \sqrt{3x}$ $G(x) = (g \circ f)(x)$
25. $F(x) = 9x^2 + 2$ $F(x) = (h \circ f)(x)$
26. $H(x) = 3x^2 + 6$ $H(x) = (f \circ h)(x)$
27. $G(x) = 3\sqrt{x}$ $G(x) = (f \circ g)(x)$
28. $F(x) = x + 2$ $F(x) = (h \circ g)(x)$

Find $f(x)$ *and* $g(x)$ *so that the given function* $h(x) = (f \circ g)(x)$.

29. $h(x) = (x + 2)^2$ answers may vary; for example,
    $g(x) = x + 2$ and $f(x) = x^2$
30. $h(x) = |x - 1|$ answers may vary; for example,
    $g(x) = x - 1$ and $f(x) = |x|$
31. $h(x) = \sqrt{x + 5} + 2$ answers may vary; for example,
    $g(x) = x + 5$ and $f(x) = \sqrt{x} + 2$

19. $(f \circ g)(x) = -8x^3 - 2x - 2$; $(g \circ f)(x) = -2x^3 - 2x + 4$

32. $h(x) = (3x + 4)^2 + 3$ answers may vary; for example,
$g(x) = 3x + 4$ and $f(x) = x^2 + 3$

33. $h(x) = \dfrac{1}{2x - 3}$ answers may vary; for example,
$g(x) = 2x - 3$ and $f(x) = \frac{1}{x}$

34. $h(x) = \dfrac{1}{x + 10}$ answers may vary; for example,
$g(x) = x + 10$ and $f(x) = \frac{1}{x}$

REVIEW AND PREVIEW

Solve each equation for y. See Section 1.8.

35. $x = y + 2$ $y = x - 2$ **36.** $x = y - 5$ $y = x + 5$

37. $x = 3y$ $y = \dfrac{x}{3}$ **38.** $x = -6y$ $y = -\dfrac{x}{6}$

39. $x = -2y - 7$ **40.** $x = 4y + 7$ $y = \dfrac{x - 7}{4}$

Concept Extensions

Given that $f(-1) = 4$ $g(-1) = -4$
$f(0) = 5$ $g(0) = -3$
$f(2) = 7$ $g(2) = -1$
$f(7) = 1$ $g(7) = 4$

8. a. $4x^3 - 6x$ **b.** $4x^3 + 6x$
  c. $-24x^4$ **d.** $-\dfrac{2}{3}x^2$, where $x \neq 0$

39. $y = -\dfrac{x + 7}{2}$

Find each function value.

41. $(f + g)(2)$ 6   **42.** $(f - g)(7)$ 8

43. $(f \circ g)(2)$ 4   **44.** $(g \circ f)(2)$ 4

45. $(f \cdot g)(7)$ 4   **46.** $(f \cdot g)(0)$ -15

47. $\left(\dfrac{f}{g}\right)(-1)$ -1 **48.** $\left(\dfrac{g}{f}\right)(-1)$ -1

49. If you are given $f(x)$ and $g(x)$, explain in your own words how to find $(f \circ g)(x)$, and then how to find $(g \circ f)(x)$. answers may vary

50. Given $f(x)$ and $g(x)$, describe in your own words the difference between $(f \circ g)(x)$ and $(f \cdot g)(x)$. answers may vary

Solve.

51. Business people are concerned with cost functions, revenue functions, and profit functions. Recall that the profit $P(x)$ obtained from x units of a product is equal to the revenue $R(x)$ from selling the x units minus the cost $C(x)$ of manufacturing the x units. Write an equation expressing this relationship among $C(x)$, $R(x)$, and $P(x)$. $P(x) = R(x) - C(x)$

52. Suppose the revenue $R(x)$ for x units of a product can be described by $R(x) = 25x$, and the cost $C(x)$ can be described by $C(x) = 50 + x^2 + 4x$. Find the profit $P(x)$ for x units.
$P(x) = 21x - x^2 - 50$

20. $(f \circ g)(x) = -4x^3 - 4x^2 + 24$; $(g \circ f)(x) = -64x^3 + 16x^2 - 6$

STUDY SKILLS REMINDER

How Are You Doing?

If you haven't done so yet, take a few moments and think about how you are doing in this course. Are you working toward your goal of successfully completing this course? Is your performance on homework, quizzes, and tests satisfactory? If not, you might want to see your instructor to see if he/she has any suggestions on how you can improve your performance. Let me once again remind you that, in addition to your instructor, there are many places to get help with your mathematics course. A few suggestions are below.

▶ This text has an accompanying video lesson for every section in this text.

▶ The back of this book contains answers to odd-numbered exercises and selected solutions.

▶ A tutorial software program is available with this text. It has lessons corresponding to each section in the text.

▶ There is a student solutions manual available that contains worked-out solutions to odd-numbered exercises as well as solutions to every exercise in the Integrated Reviews, Chapter Reviews, Chapter Tests, and Cumulative Reviews.

▶ Don't forget to check with your instructor for other local resources available to you, such as a tutor center.

9.2 INVERSE FUNCTIONS

Objectives

1 Determine whether a function is a one-to-one function.

2 Use the horizontal line test to decide whether a function is a one-to-one function.

3 Find the inverse of a function.

4 Find the equation of the inverse of a function.

5 Graph functions and their inverses.

6 Determine whether two functions are inverses of each other.

1 In the next section, we begin a study of two new functions: exponential and logarithmic functions. As we learn more about these functions, we will discover that they share a special relation to each other: They are inverses of each other.

Before we study these functions, we need to learn about inverses. We begin by defining one-to-one functions.

Study the following diagram.

Recall that since each degrees Fahrenheit (input) corresponds to exactly one degrees Celsius (output), this pairing of inputs and outputs does describe a function. Also notice that each output corresponds to exactly one input. This type of function is given a special name—a one-to-one function.

Does the set $f = \{(0, 1), (2, 2), (-3, 5), (7, 6)\}$ describe a one-to-one function? It is a function since each x-value corresponds to a unique y-value. For this particular function f, each y-value also corresponds to a unique x-value. Thus, this function is also a **one-to-one function.**

One-to-one Function

For a **one-to-one function,** each x-value (input) corresponds to only one y-value (output), and each y-value (output) corresponds to only one x-value (input).

EXAMPLE 1

Determine whether each function described is one-to-one.

a. $f = \{(6,2),(5,4),(-1,0),(7,3)\}$

b. $g = \{(3,9),(-4,2),(-3,9),(0,0)\}$

c. $h = \{(1,1),(2,2),(10,10),(-5,-5)\}$

d.

Mineral (Input)	Talc	Gypsum	Diamond	Topaz	Stibnite
Hardness on the Mohs Scale (Output)	1	2	10	8	2

e.

f.

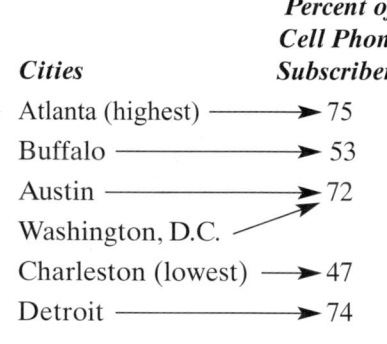

Cities	Percent of Cell Phone Subscribers
Atlanta (highest)	75
Buffalo	53
Austin	72
Washington, D.C.	
Charleston (lowest)	47
Detroit	74

Solution

a. f is one-to-one since each y-value corresponds to only one x-value.

b. g is not one-to-one because the y-value 9 in $(3,9)$ and $(-3,9)$ corresponds to two different x-values.

c. h is a one-to-one function since each y-value corresponds to only one x-value.

d. This table does not describe a one-to-one function since the output 2 corresponds to two different inputs, gypsum and stibnite.

e. This graph does not describe a one-to-one function since the y-value -1 corresponds to three different x-values, -2, -1, and 3.

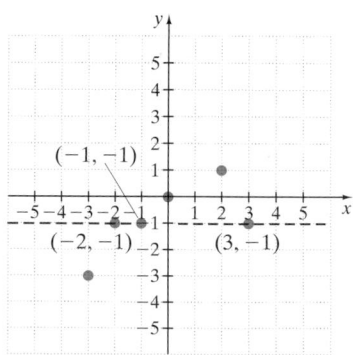

f. The mapping is not one-to-one since 72% corresponds to Austin and Washington, D.C.

2   Recall that we recognize the graph of a function when it passes the vertical line test. Since every x-value of the function corresponds to exactly one y-value, each vertical line intersects the function's graph at most once. The graph shown (left), for instance, is the graph of a function.

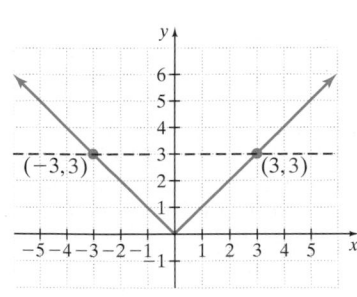

Is this function a *one-to-one* function? The answer is no. To see why not, notice that the y-value of the ordered pair $(-3, 3)$, for example, is the same as the y-value of the ordered pair $(3, 3)$. In other words, the y-value 3 corresponds to two x-values, -3 and 3. This function is therefore not one-to-one.

To test whether a graph is the graph of a one-to-one function, apply the vertical line test to see if it is a function, and then apply a similar **horizontal line test** to see if it is a one-to-one function.

Horizontal Line Test

If every horizontal line intersects the graph of a function at most once, then the function is a one-to-one function.

EXAMPLE 2

Determine whether each graph is the graph of a one-to-one function.

a.

b.

c.

d.

e.

Solution Graphs **a**, **b**, **c**, and **d** all pass the vertical line test, so only these graphs are graphs of functions. But, of these, only **b** and **c** pass the horizontal line test, so only **b** and **c** are graphs of one-to-one functions.

> **Helpful Hint**
>
> All linear equations are one-to-one functions except those whose graphs are horizontal or vertical lines. A vertical line does not pass the vertical line test and hence is not the graph of a function. A horizontal line is the graph of a function but does not pass the horizontal line test and hence is not the graph of a one-to-one function.

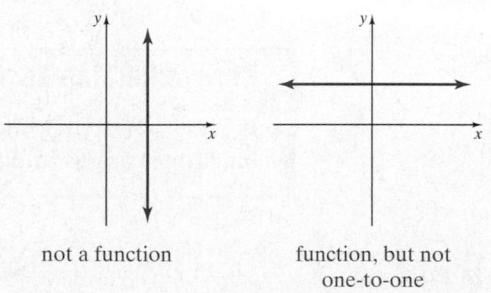

not a function   function, but not one-to-one

f

Input: degrees Fahrenheit

Output: degrees Celsius

212 → 100

68 → 20

32 → 0

−13 → −25

f^{-1}

3 One-to-one functions are special in that their graphs pass both the vertical and horizontal line tests. They are special, too, in another sense: For each one-to-one function, we can find its **inverse function** by switching the coordinates of the ordered pairs of the function, or the inputs and the outputs. For example, the inverse of the one-to-one function $f = \{(2, -3), (5, 10), (9, 1)\}$ is $\{(-3, 2), (10, 5), (1, 9)\}$. For a function f, we use the notation f^{-1}, read "f inverse," to denote its inverse function. Notice that since the coordinates of each ordered pair have been switched, the domain (set of inputs) of f is the range (set of outputs) of f^{-1}, and the range of f is the domain of f^{-1}.

The diagram to the left shows the inverse of the one-to-one function f with ordered pairs of the form (degrees Fahrenheit, degrees Celsuis) is the function f^{-1} with ordered pairs of the form (degrees Celsuis, degrees Fahrenheit). Notice that the ordered pair $(-13, -25)$ of the function, for example, becomes the ordered pair $(-25, -13)$ of its inverse.

> **Inverse Function**
>
> The inverse of a one-to-one function f is the one-to-one function f^{-1} that consists of the set of all ordered pairs (y, x) where (x, y) belongs to f.

> **Helpful Hint**
>
> If a function is not one-to-one, it does not have an inverse function.

EXAMPLE 3

Find the inverse of the one-to-one function.

$$f = \{(0, 1), (-2, 7), (3, -6), (4, 4)\}$$

Solution $f^{-1} = \{(1,0),(7,-2),(-6,3),(4,4)\}$

↑   ↑   ↑   ↑   Switch coordinates
of each ordered pair.

> **Helpful Hint**
> The symbol f^{-1} is the single symbol used to denote the inverse of the function f.
> It is read as "f inverse." This symbol *does not mean* $\dfrac{1}{f}$.

✔ **CONCEPT CHECK**

Suppose that f is a one-to-one function. If the ordered pair $(1, 5)$ belongs to f, name one point that we know must belong to the inverse function, f^{-1}.

4   If a one-to-one function f is defined as a set of ordered pairs, we can find f^{-1} by interchanging the x- and y-coordinates of the ordered pairs. If a one-to-one function f is given in the form of an equation, we can find f^{-1} by using a similar procedure.

Finding the Inverse of a One-to-One Function $f(x)$

Step 1: Replace $f(x)$ with y.
Step 2: Interchange x and y.
Step 3: Solve the equation for y.
Step 4: Replace y with the notation $f^{-1}(x)$.

EXAMPLE 4

Find an equation of the inverse of $f(x) = x + 3$.

Solution $f(x) = x + 3$

Step 1: $y = x + 3$ Replace $f(x)$ with y.
Step 2: $x = y + 3$ Interchange x and y.
Step 3: $x - 3 = y$ Solve for y.
Step 4: $f^{-1}(x) = x - 3$ Replace y with $f^{-1}(x)$.

The inverse of $f(x) = x + 3$ is $f^{-1}(x) = x - 3$. Notice that, for example,

$$f(1) = 1 + 3 = 4 \quad \text{and} \quad f^{-1}(4) = 4 - 3 = 1$$

Ordered pair: $(1, 4)$ Ordered pair: $(4, 1)$

The coordinates are switched, as expected.

EXAMPLE 5

Find the equation of the inverse of $f(x) = 3x - 5$.

Solution

$f(x) = 3x - 5$

Step 1: $y = 3x - 5$ Replace $f(x)$ with y.

Step 2: $x = 3y - 5$ Interchange x and y.

Step 3: $3y = x + 5$ Solve for y.

$$y = \frac{x + 5}{3}$$

Step 4: $f^{-1}(x) = \dfrac{x + 5}{3}$ Replace y with $f^{-1}(x)$.

The inverse of $f(x) = 3x - 5$ is $f^{-1}(x) = \dfrac{x + 5}{3}$.

CLASSROOM EXAMPLE

Find the equation of the inverse of $f(x) = 2x + 3$. Graph f and f^{-1} on the same set of axes.

answer:

TEACHING TIP

After discussing the Discover the Concept, have students find $(f \circ f^{-1})(7)$ and $(f^{-1} \circ f)(7)$. Then have them find $(f \circ f^{-1})(x)$ and $(f^{-1} \circ f)(x)$.

Discover the Concept

a. Use your graphing utility to graph both $f(x) = 3x - 5$ as $y_1 = 3x - 5$ and $f^{-1}(x) = \frac{x + 5}{3}$ as $y_2 = (x + 5)/3$ using a square window.

b. Trace to the point $(2, 1)$ on the graph of y_1. Is there a point $(1, 2)$ on y_2? Try this for a point on y_2. If (a, b) is a point of y_2, is (b, a) a point of y_1?

c. Next include the graph of $y_3 = x$ in the display. How are the graphs of $f(x) = 3x - 5$ and $f^{-1}(x) = \frac{x + 5}{3}$ related to the graph of $y = x$?

TEACHING TIP

Before discussing the solution to Example 5 ask, "What does f do to an input? What would you expect the inverse of f to do to an input?" After discussing the solution, state "Notice that the inverse function does the inverse operations of the function in reverse order."

TEACHING TIP

After discussing Example 5, have students find $(f \circ f^{-1})(7)$ and $(f^{-1} \circ f)(7)$. Then have them find $(f \circ f^{-1})(x)$ and $(f^{-1} \circ f)(x)$.

⑤   Notice in the Discover the Concept that the graphs of f and f^{-1} are mirror images of each other, and the "mirror" is the dashed line $y = x$. This is true for every function and its inverse. For this reason, we say that *the graphs of f and f^{-1} are symmetric about the line* $y = x$.

To see why this happens, study the graph of a few ordered pairs and their switched coordinates in the diagram below.

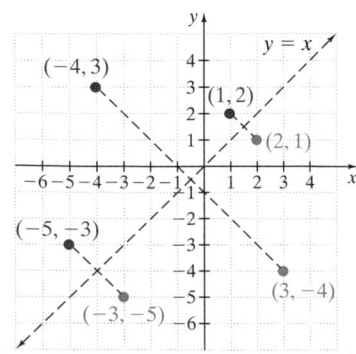

EXAMPLE 6

Graph the inverse of each function.

Solution   The function is graphed in blue and the inverse is graphed in red.

a.

b.

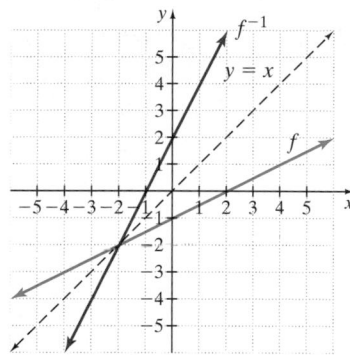

TEACHING TIP

Ask, "Where will the graphs of inverses intersect? Explain."

6   Next, study the table of values below from Example 5.

X	Y1
0	-5
1	-2
2	1

Y1 ▊3X−5

X	Y2
-5	0
-2	1
1	2

Y2 ▊(X+5)/3

Notice in the table of values above that $f(0) = -5$ and $f^{-1}(-5) = 0$, as expected. Also, for example, $f(1) = -2$ and $f^{-1}(-2) = 1$. In words, we say that for some input x, the function f^{-1} takes the output of x, called $f(x)$, back to x.

TECHNOLOGY NOTE

The inverse of a function in Y_1 can be drawn on a graph using the Draw menu and the Draw Inv Y_1 option. Notice you cannot trace on the graph of the inverse, as it has not been plotted.

$$\begin{array}{ccc} x \rightarrow f(x) & \text{and} & f^{-1}(f(x)) \rightarrow x \\ \downarrow \quad \downarrow & & \downarrow \qquad \downarrow \\ f(0) = -5 & \text{and} & f^{-1}(-5) = 0 \\ f(1) = -2 & \text{and} & f^{-1}(-2) = 1 \end{array}$$

In general,

If f is a one-to-one function, then the inverse of f is the function f^{-1} such that
$$(f^{-1} \circ f)(x) = x \quad \text{and} \quad (f \circ f^{-1})(x) = x$$

EXAMPLE 7

Show that if $f(x) = 3x + 2$, then $f^{-1}(x) = \dfrac{x - 2}{3}$.

Solution   See that $(f^{-1} \circ f)(x) = x$ and $(f \circ f^{-1})(x) = x$.

$$\begin{aligned} (f^{-1} \circ f)(x) &= f^{-1}(f(x)) \\ &= f^{-1}(3x + 2) \qquad \text{Replace } f(x) \text{ with } 3x + 2. \\ &= \frac{3x + 2 - 2}{3} \\ &= \frac{3x}{3} \\ &= x \end{aligned}$$

$$
\begin{aligned}
(f \circ f^{-1})(x) &= f(f^{-1}(x)) \\
&= f\left(\frac{x-2}{3}\right) \qquad \text{Replace } f^{-1}(x) \text{ with } \frac{x-2}{3}. \\
&= 3\left(\frac{x-2}{3}\right) + 2 \\
&= x - 2 + 2 \\
&= x
\end{aligned}
$$

To visually see these results, graph $y_1 = 3x + 2$, $y_2 = \frac{x-2}{3}$, and $y_3 = x$ using a square window, as shown below. Graphically, we see that the graphs of f and f^{-1} are mirror images of each other across the line $y = x$.

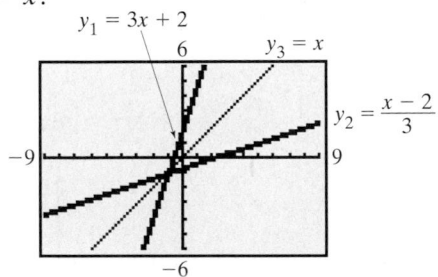

1. one-to-one; $f^{-1} = \{(-1,-1),(1,1),(2,0),(0,2)\}$
2. not one-to-one
3. one-to-one; $h^{-1} = \{(10,10)\}$
4. one-to-one; $r^{-1} = \{(2,1),(4,3),(6,5),(7,6)\}$
5. one-to-one; $f^{-1} = \{(12,11),(3,4),(4,3),(6,6)\}$
6.–8. not one-to-one

We can also make a table of values for f and f^{-1} and see that if $f(a) = b$ [the ordered pair (a, b)] then $f^{-1}(b) = a$ [the ordered pair (b, a)].

Notice that $f(0) = 2$

Notice that $f^{-1}(2) = 0$

TEACHING TIP

A Group Activity for this section is available in the Instructor's Resource Manual.

EXERCISE SET 9.2

STUDY GUIDE/SSM   CD/ VIDEO   PH MATH TUTOR CENTER   MathXL®Tutorials ON CD   MathXL®   MyMathLab®

Determine whether each function is a one-to-one function. If it is one-to-one, list the inverse function by switching coordinates, or inputs and outputs. See Examples 1 and 3.

1. $f = \{(-1,-1),(1,1),(0,2),(2,0)\}$
2. $g = \{(8,6),(9,6),(3,4),(-4,4)\}$
3. $h = \{(10,10)\}$
4. $r = \{(1,2),(3,4),(5,6),(6,7)\}$
5. $f = \{(11,12),(4,3),(3,4),(6,6)\}$
6. $g = \{(0,3),(3,7),(6,7),(-2,-2)\}$

7.

Month of 2003 (Input)	January	March	May	July	September
Unemployment rate in percent	5.7	5.8	6.1	6.2	6.1

(*Source:* U.S. Department of Housing and Urban Development)

8.

State (Input)	Wisconson	Ohio	Georgia	Colorado	California	Arizona
Electoral Votes (Output)	10	20	15	9	55	10

9.

State (Input)	California	Vermont	Virginia	Texas	South Dakota
Rank in Population (Output)	1	49	12	2	46

(*Source:* U.S. Bureau of the Census)

31. $f^{-1}(x) = \dfrac{x-2}{5}$ **32.** $f^{-1}(x) = \dfrac{x+1}{6}$ **33.** $f^{-1}(x) = 5x + 2$ **34.** $f^{-1}(x) = \dfrac{2x+3}{4}$ INVERSE FUNCTIONS   SECTION 9.2   **691**

△ **10.**

Shape (Input)	Triangle	Pentagon	Quadrilateral	Hexagon	Decagon
Number of Sides (Output)	3	5	4	6	10

(*Source:* U.S. Bureau of the Census)   **10.** one-to-one

Given the one-to-one function $f(x) = x^3 + 2$, find the following.
[*Hint: You do not need to find the equation for $f^{-1}(x)$.*]

11. **a.** $f(1)$ 3
   b. $f^{-1}(3)$ 1

12. **a.** $f(0)$ 2
   b. $f^{-1}(2)$ 0

13. **a.** $f(-1)$ 1
   b. $f^{-1}(1)$ −1

14. **a.** $f(-2)$ −6
   b. $f^{-1}(-6)$ −2

Determine whether the graph of each function is the graph of a one-to-one function. See Example 2.

15. one-to-one

16. not one-to-one

 17. not one-to-one

18. one-to-one

19. not one-to-one

20. not one-to-one

21. not one-to-one

22. one-to-one

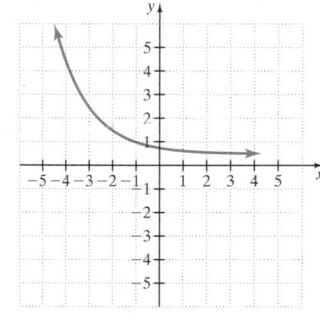

Each of the following functions is one-to-one. Find the inverse of each function and graph the function and its inverse on the same set of axes. See Examples 4 and 5.   **23.–30.** See graphing answer section.

23. $f(x) = x + 4$
   $f^{-1}(x) = x - 4$

24. $f(x) = x - 5$
   $f^{-1}(x) = x + 5$

25. $f(x) = 2x - 3$
   $f^{-1}(x) = \dfrac{x+3}{2}$

26. $f(x) = 4x + 9$
   $f^{-1}(x) = \dfrac{x-9}{4}$

27. $f(x) = \dfrac{1}{2}x - 1$
   $f^{-1}(x) = 2x + 2$

28. $f(x) = -\dfrac{1}{2}x + 2$
   $f^{-1}(x) = 4 - 2x$

29. $f(x) = x^3$
   $f^{-1}(x) = \sqrt[3]{x}$

30. $f(x) = x^3 - 1$
   $f^{-1}(x) = \sqrt[3]{x+1}$

Find the inverse of each one-to-one function. See Examples 4 and 5.

31. $f(x) = 5x + 2$

32. $f(x) = 6x - 1$

33. $f(x) = \dfrac{x-2}{5}$

34. $f(x) = \dfrac{4x-3}{2}$

35. $f(x) = \sqrt[3]{x}$

36. $f(x) = \sqrt[3]{x+1}$

37. $f(x) = \dfrac{5}{3x+1}$

38. $f(x) = \dfrac{7}{2x+4}$

39. $f(x) = (x+2)^3$

40. $f(x) = (x-5)^3$

35. $f^{-1}(x) = x^3$

36. $f^{-1}(x) = x^3 - 1$

37. $f^{-1}(x) = \dfrac{5-x}{3x}$

38. $f^{-1}(x) = \dfrac{7-4x}{2x}$

39. $f^{-1}(x) = \sqrt[3]{x} - 2$

40. $f^{-1}(x) = \sqrt[3]{x} + 5$

9.

Rank in Pop. (Input)	1	49	12	2	46
State (Output)	CA	VT	VA	TX	SD

10.

No. of Sides (Input)	3	5	4	6	10
Shape (Output)	Triangle	Pentagon	Quadrilateral	Hexagon	Decagon

Graph the inverse of each function on the same set of axes. See Example 6.

41.

42.

43.

44.

45.

46.

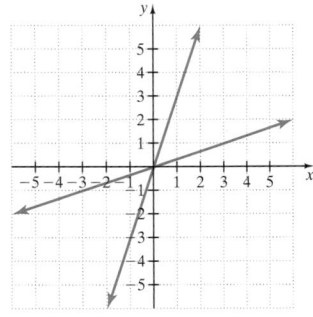

Solve. See Example 7. **47.** $(f \circ f^{-1})(x) = x; (f^{-1} \circ f)(x) = x$

47. If $f(x) = 2x + 1$, show that $f^{-1}(x) = \dfrac{x - 1}{2}$.

48. If $f(x) = 3x - 10$, show that $f^{-1}(x) = \dfrac{x + 10}{3}$.

49. If $f(x) = x^3 + 6$, show that $f^{-1}(x) = \sqrt[3]{x - 6}$.

50. If $f(x) = x^3 - 5$, show that $f^{-1}(x) = \sqrt[3]{x + 5}$.

48. $(f \circ f^{-1})(x) = x; (f^{-1} \circ f)(x) = x$

REVIEW AND PREVIEW

Evaluate each of the following. See Section 7.2.

51. $25^{1/2}$ 5

52. $49^{1/2}$ 7

53. $16^{3/4}$ 8

54. $27^{2/3}$ 9

55. $9^{-3/2}$ $\frac{1}{27}$

56. $81^{-3/4}$ $\frac{1}{27}$

49. $(f \circ f^{-1})(x) = x; (f^{-1} \circ f)(x) = x$

50. $(f \circ f^{-1})(x) = x; (f^{-1} \circ f)(x) = x$

If $f(x) = 3^x$, find the following. In Exercises 59 and 60, give an exact answer and a two-decimal-place approximation. See Sections 2.2 and 5.1.

57. $f(2)$ 9

58. $f(0)$ 1

59. $f\left(\frac{1}{2}\right)$ $3^{1/2} \approx 1.73$

60. $f\left(\frac{2}{3}\right)$ $3^{2/3} \approx 2.08$

Concept Extensions

For Exercises 61 and 62,

    a. Write the ordered pairs for $f(x)$ whose points are highlighted. (Include the points whose coordinates are given.)

    b. Write the corresponding ordered pairs for the inverse of f, f^{-1}.

    c. Graph the ordered pairs for f^{-1} found in part **b.**

    d. Graph $f^{-1}(x)$ by drawing a smooth curve through the plotted points.

61.

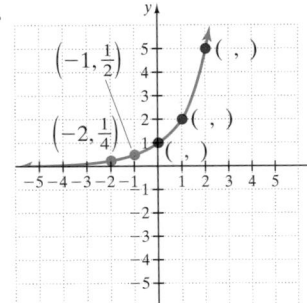

    a. $\left(-2, \frac{1}{4}\right), \left(-1, \frac{1}{2}\right)(0, 1), (1, 2), (2, 5)$

    b. $\left(\frac{1}{4}, -2\right), \left(\frac{1}{2}, -1\right), (1, 0), (2, 1), (5, 2)$

    c., d. See graphing answer section.

62.

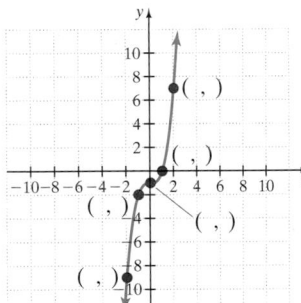

    a. $(-2, -9), (-1, -2), (0, -1), (1, 0), (2, 7)$

    b. $(-9, -2), (-2, -1), (-1, 0), (0, 1), (7, 2)$

    c., d. See graphing answer section.

63. If you are given the graph of a function, describe how you can tell from the graph whether a function has an inverse. answers may vary

64. Describe the appearance of the graphs of a function and its inverse. answers may vary

Find the inverse of each given one-to-one function. Then graph the function and its inverse on a square window.

65. $f(x) = 3x + 1$

66. $f(x) = -2x - 6$

67. $f(x) = \sqrt[3]{x + 1}$

68. $f(x) = x^3 - 3$

65. See graphing answer section.

66. $f^{-1}(x) = \dfrac{-x - 6}{2}$ See graphing answer section.

67. $f^{-1}(x) = x^3 - 1$ See graphing answer section.

68. $f^{-1}(x) = \sqrt[3]{x + 3}$ See graphing answer section.

9.3   EXPONENTIAL FUNCTIONS

Objectives

1. Graph exponential functions.
2. Solve equations of the form $b^x = b^y$.
3. Solve problems modeled by exponential equations.

1   In earlier chapters, we gave meaning to exponential expressions such as 2^x, where x is a rational number. For example,

$$2^3 = 2 \cdot 2 \cdot 2 \qquad \text{Three factors, each factor is } 2$$

$$2^{3/2} = (2^{1/2})^3 = \sqrt{2} \cdot \sqrt{2} \cdot \sqrt{2} \qquad \text{Three factors, each factor is } \sqrt{2}$$

When x is an irrational number (for example, $\sqrt{3}$), what meaning can we give to $2^{\sqrt{3}}$?

It is beyond the scope of this book to give precise meaning to 2^x if x is irrational. We can confirm your intuition and say that $2^{\sqrt{3}}$ is a real number, and since $1 < \sqrt{3} < 2$, then $2^1 < 2^{\sqrt{3}} < 2^2$. We can also use a calculator and approximate $2^{\sqrt{3}}$: $2^{\sqrt{3}} \approx 3.321997$. In fact, as long as the base b is positive, b^x is a real number for all real numbers x. Finally, the rules of exponents apply whether x is rational or irrational, as long as b is positive. In this section, we are interested in functions of the form $f(x) = b^x$, where $b > 0$. A function of this form is called an **exponential function.**

Exponential Function

A function of the form

$$f(x) = b^x$$

is called an **exponential function** if $b > 0$, b is not 1, and x is a real number.

Next, we practice graphing exponential functions.

 EXAMPLE 1

CLASSROOM EXAMPLE

Graph: $f(x) = 4^x$

answer:

Graph the exponential functions defined by $f(x) = 2^x$ and $g(x) = 3^x$ on the same set of axes.

Solution   Graph each function by plotting points. Set up a table of values for each of the two functions.

$f(x) = 2^x$

x	0	1	2	3	-1	-2
$f(x)$	1	2	4	8	$\dfrac{1}{2}$	$\dfrac{1}{4}$

$g(x) = 3^x$

x	0	1	2	3	-1	-2
$g(x)$	1	3	9	27	$\dfrac{1}{3}$	$\dfrac{1}{9}$

If each set of points is plotted and connected with a smooth curve, the following graphs result.

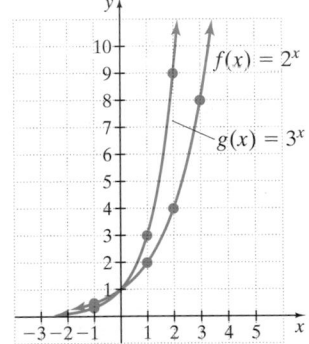

CLASSROOM EXAMPLE

Graph: $f(x) = \left(\dfrac{1}{5}\right)^x$

answer:

A number of things should be noted about the two graphs of exponential functions in Example 1. First, the graphs show that $f(x) = 2^x$ and $g(x) = 3^x$ are one-to-one functions since each graph passes the vertical and horizontal line tests. The y-intercept of each graph is $(0, 1)$, but neither graph has an x-intercept. From the graph, we can also see that the domain of each function is all real numbers and that the range is $(0, \infty)$. We can also see that as x-values are increasing, y-values are increasing also.

EXAMPLE 2

Graph the exponential functions $y = \left(\dfrac{1}{2}\right)^x$ and $y = \left(\dfrac{1}{3}\right)^x$ on the same set of axes.

Solution As before, plot points and connect them with a smooth curve.

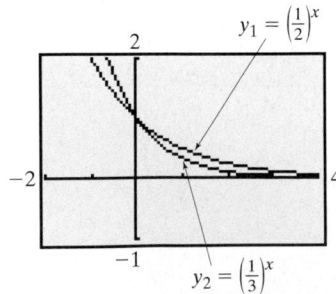

$y = \left(\dfrac{1}{2}\right)^x$

x	0	1	2	3	-1	-2
y	1	$\dfrac{1}{2}$	$\dfrac{1}{4}$	$\dfrac{1}{8}$	2	4

$y = \left(\dfrac{1}{3}\right)^x$

x	0	1	2	3	-1	-2
y	1	$\dfrac{1}{3}$	$\dfrac{1}{9}$	$\dfrac{1}{27}$	3	9

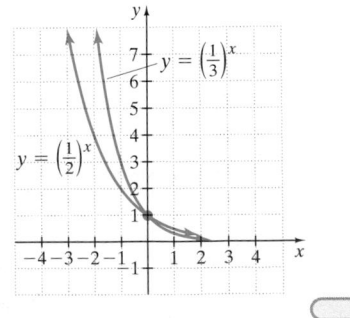

To check, graph $y_1 = (1/2)^x$ and $y_2 = (1/3)^x$ using the same window. Compare the screens to the calculations done by hand.

Each function in Example 2 again is a one-to-one function. The y-intercept of both is $(0, 1)$. The domain is the set of all real numbers, and the range is $(0, \infty)$.

Notice the difference between the graphs of Example 1 and the graphs of Example 2. An exponential function is always increasing if the base is greater than 1.

When the base is between 0 and 1, the graph is always decreasing. The following figures summarize these characteristics of exponential functions.

$$f(x) = b^x, \quad b > 0, \quad b \neq 1$$

- one-to-one function
- y-intercept $(0, 1)$
- no x-intercept
- domain: $(-\infty, \infty)$
- range: $(0, \infty)$

EXAMPLE 3

Graph the exponential function $f(x) = 3^{x+2}$.

Solution

As before, we find and plot a few ordered pair solutions. Then we connect the points with a smooth curve.

$y = 3^{x+2}$

x	0	-1	-2	-3	-4
y	9	3	1	$\dfrac{1}{3}$	$\dfrac{1}{9}$

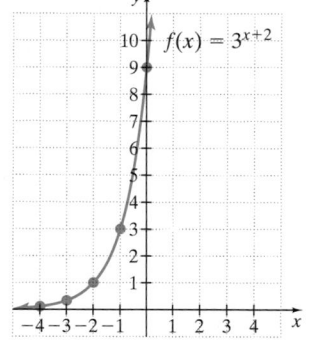

✔ **CONCEPT CHECK**

Which functions are exponential functions?

a. $f(x) = x^3$

b. $g(x) = \left(\dfrac{2}{3}\right)^x$

c. $h(x) = 5^{x-2}$

d. $w(x) = (2x)^2$

2   We have seen that an exponential function $y = b^x$ is a one-to-one function. Another way of stating this fact is a property that we can use to solve exponential equations.

Uniqueness of b^x

Let $b > 0$ and $b \neq 1$. Then $b^x = b^y$ is equivalent to $x = y$.

EXAMPLE 4

Solve each equation for x.

a. $2^x = 16$ **b.** $9^x = 27$ **c.** $4^{x+3} = 8^x$ **d.** $5^x = 10$

Solution   **a.** We write 16 as a power of 2 and then use the uniqueness of b^x to solve.

$$2^x = 16$$
$$2^x = 2^4$$

Since the bases are the same and are nonnegative, by the uniqueness of b^x, we then have that the exponents are equal. Thus,

$$x = 4$$

The solution is 4, or the solution set is $\{4\}$.

Algebraic Solution:

b. Notice that both 9 and 27 are powers of 3.

$$9^x = 27$$
$$(3^2)^x = 3^3 \qquad \text{Write 9 and 27 as powers of 3.}$$
$$3^{2x} = 3^3$$
$$2x = 3 \qquad \text{Apply the uniqueness of } b^x.$$
$$x = \frac{3}{2} \qquad \text{Divide by 2.}$$

Graphical Solution:

Graph $y_1 = 9^x$ and $y_2 = 27$.

The x-value of the point of intersection is 1.5.

To check, replace x with $\frac{3}{2}$ in the original expression, $9^x = 27$. The solution is $\frac{3}{2}$.

Algebraic Solution:

c. Write both 4 and 8 as powers of 2.

$$4^{x+3} = 8^x$$
$$(2^2)^{x+3} = (2^3)^x$$
$$2^{2x+6} = 2^{3x}$$
$$2x + 6 = 3x \qquad \text{Apply the uniqueness of } b^x.$$
$$6 = x \qquad \text{Subtract } 2x \text{ from both sides.}$$

The solution is 6.

Graphical Solution:

Graph $y_1 = 4^{x+3}$ and $y_2 = 8^x$.

The x-value of the point of intersection is 6.

d. For $5^x = 10$, notice that 5 and 10 cannot be easily written as powers of a common base. To solve this equation, we will approximate the solution graphically. Graph $y_1 = 5^x$ and $y_2 = 10$.

A calculator check for Example 4d.

The point of intersection is approximately $(1.431, 10)$, which indicates that the solution of the equation is $x \approx 1.431$. To check, replace x with 1.431 in the original equation.

$$5^x = 10$$

$$5^{1.431} = 10 \qquad \text{Let } x = 1.431.$$

$$10.00520695 \approx 10 \qquad \text{Approximately true}$$

The solution is approximately 1.431.

As we see in Example 4d, often the two sides of an equation cannot easily be written as powers of a common base. We explore how to find exact solutions to an equation such as $5^x = 10$ with the help of logarithms later.

3   The bar graph below shows the increase in the number of cellular phone users. Notice that the graph of the exponential function $y = 34(1.254)^x$ approximates the heights of the bars.

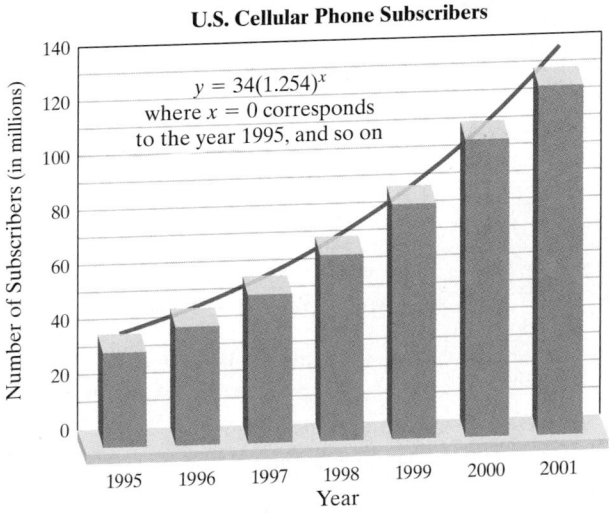

U.S. Cellular Phone Subscribers

$y = 34(1.254)^x$
where $x = 0$ corresponds
to the year 1995, and so on

Number of Subscribers (in millions)

Year

Source: The CTIA Semi-Annual Wireless Survey

The graph above shows just one example of how the world abounds with patterns that can be modeled by exponential functions. To make these applications realistic, we use numbers that warrant a calculator. Another application of an exponential function has to do with interest rates on loans.

The exponential function defined by $A = P\left(1 + \dfrac{r}{n}\right)^{nt}$ models the dollars A accrued (or owed) after P dollars are invested (or loaned) at an annual rate of interest r compounded n times each year for t years. This function is known as the compound interest formula.

EXAMPLE 5

USING THE COMPOUND INTEREST FORMULA

Find the amount owed at the end of 5 years if $1600 is loaned at a rate of 9% compounded monthly.

Solution   We use the formula $A = P\left(1 + \dfrac{r}{n}\right)^{nt}$, with the following values.

$$P = \$1600 \text{ (the amount of the loan)}$$

$$r = 9\% = 0.09 \text{ (the annual rate of interest)}$$

$$n = 12 \text{ (the number of times interest is compounded each year)}$$

$$t = 5 \text{ (the duration of the loan, in years)}$$

$$A = P\left(1 + \frac{r}{n}\right)^{nt} \qquad \text{Compound interest formula}$$

$$= 1600\left(1 + \frac{0.09}{12}\right)^{12(5)} \qquad \text{Substitute known values.}$$

$$= 1600(1.0075)^{60}$$

Use your calculator to approximate A. Two ways to do this are shown below.

Home Screen

Finance Menu (See Appendix D)

Thus, the amount A owed is approximately \$2505.09.

EXAMPLE 6

ESTIMATING PERCENT OF RADIOACTIVE MATERIAL

As a result of the Chernobyl nuclear accident, radioactive debris was carried through the atmosphere. One immediate concern was the impact that the debris had on the milk supply. The percent y of radioactive material in raw milk after t days is estimated by $y = 100(2.7)^{-0.1t}$. Estimate the expected percent of radioactive material in the milk after 30 days.

Solution Replace t with 30 in the given equation.

$$y = 100(2.7)^{-0.1t}$$

$$= 100(2.7)^{-0.1(30)} \qquad \text{Let } t = 30.$$

$$= 100(2.7)^{-3}$$

$$\approx 5.0805$$

Thus, approximately 5% of the radioactive material still remained in the milk supply after 30 days.

We can use a graphing calculator and its TRACE feature to solve Example 6 graphically.

To estimate the expected percent of radioactive material in the milk after 30 days, enter $Y_1 = 100(2.7)^{-0.1x}$. (The variable t in Example 6 is changed to x here to better accommodate our work on the graphing calculator.) The graph does not appear on a standard viewing window, so we need to determine an appropriate viewing window. Because it doesn't make sense to look at radioactivity *before* the Chernobyl nuclear accident, we use Xmin = 0. We are interested in finding the percent of radioactive material in the milk when $x = 30$, so we choose Xmax = 35 to leave enough space to see the graph at $x = 30$. Because the values of y are percents, it seems appropriate that $0 \leq y \leq 100$. (We also use Xscl = 5 and Yscl = 10.) Now we graph the function.

Notice from the graph that as the number of days, x, increases, the percent of radioactive material, y, decreases.

We can use the TRACE feature to obtain an approximation of the expected percent of radioactive material in the milk when $x = 30$. (A TABLE feature may also be used to approximate the percent.)

X	Y1
0	100
10	37.037
20	13.717
30	5.0805
40	1.8817
50	.69692
60	.25812

Y1 ■ 100(2.7)^(-....

The percent of radioactive material in the milk 30 days after the Chernobyl accident was 5.08%, accurate to two decimal places.

EXAMPLE 7

Use a graphing calculator to find each percent. Approximate your solutions so that they are accurate to two decimal places.

a. Estimate the expected percent of radioactive material in the milk 2 days after the Chernobyl nuclear accident.

b. Estimate the expected percent of radioactive material in the milk 10 days after the Chernobyl nuclear accident.

c. Estimate the expected percent of radioactive material in the milk 15 days after the Chernobyl nuclear accident.

d. Estimate the expected percent of radioactive material in the milk 25 days after the Chernobyl nuclear accident.

Solution Use a table similar to the one above with $y_1 = 100(2.7)^{-0.1x}$. In Ask mode, let $x = 2, 10, 15,$ and 25. The results are

a. 81.98%   **b.** 37.04%   **c.** 22.54%   **d.** 8.35%

1.–16. See graphing answer section.

EXERCISE SET 9.3

| STUDY GUIDE/SSM | CD/ VIDEO | PH MATH TUTOR CENTER | MathXL®Tutorials ON CD | MathXL® | MyMathLab® |

Graph each exponential function. See Examples 1 through 3.

1. $y = 4^x$

2. $y = 5^x$

3. $y = 2^x + 1$

4. $y = 3^x - 1$

5. $y = \left(\frac{1}{4}\right)^x$

6. $y = \left(\frac{1}{5}\right)^x$

7. $y = \left(\frac{1}{2}\right)^x - 2$

8. $y = \left(\frac{1}{3}\right)^x + 2$

9. $y = -2^x$

10. $y = -3^x$

11. $y = -\left(\frac{1}{4}\right)^x$

12. $y = -\left(\frac{1}{5}\right)^x$

13. $f(x) = 2^{x+1}$

14. $f(x) = 3^{x-1}$

15. $f(x) = 4^{x-2}$

16. $f(x) = 2^{x+3}$

Match each exponential equation with its graph below or in the next column.

17. $f(x) = \left(\frac{1}{2}\right)^x$ C

18. $f(x) = 2^x$ B

19. $f(x) = \left(\frac{1}{4}\right)^x$ D

20. $f(x) = 3^x$ A

A

B

C

D

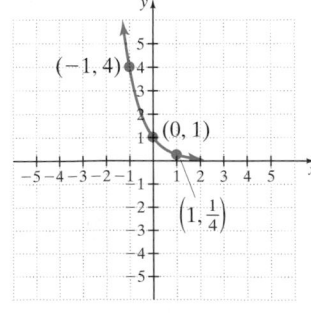

Solve each equation for x. See Example 4.

21. $3^x = 27$ 3

22. $6^x = 36$ 2

23. $16^x = 8$ $\frac{3}{4}$

24. $64^x = 16$ $\frac{2}{3}$

25. $32^{2x-3} = 2$ $\frac{8}{5}$

26. $9^{2x+1} = 81$ $\frac{1}{2}$

27. $\frac{1}{4} = 2^{3x}$ $-\frac{2}{3}$

28. $\frac{1}{27} = 3^{2x}$ $-\frac{3}{2}$

29. $5^x = 625$ 4

30. $2^x = 64$ 6

31. $4^x = 8$ $\frac{3}{2}$

32. $32^x = 4$ $\frac{2}{5}$

33. $27^{x+1} = 9$ $-\frac{1}{3}$

34. $125^{x-2} = 25$ $\frac{8}{3}$

35. $81^{x-1} = 27^{2x}$ -2

36. $4^{3x-7} = 32^{2x}$ $-\frac{7}{2}$

Solve. Unless otherwise indicated, round results to one decimal place. See Examples 6 and 7.

37. One type of uranium has a daily radioactive decay rate of 0.4%. If 30 pounds of this uranium is available today, find how much will still remain after 50 days. Use $y - 30(2\,7)^{-0.004t}$, and let t be 50.   24.6 lb

38. The nuclear waste from an atomic energy plant decays at a rate of 3% each century. If 150 pounds of nuclear waste is disposed of, find how much of it will still remain after 10 centuries. Use $y = 150(2.7)^{-0.03t}$, and let t be 10.   111.3 lb

39. National Park Service personnel are trying to increase the size of the bison population of Theodore Roosevelt National Park. If 260 bison currently live in the park, and if the population's rate of growth is 2.5% annually, find how many bison (rounded to the nearest whole) there should be in 10 years. Use $y = 260(2.7)^{0.025t}$.   333 bison

40. The size of the rat population of a wharf area grows at a rate of 8% monthly. If there are 200 rats in January, find how many rats (rounded to the nearest whole) should be expected by next January. Use $y = 200(2.7)^{0.08t}$.   519 rats

41. A rare isotope of a nuclear material is very unstable, decaying at a rate of 15% each second. Find how much isotope remains 10 seconds after 5 grams of the isotope is created. Use $y = 5(2.7)^{-0.15t}$.   1.1 g

42. An accidental spill of 75 grams of radioactive material in a local stream has led to the presence of radioactive debris decaying at a rate of 4% each day. Find how much debris still remains after 14 days. Use $y = 75(2.7)^{-0.04t}$.   43.0 g

43. Retail revenue from shopping on the Internet is currently growing at a rate of 56% per year. In 2000, a total of $42.1 billion in revenue was collected through Internet retail sales. Answer the following questions using $y = 42.1(1.56)^t$ where y is Internet revenue in billions of dollars and t is the number of years after 2000. Round answers to the nearest tenth of a billion dollars. (*Source:* Based on data from eMarketer)

  a. According to the model, what level of retail revenues from Internet shopping was expected in 2001? $65.7 billion

  b. Predict the level of Internet shopping revenues in 2009. $2303.6 billion

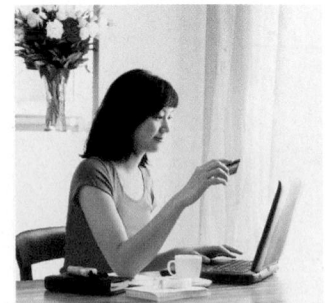

44. An unusually wet spring has caused the size of the Cape Cod mosquito population to increase by 8% each day. If an estimated 200,000 mosquitoes are on Cape Cod on May 12, find how many mosquitoes will inhabit the Cape on May 25. Use $y = 200,000(2.7)^{0.08t}$. Round to the nearest thousand. 562,000 mosquitoes

45. The equation $y = 120.882(1.012)^x$ models the population of the United States from 1930 through 2000. In the equation, y is the population in millions and x represents the number of years after 1930. Round answers to the nearest tenth of a million.(*Source:* Based on data from the U.S. Bureau of the Census)

   a. Estimate the population of the United States in 1970. 194.8 million people
   b. Assuming this equation continues to be valid in the future, use the equation to predict the population of the United States in 2020. 353.7 million people

46. Carbon dioxide (CO_2) is a greenhouse gas that contributes to global warming. Partially due to the combustion of fossil fuels, the amount of CO_2 in Earth's atmosphere has been increasing by 0.4% annually over the past century. In 2000, the concentration of CO_2 in the atmosphere was 369.4 parts per million by volume. To make the following predictions, use $y = 369.4(1.004)^t$ where y is the concentration of CO_2 in parts per million and t is the number of years after 2000. Round answers to the nearest tenth. (*Sources:* Based on data from the United Nations Environment Programme and the Carbon Dioxide Information Analysis Center)

   a. Predict the concentration of CO_2 in the atmosphere in the year 2006. 378.4 parts per million
   b. Predict the concentration of CO_2 in the atmosphere in the year 2030. 416.4 parts per million

Solve. Use $A = P\left(1 + \dfrac{r}{n}\right)^{nt}$. *Round answers to two decimal places. See Example 5.*

47. Find the amount Erica owes at the end of 3 years if $6000 is loaned to her at a rate of 8% compounded monthly. $7621.42

48. Find the amount owed at the end of 5 years if $3000 is loaned at a rate of 10% compounded quarterly. $4915.85

49. Find the total amount Janina has in a college savings account if $2000 was invested and earned 6% compounded semiannually for 12 years. $4065.59

50. Find the amount accrued if $500 is invested and earns 7% compounded monthly for 4 years. $661.03

The formula $y = 34(1.254)^x$ gives the number of cellular phone users y (in millions) in the United States for the years 1995 through 2001. In this formula, $x = 0$ corresponds to 1995, $x = 1$ corresponds to 1996, and so on. Use this formula to solve Exercises 51 and 52. Round results to the nearest whole million.

51. Use this model to predict the number of cellular phone users in the year 2008. 645 million cellular phone users

52. Use this model to predict the number of cellular phone users in the year 2010. 1014 million cellular phone users

REVIEW AND PREVIEW
Solve each equation. See Sections 1.5 and 5.8.

53. $5x - 2 = 18$ 4   **54.** $3x - 7 = 11$ 6

55. $3x - 4 = 3(x + 1)$ $\varnothing$ **56.** $2 - 6x = 6(1 - x)$ $\varnothing$

57. $x^2 + 6 = 5x$ 2, 3   **58.** $18 = 11x - x^2$ 2, 9

By inspection, find the value for x that makes each statement true.

59. $2^x = 8$ 3   **60.** $3^x = 9$ 2

61. $5^x = \dfrac{1}{5}$ -1 **62.** $4^x = 1$ 0

Concept Extensions

63. Explain why the graph of an exponential function $y = b^x$ contains the point $(1, b)$.   answers may vary

64. Explain why an exponential function $y = b^x$ has a y-intercept of $(0, 1)$.   answers may vary

Graph.   **65.–68.** See graphing answer section.

65. $y = |3^x|$ **66.** $y = \left|\left(\dfrac{1}{3}\right)^x\right|$

67. $y = 3^{|x|}$ **68.** $y = \left(\dfrac{1}{3}\right)^{|x|}$

69. Graph $y = 2^x$ and $y = \left(\dfrac{1}{2}\right)^{-x}$ on the same set of axes. Describe what you see and why.

70. Graph $y = 2^x$ and $x = 2^y$ on the same set of axes. Describe what you see.   answers may vary

69. The graphs are the same since $\left(\dfrac{1}{2}\right)^{-x} = 2^x$.

9.4   LOGARITHMIC FUNCTIONS

Objectives

① Write exponential equations with logarithmic notation and write logarithmic equations with exponential notation.

② Solve logarithmic equations by using exponential notation.

③ Identify and graph logarithmic functions.

① Since the exponential function $f(x) = 2^x$ is a one-to-one function, it has an inverse.

We can create a table of values for f^{-1} by switching the coordinates in the accompanying table of values for $f(x) = 2^x$.

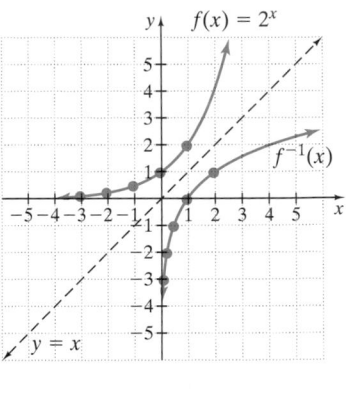

x	$y = f(x)$	x	$y = f^{-1}(x)$
-3	$\dfrac{1}{8}$	$\dfrac{1}{8}$	-3
-2	$\dfrac{1}{4}$	$\dfrac{1}{4}$	-2
-1	$\dfrac{1}{2}$	$\dfrac{1}{2}$	-1
0	1	1	0
1	2	2	1
2	4	4	2
3	8	8	3

The graphs of $f(x)$ and its inverse are shown above. Notice that the graphs of f and f^{-1} are symmetric about the line $y = x$, as expected.

Now we would like to be able to write an equation for f^{-1}. To do so, we follow the steps for finding an inverse.

$$f(x) = 2^x$$

Step 1: Replace $f(x)$ by y. $y = 2^x$

Step 2: Interchange x and y. $x = 2^y$

Step 3: Solve for y.

At this point, we are stuck. To solve this equation for y, a new notation, the **logarithmic notation,** is needed. The symbol $\log_b x$ means "the power to which b is raised in order to produce a result of x."

$$\log_b x = y \quad \text{means} \quad b^y = x$$

We say that $\log_b x$ is "the logarithm of x to the base b" or "the log of x to the base b."

TEACHING TIP

Make sure that students are comfortable with writing logarithm equations as exponential equations and vice versa before continuing.

Logarithmic Definition

If $b > 0$ and $b \neq 1$, then

$$y = \log_b x \text{ means } x = b^y$$

for every $x > 0$ and every real number y.

Before returning to the function $x = 2^y$ and solving it for y in terms of x, let's practice using the new notation $\log_b x$.

Helpful Hint

Notice that a *logarithm* is an *exponent*. In other words, $\log_3 9$ is the *power* that we raise 3 to in order to get 9.

It is important to be able to write exponential equations from logarithmic notation, and vice versa. The following table shows examples of both forms.

Logarithmic Equation	*Corresponding Exponential Equation*
$\log_3 9 = 2$	$3^2 = 9$
$\log_6 1 = 0$	$6^0 = 1$
$\log_2 8 = 3$	$2^3 = 8$
$\log_4 \dfrac{1}{16} = -2$	$4^{-2} = \dfrac{1}{16}$
$\log_8 2 = \dfrac{1}{3}$	$8^{1/3} = 2$

EXAMPLE 1

CLASSROOM EXAMPLE

Write as an exponential equation.

a. $\log_7 49 = 2$

b. $\log_8 \dfrac{1}{8} = -1$

c. $\log_3 \sqrt{3} = \dfrac{1}{2}$

answer: a. $7^2 = 49$ b. $8^{-1} = \dfrac{1}{8}$
c. $3^{1/2} = \sqrt{3}$

Write as an exponential equation.

a. $\log_5 25 = 2$ **b.** $\log_6 \dfrac{1}{6} = -1$ **c.** $\log_2 \sqrt{2} = \dfrac{1}{2}$

Solution **a.** $\log_5 25 = 2$ means $5^2 = 25$

b. $\log_6 \dfrac{1}{6} = -1$ means $6^{-1} = \dfrac{1}{6}$

c. $\log_2 \sqrt{2} = \dfrac{1}{2}$ means $2^{1/2} = \sqrt{2}$

EXAMPLE 2

CLASSROOM EXAMPLE

Write as a logarithmic equation.

a. $3^4 = 81$

b. $2^{-3} = \dfrac{1}{8}$

c. $7^{1/3} = \sqrt[3]{7}$

answer: a. $\log_3 81 = 4$ b. $\log_2 \dfrac{1}{8} = -3$
c. $\log_7 \sqrt[3]{7} = \dfrac{1}{3}$

Write as a logarithmic equation.

a. $9^3 = 729$ **b.** $6^{-2} = \dfrac{1}{36}$ **c.** $5^{1/3} = \sqrt[3]{5}$

Solution **a.** $9^3 = 729$ means $\log_9 729 = 3$

b. $6^{-2} = \dfrac{1}{36}$ means $\log_6 \dfrac{1}{36} = -2$

c. $5^{1/3} = \sqrt[3]{5}$ means $\log_5 \sqrt[3]{5} = \dfrac{1}{3}$

EXAMPLE 3

Find the value of each logarithmic expression.

a. $\log_4 16$ **b.** $\log_{10} \dfrac{1}{10}$ **c.** $\log_9 3$

Solution **a.** $\log_4 16 = 2$ because $4^2 = 16$

b. $\log_{10} \dfrac{1}{10} = -1$ because $10^{-1} = \dfrac{1}{10}$

c. $\log_9 3 = \dfrac{1}{2}$ because $9^{1/2} = \sqrt{9} = 3$

CLASSROOM EXAMPLE

Find each value.

a. $\log_2 8$ **b.** $\log_3 \dfrac{1}{9}$ **c.** $\log_{25} 5$

answer: **a.** 3 **b.** -2 **c.** $\dfrac{1}{2}$

TEACHING TIP

For Example 3, suggest that students ask themselves, "What power of b is x? For instance, 4 raised to what power is 16? 4 raised to the *second* power is 16."

TECHNOLOGY NOTE

A change of base formula, such as

$$\log_a b = \frac{\log b}{\log a}$$

can be used to evaluate logarithms on a calculator. (See Section 9.6.)

> **Helpful Hint**
>
> Another method for evaluating logarithms such as those in Example 3 is to set the expression equal to x and then write them in exponential form to find x. For example:
>
> **a.** $\log_4 16 = x$ means $4^x = 16$. Since $4^2 = 16$, $x = 2$ or $\log_4 16 = 2$.
>
> **b.** $\log_{10} \dfrac{1}{10} = x$ means $10^x = \dfrac{1}{10}$. Since $10^{-1} = \dfrac{1}{10}$, $x = -1$ or $\log_{10} \dfrac{1}{10} = -1$.
>
> **c.** $\log_9 3 = x$ means $9^x = 3$. Since $9^{1/2} = 3$, $x = \dfrac{1}{2}$ or $\log_9 3 = \dfrac{1}{2}$.

2   The ability to interchange the logarithmic and exponential forms of a statement is often the key to solving logarithmic equations.

EXAMPLE 4

Solve each equation for x.

a. $\log_4 \dfrac{1}{4} = x$ **b.** $\log_5 x = 3$ **c.** $\log_x 25 = 2$

d. $\log_3 1 = x$ **e.** $\log_b 1 = x$

Solution **a.** $\log_4 \dfrac{1}{4} = x$ means $4^x = \dfrac{1}{4}$. Solve $4^x = \dfrac{1}{4}$ for x.

$$4^x = \frac{1}{4}$$
$$4^x = 4^{-1}$$

Since the bases are the same, by the uniqueness of b^x, we have that

$$x = -1$$

The solution is -1 or the solution set is $\{-1\}$. To check, see that $\log_4 \dfrac{1}{4} = -1$, since $4^{-1} = \dfrac{1}{4}$.

b. $\log_5 x = 3$ means $5^3 = x$ or

$$x = 125$$

The solution is 125.

CLASSROOM EXAMPLE

Solve.

a. $\log_2 x = 4$ **b.** $\log_x 9 = 2$

c. $\log_7 \dfrac{1}{7} = x$ **d.** $\log_2 1 = x$

e. $\log_a 1 = x$

answer:
a. 16 **b.** 3 **c.** -1 **d.** 0 **e.** 0

TEACHING TIP

For Property 1, ask, "*b* raised to what power is 1?"

c. $\log_x 25 = 2$ means $x^2 = 25$ and $x > 0$ and $x \neq 1$. Thus,

$$x = 5$$

Even though $(-5)^2 = 25$, the base b of a logarithm must be positive. The solution is 5.

TEACHING TIP

For Property 2 ask, "*b* raised to what power is *b*?" Then give the example, "What power of 2 is 2^5? In other words, what power of 2 is 32?"

d. $\log_3 1 = x$ means $3^x = 1$. Either solve this equation by inspection or solve by writing 1 as 3^0 as shown.

$$3^x = 3^0 \quad \text{Write 1 as } 3^0.$$

$$x = 0 \quad \text{Apply the uniqueness of } b^x.$$

The solution is 0.

e. $\log_b 1 = x$ means $b^x = 1$ and $b > 0$ and $b \neq 1$.

$$b^x = b^0 \quad \text{Write 1 as } b^0.$$

$$x = 0 \quad \text{Apply the uniqueness of } b^x.$$

TEACHING TIP

For Property 3, have students convert the equation to logarithmic form.

The solution is 0.

In Example 4e we proved an important property of logarithms. That is, $\log_b 1$ is always 0. This property as well as two important others are given next.

> ## Properties of Logarithms
>
> If b is a real number, $b > 0$, and $b \neq 1$, then
>
> **1.** $\log_b 1 = 0$
> **2.** $\log_b b^x = x$
> **3.** $b^{\log_b x} = x$

To see that **2.** $\log_b b^x = x$, change the logarithmic form to exponential form. Then, $\log_b b^x = x$ means $b^x = b^x$. In exponential form, the statement is true, so in logarithmic form, the statement is also true. To understand **3.** $b^{\log_b x} = x$, write this exponential equation as an equivalent logarithm.

EXAMPLE 5

Simplify.

a. $\log_3 3^2$ **b.** $\log_7 7^{-1}$ **c.** $5^{\log_5 3}$ **d.** $2^{\log_2 6}$

Solution **a.** From Property 2, $\log_3 3^2 = 2$.

b. From Property 2, $\log_7 7^{-1} = -1$.

c. From Property 3, $5^{\log_5 3} = 3$.

d. From Property 3, $2^{\log_2 6} = 6$.

CLASSROOM EXAMPLE

Simplify.

a. $\log_6 6^3$ **b.** $\log_{11} 11^{-4}$

c. $7^{\log_7 5}$ **d.** $3^{\log_3 10}$

answer: a. 3 **b.** -4 **c.** 5 **d.** 10

3 Let us now return to the function $f(x) = 2^x$ and write an equation for its inverse, $f^{-1}(x)$. Recall our earlier work.

$$f(x) = 2^x$$

Step 1: Replace $f(x)$ by y. $y = 2^x$

Step 2: Interchange x and y. $x = 2^y$

Having gained proficiency with the notation $\log_b x$, we can now complete the steps for writing the inverse equation by writing $x = 2^y$ as an equivalent logarithm.

Step 1: Solve for y. $y = \log_2 x$

Step 2: Replace y with $f^{-1}(x)$. $f^{-1}(x) = \log_2 x$

Thus, $f^{-1}(x) = \log_2 x$ defines a function that is the inverse function of the function $f(x) = 2^x$. The function $f^{-1}(x)$ or $y = \log_2 x$ is called a **logarithmic function**.

Logarithmic Function

If x is a positive real number, b is a constant positive real number, and b is not 1, then a **logarithmic function** is a function that can be defined by

$$f(x) = \log_b x$$

The domain of f is the set of positive real numbers, and the range of f is the set of real numbers.

✔ **CONCEPT CHECK**

Let $f(x) = \log_3 x$ and $g(x) = 3^x$. These two functions are inverses of each other. Since $(2, 9)$ is an ordered pair solution of $g(x)$ or $g(2) = 9$, what ordered pair do we know to be a solution of $f(x)$? Also, find $f(9)$. Explain why.

We can explore logarithmic functions by graphing them.

EXAMPLE 6

Graph the logarithmic function $y = \log_2 x$.

Solution First we write the equation with exponential notation as $2^y = x$. Then we find some ordered pair solutions that satisfy this equation. Finally, we plot the points and connect them with a smooth curve. The domain of this function is $(0, \infty)$, and the range is all real numbers.

Since $x = 2^y$ is solved for x, we choose y-values and compute corresponding x-values.

If $y = 0, x = 2^0 = 1$

If $y = 1, x = 2^1 = 2$

If $y = 2, x = 2^2 = 4$

If $y = -1, x = 2^{-1} = \dfrac{1}{2}$

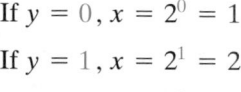

$x = 2^y$	y
1	0
2	1
4	2
$\dfrac{1}{2}$	-1

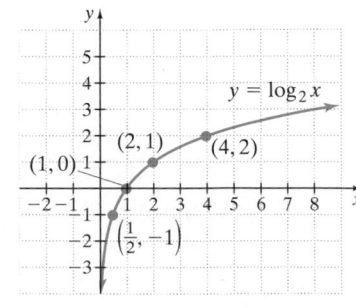

CLASSROOM EXAMPLE

Graph: $y = \log_4 x$

answer:

$y = \log_4 x$

Concept Check Answer:

$(9, 2)$; $f(9) = 2$; answers may vary

EXAMPLE 7

Graph the logarithmic function $f(x) = \log_{1/3} x$.

Solution

Replace $f(x)$ with y, and write the result with exponential notation.

$$f(x) = \log_{1/3} x$$

$$y = \log_{1/3} x \qquad \text{Replace } f(x) \text{ with } y.$$

$$\left(\frac{1}{3}\right)^y = x \qquad \text{Write in exponential form.}$$

Now we can find ordered pair solutions that satisfy $\left(\frac{1}{3}\right)^y = x$, plot these points, and connect them with a smooth curve.

If $y = 0$, $x = \left(\frac{1}{3}\right)^0 = 1$

If $y = 1$, $x = \left(\frac{1}{3}\right)^1 = \frac{1}{3}$

If $y = -1$, $x = \left(\frac{1}{3}\right)^{-1} = 3$

If $y = -2$, $x = \left(\frac{1}{3}\right)^{-2} = 9$

$x = \left(\frac{1}{3}\right)^y$	y
1	0
$\frac{1}{3}$	1
3	-1
9	-2

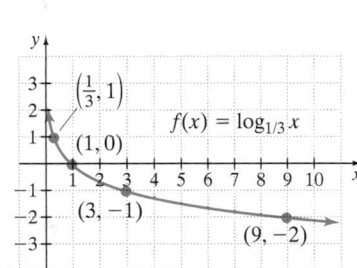

The domain of this function is $(0, \infty)$, and the range is the set of all real numbers.

The following figures summarize characteristics of logarithmic functions.

$$f(x) = \log_b x, b > 0, b \neq 1$$

- one-to-one function
- x-intercept $(1, 0)$
- no y-intercept

- domain: $(0, \infty)$
- range: $(-\infty, \infty)$

$f(x) = \log_b x,$
if $b > 1$

$f(x) = \log_b x,$
if $0 < b < 1$

Spotlight on
DECISION
∽ MAKING

Suppose you are the Webmaster for a small but growing company. One of your duties is to ensure that your company's newly established Web site can adequately handle the number of visitors to it. You decide to find a mathematical model for recent Web site usage statistics to help predict future numbers of visitors. Ultimately, you would like to use this model to predict when your Web site's server capacity must be expanded.

The first step in finding a model for the usage statistics is to decide what type of mathematical model to use: linear, quadratic, exponential, or logarithmic. The graph shows the number of visitors to your company's Web site in each of the first five months since it was established. Use the graph to decide which type of mathematical model to use. Explain your reasoning.

Month	1	2	3	4	5
Visitors	166,511	1,320,978	1,996,298	2,475,445	2,847,100

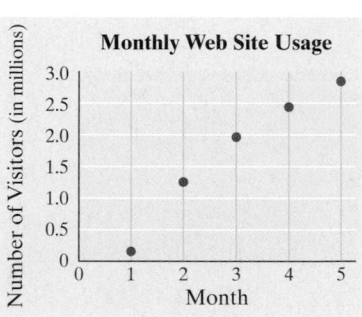

Monthly Web Site Usage

EXERCISE SET 9.4

5. $10^3 = 1000$ **12.** $11^{1/4} = \sqrt[4]{11}$
16. $\log_{10} 10,000 = 4$

STUDY GUIDE/SSM   CD/VIDEO   PH MATH TUTOR CENTER   MathXL®Tutorials ON CD   MathXL®   MyMathLab®

19. $\log_{10} \dfrac{1}{10} = -1$

Write each as an exponential equation. See Example 1.

1. $\log_6 36 = 2$ $6^2 = 36$ **2.** $\log_2 32 = 5$ $2^5 = 32$

3. $\log_3 \dfrac{1}{27} = -3$ $3^{-3} = \dfrac{1}{27}$ **4.** $\log_5 \dfrac{1}{25} = -2$ $5^{-2} = \dfrac{1}{25}$

5. $\log_{10} 1000 = 3$ **6.** $\log_{10} 10 = 1$ $10^1 = 10$

7. $\log_e x = 4$ $e^4 = x$ **8.** $\log_e \dfrac{1}{e} = -1$ $e^{-1} = \dfrac{1}{e}$

9. $\log_e \dfrac{1}{e^2} = -2$ $e^{-2} = \dfrac{1}{e^2}$ **10.** $\log_e y = 7$ $e^7 = y$

11. $\log_7 \sqrt{7} = \dfrac{1}{2}$ $7^{1/2} = \sqrt{7}$ **12.** $\log_{11} \sqrt[4]{11} = \dfrac{1}{4}$

Write each as a logarithmic equation. See Example 2.

13. $2^4 = 16$ $\log_2 16 = 4$ **14.** $5^3 = 125$ $\log_5 125 = 3$

15. $10^2 = 100$ $\log_{10} 100 = 2$ **16.** $10^4 = 10,000$

17. $e^3 = x$ $\log_e x = 3$ **18.** $e^5 = y$ $\log_e y = 5$

19. $10^{-1} = \dfrac{1}{10}$ **20.** $10^{-2} = \dfrac{1}{100}$ $\log_{10} \dfrac{1}{100} = -2$

21. $4^{-2} = \dfrac{1}{16}$ $\log_4 \dfrac{1}{16} = -2$ **22.** $3^{-4} = \dfrac{1}{81}$ $\log_3 \dfrac{1}{81} = -4$

23. $5^{1/2} = \sqrt{5}$ $\log_5 \sqrt{5} = \dfrac{1}{2}$ **24.** $4^{1/3} = \sqrt[3]{4}$ $\log_4 \sqrt[3]{4} = \dfrac{1}{3}$

Find the value of each logarithmic expression. See Examples 3 and 5.

 25. $\log_2 8$ 3   **26.** $\log_3 9$ 2

27. $\log_3 \dfrac{1}{9}$ -2 **28.** $\log_2 \dfrac{1}{32}$ -5

29. $\log_{25} 5$ $\dfrac{1}{2}$ **30.** $\log_8 \dfrac{1}{2}$ $-\dfrac{1}{3}$

 31. $\log_{1/2} 2$ -1 **32.** $\log_{2/3} \dfrac{4}{9}$ 2

33. $\log_7 1$ 0   **34.** $\log_9 9$ 1

35. $\log_2 2^4$ 4   **36.** $\log_6 6^{-2}$ -2

37. $\log_{10} 100$ 2   **38.** $\log_{10} \dfrac{1}{10}$ -1

39. $3^{\log_3 5}$ 5   **40.** $5^{\log_5 7}$ 7

41. $\log_3 81$ 4   **42.** $\log_2 16$ 4

43. $\log_4 \dfrac{1}{64}$ -3 **44.** $\log_3 \dfrac{1}{9}$ -2

45. Explain why negative numbers are not included as logarithmic bases.   answers may vary

46. Explain why 1 is not included as a logarithmic base.   answers may vary

Solve each equation for x. See Example 4.

47. $\log_3 9 = x$ 2   **48.** $\log_2 8 = x$ 3

49. $\log_3 x = 4$ 81   **50.** $\log_2 x = 3$ 8

51. $\log_x 49 = 2$ 7   **52.** $\log_x 8 = 3$ 2

53. $\log_2 \dfrac{1}{8} = x$ -3 **54.** $\log_3 \dfrac{1}{81} = x$ -4

55. $\log_3 \dfrac{1}{27} = x$ -3

56. $\log_5 \dfrac{1}{125} = x$ -3

🔒 **57.** $\log_8 x = \dfrac{1}{3}$ 2

58. $\log_9 x = \dfrac{1}{2}$ 3

59. $\log_4 16 = x$ 2

60. $\log_2 16 = x$ 4

61. $\log_{3/4} x = 3$ $\dfrac{27}{64}$

62. $\log_{2/3} x = 2$ $\dfrac{4}{9}$

🔒 **63.** $\log_x 100 = 2$ 10

64. $\log_x 27 = 3$ 3

Simplify. See Example 5.

65. $\log_5 5^3$ 3

66. $\log_6 6^2$ 2

🔒 **67.** $2^{\log_2 3}$ 3

68. $7^{\log_7 4}$ 4

🔒 **69.** $\log_9 9$ 1

70. $\log_8 (8)^{-1}$ -1

Graph each logarithmic function. Label any intercepts. See Examples 6 and 7.   **71.–78.** See graphing answer section.

🔒 **71.** $y = \log_3 x$

72. $y = \log_2 x$

73. $f(x) = \log_{1/4} x$

74. $f(x) = \log_{1/2} x$

75. $f(x) = \log_5 x$

76. $f(x) = \log_6 x$

77. $f(x) = \log_{1/6} x$

78. $f(x) = \log_{1/5} x$

REVIEW AND PREVIEW

Simplify each rational expression. See Section 6.1.

79. $\dfrac{x+3}{3+x}$ 1

80. $\dfrac{x-5}{5-x}$ -1

81. $\dfrac{x^2 - 8x + 16}{2x - 8}$ $\dfrac{x-4}{2}$

82. $\dfrac{x^2 - 3x - 10}{2 + x}$ $x - 5$

Add or subtract as indicated. See Section 6.2.

83. $\dfrac{2}{x} + \dfrac{3}{x^2}$ $\dfrac{2x+3}{x^2}$

84. $\dfrac{3x}{x+3} + \dfrac{9}{x+3}$ 3

85. $\dfrac{m^2}{m+1} - \dfrac{1}{m+1}$ $m-1$

86. $\dfrac{5}{y+1} - \dfrac{4}{y-1}$ $\dfrac{y-9}{y^2-1}$

Concept Extensions

Solve by first writing as an exponential.

87. $\log_7(5x - 2) = 1$ $\dfrac{9}{5}$

88. $\log_3(2x + 4) = 2$ $\dfrac{5}{2}$

89. Simplify: $\log_3(\log_5 125)$ 1 **90.** Simplify: $\log(\log_4(\log_2 16)$ 0

Graph each function and its inverse function on the same set of axes. Label any intercepts.   **91.–94.** See graphing answer section.

91. $y = 4^x; y = \log_4 x$

92. $y = 3^x; y = \log_3 x$

93. $y = \left(\dfrac{1}{3}\right)^x; y = \log_{1/3} x$

94. $y = \left(\dfrac{1}{2}\right)^x; y = \log_{1/2} x$

95. The formula $\log_{10}(1 - k) = \dfrac{-0.3}{H}$ models the relationship between the half-life H of a radioactive material and its rate of decay k. Find the rate of decay of the iodine isotope I-131 if its half-life is 8 days. Round to 4 decimal places. 0.0827

96. Explain why the graph of the function $y = \log_b x$ contains the point $(1, 0)$ no matter what b is.   answers may vary

97. Log$_3$ 10 is between which two integers? Explain your answer.   2 and 3; answers may vary

---

9.5   PROPERTIES OF LOGARITHMS

Objectives

① Use the product property of logarithms.

② Use the quotient property of logarithms.

③ Use the power property of logarithms.

④ Use the properties of logarithms together.

In the previous section we explored some basic properties of logarithms. We now introduce and explore additional properties. Because a logarithm is an exponent, logarithmic properties are just restatements of exponential properties.

① The first of these properties is called the **product property of logarithms,** because it deals with the logarithm of a product.

Product Property of Logarithms

If x, y, and b are positive real numbers and $b \neq 1$, then

$$\log_b xy = \log_b x + \log_b y$$

To prove this, let $\log_b x = M$ and $\log_b y = N$. Now write each logarithm with exponential notation.

$$\log_b x = M \qquad \text{is equivalent to} \qquad b^M = x$$

$$\log_b y = N \qquad \text{is equivalent to} \qquad b^N = y$$

Multiply the left sides and the right sides of the exponential equations, and we have that

$$xy = (b^M)(b^N) = b^{M+N}$$

If we write the equation $xy = b^{M+N}$ in equivalent logarithmic form, we have

$$\log_b xy = M + N$$

But since $M = \log_b x$ and $N = \log_b y$, we can write

$$\log_b xy = \log_b x + \log_b y \qquad \text{Let } M = \log_b x \text{ and } N = \log_b y.$$

In other words, the logarithm of a product is the sum of the logarithms of the factors. This property is sometimes used to simplify logarithmic expressions.

In the examples that follow, assume that variables represent positive numbers.

EXAMPLE 1

Write each sum as a single logarithm.

a. $\log_{11} 10 + \log_{11} 3$ **b.** $\log_3 \dfrac{1}{2} + \log_3 12$ **c.** $\log_2(x + 2) + \log_2 x$

Solution   In each case, both terms have a common logarithmic base.

a. $\log_{11}10 + \log_{11}3 = \log_{11}(10 \cdot 3)$ *Apply the product property.*

$$= \log_{11} 30$$

> **Helpful Hint**
>
> Check your logarithm properties. Make sure you understand that $\log_2(x + 2)$ *is not* $\log_2 x + \log_2 2$.

b. $\log_3 \dfrac{1}{2} + \log_3 12 = \log_3\left(\dfrac{1}{2} \cdot 12\right) = \log_3 6$

c. $\log_2(x + 2) + \log_2 x = \log_2[(x + 2) \cdot x] = \log_2(x^2 + 2x)$

2   The second property is the **quotient property of logarithms.**

CLASSROOM EXAMPLE

Write as a single logarithm.

a. $\log_2 7 + \log_2 5$

b. $\log_6 \dfrac{1}{3} + \log_6 15$

c. $\log_3 x + \log_3(x - 9)$

answer: **a.** $\log_2 35$ **b.** $\log_6 5$

c. $\log_3 x^2 - 9x$

Quotient Property of Logarithms

If x, y, and b are positive real numbers and $b \neq 1$, then

$$\log_b \frac{x}{y} = \log_b x - \log_b y$$

The proof of the quotient property of logarithms is similar to the proof of the product property. Notice that the quotient property says that the logarithm of a quotient is the difference of the logarithms of the dividend and divisor.

✔ **CONCEPT CHECK**

Which of the following is the correct way to rewrite $\log_5 \frac{7}{2}$?

a. $\log_5 7 - \log_5 2$ **b.** $\log_5(7 - 2)$ **c.** $\dfrac{\log_5 7}{\log_5 2}$ **d.** $\log_5 14$

EXAMPLE 2

Write each difference as a single logarithm.

a. $\log_{10} 27 - \log_{10} 3$ **b.** $\log_5 8 - \log_5 x$ **c.** $\log_3(x^2 + 5) - \log_3(x^2 + 1)$

Solution All terms have a common logarithmic base.

a. $\log_{10} 27 - \log_{10} 3 = \log_{10} \dfrac{27}{3} = \log_{10} 9$

b. $\log_5 8 - \log_5 x = \log_5 \dfrac{8}{x}$

c. $\log_3(x^2 + 5) - \log_3(x^2 + 1) = \log_3 \dfrac{x^2 + 5}{x^2 + 1}$ *Apply the quotient property.*

CLASSROOM EXAMPLE

Write as a single logarithm.

a. $\log_7 40 - \log_7 8$

b. $\log_9 x - \log_9 5$

c. $\log_3(x^3 + 4) - \log_3(x^2 + 2)$

answer: **a.** $\log_7 5$ **b.** $\log_9 \dfrac{x}{5}$

c. $\log_3 \dfrac{x^3 + 4}{x^2 + 2}$

---

3   The third and final property we introduce is the **power property of logarithms.**

TEACHING TIP

Show how the Power Property of Logarithms is related to the product property of logarithms. For instance, $\log_2 5^3 = \log_2(5 \cdot 5 \cdot 5) = \log_2 5 + \log_2 5 + \log_2 5 = 3 \log_2 5.$

Power Property of Logarithms

If x and b are positive real numbers, $b \neq 1$, and r is a real number, then

$$\log_b x^r = r \log_b x$$

EXAMPLE 3

CLASSROOM EXAMPLE

Use the power property to rewrite.

a. $\log_3 x^5$

b. $\log_7 \sqrt[3]{4}$

answer: **a.** $5 \log_3 x$ **b.** $\frac{1}{3}\log_7 4$

Use the power property to rewrite each expression.

a. $\log_5 x^3$ **b.** $\log_4 \sqrt{2}$

Solution   **a.** $\log_5 x^3 = 3 \log_5 x$

b. $\log_4 \sqrt{2} = \log_4 2^{1/2} = \dfrac{1}{2}\log_4 2$

Concept Check Answer:
a

4   Many times we must use more than one property of logarithms to simplify a logarithmic expression.

EXAMPLE 4

Write as a single logarithm.

a. $2 \log_5 3 + 3 \log_5 2$ **b.** $3 \log_9 x - \log_9(x + 1)$ **c.** $\log_4 25 + \log_4 3 - \log_4 5$

Solution   In each case, all terms have a common logarithmic base.

a. $2 \log_5 3 + 3 \log_5 2 = \log_5 3^2 + \log_5 2^3$ Apply the power property.

$= \log_5 9 + \log_5 8$

$= \log_5(9 \cdot 8)$ Apply the product property.

$= \log_5 72$

b. $3 \log_9 x - \log_9(x + 1) = \log_9 x^3 - \log_9(x + 1)$ Apply the power property.

$= \log_9 \dfrac{x^3}{x + 1}$ Apply the quotient property.

c. Use both the product and quotient properties.

$\log_4 25 + \log_4 3 - \log_4 5 = \log_4(25 \cdot 3) - \log_4 5$ Apply the product property.

$= \log_4 75 - \log_4 5$ Simplify.

$= \log_4 \dfrac{75}{5}$ Apply the quotient property.

$= \log_4 15$ Simplify.

EXAMPLE 5

Write each expression as sums or differences of multiples of logarithms.

a. $\log_3 \dfrac{5 \cdot 7}{4}$ **b.** $\log_2 \dfrac{x^5}{y^2}$

Solution **a.** $\log_3 \dfrac{5 \cdot 7}{4} = \log_3(5 \cdot 7) - \log_3 4$ Apply the quotient property.

$= \log_3 5 + \log_3 7 - \log_3 4$ Apply the product property.

b. $\log_2 \dfrac{x^5}{y^2} = \log_2(x^5) - \log_2(y^2)$ Apply the quotient property.

$= 5 \log_2 x - 2 \log_2 y$ Apply the power property.

> **Helpful Hint**
>
> Notice that we are not able to simplify further a logarithmic expression such as $\log_5(2x - 1)$. None of the basic properties gives a way to write the logarithm of a difference in some equivalent form.

✔ **CONCEPT CHECK**

What is **wrong** with the following?

$$\log_{10}(x^2 + 5) = \log_{10} x^2 + \log_{10} 5$$
$$= 2\log_{10} x + \log_{10} 5$$

Use a numerical example to demonstrate that the result is incorrect.

EXAMPLE 6

If $\log_b 2 = 0.43$ and $\log_b 3 = 0.68$, use the properties of logarithms to evaluate.

a. $\log_b 6$ **b.** $\log_b 9$ **c.** $\log_b \sqrt{2}$

Solution   **a.** $\log_b 6 = \log_b(2 \cdot 3)$ Write 6 as $2 \cdot 3$.

$\qquad\qquad = \log_b 2 + \log_b 3$ Apply the product property.

$\qquad\qquad = 0.43 + 0.68$ Substitute given values.

$\qquad\qquad = 1.11$ Simplify.

b. $\log_b 9 = \log_b 3^2$ Write 9 as 3^2.

$\qquad\quad = 2\log_b 3$

$\qquad\quad = 2(0.68)$ Substitute 0.68 for $\log_b 3$.

$\qquad\quad = 1.36$ Simplify.

c. First, recall that $\sqrt{2} = 2^{1/2}$. Then

$\log_b \sqrt{2} = \log_b 2^{1/2}$ Write $\sqrt{2}$ as $2^{1/2}$.

$\qquad\quad = \dfrac{1}{2}\log_b 2$ Apply the power property.

$\qquad\quad = \dfrac{1}{2}(0.43)$ Substitute the given value.

$\qquad\quad = 0.215$ Simplify.

A summary of the basic properties of logarithms that we have developed so far is given next.

TEACHING TIP

Caution students to be careful and make sure that they use these properties exactly as they are written. You may want to make your point by asking them if there is a property they can use to rewrite $\log_7(x + y)$ as $\log_7 x + \log_7 y$.

Concept Check Answer:
The properties do not give any way to simplify the logarithm of a sum; answers may vary.

Properties of Logarithms

If x, y, and b are positive real numbers, $b \neq 1$, and r is a real number, then

1. $\log_b 1 = 0$

2. $\log_b b^x = x$

3. $b^{\log_b x} = x$

4. $\log_b xy = \log_b x + \log_b y$ Product property.

5. $\log_b \dfrac{x}{y} = \log_b x - \log_b y$ Quotient property.

6. $\log_b x^r = r\log_b x$ Power property.

Spotlight on

DECISION *MAKING*

Suppose you are a quality assurance inspector for an electronics manufacturer. Your department has conducted reliability studies of a new model of DVD player. Your studies show that the DVD player's reliability can be described by the exponential function $R(t) = 2.7^{-(1/3)t}$, where the reliability R is the probability that the DVD player is still working t years after it is manufactured.

The marketing department asks for your input in choosing a warranty period for the DVD player. Popular warranty periods for similar competing DVD players are 1 year, 2 years, and 3 years. Using the graph of the reliability for this DVD player, which warranty period would you recommend? Explain your reasoning. What other factors would you want to consider?

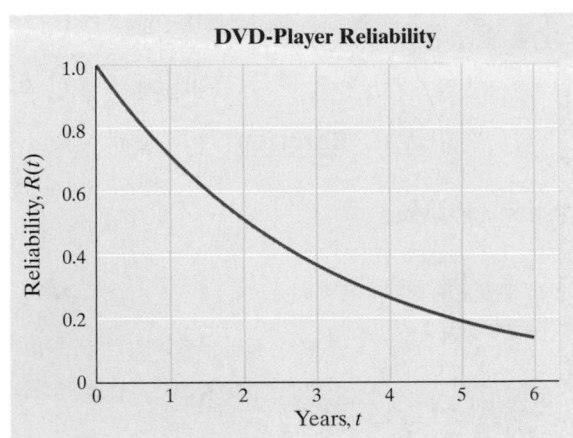

MENTAL MATH

Select the correct choice.

1. $\log 12 + \log 3 = \log$ _____ a

a. 36   **b.** 15   **c.** 4   **d.** 9

2. $\log 12 - \log 3 = \log$ _____ c

a. 36   **b.** 15   **c.** 4   **d.** 9

3. $7 \log 2 =$ _____ b

a. $\log 14$ **b.** $\log 2^7$ **c.** $\log 7^2$ **d.** $(\log 2)^7$

4. The base of $\log 7$ is _____ c

a. e   **b.** 7   **c.** 10   **d.** no base

5. The base of $\ln 7$ is _____ a

a. e   **b.** 7   **c.** 10   **d.** no base

6. $\log_5 5^2 =$ _____ b

a. 25   **b.** 2   **c.** 5^{5^2} **d.** 32

39. $\log_3 4 + \log_3 y - \log_3 5$ **40.** $\log_4 2 - \log_4 9 - \log_4 z$

43. $\frac{1}{2}\log_b 7 + \frac{1}{2}\log_b x$ **45.** $\log_7 5 + \log_7 x - \log_7 4$

EXERCISE SET 9.5

STUDY GUIDE/SSM   CD/VIDEO   PH MATH TUTOR CENTER   MathXL®Tutorials ON CD   MathXL®   MyMathLab®

47. $3\log_5 x + \log_5(x + 1)$

11. $\log_4 4$, or 1 **12.** $\log_6 4$ **15.** $-1\log_4 5 = -\log_4 5$

Write each sum as the logarithm of a single expression. Assume that variables represent positive numbers. See Example 1.

 1. $\log_5 2 + \log_5 7$ $\log_5 14$ **2.** $\log_3 8 + \log_3 4$ $\log_3 32$

3. $\log_4 9 + \log_4 x$ $\log_4 9x$ **4.** $\log_2 x + \log_2 y$ $\log_2 xy$

5. $\log_{10} 5 + \log_{10} 2 + \log_{10}(x^2 + 2)$ $\log_{10}(10x^2 + 20)$

6. $\log_6 3 + \log_6(x + 4) + \log_6 5$ $\log_6(15x + 60)$

Write each as the logarithm of a single expression. Assume that variables represent positive numbers. See Examples 2 and 4.

7. $\log_5 12 - \log_5 4$ $\log_5 3$ **8.** $\log_7 20 - \log_7 4$ $\log_7 5$

9. $\log_2 x - \log_2 y$ $\log_2 \frac{x}{y}$ **10.** $\log_3 12 - \log_3 z$ $\log_3 \frac{12}{z}$

11. $\log_4 2 + \log_4 10 - \log_4 5$ **12.** $\log_6 18 + \log_6 2 - \log_6 9$

Use the power property to rewrite each expression. See Example 3.

13. $\log_3 x^2$ $2\log_3 x$ **14.** $\log_2 x^5$ $5\log_2 x$

15. $\log_4 5^{-1}$ **16.** $\log_6 7^{-2}$ $-2\log_6 7$

17. $\log_5 \sqrt{y}$ $\frac{1}{2}\log_5 y$ **18.** $\log_5 \sqrt[3]{x}$ $\frac{1}{3}\log_5 x$

MIXED PRACTICE   **21.** $\log_5 x^3 z^6$ **29.** $\log_7 \frac{9}{2}$ **30.** $\log_8 \frac{15}{4}$

Write each as a single logarithm. Assume that variables represent positive numbers. See Example 4.

19. $2\log_2 5$ $\log_2 25$ **20.** $3\log_5 2$ $\log_5 8$

21. $3\log_5 x + 6\log_5 z$ **22.** $2\log_7 y + 6\log_7 z$ $\log_7 y^2 z^6$

23. $\log_{10} x - \log_{10}(x + 1) + \log_{10}(x^2 - 2)$ $\log_{10}\frac{x^3 - 2x}{x + 1}$

24. $\log_9(4x) - \log_9(x - 3) + \log_9(x^3 + 1)$ $\log_9\frac{4x^4 + 4x}{x - 3}$

25. $\log_4 5 + \log_4 7$ $\log_4 35$ **26.** $\log_3 2 + \log_3 5$ $\log_3 10$

27. $\log_3 8 - \log_3 2$ $\log_3 4$ **28.** $\log_5 12 - \log_5 3$ $\log_5 4$

29. $\log_7 6 + \log_7 3 - \log_7 4$ **30.** $\log_8 5 + \log_8 15 - \log_8 20$

31. $3\log_4 2 + \log_4 6$ $\log_4 48$ **32.** $2\log_3 5 + \log_3 2$ $\log_3 50$

33. $3\log_2 x + \frac{1}{2}\log_2 x - 2\log_2(x + 1)$ $\log_2\frac{x^{7/2}}{(x + 1)^2}$

34. $2\log_5 x + \frac{1}{3}\log_5 x - 3\log_5(x + 5)$ $\log_5\frac{x^{7/3}}{(x + 5)^3}$

35. $2\log_8 x - \frac{2}{3}\log_8 x + 4\log_8 x$ $\log_8 x^{16/3}$

36. $5\log_6 x - \frac{3}{4}\log_6 x + 3\log_6 x$ $\log_6 x^{29/4}$

Write each expression as a sum or difference of multiples of logarithms. Assume that variables represent positive numbers. See Example 5.   **37.** $\log_2 7 + \log_2 11 - \log_2 3$

37. $\log_2\frac{7 \cdot 11}{3}$ **38.** $\log_5\frac{2 \cdot 9}{13}$

38. $\log_5 2 + \log_5 9 - \log_5 13$

39. $\log_3\frac{4y}{5}$ **40.** $\log_4\frac{2}{9z}$

41. $\log_2\frac{x^3}{y}$ $3\log_2 x - \log_2 y$ **42.** $\log_5\frac{x}{y^4}$ $\log_5 x - 4\log_5 y$

43. $\log_b\sqrt{7x}$ **44.** $\log_b\sqrt{\frac{3}{y}}$ $\frac{1}{2}\log_b 3 - \frac{1}{2}\log_b y$

45. $\log_7\frac{5x}{4}$ **46.** $\log_9\frac{7}{y}$ $\log_9 7 - \log_9 y$

47. $\log_5 x^3(x + 1)$ **48.** $\log_2 y^3 z$ $3\log_2 y + \log_2 z$

49. $\log_6\frac{x^2}{x + 3}$ **50.** $\log_3\frac{(x + 5)^2}{x}$

49. $2\log_6 x - \log_6(x + 3)$ **50.** $2\log_3(x + 5) - \log_3 x$

If $\log_b 3 = 0.5$ and $\log_b 5 = 0.7$, evaluate the following. See Example 6. If necessary, round to three decimal places.

51. $\log_b\frac{5}{3}$ 0.2 **52.** $\log_b 25$ 1.4

53. $\log_b 15$ 1.2 **54.** $\log_b\frac{3}{5}$ -0.2

55. $\log_b\sqrt[3]{5}$ 0.233 **56.** $\log_b\sqrt[4]{3}$ 0.125

If $\log_b 2 = 0.43$ and $\log_b 3 = 0.68$, evaluate the following.

57. $\log_b 8$ 1.29 **58.** $\log_b 81$ 2.72

59. $\log_b\frac{3}{9}$ -0.68 **60.** $\log_b\frac{4}{32}$ -1.29

61. $\log_b\sqrt{\frac{2}{3}}$ -0.125 **62.** $\log_b\sqrt{\frac{3}{2}}$ 0.125

REVIEW AND PREVIEW

63. Graph the functions $y = 10^x$ and $y = \log_{10} x$ on the same set of axes. See Section 9.4.   See graphing answer section.

Evaluate each expression. See Section 9.4.

64. $\log_{10} 100$ 2 **65.** $\log_{10}\frac{1}{10}$ -1

66. $\log_7 7^2$ 2 **67.** $\log_7\sqrt{7}$ $\frac{1}{2}$

Concept Extensions

Answer the following true or false. Study your logarithm properties carefully before answering.   **71.** true

68. $\log_2 x^3 = 3\log_2 x$ true

69. $\log_3(x + y) = \log_3 x + \log_3 y$ false

70. $\frac{\log_7 10}{\log_7 5} = \log_7 2$ false   **71.** $\log_7\frac{14}{8} = \log_7 14 - \log_7 8$

72. $\frac{\log_7 x}{\log_7 y} = (\log_7 x) - (\log_7 y)$ false

73. $(\log_3 6) \cdot (\log_3 4) = \log_3 24$ false

74. It is true that $\log 8 = \log(8 \cdot 1) = \log 8 + \log 1$. Explain how $\log 8$ can equal $\log 8 + \log 1$.   because $\log 1 = 0$

INTEGRATED REVIEW · FUNCTIONS AND PROPERTIES OF LOGARITHMS

If $f(x) = x - 6$ and $g(x) = x^2 + 1$, find each value.

1. $(f + g)(x)$

2. $(f - g)(x)$

3. $(f \cdot g)(x)$

4. $\left(\dfrac{f}{g}\right)(x)$

$x^2 + x - 5$ $\qquad$ $-x^2 + x - 7$ $\qquad$ $x^3 - 6x^2 + x - 6$ $\qquad$ $\dfrac{x - 6}{x^2 + 1}$

If $f(x) = \sqrt{x}$ and $g(x) = 3x - 1$, find each function.

5. $(f \circ g)(x)$

6. $(g \circ f)(x)$

$\sqrt{3x - 1}$ $\qquad\qquad$ $3\sqrt{x} - 1$

Determine whether each is a one-to-one function. If it is, find its inverse.

7. $f = \{(-2, 6), (4, 8), (2, -6), (3, 3)\}$

8. $g = \{(4, 2), (-1, 3), (5, 3), (7, 1)\}$

one-to-one; $\{(6, -2), (8, 4), (-6, 2), (3, 3)\}$ $\qquad$ not one-to-one

Determine whether the graph of each function is one-to-one.

9.

not one-to-one

10.

one-to-one

11.

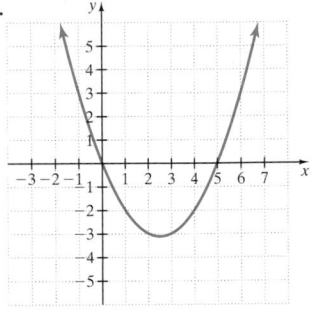

not one-to-one

Each function listed is one-to-one. Find the inverse of each function.

12. $f(x) = 3x$

$f^{-1}(x) = \dfrac{x}{3}$

13. $f(x) = x + 4$

$f^{-1}(x) = x - 4$

14. $f(x) = 5x - 1$

$f^{-1}(x) = \dfrac{x + 1}{5}$

15. $f(x) = 3x + 2$

$f^{-1}(x) = \dfrac{x - 2}{3}$

Graph each function.

16.–19. See graphing answer section.

16. $y = \left(\dfrac{1}{2}\right)^x$

17. $y = 2^x + 1$

18. $y = \log_3 x$

19. $y = \log_{1/3} x$

Solve.

20. $2^x = 8$
3

21. $9 = 3^{x-5}$
7

22. $4^{x-1} = 8^{x+2}$
−8

23. $25^x = 125^{x-1}$
3

24. $\log_4 16 = x$
2

25. $\log_{49} 7 = x$
$\dfrac{1}{2}$

26. $\log_2 x = 5$
32

27. $\log_x 64 = 3$
4

28. $\log_x \dfrac{1}{125} = -3$
5

29. $\log_3 x = -2$
$\dfrac{1}{9}$

Write each as a single logarithm.

30. $5 \log_2 x$
$\log_2 x^5$

31. $x \log_2 5$
$\log_2 5^x$

32. $3 \log_5 x - 5 \log_5 y$
$\log_5 \dfrac{x^3}{y^5}$

33. $9 \log_5 x + 3 \log_5 y$
$\log_5 x^9 y^3$

34. $\log_2 x + \log_2(x - 3) - \log_2(x^2 + 4)$
$\log_2 \dfrac{x^2 - 3x}{x^2 + 4}$

35. $\log_3 y - \log_3(y + 2) + \log_3(y^3 + 11)$
$\log_3 \dfrac{y^4 + 11y}{y + 2}$

Write each expression as a sum or difference of multiples of logarithms.

36. $\log_7 \dfrac{9x^2}{y}$
$\log_7 9 + 2 \log_7 x - \log_7 y$

37. $\log_6 \dfrac{5y}{z^2}$
$\log_6 5 + \log_6 y - 2 \log_6 z$

9.6   COMMON LOGARITHMS, NATURAL LOGARITHMS, AND CHANGE OF BASE

Objectives

1 Identify common logarithms and approximate them by calculator.

2 Evaluate common logarithms of powers of 10.

3 Identify natural logarithms and approximate them by calculator.

4 Evaluate natural logarithms of powers of e.

5 Use the change of base formula.

In this section we look closely at two particular logarithmic bases. These two logarithmic bases are used so frequently that logarithms to their bases are given special names. **Common logarithms** are logarithms to base 10. **Natural logarithms** are logarithms to base e, which we introduce in this section. The work in this section is based on the use of the calculator, which has both the common "log" $\boxed{\text{LOG}}$ and the natural "log" $\boxed{\text{LN}}$ keys.

1 Logarithms to base 10, common logarithms, are used frequently because our number system is a base 10 decimal system. The notation $\log x$ means the same as $\log_{10} x$.

Common Logarithms

$$\log x \text{ means } \log_{10} x$$

EXAMPLE 1

a. Use a calculator to approximate $\log 7$ to four decimal places.

b. Use a graphing utility to graph $f(x) = \log_{10} x$ in an appropriate window. From the graph, complete the ordered pair $(4,\ \ )$.

Solution

a. See the screen to the left. To four decimal places, $\log 7 \approx 0.8451$.

b. We know from Section 9.4 that the domain of f is the positive real numbers. Define the window to be $[-1, 9, 1]$ by $[-3, 3, 1]$.

The approximate ordered pair $(4, 0.602)$ is on the graph of $f(x) = \log_{10} x$.

2 To evaluate the common log of a power of 10, a calculator is not needed. According to the property of logarithms,

$$\log_b b^x = x$$

It follows that if b is replaced with 10, we have

$$\log 10^x = x$$

> ### Helpful Hint
> Remember that $\log 10^x$ means $\log_{10} 10^x = x$.

EXAMPLE 2

Find the exact value of each logarithm.

a. $\log 10$ **b.** $\log 1000$ **c.** $\log \dfrac{1}{10}$ **d.** $\log \sqrt{10}$

Solution

a. $\log 10 = \log 10^1 = 1$ **b.** $\log 1000 = \log 10^3 = 3$

c. $\log \dfrac{1}{10} = \log 10^{-1} = -1$ **d.** $\log \sqrt{10} = \log 10^{1/2} = \dfrac{1}{2}$

As we will soon see, equations containing common logs are useful models of many natural phenomena.

EXAMPLE 3

Solve $\log x = 1.2$ for x. Give an exact solution, and then approximate the solution to four decimal places.

Solution   Remember that the base of a common log is understood to be 10.

> **Helpful Hint**
> The understood base is 10.

$$\log x = 1.2$$

$$10^{1.2} = x \qquad \textit{Write with exponential notation}$$

The exact solution is $10^{1.2}$. To four decimal places, $x \approx 15.8489$.

We can check this solution with a graphing utility by graphing the left side and the right side of the original equation and finding the point of intersection.

The point of intersection of $y_1 = \log x$ and $y_2 = 1.2$ is approximately (15.8489, 1.2). Thus, the solution of the equation $\log x = 1.2$ is $x \approx 15.8489$.

The Richter scale measures the intensity, or magnitude, of an earthquake. The formula for the magnitude R of an earthquake is $R = \log\left(\dfrac{a}{T}\right) + B$, where a is the amplitude in micrometers of the vertical motion of the ground at the recording station, T is the number of seconds between successive seismic waves, and B is an adjustment factor that takes into account the weakening of the seismic wave as the distance increases from the epicenter of the earthquake.

EXAMPLE 4

FINDING THE MAGNITUDE OF AN EARTHQUAKE

Find an earthquake's magnitude on the Richter scale if a recording station measures an amplitude of 300 micrometers and 2.5 seconds between waves. Assume that B is 4.2. Approximate the solution to the nearest tenth.

Solution   Substitute the known values into the formula for earthquake intensity.

$$R = \log\left(\frac{a}{T}\right) + B \qquad \textit{Richter scale formula}$$

$$= \log\left(\frac{300}{2.5}\right) + 4.2 \qquad \textit{Let } a = 300, T = 2.5, \textit{ and } B = 4.2.$$

$$= \log(120) + 4.2$$

$$\approx 2.1 + 4.2 \qquad \textit{Approximate } \log 120 \textit{ by 2.1.}$$

$$= 6.3$$

This earthquake had a magnitude of 6.3 on the Richter scale.

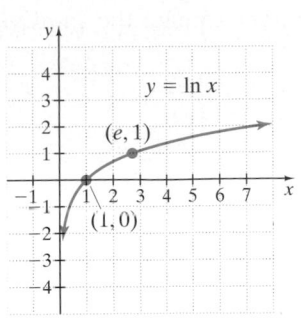

3 **Natural logarithms** are also frequently used, especially to describe natural events; hence the label "natural logarithm." Natural logarithms are logarithms to the base e, which is a constant approximately equal to 2.7183. The number e is an irrational number, as is π. The notation $\log_e x$ is usually abbreviated to $\ln x$. (The abbreviation ln is read "el en.")

Natural Logarithms

$$\ln x \text{ means } \log_e x$$

The graph of $y = \ln x$ is shown to the left.

The table below shows some values of the natural logarithm function.

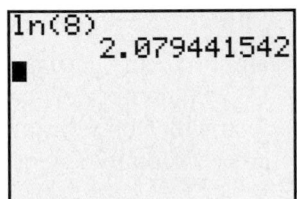

EXAMPLE 5

Use a calculator to approximate ln 8 to four decimal places.

Solution See the screen below.

```
ln(8)
        2.079441542
■
```

To four decimal places,

$$\ln 8 \approx 2.0794$$

4 As a result of the property $\log_b b^x = x$, we know that $\log_e e^x = x$, or $\ln e^x = x$.

Since $\ln e^x = x$, then $\ln e^5 = 5$, $\ln e^{22} = 22$ and so on. Also,

$$\ln e^1 = 1 \text{ or simply } \ln e = 1.$$

That is why the graph of $y = \ln x$ shown on the previous page passes through $(e, 1)$. If $x = e$, then $y = \ln e = 1$, thus the ordered pair $(e, 1)$.

EXAMPLE 6

Find the exact value of each natural logarithm.

a. $\ln e^3$ **b.** $\ln \sqrt[5]{e}$

Solution   **a.** $\ln e^3 = 3$ **b.** $\ln \sqrt[5]{e} = \ln e^{1/5} = \dfrac{1}{5}$

EXAMPLE 7

Solve $\ln 3x = 5$. Give an exact solution, and then approximate the solution to four decimal places.

Solution   Remember that the base of a natural logarithm is understood to be e.

$$\ln 3x = 5$$

> **Helpful Hint**
> The understood base is e.

$$e^5 = 3x \qquad \text{Write with exponential notation.}$$

$$\frac{e^5}{3} = x \qquad \text{Solve for } x.$$

CLASSROOM EXAMPLE
Solve: $\ln 7x = 10$. Give an exact solution and then approximate the solution to four decimal places.

answer: $x = \dfrac{e^{10}}{7} \approx 3146.6380$

The exact solution is $\dfrac{e^5}{3}$. To four decimal places,

$$x \approx 49.4711.$$

The point of intersection of $y_1 = \ln(3x)$ and $y_2 = 5$ is approximately $(49.4711, 5)$. Thus, the solution of the equation $\ln 3x = 5$ is $x \approx 49.4711$.

To check this solution with a graphing utility, see the screen to the left.

Recall from Section 9.3 the formula $A = P\left(1 + \dfrac{r}{n}\right)^{nt}$ for compound interest, where n represents the number of compoundings per year. When interest is compounded continuously, the formula $A = Pe^{rt}$ is used, where r is the annual interest rate and interest is compounded continuously for t years.

EXAMPLE 8

FINDING FINAL LOAN PAYMENT

Find the amount owed at the end of 5 years if $1600 is loaned at a rate of 9% compounded continuously.

Solution   Use the formula $A = Pe^{rt}$, where

$$P = \$1600 \text{ (the size of the loan)}$$

$$r = 9\% = 0.09 \text{ (the rate of interest)}$$

$$t = 5 \text{ (the 5-year duration of the loan)}$$

$$A = Pe^{rt}$$

$$= 1600e^{0.09(5)} \qquad \text{Substitute in known values.}$$

$$= 1600e^{0.45}$$

Now we can use a calculator to approximate the solution.

$$A \approx 2509.30$$

The total amount of money owed is $2509.30.

CLASSROOM EXAMPLE
Find the amount owed at the end of 3 years if $1200 is loaned at a rate of 8% compounded continuously.
answer: $1525.50

TEACHING TIP

Have students repeat Example 8 assuming that interest is compounded ...

a. annually.
b. quarterly.
c. monthly.
d. daily.

5   Calculators are handy tools for approximating natural and common logarithms. Unfortunately, some calculators cannot be used to approximate logarithms to bases other than e or 10—at least not directly. In such cases, we use the change of base formula.

Change of Base

If a, b, and c are positive real numbers and neither b nor c is 1, then

$$\log_b a = \frac{\log_c a}{\log_c b}$$

EXAMPLE 9

Approximate $\log_5 3$ to four decimal places.

Solution   Use the change of base property to write $\log_5 3$ as a quotient of logarithms to base 10.

$$\log_5 3 = \frac{\log 3}{\log 5} \qquad \text{Use the change of base property. In the change of base property, we let } a = 3, b = 5, \text{ and } c = 10.$$

$$\approx \frac{0.4771213}{0.69897} \qquad \text{Approximate logarithms by calculator.}$$

$$\approx 0.6826062 \qquad \text{Simplify by calculator.}$$

To four decimal places, $\log_5 3 \approx 0.6826$.

CLASSROOM EXAMPLE

Approximate $\log_7 5$ to four decimal places.

answer: 0.8271

 CONCEPT CHECK

If a graphing calculator cannot directly evaluate logarithms to base 5, describe how you could use the graphing calculator to graph the function $f(x) = \log_5 x$.

STUDY SKILLS REMINDER

Do You Remember What to Do the Day of an Exam?

On the day of an exam, don't forget to try the following:

▶ Allow yourself plenty of time to arrive.

▶ Read the directions on the test carefully.

▶ Read each problem carefully as you take your test. Make sure that you answer the question asked.

▶ Watch your time and pace yourself so that you may attempt each problem on your test.

▶ If you have time, check your work and answers.

▶ Do not turn your test in early. If you have extra time, spend it double-checking your work.

Good luck!

Concept Check Answer:

$$f(x) = \frac{\log x}{\log 5}$$

35. $10^{1.3} \approx 19.9526$ **36.** $10^{2.1} \approx 125.8925$

EXERCISE SET 9.6

STUDY
GUIDE/SSM   CD/
VIDEO   PH MATH
TUTOR CENTER   MathXL®Tutorials
ON CD   MathXL®   MyMathLab®

37. $\dfrac{10^{1.1}}{2} \approx 6.2946$ **38.** $\dfrac{10^{1.3}}{3} \approx 6.6509$

Use a calculator to approximate each logarithm to four decimal places. See Examples 1 and 5.

1. $\log 8$ 0.9031
2. $\log 6$ 0.7782

3. $\log 2.31$ 0.3636
4. $\log 4.86$ 0.6866

5. $\ln 2$ 0.6931
6. $\ln 3$ 1.0986

7. $\ln 0.0716$ -2.6367
8. $\ln 0.0032$ -5.7446

9. $\log 12.6$ 1.1004
10. $\log 25.9$ 1.4133

11. $\ln 5$ 1.6094
12. $\ln 7$ 1.9459

13. $\log 41.5$ 1.6180
14. $\ln 41.5$ 3.7257

15. Use a calculator and try to approximate log 0. Describe what happens and explain why.   answers may vary

16. Use a calculator and try to approximate ln 0. Describe what happens and explain why.   answers may vary

Find the exact value. See Examples 2 and 6.

 17. $\log 100$ 2
18. $\log 10,000$ 4

19. $\log\left(\dfrac{1}{1000}\right)$ -3
20. $\log\left(\dfrac{1}{100}\right)$ -2

 21. $\ln e^2$ 2
22. $\ln e^4$ 4

23. $\ln \sqrt[4]{e}$ $\frac{1}{4}$
24. $\ln \sqrt[5]{e}$ $\frac{1}{5}$

25. $\log 10^3$ 3
26. $\ln e^5$ 5

27. $\ln e^2$ 2
28. $\log 10^7$ 7

29. $\log 0.0001$ -4
30. $\log 0.001$ -3

31. $\ln \sqrt{e}$ $\frac{1}{2}$
32. $\log \sqrt{10}$ $\frac{1}{2}$

33. Without using a calculator, explain which of log 50 or ln 50 must be larger.   answers may vary

34. Without using a calculator, explain which of $\log 50^{-1}$ or $\ln 50^{-1}$ must be larger.   answers may vary

Solve each equation for x. Give an exact solution and a four-decimal-place approximation. See Examples 3 and 7.

35. $\log x = 1.3$
36. $\log x = 2.1$

37. $\log 2x = 1.1$
38. $\log 3x = 1.3$

39. $\ln x = 1.4$
40. $\ln x = 2.1$

41. $\ln(3x - 4) = 2.3$
42. $\ln(2x + 5) = 3.4$

43. $\log x = 2.3$
44. $\log x = 3.1$

45. $\ln x = -2.3$
46. $\ln x = -3.7$

47. $\log(2x + 1) = -0.5$
48. $\log(3x - 2) = -0.8$

49. $\ln 4x = 0.18$
50. $\ln 3x = 0.76$

Approximate each logarithm to four decimal places. See Example 9.

51. $\log_2 3$ 1.5850
52. $\log_3 2$ 0.6309

53. $\log_{1/2} 5$ -2.3219
54. $\log_{1/3} 2$ -0.6309

55. $\log_4 9$ 1.5850
56. $\log_9 4$ 0.6309

57. $\log_3 \dfrac{1}{6}$ -1.6309
58. $\log_6 \dfrac{2}{3}$ -0.2263

59. $\log_8 6$ 0.8617
60. $\log_6 8$ 1.1606

Use the formula $R = \log\left(\dfrac{a}{T}\right) + B$ to find the intensity R on the Richter scale of the earthquakes that fit the descriptions given. Round answers to one decimal place. See Example 4.

61. Amplitude a is 200 micrometers, time T between waves is 1.6 seconds, and B is 2.1. 4.2

62. Amplitude a is 150 micrometers, time T between waves is 3.6 seconds, and B is 1.9. 3.5

63. Amplitude a is 400 micrometers, time T between waves is 2.6 seconds, and B is 3.1. 5.3

64. Amplitude a is 450 micrometers, time T between waves is 4.2 seconds, and B is 2.7. 4.7

Use the formula $A = Pe^{rt}$ to solve. See Example 8.

65. Find how much money Dana Jones has after 12 years if $1400 is invested at 8% interest compounded continuously. $3656.38

66. Determine the size of an account in which $3500 earns 6% interest compounded continuously for 1 year. $3716.43

67. Find the amount of money Barbara Mack owes at the end of 4 years if 6% interest is compounded continuously on her $2000 debt. $2542.50

68. Find the amount of money for which a $2500 certificate of deposit is redeemable if it has been paying 10% interest compounded continuously for 3 years. $3374.65

REVIEW AND PREVIEW

Solve each equation for x. See Sections 1.5 and 5.8.

69. $6x - 3(2 - 5x) = 6$ $\frac{4}{7}$ **70.** $2x + 3 = 5 - 2(3x - 1)$ $\frac{1}{2}$

71. $2x + 3y = 6x$ $x = \dfrac{3y}{4}$ **72.** $4x - 8y = 10x$ $x = -\dfrac{4y}{3}$

73. $x^2 + 7x = -6$ $-6, -1$ **74.** $x^2 + 4x = 12$ $-6, 2$

Solve each system of equations. See Section 4.1.

75. $\begin{cases} x + 2y = -4 \\ 3x - y = 9 \end{cases}$ $(2, -3)$ **76.** $\begin{cases} 5x + y = 5 \\ -3x - 2y = -10 \end{cases}$ $(0, 5)$

39. $e^{1.4} \approx 4.0552$ **40.** $e^{2.1} \approx 8.1662$ **41.** $\dfrac{4 + e^{2.3}}{3} \approx 4.6581$ **42.** $\dfrac{e^{3.4} - 5}{2} \approx 12.4821$ **43.** $10^{2.3} \approx 199.5262$

Concept Extensions

77. Without using a calculator, explain which of log 50 or ln 50 must be larger.   answers may vary

78. Without using a calculator, explain which of $\log 50^{-1}$ or $\ln 50^{-1}$ must be larger.   answers may vary

Graph each function by finding ordered pair solutions, plotting the solutions, and then drawing a smooth curve through the plotted points.

79. $f(x) = e^x$

80. $f(x) = e^{2x}$

81. $f(x) = e^{-3x}$

82. $f(x) = e^{-x}$

83. $f(x) = e^x + 2$

84. $f(x) = e^x - 3$

85. $f(x) = e^{x-1}$

86. $f(x) = e^{x+4}$

87. $f(x) = 3e^x$

88. $f(x) = -2e^x$

89. $f(x) = \ln x$

90. $f(x) = \log x$

91. $f(x) = -2 \log x$

92. $f(x) = 3 \ln x$

93. $f(x) = \log (x + 2)$

94. $f(x) = \log (x - 2)$

95. $f(x) = \ln x - 3$

96. $f(x) = \ln x + 3$

97. Graph $f(x) = e^x$ (Exercise 79), $f(x) = e^x + 2$ (Exercise 83), and $f(x) = e^x - 3$ (Exercise 84) on the same screen. Discuss any trends shown on the graphs.   answers may vary

98. Graph $f(x) = \ln x$ (Exercise 89), $f(x) = \ln x - 3$ (Exercise 95), and $f(x) = \ln x + 3$ (Exercise 96). Discuss any trends shown on the graphs.   answers may vary

44. $10^{3.1} \approx 1258.9254$ 45. $e^{-2.3} \approx 0.1003$ 46. $e^{-3.7} \approx 0.0247$ 47. $\dfrac{10^{-0.5} - 1}{2} \approx -0.3419$ 48. $\dfrac{10^{-0.8} + 2}{3} \approx 0.7195$

49. $\dfrac{e^{0.18}}{4} \approx 0.2993$ 50. $\dfrac{e^{0.76}}{3} \approx 0.7128$ 79.–98. See graphing answer section.

9.7 EXPONENTIAL AND LOGARITHMIC EQUATIONS AND APPLICATIONS

Objectives

1 Solve exponential equations.

2 Solve logarithmic equations.

3 Solve problems that can be modeled by exponential and logarithmic equations.

1   In Section 9.3 we solved exponential equations such as $2^x = 16$ by writing 16 as a power of 2 and applying the uniqueness of b^x.

TEACHING TIP

Point out that logarithms are used to solve equations in which the variable is in the exponent.

$$2^x = 16$$

$$2^x = 2^4 \qquad \text{Write 16 as } 2^4.$$

$$x = 4 \qquad \text{Use the uniqueness of } b^x.$$

Solving the equation in this manner is possible since 16 is a power of 2. If solving an equation such as $2^x = a\ number$, where the number is not a power of 2, we use logarithms. For example, to solve an equation such as $3^x = 7$, we use the fact that $f(x) = \log_b x$ is a one-to-one function. Another way of stating this fact is as a property of equality.

Logarithm Property of Equality

Let a, b, and c be real numbers such that $\log_b a$ and $\log_b c$ are real numbers and b is not 1. Then

$$\log_b a = \log_b c \text{ is equivalent to } a = c$$

EXAMPLE 1

Solution

Solve $3^x = 7$.

To solve, we use the logarithm property of equality and take the logarithm of both sides. For this example, we use the common logarithm.

$$3^x = 7$$

$$\log 3^x = \log 7 \qquad \text{Take the common log of both sides.}$$

$$x \log 3 = \log 7 \qquad \text{Apply the power property of logarithms.}$$

$$x = \frac{\log 7}{\log 3} \qquad \text{Divide both sides by } \log 3.$$

The exact solution is $\dfrac{\log 7}{\log 3}$. If a decimal approximation is preferred,

$$\frac{\log 7}{\log 3} \approx \frac{0.845098}{0.4771213} \approx 1.7712 \text{ to four decimal places.}$$

The solution is $\dfrac{\log 7}{\log 3}$, or *approximately* 1.7712.

2   By applying the appropriate properties of logarithms, we can solve a broad variety of logarithmic equations.

EXAMPLE 2

Solution

Solve $\log_4(x - 2) = 2$.

Notice that $x - 2$ must be positive, so x must be greater than 2. With this in mind, we first write the equation with exponential notation.

$$\log_4(x - 2) = 2$$

$$4^2 = x - 2$$

$$16 = x - 2$$

$$18 = x \qquad \text{Add 2 to both sides.}$$

Check

To check numerically, we replace x with 18 in the original equation.

$$\log_4(x - 2) = 2$$

$$\log_4(18 - 2) \overset{?}{=} 2 \qquad \text{Let } x = 18.$$

$$\log_4 16 \overset{?}{=} 2$$

$$4^2 = 16 \qquad \text{True.}$$

The solution is 18.

To check graphically,

EXAMPLE 3

Solve $\log_2 x + \log_2(x - 1) = 1$.

Solution

Notice that $x - 1$ must be positive, so x must be greater than 1. We use the product property on the left side of the equation.

$$\log_2 x + \log_2(x - 1) = 1$$

$$\log_2 x(x - 1) = 1 \qquad \text{Apply the product property.}$$

$$\log_2(x^2 - x) = 1$$

Next we write the equation with exponential notation and solve for x.

$$2^1 = x^2 - x$$

$$0 = x^2 - x - 2 \qquad \text{Subtract 2 from both sides.}$$

$$0 = (x - 2)(x + 1) \qquad \text{Factor.}$$

$$0 = x - 2 \quad \text{or} \quad 0 = x + 1 \qquad \text{Set each factor equal to 0.}$$

$$2 = x \qquad\qquad -1 = x$$

Recall that -1 cannot be a solution because x must be greater than 1. If we forgot this, we would still reject -1 after checking. To see this, we replace x with -1 in the original equation.

$$\log_2 x + \log_2(x - 1) = 1$$

$$\log_2(-1) + \log_2(-1 - 1) \stackrel{?}{=} 1 \quad \text{Let } x = -1.$$

Because the logarithm of a negative number is undefined, -1 is rejected. Check to see that the solution is 2.

CLASSROOM EXAMPLE

Solve: $\log_6 x + \log_6(x + 1) = 1$

answer: 2

A graphic check for Example 3. Use the change of base formula to enter $y_1 = \log x/\log 2 + \log(x - 1)/\log 2$ and $y_2 = 1$.

EXAMPLE 4

Solve $\log(x + 2) - \log x = 2$.

Solution

We use the quotient property of logarithms on the left side of the equation.

$$\log(x + 2) - \log x = 2$$

$$\log\frac{x + 2}{x} = 2 \qquad \text{Apply the quotient property.}$$

$$10^2 = \frac{x + 2}{x} \qquad \text{Write using exponential notation.}$$

$$100 = \frac{x + 2}{x} \qquad \text{Simplify.}$$

$$100x = x + 2 \qquad \text{Multiply both sides by } x.$$

$$99x = 2 \qquad \text{Subtract } x \text{ from both sides.}$$

$$x = \frac{2}{99} \qquad \text{Divide both sides by 99.}$$

CLASSROOM EXAMPLE

Solve: $\log(x - 1) - \log x = 1$

answer: ∅

```
2/99→X:log(X+2)-
log(X)
                   2
▮
```

A numerical check for Example 4.

Verify that the solution is $\dfrac{2}{99}$. See a numerical check to the left.

3   Logarithmic and exponential functions are used in a variety of scientific, technical, and business settings. A few examples follow.

EXAMPLE 5

ESTIMATING POPULATION SIZE

The population size y of a community of lemmings varies according to the relationship $y = y_0 e^{0.15t}$. In this formula, t is time in months, and y_0 is the initial population at time 0. Estimate the population after 6 months if there were originally 5000 lemmings.

Solution   We substitute 5000 for y_0 and 6 for t.

$$y = y_0 e^{0.15t}$$

$$= 5000 e^{0.15(6)} \qquad \text{Let } t = 6 \text{ and } y_0 = 5000.$$

$$= 5000 e^{0.9} \qquad \text{Multiply.}$$

Using a calculator, we find that $y \approx 12{,}298.016$. In 6 months the population will be approximately 12,300 lemmings.

EXAMPLE 6

DOUBLING AN INVESTMENT

How long does it take an investment of \$2000 to double if it is invested at 5% interest compounded quarterly? The necessary formula is $A = P\left(1 + \dfrac{r}{n}\right)^{nt}$, where A is the accrued (or owed) amount, P is the principal invested, r is the annual rate of interest, n is the number of compounding periods per year, and t is the number of years.

Solution   We are given that $P = \$2000$ and $r = 5\% = 0.05$. Compounding quarterly means 4 times a year, so $n = 4$. The investment is to double, so A must be \$4000. Substitute these values and solve for t.

$$A = P\left(1 + \frac{r}{n}\right)^{nt}$$

$$4000 = 2000\left(1 + \frac{0.05}{4}\right)^{4t} \qquad \text{Substitute in known values.}$$

$$4000 = 2000(1.0125)^{4t} \qquad \text{Simplify } 1 + \frac{0.05}{4}.$$

$$2 = (1.0125)^{4t} \qquad \text{Divide both sides by 2000.}$$

$$\log 2 = \log 1.0125^{4t} \qquad \text{Take the logarithm of both sides.}$$

$$\log 2 = 4t(\log 1.0125) \qquad \text{Apply the power property.}$$

$$\frac{\log 2}{4 \log 1.0125} = t \qquad \text{Divide both sides by } 4 \log 1.0125.$$

$$13.949408 \approx t \qquad \text{Approximate by calculator.}$$

Thus, it takes nearly 14 years for the money to double in value.

EXAMPLE 7

TRACKING AN INVESTIMENT

Suppose that you invest $1500 at an annual rate of 8% compounded monthly.

a. Make a table giving the value of the investment at the end of each year for the next 14 years.

b. How long does it take the investment to triple?

Solution **a.** First, let $P = \$1500$, $r = 0.08$, and $n = 12$ (for monthly compounding) in the formula $A = P\left(1 + \frac{r}{n}\right)^{nt}$. Using the function $A = 1500\left(1 + \frac{0.08}{12}\right)^{12x}$, we can use a graphing utility to make the following table shown below in two screens:

X	Y1
1	1624.5
2	1759.3
3	1905.4
4	2063.5
5	2234.8
6	2420.3
7	2621.1

Y1☰1500(1+.08/1...

X	Y1
8	2838.7
9	3074.3
10	3329.5
11	3605.8
12	3905.1
13	4229.2
14	4580.2

Y1☰1500(1+.08/1...

b. Notice that when the investment has tripled the accrued amount A is $4500. Use a graphing utility to approximate the value of x that gives $A = 4500$. We can see from the table in Part **a** that the value of the investment is $4500 between $x = 13$ years and $x = 14$ years. Using this information, we can determine an appropriate viewing window, such as $[0, 20, 1]$ by $[0, 5000, 1000]$. Graph $y_1 = 1500(1 + 0.08/12)^{12x}$ and $y_2 = 4500$ in the same window. As usual, the point of intersection of the two curves is the solution.

$$y_1 = 1500\left(1 + \frac{0.08}{12}\right)^{12x}$$

The screen above shows that it takes $x \approx 13.78$ years, or 13 years and 9 months (rounded to the nearest month), for the original investment to triple in value to $4500.

STUDY SKILLS REMINDER

Continue your outline started in Section 1.5. Write how to recognize and how to solve exponential and logarithmic equations in your own words. For example:

Solving Equations and Inequalities

I. Equations
   A. Linear equations (Sec. 1.5)
   B. Absolute value equations (Sec. 3.4)
   C. Quadratic and higher degree equations (Sec. 5.8 and Chapter 8)
   D. Equations with rational expressions (Sec. 6.6)
   E. Equations with radicals (Sec. 7.6)
   F. **Exponential equations**—Recognize: *equations where the variable is in the exponent*—Solve:
      1. If we can write the bases the same, then set the exponents equal to each other and solve.
      2. If we can't write the bases the same, then solve using logarithms.
   G. **Logarithmic equations**—Recognize: *equations with logarithms*—Solve: Write the equation so that there is a single logarithm on one side and a constant on the other side. Then use the definition of a logarithm to write the logarithm as an equivalent exponential and solve.

II. Inequalities
   A. Linear inequalities (Sec. 3.2)
   B. Compound inequalities (Sec. 3.3)
   C. Absolute value inequalities (Sec. 3.5)
   D. Nonlinear inequalities (Sec. 8.4)
      1. Polynomial inequalities
      2. Rational inequalities

EXERCISE SET 9.7

STUDY GUIDE/SSM | CD/VIDEO | PH MATH TUTOR CENTER | MathXL®Tutorials ON CD | MathXL® | MyMathLab®

Solve each equation. Give an exact solution, and also approximate the solution to four decimal places. See Example 1.

 1. $3^x = 6$ $\dfrac{\log 6}{\log 3}$; 1.6309

2. $4^x = 7$ $\dfrac{\log 7}{\log 4}$; 1.4037

3. $3^{2x} = 3.8$ $\dfrac{\log 3.8}{2 \log 3}$; 0.6076

4. $5^{3x} = 5.6$ $\dfrac{\log 5.6}{3 \log 5}$; 0.3568

 5. $2^{x-3} = 5$

6. $8^{x-2} = 12$

7. $9^x = 5$ $\dfrac{\log 5}{\log 9}$; 0.7325

8. $3^x = 11$ $\dfrac{\log 11}{\log 3}$; 2.1827

9. $4^{x+7} = 3$

10. $6^{x+3} = 2$

MIXED PRACTICE

 Solve each equation. See Examples 1 through 4.

11. $7^{3x-4} = 11$

12. $5^{2x-6} = 12$

13. $e^{6x} = 5$ $\dfrac{\ln 5}{6}$; 0.2682

14. $e^{2x} = 8$ $\dfrac{\ln 8}{2}$; 1.0397

15. $\log_2(x + 5) = 4$ 11

16. $\log_6(x^2 - x) = 1$ $-2, 3$

17. $\log_3 x^2 = 4$ $9, -9$

18. $\log_2 x^2 = 6$ $-8, 8$

19. $\log_4 2 + \log_4 x = 0$ $\frac{1}{2}$

20. $\log_3 5 + \log_3 x = 1$ $\frac{3}{5}$

21. $\log_2 6 - \log_2 x = 3$ $\frac{3}{4}$

22. $\log_4 10 - \log_4 x = 2$ $\frac{5}{8}$

23. $\log_4 x + \log_4(x + 6) = 2$

24. $\log_3 x + \log_3(x + 6) = 3$

25. $\log_5(x + 3) - \log_5 x = 2$

26. $\log_6(x + 2) - \log_6 x = 2$

27. $\log_3(x - 2) = 2$ 11

28. $\log_2(x - 5) = 3$ 13

29. $\log_4(x^2 - 3x) = 1$

30. $\log_8(x^2 - 2x) = 1$ $-2, 4$

31. $\ln 5 + \ln x = 0$ $\frac{1}{5}$

32. $\ln 3 + \ln (x - 1) = 0$ $\frac{4}{3}$

33. $3 \log x - \log x^2 = 2$

34. $2 \log x - \log x = 3$ 1000

35. $\log_2 x + \log_2(x + 5) = 1$

36. $\log_4 x + \log_4(x + 7) = 1$

37. $\log_4 x - \log_4(2x - 3) = 3$

38. $\log_2 x - \log_2(3x + 5) = 4$

39. $\log_2 x + \log_2(3x + 1) = 1$ **40.** $\log_3 x + \log_3(x - 8) = 2$

Solve. See Example 5.

41. The size of the wolf population at Isle Royale National Park increases at a rate of 4.3% per year. If the size of the current population is 83 wolves, find how many there should be in 5 years. Use $y = y_0 e^{0.043t}$ and round to the nearest whole.   103 wolves

42. The number of victims of a flu epidemic is increasing at a rate of 7.5% per week. If 20,000 persons are currently infected, find in how many days we can expect 45,000 to have the flu. Use $y = y_0 e^{0.075t}$ and round to the nearest whole. (Hint: Don't forget to convert your answer to days.)   76 days

43. The size of the population of Senegal is increasing at a rate of 2.6% per year. If 10,589,571 people lived in Senegal in 2002, find how many inhabitants there will be by 2008. Round to the nearest ten-thousand. Use $y = y_0 e^{0.026t}$.   12,380,000 inhabitants

44. In 2002, 1046 million people were citizens of India. Find how long it will take India's population to reach a size of 1500 million (that is, 1.5 billion) if the population size is growing at a rate of 1.7% per year. Use $y = y_0 e^{0.017t}$ and round to the nearest tenth. (*Source:* U.S. Bureau of the Census, International Data Base)   21.2 yr

45. In 2002, Russia had a population of 144,979 thousand. At that time, Russia's population was declining at a rate of 1.8% per year. If this continues, how long will it take for Russia's population to reach 120,000 thousand? Use $y = y_0 e^{-0.018t}$ and round to the nearest tenth. (*Source:* U.S. Bureau of the Census, International Data Base)   10.5 yr

46. The population of Italy has been decreasing at a rate of 0.1% per year. If there were 57,715,625 people living in Italy in 2002, how many inhabitants will there be by 2020? Use $y = y_0 e^{-0.001t}$ and round to the nearest hundred. (*Source:* U.S. Bureau of the Census, International Data Base)   56,686,000 inhabitants

Use the formula $A = P\left(1 + \dfrac{r}{n}\right)^{nt}$ to solve these compound interest problems. Round to the nearest tenth. See Example 6.

47. Find how long it takes $600 to double if it is invested at 7% interest compounded monthly.   9.9 yr

48. Find how long it takes $600 to double if it is invested at 12% interest compounded monthly.   5.8 yr

49. Find how long it takes a $1200 investment to earn $200 interest if it is invested at 9% interest compounded quarterly.   1.7 yr

50. Find how long it takes a $1500 investment to earn $200 interest if it is invested at 10% compounded semiannually.   1.3 yr

51. Find how long it takes $1000 to double if it is invested at 8% interest compounded semiannually.   8.8 yr

52. Find how long it takes $1000 to double if it is invested at 8% interest compounded monthly.   8.7 yr

The formula $w = 0.00185h^{2.67}$ is used to estimate the normal weight w of a boy h inches tall. Use this formula to solve the height-weight problems. Round to the nearest tenth.

53. Find the expected weight of a boy who is 35 inches tall.   24.5 lb

54. Find the expected weight of a boy who is 43 inches tall.   42.5 lb

55. Find the expected height of a boy who weighs 85 pounds.

56. Find the expected height of a boy who weighs 140 pounds.   67.2 in.

The formula $P = 14.7e^{-0.21x}$ gives the average atmospheric pressure P, in pounds per square inch, at an altitude x, in miles above sea level. Use this formula to solve these pressure problems. Round answers to the nearest tenth.

57. Find the average atmospheric pressure of Denver, which is 1 mile above sea level.   11.9 lb/sq. in.

58. Find the average atmospheric pressure of Pikes Peak, which is 2.7 miles above sea level.   8.3 lb/sq. in.

59. Find the elevation of a Delta jet if the atmospheric pressure outside the jet is 7.5 lb/in.^{2}.   3.2 mi

60. Find the elevation of a remote Himalayan peak if the atmospheric pressure atop the peak is 6.5 lb/in.^{2}.   3.9 mi

Psychologists call the graph of the formula $t = \dfrac{1}{c}\ln\left(\dfrac{A}{A - N}\right)$ the learning curve, since the formula relates time t passed, in weeks, to a measure N of learning achieved, to a measure A of maximum learning possible, and to a measure c of an individual's learning style. Round to the nearest week.

61. Norman is learning to type. If he wants to type at a rate of 50 words per minute (N is 50) and his expected maximum rate is 75 words per minute (A is 75), find how many weeks it should take him to achieve his goal. Assume that c is 0.09.   12 weeks

62. An experiment with teaching chimpanzees sign language shows that a typical chimp can master a maximum of 65 signs. Find how many weeks it should take a chimpanzee to master 30 signs if c is 0.03.   21 weeks

63. Janine is working on her dictation skills. She wants to take dictation at a rate of 150 words per minute and believes that the maximum rate she can hope for is 210 words per minute. Find how many weeks it should take her to achieve the 150 words per minute level if c is 0.07.   18 weeks

64. A psychologist is measuring human capability to memorize nonsense syllables. Find how many weeks it should take a subject to learn 15 nonsense syllables if the maximum possible to learn is 24 syllables and c is 0.17.   6 weeks

REVIEW AND PREVIEW

If $x = -2$, $y = 0$, and $z = 3$, find the value of each expression. See Section 1.3.

65. $\dfrac{x^2 - y + 2z}{3x}$ $-\dfrac{5}{3}$

66. $\dfrac{x^3 - 2y + z}{2z}$ $-\dfrac{5}{6}$

67. $\dfrac{3z - 4x + y}{x + 2z}$ $\dfrac{17}{4}$

68. $\dfrac{4y - 3x + z}{2x + y}$ $-\dfrac{9}{4}$

Find the inverse function of each one-to-one function. See Section 9.2.

69. $f(x) = 5x + 2$

70. $f(x) = \dfrac{x - 3}{4}$

Concept Extensions

The formula $y = y_0 e^{kt}$ gives the population size y of a population that experiences an annual rate of population growth k (given as a decimal). In this formula, t is time in years and y_0 is the initial population at time 0. Use this formula to solve Exercises 71 and 72.

71. In 1990, the population of Arizona was 3,665,228. By 2000, the population had grown to 5,130,632. Find the annual rate of population growth over this period. Round your answer to the nearest tenth of a percent. (*Source:* U.S. Bureau of the Census)   3.4%

72. In 1990, the population of Nevada was 1,201,833. By 2000, the population had grown to 1,998,257. Find the annual rate of population growth over this period. Round your answer to the nearest tenth of a percent. (*Source:* U.S. Bureau of the Census)   5.1%

73. When solving a logarithmic equation, explain why you must check possible solutions in the original equation.   answers may vary

74. Solve $5^x = 9$ by taking the common logarithm of both sides of the equation. Next, solve this equation by taking the natural logarithm of both sides. Compare your solutions. Are they the same? Why or why not?   answers may vary

5. $3 + \dfrac{\log 5}{\log 2}$; 5.3219   **6.** $\dfrac{\log 12}{\log 8} + 2$; 3.1950

9. $\dfrac{\log 3}{\log 4} - 7$; −6.2075   **10.** $\dfrac{\log 2}{\log 6} - 3$; −2.6131

11. $\dfrac{1}{3}\left(4 + \dfrac{\log 11}{\log 7}\right)$; 1.7441   **12.** $\dfrac{1}{2}\left(6 + \dfrac{\log 12}{\log 5}\right)$; 3.7720

23. 2   **24.** 3   **25.** $\frac{1}{8}$ **26.** $\frac{2}{35}$ **29.** 4, −1   **33.** 100

35. $\frac{-5 + \sqrt{33}}{2}$ **36.** $\frac{-7 + \sqrt{65}}{2}$ **37.** $\frac{192}{127}$ **38.** ∅   **39.** $\frac{2}{3}$ **40.** 9

55. 55.7 in.   **69.** $f^{-1}(x) = \dfrac{x - 2}{5}$ **70.** $f^{-1}(x) = 4x + 3$

Solve using exponential regression features on a graphing calculator.

75. The number of personal computers manufactured in the United States has risen sharply since 1980.

  a. Use the data in the table below to find and graph the exponential regression equation for the data with y representing the number of computer shipments in the United States in the thousands. (Let $x = 0$ represent the year 1980)
  $y = 942.42(1.235)^x$; See graphing answer section.

Year (0 represents 1980)	U.S. Shipments (in thousands)
0	490
5	6072
10	9430
15	22,583
16	25,650
17	30,989
18	35,550

  b. Find the approximate number of shipments for the year 1987.   4136 thousand

  c. Predict the number of shipments to be made in the year 2005.   185,548 thousand

76. Using the exponential regression equation from Exercise 75, find the number of shipments in the year 1983. Predict the number of shipments to be made in the year 2002.
1,776,000 in 1983 98,435,000 in 2002

77. The revenue for sales of computers manufactured in the U.S. is shown in the table below.

Year (0 represents 1980)	U.S. Revenue (in millions of U.S. dollars)
5	11,980
10	18,898
15	47,749
16	60,129
17	70,086
18	69,698

  a. Use the data in this table to find and graph the exponential regression equation for the data with x representing the revenue of sales of computers manufactured in the U.S. in millions of U.S. dollars. (Let $x = 0$ represent the year 1980)
  $y = 5173.90(1.160)^x$; see graphing answer section

  b. Find the approximate revenue for the year 1992. $30,689 million for 1992

  c. Predict the approximate revenue in the year 2005. $211.144 million for 2005

78. Using the exponential regression equation from Exercise 77, find the revenue in the year 1987. Predict the revenue in the year 2002. $14,616 million for 1987; $135,297 million for 2002

79. The revenue for sales of computers manufactured worldwide is shown in the table below.

  a. Use the data in this table to find and graph the exponential regression equation for the data with y representing the revenue from sales of computers manufactured worldwide in millions of U.S. dollars. (Let $x = 0$ represent the year 1980) $y = 9500.669(1.182)^x$; see graphing answer section

Year (0 represents 1980)	Worldwide Revenue (in millions of U.S. dollars)
5	22,765
10	46,000
15	123,643
16	150,414
17	162,834
18	181,544

  b. Find the approximate revenue for the year 1992. $70,940 million for 1992

  c. Predict the revenue in the year 2005. $626,315 million for 2005

80. Using the exponential regression equation from Exercise 79, find the revenue in the year 1987. Predict the revenue in the year 2002. $30,696 million for 1987; $378,886 million for 2002

CHAPTER 9 PROJECT

Modeling Temperature

When a cold object is placed in a warm room, the object's temperature gradually rises until it becomes, or nearly becomes, room temperature. Similarly, if a hot object is placed in a cooler room, the object's temperature gradually falls to room temperature. The way in which a cold or hot object warms up or cools off is modeled by an exact mathematical relationship, known as Newton's law of cooling. This law relates the temperature of an object to the time elapsed since its warming or cooling began. In this project, you will have the opportunity to investigate this model of cooling and warming. This project may be completed by working in groups or individually.

To investigate Newton's law of cooling in this project, you will collect experimental data in one of two methods: Method 1, using a stopwatch and thermometer, or Method 2, using Texas Instruments' Calculator-Based Laboratory (CBL™) or Second Generation Calculator-Based Laboratory (CBL 2™).

Method 1 Materials

- Container of either cold or hot liquid
- Thermometer
- Stopwatch
- Graphing calculator with regression capabilities

Method 2 Materials

- Container of either cold or hot liquid
- A TI-82, TI-83, or TI-85 graphing calculator with unit-to-unit link cable
- CBL™ or CBL 2™ unit with temperature probe

DATA TABLE

Time, t	Temperature, T
0	

Steps for Collecting Data with Method 1:

a. Insert the thermometer into the liquid and allow a thermometer reading to register. Take a temperature reading T as you start the stopwatch (at $t = 0$) and record it in the accompanying data table.

b. Continue taking temperature readings at uniform intervals anywhere between 5 and 10 minutes long. At each reading use the stopwatch to measure the length of time that has elapsed since the temperature readings started with your first reading at $t = 0$. Record your time t and liquid temperature T in the data table. Gather data for six to twelve readings.

c. Plot the data from the data table. Plot t on the horizontal axis and T on the vertical axis.

Steps for Collecting Data with Method 2:

a. Enter the HEAT program appropriate for your calculator.

b. Prepare the CBL or CBL 2 and the graphing calculator. Insert the temperature probe into the liquid.

c. Start the HEAT program on the graphing calculator and follow its instructions to begin collecting data. The program will collect 36 temperature readings in degrees Celsius and plot them in real time with t on the horizontal axis and T on the vertical axis.

1. Which of the following mathematical models best fits the data you collected? Explain your reasoning. (Assume $a > 0$.)

    a. $T = ab^t + c$

    b. $T = ab^{-t} + c$

    c. $T = -ab^{-t} + c$

    d. $T = \ln(-ax + b) + c$

    e. $T = -\ln(-ax + b) + c$

2. What does the constant c represent in the model you chose? What is the value of c in this activity?

3. (Optional) Subtract the value of c from each of your observations of T. Enter the new ordered pairs $(t, T - c)$ into a graphing calculator. Use the exponential or logarithmic regression feature to find a model for your experimental data. Graph the ordered pairs $(t, T - c)$ with the model you found. How well does the model fit the data? How does the model compare with your selection from Question 1?

STUDY SKILLS REMINDER

Are You Preparing for a Test on Chapter 9?

Below I have listed some common trouble areas for students in Chapter 9. After studying for your test—but before taking your test—read these.

▶ Don't forget how to find the composition of two functions.

If $f(x) = x^2 + 5$ and $g(x) = 3x$, then
$(f \circ g)(x) = f[g(x)] = f(3x) = (3x)^2 + 5 = 9x^2 + 5$
$(g \circ f)(x) = g[f(x)] = g(x^2 + 5) = 3(x^2 + 5) = 3x^2 + 15$

▶ Don't forget that f^{-1} is a special notation used to denote the inverse of a function.

Let's find the inverse of the invertible function $f(x) = 3x - 5$.

$$f(x) = 3x - 5$$
$$y = 3x - 5 \qquad \text{Replace } f(x) \text{ with } y.$$
$$x = 3y - 5 \qquad \text{Switch variables.}$$
$$x + 5 = 3y$$
$$\frac{x + 5}{3} = y \qquad \text{Solve for } y.$$
$$f^{-1}(x) = \frac{x + 5}{3} \qquad \text{Replace } y \text{ with } f^{-1}(x).$$

▶ Don't forget that $y = \log_b x$ means $b^y = x$.

Thus, $3 = \log_5 125$ means $5^3 = 125$.

▶ Remember rules for logarithms.
$$\log_b 3x = \log_b 3 + \log_b x$$
$$\log_b(3 + x) \quad \text{cannot be simplified in the same manner.}$$

Remember: This is simply a checklist of common trouble areas. For a review of Chapter 9, see the Highlights and Chapter Review at the end of this chapter.

CHAPTER VOCABULARY CHECK

Fill in each blank with one of the words or phrases listed below.

| inverse | common | composition | symmetric | exponential |
| vertical | logarithmic | natural | horizontal | |

1. For each one-to-one function, we can find its <u>inverse</u> function by switching the coordinates of the ordered pairs of the function.

2. The <u>composition</u> of functions f and g is $(f \circ g)(x) = f(g(x))$.

3. A function of the form $f(x) = b^x$ is called an <u>exponential</u> function if $b > 0$, b is not 1, and x is a real number.

4. The graphs of f and f^{-1} are <u>symmetric</u> about the line $y = x$.

5. <u>Natural</u> logarithms are logarithms to base e.

6. <u>Common</u> logarithms are logarithms to base 10.

7. To see whether a graph is the graph of a one-to-one function, apply the <u>vertical</u> line test to see if it is a function, and then apply the <u>horizontal</u> line test to see if it is a one-to-one function.

8. A <u>logarithmic</u> function is a function that can be defined by $f(x) = \log_b x$ where x is a positive real number, b is a constant positive real number, and b is not 1.

CHAPTER 9 HIGHLIGHTS

Definitions and Concepts	**Examples**

Section 9.1 The Algebra of Functions; Composite Functions

Algebra of Functions

Sum $\qquad (f + g)(x) = f(x) + g(x)$

Difference $\quad (f - g)(x) = f(x) - g(x)$

Product $\qquad (f \cdot g)(x) = f(x) \cdot g(x)$

Quotient $\qquad \left(\dfrac{f}{g}\right)(x) = \dfrac{f(x)}{g(x)}, g(x) \neq 0$

Composite Functions

The notation $(f \circ g)(x)$ means "f composed with g."

$$(f \circ g)(x) = f(g(x))$$

$$(g \circ f)(x) = g(f(x))$$

If $f(x) = 7x$ and $g(x) = x^2 + 1$,

$$(f + g)(x) = f(x) + g(x) = 7x + x^2 + 1$$

$$(f - g)(x) = f(x) - g(x) = 7x - (x^2 + 1)$$

$$= 7x - x^2 - 1$$

$$(f \cdot g)(x) = f(x) \cdot g(x) = 7x(x^2 + 1)$$

$$= 7x^3 + 7x$$

$$\left(\frac{f}{g}\right)(x) = \frac{f(x)}{g(x)} = \frac{7x}{x^2 + 1}$$

If $f(x) = x^2 + 1$ and $g(x) = x - 5$, find $(f \circ g)(x)$.

$$(f \circ g)(x) = f(g(x))$$

$$= f(x - 5)$$

$$= (x - 5)^2 + 1$$

$$= x^2 - 10x + 26$$

Section 9.2 Inverse Functions

If f is a function, then f is a **one-to-one function** only if each y-value (output) corresponds to only one x-value (input).

Horizontal Line Test

If every horizontal line intersects the graph of a function at most once, then the function is a one-to-one function.

Determine whether each graph is a one-to-one function.

A.

(continued)

Definitions and Concepts	Examples

Section 9.2   Inverse Functions

B.   **C.**

Graphs **A** and **C** pass the vertical line test, so only these are graphs of functions. Of graphs **A** and **C**, only graph **A** passes the horizontal line test, so only graph **A** is the graph of a one-to-one function.

The **inverse** of a one-to-one function f is the one-to-one function f^{-1} that is the set of all ordered pairs (b, a) such that (a, b) belongs to f.

To Find the Inverse of a One-to-One Function f(x)

Step 1: Replace $f(x)$ with y.

Step 2: Interchange x and y.

Step 3: Solve for y.

Step 4: Replace y with $f^{-1}(x)$.

Find the inverse of $f(x) = 2x + 7$.

$$y = 2x + 7 \qquad \text{Replace } f(x) \text{ with } y.$$
$$x = 2y + 7 \qquad \text{Interchange } x \text{ and } y.$$
$$2y = x - 7 \qquad \text{Solve for } y.$$
$$y = \frac{x - 7}{2}$$
$$f^{-1}(x) = \frac{x - 7}{2} \qquad \text{Replace } y \text{ with } f^{-1}(x).$$

The inverse of $f(x) = 2x + 7$ is $f^{-1}(x) = \dfrac{x - 7}{2}$.

Section 9.3   Exponential Functions

A function of the form $f(x) = b^x$ is an **exponential function,** where $b > 0$, $b \neq 1$, and x is a real number.

Graph the exponential function $y = 4^x$.

x	y
-2	$\frac{1}{16}$
-1	$\frac{1}{4}$
0	1
1	4
2	16

Uniqueness of b^x

If $b > 0$ and $b \neq 1$, then $b^x = b^y$ is equivalent to $x = y$.

Solve $2^{x+5} = 8$.
$$2^{x+5} = 2^3 \qquad \text{Write 8 as } 2^3.$$
$$x + 5 = 3 \qquad \text{Use the uniqueness of } b^x.$$
$$x = -2 \qquad \text{Subtract 5 from both sides.}$$

Definitions and Concepts	**Examples**

Section 9.4 Logarithmic Functions

Logarithmic Definition

If $b > 0$ and $b \neq 1$, then

$$y = \log_b x \quad \text{means} \quad x = b^y$$

for any positive number x and real number y.

Properties of Logarithms

If b is a real number, $b > 0$ and $b \neq 1$, then

$$\log_b 1 = 0, \quad \log_b b^x = x, \quad b^{\log_b x} = x$$

Logarithmic Function

If $b > 0$ and $b \neq 1$, then a **logarithmic function** is a function that can be defined as

$$f(x) = \log_b x$$

The domain of f is the set of positive real numbers, and the range of f is the set of real numbers.

Logarithmic Form	Corresponding Exponential Statement
$\log_5 25 = 2$	$5^2 = 25$
$\log_9 3 = \dfrac{1}{2}$	$9^{1/2} = 3$

$$\log_5 1 = 0, \quad \log_7 7^2 = 2, \quad 3^{\log_3 6} = 6$$

Graph $y = \log_3 x$.

Write $y = \log_3 x$ as $3^y = x$. Plot the ordered pair solutions listed in the table, and connect them with a smooth curve.

x	y
3	1
1	0
$\frac{1}{3}$	-1
$\frac{1}{9}$	-2

Section 9.5 Properties of Logarithms

Let x, y, and b be positive numbers and $b \neq 1$.

Product Property

$$\log_b xy = \log_b x + \log_b y$$

Quotient Property

$$\log_b \frac{x}{y} = \log_b x - \log_b y$$

Power Property

$$\log_b x^r = r \log_b x$$

Write as a single logarithm.

$2 \log_5 6 + \log_5 x - \log_5(y + 2)$

$= \log_5 6^2 + \log_5 x - \log_5(y + 2)$ Power property

$= \log_5 36 \cdot x - \log_5(y + 2)$ Product property

$= \log_5 \dfrac{36x}{y + 2}$ Quotient property

Definitions and Concepts	**Examples**

Section 9.6 Common Logarithms, Natural Logarithms, and Change of Base

Common Logarithms $\log x$ means $\log_{10} x$ ***Natural Logarithms*** $\ln x$ means $\log_e x$ ***Continuously Compounded Interest Formula*** $$A = Pe^{rt}$$ where r is the annual interest rate for P dollars invested for t years.	$\log 5 = \log_{10} 5 \approx 0.69897$ $\ln 7 = \log_e 7 \approx 1.94591$ Find the amount in an account at the end of 3 years if \$1000 is invested at an interest rate of 4% compounded continuously. Here, $t = 3$ years, $P = \$1000$, and $r = 0.04$. $$A = Pe^{rt}$$ $$= 1000e^{0.04(3)}$$ $$\approx \$1127.50$$

Section 9.7 Exponential and Logarithmic Equations and Applications

Logarithm Property of Equality Let $\log_b a$ and $\log_b c$ be real numbers and $b \neq 1$. Then $\log_b a = \log_b c$ is equivalent to $a = c$	Solve $2^x = 5$. $\log 2^x = \log 5$ Log property of equality $x \log 2 = \log 5$ Power property $x = \dfrac{\log 5}{\log 2}$ Divide both sides by $\log 2$. $x \approx 2.3219$ Use a calculator.

CHAPTER REVIEW

(9.1) *If $f(x) = x - 5$ and $g(x) = 2x + 1$, find*

1. $(f + g)(x)$ $3x - 4$ **2.** $(f - g)(x)$ $-x - 6$

3. $(f \cdot g)(x)$ $2x^2 - 9x - 5$ **4.** $\left(\dfrac{g}{f}\right)(x)$ $\dfrac{2x + 1}{x - 5}, x \neq 5$

If $f(x) = x^2 - 2$, $g(x) = x + 1$, and $h(x) = x^3 - x^2$, find each composition.

5. $(f \circ g)(x)$ $x^2 + 2x - 1$ **6.** $(g \circ f)(x)$ $x^2 - 1$

7. $(h \circ g)(2)$ 18 **8.** $(f \circ f)(x)$ $x^4 - 4x^2 + 2$

9. $(f \circ g)(-1)$ -2 **10.** $(h \circ h)(2)$ 48

(9.2) *Determine whether each function is a one-to-one function. If it is one-to-one, list the elements of its inverse.*

11. $h = \{(-9, 14), (6, 8), (-11, 12), (15, 15)\}$ one-to-one;
      $h^{-1} = \{(14, -9), (8, 6), (12, -11), (15, 15)\}$

12. $f = \{(-5, 5), (0, 4), (13, 5), (11, -6)\}$ not one-to-one

13.

U.S. Region (Input)	West	Midwest	South	Northeast
Rank in Automobile Thefts (Output)	2	4	1	3

one-to-one

△ **14.**

Shape (Input)	Square	Triangle	Parallelogram	Rectangle
Number of Sides (Output)	4	3	4	4

not one-to-one

Given that $f(x) = \sqrt{x+2}$ *is a one-to-one function, find the following.*

15. a. $f(7)$ 3
   b. $f^{-1}(3)$ 7

16. a. $f(-1)$ 1
   b. $f^{-1}(1)$ −1

Determine whether each function is a one-to-one function.

17.

not one-to-one

18.

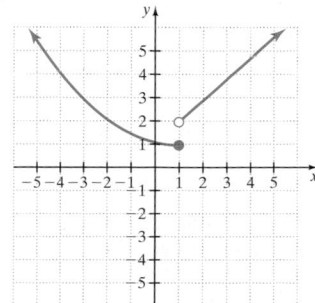

not one-to-one

13.

Rank in Automobile Thefts (Input)	2	4	1	3
U.S. Region (Output)	W	Midwest	S	NE

19.

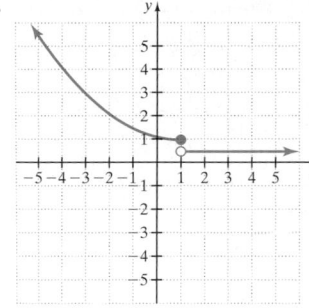

not one-to-one

21. $f^{-1}(x) = x + 9$

22. $f^{-1}(x) = x - 8$

23. $f^{-1}(x) = \dfrac{x - 11}{6}$

24. $f^{-1}(x) = \dfrac{x}{12}$

25. $f^{-1}(x) = \sqrt[3]{x + 5}$

20.

one-to-one

Find an equation defining the inverse function of the given one-to-one function.

21. $f(x) = x - 9$

22. $f(x) = x + 8$

23. $f(x) = 6x + 11$

24. $f(x) = 12x$

25. $f(x) = x^3 - 5$

26. $f(x) = \sqrt[3]{x + 2}$

27. $g(x) = \dfrac{12x - 7}{6}$

28. $r(x) = \dfrac{13}{2}x - 4$

On the same set of axes, graph the given one-to-one function and its inverse. **29.–31.** See graphing answer section.

29. $g(x) = \sqrt{x}$

30. $h(x) = 5x - 5$

31. Find the inverse of the one-to-one function $f(x) = 2x - 3$. Then graph both $f(x)$ and $f^{-1}(x)$ on a graphing utility using a square window.

(9.3) Solve each equation for x.

32. $4^x = 64$ 3

33. $3^x = \dfrac{1}{9}$ −2

34. $2^{3x} = \dfrac{1}{16}$ $-\dfrac{4}{3}$

35. $5^{2x} = 125$ $\dfrac{3}{2}$

36. $9^{x+1} = 243$ $\dfrac{3}{2}$

37. $8^{3x-2} = 4$ $\dfrac{8}{9}$

26. $f^{-1}(x) = x^3 - 2$ **27.** $g^{-1}(x) = \dfrac{6x + 7}{12}$

28. $r^{-1}(x) = \dfrac{2(x + 4)}{13}$ **31.** $f^{-1}(x) = \dfrac{x + 3}{2}$

Graph each exponential function. **38.–41.** See graphing answer section.

38. $y = 3^x$

39. $y = \left(\dfrac{1}{3}\right)^x$

40. $y = 4 \cdot 2^x$

41. $y = 2^x + 4$

Use the formula $A = P\left(1 + \dfrac{r}{n}\right)^{nt}$ *to solve the interest problems. In this formula,*

A = amount accrued (or owed)

P = principal invested (or loaned)

r = rate of interest

n = number of compounding periods per year

t = time in years

42. Find the amount accrued if $1600 is invested at 9% interest compounded semiannually for 7 years. $2963.11

43. A total of $800 is invested in a 7% certificate of deposit for which interest is compounded quarterly. Find the value that this certificate will have at the end of 5 years. $1131.82

44. Use a graphing calculator to verify the results of Exercise 40. See graphing answer section.

(9.4) Write each equation with logarithmic notation.

45. $49 = 7^2$ $\log_7 49 = 2$ **46.** $2^{-4} = \dfrac{1}{16}$ $\log_2 \dfrac{1}{16} = -4$

Write each logarithmic equation with exponential notation.

47. $\log_{1/2} 16 = -4$ **48.** $\log_{0.4} 0.064 = 3$

$\left(\dfrac{1}{2}\right)^{-4} = 16$ $0.4^3 = 0.064$

Solve for x.

49. $\log_4 x = -3$ $\dfrac{1}{64}$ **50.** $\log_3 x = 2$ 9

51. $\log_3 1 = x$ 0 **52.** $\log_4 64 = x$ 3

53. $\log_x 64 = 2$ 8 **54.** $\log_x 81 = 4$ 3

55. $\log_4 4^5 = x$ 5 **56.** $\log_7 7^{-2} = x$ -2

57. $5^{\log_5 4} = x$ 4 **58.** $2^{\log_2 9} = x$ 9

59. $\log_2(3x - 1) = 4$ $\dfrac{17}{3}$ **60.** $\log_3(2x + 5) = 2$ 2

61. $\log_4(x^2 - 3x) = 1$ $-1, 4$ **62.** $\log_8(x^2 + 7x) = 1$ $-8, 1$

Graph each pair of equations on the same coordinate system.

63. $y = 2^x$ and $y = \log_2 x$ **64.** $y = \left(\dfrac{1}{2}\right)^x$ and $y = \log_{1/2} x$

63.–64. See graphing answer section.

(9.5) Write each of the following as single logarithms.

65. $\log_3 8 + \log_3 4$ $\log_3 32$ **66.** $\log_2 6 + \log_2 3$ $\log_2 18$

67. $\log_7 15 - \log_7 20$ $\log_7 \dfrac{3}{4}$ **68.** $\log 18 - \log 12$ $\log \dfrac{3}{2}$

73. $3 \log_3 x - \log_3(x + 2)$ **74.** $\log_4(x + 5) - 2 \log_4 x$ **75.** $\log_2 3 + 2 \log_2 x + \log_2 y - \log_2 z$

76. $\log_7 y + 3 \log_7 z - \log_7 x$ **89.** $\dfrac{e^{-1} + 3}{2}$

69. $\log_{11} 8 + \log_{11} 3 - \log_{11} 6$ $\log_{11} 4$

70. $\log_5 14 + \log_5 3 - \log_5 21$ $\log_5 2$

71. $2 \log_5 x - 2 \log_5(x + 1) + \log_5 x$ $\log_5 \dfrac{x^3}{(x + 1)^2}$

72. $4 \log_3 x - \log_3 x + \log_3(x + 2)$ $\log_3(x^4 + 2x^3)$

Use properties of logarithms to write each expression as a sum or difference of multiples of logarithms.

73. $\log_3 \dfrac{x^3}{x + 2}$ **74.** $\log_4 \dfrac{x + 5}{x^2}$

75. $\log_2 \dfrac{3x^2 y}{z}$ **76.** $\log_7 \dfrac{yz^3}{x}$

If $\log_b 2 = 0.36$ *and* $\log_b 5 = 0.83$, *find the following.*

77. $\log_b 50$ 2.02 **78.** $\log_b \dfrac{4}{5}$ -0.11

(9.6) Use a calculator to approximate the logarithm to four decimal places.

79. $\log 3.6$ 0.5563 **80.** $\log 0.15$ -0.8239

81. $\ln 1.25$ 0.2231 **82.** $\ln 4.63$ 1.5326

Find the exact value.

83. $\log 1000$ 3 **84.** $\log \dfrac{1}{10}$ -1

85. $\ln \dfrac{1}{e}$ -1 **86.** $\ln e^4$ 4

Solve each equation for x.

87. $\ln(2x) = 2$ $\dfrac{e^2}{2}$ **88.** $\ln(3x) = 1.6$ $\dfrac{e^{1.6}}{3}$

89. $\ln(2x - 3) = -1$ **90.** $\ln(3x + 1) = 2$ $\dfrac{e^2 - 1}{3}$

Use the formula $\ln \dfrac{I}{I_0} = -kx$ *to solve radiation problems. In this formula,*

x = depth in millimeters

I = intensity of radiation

I_0 = initial intensity

k = a constant measure dependent on the material

Round answers to two decimal places.

91. Find the depth at which the intensity of the radiation passing through a lead shield is reduced to 3% of the original intensity if the value of k is 2.1.   1.67 mm

92. If k is 3.2, find the depth at which 2% of the original radiation will penetrate.   1.22 mm

Approximate the logarithm to four decimal places.

93. $\log_5 1.6$ 0.2920   **94.** $\log_3 4$ 1.2619

Use the formula $A = Pe^{rt}$ to solve the interest problems in which interest is compounded continuously. In this formula,   **96.** $1307.51

$A = $ amount accrued (or owed)

$P = $ principal invested (or loaned)

$r = $ rate of interest

$t = $ time in years

95. Bank of New York offers a 5-year, 6% continuously compounded investment option. Find the amount accrued if $1450 is invested. $1957.30

96. Find the amount to which a $940 investment grows if it is invested at 11% compounded continuously for 3 years.

(9.7) *Solve each exponential equation for x. Give an exact solution and also approximate the solution to four decimal places.*

97. $3^{2x} = 7$

98. $6^{3x} = 5$

99. $3^{2x+1} = 6$

100. $4^{3x+2} = 9$

101. $5^{3x-5} = 4$

102. $8^{4x-2} = 3$

103. $2 \cdot 5^{x-1} = 1$

104. $3 \cdot 4^{x+5} = 2$

97. $\dfrac{\log 7}{2 \log 3}$; 0.8856

98. $\dfrac{\log 5}{3 \log 6}$; 0.2994

99. $\dfrac{1}{2}\left(\dfrac{\log 6}{\log 3} - 1\right)$; 0.3155

100. $\dfrac{1}{3}\left(\dfrac{\log 9}{\log 4} - 2\right)$; -0.1383

101. $\dfrac{1}{3}\left(\dfrac{\log 4}{\log 5} + 5\right)$; 1.9538

102. $\dfrac{1}{4}\left(\dfrac{\log 3}{\log 8} + 2\right)$; 0.6321

103. $-\dfrac{\log 2}{\log 5} + 1$; 0.5693

Solve the equation for x.

105. $\log_5 2 + \log_5 x = 2$ $\dfrac{25}{2}$

106. $\log_3 x + \log_3 10 = 2$ $\dfrac{9}{10}$

107. $\log (5x) - \log (x + 1) = 4$ $\varnothing$

108. $\ln (3x) - \ln (x - 3) = 2$ $\dfrac{3e^2}{e^2 - 3}$

109. $\log_2 x + \log_2 2x - 3 = 1$ $2\sqrt{2}$

110. $-\log_6(4x + 7) + \log_6 x = 1$ $\varnothing$

Use the formula $y = y_0 e^{kt}$ to solve the population growth problems. In this formula,

$y = $ size of population

$y_0 = $ initial count of population

$k = $ rate of growth written as a decimal

$t = $ time

104. $\dfrac{\log \dfrac{2}{3}}{\log 4} - 5$; -5.2925 **116.** 8.8 y

Round each answer to the nearest whole.

111. The population of mallard ducks in Nova Scotia is expected to grow at a rate of 6% per week during the spring migration. If 155,000 ducks are already in Nova Scotia, find how many are expected by the end of 4 weeks.   197,044 ducks

112. The population of Indonesia is growing at a rate of 1.5% per year. If the population in 2002 was 232,073,071, find the expected population by the year 2010. (*Source:* U.S. Bureau of the Census, International Data Base)   261,661,657

113. Japan is experiencing an annual growth rate of 0.1%. In 2002, the population of Japan was 126,975,000. How long will it take for the population to be 130,000,000? (*Source:* U.S. Bureau of the Census, International Data Base)   24 yr

114. In 2002, Canada had a population of 31,902,268. How long will it take Canada to double in population if its growth rate is 0.8% annually? (*Source:* U.S. Bureau of the Census, International Data Base)   87 yr

115. Egypt's population is increasing at a rate of 1.6% per year. How long will it take for its 2002 population of 70,712,345 to double in size? (*Source:* U.S. Bureau of the Census, International Data Base)   43 yr

Use the compound interest equation $A = P\left(1 + \dfrac{r}{n}\right)^{nt}$ to solve the following. (See the directions for Exercises 42 and 43 for a reminder of this formula. Round answers to the nearest tenth.)

116. Find how long it will take a $5000 investment to grow to $10,000 if it is invested at 8% interest compounded quarterly.

117. An investment of $6000 has grown to $10,000 while the money was invested at 6% interest compounded monthly. Find how long it was invested.   8.5 yr

Use a graphing calculator to solve each equation. Round all solutions to two decimal places.   **118.–119.** See graphing answer section.

118. $e^x = 2$ 0.69   **119.** $10^{0.3x} = 7$ 2.82

Solve using exponential regression features on a graphing calculator.

120. Classroom costs have risen sharply since 1940. The chart below lists the costs for public elementary and secondary schools in billions of dollars. Use the chart to find the exponential regression equation and predict the amount of cost for the year 2005. (Let $x = 40$ represent the year 1940) $y = 0.219(1.078)^x$
562.54 billion for 2005

Year (40 represents 1940)	Cost $ (in billions)
40	3
50	10
60	25
70	48
80	100
90	200
100	275

CHAPTER 9 TEST

Remember to use your Chapter Test Prep Video CD to help you study and view solutions to the test questions you need help with.

If $f(x) = x$, $g(x) = x - 7$, and $h(x) = x^2 - 6x + 5$, find the following.

1. $(f \circ h)(0)$ 5

2. $(g \circ f)(x)$ $x - 7$

3. $(g \circ h)(x)$ $x^2 - 6x - 2$

13. -1.53

16. $\dfrac{1}{2}\left(\dfrac{\log 4}{\log 3} - 5\right)$; -1.8691

On the same set of axes, graph the given one-to-one function and its inverse.

4. $f(x) = 7x - 14$ See graphing answer section.

Determine whether the given graph is the graph of a one-to-one function.

5.

one-to-one

6.

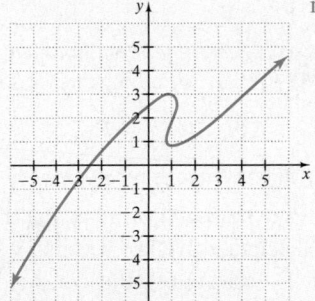

not one-to-one

Determine whether each function is one-to-one. If it is one-to-one, find an equation or a set of ordered pairs that defines the inverse function of the given function.

7. $y = 6 - 2x$ one-to-one; $f^{-1}(x) = \dfrac{-x + 6}{2}$

8. $f = \{(0,0), (2,3), (-1,5)\}$ one-to-one; $f^{-1} = \{(0,0), (3,2), (5,-1)\}$

9.   not one-to-one

Word (Input)	Dog	Cat	House	Desk	Circle
First Letter of Word (Output)	d	c	h	d	c

Use the properties of logarithms to write each expression as a single logarithm.

10. $\log_3 6 + \log_3 4$ $\log_3 24$

11. $\log_5 x + 3 \log_5 x - \log_5(x + 1)$ $\log_5 \dfrac{x^4}{x + 1}$

12. Write the expression $\log_6 \dfrac{2x}{y^3}$ as the sum or difference of multiples of logarithms. $\log_6 2 + \log_6 x - 3 \log_6 y$

13. If $\log_b 3 = 0.79$ and $\log_b 5 = 1.16$, find the value of $\log_b \dfrac{3}{25}$.

14. Approximate $\log_7 8$ to four decimal places.   1.0686

15. Solve $8^{x-1} = \dfrac{1}{64}$ for x. Give an exact solution. -1

16. Solve $3^{2x+5} = 4$ for x. Give an exact solution, and also approximate the solution to four decimal places.

Solve each logarithmic equation for x. Give an exact solution.

17. $\log_3 x = -2$ $\dfrac{1}{9}$

18. $\ln \sqrt{e} = x$ $\dfrac{1}{2}$

19. $\log_8(3x - 2) = 2$ 22

20. $\log_5 x + \log_5 3 = 2$ $\dfrac{25}{3}$

21. $\log_4(x + 1) - \log_4(x - 2) = 3$ $\dfrac{43}{21}$

22. Solve $\ln (3x + 7) = 1.31$ accurate to four decimal places. -1.0979

23. Graph $y = \left(\dfrac{1}{2}\right)^x + 1$.   See graphing answer section.

24. Graph the functions $y = 3^x$ and $y = \log_3 x$ on the same coordinate system.   See graphing answer section.

Use the formula $A = P\left(1 + \dfrac{r}{n}\right)^{nt}$ to solve Exercises 25 and 26.

25. Find the amount in the account if $4000 is invested for 3 years at 9% interest compounded monthly. $5234.58

26. Find how long it will take $2000 to grow to $3000 if the money is invested at 7% interest compounded semiannually. Round to the nearest whole.   6 yr

Use the population growth formula $y = y_0 e^{kt}$ to solve Exercises 27 and 28.

27. The prairie dog population of the Grand Rapids area now stands at 57,000 animals. If the population is growing at a rate of 2.6% annually, find how many prairie dogs there will be in that area 5 years from now.   64,913 prairie dogs

28. In an attempt to save an endangered species of wood duck, naturalists would like to increase the wood duck population from 400 to 1000 ducks. If the annual population growth rate is 6.2%, find how long it will take the naturalists to reach their goal. Round to the nearest whole year.   15 yr

29. The formula $\log(1 + k) = \dfrac{0.3}{D}$ relates the doubling time D, in days, and the growth rate k for a population of mice. Find the rate at which the population is increasing if the doubling time is 56 days. Round to the nearest tenth of a percent.   1.2%

30. Use a graphing calculator to approximate the solution of
$$e^{0.2x} = e^{-0.4x} + 2$$
to two decimal places.   3.95

31. $y = 247.71 \, (1.138)^x$; $1971.1 billion in 2006

Solve using exponential regression features on a graphing calculator.

31. The growth in the amount of money charged to two major credit cards is increasing rapidly. The chart below shows the volume of billions of dollars charged in the years 1990 to 2002. Graph the data and find the exponential regression equation to predict the amount charged in the year 2006. (Let $x = 0$ represent the year 1990)

Year (0 represents 1990)	Amount $ charged (in billions)
0	243
9	839
10	938
12	1095

CHAPTER CUMULATIVE REVIEW

1. Multiply.   (Sec. 1.3, Ex. 4)

a. $(-8)(-1)$ 8

b. $(-2)\dfrac{1}{6}$ $-\dfrac{1}{3}$

c. $3(-3)$ -9

d. $0(11)$ 0

e. $\left(\dfrac{1}{5}\right)\left(-\dfrac{10}{11}\right)$ $-\dfrac{2}{11}$

f. $(7)(1)(-2)(-3)$ 42

g. $8(-2)(0)$ 0

2. Solve $\dfrac{1}{3}(x - 2) = \dfrac{1}{4}(x + 1)$.   11 (Sec. 1.5)

3. Graph $y = x^2$.   See graphing answer section. (Sec. 2.1, Ex. 7)

4. Find the equation of a line through $(-2, 6)$ and perpendicular to $f(x) = -3x + 4$. Write the equation using function notation.
   4. $f(x) = \dfrac{x + 20}{3}$ (Sec. 2.5)

5. Solve the system.   (Sec. 4.2, Ex. 4)
$$\begin{cases} x - 5y - 2z = 6 \\ -2x + 10y + 4z = -12 \\ \dfrac{1}{2}x - \dfrac{5}{2}y - z = 3 \end{cases}$$
   $\{(x, y, z)\,|\,x - 5y - 2z = 6\}$

6. Line l and line m are parallel lines cut by transversal t. Find the values of x and y. $x = 110$ $y = 70$ (Sec. 4.3)

7. Use the quotient rule to simplify.   (Sec. 5.1, Ex. 4)

a. $\dfrac{x^7}{x^4}$ x^3

b. $\dfrac{5^8}{5^2}$ 5^6

c. $\dfrac{20x^6}{4x^5}$ $5x$

d. $\dfrac{12y^{10}z^7}{14y^8z^7}$ $\dfrac{6y^2}{7}$

8. Use the power rules to simplify the following. Use positive exponents to write all results.

a. $(4a^3)^2$ $16a^6$

b. $\left(-\dfrac{2}{3}\right)^3$ $-\dfrac{8}{27}$

c. $\left(\dfrac{4a^5}{b^3}\right)^3$ $\dfrac{64a^{15}}{b^9}$

d. $\left(\dfrac{3^{-2}}{x}\right)^{-3}$ $729x^3$ (Sec. 5.2)

e. $(a^{-2}b^3c^{-4})^{-2}$ $\dfrac{a^4c^8}{b^6}$

9. For the ICL Production Company, the rational function
$$C(x) = \dfrac{2.6x + 10,000}{x}$$
describes the company's cost per disc of pressing x compact discs. Find the cost per disc for pressing:
   9. (Sec. 6.1, Ex. 9)   a. $102.60

a. 100 compact discs

b. 1000 compact discs $12.60

10. Multiply.

a. $(3x - 1)^2$ $9x^2 - 6x + 1$

b. $\left(\dfrac{1}{2}x + 3\right)\left(\dfrac{1}{2}x - 3\right)$ $\dfrac{1}{4}x^2 - 9$

c. $(2x - 5)(6x + 7)$ $12x^2 - 16x - 35$ (Sec. 5.4)

11. Add or subtract.   (Sec. 6.2, Ex. 1)

a. $\dfrac{x}{4} + \dfrac{5x}{4} - \dfrac{3x}{2}$ **b.** $\dfrac{x^2}{x + 7} - \dfrac{49}{x + 7}$ $x - 7$

c. $\dfrac{x}{3y^2} - \dfrac{x + 1}{3y^2} - \dfrac{1}{3y^2}$

12. Perform the indicated operation and simplify if possible.

$\dfrac{5}{x - 2} + \dfrac{3}{x^2 + 4x + 4} - \dfrac{6}{x + 2}$ $\dfrac{-x^2 + 23x + 38}{(x - 2)(x + 2)^2}$ (Sec. 6.2)

13. Divide $3x^4 + 2x^3 - 8x + 6$ by $x^2 - 1$.

14. Simplify each complex fraction.

a. $\dfrac{\dfrac{a}{5}}{\dfrac{a - 1}{10}}$ **b.** $\dfrac{\dfrac{3}{2 + a} + \dfrac{6}{2 - a}}{\dfrac{5}{a + 2} - \dfrac{1}{a - 2}}$ **c.** $\dfrac{x^{-1} + y^{-1}}{xy}$

15. Solve: $\dfrac{2x}{2x - 1} + \dfrac{1}{x} = \dfrac{1}{2x - 1}$ -1 (Sec. 6.6, Ex. 4)

16. Divide $x^3 - 8$ by $x - 2$. $x^2 + 2x + 4$ (Sec. 6.4)

17. Steve Deitmer takes $1\dfrac{1}{2}$ times as long to go 72 miles upstream in his boat as he does to return. If the boat cruises at 30 mph in still water, what is the speed of the current?   6 mph (Sec. 6.7, Ex. 5)

18. Use synthetic division to divide: $(8x^2 - 12x - 7) \div (x - 2)$

19. Simplify the following expressions.   (Sec. 7.1, Ex. 4)

a. $\sqrt[4]{81}$ 3   **b.** $\sqrt[5]{-243}$ -3 **c.** $-\sqrt{25}$ -5

d. $\sqrt[4]{-81}$ **e.** $\sqrt[3]{64x^3}$ $4x$

20. Solve $\dfrac{1}{a + 5} = \dfrac{1}{3a + 6} - \dfrac{a + 2}{a^2 + 7x + 10}$ $-\dfrac{7}{5}$ (Sec. 6.6)

21. Use rational exponents to write as a single radical.

a. $\sqrt{x} \cdot \sqrt[4]{x}$ $\sqrt[4]{x^3}$ **b.** $\dfrac{\sqrt{x}}{\sqrt[3]{x}}$ $\sqrt[6]{x}$ **c.** $\sqrt[3]{3} \cdot \sqrt{2}$ $\sqrt[6]{72}$

22. Suppose that y varies directly as x. If $y = \dfrac{1}{2}$ when $x = 12$, find the constant of variation and the direct variation equation.

23. Multiply.   (Sec. 7.4, Ex. 4)

a. $\sqrt{3}(5 + \sqrt{30})$ **b.** $(\sqrt{5} - \sqrt{6})(\sqrt{7} + 1)$

c. $(7\sqrt{x} + 5)(3\sqrt{x} - \sqrt{5})$ **d.** $(4\sqrt{3} - 1)^2$

e. $(\sqrt{2x} - 5)(\sqrt{2x} + 5)$ **f.** $(\sqrt{x - 3} + 5)^2$

24. Find each root. Assume that all variables represent nonnegative real numbers.

a. $\sqrt[4]{81}$ 3   **b.** $\sqrt[3]{-27}$ -3 **c.** $\sqrt{\dfrac{9}{64}}$ $\dfrac{3}{8}$

d. $\sqrt[4]{x^{12}}$ x^3 **e.** $\sqrt[3]{-125y^6}$ $-5y^2$ (Sec. 7.1)

25. Rationalize the denominator of $\dfrac{\sqrt[4]{x}}{\sqrt[4]{81y^5}}$.

26. Multiply.

a. $a^{1/4}(a^{3/4} - a^8)$ $a - a^{33/4}$ **b.** $(x^{1/2} - 3)(x^{1/2} + 5)$

27. Solve $\sqrt{4 - x} = x - 2$.   3 (Sec. 7.6, Ex. 4)

28. Use the quotient rule to divide and simplify if possible.

a. $\dfrac{\sqrt{54}}{\sqrt{6}}$ 3   **b.** $\dfrac{\sqrt{108a^2}}{3\sqrt{3}}$ $2a$ **c.** $\dfrac{3\sqrt[3]{81a^5b^{10}}}{\sqrt[3]{3b^4}}$

29. Solve $3x^2 - 9x + 8 = 0$ by completing the square.

30. Add or subtract as indicated.

a. $\dfrac{\sqrt{20}}{3} + \dfrac{\sqrt{5}}{4}$ $\dfrac{11\sqrt{5}}{12}$ **b.** $\sqrt[3]{\dfrac{24x}{27}} - \dfrac{\sqrt[3]{3x}}{2}$ $\dfrac{\sqrt[3]{3x}}{6}$ (Sec. 7.4)

31. Solve $\dfrac{3x}{x - 2} - \dfrac{x + 1}{x} = \dfrac{6}{x(x - 2)}$.

32. Rationalize the denominator. $\sqrt[3]{\dfrac{27}{m^4 n^8}}$ $\dfrac{3\sqrt[3]{m^2 n}}{m^2 n^3}$ (Sec. 7.5)

33. Solve $x^2 - 4x \le 0$. $[0, 4]$ (Sec. 8.4, Ex. 2)

34. Find the length of the unknown side of the triangle. $4\sqrt{3}$ in. (Sec. 7.6)

8 in.   4 in.

35. Graph $F(x) = (x - 3)^2 + 1$.

36. Find the followings powers of i.

   a. i^8 1

   b. i^{21} i

   c. i^{42} -1

   d. i^{-13} $-i$ (Sec. 7.7)

37. If $f(x) = x - 1$ and $g(x) = 2x - 3$, find   (Sec. 9.1, Ex. 1)

   a. $(f + g)(x)$ $3x - 4$ **b.** $(f - g)(x)$ $-x + 2$

   c. $(f \cdot g)(x)$ $2x^2 - 5x + 3$ **d.** $\left(\dfrac{f}{g}\right)(x)$

38. Solve $4x^2 + 8x - 1 = 0$ by completing the square.

39. Find an equation of the inverse of $f(x) = x + 3$.

40. Solve by using the quadratic formula. $\left(x - \dfrac{1}{2}\right)^2 = \dfrac{x}{2}$

41. Find the value of each logarithmic expression.

   a. $\log_4 16$ 2 **b.** $\log_{10}\dfrac{1}{10}$ -1 **c.** $\log_9 3$ $\dfrac{1}{2}$

42. Graph $f(x) = -(x + 1)^2 + 1$. Find the vertex and the axis of symmetry.

13. $3x^2 + 2x + 3 + \dfrac{-6x + 9}{x^2 - 1}$ (Sec. 6.4, Ex. 6)   **14. a.** $\dfrac{2a}{a - 1}$ **b.** $\dfrac{-3a - 18}{4a - 12}$ **c.** $\dfrac{y + x}{x^2 y^2}$ (Sec. 6.3)   **18.** $8x + 4 + \dfrac{1}{x - 2}$ (Sec. 6.5)

19. d. not a real number   **21.** (Sec. 7.2, Ex. 8)   **22.** $k = \dfrac{1}{24}; y = \dfrac{1}{24}x$ (Sec. 6.8)   **23 a.** $5\sqrt{3} + 3\sqrt{10}$

b. $\sqrt{35} + \sqrt{5} - \sqrt{42} - \sqrt{6}$ **c.** $21x - 7\sqrt{5x} + 15\sqrt{x} - 5\sqrt{5}$ **d.** $49 - 8\sqrt{3}$ **e.** $2x - 25$ **f.** $x + 22 + 10\sqrt{x - 3}$

25. $\dfrac{\sqrt[4]{xy^3}}{3y^2}$ (Sec. 7.5, Ex. 3)   **26. b.** $x + 2x^{\frac{1}{2}} - 15$ (Sec. 7.2)   **28. c.** $9ab^2\sqrt[3]{a^2}$ (Sec. 7.3)   **29.** $\dfrac{9 + i\sqrt{15}}{6}, \dfrac{9 - i\sqrt{15}}{6}$ (Sec. 8.1, Ex. 8)

31. $\dfrac{-1 + \sqrt{33}}{4}, \dfrac{-1 - \sqrt{33}}{4}$ (Sec. 8.3, Ex. 2)   **35.** See graphing answer section. (Sec. 8.5, Ex. 3)   **37. d.** $\dfrac{x - 1}{2x - 3}$, where $x \ne \dfrac{3}{2}$

38. $\dfrac{-2 + \sqrt{5}}{2}, \dfrac{-2 - \sqrt{5}}{2}$ (Sec. 8.1)   **39.** $f^{-1}(x) = x - 3$ (Sec. 9.2, Ex. 4)   **40.** $\dfrac{3 - \sqrt{5}}{4}, \dfrac{3 + \sqrt{5}}{4}$ (Sec. 8.3)   **41.** (Sec. 9.4, Ex. 3)

42. See graphing answer section; vertex: $(-1, 1)$; axis of symmetry: $x = -1$

Conic Sections

In Chapter 8, we analyzed some of the important connections between a parabola and its equation. Parabolas are interesting in their own right but are more interesting still because they are part of a collection of curves known as conic sections. This chapter is devoted to quadratic equations in two variables and their conic section graphs: the parabola, circle, ellipse, and hyperbola.

Schools, homes, hospitals, airports, auditoriums, community centers, jails, theaters, day-care centers, and office buildings are just some of the types of structures designed by architects to be safe, economical, and functional.

To become a licensed architect, a person must have a professional degree in architecture, complete a three-year internship, and pass the Architect Registration Examination. An architecture degree typically includes courses in building design, computer-aided design and drafting (CADD), physics and other physical sciences, architectural history, and mathematics. Architects need solid computer and communication skills. They also need a good understanding of geometry and spatial relationships to visualize a building during the design process.

In the Spotlight on Decision Making feature on page 778, you will have the opportunity to make a decision as an architect about redesigning a bridge arch in the shape of a half-ellipse.

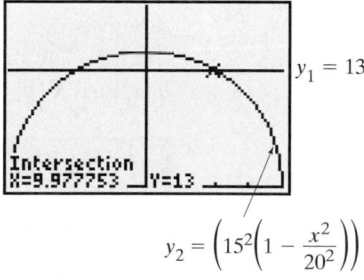

$$y_1 = 13$$

$$y_2 = \left(15^2\left(1 - \frac{x^2}{20^2}\right)\right)$$

The portion of the ellipse represents the tunnel in the Spotlight on Decision Making feature on page 778.

Link: The American Institute of Architects website.

10.1 THE PARABOLA AND THE CIRCLE

Objectives

1. Graph parabolas of the form $x = a(y - k)^2 + h$ and $y = a(x - h)^2 + k$.
2. Use the distance formula and the midpoint formula.
3. Graph circles of the form $(x - h)^2 + (y - k)^2 = r^2$.
4. Write the equation of a circle, given its center and radius.
5. Find the center and the radius of a circle, given its equation.

Conic sections derive their name because each conic section is the intersection of a right circular cone and a plane. The circle, parabola, ellipse, and hyperbola are the conic sections.

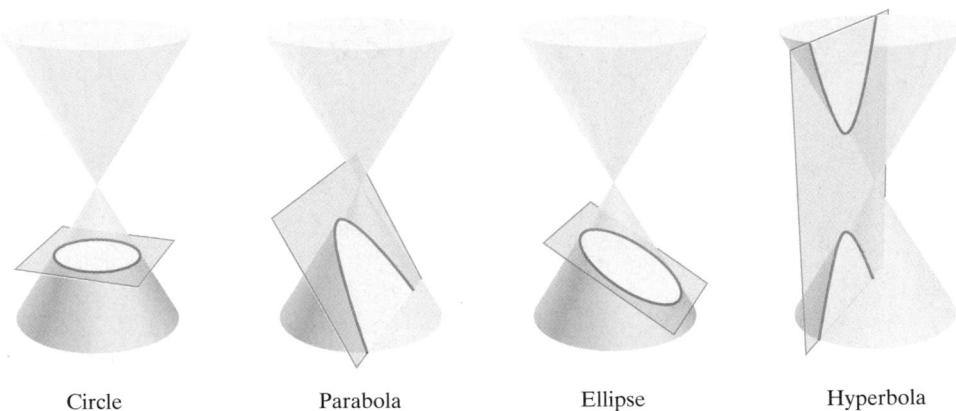

Circle   Parabola   Ellipse   Hyperbola

TEACHING TIP

After discussing Example 1, ask, "What would we need to do in order to graph this equation using a graphing calculator?" Solve the equation for y with the class, pointing out that the equation turns into two equations when it is solved for y. Then have them graph the equations using graphing calculators.

1. Thus far, we have seen that $f(x)$ or $y = a(x - h)^2 + k$ is the equation of a parabola that opens upward if $a > 0$ or downward if $a < 0$. Parabolas can also open left or right, or even on a slant. Equations of these parabolas are not functions of x, of course, since a parabola opening any way other than upward or downward fails the vertical line test. In this section, we introduce parabolas that open to the left and to the right. Parabolas opening on a slant will not be developed in this book.

Just as $y = a(x - h)^2 + k$ is the equation of a parabola that opens upward or downward, $x = a(y - k)^2 + h$ is the equation of a parabola that opens to the right or to the left. The parabola opens to the right if $a > 0$ and to the left if $a < 0$. The parabola has vertex (h, k), and its axis of symmetry is the line $y = k$.

Parabolas

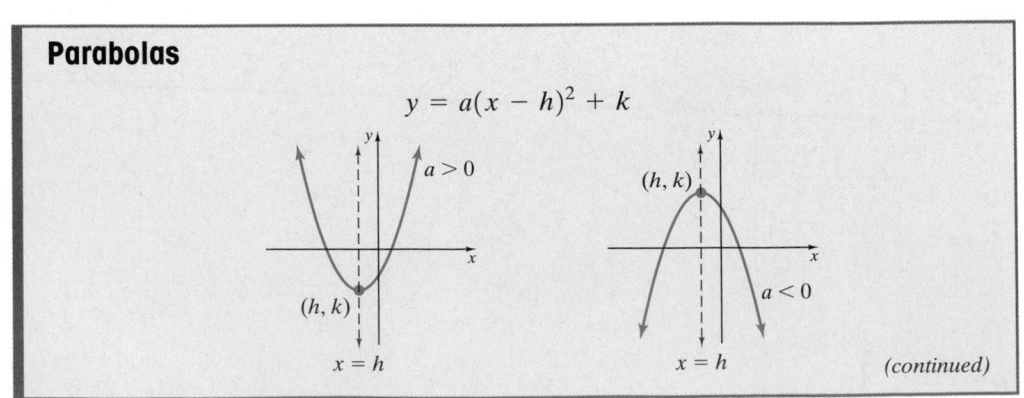

$$y = a(x - h)^2 + k$$

(continued)

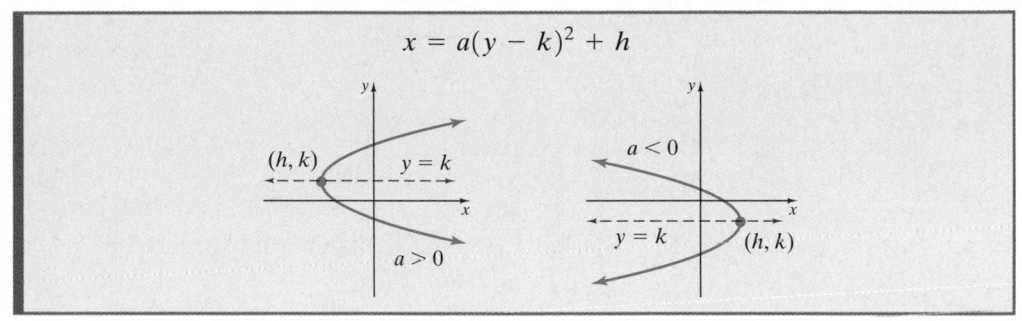

$$x = a(y - k)^2 + h$$

The equations $y = a(x - h)^2 + k$ and $x = a(y - k)^2 + h$ are called **standard forms.**

In this section, we graph parabolas and circles by hand and by using a graphing utility. Often a hand-drawn sketch is sufficient, but sometimes an exact graph drawn by a graphing utility is needed to further analyze the graph.

✔ **CONCEPT CHECK**

Does the graph of the parabola given by the equation $x = -3y^2$ open to the left, to the right, upward, or downward?

EXAMPLE 1

Graph the parabola $x = 2y^2$.

Solution   Written in standard form, the equation $x = 2y^2$ is $x = 2(y - 0)^2 + 0$ with $a = 2, h = 0$, and $k = 0$. Its graph is a parabola with vertex $(0, 0)$, and its axis of symmetry is the line $y = 0$. Since $a > 0$, this parabola opens to the right. The table shows a few more ordered pair solutions of $x = 2y^2$. Its graph is also shown.

CLASSROOM EXAMPLE

Graph $x = 3y^2$.
answer:

x	y
8	-2
2	-1
0	0
2	1
8	2

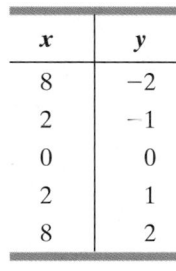

Notice that this parabola does not pass the vertical line test and is not the graph of a function. This means that we cannot enter a single function into the Y= editor to graph $x = 2y^2$. Instead, we solve for y and enter two functions separately, as shown below.

To graph $x = 2y^2$ using a graphing utility, solve the equation for y.

$$x = 2y^2$$

$$\frac{x}{2} = y^2$$

$$\pm\sqrt{\frac{x}{2}} = y$$

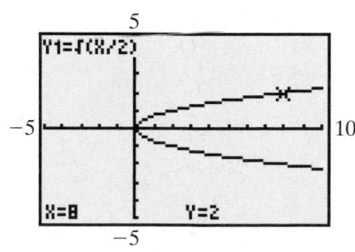

The graph of $y_1 = \sqrt{\frac{x}{2}}$ and $y_2 = -y_1$.

Both $y = \sqrt{\frac{x}{2}}$ and $y = -\sqrt{\frac{x}{2}}$ describe functions. Graph $y_1 = \sqrt{\frac{x}{2}}$ and $y_2 = -\sqrt{\frac{x}{2}}$ (or $y_1 = \sqrt{\frac{x}{2}}$ and $y_2 = -y_1$). The graph of $y = \sqrt{\frac{x}{2}}$ is the top half of the parabola and the graph of $y = -\sqrt{\frac{x}{2}}$ is the bottom half of the parabola. The two graphs together form the graph of $x = 2y^2$, as shown to the left.

EXAMPLE 2

Graph the parabola $x = -3(y - 1)^2 + 2$.

Solution   The equation $x = -3(y - 1)^2 + 2$ is in the form $x = a(y - k)^2 + h$ with $a = -3$, $k = 1$, and $h = 2$. Since $a < 0$, the parabola opens to the left. The vertex (h, k) is $(2, 1)$, and the axis of symmetry is the line $y = 1$. When $y = 0$, $x = -1$, so the x-intercept is $(-1, 0)$. Again, we obtain a few ordered pair solutions and then graph the parabola.

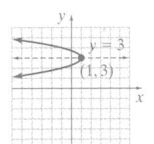

x	y
2	1
-1	0
-1	2
-10	3
-10	-1

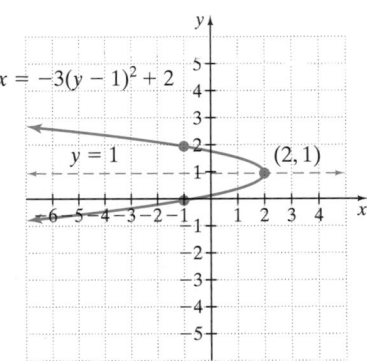

EXAMPLE 3

Graph $y = -x^2 - 2x + 15$.

Solution   Complete the square on x to write the equation in standard form.

$$y - 15 = -x^2 - 2x \qquad \text{Subtract 15 from both sides.}$$
$$y - 15 = -1(x^2 + 2x) \qquad \text{Factor } -1 \text{ from the terms } -x^2 - 2x.$$

The coefficient of x is 2. Find the square of half of 2.

$$\frac{1}{2}(2) = 1 \quad \text{and} \quad 1^2 = 1$$

$$y - 15 - 1(1) = -1(x^2 + 2x + 1) \qquad \text{Add } -1(1) \text{ to both sides.}$$

$$y - 16 = -1(x + 1)^2 \qquad \begin{array}{l}\text{Simplify the left side and} \\ \text{factor the right side.}\end{array}$$

$$y = -(x + 1)^2 + 16 \qquad \text{Add 16 to both sides.}$$

The equation is now in standard form $y = a(x - h)^2 + k$ with $a = -1$, $h = -1$, and $k = 16$.

The vertex is then (h, k), or $(-1, 16)$.

A second method for finding the vertex is by using the formula $\dfrac{-b}{2a}$.

$$x = \frac{-(-2)}{2(-1)} = \frac{2}{-2} = -1$$

$$y = -(-1)^2 - 2(-1) + 15 = -1 + 2 + 15 = 16$$

Again, we see that the vertex is $(-1, 16)$, and the axis of symmetry is the vertical line $x = -1$. The y-intercept is $(0, 15)$. Now we can use a few more ordered pair solutions to graph the parabola.

x	y
-1	16
0	15
-2	15
1	12
-3	12
3	0
-5	0

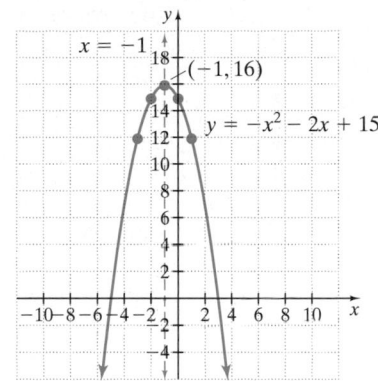

EXAMPLE 4

Graph $x = 2y^2 + 4y + 5$.

Solution Notice that this equation is quadratic in y, so its graph is a parabola that opens to the left or the right. We can complete the square on y or we can use the formula $\dfrac{-b}{2a}$ to find the vertex.

Since the equation is quadratic in y, the formula gives us the y-value of the vertex.

$$y = \frac{-4}{2 \cdot 2} = \frac{-4}{4} = -1$$

$$x = 2(-1)^2 + 4(-1) + 5 = 2 \cdot 1 - 4 + 5 = 3$$

The vertex is $(3, -1)$, and the axis of symmetry is the line $y = -1$. The parabola opens to the right since $a > 0$. The x-intercept is $(5, 0)$.

EXAMPLE 5

Use a graphing utility to graph $x = -y^2 - 4y - 1$.

Solution Solve the equation for y. To do so, write the equation so that one side is 0 and then use the quadratic formula.

$$x = -y^2 - 4y - 1$$

$$y^2 + 4y + (x + 1) = 0$$

Since we are solving for $y, a = 1, b = 4$, and $c = x + 1$. By the quadratic formula,

$$y = \frac{-b \pm \sqrt{b^2 - 4ac}}{2a}$$

or

$$y = \frac{-4 \pm \sqrt{4^2 - 4(1)(x + 1)}}{2(1)}.$$

To enter these expressions on a graphing utility, it often is convenient to store the expression under the radical, $4^2 - 4(1)(x + 1)$, in y_1. Then deselect y_1, but use it to form y_2 and y_3 as shown below.

$$y_2 = \frac{-4 + \sqrt{y_1}}{2}$$

$$y_3 = \frac{-4 - \sqrt{y_1}}{2}$$

Enter y_2 and y_3 and graph these two equations in the standard window as shown below. (Don't forget to deselect y_1 so that its graph is not shown. The purpose of entering y_1 is to simplify the expressions for y_2 and y_3.)

2   Another conic section is the circle. Before we review the circle, we need a formula to calculate the distance between points of the Cartesian coordinate system. To find the distance between two points, we use the distance formula, which is derived from the Pythagorean theorem.

To find the distance d between two points (x_1, y_1) and (x_2, y_2) as shown to the left, notice that the length of leg a is $x_2 - x_1$ and that the length of leg b is $y_2 - y_1$.

Thus, the Pythagorean theorem tells us that

$$d^2 = a^2 + b^2$$

or

$$d^2 = (x_2 - x_1)^2 + (y_2 - y_1)^2$$

or

$$d = \sqrt{(x_2 - x_1)^2 + (y_2 - y_1)^2}$$

This formula gives us the distance between any two points on the real plane.

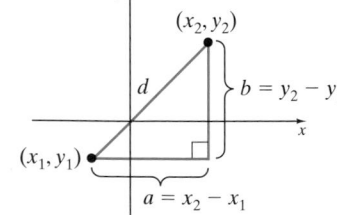

TEACHING TIP

Point out that when solving $d^2 = 3^2 + 4^2$, for example,

$d = \pm\sqrt{3^2 + 4^2}$. Then ask, "Why is the distance formula defined only as the principle square root?"

Distance Formula

The distance d between two points (x_1, y_1) and (x_2, y_2) is given by

$$d = \sqrt{(x_2 - x_1)^2 + (y_2 - y_1)^2}$$

EXAMPLE 6

Find the distance between $(2, -5)$ and $(1, -4)$. Give an exact distance and a three-decimal-place approximation.

Solution   To use the distance formula, it makes no difference which point we call (x_1, y_1) and which point we call (x_2, y_2). We will let $(x_1, y_1) = (2, -5)$ and $(x_2, y_2) = (1, -4)$.

$$
\begin{aligned}
d &= \sqrt{(x_2 - x_1)^2 + (y_2 - y_1)^2} \\
&= \sqrt{(1 - 2)^2 + [-4 - (-5)]^2} \\
&= \sqrt{(-1)^2 + (1)^2} \\
&= \sqrt{1 + 1} \\
&= \sqrt{2} \approx 1.414
\end{aligned}
$$

The distance between the two points is exactly $\sqrt{2}$ units, or approximately 1.414 units.

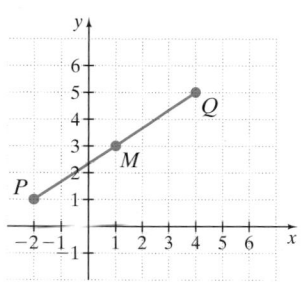

The **midpoint** of a line segment is the **point** located exactly halfway between the two endpoints of the line segment. On the graph to the left, the point M is the midpoint of line segment PQ. Thus, the distance between M and P equals the distance between M and Q.

The x-coordinate of M is at half the distance between the x-coordinates of P and Q, and the y-coordinate of M is at half the distance between the y-coordinates of P and Q. That is, the x-coordinate of M is the average of the x-coordinates of P and Q; the y-coordinate of M is the average of the y-coordinates of P and Q.

Midpoint Formula

The midpoint of the line segment whose endpoints are (x_1, y_1) and (x_2, y_2) is the point with coordinates

$$
\left(\frac{x_1 + x_2}{2}, \frac{y_1 + y_2}{2} \right)
$$

EXAMPLE 7

Find the midpoint of the line segment that joins points $P(-3, 3)$ and $Q(1, 0)$.

Solution   Use the midpoint formula. It makes no difference which point we call (x_1, y_1) or which point we call (x_2, y_2). Let $(x_1, y_1) = (-3, 3)$ and $(x_2, y_2) = (1, 0)$.

$$
\begin{aligned}
\text{midpoint} &= \left(\frac{x_1 + x_2}{2}, \frac{y_1 + y_2}{2} \right) \\
&= \left(\frac{-3 + 1}{2}, \frac{3 + 0}{2} \right) \\
&= \left(\frac{-2}{2}, \frac{3}{2} \right) \\
&= \left(-1, \frac{3}{2} \right)
\end{aligned}
$$

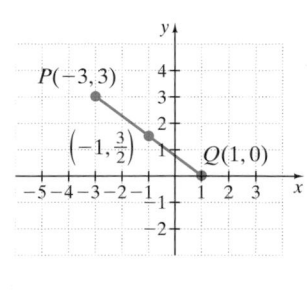

The midpoint of the segment is $\left(-1, \dfrac{3}{2} \right)$.

3 Let's now review a **circle**; which is also a conic section. A circle is the set of all points in a plane that are the same distance from a fixed point called the **center.** The distance is called the **radius** of the circle. To find a standard equation for a circle, let (h, k) represent the center of the circle, and let (x, y) represent any point on the circle. The distance between (h, k) and (x, y) is defined to be the circle's radius, r units. We can find this distance r by using the distance formula.

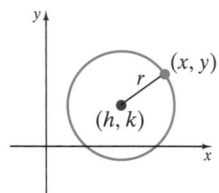

$$r = \sqrt{(x - h)^2 + (y - k)^2}$$

$$r^2 = (x - h)^2 + (y - k)^2 \qquad \text{Square both sides.}$$

Circle

The graph of $(x - h)^2 + (y - k)^2 = r^2$ is a circle with center (h, k) and radius r.

The equation $(x - h)^2 + (y - k)^2 = r^2$ is called **standard form.**
    If an equation can be written in the standard form

$$(x - h)^2 + (y - k)^2 = r^2$$

then its graph is a circle, which we can draw by graphing the center (h, k) and using the radius r.

EXAMPLE 8

Graph $x^2 + y^2 = 4$.

Solution The equation can be written in standard form as

$$(x - 0)^2 + (y - 0)^2 = 2^2$$

The center of the circle is $(0, 0)$, and the radius is 2. Its graph is shown.

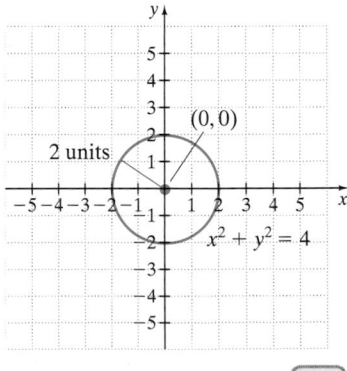

CLASSROOM EXAMPLE
Graph $x^2 + y^2 = 9$.
answer:

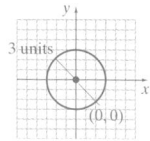

▶ **Helpful Hint**
Notice the difference between the equation of a circle and the equation of a parabola. The equation of a circle contains both x^2 and y^2 terms on the same side of the equation with equal coefficients. The equation of a parabola has either an x^2 term or a y^2 term but not both.

EXAMPLE 9

Graph $(x + 1)^2 + y^2 = 8$. Use a graphing utility to check.

Solution   The equation can be written as $(x + 1)^2 + (y - 0)^2 = 8$ with $h = -1, k = 0,$ and $r = \sqrt{8}$. The center is $(-1, 0)$, and the radius is $\sqrt{8} = 2\sqrt{2} \approx 2.8$.

Algebraic Solution:

The equation can be written as $(x + 1)^2 + (y - 0)^2 = 8$ with $h = -1, k = 0,$ and $r = \sqrt{8}$. The center is $(-1, 0)$, and the radius is $\sqrt{8} = 2\sqrt{2} \approx 2.8$.

Graphical Solution:

Solve the equation for y.

$$(x + 1)^2 + y^2 = 8$$
$$y^2 = 8 - (x + 1)^2$$
$$y = \pm\sqrt{8 - (x + 1)^2}$$

Graph $y_1 = \sqrt{8 - (x + 1)^2}$ and $y_2 = -y_1$ in a decimal window as shown below.

TECHNOLOGY NOTE

For the graph to appear circular, a square window must be used. Recall that a square window is one whose tick marks are equally spaced on the x- and y-axes.

✔ **CONCEPT CHECK**

In the graph of the equation $(x - 3)^2 + (y - 2)^2 = 5$, what is the distance between the center of the circle and any point on the circle?

4   Since a circle is determined entirely by its center and radius, this information is all we need to write the equation of a circle.

EXAMPLE 10

Find an equation of the circle with center $(-7, 3)$ and radius 10.

Solution   Using the given values $h = -7, k = 3,$ and $r = 10$, we write the equation

$$(x - h)^2 + (y - k)^2 = r^2$$

or

$$[x - (-7)]^2 + (y - 3)^2 = 10^2 \qquad \text{Substitute the given values.}$$

or

$$(x + 7)^2 + (y - 3)^2 = 100$$

5 To find the center and the radius of a circle from its equation, write the equation in standard form. To write the equation of a circle in standard form, we complete the square on both x and y.

EXAMPLE 11

Graph $x^2 + y^2 + 4x - 8y = 16$.

Solution Since this equation contains x^2 and y^2 terms on the same side of the equation with equal coefficients, its graph is a circle. To write the equation in standard form, group the terms involving x and the terms involving y, and then complete the square on each variable.

$$(x^2 + 4x) + (y^2 - 8y) = 16$$

CLASSROOM EXAMPLE

Graph $x^2 + y^2 - 2x + 6y = 6$.

answer:

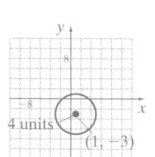

Thus, $\frac{1}{2}(4) = 2$ and $2^2 = 4$. Also, $\frac{1}{2}(-8) = -4$ and $(-4)^2 = 16$. Add 4 and then 16 to both sides.

$$(x^2 + 4x + 4) + (y^2 - 8y + 16) = 16 + 4 + 16$$
$$(x + 2)^2 + (y - 4)^2 = 36 \qquad \text{Factor.}$$

This circle has the center $(-2, 4)$ and radius 6, as shown.

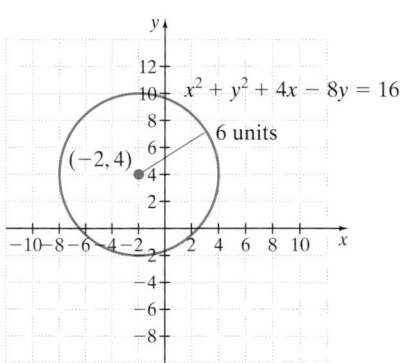

MENTAL MATH

The graph of each equation is a parabola. Determine whether the parabola opens upward, downward, to the left, or to the right.

1. $y = x^2 - 7x + 5$ upward

2. $y = -x^2 + 16$ downward

3. $x = -y^2 - y + 2$ to the left

4. $x = 3y^2 + 2y - 5$ to the right

5. $y = -x^2 + 2x + 1$ downward

6. $x = -y^2 + 2y - 6$ to the left

34. $\left(\dfrac{5\sqrt{2}}{2}, \dfrac{5\sqrt{3}}{2}\right)$ **35.** $(6.2, -6.65)$

TEACHING TIP
A Group Activity for this section is available in the Instructor's Resource Manual.

EXERCISE SET 10.1

STUDY GUIDE/SSM   CD/VIDEO   PH MATH TUTOR CENTER   MathXL®Tutorials ON CD   MathXL®   MyMathLab®

15. $\sqrt{41}$ units   **16.** $\sqrt{58}$ units
17. $\sqrt{10}$ units   **18.** $\sqrt{17}$ units   **19.** $\sqrt{5}$ units   **20.** $2\sqrt{3}$ units

The graph of each equation is a parabola. Find the vertex of the parabola and sketch its graph. See Examples 1 through 5.

1. $x = 3y^2$
2. $x = -2y^2$
3. $x = (y - 2)^2 + 3$
4. $x = (y - 4)^2 - 1$
5. $y = 3(x - 1)^2 + 5$
6. $x = -4(y - 2)^2 + 2$
7. $x = y^2 + 6y + 8$
8. $x = y^2 - 6y + 6$
9. $y = x^2 + 10x + 20$
10. $y = x^2 + 4x - 5$
11. $x = -2y^2 + 4y + 6$
12. $x = 3y^2 + 6y + 7$

Find the distance between each pair of points. Approximate the distance in Exercises 21 and 22 to two decimal places. See Example 6.

13. $(5, 1)$ and $(8, 5)$ 5 units
14. $(2, 3)$ and $(14, 8)$ 13 units
15. $(-3, 2)$ and $(1, -3)$
16. $(3, -2)$ and $(-4, 1)$
17. $(-9, 4)$ and $(-8, 1)$
18. $(-5, -2)$ and $(-6, -6)$
19. $\left(0, -\sqrt{2}\right)$ and $\left(\sqrt{3}, 0\right)$
20. $\left(-\sqrt{5}, 0\right)$ and $\left(0, \sqrt{7}\right)$
21. $(1.7, -3.6)$ and $(-8.6, 5.7)$
22. $(9.6, 2.5)$ and $(-1.9, -3.7)$ **27.** $\left(-5, \dfrac{5}{2}\right)$ **28.** $\left(\dfrac{3}{2}, -6\right)$
23. $\left(2\sqrt{3}, \sqrt{6}\right)$ and $\left(-\sqrt{3}, 4\sqrt{6}\right)$ 9 units
24. $\left(5\sqrt{2}, -4\right)$ and $\left(-3\sqrt{2}, -8\right)$ 12 units

Find the midpoint of the line segment whose endpoints are given. See Example 7.   **50.** $(x + 7)^2 + (y - 6)^2 = 4$

25. $(6, -8), (2, 4)$ $(4, -2)$
26. $(3, 9), (7, 11)$ $(5, 10)$
27. $(-2, -1), (-8, 6)$
28. $(-3, -4), (6, -8)$
29. $(7, 3), (-1, -3)$ $(3, 0)$
30. $(-2, 5), (-1, 6)$
31. $\left(\dfrac{1}{2}, \dfrac{3}{8}\right), \left(-\dfrac{3}{2}, \dfrac{5}{8}\right)$
32. $\left(-\dfrac{2}{5}, \dfrac{7}{15}\right), \left(-\dfrac{2}{5}, -\dfrac{4}{15}\right)$
33. $\left(\sqrt{2}, 3\sqrt{5}\right), \left(\sqrt{2}, -2\sqrt{5}\right)$ **49.** $(x - 2)^2 + (y - 3)^2 = 36$
34. $\left(\sqrt{8}, -\sqrt{12}\right), \left(3\sqrt{2}, 7\sqrt{3}\right)$ **52.** $x^2 + (y + 6)^2 = 2$
35. $(4.6, -3.5), (7.8, -9.8)$
36. $(-4.6, 2.1), (-6.7, 1.9)$

The graph of each equation is a circle. Find the center and the radius, and then sketch. See Examples 8, 9, and 11.

37. $x^2 + y^2 = 9$
38. $x^2 + y^2 = 25$
39. $x^2 + (y - 2)^2 = 1$
40. $(x - 3)^2 + y^2 = 9$
41. $(x - 5)^2 + (y + 2)^2 = 1$
42. $(x + 3)^2 + (y + 3)^2 = 4$
43. $x^2 + y^2 + 6y = 0$
44. $x^2 + 10x + y^2 = 0$
45. $x^2 + y^2 + 2x - 4y = 4$
46. $x^2 + 6x - 4y + y^2 = 3$

1.–12. See graphing answer section.   **21.** 13.88 units   **22.** 13.06 units   **37.–48.** See graphing answer section.

47. $x^2 + y^2 - 4x - 8y - 2 = 0$
48. $x^2 + y^2 - 2x - 6y - 5 = 0$ **31.** $\left(-\dfrac{1}{2}, \dfrac{1}{2}\right)$ **32.** $\left(-\dfrac{2}{5}, \dfrac{1}{10}\right)$

Write an equation of the circle with the given center and radius. See Example 10.   **36.** $(-5.65, 2)$ **53.** $(x + 5)^2 + (y - 4)^2 = 45$

49. $(2, 3); 6$
50. $(-7, 6); 2$
51. $(0, 0); \sqrt{3}$ $x^2 + y^2 = 3$
52. $(0, -6); \sqrt{2}$
53. $(-5, 4); 3\sqrt{5}$
54. the origin; $4\sqrt{7}$

55. If you are given a list of equations of circles and parabolas and none are in standard form, explain how you would determine which is an equation of a circle and which is an equation of a parabola. Explain also how you would distinguish the upward or downward parabolas from the left-opening or right-opening parabolas.   answers may vary

MIXED PRACTICE   **30.** $\left(-\dfrac{3}{2}, \dfrac{11}{2}\right)$ **33.** $\left(\sqrt{2}, \dfrac{\sqrt{5}}{2}\right)$

Sketch the graph of each equation. If the graph is a parabola, find its vertex. If the graph is a circle, find its center and radius.
56.–81. See graphing answer section.   **54.** $x^2 + y^2 = 112$

56. $x = y^2 + 2$
57. $x = y^2 - 3$
58. $y = (x + 3)^2 + 3$
59. $y = (x - 2)^2 - 2$
60. $x^2 + y^2 = 49$
61. $x^2 + y^2 = 1$
62. $x = (y - 1)^2 + 4$
63. $x = (y + 3)^2 - 1$
64. $(x + 3)^2 + (y - 1)^2 = 9$
65. $(x - 2)^2 + (y - 2)^2 = 16$
66. $x = -2(y + 5)^2$
67. $x = -(y - 1)^2$
68. $x^2 + (y + 5)^2 = 5$
69. $(x - 4)^2 + y^2 = 7$
70. $y = 3(x - 4)^2 + 2$
71. $y = 5(x + 5)^2 + 3$
72. $2x^2 + 2y^2 = \dfrac{1}{2}$
73. $\dfrac{x^2}{8} + \dfrac{y^2}{8} = 2$
74. $y = x^2 - 2x - 15$
75. $y = x^2 + 7x + 6$
76. $x^2 + y^2 + 6x + 10y - 2 = 0$
77. $x^2 + y^2 + 2x + 12y - 12 = 0$
78. $x = y^2 + 6y + 2$
79. $x = y^2 + 8y - 4$
80. $x^2 + y^2 - 8y + 5 = 0$
81. $x^2 - 10y + y^2 + 4 = 0$
82. $x = -2y^2 - 4y$
83. $x = -3y^2 + 30y$
84. $\dfrac{x^2}{3} + \dfrac{y^2}{3} = 2$
85. $5x^2 + 5y^2 = 25$
86. $y = 4x^2 - 40x + 105$
87. $y = 5x^2 - 20x + 16$

82.–87. See graphing answer section.

Use a graphing utility to graph each conic section. If the conic section is a circle, graph using a square window. See Examples 5 and 9. **88.–95.** See graphing answer section.

88. $x^2 + y^2 = 55$

89. $x^2 + y^2 = 20$

90. $x = 2 - 2y - y^2$

91. $x = -7 - 6y - y^2$

92. $(x - 4)^2 + (y + 2)^2 = 10$

93. $(x + 3)^2 + (y - 1)^2 = 15$

94. $x = 4y^2 - 16y + 11$

95. $x = 9y^2 - 6y + 4$

REVIEW AND PREVIEW **96.–99.** See graphing answer section.

Graph each equation. See Section 2.1.

96. $y = 2x + 5$

97. $y = -3x + 3$

98. $y = 3$

99. $x = -2$

Rationalize each denominator and simplify if possible. See Section 7.5.

100. $\dfrac{1}{\sqrt{3}}$ $\dfrac{\sqrt{3}}{3}$

101. $\dfrac{\sqrt{5}}{\sqrt{8}}$ $\dfrac{\sqrt{10}}{4}$

102. $\dfrac{4\sqrt{7}}{\sqrt{6}}$ $\dfrac{2\sqrt{42}}{3}$

103. $\dfrac{10}{\sqrt{5}}$ $2\sqrt{5}$

Concept Extensions

△ **104.** Determine whether the triangle with vertices $(2, 6)$, $(0, -2)$, and $(5, 1)$ is an isosceles triangle. yes, it is

△ **105.** In 1893, Pittsburgh bridge builder George Ferris designed and built a gigantic revolving steel wheel whose height was 264 feet and diameter was 250 feet. This Ferris wheel opened at the 1893 exposition in Chicago. It had 36 wooden cars, each capable of holding 60 passengers. (*Source:* The Handy Science Answer Book)

a. What was the radius of this Ferris wheel? 125 ft

b. How close is the wheel to the ground? 14 ft

c. How high is the center of the wheel from the ground?

d. Using the axes in the drawing, what are the coordinates of the center of the wheel? $(0, 139)$

e. Use parts **a** and **d** to write the equation of the wheel. $x^2 + (y - 139)^2 = 125^2$ **105. c.** 139 ft

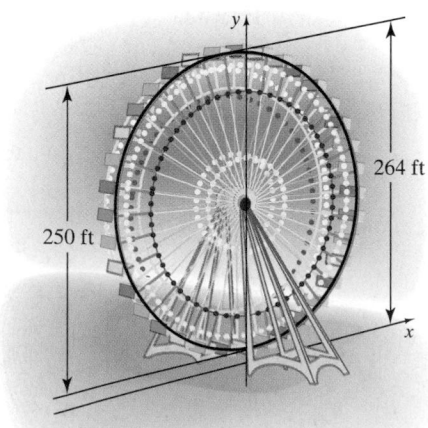

△ **106.** The world's largest-diameter Ferris wheel currently operating is the Cosmoclock 21 at Yokohama City, Japan. It has a

60-armed wheel, its diameter is 100 meters and it has a height of 105 meters. (*Source:* The Handy Science Answer Book)

a. What is the radius of this Ferris wheel? 50 m

b. How close is the wheel to the ground? 5 m

c. How high is the center of the wheel from the ground?

d. Using the axes in the drawing, what are the coordinates of the center of the wheel? $(0, 55)$

e. Use parts **a** and **d** to write the equation of the wheel. $x^2 + (y - 55)^2 = 50^2$ **106. c.** 55 m

△ **107.** Two surveyors need to find the distance across a lake. They place a reference pole at point A in the diagram. Point B is 3 meters east and 1 meter north of the reference point A. Point C is 19 meters east and 13 meters north of point A. Find the distance across the lake, from B to C. 20 m

108. a. See graphing answer section.

△ **108.** Cindy Brown, an architect, is drawing plans on grid paper for a circular pool with a fountain in the middle. The paper is marked off in centimeters, and each centimeter represents 1 foot. On the paper, the diameter of the "pool" is 20 centimeters, and "fountain" is the point $(0, 0)$. **108. b.** $x^2 + y^2 = 100$

a. Sketch the architect's drawing. Be sure to label the axes.

b. Write an equation that describes the circular pool.

c. Cindy plans to place a circle of lights around the fountain such that each light is 5 feet from the fountain. Write an equation for the circle of lights and sketch the circle on your drawing. $x^2 + y^2 = 25$; See graphing answer section.

△ **109.** A bridge constructed over a bayou has a supporting arch in the shape of a parabola. Find an equation of the parabolic arch if the length of the road over the arch is 100 meters and the maximum height of the arch is 40 meters.

Use a graphing calculator to verify each exercise. Use a square viewing window.   **110.–113.** See graphing answer section.

110. Exercise 84.

111. Exercise 85.

112. Exercise 86.

113. Exercise 87.

109. $y = -\dfrac{2}{125}x^2 + 40$

STUDY SKILLS REMINDER

How Are You Doing?

If you haven't done so yet, take a few moments and think about how you are doing in this course. Are you working toward your goal of successfully completing this course? Is your performance on homework, quizzes, and tests satisfactory? If not, you might want to see your instructor to see if he/she has any suggestions on how you can improve your performance. Let me once again remind you that, in addition to your instructor, there are many places to get help with your mathematics course. A few suggestions are below.

► This text has an accompanying video lesson for every section in this text.

► The back of this book contains answers to odd-numbered exercises.

► A tutorial software program is available with this text. It contains lessons corresponding to each section in the text.

► There is a student solutions manual available that contains worked-out solutions to odd-numbered exercises as well as solutions to every exercise in the Integrated Reviews, Chapter Reviews, Chapter Tests, and Cumulative Reviews.

► Don't forget to check with your instructor for other local resources available to you, such as a tutor center.

10.2   THE ELLIPSE AND THE HYPERBOLA

Objectives

1   Define and graph an ellipse.

2   Define and graph a hyperbola.

1   An **ellipse** can be thought of as the set of points in a plane such that the sum of the distances of those points from two fixed points is constant. Each of the two fixed points is called a **focus.** (The plural of focus is **foci.**) The point midway between the foci is called the **center.**

An ellipse may be drawn by hand by using two thumbtacks, a piece of string, and a pencil. Secure the two thumbtacks in a piece of cardboard, for example, and tie each end of the string to a tack. Use your pencil to pull the string tight and draw the ellipse. The two thumbtacks are the foci of the drawn ellipse.

  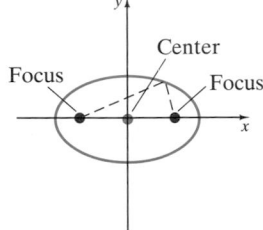

Ellipse with Center $(0, 0)$

The graph of an equation of the form $\dfrac{x^2}{a^2} + \dfrac{y^2}{b^2} = 1$ is an ellipse with center $(0, 0)$.

The x-intercepts are $(a, 0)$ and $(-a, 0)$, and the y-intercepts are $(0, b)$, and $(0, -b)$.

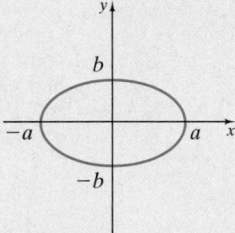

The **standard form** of an ellipse with center $(0, 0)$ is $\dfrac{x^2}{a^2} + \dfrac{y^2}{b^2} = 1$.

EXAMPLE 1

Graph $\dfrac{x^2}{9} + \dfrac{y^2}{16} = 1$.

Solution   The equation is of the form $\dfrac{x^2}{a^2} + \dfrac{y^2}{b^2} = 1$, with $a = 3$ and $b = 4$, so its graph is an ellipse with center $(0, 0)$, x-intercepts $(3, 0)$ and $(-3, 0)$, and y-intercepts $(0, 4)$ and $(0, -4)$.

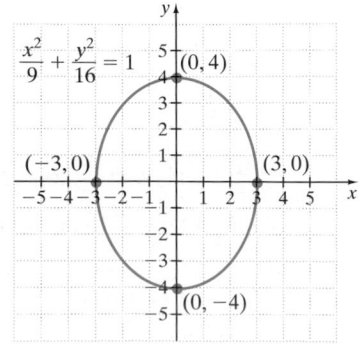

CLASSROOM EXAMPLE

Graph $\dfrac{x^2}{4} + \dfrac{y^2}{9} = 1$.

answer:

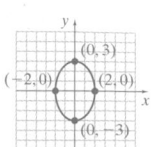

EXAMPLE 2

Graph $4x^2 + 16y^2 = 64$. Use a graphing utility to check.

Solution

Although this equation contains a sum of squared terms in x and y on the same side of an equation, this is not the equation of a circle since the coefficients of x^2 and y^2 are not the same. The graph of this equation is an ellipse. Since the standard form of the equation of an ellipse has 1 on one side, divide both sides of this equation by 64.

$$4x^2 + 16y^2 = 64$$

$$\frac{4x^2}{64} + \frac{16y^2}{64} = \frac{64}{64} \qquad \text{Divide both sides by 64.}$$

$$\frac{x^2}{16} + \frac{y^2}{4} = 1 \qquad \text{Simplify.}$$

We now recognize the equation of an ellipse with $a = 4$ and $b = 2$. This ellipse has center $(0, 0)$, x-intercepts $(4, 0)$ and $(-4, 0)$, and y-intercepts $(0, 2)$ and $(0, -2)$.

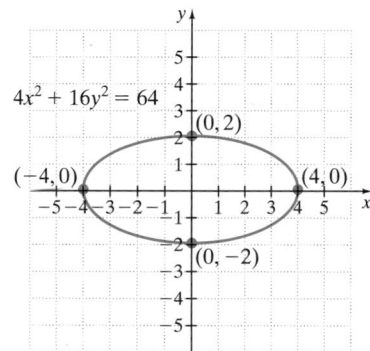

To check using a graphing utility, use the same procedure as for graphing a circle. First, solve the equation for y.

$$4x^2 + 16y^2 = 64$$

$$16y^2 = 64 - 4x^2 \qquad \text{Subtract } 4x^2 \text{ from both sides.}$$

$$y^2 = \frac{64 - 4x^2}{16} \qquad \text{Divide both sides by 16.}$$

$$y = \pm\sqrt{\frac{64 - 4x^2}{16}} \qquad \text{Solve for y.}$$

$$y = \pm\frac{\sqrt{64 - 4x^2}}{4} \qquad \text{Simplify.}$$

Graph $y_1 = \frac{\sqrt{64 - 4x^2}}{4}$ and $y_2 = -y_1$. (Insert two sets of parentheses in the radicand as $\sqrt{((64 - 4x^2)/4)}$ so that the desired graph is obtained.) The graph is shown to the left.

The center of an ellipse is not always $(0, 0)$, as shown in the next example.

CLASSROOM EXAMPLE

Graph $4x^2 + 25y^2 = 100$.

answer:

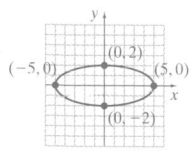

TEACHING TIP

After discussing Example 2, ask, "What would we need to do in order to graph this equation using a graphing calculator?" Solve the equation for y with the class, pointing out that the equation turns into 2 equations when it is solved for y. Then show students how to graph the equations using graphing calculators.

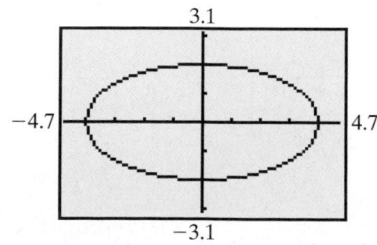

EXAMPLE 3

Graph $\dfrac{(x + 3)^2}{25} + \dfrac{(y - 2)^2}{36} = 1$.

Solution The center of this ellipse is found in a way that is similar to finding the center of a circle. This ellipse has center $(-3, 2)$. Notice that $a = 5$ and $b = 6$. To find four points on the graph of the ellipse, first graph the center, $(-3, 2)$. Since $a = 5$, count 5 units right and then 5 units left of the point with coordinates $(-3, 2)$. Next, since $b = 6$, start at $(-3, 2)$ and count 6 units up and then 6 units down to find two more points on the ellipse.

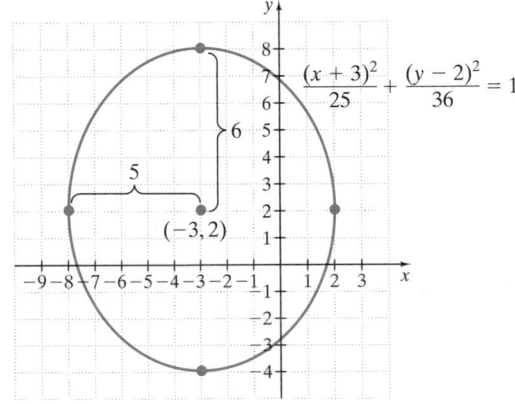

$$\frac{(x + 3)^2}{25} + \frac{(y - 2)^2}{36} = 1$$

CLASSROOM EXAMPLE

Graph
$\dfrac{(x - 1)^2}{9} + \dfrac{(y - 3)^2}{16} = 1$.

answer:

TEACHING TIP

After discussing Example 3, have students expand the equation by clearing the fractions, multiplying the binomials, and collecting like terms. Then have them work in pairs to answer, "If an equation was in this form, what steps would you use to write it in the standard ellipse form?"

✔ **CONCEPT CHECK**

In the graph of the equation $\dfrac{x^2}{64} + \dfrac{y^2}{36} = 1$, which distance is longer: the distance between the x-intercepts or the distance between the y-intercepts? How much longer? Explain.

2 The final conic section is the **hyperbola.** A hyperbola is the set of points in a plane such that the absolute value of the difference of the distances from two fixed points is constant. Each of the two fixed points is called a **focus.** The point midway between the foci is called the **center.**

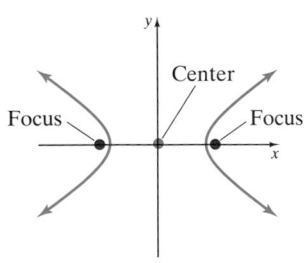

Using the distance formula, we can show that the graph of $\dfrac{x^2}{a^2} - \dfrac{y^2}{b^2} = 1$ is a hyperbola with center $(0, 0)$ and x-intercepts $(a, 0)$ and $(-a, 0)$. Also, the graph of $\dfrac{y^2}{b^2} - \dfrac{x^2}{a^2} = 1$ is a hyperbola with center $(0, 0)$ and y-intercepts $(0, b)$ and $(0, -b)$.

Hyperbola with Center $(0, 0)$

The graph of an equation of the form

$\dfrac{x^2}{a^2} - \dfrac{y^2}{b^2} = 1$ is a hyperbola with center $(0, 0)$

and x-intercepts $(a, 0)$ and $(-a, 0)$.

(continued)

Concept Check Answer:
x-intercepts, by 4 units

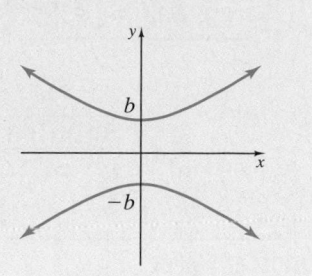

The graph of an equation of the form

$$\frac{y^2}{b^2} - \frac{x^2}{a^2} = 1 \text{ is a hyperbola with center } (0,0)$$

and y-intercepts $(0, b)$ and $(0, -b)$.

The equations $\dfrac{x^2}{a^2} - \dfrac{y^2}{b^2} = 1$ and $\dfrac{y^2}{b^2} - \dfrac{x^2}{a^2} = 1$ are the **standard forms** for the equation of a hyperbola.

> **Helpful Hint**
>
> Notice the difference between the equation of an ellipse and a hyperbola. The equation of the ellipse contains x^2 and y^2 terms on the same side of the equation with same-sign coefficients. For a hyperbola, the coefficients on the same side of the equation have different signs.

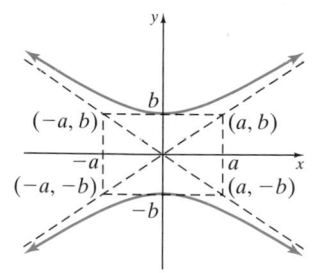

Graphing a hyperbola such as $\dfrac{y^2}{b^2} - \dfrac{x^2}{a^2} = 1$ is made easier by recognizing one of its important characteristics. Examining the figure to the left, notice how the sides of the branches of the hyperbola extend indefinitely and seem to approach the dashed lines in the figure. These dashed lines are called the **asymptotes** of the hyperbola.

To sketch these lines, or asymptotes, draw a rectangle with vertices (a, b), $(-a, b)$, $(a, -b)$, and $(-a, -b)$. The asymptotes of the hyperbola are the extended diagonals of this rectangle.

EXAMPLE 4

Graph $\dfrac{x^2}{16} - \dfrac{y^2}{25} = 1$.

Solution   This equation has the form $\dfrac{x^2}{a^2} - \dfrac{y^2}{b^2} = 1$, with $a = 4$ and $b = 5$. Thus, its graph is a hyperbola that opens to the left and right. It has center $(0, 0)$ and x-intercepts $(4, 0)$ and $(-4, 0)$. To aid in graphing the hyperbola, we first sketch its asymptotes. The extended diagonals of the rectangle with corners $(4, 5)$, $(4, -5)$, $(-4, 5)$, and $(-4, -5)$ are the asymptotes of the hyperbola. Then we use the asymptotes to aid in sketching the hyperbola.

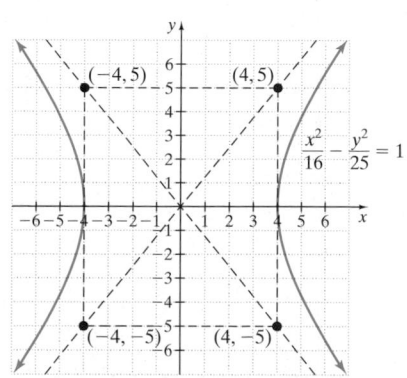

EXAMPLE 5

Graph $4y^2 - 9x^2 = 36$. Use a graphing utility to check.

Solution   Since this is a difference of squared terms in x and y on the same side of the equation, its graph is a hyperbola, as opposed to an ellipse or a circle. The standard form of the equation of a hyperbola has a 1 on one side, so divide both sides of the equation by 36.

CLASSROOM EXAMPLE

Graph $9y^2 - 16x^2 = 144$.

answer:

$$4y^2 - 9x^2 = 36$$

$$\frac{4y^2}{36} - \frac{9x^2}{36} = \frac{36}{36} \qquad \text{Divide both sides by 36.}$$

$$\frac{y^2}{9} - \frac{x^2}{4} = 1 \qquad \text{Simplify.}$$

The equation is of the form $\dfrac{y^2}{b^2} - \dfrac{x^2}{a^2} = 1$, with $a = 2$ and $b = 3$, so the hyperbola is centered at $(0, 0)$ with y-intercepts $(0, 3)$ and $(0, -3)$. The sketch of the hyperbola is shown.

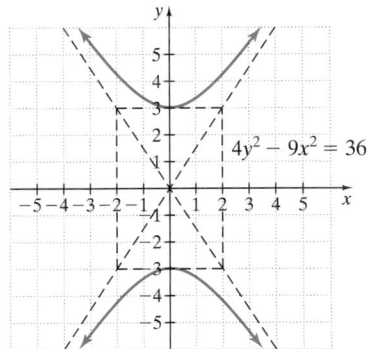

To check, solve the equation for y.

$$4y^2 - 9x^2 = 36$$

$$4y^2 = 36 + 9x^2 \qquad \text{Add } 9x^2 \text{ to both sides.}$$

$$y^2 = \frac{36 + 9x^2}{4} \qquad \text{Divide both sides by 4.}$$

$$y = \pm\sqrt{\frac{36 + 9x^2}{4}} \qquad \text{Solve for y.}$$

$$y = \pm\frac{\sqrt{36 + 9x^2}}{2} \qquad \text{Simplify.}$$

Graph $y_1 = \frac{\sqrt{36 + 9x^2}}{2}$ and $y_2 = -y_1$. The graph is shown below.

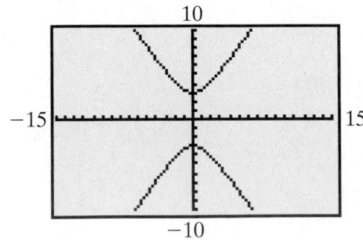

Spotlight on
DECISION
MAKING

Suppose you are an astronomer. You know that the orbits of stars, planets, comets, asteroids, and satellites all have the shape of one of the conic sections. *Eccentricity* is a measure used to describe the shape and elongation of an orbital path. The table shows ranges of eccentricities for the different types of conic sections.

For each of the following comets known to pass through our solar system, decide what type of orbit the comet has based on its eccentricity e. Describe how the shape of the comet's orbit affects how often it passes through our solar system. (For more exercises on eccentricity, see Exercise Set 10.2.)

a. Spacewatch (1997 P2), $e = 1.02851919$

b. Whipple, $e = 0.25871336$

c. Lee (1999 H1), $e = 0.99973749$

d. Giacobini-Zinner, $e = 0.70647162$

e. Tabur (1997 N1), $e = 1.00004712$

Conic Section	Eccentricity e
Circle	$e = 0$
Ellipse	$0 < e < 1$
Parabola	$e = 1$
Hyperbola	$e > 1$

59. A: $\frac{7}{6}$; B: 0; C: $\frac{3}{5}$; D: $\frac{8}{5}$; E: $\frac{8}{9}$; F: 0; G: $\frac{9}{4}$; H: $\frac{1}{6}$ **65.** $\dfrac{x^2}{1.69 \cdot 10^{16}} + \dfrac{y^2}{1.5625 \cdot 10^{16}} = 1$

MENTAL MATH

Identify the graph of each equation as an ellipse or a hyperbola.

1. $\dfrac{x^2}{16} + \dfrac{y^2}{4} = 1$ ellipse

2. $\dfrac{x^2}{16} - \dfrac{y^2}{4} = 1$ hyperbola

3. $x^2 - 5y^2 = 3$ hyperbola

4. $-x^2 + 5y^2 = 3$ hyperbola

5. $-\dfrac{y^2}{25} + \dfrac{x^2}{36} = 1$ hyperbola

6. $\dfrac{y^2}{25} + \dfrac{x^2}{36} = 1$ ellipse

1.–20. See graphing answer section.

EXERCISE SET 10.2

STUDY GUIDE/SSM   CD/VIDEO   PH MATH TUTOR CENTER   MathXL®Tutorials ON CD   MathXL®   MyMathLab®

Sketch the graph of each equation. See Examples 1 and 2.

1. $\dfrac{x^2}{4} + \dfrac{y^2}{25} = 1$

2. $\dfrac{x^2}{9} + y^2 = 1$

3. $\dfrac{x^2}{16} + \dfrac{y^2}{9} = 1$

4. $x^2 + \dfrac{y^2}{4} = 1$

 5. $9x^2 + 4y^2 = 36$

6. $x^2 + 4y^2 = 16$

7. $4x^2 + 25y^2 = 100$

8. $36x^2 + y^2 = 36$

Sketch the graph of each equation. See Example 3.

9. $\dfrac{(x+1)^2}{36} + \dfrac{(y-2)^2}{49} = 1$ **10.** $\dfrac{(x-3)^2}{9} + \dfrac{(y+3)^2}{16} = 1$

11. $\dfrac{(x-1)^2}{4} + \dfrac{(y-1)^2}{25} = 1$ **12.** $\dfrac{(x+3)^2}{16} + \dfrac{(y+2)^2}{4} = 1$

Sketch the graph of each equation. See Examples 4 and 5.

13. $\dfrac{x^2}{4} - \dfrac{y^2}{9} = 1$

14. $\dfrac{x^2}{36} - \dfrac{y^2}{36} = 1$

15. $\dfrac{y^2}{25} - \dfrac{x^2}{16} = 1$

16. $\dfrac{y^2}{25} - \dfrac{x^2}{49} = 1$

17. $x^2 - 4y^2 = 16$

18. $4x^2 - y^2 = 36$

19. $16y^2 - x^2 = 16$

20. $4y^2 - 25x^2 = 100$

21. If you are given a list of equations of circles, parabolas, ellipses, and hyperbolas, explain how you could distinguish the different conic sections from their equations. answers may vary

MIXED PRACTICE **23.–24.** parabola. See graphing answer section.
22. circle. See graphing answer section.
Identify whether each equation, when graphed, will be a parabola, circle, ellipse, or hyperbola. Sketch the graph of each equation.

22. $(x - 7)^2 + (y - 2)^2 = 4$ **23.** $y = x^2 + 4$

24. $y = x^2 + 12x + 36$ **25.** $\dfrac{x^2}{4} + \dfrac{y^2}{9} = 1$

26. $\dfrac{y^2}{9} \quad \dfrac{x^2}{9} - 1$ **27.** $\dfrac{x^2}{16} - \dfrac{y^2}{4} = 1$

28. $\dfrac{x^2}{16} + \dfrac{y^2}{4} = 1$ **29.** $x^2 + y^2 = 16$

30. $x = y^2 + 4y - 1$ **31.** $x = -y^2 + 6y$

32. $9x^2 - 4y^2 = 36$ **33.** $9x^2 + 4y^2 = 36$

34. $\dfrac{(x - 1)^2}{49} + \dfrac{(y + 2)^2}{25} = 1$ **26.–27.** hyperbola. See graphing answer section.

35. $y^2 = x^2 + 16$ **25.** ellipse. See graphing answer section.

36. $\left(x + \dfrac{1}{2}\right)^2 + \left(y - \dfrac{1}{2}\right)^2 = 1$ **28.** ellipse. See graphing answer section.

37. $y = -2x^2 + 4x - 3$ **33.–34.** ellipse. See graphing answer section.

38.–45. See graphing answer section.

Use a graphing utility to graph each conic section.

38. $10x^2 + y^2 = 32$ **39.** $20x^2 + 5y^2 = 100$

40. $2y^2 - 5x^2 = 10$ **41.** $7y^2 - 3x^2 = 21$

42. $7.3x^2 + 15.5y^2 = 95.2$ **43.** $18.8x^2 + 36.1y^2 = 205.8$

44. $4.6x^2 - 3.7y^2 = 70.2$ **45.** $4.5x^2 - 6.7y^2 = 50.7$

32.&35. hyperbola. See graphing answer section.

REVIEW AND PREVIEW

Solve each inequality. See Section 3.3.

30.–31., 37. parabola. See graphing answer section.

46. $x < 5$ and $x < 1$ $(-\infty, 1)$

47. $x < 5$ or $x < 1$ $(-\infty, 5)$

48. $2x - 1 \geq 7$ or $-3x \leq -6$ $[2, \infty)$

54. $\dfrac{x^2}{25} + \dfrac{y^2}{25} = 1$;

49. $2x - 1 \geq 7$ and $-3x \leq -6$ $[4, \infty)$ ellipse; when $a = b$

Perform the indicated operations. See Sections 5.3 and 5.4.

50. $(2x^3)(-4x^2)$ $-8x^5$ **51.** $2x^3 - 4x^3$ $-2x^3$

52. $-5x^2 + x^2$ $-4x^2$ **53.** $(-5x^2)(x^2)$ $-5x^4$

Concept Extensions

29., 36. circle. See graphing answer section.

54. We know that $x^2 + y^2 = 25$ is the equation of a circle. Rewrite the equation so that the right side is equal to 1. Which type of conic section does this equation form resemble? In fact, the circle is a special case of this type of conic section. Describe the conditions under which this type of conic section is a circle. **67.–68.** See graphing answer section.

The orbits of stars, planets, comets, asteroids, and satellites all have the shape of one of the conic sections. Astronomers use a measure called eccentricity to describe the shape and elongation of an orbital

path. For the circle and ellipse, eccentricity e is calculated with the formula $e = \dfrac{c}{d}$, *where* $c^2 = |a^2 - b^2|$ *and d is the larger value of a or b. For a hyperbola, eccentricity e is calculated with the formula* $e = \dfrac{c}{d}$, *where* $c^2 = a^2 + b^2$ *and the value of d is equal to a if the hyperbola has x-intercepts or equal to b if the hyperbola has y-intercepts. Use equations A–H to answer Exercises 55–64.*

A $\dfrac{x^2}{36} - \dfrac{y^2}{13} = 1$ **B** $\dfrac{x^2}{4} + \dfrac{y^2}{4} = 1$ **C** $\dfrac{x^2}{25} + \dfrac{y^2}{16} = 1$

D $\dfrac{y^2}{25} - \dfrac{x^2}{39} = 1$ **E** $\dfrac{x^2}{17} + \dfrac{y^2}{81} = 1$ **F** $\dfrac{x^2}{36} + \dfrac{y^2}{36} = 1$

G $\dfrac{x^2}{16} - \dfrac{y^2}{65} = 1$ **H** $\dfrac{x^2}{144} + \dfrac{y^2}{140} = 1$

55. Identify the type of conic section represented by each of the equations A–H. **58.** A: 6; B: 2; C: 5; D: 5; E: 9; F: 6; G: 4; H: 12

56. For each of the equations A–H, identify the values of a^2 and b^2.

57. For each of the equations A–H, calculate the value of c^2 and c.

58. For each of the equations A–H, find the value of d.

59. For each of the equations A–H, calculate the eccentricity e.

60. What do you notice about the values of e for the equations you identified as ellipses?   greater than 0 and less than 1

61. What do you notice about the values of e for the equations you identified as circles?   equal to 0

62. What do you notice about the values of e for the equations you identified as hyperbolas?   greater than 1

63. The eccentricity of a parabola is exactly 1. Use this information and the observations you made in Exercises 31, 32, and 33 to describe a way that could be used to identify the type of conic section based on its eccentricity value.   answers may vary

64. Graph each of the conic sections given in equations A–H. What do you notice about the shape of the ellipses for increasing values of eccentricity? Which is the most elliptical? Which is the least elliptical, that is, the most circular?

65. A planet's orbit about the Sun can be described as an ellipse. Consider the Sun as the origin of a rectangular coordinate system. Suppose that the x-intercepts of the elliptical path of the planet are ±130,000,000 and that the y-intercepts are ±125,000,000. Write the equation of the elliptical path of the planet.   **64.** answers may vary. See graphing answer section.

66. Comets orbit the Sun in elongated ellipses. Consider the Sun as the origin of a rectangular coordinate system. Suppose that the equation of the path of the comet is
(1,782,000,000, 356,400,000)

$$\dfrac{(x - 1,782,000,000)^2}{3.42 \cdot 10^{23}} + \dfrac{(y - 356,400,000)^2}{1.368 \cdot 10^{22}} = 1$$

Find the center of the path of the comet.

67. Use a graphing calculator to verify Exercise 5.

68. Use a graphing calculator to verify Exercise 6.

55. ellipses: C, E, H; circles: B, F; hyperbolas: A, D, G   **56.** A: 36, 13; B: 4, 4; C: 25, 16; D: 39, 25; E: 17, 81; F: 36, 36; G: 16, 65; H: 144, 140
57. A: 49, 7; B: 0, 0; C: 9, 3; D: 64, 8; E: 64, 8; F: 0, 0; G: 81, 9; H: 4, 2

For Exercises 69 through 74, see the example below.

Example

Sketch the graph of $\dfrac{(x-2)^2}{25} - \dfrac{(y-1)^2}{9} = 1.$

Solution

This hyperbola has center $(2, 1)$. Notice that $a = 5$ and $b = 3$.

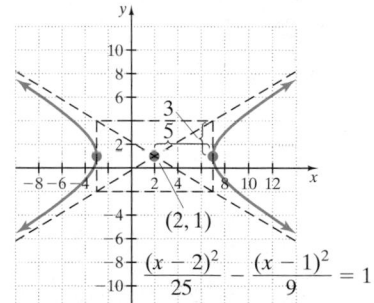

Sketch the graph of each equation. **69.–74.** See graphing answer section.

69. $\dfrac{(x-1)^2}{4} - \dfrac{(y+1)^2}{25} = 1$ **70.** $\dfrac{(x+2)^2}{9} - \dfrac{(y-1)^2}{4} = 1$

71. $\dfrac{y^2}{16} - \dfrac{(x+3)^2}{9} = 1$ **72.** $\dfrac{(y+4)^2}{4} - \dfrac{x^2}{25} = 1$

73. $\dfrac{(x+5)^2}{16} - \dfrac{(y+2)^2}{25} = 1$ **74.** $\dfrac{(x-3)^2}{9} - \dfrac{(y-2)^2}{4} = 1$

INTEGRATED REVIEW   GRAPHING CONIC SECTIONS

Following is a summary of conic sections.

Conic Sections

	Standard Form	**Graph**
Parabola	$y = a(x-h)^2 + k$	
Parabola	$x = a(y-k)^2 + h$	
Circle	$(x-h)^2 + (y-k)^2 = r^2$	
Ellipse	$\dfrac{x^2}{a^2} + \dfrac{y^2}{b^2} = 1$	
Hyperbola	$\dfrac{x^2}{a^2} - \dfrac{y^2}{b^2} = 1$	

(continued)

1., 8., 15. circle   2.–3. parabola   4.,7. ellipse
5.–6., 11. hyperbola 9.–10. parabola 12.–13. ellipse   14. hyperbola

Hyperbola	$\dfrac{y^2}{b^2} - \dfrac{x^2}{a^2} = 1$	

Identify whether each equation, when graphed, will be a parabola, circle, ellipse, or hyperbola. Then graph each equation.   See graphing answer section.

1. $(x - 7)^2 + (y - 2)^2 = 4$ **2.** $y = x^2 + 4$ **3.** $y = x^2 + 12x + 36$

4. $\dfrac{x^2}{4} + \dfrac{y^2}{9} = 1$ **5.** $\dfrac{y^2}{9} - \dfrac{x^2}{9} = 1$ **6.** $\dfrac{x^2}{16} - \dfrac{y^2}{4} = 1$

7. $\dfrac{x^2}{16} + \dfrac{y^2}{4} = 1$ **8.** $x^2 + y^2 = 16$ **9.** $x = y^2 + 4y - 1$

10. $x = -y^2 + 6y$ **11.** $9x^2 - 4y^2 = 36$ **12.** $9x^2 + 4y^2 = 36$

13. $\dfrac{(x - 1)^2}{49} + \dfrac{(y + 2)^2}{25} = 1$ **14.** $y^2 = x^2 + 16$ **15.** $\left(x + \dfrac{1}{2}\right)^2 + \left(y - \dfrac{1}{2}\right)^2 = 1$

10.3   SOLVING NONLINEAR SYSTEMS OF EQUATIONS

Objectives

1 Solve a nonlinear system by substitution.

2 Solve a nonlinear system by elimination.

   In Section 4.1, we used graphing, substitution, and elimination methods to find solutions of systems of linear equations in two variables. We now apply these same methods to nonlinear systems of equations in two variables. A **nonlinear system of equations** is a system of equations at least one of which is not linear. Since we will be graphing the equations in each system, we are interested in real number solutions only.

1   First, nonlinear systems are solved by the substitution method.

EXAMPLE 1

Solve the system.

$$\begin{cases} y = \sqrt{x} \\ x^2 + y^2 = 6 \end{cases}$$

This system is ideal for substitution since y is expressed in terms of x in the first equation. Notice that if $y = \sqrt{x}$, then both x and y must be nonnegative if they are real numbers. Substitute $\sqrt{x}$ for y in the second equation, and solve for x.

$$x^2 + y^2 = 6$$
$$x^2 + \left(\sqrt{x}\right)^2 = 6 \qquad \text{Let } y = \sqrt{x}.$$
$$x^2 + x = 6$$
$$x^2 + x - 6 = 0$$
$$(x + 3)(x - 2) = 0$$
$$x = -3 \quad \text{or} \quad x = 2$$

The solution -3 is discarded because we have noted that x must be nonnegative. To see this, let $x = -3$ in the first equation. Then let $x = 2$ in the first equation to find a corresponding y-value.

$$\text{Let } x = -3. \qquad\qquad\qquad \text{Let } x = 2.$$
$$y = \sqrt{x} \qquad\qquad\qquad\qquad y = \sqrt{x}$$
$$y = \sqrt{-3} \quad \text{Not a real number} \qquad y = \sqrt{2}$$

Since we are interested only in real number solutions, the only solution is $\left(2, \sqrt{2}\right)$. Check to see that this solution satisfies both equations. The graph of each equation in

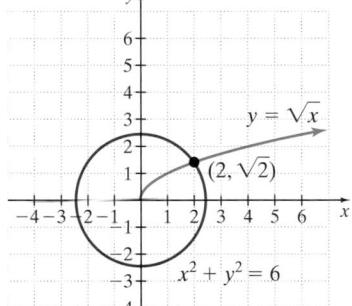

the system is shown next.

EXAMPLE 2

Solve the system. Use a graphical approach to check.
$$\begin{cases} x^2 - 3y = 1 \\ x - y = 1 \end{cases}$$

Solution

We can solve this system by substitution if we solve one equation for one of the variables. Solving the first equation for x is not the best choice since doing so introduces a radical. Also, solving for y in the first equation introduces a fraction. We solve the second equation for y.

$$x - y = 1 \qquad \text{Second equation}$$
$$x - 1 = y \qquad \text{Solve for } y.$$

Replace y with $x - 1$ in the first equation, and then solve for x.

$$x^2 - 3y = 1 \qquad \text{First equation}$$

$$x^2 - 3\overbrace{(x - 1)} = 1 \qquad \text{Replace } y \text{ with } x - 1.$$

$$x^2 - 3x + 3 = 1$$

$$x^2 - 3x + 2 = 0$$

$$(x - 2)(x - 1) = 0$$

$$x = 2 \quad \text{or} \quad x = 1$$

Let $x = 2$ and then let $x = 1$ in the equation $y = x - 1$ to find corresponding y-values.

$$\text{Let } x = 2. \qquad\qquad \text{Let } x = 1.$$
$$y = x - 1 \qquad\qquad y = x - 1$$
$$y = 2 - 1 = 1 \qquad\qquad y = 1 - 1 = 0$$

The solutions are $(2, 1)$ and $(1, 0)$ or the solution set is $\{(2, 1), (1, 0)\}$.
To check these solutions, graph each equation of the system. Since we are using a graphing utility, we first solve each equation for y.

$$x^2 - 3y = 1 \qquad\qquad\qquad\qquad x - y = 1$$

$$x^2 - 1 = 3y$$

$$y = \frac{x^2 - 1}{3} \quad \text{Solve for y.} \qquad\qquad y = x - 1 \quad \text{Solve for y.}$$

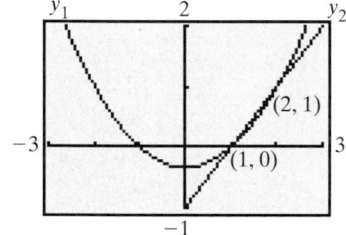

The graph of $y_1 = \frac{x^2 - 1}{3}$ and $y_2 = x - 1$ is shown to the left with the points of inter-section noted. Since the coordinates of the points of intersection are the same as the ordered pair solutions found above, the solutions check.

EXAMPLE 3

Solve the system.

$$\begin{cases} x^2 + y^2 = 4 \\ x + y = 3 \end{cases}$$

Solution We use the substitution method and solve the second equation for x.

$$x + y = 3 \qquad \text{Second equation}$$
$$x = 3 - y$$

Now we let $x = 3 - y$ in the first equation.

$$x^2 + y^2 = 4 \qquad \text{First equation}$$

$$(3 - y)^2 + y^2 = 4 \qquad \text{Let } x = 3 - y.$$

$$9 - 6y + y^2 + y^2 = 4$$

$$2y^2 - 6y + 5 = 0$$

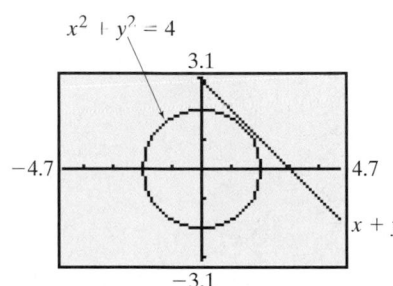

$x^2 + y^2 = 4$

By the quadratic formula, where $a = 2$, $b = -6$, and $c = 5$, we have

$$y = \frac{6 \pm \sqrt{(-6)^2 - 4 \cdot 2 \cdot 5}}{2 \cdot 2} = \frac{6 \pm \sqrt{-4}}{4}$$

Since $\sqrt{-4}$ is not a real number, there is no real solution, or $\emptyset$.

To check graphically, notice that $x^2 + y^2 = 4$ solved for y is $y = \pm\sqrt{4 - x^2}$, and $x + y = 3$ solved for y is $y = 3 - x$. Graph $y_1 = \sqrt{4 - x^2}$, $y_2 = -y_1$, and $y_3 = 3 - x$ in a decimal window.

The graph of the circle and the line do not intersect, as expected. The system has no solution.

✔ CONCEPT CHECK

Without solving, how can you tell that $x^2 + y^2 = 9$ and $x^2 + y^2 = 16$ do not have any points of intersection?

 2   Some nonlinear systems may be solved by the elimination method.

EXAMPLE 4

Solve the system.

$$\begin{cases} x^2 + 2y^2 = 10 \\ x^2 - y^2 = 1 \end{cases}$$

Solution

We will use the elimination, or addition, method to solve this system. To eliminate x^2 when we add the two equations, multiply both sides of the second equation by -1. Then

$$\begin{cases} x^2 + 2y^2 = 10 \\ (-1)(x^2 - y^2) = -1 \cdot 1 \end{cases} \text{ is equivalent to } \begin{cases} x^2 + 2y^2 = 10 \\ -x^2 + y^2 = -1 \end{cases}$$

$$\begin{aligned} 3y^2 &= 9 \quad \text{Add.} \\ y^2 &= 3 \quad \text{Divide both} \\ y &= \pm\sqrt{3} \quad \text{sides by 3.} \end{aligned}$$

To find the corresponding x-values, we let $y = \sqrt{3}$ and $y = -\sqrt{3}$ in either original equation. We choose the second equation.

Let $y = \sqrt{3}$.	Let $y = -\sqrt{3}$.
$x^2 - y^2 = 1$	$x^2 - y^2 = 1$
$x^2 - \left(\sqrt{3}\right)^2 = 1$	$x^2 - \left(-\sqrt{3}\right)^2 = 1$
$x^2 - 3 = 1$	$x^2 - 3 = 1$
$x^2 = 4$	$x^2 = 4$
$x = \pm\sqrt{4} = \pm 2$	$x = \pm\sqrt{4} = \pm 2$

TEACHING TIP

After discussing the solution to Example 4, have students work in pairs to solve it using substitution.

The solutions are $\left(2, \sqrt{3}\right), \left(-2, \sqrt{3}\right), \left(2, -\sqrt{3}\right),$ and $\left(-2, -\sqrt{3}\right).$ Check all four ordered pairs in both equations of the system. The graph of each equation in this system is shown below.

1. $(3, -4), (-3, 4)$
2. $(-4, 3), (4, -3)$
3. $\left(\sqrt{2}, \sqrt{2}\right), \left(-\sqrt{2}, -\sqrt{2}\right)$
4. $\left(-\sqrt{2}, -\sqrt{2}\right), \left(\sqrt{2}, \sqrt{2}\right)$
5. $(4, 0), (0, -2)$ 6. $(-2, 0), (0, -2)$
11. $(1, -2), (3, 6)$ 12. $(-1, 0)$
7. $\left(-\sqrt{5}, -2\right), \left(-\sqrt{5}, 2\right), \left(\sqrt{5}, -2\right), \left(\sqrt{5}, 2\right)$ 8. $\left(2\sqrt{2}, 1\right), \left(-2\sqrt{2}, 1\right), \left(2\sqrt{2}, -1\right), \left(-2\sqrt{2}, -1\right)$ 38. $(8x^2 + 4x + 8)$ ft

EXERCISE SET 10.3

STUDY GUIDE/SSM   CD/VIDEO   PH MATH TUTOR CENTER   MathXL®Tutorials ON CD   MathXL®   MyMathLab®

13. $(2, 4), (-5, 25)$ 14. $\left(-\dfrac{1}{6}, -6\right), (1, 1)$

25. $\left(\sqrt{3}, 0\right), \left(-\sqrt{3}, 0\right)$

MIXED PRACTICE

19. $(-1, -2), (-1, 2), (1, -2), (1, 2)$

Solve each nonlinear system of equations for real solutions. See Examples 1 through 4.   24. $(-4, -1), (-4, 1)$

1. $\begin{cases} x^2 + y^2 = 25 \\ 4x + 3y = 0 \end{cases}$ 2. $\begin{cases} x^2 + y^2 = 25 \\ 3x + 4y = 0 \end{cases}$

3. $\begin{cases} x^2 + 4y^2 = 10 \\ y = x \end{cases}$ 4. $\begin{cases} 4x^2 + y^2 = 10 \\ y = x \end{cases}$

5. $\begin{cases} y^2 = 4 - x \\ x - 2y = 4 \end{cases}$ 6. $\begin{cases} x^2 + y^2 = 4 \\ x + y = -2 \end{cases}$

7. $\begin{cases} x^2 + y^2 = 9 \\ 16x^2 - 4y^2 = 64 \end{cases}$ 8. $\begin{cases} 4x^2 + 3y^2 = 35 \\ 5x^2 + 2y^2 = 42 \end{cases}$

9. $\begin{cases} x^2 + 2y^2 = 2 \\ x - y = 2 \end{cases} \varnothing$ 10. $\begin{cases} x^2 + 2y^2 = 2 \\ x^2 - 2y^2 = 6 \end{cases} \varnothing$

11. $\begin{cases} y = x^2 - 3 \\ 4x - y = 6 \end{cases}$ 12. $\begin{cases} y = x + 1 \\ x^2 - y^2 = 1 \end{cases}$

13. $\begin{cases} y = x^2 \\ 3x + y = 10 \end{cases}$ 14. $\begin{cases} 6x - y = 5 \\ xy = 1 \end{cases}$

15. $\begin{cases} y = 2x^2 + 1 \\ x + y = -1 \end{cases} \varnothing$ 16. $\begin{cases} x^2 + y^2 = 9 \\ x + y = 5 \end{cases} \varnothing$

17. $\begin{cases} y = x^2 - 4 \\ y = x^2 - 4x \end{cases} (1, -3)$ 18. $\begin{cases} x = y^2 - 3 \\ x = y^2 - 3y \end{cases} (-2, 1)$

19. $\begin{cases} 2x^2 + 3y^2 = 14 \\ -x^2 + y^2 = 3 \end{cases}$ 20. $\begin{cases} 4x^2 - 2y^2 = 2 \\ -x^2 + y^2 = 2 \end{cases}$

21. $\begin{cases} x^2 + y^2 = 1 \\ x^2 + (y + 3)^2 = 4 \end{cases}$ 22. $\begin{cases} x^2 + 2y^2 = 4 \\ x^2 - y^2 = 4 \end{cases}$

23. $\begin{cases} y = x^2 + 2 \\ y = -x^2 + 4 \end{cases}$ 24. $\begin{cases} x = -y^2 - 3 \\ x = y^2 - 5 \end{cases}$

20. $\left(\sqrt{3}, \sqrt{5}\right), \left(\sqrt{3}, -\sqrt{5}\right), \left(-\sqrt{3}, \sqrt{5}\right), \left(-\sqrt{3}, -\sqrt{5}\right)$

25. $\begin{cases} 3x^2 + y^2 = 9 \\ 3x^2 - y^2 = 9 \end{cases}$ 26. $\begin{cases} x^2 + y^2 = 25 \\ x = y^2 - 5 \end{cases}$

27. $\begin{cases} x^2 + 3y^2 = 6 \\ x^2 - 3y^2 = 10 \end{cases} \varnothing$ 28. $\begin{cases} x^2 + y^2 = 1 \\ y = x^2 - 9 \end{cases} \varnothing$

29. $\begin{cases} x^2 + y^2 = 36 \\ y = \dfrac{1}{6}x^2 - 6 \end{cases}$ 30. $\begin{cases} x^2 + y^2 = 16 \\ y = -\dfrac{1}{4}x^2 + 4 \end{cases}$

21. $(0, -1)$ 22. $(-2, 0), (2, 0)$

REVIEW AND PREVIEW   23. $(-1, 3), (1, 3)$

Graph each inequality in two variables. See Section 3.6.

26. $(-5, 0), (4, -3), (4, 3)$

31. $x > -3$ 32. $y \le 1$

33. $y < 2x - 1$ 34. $3x - y \le 4$

Find the perimeter of each geometric figure. See Section 5.3.

△ 35.

x inches
$(2x - 5)$ inches
$(5x - 20)$ inches

△ 36.

$(3x + 2)$ centimeters

31.–34. See graphing answer section

△ 37. $(x^2 + 3x + 1)$ meters

x^2 meters

37. $(4x^2 + 6x + 2)$ m

△ 38. $2x^2$ feet

$4x$ feet $(3x^2 + 1)$ feet

$(3x^2 + 7)$ feet

35. $(8x - 25)$ in.   36. $(12x + 8)$ cm

Concept Extensions

39. How many real solutions are possible for a system of equations whose graphs are a circle and a parabola? Draw diagrams to illustrate each possibility.   0, 1, 2, 3, or 4; answers may vary

29. $(-6, 0), (6, 0), (0, -6)$ 30. $(0, 4), (-4, 0), (4, 0)$

40. How many real solutions are possible for a system of equations whose graphs are an ellipse and a line? Draw diagrams to illustrate each possibility.   0, 1, or 2; answers may vary

41. The sum of the squares of two numbers is 130. The difference of the squares of the two numbers is 32. Find the two numbers.   9 and 7; 9 and -7; -9 and 7; -9 and -7

42. The sum of the squares of two numbers is 20. Their product is 8. Find the two numbers.   2 and 4; -2 and -4

△ **43.** During the development stage of a new rectangular keypad for a security system, it was decided that the area of the rectangle should be 285 square centimeters and the perimeter should be 68 centimeters. Find the dimensions of the keypad.

△ **44.** A rectangular holding pen for cattle is to be designed so that its perimeter is 92 feet and its area is 525 feet. Find the dimensions of the holding pen.   21 ft by 25 ft

Recall that in business, a demand function expresses the quantity of a commodity demanded as a function of the commodity's unit

43. 15 cm by 19 cm

*price. A supply function expresses the quantity of a commodity supplied as a function of the commodity's unit price. When the quantity produced and supplied is equal to the quantity demanded, then we have what is called **market equilibrium**.*

45. The demand function for a certain compact disc is given by the function   15 thousand compact discs; price, $3.75

$$p = -0.01x^2 - 0.2x + 9$$

and the corresponding supply function is given by

$$p = 0.01x^2 - 0.1x + 3$$

where p is in dollars and x is in thousands of units. Find the equilibrium quantity and the corresponding price by solving the system consisting of the two given equations.

46. The demand function for a certain style of picture frame is given by the function

$$p = -2x^2 + 90$$

and the corresponding supply function is given by

$$p = 9x + 34$$

where p is in dollars and x is in thousands of units. Find the equilibrium quantity and the corresponding price by solving the system consisting of the two given equations.   3.5 thousand frames; price, $65.50

Use a graphing calculator to verify the results of each exercise.
47.–50. See graphing answer section.

47. Exercise 3.	**48.** Exercise 4.
49. Exercise 23.	**50.** Exercise 24.

STUDY SKILLS REMINDER

Are You Satisfied with Your Performance on a Particular Quiz or Exam?

If not, don't forget to analyze your quiz or exam and look for common errors.

Were most of your errors a result of

▶ *Carelessness?* If your errors were careless, did you turn in your work before the allotted time expired? If so, resolve next time to use the entire time allotted. Any extra time can be spent checking your work.

▶ *Running out of time?* If so, make a point to better manage your time on your next exam. A few suggestions are to work any questions that you are unsure of last and to check your work after all questions have been answered.

▶ *Not understanding a concept?* If so, review that concept and correct your work. Remember next time to make sure that all concepts on a quiz or exam are understood before the exam.

Objectives

1 Graph a nonlinear inequality.

2 Graph a system of nonlinear inequalities.

Discover the Concept

Graph the circle defined by $x^2 + y^2 = 100$ in an integer window by graphing $y_1 = \sqrt{100 - x^2}$ and $y_2 = -\sqrt{100 - x^2}$ or $y_2 = -y_1$.

a. Move the cursor to several points in the interior of the circle. For each ordered pair of numbers displayed, compare the value of the expression $x^2 + y^2$ to 100.

b. Move the cursor to several points in the exterior of the circle. For each ordered pair of numbers displayed, compare the value of the expression $x^2 + y^2$ to 100.

c. What region do you think corresponds to ordered pair solutions of $x^2 + y^2 < 100$, and what region do you think corresponds to ordered pair solutions of $x^2 + y^2 > 100$?

1 From the Discover the Concept above, we see that the circle divides the plane into two regions. All points in the interior of the circle correspond to ordered pair solutions of $x^2 + y^2 < 100$. All points in the exterior of the circle correspond to ordered pair solutions of $x^2 + y^2 > 100$. Examples are shown on the screens below. Don't forget that all points of the circle correspond to ordered pair solutions of $x^2 + y^2 = 100$.

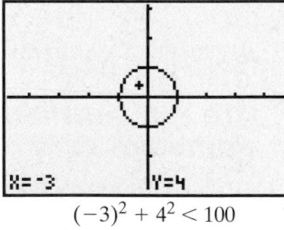

$(-3)^2 + 4^2 < 100$

$13^2 + (-6)^2 > 100$

In general, we can graph a nonlinear equality in two variables in a way similar to the way we graphed a linear inequality in two variables in Section 3.6. First, we graph the related equation. The graph of this equation is our boundary. Then, using test points, we determine and shade the region whose points satisfy the inequality.

EXAMPLE 1

Graph $\dfrac{x^2}{9} + \dfrac{y^2}{16} \leq 1$.

Solution First, graph the equation $\dfrac{x^2}{9} + \dfrac{y^2}{16} = 1$. Sketch a solid curve since the graph of $\dfrac{x^2}{9} + \dfrac{y^2}{16} \leq 1$ includes the graph of $\dfrac{x^2}{9} + \dfrac{y^2}{16} = 1$. The graph is an ellipse, and it divides the plane into two regions, the "inside" and the "outside" of the ellipse. To determine

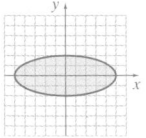

which region contains the solutions, select a test point in either region and determine whether the coordinates of the point satisfy the inequality. We choose $(0, 0)$ as the test point.

$$\frac{x^2}{9} + \frac{y^2}{16} \le 1$$

$$\frac{0^2}{9} + \frac{0^2}{16} \le 1 \qquad \text{Let } x = 0 \text{ and } y = 0.$$

$$0 \le 1 \qquad \text{True.}$$

Since this statement is true, the solution set is the region containing $(0, 0)$. The graph of the solution set includes the points on and inside the ellipse, as shaded in the figure.

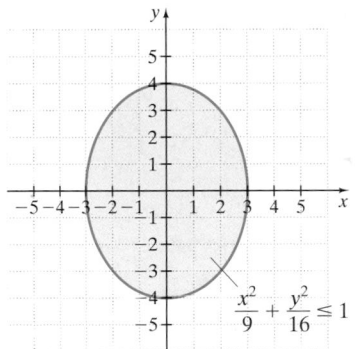

EXAMPLE 2

Use a graphing utility to graph $y \ge x^2 - 3$.

Solution   Graph the related equation $y_1 = x^2 - 3$ and shade the region above the boundary curve since the inequality symbol is $\ge$.

On some graphing calculators, the graph style can be changed in the Y= editor.

The solution region of the inequality $y \ge x^2 - 3$ is the region consisting of all points on the parabola itself and above the parabola.

EXAMPLE 3

Graph $4y^2 > x^2 + 16$.

Solution   The related equation is $4y^2 = x^2 + 16$. Subtract x^2 from both sides and divide both sides by 16, and we have $\dfrac{y^2}{4} - \dfrac{x^2}{16} = 1$, which is a hyperbola. Graph the hyperbola as a dashed curve since the graph of $4y^2 > x^2 + 16$ does *not* include the graph of $4y^2 = x^2 + 16$. The hyperbola divides the plane into three regions. Select a test point in each region—not on a boundary line—to determine whether that region contains solutions of the inequality.

Test region A with (0, 4)	**Test region B with (0, 0)**	**Test region C with (0, −4)**
$4y^2 > x^2 + 16$	$4y^2 > x^2 + 16$	$4y^2 > x^2 + 16$
$4(4)^2 > 0^2 + 16$	$4(0)^2 > 0^2 + 16$	$4(-4)^2 > 0^2 + 16$
$64 > 16$ True.	$0 > 16$ False.	$64 > 16$ True.

The graph of the solution set includes the shaded regions A and C only, not the boundary.

The solution to Example 3 using a graphing calculator.

2   In Section 3.6 we graphed systems of linear inequalities. Recall that the graph of a system of inequalities is the intersection of the graphs of the inequalities.

EXAMPLE 4

Graph the system.

$$\begin{cases} x \le 1 - 2y \\ y \le x^2 \end{cases}$$

Solution We graph each inequality on the same set of axes. The intersection is shown in the third graph on the following page. It is the darkest shaded (appears purple) region along with its boundary lines. The coordinates of the points of intersection can be found by solving the related system.

$$\begin{cases} x = 1 - 2y \\ y = x^2 \end{cases}$$

The solution to Example 4 using a graphing calculator.

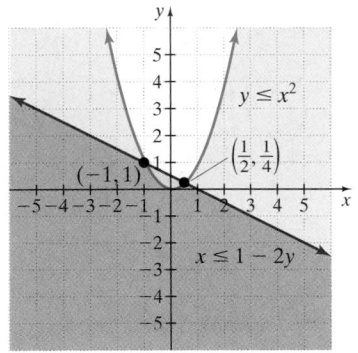

EXAMPLE 5

Graph the system.

$$\begin{cases} x^2 + y^2 < 25 \\ \dfrac{x^2}{9} - \dfrac{y^2}{25} < 1 \\ y < x + 3 \end{cases}$$

Solution   We graph each inequality. The graph of $x^2 + y^2 < 25$ contains points "inside" the circle that has center $(0, 0)$ and radius 5. The graph of $\dfrac{x^2}{9} - \dfrac{y^2}{25} < 1$ is the region between the two branches of the hyperbola with x-intercepts -3 and 3 and center $(0, 0)$. The graph of $y < x + 3$ is the region "below" the line with slope 1 and y-intercept $(0, 3)$. The graph of the solution set of the system is the intersection of all the graphs, the darkest shaded region shown. The boundary of this region is not part of the solution.

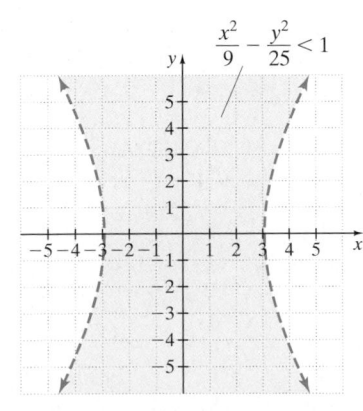

TEACHING TIP

After discussing the solution to Example 5, have students find an inequality which could be added to the system which …
a. would not change the solution.
b. would change the solution to the empty set.

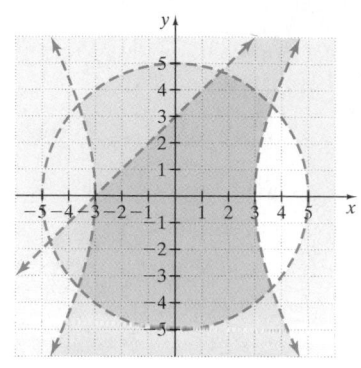

Spotlight on
DECISION
& MAKING

Suppose you are an architect. You have just designed a bridge over a two-lane road. Spanning the road is an arch in the shape of a half-ellipse that is 40 feet wide at the base of the arch and is 15 feet tall at the center of the arch. A colleague has just pointed out that the bridge must have a 13-foot clearance for vehicles on the road. The road is 22 feet wide, and its center line falls directly beneath the highest point of the arch. Decide whether your current bridge design will allow 13-foot-tall vehicles to pass on the road beneath it or if your bridge must be redesigned. Explain your reasoning. (*Hint:* Envision a 13-foot-tall semi truck passing under the bridge in the right-hand lane of the road. Try checking whether points along the top of the truck would fall within the ellipse that defines the arch of the bridge. Remember that the truck can drive within any portion of the right-hand lane.)

EXERCISE SET 10.4

STUDY GUIDE/SSM   CD/VIDEO   PH MATH TUTOR CENTER   MathXL®Tutorials ON CD   MathXL®   MyMathLab®

Graph each inequality. See Examples 1 through 3.

1.–16. See graphing answer section.

1. $y < x^2$

2. $y < -x^2$

3. $x^2 + y^2 \geq 16$

4. $x^2 + y^2 < 36$

5. $\dfrac{x^2}{4} - y^2 < 1$

6. $x^2 - \dfrac{y^2}{9} \geq 1$

7. $y > (x - 1)^2 - 3$

8. $y > (x + 3)^2 + 2$

9. $x^2 + y^2 \leq 9$

10. $x^2 + y^2 > 4$

11. $y > -x^2 + 5$

12. $y < -x^2 + 5$

13. $\dfrac{x^2}{4} + \dfrac{y^2}{9} \leq 1$

14. $\dfrac{x^2}{25} + \dfrac{y^2}{4} \geq 1$

15. $\dfrac{y^2}{4} - x^2 \leq 1$

16. $\dfrac{y^2}{16} - \dfrac{x^2}{9} > 1$

17. $y < (x - 2)^2 + 1$

18. $y > (x - 2)^2 + 1$

19. $y \leq x^2 + x - 2$

20. $y > x^2 + x - 2$

Graph each system. See Examples 4 and 5.

21. $\begin{cases} 2x - y < 2 \\ \quad\quad y \leq -x^2 \end{cases}$

22. $\begin{cases} x - 2y > 4 \\ \quad\quad y > -x^2 \end{cases}$

23. $\begin{cases} 4x + 3y \geq 12 \\ x^2 + y^2 < 16 \end{cases}$

24. $\begin{cases} 3x - 4y \leq 12 \\ x^2 + y^2 < 16 \end{cases}$

25. $\begin{cases} x^2 + y^2 \leq 9 \\ x^2 + y^2 \geq 1 \end{cases}$

26. $\begin{cases} x^2 + y^2 \geq 9 \\ x^2 + y^2 \geq 16 \end{cases}$

27. $\begin{cases} y > x^2 \\ y \geq 2x + 1 \end{cases}$

28. $\begin{cases} y \leq -x^2 + 3 \\ y \leq 2x - 1 \end{cases}$

29. $\begin{cases} x > y^2 \\ y > 0 \end{cases}$

30. $\begin{cases} x < (y + 1)^2 + 2 \\ x + y \geq 3 \end{cases}$

🔒 **31.** $\begin{cases} x^2 + y^2 > 9 \\ \quad\quad y > x^2 \end{cases}$

32. $\begin{cases} x^2 + y^2 \leq 9 \\ \quad\quad y < x^2 \end{cases}$

33. $\begin{cases} \dfrac{x^2}{4} + \dfrac{y^2}{9} \geq 1 \\ x^2 + y^2 \geq 4 \end{cases}$

34. $\begin{cases} x^2 + (y - 2)^2 \geq 9 \\ \dfrac{x^2}{4} + \dfrac{y^2}{25} < 1 \end{cases}$

35. $\begin{cases} x^2 - y^2 \geq 1 \\ \quad\quad y \geq 0 \end{cases}$

36. $\begin{cases} x^2 - y^2 \geq 1 \\ \quad\quad x \geq 0 \end{cases}$

37. $\begin{cases} x + y \geq 1 \\ 2x + 3y < 1 \\ \quad\quad x > -3 \end{cases}$

38. $\begin{cases} x - y < -1 \\ 4x - 3y > 0 \\ \quad\quad y > 0 \end{cases}$

39. $\begin{cases} x^2 - y^2 < 1 \\ \dfrac{x^2}{16} + y^2 \leq 1 \\ \quad\quad x \geq -2 \end{cases}$

40. $\begin{cases} x^2 - y^2 \geq 1 \\ \dfrac{x^2}{16} + \dfrac{y^2}{4} \leq 1 \\ \quad\quad y \geq 1 \end{cases}$

17.–40., 51. See graphing answer section.

REVIEW AND PREVIEW

Determine which graph is the graph of a function. See Section 3.2.

41.

not a function

42.

function

43.

function

44.

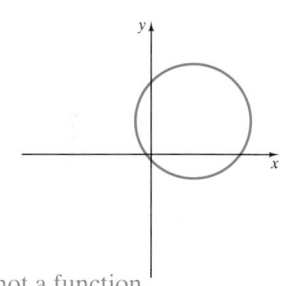

not a function

Find each function value if $f(x) = 3x^2 - 2$. See Section 3.2.

45. $f(-1)$ 1

46. $f(-3)$ 25

47. $f(a)$ $3a^2 - 2$

48. $f(b)$ $3b^2 - 2$

Concept Extensions

49. Discuss how graphing a linear inequality such as $x + y < 9$ is similar to graphing a nonlinear inequality such as $x^2 + y^2 < 9$. answers may vary

50. Discuss how graphing a linear inequality such as $x + y < 9$ is different from graphing a nonlinear inequality such as $x^2 + y^2 < 9$. answers may vary

51. Graph the system. $\begin{cases} y \leq x^2 \\ y \geq x + 2 \\ x \geq 0 \\ y \geq 0 \end{cases}$

CHAPTER 10 PROJECT

Modeling Conic Sections

In this project, you will have the opportunity to construct and investigate a model of an ellipse. You will need two thumbtacks or nails, graph paper, cardboard, tape, string, a pencil, and a ruler. This project may be completed by working in groups or individually.

Follow these steps, answering any questions as you go.

1. Draw an *x*-axis and a *y*-axis on the graph paper as shown in Figure 1.

2. Place the graph paper on the cardboard and attach it with tape.

3. Locate two points on the *x*-axis, each about $1\frac{1}{2}$ inches from the origin and on opposite sides of the origin (see Figure 1). Insert thumbtacks (or nails) at each of these locations.

4. Fasten a 9-inch piece of string to the thumbtacks as shown in Figure 2. Use your pencil to draw and keep the string taut while you carefully move the pencil in a path all around the thumbtacks.

5. Using the grid of the graph paper as a guide, find an approximate equation of the ellipse you drew.

6. Experiment by moving the tacks closer together or farther apart and drawing new ellipses. What do you observe?

7. Write a paragraph explaining why the figure drawn by the pencil is an ellipse. How might you use the same materials to draw a circle?

8. (Optional) Choose one of the ellipses you drew with the string and pencil. Use a ruler to draw any six tangent lines to the ellipse. (A line is tangent to the ellipse if it intersects, or just touches, the ellipse at only one point. See Figure 3.) Extend the tangent lines to yield six points of intersection among the tangents. Use a straightedge to draw a line connecting each pair of opposite points of intersection. What do you observe? Repeat with a different ellipse. Can you make a conjecture about the relationship among the lines that connect opposite points of intersection?

Figure 1

Figure 2

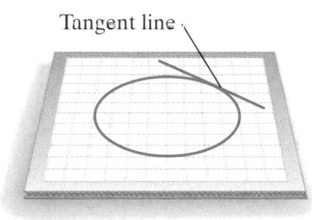

Figure 3

STUDY SKILLS REMINDER

Are You Preparing for a Test on Chapter 10?

Below I have listed some common trouble areas for students in Chapter 10. After studying for your test—but before taking your test—read these.

▶ Don't forget to review all the standard forms for the conic sections.

▶ Remember that the midpoint of a segment is a *point*. The x-coordinate is the average of the x-coordinates of the endpoints of the segment and the y-coordinate is the average of the y-coordinates of the endpoints of the segment.

The midpoint of the segment joining $(-1, 5)$ and $(3, 4)$ is $\left(\dfrac{-1 + 3}{2}, \dfrac{5 + 4}{2}\right)$ or $\left(1, \dfrac{9}{2}\right)$.

▶ Remember that the distance formula gives the *distance* between two points.

The distance between $(-1, 5)$ and $(3, 4)$ is

$$\sqrt{(3 - (-1))^2 + (4 - 5)^2} = \sqrt{4^2 + (-1)^2}$$
$$= \sqrt{16 + 1} = \sqrt{17} \text{ units}$$

▶ Don't forget that both methods, substitution and elimination, are available for solving nonlinear systems of equations.

$$\begin{cases} x^2 + y^2 = 7 \\ 2x^2 - 3y^2 = 4 \end{cases} \text{ is equivalent to } \begin{cases} 3x^2 + 3y^2 = 21 \\ 2x^2 - 3y^2 = 4 \end{cases}$$
$$\overline{ 5x^2 = 25}$$
$$x^2 = 5$$
$$x = \pm\sqrt{5}$$

Let $x = \pm\sqrt{5}$ in either original equation, and $y = \pm\sqrt{2}$, the solution set is $\{(\sqrt{5}, \sqrt{2}), (-\sqrt{5}, \sqrt{2}), (\sqrt{5}, -\sqrt{2}), (-\sqrt{5}, -\sqrt{2})\}$.

Remember: This is simply a checklist of common trouble areas. For a review of Chapter 10, see the Highlights and Chapter Review at the end of this chapter.

10 CHAPTER VOCABULARY CHECK

Fill in each blank with one of the words or phrases listed below.

circle	midpoint	radius	distance
center	ellipse	hyperbola	nonlinear system of equations

1. The <u>distance</u> formula is $d = \sqrt{(x_2 - x_1)^2 + (y_2 - y_1)^2}$.

2. A(n) <u>circle</u> is the set of all points in a plane that are the same distance from a fixed point, called the <u>center</u>.

3. A <u>nonlinear system of equations</u> is a system of equations at least one of which is not linear.

4. A(n) <u>ellipse</u> is the set of points on a plane such that the sum of the distances of those points from two fixed points is a constant.

5. In a circle, the distance from the center to a point of the circle is called its <u>radius</u>.

6. A(n) <u>hyperbola</u> is the set of points in a plane such that the absolute value of the difference of the distance from two fixed points is constant.

7. The <u>midpoint</u> formula is $\left(\dfrac{x_1 + x_2}{2}, \dfrac{y_1 + y_2}{2} \right)$.

CHAPTER 10 HIGHLIGHTS

Definitions and Concepts	**Examples**
Section 10.1 The Parabola and the Circle	

Parabolas $y = a(x - h)^2 + k$ 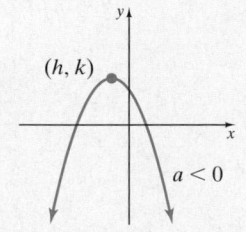	Graph $$x = 3y^2 - 12y + 13.$$ $$x - 13 = 3y^2 - 12y$$ $$x - 13 + 3(4) = 3(y^2 - 4y + 4) \quad \text{Add } 3(4)$$ $$x = 3(y - 2)^2 + 1 \quad \text{to both sides.}$$ Since $a = 3$, this parabola opens to the right with vertex $(1, 2)$. Its axis of symmetry is $y = 2$. The x-intercept is $(13, 0)$. *(continued)*

Definitions and Concepts	Examples

Section 10.1   The Parabola and the Circle

$$x = a(y - k)^2 + h$$

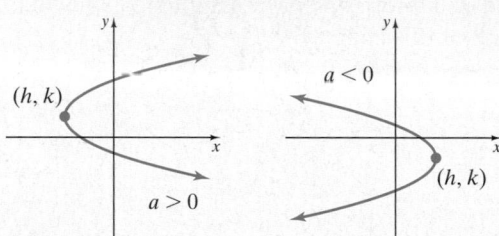

To use a graphing utility to graph an equation, we solve the equation for y.

Graph $x = 3y^2 - 12y + 13$ using a graphing utility.

$$0 = 3y^2 - 12y + (13 - x)$$

$$a = 3, \quad b = -12, \quad c = 13 - x$$

Substitute these values in the quadratic formula.

$$y = \frac{12 \pm \sqrt{144 - 4(3)(13 - x)}}{2(3)}$$

Let y_1 be the expression under the radical. Then *deselect* y_1 and graph y_2 and y_3 as shown.

$$y_1 = 144 - 4(3)(13 - x)$$

$$y_2 = \frac{12 + \sqrt{y_1}}{6}, \; y_3 = \frac{12 - \sqrt{y_1}}{6}$$

Distance formula

The distance d between two points (x_1, y_1) and (x_2, y_2) is given by

$$d = \sqrt{(x_2 - x_1)^2 + (y_2 - y_1)^2}$$

Find the distance between points $(-1, 6)$ and $(-2, -4)$. Let $(x_1, y_1) = (-1, 6)$ and $(x_2, y_2) = (-2, -4)$.

$$\begin{aligned} d &= \sqrt{(x_2 - x_1)^2 + (y_2 - y_1)^2} \\ &= \sqrt{(-2 - (-1))^2 + (-4 - 6)^2} \\ &= \sqrt{1 + 100} = \sqrt{101} \end{aligned}$$

Midpoint formula

The midpoint of the line segment whose endpoints are (x_1, y_1) and (x_2, y_2) is the point with coordinates

$$\left(\frac{x_1 + x_2}{2}, \frac{y_1 + y_2}{2} \right)$$

Find the midpoint of the line segment whose endpoints are $(-1, 6)$ and $(-2, -4)$.

$$\left(\frac{-1 + (-2)}{2}, \frac{6 + (-4)}{2} \right)$$

The midpoint is

$$\left(-\frac{3}{2}, 1 \right)$$

Definitions and Concepts	**Examples**
Circle The graph of $(x - h)^2 + (y - k)^2 = r^2$ is a circle with center (h, k) and radius r. 	Graph $x^2 + (y + 3)^2 = 5$. This equation can be written as $(x - 0)^2 + (y + 3)^2 = 5$ with $h = 0, k = -3$, and $r = \sqrt{5}$. The center of this circle is $(0, -3)$, and the radius is $\sqrt{5}$.

Section 10.2 The Ellipse and the Hyperbola

Ellipse with center $(0, 0)$ The graph of an equation of the form $\dfrac{x^2}{a^2} + \dfrac{y^2}{b^2} = 1$ is an ellipse with center $(0, 0)$. The x-intercepts are $(a, 0)$ and $(-a, 0)$, and the y-intercepts are $(0, b)$ and $(0, -b)$. 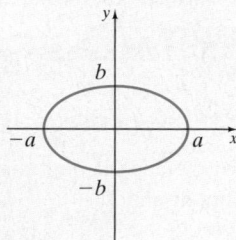	Graph $4x^2 + 9y^2 = 36$. $$\frac{x^2}{9} + \frac{y^2}{4} = 1 \qquad \text{Divide by 36.}$$ $$\frac{x^2}{3^2} + \frac{y^2}{2^2} = 1$$ The ellipse has center $(0, 0)$, x-intercepts $(3, 0)$ and $(-3, 0)$, and y-intercepts $(0, 2)$ and $(0, -2)$. 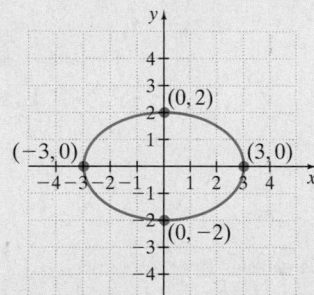
Hyperbola with center $(0, 0)$ The graph of an equation of the form $\dfrac{x^2}{a^2} - \dfrac{y^2}{b^2} = 1$ is a hyperbola with center $(0, 0)$ and x-intercepts $(a, 0)$ and $(-a, 0)$. 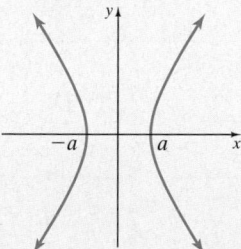 The graph of an equation of the form $\dfrac{y^2}{b^2} - \dfrac{x^2}{a^2} = 1$ is a hyperbola with center $(0, 0)$ and y-intercepts $(0, b)$ and $(0, -b)$. 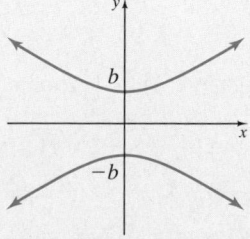	Graph $\dfrac{x^2}{9} - \dfrac{y^2}{4} = 1$. Here $a = 3$ and $b = 2$. 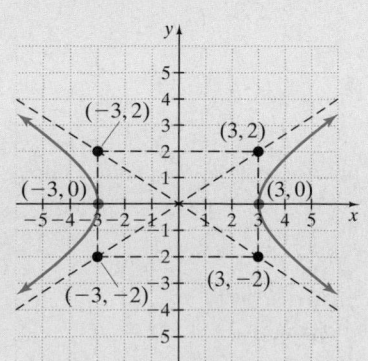

Definitions and Concepts	Examples

Section 10.3 Solving Nonlinear Systems of Equations

A **nonlinear system of equations** is a system of equations at least one of which is not linear. Both the substitution method and the elimination method may be used to solve a nonlinear system of equations.

Solve the nonlinear system $\begin{cases} y = x + 2 \\ 2x^2 + y^2 = 3 \end{cases}$

Substitute $x + 2$ for y in the second equation.

$$2x^2 + y^2 = 3$$
$$2x^2 + (x + 2)^2 = 3$$
$$2x^2 + x^2 + 4x + 4 = 3$$
$$3x^2 + 4x + 1 = 0$$
$$(3x + 1)(x + 1) = 0$$
$$x = -\frac{1}{3}, x = -1$$

If $x = -\frac{1}{3}, y = x + 2 = -\frac{1}{3} + 2 = \frac{5}{3}$.

If $x = -1, y = x + 2 = -1 + 2 = 1$.

The solutions are $\left(-\frac{1}{3}, \frac{5}{3}\right)$ and $(-1, 1)$.

Section 10.4 Nonlinear Inequalities and Systems of Inequalities

The graph of a system of inequalities is the intersection of the graphs of the inequalities.

Graph the system $\begin{cases} x \geq y^2 \\ x + y \leq 4 \end{cases}$

The graph of the system is the pink shaded region along with its boundary lines.

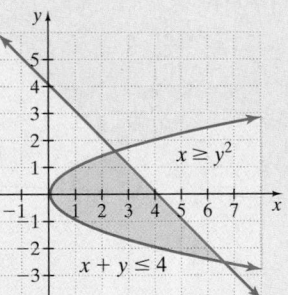

10 CHAPTER REVIEW

(10.1) *Find the distance between each pair of points. For Exercises 7 and 8, round the distance to two decimal places.*

1. $(-6, 3)$ and $(8, 4)$ $\sqrt{197}$ units

2. $(3, 5)$ and $(8, 9)$ $\sqrt{41}$ units

3. $(-4, -6)$ and $(-1, 5)$ $\sqrt{130}$ units

4. $(-1, 5)$ and $(2, -3)$ $\sqrt{73}$ units

5. $\left(-\sqrt{2}, 0\right)$ and $\left(0, -4\sqrt{6}\right)$ $7\sqrt{2}$ units

6. $\left(-\sqrt{5}, -\sqrt{11}\right)$ and $\left(-\sqrt{5}, -3\sqrt{11}\right)$ $2\sqrt{11}$ units

7. $(7.4, -8.6)$ and $(-1.2, 5.6)$ 16.60 units

8. $(2.3, 1.8)$ and $(10.7, -9.2)$ 13.84 units

Find the midpoint of the line segment whose endpoints are given.

9. $(2, 6)$ and $(-12, 4)$ $(-5, 5)$ **10.** $(-3, 8)$ and $(11, 24)$

11. $(-6, -5)$ and $(-9, 7)$ **12.** $(4, -6)$ and $(-15, 2)$

13. $\left(0, -\dfrac{3}{8}\right)$ and $\left(\dfrac{1}{10}, 0\right)$ **14.** $\left(\dfrac{3}{4}, -\dfrac{1}{7}\right)$ and $\left(-\dfrac{1}{4}, -\dfrac{3}{7}\right)$

15. $\left(\sqrt{3}, -2\sqrt{6}\right)$ and $\left(\sqrt{3}, -4\sqrt{6}\right)$ $\left(\sqrt{3}, -3\sqrt{6}\right)$

16. $\left(-5\sqrt{3}, 2\sqrt{7}\right)$ and $\left(-3\sqrt{3}, 10\sqrt{7}\right)$ $\left(-4\sqrt{3}, 6\sqrt{7}\right)$

10. $(4, 16)$ **54.** ellipse **60.** hyperbola

Write an equation of the circle with the given center and radius.

17. center $(-4, 4)$ radius 3 $(x + 4)^2 + (y - 4)^2 = 9$

18. center $(5, 0)$, radius 5 $(x - 5)^2 + y^2 = 25$

19. center $(-7, -9)$, radius $\sqrt{11}$ $(x + 7)^2 + (y + 9)^2 = 11$

20. center $(0, 0)$, radius $\dfrac{7}{2}$ $x^2 + y^2 = \dfrac{49}{4}$ **11.** $\left(-\dfrac{15}{2}, 1\right)$

21.–34. See graphing answer section.

Sketch the graph of the equation. If the graph is a circle, find its center. If the graph is a parabola, find its vertex.

21. $x^2 + y^2 = 7$ **22.** $x = 2(y - 5)^2 + 4$

23. $x = -(y + 2)^2 + 3$ **24.** $(x - 1)^2 + (y - 2)^2 = 4$

25. $y = -x^2 + 4x + 10$ **26.** $x = -y^2 - 4y + 6$

27. $x = \dfrac{1}{2}y^2 + 2y + 1$ **28.** $y = -3x^2 + \dfrac{1}{2}x + 4$

29. $x^2 + y^2 + 2x + y = \dfrac{3}{4}$ **30.** $x^2 + y^2 + 3y = \dfrac{7}{4}$

31. $4x^2 + 4y^2 + 16x + 8y = 1$ **32.** $3x^2 + 6x + 3y^2 = 9$

33. $y = x^2 + 6x + 9$ **34.** $x = y^2 + 6y + 9$

35. Write an equation of the circle centered at $(5.6, -2.4)$ with diameter 6.2. $(x - 5.6)^2 + (y + 2.4)^2 = 9.61$

36.–50. See graphing answer section.

(10.2) Sketch the graph of each equation.

36. $x^2 + \dfrac{y^2}{4} = 1$ **37.** $x^2 - \dfrac{y^2}{4} = 1$

38. $\dfrac{y^2}{4} - \dfrac{x^2}{16} = 1$ **39.** $\dfrac{y^2}{4} + \dfrac{x^2}{16} = 1$

40. $\dfrac{x^2}{5} + \dfrac{y^2}{5} = 1$ **41.** $\dfrac{x^2}{5} - \dfrac{y^2}{5} = 1$

42. $-5x^2 + 25y^2 = 125$ **43.** $4y^2 + 9x^2 = 36$

44. $\dfrac{(x - 2)^2}{4} + (y - 1)^2 = 1$ **45.** $\dfrac{(x + 3)^2}{9} + \dfrac{(y - 4)^2}{25} = 1$

46. $x^2 - y^2 = 1$ **47.** $36y^2 - 49x^2 = 1764$

48. $y^2 = x^2 + 9$ **49.** $x^2 = 4y^2 - 16$

50. $100 - 25x^2 = 4y^2$

Sketch the graph of each equation. Identify whether each equation, when graphed, will be a parabola, circle, ellipse, or hyperbola.
See graphing answer section.

51. $y = x^2 + 4x + 6$ parabola **52.** $y^2 = x^2 + 6$ hyperbola

53. $y^2 + x^2 = 4x + 6$ circle **54.** $y^2 + 2x^2 = 4x + 6$

55. $x^2 + y^2 - 8y = 0$ circle **56.** $x - 4y = y^2$ parabola

57. $x^2 - 4 = y^2$ hyperbola **58.** $x^2 = 4 - y^2$ circle

59. $6(x - 2)^2 + 9(y + 5)^2 = 36$ ellipse

60. $36y^2 = 576 + 16x^2$ **61.** $\dfrac{x^2}{16} - \dfrac{y^2}{25} = 1$ hyperbola

62. $3(x - 7)^2 + 3(y + 4)^2 = 1$ circle

Use a graphing calculator to verify the results of each exercise.
See graphing answer section.

63. Exercise 39. **64.** Exercise 40.

65. Exercise 51. **66.** Exercise 58.

(10.3) Solve each system of equations. **74.** $\left(2, \dfrac{5}{2}\right), (-7, -20)$

67. $\begin{cases} y = 2x - 4 \\ y^2 = 4x \end{cases}$ **68.** $\begin{cases} x^2 + y^2 = 4 \\ x - y = 4 \end{cases}$ $\varnothing$

69. $\begin{cases} y = x + 2 \\ y = x^2 \end{cases}$ $(-1, 1), (2, 4)$ **70.** $\begin{cases} y = x^2 - 5x + 1 \\ y = -x + 6 \end{cases}$

71. $\begin{cases} 4x - y^2 = 0 \\ 2x^2 + y^2 = 16 \end{cases}$ **72.** $\begin{cases} x^2 + 4y^2 = 16 \\ x^2 + y^2 = 4 \end{cases}$

73. $\begin{cases} x^2 + y^2 = 10 \\ 9x^2 + y^2 = 18 \end{cases}$ **74.** $\begin{cases} x^2 + 2y = 9 \\ 5x - 2y = 5 \end{cases}$

75. $\begin{cases} y = 3x^2 + 5x - 4 \\ y = 3x^2 - x + 2 \end{cases}$ $(1, 4)$ **76.** $\begin{cases} x^2 - 3y^2 = 1 \\ 4x^2 + 5y^2 = 21 \end{cases}$

△ **77.** Find the length and the width of a room whose area is 150 square feet and whose perimeter is 50 feet. 15 ft by 10 ft

78. What is the greatest number of real solutions possible for a system of two equations whose graphs are an ellipse and a hyperbola? 4

(10.4) Graph the inequality or system of inequalities.
79.–88. See graphing answer section.

79. $y \leq -x^2 + 3$ **80.** $x^2 + y^2 < 9$

81. $x^2 - y^2 < 1$ **82.** $\dfrac{x^2}{4} + \dfrac{y^2}{9} \geq 1$

83. $\begin{cases} 2x \leq 4 \\ x + y \geq 1 \end{cases}$ **84.** $\begin{cases} 3x + 4y \leq 12 \\ x - 2y > 6 \end{cases}$

85. $\begin{cases} y > x^2 \\ x + y \geq 3 \end{cases}$ **86.** $\begin{cases} x^2 + y^2 \leq 16 \\ x^2 + y^2 \geq 4 \end{cases}$

87. $\begin{cases} x^2 + y^2 < 4 \\ x^2 - y^2 \leq 1 \end{cases}$ **88.** $\begin{cases} x^2 + y^2 < 4 \\ y \geq x^2 - 1 \\ x \geq 0 \end{cases}$

12. $\left(-\dfrac{11}{2}, -2\right)$ **13.** $\left(\dfrac{1}{20}, -\dfrac{3}{16}\right)$ **14.** $\left(\dfrac{1}{4}, -\dfrac{2}{7}\right)$ **67.** $(1, -2), (4, 4)$ **70.** $(5, 1), (-1, 7)$ **71.** $\left(2, 2\sqrt{2}\right), \left(2, -2\sqrt{2}\right)$

72. $(0, 2), (0, -2)$ **73.** $(-1, 3), (-1, -3), (1, 3), (1, -3)$ **76.** $(-2, -1), (-2, 1), (2, -1), (2, 1)$

CHAPTER 10 TEST

Remember to use the Chapter Test Prep Video CD to help you study and view solutions to the test questions you need help with.

1. Find the distance between the points $(-6, 3)$ and $(-8, -7)$. $2\sqrt{26}$ units

2. Find the distance between the points $\left(-2\sqrt{5}, \sqrt{10}\right)$ and $\left(-\sqrt{5}, 4\sqrt{10}\right)$. $\sqrt{95}$ units

3. Find the midpoint of the line segment whose endpoints are $(-2, -5)$ and $(-6, 12)$. $\left(-4, \dfrac{7}{2}\right)$

Sketch the graph of each equation. 4.–11. See graphing answer section.

4. $x^2 + y^2 = 36$

5. $x^2 - y^2 = 36$

6. $16x^2 + 9y^2 = 144$

7. $y = x^2 - 8x + 16$

8. $x^2 + y^2 + 6x = 16$

9. $x = y^2 + 8y - 3$

10. $\dfrac{(x-4)^2}{16} + \dfrac{(y-3)^2}{9} = 1$ 11. $y^2 - x^2 = 1$

Solve each system.

12. $\begin{cases} x^2 + y^2 = 26 \\ x^2 - 2y^2 = 23 \end{cases}$ $(-5, -1), (-5, 1), (5, -1), (5, 1)$

13. $\begin{cases} y = x^2 - 5x + 6 \\ y = 2x \end{cases}$ $(6, 12), (1, 2)$

Graph the solution of each system.   14.–17. See graphing answer section.

14. $\begin{cases} 2x + 5y \geq 10 \\ y \geq x^2 + 1 \end{cases}$

15. $\begin{cases} \dfrac{x^2}{4} + y^2 \leq 1 \\ x + y > 1 \end{cases}$

16. $\begin{cases} x^2 + y^2 \geq 4 \\ x^2 + y^2 < 16 \\ y \geq 0 \end{cases}$

17. A bridge has an arch in the shape of a half-ellipse. If the equation of the ellipse, measured in feet, is $100x^2 + 225y^2 = 22{,}500$, find the height of the arch from the road and the width of the arch.   height, 10 ft; width, 30 ft

CHAPTER CUMULATIVE REVIEW

3. See graphing answer section. (Sec. 2.3) (Ex. 3)

4. 3 (Sec. 2.4)   7. e. $\dfrac{y^{14}}{x^{35}z^7}$

1. Use the associative property of multiplication to write an expression equivalent to $4 \cdot (9y)$. Then simplify the equivalent expression. $(4 \cdot 9)y = 36y$ (Sec. 1.4, Ex. 7)

2. Solve $3x + 4 > 1$ and $2x - 5 \leq 9$. Write the solution in interval notation. $(-1, 7]$ (Sec. 3.3)

3. Graph $x - 3y = 6$ by plotting intercepts.

4. Find the slope of the line that goes through $(3, 2)$ and $(1, -4)$.

5. Use the elimination method to solve the system:
$\begin{cases} 3x + \dfrac{y}{2} = 2 \\ 6x + y = 5 \end{cases}$ $\varnothing$ (Sec. 4.1, Ex. 8)

6. Two planes leave Greensboro, one traveling north and the other south. After 2 hours they are 650 miles apart. If one plane is flying 25 mph faster than the other, what is the speed of each?   150 mph, 175 mph   (Sec. 4.3)

7. Use the power rules to simplify the following. Use positive exponents to write all results.   (Sec. 5.2, Ex. 2)

a. $(5x^2)^3$ $125x^6$

b. $\left(\dfrac{2}{3}\right)^3$ $\dfrac{8}{27}$

c. $\left(\dfrac{3p^4}{q^5}\right)^2$ $\dfrac{9p^8}{q^{10}}$

d. $\left(\dfrac{2^{-3}}{y}\right)^{-2}$ $64y^2$

e. $(x^{-5}y^2z^{-1})^7$

8. Use the quotient rule to simplify.

a. $\dfrac{4^8}{4^3}$ 4^5 b. $\dfrac{y^{11}}{y^5}$ y^6

c. $\dfrac{32x^7}{4x^6}$ $8x$ d. $\dfrac{18a^{12}b^6}{12a^8b^6}$ $\dfrac{3a^4}{2}$ (Sec. 5.1)

9. Solve $2x^2 = \dfrac{17}{3}x + 1$. $-\dfrac{1}{6}, 3$ (Sec. 5.8, Ex. 5)

10. Factor.

a. $3y^2 + 14y + 15$ $(3y + 5)(y + 3)$

b. $20a^5 + 54a^4 + 10a^3$ $2a^3(2a + 5)(5a + 1)$

c. $(y - 3)^2 - 2(y - 3) - 8$ $(y - 1)(y - 7)$ (Sec. 5.6)

11. Perform each indicated operation.

$\dfrac{7}{x - 1} + \dfrac{10x}{x^2 - 1} - \dfrac{5}{x + 1}$ $\dfrac{12}{x - 1}$ (Sec. 6.2, Ex. 6)

12. Perform the indicated operation and simplify if possible.

$$\frac{2}{3a - 15} - \frac{a}{25 - a^2} \quad \frac{5a + 10}{3(a + 5)(a - 5)} \quad \text{(Sec. 6.2)}$$

13. Simplify each complex fraction.   (Sec. 6.3, Ex. 1)

a. $\dfrac{\dfrac{2x}{27y^2}}{\dfrac{6x^2}{9}} \quad \dfrac{1}{9xy^2}$ **b.** $\dfrac{\dfrac{5x}{x + 2}}{\dfrac{10}{x - 2}} \quad \dfrac{x(x - 2)}{2(x + 2)}$ **c.** $\dfrac{\dfrac{x}{y^2} + \dfrac{1}{y}}{\dfrac{y}{x^2} + \dfrac{1}{x}} \quad \dfrac{x^2}{y^2}$

14. Simplify each complex fraction.

a. $(a^{-1} - b^{-1})^{-1} \quad \dfrac{ab}{b - a}$

b. $\dfrac{2 - \dfrac{1}{x}}{4x - \dfrac{1}{x}} \quad \dfrac{1}{2x + 1}$ (Sec. 6.3)

15. Divide $2x^2 - x - 10$ by $x + 2$. $2x - 5$ (Sec. 6.4, Ex. 3)

16. Solve $\dfrac{2}{x + 3} = \dfrac{1}{x^2 - 9} - \dfrac{1}{x - 3} \quad \dfrac{4}{3}$ (Sec. 6.6)

17. Use the remainder theorem and synthetic division to find $P(4)$ if
$P(x) = 4x^6 - 25x^5 + 35x^4 + 17x^2$.   (Sec. 6.5, Ex. 4) 16

18. Suppose that y varies inversely as x. If $y = 3$ when $x = \frac{2}{3}$, find the constant of variation and the direct variation equation.

19. Solve: $\dfrac{2x}{x - 3} + \dfrac{6 - 2x}{x^2 - 9} = \dfrac{x}{x + 3}$. $-6, -1$ (Sec. 6.6, Ex. 5)

20. Simplify the following expressions. Assume that all variables represent nonnegative real numbers.

a. $\sqrt[5]{-32} \quad -2$ **b.** $\sqrt[4]{625} \quad 5$
c. $-\sqrt{36} \quad -6$ **d.** $-\sqrt[3]{-27x^3} \quad 3x$
e. $\sqrt{144y^2} \quad 12y$ (Sec. 7.3)

21. Melissa Scarlatti can clean the house in 4 hours, whereas her husband, Zack, can do the same job in 5 hours. They have agreed to clean together so that they can finish in time to watch a movie on TV that starts in 2 hours. How long will it take them to clean the house together? Can they finish before the movie starts? $2\frac{2}{9}$ hr;   no (Sec. 6.7, Ex. 4)

22. Use the quotient rule to simplify.

a. $\dfrac{\sqrt{32}}{\sqrt{4}} \quad 2\sqrt{2}$ **b.** $\dfrac{\sqrt[3]{240y^2}}{5\sqrt[3]{3y^{-4}}} \quad \dfrac{2y^2\sqrt[3]{10}}{5}$

c. $\dfrac{\sqrt[5]{64x^9y^2}}{\sqrt[5]{2x^2y^{-8}}} \quad 2xy^2\sqrt[5]{x^2}$ (Sec. 7.3)

23. Find the cube roots.   (Sec. 7.1, Ex. 3)

a. $\sqrt[3]{1} \quad 1$ **b.** $\sqrt[3]{-64} \quad -4$

c. $\sqrt[3]{\dfrac{8}{125}} \quad \dfrac{2}{5}$ **d.** $\sqrt[3]{x^6} \quad x^2$

e. $\sqrt[3]{-27x^9} \quad -3x^3$

24. Multiply and simplify if possible.

a. $\sqrt{5}(2 + \sqrt{15})$

b. $(\sqrt{3} - \sqrt{5})(\sqrt{7} - 1)$

c. $(2\sqrt{5} - 1)^2 \quad 21 - 4\sqrt{5}$

d. $(3\sqrt{2} + 5)(3\sqrt{2} - 5) \quad -7$ (Sec. 7.4)

25. Multiply.   (Sec. 7.2, Ex. 5)

a. $z^{2/3}(z^{1/3} - z^5) \quad z - z^{17/3}$ **b.** $(x^{1/3} - 5)(x^{1/3} + 2)$

26. Rationalize the denominator. $\dfrac{-2}{\sqrt{3} + 3} \quad \dfrac{\sqrt{3} - 3}{3}$ (Sec. 7.5)

27. Use the quotient rule to divide, and simplify if possible.

a. $\dfrac{\sqrt{20}}{\sqrt{5}} \quad 2$ (Sec. 7.3, Ex. 5)   **b.** $\dfrac{\sqrt{50x}}{2\sqrt{2}} \quad \dfrac{5}{2}\sqrt{x}$

c. $\dfrac{7\sqrt[3]{48x^4y^8}}{\sqrt[3]{6y^2}} \quad 14xy^2\sqrt[3]{x}$ **d.** $\dfrac{2\sqrt[4]{32a^8b^6}}{\sqrt[4]{a^{-1}b^2}}$

28. Solve $\sqrt{2x - 3} = x - 3$.   6 (Sec. 7.6)

29. Add or subtract as indicated.   (Sec. 7.4, Ex. 3)

a. $\dfrac{\sqrt{45}}{4} - \dfrac{\sqrt{5}}{3} \quad \dfrac{5\sqrt{5}}{12}$ **b.** $\sqrt[3]{\dfrac{7x}{8}} + 2\sqrt[3]{7x} \quad \dfrac{5\sqrt[3]{7x}}{2}$

30. Use the discriminant to determine the number and type of solutions for $9x^2 - 6x = -4$.

31. Rationalize the denominator of $\sqrt{\dfrac{7x}{3y}}$.

32. Solve $\dfrac{4}{x - 2} - \dfrac{x}{x + 2} = \dfrac{16}{x^2 - 4}$ 4 (Sec. 6.6)

33. Solve $\sqrt{2x - 3} = 9$.   42 (Sec. 7.6, Ex. 1)

34. Solve $x^3 + 2x^2 - 4x \geq 8$ $[2, \infty)$ (Sec. 8.4)

35. Find the following powers of i.   (Sec. 7.7, Ex. 6)

a. $i^7 \quad -i$ **b.** $i^{20} \quad 1$
c. $i^{46} \quad -1$ **d.** $i^{-12} \quad 1$

36. Graph $f(x) = (x + 2)^2 - 1$

37. Solve $p^2 + 2p = 4$ by completing the square.

38. Find the maximum value of $f(x) = -x^2 - 6x + 4$.

39. Solve $\dfrac{1}{4}m^2 - m + \dfrac{1}{2} = 0$.

40. Find the inverse of $f(x) = \dfrac{x + 1}{2}$.

41. Solve $p^4 - 3p^2 - 4 = 0$. $2, -2, i, -i$ (Sec. 8.3, Ex. 3)

18. $k = 2; y = \dfrac{2}{x}$ (Sec. 6.8)   **24. a.** $2\sqrt{5} + 5\sqrt{3}$

24. b. $\sqrt{21} - \sqrt{3} - \sqrt{35} + \sqrt{5}$ **25. b.** $x^{2/3} - 3x^{1/3} - 10$

27d. $4a^2b\sqrt[4]{2a}$ **30.** two complex but not real solutions (Sec. 8.3)   **31.** $\dfrac{\sqrt{21xy}}{3y}$ (Sec. 7.5, Ex. 2)

36. See graphing answer section. (Sec. 8.5)   **37.** $-1 + \sqrt{5}, -1 - \sqrt{5}$ (Sec. 8.1, Ex. 5)   **38.** 13 (Sec. 8.6)
39. $2 + \sqrt{2}, 2 - \sqrt{2}$ (Sec. 8.2, Ex. 3)   **40.** $f^{-1}(x) = 2x - 1$ (Sec. 9.2)

42. If $f(x) = x^2 - 3x + 2$ and $g(x) = -3x + 5$ find:
    a. $(f \circ g)(x)$ $9x^2 - 21x + 12$
    b. $(f \circ g)(-2)$ 90
    c. $(g \circ f)(x)$ $-3x^2 + 9x - 1$
    d. $(g \circ f)(5)$ -31 (Sec. 9.1)

43. Solve $\dfrac{x + 2}{x - 3} \le 0$. $[-2, 3)$ (Sec. 8.4, Ex. 6)

44. Graph $4x^2 + 9y^2 = 36$.

45. Graph $g(x) = \dfrac{1}{2}(x + 2)^2 + 5$. Find the vertex and the axis
    of symmetry.   See graphing answer section. (Sec. 8.5, Ex. 5)

46. Solve each equation for x.
    a. $64^x = 4$ $\frac{1}{3}$ **b.** $125^{x-3} = 25$ $\frac{11}{3}$
    c. $\dfrac{1}{81} = 3^{2x}$ -2 (Sec. 9.3)

47. Find the vertex of the graph of $f(x) = x^2 - 4x - 12$.

48. Graph the system: $\begin{cases} x + 2y < 8 \\ \quad y \ge x^2 \end{cases}$

49. Find the distance between $(2, -5)$ and $(1, -4)$. Give an exact distance and a three decimal-place approximation.

50. Solve the system $\begin{cases} x^2 + y^2 = 36 \\ \quad y = x + 6 \end{cases}$ (Sec. 10.3)

44. See graphing answer section. (Sec. 10.2)   **47.** $(2, -16)$ (Sec. 8.6, Ex. 4)   **48.** See graphing answer section. (Sec. 10.4)
49. $\sqrt{2} \approx 1.414$ (Sec. 10.1, Ex. 6)   **50.** $(-6, 0), (0, 6)$

Sequences, Series, and the Binomial Theorem

Having explored in some depth the concept of function, we turn now in this final chapter to *sequences.* In one sense, a sequence is simply an ordered list of numbers. In another sense, a sequence is itself a function. Phenomena modeled by such functions are everywhere around us. The starting place for all mathematics is the sequence of natural numbers: 1, 2, 3, 4, and so on.

Sequences lead us to *series*, which are a sum of ordered numbers. Through series we gain new insight, for example about the expansion of a binomial $(a + b)^n$, the concluding topic of this book.

As Baby Boomers approach retirement age and life spans lengthen overall, more and more people will seek professional assistance with financial planners.

Financial planners work for investment firms, accounting firms, insurance companies, banks, credit counseling organizations, law firms, or in private practice. They give advice on retirement planning, insurance needs, investment options, estate planning, tax strategies, and employee benefits. Certified financial planners use math and problem-solving skills in such tasks as analyzing clients' current cash flow, estimating cash needs for future goals, and calculating investment returns.

In the Spotlight on Decision Making feature on page 812, you will have the opportunity to make a decision about reaching a client's retirement goals as a certified financial planner.

Use the Finance menu on the TI-83 in the Spotlight on Decision Making feature, page 812. For instructions on using the Finance menu, see Appendix D.

Source: Certified Financial Planner Board of Standards website

11.1   SEQUENCES

Objectives

1. Write the terms of a sequence given its general term.
2. Find the general term of a sequence.
3. Solve applications that involve sequences.

Suppose that a town's present population of 100,000 is growing by 5% each year. After the first year, the town's population will be

$$100{,}000 + 0.05(100{,}000) = 105{,}000$$

After the second year, the town's population will be

$$105{,}000 + 0.05(105{,}000) = 110{,}250$$

After the third year, the town's population will be

$$110{,}250 + 0.05(110{,}250) \approx 115{,}763$$

If we continue to calculate, the town's yearly population can be written as the **infinite sequence** of numbers

$$105{,}000, 110{,}250, 115{,}763, \ldots$$

If we decide to stop calculating after a certain year (say, the fourth year), we obtain the **finite sequence**

$$105{,}000, 110{,}250, 115{,}763, 121{,}551$$

Sequences

An infinite sequence is a function whose domain is the set of natural numbers $\{1, 2, 3, 4, \ldots\}$.

   A finite sequence is a function whose domain is the set of natural numbers $\{1, 2, 3, 4, \ldots, n\}$, where n is some natural number.

---

1   Given the sequence $2, 4, 8, 16, \ldots$, we say that each number is a **term** of the sequence. Because a sequence is a function, we could describe it by writing $f(n) = 2^n$, where n is a natural number. Instead, we use the notation

$$a_n = 2^n$$

Some function values are

$a_1 = 2^1 = 2$	First term of the sequence
$a_2 = 2^2 = 4$	Second term
$a_3 = 2^3 = 8$	Third term
$a_4 = 2^4 = 16$	Fourth term
$a_{10} = 2^{10} = 1024$	Tenth term

The nth term of the sequence a_n is called the **general term.**

> **Helpful Hint**
>
> If it helps, think of a sequence as simply a list of values in which a position is assigned. For the first sequence in this section,
>
> Value: 2, 4, 8, 16, ..., 1024
>
> $\quad\quad\quad\quad \uparrow \quad \uparrow \quad \uparrow \quad \uparrow \quad\quad\quad \uparrow$
>
> Position   1st 2nd 3rd 4th   10th

EXAMPLE 1

Write the first five terms of the sequence whose general term is given by

$$a_n = n^2 - 1$$

Solution Evaluate a_n, where n is 1, 2, 3, 4, and 5.

$$a_n = n^2 - 1$$

$a_1 = 1^2 - 1 = 0$ Replace n with **1**.

$a_2 = 2^2 - 1 = 3$ Replace n with **2**.

$a_3 = 3^2 - 1 = 8$ Replace n with **3**.

$a_4 = 4^2 - 1 = 15$ Replace n with **4**.

$a_5 = 5^2 - 1 = 24$ Replace n with **5**.

n	$u(n)$
1	0
2	3
3	8
4	15
5	24
6	35
7	48

$u(n) = n^2 - 1$

Here we use a sequence mode and list the terms of the sequence in a table.

Thus, the first five terms of the sequence $a_n = n^2 - 1$ are 0, 3, 8, 15, and 24.

EXAMPLE 2

If the general term of a sequence is given by $a_n = \dfrac{(-1)^n}{3n}$, find

a. the first term of the sequence

b. a_8

c. the one-hundredth term of the sequence

d. a_{15}

Solution **a.** $a_1 = \dfrac{(-1)^1}{3(1)} = -\dfrac{1}{3}$ Replace n with 1.

b. $a_8 = \dfrac{(-1)^8}{3(8)} = \dfrac{1}{24}$ Replace n with 8.

c. $a_{100} = \dfrac{(-1)^{100}}{3(100)} = \dfrac{1}{300}$ Replace n with 100.

d. $a_{15} = \dfrac{(-1)^{15}}{3(15)} = -\dfrac{1}{45}$ Replace n with 15.

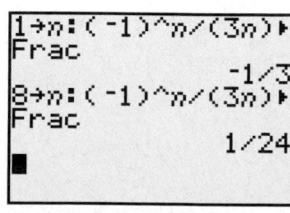

```
1→n:(-1)^n/(3n)►
Frac
                -1/3
8→n:(-1)^n/(3n)►
Frac
                1/24
■
```

Single terms of a sequence may be found by evaluating the general term expression at given values.

2   Suppose we know the first few terms of a sequence and want to find a general term that fits the pattern of the first few terms.

EXAMPLE 3

Find a general term a_n of the sequence whose first few terms are given.

a. $1, 4, 9, 16, \ldots$

b. $\dfrac{1}{1}, \dfrac{1}{2}, \dfrac{1}{3}, \dfrac{1}{4}, \dfrac{1}{5}, \ldots$

c. $-3, -6, -9, -12, \ldots$

d. $\dfrac{1}{2}, \dfrac{1}{4}, \dfrac{1}{8}, \dfrac{1}{16}, \ldots$

Solution

a. These numbers are the squares of the first four natural numbers, so a general term might be $a_n = n^2$.

b. These numbers are the reciprocals of the first five natural numbers, so a general term might be $a_n = \dfrac{1}{n}$.

c. These numbers are the product of -3 and the first four natural numbers, so a general term might be $a_n = -3n$.

d. Notice that the denominators double each time.

$$\frac{1}{2}, \quad \frac{1}{2 \cdot 2}, \quad \frac{1}{2(2 \cdot 2)}, \quad \frac{1}{2(2 \cdot 2 \cdot 2)}$$

or

$$\frac{1}{2^1}, \quad \frac{1}{2^2}, \quad \frac{1}{2^3}, \quad \frac{1}{2^4}$$

We might then suppose that the general term is $a_n = \dfrac{1}{2^n}$.

3   Sequences model many phenomena of the physical world, as illustrated by the following example.

EXAMPLE 4

FINDING A PUPPY'S WEIGHT GAIN

The amount of weight, in pounds, a puppy gains in each month of its first year is modeled by a sequence whose general term is $a_n = n + 4$, where n is the number of the month. Write the first five terms of the sequence, and find how much weight the puppy should gain in its fifth month.

Solution

Evaluate $a_n = n + 4$ when n is $1, 2, 3, 4,$ and 5.

$$a_1 = 1 + 4 = 5$$
$$a_2 = 2 + 4 = 6$$
$$a_3 = 3 + 4 = 7$$
$$a_4 = 4 + 4 = 8$$
$$a_5 = 5 + 4 = 9$$

The puppy should gain 9 pounds in its fifth month.

Spotlight on
DECISION MAKING

Suppose you are considering two job offers. The first job offer pays $11.50 per hour and guarantees a $0.65-per-hour raise each year. The second job offer pays $10.75 per hour and guarantees a $1.10-per-hour raise each year. If one of your goals is to be earning at least $15 per hour in 5 years, which job offer would you accept? Explain. What other factors would you want to consider?

8. $-6, -12, -18, -24, -30$ **9.** $-1, -4, -9, -16, -25$ **10.** $3, 6, 11, 18, 27$ **14.** $-2, -5, -8, -11, -14$ **15.** $-1, 4, -9, 16, -25$

EXERCISE SET 11.1

3. $-1, 1, -1, 1, -1$ **4.** $-2, 4, -8, 16, -32$ STUDY GUIDE/SSM   CD/ VIDEO   PH MATH TUTOR CENTER   MathXL®Tutorials ON CD   MathXL®   MyMathLab®

Write the first five terms of each sequence whose general term is given. See Example 1.

1. $a_n = n + 4$ $5, 6, 7, 8, 9$
2. $a_n = 5 - n$ $4, 3, 2, 1, 0$
3. $a_n = (-1)^n$
4. $a_n = (-2)^n$
5. $a_n = \dfrac{1}{n + 3}$ $\dfrac{1}{4}, \dfrac{1}{5}, \dfrac{1}{6}, \dfrac{1}{7}, \dfrac{1}{8}$
6. $a_n = \dfrac{1}{7 - n}$ $\dfrac{1}{6}, \dfrac{1}{5}, \dfrac{1}{4}, \dfrac{1}{3}, \dfrac{1}{2}$
7. $a_n = 2n$ $2, 4, 6, 8, 10$
8. $a_n = -6n$
9. $a_n = -n^2$
10. $a_n = n^2 + 2$
11. $a_n = 2^n$ $2, 4, 8, 16, 32$
12. $a_n = 3^{n-2}$ $\dfrac{1}{3}, 1, 3, 9, 27$
13. $a_n = 2n + 5$ $7, 9, 11, 13, 15$
14. $a_n = 1 - 3n$
15. $a_n = (-1)^n n^2$
16. $a_n = (-1)^{n+1}(n - 1)$ $0, -1, 2, -3, 4$

Find the indicated term for each sequence whose general term is given. See Example 2.

17. $a_n = 3n^2$; a_5 75
18. $a_n = -n^2$; a_{15} -225
19. $a_n = 6n - 2$; a_{20} 118
20. $a_n = 100 - 7n$; a_{50} -250
21. $a_n = \dfrac{n + 3}{n}$; a_{15} $\dfrac{6}{5}$
22. $a_n = \dfrac{n}{n + 4}$; a_{24} $\dfrac{6}{7}$
23. $a_n = (-3)^n$; a_6 729
24. $a_n = 5^{n+1}$; a_3 625
25. $a_n = \dfrac{n - 2}{n + 1}$; a_6 $\dfrac{4}{7}$
26. $a_n = \dfrac{n + 3}{n + 4}$; a_8 $\dfrac{11}{12}$
27. $a_n = \dfrac{(-1)^n}{n}$; a_8 $\dfrac{1}{8}$
28. $a_n = \dfrac{(-1)^n}{2n}$; a_{100} $\dfrac{1}{200}$
29. $a_n = -n^2 + 5$; a_{10} -95
30. $a_n = 8 - n^2$; a_{20} -392
31. $a_n = \dfrac{(-1)^n}{n + 6}$; a_{19} $-\dfrac{1}{25}$
32. $a_n = \dfrac{n - 4}{(-2)^n}$; a_6 $\dfrac{1}{32}$

Find a general term a_n for each sequence whose first four terms are given. See Example 3.

33. $3, 7, 11, 15$ $a_n = 4n - 1$
34. $2, 7, 12, 17$ $a_n = 5n - 3$
35. $-2, -4, -8, -16$
36. $-4, 16, -64, 256$
37. $\dfrac{1}{3}, \dfrac{1}{9}, \dfrac{1}{27}, \dfrac{1}{81}$ $a_n = \dfrac{1}{3^n}$
38. $\dfrac{2}{5}, \dfrac{2}{25}, \dfrac{2}{125}, \dfrac{2}{625}$ $a_n = \dfrac{2}{5^n}$

35. $a_n = -2^n$ **36.** $a_n = (-4)^n$

Solve. See Example 4.

39. The distance, in feet, that a Thermos dropped from a cliff falls in each consecutive second is modeled by a sequence whose general term is $a_n = 32n - 16$, where n is the number of seconds. Find the distance the Thermos falls in the second, third, and fourth seconds.   48 ft, 80 ft, and 112 ft

40. The population size of a culture of bacteria triples every hour such that its size is modeled by the sequence $a_n = 50(3)^{n-1}$, where n is the number of the hour just beginning. Find the size of the culture at the beginning of the fourth hour and the size of the culture at the beginning of the first hour.   1350 bacteria; 50 bacteria

41. Mrs. Laser agrees to give her son Mark an allowance of $0.10 on the first day of his 14-day vacation, $0.20 on the second day, $0.40 on the third day, and so on. Write an equation of a sequence whose terms correspond to Mark's allowance. Find the allowance Mark will receive on the last day of his vacation. $a_n = 0.10(2)^{n-1}$; $819.20

42. A small theater has 10 rows with 12 seats in the first row, 15 seats in the second row, 18 seats in the third row, and so on. Write an equation of a sequence whose terms correspond to the seats in each row. Find the number of seats in the eighth row. $a_n = 12 + 3(n - 1)$;   33 seats

43. The number of cases of a new infectious disease is doubling every year such that the number of cases is modeled by a sequence whose general term is $a_n = 75(2)^{n-1}$, where n is the number of the year just beginning. Find how many cases there will be at the beginning of the sixth year. Find how many cases there were at the beginning of the first year.   2400 cases; 75 cases

44. A new college had an initial enrollment of 2700 students in 2000, and each year the enrollment increases by 150 students. Find the enrollment for each of 5 years, beginning with 2000.   2700, 2850, 3000, 3150, 3300 students

45. An endangered species of sparrow had an estimated population of 800 in 2000, and scientists predict that its population will decrease by half each year. Estimate the population in 2004. Estimate the year the sparrow will be extinct.   50 sparrows in 2004; extinct in 2010

46. A **Fibonacci sequence** is a special type of sequence in which the first two terms are 1 and each term thereafter is the sum of the two previous terms: $1, 1, 2, 3, 5, 8, \ldots$. Many plants and animals seem to grow according to a Fibonacci sequence, including pine cones, pineapple scales, nautilus shells, and certain flowers. Write the first 15 terms of the Fibonacci sequence. $1, 1, 2, 3, 5, 8, 13, 21, 34, 55, 89, 144, 233, 377, 610$

REVIEW AND PREVIEW

Sketch the graph of each quadratic function. See Section 8.5.

47. $f(x) = (x - 1)^2 + 3$ See graphing answer section.

48. $f(x) = (x - 2)^2 + 1$ See graphing answer section.

49. $f(x) = 2(x + 4)^2 + 2$ See graphing answer section.

50. $f(x) = 3(x - 3)^2 + 4$ See graphing answer section.

Find the distance between each pair of points. See Section 10.1.

51. $(-4, -1)$ and $(-7, -3)$ $\sqrt{13}$ units

52. $(-2, -1)$ and $(-1, 5)$ $\sqrt{37}$ units

53. $(2, -7)$ and $(-3, -3)$ $\sqrt{41}$ units

54. $(10, -14)$ and $(5, -11)$ $\sqrt{34}$ units

Concept Extensions

Use a graphing utility to find the first five terms of each sequence. Round each term after the first to four decimal places.

55. $a_n = \dfrac{1}{\sqrt{n}}$ $1, 0.7071, 0.5774, 0.5, 0.4472$

56. $\dfrac{\sqrt{n}}{\sqrt{n} + 1}$ $0.5, 0.5858, 0.6340, 0.6667, 0.6910$

57. $a_n = \left(1 + \dfrac{1}{n}\right)^n$ $2, 2.25, 2.3704, 2.4414, 2.4883$

58. $a_n = \left(1 + \dfrac{0.05}{n}\right)^n$ $1.05, 1.0506, 1.0508, 1.0509, 1.0510$

11.2   ARITHMETIC AND GEOMETRIC SEQUENCES

Objectives

1 Identify arithmetic sequences and their common differences.

2 Identify geometric sequences and their common ratios.

1   Find the first four terms of the sequence whose general term is $a_n = 5 + (n - 1)3$.

$$a_1 = 5 + (1 - 1)3 = 5 \qquad \text{Replace } n \text{ with } 1.$$
$$a_2 = 5 + (2 - 1)3 = 8 \qquad \text{Replace } n \text{ with } 2.$$
$$a_3 = 5 + (3 - 1)3 = 11 \qquad \text{Replace } n \text{ with } 3.$$
$$a_4 = 5 + (4 - 1)3 = 14 \qquad \text{Replace } n \text{ with } 4.$$

The first four terms are $5, 8, 11,$ and 14. Notice that the difference of any two successive terms is 3.

$$8 - 5 = 3$$
$$11 - 8 = 3$$
$$14 - 11 = 3$$
$$\vdots$$
$$a_n - a_{n-1} = 3$$

$\uparrow$ *n*th term $\uparrow$ previous term

Because the difference of any two successive terms is a constant, we call the sequence an **arithmetic sequence,** or an **arithmetic progression.** The constant difference d in successive terms is called the **common difference.** In this example, d is 3.

> ## Arithmetic Sequence and Common Difference
>
> An **arithmetic sequence** is a sequence in which each term (after the first) differs from the preceding term by a constant amount d. The constant d is called the **common difference** of the sequence.

The sequence $2, 6, 10, 14, 18, \ldots$ is an arithmetic sequence. Its common difference is 4. Given the first term a_1 and the common difference d of an arithmetic sequence, we can find any term of the sequence.

EXAMPLE 1

Write the first five terms of the arithmetic sequence whose first term is 7 and whose common difference is 2.

Solution

$$a_1 = 7$$
$$a_2 = 7 + 2 = 9$$
$$a_3 = 9 + 2 = 11$$
$$a_4 = 11 + 2 = 13$$
$$a_5 = 13 + 2 = 15$$

The first five terms are $7, 9, 11, 13, 15$.

CLASSROOM EXAMPLE
Write the first six terms of the arithmetic sequence whose first term is 5 and whose common difference is 3.
answer: $5, 8, 11, 14, 17, 20$

Notice the general pattern of the terms in Example 1.

$$a_1 = 7$$
$$a_2 = 7 + 2 = 9 \quad \text{or} \quad a_2 = a_1 + d$$
$$a_3 = 9 + 2 = 11 \quad \text{or} \quad a_3 = a_2 + d = (a_1 + d) + d = a_1 + 2d$$
$$a_4 = 11 + 2 = 13 \quad \text{or} \quad a_4 = a_3 + d = (a_1 + 2d) + d = a_1 + 3d$$
$$a_5 = 13 + 2 = 15 \quad \text{or} \quad a_5 = a_4 + d = (a_1 + 3d) + d = a_1 + 4d$$

$$\text{(subscript} - 1) \text{ is multiplier}$$

CLASSROOM EXAMPLE
Consider the arithmetic sequence whose first term is 4 and common difference is -7.
a. Write an expression for the general term a_n
b. Find the fifteenth term of the sequence
answer: **a.** $a_n = 11 - 7n$ **b.** -94

The pattern on the right suggests that the general term a_n of an arithmetic sequence is given by

$$a_n = a_1 + (n - 1)d$$

> ## General Term of an Arithmetic Sequence
>
> The general term a_n of an arithmetic sequence is given by
> $$a_n = a_1 + (n - 1)d$$
> where a_1 is the first term and d is the common difference.

EXAMPLE 2

Consider the arithmetic sequence whose first term is 3 and common difference is -5.

a. Write an expression for the general term a_n.

b. Find the twentieth term of this sequence.

Solution

a. Since this is an arithmetic sequence, the general term a_n is given by $a_n = a_1 + (n - 1)d$. Here, $a_1 = 3$ and $d = -5$, so

$$a_n = 3 + (n - 1)(-5) \qquad \text{Let } a_1 = 3 \text{ and } d = -5.$$
$$= 3 - 5n + 5 \qquad \text{Multiply.}$$
$$= 8 - 5n \qquad \text{Simplify.}$$

b. $a_n = 8 - 5n$

$$a_{20} = 8 - 5 \cdot 20 \qquad \text{Let } n = 20.$$
$$= 8 - 100 = -92$$

EXAMPLE 3

Find the eleventh term of the arithmetic sequence whose first three terms are 2, 9, and 16.

Solution

Since the sequence is arithmetic, the eleventh term is

$$a_{11} = a_1 + (11 - 1)d = a_1 + 10d$$

We know a_1 is the first term of the sequence, so $a_1 = 2$. Also, d is the constant difference of terms, so $d = a_2 - a_1 = 9 - 2 = 7$. Thus,

$$a_{11} = a_1 + 10d$$
$$= 2 + 10 \cdot 7 \qquad \text{Let } a_1 = 2 \text{ and } d = 7.$$
$$= 72$$

EXAMPLE 4

If the third term of an arithmetic sequence is 12 and the eighth term is 27, find the fifth term.

Solution

We need to find a_1 and d to write the general term, which then enables us to find a_5, the fifth term. The given facts about terms a_3 and a_8 lead to a system of linear equations.

$$\begin{cases} a_3 = a_1 + (3 - 1)d \\ a_8 = a_1 + (8 - 1)d \end{cases} \quad \text{or} \quad \begin{cases} 12 = a_1 + 2d \\ 27 = a_1 + 7d \end{cases}$$

Next, we solve the system $\begin{cases} 12 = a_1 + 2d \\ 27 = a_1 + 7d \end{cases}$ by elimination. Multiply both sides of the second equation by -1 so that

$$\begin{cases} 12 = a_1 + 2d \\ -1(27) = -1(a_1 + 7d) \end{cases} \quad \begin{array}{c} \text{simplifies} \\ \text{to} \end{array} \quad \begin{cases} 12 = a_1 + 2d \\ \underline{-27 = -a_1 - 7d} \end{cases}$$
$$-15 = -5d \qquad \text{Add the equations.}$$
$$3 = d \qquad \text{Divide both sides by } -5.$$

To find a_1, let $d = 3$ in $12 = a_1 + 2d$. Then

$$12 = a_1 + 2(3)$$
$$12 = a_1 + 6$$
$$6 = a_1$$

Thus, $a_1 = 6$ and $d = 3$, so

$$a_n = 6 + (n - 1)(3)$$
$$= 6 + 3n - 3$$
$$= 3 + 3n$$

and

$$a_5 = 3 + 3 \cdot 5 = 18$$

EXAMPLE 5

FINDING SALARY

Donna Theime has an offer for a job starting at $40,000 per year and guaranteeing her a raise of $1600 per year for the next 5 years. Write the general term for the arithmetic sequence that models Donna's potential annual salaries, and find her salary for the fourth year.

Solution   The first term, a_1, is 40,000, and d is 1600. So

$$a_n = 40,000 + (n-1)(1600) = 38,400 + 1600n$$
$$a_4 = 38,400 + 1600 \cdot 4 = 44,800$$

Her salary for the fourth year will be $44,800.

Verify that her salary for the fourth year is $44,800.

2   We now investigate a **geometric sequence,** also called a **geometric progression.** In the sequence $5, 15, 45, 135, \ldots$, each term after the first is the *product* of 3 and the preceding term. This pattern of multiplying by a constant to get the next term defines a geometric sequence. The constant is called the **common ratio** because it is the ratio of any term (after the first) to its preceding term.

$$\frac{15}{5} = 3$$
$$\frac{45}{15} = 3$$
$$\frac{135}{45} = 3$$
$$\vdots$$

nth term $\longrightarrow$ $\dfrac{a_n}{a_{n-1}} = 3$
previous term $\longrightarrow$

Geometric Sequence and Common Ratio

A **geometric sequence** is a sequence in which each term (after the first) is obtained by multiplying the preceding term by a constant r. The constant r is called the **common ratio** of the sequence.

The sequence $12, 6, 3, \dfrac{3}{2}, \ldots$ is geometric since each term after the first is the product of the previous term and $\dfrac{1}{2}$.

EXAMPLE 6

Write the first five terms of a geometric sequence whose first term is 7 and whose common ratio is 2.

Solution

$$a_1 = 7$$
$$a_2 = 7(2) = 14$$
$$a_3 = 14(2) = 28$$
$$a_4 = 28(2) = 56$$
$$a_5 = 56(2) = 112$$

The first five terms are $7, 14, 28, 56$, and 112.

Notice the general pattern of the terms in Example 6.

$$a_1 = 7$$

$$a_2 = 7(2) = 14 \quad \text{or} \quad a_2 = a_1(r)$$

$$a_3 = 14(2) = 28 \quad \text{or} \quad a_3 = a_2(r) = (a_1 \cdot r) \cdot r = a_1 r^2$$

$$a_4 = 28(2) = 56 \quad \text{or} \quad a_4 = a_3(r) = (a_1 \cdot r^2) \cdot r = a_1 r^3$$

$$a_5 = 56(2) = 112 \quad \text{or} \quad a_5 = a_4(r) = (a_1 \cdot r^3) \cdot r = a_1 r^4$$

$\longrightarrow$ (subscript -1) is power

TEACHING TIP

Ask students whether a sequence can be both arithmetic and geometric. Help them write a sequence that is both. For example, $3, 3, 3, 3, \ldots$.

The pattern on the right above suggests that the general term of a geometric sequence is given by $a_n = a_1 r^{n-1}$.

General Term of a Geometric Sequence

The general term a_n of a geometric sequence is given by

$$a_n = a_1 r^{n-1}$$

where a_1 is the first term and r is the common ratio.

EXAMPLE 7

Find the eighth term of the geometric sequence whose first term is 12 and whose common ratio is $\dfrac{1}{2}$.

Solution   Since this is a geometric sequence, the general term a_n is given by

$$a_n = a_1 r^{n-1}$$

CLASSROOM EXAMPLE

Find the sixth term of the geometric sequence whose first term is 21 and whose common ratio is $\dfrac{1}{3}$.

answer: $\dfrac{7}{81}$

Here $a_1 = 12$ and $r = \dfrac{1}{2}$, so $a_n = 12\left(\dfrac{1}{2}\right)^{n-1}$. Evaluate a_n for $n = 8$.

$$a_8 = 12\left(\dfrac{1}{2}\right)^{8-1} = 12\left(\dfrac{1}{2}\right)^{7} = 12\left(\dfrac{1}{128}\right) = \dfrac{3}{32}$$

EXAMPLE 8

Find the fifth term of the geometric sequence whose first three terms are $2, -6$, and 18.

Solution   Since the sequence is geometric and $a_1 = 2$, the fifth term must be $a_1 r^{5-1}$, or $2r^4$. We know that r is the common ratio of terms, so r must be $\dfrac{-6}{2}$, or -3. Thus,

CLASSROOM EXAMPLE

Find the sixth term of the geometric sequence whose first three terms are $-2, 8, -32 \ldots$.

answer: 2048

$$a_5 = 2r^4$$

$$a_5 = 2(-3)^4 = 162$$

EXAMPLE 9

If the second term of a geometric sequence is $\frac{5}{4}$ and the third term is $\frac{5}{16}$, find the first term and the common ratio.

Solution

Notice that $\frac{5}{16} \div \frac{5}{4} = \frac{1}{4}$, so $r = \frac{1}{4}$. Then

$$a_2 = a_1\left(\frac{1}{4}\right)^{2-1}$$

$$\frac{5}{4} = a_1\left(\frac{1}{4}\right)^1, \text{ or } a_1 = 5 \qquad \text{Replace } a_2 \text{ with } \frac{5}{4}.$$

The first term is 5.

EXAMPLE 10

PREDICTING POPULATION OF A BACTERIAL CULTURE

The population size of a bacterial culture growing under controlled conditions is doubling each day. Predict how large the culture will be at the beginning of day 7 if it measures 10 units at the beginning of day 1.

Solution

Since the culture doubles in size each day, the population sizes are modeled by a geometric sequence. Here $a_1 = 10$ and $r = 2$. Thus,

$$a_n = a_1 r^{n-1} = 10(2)^{n-1} \quad \text{and} \quad a_7 = 10(2)^{7-1} = 640$$

The bacterial culture should measure 640 units at the beginning of day 7.

n	$u(n)$
1	10
2	20
3	40
4	80
5	160
6	320
7	640

$u(n)\text{\small■}10*2\char`^(n-1)$

The bacterial culture measures 640 units at the beginning of day 7.

Spotlight on
DECISION
♂ MAKING

Suppose you are a research biologist studying a particular strain of bacteria that grows at a rate of 1.5 times per hour. For a particular experiment, you will start with a culture of 200 units of bacteria and will allow the culture to grow for 7 hours. Decide whether a culture dish that holds 5000 units will be large enough for this experiment. If not, would a dish that holds 10,000 units be a better choice?

EXERCISE SET 11.2

STUDY GUIDE/SSM | CD/VIDEO | PH MATH TUTOR CENTER | MathXL®Tutorials ON CD | MathXL® | MyMathLab®

Write the first five terms of the arithmetic or geometric sequence whose first term, a_1, and common difference, d, or common ratio, r, are given. See Examples 1 and 6.

1. $a_1 = 4; d = 2$ $4, 6, 8, 10, 12$

2. $a_1 = 3; d = 10$ $3, 13, 23, 33, 43$

 3. $a_1 = 6; d = -2$ $6, 4, 2, 0, -2$

4. $a_1 = -20; d = 3$ $-20, -17, -14, -11, -8$

5. $a_1 = 1; r = 3$ $1, 3, 9, 27, 81$

6. $a_1 = -2; r = 2$ $-2, -4, -8, -16, -32$

7. $a_1 = 48; r = \dfrac{1}{2}$ $48, 24, 12, 6, 3$

8. $a_1 = 1; r = \dfrac{1}{3}$ $1, \dfrac{1}{3}, \dfrac{1}{9}, \dfrac{1}{27}, \dfrac{1}{81}$

Find the indicated term of each sequence. See Examples 2 and 7.

9. The eighth term of the arithmetic sequence whose first term is 12 and whose common difference is 3   33

10. The twelfth term of the arithmetic sequence whose first term is 32 and whose common difference is -4 -12

11. The fourth term of the geometric sequence whose first term is 7 and whose common ratio is -5 -875

12. The fifth term of the geometric sequence whose first term is 3 and whose common ratio is 3   243

13. The fifteenth term of the arithmetic sequence whose first term is -4 and whose common difference is -4 -60

14. The sixth term of the geometric sequence whose first term is 5 and whose common ratio is -4 -5120

16. 33   19. 1250

Find the indicated term of each sequence. See Examples 3 and 8.

15. The ninth term of the arithmetic sequence $0, 12, 24, \ldots$ 96

16. The thirteenth term of the arithmetic sequence $-3, 0, 3, \ldots$

17. The twenty-fifth term of the arithmetic sequence $20, 18, 16, \ldots$ -28

18. The ninth term of the geometric sequence $5, 10, 20, \ldots$ 1280

19. The fifth term of the geometric sequence $2, -10, 50, \ldots$

20. The sixth term of the geometric sequence $\dfrac{1}{2}, \dfrac{3}{2}, \dfrac{9}{2}, \ldots$ $\dfrac{243}{2}$

Find the indicated term of each sequence. See Examples 4 and 9.

21. The eighth term of the arithmetic sequence whose fourth term is 19 and whose fifteenth term is 52   31

22. If the second term of an arithmetic sequence is 6 and the tenth term is 30, find the twenty-fifth term.   75

23. If the second term of an arithmetic sequence is -1 and the fourth term is 5, find the ninth term.   20

24. If the second term of a geometric sequence is 15 and the third term is 3, find a_1 and r. $a_1 = 75; r = \dfrac{1}{5}$

25. If the second term of a geometric sequence is $-\dfrac{4}{3}$ and the third term is $\dfrac{8}{3}$, find a_1 and r. $a_1 = \dfrac{2}{3}; r = -2$

26. If the third term of a geometric sequence is 4 and the fourth term is -12, find a_1 and r. $a_1 = \dfrac{4}{9}; r = -3$

27. Explain why 14, 10, and 6 may be the first three terms of an arithmetic sequence when it appears we are subtracting instead of adding to get the next term.   answers may vary

28. Explain why 80, 20, and 5 may be the first three terms of a geometric sequence when it appears we are dividing instead of multiplying to get the next term.   answers may vary

37. $a_1 = p; d = 4$ 38. $a_1 = t; d = -1$ 44. $-\dfrac{4}{3}$

MIXED PRACTICE

Given are the first three terms of a sequence that is either arithmetic or geometric. If the sequence is arithmetic, find a_1 and d. If a sequence is geometric, find a_1 and r.

29. $2, 4, 6$ $a_1 = 2; d = 2$
30. $8, 16, 24$ $a_1 = 8; d = 8$

31. $5, 10, 20$ $a_1 = 5; r = 2$
32. $2, 6, 18$ $a_1 = 2; r = 3$

33. $\dfrac{1}{2}, \dfrac{1}{10}, \dfrac{1}{50}$ $a_1 = \dfrac{1}{2}; r = \dfrac{1}{5}$
34. $\dfrac{2}{3}, \dfrac{4}{3}, 2$ $a_1 = \dfrac{2}{3}; d = \dfrac{2}{3}$

35. $x, 5x, 25x$ $a_1 = x; r = 5$
36. $y, -3y, 9y$ $a_1 = y; r = -3$

37. $p, p + 4, p + 8$
38. $t, t - 1, t - 2$

Find the indicated term of each sequence.

39. The twenty-first term of the arithmetic sequence whose first term is 14 and whose common difference is $\dfrac{1}{4}$ 19

40. The fifth term of the geometric sequence whose first term is 8 and whose common ratio is -3 648

41. The fourth term of the geometric sequence whose first term is 3 and whose common ratio is $-\dfrac{2}{3}$ $-\dfrac{8}{9}$

42. The fourth term of the arithmetic sequence whose first term is 9 and whose common difference is 5   24

43. The fifteenth term of the arithmetic sequence $\dfrac{3}{2}, 2, \dfrac{5}{2}, \ldots$ $\dfrac{17}{2}$

44. The eleventh term of the arithmetic sequence $2, \dfrac{5}{3}, \dfrac{4}{3}, \ldots$

45. The sixth term of the geometric sequence $24, 8, \dfrac{8}{3}, \ldots$ $\dfrac{8}{81}$

46. The eighteenth term of the arithmetic sequence $5, 2, -1, \ldots$

47. If the third term of an arithmetic sequence is 2 and the seventeenth term is -40, find the tenth term. -19

48. If the third term of a geometric sequence is -28 and the fourth term is -56, find a_1 and r. $a_1 = -7; r = 2$

Solve. See Examples 5 and 10.

49. An auditorium has 54 seats in the first row, 58 seats in the second row, 62 seats in the third row, and so on. Find the general term of this arithmetic sequence and the number of seats in the twentieth row. $a_n = 4n + 50$; 130 seats

50. A triangular display of cans in a grocery store has 20 cans in the first row, 17 cans in the next row, and so on, in an arithmetic sequence. Find the general term and the number of cans in the fifth row. Find how many rows there are in the display and how many cans are in the top row.

51. The initial size of a virus culture is 6 units, and it triples its size every day. Find the general term of the geometric sequence that models the culture's size. $a_n = 6(3)^{n-1}$

52. A real estate investment broker predicts that a certain property will increase in value 15% each year. Thus, the yearly property values can be modeled by a geometric sequence whose common ratio r is 1.15. If the initial property value was $500,000, write the first four terms of the sequence and predict the value at the end of the third year.

53. A rubber ball is dropped from a height of 486 feet, and it continues to bounce one-third the height from which it last fell. Write out the first five terms of this geometric sequence and find the general term. Find how many bounces it takes for the ball to rebound less than 1 foot.

54. On the first swing, the length of the arc through which a pendulum swings is 50 inches. The length of each successive swing is 80% of the preceding swing. Determine whether this sequence is arithmetic or geometric. Find the length of the fourth swing.   geometric; 25.6 in.

55. Jose takes a job that offers a monthly starting salary of $4000 and guarantees him a monthly raise of $125 during his first year of training. Find the general term of this arithmetic sequence and his monthly salary at the end of his training.

56. At the beginning of Claudia Schaffer's exercise program, she rides 15 minutes on the Lifecycle. Each week she increases her riding time by 5 minutes. Write the general term of this arithmetic sequence, and find her riding time after 7 weeks. Find how many weeks it takes her to reach a riding time of 1 hour.

57. If a radioactive element has a half-life of 3 hours, then x grams of the element dwindles to $\dfrac{x}{2}$ grams after 3 hours. If a nuclear reactor has 400 grams of that radioactive element, find the amount of radioactive material after 12 hours.   25 g

REVIEW AND PREVIEW

Evaluate. See Section 1.3.

58. $5(1) + 5(2) + 5(3) + 5(4)$ 50

59. $\dfrac{1}{3(1)} + \dfrac{1}{3(2)} + \dfrac{1}{3(3)}$ $\dfrac{11}{18}$

60. $2(2 - 4) + 3(3 - 4) + 4(4 - 4)$ -7

61. $3^0 + 3^1 + 3^2 + 3^3$ 40

62. $\dfrac{1}{4(1)} + \dfrac{1}{4(2)} + \dfrac{1}{4(3)}$ $\dfrac{11}{24}$

63. $\dfrac{8 - 1}{8 + 1} + \dfrac{8 - 2}{8 + 2} + \dfrac{8 - 3}{8 + 3}$ $\dfrac{907}{495}$

Concept Extensions

Use a graphing utility to write the first four terms of the arithmetic or geometric sequence whose first term, a_1, and common difference, d, or common ratio, r, are given.

64. $a_1 = \$3720, d = -\$268.50$ $\$3720, \$3451.50, \$3183, \$2914.50$

65. $a_1 = \$11{,}782.40, r = 0.5$

66. $a_1 = 26.8, r = 2.5$ $26.8, 67, 167.5, 418.75$

67. $a_1 = 19.652; d = -0.034$ $19.652, 19.618, 19.584, 19.55$

68. Describe a situation in your life that can be modeled by a geometric sequence. Write an equation for the sequence.

69. Describe a situation in your life that can be modeled by an arithmetic sequence. Write an equation for the sequence.

---

46. -46 **50.** $a_n = 20 + (n - 1)(-3)$ or $a_n = 23 - 3n$; 8 cans in the fifth row; 7 rows in the display; top row, 2 cans

52. $500,000, $575,000, $661,250, $760,437.50; $760,437.50   **53.** $486, 162, 54, 18, 6$; $a_n = \dfrac{486}{3^{n-1}}$; 6 bounces

55. $a_n = 4000 + 125(n - 1)$ or $a_n = 3875 + 125n$; $5375   **56.** $a_n = 15 + (n - 1)(5)$ or $a_n = 10 + 5n$; 45 min; 10 weeks

65. $11,782.40, $5891.20, $2945.60, $1472.80   **68.** answers may vary   **69.** answers may vary

11.3   *SERIES*

Objectives

1 Identify finite and infinite series and use summation notation.

2 Find partial sums.

1 A person who conscientiously saves money by saving first $100 and then saving $10 more each month than he saved the preceding month is saving money according to the arithmetic sequence

$$a_n = 100 + 10(n - 1)$$

Following this sequence, he can predict how much money he should save for any particular month. But if he also wants to know how much money *in total* he has saved, say, by the fifth month, he must find the *sum* of the first five terms of the sequence

$$\underbrace{100}_{a_1} + \underbrace{100 + 10}_{a_2} + \underbrace{100 + 20}_{a_3} + \underbrace{100 + 30}_{a_4} + \underbrace{100 + 40}_{a_5}$$

A sum of the terms of a sequence is called a **series** (the plural is also "series"). As our example here suggests, series are frequently used to model financial and natural phenomena.

A series is a **finite series** if it is the sum of a finite number of terms. A series is an **infinite series** if it is the sum of all the terms of the sequence. For example,

Sequence	*Series*	
$5, 9, 13$	$5 + 9 + 13$	Finite; sum of **3** terms.
$5, 9, 13, \ldots$	$5 + 9 + 13 + \cdots$	Infinite
$4, -2, 1, -\dfrac{1}{2}, \dfrac{1}{4}$	$4 + (-2) + 1 + \left(-\dfrac{1}{2}\right) + \left(\dfrac{1}{4}\right)$	Finite; sum of **5** terms.
$4, -2, 1, \ldots$	$4 + (-2) + 1 + \cdots$	Infinite
$3, 6, \ldots, 99$	$3 + 6 + \cdots + 99$	Finite; sum of **33** terms.

TECHNOLOGY NOTE

You can use a graphing utility to evaluate summations. Use a sum command to add terms of a specified sequence.

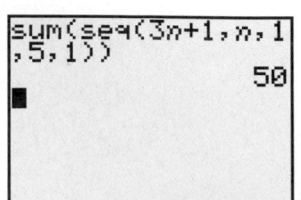

A shorthand notation for denoting a series when the general term of the sequence is known is called **summation notation.** The Greek uppercase letter **sigma**, Σ, is used to mean "sum." The expression $\displaystyle\sum_{n=1}^{5}(3n + 1)$ is read "the sum of $3n + 1$ as n goes from 1 to 5"; this expression means the sum of the first five terms of the sequence whose general term is $a_n = 3n + 1$. Often, the variable i is used instead of n in summation notation: $\displaystyle\sum_{i=1}^{5}(3i + 1)$. Whether we use $n, i, k,$ or some other variable, the variable is called the **index of summation.** The notation $i = 1$ below the symbol Σ indicates the beginning value of i, and the number 5 above the symbol Σ indicates the ending value of i. Thus, the terms of the sequence are found by successively replacing i with the natural numbers $1, 2, 3, 4, 5$. To find the sum, we write out the terms and then add.

$$\sum_{i=1}^{5}(3i + 1) = (3 \cdot 1 + 1) + (3 \cdot 2 + 1) + (3 \cdot 3 + 1)$$
$$+ (3 \cdot 4 + 1) + (3 \cdot 5 + 1)$$
$$= 4 + 7 + 10 + 13 + 16 = 50$$

EXAMPLE 1

Evaluate.

a. $\displaystyle\sum_{i=0}^{6} \frac{i-2}{2}$ **b.** $\displaystyle\sum_{i=3}^{5} 2^i$

Solution   **a.** $\displaystyle\sum_{i=0}^{6} \frac{i-2}{2} = \frac{0-2}{2} + \frac{1-2}{2} + \frac{2-2}{2} + \frac{3-2}{2} + \frac{4-2}{2} + \frac{5-2}{2} + \frac{6-2}{2}$

$$= (-1) + \left(-\frac{1}{2}\right) + 0 + \frac{1}{2} + 1 + \frac{3}{2} + 2$$

$$= \frac{7}{2}, \text{ or } 3\frac{1}{2}$$

b. $\displaystyle\sum_{i=3}^{5} 2^i = 2^3 + 2^4 + 2^5$

$$= 8 + 16 + 32$$

$$= 56$$

EXAMPLE 2

Write each series with summation notation.

a. $3 + 6 + 9 + 12 + 15$

b. $\dfrac{1}{2} + \dfrac{1}{4} + \dfrac{1}{8} + \dfrac{1}{16}$

Solution   **a.** Since the *difference* of each term and the preceding term is 3, the terms correspond to the first five terms of the arithmetic sequence $a_n = a_1 + (n-1)d$ with $a_1 = 3$ and $d = 3$. So $a_n = 3 + (n-1)3 = 3n$, when simplified. Thus, in summation notation,

$$3 + 6 + 9 + 12 + 15 = \sum_{i=1}^{5} 3i.$$

b. Since each term is the *product* of the preceding term and $\dfrac{1}{2}$, these terms correspond to the first four terms of the geometric sequence $a_n = a_1 r^{n-1}$. Here $a_1 = \dfrac{1}{2}$ and $r = \dfrac{1}{2}$, so $a_n = \left(\dfrac{1}{2}\right)\left(\dfrac{1}{2}\right)^{n-1} = \left(\dfrac{1}{2}\right)^{1+(n-1)} = \left(\dfrac{1}{2}\right)^{n}$. In summation notation,

$$\frac{1}{2} + \frac{1}{4} + \frac{1}{8} + \frac{1}{16} = \sum_{i=1}^{4}\left(\frac{1}{2}\right)^i$$

2   The sum of the first n terms of a sequence is a finite series known as a **partial sum**, S_n. Thus, for the sequence $a_1, a_2, \ldots, a_n$, the first three partial sums are

$$S_1 = a_1$$
$$S_2 = a_1 + a_2$$
$$S_3 = a_1 + a_2 + a_3$$

In general, S_n is the sum of the first n terms of a sequence.

$$S_n = \sum_{i=1}^{n} a_n$$

EXAMPLE 3

Find the sum of the first three terms of the sequence whose general term is $a_n = \dfrac{n + 3}{2n}$.

Solution

$$S_3 = \sum_{i=1}^{3} \frac{i + 3}{2i} = \frac{1 + 3}{2 \cdot 1} + \frac{2 + 3}{2 \cdot 2} + \frac{3 + 3}{2 \cdot 3}$$

$$= 2 + \frac{5}{4} + 1 = 4\frac{1}{4}$$

The next example illustrates how these sums model real-life phenomena.

EXAMPLE 4

NUMBER OF BABY GORILLAS BORN

The number of baby gorillas born at the San Diego Zoo is a sequence defined by $a_n = n(n - 1)$, where n is the number of years the zoo has owned gorillas. Find the *total* number of baby gorillas born in the *first 4 years*.

Solution

To solve, find the sum

$$S_4 = \sum_{i=1}^{4} i(i - 1)$$

$$= 1(1 - 1) + 2(2 - 1) + 3(3 - 1) + 4(4 - 1)$$

$$= 0 + 2 + 6 + 12 = 20$$

There were 20 gorillas born in the first 4 years.

```
sum(seq(n(n-1),n
,1,4,1))
                20
■
```

19. $\sum_{i=1}^{5}(2i - 1)$ **20.** $\sum_{i=1}^{4}(3i + 1)$ **21.** $\sum_{i=1}^{4}4(3)^{i-1}$ **22.** $\sum_{i=1}^{6}5(2)^{i-1}$ **23.** $\sum_{i=1}^{6}(-3i + 15)$ **24.** $\sum_{i=1}^{4}(-4i + 9)$

EXERCISE SET 11.3

STUDY GUIDE/SSM   CD/VIDEO   PH MATH TUTOR CENTER   MathXL®Tutorials ON CD   MathXL®   MyMathLab®

Evaluate. See Example 1.

1. $\sum_{i=1}^{4}(i - 3)$ -2

2. $\sum_{i=1}^{5}(i + 6)$ 45

 3. $\sum_{i=4}^{7}(2i + 4)$ 60

4. $\sum_{i=2}^{3}(5i - 1)$ 23

5. $\sum_{i=2}^{4}(i^2 - 3)$ 20

6. $\sum_{i=3}^{5}i^3$ 216

7. $\sum_{i=1}^{3}\left(\frac{1}{i + 5}\right)$ $\frac{73}{168}$

8. $\sum_{i=2}^{4}\left(\frac{2}{i + 3}\right)$ $\frac{107}{105}$

9. $\sum_{i=1}^{3}\frac{1}{6i}$ $\frac{11}{36}$

10. $\sum_{i=1}^{3}\frac{1}{3i}$ $\frac{11}{18}$

11. $\sum_{i=2}^{6}3i$ 60

12. $\sum_{i=3}^{6}-4i$ -72

13. $\sum_{i=3}^{5}i(i + 2)$ 74

14. $\sum_{i=2}^{4}i(i - 3)$ 2

15. $\sum_{i=1}^{5}2^i$ 62

16. $\sum_{i=1}^{4}3^{i-1}$ 40

17. $\sum_{i=1}^{4}\frac{4i}{i + 3}$ $\frac{241}{35}$

18. $\sum_{i=2}^{5}\frac{6 - i}{6 + i}$ $\frac{371}{330}$

Write each series with summation notation. See Example 2.

19. $1 + 3 + 5 + 7 + 9$

20. $4 + 7 + 10 + 13$

21. $4 + 12 + 36 + 108$

22. $5 + 10 + 20 + 40 + 80 + 160$

23. $12 + 9 + 6 + 3 + 0 + (-3)$

24. $5 + 1 + (-3) + (-7)$

25. $12 + 4 + \frac{4}{3} + \frac{4}{9}$ $\sum_{i=1}^{4}\frac{4}{3^{i-2}}$

26. $80 + 20 + 5 + \frac{5}{4} + \frac{5}{16}$ $\sum_{i=1}^{5}\frac{5}{4^{i-3}}$

41. $a_n = 6(2)^{n-1}$; 96 units   **42.** $a_n = 100(2)^n$; n represents the number of 6-hr periods; 1600 bacteria
43. $a_n = 50(2)^n$; n represents the number of 12-hr periods; 800 bacteria

27. $1 + 4 + 9 + 16 + 25 + 36 + 49$ $\sum\limits_{i=1}^{7} i^2$

28. $1 + (-4) + 9 + (-16)$ $\sum\limits_{i=1}^{4} (-1)^{i-1} i^2$

Find each partial sum. See Example 3.

29. Find the sum of the first two terms of the sequence whose general term is $a_n = (n + 2)(n - 5)$. -24

30. Find the sum of the first six terms of the sequence whose general term is $a_n = (-1)^n$. 0

31. Find the sum of the first two terms of the sequence whose general term is $a_n = n(n - 6)$. -13

32. Find the sum of the first seven terms of the sequence whose general term is $a_n = (-1)^{n-1}$. 1

33. Find the sum of the first four terms of the sequence whose general term is $a_n = (n + 3)(n + 1)$. 82

34. Find the sum of the first five terms of the sequence whose general term is $a_n = \dfrac{(-1)^n}{2n}$. $-\dfrac{47}{120}$

35. Find the sum of the first four terms of the sequence whose general term is $a_n = -2n$. -20

36. Find the sum of the first five terms of the sequence whose general term is $a_n = (n - 1)^2$. 30

37. Find the sum of the first three terms of the sequence whose general term is $a_n = -\dfrac{n}{3}$. -2

38. Find the sum of the first three terms of the sequence whose general term is $a_n = (n + 4)^2$. 110

Solve. See Example 4.

39. A gardener is making a triangular planting with 1 tree in the first row, 2 trees in the second row, 3 trees in the third row, and so on for 10 rows. Write the sequence that describes the number of trees in each row. Find the total number of trees planted. $1, 2, 3, \ldots, 10$; 55 trees

45. 30 opossums; 68 opossums   **46.** 12,000 fish; 48,000 fish
48. Job B pays $3000 more over the next 5 yrs

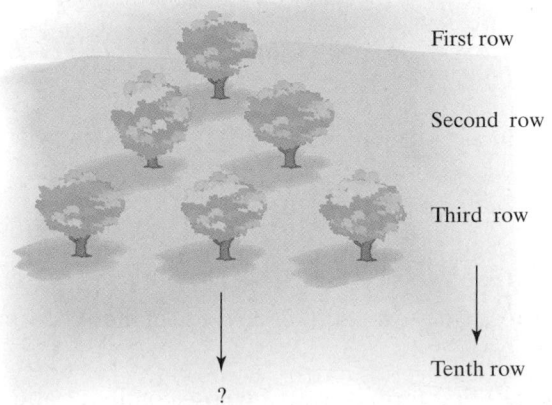

First row

Second row

Third row

Tenth row

?

40. Some surfers at the beach form a human pyramid with 2 surfers in the top row, 3 surfers in the second row, 4 surfers in the third row, and so on. If there are 6 rows in the pyramid, write the sequence that describes the number of surfers in each row of the pyramid. Find the total number of surfers. $2, 3, 4, 5, 6, 7$; 27 surfers

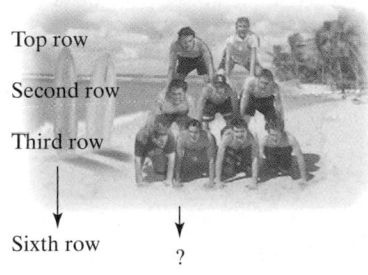

Top row

Second row

Third row

Sixth row   ?

41. A culture of fungus starts with 6 units and doubles every day. Write the general term of the sequence that describes the growth of this fungus. Find the number of fungus units there will be at the beginning of the fifth day.

42. A bacterial colony begins with 100 bacteria and doubles every 6 hours. Write the general term of the sequence describing the growth of the bacteria. Find the number of bacteria there will be after 24 hours.

43. A bacterial colony begins with 50 bacteria and doubles every 12 hours. Write the sequence that describes the growth of the bacteria. Find the number of bacteria there will be after 48 hours.

44. The number of otters born each year in a new aquarium forms a sequence whose general term is $a_n = (n - 1)(n + 3)$. Find the number of otters born in the third year, and find the total number of otters born in the first three years.   12 otters; 17 otters

45. The number of opossums killed each month on a new highway forms the sequence whose general term is $a_n = (n + 1)(n + 2)$, where n is the number of the months. Find the number of opossums killed in the fourth month, and find the total number killed in the first four months.

46. In 2003 the population of an endangered fish was estimated by environmentalists to be decreasing each year. The size of the population in a given year is $24 - 4n$ thousand fish fewer than the previous year. Find the decrease in population in 2005, if year 1 is 2003. Find the total decrease in the fish population for the years 2003 through 2005.

47. The amount of decay in pounds of a radioactive isotope each year is given by the sequence whose general term is $a_n = 100(0.5)^n$, where n is the number of the year. Find the amount of decay in the fourth year, and find the total amount of decay in the first four years.   6.25 lb; 93.75 lb

48. Susan has a choice between two job offers. Job A has an annual starting salary of $20,000 with guaranteed annual raises of $1200 for the next four years, whereas job B has an annual starting salary of $18,000 with guaranteed annual raises of $2500 for the next four years. Compare the fifth partial sums for each sequence to determine which job would pay Susan more money over the next 5 years.

49. A pendulum swings a length of 40 inches on its first swing. Each successive swing is $\frac{4}{5}$ of the preceding swing. Find the length of the fifth swing and the total length swung during the first five swings. (Round to the nearest tenth of an inch.)

50. Explain the difference between a sequence and a series. Answers may vary

REVIEW AND PREVIEW
49. 16.4 in.; 134.5 in.

Evaluate. See Section 1.3.

51. $\dfrac{5}{1 - \dfrac{1}{2}}$ 10

52. $\dfrac{-3}{1 - \dfrac{1}{7}}$ $-\dfrac{7}{2}$

53. $\dfrac{\dfrac{1}{3}}{1 - \dfrac{1}{10}}$ $\dfrac{10}{27}$

54. $\dfrac{\dfrac{6}{11}}{1 - \dfrac{1}{10}}$ $\dfrac{20}{33}$

55. $\dfrac{3(1 - 2^4)}{1 - 2}$ 45

56. $\dfrac{2(1 - 5^3)}{1 - 5}$ 62

57. $\dfrac{10}{2}(3 + 15)$ 90

58. $\dfrac{12}{2}(2 + 19)$ 126

59. a. $2 + 6 + 12 + 20 + 30 + 42 + 56$ **59. b.** $1 + 2 + 3 + 4 + 5 + 6 + 7 + 1 + 4 + 9 + 16 + 25 + 36 + 49$
60. a. $5 + 40 + 135 + 320 + 625 + 1080$

Concept Extensions

59. a. Write the sum $\sum_{i=1}^{7} (i + i^2)$ without summation notation.

b. Write the sum $\sum_{i=1}^{7} i + \sum_{i=1}^{7} i^2$ without summation notation.

c. Compare the results of parts **a** and **b.** Answers may vary

d. Do you think the following is true or false? Explain your answer.   true; answers may vary

$$\sum_{i=1}^{n} (a_n + b_n) = \sum_{i=1}^{n} a_n + \sum_{i=1}^{n} b_n$$

60. a. Write the sum $\sum_{i=1}^{6} 5i^3$ without summation notation.

b. Write the expression $5 \cdot \sum_{i=1}^{6} i^3$ without summation notation. $5(1 + 8 + 27 + 64 + 125 + 216)$

c. Compare the results of parts **a** and **b.**   Answers may vary

d. Do you think the following is true or false? Explain your answer.   true; answers may vary

$$\sum_{i=1}^{n} c \cdot a_n = c \cdot \sum_{i=1}^{n} a_n \text{ where } c \text{ is a constant}$$

INTEGRATED REVIEW   SEQUENCES AND SERIES

Write the first five terms of each sequence whose general term is given.

1. $a_n = n - 3$ $-2, -1, 0, 1, 2$

2. $a_n = \dfrac{7}{1 + n}$ $\dfrac{7}{2}, \dfrac{7}{3}, \dfrac{7}{4}, \dfrac{7}{5}, \dfrac{7}{6}$

3. $a_n = 3^{n-1}$ $1, 3, 9, 27, 81$

4. $a_n = n^2 - 5$ $-4, -1, 4, 11, 20$

Find the indicated term for each sequence.

5. $(-2)^n; a_6$ 64

6. $-n^2 + 2; a_4$ -14

7. $\dfrac{(-1)^n}{n}; a_{40}$ $\dfrac{1}{40}$

8. $\dfrac{(-1)^n}{2n}; a_{41}$ $-\dfrac{1}{82}$

Write the first five terms of the arithmetic or geometric sequence whose first term is a_1, and common difference, d, or common ratio, r, are given.

9. $a_1 = 7; d = -3$ $7, 4, 1, -2, -5$

10. $a_1 = -3; r = 5$ $-3, -15, -75, -375, -1875$

11. $a_1 = 45; r = \dfrac{1}{3}$ $45, 15, 5, \dfrac{5}{3}, \dfrac{5}{9}$

12. $a_1 = -12; d = 10$ $-12, -2, 8, 18, 28$

Find the indicated term of each sequence.

14. $\dfrac{243}{16}$

13. The tenth term of the arithmetic sequence whose first term is 20 and whose common difference is 9.   101

14. The sixth term of the geometric sequence whose first term is 64 and whose common ratio is $\dfrac{3}{4}$.

15. The seventh term of the geometric sequence $6, -12, 24, \ldots$ 384

16. The twentieth term of the arithmetic sequence $-100, -85, -70, \ldots$ 185

17. −10

17. The fifth term of the arithmetic sequence whose fourth term is -5 and whose tenth term is -35.

18. The fifth term of the geometric sequence whose fourth term is 1 and whose seventh term is $\dfrac{1}{125}$.

18. $\dfrac{1}{5}$

Evaluate.

19. $\displaystyle\sum_{i=1}^{4} 5i$ 50
20. $\displaystyle\sum_{i=1}^{7} (3i + 2)$ 98
21. $\displaystyle\sum_{i=3}^{7} 2^{i-4}$ $\dfrac{31}{2}$
22. $\displaystyle\sum_{i=2}^{5} \dfrac{i}{i+1}$ $\dfrac{61}{20}$

Find each partial sum.

23. Find the sum of the first three terms of the sequence whose general term is $a_n = n(n - 4)$.
−10

24. Find the sum of the first ten terms of the sequence whose general term is $a_n = (-1)^n(n + 1)$.
5

11.4   *PARTIAL SUMS OF ARITHMETIC AND GEOMETRIC SEQUENCES*

Objectives

1   Find the partial sum of an arithmetic sequence.

2   Find the partial sum of a geometric sequence.

3   Find the sum of the terms of an infinite geometric sequence.

1   Partial sums S_n are relatively easy to find when n is small—that is, when the number of terms to add is small. But when n is large, finding S_n can be tedious. For a large n, S_n is still relatively easy to find if the addends are terms of an arithmetic sequence or a geometric sequence.

For an arithmetic sequence, $a_n = a_1 + (n - 1)d$ for some first term a_1 and some common difference d. So S_n, the sum of the first n terms, is

$$S_n = a_1 + (a_1 + d) + (a_1 + 2d) + \cdots + (a_1 + (n - 1)d)$$

We might also find S_n by "working backward" from the nth term a_n, finding the preceding term a_{n-1}, by subtracting d each time.

$$S_n = a_n + (a_n - d) + (a_n - 2d) + \cdots + (a_n - (n - 1)d)$$

Now add the left sides of these two equations and add the right sides.

$$2S_n = (a_1 + a_n) + (a_1 + a_n) + (a_1 + a_n) + \cdots + (a_1 + a_n)$$

The d terms subtract out, leaving n sums of the first term, a_1, and last term, a_n. Thus, we write

$$2S_n = n(a_1 + a_n)$$

or

$$S_n = \frac{n}{2}(a_1 + a_n)$$

> **Partial Sum S_n of an Arithmetic Sequence**
>
> The partial sum S_n of the first n terms of an arithmetic sequence is given by
>
> $$S_n = \frac{n}{2}(a_1 + a_n)$$
>
> where a_1 is the first term of the sequence and a_n is the nth term.

EXAMPLE 1

Use the partial sum formula to find the sum of the first six terms of the arithmetic sequence $2, 5, 8, 11, 14, 17, \ldots$.

Solution   Use the formula for S_n of an arithmetic sequence, replacing n with 6, a_1 with 2, and a_n with 17.

$$S_n = \frac{n}{2}(a_1 + a_n) = \frac{6}{2}(2 + 17) = 3(19) = 57$$

EXAMPLE 2

Find the sum of the first 30 positive integers.

Solution   Because $1, 2, 3, \ldots, 30$ is an arithmetic sequence, use the formula for S_n with $n = 30$, $a_1 = 1$, and $a_n = 30$. Thus,

$$S_n = \frac{n}{2}(a_1 + a_n) = \frac{30}{2}(1 + 30) = 15(31) = 465$$

EXAMPLE 3

STACKING ROLLS OF CARPET

Rolls of carpet are stacked in 20 rows with 3 rolls in the top row, 4 rolls in the next row, and so on, forming an arithmetic sequence. Find the total number of carpet rolls if there are 22 rolls in the bottom row.

3 rolls

4 rolls

5 rolls

Solution   The list $3, 4, 5, \ldots, 22$ is the first 20 terms of an arithmetic sequence. Use the formula for S_n with $a_1 = 3$, $a_n = 22$, and $n = 20$ terms. Thus,

$$S_{20} = \frac{20}{2}(3 + 22) = 10(25) = 250$$

There are a total of 250 rolls of carpet.

2   We can also derive a formula for the partial sum S_n of the first n terms of a geometric series. If $a_n = a_1 r^{n-1}$, then

$$S_n = a_1 + a_1 r + a_1 r^2 + \cdots + a_1 r^{n-1}$$

$\uparrow$ $\uparrow$ $\uparrow$ $\uparrow$
1st **2nd** **3rd** **nth**
term term term term

Multiply each side of the equation by $-r$.

$$-rS_n = -a_1 r - a_1 r^2 - a_1 r^3 - \cdots - a_1 r^n$$

Add the two equations.

$$S_n - rS_n = a_1 + (a_1 r - a_1 r) + (a_1 r^2 - a_1 r^2) + (a_1 r^3 - a_1 r^3) + \cdots - a_1 r^n$$
$$S_n - rS_n = a_1 - a_1 r^n$$

Now factor each side.

$$S_n(1 - r) = a_1(1 - r^n)$$

Solve for S_n by dividing both sides by $1 - r$. Thus,

$$S_n = \frac{a_1(1 - r^n)}{1 - r}$$

as long as r is not 1.

Partial Sum S_n of a Geometric Sequence

The partial sum S_n of the first n terms of a geometric sequence is given by

$$S_n = \frac{a_1(1 - r^n)}{1 - r}$$

where a_1 is the first term of the sequence, r is the common ratio, and $r \neq 1$.

EXAMPLE 4

Find the sum of the first six terms of the geometric sequence $5, 10, 20, 40, 80, 160$.

Solution   Use the formula for the partial sum S_n of the terms of a geometric sequence. Here, $n = 6$, the first term $a_1 = 5$, and the common ratio $r = 2$.

$$S_n = \frac{a_1(1 - r^n)}{1 - r}$$

$$S_6 = \frac{5(1 - 2^6)}{1 - 2} = \frac{5(-63)}{-1} = 315$$

CLASSROOM EXAMPLE

Find the sum of the first seven terms of the geometric sequence. $16, 8, 4, 2, 1, \frac{1}{2}, \frac{1}{4}$.

answer: $\frac{127}{4}$ or $31\frac{3}{4}$

EXAMPLE 5

FINDING AMOUNT OF DONATION

A grant from an alumnus to a university specified that the university was to receive $800,000 during the first year and 75% of the preceding year's donation during each of the following 5 years. Find the total amount donated during the 6 years.

Solution The donations are modeled by the first six terms of a geometric sequence. Evaluate S_n when $n = 6$, $a_1 = 800,000$, and $r = 0.75$.

$$S_6 = \frac{800,000[1 - (0.75)^6]}{1 - 0.75}$$

$$= \$2,630,468.75$$

The total amount donated during the 6 years is $2,630,468.75.

3 Is it possible to find the sum of all the terms of an infinite sequence? Examine the partial sums of the geometric sequence $\frac{1}{2}, \frac{1}{4}, \frac{1}{8}, \dots$.

$$S_1 = \frac{1}{2}$$

$$S_2 = \frac{1}{2} + \frac{1}{4} = \frac{3}{4}$$

$$S_3 = \frac{1}{2} + \frac{1}{4} + \frac{1}{8} = \frac{7}{8}$$

$$S_4 = \frac{1}{2} + \frac{1}{4} + \frac{1}{8} + \frac{1}{16} = \frac{15}{16}$$

$$S_5 = \frac{1}{2} + \frac{1}{4} + \frac{1}{8} + \frac{1}{16} + \frac{1}{32} = \frac{31}{32}$$

$$\vdots$$

$$S_{10} = \frac{1}{2} + \frac{1}{4} + \frac{1}{8} + \cdots + \frac{1}{2^{10}} = \frac{1023}{1024}$$

Even though each partial sum is larger than the preceding partial sum, we see that each partial sum is closer to 1 than the preceding partial sum. If n gets larger and larger, then S_n gets closer and closer to 1. We say that 1 is the **limit** of S_n and also that 1 is the sum of the terms of this infinite sequence. In general, if $|r| < 1$, the following formula gives the sum of the terms of an infinite geometric sequence.

Sum of the Terms of an Infinite Geometric Sequence

The sum S_∞ of the terms of an infinite geometric sequence is given by

$$S_\infty = \frac{a_1}{1 - r}$$

where a_1 is the first term of the sequence, r is the common ratio, and $|r| < 1$. If $|r| \geq 1$, S_∞ does not exist.

What happens for other values of r? For example, in the following geometric sequence, $r = 3$.

$$6, 18, 54, 162, \dots$$

Here, as n increases, the sum S_n increases also. This time, though, S_n does not get closer and closer to a fixed number but instead increases without bound.

EXAMPLE 6

Find the sum of the terms of the geometric sequence $2, \dfrac{2}{3}, \dfrac{2}{9}, \dfrac{2}{27}, \ldots$.

Solution For this geometric sequence, $r = \dfrac{1}{3}$. Since $|r| < 1$, we may use the formula for S_∞ of a geometric sequence with $a_1 = 2$ and $r = \dfrac{1}{3}$.

$$S_\infty = \frac{a_1}{1 - r} = \frac{2}{1 - \dfrac{1}{3}} = \frac{2}{\dfrac{2}{3}} = 3$$

The formula for the sum of the terms of an infinite geometric sequence can be used to write a repeating decimal as a fraction. For example,

$$0.33\overline{3} = \frac{3}{10} + \frac{3}{100} + \frac{3}{1000} + \cdots$$

This sum is the sum of the terms of an infinite geometric sequence whose first term a_1 is $\dfrac{3}{10}$ and whose common ratio r is $\dfrac{1}{10}$. Using the formula for S_∞,

$$S_\infty = \frac{a_1}{1 - r} = \frac{\dfrac{3}{10}}{1 - \dfrac{1}{10}} = \frac{1}{3}$$

So, $0.33\overline{3} = \dfrac{1}{3}$.

EXAMPLE 7

DISTANCE TRAVELED BY A PENDULUM

On its first pass, a pendulum swings through an arc whose length is 24 inches. On each pass thereafter, the arc length is 75% of the arc length on the preceding pass. Find the total distance the pendulum travels before it comes to rest.

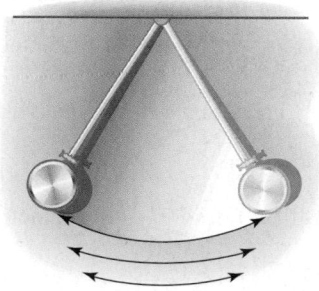

Solution We must find the sum of the terms of an infinite geometric sequence whose first term, a_1, is 24 and whose common ratio, r, is 0.75. Since $|r| < 1$, we may use the formula for S_∞.

$$S_\infty = \frac{a_1}{1 - r} = \frac{24}{1 - 0.75} = \frac{24}{0.25} = 96$$

The pendulum travels a total distance of 96 inches before it comes to rest.

Spotlight on
DECISION
⌇MAKING

Suppose you are a certified financial planner. You are working with a 30-year-old client whose goal is to retire at age 65 with a sum of $500,000 to live on. This year she just started making the maximum annual contribution of $2000 to a Roth IRA (Individual Retirement Account) that pays 8% interest compounded annually.

You know that if she continues to make a $2000 contribution at the beginning of each year, by the end of the nth year, her account increases in value by the nth term of the geometric sequence $a_n = 2000(1.08)^n$, which considers both the annual $2000 contribution and her earned interest. Decide whether your client will be able to reach her retirement goal by making only the maximum Roth IRA contribution each year until she retires, or if you should suggest an additional investment to help her reach her goal. (Note: Use a partial sum to find the value of the Roth IRA at the end of 35 years. To find a_1 of the geometric sequence, be sure to evaluate the equation for the nth term at $n = 1$.)

EXERCISE SET 11.4

STUDY GUIDE/SSM   CD/VIDEO   PH MATH TUTOR CENTER   MathXL®Tutorials ON CD   MathXL®   MyMathLab®

Use the partial sum formula to find the partial sum of the given arithmetic or geometric sequence. See Examples 1 and 4.

1. Find the sum of the first six terms of the arithmetic sequence $1, 3, 5, 7, \ldots$.   36

2. Find the sum of the first seven terms of the arithmetic sequence $-7, -11, -15, \ldots$. -133

 3. Find the sum of the first five terms of the geometric sequence $4, 12, 36, \ldots$.   484

4. Find the sum of the first eight terms of the geometric sequence $-1, 2, -4, \ldots$.   85

 5. Find the sum of the first six terms of the arithmetic sequence $3, 6, 9, \ldots$.   63

6. Find the sum of the first four terms of the arithmetic sequence $-4, -8, -12, \ldots$. -40

7. Find the sum of the first four terms of the geometric sequence $2, \dfrac{2}{5}, \dfrac{2}{25}, \ldots$.   2.496

8. Find the sum of the first five terms of the geometric sequence $\dfrac{1}{3}, -\dfrac{2}{3}, \dfrac{4}{3}, \ldots$. $\dfrac{11}{3}$

Solve. See Example 2.

9. Find the sum of the first ten positive integers.   55

10. Find the sum of the first eight negative integers. -36

11. Find the sum of the first four positive odd integers.   16

12. Find the sum of the first five negative odd integers. -25

Find the sum of the terms of each infinite geometric sequence. See Example 6.

13. $12, 6, 3, \ldots$ 24

14. $45, 15, 5, \ldots$ 67.5

15. $\dfrac{1}{10}, \dfrac{1}{100}, \dfrac{1}{1000}, \ldots$ $\dfrac{1}{9}$

16. $\dfrac{3}{5}, \dfrac{3}{20}, \dfrac{3}{80}, \ldots$ $\dfrac{4}{5}$

17. $-10, -5, -\dfrac{5}{2}, \ldots$ -20

18. $-16, -4, -1, \ldots$ $-\dfrac{64}{3}$

19. $2, -\dfrac{1}{4}, \dfrac{1}{32}, \ldots$ $\dfrac{16}{9}$

20. $-3, \dfrac{3}{5}, -\dfrac{3}{25}, \ldots$ $-\dfrac{5}{2}$

21. $\dfrac{2}{3}, -\dfrac{1}{3}, \dfrac{1}{6}, \ldots$ $\dfrac{4}{9}$

22. $6, -4, \dfrac{8}{3}, \ldots$ $\dfrac{18}{5}$

MIXED PRACTICE

Solve.   24. -696

23. Find the sum of the first ten terms of the sequence $-4, 1, 6, \ldots, 41$ where 41 is the tenth term.   185

24. Find the sum of the first twelve terms of the sequence $-3, -13, -23, \ldots, -113$ where -113 is the twelfth term.

25. Find the sum of the first seven terms of the sequence $3, \dfrac{3}{2}, \dfrac{3}{4}, \ldots$. $\dfrac{381}{64}$

26. Find the sum of the first five terms of the sequence $-2, -6, -18, \ldots$. -242

27. Find the sum of the first five terms of the sequence $-12, 6, -3, \ldots$. $-\dfrac{33}{4}$, or -8.25

28. Find the sum of the first four terms of the sequence $-\dfrac{1}{4}, -\dfrac{3}{4}, -\dfrac{9}{4}, \ldots$ -10

29. Find the sum of the first twenty terms of the sequence $\dfrac{1}{2}, \dfrac{1}{4}, 0, \ldots, -\dfrac{17}{4}$ where $-\dfrac{17}{4}$ is the twentieth term. $-\dfrac{75}{2}$

30. Find the sum of the first fifteen terms of the sequence $-5, -9, -13, \ldots, -61$ where -61 is the fifteenth term. -495

31. If a_1 is 8 and r is $-\dfrac{2}{3}$, find S_3. $\dfrac{56}{9}$

32. If a_1 is 10, a_{18} is $\dfrac{3}{2}$, and d is $-\dfrac{1}{2}$, find S_{18}. $\dfrac{207}{2}$

Solve. See Example 3.

33. 4000, 3950, 3900, 3850, 3800; 3450 cars; 44,700 cars

33. Modern Car Company has come out with a new car model. Market analysts predict that 4000 cars will be sold in the first month and that sales will drop by 50 cars per month after that during the first year. Write out the first five terms of the sequence, and find the number of sold cars predicted for the twelfth month. Find the total predicted number of sold cars for the first year.

34. 3, 2.90, 2.80, 2.70, 2.60; $23.40

34. A company that sends faxes charges $3 for the first page sent and $0.10 less than the preceding page for each additional page sent. The cost per page forms an arithmetic sequence. Write the first five terms of this sequence, and use a partial sum to find the cost of sending a nine-page document.

35. Sal has two job offers: Firm A starts at $22,000 per year and guarantees raises of $1000 per year, whereas Firm B starts at $20,000 and guarantees raises of $1200 per year. Over a 10-year period, determine the more profitable offer.

36. The game of pool uses 15 balls numbered 1 to 15. In the variety called rotation, a player who sinks a ball receives as many points as the number on the ball. Use an arithmetic series to find the score of a player who sinks all 15 balls.   120 points

35. Firm A (Firm A, $265,000; Firm B, $254,000)

Solve. See Example 5.

38. 240 ft; 1024 ft

37. A woman made $30,000 during the first year she owned her business and made an additional 10% over the previous year in each subsequent year. Find how much she made during her fourth year of business. Find her total earnings during the first four years. $39,930; $139,230

38. In free fall, a parachutist falls 16 feet during the first second, 48 feet during the second second, 80 feet during the third second, and so on. Find how far she falls during the eighth second. Find the total distance she falls during the first 8 seconds.

39. A trainee in a computer company takes 0.9 times as long to assemble each computer as he took to assemble the preceding computer. If it took him 30 minutes to assemble the first computer, find how long it takes him to assemble the fifth computer. Find the total time he takes to assemble the first five computers (round answers to the nearest minute).   20 min; 123 min

40. On a gambling trip to Reno, Carol doubled her bet each time she lost. If her first losing bet was $5 and she lost six consecutive bets, find how much she lost on the sixth bet. Find the total amount lost on these six bets. $160; $315

Solve. See Example 7.

41. A ball is dropped from a height of 20 feet and repeatedly rebounds to a height that is $\dfrac{4}{5}$ of its previous height. Find the total distance the ball covers before it comes to rest.   180 ft

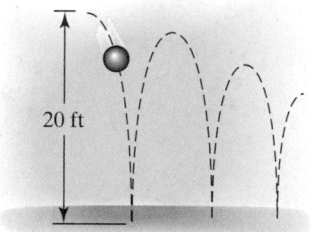

42. A rotating flywheel coming to rest makes 300 revolutions in the first minute and in each minute thereafter makes $\dfrac{2}{5}$ as many revolutions as in the preceding minute. Find how many revolutions the wheel makes before it comes to rest.   500 revolutions

43. Player A, 45 points; Player B, 75 points
44. $1300; $17,050   **46.** 8525 weevils

Solve.

43. In the pool game of rotation, player A sinks balls numbered 1 to 9, and player B sinks the rest of the balls. Use arithmetic series to find each player's score (see Exercise 36).

44. A godfather deposited $250 in a savings account on the day his godchild was born. On each subsequent birthday he deposited $50 more than he deposited the previous year. Find how much money he deposited on his godchild's twenty-first birthday. Find the total amount deposited over the 21 years.

45. During the holiday rush a business can rent a computer system for $200 the first day, with the rental fee decreasing $5 for each additional day. Find the fee paid for 20 days during the holiday rush. $3050

46. The spraying of a field with insecticide killed 6400 weevils the first day, 1600 the second day, 400 the third day, and so on. Find the total number of weevils killed during the first 5 days.

47. A college student humorously asks his parents to charge him room and board according to this geometric sequence: $0.01 for the first day of the month, $0.02 for the second day, $0.04 for the third day, and so on. Find the total room and board he would pay for 30 days. $10,737,418.23

48. Following its television advertising campaign, a bank attracted 80 new customers the first day, 120 the second day, 160 the third day, and so on, in an arithmetic sequence. Find how many new customers were attracted during the first 5 days following its television campaign.   800 customers

REVIEW AND PREVIEW

Evaluate. See Section 1.3.

49. $6 \cdot 5 \cdot 4 \cdot 3 \cdot 2 \cdot 1$ 720

50. $8 \cdot 7 \cdot 6 \cdot 5 \cdot 4 \cdot 3 \cdot 2 \cdot 1$ 40,320

51. $\dfrac{3 \cdot 2 \cdot 1}{2 \cdot 1}$ 3

52. $\dfrac{5 \cdot 4 \cdot 3 \cdot 2 \cdot 1}{3 \cdot 2 \cdot 1}$ 20

Multiply. See Section 5.4.

53. $(x + 5)^2$ $x^2 + 10x + 25$ **54.** $(x - 2)^2$ $x^2 - 4x + 4$

55. $(2x - 1)^3$ $8x^3 - 12x^2 + 6x - 1$

56. $(3x + 2)^3$ $27x^3 + 54x^2 + 36x + 8$

Concept Extensions

57. Write $0.88\overline{8}$ as an infinite geometric series and use the formula for S_∞ to write it as a rational number.

58. Write $0.54\overline{54}$ as an infinite geometric series and use the formula S_∞ to write it as a rational number.

59. Explain whether the sequence $5, 5, 5, \ldots$ is arithmetic, geometric, neither, or both.   answers may vary

60. Describe a situation in everyday life that can be modeled by an infinite geometric series.   answers may vary

57. $\dfrac{8}{10} + \dfrac{8}{100} + \dfrac{8}{1000} + \cdots ; \dfrac{8}{9}$

58. $\dfrac{54}{100} + \dfrac{54}{10,000} + \dfrac{54}{1,000,000} + \cdots ; \dfrac{6}{11}$

11.5 THE BINOMIAL THEOREM

Objectives

1 Use Pascal's triangle to expand binomials.

2 Evaluate factorials.

3 Use the binomial theorem to expand binomials.

4 Find the nth term in the expansion of a binomial raised to a positive power.

In this section, we learn how to **expand** binomials of the form $(a + b)^n$ easily. Expanding a binomial such as $(a + b)^n$ means to write the factored form as a sum. First, we review the patterns in the expansions of $(a + b)^n$.

$$(a + b)^0 = 1 \qquad \textbf{1 term}$$

$$(a + b)^1 = a + b \qquad \textbf{2 terms}$$

$$(a + b)^2 = a^2 + 2ab + b^2 \qquad \textbf{3 terms}$$

$$(a + b)^3 = a^3 + 3a^2b + 3ab^2 + b^3 \qquad \textbf{4 terms}$$

$$(a + b)^4 = a^4 + 4a^3b + 6a^2b^2 + 4ab^3 + b^4 \qquad \textbf{5 terms}$$

$$(a + b)^5 = a^5 + 5a^4b + 10a^3b^2 + 10a^2b^3 + 5ab^4 + b^5 \qquad \textbf{6 terms}$$

Notice the following patterns.

1. The expansion of $(a + b)^n$ contains $n + 1$ terms. For example, for $(a + b)^3$, $n = 3$, and the expansion contains $3 + 1$ terms, or 4 terms.

2. The first term of the expansion of $(a + b)^n$ is a^n, and the last term is b^n.

3. The powers of a decrease by 1 for each term, whereas the powers of b increase by 1 for each term.

4. For each term of the expansion of $(a + b)^n$, the sum of the exponents of a and b is n. (For example, the sum of the exponents of $5a^4b$ is $4 + 1$, or 5, and the sum of the exponents of $10a^3b^2$ is $3 + 2$, or 5.)

1 There are patterns in the coefficients of the terms as well. Written in a triangular array, the coefficients are called **Pascal's triangle.**

$(a + b)^0$:						1						$n = 0$
$(a + b)^1$:					1		1					$n = 1$
$(a + b)^2$:				1		2		1				$n = 2$
$(a + b)^3$:			1		3		3		1			$n = 3$
$(a + b)^4$:		1		4		6		4		1		$n = 4$
$(a + b)^5$:	1		5		10		10		5		1	$n = 5$

Each row in Pascal's triangle begins and ends with 1. Any other number in a row is the sum of the two closest numbers above it. Using this pattern, we can write the next row, for $n = 6$, by first writing the number 1. Then we can add the consecutive numbers in the row for $n = 5$ and write each sum "between and below" the pair. We complete the row by writing a 1.

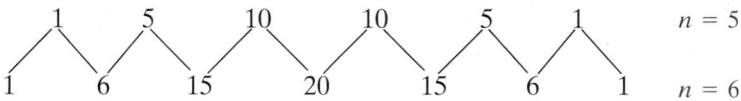

We can use Pascal's triangle and the patterns noted to expand $(a + b)^n$ without actually multiplying any terms.

EXAMPLE 1

Expand $(a + b)^6$.

Solution Using the $n = 6$ row of Pascal's triangle as the coefficients and following the patterns noted, $(a + b)^6$ can be expanded as

$$a^6 + 6a^5b + 15a^4b^2 + 20a^3b^3 + 15a^2b^4 + 6ab^5 + b^6$$

CLASSROOM EXAMPLE
Expand $(x + y)^5$.
answer:
$x^5 + 5x^4y + 10x^3y^2 + 10x^2y^3 + 5xy^4 + y^5$

2 For a large n, the use of Pascal's triangle to find coefficients for $(a + b)^n$ can be tedious. An alternative method for determining these coefficients is based on the concept of a **factorial**.

The **factorial of n**, written $n!$ (read "n factorial"), is the product of the first n consecutive natural numbers.

Factorial of n: $n!$

If n is a natural number, then $n! = n(n - 1)(n - 2)(n - 3) \cdots \cdot 3 \cdot 2 \cdot 1$. The factorial of 0, written $0!$, is defined to be 1.

For example, $3! = 3 \cdot 2 \cdot 1 = 6$, $\quad 5! = 5 \cdot 4 \cdot 3 \cdot 2 \cdot 1 = 120$, $\quad$ and $0! = 1$.

EXAMPLE 2

CLASSROOM EXAMPLE
Evaluate.
a. $\dfrac{7!}{8!}$ **b.** $\dfrac{9!}{5!4!}$ **c.** $\dfrac{4!}{3!1!}$ **d.** $\dfrac{8!}{8!0!}$
answer:
a. $\dfrac{1}{8}$ **b.** 126 **c.** 4 **d.** 1

Evaluate each expression.

a. $\dfrac{5!}{6!}$ **b.** $\dfrac{10!}{7!3!}$

c. $\dfrac{3!}{2!1!}$ **d.** $\dfrac{7!}{7!0!}$

Solution   **a.** $\dfrac{5!}{6!} = \dfrac{5 \cdot 4 \cdot 3 \cdot 2 \cdot 1}{6 \cdot 5 \cdot 4 \cdot 3 \cdot 2 \cdot 1} = \dfrac{1}{6}$

b. $\dfrac{10!}{7!3!} = \dfrac{10 \cdot 9 \cdot 8 \cdot 7!}{7! \cdot 3 \cdot 2 \cdot 1} = \dfrac{10 \cdot 9 \cdot 8}{3 \cdot 2 \cdot 1} = 10 \cdot 3 \cdot 4 = 120$

c. $\dfrac{3!}{2!1!} = \dfrac{3 \cdot 2 \cdot 1}{2 \cdot 1 \cdot 1} = 3$

d. $\dfrac{7!}{7!0!} = \dfrac{7!}{7! \cdot 1} = 1$

TECHNOLOGY NOTE

When evalutating factorials using a graphing utility, notice the need for parentheses grouping the denominator.

Helpful Hint

We can use a calculator with a factorial key to evaluate a factorial. A calculator uses scientific notation for large results.

3   It can be proved, although we won't do so here, that the coefficients of terms in the expansion of $(a + b)^n$ can be expressed in terms of factorials. Following patterns 1 through 4 given earlier and using the factorial expressions of the coefficients, we have what is known as the **binomial theorem.**

Binomial Theorem

If n is a positive integer, then

$$(a + b)^n = a^n + \dfrac{n}{1!}a^{n-1}b^1 + \dfrac{n(n-1)}{2!}a^{n-2}b^2$$

$$+ \dfrac{n(n-1)(n-2)}{3!}a^{n-3}b^3 + \cdots + b^n$$

We call the formula for $(a + b)^n$ given by the binomial theorem the **binomial formula.**

EXAMPLE 3

Use the binomial theorem to expand $(x + y)^{10}$.

Solution   Let $a = x$, $b = y$, and $n = 10$ in the binomial formula.

$$(x + y)^{10} = x^{10} + \dfrac{10}{1!}x^9y + \dfrac{10 \cdot 9}{2!}x^8y^2 + \dfrac{10 \cdot 9 \cdot 8}{3!}x^7y^3 + \dfrac{10 \cdot 9 \cdot 8 \cdot 7}{4!}x^6y^4$$

$$+ \dfrac{10 \cdot 9 \cdot 8 \cdot 7 \cdot 6}{5!}x^5y^5 + \dfrac{10 \cdot 9 \cdot 8 \cdot 7 \cdot 6 \cdot 5}{6!}x^4y^6$$

$$+ \dfrac{10 \cdot 9 \cdot 8 \cdot 7 \cdot 6 \cdot 5 \cdot 4}{7!}x^3y^7$$

$$+ \dfrac{10 \cdot 9 \cdot 8 \cdot 7 \cdot 6 \cdot 5 \cdot 4 \cdot 3}{8!}x^2y^8$$

$$+ \dfrac{10 \cdot 9 \cdot 8 \cdot 7 \cdot 6 \cdot 5 \cdot 4 \cdot 3 \cdot 2}{9!}xy^9 + y^{10}$$

$$= x^{10} + 10x^9y + 45x^8y^2 + 120x^7y^3 + 210x^6y^4 + 252x^5y^5 + 210x^4y^6$$

$$+ 120x^3y^7 + 45x^2y^8 + 10xy^9 + y^{10}$$

EXAMPLE 4

Use the binomial theorem to expand $(x + 2)^5$.

Solution   Let $a = x$ and $b = 2$ in the binomial formula.

$$(x + 2)^5 = x^5 + \frac{5}{1!}x^4(2) + \frac{5 \cdot 4}{2!}x^3(2)^2 + \frac{5 \cdot 4 \cdot 3}{3!}x^2(2)^3$$

$$+ \frac{5 \cdot 4 \cdot 3 \cdot 2}{4!}x(2)^4 + (2)^5$$

$$= x^5 + 10x^4 + 40x^3 + 80x^2 + 80x + 32$$

Technology Note

To visually check Example 4, graph $y_1 = (x + 2)^5$ and $y_2 = x^5 + 10x^4 + 40x^3 + 80x^2 + 80x + 32$ in the same viewing window. If the graphs do not coincide, you should double-check your binomial expansion. In this case, the graphs appear to coincide, reinforcing the conclusion of Example 4.

EXAMPLE 5

Use the binomial theorem to expand $(3m - n)^4$.

Solution   Let $a = 3m$ and $b = -n$ in the binomial formula.

$$(3m - n)^4 = (3m)^4 + \frac{4}{1!}(3m)^3(-n) + \frac{4 \cdot 3}{2!}(3m)^2(-n)^2$$

$$+ \frac{4 \cdot 3 \cdot 2}{3!}(3m)(-n)^3 + (-n)^4$$

$$= 81m^4 - 108m^3n + 54m^2n^2 - 12mn^3 + n^4$$

4   Sometimes it is convenient to find a specific term of a binomial expansion without writing out the entire expansion. By studying the expansion of binomials, a pattern forms for each term. This pattern is most easily stated for the $(r + 1)$st term.

$(r + 1)st$ Term in a Binomial Expansion

The $(r + 1)$st term of the expansion of $(a + b)^n$ is $\dfrac{n!}{r!(n - r)!}a^{n-r}b^r$.

EXAMPLE 6

Find the eighth term in the expansion of $(2x - y)^{10}$.

Solution   Use the formula, with $n = 10, a = 2x, b = -y$, and $r + 1 = 8$. Notice that, since $r + 1 = 8, r = 7$.

$$\frac{n!}{r!(n - r)!}a^{n-r}b^r = \frac{10!}{7!3!}(2x)^3(-y)^7$$

$$= 120(8x^3)(-y^7)$$

$$= -960x^3y^7$$

EXERCISE SET 11.5

STUDY GUIDE/SSM   CD/VIDEO   PH MATH TUTOR CENTER   MathXL®Tutorials ON CD   MathXL®   MyMathLab®

Use Pascal's triangle to expand the binomial. See Example 1.

1. $(m + n)^3$ $m^3 + 3m^2n + 3mn^2 + n^3$

2. $(x + y)^4$ $x^4 + 4x^3y + 6x^2y^2 + 4xy^3 + y^4$

3. $(c + d)^5$ $c^5 + 5c^4d + 10c^3d^2 + 10c^2d^3 + 5cd^4 + d^5$

4. $(a + b)^6$

5. $(y - x)^5$ $y^5 - 5y^4x + 10y^3x^2 - 10y^2x^3 + 5yx^4 - x^5$

6. $(q - r)^7$

7. Explain how to generate a row of Pascal's triangle.

8. Write the $n = 8$ row of Pascal's triangle.

Evaluate each expression. See Example 2.

9. $\dfrac{8!}{7!}$ 8

10. $\dfrac{6!}{0!}$ 720

11. $\dfrac{7!}{5!}$ 42

12. $\dfrac{8!}{5!}$ 336

13. $\dfrac{10!}{7!2!}$ 360

14. $\dfrac{9!}{5!3!}$ 504

15. $\dfrac{8!}{6!0!}$ 56

16. $\dfrac{10!}{4!6!}$ 210

Use the binomial formula to expand each binomial. See Examples 3 through 5.

17. $(a + b)^7$

18. $(x + y)^8$

19. $(a + 2b)^5$

20. $(x + 3y)^6$

21. $(q + r)^9$

22. $(b + c)^6$

23. $(4a + b)^5$

24. $(3m + n)^4$

25. $(5a - 2b)^4$

26. $(m - 4)^6$

27. $(2a + 3b)^3$

28. $(4 - 3x)^5$

29. $(x + 2)^5$

30. $(3 + 2a)^4$

Find the indicated term. See Example 6.

31. The fifth term of the expansion of $(c - d)^5$ $5cd^4$

32. The fourth term of the expansion of $(x - y)^6$ $-20x^3y^3$

33. The eighth term of the expansion of $(2c + d)^7$ d^7

34. The tenth term of the expansion of $(5x - y)^9$ $-y^9$

35. The fourth term of the expansion of $(2r - s)^5$ $-40r^2s^3$

36. The first term of the expansion of $(3q - 7r)^6$ $729q^6$

37. The third term of the expansion of $(x + y)^4$ $6x^2y^2$

38. The fourth term of the expansion of $(a + b)^8$ $56a^5b^3$

39. The second term of the expansion of $(a + 3b)^{10}$ $30a^9b$

40. The third term of the expansion of $(m + 5n)^7$ $525m^5n^2$

REVIEW AND PREVIEW

Sketch the graph of each function. Decide whether each function is one-to-one. See Sections 2.3 and 8.5.

41. $f(x) = |x|$

42. $g(x) = 3(x - 1)^2$

43. $H(x) = 2x + 3$

44. $F(x) = -2$

45. $f(x) = x^2 + 3$

46. $h(x) = -(x + 1)^2 - 4$

41.–46. See graphing answer section.

Concept Extensions

47. Expand the expression $\left(\sqrt{x} + \sqrt{3}\right)^5$

48. Find the term containing x^2 in the expansion of $\left(\sqrt{x} - \sqrt{5}\right)^6$ $75x^2$

Evaluate the following.

The notation $\dbinom{n}{r}$ means $\dfrac{n!}{r!(n - r)!}$. For example,

$$\binom{5}{3} = \frac{5!}{3!(5 - 3)!} = \frac{5!}{3!2!} = \frac{5 \cdot 4 \cdot 3 \cdot 2 \cdot 1}{(3 \cdot 2 \cdot 1) \cdot (2 \cdot 1)} = 10.$$

49. $\dbinom{9}{5}$ 126

50. $\dbinom{4}{3}$ 4

51. $\dbinom{8}{2}$ 28

52. $\dbinom{12}{11}$ 12

53. Show that $\dbinom{n}{n} = 1$ for any whole number n.

   answers may vary

---

4. $a^6 + 6a^5b + 15a^4b^2 + 20a^3b^3 + 15a^2b^4 + 6ab^5 + b^6$ **6.** $q^7 - 7q^6r + 21q^5r^2 - 35q^4r^3 + 35q^3r^4 - 21q^2r^5 + 7qr^6 - r^7$
7. answers may vary **8.** 1 8 28 56 70 56 28 8 1 **17.** $a^7 + 7a^6b + 21a^5b^2 + 35a^4b^3 + 35a^3b^4 + 21a^2b^5 + 7ab^6 + b^7$
18. $x^8 + 8x^7y + 28x^6y^2 + 56x^5y^3 + 70x^4y^4 + 56x^3y^5 + 28x^2y^6 + 8xy^7 + y^8$
19. $a^5 + 10a^4b + 40a^3b^2 + 80a^2b^3 + 80ab^4 + 32b^5$ **20.** $x^6 + 18x^5y + 135x^4y^2 + 540x^3y^3 + 1215x^2y^4 + 1458xy^5 + 729y^6$
21. $q^9 + 9q^8r + 36q^7r^2 + 84q^6r^3 + 126q^5r^4 + 126q^4r^5 + 84q^3r^6 + 36q^2r^7 + 9qr^8 + r^9$
22. $b^6 + 6b^5c + 15b^4c^2 + 20b^3c^3 + 15b^2c^4 + 6bc^5 + c^6$ **23.** $1024a^5 + 1280a^4b + 640a^3b^2 + 160a^2b^3 + 20ab^4 + b^5$
24. $81m^4 + 108m^3n + 54m^2n^2 + 12mn^3 + n^4$ **25.** $625a^4 - 1000a^3b + 600a^2b^2 - 160ab^3 + 16b^4$
26. $m^6 - 24m^5 + 240m^4 - 1280m^3 + 3840m^2 - 6144m + 4096$ **27.** $8a^3 + 36a^2b + 54ab^2 + 27b^3$
28. $1024 - 3840x + 5760x^2 - 4320x^3 + 1620x^4 - 243x^5$ **29.** $x^5 + 10x^4 + 40x^3 + 80x^2 + 80x + 32$
30. $81 + 216a + 216a^2 + 96a^3 + 16a^4$ **47.** $x^2\sqrt{x} + 5\sqrt{3}x^2 + 30x\sqrt{x} + 30\sqrt{3}x + 45\sqrt{x} + 9\sqrt{3}$

STUDY SKILLS REMINDER

Are You Prepared for Your Final Exam?

To prepare for your final exam, try the following study techniques.

▶ Review the material that you will be responsible for on your exam. Also check your notebook for any lecture notes that you highlighted.

▶ Review any formulas that you may need to memorize.

▶ Check to see if your instructor or math department will be conducting a final exam review.

▶ Check with your instructor to see whether there are final exams from previous semesters/quarters that are available to students for study.

▶ Use your previously taken tests as a practice final exam. To do so, rewrite the test questions in mixed order on blank sheets of paper. This will help you prepare for exam conditions.

▶ If you are unsure of a few topics, see your instructor or visit a learning lab for further assistance. Also, viewing the video segment of a troublesome section will help.

▶ If you need further exercises to work, try the chapter tests at the end of appropriate chapters.

Good luck! I hope you have enjoyed this textbook and your intermediate algebra course.

CHAPTER 11 PROJECT

Modeling College Tuition

Annual college tuition has steadily increased since 1970. According to the College Board, by the 2000–2001 academic year, the average annual tuition at a public four-year university had increased to $3506. Similarly, average annual tuition at private four-year universities had grown to $15,531.

     Over the past few years, annual tuition at four-year public universities has been increasing at an average rate of 4.5% per year. Over the same time period, annual tuition at four-year private universities has been increasing at an average rate of $804.30 per year. In this project, you will have the opportunity to model and investigate the trend in increasing tuition at public and private universities. This project may be completed by working in groups or individually.

1. Using the information given in the introductory paragraphs, decide whether the sequence of public university tuitions is arithmetic or geometric.

2. Using the information given in the introductory paragraphs, decide whether the sequence of private university tuitions is arithmetic or geometric.

3. Find the general term of the sequence that describes the pattern of average annual tuition for four-year public universities. Let $n = 1$ represent the 2000–2001 academic year.

4. Find the general term of the sequence that describes the pattern of average annual tuition for four-year private universities. Let $n = 1$ represent the 2000–2001 academic year.

5. Assuming that the rate of tuition increase remains the same, use the general term equation from Question 3 to find the average annual tuition at a four-year public university for the 2003–2004 academic year.

6. Assuming that the rate of tuition increase remains the same, use the general term equation from Question 4 to find the average annual tuition at a four-year private university for the 2004–2005 academic year.

7. Use partial sums to find the average cost of a four-year college education at a public university for a student who started college in the 2000–2001 academic year.

8. Use partial sums to find the average cost of a four-year college education at a private university for a student who starts college in the 2003–2004 academic year. (Hint: One way to do this is to find S_7 and subtract S_3 from it. If you use this method, explain why this gives the desired sum.)

9. (Optional) Use newspapers or news magazines to find a situation that can be modeled by a sequence. Briefly describe the situation, and decide whether it is an arithmetic or a geometric sequence. Find an equation of the general term of the sequence.

Academic Year	*n*
2000–2001	$n = 1$
2001–2002	$n = 2$
2002–2003	$n = 3$
2003–2004	$n = 4$
2004–2005	$n = 5$
2005–2006	$n = 6$
2006–2007	$n = 7$

CHAPTER VOCABULARY CHECK

Fill in each blank with one of the words or phrases listed below.

general term	common difference	finite sequence	common ratio
Pascal's triangle	infinite sequence	factorial of n	arithmetic sequence
geometric sequence	series		

1. A(n) <u>finite sequence</u> is a function whose domain is the set of natural numbers $\{1, 2, 3, \ldots, n\}$, where n is some natural number.

2. The <u>factorial of n</u>, written $n!$, is the product of the first n consecutive natural numbers.

3. A(n) <u>infinite sequence</u> is a function whose domain is the set of natural numbers.

4. A(n) <u>geometric sequence</u> is a sequence in which each term (after the first) is obtained by multiplying the preceding term by a constant amount r. The constant r is called the <u>common ratio</u> of the sequence.

5. A sum of the terms of a sequence is called a <u>series</u>.

6. The nth term of the sequence a_n is called the <u>general term</u>.

7. A(n) <u>arithmetic sequence</u> is a sequence in which each term (after the first) differs from the preceding term by a constant amount d. The constant d is called the <u>common difference</u> of the sequence.

8. A(n) triangle array of the coefficients of the terms of the expansions of $(a + b)^n$ is called <u>Pascal's triangle</u>.

CHAPTER 11 HIGHLIGHTS

Definitions and Concepts	**Examples**

Section 11.1 Sequences

An **infinite sequence** is a function whose domain is the set of natural numbers $\{1, 2, 3, 4, \ldots\}$.

Infinite Sequence

$$2, 4, 6, 8, 10, \ldots$$

A **finite sequence** is a function whose domain is the set of natural numbers $\{1, 2, 3, 4, \ldots, n\}$, where n is some natural number.

The notation a_n, where n is a natural number, is used to denote a sequence.

Finite Sequence

$$1, -2, 3, -4, 5, -6$$

Write the first four terms of the sequence whose general term is $a_n = n^2 + 1$.

$$a_1 = 1^2 + 1 = 2$$

$$a_2 = 2^2 + 1 = 5$$

$$a_3 = 3^2 + 1 = 10$$

$$a_4 = 4^2 + 1 = 17$$

Section 11.2 Arithmetic and Geometric Sequences

An **arithmetic sequence** is a sequence in which each term differs from the preceding term by a constant amount d, called the **common difference.**

The **general term** a_n of an arithmetic sequence is given by

$$a_n = a_1 + (n - 1)d$$

where a_1 is the first term and d is the common difference.

A **geometric sequence** is a sequence in which each term is obtained by multiplying the preceding term by a constant r, called the **common ratio.**

The **general term** a_n of a geometric sequence is given by

$$a_n = a_1 r^{n-1}$$

where a_1 is the first term and r is the common ratio.

Arithmetic Sequence

$$5, 8, 11, 14, 17, 20, \ldots$$

Here, $a_1 = 5$ and $d = 3$.

The general term is

$$a_n = a_1 + (n - 1)d \text{ or}$$

$$a_n = 5 + (n - 1)3$$

Geometric Sequence

$$12, -6, 3, -\frac{3}{2}, \ldots$$

Here $a_1 = 12$ and $r = -\frac{1}{2}$.

The general term is

$$a_n = a_1 r^{n-1} \text{ or}$$

$$a_n = 12\left(-\frac{1}{2}\right)^{n-1}$$

Definitions and Concepts	**Examples**

Section 11.3   Series

A sum of the terms of a sequence is called a **series**.

A shorthand notation for denoting a series is called **summation notation:**

$$\text{index of summation} \rightarrow \underset{i=1}{\overset{4}{\sum}} \rightarrow \begin{array}{l}\text{Greek letter sigma}\\ \text{used to mean sum}\end{array}$$

Sequence	Series	
$3, 7, 11, 15$	$3 + 7 + 11 + 15$	finite
$3, 7, 11, 15, \ldots$	$3 + 7 + 11 + 15 + \cdots$	infinite

$$\sum_{i=1}^{4} 3^i = 3^1 + 3^2 + 3^3 + 3^4$$
$$= 3 + 9 + 27 + 81$$
$$= 120$$

Section 11.4   Partial Sums of Arithmetic and Geometric Sequences

Partial sum, S_n, of the first n terms of an arithmetic sequence:

$$S_n = \frac{n}{2}(a_1 + a_n)$$

where a_1 is the first term and a_n is the nth term.

Partial sum, S_n, of the first n terms of a geometric sequence:

$$S_n = \frac{a_1(1 - r^n)}{1 - r}$$

where a_1 is the first term, r is the common ratio, and $r \neq 1$.

Sum of the terms of an infinite geometric sequence:

$$S_\infty = \frac{a_1}{1 - r}$$

where a_1 is the first term, r is the common ratio, and $|r| < 1$. (If $|r| \geq 1$, S_∞ does not exist.)

The sum of the first five terms of the arithmetic sequence

$$12, 24, 36, 48, 60, \ldots \text{ is}$$

$$S_n = \frac{5}{2}(12 + 60) = 180$$

The sum of the first five terms of the geometric sequence

$$15, 30, 60, 120, 240, \ldots \text{ is}$$

$$S_5 = \frac{15(1 - 2^5)}{1 - 2} = 465$$

The sum of the terms of the infinite geometric sequence

$$1, \frac{1}{3}, \frac{1}{9}, \frac{1}{27}, \ldots \text{ is}$$

$$S_\infty = \frac{1}{1 - \dfrac{1}{3}} = \frac{3}{2}$$

Section 11.5   The Binomial Theorem

The **factorial of n**, written $n!$, is the product of the first n consecutive natural numbers.

Binomial Theorem

If n is a positive integer, then

$$(a + b)^n = a^n + \frac{n}{1!}a^{n-1}b^1 + \frac{n(n-1)}{2!}a^{n-2}b^2$$
$$+ \frac{n(n-1)(n-2)}{3!}a^{n-3}b^3 + \cdots + b^n$$

$$5! = 5 \cdot 4 \cdot 3 \cdot 2 \cdot 1 = 120$$

Expand $(3x + y)^4$.

$$(3x + y)^4 = (3x)^4 + \frac{4}{1!}(3x)^3(y)^1$$
$$+ \frac{4 \cdot 3}{2!}(3x)^2(y)^2 + \frac{4 \cdot 3 \cdot 2}{3!}(3x)^1 y^3 + y^4$$
$$= 81x^4 + 108x^3 y + 54x^2 y^2 + 12xy^3 + y^4$$

1. $-3, -12, -27, -48, -75$ 2. $3, 8, 15, 24, 35$ 6. $a_n = (-1)^n n^2$ 7. 144 ft, 176 ft, 208 ft
10. $50, 58, 66, 74, 82, 90, 98, 106, 114, 122$; 122 seats 12. $12, 10.5, 9, 7.5, 6$

11 CHAPTER REVIEW

(11.1) *Find the indicated term(s) of the given sequence.*

1. The first five terms of the sequence $a_n = -3n^2$

2. The first five terms of the sequence $a_n = n^2 + 2n$

3. The one-hundredth term of the sequence $a_n = \dfrac{(-1)^n}{100}$ $\frac{1}{100}$

4. The fiftieth term of the sequence $a_n = \dfrac{2n}{(-1)^2}$ 100

5. The general term a_n of the sequence $\dfrac{1}{6}, \dfrac{1}{12}, \dfrac{1}{18}, \ldots$ $a_n = \frac{1}{6n}$

6. The general term a_n of the sequence $-1, 4, -9, 16, \ldots$.

Solve the following applications.

7. The distance in feet that an olive falling from rest in a vacuum will travel during each second is given by an arithmetic sequence whose general term is $a_n = 32n - 16$, where n is the number of the second. Find the distance the olive will fall during the fifth, sixth, and seventh seconds.

8. A culture of yeast doubles every day in a geometric progression whose general term is $a_n = 100(2)^{n-1}$, where n is the number of the day just ending. Find how many days it takes the yeast culture to measure at least 10,000. Find the original measure of the yeast culture.   8 days; 50

9. The Centers for Disease Control and Prevention (CDC) reported that a new type of virus infected approximately 450 people during 2003, the year it was first discovered. The CDC predicts that during the next decade the virus will infect three times as many people each year as the year before. Write out the first five terms of this geometric sequence, and predict the number of infected people there will be in 2007.
   450, 1350, 4050, 12,150, 36,450; 36,450 infected people in 2007

10. The first row of an amphitheater contains 50 seats, and each row thereafter contains 8 additional seats. Write the first ten terms of this arithmetic progression, and find the number of seats in the tenth row.

(11.2)

11. $-2, -\frac{4}{3}, -\frac{8}{9}, -\frac{16}{27}, -\frac{32}{81}$

11. Find the first five terms of the geometric sequence whose first term is -2 and whose common ratio is $\dfrac{2}{3}$.

12. Find the first five terms of the arithmetic sequence whose first term is 12 and whose common difference is -1.5.

13. Find the thirtieth term of the arithmetic sequence whose first term is -5 and whose common difference is 4.   111

14. Find the eleventh term of the arithmetic sequence whose first term is 2 and whose common difference is $\dfrac{3}{4}$. $\frac{19}{2}$

15. Find the twentieth term of the arithmetic sequence whose first three terms are 12, 7, and 2. -83

16. Find the sixth term of the geometric sequence whose first three terms are 4, 6, and 9. $\frac{243}{8}$

17. If the fourth term of an arithmetic sequence is 18 and the twentieth term is 98, find the first term and the common difference. $a_1 = 3; d = 5$

18. If the third term of a geometric sequence is -48 and the fourth term is 192, find the first term and the common ratio. $a_1 = -3; r = -4$

19. Find the general term of the sequence $\dfrac{3}{10}, \dfrac{3}{100}, \dfrac{3}{1000}, \ldots$.

20. Find a general term that satisfies the terms shown for the sequence $50, 58, 66, \ldots$.   19. $a_n = \dfrac{3}{10^n}$

20. $a_n = 50 + (n-1)(8)$ or $a_n = 42 + 8n$

Determine whether each of the following sequences is arithmetic, geometric, or neither. If a sequence is arithmetic, find a_1 and d. If a sequence is geometric, find a_1 and r.

21. $\dfrac{8}{3}, 4, 6, \ldots$ $a_1 = \frac{8}{3}, r = \frac{3}{2}$

22. $-10.5, -6.1, -1.7$ $a_1 = -10.5, d = 4.4$

23. $7x, -14x, 28x$ $a_1 = 7x, r = -2$

24. $3x^2, 9x^4, 81x^8, \ldots$ neither

Solve the following applications.   25. $8, 6, 4.5, 3.4, 2.5, 1.9$; good

25. To test the bounce of a racquetball, the ball is dropped from a height of 8 feet. The ball is judged "good" if it rebounds at least 75% of its previous height with each bounce. Write out the first six terms of this geometric sequence (round to the nearest tenth). Determine if a ball is "good" that rebounds to a height of 2.5 feet after the fifth bounce.

26. A display of oil cans in an auto parts store has 25 cans in the bottom row, 21 cans in the next row, and so on, in an arithmetic progression. Find the general term and the number of cans in the top row.
   $a_n = 25 + (n-1)(-4)$ or $a_n = 29 - 4n$; 1 can

27. Suppose that you save \$1 the first day of a month, \$2 the second day, \$4 the third day, continuing to double your savings each day. Write the general term of this geometric sequence and find the amount you will save on the tenth day. Estimate the amount you will save on the thirtieth day of the month, and check your estimate with a calculator.
   $a_n = 2^{n-1}$, \$512, \$536,870,912

28. On the first swing, the length of an arc through which a pendulum swings is 30 inches. The length of the arc for each successive swing is 70% of the preceding swing. Find the length of the arc for the fifth swing.   7.203 in.

29. Rosa takes a job that has a monthly starting salary of $900 and guarantees her a monthly raise of $150 during her 6-month training period. Find the general term of this sequence and her salary at the end of her training.
$a_n = 900 + (n - 1)150$ or $a_n = 150n + 750$; $1650/month

30. A sheet of paper is $\frac{1}{512}$-inch thick. By folding the sheet in half, the total thickness will be $\frac{1}{256}$ inch. A second fold produces a total thickness of $\frac{1}{128}$ inch. Estimate the thickness of the stack after 15 folds, and then check your estimate with a calculator. 64 in.

31. $1 + 3 + 5 + 7 + 9 = 25$
32. $3 + 8 + 15 + 24 + 35 = 85$ **34.** $5 - 5 + 5 = 5$

(11.3) Write out the terms and find the sum for each of the following.

31. $\displaystyle\sum_{i=1}^{5}(2i - 1)$ **32.** $\displaystyle\sum_{i=1}^{5}i(i + 2)$

33. $\displaystyle\sum_{i=2}^{4}\frac{(-1)^i}{2i}$ $\frac{1}{4} - \frac{1}{6} + \frac{1}{8} = \frac{5}{24}$ **34.** $\displaystyle\sum_{i=3}^{5}5(-1)^{i-1}$

Find the partial sum of the given sequence.

35. S_4 of the sequence $a_n = (n - 3)(n + 2)$ -4

36. S_6 of the sequence $a_n = n^2$ 91

37. S_5 of the sequence $a_n = -8 + (n - 1)3$ -10

38. S_3 of the sequence $a_n = 5(4)^{n-1}$ 105

Write the sum with Σ notation.

39. $1 + 3 + 9 + 27 + 81 + 243$ $\displaystyle\sum_{j=1}^{6}3^{i-1}$

40. $6 + 2 + (-2) + (-6) + (-10) + (-14) + (-18)$

41. $\frac{1}{4} + \frac{1}{16} + \frac{1}{64} + \frac{1}{256}$ $\displaystyle\sum_{i=1}^{4}\frac{1}{4^i}$ **40.** $\displaystyle\sum_{i=1}^{7}[6 + (i - 1)(-4)]$

42. $1 + \left(-\frac{3}{2}\right) + \frac{9}{4}$ $\displaystyle\sum_{i=1}^{3}\left(-\frac{3}{2}\right)^{i-1}$

43. $a_n = 20(2)^n$; n represents the number of 8-hour periods; 1280 yeast

Solve.

44. 23 cranes; 46 cranes **45.** Job A, $48,300; Job B, $46,600

43. A yeast colony begins with 20 yeast and doubles every 8 hours. Write the sequence that describes the growth of the yeast, and find the total yeast after 48 hours.

44. The number of cranes born each year in a new aviary forms a sequence whose general term is $a_n = n^2 + 2n - 1$. Find the number of cranes born in the fourth year and the total number of cranes born in the first four years.

45. Harold has a choice between two job offers. Job A has an annual starting salary of $39,500 with guaranteed annual raises of $2200 for the next four years, whereas job B has an annual starting salary of $41,000 with guaranteed annual raises of $1400 for the next four years. Compare the salaries for the fifth year under each job offer.

46. 25 kg; 175 kg

46. A sample of radioactive waste is decaying such that the amount decaying in kilograms during year n is $a_n = 200(0.5)^n$. Find the amount of decay in the third year, and the total amount of decay in the first three years.

(11.4) Find the partial sum of the given sequence.

47. The sixth partial sum of the sequence $15, 19, 23, \ldots$. 150

48. The ninth partial sum of the sequence $5, -10, 20, \ldots$. 855

49. The sum of the first 30 odd positive integers 900

50. The sum of the first 20 positive multiples of 7 1470

51. The sum of the first 20 terms of the sequence $8, 5, 2, \ldots$

52. The sum of the first eight terms of the sequence $\frac{3}{4}, \frac{9}{4}, \frac{27}{4}, \ldots$ 2460

53. S_4 if $a_1 = 6$ and $r = 5$ 936

54. S_{100} if $a_1 = -3$ and $d = -6$ $-30,000$

Find the sum of each infinite geometric sequence.

55. $5, \frac{5}{2}, \frac{5}{4}, \ldots$ 10

56. $18, -2, \frac{2}{9}, \ldots$ $\frac{81}{5}$

57. $-20, -4, -\frac{4}{5}, \ldots$ -25

58. $0.2, 0.02, 0.002, \ldots$ $\frac{2}{9}$

51. -410
59. $30,418; $99,868

Solve.

59. A frozen yogurt store owner cleared $20,000 the first year he owned his business and made an additional 15% over the previous year in each subsequent year. Find how much he made during his fourth year of business. Find his total earnings during the first 4 years (round to the nearest dollar).

60. On his first morning in a television assembly factory, a trainee takes 0.8 times as long to assemble each television as he took to assemble the one before. If it took him 40 minutes to assemble the first television, find how long it takes him to assemble the fourth television. Find the total time he takes to assemble the first four televisions (round answers to the nearest minute). 20 min; 118 min

61. During the harvest season a farmer can rent a combine machine for $100 the first day, with the rental fee decreasing $7 for each additional day. Find how much the farmer pays for the rental on the seventh day. Find how much total rent the farmer pays for 7 days. $58; $553

62. A rubber ball is dropped from a height of 15 feet and rebounds 80% of its previous height after each bounce. Find the total distance the ball travels before it comes to rest. 135 ft

63. After a pond was sprayed once with insecticide, 1800 mosquitoes were killed the first day, 600 the second day, 200 the third day, and so on. Find the total number of mosquitoes killed during the first 6 days after the spraying (round to the nearest unit). 2696 mosquitoes

64. See Exercise 63. Find the day on which the insecticide is no longer effective, and find the total number of mosquitoes killed (round to the nearest mosquito). eighth day; 2700 mosquitoes

65. Use the formula S_∞ to write $0.5\overline{55}$ as a fraction. $\frac{5}{9}$

66. A movie theater has 27 seats in the first row, 30 seats in the second row, 33 seats in the third row, and so on. Find the total number of seats in the theater if there are 20 rows.

(11.5) Use Pascal's triangle to expand each binomial.

67. $(x + z)^5$

68. $(y - r)^6$

69. $(2x + y)^4$

70. $(3y - z)^4$

Use the binomial formula to expand the following.

71. $(b + c)^8$

72. $(x - w)^7$

73. $(4m - n)^4$

74. $(p - 2r)^5$

Find the indicated term.

75. The fourth term of the expansion of $(a + b)^7$ $35a^4b^3$

76. The eleventh term of the expansion of $(y + 2z)^{10}$ $1024z^{10}$

66. 1110 seats   **67.** $x^5 + 5x^4z + 10x^3z^2 + 10x^2z^3 + 5xz^4 + z^5$ **68.** $y^6 - 6y^5r + 15y^4r^2 - 20y^3r^3 + 15y^2r^4 - 6yr^5 + r^6$
69. $16x^4 + 32x^3y + 24x^2y^2 + 8xy^3 + y^4$ **70.** $81y^4 - 108y^3z + 54y^2z^2 - 12yz^3 + z^4$
71. $b^8 + 8b^7c + 28b^6c^2 + 56b^5c^3 + 70b^4c^4 + 56b^3c^5 + 28b^2c^6 + 8bc^7 + c^8$
72. $x^7 - 7x^6w + 21x^5w^2 - 35x^4w^3 + 35x^3w^4 - 21x^2w^5 + 7xw^6 - w^7$ **73.** $256m^4 - 256m^3n + 96m^2n^2 - 16mn^3 + n^4$
74. $p^5 - 10p^4r + 40p^3r^2 - 80p^2r^3 + 80pr^4 - 32r^5$

CHAPTER 11 TEST

Remember to use your Chapter Test Prep Video CD to help you study and view solutions to the test questions you need help with.

Find the indicated term(s) of the given sequence.

1. The first five terms of the sequence $a_n = \dfrac{(-1)^n}{n + 4}$

2. The eightieth term of the sequence $a_n = 10 + 3(n - 1)$ 247

3. The general term of the sequence $\dfrac{2}{5}, \dfrac{2}{25}, \dfrac{2}{125}, \ldots$

4. The general term of the sequence $-9, 18, -27, 36, \ldots$

Find the partial sum of the given sequence.

5. S_5 of the sequence $a_n = 5(2)^{n-1}$ 155

6. S_{30} of the sequence $a_n = 18 + (n - 1)(-2)$ -330

7. S_∞ of the sequence $a_1 = 24$ and $r = \dfrac{1}{6}$ $\dfrac{144}{5}$

8. S_∞ of the sequence $\dfrac{3}{2}, -\dfrac{3}{4}, \dfrac{3}{8}, \ldots$ 1

9. $\displaystyle\sum_{i=1}^{4} i(i - 2)$ 10

10. $\displaystyle\sum_{i=2}^{4} 5(2)^i (-1)^{i-1}$ -60

Expand each binomial.

11. $(a - b)^6$

12. $(2x + y)^5$

Solve the following applications.

13. The population of a small town is growing yearly according to the sequence defined by $a_n = 250 + 75(n - 1)$, where n is the number of the year just beginning. Predict the population at the beginning of the tenth year. Find the town's initial population.   925 people; 250 people initially

14. A gardener is making a triangular planting with one shrub in the first row, three shrubs in the second row, five shrubs in the third row, and so on, for eight rows. Write the finite series of this sequence, and find the total number of shrubs planted.

15. A pendulum swings through an arc of length 80 centimeters on its first swing. On each successive swing, the length of the arc is $\dfrac{3}{4}$ the length of the arc on the preceding swing. Find the length of the arc on the fourth swing, and find the total arc length for the first four swings.

16. See Exercise 15. Find the total arc length before the pendulum comes to rest.   320 cm

17. A parachutist in free-fall falls 16 feet during the first second, 48 feet during the second second, 80 feet during the third second, and so on. Find how far he falls during the tenth second. Find the total distance he falls during the first 10 seconds.   304 ft; 1600 ft

18. Use the formula S_∞ to write $0.42\overline{42}$ as a fraction. $\frac{14}{33}$

1. $-\frac{1}{5}, \frac{1}{6}, -\frac{1}{7}, \frac{1}{8}, -\frac{1}{9}$ **3.** $a_n = \frac{2}{5}\left(\frac{1}{5}\right)^{n-1}$ **4.** $a_n = (-1)^n 9n$ **11.** $a^6 - 6a^5b + 15a^4b^2 - 20a^3b^3 + 15a^2b^4 - 6ab^5 + b^6$
12. $32x^5 + 80x^4y + 80x^3y^2 + 40x^2y^3 + 10xy^4 + y^5$ **14.** $1 + 3 + 5 + 7 + 9 + 11 + 13 + 15$; 64 shrubs   **15.** 33.75 cm, 218.75 cm

4. $325 (Sec. 1.6)

11 CHAPTER CUMULATIVE REVIEW

9. $x^3 - 4x^2 - 3x + 11 + \dfrac{12}{x + 2}$ (Sec. 6.5, Ex. 2)

1. Divide.   (Sec. 1.3, Ex. 5)

  a. $\dfrac{20}{-4}$ -5 **b.** $\dfrac{-9}{-3}$ 3

  c. $-\dfrac{3}{8} \div 3$ $-\dfrac{1}{8}$ **d.** $\dfrac{-40}{10}$ -4

  e. $\dfrac{-1}{10} \div \dfrac{-2}{5}$ $\dfrac{1}{4}$ **f.** $\dfrac{8}{0}$ undefined

2. Simplify each expression.

  a. $3a - (4a + 3)$ $-a - 3$

  b. $(5x - 3) + (2x + 6)$ $7x + 3$

  c. $4(2x - 5) - 3(5x + 1)$ $-7x - 23$ (Sec. 1.4)

3. Suppose that a computer store just announced an 8% decrease in the price of a particular computer model. If this computer sells for $2162 after the decrease, find the original price of this computer. $2350 (Sec. 1.6, Ex. 5)

4. Sara bought a digital camera for $344.50 including tax. If the tax rate is 6%, what was the price of the camera before taxes.

5. Use Cramer's rule to solve each system.   (Sec. 4.5, Ex. 2)

  a. $\begin{cases} 3x + 4y = -7 \\ x - 2y = -9 \end{cases}$ $(-5, 2)$

  6. $f(x) = \dfrac{3}{2}x - \dfrac{13}{2}$,

  b. $\begin{cases} 5x + y = 5 \\ -7x - 2y = -7 \end{cases}$ $(1, 0)$ (Sec. 3.5)

6. Find an equation of a line through $(3, -2)$ and parallel to $3x - 2y = 6$. Write the equation using function notation.

7. Use the product rule to multiply.   (Sec. 5.1, Ex. 2)

  a. $(3x^6)(5x)$ $15x^7$

  b. $(-2x^3p^2)(4xp^{10})$ $-8x^4p^{12}$

8. Solve $y^3 + 5y^2 - y = 5$ $-5, -1, 1$, (Sec. 5.8)

9. Use synthetic division to divide $(x^4 - 2x^3 - 11x^2 + 5x + 34)$ by $(x + 2)$.

10. Perform the indicated operation and simplify if possible.

  $\dfrac{5}{3a - 6} - \dfrac{a}{a - 2} + \dfrac{3 + 2a}{5a - 10}$ $\dfrac{34 - 9a}{15(a - 2)}$ (Sec. 6.2)

11. Simplify the following.   (Sec. 7.3, Ex. 3)

  a. $\sqrt{50}$ $5\sqrt{2}$

  b. $\sqrt[3]{24}$ $2\sqrt[3]{3}$

  c. $\sqrt{26}$ $\sqrt{26}$

  d. $\sqrt[4]{32}$ $2\sqrt[4]{2}$

12. Solve $\sqrt{3x + 6} - \sqrt{7x - 6} = 0$ 3 (Sec. 7.6)

13. Find the interest rate r if $2000 compounded annually grows to $2420 in 2 years. 10% (Sec. 8.1, Ex. 9)

14. Rationalize each denominator.

  a. $\sqrt[3]{\dfrac{4}{3x}}$ $\dfrac{\sqrt[3]{36x^2}}{3x}$ **18.** vertex: $(-2, -6)$ axis of symmetry: $x = -z$. See graphing answer section.

  b. $\dfrac{\sqrt{2} + 1}{\sqrt{2} - 1}$ $3 + 2\sqrt{2}$ (Sec. 7.5)

15. Solve $(x - 3)^2 - 3(x - 3) - 4 = 0$. $2, 7$ (Sec. 8.3, Ex. 4)

16. Solve $\dfrac{10}{(2x + 4)^2} - \dfrac{1}{2x + 4} = 3$ $\dfrac{-7}{6}, -3$ (Sec. 6.6)

17. Solve $\dfrac{5}{x + 1} < -2$. $\left(-\dfrac{7}{2}, -1\right)$ (Sec. 8.4, Ex. 7)

18. Graph $f(x) = (x + 2)^2 - 6$. Find the vertex and axis of symmetry.   See graphing answer section. (Sec. 8.5)

19. A rock is thrown upward from the ground. Its height in feet above ground after t seconds is given by the function $f(t) = -16t^2 + 20t$. Find the maximum height of the rock and the number of seconds it took for the rock to reach its maximum height. $\dfrac{25}{4}$ ft; $\dfrac{5}{8}$ sec (Sec. 8.6, Ex. 6)

20. Find the vertex of $f(x) = x^2 + 3x - 18$

21. If $f(x) = x^2$ and $g(x) = x + 3$, find each composition.

  a. $(f \circ g)(2)$ and $(g \circ f)(2)$ $25; 7$ (Sec. 9.1, Ex. 2)

  b. $(f \circ g)(x)$ and $(g \circ f)(x)$ $x^2 + 6x + 9; x^2 + 3$

22. Find the inverse of $f(x) = -2x + 3$

23. Find the inverse of the one-to-one function. $f = \{(0, 1), (-2, 7), (3, -6), (4, 4)\}$

24. If $f(x) = x^2 - 2$ and $g(x) = x + 1$, find each composition.

  a. $(f \circ g)(2)$ and $(g \circ f)(2)$ $7; 3$

  b. $(f \circ g)(x)$ and $(g \circ f)(x)$ $x^2 + 2x - 1; x^2 - 1$ (Sec. 9.1)

25. Solve each equation for x.   (Sec. 9.3, Ex. 4)

  a. $2^x = 16$ 4 **b.** $9^x = 27$ $\dfrac{3}{2}$

  c. $4^{x+3} = 8^x$ 6 **d.** $5^x = 10$ 1.431

26. Solve each equation for x.

  a. $\log_2 32 = x$ 5 **20.** $\left(-\dfrac{3}{2}, \dfrac{-81}{4}\right)$ (Sec. 8.6)

  b. $\log_4 \dfrac{1}{64} = x$ -3

  c. $\log_{\frac{1}{2}} x = 5$ $\dfrac{1}{32}$ (Sec. 9.4)   **22.** $f^{-1}(x) = \dfrac{3 - x}{2}$ (Sec. 9.2)

27. Simplify.   (Sec. 9.4, Ex. 5)

  a. $\log_3 3^2$ 2 **23.** $f^{-1} = \{(1, 0), (7, -2), (-6, 3), (4, 4)\}$

  b. $\log_7 7^{-1}$ -1 (Sec. 9.2, Ex. 3)

  c. $5^{\log_5 3}$ 3

  d. $2^{\log_2 6}$ 6

28. Solve each equation for x.

   a. $4^x = 64$ 3

   b. $8^x = 32$ $\frac{5}{3}$

   c. $9^{x+4} = 243^x$ $\frac{8}{3}$ (Sec. 9.3)

29. Write each sum as a single logarithm.   (Sec. 9.5, Ex. 1)

   a. $\log_{11} 10 + \log_{11} 3$ $\log_{11} 30$

   b. $\log_3 \frac{1}{2} + \log_3 12$ $\log_3 6$

   c. $\log_2(x + 2) + \log_2 x$ $\log_2(x^2 + 2x)$

30. Find the exact value.

   a. $\log 100{,}000$ 5

   b. $\log 10^{-3}$ -3

   c. $\ln \sqrt[5]{e}$ $\frac{1}{5}$

   d. $\ln e^4$ 4 (Sec. 9.6)

31. Find the amount owed at the end of 5 years if $1600 is loaned at a rate of 9% compounded continuously.
   $2509.30 (Sec. 9.6, Ex. 8)

32. Write each expression as a single logarithm.

   a. $\log_6 5 + \log_6 4$ $\log_6 20$

   b. $\log_8 12 - \log_8 4$ $\log_8 3$ (Sec. 9.5)

   c. $2 \log_2 x + 3 \log_2 x - 2 \log_2(x - 1)$ $\log_2 \dfrac{x^5}{(x - 1)^2}$

33. Solve $3^x = 7$.

34. Using $A = P\left(1 + \dfrac{r}{n}\right)^{nt}$, find how long it takes $5000 to double if it is invested at 2% interest compounded quarterly. Round to the nearest tenth.   34.7 yr. (Sec. 9.6)

35. Solve $\log_4(x - 2) = 2$.   18 (Sec. 9.7, Ex. 2)

36. Solve $\log_4 10 - \log_4 x = 2$ $\frac{5}{8}$ (Sec. 9.7)

37. Graph $\dfrac{x^2}{16} - \dfrac{y^2}{25} = 1$.

38. Find the distance between $(8, 5)$ and $(-2, 4)$. $\sqrt{101}$ units (Sec. 10.1)

39. Solve the system. $\begin{cases} y = \sqrt{x} \\ x^2 + y^2 = 6 \end{cases}$.

40. Solve the system. $\begin{cases} x^2 + y^2 = 36 \\ x - y = 6 \end{cases}$ $(0, -6); (6, 0)$ (Sec. 10.3)

41. Graph $\dfrac{x^2}{9} + \dfrac{y^2}{16} \leq 1$.

42. Graph $\begin{cases} y \geq x^2 \\ y \leq 4 \end{cases}$ See graphing answer section. (Sec. 10.4)

43. Write the first five terms of the sequence whose general term is given by $a_n = n^2 - 1$. $0, 3, 8, 15, 24$ (Sec. 11.1, Ex. 1)

44. If the general term of a sequence is $a_n = \dfrac{n}{n + 4}$, find a_8.

45. Find the eleventh term of the arithmetic sequence whose first three terms are $2, 9,$ and 16.   72 (Sec. 11.2, Ex. 3)

46. Find the sixth term of the geometric sequence $2, 10, 50, \ldots$

47. Evaluate.   (Sec. 11.3, Ex. 1)

   a. $\displaystyle\sum_{i=0}^{6} \dfrac{i - 2}{2}$ $\frac{7}{2}$

   b. $\displaystyle\sum_{i=3}^{5} 2^i$ 56

48. Evaluate.

   a. $\displaystyle\sum_{i=0}^{4} i(i + 1)$ 40

   b. $\displaystyle\sum_{i=0}^{3} 2^i$ 15 (Sec. 11.3)

49. Find the sum of the first 30 positive integers.

50. Find the third term of the expansion of $(x - y)^6$. $15x^4 y^2$ (Sec. 11.5)

33. $\dfrac{\log 7}{\log 3} \approx 1.7712$ (Sec. 9.7, Ex. 1)

37. See graphing answer section. (Sec. 10.2, Ex. 4)

39. (Sec. 10.3, Ex. 1) $\left(2, \sqrt{2}\right)$

41. See graphing answer section. (Sec. 10.4, Ex. 1)

44. $\dfrac{2}{3}$ (Sec. 11.1)

46. 6250 (Sec. 11.2)

49. 465 (Sec. 11.4, Ex. 2)

Solving Equations and Inequalities— Study Skills Practice

Each set of exercises corresponds to the Study Skills Reminder in the section noted. After reviewing the outline in that section's Study Skills Reminder, solve each equation or inequality.

APPENDIX A EXERCISE SET

Section 1.5

Solve.

1. $3x - 4 = 3(2x - 1) + 7$ $-\dfrac{8}{3}$

2. $\dfrac{7}{5} + \dfrac{y}{10} = 2$ 6

3. $5 + 2x = 5(x + 1)$ 0

4. $5 + 2x = 2(x + 1)$ $\varnothing$

Section 3.2

Solve. Write inequality solutions in interval notation.

1. $\dfrac{x + 3}{2} > 1$ $(-1, \infty)$

2. $4(x - 2) + 3x \geq 9(x - 1) - 2$ $\left(-\infty, \frac{3}{2}\right]$

3. $\dfrac{x - 2}{2} - \dfrac{x - 4}{3} = \dfrac{5}{6}$ 3

4. $6(x + 1) - 2 = 6x + 4$ all real numbers, or $(-\infty, \infty)$

Section 3.3

Solve. Write inequality solutions in interval notation.   **1.** $[-1, 3]$

1. $x - 2 \leq 1$ and $3x - 1 \geq -4$

2. $-2 < x - 1 < 5$ $(-1, 6)$

3. $-2x + 2.5 = -7.7$ 5.1

4. $-5x > 20$ $(-\infty, -4)$

5. $x \leq -3$ or $x \leq -5$ $(-\infty, -3]$

Section 3.4

Solve. Write inequality solutions in interval notation.   **2.** $(-\infty, -2) \cup (2, \infty)$

1. $|2 + 3x| = 7$ $\frac{5}{3}, -3$

2. $5x < -10$ or $3x - 4 > 2$

3. $\dfrac{5t}{2} - \dfrac{3t}{4} = 7$ 4

4. $|x - 2| = |x + 1|$ $\frac{1}{2}$

5. $5(x - 3) + x + 2 \geq 3(x + 2) + 2x$ $[19, \infty)$

Section 3.5

Solve. Write inequality solutions in interval notation.

1. $|x - 11| \geq 7$ $(-\infty, 4] \cup [18, \infty)$ **2.** $|x - 11| = 7$ $4, 18$

3. $-5 < x - (2x + 3) < 0$ $(-3, 2)$ **4.** $|9x| - 8 = -1$ $-\dfrac{7}{9}, \dfrac{7}{9}$

5. $\dfrac{4x}{5} - 1 = \dfrac{x}{2} + 2$ 10

Section 5.8

Solve. Write inequality solutions in interval notation.   **5.** $(-\infty, 2)$

1. $2x^2 - 17x = 9$ $9, -\frac{1}{2}$ **2.** $14x - 17x + 6 = 6(x + 1)$ 0

3. $|4x + 7| = |-35|$ $7, -\dfrac{21}{2}$ **4.** $\left|\dfrac{3x - 5}{4}\right| > 1$ $\left(-\infty, \frac{1}{3}\right) \cup (3, \infty)$

5. $3(2x - 1) < 9$ and $-4x > -12$ **6.** $\frac{2}{3}x \leq \frac{5}{6}$ $\left(-\infty, \frac{5}{4}\right]$

Section 6.6

Solve. Write inequality solutions in interval notation.   **4.** all real numbers or $(-\infty, \infty)$

1. $\dfrac{x}{10} - \dfrac{1}{2} = \dfrac{7}{5x}$ $7, -2$ **2.** $|1 - 5x| = 9$ $-\frac{8}{5}, 2$

3. $x + 2 \leq 0$ or $5x \leq 0$ $(-\infty, 0]$ **4.** $5(x - 3) + 2x = 7(x + 1) - 22$

5. $-8 + |2x - 4| \leq -2$ $[-1, 5]$ **6.** $x^3 = 25x$ $0, 5, -5$

Section 7.6

Solve. Write inequality solutions in interval notation.   **5.** $(-\infty, -4]$

1. $x(3x + 14) = 5$ $\frac{1}{3}, -5$ **2.** $\dfrac{2}{x - 2} + \dfrac{3}{x + 2} = \dfrac{7}{x^2 - 4}$ $\dfrac{9}{5}$

3. $|5x - 4| = |4x + 1|$ $5, \dfrac{1}{3}$ **4.** $\sqrt{6x - 1} = 7$ $\dfrac{25}{3}$

5. $-2(x - 4) + 3x \leq -3(x + 2) - 2$ **6.** $|-x + 3| > 5$ $(-\infty, -2) \cup (8, \infty)$

Section 8.2

Solve. Write inequality solutions in interval notation.

1. $(x - 2)^2 = 17$ $2 \pm \sqrt{17}$ **2.** $x^2 - 5x + 2 = 0$ $\dfrac{5 \pm \sqrt{17}}{2}$

3. $x^2 - 5x + 6 = 0$ $3, 2$ **4.** $\dfrac{x}{4} - \dfrac{3}{2} > 1$ $(10, \infty)$

5. $\sqrt{2x + 30} = x + 3$ 3 **6.** $|3x + 11| - 12 = -1$ $0, -\dfrac{22}{3}$

7. $\dfrac{3x^2 - 7}{3x^2 - 8x - 3} = \dfrac{1}{x - 3} + \dfrac{2}{3x + 1}$ 2 **8.** $\left|\dfrac{x - 4}{3}\right| < 10$ $(-26, 34)$

Section 8.4

Solve. Write inequality solutions in interval notation.

1. $x^2 - 3x - 10 = 0$ $5, -2$

2. $x^2 - 3x - 10 < 0$ $(-2, 5)$

3. $\dfrac{x + 4}{x - 10} = 0$ -4

4. $\dfrac{x + 4}{x - 10} \geq 0$ $(-\infty, -4] \cup (10, \infty)$

5. $\sqrt{x - 7} - 12 = -8$ 23

6. $(5x - 2)^2 = 10$ $\dfrac{2 \pm \sqrt{10}}{5}$

7. $\left|\dfrac{3x + 5}{2}\right| = -9$ $\varnothing$

8. $2(3x - 6) + 4 - 7(x + 1) - 2x$ 15

9. $-4(x - 3) + 2x < 6x + 4$ $(1, \infty)$

10. $x \leq 7$ and $2(x + 1) + 3 \leq 7$ $(-\infty, 1]$

Section 9.7

Solve. Write inequality solutions in interval notation.

1. $4^x = 8^{x-1}$ 3

2. $2^{x+7} = 5$ $\dfrac{\log 5 - 7 \log 2}{\log 2}$

3. $x(x - 9) > 0$ $(-\infty, 0) \cup (9, \infty)$

4. $\sqrt[3]{6x + 1} - 2 = 0$ $\dfrac{7}{6}$

5. $\log_4(x^2 - 3x) = 1$ $4, -1$

6. $3x^2 + 2x = 3$ $\dfrac{-1 \pm \sqrt{10}}{3}$

7. $\dfrac{6}{x - 2} \geq 3$ $(2, 4]$

8. $\dfrac{x + 3}{5} - \dfrac{7 - x}{3} = \dfrac{3}{10}$ $\dfrac{61}{16}$

9. $\log_3(2x + 1) - \log_3 x = 1$ 1

10. $|6x - 5| > 1$ $\left(-\infty, \dfrac{2}{3}\right) \cup (1, \infty)$

Review of Geometric Figures

Plane figures have length and width but no thickness or depth

Name	Description	Figure
Polygon	Union of three or more coplanar line segments that intersect with each other only at each endpoint, with each endpoint shared by two segments.	
Triangle	Polygon with three sides (sum of measures of three angles is 180°).	
Scalene Triangle	Triangle with no sides of equal length.	
Isosceles Triangle	Triangle with two sides of equal length.	
Equilateral Triangle	Triangle with all sides of equal length.	
Right Triangle	Triangle that contains a right angle.	hypotenuse leg leg
Quadrilateral	Polygon with four sides (sum of measures of four angles is 360°).	

Plane figures have length and width but no thickness or depth.

Name	Description	Figure
Trapezoid	Quadrilateral with exactly one pair of opposite sides parallel.	
Isosceles Trapezoid	Trapezoid with legs of equal length.	
Parallelogram	Quadrilateral with both pairs of opposite sides parallel and equal in length.	
Rhombus	Parallelogram with all sides of equal length.	
Rectangle	Parallelogram with four right angles.	
Square	Rectangle with all sides of equal length.	
Circle	All points in a plane the same distance from a fixed point called the **center**.	

Solids have length, width, and depth.

Name	Description	Figure
Rectangular Solid	A solid with six sides, all of which are rectangles.	
Cube	A rectangular solid whose six sides are squares.	
Sphere	All points the same distance from a fixed point, called the center.	
Right Circular Cylinder	A cylinder with two circular bases that are perpendicular to its altitude.	
Right Circular Cone	A cone with a circular base that is perpendicular to its altitude.	

Review of Volume and Surface Area

A **convex solid** is a set of points, S, not all in one plane, such that for any two points A and B in S, all points between A and B are also in S. In this appendix, we will find the volume and surface area of special types of solids called polyhedrons. A solid formed by the intersection of a finite number of planes is called a **polyhedron**. The box below is an example of a polyhedron.

Each of the plane regions of the polyhedron is called a **face** of the polyhedron. If the intersection of two faces is a line segment, this line segment is an **edge** of the polyhedron. The intersections of the edges are the **vertices** of the polyhedron.

Volume is a measure of the space of a solid. The volume of a box or can, for example, is the amount of space inside. Volume can be used to describe the amount of juice in a pitcher or the amount of concrete needed to pour a foundation for a house.

The volume of a solid is the number of **cubic units** in the solid. A cubic centimeter and a cubic inch are illustrated.

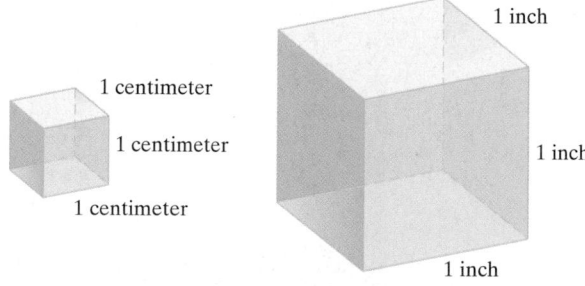

The **surface area** of a polyhedron is the sum of the areas of the faces of the polyhedron. For example, each face of the cube to the left above has an area of 1 square centimeter. Since there are 6 faces of the cube, the sum of the areas of the faces is 6 square centimeters. Surface area can be used to describe the amount of material needed to cover or form a solid. Surface area is measured in square units.

Formulas for finding the volumes, V, and surface areas, SA, of some common solids are given next.

Volume and Surface Area Formulas of Common Solids

Solid	*Formulas*

RECTANGULAR SOLID

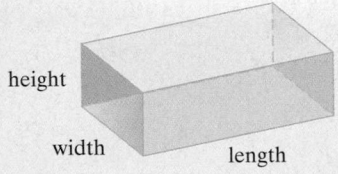

$V = lwh$
$SA = 2lh + 2wh + 2lw$
where h = height, w = width, l = length

CUBE

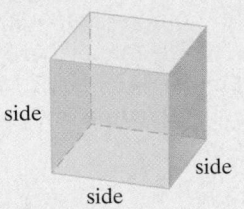

$V = s^3$
$SA = 6s^2$
where s = side

SPHERE

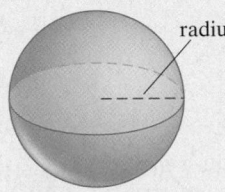

$V = \dfrac{4}{3}\pi r^3$
$SA = 4\pi r^2$
where r = radius

CIRCULAR CYLINDER

$V = \pi r^2 h$
$SA = 2\pi rh + 2\pi r^2$
where h = height, r = radius

CONE

$V = \dfrac{1}{3}\pi r^2 h$
$SA = \pi r\sqrt{r^2 + h^2} + \pi r^2$
where h = height, r = radius

SQUARE-BASED PYRAMID

$V = \dfrac{1}{3}s^2 h$

$SA = B + \dfrac{1}{2}pl$

where B = area of base, p = perimeter of base, h = height, s = side, l = slant height

> **Helpful Hint**
> Volume is measured in cubic units. Surface area is measured in square units.

EXAMPLE 1

Find the volume and surface area of a rectangular box that is 12 inches long, 6 inches wide, and 3 inches high.

3 in.

6 in.   12 in.

Solution   Let $h = 3$ in., $l = 12$ in., and $w = 6$ in.

$$V = lwh$$
$$V = 12 \text{ inches} \cdot 6 \text{ inches} \cdot 3 \text{ inches} = 216 \text{ cubic inches}$$

The volume of the rectangular box is 216 cubic inches.

$$SA = 2lh + 2wh + 2lw$$
$$= 2(12 \text{ in.})(3 \text{ in.}) + 2(6 \text{ in.})(3 \text{ in.}) + 2(12 \text{ in.})(6 \text{ in.})$$
$$= 72 \text{ sq. in.} + 36 \text{ sq. in.} + 144 \text{ sq. in.}$$
$$= 252 \text{ sq. in.}$$

The surface area of the rectangular box is 252 square inches.

EXAMPLE 2

Find the volume and surface area of a ball of radius 2 inches. Give the exact volume and surface area and then use the approximation $\dfrac{22}{7}$ for π.

Solution

2 in.

$$V = \frac{4}{3}\pi r^3 \qquad \text{Formula for volume of a sphere.}$$

$$V = \frac{4}{3}\pi (2 \text{ in.})^3 \qquad \text{Let } r = 2 \text{ inches.}$$

$$= \frac{32}{3}\pi \text{ cu. in.} \qquad \text{Simplify.}$$

$$\approx \frac{32}{3} \cdot \frac{22}{7} \text{cu. in.} \qquad \text{Approximate } \pi \text{ with } \frac{22}{7}.$$

$$= \frac{704}{21} \text{ or } 33\frac{11}{21}\text{cu. in.}$$

The volume of the sphere is exactly $\dfrac{32}{3}\pi$ cubic inches or approximately $33\dfrac{11}{21}$ cubic inches.

$$SA = 4\pi r^2 \qquad \text{Formula for surface area.}$$
$$SA = 4\pi (2 \text{ in.})^2 \qquad \text{Let } r = 2 \text{ inches.}$$
$$= 16\pi \text{ sq. in.} \qquad \text{Simplify.}$$
$$\approx 16 \cdot \frac{22}{7}\text{sq. in.} \qquad \text{Approximate } \pi \text{ with } \frac{22}{7}.$$
$$= \frac{352}{7} \text{ or } 50\frac{2}{7}\text{sq. in.}$$

The surface area of the sphere is exactly 16π square inches or approximately $50\dfrac{2}{7}$ square inches.

APPENDIX C EXERCISE SET

Find the volume and surface area of each solid. See Examples 1 and 2. For formulas that contain π, give an exact answer and then approximate using $\frac{22}{7}$ for π.

1.

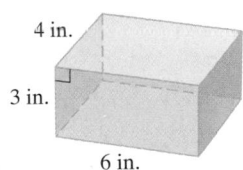
4 in.
3 in.
6 in.

2.

3 mi

3.

8 cm
8 cm
8 cm

4.

4 cm
4 cm
8 cm

5. (For surface area, use 3.14 for π and approximate to two decimal places.)

3 yd
2 yd

6.

10 ft
6 ft

7.

10 in.

8. Find the volume only.

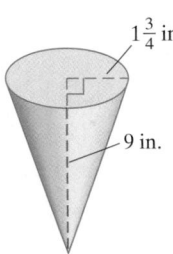
$1\frac{3}{4}$ in.
9 in.

9.

5 cm
4 cm
6 cm

9. $V = 48$ cu. cm; $SA = 96$ sq. cm

10.

1 ft

10. $V = 18$ cu. ft; $SA = 42$ sq. in.

Solve.

11. Find the volume of a cube with edges of $1\frac{1}{3}$ inches.

$2\frac{10}{27}$ cu. in.

$1\frac{1}{3}$ in.

12. A water storage tank is in the shape of a cone with the pointed end down. If the radius is 14 ft and the depth of the tank is 15 ft, approximate the volume of the tank in cubic feet. Use $\frac{22}{7}$ for π. 3080 cu. ft

14 ft
15 ft

1. $V = 72$ cu. in; $SA = 108$ sq. in **2.** $V = 36\pi$ cu. mi $\approx 113\frac{1}{7}$ cu. mi; $SA = 36\pi$ sq. mi $\approx 113\frac{1}{7}$ sq. mi

3. $V = 512$ cu. cm; $SA = 384$ sq. cm **4.** $V = 128$ cu. cm; $SA = 160$ sq. cm

5. $V = 4\pi$ cu. yd ≈ 12.56 cu. yd; $SA = (2\sqrt{13}\pi + 4\pi)$ sq. yd ≈ 35.20 sq. yd

6. $V = 150\pi$ cu. ft $\approx 471\frac{3}{7}$ cu. ft; $SA = 110\pi$ sq. ft; $\approx 345\frac{5}{7}$ sq. ft

7. $V = \frac{500}{3}\pi$ cu. in. $\approx 523\frac{17}{21}$ cu. in.; $SA = 100\pi$ sq. in. $\approx 314\frac{2}{7}$ sq. in. **8.** $V = \frac{147}{16}\pi$ cu. in. $\approx 28\frac{7}{8}$ cu. in.

13. Find the surface area of a rectangular box 2 ft by 1.4 ft by 3 ft.   26 sq. ft

14. Find the surface area of a box in the shape of a cube that is 5 ft on each side.   150 sq. ft

15. Find the volume of a pyramid with a square base 5 in. on a side and a height of 1.3 in.

16. Approximate to the nearest hundredth the volume of a sphere with a radius of 2 cm. Use 3.14 for π.   33.49 cu. cm

17. A paperweight is in the shape of a square-based pyramid 20 cm tall. If an edge of the base is 12 cm, find the volume of the paperweight.   960 cu. cm

18. A bird bath is made in the shape of a hemisphere (half-sphere). If its radius is 10 in., approximate the volume. Use $\frac{22}{7}$ for π. $2095\frac{5}{21}$ cu. in.

10 in.

19. Find the exact surface area of a sphere with a radius of 7 in. 196π sq. in.

20. A tank is in the shape of a cylinder 8 ft tall and 3 ft in radius. Find the exact surface area of the tank. 66π sq. ft

21. Find the volume of a rectangular block of ice 2 ft by $2\frac{1}{2}$ ft by $1\frac{1}{2}$ ft. $7\frac{1}{2}$ cu. ft

22. Find the capacity (volume in cubic feet) of a rectangular ice chest with inside measurements of 3 ft by $1\frac{1}{2}$ ft by $1\frac{3}{4}$ ft.

23. An ice cream cone with a 4-cm diameter and 3-cm depth is filled exactly level with the top of the cone. Approximate how much ice cream (in cubic centimeters) is in the cone. Use $\frac{22}{7}$ for π. $12\frac{4}{7}$ cu. cm

24. A child's toy is in the shape of a square-based pyramid 10 in. tall. If an edge of the base is 7 in., find the volume of the toy. $163\frac{1}{3}$ cu. in.

15. $10\frac{5}{6}$ cu. in.   **22.** $7\frac{7}{8}$ cu. ft

An Introduction to Using a Graphing Utility

THE VIEWING WINDOW AND INTERPRETING WINDOW SETTINGS

In this appendix, we will use the term **graphing utility** to mean a graphing calculator or a computer software graphing package. All graphing utilities graph equations by plotting points on a screen. While plotting several points can be slow and sometimes tedious for us, a graphing utility can quickly and accurately plot hundreds of points. How does a graphing utility show plotted points? A computer or calculator screen is made up of a grid of small rectangular areas called **pixels**. If a pixel contains a point to be plotted, the pixel is turned "on"; otherwise, the pixel remains "off." The graph of an equation is then a collection of pixels turned "on." The graph of $y = 3x + 1$ from a graphing calculator is shown in Figure A–1. Notice the irregular shape of the line caused by the rectangular pixels.

The portion of the coordinate plane shown on the screen in Figure A-1 is called the **viewing window** or the **viewing rectangle**. Notice the x-axis and the y-axis on the graph. While tick marks are shown on the axes, they are not labeled. This means that from this screen alone, we do not know how many units each tick mark represents. To see what each tick mark represents and the minimum and maximum values on the axes, check the window setting of the graphing utility. It defines the viewing window. The window of the graph of $y = 3x + 1$ shown in Figure A-1 has the following setting (Figure A-2):

$y = 3x + 1$

Figure A-1

(Ymax) 10

Yscl = 1

−10 (Xmin)   10 (Xmax)

Xscl = 1

−10 (Ymin)

Figure A-2

$Xmin = -10$	The minimum x-value is -10.
$Xmax = 10$	The maximum x-value is 10.
$Xscl = 1$	The x-axis scale is 1 unit per tick mark.
$Ymin = -10$	The minimum y-value is -10.
$Ymax = 10$	The maximum y-value is 10.
$Yscl = 1$	The y-axis scale is 1 unit per tick mark.

By knowing the scale, we can find the minimum and the maximum values on the axes simply by counting tick marks. For example, if both the Xscl (x-axis scale) and the

Yscl are 1 unit per tick mark on the graph in Figure A-3, we can count the tick marks and find that the minimum *x*-value is −10 and the maximum *x*-value is 10. Also, the minimum *y*-value is −10 and the maximum *y*-value is 10. If the Xscl (*x*-axis scale) changes to 2 units per tick mark (shown in Figure A-4), by counting tick marks, we see that the minimum *x*-value is now −20 and the maximum *x*-value is now 20.

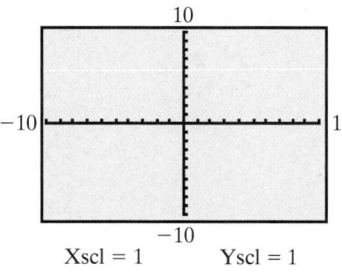

Xscl = 1   Yscl = 1

Figure A-3

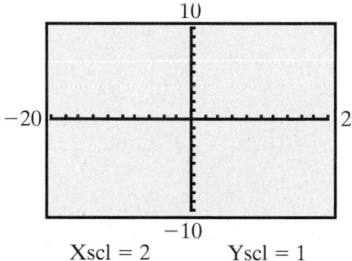

Xscl = 2   Yscl = 1

Figure A-4

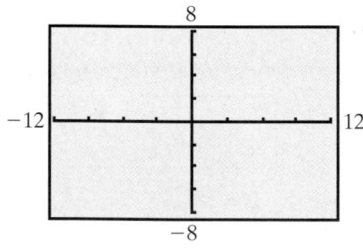

Figure A-5

It is also true that if we know the Xmin and the Xmax values, we can calculate the Xscl by the displayed axes. For example, the Xscl of the graph in Figure A-5 must be 3 units per tick mark for the maximum and minimum *x*-values to be as shown. Also, the Yscl of that graph must be 2 units per tick mark for the maximum and minimum *y*-values to be as shown.

We will call the viewing window in Figure A-3 a *standard* viewing window or rectangle. Although a standard viewing window is sufficient for much of this text, special care must be taken to ensure that all key features of a graph are shown. Figures A-6, A-7, and A-8 show the graph of $y = x^2 + 11x - 1$ on three different viewing windows. Note that certain viewing windows for this equation are misleading.

Figure A-6

Figure A-7

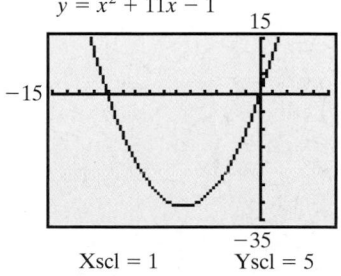

Xscl = 1   Yscl = 5

Figure A-8

How do we ensure that all distinguishing features of the graph of an equation are shown? It helps to know about the equation that is being graphed. For example, the equation $y = x^2 + 11x - 1$ is not a linear equation and its graph is not a line. This equation is a quadratic equation and, therefore, its graph is a parabola. By knowing this information, we know that the graph shown in Figure A-6, although correct, is misleading. Of the three viewing rectangles shown, the graph in Figure A-8 is best because it shows more of the distinguishing features of the parabola. Properties of equations needed for graphing will be studied in this text.

VIEWING WINDOW AND INTERPRETING WINDOW SETTINGS EXERCISE SET

In Exercises 1–4, determine whether all ordered pairs listed will lie within a standard viewing rectangle.

1. $(-9, 0), (5, 8), (1, -8)$ yes

2. $(4, 7), (0, 0), (-8, 9)$ yes

3. $(-11, 0), (2, 2), (7, -5)$ no

4. $(3, 5), (-3, -5), (15, 0)$ no

In Exercises 5–10, choose an Xmin, Xmax, Ymin, and Ymax so that all ordered pairs listed will lie within the viewing rectangle.

5.–10. answers may vary

5. $(-90, 0), (55, 80), (0, -80)$

6. $(4, 70), (20, 20), (-18, 90)$

7. $(-11, 0), (2, 2), (7, -5)$

8. $(3, 5), (-3, -5), (15, 0)$

9. $(200, 200), (50, -50), (70, -50)$

10. $(40, 800), (-30, 500), (15, 0)$

Write the window setting for each viewing window shown. Use the following format:

Xmin =	Ymin =
Xmax =	Ymax =
Xscl =	Yscl =

11. Xmin $= -12$ Ymin $= -12$
Xmax $= 12$ Ymax $= 12$
Xscl $= 3$ Yscl $= 3$

12. Xmin $= -20$ Ymin $= -20$
Xmax $= 20$ Ymax $= 20$
Xscl $= 5$ Yscl $= 5$

13. Xmin $= -9$ Ymin $= -12$
Xmax $= 9$ Ymax $= 12$
Xscl $= 1$ Yscl $= 2$

14. Xmin $= -27$ Ymin $= -6$
Xmax $= 27$ Ymax $= 6$
Xscl $= 3$ Yscl $= 1$

15. Xmin $= -10$ Ymin $= -25$
Xmax $= 10$ Ymax $= 25$
Xscl $= 2$ Yscl $= 5$

16. Xmin $= -50$ Ymin $= -20$
Xmax $= 50$ Ymax $= 20$
Xscl $= 10$ Yscl $= 4$

17. Xmin $= -10$ Ymin $= -30$
Xmax $= 10$ Ymax $= 30$
Xscl $= 1$ Yscl $= 3$

18. Xmin $= -100$ Ymin $= -20$
Xmax $= 100$ Ymax $= 20$
Xscl $= 10$ Yscl $= 2$

19. Xmin $= -20$ Ymin $= -30$
Xmax $= 30$ Ymax $= 50$
Xscl $= 5$ Yscl $= 10$

20. Xmin $= -500$ Ymin $= -800$
Xmax $= 700$ Ymax $= 400$
Xscl $= 100$ Yscl $= 200$

11. 12, −12, 12, −12

12. 20, −20, 20, −20

13. 12, −9, 9, −12

14. 6, −27, 27, −6

15. 25, −10, 10, −25

16. 20, −50, 50, −20

17. Xscl = 1, Yscl = 3

18. Xscl = 10, Yscl = 2

19. Xscl = 5, Yscl = 10

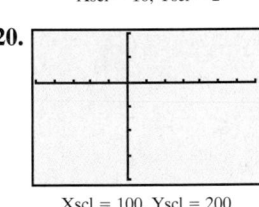

20. Xscl = 100, Yscl = 200

GRAPHING EQUATIONS AND SQUARE VIEWING WINDOW

In general, the following steps may be used to graph an equation on a standard viewing window.

Graphing an Equation in X and Y with a Graphing Utility on a Standard Viewing Window

Step 1: Solve the equation for y.

Step 2: Using your graphing utility and enter the equation in the form
$Y = $ *expression involving x.*

Step 3: Activate the graphing utility.

Special care must be taken when entering the *expression involving x* in Step 2. You must be sure that the graphing utility you are using interprets the expression as you want it to. For example, let's graph $3y = 4x$. To do so,

Step 1: Solve the equation for y.

$$3y = 4x$$

$$\frac{3y}{3} = \frac{4x}{3}$$

$$y = \frac{4}{3}x$$

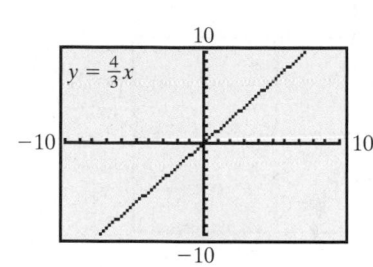

Figure A-9

Step 2: Using your graphing utility, enter the expression $\frac{4}{3}x$ after the Y = prompt. In order for your graphing utility to correctly interpret the expression, you may need to enter $(4/3)x$ or $(4 \div 3)x$.

Step 3: Activate the graphing utility. The graph should appear as in Figure A-9.

Distinguishing features of the graph of a line include showing all the intercepts of the line. For example, the window of the graph of the line in Figure A-10 does not show both intercepts of the line, but the window of the graph of the same line in Figure A-11 does show both intercepts. Notice the notation below each graph. This is a shorthand notation of the range setting of the graph. This notation means [Xmin, Xmax] by [Ymin, Ymax].

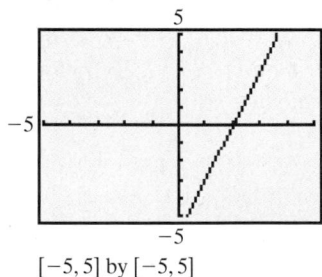

$[-5, 5]$ by $[-5, 5]$

Figure A-10

$[-4, 8]$ by $[-8, 5]$

Figure A-11

On a standard viewing window, the tick marks on the y-axis are closer together than the tick marks on the x-axis. This happens because the viewing window is a rectangle, and so 10 equally spaced tick marks on the positive y-axis will be closer together than 10 equally spaced tick marks on the positive x-axis. This causes the appearance of graphs to be distorted.

For example, notice the different appearances of the same line graphed using different viewing windows. The line in Figure A-12 is distorted because the tick marks along the x-axis are farther apart than the tick marks along the y-axis. The graph of the same line in Figure A-13 is not distorted because the viewing rectangle has been selected so that there is equal spacing between tick marks on both axes.

Figure A-12

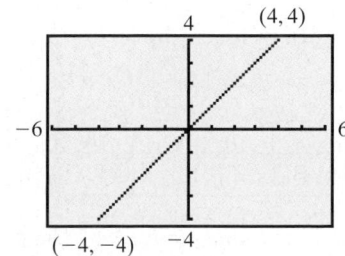

Figure A-13

We say that the line in Figure A-13 is graphed on a *square* setting. Some graphing utilities have a built-in program that, if activated, will automatically provide a square setting. A square setting is especially helpful when we are graphing perpendicular lines, circles, or when a true geometric perspective is desired. Some examples of square screens are shown in Figures A-14 and A-15.

Figure A-14

Figure A-15

Other features of a graphing utility such as Trace, Zoom, Intersect, and Table are discussed in appropriate Graphing Calculator Explorations in this text.

GRAPHING EQUATIONS AND SQUARE VIEWING WINDOW EXERCISE SET

Graph each linear equation in two variables, using the two different range settings given. Determine which setting shows all intercepts of a line.

1. $y = 2x + 12$ Setting B

  Setting A: $[-10, 10]$ by $[-10, 10]$
  Setting B: $[-10, 10]$ by $[-10, 15]$

2. $y = -3x + 25$ Setting B

  Setting A: $[-5, 5]$ by $[-30, 10]$
  Setting B: $[-10, 10]$ by $[-10, 30]$

3. $y = -x - 41$ Setting B

  Setting A: $[-50, 10]$ by $[-10, 10]$
  Setting B: $[-50, 10]$ by $[-50, 15]$

4. $y = 6x - 18$ Setting A

  Setting A: $[-10, 10]$ by $[-20, 10]$
  Setting B: $[-10, 10]$ by $[-10, 10]$

5. $y = \frac{1}{2}x - 15$ Setting B

  Setting A: $[-10, 10]$ by $[-20, 10]$
  Setting B: $[-10, 35]$ by $[-20, 15]$

6. $y = -\frac{2}{3}x - \frac{29}{3}$ Setting B

  Setting A: $[-10, 10]$ by $[-10, 10]$
  Setting B: $[-15, 5]$ by $[-15, 5]$

The graph of each equation is a line. Use a graphing utility and a standard viewing window to graph each equation.

7.–22. See graphing answer section.

7. $3x = 5y$

8. $7y = -3x$

9. $9x - 5y = 30$

10. $4x + 6y = 20$

11. $y = -7$

12. $y = 2$

13. $x + 10y = -5$

14. $x - 5y = 9$

Graph the following equations using the square setting given. Some keystrokes that may be helpful are given.

15. $y = \sqrt{x}$ $[-12, 12]$ by $[-8, 8]$

  Suggested keystrokes: $\sqrt{\ } x$

16. $y = \sqrt{2x}$ $[-12, 12]$ by $[-8, 8]$

  Suggested keystrokes: $\sqrt{\ } (2x)$

17. $y = x^2 + 2x + 1$ $[-15, 15]$ by $[-10, 10]$

  Suggested keystrokes: $x^\wedge 2 + 2x + 1$

18. $y = x^2 - 5$ $[-15, 15]$ by $[-10, 10]$

  Suggested keystrokes: $x^\wedge 2 - 5$

19. $y = |x|$ $[-9, 9]$ by $[-6, 6]$

  Suggested keystrokes: $ABS(x)$

20. $y = |x - 2|$ $[-9, 9]$ by $[-6, 6]$

  Suggested keystrokes: $ABS(x - 2)$

Graph each line. Use a standard viewing window; then, if necessary, change the viewing window so that all intercepts of each line show.

21. $x + 2y = 30$

22. $1.5x - 3.7y = 40.3$

CHAPTER 1   REAL NUMBERS, ALGEBRAIC EXPRESSIONS, AND EQUATIONS

Exercise Set 1.2   **1.** 35   **3.** 30.38   **5.** $\dfrac{3}{8}$ **7.** 22   **9.** 2000 mi   **11.** 20.4 sq. ft   **13.** $10,612.80   **15.** $\{1, 2, 3, 4, 5\}$

17. $\{11, 12, 13, 14, 15, 16\}$ **19.** $\{0\}$ **21.** $\{0, 2, 4, 6, 8\}$ **23.**   **25.**

27.   **29.** answers may vary   **31.** $\{3, 0, \sqrt{36}\}$ **33.** $\{3, \sqrt{36}\}$

35. $\{\sqrt{7}\}$ **37.** $\in$ **39.** $\notin$ **41.** $\notin$ **43.** $\notin$ **45.** true   **47.** true   **49.** false   **51.** false   **53.** true   **55.** false   **57.** answers

may vary   **59.** -2 **61.** 4   **63.** 0   **65.** -3 **67.** answers may vary   **69.** 6.2   **71.** $-\dfrac{4}{7}$ **73.** $\dfrac{2}{3}$ **75.** 0   **77.** $2x$

79. $2x + 5$ **81.** $x - 10$ **83.** $x + 2$ **85.** $\dfrac{x}{11}$ **87.** $3x + 12$ **89.** $x - 17$ **91.** $2(x + 3)$ **93.** $\dfrac{5}{4 - x}$ **95.** 137; 100; 91; 69; 56

97. answers may vary

Mental Math   **1.** B, C   **3.** B, D   **5.** B

Exercise Set 1.3   **1.** 5   **3.** -24 **5.** -11 **7.** -4 **9.** $\dfrac{4}{3}$ **11.** -2 **13.** -60 **15.** 80   **17.** 3   **19.** 0   **21.** -8 **23.** $-\dfrac{3}{7}$

25. $\dfrac{1}{21}$ **27.** -49 **29.** 36   **31.** -8 **33.** 7   **35.** $-\dfrac{1}{3}$ **37.** 4   **39.** 3   **41.** not a real number   **43.** 48   **45.** -1 **47.** -3

49. 14.4   **51.** -2.1 **53.** $-\dfrac{1}{3}$ **55.** 17   **57.** 40   **59.** 11   **61.** $-\dfrac{3}{4}$ **63.** 7   **65.** -11 **67.** $-\dfrac{4}{5}$ **69.** $-\dfrac{79}{15}$ **71.** $-\dfrac{5}{14}$

73. 13   **75.** 65   **77.** -2 **79.** 16.6   **81.** -0.61 **83.** -121.2 **85.** $x = -2.7, y = 0.9; x^2 + 2y = 9.09$ **87.** $A = 4, B = -2.5;$

$A^2 + B^2 = 22.25$ **89. a.** $(1/2)x = 1; 1/2x = 1$ **b.** same; answers may vary   **91. a.** 18; 22; 28; 208   **b.** increase   **93. a.** 600; 150; 105

b. decrease   **95.** $\dfrac{13}{35}$ **97.** 4205 m   **99.** 5.4; 2.1; 1.7; 2.6; 1.4; 0.7   **101.** $6 - (5 \cdot 2 + 2)$ **103.** answers may vary   **105.** 16.5227

107. 4.4272   **109.** 1.4187   **111.** 13.2%   **113.** 10.8%   **115.** b   **117.** d   **119.** Yes.   Two players have 6 points each (the third player

has 0 points) or two players have 5 points each (the third has 2 points).

Exercise Set 1.4   **1.** $>$ **3.** $=$ **5.** $<$ **7.** $2x + 5 = -14$ **9.** $3(x + 1) = 7$ **11.** $\dfrac{n}{5} = 4n$ **13.** $z - 2 = 2z$ **15.** $7x \le -21$

17. $-2 + x \ne 10$ **19.** $2(x - 6) > \dfrac{1}{11}$ **21.** $5y - 7 = 6$ **23.** $2(x - 6) = -27$ **25.** $-5; \dfrac{1}{5}$ **27.** $8; -\dfrac{1}{8}$ **29.** $\dfrac{1}{7}; -7$

31. 0; undefined   **33.** $\dfrac{7}{8}; -\dfrac{7}{8}$ **35.** Zero. For every real number $x, 0 \cdot x \ne 1$, so 0 has no reciprocal. It is the only real number that has no recipro-

cal because if $x \ne 0$, then $x \cdot \dfrac{1}{x} = 1$ by definition.   **37.** $y + 7x$ **39.** $w \cdot z$ **41.** $\dfrac{x}{5} \cdot \dfrac{1}{3}$ **43.** no; answers may vary   **45.** $(5 \cdot 7)x$

47. $x + (1.2 + y)$ **49.** $14(z \cdot y)$ **51.** 10 and 4. Subtraction is not associative.   **53.** $3x + 15$ **55.** $-2a - b$ **57.** $12x + 10y + 4z$

59. $-4x + 8y - 28$ **61.** $3xy - 1.5x$ **63.** $6 + 3x$ **65.** 0   **67.** 7   **69.** $(10 \cdot 2)y$ **71.** Associative Property of Addition.   **73.** Com-

mutation Property of Multiplication.   **75.** $a(b + c) = ab + ac$ **77.** $0.1d$ **79.** $112 - x$ **81.** $90 - 5x$ **83.** $35.61y$ **85.** $2x + 2$

87. $-8y - 14$ **89.** $-9c - 4$ **91.** $4 - 8y$ **93.** $y^2 - 11yz - 11$ **95.** $3t - 14$ **97.** 0   **99.** $13n - 20$ **101.** $8.5y - 20.8$

103. $\dfrac{3}{8}a - \dfrac{1}{12}$ **105.** $-x - 6y - \dfrac{11}{24}$ **107.** $-180.96y - 74.33$ **109.** $6.5y - 7.92x + 25.47$

111. no   **113.** 80 million   **115.** 35 million   **117.** 20.25%

Integrated Review   **1.** 16   **2.** -16 **3.** 0   **4.** -11 **5.** -5 **6.** $-\dfrac{1}{60}$ **7.** undefined   **8.** -2.97 **9.** 4   **10.** -50

11. 35   **12.** 92   **13.** $-15 - 2x$ **14.** $3x + 5$ **15.** 0   **16.** true   **17.** $-45x$ **18.** $-x - 8$ **19.** $7.6a + 7.2b$ **20.** $\dfrac{14}{9}y + \dfrac{7}{30}$

Mental Math   **1.** $8x + 21$ **3.** $6n - 7$ **5.** $-4x - 1$ **7.** expression   **9.** equation   **11.** all real numbers   **13.** no solution

Exercise Set 1.5 **1.** -12 **3.** -0.9 **5.** 6 **7.** -5 **9.** -1.1 **11.** -5 **13.** 0 **15.** 2 **17.** -9 **19.** $-\dfrac{10}{7}$ **21. a.** $4x + 5$

b. -3 **c.** answers may vary **23.** $\dfrac{1}{6}$ **25.** 4 **27.** 1 **29.** 5 **31.** all real numbers **33.** $\emptyset$ **35.** answers may vary

37. $x = -8$; $6x - 5 = 4x - 21$ **39.** $N = 3$; $3(N - 3) + 2 = -1 + N$ **41.** 7 **43.** 4.2 **45.** 2 **47.** -2 **49.** 0 **51.** 5 **53.** $\emptyset$

55. $\dfrac{1}{8}$ **57.** 0 **59.** 29 **61.** -8 **63.** all real numbers **65.** 4 **67.** -2 **69.** all real numbers **71.** $\dfrac{40}{3}$ **73.** 17 **75.** $\dfrac{3}{5}$

77. $\dfrac{103}{5}$ or 20.6 **79.** $K = -11$ **81.** $K = 24$ **83.** 1 **85.** 3 **87.** -4.86 **89.** 1.53 **91.** not a fair game

Exercise Set 1.6 **1.** $4y$ **3.** $3z + 3$ **5.** $(15x + 30)$ cents **7.** $10x + 3$ **9.** -5 **11.** 45,225 **13.** 78 **15.** 1.92 **17.** 1612.41 million acres **19.** 1991 earthquakes **21.** 860 shoppers **23.** 17% **25.** 6750 users **27.** Los Angeles: 61.6 million; Atlanta: 76.9 million; Chicago: 67.4 million **29.** $737 - 200$: 113 seats; $737 - 300$: 134 seats; $757 - 200$: 190 seats **31.** $430.00 **33.** Mile High Stadium: 76,125 seats; Heinz Field Stadium: 64,450 seats **35.** AOL: 23,700,000; Earthlink: 4,700,000; MSN: 4,000,000 **37.** 28.6 million **39. a.** 214,866 operators **b.** answers may vary **41.** 75, 76, 77 **43.** $64°, 32°, 84°$ **45.** height: 48 in.; width: 108 in. **47.** length: 14 cm; width: 6 cm **49.** $80°, 100°$ **51.** $15°, 75°$ **53.** $40°, 140°$ **55.** width: 8.4 m; height: 47 m **57.** incandescent: 1500 bulb hours; fluorescent: 100,000 bulb hours; halogen: 4000 bulb hours **59.** 21.1 million returns **61.** Thome: 49; Palmeiro: 47; Sexson: 45 **63.** 11 million trees **65. a.** during the year 2033 **b.** 1828.75 **c.** 5; no: this is the average daily number of cigarettes for all American adults—smokers and nonsmokers **67.** no such odd integers exist **69.** 500 boards; $30,000 **71.** company makes a profit

Mental Math **1.** f **3.** b **5.** d

Exercise Set 1.7 **1.** 343; 512; 729; 1000; 1331 **a.** 729 cu. cm **b.** 1331 cu. ft **c.** 7 in. **3.** 298; 301.50; 305; 308.50 **a.** 44 hr **b.** 35 hr **c.** 37 hr **5. a.** $y_1 = 5.25x$ **b.** $y_2 = 7x$ **c.** Rough draft: 5.25, 6.56, 7.88, 9.19, 10.50, 11.81, 13.13, 14.44, 15.75; Manuscript: 7, 8.75, 10.50, 12.25, 14, 15.75, 17.50, 19.25, 21 **d.** $6.56 **e.** $75.25 **7.** Circumference: 12.57, 31.42, 131.95, 591.88; Area: 12.57, 78.54, 1385.44, 27,877.36 **a.** 188.5 yd; 2827.43 sq. yd **b.** 493.23 mm; 19,359.28 sq. mm **9. a.** 5500, 8500, 10,500, 14,500, 16,500 **b.** $5500 and $16,500 **11. a.** $y_1 = 12.96 + 1.10x$ **b.** 12.96, 14.06, 15.16, 16.26, 17.36, 18.46, 19.56 **13. a.** $y_1 = 500 - 8x$ **b.** 500; 420; 340; 260; 180; 100; 20; -60 **c.** 30 min **d.** 62.5 min **15. a.** First job: $18,500, $23,500, $28,500, $33,500; Second job: $22,300, $25,175, $28,050, $30,925 **b.** answers may vary **17. a.** $y_1 = 15 + 0.08924x$ **b.** 15; 59.62; 104.24; 148.86; 193.48 **19. a.** $y_1 = 150 - x(42/30)$ **b.** 150; 108; 66; 24 **c.** 107 min or 6:47 pm **21. a.** $y_1 = 24 - x/15$ **b.** $y_2 = 24 - x/25$ **c.** 24; 17.3, 10.7, 4, -2.7, -9.3, -16; 24, 20, 16, 12, 8, 4, 0 **d.** 17.3 gal **e.** 20 gal **f.** 600 mi **g.** 360 mi **h.** answers may vary **23. a.** $y_1 = 0.65x$ **b.** $19.47; $23.37; $12.51; $25.97; $11.67; $6.47; $16.74 **c.** $116.20 **d.** $124.33 **e.** $62.55 **25.** Cost: 9000, 10.375, 11,750, 13,125, 14,500; Revenue: 7500, 9375, 11,250, 13,125, 15,000 **27.** When 200 calculators are produced and sold, the profit is $500. **29.** 6 in.: 28.27, 18.85, $1.41, $4.71; 12 in.: 113.10, 37.70, $5.65, $9.42; 18 in.: 254.47, 56.55, $12.72, $14.14; 24 in.: 452.39, 75.40, $22.62, $18.85; 30 in.: 706.86, 94.25, $35.34, $23.56; 36 in.: 1017.88, 113.10, $50.89, $28.27 **31. a.** 1569 ft, 1606 ft, 1611 ft, 1584 ft, 1525 ft **b.** 1613 ft

33. a. Anne: $49.00; Michelle: $50.00 **b.** distances less than $333\dfrac{1}{3}$ mi **35. a.** $C = 90 + 75x$ **b.** 240; 390; 540; 690 **37. a.** 3.5 ft

b. 5 ft **c.** 10 ft **39. a.** 6.73 or 6:44 A.M. **b.** 4; July 1. **c.** 7:14 A.M. **d.** 6:17 A.M. **e.** answers may vary

Mental Math **1.** $y = 5 - 2x$ **3.** $a = 5b + 8$ **5.** $k = h - 5j + 6$

Exercise Set 1.8 **1.** $t = \dfrac{D}{r}$ **3.** $R = \dfrac{I}{PT}$ **5.** $y = \dfrac{9x - 16}{4}$ **7.** $W = \dfrac{P - 2L}{2}$ **9.** $A = \dfrac{J + 3}{C}$ **11.** $g = \dfrac{W}{h - 3t^2}$

13. $B = \dfrac{T - 2C}{AC}$ **15.** $r = \dfrac{C}{2\pi}$ **17.** $r = \dfrac{E - IR}{I}$ **19.** $L = \dfrac{2s - an}{n}$ **21.** $v = \dfrac{3st^4 - N}{5s}$ **23.** $H = \dfrac{S - 2LW}{2L + 2W}$ **25.** $4703.71; $4713.99; $4719.22; $4722.74; $4724.45 **27. a.** $7313.97 **b.** $7321.14 **c.** $7325.98 **29.** $40°C$ **31.** 3.6 hr, or 3 hr and 36 min

33. 171 packages **35.** 9 ft **37.** 2 gal **39. a.** 1174.86 cu. m **b.** 310.34 cu. m **c.** 1485.20 cu. m **41.** 164,921 mi **43.** 0.42 ft

45. 41.125π ft ≈ 129.1325 ft **47.** $1831.96 **49.** $f = \dfrac{C - 4h - 4p}{9}$ **51.** 178 cal **53.** 1.5 g **55.** 0.388; 0.723; 1.00; 1.523; 5.202; 9.538; 19.193;

30.065; 39.505 **57.** $6.80 per person **59.** answers may vary **61.** 0.25 sec **63.** $\dfrac{1}{4}$ **65.** $\dfrac{3}{8}$ **67.** $\dfrac{3}{8}$ **69.** $\dfrac{3}{4}$ **71.** 1 **73.** 1

Exercise Set 1.9 **1.** 18.6 **3.** $5.33 **5.** 116.7 million **7. a.** 5, 14, 19, 16, 22, 18, 17 **b.** 13.2 thousand **9.** 85 **11.** Growth/Income: $5250; Small Company/Aggressive Growth: $3750; International: $3000; Bonds: $3000 **13.** answers may vary **15.** 422.1 m

17. 418 m   **19.** 6.8 sec   **21.** 6.9 sec   **23.** 85.5   **25.** 73   **27.** 70 and 71   **29.** 9   **31.** 21, 21, 20   **33.** Police officer   **35.** 940 mi
37. answers may vary   **39.** Democrat   **41.** 1964   **43.** Democrat: 50%; Republican: 38%; Independent; 10%   **45.** answers may vary
47. $210.54   **49.** $33.69   **51.** 24.3 million   **53.** 17.6 million   **55.**

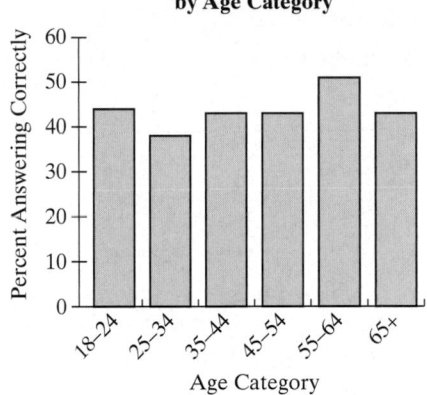

Correct Answer to History Question by Age Category

57. answers may vary
59. no
61. The mean is lowered; the median and mode are unchanged.

Chapter 1 Review   **1.** 21   **3.** 324,000   **5.** $\{-2, 0, 2, 4, 6\}$ **7.** $\varnothing$ **9.** $\{\dots, -1, 0, 1, 2\}$ **11.** false   **13.** true   **15.** true

17. true   **19.** true   **21.** true   **23.** true   **25.** true   **27.** true   **29.** $\left\{5, \frac{8}{2}, \sqrt{9}\right\}$ **31.** $\left\{\sqrt{7}, \pi\right\}$ **33.** $\left\{5, \frac{8}{2}, \sqrt{9}, -1\right\}$ **35.** -0.6

37. -1 **39.** $\frac{1}{0.6}$ **41.** 1   **43.** -35 **45.** 0.31   **47.** 13.3   **49.** 0   **51.** 0   **53.** -5 **55.** 4   **57.** 9   **59.** 3   **61.** $-\frac{32}{135}$

63. $-\frac{5}{4}$ **65.** $\frac{5}{8}$ **67.** -1 **69.** 1   **71.** -4 **73.** $\frac{5}{7}$ **75.** $\frac{1}{5}$ **77.** -5 **79.** 5   **81. a.** 6.28; 62.8; 628   **b.** increase

83. $-5x - 9$ **85.** $-15x^2 + 6$ **87.** $5.7x + 1.1$ **89.** $n + 2n = -15$ **91.** $6(t - 5) = 4$ **93.** $9x - 10 = 5$ **95.** $-4 < 7y$

97. $t + 6 \le -12$ **99.** distributive property   **101.** commutative property of addition   **103.** multiplicative inverse property

105. associative property of multiplication   **107.** multiplicative identity property   **109.** $(3 + x) + (7 + y)$ **111.** $2 \cdot \frac{1}{2}$, for example

113. $7 + 0$ **115.** $>$ **117.** $=$ **119.** 3   **121.** $-\frac{45}{14}$ **123.** 6   **125.** all real numbers   **127.** $\varnothing$ **129.** -3 **131** $\frac{96}{5}$ **133.** 8

135. 2   **137.** -7 **139.** 52   **141.** 55 million viewers   **143.** No such odd integers exist.   **145.** 358 mi   **147.** 5 plants, $200

149. 34; 39.36; 49; 55.36   **151.** Coast charge: 9.1, 10.2, 11.3, 12.4; Cross Gates charge: 12, 13.5, 15, 16.5   **153.** $w = \frac{V}{lh}$ **155.** $y = \frac{5x + 12}{4}$

157. $m = \frac{y - y_1}{x - x_1}$ **159.** $r = \frac{E - IR}{I}$ **161.** $g = \frac{T}{r + vt}$ **163.** $B = \frac{2A - hb}{h}$ **165.** $r_1 = 2R - r_2$ **167. a.** $3695.27   **b.** $3700.81

169. a. -40; 5; 50; 140   **b.** 212 iF   **c.** 32 iF   **171.** 16 packages   **173.** 58 mph   **175.** 23.56; 46.18; 79.19; 94.25; 128.28   **177.** mean: 40.6;
median: 40; mode: none   **179.** mean: 5.97; median: 6.05; mode: 4.9 and 6.8   **181.** mean: 0.53; median: 0.6; mode: 0.6 and 0.8   **183.** $ 8.55 billion
185. less than 10; 29%   **187.** answers may vary

Chapter 1 Test   **1.** true   **2.** false   **3.** false   **4.** false   **5.** true   **6.** false   **7.** -3 **8.** -56 **9.** -225 **10.** 3   **11.** 1
12. $-\frac{3}{2}$ **13.** 12   **14.** 1   **15. a.** 5.75; 17.25; 57.50; 115.00   **b.** increase   **16.** $2(x + 5) = 30$ **17.** $\frac{(6 - y)^2}{7} < -2$

18. $\frac{9z}{|-12|} \ne 10$ **19.** $3\left(\frac{n}{5}\right) = -n$ **20.** $20 = 2x - 6$ **21.** $-2 = \frac{x}{x + 5}$ **22.** distributive property   **23.** associative property of addition

24. additive inverse property   **25.** multiplication property of zero   **26.** $0.05n + 0.1d$ **27.** $-6x - 14$ **28.** $\frac{1}{2}a - \frac{9}{8}$ **29.** $2y - 10$

30. $-1.3x + 1.9$ **31.** 64.3118   **32.** 6.28; 11.94; 31.42; 46.81; 3.14; 11.34; 78.54; 174.37   **33.** 10   **34.** 1   **35.** $\varnothing$ **36.** all real numbers

37. 12   **38.** $y = \frac{3x - 8}{4}$ **39.** $n = \frac{9}{7}m$ **40.** $g = \frac{S}{t^2 + vt}$ **41.** $[-3, -1)$ **42.** $(-\infty, \infty)$ **43.** 9.6   **44.** 211,468 people

45. approximately 8 dogs   **46. a.** 1900; 1950; 2000; 2050; 2100   **b.** $24,600   **c.** $14,000   **47. a.** 2 and 3 sec   **b.** 120 ft   **c.** 2.5 sec
d. 3.0 sec   **48.** 4.75   **49.** 5   **50.** 5   **51.** 80 lb   **52.** 119 lb   **53.** 2000   **54.** answers may vary

CHAPTER 2   GRAPHS AND FUNCTIONS

Mental Math   **1.** $(5,2)$ **3.** $(3,0)$ **5.** $(-5,-2)$ **7.** $(-1,0)$ **9.** QI   **11.** QII   **13.** QIII   **15.** y-axis   **17.** QIII   **19.** x-axis

Exercise Set 2.1

1. Quadrant I   **3.** Quadrant II   **5.** Quadrant IV   **7.** y-axis   **9.** Quadrant III

    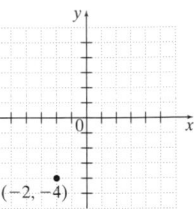

11. Quadrant IV   **13.** x-axis   **15.** Quadrant III   **17.** $(2,8)$; quadrant I   **19.** $(1,-4)$; quadrant IV   **21.** Possible answer: $[-10, 10, 1]$ by $[0, 20, 1]$
23. Possible answer: $[-100, 100, 10]$ by $[-100, 100, 10]$ **25.** c   **27.** d   **29.** no; yes   **31.** yes; yes   **33.** yes; yes   **35.** linear, line; linear, line; non-linear, parabola; linear, line; non-linear, V-shaped; non-linear, parabola; linear, line; non-linear, V-shaped; non-linear, cubic
37. $y = -2x + 10$ **39.** $y = (-7x - 4)/3$

41.

     43. d   **45.** C   **47.** A   **49.** D
     51. C   **53.** B   **55.** C

y-intercept: $(0, -5.6)$
x-intercepts: $(-2, 0)$ and $(1.4, 0)$

57.   **59.**   **61.**   **63.**

65.   **67.**   **69.**   **71.**

73.   **75.**   **77.**   **79.**

81.

$y = x^3$

83.

$y = -|x|$

85.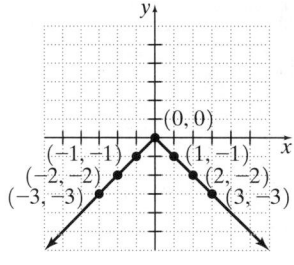

$y = -|x|$

87. a. parabola   **b.** line
89. a. line   **b.** parabola

91. a.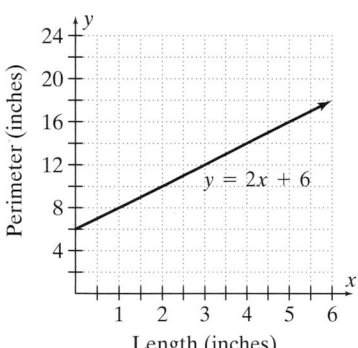

$y = 2x + 6$

b. 14 in.   **93.** -5 **95.** $-\dfrac{1}{10}$ **97.** b   **99.** c   **101. a.** 1991   **b.** 1997
c. answers may vary   **d.** $4.90   **103.**

$y = 3x + 5$

105.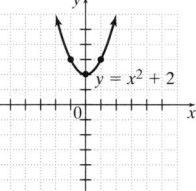

$y = x^2 + 2$

Exercise Set 2.2   **1.** domain: $\{-1, 0, -2, 5\}$; range: $\{7, 6, 2\}$; function   **3.** domain: $\{-2, 6, -7\}$; range: $\{4, -3, -8\}$; not a function

5. domain: $\{1\}$; range: $\{1, 2, 3, 4\}$; not a function   **7.** domain: $\left\{\dfrac{3}{2}, 0\right\}$; range: $\left\{\dfrac{1}{2}, -7, \dfrac{4}{5}\right\}$; not a function   **9.** domain: $\{-3, 0, 3\}$; range:

$\{-3, 0, 3\}$; function   **11.** domain: $\{-1, 1, 2, 3\}$; range: $\{2, 1\}$; function   **13.** domain: {Colorado, Alaska, Delaware, Illinois, Connecticut, Texas}; range: $\{6, 1, 20, 30\}$; function   **15.** domain: $\{32°, 104°, 212°, 50°\}$; range: $\{0°, 40°, 10°, 100°\}$; function   **17.** domain: $\{0\}$; range: $\{2, -1, 5, 100\}$; function   **19.** function   **21.** not a function   **23.** function   **25.** not a function   **27.** function   **29.** function
31. not a function   **33.** domain: $[0, \infty)$; range: $(-\infty, \infty)$; not a function   **35.** domain: $[-1, 1]$; range: $(-\infty, \infty)$; not a function   **37.** domain: $(-\infty, \infty)$; range: $(-\infty, -3] \cup [3, \infty)$; not a function   **39.** domain: $[2, 7]$; range $[1, 6]$; not a function   **41.** domain: $\{-2\}$; range: $(-\infty, \infty)$; not a function   **43.** domain: $(-\infty, \infty)$; range: $(-\infty, 3]$; function   **45.** domain: $(-\infty, \infty)$; range: $(-\infty, \infty)$ **47.** answers may vary
49. yes   **51.** no   **53.** yes   **55.** yes   **57.** yes   **59.** no   **61.** 15   **63.** 38   **65.** 7   **67.** 3   **69. a.** 0   **b.** 1   **c.** -1

71. a. 246   **b.** 6   **c.** $\dfrac{9}{2}$ **73. a.** -5 **b.** -5 **c.** -5 **75. a.** 5.1   **b.** 15.5   **c.** 9.533   **77. a.** 23   **b.** 23   **c.** 33   **79.** 7

81. $(1, -10)$ **83.** $(4, 56)$ **85.** $f(-1) = -2$ **87.** $g(2) = 0$ **89.** $-4, 0$ **91.** -4 **93.** infinite number   **95.** 55; 90; 125; 160; 195
97. a. 36   **b.** 64   **c.** 84   **d.** 96   **e.** 100   **f.** 0   **99. a.** $17.1 billion   **b.** $16.21 billion   **101.** $36.464 billion   **103.** $f(x) = x + 7$
105. 25π sq. cm   **107.** 2744 cu. in.   **109.** 166.38 cm   **111.** 163.2 mg   **113. a.** 95.99; per capita consumption of poultry was 95.99 lb in 2000.
b. 99.37 lb

115. $5, -5, 6$

117. $2, \dfrac{8}{7}, \dfrac{12}{7}$

119. $0, 0, -6$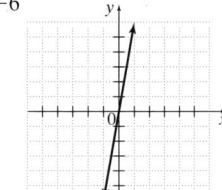

121. yes; 170 m

123. a. $-3s + 12$

b. $-3r + 12$

125. a. 132   **b.** $a^2 - 12$

127. answers may vary

Exercise Set 2.3   **1.**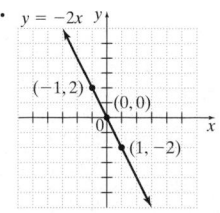

$y = -2x$ $(-1, 2)$ $(0, 0)$ $(1, -2)$

3.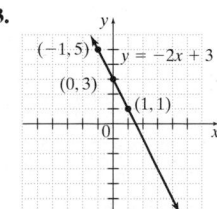

$(-1, 5)$ $y = -2x + 3$ $(0, 3)$ $(1, 1)$

5.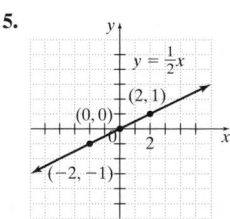

$y = \dfrac{1}{2}x$ $(2, 1)$ $(0, 0)$ $(-2, -1)$

7.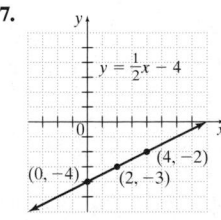

$y = \dfrac{1}{2}x - 4$ $(4, -2)$ $(0, -4)$ $(2, -3)$

9. C **11.** D **13.**   **15.**   **17.**   **19.**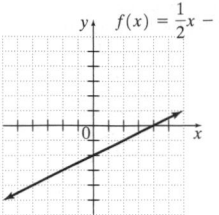

21. answers may vary **23.**   **25.**   **27.**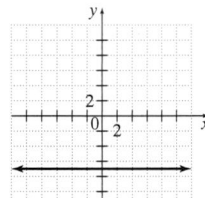

29. C **31.** A **33.** The vertical line $x = 0$ has y-intercepts.

35.   **37.**   **39.**   **41.**   **43.**

45.   **47.**   **49.**   **51.**   **53.**

55.   **57.**   **59.** 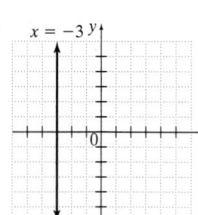

61. b, d **63.** 677.25; 706.78; 807.19; 860.34

65. $\dfrac{3}{2}$ **67.** 6 **69.** $-\dfrac{6}{5}$

71. a. $(0, 500)$; if no tables are produced, 500 chairs can be produced **b.** $(750, 0)$; if no chairs are produced, 750 tables can be produced **c.** 466 chairs

73. a. \$64 **b.** 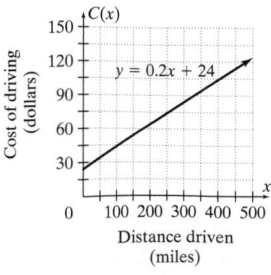 **c.** The line moves upward from left to right. **75. a.** \$1921.88 **b.** 2012 **c.** answers may vary

77. a. a line parallel to $-y = -4x$ but with x-intercept $(0, 2)$ **b.** a line parallel to $y = -4x$ but with x-intercept $(0, -5)$ **79.** b **81.** a

Mental Math **1.** upward **3.** horizontally

Exercise Set 2.4 **1.** $\frac{9}{5}$ **3.** $-\frac{5}{6}$ **5.** $\frac{1}{3}$ **7.** 0   **9.** undefined   **11.** -1 **13.** $-\frac{2}{3}$ **15.** 1.5   **17.** 0.2   **19.** l_2 **21.** l_2 **23.** l_2

25. $m = 5, b = -2$ **27.** $m = -2, b = 7$ **29.** $m = \frac{2}{3}, b = -\frac{10}{3}$ **31.** $m = \frac{1}{2}, b = 0$ **33.** A   **35.** B   **37.** undefined   **39.** 0

41. undefined   **43.** answers may vary   **45.** $m = -1, b = 5$ **47.** $m = \frac{6}{5}, b = 6$ **49.** $m = 3, b = 9$ **51.** $m = 0, b = 4$ **53.** $m = 7, b = 0$

55. $m = 0, b = -6$ **57.** slope is undefined, no y-intercept   **59.** neither   **61.** parallel   **63.** perpendicular   **65.** answers may vary   **67.** $\frac{3}{2}$

69. $-\frac{1}{2}$ **71.** $\frac{2}{3}$ **73.** approximately -0.12 **75. a.** \$46,221.60   **b.** $m = 1545.4$; The annual income increases \$1545.40 every year.

c. $b = 33,858.4$; At year $x = 0$, or 1997, the annual average income was \$33,858.40.   **77. a.** $m = 24.5, b = 5.9$ **b.** The number of internet access points increases by 24.5 thousand for every 1 year.   **c.** There were 5.9 thousand internet access points in 2002.   **79. a.** The yearly cost of tuition increases \$174.40 every 1 year.   **b.** The yearly cost of tuition in 1990 was \$2074.38.   **81.** $-\frac{7}{2}$ **83.** $\frac{2}{7}$ **85.** $\frac{5}{2}$ **87.** $-\frac{2}{5}$ **89.** $\frac{2}{11}$

91. $\frac{3}{11}$ **93.** $\frac{4}{11}$ **95.** $y = -3x - 30$ **97.** $y = -8x - 23$ **99. a.** $(6, 20)$ **b.** $(10, 13)$ **c.** $-\frac{7}{4}$ or -1.75 yd per sec   **d.** $\frac{3}{2}$ or 1.5 yd per sec

101.   **103. a.**   **b.** 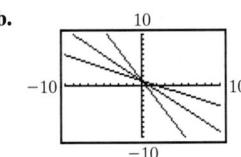   **c.** true

Mental Math **1.** $m = -4, b = 12$ **3.** $m = 5, b = 0$ **5.** $m = \frac{1}{2}, b = 6$ **7.** parallel   **9.** neither

Exercise Set 2.5 **1.** $y = -x + 1$ **3.** $y = 2x + \frac{3}{4}$ **5.** $y = \frac{2}{7}x$

7.   **9.**   **11.**   **13.** $y = 3x - 1$ **15.** $y = -2x - 1$ **17.** $y = \frac{1}{2}x + 5$

19. $y = -\frac{9}{10}x - \frac{27}{10}$ **21.** $2x + y = 3$

23. $2x - 3y = -7$ **25.** $f(x) = 3x - 6$

27. $f(x) = -2x + 1$ **29.** $f(x) = -\frac{1}{2}x - 5$

31. $f(x) = \frac{1}{3}x - 7$ **33.** answers may vary   **35.** -2 **37.** 2

39. -2 **41.** $y = -4$ **43.** $x = 4$ **45.** $y = 5$ **47.** $f(x) = 4x - 4$ **49.** $f(x) = -3x + 1$ **51.** $f(x) = -\frac{3}{2}x - 6$ **53.** $2x - y = -7$

55. $f(x) = -x + 7$ **57.** $x + 2y = 22$ **59.** $2x + 7y = -42$ **61.** $4x + 3y = -20$ **63.** $x = -2$ **65.** $x + 2y = 2$ **67.** $y = 12$

69. $8x - y = 47$ **71.** $x = 5$ **73.** $f(x) = -\frac{3}{8}x - \frac{29}{4}$ **75. a.** $P(x) = 12,000x + 18,000$ **b.** \$102,000   **c.** end of the ninth yr

77. a. $y = -1000x + 13,000$ **b.** 9500 Fun Noodles   **79. a.** $4834x + 133,300$ **b.** \$176,806   **c.** every year, the median price of a home increases by \$4834.   **81. a.** $y = 29.5x + 757$ **b.** 875 thousand people

83.   **85.** 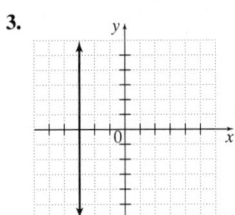   **87.** 14   **89.** $\frac{7}{2}$ **91.** $-\frac{1}{4}$

93. $-4x + y = 4$ **95.** $2x + y = -23$

97. $3x - 2y = -13$ **99.** true

Integrated Review **1.**   **2.**   **3.**   **4.**

5. 0 **6.** $-\dfrac{3}{5}$ **7.** $m = 3; (0, -5)$ **8.** $m = \dfrac{5}{2}; \left(0, -\dfrac{7}{2}\right)$ **9.** parallel **10.** perpendicular **11.** $y = -x + 7$ **12.** $x = -2$

13. $y = 0$ **14.** $y = -\dfrac{3}{8}x - \dfrac{29}{4}$ **15.** $y = -5x - 6$ **16.** $y = -4x + \dfrac{1}{3}$ **17.** $y = \dfrac{1}{2}x - 1$ **18.** $y = 3x - \dfrac{3}{2}$ **19.** $y = 3x - 2$

20. $y = -\dfrac{5}{4}x + 4$ **21.** $y = \dfrac{1}{4}x - \dfrac{7}{2}$ **22.** $y = -\dfrac{5}{2}x - \dfrac{5}{2}$ **23.** $x = -1$ **24.** $y = 3$

Exercise Set 2.6 **1.** 38.022 billion **3.** 1995 **5. a.** $y = 11{,}575.003x + 58{,}440.426$ **b.** 637,191 thousand or 637,191,000
c. 11,575 thousand or 11,575,000 visits per year **7. a.** $y = -0.411x + 36.836$ **b.** 15.9% **c.** 0.411% per year **9. a.** $y = 3.82x + 6.6$
b. 117 thousand or 117,000 **c.** 4 thousand or 4000 **11.** $y = 114.453x - 81{,}378.675$ **13.** \$170,418 **15.** $y = 0.08x + 700$
17. a. $y - 0.475x - 0.027$ **b.** \$4.723 billion **c.** \$0.475 or \$475,000,000 per year **19. a.** $y = 0.262x + 48.740$ **b.** 0.262 years annually
c. 77.6 years **21. a.** $y = 8.209x + 41.170$ **b.** 205.36 million or 205,360,000 members **c.** 2015 **d.** The number of PPO's is increasing at
the rate of 8.209 million per year. **23. a.** $y = 3.933x - 11.025$ **b.** \$158.09 **c.** \$3.93 per year
25. a.

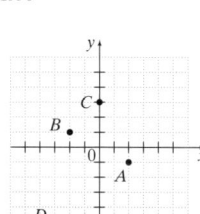

Academy Award
Commercials
Millions of dollars
Years since 1990

b. $y = 0.097x + 0.163$ **27.** 8 **29.** $\dfrac{1}{10}$ **31.** dining in: $y = 0.0582x + 65.790$; take out:
$y = 1.715x + 37.398$; in the year 1997 **33.** answers may vary

Chapter 2 Review **1.**

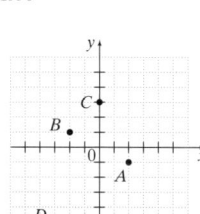

3. no, yes **5.** yes, yes **7.** linear

9. linear

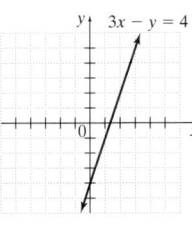

$3x - y = 4$

11. nonlinear

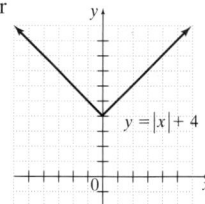

$y = |x| + 4$

13. linear

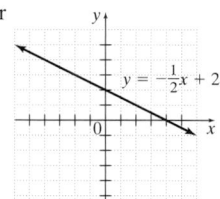

$y = -\dfrac{1}{2}x + 2$

15. D **17.** C

19. domain: $\left\{-\dfrac{1}{2}, 6, 0, 25\right\}$; range: $\left\{\dfrac{3}{4} \text{ or } 0.75, -12, 25\right\}$; function

21. domain: $\{2, 4, 6, 8\}$; range: $\{2, 4, 5, 6\}$; not a function
23. domain: $(-\infty, \infty)$; range: $(-\infty, -1] \cup [1, \infty)$; not a function
25. domain: $(-\infty, \infty)$; range: $\{4\}$; function **27.** -3 **29.** 18
31. -3 **33.** 381 lb **35.** 0 **37.** $-2, 4$

39.

$y = x$

41.

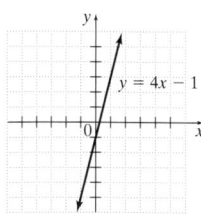

$y = 4x - 1$

43. A **45.** D **47.**

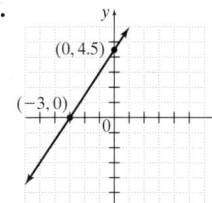

$(0, 4.5)$
$(-3, 0)$

49.

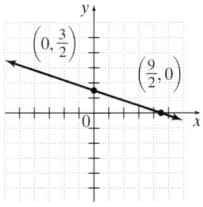

$\left(0, \dfrac{3}{2}\right)$
$\left(\dfrac{9}{2}, 0\right)$

51.

$x = -2$

53.

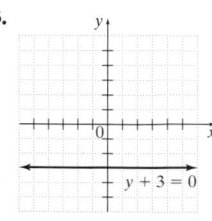

$y + 3 = 0$

55. -3 **57.** $\dfrac{5}{2}$ **59.** $m = \dfrac{2}{5}, b = -\dfrac{4}{3}$ **61.** 0 **63.** l_2 **65.** l_2

67. a. $m = 0.3$; The cost increases by \$0.30 for each additional mile driven.
    b. $b = 42$; The cost for 0 miles driven is \$42.
69. parallel

71.

73.

75. $x = -2$ **77.** $y = 5$ **79.** $2x - y = 12$ **81.** $11x + y = -52$ **83.** $y = -5$

85. $f(x) = -x - 2$ **87.** $f(x) = -\dfrac{3}{2}x - 8$ **89.** $f(x) = -\dfrac{3}{2}x - 1$

91. a. $y = \dfrac{17}{22}x + 43$ **b.** 52 million

Chapter 2 Test   **1.** A: quadrant IV   **2.** $(-6, -3)$ **3.**   **4.**
B: no quadrant
C: quadrant II

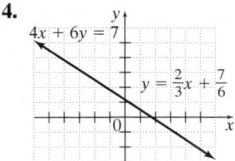

5.   **6.**   **7.** $-\dfrac{3}{2}$ **8.** $m = -\dfrac{1}{4}, b = \dfrac{2}{3}$ **9.** $y = -8$ **10.** $x = -4$ **11.** $y = -2$

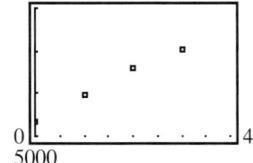

12. $3x + y = 11$ **13.** $5x - y - 2$ **14.** $f(x) = -\dfrac{1}{2}x$ **15.** $f(x) = -\dfrac{1}{3}x + \dfrac{5}{3}$

16. $f(x) = -\dfrac{1}{2}x - \dfrac{1}{2}$ **17.** neither   **18.** B   **19.** A   **20.** D   **21.** C

22. domain: $(-\infty, \infty)$; range: $\{5\}$; function   **23.** domain: $\{-2\}$; range: $(-\infty, \infty)$; not a function   **24.** domain: $(-\infty, \infty)$; range: $[0, \infty)$; function   **25.** domain: $(-\infty, \infty)$; range: $(-\infty, \infty)$; function   **26. a.** $22,892   **b.** $28,016   **c.** 2008

d. The average yearly earnings for high school graduates increases $732 per year.   **e.** The average yearly earnings for a high school graduate in 1996 was $21,428.   **27. a.** 20,000   **b.** $y = 285.18x + 6899.3$; 20,000   **c.** 18,306.5 thousand

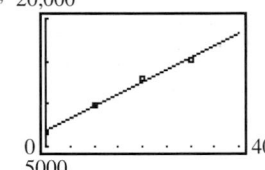

Chapter 2 Cumulative Review   **1.** 41; Sec. 1.2, Ex. 2   **2. a.** -7 **b.** $\dfrac{5}{6}$ **c.** -13; Sec. 1.3   **3. a.** true   **b.** false   **c.** false

d. false; Sec. 1.2, Ex. 5   **4. a.** 7   **b.** 0   **c.** $-\dfrac{1}{4}$; Sec. 1.2   **5. a.** -6 **b.** -7 **c.** -16 **d.** 20.5   **e.** $\dfrac{1}{6}$ **f.** 0.94

g. -3; Sec. 1.3, Ex. 2   **6. a.** 7   **b.** 0   **c.** -10; Sec. 1.3   **7. a.** 9   **b.** $\dfrac{1}{16}$ **c.** -25 **d.** 25   **e.** -125 **f.** -125; Sec. 1.3, Ex. 6

8. a. distributive property   **b.** commutative property for addition; Sec. 1.4   **9. a.** $>$ **b.** $=$ **c.** $<$ **d.** $<$; Sec. 1.4, Ex. 2

10. a. 98   **b.** 98; Sec. 1.3   **11. a.** $\dfrac{1}{11}$ **b.** $-\dfrac{1}{9}$ **c.** $\dfrac{4}{7}$; Sec. 1.4, Ex. 5   **12.** 22; Sec. 1.3   **13.** 0.4; Sec. 1.5, Ex. 2   **14.** -17; Sec. 1.5

15. $\{\ \ \}$ or $\varnothing$; Sec. 1.5, Ex. 8   **16.** all real numbers; Sec. 1.5   **17.** 4; Sec. 1.6, Ex. 4   **18.** 4; Sec. 1.6   **19.** 86, 88 and 90; Sec. 1.6, Ex. 7

20. 69, 71, 73; Sec. 1.6   **21.** $\dfrac{V}{lw} = h$; Sec. 1.8, Ex. 1   **22.** $y = \dfrac{-7x + 21}{3}$ or $y = -\dfrac{7}{3}x + 7$; Sec. 1.8   **23.** median: 83.5; mode: 89 (Sec. 1.9, Ex. 2)

24. median: 102; mode: 118   **25. a.** QIV   **b.** no quadrant   **c.** QII   **d.** no quadrant   **e.** QIII   **f.** QI (Sec. 2.1, Ex. 1)   **26. a.** y-axis

b. QIII   **c.** x-axis   **d.** QIV   **27.** solutions: $(2, -6), (0, -12)$; not a solution $(1, 9)$; Sec. 3.1, Ex. 2   **28.** $m = -\dfrac{7}{2}$, y-intercept $(0, 5)$; Sec. 3.1

29. yes; Sec. 2.2, Ex. 3   **30.** no; Sec. 2.2   **31. a.** $\left(0, \dfrac{3}{7}\right)$ **b.** $(0, -3.2)$; Sec. 2.3, Ex. 3   **32.** $m = 3$; Sec. 2.4   **33.** $y = \dfrac{1}{4}x - 3$; Sec. 2.5 Ex. 1

34. $x = -2$; Sec. 2.3   **35.** $y = -\dfrac{3}{4}$; Sec. 2.3   **36.** $x + y - 3$; Sec. 2.5

CHAPTER 3 EQUATIONS AND INEQUALITIES

Mental Math   **1.** 6   **3.** 5   **5.** 0

Exercise Set 3.1   **1.** 2   **3.** -1 **5.** -2 **7.** $3x - 13 = 2(x - 5) + 3$; 6   **9.** $-(2x + 3) + 2 = 5x - 1 - 7x$; all real numbers   **11.** $\varnothing$
13. all real numbers   **15.** $\varnothing$ **17.** all real numbers   **19.** 25.8   **21.** 3.5   **23.** -3.28 **25.** $\varnothing$ **27.** 4   **29.** 8   **31.** all real numbers
33. -0.32 **35.** 0.6   **37.** 17   **39.** 0.8   **41. a.** second consultant   **b.** first consultant   **c.** 6-hr job   **43. a.** first agency   **b.** second
agency   **c.** 60 mi   **45.** $\{-3, -2, -1\}$ **47.** $\{-3, -2, -1, 0, 1\}$ **49.** answers may vary   **51.** $(12, 22)$ **53.** greater than   **55.** 1.43
57. 1.51   **59.** 0.91   **61.** 22.87   **63.** The equation has no solution.

Mental Math   **1.** $\{x | x < 6\}$ **3.** $\{x | x \geq 10\}$ **5.** $\{x | x > 4\}$ **7.** $\{x | x \leq 2\}$

Exercise Set 3.2   **1.** $\xrightarrow{\hspace{1cm})\hspace{1cm}}$; $(-\infty, -3)$ **3.** $\xleftarrow{\hspace{1cm}[\hspace{1cm}}$; $[0.3, \infty)$ **5.** $\xleftarrow{\hspace{1cm}(\hspace{1cm}}$; $\left(\dfrac{5}{9}, \infty\right)$

7. $\xleftarrow{(\hspace{0.5cm})}$; $(-2, 5)$ **9.** $\xleftarrow{(\hspace{0.5cm})}$; $(-1, 5)$ **11.** answers may vary   **13.** d   **15.** b   **17.** $(-\infty, 4)$ **19.** $[-3, \infty)$

21. ∅ **23.** ;$(-\infty, 1)$ **25.** ;$(-\infty, 2]$ **27.** ;$\left[\frac{8}{3}, \infty\right)$

29. ;$(-\infty, -4.7)$ **31.** ;$(-\infty, -3]$ **33.** ;$(4, \infty)$ **35.** $(-\infty, -1]$ **37.** $(-\infty, 11]$

39. $(-13, \infty)$ **41.** $(-\infty, 7]$ **43.** $(-\infty, \infty)$ **45.** ∅ **47** $(0, \infty)$ **49.** $\left[-\frac{3}{5}, \infty\right]$ **51.** $[-9.6, \infty)$

53. answers may vary **55.** $[0, \infty)$ **57.** $(-\infty, -5]$ **59.** $\left(-\infty, \frac{1}{4}\right)$ **61.** $(-\infty, -1]$ **63.** $\left[-\frac{79}{3}, \infty\right)$ **65.** $(-\infty, -15)$ **67.** $[3, \infty)$

69. $\left[-\frac{37}{3}, \infty\right)$ **71.** $(-\infty, 5)$ **73.** 30 **75.** 1040 lb **77.** 16 oz **79.** more than 200 calls **81.** $F \geq 932°$ **83. a.** the end of 2004
b. answers may vary **85.** decreasing **87.** 6.57 gal **89.** 2004 **91.** answers may vary **93.** 2, 3, 4 **95.** 2, 3, 4…

97. ;$[0, 5]$ **99.** ;$\left(-\frac{1}{2}, \frac{3}{2}\right)$ **101.** $(-\infty, \infty)$ **103.** ∅ **105. a.** $y = 0.098x + 1.918$

b. 0.098% per year **c.** 2021 and later

Integrated Review **1.** -5 **2.** $(-5, \infty)$ **3.** $\left[\frac{8}{3}, \infty\right)$ **4.** $[-1, \infty)$ **5.** 0 **6.** $\left(-\frac{1}{10}, \infty\right)$ **7.** $\left(-\infty, -\frac{1}{6}\right]$ **8.** 0 **9.** ∅

10. $\left[-\frac{3}{5}, \infty\right)$ **11.** 4.2 **12.** 6 **13.** -8 **14.** $(-\infty, -16)$ **15.** $\frac{20}{11}$ **16.** 1 **17.** $(38, \infty)$ **18.** $-5, 5$ **19.** $\frac{3}{5}$ **20.** $(-\infty, \infty)$

21. 29 **22.** all real numbers **23.** $(-\infty, 5)$ **24.** $\frac{9}{13}$ **25.** $(23, \infty)$ **26.** $(-\infty, 6]$ **27.** $\left(-\infty, \frac{3}{5}\right]$ **28.** $\left(-\infty, -\frac{19}{32}\right)$

Exercise Set 3.3 **1.** $\{2, 3, 4, 5, 6, 7\}$ **3.** $\{4, 6\}$ **5.** $\{\ldots, -2, -1, 0, 1, \ldots\}$ **7.** $\{5, 7\}$ **9.** $\{x \mid x \text{ is an odd integer or } x = 2 \text{ or } x = 4\}$

11. $\{2, 4\}$ **13.** ;$(-2, 5)$ **15.** ;$[6, \infty)$ **17.** ;$(-\infty, -3]$ **19.** ;$(11, 17)$

21. ;$[1, 4]$ **23.** ;$\left[-3, \frac{3}{2}\right]$ **25.** ;$[-21, -9]$ **27.** ;$(-\infty, -1) \cup (0, \infty)$

29. ;$[2, \infty)$ **31.** ;$(-\infty, \infty)$ **33.** answers may vary **35.** ;$(-1, 2)$

37. ;$(-\infty, \infty)$ **39.** ;$[-1, \infty)$ **41.** ;$[-5, \infty)$ **43.** ;$\left[\frac{3}{2}, 6\right]$

45. ;$\left(\frac{5}{4}, \frac{11}{4}\right)$ **47.** ;∅ **49.** ;$(-7, \infty)$ **51.** ;$\left(-5, \frac{5}{2}\right)$

53. ;$\left(0, \frac{14}{3}\right]$ **55.** ;$(-\infty, -3]$ **57.** ;$(-\infty, 1] \cup \left(\frac{29}{7}, \infty\right)$ **59.** ;∅

61. ;$\left[-\frac{1}{2}, \frac{3}{2}\right)$ **63.** ;$\left(-\frac{4}{3}, \frac{7}{3}\right)$ **65.** ;$(6, 12)$ **67. a.** $(-5, 2.5)$ **b.** $(-\infty, -5) \cup (2.5, \infty)$
69. a. $[2, 9]$ **b.** $(-\infty, 2] \cup [9, \infty)$ **71.** -12 **73.** -4 **75.** $-7, 7$ **77.** 0

79. $-20.2° \leq F \leq 95°$ **81.** $67 \leq \text{final score} \leq 94$ **83.** 1994–1995; 1998–1999 **85.** ;$(6, \infty)$

87. ;$[3, 7]$ **89.** ;$(-\infty, -1)$

Mental Math **1.** 7 **3.** -5 **5.** -6 **7.** 12

Exercise Set 3.4 **1.** $7, -7$ **3.** $4.2, -4.2$ **5.** $7, -2$ **7.** $8, 4$ **9.** $5, -5$ **11.** $3, -3$ **13.** 0 **15.** ∅ **17.** $\frac{1}{5}$ **19.** $|x| = 5$

21. $9, -\frac{1}{2}$ **23.** $-\frac{5}{2}$ **25.** answers may vary **27.** $\{-1, 4\}$ **29.** $\{2.5\}$ **31.** $4, -4$ **33.** ∅ **35.** $0, \frac{14}{3}$ **37.** $2, -2$ **39.** $7, -1$

41. ∅ **43.** ∅ **45.** $-\frac{1}{8}$ **47.** $\frac{1}{2}, -\frac{5}{6}$ **49.** $2, -\frac{12}{5}$ **51.** $3, -2$ **53.** $-8, \frac{2}{3}$ **55.** ∅ **57.** 4 **59.** $13, -8$ **61.** $3, -3$

63. $8, -7$ **65.** $2, 3$ **67.** $2, -\frac{10}{3}$ **69.** $\frac{3}{2}$ **71.** ∅ **73.** answers may vary **75.** 33% **77.** 38.4 lb **79.** answers may vary

81. no solution **83.** $|x| = 2$ **85.** $|2x - 1| = 4$ **87. a.** $c = 0$ **b.** c is a negative number **c.** c is a positive number **89.** $0.09, 0.59$
91. $-4, -1.37$

Mental Math   **1.** D   **3.** C   **5.** A

Exercise Set 3.5   **1.** ⟵[▬▬]⟶; $[-4, 4]$ **3.** ⟵(▬▬)⟶; $(1, 5)$ **5.** ⟵(▬▬)⟶; $(-5, -1)$ **7.** ⟵[▬▬]⟶; $[-10, 3]$

9. ⟵[▬▬]⟶; $[-5, 5]$ **11.** ⟵⟶; $\varnothing$ **13.** ⟵[▬▬]⟶; $[0, 12]$ **15.** ⟵▬)(▬⟶; $(-\infty, -3) \cup (3, \infty)$

17. ⟵▬] [▬⟶; $(-\infty, -24] \cup [4, \infty)$ **19.** ⟵▬) (▬⟶; $(-\infty, -4) \cup (4, \infty)$ **21.** ⟵▬▬⟶; $(-\infty, \infty)$

23. ⟵▬) (▬⟶; $\left(-\infty, \dfrac{2}{3}\right) \cup (2, \infty)$ **25.** ⟵●⟶; $\{0\}$ **27.** ⟵▬✕▬⟶; $\left(-\infty, -\dfrac{3}{8}\right) \cup \left(-\dfrac{3}{8}, \infty\right)$

29. ⟵[▬▬]⟶; $[-2, 2]$ **31.** ⟵▬)(▬⟶; $(-\infty, -1) \cup (1, \infty)$ **33.** ⟵(▬▬)⟶; $(-5, 11)$

35. ⟵▬)(▬⟶; $(-\infty, 4) \cup (6, \infty)$ **37.** ⟵⟶; $\varnothing$ **39.** ⟵▬▬⟶; $(-\infty, \infty)$ **41.** ⟵[▬▬]⟶; $[-2, 9]$

43. ⟵▬] [▬⟶; $(-\infty, -11] \cup [1, \infty)$ **45.** ⟵▬✕▬⟶; $(-\infty, 0) \cup (0, \infty)$ **47.** ⟵▬▬⟶; $(-\infty, \infty)$

49. ⟵[▬▬]⟶; $\left[-\dfrac{1}{2}, 1\right]$ **51.** ⟵▬)(▬⟶; $(-\infty, -3) \cup (0, \infty)$ **53.** ⟵⟶; $\varnothing$ **55.** ⟵▬▬⟶; $(-\infty, \infty)$

57. ⟵(▬▬)⟶; $\left(-\dfrac{2}{3}, 0\right)$ **59.** ⟵▬)(▬⟶; $(-\infty, -12) \cup (0, \infty)$ **61.** ⟵[▬▬]⟶; $[-1, 8]$ **63.** ⟵[▬▬]⟶; $\left[-\dfrac{23}{8}, \dfrac{17}{8}\right]$

65. $(-2, 5)$ **67.** $5, -2$ **69.** $(-\infty, -7] \cup [17, \infty)$ **71.** $-\dfrac{9}{4}$ **73.** $(-2, 1)$ **75.** $2, \dfrac{4}{3}$ **77.** $\varnothing$ **79.** $\dfrac{19}{2}, -\dfrac{17}{2}$

81. $\left(-\infty, -\dfrac{25}{3}\right) \cup \left(\dfrac{35}{3}, \infty\right)$ **83. a.** $\{-5, 11\}$ **b.** $(-5, 11)$ **c.** $(-\infty, -5] \cup [11, \infty)$ **85. a.** $\{-8, 4\}$ **b.** $[-8, 4]$ **c.** $(-\infty, -8] \cup (4, \infty)$

87. $\dfrac{1}{6}$ **89.** 0   **91.** $\dfrac{1}{3}$ **93.** -1.5 **95.** 0   **97.** $|x| < 7$ **99.** $|x| \le 5$ **101.** answers may vary   **103.** $3.45 < x < 3.55$

Exercise Set 3.6   **1.**   **3.**   **5.**   **7.**

     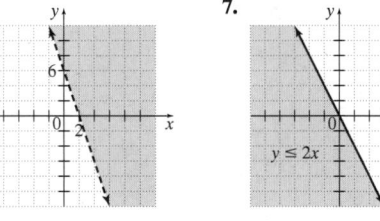

9.   **11.**   **13.** answers may vary   **15.**   **17.**

        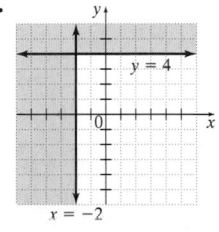

19.   **21.**   **23.**   **25.**   **27.**

           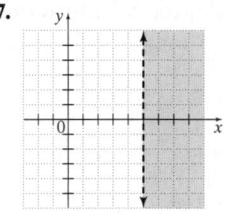

29.   **31.**   **33.**   **35.**   **37.**

39. **41.** **43.** **45.**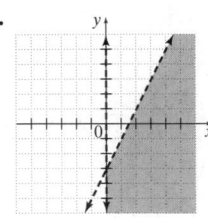

47. D **49.** A

51. $x \geq 2$ **53.** $y \leq -3$

55. $y > 4$ **57.** $x < 1$

59. 8 **61.** -25

63. 16 **65.** $\dfrac{27}{125}$

67. domain: $[1, 5]$; range: $[1, 3]$; no **69.** $x \leq 20$ and $y \geq 10$.

71.

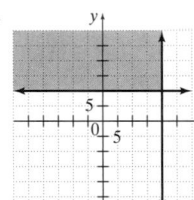

Chapter 3 Review **1.** 12 **3.** $\emptyset$ **5.** $\dfrac{17}{5}$ **7.** -1.51 **9.** $(3, \infty)$ **11.** $(-4, \infty)$ **13.** $(-\infty, 7]$ **15.** $\left(\dfrac{1}{2}, \infty\right)$ **17.** $[-19, \infty)$

19. more than 30 pounds per week **21.** at least 9.6 **23.** $\left[2, \dfrac{5}{2}\right]$ **25.** $\left(\dfrac{1}{8}, 2\right)$ **27.** $\left(\dfrac{7}{8}, \dfrac{27}{20}\right]$ **29.** $(-5, 2]$ **31.** $\left(\dfrac{11}{3}, \infty\right)$ **33.** $16, -2$

35. $0, -9$ **37.** $2, -\dfrac{2}{3}$ **39.** $\emptyset$ **41.** $3, -3$ **43.** $5, -\dfrac{1}{3}$ **45.** $7, -\dfrac{8}{5}$ **47.** $\left(\dfrac{8}{5}, 2\right)$ **49.** $(-\infty, -3) \cup (3, \infty)$ **51.** $\emptyset$ **53.** $\left(-\infty, \dfrac{22}{15}\right] \cup \left[\dfrac{6}{5}, \infty\right)$

55. $(-\infty, -27) \cup (-9, \infty)$ **57.** **59.** **61.** **63.**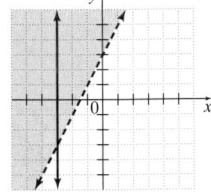

Chapter 3 Test **1.** $-\dfrac{29}{17}$ **2.** 0.15 **3.** $1, \dfrac{2}{3}$ **4.** $\emptyset$ **5.** $(5, \infty)$ **6.** $(-\infty, 2]$ **7.** $\left(\dfrac{3}{2}, 5\right]$ **8.** $(-\infty, -2) \cup \left(\dfrac{4}{3}, \infty\right)$ **9.** $(-\infty, -5]$

10. $(-\infty, -2]$ **11.** $[-3, -1]$ **12.** $(-\infty, \infty)$ **13.** $(3, 7)$ **14.** $3, -\dfrac{1}{5}$ **15.** $(-\infty, \infty)$ **16.** $\dfrac{3}{2}$

17. **18.** **19.** 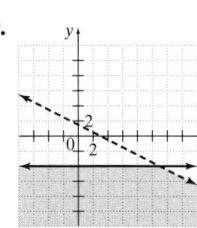 **20.** $(-\infty, -3)$ **21.** $(-3, \infty)$

Chapter 3 Cumulative Review **1. a.** $\{2, 3, 4, 5\}$ **b.** $\{101, 102, 103, \dots\}$; Sec. 1.2, Ex. 3 **2. a.** $\{-2, -1, 0, 1, 2, 3, 4\}$ **b.** $\{4\}$; Sec. 1.2

3. a. 3 **b.** 5 **c.** -2 **d.** -8 **e.** 0; Sec. 1.2, Ex. 6 **4. a.** $-\dfrac{2}{3}$ **b.** 9 **c.** -1.5; Sec. 1.2 **5. a.** -14 **b.** -4 **c.** 5 **d.** -10.2

e. $\dfrac{1}{4}$ **f.** $-\dfrac{5}{21}$; Sec. 1.3, Ex. 1 **6. a.** 8 **b.** -7.2 **c.** $-\dfrac{3}{4}$; Sec. 1.3 **7. a.** 3 **b.** 5 **c.** $\dfrac{1}{2}$ Sec. 1.3, Ex. 7 **d.** -6 **e.** not a real number

8. a. 6 **b.** $\dfrac{3}{7}$ **c.** 0 **d.** 10; Sec. 1.3 **9. a.** -2 **b.** 9 **c.** -1 **d.** -12; Sec. 1.3, Ex. 11 **10. a.** 1 **b.** 2 **c.** 3; Sec. 1.3

11. a. $x + 5 = 20$ **b.** $2(3 + y) = 4$ **c.** $x - 8 = 2x$ **d.** $\dfrac{z}{9} = 3(z - 5)$; Sec. 1.4, Ex. 1 **12. a.** $>$ **b.** $=$ **c.** $>$; Sec. 1.4

13. $5 + 7x$; Sec. 1.4, Ex. 6 **14.** $5 \cdot (7x) = (5 \cdot 7)x = 35x$; Sec. 1.4 **15.** 2; Sec. 1.5, Ex. 1 **16.** -2; Sec. 1.5 **17.** all real numbers; Sec. 1.5, Ex. 9

18. -4.25; Sec. 1.5 **19. a.** $3x + 3$ **b.** $12x - 3$; Sec. 1.6, Ex. 1 **20. a.** $3x + 3$ **b.** $12x + 4$; Sec. 1.6 **21.** $23, 49$; Sec. 1.6, Ex. 3

22. $11, 35$; Sec. 1.6 **23.** $y = \dfrac{2x + 7}{3}$ or $y = \dfrac{2x}{3} + \dfrac{7}{3}$; Sec. 1.8, Ex. 2 **24.** $x = \dfrac{10 + 4y}{7}$; Sec. 1.8 **25.** $b = \dfrac{2A - Bh}{h}$; Sec. 1.8, Ex. 3

26. $l = \dfrac{P - 2w}{2}$; Sec. 1.8 **27. a.** $[2, \infty)$ **b.** $(-\infty, -1)$ **c.** $(0.5, 3]$; Sec. 3.2, Ex. 1 **28. a.** $(-\infty, -3]$ **b.** $[-2, 0.1)$; Sec. 3.2

29. $\left[\dfrac{5}{2}, \infty\right)$; Sec. 3.2, Ex. 5 **30.** $(3, \infty)$; Sec. 3.2 **31. a.** $(-\infty, \infty)$ **b.** $\emptyset$; Sec. 3.2, Ex. 7 **32.** $\emptyset$; Sec. 3.2 **33.** $\{4, 6\}$; Sec. 3.3, Ex. 1

34. $\{-2, -1, 0, 1, 2, 3, 4, 5\}$; Sec. 3.3 **35.** $(-\infty, 4)$; Sec. 3.3, Ex. 2 **36.** $(-\infty, 3)$; Sec. 3.3 **37.** $\{2, 3, 4, 5, 6, 8\}$; Sec. 3.3, Ex. 6 **38.** $\emptyset$; Sec. 3.3 **39.** $(-\infty, \infty)$; Sec. 3.3, Ex. 8 **40.** $(-1, 0)$; Sec. 3.3 **41.** $2, -2$; Sec. 3.4, Ex. 1 **42.** $-5, 5$; Sec. 3.4 **43.** $24, -20$; Sec. 3.4, Ex. 3 **44.** $24, -36$; Sec. 3.4 **45.** 4; Sec. 3.4, Ex. 9 **46.** 2; Sec. 3.4 **47.** $[-3, 3]$; Sec. 3.5, Ex. 1 **48.** $(-\infty, -1) \cup (1, \infty)$; Sec. 3.5 **49.** $(-\infty, \infty)$; Sec. 3.5, Ex. 6 **50.** $\emptyset$; Sec. 3.5

CHAPTER 4 SYSTEMS OF EQUATIONS

Mental Math 1. B **3.** A

Exercise Set 4.1 1. yes **3.** $(1.5, 1)$ **5.** $(-2, 0.75)$
7. $(2, -1)$

9. $(1, 2)$

11. $\emptyset$

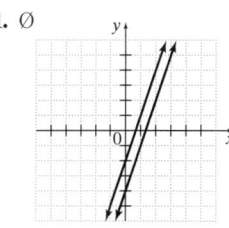

13. No; answers may vary **15.** $(2, 8)$
17. $(0, -9)$ **19.** $(1, -1)$ **21.** $(-5, 3)$
23. $\left(\dfrac{5}{2}, \dfrac{5}{4}\right)$ **25.** $(1, -2)$ **27.** $(9, 9)$
29. $(7, 2)$ **31.** $\emptyset$
33. $\{(x, y) | 3x + y = 1\}$

35. $\left(\dfrac{3}{2}, 1\right)$ **37.** $(2, -1)$ **39.** $(-5, 3)$ **41.** $\{(x, y) | 3x + 9y = 12\}$ **43.** $\emptyset$ **45.** $\left(\dfrac{1}{2}, \dfrac{1}{5}\right)$ **47.** $(8, 2)$ **49.** $\{(x, y) | x = 3y + 2\}$

51. $\left(-\dfrac{1}{4}, \dfrac{1}{2}\right)$ **53.** $(3, 2)$ **55.** $(7, -3)$ **57.** $\emptyset$ **59.** $(3, 4)$ **61.** $(-2, 1)$ **63.** $(1.2, -3.6)$ **65.** $(2.11, 0.17)$ **67.** $(0.57, -1.97)$

69. true **71.** false **73.** $6y - 4z = 25$ **75.** $x + 10y = 2$ **77.** 5000 DVDs; \$21 **79.** supply greater than demand **81.** $(1875; 4687.5)$

83. makes money **85.** for x-values greater than 1875 **87.** answers may vary; One possibility: $\begin{cases} -2x + y = 1 \\ x - 2y = -8 \end{cases}$ **89. a.** Consumption of red meat is decreasing while consumption of poultry is increasing. **b.** $(17, 108)$ **c.** In the year 2015, red meat and poultry consumption will each be about 108 pounds per person. **91.** $\left(\dfrac{1}{4}, 8\right)$ **93.** $\left(\dfrac{1}{3}, \dfrac{1}{2}\right)$ **95.** $\left(\dfrac{1}{4}, -\dfrac{1}{3}\right)$ **97.** $\emptyset$

Exercise Set 4.2 1. a, b, c, d **3.** Yes; answers may vary. **5.** $(-1, 5, 2)$ **7.** $(-2, 5, 1)$ **9.** $(-2, 3, -1)$
11. $\{(x, y, z) | x - 2y + z = -5\}$ **13.** $\emptyset$ **15.** $(0, 0, 0)$ **17.** $(-3, -35, -7)$ **19.** $(6, 22, -20)$ **21.** $\emptyset$ **23.** $(3, 2, 2)$
25. $\{(x, y, z) | x + 2y - 3z = 4\}$ **27.** $(-3, -4, -5)$ **29.** $\left(0, \dfrac{1}{2}, -4\right)$ **31.** $(12, 6, 4)$ **33.** 15 and 30 **35.** 5 **37.** $-\dfrac{5}{3}$
39. answers may vary **41.** answers may vary **43.** $(1, 1, -1)$ **45.** $(1, 1, 0, 2)$ **47.** $(1, -1, 2, 3)$ **49.** answers may vary

Exercise Set 4.3 1. 10 and 8 **3. a.** Enterprise class: 1101 ft; Nimitz class: 1092 ft **b.** 3.67 foot ball fields **5.** plane: 520 mph; wind: 40 mph **7.** 20 qt of 4%; 40 qt of 1% **9.** United Kingdom; 27,720 students; Spain: 12,292 students **11.** 9 large frames; 13 small frames
13. -10 and -8 **15.** 2005 **17.** tablets: \$0.80; pens: \$0.20 **19.** speed of plane: 630 mph; speed of wind: 90 mph **21. a.** answers may vary but notice the slope of each function **b.** 2006 **23.** 28 cm; 28 cm; 37 cm **25.** 600 mi **27.** $x = 75; y = 105$ **29.** 625 units
31. 3000 units **33.** 1280 units **35. a.** $R(x) = 450x$ **b.** $C(x) = 200x + 6000$ **c.** 24 desks **37.** 2 units of Mix A; 3 units of Mix B; 1 unit of Mix C **39.** 5 in.; 7 in.; 10 in. **41.** 18, 13, and 9 **43.** 151 free throws; 215 two-point field goals; 39 three-point field goals
45. $x = 60; y = 55; z = 65$ **47.** $5x + 5z = 10$ **49.** $-5y + 2z = 2$ **51.** 1980: 300,000; 2001: 1,400,000 **53.** $a = 3, b = 4, c = -1$
55. $a = 15\dfrac{5}{6}, b = -10\dfrac{5}{6}, c = 1065$; 2250 students in 2009

Integrated Review 1. C **2.** D **3.** A **4.** B **5.** $(1, 3)$ **6.** $\left(\dfrac{4}{3}, \dfrac{16}{3}\right)$ **7.** $(2, -1)$ **8.** $(5, 2)$ **9.** $\left(\dfrac{3}{2}, 1\right)$ **10.** $\left(-2, \dfrac{3}{4}\right)$

11. $\emptyset$ **12.** $\{(x, y) | 2x - 5y = 3\}$ **13.** $(-1, 3, 2)$ **14.** $(1, -3, 0)$ **15.** $\emptyset$ **16.** $\{(x, y, z) | x - y + 3z = 2\}$ **17.** $\left(2, 5, \dfrac{1}{2}\right)$
18. $\left(1, 1, \dfrac{1}{3}\right)$ **19.** 19 and 27 **20.** $70°; 70°; 100°; 120°$

Exercise Set 4.4 1. $(2, -1)$ **3.** $(-4, 2)$ **5.** $\emptyset$ **7.** $\{(x, y) | 3x - 3y = 9\}$ **9.** $(-2, 5, -2)$ **11.** $(1, -2, 3)$ **13.** $(4, -3)$
15. $(2, 1, -1)$ **17.** $(9, 9)$ **19.** $\emptyset$ **21.** $\emptyset$ **23.** $(1, -4, 3)$ **25.** function **27.** not a function **29.** -13 **31.** -36 **33.** 0
35. a. end of 1984 **b.** black-and-white sets; microwave ovens; The percent of households owning black-and-white television sets is decreasing and the percent of households owning microwave overs is increasing; answers may vary **c.** in 2002 **d.** no; answers may vary **37.** answers may vary

Exercise Set 4.5 **1.** 26 **3.** −19 **5.** 0 **7.** (1, 2) **9.** $\{(x, y)|3x + y = 1\}$ **11.** (9, 9) **13.** 8 **15.** 0 **17.** 54 **19.** (−2, 0, 5) **21.** (6, −2, 4) **23.** 16 **25.** 15 **27.** $\dfrac{13}{6}$ **29.** 0 **31.** 56 **33.** (−3, −2) **35.** ∅ **37.** (−2, 3, −1) **39.** (3, 4) **41.** (−2, 1) **43.** $\{(x, y, z)|x − 2y + z = −3\}$ **45.** (0, 2, −1) **47.** $6x − 18$ **49.** $9x − 15$

51. **53.** **55.** 5 **57.** 0 **59.** **61.** −125 **63.** 24

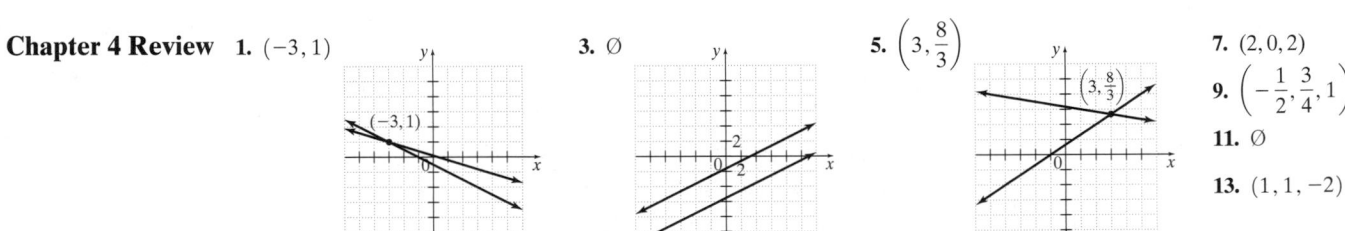

$$\begin{array}{cccc} + & - & + & - \\ - & + & - & + \\ + & - & + & - \\ - & + & - & + \end{array}$$

Chapter 4 Review **1.** (−3, 1) **3.** ∅ **5.** $\left(3, \dfrac{8}{3}\right)$ **7.** (2, 0, 2) **9.** $\left(-\dfrac{1}{2}, \dfrac{3}{4}, 1\right)$ **11.** ∅ **13.** (1, 1, −2)

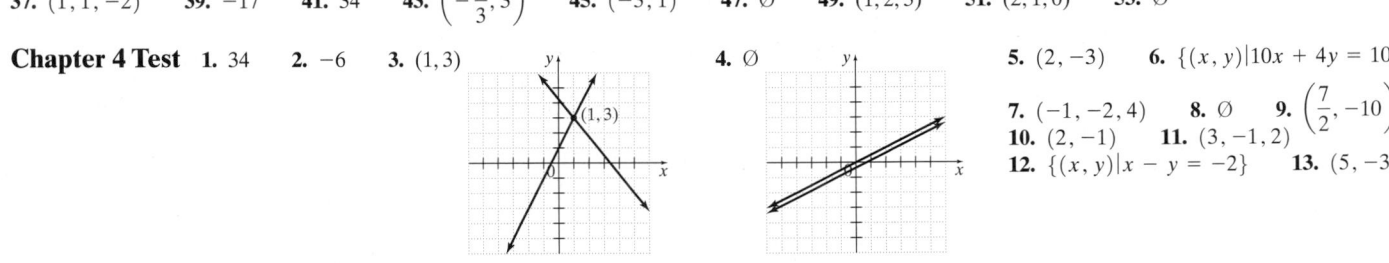

15. 10, 40, and 48 **17.** 58 mph, 65 mph **19.** 20 liters of 10% solution, 30 liters of 60% solution **21.** 17 pennies, 20 nickels, and 16 dimes

23. Two sides are 22 cm each; third side is 29 cm. **25.** (−3, 1) **27.** $\left(-\dfrac{2}{3}, 3\right)$ **29.** $\left(\dfrac{5}{4}, \dfrac{5}{8}\right)$ **31.** (1, 3) **33.** (1, 2, 3) **35.** (3, −2, 5)

37. (1, 1, −2) **39.** −17 **41.** 34 **43.** $\left(-\dfrac{2}{3}, 3\right)$ **45.** (−3, 1) **47.** ∅ **49.** (1, 2, 3) **51.** (2, 1, 0) **53.** ∅

Chapter 4 Test **1.** 34 **2.** −6 **3.** (1, 3) **4.** ∅ **5.** (2, −3) **6.** $\{(x, y)|10x + 4y = 10\}$ **7.** (−1, −2, 4) **8.** ∅ **9.** $\left(\dfrac{7}{2}, -10\right)$ **10.** (2, −1) **11.** (3, −1, 2) **12.** $\{(x, y)|x − y = −2\}$ **13.** (5, −3)

14. (−1, −1, 0) **15.** 53 double rooms and 27 single rooms **16.** 5 gal of 10%, 15 gal of 20% **17.** 800 packages **18.** 23°, 45°, 112°

Chapter 4 Cumulative Review **1. a.** true **b.** true; Sec. 1.2, Ex. 4 **2. a.** false **b.** true; Sec. 1.2 **3. a.** 6 **b.** −7; Sec. 1.3, Ex. 3

4. a. −5 **b.** −24; Sec. 1.3 **5. a.** −8 **b.** $-\dfrac{1}{5}$ **c.** 9.6; Sec. 1.4, Ex. 4 **6. a.** $\dfrac{1}{5}$ **b.** $-\dfrac{3}{2}$; Sec. 1.4 **7. a.** $6x + 3y$

b. $−3x + 1$ **c.** $0.7ab − 1.4a$; Sec. 1.4, Ex. 8 **8. a.** $21x − 14y + 28$ **b.** $2s + 3t$; Sec. 1.4 **9. a.** $−2x + 4$ **b.** $8yz$ **c.** $4z + 6.1$; Sec. 1.4, Ex. 12 **10. a.** $7y^2 + 3$ **b.** $2.2x + 1.2$; Sec. 1.4 **11.** −4; Sec. 1.5, Ex. 3 **12.** 0; Sec. 1.5 **13.** −4; Sec. 1.5, Ex. 7 **14.** all real numbers; Sec. 1.5 **15.** 25 cm, 62 cm, 62 cm; Sec. 1.6, Ex. 6 **16.** 100°, 100°, 110°, 50°; Sec. 1.6 **17.** $\{x|x \geq −10\}$; Sec. 3.2, Ex. 3 **18.** (0, ∞); Sec. 3.2 **19.** (−3, 2); Sec. 3.3, Ex. 4 **20.** (−2, 1); Sec. 3.3 **21.** 1, −1; Sec. 3.4, Ex. 4 **22.** 1, 9; Sec. 3.4 **23.** (4, 8); Sec. 3.5, Ex. 2 **24.** $(−∞, −3) \cup (2, ∞)$; Sec. 3.5 **25. a.** IV **b.** y-axis **c.** II **d.** x-axis **e.** III

f. I; ; Sec. 2.1, Ex. 1 **26. a.** III **b.** IV **c.** y-axis; Sec. 2.1 **27.** yes; Sec. 2.2, Ex. 3 **28.** ; Sec. 2.1

29. a. 5 **b.** 7 **c.** 4 **d.** −2; Sec 2.2, Ex. 7 **30. a.** 75 **b.** 12; Sec. 2.2

31. a. $\left(0, \dfrac{3}{7}\right)$ **b.** (0, −3.2); Sec. 2.3, Ex. 4 **32.** $m = \dfrac{3}{2}$; Sec. 2.4 **33.** slope: $\dfrac{3}{4}$; y-intercept: (0, −1); Sec 2.4, Ex. 4 **34.** $m = 0$, y-intercept: (0, 2); Sec. 2.4 **35. a.** parallel **b.** neither; Sec. 2.4, Ex. 7 **36.** $y = \dfrac{1}{5}x − 9$; Sec. 2.5 **37.** $f(x) = \dfrac{5}{8}x − \dfrac{5}{2}$; Sec. 2.5, Ex. 5

38. $y = −2x + 2$; Sec. 2.5

39. ; Sec. 3.6, Ex. 2   **40.** ; Sec. 3.6   **41. a.** yes   **b.** no; Sec 4.1, Ex. 1   **42.** $(-1, 3)$; Sec. 4.2

43. $(-4, 2, -1)$; Sec. 4.2, Ex. 1   **44.** $(-1, 2, 5)$; Sec. 4.3

45. $(-1, 2)$; Sec. 4.4, Ex. 1   **46.** $\varnothing$; Sec. 4.2

CHAPTER 5 EXPONENTS, POLYNOMIALS, AND POLYNOMIAL FUNCTIONS

Mental Math 1. $\dfrac{5}{xy^2}$ **3.** $\dfrac{a^2}{bc^5}$ **5.** $\dfrac{x^4}{y^2}$

Exercise Set 5.1 1. 4^5 **3.** x^8 **5.** $-140x^{12}$ **7.** $-20x^2y$ **9.** $-16x^6y^3p^2$ **11.** -1 **13.** 1 **15.** 6 **17.** answers may vary

19. a^3 **21.** x **23.** $-13z^4$ **25.** $-6a^4b^4c^6$ **27.** $\dfrac{1}{16}$ **29.** $\dfrac{1}{x^8}$ **31.** $\dfrac{5}{a^4}$ **33.** $\dfrac{1}{x^7}$ **35.** $4r^8$ **37.** 1 **39.** $\dfrac{13}{36}$ **41.** 9 **43.** x^{16}

45. $10x^{10}$ **47.** $\dfrac{1}{z^3}$ **49.** y^4 **51.** $\dfrac{3}{x}$ **53.** -2 **55.** r^8 **57.** $\dfrac{1}{x^9y^4}$ **59.** $\dfrac{b^7}{9a^7}$ **61.** $\dfrac{6x^{16}}{5}$ **63.** 3.125×10^7 **65.** 1.6×10^{-2}

67. 6.7413×10^4 **69.** 1.25×10^{-2} **71.** 5.3×10^{-5} **73.** 7.783×10^8 **75.** 7.37×10^5 **77.** 1.41×10^9 **79.** 1.0×10^{-3}
81. 0.0000000036 **83.** $93,000,000$ **85.** $1,278,000$ **87.** $7,350,000,000,000$ **89.** 0.000000403 **91.** $200,000,000$ **93.** $4,900,000,000$

95. 100 **97.** $\dfrac{27}{64}$ **99.** 64 **101.** $\dfrac{1}{16}$ **103.** answers may vary   **105. a.** x^{2a} **b.** $2x^a$ **c.** x^{a-b} **d.** x^{a+b} **e.** $x^a + x^b$ **107.** 7^{13}

109. 7^{-11} **111.** x^{7a+5} **113.** x^{2t-1} **115.** x^{4a+7} **117.** z^{6x-7} **119.** x^{6t-1} **121.** x^{3a+9}

Mental Math 1. x^{20} **3.** x^9 **5.** y^{42} **7.** z^{36} **9.** z^{18}

Exercise Set 5.2 1. $\dfrac{1}{9}$ **3.** $\dfrac{1}{x^{36}}$ **5.** $\dfrac{1}{y^5}$ **7.** $9x^4y^6$ **9.** $16x^{20}y^{12}$ **11.** $\dfrac{c^{18}}{a^{12}b^6}$ **13.** $\dfrac{y^{15}}{x^{35}z^{20}}$ **15.** $\dfrac{1}{125}$ **17.** x^{15} **19.** $\dfrac{8}{x^{12}y^{18}}$

21. $\dfrac{y^{16}}{64x^5}$ **23.** $\dfrac{64}{p^9}$ **25.** $-\dfrac{1}{x^9a^9}$ **27.** $\dfrac{x^5y^{10}}{5^{15}}$ **29.** $\dfrac{1}{x^{63}}$ **31.** $\dfrac{343}{512}$ **33.** $16x^4$ **35.** $-\dfrac{y^3}{64}$ **37.** $4^8x^2y^6$ **39.** 64 **41.** $\dfrac{x^4}{16}$ **43.** $\dfrac{1}{y^{15}}$

45. $\dfrac{x^9}{8y^3}$ **47.** $\dfrac{16a^2b^9}{9}$ **49.** $\dfrac{3}{8x^8y^7}$ **51.** $\dfrac{1}{x^{30}b^6c^6}$ **53.** $\dfrac{25}{8x^5y^4}$ **55.** $\dfrac{2}{x^4y^{10}}$ **57.** 1.45×10^9 **59.** 8×10^{15} **61.** 4×10^{-7} **63.** 3×10^{-1}

65. 2×10^1 **67.** 1×10^1 **69.** 8×10^{-5} **71.** 1.1×10^7 **73.** $8.877840909 \times 10^{20}$ **75.** 2×10^{-3} sec   **77.** 6.232×10^{-11} cu. m

79. $-3m - 15$ **81.** $-3y - 5$ **83.** $-3x + 5$ **85.** x^{4b+14} **87.** x^{-3y+1} **89.** c^{6a+9} **91.** y^{26a+1} **93.** $9y^{12a-2}$ **95.** y^{3b-a}

97. $x^{-3a-3b}y^{b-a}$ **99.** $\dfrac{15y^3}{x^8}$ sq. ft   **101.** 1.331928×10^{13} **103.** no   **105.** 83 people per sq.mi   **107.** 3.9   **109.** 4.5 times

Exercise Set 5.3 1. 0 **3.** 2 **5.** 3 **7.** degree 1; binomial   **9.** degree 2; trinomial   **11.** degree 3; monomial   **13.** degree 3;

none of these   **15.** answers may vary   **17.** 57   **19.** 499   **21.** $-\dfrac{11}{16}$ **23. a.** 2 and 3   **b.** 2 and 3   **25.** 5.7 sec.   **27. a.** the second egg

b. 1 sec.   **29.** $6y$ **31.** $11x - 3$ **33.** $xy + 2x - 1$ **35.** $6x^2 - xy + 16y^2$ **37.** $18y^2 - 17$ **39.** $3x^2 - 3xy + 6y^2$ **41.** $x^2 - 4x + 8$

43. $y^2 + 3$ **45.** $-2x^2 + 5x$ **47.** $-2x^2 - 4x + 15$ **49.** $4x - 13$ **51.** $x^2 + 2$ **53.** $12x^3 + 8x + 8$ **55.** $7x^3 + 4x^2 + 8x - 10$

57. $-18y^2 + 11yx + 14$ **59.** $-x^3 + 8a - 12$ **61.** $5x^2 - 9x - 3$ **63.** $-3x^2 + 3$ **65.** $8xy^2 + 2x^3 + 3x^2 - 3$ **67.** $7y^2 - 3$

69. $5x^2 + 22x + 16$ **71.** $\dfrac{3}{4}x^2 - \dfrac{1}{3}x^2y - \dfrac{8}{3}x^2y^2 + \dfrac{3}{2}y^3$ **73.** $-q^4 + q^2 - 3q + 5$ **75.** $15x^2 + 8x - 6$ **77.** $x^4 - 7x^2 + 5$

79. $\dfrac{1}{3}x^2 - x + 1$ **81.** 202 sq. in.   **83. a.** 284 ft   **b.** 536 ft   **c.** 756 ft   **d.** 944 ft   **e.** answers may vary   **f.** 19 sec   **85.** $80,000

87. $40,000   **89.** A   **91.** D   **93.** $15x - 10$ **95.** $-2x^2 + 10x - 12$ **97.** $3x^{2a} + 2x^a + 0.7$ **99.** $4x^{2y} + 2x^y - 11$
101. $(6x^2 + 14y)$ units   **103.** $4x^2 - 3x + 6$ **105.** $-x^2 - 6x + 10$ **107.** $3x^2 - 12x + 13$ **109.** $15x^2 + 12x - 9$ **111. a.** $2a - 3$
b. $-2x - 3$ **c.** $2x + 2h - 3$ **113. a.** $4a$ **b.** $-4x$ **c.** $4x + 4h$ **115. a.** $4a - 1$ **b.** $-4x - 1$ **c.** $4x + 4h - 1$ **117. a.** 1017
stations   **b.** 3806 stations   **c.** 331 stations   **d.** answers may vary   **119. a.** 3.1 million SUVs   **b.** 5.8 million SUVs   **121. a.** $1622
b. $3198   **c.** $6087   **d.** No, $f(x)$ is not linear.

Exercise Set 5.4 1. $-12x^5$ **3.** $12x^2 + 21x$ **5.** $-24x^2y - 6xy^2$ **7.** $-4a^3bx - 4a^3by + 12ab$ **9.** $2x^2 - 2x - 12$
11. $2x^4 + 3x^3 - 2x^2 + x + 6$ **13.** $15x^2 - 7x - 2$ **15.** $15m^3 + 16m^2 - m - 2$ **17.** answers may vary   **19.** $x^2 + x - 12$

21. $10x^2 + 11xy - 8y^2$ **23.** $3x^2 + 8x - 3$ **25.** $9x^2 - \dfrac{1}{4}$ **27.** $x^2 + 8x + 16$ **29.** $36y^2 - 1$ **31.** $9x^2 - 6xy + y^2$ **33.** $9b^2 - 36y^2$

35. $16b^2 + 32b + 16$ **37.** $4s^2 - 12s + 8$ **39.** $x^2y^2 - 4xy + 4$ **41.** answers may vary **43.** $2x^3 + 2x^2y + x^2 + xy - x - y$

45. $x^4 - 8x^3 + 24x^2 - 32x + 16$ **47.** $x^4 - 625$ **49.** $9x^2 + 18x + 5$ **51.** $10x^5 + 8x^4 + 2x^3 + 25x^2 + 20x + 5$ **53.** $49x^2 - 9$

55. $9x^3 + 30x^2 + 12x - 24$ **57.** $16x^2 - \frac{2}{3}x - \frac{1}{6}$ **59.** $36x^2 + 12x + 1$ **61.** $x^4 - 4y^2$ **63.** $-30a^4b^4 + 36a^3b^2 + 36a^2b^3$

65. $2a^2 - 12a + 16$ **67.** $49a^2b^2 - 9c^2$ **69.** $m^2 - 8m + 16$ **71.** $9x^2 + 6x + 1$ **73.** $y^2 - 7y + 12$
75. $2x^3 + 2x^2y + x^2 + xy - x - y$ **77.** $9x^4 + 12x^3 - 2x^2 - 4x + 1$ **79.** $12x^3 - 2x^2 + 13x + 5$ **81.** $a^2 - 3a$

83. $a^2 + 2ah + h^2 - 3a - 3h$ **85.** $b^2 - 7b + 10$ **87.** -2 **89.** $\frac{3}{5}$ **91.** function **93. a.** $a^2 + 2ah + h^2 + 3a + 3h + 2$
b. $a^2 + 3a + 2$ **c.** $2ah + h^2 + 3h$ **95.** $30x^2y^{2n+1} - 10x^2y^n$ **97.** $x^{3a} + 5x^{2a} - 3x^a - 15$ **99.** $\pi(25x^2 - 20x + 4)$ sq. km
101. $(8x^2 - 12x + 4)$ sq. in. **103. a.** $6x + 12$ **b.** $9x^2 + 36x + 35$; one operation is addition, the other is multiplication. **105.** $5x^2 + 25x$
107. $x^4 - 4x^2 + 4$ **109.** $x^3 + 5x^2 - 2x - 10$

Mental Math 1. 6 **3.** 5 **5.** x **7.** $7x$

Exercise Set 5.5 1. a^3 **3.** y^2z^2 **5.** $3x^2y$ **7.** $5xz^3$ **9.** $6(3x - 2)$ **11.** $4y^2(1 - 4xy)$ **13.** $2x^3(3x^2 - 4x + 1)$
15. $4ab(2a^2b^2 - ab + 1 + 4b)$ **17.** $(x + 3)(6 + 5a)$ **19.** $(z + 7)(2x + 1)$ **21.** $(x^2 + 5)(3x - 2)$ **23.** answers may vary
25. $(a + 2)(b + 3)$ **27.** $(a - 2)(c + 4)$ **29.** $(x - 2)(2y - 3)$ **31.** $(4x - 1)(3y - 2)$ **33.** $3(2x^3 + 3)$ **35.** $x^2(x + 3)$
37. $4a(2a^2 - 1)$ **39.** $-4xy(5x - 4y^2)$ **41.** $5ab^2(2ab + 1 - 3b)$ **43.** $3b(3ac^2 + 2a^2c - 2a + c)$ **45.** $(y - 2)(4x - 3)$
47. $(2x + 3)(3y + 5)$ **49.** $(x + 3)(y - 5)$ **51.** $(2a - 3)(3b - 1)$ **53.** $(6x + 1)(2y + 3)$ **55.** $(n - 8)(2m - 1)$ **57.** $3x^2y^2(5x - 6)$
59. $(2x + 3y)(x + 2)$ **61.** $(5x - 3)(x + y)$ **63.** $(x^2 + 4)(x + 3)$ **65.** $(x^2 - 2)(x - 1)$ **67.** $55x^7$ **69.** $125x^6$ **71.** $x^2 - 3x - 10$
73. $x^2 + 5x + 6$ **75.** $y^2 - 4y + 3$ **77.** none **79.** a **81.** $I(R_1 + R_2) = E$ **83.** $x(x + 40)$ sq. in. **85. a.** $h(t) = -16(t^2 - 14)$
b. 160 ft **c.** answers may vary **87.** $y^n(3 + 3y^n + 5y^{7n})$ **89.** $3x^{2a}(x^{3a} - 2x^a + 3)$

Mental Math 1. 5 and 2 **3.** 8 and 3

Exercise Set 5.6 1. $(x + 3)(x + 6)$ **3.** $(x - 8)(x - 4)$ **5.** $(x + 12)(x - 2)$ **7.** $(x - 6)(x + 4)$ **9.** $3(x - 2)(x - 4)$
11. $4z(x + 2)(x + 5)$ **13.** $2(x + 18)(x - 3)$ **15.** $\pm5, \pm7$ **17.** $(x + 5), (x - 3)$ **19.** $(x - 2), (x - 6)$ **21.** $(2x - 3)(x - 4)$
23. prime polynomial **25.** $2(3x - 5)(2x + 5)$ **27.** $y^2(3y + 5)(y - 2)$ **29.** $2x(3x^2 + 4x + 12)$ **31.** $(x + 7z)(x + z)$
33. $(2x + y)(x - 3y)$ **35.** $(x - 4)(x + 3)$ **37.** $2(7y + 2)(2y + 1)$ **39.** $(2x - 3)(x + 9)$ **41.** $\pm8, \pm16$ **43.** $(x^2 + 3)(x^2 - 2)$
45. $(5x + 8)(5x + 2)$ **47.** $(x^3 - 4)(x^3 - 3)$ **49.** $(a - 3)(a + 8)$ **51.** $x(3x + 4)(x - 2)$ **53.** $(x - 27)(x + 3)$
55. $(x - 18)(x + 3)$ **57.** $3(x - 1)^2$ **59.** $(3x + 1)(x - 2)$ **61.** $(4x - 3)(2x - 5)$ **63.** $3x^2(2x + 1)(3x + 2)$ **65.** $3(a + 2b)^2$
67. prime polynomial **69.** $(2x + 13)(x + 3)$ **71.** $(3x - 2)(2x - 15)$ **73.** $(x^2 - 6)(x^2 + 1)$ **75.** $x(3x + 1)(2x - 1)$
77. $(4a - 3b)(3a - 5b)$ **79.** $(3x + 5)^2$ **81.** $y(3x - 8)(x - 1)$ **83.** $2(x + 3)(x - 2)$ **85.** $(x + 2)(x - 7)$ **87.** $(2x^3 - 3)(x^3 + 3)$
89. $2x(6y^2 - z)^2$ **91.** $x^2 - 9$ **93.** $4x^2 + 4x + 1$ **95.** $x^3 - 8$ **97. a.** 576 ft; 672 ft; 640 ft; 480 ft **b.** answers may vary
c. $-16(t + 4)(t - 9)$ **99.** $(x^n + 8)(x^n + 2)$ **101.** $(x^n - 6)(x^n + 3)$ **103.** $(2x^n + 1)(x^n + 5)$ **105.** $(2x^n - 3)^2$
107. $x^2(x + 5)(x + 1)$ **109.** $3x(5x - 1)(2x + 1)$

Exercise Set 5.7 1. $(x + 3)^2$ **3.** $(2x - 3)^2$ **5.** $3(x - 4)^2$ **7.** $x^2(3y + 2)^2$ **9.** $(x + 5)(x - 5)$ **11.** $(3 + 2z)(3 - 2z)$
13. $(y + 9)(y - 5)$ **15.** $4(4x + 5)(4x - 5)$ **17.** $(x + 3)(x^2 - 3x + 9)$ **19.** $(z - 1)(z^2 + z + 1)$ **21.** $(m + n)(m^2 - mn + n^2)$
23. $y^2(x - 3)(x^2 + 3x + 9)$ **25.** $b(a + 2b)(a^2 - 2ab + 4b^2)$ **27.** $(5y - 2x)(25y^2 + 10yx + 4x^2)$ **29.** $(x + 3 + y)(x + 3 - y)$
31. $(x - 5 + y)(x - 5 - y)$ **33.** $(2x + 1 + z)(2x + 1 - z)$ **35.** $(3x + 7)(3x - 7)$ **37.** $(x - 6)^2$ **39.** $(x^2 + 9)(x + 3)(x - 3)$
41. $(x + 4 + 2y)(x + 4 - 2y)$ **43.** $(x + 2y + 3)(x + 2y - 3)$ **45.** $(x - 6)(x^2 + 6x + 36)$ **47.** $(x + 5)(x^2 - 5x + 25)$
49. prime polynomial **51.** $(2a + 3)^2$ **53.** $2y(3x + 1)(3x - 1)$ **55.** $(2x + y)(4x^2 - 2xy + y^2)$ **57.** $(x^2 - y)(x^4 + x^2y + y^2)$
59. $(x + 8 + x^2)(x + 8 - x^2)$ **61.** $3y^2(x^2 + 3)(x^4 - 3x^2 + 9)$ **63.** $(x + y + 5)(x^2 + 2xy + y^2 - 5x - 5y + 25)$

65. $(2x - 1)(4x^2 + 20x + 37)$ **67.** 5 **69.** $-\frac{1}{3}$ **71.** 0 **73.** 5 **75.** $\pi R^2 - \pi r^2 = \pi(R + r)(R - r)$

77. $x^3 - y^2x = x(x + y)(x - y)$ **79.** $c = 9$ **81.** $c = 49$ **83.** $c = \pm8$ **85. a.** $(x + 1)(x^2 - x + 1)(x - 1)(x^2 + x + 1)$
b. $(x + 1)(x - 1)(x^4 + x^2 + 1)$ **c.** answers may vary **87.** $(x^n + 6)(x^n - 6)$ **89.** $(5x^n + 9)(5x^n - 9)$ **91.** $(x^{2n} + 25)(x^n + 5)(x^n - 5)$

Integrated Review 1. $2y^2 + 2y - 11$ **2.** $-2z^4 - 6z^2 + 3z$ **3.** $x^2 - 7x + 7$ **4.** $7x^2 - 4x - 5$ **5.** $25x^2 - 30x + 9$ **6.** $x - 3$

7. $2x^3 - 4x^2 + 5x - 5 + \frac{8}{x + 2}$ **8.** $4x^3 - 13x^2 - 5x + 2$ **9.** $(x - 4 + y)(x - 4 - y)$ **10.** $2(3x + 2)(2x - 5)$

11. $x(x - 1)(x^2 + x + 1)$ **12.** $2x(2x - 1)$ **13.** $2xy(7x - 1)$ **14.** $6ab(4b - 1)$ **15.** $4(x + 2)(x - 2)$ **16.** $9(x + 3)(x - 3)$
17. $(3x - 11)(x + 1)$ **18.** $(5x + 3)(x - 1)$ **19.** $4(x + 3)(x - 1)$ **20.** $6(x + 1)(x - 2)$ **21.** $(2x + 9)^2$ **22.** $(5x + 4)^2$
23. $(2x + 5y)(4x^2 - 10xy + 25y^2)$ **24.** $(3x - 4y)(9x^2 + 12xy + 16y^2)$ **25.** $8x^2(2y - 1)(4y^2 + 2y + 1)$
26. $27x^2y(xy - 2)(x^2y^2 + 2xy + 4)$ **27.** $(x + 5 + y)(x^2 + 10x - xy - 5y + y^2 + 25)$ **28.** $(y - 1 + 3x)(y^2 - 2y + 1 - 3xy + 3x + 9x^2)$
29. $(5a - 6)^2$ **30.** $(4r + 5)^2$ **31.** $7x(x - 9)$ **32.** $(4x + 3)(5x + 2)$ **33.** $(a + 7)(b - 6)$ **34.** $20(x - 6)(x - 5)$
35. $(x^2 + 1)(x - 1)(x + 1)$ **36.** $5x(3x - 4)$ **37.** $(5x - 11)(2x + 3)$ **38.** $9m^2n^2(5mn - 3)$ **39.** $5a^3b(b^2 - 10)$
40. $x(x + 1)(x^2 - x + 1)$ **41.** prime **42.** $20(x + y)(x^2 - xy + y^2)$ **43.** $10x(x - 10)(x - 11)$ **44.** $(3y - 7)^2$
45. $a^3b(4b - 3)(16b^2 + 12b + 9)$ **46.** $(y^2 + 4)(y + 2)(y - 2)$ **47.** $2(x - 3)(x^2 + 3x + 9)$ **48.** $(2s - 1)(r + 5)$ **49.** $(y^4 + 2)(3y - 5)$
50. prime **51.** $100(z + 1)(z^2 - z + 1)$ **52.** $2x(5x - 2)(25x^2 + 10x + 4)$ **53.** $(2b - 9)^2$ **54.** $(a^4 + 3)(2a - 1)$ **55.** $(y - 4)(y - 5)$
56. $(c - 3)(c + 1)$ **57.** $A = 9 - 4x^2 = (3 + 2x)(3 - 2x)$

Mental Math 1. $3, -5$ **3.** $3, -7$ **5.** $0, 9$

Exercise Set 5.8 **1.** $-3, \frac{4}{3}$ **3.** $\frac{5}{2}, -\frac{3}{4}$ **5.** $-3, -8$ **7.** $\frac{1}{4}, -\frac{2}{3}$ **9.** $1, 9$ **11.** $\frac{3}{5}, -1$ **13.** 0 **15.** $6, -3$ **17.** $\frac{2}{5}, -\frac{1}{2}$

19. $\frac{3}{4}, -\frac{1}{2}$ **21.** $-2, 7, \frac{8}{3}$ **23.** $0, 3, -3$ **25.** $2, 1, -1$ **27.** answers may vary **29.** $-\frac{7}{2}, 10$ **31.** $0, 5$ **33.** $-3, 5$ **35.** $-\frac{1}{2}, \frac{1}{3}$

37. $-4, 9$ **39.** $\frac{4}{5}$ **41.** $-5, 0, 2$ **43.** $-3, 0, \frac{4}{5}$ **45.** $\varnothing$ **47.** $-7, 4$ **49.** $4, 6$ **51.** $-\frac{1}{2}$ **53.** $-4, -3, 3$ **55.** $-5, 0, 5$

57. $-6, 5$ **59.** $-\frac{1}{3}, 0, 1$ **61.** $-\frac{1}{3}, 0$ **63.** $-\frac{7}{8}$ **65.** $\frac{31}{4}$ **67.** 1 **69. a.** incorrect **b.** correct **c.** correct **d.** incorrect

71. -11 and -6 or 6 and 11 **73.** 75 ft **75.** 105 units **77.** 12 cm and 9 cm **79.** 2 in. **81.** 10 sec **83.** width: $7\frac{1}{2}$ ft; length: 12 ft

85. 10 in. sq. tier **87.** 9 sec **89.** E **91.** F **93.** B **95.** $(-3, 0), (0, 2)$; function **97.** $(-4, 0), (0, 2), (4, 0), (0, -2)$; not a function

99. answers may vary **101.** $\left\{-3, -\frac{1}{3}, 2, 5\right\}$ **103.** No; answers may vary. **105.** answers may vary. Ex.: $f(x) = x^2 - 13x + 42$

107. answers may vary. Ex.: $f(x) = x^2 - x - 12$

Chapter 5 Review **1.** 4 **3.** -4 **5.** 1 **7.** $-\frac{1}{16}$ **9.** $-x^2y^7z$ **11.** $\frac{1}{a^9}$ **13.** $\frac{1}{x^{11}}$ **15.** $\frac{1}{y^5}$ **17.** -3.62×10^{-4} **19.** $410{,}000$

21. $\frac{a^2}{16}$ **23.** $\frac{1}{16x^2}$ **25.** $\frac{1}{8^{18}}$ **27.** $-\frac{1}{8x^9}$ **29.** $\frac{-27y^6}{x^6}$ **31.** $\frac{xz}{4}$ **33.** $\frac{2}{27z^3}$ **35.** $2y^{x-7}$ **37.** -2.21×10^{-11} **39.** $\frac{x^3y^{10}}{3z^{12}}$ **41.** 5

43. $12x - 6x^2 - 6x^2y$ **45.** $4x^2 + 8y + 6$ **47.** $8x^2 + 2b - 22$ **49.** $12x^2y - 7xy + 3$ **51.** $x^3 + x - 2xy^2 - y - 7$ **53.** 58

55. $x^2 + 4x - 6$ **57.** $(6x^2y - 12x + 12)$ cm **59.** $-12a^2b^5 - 28a^2b^3 - 4ab^2$ **61.** $9x^2a^2 - 24xab + 16b^2$ **63.** $15x^2 + 18xy - 81y^2$

65. $x^4 + 18x^3 + 83x^2 + 18x + 1$ **67.** $16x^2 + 72x + 81$ **69.** $16 - 9a^2 + 6ab - b^2$ **71.** $(9y^2 - 49z^2)$ sq. units

73. $16x^2y^{2z} - 8xy^zb + b^2$ **75.** $8x^2(2x - 3)$ **77.** $2ab(3b + 4 - 2ab)$ **79.** $(a + 3b)(6a - 5)$ **81.** $(x - 6)(y + 3)$

83. $(p - 5)(q - 3)$ **85.** $x(2y - x)$ **87.** $(x - 4)(x + 20)$ **89.** $3(x + 2)(x + 9)$ **91.** $(3x + 8)(x - 2)$ **93.** $(15x - 1)(x - 6)$

95. $3(x - 2)(3x + 2)$ **97.** $(x + 7)(x + 9)$ **99.** $(x^2 - 2)(x^2 + 10)$ **101.** $(x + 9)(x - 9)$ **103.** $6(x + 3)(x - 3)$

105. $(4 + y^2)(2 + y)(2 - y)$ **107.** $(x - 7)(x + 1)$ **109.** $(y + 8)(y^2 - 8y + 64)$ **111.** $(1 - 4y)(1 + 4y + 16y^2)$

113. $2x^2(x + 2y)(x^2 - 2xy + 4y^2)$ **115.** $(x - 3 - 2y)(x - 3 + 2y)$ **117.** $(4a - 5b)^2$ **119.** $\frac{1}{3}, -7$ **121.** $0, 4, \frac{9}{2}$ **123.** $0, 6$

125. $-\frac{1}{3}, 2$ **127.** $-4, 1$ **129.** $0, 6, -3$ **131.** $0, -2, 1$ **133.** $-\frac{15}{2}, 7$ **135.** 5 sec

Chapter 5 Test **1.** $\frac{1}{81x^2}$ **2.** $-12x^2z$ **3.** $\frac{3a^7}{2b^5}$ **4.** $-\frac{y^{40}}{z^5}$ **5.** 6.3×10^8 **6.** 1.2×10^{-2} **7.** 0.000005 **8.** 0.0009

9. $-5x^3y - 11x - 9$ **10.** $-12x^2y - 3xy^2$ **11.** $12x^2 - 5x - 28$ **12.** $25a^2 - 4b^2$ **13.** $36m^2 + 12mn + n^2$ **14.** $2x^3 - 13x^2 + 14x - 4$

15. $4x^2y(4x - 3y^3)$ **16.** $(x - 15)(x + 2)$ **17.** $(2y + 5)^2$ **18.** $3(2x + 1)(x - 3)$ **19.** $(2x + 5)(2x - 5)$ **20.** $(x + 4)(x^2 - 4x + 16)$

21. $3y(x + 3y)(x - 3y)$ **22.** $6(x^2 + 4)$ **23.** $2(2y - 1)(4y^2 + 2y + 1)$ **24.** $(x + 3)(x - 3)(y - 3)$ **25.** $4, -\frac{8}{7}$ **26.** $-3, 8$

27. $-\frac{5}{2}, -2, 2$ **28.** $(x + 2y)(x - 2y)$ **29. a.** 960 ft **b.** 953.44 ft **c.** 11 sec

Chapter 5 Cumulative Review **1. a.** 3 **b.** 1 **c.** 2; Sec. 1.3, Ex. 8 **2. a.** 4 **b.** 3 **c.** 2; Sec. 1.3 **3.** 1; Sec. 1.5, Ex. 4

4. 8; Sec. 1.5 **5.** $11,607.55$; Sec. 1.8, Ex. 5 **6.** 3 gals; Sec. 1.8 **7. a.** $\left\{x \mid x \le \frac{3}{2}\right\}$ **b.** $\{x \mid x > -3\}$; Sec. 3.2, Ex. 4

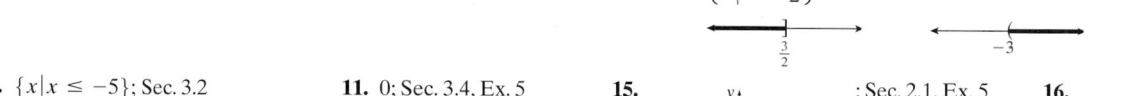

8. $\{x \mid x \le -5\}$; Sec. 3.2   **11.** 0; Sec. 3.4, Ex. 5   **15.** ; Sec. 2.1, Ex. 5   **16.** ; Sec. 2.1

12. $-4, 4$; Sec. 3.4

9. $\left[-9, -\frac{9}{2}\right]$; Sec. 3.3, Ex. 5   **13.** $\varnothing$; Sec. 3.5, Ex. 4

14. $(-\infty, \infty)$; Sec. 3.5

10. $\left(-1, \frac{1}{3}\right)$; Sec. 2.5

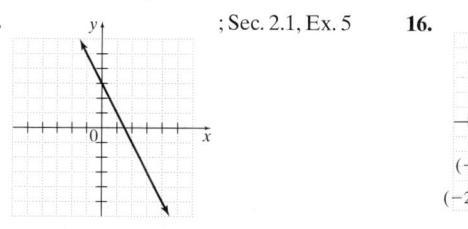

17. no; Sec. 2.2, Ex. 4 **19.** ; Sec. 2.3, Ex. 6 **20.** ; Sec. 2.3 **21.** 0; Sec. 2.4, Ex. 6 **22.** -2; Sec. 2.4

18. 24; Sec. 2.2   **23.** $y = 3$; Sec. 2.5, Ex. 7 **24.** $x = -3$; Sec. 2.5

25. 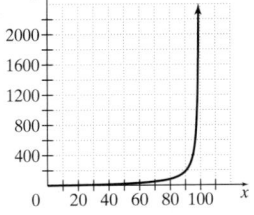 ; Sec. 3.6, Ex. 4   **26.** $y = 3$; Sec. 3.6   **27.** $(1, 6)$ Sec. 4.1, Ex. 4   **28.** $(2, 0)$; Sec. 4.1   **29.** $\left(\frac{1}{2}, 0, \frac{3}{4}\right)$; Sec. 4.2, Ex. 3

30. $(-3, -4, -5)$; Sec. 4.2   **31.** $7, 11$; Sec. 4.3, Ex. 2   **32.** 37.5 oz. of 20% solution; 12.5 oz of 60% solution; Sec. 4.3

33. $\varnothing$; Sec. 4.4, Ex. 2   **34.** $(5, 2)$ **35.** $(1, -2, -1)$; Sec. 4.5, Ex. 4   **36.** $(-3, -2, 5)$; Sec. 4.5   **37. a.** 7.3×10^5

b. 1.04×10^{-6}; Sec. 5.1, Ex. 8   **38. a.** 8.25×10^6 **b.** 3.46×10^{-5}; Sec. 5.1   **39. a.** $\frac{y^6}{4}$ **b.** x^9 **c.** $\frac{49}{4}$

d. $\frac{y^{16}}{25x^5}$; Sec. 5.2, Ex. 3   **40. a.** $\frac{a^3}{64}$ **b.** $\frac{1}{a^4}$ **c.** $\frac{27}{8}$ **d.** $\frac{b^{17}}{9a^6}$; Sec. 5.2   **41.** 4; Sec. 5.3, Ex. 3

42. $-2x^2 - 5x$; Sec. 5.3   **43. a.** $10x^9$ **b.** $-7xy^{15}z^9$; Sec. 5.4, Ex. 1   **44. a.** $12y^8$ **b.** $-6a^3b^3c^4$; Sec. 5.4   **45.** $17x^3y^2(1 - 2x)$; Sec. 5.5,

Ex. 3   **46.** $3xy(2x + y)(2x - y)$; Sec. 5.7   **47.** $(x + 2)(x + 8)$; Sec. 5.6, Ex. 1   **48.** $(5a - 1)(a + 3)$; Sec. 5.6   **49.** $-5, \frac{1}{2}$; Sec. 5.8, Ex. 2

50. $4, -\frac{2}{3}$; Sec. 5.8

CHAPTER 6   RATIONAL EXPRESSIONS

Exercise Set 6.1   **1.** $\frac{10}{3}, -8, -\frac{7}{3}$ **3.** $-\frac{17}{48}, \frac{2}{7}, -\frac{3}{8}$ **5.** $\{x \mid x \text{ is a real number}\}$ **7.** $\{t \mid t \text{ is a real number and } t \neq 0\}$

9. $\{x \mid x \text{ is a real number and } x \neq 7\}$ **11.** $\left\{x \mid x \text{ is a real number and } x \neq \frac{1}{3}\right\}$ **13.** $\{x \mid x \text{ is a real number and } x \neq -2, x \neq 0, x \neq 1\}$

15. $\{x \mid x \text{ is a real number and } x \neq 2, x \neq -2\}$ **17.** answers may vary   **19.** $\frac{4}{3}$ **21.** -2 **23.** $\frac{x + 1}{x - 3}$ **25.** $\frac{2(x + 3)}{x - 3}$ **27.** $\frac{3}{x}$

29. $\frac{x + 1}{x^2 + 1}$ **31.** $\frac{1}{2(q - 1)}$ **33.** $x - 4$ **35.** $-x^2 - 5x - 25$ **37.** $\frac{4x^2 + 6x + 9}{2}$ **39.** D   **41.** $-\frac{2}{3x^3y^2}$ **43.** $\frac{4}{ab^6}$ **45.** $\frac{1}{4a(a - b)}$

47. $\frac{(x + 2)(x + 3)}{4}$ **49.** $\frac{3}{2(x - 1)}$ **51.** $\frac{4a^2}{a - b}$ **53.** $\frac{2(x + 3)(x - 3)}{5(x^2 - 8x - 15)}$ **55.** $\frac{x + 2}{x + 3}$ **57.** $\frac{3b}{a - b}$ **59.** $\frac{3a}{a - b}$ **61.** $\frac{1}{4}$ **63.** -1

65. $\frac{8}{3}$ **67.** $\frac{8(a - 2)}{3(a + 2)}$ **69.** $\frac{8}{x^2 y}$ **71.** $\frac{(y + 5)(2x - 1)}{(y + 2)(5x + 1)}$ **73.** $\frac{5(3a + 2)}{a}$ **75.** $\frac{5x^2 - 2}{(x - 1)^2}$ **77.** $\frac{7}{5}$ **79.** $\frac{1}{12}$ **81.** $\frac{11}{16}$

83. $\frac{5}{x - 2}$ sq. m   **85. a.** $\{x \mid 0 \le x < 100\}$ **b.** \$42,857.14   **c.** \$150,000; \$400,000   **d.** \$900,000; \$1,900,000; \$9,900,000; answers may vary

87. answers may vary   **89.** $\frac{(x + 2)(x - 1)^2}{x^5}$ ft.   **91.** $0, \frac{20}{9}, \frac{60}{7}, 20, \frac{140}{3}, 180, 380, 1980;$

93. $2x^2(x^n + 2)$ **95.** $\frac{1}{10y(y^n + 3)}$ **97.** $\frac{y^n + 1}{2(y^n - 1)}$

Mental Math   **1.** A, B   **3.** C   **5.** $\frac{12}{y}$ **7.** $\frac{35}{y^2}$

Exercise Set 6.2   **1.** $-\frac{3}{x}$ **3.** $\frac{x + 2}{x - 2}$ **5.** $x - 2$ **7.** $\frac{1}{2 - x}$ **9.** $\frac{4x}{x + 5}$ ft; $\frac{x^2}{x^2 + 10x + 25}$ sq. ft   **11.** $35x$ **13.** $x(x + 1)$

15. $(x + 7)(x - 7)$ **17.** $6(x + 2)(x - 2)$ **19.** $(3x - 1)(x + 2)$ **21.** $(a + b)(a - b)^2$ **23.** $-4x(x + 3)(x - 3)$

25. answers may vary   **27.** $\frac{17}{6x}$ **29.** $\frac{35 - 4y}{14y^2}$ **31.** $\frac{-13x + 4}{(x + 4)(x - 4)}$ **33.** $\frac{2x + 4}{(x - 5)(x + 4)}$ **35.** 0 **37.** $-\frac{x}{x - 1}$ **39.** $\frac{-x + 1}{x - 2}$

41. $\frac{y^2 + 2y + 10}{(y + 4)(y - 4)(y - 2)}$ **43.** $\frac{5(x^2 + x - 4)}{(3x + 2)(x + 3)(2x - 5)}$ **45.** $\frac{x^2 + 5x + 21}{(x - 2)(x + 1)(x + 3)}$ **47.** $\frac{-2x + 5}{2(x + 1)}$ **49.** $\frac{2(x^2 + x - 21)}{(x + 3)^2(x - 3)}$

51. $\frac{3}{x^2 y^3}$ **53.** $-\frac{5}{x}$ **55.** $\frac{25}{6(x + 5)}$ **57.** $\frac{-2x - 1}{x^2(x - 3)}$ **59.** $\frac{2ab - b^2}{(a + b)(a - b)}$ **61.** $\frac{2x + 16}{(x + 2)^2(x - 2)}$ **63.** $\frac{5a + 1}{(a + 1)^2(a - 1)}$

65. answers may vary   **67.** answers may vary   **69.** $\frac{2x^2 + 9x - 18}{6x^2}$ **71.** $\frac{4}{3}$ **73.** $\frac{4a^2}{9(a - 1)}$ **75.** 4 **77.** $\frac{6x}{(x + 3)(x - 3)^2}$

79. $-\frac{4}{x - 1}$ **81.** $-\frac{32}{x(x + 2)(x - 2)}$ **83.** 10 **85.** $4 + x^2$ **87.** 10 **89.** 2 **91.** 3 **93.** 5 m   **95.** $\frac{3}{2x}$ **97.** $\frac{4 - 3x}{x^2}$

99. $\dfrac{1-3x}{x^3}$ **101.**   **103.**

Exercise Set 6.3 **1.** 4 **3.** $\dfrac{7}{13}$ **5.** $\dfrac{4}{x}$ **7.** $\dfrac{9(x-2)}{9x^2+4}$ **9.** $2x+y$ **11.** $\dfrac{2(x+1)}{2x-1}$ **13.** $\dfrac{2x+3}{4-9x}$ **15.** $\dfrac{1}{x^2-2x+4}$ **17.** $\dfrac{x}{5x-10}$

19. $\dfrac{x-2}{2x-1}$ **21.** $\dfrac{x}{2-3x}$ **23.** $-\dfrac{y}{x+y}$ **25.** $-\dfrac{2x^3}{y(x-y)}$ **27.** $\dfrac{2x+1}{y}$ **29.** $\dfrac{x-3}{9}$ **31.** $\dfrac{1}{x+2}$ **33.** $\dfrac{xy^2}{x^2+y^2}$ **35.** $\dfrac{2b^2+3a}{b(b-a)}$

37. $\dfrac{x}{(x+1)(x-1)}$ **39.** $\dfrac{1+a}{1-a}$ **41.** $\dfrac{x(x+6y)}{2y}$ **43.** $\dfrac{5a}{2a+4}$ **45.** $5xy^2+2x^2y$ **47.** $\dfrac{xy}{2x+5y}$ **49.** $\dfrac{x^2y^2}{4}$ **51.** $-9x^3y^4$

53. $-4, 14$ **55.** $\dfrac{770a}{770-s}$ **57.** a, b **59.** $\dfrac{1+x}{2+x}$ **61.** x^2+x **63.** $\dfrac{x-3y}{x+3y}$ **65.** $3a^2+4a+4$ **67. a.** $\dfrac{1}{a+h}$ **b.** $\dfrac{1}{a}$

c. $\dfrac{\dfrac{1}{a+h}-\dfrac{1}{a}}{h}$ **d.** $\dfrac{-1}{a(a+h)}$ **69. a.** $\dfrac{3}{a+h+1}$ **b.** $\dfrac{3}{a+1}$ **c.** $\dfrac{\dfrac{3}{a+h+1}-\dfrac{3}{a+1}}{h}$ **d.** $\dfrac{-3}{(a+h+1)(a+1)}$

Exercise Set 6.4 **1.** $2a+4$ **3.** $3ab+4$ **5.** $2y+\dfrac{3y}{x}-\dfrac{2y}{x^2}$ **7.** x^2+2x+1 **9.** $(x^4+2x^2-6)\,m$ **11.** $x+1$ **13.** $2x-8+\dfrac{1}{x+1}$

15. $x-\dfrac{1}{2}$ **17.** $2x^2-\dfrac{1}{2}x+5$ **19.** $(3x-7)\,\text{in.}$ **21.** $\dfrac{5b^5}{2a^3}$ **23.** x^3y^3-1 **25.** $a+3$ **27.** $2x+5$ **29.** $4y-6y^2$

31. $2x+23+\dfrac{130}{x-5}$ **33.** $10x+3y-6x^2y^2$ **35.** $2x+4$ **37.** $y+5$ **39.** $2x+3$ **41.** $2x^2-8x+38-\dfrac{156}{x+4}$ **43.** $3x+3-\dfrac{1}{x-1}$

45. $-2x^3+3x^2-x+4$ **47.** $3x^3+5x+4-\dfrac{2x}{x^2-2}$ **49.** $x-\dfrac{5}{3x^2}$ **51.** $=$ **53.** $=$ **55.** $(-9,-1)$ **57.** $(-\infty,-8]\cup[1,\infty)$

59. 4 **61.** 372 **63.** answers may vary **65.** $x^3+\dfrac{5}{3}x^2+\dfrac{5}{3}x+\dfrac{8}{3}+\dfrac{8}{3(x-1)}$ **67.** $\dfrac{3}{2}x^3+\dfrac{1}{4}x^2+\dfrac{1}{8}x-\dfrac{7}{16}+\dfrac{1}{16(2x-1)}$

69. $x^3-\dfrac{2}{5}x$ **71.** $5x-1+\dfrac{6}{x}; x\neq 0$ **73.** $7x^3+14x^2+25x+50+\dfrac{102}{x-2}; x\neq 2$ **75.** answers may vary

Exercise Set 6.5 **1.** $x+8$ **3.** $x-1$ **5.** $x^2-5x-23-\dfrac{41}{x-2}$ **7.** $4x+8+\dfrac{7}{x-2}$ **9.** 3 **11.** 73 **13.** -8

15. $x^2+\dfrac{2}{x-3}$ **17.** $6x+7+\dfrac{1}{x+1}$ **19.** $2x^3-3x^2+x-4$ **21.** $3x-9+\dfrac{12}{x+3}$ **23.** $3x^2-\dfrac{9}{2}x+\dfrac{7}{4}+\dfrac{47}{8\left(x-\frac{1}{2}\right)}$

25. $3x^2+3x-3$ **27.** $3x^2+4x-8+\dfrac{20}{x+1}$ **29.** x^2+x+1 **31.** $x-6$ **33.** 1 **35.** -133 **37.** 3 **39.** $-\dfrac{187}{81}$ **41.** $\dfrac{95}{32}$

43. answers may vary **45.** $-\dfrac{5}{6}$ **47.** 2 **49.** 54 **51.** $(x-1)(x^2+x+1)$ **53.** $(5z+2)(25z^2-10z+4)$ **55.** $(y+2)(x+3)$

57. $x(x+3)(x-3)$ **59.** $(x+3)(x^2+4)=x^3+3x^2+4x+12$ **61.** 0 **63.** $x^3+2x^2+7x+28$ **65.** $(x-1)\,m$

Exercise Set 6.6 **1.** 72 **3.** 2 **5.** 6 **7.** $2,-2$ **9.** $\varnothing$ **11.** $-\dfrac{28}{3}$ **13.** 3 **15.** -8 **17.** 3 **19.** $\varnothing$ **21.** 1 **23.** 3

25. -1 **27.** 6 **29.** $\dfrac{1}{3}$ **31.** $-5,5$ **33.** 3 **35.** 7 **37.** $\varnothing$ **39.** $\dfrac{4}{3}$ **41.** -12 **43.** $1,\dfrac{11}{4}$ **45.** $-5,-1$ **47.** $-\dfrac{7}{5}$ **49.** 5

51. length, 15 in.; width, 10 in. **53.** 10% **55.** 25–29 and 30–34 **57.** 6785 inmates **59.** 800 pencil sharpeners **61.** $\dfrac{1}{9},-\dfrac{1}{4}$ **63.** $3,2$

65. 1.39 **67.** -0.08 **69.** $1,2$ **71.** $-3,-\dfrac{3}{4}$ **73.**   **75.**

Integrated Review **1.** $\dfrac{1}{2}$ **2.** 10 **3.** $\dfrac{1+2x}{8}$ **4.** $\dfrac{15+x}{10}$ **5.** $\dfrac{2(x-4)}{(x+2)(x-1)}$ **6.** $-\dfrac{5(x-8)}{(x-2)(x+4)}$ **7.** 4 **8.** 8 **9.** -5

10. $-\dfrac{2}{3}$ **11.** $\dfrac{2x+5}{x(x-3)}$ **12.** $\dfrac{5}{2x}$ **13.** -2 **14.** $-\dfrac{y}{x}$ **15.** $\dfrac{(a+3)(a+1)}{a+2}$ **16.** $\dfrac{-a^2+31a+10}{5(a-6)(a+1)}$ **17.** $-\dfrac{1}{5}$ **18.** $-\dfrac{3}{13}$

19. $\dfrac{4a + 1}{(3a + 1)(3a - 1)}$ **20.** $\dfrac{-a - 8}{4a(a - 2)}$ or $-\dfrac{a + 8}{4a(a - 2)}$ **21.** $-1, \dfrac{3}{2}$ **22.** $\dfrac{x^2 - 3x + 10}{2(x + 3)(x - 3)}$ **23.** $\dfrac{3}{x + 1}$ **24.** $\{x \mid x \text{ is a real number and}$

$x \neq 2, x \neq -1\}$ **25.** -1 **26.** $\dfrac{22z - 45}{3z(z - 3)}$ **27. a.** $\dfrac{x}{5} - \dfrac{x}{4} + \dfrac{1}{10}$ **b.** Write each rational expression term so that the denominator is the

LCD, 20. **c.** $\dfrac{-x + 2}{20}$ **28. a.** $\dfrac{x}{5} - \dfrac{x}{4} = \dfrac{1}{10}$ **b.** Clear the equation of fractions by multiplying each term by the LCD, 20. **c.** -2 **29.** b

30. d **31.** d **32.** a **33.** d

Exercise Set 6.7

1. $C = \dfrac{5}{9}(F - 32)$ **3.** $I = A\ QL$ **5.** $R = \dfrac{R_1 R_2}{R_1 + R_2}$ **7.** $n = \dfrac{2S}{a + L}$ **9.** $b = \dfrac{2A - ah}{h}$ **11.** $T_2 = \dfrac{P_2 V_2 T_1}{P_1 V_1}$

13. $f_2 = \dfrac{f_1 f}{f_1 - f}$ **15.** $L = \dfrac{n\lambda}{2}$ **17.** $c = \dfrac{2L\omega}{\theta}$ **19.** 1 and 5 **21.** 5 **23.** 4.5 gal **25.** 3643 women **27.** 15.6 hr **29.** 10 min

31. 200 mph **33.** 15 mph **35.** -8 and -7 **37.** 36 min **39.** 45 mph; 60 mph **41.** 5.9 hr **43.** 2 hr **45.** 135 mph **47.** 12 mi

49. $2\dfrac{2}{9}$ hr **51.** $\dfrac{7}{8}$ **53.** $1\dfrac{1}{2}$ min **55.** 63 mph **57.** 1 hr **59.** 2 hr **61.** $\{6\}$ **63.** $\{22\}$ **65.** answers may vary; 60 in. or 5 ft

67. 6 ohms **69.** $\dfrac{1}{R} = \dfrac{1}{R_1} + \dfrac{1}{R_2} + \dfrac{1}{R_3}$; $R = \dfrac{15}{13}$ ohms

Mental Math

1. direct **3.** joint **5.** inverse **7.** direct

Exercise Set 6.8

1. $y = kx$ **3.** $a = \dfrac{k}{b}$ **5.** $y = kxz$ **7.** $y = \dfrac{k}{x^3}$ **9.** $y = \dfrac{kx}{p^2}$ **11.** $k = \dfrac{1}{5}$; $y = \dfrac{1}{5}x$ **13.** $k = \dfrac{3}{2}$; $y = \dfrac{3}{2}x$

15. $k = 14$; $y = 14x$ **17.** $k = 0.25$; $y = 0.25x$ **19.** 4.05 lb **21.** $P = 566{,}222$ tons **23.** $k = 30$; $y = \dfrac{30}{x}$ **25.** $k = 700$; $y = \dfrac{700}{x}$

27. $k = 2$; $y = \dfrac{2}{x}$ **29.** $k = 0.14$; $y = \dfrac{0.14}{x}$ **31.** 54 mph **33.** 72 amps **35.** divided by 4 **37.** $x = kyz$ **39.** $r = kst^3$

41. $k = \dfrac{1}{3}$; $y = \dfrac{1}{3}x^3$ **43.** $k = 0.2$; $y = 0.2\sqrt{x}$ **45.** $k = 1.3$; $y = \dfrac{1.3}{x^2}$ **47.** $k = 3$; $y = 3xz^3$ **49.** 22.5 tons **51.** 15π cu. in.

53. 90 hp **55.** 800 millibars **57.** $C = 12\pi$ cm; $A = 36\pi$ sq. cm **59.** $C = 14\pi$ m; $A = 49\pi$ sq. m **61.** 6 **63.** 2 **65.** $\dfrac{1}{5}$ **67.** $\dfrac{5}{11}$

69. multiplied by 2 **71.** multiplied by 4 **73.** 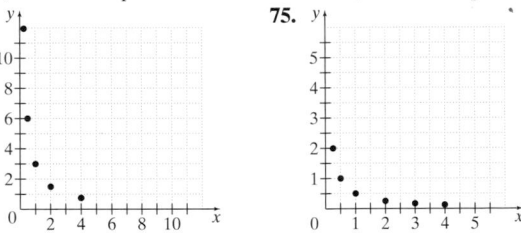 **75.**

Chapter 6 Review

1. $\{x \mid x \text{ is a real number}\}$ **3.** $\{x \mid x \text{ is a real number and } x \neq 5\}$ **5.** $\{x \mid x \text{ is a real number and } x \neq 0, x \neq -8\}$

7. $\dfrac{x^2}{3}$ **9.** $\dfrac{9m^2 p}{5}$ **11.** $\dfrac{1}{5}$ **13.** $\dfrac{1}{x - 1}$ **15.** $\dfrac{2(x - 3)}{x - 4}$ **17. a.** \$119 **b.** \$77 **c.** decrease **19.** $\dfrac{2x^3}{z^3}$ **21.** $\dfrac{2}{5}$ **23.** $\dfrac{1}{6}$

25. $\dfrac{3x}{16}$ **27.** $\dfrac{3c^2}{14a^2 b}$ **29.** $\dfrac{(x + 4)(x + 5)}{3}$ **31.** $\dfrac{7(x - 4)}{2(x - 2)}$ **33.** $-\dfrac{1}{x}$ **35.** $\dfrac{8}{9a^2}$ **37.** $\dfrac{6}{a}$ **39.** $60x^2 y^5$ **41.** $5x(x - 5)$ **43.** $\dfrac{2}{5}$

45. $\dfrac{2}{x^2}$ **47.** $\dfrac{1}{x - 2}$ **49.** $\dfrac{5x^2 - 3y^2}{15x^4 y^3}$ **51.** $\dfrac{-x + 5}{(x + 1)(x - 1)}$ **53.** $\dfrac{2x^2 - 5x - 4}{x - 3}$ **55.** $\dfrac{3x^2 - 7x - 4}{(3x - 4)(9x^2 + 12x + 16)}$

57. $-\dfrac{12}{x(x + 1)(x - 3)}$ **59.** $\dfrac{14x - 40}{(x + 4)^2(x - 4)}$ **61.** $\dfrac{2}{3}$ **63.** $\dfrac{2}{15 - 2x}$ **65.** $\dfrac{y}{2}$ **67.** $\dfrac{20x - 15}{10x^2 - 4}$ **69.** $\dfrac{5xy + x}{3y}$ **71.** $\dfrac{1 + x}{1 - x}$

73. $\dfrac{x - 1}{3x - 1}$ **75.** $-\dfrac{x^2 + 9}{6x}$ **77. a.** $\dfrac{3}{a + h}$ **b.** $\dfrac{3}{a}$ **c.** $\dfrac{\dfrac{3}{a + h} - \dfrac{3}{a}}{h}$ **d.** $\dfrac{-3}{a(a + h)}$ **79.** $\dfrac{9b^2 z^3}{4a}$ **81.** $\dfrac{3}{b} + 4b$

83. $2x^3 - 4x^2 + 7x - 9 + \dfrac{6}{x + 2}$ **85.** $x^2 - 1 + \dfrac{5}{2x + 3}$ **87.** $3x^2 + 6$ **89.** $3x^2 - \dfrac{5}{2}x - \dfrac{1}{4} - \dfrac{5}{8\left(x + \dfrac{3}{2}\right)}$ **91.** $x^2 + 3x + 9 - \dfrac{54}{x - 3}$

93. $3x^3 - 6x^2 + 10x - 20 + \dfrac{50}{x + 2}$ **95.** -9323 **97.** $\dfrac{365}{32}$ **99.** 6 **101.** 2 **103.** $\dfrac{3}{2}$ **105.** $\dfrac{5}{3}$ **107.** $-\dfrac{1}{3}, 2$ **109.** $a = \dfrac{2A - hb}{h}$

111. $R = \dfrac{E - Ir}{I}$ **113.** $A = \dfrac{HL}{k(T_1 - T_2)}$ **115.** 7 **117.** -10 and -8 **119.** 12 hr **121.** 490 mph **123.** 8 mph **125.** 4 mph

127. 9 **129.** 3.125 cu. ft

Chapter 6 Test   **1.** $\{x \mid x \text{ is a real number and } x \neq 1\}$ **2.** $\{x \mid x \text{ is a real number and } x \neq -3, x \neq -1\}$ **3.** $-\dfrac{7}{8}$ **4.** $\dfrac{x}{x+9}$ **5.** $\dfrac{5}{3x}$
6. $\dfrac{x+2}{2(x+3)}$ **7.** $-\dfrac{4(2x+9)}{5}$ **8.** -1 **9.** $\dfrac{5x-2}{(x-3)(x+2)(x-2)}$ **10.** $\dfrac{-x+30}{6(x-7)}$ **11.** $\dfrac{3}{2}$ **12.** $\dfrac{1}{5}$ **13.** $\dfrac{64}{3}$ **14.** $\dfrac{4xy}{3z} + \dfrac{3}{z} + 1$
15. $2x^2 - x - 2 + \dfrac{2}{2x+1}$ **16.** $4x^3 - 15x^2 + 45x - 136 + \dfrac{407}{x+3}$ **17.** 91 **18.** $\dfrac{2}{7}$ **19.** 3 **20.** $x = \dfrac{7a^2 + b^2}{4a - b}$ **21.** 5 **22.** $\dfrac{6}{7}$ hr
23. 16 **24.** 9 **25.** 256 ft

Chapter 6 Cumulative Review   **1. a.** $8x$ **b.** $8x + 3$ **c.** $x \div -7$ or $\dfrac{x}{-7}$ **d.** $2x - 1.6$; Sec. 1.2, Ex. 8
2. a. $x - \dfrac{1}{3}$ **b.** $5x - 6$ **c.** $8x + 3$ **d.** $\dfrac{7}{2-x}$; Sec. 1.2   **3.** 2; Sec. 1.5, Ex. 5   **4.** 7; Sec. 1.5   **5.** 2013 and after; Sec. 3.2, Ex. 9
6. 82; Sec. 3.2   **7.** $\varnothing$; Sec. 3.4, Ex. 7   **8.** $\varnothing$; Sec. 3.4   **9.** -1; Sec. 3.5, Ex. 8   **10.** $\left(\infty, -\frac{5}{3}\right] \cup \left[\frac{11}{3}, \infty\right)$; Sec. 3.5

11. ; Sec. 2.1, Ex. 5   **12.** ; Sec. 2.1   **13. a.** function
b. not a function
c. function; Sec. 2.2, Ex. 2

14. a. -2 **b.** -20 **c.** $-\dfrac{10}{9}$; Sec. 2.2

15. 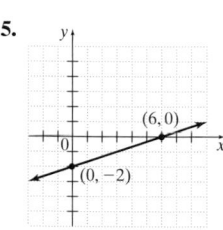 ; Sec. 2.3, Ex. 3   **16.** 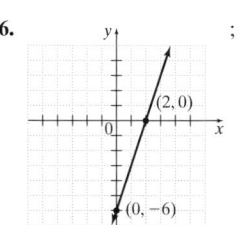 ; Sec. 2.3   **17.** $y = -3x - 2$; Sec. 2.5, Ex. 4

18. $f(x) = \dfrac{1}{2}x + \dfrac{7}{2}$; Sec. 2.5

19. ; Sec. 3.6, Ex. 3   **20.** 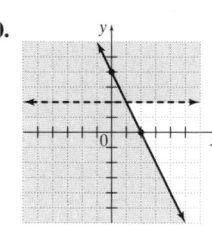 ; Sec. 3.6   **21.** $(0, -5)$; Sec. 4.1, Ex. 7   **22.** $(3, 4)$; Sec. 4.1   **23.** $\varnothing$;
Sec. 4.2, Ex. 2   **24.** $(2, 1, 1)$; Sec. 4.2   **25.** $30°, 110°, 40°$;
Sec. 4.3, Ex. 6   **26.** Paper, $3.80; folders, $5.25; Sec. 4.3
27. $(1, -1, 3)$; Sec. 4.4, Ex. 4   **28.** $(0, 5, 4)$; Sec. 4.4

29. a. 1 **b.** -1 **c.** 1 **d.** 2; Sec. 5.1, Ex. 3   **30. a.** $\dfrac{7}{12}$
b. -6 **c.** $\dfrac{1}{x^3}$; Sec. 5.1   **31. a.** $4x^b$ **b.** y^{5a+6}; Sec. 5.2, Ex. 5
32. a. $48a^{2a}$ **b.** y^{10b+3}; Sec. 5.2   **33. a.** 2 **b.** 5 **c.** 1

d. 6 **e.** 0; Sec. 5.3, Ex. 1   **34.** $2x^2 + 6x + 4$; Sec. 5.3   **35.** $9 + 12a + 6b + 4a^2 + 4ab + b^2$; Sec. 5.4, Ex. 9
36. $16 + 24x - 8y + 9x^2 - 6xy + y^2$; Sec. 5.4   **37.** $(b - 6)(a + 2)$; Sec 5.5, Ex. 7   **38.** $(x - 5)(y + 2)$; Sec. 5.5   **39.** $2(n^2 - 19n + 40)$;
Sec. 5.6, Ex. 4   **40.** $(2x - 5)(3x + 7)$; Sec. 5.6   **41.** $(x + 2 + y)(x + 2 - y)$; Sec. 5.7, Ex. 5   **42.** $(2x + 3y - 1)(2x - 3y - 1)$; Sec. 5.7
43. $-2, 6$; Sec. 5.8, Ex. 1   **44.** $0, -\frac{1}{3}, 3$; Sec. 5.8   **45.** $\dfrac{1}{5x-1}$; Sec. 6.1, Ex. 2   **46. a.** domain: $(-\infty, \infty)$ range: $[-4, \infty)$
b. x-intercepts: $(-2, 0), (2, 0)$ y-intercepts: $(0, -4)$ **c.** there is no such point   **d.** $(0, -4)$ **e.** $-2, 2$ **f.** between $x = -2$ and $x = 2$
g. $-2, 2$; Sec. 2.2   **47.** $\dfrac{5k^2 - 7k + 4}{(k+2)(k-2)(k-1)}$; Sec. 6.2, Ex. 4   **48.** $\dfrac{8a+6}{(a+2)(a-2)}$; Sec. 6.2   **49.** -2; Sec. 6.6, Ex. 2   **50.** -4; Sec. 6.6

CHAPTER 7   RATIONAL EXPONENTS, RADICALS, AND COMPLEX NUMBERS

Mental Math   **1.** D   **3.** D

Exercise Set 7.1   **1.** 10 **3.** $\dfrac{1}{2}$ **5.** 0.01 **7.** -6 **9.** x^5 **11.** $4y^3$ **13.** 2.646 **15.** 6.164 **17.** 14.142 **19.** 4
21. $\dfrac{1}{2}$ **23.** -1 **25.** x^4 **27.** $-3x^3$ **29.** -2 **31.** not a real number   **33.** -2 **35.** x^4 **37.** $2x^2$ **39.** $9x^2$ **41.** $4x^2$
43. 8 **45.** -8 **47.** $2|x|$ **49.** x **51.** $|x - 5|$ **53.** $|x + 2|$ **55.** -11 **57.** $2x$ **59.** y^6 **61.** $5ab^{10}$ **63.** $-3x^4y^3$
65. a^4b **67.** $-2x^2y$ **69.** $\dfrac{5}{7}$ **71.** $\dfrac{x}{2y}$ **73.** $-\dfrac{z^7}{3x}$ **75.** $\dfrac{x}{2}$ **77.** $\sqrt{3}$ **79.** -1 **81.** -3 **83.** $\sqrt{7}$

85. $[0, \infty)$;

87. $[3, \infty)$; $0, 1, 2, 3$

89. D **91.** D

93. A **95.** B

97. $(-\infty, \infty)$;

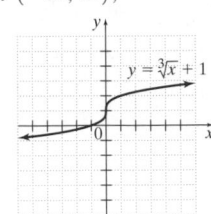

99. $(-\infty, \infty)$; $0, 1, -1, 2, -2$

101. $-32x^{15}y^{10}$ **103.** $-60x^7y^{10}z^5$ **105.** $\dfrac{x^9y^5}{2}$ **107.** answers may vary **109.** 13 **111.** 18 **113.** 1.69 sq. m **115.** answers may vary

Mental Math **1.** A **3.** C **5.** B **7.** B **9.** B

Exercise Set 7.2 **1.** 7 **3.** 3 **5.** $\dfrac{1}{2}$ **7.** 13 **9.** $2\sqrt[3]{m}$ **11.** $3x^2$ **13.** -3 **15.** -2 **17.** 8 **19.** 16 **21.** not a real number

23. $\sqrt[5]{(2x)^3}$ **25.** $\sqrt[3]{(7x+2)^2}$ **27.** $\dfrac{64}{27}$ **29.** $\dfrac{1}{16}$ **31.** $\dfrac{1}{16}$ **33.** not a real number **35.** $\dfrac{1}{x^{1/4}}$ **37.** $a^{2/3}$ **39.** $\dfrac{5x^{3/4}}{7}$

41. answers may vary **43.** $a^{7/3}$ **45.** x **47.** $3^{5/8}$ **49.** $y^{1/6}$ **51.** $8u^3$ **53.** $-b$ **55.** $27x^{2/3}$ **57.** $y - y^{7/6}$ **59.** $2x^{5/3} - 2x^{2/3}$

61. $4x^{2/3} - 9$ **63.** $x^{8/3}(1 + x^{2/3})$ **65.** $x^{1/5}(x^{1/5} - 3)$ **67.** $x^{-1/3}(5 + x)$ **69.** $\sqrt{x}$ **71.** $\sqrt[3]{2}$ **73.** $2\sqrt{x}$ **75.** $\sqrt{xy}$ **77.** $\sqrt[15]{y^{11}}$

79. $\sqrt[12]{b^5}$ **81.** $\sqrt{a}$ **83.** $\sqrt[3]{432}$ **85.** $\sqrt[15]{343y^5}$ **87.** $25 \cdot 3$ **89.** $16 \cdot 3$ or $4 \cdot 12$ **91.** $8 \cdot 2$ **93.** $27 \cdot 2$ **95.** 1509 calories

97. 97.2 million **99.** $a^{1/3}$ **101.** $x^{1/5}$ **103.** 1.6818 **105.** 5.6645 **107.** $\dfrac{t^{1/2}}{u^{1/2}}$

Exercise Set 7.3 **1.** $\sqrt{14}$ **3.** 2 **5.** $\sqrt[3]{36}$ **7.** $\sqrt{6x}$ **9.** $\sqrt{\dfrac{14}{xy}}$ **11.** $\sqrt[4]{20x^3}$ **13.** $\dfrac{\sqrt{6}}{7}$ **15.** $\dfrac{\sqrt{2}}{7}$ **17.** $\dfrac{\sqrt[4]{x^3}}{2}$ **19.** $\dfrac{\sqrt[3]{4}}{3}$

21. $\dfrac{\sqrt[4]{8}}{x^2}$ **23.** $\dfrac{\sqrt[3]{2x}}{3y^4\sqrt[3]{3}}$ **25.** $\dfrac{x\sqrt{y}}{10}$ **27.** $\dfrac{\sqrt{5x}}{2y}$ **29.** $-\dfrac{z^2\sqrt[3]{z}}{3x}$ **31.** $4\sqrt{2}$ **33.** $4\sqrt[3]{3}$ **35.** $25\sqrt{3}$ **37.** $2\sqrt{6}$ **39.** $10x^2\sqrt{x}$

41. $2y^2\sqrt[3]{2y}$ **43.** $a^2b\sqrt[4]{b^3}$ **45.** $y^2\sqrt{y}$ **47.** $5ab\sqrt{b}$ **49.** $-2x^2\sqrt[5]{y}$ **51.** $x^4\sqrt[3]{50x^2}$ **53.** $-4a^4b^3\sqrt{2b}$ **55.** $3x^3y^4\sqrt{xy}$ **57.** $5r^3s^4$

59. $\sqrt{2}$ **61.** 2 **63.** 10 **65.** x^2y **67.** $24m^2$ **69.** $\dfrac{15x\sqrt{2x}}{2}$ or $\dfrac{15x}{2}\sqrt{2x}$ **71.** $2a^2\sqrt[4]{2}$ **73.** $14x$ **75.** $2x^2 - 7x - 15$ **77.** y^2

79. $-3x - 15$ **81.** $x^2 - 8x - 16$ **83. a.** 20π sq. cm **b.** 211.57 sq. ft **85. a.** 3.8 times **b.** 2.9 times **c.** answers may vary

Mental Math **1.** $6\sqrt{3}$ **3.** $3\sqrt{x}$ **5.** $12\sqrt[3]{x}$ **7.** $2\sqrt{11}$ **9.** $8\sqrt{13}$ **11.** $10\sqrt[3]{2x}$

Exercise Set 7.4 **1.** $-2\sqrt{2}$ **3.** $10x\sqrt{2x}$ **5.** $17\sqrt{2} - 15\sqrt{5}$ **7.** $-\sqrt[3]{2x}$ **9.** $5b\sqrt{b}$ **11.** $\dfrac{31\sqrt{2}}{15}$ **13.** $\dfrac{\sqrt[3]{11}}{3}$ **15.** $\dfrac{5\sqrt{5x}}{9}$

17. $14 + \sqrt{3}$ **19.** $7 - 3y$ **21.** $6\sqrt{3} - 6\sqrt{2}$ **23.** $-23\sqrt[3]{5}$ **25.** $2b\sqrt{b}$ **27.** $20y\sqrt{2y}$ **29.** $2y\sqrt[3]{2x}$ **31.** $6\sqrt[3]{11} - 4\sqrt{11}$

33. $4x\sqrt[4]{x^3}$ **35.** $\dfrac{2\sqrt{3}}{3}$ **37.** $\dfrac{5x\sqrt[3]{x}}{7}$ **39.** $\dfrac{5\sqrt{7}}{2x}$ **41.** $\dfrac{\sqrt[3]{2}}{6}$ **43.** $\dfrac{14x\sqrt[3]{2x}}{9}$ **45.** $15\sqrt{3}$ in. **47.** $\sqrt{35} + \sqrt{21}$ **49.** $7 - 2\sqrt{10}$

51. $3\sqrt{x} - x\sqrt{3}$ **53.** $6x - 13\sqrt{x} - 5$ **55.** $\sqrt[3]{a^2} + \sqrt[3]{a} - 20$ **57.** $6\sqrt{2} - 12$ **59.** $2 + 2x\sqrt{3}$ **61.** $-16 - \sqrt{35}$ **63.** $x - y^2$

65. $3 + 2x\sqrt{3} + x^2$ **67.** $5x - 3\sqrt{15x} - 3\sqrt{10x} + 9\sqrt{6}$ **69.** $2\sqrt[3]{2} - \sqrt[3]{4}$ **71.** $-4\sqrt[6]{x^5} + \sqrt[3]{x^2} + 8\sqrt[3]{x} - 4\sqrt{x} + 7$

73. $x + 24 + 10\sqrt{x - 1}$ **75.** $2x + 6 - 2\sqrt{2x + 5}$ **77.** $x - 7$ **79.** $\dfrac{7}{x + y}$ **81.** $2a - 3$ **83.** $\dfrac{-2 + \sqrt{3}}{3}$ **85.** $22\sqrt{5}$ ft; 150 sq. ft

87. a. $2\sqrt{3}$ **b.** 3 **c.** answers may vary **89.** answers may vary

Mental Math **1.** $\sqrt{2} - x$ **3.** $5 + \sqrt{a}$ **5.** $7\sqrt{5} - 8\sqrt{x}$

Exercise Set 7.5 **1.** $\dfrac{\sqrt{14}}{7}$ **3.** $\dfrac{\sqrt{5}}{5}$ **5.** $\dfrac{\sqrt[3]{6}}{2}$ **7.** $\dfrac{4\sqrt[3]{9}}{3}$ **9.** $\dfrac{3\sqrt{2x}}{4x}$ **11.** $\dfrac{3\sqrt[3]{2x}}{2x}$ **13.** $\dfrac{2\sqrt{x}}{x}$ **15.** $\dfrac{3\sqrt{3a}}{a}$ **17.** $\dfrac{3\sqrt[3]{4}}{2}$

19. $\dfrac{2\sqrt{21}}{7}$ **21.** $\dfrac{\sqrt{10xy}}{5y}$ **23.** $\dfrac{3\sqrt[4]{2}}{2}$ **25.** $\dfrac{2\sqrt[4]{9x}}{3x^2}$ **27.** $\dfrac{5a\sqrt[5]{4ab^4}}{2a^2b^3}$ **29.** $-2(2 + \sqrt{7})$ **31.** $\dfrac{7(3 + \sqrt{x})}{9 - x}$ **33.** $-5 + 2\sqrt{6}$

35. $\dfrac{2a + 2\sqrt{a} + \sqrt{ab} + \sqrt{b}}{4a - b}$ **37.** $-\dfrac{8(1 - \sqrt{10})}{9}$ **39.** $\dfrac{x - \sqrt{xy}}{x - y}$ **41.** $\dfrac{5 + 3\sqrt{2}}{7}$ **43.** $\dfrac{5}{\sqrt{15}}$ **45.** $\dfrac{6}{\sqrt{10}}$ **47.** $\dfrac{2x}{7\sqrt{x}}$

49. $\dfrac{5y}{\sqrt[3]{100xy}}$ **51.** $\dfrac{2}{\sqrt{10}}$ **53.** $\dfrac{2x}{11\sqrt{2x}}$ **55.** $\dfrac{7}{2\sqrt[3]{49}}$ **57.** $\dfrac{3x^2}{10\sqrt[3]{9x}}$ **59.** $\dfrac{6x^2y^3}{\sqrt{6z}}$ **61.** answers may vary **63.** $\dfrac{-7}{12 + 6\sqrt{11}}$

65. $\dfrac{3}{10 + 5\sqrt{7}}$ **67.** $\dfrac{x - 9}{x - 3\sqrt{x}}$ **69.** $\dfrac{1}{3 + 2\sqrt{2}}$ **71.** $\dfrac{x - 1}{x - 2\sqrt{x} + 1}$ **73.** 5 **75.** $-\dfrac{1}{2}, 6$ **77.** 2, 6 **79.** $r = \dfrac{\sqrt{A\pi}}{2\pi}$

81. answers may vary

Integrated Review **1.** 9 **2.** -2 **3.** $\dfrac{1}{2}$ **4.** x^3 **5.** y^3 **6.** $2y^5$ **7.** $-2y$ **8.** $3b^3$ **9.** 6 **10.** $\sqrt[4]{3y}$ **11.** $\dfrac{1}{16}$

12. $\sqrt[5]{(x + 1)^3}$ **13.** y **14.** $16x^{1/2}$ **15.** $x^{5/4}$ **16.** $4^{11/15}$ **17.** $2x^2$ **18.** $\sqrt[4]{a^3b^2}$ **19.** $\sqrt[4]{x^3}$ **20.** $\sqrt[6]{500}$ **21.** $2\sqrt{10}$

22. $2xy^2\sqrt[4]{x^3y^2}$ **23.** $3x\sqrt[3]{2x}$ **24.** $-2b^2\sqrt[5]{2}$ **25.** $\sqrt{5x}$ **26.** $4x$ **27.** $7y^2\sqrt{y}$ **28.** $2a^2\sqrt[4]{3}$ **29.** $2\sqrt{5} - 5\sqrt{3} + 5\sqrt{7}$

30. $y\sqrt[3]{2y}$ **31.** $\sqrt{15} - \sqrt{6}$ **32.** $10 + 2\sqrt{21}$ **33.** $4x^2 - 5$ **34.** $x + 2 - 2\sqrt{x + 1}$ **35.** $\dfrac{\sqrt{21}}{3}$ **36.** $\dfrac{5\sqrt[3]{4x}}{2x}$ **37.** $\dfrac{13 - 3\sqrt{21}}{5}$

38. $\dfrac{7}{\sqrt{21}}$ **39.** $\dfrac{3y}{\sqrt[3]{33y^2}}$ **40.** $\dfrac{x - 4}{x + 2\sqrt{x}}$

Exercise Set 7.6 **1.** 8 **3.** 7 **5.** $\varnothing$ **7.** 7 **9.** 6 **11.** $-\dfrac{9}{2}$ **13.** 29 **15.** 4 **17.** -4 **19.** $\varnothing$ **21.** 7 **23.** 9 **25.** 50

27. $\varnothing$ **29.** $\dfrac{15}{4}$ **31.** 13 **33.** 5 **35.** -12 **37.** 9 **39.** -3 **41.** 1 **43.** 1 **45.** $\dfrac{1}{2}$ **47.** 0, 4 **49.** $\dfrac{37}{4}$ **51.** $3\sqrt{5}$ ft

53. $2\sqrt{10}$ m **55.** $2\sqrt{131}$ m ≈ 22.9 m **57.** $\sqrt{100.84}$ mm ≈ 10.0 mm **59.** 17 ft **61.** 13 ft **63.** 14,657,415 sq. mi **65.** 100 ft

67. 100 **69.** $\dfrac{\pi}{2}$ sec ≈ 1.57 sec **71.** 12.97 ft **73.** answers may vary **75.** $15\sqrt{3}$ sq. mi ≈ 25.98 sq. mi **77.** answers may vary

79. 0.51 km **81.** function **83.** function **85.** not a function **87.** $\dfrac{x}{4x + 3}$ **89.** $-\dfrac{4z + 2}{3z}$ **91.** 1 **93.** 2743 deliveries

95. $-1, 0, 8, 9$ **97.** $-1, 4$

Mental Math **1.** $9i$ **3.** $i\sqrt{7}$ **5.** -4 **7.** $8i$

Exercise Set 7.7 **1.** $2i\sqrt{6}$ **3.** $-6i$ **5.** $24i\sqrt{7}$ **7.** $-3\sqrt{6}$ **9.** $-\sqrt{14}$ **11.** $-5\sqrt{2}$ **13.** $4i$ **15.** $i\sqrt{3}$ **17.** $2\sqrt{2}$

19. $6 - 4i$ **21.** $-2 + 6i$ **23.** $-2 - 4i$ **25.** $18 + 12i$ **27.** 7 **29.** $12 - 16i$ **31.** $-4i$ **33.** $\dfrac{28}{25} - \dfrac{21}{25}i$ **35.** $4 + i$

37. $\dfrac{17}{13} + \dfrac{7}{13}i$ **39.** 63 **41.** $2 - i$ **43.** 20 **45.** 10 **47.** 2 **49.** $-5 + \dfrac{16}{3}i$ **51.** $17 + 144i$ **53.** $\dfrac{3}{5} - \dfrac{1}{5}i$ **55.** $5 - 10i$

57. $\dfrac{1}{5} - \dfrac{8}{5}i$ **59.** $8 - i$ **61.** 1 **63.** i **65.** $-i$ **67.** -1 **69.** -64 **71.** $-243i$ **73.** $40°$ **75.** $x^2 - 5x - 2 - \dfrac{6}{x - 1}$

77. 5 people **79.** 14 people **81.** 16.7% **83.** $1 - i$ **85.** 0 **87.** $2 + 3i$ **89.** $2 + i\sqrt{2}$ **91.** $\dfrac{1}{2} - \dfrac{\sqrt{3}}{2}i$

93. answers may vary **95.** $6 - 6i$ **97.** yes

Chapter 7 Review **1.** 9 **3.** -2 **5.** $-\dfrac{1}{7}$ **7.** -6 **9.** $-a^2b^3$ **11.** $2ab^2$ **13.** $\dfrac{x^6}{6y}$ **15.** $|-x|$ **17.** -27 **19.** $-x$

21. $5|(x - y)^5|$ **23.** $-x$ **25.** $(-\infty, \infty); -2, -1, 0, 1, 2$ **27.** $-\dfrac{1}{3}$ **29.** $-\dfrac{1}{4}$ **31.** $\dfrac{1}{4}$ **33.** $\dfrac{343}{125}$ **35.** not a real number

37. $5^{1/5}x^{2/5}y^{3/5}$ **39.** $5\sqrt[3]{xy^2z^5}$ **41.** $a^{13/6}$ **43.** $\dfrac{1}{a^{9/2}}$ **45.** a^4b^6

47. $\dfrac{b^{5/6}}{49a^{1/4}c^{5/3}}$ **49.** 4.472 **51.** 5.191 **53.** -26.246 **55.** $\sqrt[6]{1372}$

57. $2\sqrt{6}$ **59.** $2x$ **61.** $2\sqrt{15}$ **63.** $3\sqrt[3]{6}$ **65.** $6x^3\sqrt{x}$ **67.** $\dfrac{p^8\sqrt{p}}{11}$

69. $\dfrac{y\sqrt[4]{xy^2}}{3}$ **71. a.** $\dfrac{5}{\sqrt{\pi}}$ m or $\dfrac{5\sqrt{\pi}}{\pi}$ m **b.** 5.75 in. **73.** $xy\sqrt{2y}$

75. $3a\sqrt[4]{2a}$ **77.** $\dfrac{3\sqrt{2}}{4x}$ **79.** $-4ab\sqrt[4]{2b}$ **81.** $x - 6\sqrt{x} + 9$ **83.** $4x - 9y$ **85.** $\sqrt[3]{a^2} + 4\sqrt[3]{a} + 4$ **87.** $a + 64$ **89.** $\dfrac{\sqrt{3x}}{6}$

91. $\dfrac{2x^2\sqrt{2x}}{y}$ **93.** $-\dfrac{10 + 5\sqrt{7}}{3}$ **95.** $-5 + 2\sqrt{6}$ **97.** $\dfrac{6}{\sqrt{2y}}$ **99.** $\dfrac{4x^3}{y\sqrt{2x}}$ **101.** $\dfrac{x - 25}{-3\sqrt{x} + 15}$ **103.** $\varnothing$ **105.** $\varnothing$ **107.** 16

109. $\sqrt{241}$ **111.** 4.24 ft **113.** $-i\sqrt{6}$ **115.** $-\sqrt{10}$ **117.** $-13 - 3i$ **119.** $10 + 4i$ **121.** $1 + 5i$ **123.** 87 **125.** $-\dfrac{1}{3} + \dfrac{1}{3}i$

Chapter 7 Test   **1.** $6\sqrt{6}$ **2.** $-x^{16}$ **3.** $\dfrac{1}{5}$ **4.** 5   **5.** $\dfrac{4x^2}{9}$ **6.** $-a^6b^3$ **7.** $\dfrac{8a^{1/3}c^{2/3}}{b^{5/12}}$ **8.** $a^{7/12} - a^{7/3}$ **9.** $|4xy|$ or $4|xy|$

10. -27 **11.** $\dfrac{3\sqrt{y}}{y}$ **12.** $\dfrac{8 - 6\sqrt{x} + x}{8 - 2x}$ **13.** $\dfrac{\sqrt[3]{b^2}}{b}$ **14.** $\dfrac{6 - x^2}{8(\sqrt{6} - x)}$ **15.** $-x\sqrt{5x}$ **16.** $4\sqrt{3} - \sqrt{6}$ **17.** $x + 2\sqrt{x} + 1$

18. $\sqrt{6} - 4\sqrt{3} + \sqrt{2} - 4$ **19.** -20 **20.** 23.685   **21.** 0.019   **22.** 2, 3   **23.** $\varnothing$ **24.** 6   **25.** $i\sqrt{2}$ **26.** $-2i\sqrt{2}$ **27.** $-3i$

28. 40   **29.** $7 + 24i$ **30.** $-\dfrac{3}{2} + \dfrac{5}{2}i$ **31.** $\dfrac{5\sqrt{2}}{2}$ **32.** $[-2, \infty)$;   **33.** 27 mph   **34.** 360 ft

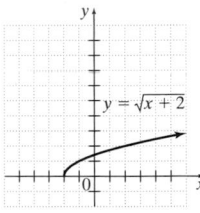

Chapter 7 Cumulative Review   **1. a.** $2xy - 2$ **b.** $2x^2 + 23$ **c.** $3.1x - 0.3$ **d.** $-a - 7b + \dfrac{7}{12}$; Sec. 1.4, Ex. 13   **2. a.** $7x - 3$

b. $-3x + 11$ **c.** $6x - 14$; Sec. 1.4   **3.** $\dfrac{21}{11}$; Sec. 1.5, Ex. 6   **4.** $\dfrac{13}{14}$; Sec. 1.5   **5.** \$4500 per month; Sec. 3.2, Ex. 8   **6.** 54 mph; Sec. 1.6

7. $\varnothing$; Sec. 3.4, Ex. 6   **8.** $\dfrac{2}{3}$; Sec. 3.4   **9.** $(-\infty, -3] \cup [9, \infty)$; Sec. 3.5, Ex. 7   **10.** 2; Sec. 3.5

11.   ; Sec. 2.1, Ex. 9   **12.**   ; Sec. 2.1   **13. a.** domain: $\{2, 0, 3\}$; range: $\{3, 4, -1\}$

b. domain: $\{-4, -3, -2, -1, 0, 1, 2, 3\}$; range: $\{1\}$

c. domain: {Erie, Escondido, Gary, Miami, Waco}; range: $\{104, 109, 117, 359\}$; Sec. 2.2, Ex. 1   **14. a.** domain: $(-\infty, 0]$; range: $(-\infty, \infty)$; not a function   **b.** domain: $(-\infty, \infty)$; range: $(-\infty, \infty)$; function   **c.** domain: $(-\infty, -2] \cup [2, \infty)$; range: $(-\infty, \infty)$; not a function; Sec. 2.2

15.   ; Sec. 2.3, Ex. 6   **16.**   ; Sec. 2.3   **17.** undefined; Sec. 2.4, Ex. 5   **18.** 0; Sec. 2.4

19. $\left(-\dfrac{21}{10}, \dfrac{3}{10}\right)$; Sec. 4.1, Ex. 5   **20.** $(6, 0)$; Sec. 4.1

21. a. 2^7 **b.** x^{10} **c.** y^7; Sec. 5.1, Ex. 1   **22.** 6 shirts; 3 shorts; Sec. 4.3   **23.** 6×10^{-5}; Sec. 5.2, Ex. 7   **24.** 5×10^{-2}; Sec. 5.2   **25. a.** -4 **b.** 11; Sec. 5.3, Ex. 4   **26.** $9x^2 + 8$; Sec. 5.3   **27. a.** $2x^2 + 11x + 15$ **b.** $10x^3 - 27x^2 + 32x - 21$;

Sec. 5.4, Ex. 3   **28. a.** $3y^2 - 2y - 8$ **b.** $6y^3 + 7y^2 + 6y + 1$; Sec. 5.4   **29.** $5x^2$; Sec. 5.5, Ex. 1   **30.** $(x - 1)(x^2 + 4)$; Sec. 5.5

31. a. $x^2 - 2x + 4$ **b.** $\dfrac{2}{y - 5}$; Sec. 6.1, Ex. 5   **32. a.** $-a^2 - 2a - 4$ **b.** $\dfrac{3}{a + 5}$; Sec. 6.1   **33. a.** $\dfrac{6x + 5}{3x^3y}$ **b.** $\dfrac{5x^2 - 2x}{(x + 2)(x - 2)}$

c. $\dfrac{x + 4}{x - 1}$; Sec. 6.2, Ex. 3   **34. a.** $\dfrac{9x - 2y}{3x^2y^2}$ **b.** $\dfrac{3x^2 - 21x}{(x + 3)(x - 3)}$ **c.** $\dfrac{x + 5}{x - 2}$; Sec. 6.2   **35. a.** $\dfrac{x(x - 2)}{2(x + 2)}$ **b.** $\dfrac{x^2}{y^2}$; Sec. 6.3, Ex. 2

36. a. $\dfrac{3(y - 2)}{4(2y + 3)}$ **b.** $\dfrac{x + 4}{16}$; Sec. 6.3   **37.** $2x^2 - x + 4$; Sec. 6.4, Ex. 1   **38.** $x^2 + 3$; Sec. 6.4   **39.** $2x^2 + 5x + 2 + \dfrac{7}{x - 3}$; Sec. 6.5, Ex. 1

40. $4y^2 - 1 + \dfrac{9}{y - 3}$ **41.** $\varnothing$; Sec. 6.6, Ex. 3   **42.** $-1; -5$; Sec. 6.6   **43.** $x = \dfrac{yz}{y - z}$; Sec. 6.7, Ex. 1   **44.** $a = \dfrac{2A - bh}{h}$; Sec. 6.7

45. constant of variation: 15; $u = \dfrac{15}{w}$; Sec. 6.8, Ex. 3   **46.** $k = 0.17$; $y = 0.17x$; Sec. 6.8   **47. a.** $\dfrac{1}{8}$ **b.** $\dfrac{1}{9}$; Sec. 7.2, Ex. 3   **48. a.** $\dfrac{1}{27}$ **b.** $\dfrac{1}{25}$;

Sec. 7.2   **49.** $\dfrac{x - 4}{5(\sqrt{x} - 2)}$; Sec. 7.5, Ex. 7   **50. a.** $-4a\sqrt{a}$ **b.** $3b\sqrt[3]{2a}$ **c.** $\dfrac{11\sqrt[3]{3}}{10}$; Sec. 7.4

CHAPTER 8 QUADRATIC EQUATIONS AND FUNCTIONS

Exercise Set 8.1   **1.** $-4, 4$ **3.** $-\sqrt{7}, \sqrt{7}$ **5.** $-3\sqrt{2}, 3\sqrt{2}$ **7.** $-\sqrt{10}, \sqrt{10}$ **9.** $-8, -2$ **11.** $6 - 3\sqrt{2}, 6 + 3\sqrt{2}$

13. $\dfrac{3 - 2\sqrt{2}}{2}, \dfrac{3 + 2\sqrt{2}}{2}$ **15.** $-3i, 3i$ **17.** $-\sqrt{6}, \sqrt{6}$ **19.** $-2i\sqrt{2}, 2i\sqrt{2}$ **21.** $1 - 4i, 1 + 4i$ **23.** $-7 - \sqrt{5}, -7 + \sqrt{5}$

25. $-3 - 2i\sqrt{2}, -3 + 2i\sqrt{2}$ **27.** $x^2 + 16x + 64 = (x + 8)^2$ **29.** $z^2 - 12z + 36 = (z - 6)^2$ **31.** $p^2 + 9p + \dfrac{81}{4} = \left(p + \dfrac{9}{2}\right)^2$

33. $x^2 + x + \dfrac{1}{4} = \left(x + \dfrac{1}{2}\right)^2$ **35.** $-5, -3$ **37.** $-3 - \sqrt{7}, -3 + \sqrt{7}$ **39.** $\dfrac{-1 - \sqrt{5}}{2}, \dfrac{-1 + \sqrt{5}}{2}$ **41.** $-1 - \sqrt{6}, -1 + \sqrt{6}$

43. $\dfrac{6 - \sqrt{30}}{3}, \dfrac{6 + \sqrt{30}}{3}$ **45.** $\dfrac{3 - \sqrt{11}}{2}, \dfrac{3 + \sqrt{11}}{2}$ **47.** $-4, \dfrac{1}{2}$ **49.** $-1, 5$ **51.** $-4 - \sqrt{15}, -4 + \sqrt{15}$ **53.** $\dfrac{-3 - \sqrt{21}}{3}, \dfrac{-3 + \sqrt{21}}{3}$

55. $-1, \dfrac{5}{2}$ **57.** $-1 - i, -1 + i$ **59.** $3 - \sqrt{6}, 3 + \sqrt{6}$ **61.** $-2 - i\sqrt{2}, -2 + i\sqrt{2}$ **63.** $\dfrac{-15 - 7\sqrt{5}}{10}, \dfrac{-15 + 7\sqrt{5}}{10}$

65. $\dfrac{1 - i\sqrt{47}}{4}, \dfrac{1 + i\sqrt{47}}{4}$ **67.** $-5 - i\sqrt{3}, -5 + i\sqrt{3}$ **69.** $-4, 1$ **71.** $\dfrac{2 - i\sqrt{2}}{2}, \dfrac{2 + i\sqrt{2}}{2}$ **73.** $\dfrac{-3 - \sqrt{69}}{6}, \dfrac{-3 + \sqrt{69}}{6}$

75. 2 real solutions   **77.** no real solutions   **79.** 20%   **81.** $11\frac{1}{9}\%$ **83.** answers may vary   **85.** simple   **87.** $\dfrac{7}{5}$ **89.** $\dfrac{1}{5}$

91. $5 - 10\sqrt{3}$ **93.** $\dfrac{3 - 2\sqrt{7}}{4}$ **95.** $2\sqrt{7}$ **97.** $\sqrt{13}$ **99.** $-6y, 6y$ **101.** $-x, x$ **103.** 8.11 sec   **105.** 6.73 sec   **107.** 6 in.

109. 16.2×21.6 in.   **111.** 2.828 thousand units or 2828 units

Mental Math   1. $a = 1, b = 3, c = 1$ **3.** $a = 7, b = 0, c = -4$ **5.** $a = 6, b = -1, c = 0$

Exercise Set 8.2   1. $-6, 1$ **3.** $-\dfrac{3}{5}, 1$ **5.** 3 **7.** $\dfrac{-7 - \sqrt{33}}{2}, \dfrac{-7 + \sqrt{33}}{2}$ **9.** $\dfrac{1 - \sqrt{57}}{8}, \dfrac{1 + \sqrt{57}}{8}$ **11.** $\dfrac{7 - \sqrt{85}}{6}, \dfrac{7 + \sqrt{85}}{6}$

13. $1 - \sqrt{3}, 1 + \sqrt{3}$ **15.** $-\dfrac{3}{2}, 1$ **17.** $\dfrac{3 - \sqrt{11}}{2}, \dfrac{3 + \sqrt{11}}{2}$ **19.** answers may vary   **21.** $\dfrac{3 - i\sqrt{87}}{8}, \dfrac{3 + i\sqrt{87}}{8}$

23. $-2 - \sqrt{11}, -2 + \sqrt{11}$ **25.** $\dfrac{-5 - i\sqrt{5}}{10}, \dfrac{-5 + i\sqrt{5}}{10}$ **27.** two real solutions   **29.** one real solution   **31.** two real solutions

33. two complex but not real solutions   **35.** $\dfrac{-5 - \sqrt{17}}{2}, \dfrac{-5 + \sqrt{17}}{2}$ **37.** $\dfrac{5}{2}, 1$ **39.** $\dfrac{3 - \sqrt{29}}{2}, \dfrac{3 + \sqrt{29}}{2}$ **41.** $\dfrac{-1 - \sqrt{19}}{6}, \dfrac{-1 + \sqrt{19}}{6}$

43. $-3 - 2i, -3 + 2i$ **45.** $\dfrac{-1 - i\sqrt{23}}{4}, \dfrac{-1 + i\sqrt{23}}{4}$ **47.** 1 **49.** $\dfrac{19 - \sqrt{345}}{2}, \dfrac{19 + \sqrt{345}}{2}$ **51. a.** 2 real solutions   **b.** 1 real solution

53. 14 ft   **55.** $2 + 2\sqrt{2}$ cm, $2 + 2\sqrt{2}$ cm, $4 + 2\sqrt{2}$ cm   **57.** width: $-5 + 5\sqrt{17}$ ft; length: $5 + 5\sqrt{17}$ ft   **59. a.** $50\sqrt{2}$ m   **b.** 5000 sq. m

61. 37.4 ft by 38.5 ft   **63.** $\dfrac{1 + \sqrt{5}}{2}$ **65.** 8.9 sec   **67.** 2.8 sec   **69.** $\dfrac{11}{5}$ **71.** 15   **73.** $(x^2 + 5)(x + 2)(x - 2)$

75. $(z + 3)(z - 3)(z + 2)(z - 2)$ **77.** $0.6, 2.4$ **79.** Sunday to Monday.   **81.** Wednesday   **83.** 32; yes   **85. a.** $6901.4 million   **b.** 2007

87. answers may vary   **89.** $\dfrac{\sqrt{3}}{3}$ **91.** $\dfrac{-\sqrt{2} - i\sqrt{2}}{2}, \dfrac{-\sqrt{2} + i\sqrt{2}}{2}$ **93.** $\dfrac{\sqrt{3} - \sqrt{11}}{4}, \dfrac{\sqrt{3} + \sqrt{11}}{4}$

Exercise Set 8.3   1. 2 **3.** 16 **5.** $1, 4$ **7.** $3 - \sqrt{7}, 3 + \sqrt{7}$ **9.** $\dfrac{3 - \sqrt{57}}{4}, \dfrac{3 + \sqrt{57}}{4}$ **11.** $\dfrac{1 - \sqrt{29}}{2}, \dfrac{1 + \sqrt{29}}{2}$

13. $-2, 2, -2i, 2i$ **15.** $-\dfrac{1}{2}, \dfrac{1}{2}, -i\sqrt{3}, i\sqrt{3}$ **17.** $-3, 3, -2, 2$ **19.** $125, -8$ **21.** $-\dfrac{4}{5}, 0$ **23.** $-\dfrac{1}{8}, 27$ **25.** $-\dfrac{2}{3}, \dfrac{4}{3}$ **27.** $-\dfrac{1}{125}, \dfrac{1}{8}$

29. $-\sqrt{2}, \sqrt{2}, -\sqrt{3}, \sqrt{3}$ **31.** $\dfrac{-9 - \sqrt{201}}{6}, \dfrac{-9 + \sqrt{201}}{6}$ **33.** $2, 3$ **35.** 3 **37.** $27, 125$ **39.** $1, -3i, 3i$ **41.** $\dfrac{1}{8}, -8$ **43.** $-\dfrac{1}{2}, \dfrac{1}{3}$

45. 4 **47.** -3 **49.** $-\sqrt{5}, \sqrt{5}, -2i, 2i$ **51.** $-3, \dfrac{3 - 3i\sqrt{3}}{2}, \dfrac{3 + 3i\sqrt{3}}{2}$ **53.** $6, 12$ **55.** $-\dfrac{1}{3}, \dfrac{1}{3}, -\dfrac{i\sqrt{6}}{3}, \dfrac{i\sqrt{6}}{3}$ **57.** 5 mph, then 4 mph

59. inlet pipe, 15.5 hr; hose, 16.5 hr   **61.** 55 mph, 66 mph   **63.** 8.5 hr.   **65.** 12 or -8 **67. a.** $(x - 6)$ in.   **b.** $300 = (x - 6) \cdot (x - 6) \cdot 3$
c. 16 cm by 16 cm   **69.** 22 feet   **71.** $(-\infty, 3]$ **73.** $(-5, \infty)$ **75.** domain: $\{x \mid x$ is a real number$\}$ or $(-\infty, \infty)$; range:
$\{y \mid y$ is a real number$\}$ or $(-\infty, \infty)$; function   **77.** domain: $\{x \mid x$ is a real number$\}$ or $(-\infty, \infty)$; range: $\{y \mid y \geq -1\}$ or $[-1, \infty)$; function
79. answers may vary   **81. a.** 134.05 ft per sec   **b.** 134.93 ft per sec   **c.** Fernandez: 92.0 mph; Dominguez: 91.4 mph

Integrated Review   1. $\{-\sqrt{10}, \sqrt{10}\}$ **2.** $\{-\sqrt{14}, \sqrt{14}\}$ **3.** $\{1 - 2\sqrt{2}, 1 + 2\sqrt{2}\}$ **4.** $\{-5 - 2\sqrt{3}, -5 + 2\sqrt{3}\}$

5. $\{-1 - \sqrt{13}, -1 + \sqrt{13}\}$ **6.** $\{1, 11\}$ **7.** $\left\{\dfrac{-3 - \sqrt{69}}{6}, \dfrac{-3 + \sqrt{69}}{6}\right\}$ **8.** $\left\{\dfrac{-2 - \sqrt{5}}{4}, \dfrac{-2 + \sqrt{5}}{4}\right\}$ **9.** $\left\{\dfrac{2 - \sqrt{2}}{2}, \dfrac{2 + \sqrt{2}}{2}\right\}$

10. $\{-3 - \sqrt{5}, -3 + \sqrt{5}\}$ **11.** $\{-2 + i\sqrt{3}, -2 - i\sqrt{3}\}$ **12.** $\left\{\dfrac{-1 - i\sqrt{11}}{2}, \dfrac{-1 + i\sqrt{11}}{2}\right\}$ **13.** $\left\{\dfrac{-3 + i\sqrt{15}}{2}, \dfrac{-3 - i\sqrt{15}}{2}\right\}$

14. $\{3i, -3i\}$ **15.** $\{0, -17\}$ **16.** $\left\{\dfrac{1 + \sqrt{13}}{4}, \dfrac{1 - \sqrt{13}}{4}\right\}$ **17.** $\{2 + 3\sqrt{3}, 2 - 3\sqrt{3}\}$ **18.** $\{2 + \sqrt{3}, 2 - \sqrt{3}\}$ **19.** $\left\{-2, \dfrac{4}{3}\right\}$

20. $\left\{\dfrac{-5 + \sqrt{17}}{4}, \dfrac{-5 - \sqrt{17}}{4}\right\}$ **21.** $\{1 - \sqrt{6}, 1 + \sqrt{6}\}$ **22.** $\{-\sqrt{31}, \sqrt{31}\}$ **23.** $\{-\sqrt{11}, \sqrt{11}\}$ **24.** $\{-i\sqrt{11}, i\sqrt{11}\}$

25. $\{-11, 6\}$ **26.** $\left\{\dfrac{-3 + \sqrt{19}}{5}, \dfrac{-3 - \sqrt{19}}{5}\right\}$ **27.** $\left\{\dfrac{-3 + \sqrt{17}}{4}, \dfrac{-3 - \sqrt{17}}{4}\right\}$ **28.** $10\sqrt{2}$ ft ≈ 14.1 ft   **29.** Jack: 9.1 hr; Lucy: 7.1 hr

30. 5 mph during the first part, then 6 mph

Exercise Set 8.4   **1.** $(-\infty, -5) \cup (-1, \infty)$ **3.** $[-4, 3]$ **5.** $[2, 5]$

7. $\left(-5, -\dfrac{1}{3}\right)$ **9.** $(2, 4) \cup (6, \infty)$ **11.** $(-\infty, -4] \cup [0, 1]$

13. $[-3, -2] \cup [2, 3]$ **15.** $(-7, 2)$ **17.** $(-1, \infty)$

19. $(-\infty, -1] \cup (4, \infty)$ **21.** $(-\infty, 2) \cup \left(\dfrac{11}{4}, \infty\right)$ **23.** $(0, 2] \cup [3, \infty)$

25. $(-\infty, -7) \cup (8, \infty)$ **27.** $\left[-\dfrac{5}{4}, \dfrac{3}{2}\right]$ **29.** $(-\infty, 0) \cup (1, \infty)$ **31.** $(-\infty, -4] \cup [4, 6]$

33. $\left(-\infty, -\dfrac{2}{3}\right] \cup \left[\dfrac{3}{2}, \infty\right)$ **35.** $\left(-4, -\dfrac{3}{2}\right) \cup \left(\dfrac{3}{2}, \infty\right)$

37. $(-\infty, -5] \cup [-1, 1] \cup [5, \infty)$ **39.** $\left(-\infty, -\dfrac{5}{3}\right) \cup \left(\dfrac{7}{2}, \infty\right)$ **41.** $(0, 10)$

43. $(-\infty, -4) \cup [5, \infty)$ **45.** $(-\infty, -6] \cup (-1, 0] \cup (7, \infty)$ **47.** $(-\infty, 1) \cup (2, \infty)$

49. $(-\infty, -8] \cup (-4, \infty)$ **51.** $(-\infty, 0] \cup \left(5, \dfrac{11}{2}\right]$ **53.** $(0, \infty)$

55.   **57.**   **59.**   **61.**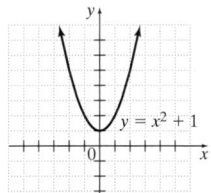

63. answers may vary   **65.** any number less than -1 or between 0 and 1   **67.** x is between 2 and 11

69.   **71.**

Mental Math   **1.** $(0, 0)$ **3.** $(2, 0)$ **5.** $(0, 3)$ **7.** $(-1, 5)$

Exercise Set 8.5 **1.**

3.

5.

7.

9.

11.

13.

15.

17.

19.

21.

23.

25.

27.

29.

31.

33.

35.

37.

39.

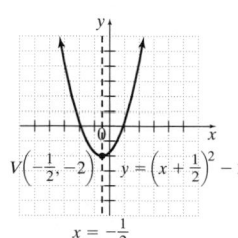

41. $y = \frac{3}{2}(x + 7)^2 + 1$

43.

45.

47.

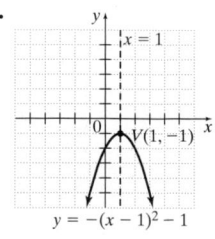

49. $y = \sqrt{3}(x + 5)^2 + \frac{3}{4}$

51.

53.

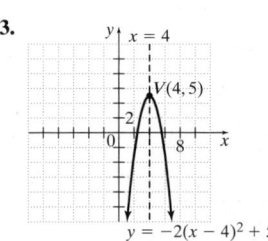

55. $x^2 + 8x + 16$ **57.** $z^2 - 16z + 64$ **59.** $y^2 + y + \dfrac{1}{4}$ **61.** $-6, 2$ **63.** $-5 - \sqrt{26}, -5 + \sqrt{26}$

65. $4 - 3\sqrt{2}, 4 + 3\sqrt{2}$ **67.** $f(x) = 5(x - 2)^2 + 3$ **69.** $f(x) = 5(x + 3)^2 + 6$

71.   **73.**   **75.**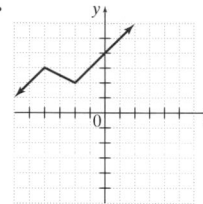

Exercise Set 8.6   **1.** $(-4, -9)$ **3.** $(5, 30)$ **5.** $(1, -2)$ **7.** $\left(\dfrac{1}{2}, \dfrac{5}{4}\right)$ **9.** D   **11.** B

13.   **15.**   **17.**   **19.**   **21.**

23.   **25.**   **27.**   **29.**   **31.**

33.   **35.**   **37.**   **39.**

41.   **43.** 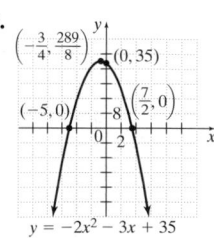   **45.** 144 ft   **47.** 16 ft   **49.** 30 and 30   **51.** $5, -5$

53. length, 20 units; width, 20 units

55.   **57.**   **59.**   **61.**

63.

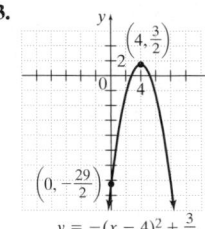

$y = -(x - 4)^2 + \frac{3}{2}$

65.

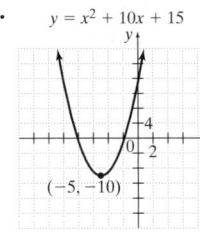

$y = x^2 + 10x + 15$

67.

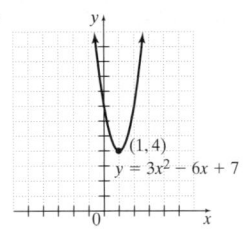

$y = 3x^2 - 6x + 7$

69. -0.84 **71.** 1.43 **73. a.** maximum; answers may vary   **b.** 2036   **c.** 32,902,500 or about 3290.3 thousands

Exercise Set 8.7   **1.** linear   **3.** neither   **5.** linear   **7.** quadratic   **9.** not linear or quadratic   **11.** $y = 9.547x + 467.466$; $658.39 or $y = 0.710x^2 + 2360x + 476.247$; $812.95   **13.** The predicted number of VCR sales is negative; answers may vary
15. $y = 0.05x^2 - 0.937x + 5.953$; $6.197 billion   **17.a.** quadratic   **b.** $y = -13.010x^2 + 1727.626x - 37,489.948$ **c.** $17,457 profit at $80, $19,332 profit at $60   **d.** $1500; no; answers may vary   **e.** $66   **19.** $y = 82.509x - 51,908.333$ **21.** $146,113   **23.** 1648.2, 1618.9, 1589.6, 1560.5; 11.2, 7.1, 19.4, 15.3   **25.** $y = 0.75x^2 - 10.774x + 86.655$; answers may vary   **27.** linear   **29.** quadratic   **31.** quadratic   **33.** quadratic   **35.** $-19, 1$

37. $-\frac{5}{2}, \frac{1}{3}$ **39.** -6 **41.** $(0, 12)$ **43.** $(0, -10)$

Chapter 8 Review   **1.** $14, 1$ **3.** $\frac{4}{5}, -\frac{1}{2}$ **5.** $-7, 7$ **7.** $-\frac{4}{9}, \frac{2}{9}$ **9.** $\frac{-3 - \sqrt{5}}{2}, \frac{-3 + \sqrt{5}}{2}$

11. $\frac{-3 - i\sqrt{7}}{8}, \frac{-3 + i\sqrt{7}}{8}$ **13.** 4.25% **15.** two complex but not real solutions   **17.** two real solutions   **19.** 8   **21.** $-i\sqrt{11}, i\sqrt{11}$

23. $\frac{5 - i\sqrt{143}}{12}, \frac{5 + i\sqrt{143}}{12}$ **25.** $\frac{21 - \sqrt{41}}{50}, \frac{21 + \sqrt{41}}{50}$ **27. a.** 20 ft   **b.** $\frac{15 + \sqrt{321}}{16}$ sec; 2.1 sec   **29.** $3, \frac{-3 + 3i\sqrt{3}}{2}, \frac{-3 - 3i\sqrt{3}}{2}$

31. $\frac{2}{3}, 5$ **33.** $-5, 5, -2i, 2i$ **35.** $1, 125$ **37.** $-1, 1, -i, i$ **39.** Jerome: 10.5 hr; Tim: 9.5 hr   **41.** $[-5, 5]$

43. $\left(-\infty, -\frac{5}{4}\right] \cup \left[\frac{3}{2}, \infty\right)$ **45.** ; $(5, 6)$ **47.** $(-\infty, -6) \cup \left(-\frac{3}{4}, 0\right) \cup (5, \infty)$

49. $(-5, -3) \cup (5, \infty)$ **51.** ; $\left(-\frac{6}{5}, 0\right) \cup \left(\frac{5}{6}, 3\right)$

53.

(graph with $(0, -4)$, $x = 0$)

55.

(graph with $(0, 0)$, $x = 0$)

57.

(graph with $(1, 0)$, $x = 1$)

59.

(graph with $(4, -2)$, $x = 4$)

61.

(graph with $(0, 25)$, $(-5, 0)$)

63.

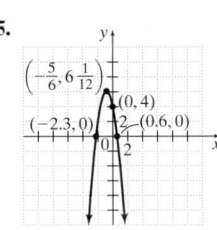

(graph with $\left(-\frac{1}{2}, 0\right)$, $\left(\frac{1}{2}, 0\right)$, $(0, -1)$)

65.

(graph with $\left(-\frac{5}{6}, 6\frac{1}{12}\right)$, $(-2.3, 0)$, $(0, 4)$, $(0.6, 0)$)

67. The numbers are both 210.   **69.** $y = 2361.58x + 37,564.2$; 285,530 thousand.

Chapter 8 Test   **1.** $\frac{7}{5}, -1$ **2.** $-1 - \sqrt{10}, -1 + \sqrt{10}$ **3.** $\frac{1 + i\sqrt{31}}{2}, \frac{1 - i\sqrt{31}}{2}$ **4.** $\frac{3 + \sqrt{29}}{2}, \frac{3 - \sqrt{29}}{2}$

5. $-2 - \sqrt{11}, -2 + \sqrt{11}$ **6.** $-1, 1, -i, i, -3$ **7.** $6, 7$ **8.** $3 - \sqrt{7}, 3 + \sqrt{7}$ **9.** $\frac{2 - i\sqrt{6}}{2}, \frac{2 + i\sqrt{6}}{2}$

10. ; $\left(-\infty, -\frac{3}{2}\right) \cup (5, \infty)$ **11.** ; $(-\infty, -5] \cup [-4, 4] \cup [5, \infty)$

12. ; $(-\infty, -3) \cup (2, \infty)$ **13.** ; $(-\infty, -3) \cup [2, 3)$

14.

15.

16.

17.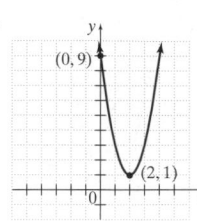

18. $\left(5 + \sqrt{17}\right)$ hr ≈ 9.12 hr
19. a. 272 ft   **b.** 5.12 sec
20. 7 ft
21. a. quadratic
b. $y = 0.034x^2$
     $- 0.563x + 16.784$;
18.26 births per 1000 people

Chapter 8 Cumulative Review **1. a.** $5 + y \geq 7$ **b.** $11 \neq z$ **c.** $20 < 5 - 2x$; Sec. 1.4, Ex. 3   **2.** $\emptyset$ (Sec. 3.4)

3. slope: 1; 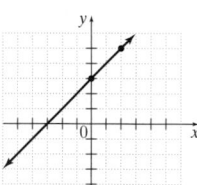 ; Sec. 2.4, Ex. 1   **4.** $(-2, -7)$; Sec 4.1   **5.** $(-2, 2)$; Sec. 4.1, Ex. 6   **6. a.** $\dfrac{a^6}{b^3c^9}$ **b.** $\dfrac{a^8c^6}{b^4}$ **c.** $\dfrac{16b^6}{a^6}$; Sec. 5.2

7. a. $6x^2 - 29x + 28$ **b.** $15x^2 - xy - 2y^2$; Sec.
5.4, Ex. 6   **8. a.** $28a^2 - 29a + 6$ **b.** $6a^2 - 7ab - 5b^2$; Sec. 5.4

9. a. $4(2x^2 + 1)$ **b.** prime polynomial   **c.** $3x^2(2 - x)$; Sec. 5.5, Ex. 2
10. a. $3x(3x^2 + 9x - 5)$ **b.** $(3y - 2)(2x - 5)$ **c.** $(2x - 1)(y + 3)$; Sec. 5.5
11. $(x - 5)(x - 7)$; Sec. 5.6, Ex. 2   **12.** $(x + 6)(x - 8)$; Sec. 5.6   **13.** $3x(a - 2b)^2$;
Sec. 5.7, Ex. 2   **14.** $2a(x - 3y)^2$; Sec. 5.7   **15.** $-\dfrac{2}{3}$; Sec. 5.8,

Ex. 4   **16.** $\dfrac{1}{2}$; Sec. 5.8   **17.** $-2, 0, 2$; Sec. 5.8, Ex. 6   **18.** vertex $\left(-\dfrac{1}{2}, -\dfrac{49}{4}\right)$; intercepts $(0, -12), (3, 0), (4, 0)$; Sec. 8.6   **19.** $\dfrac{1}{5x - 1}$; Sec. 6.1,

Ex. 2   **20.** $2 - x$; Sec. 6.1   **21.** $\dfrac{7x^2 - 9x - 13}{(2x + 1)(x - 5)(3x - 2)}$; Sec. 6.2, Ex. 5   **22.** $\dfrac{a^2 + 8a - 2}{(a - 2)(a - 4)(a + 4)}$; Sec. 6.2   **23.** $\dfrac{xy + 2x^3}{y - 1}$; Sec. 6.3, Ex. 3

24. $\dfrac{2a + b}{a + 2b}$; Sec. 6.3   **25.** $3x^3y - 15x - 1 - \dfrac{6}{xy}$; Sec. 6.4, Ex. 2   **26.** $x^2 - 6x + 8$; Sec. 6.4   **27. a.** 5   **b.** 5; Sec. 6.5, Ex. 3   **28. a.** -37

b. -37; Sec. 6.5   **29.** -3; Sec. 6.6, Ex. 1   **30.** -1; Sec. 6.6   **31.** 2; Sec. 6.7, Ex. 2   **32.** 15 hr; Sec. 6.7   **33.** $\dfrac{1}{6}$; $y = \dfrac{1}{6}x$; Sec. 6.8, Ex. 1

34. 112; $y = \dfrac{112}{x}$; Sec. 6.8   **35. a.** 3   **b.** $|x|$ **c.** $|x - 2|$ **d.** -5 **e.** $2x - 7$ **f.** $5|x|$ **g.** $|x + 1|$; Sec. 7.1, Ex. 5   **36. a.** 2

b. $|y|$ **c.** $|a - 3|$ **d.** -6 **e.** $3x - 1$; Sec. 7.1   **37. a.** $\sqrt{x}$ **b.** $\sqrt[3]{5}$ **c.** $\sqrt{rs^3}$; Sec. 7.2, Ex. 7   **38. a.** $\sqrt{5}$ **b.** $\sqrt[4]{x}$ **c.** $\sqrt[3]{xy^2}$;

Sec. 7.2   **39. a.** $5x\sqrt{x}$ **b.** $3x^2y^2\sqrt[3]{2y^2}$ **c.** $3z^2\sqrt[4]{z^3}$; Sec. 7.3, Ex. 4   **40. a.** $8a^2\sqrt{a}$ **b.** $2a^2b^3\sqrt[3]{3a}$ **c.** $2x^2\sqrt[4]{3x}$; Sec. 7.3   **41. a.** $\dfrac{2\sqrt{5}}{5}$

b. $\dfrac{8\sqrt{x}}{3x}$ **c.** $\dfrac{\sqrt[3]{4}}{2}$; Sec. 7.5, Ex. 1   **42. a.** $-2 - 6\sqrt{3}$ **b.** $5 - 2x\sqrt{5} + x^2$ **c.** $a - b^2$; Sec. 7.4   **43.** $\dfrac{2}{9}$; Sec. 7.6, Ex. 5   **44.** $2, 6$; Sec. 7.6

45. a. $\dfrac{1}{2} + \dfrac{3}{2}i$ **b.** $-\dfrac{7}{3}i$; Sec. 7.7, Ex. 5   **46. a.** $6 + 15i$ **b.** $11 - 60i$ **c.** 7; Sec. 7.7   **47.** $-1 + 2\sqrt{3}, -1 - 2\sqrt{3}$; Sec. 8.1, Ex. 3

48. $1 + 2\sqrt{6}, 1 - 2\sqrt{6}$; Sec. 8.1   **49.** 9; Sec. 8.3, Ex. 1   **50.** $2 - 2\sqrt{3}, 2 + 2\sqrt{3}$; Sec. 8.2

CHAPTER 9 EXPONENTIAL AND LOGARITHMIC FUNCTIONS

Mental Math **1.** C   **3.** F   **5.** D

Exercise Set 9.1 **1. a.** $3x - 6$ **b.** $-x - 8$ **c.** $2x^2 - 13x - 7$ **d.** $\dfrac{x - 7}{2x + 1}$, where $x \neq -\dfrac{1}{2}$ **3. a.** $x^2 + 5x + 1$ **b.** $x^2 - 5x + 1$

c. $5x^3 + 5x$ **d.** $\dfrac{x^2 + 1}{5x}$, where $x \neq 0$ **5. a.** $\sqrt{x} + x + 5$ **b.** $\sqrt{x} - x - 5$ **c.** $x\sqrt{x} + 5\sqrt{x}$ **d.** $\dfrac{\sqrt{x}}{x + 5}$, where $x \neq -5$

7. a. $5x^2 - 3x$ **b.** $-5x^2 - 3x$ **c.** $-15x^3$ **d.** $-\dfrac{3}{5x}$, where $x \neq 0$ **9.** 42   **11.** -18 **13.** 0

15. $(f \circ g)(x) = 25x^2 + 1$; $(g \circ f)(x) = 5x^2 + 5$ **17.** $(f \circ g)(x) = 2x + 11$; $(g \circ f)(x) = 2x + 4$

19. $(f \circ g)(x) = -8x^3 - 2x - 2$; $(g \circ f)(x) = -2x^3 - 2x + 4$ **21.** $(f \circ g)(x) = \sqrt{-5x + 2}$; $(g \circ f)(x) = -5\sqrt{x} + 2$

23. $H(x) = (g \circ h)(x)$ **25.** $F(x) = (h \circ f)(x)$ **27.** $G(x) = (f \circ g)(x)$ **29.** answers may vary; for example $g(x) = x + 2$ and $f(x) = x^2$

31. answers may vary; for example, $g(x) = x + 5$ and $f(x) = \sqrt{x} + 2$ **33.** answers may vary; for example, $g(x) = 2x - 3$ and $f(x) = \dfrac{1}{x}$

35. $y = x - 2$ **37.** $y = \dfrac{x}{3}$ **39.** $y = -\dfrac{x + 7}{2}$ **41.** 6   **43.** 4   **45.** 4   **47.** -1 **49.** answers may vary   **51.** $P(x) = R(x) - C(x)$

Exercise Set 9.2 **1.** one-to-one; $f^{-1} = \{(-1, -1), (1, 1), (2, 0), (0, 2)\}$ **3.** one-to-one; $h^{-1} = \{(10, 10)\}$ **5.** one-to-one;
$f^{-1} = \{(12, 11), (3, 4), (4, 3), (6, 6)\}$ **7.** not one-to-one **9.** one-to-one;

Rank in Population (Input)	1	49	12	2	46
State (Output)	CA	VT	VA	TX	SD

11. a. 3 **b.** 1 **13. a.** 1 **b.** -1 **15.** one-to-one
17. not one-to-one **19.** not one-to-one **21.** not one-to-one

23. $f^{-1}(x) = x - 4$ **25.** $f^{-1}(x) = \dfrac{x + 3}{2}$ **27.** $f^{-1}(x) = 2x + 2$ **29.** $f^{-1}(x) = \sqrt[3]{x}$

   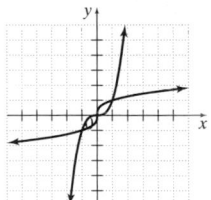

31. $f^{-1}(x) = \dfrac{x - 2}{5}$ **33.** $f^{-1}(x) = 5x + 2$ **35.** $f^{-1}(x) = x^3$ **37.** $f^{-1}(x) = \dfrac{5 - x}{3x}$ **39.** $f^{-1}(x) = \sqrt[3]{x} - 2$

41. **43.** **45.**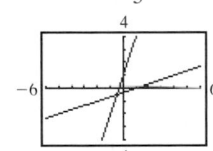

47. $(f \circ f^{-1})(x) = x; (f^{-1} \circ f)(x) = x$ **49.** $(f \circ f^{-1})(x) = x; (f^{-1} \circ f)(x) = x$ **51.** 5 **53.** 8 **55.** $\dfrac{1}{27}$ **57.** 9 **59.** $3^{1/2} \approx 1.73$

61. a. $\left(-2, \dfrac{1}{4}\right), \left(-1, \dfrac{1}{2}\right), (0, 1), (1, 2), (2, 5)$ **b.** $\left(\dfrac{1}{4}, -2\right), \left(\dfrac{1}{2}, -1\right), (1, 0), (2, 1), (5, 2)$

c. **d.** 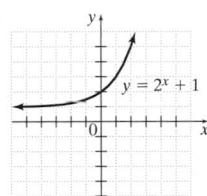 **63.** answers may vary

65. $f^{-1}(x) = \dfrac{x - 1}{3}$; 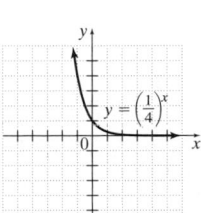 **67.** $f^{-1}(x) = x^3 - 1$;

Exercise Set 9.3 **1.** **3.** **5.** **7.**

9. **11.** **13.** **15.**

17. C **19.** D **21.** 3 **23.** $\dfrac{3}{4}$ **25.** $\dfrac{8}{5}$ **27.** $-\dfrac{2}{3}$ **29.** 4 **31.** $\dfrac{3}{2}$ **33.** $-\dfrac{1}{3}$ **35.** -2 **37.** 24.6 lb **39.** 333 bison **41.** 1.1 g

43. a. \$65.7 billion **b.** \$2303.6 billion **45. a.** 194.8 million people **b.** 353.7 million people **47.** \$7621.42 **49.** \$4065.59
51. 645 million cell phone users **53.** 4 **55.** $\varnothing$ **57.** 2, 3 **59.** 3 **61.** -1 **63.** answers may vary

65.

67.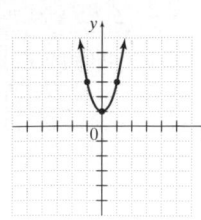

69. The graphs are the same since $\dfrac{1}{2}^{-x} = 2x$.

Exercise Set 9.4   **1.** $6^2 = 36$ **3.** $3^{-3} = \dfrac{1}{27}$ **5.** $10^3 = 1000$ **7.** $e^4 = x$ **9.** $e^{-2} = \dfrac{1}{e^2}$ **11.** $7^{1/2} = \sqrt{7}$

13. $\log_2 16 = 4$ **15.** $\log_{10} 100 = 2$ **17.** $\log_3 x = 3$ **19.** $\log_{10} \dfrac{1}{10} = -1$ **21.**

$\log_4 \dfrac{1}{16} = -2$ **23.** $\log_5 \sqrt{5} = \dfrac{1}{2}$

25. 3   **27.** -2 **29.** $\dfrac{1}{2}$ **31.** -1 **33.** 0   **35.** 4   **37.** 2   **39.** 5   **41.** 4   **43.** -3 **45.** answers may vary

47. 2   **49.** 81   **51.** 7   **53.** -3 **55.** -3 **57.** 2   **59.** 2   **61.** $\dfrac{27}{64}$ **63.** 10   **65.** 3   **67.** 3   **69.** 1

71.

73.

75.

77.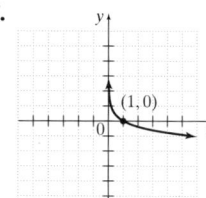

79. 1   **81.** $\dfrac{x - 4}{2}$

83. $\dfrac{2x + 3}{x^2}$

85. $m - 1$

87. $\dfrac{9}{5}$ **89.** 1

91.

93.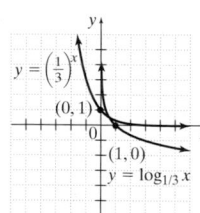

95. 0.0827   **97.** 2 and 3; answers may vary

Mental Math   **1.** a   **3.** b   **5.** a

Exercise Set 9.5   **1.** $\log_5 14$ **3.** $\log_4 9x$ **5.** $\log_{10}(10x^2 + 20)$ **7.** $\log_5 3$ **9.** $\log_2 \dfrac{x}{y}$ **11.** $\log_4 4$, or 1

13. $2 \log_3 x$ **15.** $-1 \log_4 5 = -\log_4 5$ **17.** $\dfrac{1}{2} \log_5 y$ **19.** $\log_2 25$ **21.** $\log_5 x^3 z^6$ **23.** $\log_{10} \dfrac{x^3 - 2x}{x + 1}$ **25.** $\log_4 35$

27. $\log_3 4$ **29.** $\log_7 \dfrac{9}{2}$ **31.** $\log_4 48$ **33.** $\log_2 \dfrac{x^{7/2}}{(x + 1)^2}$ **35.** $\log_8 x^{16/3}$ **37.** $\log_2 7 + \log_2 11 - \log_2 3$

39. $\log_3 4 + \log_5 y - \log_3 5$ **41.** $3 \log_2 x - \log_2 y$ **43.** $\dfrac{1}{2} \log_b 7 + \dfrac{1}{2} \log_b x$ **45.** $\log_7 5 + \log_7 x - \log_7 4$

47. $3 \log_5 x + \log_5(x + 1)$ **49.** $2 \log_6 x - \log_6(x + 3)$ **51.** 0.2   **53.** 1.2   **55.** 0.233   **57.** 1.29   **59.** -0.68 **61.** -0.125

63.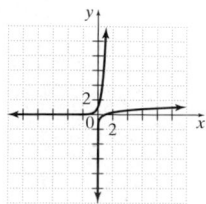

65. -1 **67.** $\dfrac{1}{2}$ **69.** false   **71.** true   **73.** false

Integrated Review **1.** $x^2 + x - 5$ **2.** $-x^2 + x - 7$ **3.** $x^3 - 6x^2 + x - 6$ **4.** $\dfrac{x - 6}{x^2 + 1}$ **5.** $\sqrt{3x - 1}$

6. $3\sqrt{x} - 1$ **7.** one-to-one; $\{(6, -2), (8, 4), (-6, 2), (3, 3)\}$ **8.** not one-to-one **9.** not one-to-one **10.** one-to-one

11. not one-to-one **12.** $f^{-1}(x) = \dfrac{x}{3}$ **13.** $f^{-1}(x) = x - 4$ **14.** $f^{-1}(x) = \dfrac{x + 1}{5}$ **15.** $f^{-1}(x) = \dfrac{x - 2}{3}$

16.

$y = \left(\dfrac{1}{2}\right)^x$

17.
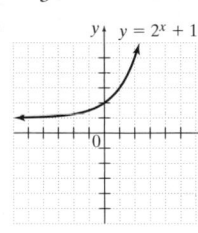
$y = 2^x + 1$

18.
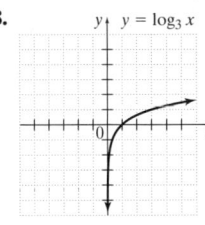
$y = \log_3 x$

19.
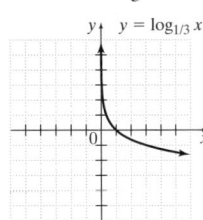
$y = \log_{1/3} x$

20. 3 **21.** 7 **22.** -8 **23.** 3 **24.** 2 **25.** $\dfrac{1}{2}$ **26.** 32 **27.** 4 **28.** 5 **29.** $\dfrac{1}{9}$ **30.** $\log_2 x^5$ **31.** $\log_2 5^x$ **32.** $\log_5 \dfrac{x^3}{y^5}$

33. $\log_5 x^9 y^3$ **34.** $\log_2 \dfrac{x^2 - 3x}{x^2 + 4}$ **35.** $\log_3 \dfrac{y^4 + 11y}{y + 2}$ **36.** $\log_7 9 + 2 \log_7 x - \log_7 y$ **37.** $\log_6 5 + \log_6 y - 2 \log_6 z$

Exercise Set 9.6 **1.** 0.9031 **3.** 0.3636 **5.** 0.6931 **7.** -2.6367 **9.** 1.1004 **11.** 1.6094 **13.** 1.6180 **15.** answers may vary

17. 2 **19.** -3 **21.** 2 **23.** $\dfrac{1}{4}$ **25.** 3 **27.** 2 **29.** -4 **31.** $\dfrac{1}{2}$ **33.** answers may vary **35.** $10^{1.3} \approx 19.9526$

37. $\dfrac{10^{1.1}}{2} \approx 6.2946$ **39.** $e^{1.4} \approx 4.0552$ **41.** $\dfrac{4 + e^{2.3}}{3} \approx 4.6581$ **43.** $10^{2.3} \approx 199.5262$ **45.** $e^{-2.3} \approx 0.1003$ **47.** $\dfrac{10^{-0.5} - 1}{2} \approx -0.3419$

49. $\dfrac{e^{0.18}}{4} \approx 0.2993$ **51.** 1.5850 **53.** -2.3219 **55.** 1.5850 **57.** -1.6309 **59.** 0.8617 **61.** 4.2 **63.** 5.3

65. \$3656.38 **67.** \$2542.50 **69.** $\dfrac{4}{7}$ **71.** $x = \dfrac{3y}{4}$ **73.** $-6, -1$ **75.** $(2, -3)$ **77.** answers may vary

79.

$f(x) = e^x$

81.

$f(x) = e^{-3x}$

83.
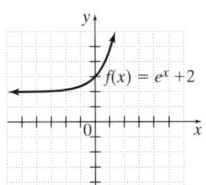
$f(x) = e^x + 2$

85.
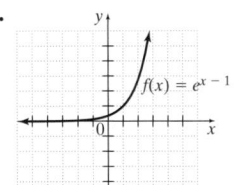
$f(x) = e^{x - 1}$

87.

$f(x) = 3e^x$

89.
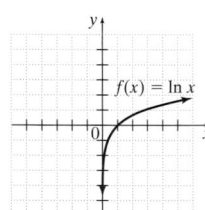
$f(x) = \ln x$

91.
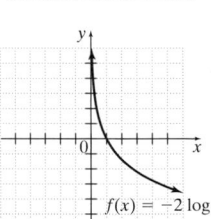
$f(x) = -2 \log x$

93.
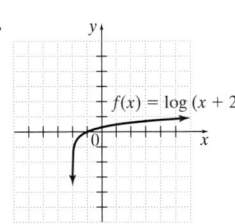
$f(x) = \log(x + 2)$

95.
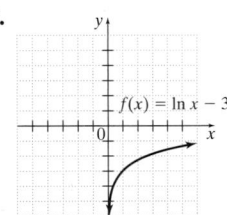
$f(x) = \ln x - 3$

97. answers may vary
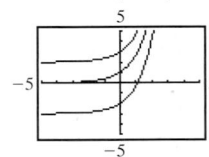

Exercise Set 9.7 **1.** $\dfrac{\log 6}{\log 3}$; 1.6309 **3.** $\dfrac{\log 3.8}{2 \log 3}$; 0.6076 **5.** $3 + \dfrac{\log 5}{\log 2}$; 5.3219 **7.** $\dfrac{\log 5}{\log 9}$; 0.7325 **9.** $\dfrac{\log 3}{\log 4} - 7$; -6.2075

11. $\dfrac{1}{3}\left(4 + \dfrac{\log 11}{\log 7}\right)$; 1.7441 **13.** $\dfrac{\ln 5}{6}$; 0.2682 **15.** 11 **17.** 9, -9 **19.** $\dfrac{1}{2}$ **21.** $\dfrac{3}{4}$ **23.** 2 **25.** $\dfrac{1}{8}$ **27.** 11 **29.** 4, -1 **31.** $\dfrac{1}{5}$

33. 100 **35.** $\dfrac{-5 + \sqrt{33}}{2}$ **37.** $\dfrac{192}{127}$ **39.** $\dfrac{2}{3}$ **41.** 103 wolves **43.** 12,380,000 inhabitants **45.** 10.5 yr **47.** 9.9 yr **49.** 1.7 yr

51. 8.8 yr **53.** 24.5 lb **55.** 55.7 in. **57.** 11.9 lb/sq. in. **59.** 3.2 mi **61.** 12 weeks **63.** 18 weeks **65.** $-\dfrac{5}{3}$ **67.** $\dfrac{17}{4}$

69. $f^{-1}(x) = \dfrac{x - 2}{5}$ **71.** 3.4% **73.** answers may vary

75. **a.** $y = 942.42(1.235)^x$　**b.** 4136 thousand
c. 185,548 thousand

77. **a.** $y = 5173.90(1.160)^x$　**b.** \$30,689 million
c. \$211,144 million

79. **a.** $y = 9500.669(1.182)^x$　**b.** \$70,940 million
c. \$626,315 million

Chapter 9 Review　**1.** $3x - 4$　**3.** $2x^2 - 9x - 5$　**5.** $x^2 + 2x - 1$　**7.** 18　**9.** -2　**11.** one-to-one;

$h^{-1} = \{(14, -9), (8, 6), (12, -11), (15, 15)\}$　**13.** one-to-one;

Rank in Automobile Thefts (Input)	2	4	1	3
US Region (Output)	W	Midwest	S	NE

15. **a.** 3　**b.** 7

17. not one-to-one

19. not one-to-one　**21.** $f^{-1}(x) = x + 9$　**23.** $f^{-1}(x) = \dfrac{x - 11}{6}$　**25.** $f^{-1}(x) = \sqrt[3]{x + 5}$　**27.** $g^{-1}(x) = \dfrac{6x + 7}{12}$

29.

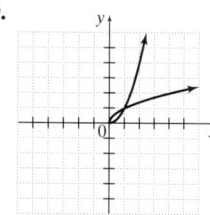

31. $f^{-1}(x) = \dfrac{x + 3}{2}$;

33. -2　**35.** $\dfrac{3}{2}$　**37.** $\dfrac{8}{9}$

39.

41.

43. \$1131.82　**45.** $\log_7 49 = 2$　**47.** $\left(\dfrac{1}{2}\right)^{-4} = 16$　**49.** $\dfrac{1}{64}$

51. 0　**53.** 8　**55.** 5　**57.** 4　**59.** $\dfrac{17}{3}$　**61.** $-1, 4$

63.

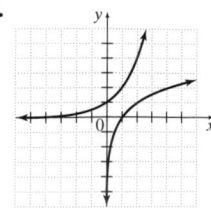

65. $\log_3 32$　**67.** $\log_7 \dfrac{3}{4}$　**69.** $\log_{11} 4$　**71.** $\log_5 \dfrac{x^3}{(x + 1)^2}$　**73.** $3 \log_3 x - \log_3(x + 2)$

75. $\log_2 3 + 2 \log_2 x + \log_2 y - \log_2 z$　**77.** 2.02　**79.** 0.5563　**81.** 0.2231　**83.** 3　**85.** -1

87. $\dfrac{e^2}{2}$　**89.** $\dfrac{e^{-1} + 3}{2}$　**91.** 1.67 mm　**93.** 0.2920　**95.** \$1957.30　**97.** $\dfrac{\log 7}{2 \log 3}$; 0.8856

99. $\dfrac{1}{2}\left(\dfrac{\log 6}{\log 3} - 1\right)$; 0.3155　**101.** $\dfrac{1}{3}\left(\dfrac{\log 4}{\log 5} + 5\right)$; 1.9538　**103.** $-\dfrac{\log 2}{\log 5} + 1$; 0.5693　**105.** $\dfrac{25}{2}$

107. $\varnothing$　**109.** $2\sqrt{2}$　**111.** 197,044 ducks　**113.** 24 yr　**115.** 43 yr　**117.** 8.5 yr　**119.** 2.82

Chapter 9 Test **1.** 5   **2.** $x - 7$ **3.** $x^2 - 6x - 2$

4.

5. one-to-one   **6.** not one-to-one   **7.** one-to-one; $f^{-1}(x) = \dfrac{-x + 6}{2}$ **8.** one-to-one; $f^{-1} = \{(0,0), (3,2), (5,-1)\}$

9. not one-to-one   **10.** $\log_3 24$ **11.** $\log_5 \dfrac{x^4}{x + 1}$ **12.** $\log_6 2 + \log_6 x - 3\log_6 y$ **13.** -1.53 **14.** 1.0686

15. -1 **16.** $\dfrac{1}{2}\left(\dfrac{\log 4}{\log 3} - 5\right)$; -1.8691 **17.** $\dfrac{1}{9}$ **18.** $\dfrac{1}{2}$ **19.** 22   **20.** $\dfrac{25}{3}$ **21.** $\dfrac{43}{21}$ **22.** -1.0979

23.

24.

25. $5234.58   **26.** 6 yr   **27.** 64,913 prairie dogs   **28.** 15 yr   **29.** 1.2%
30. 3.95   **31.** $y = 247.71(1.138)^x$; $1971.1 billion in 2006

Chapter 9 Cumulative Review **1. a.** 8   **b.** $-\dfrac{1}{3}$ **c.** -9 **d.** 0   **e.** $-\dfrac{2}{11}$ **f.** 42   **g.** 0; Sec. 1.3, Ex. 4   **2.** 11; Sec. 1.5

3. ; Sec. 2.1, Ex. 7   **4.** $f(x) = \dfrac{1}{3}x + \dfrac{20}{3}$; Sec. 2.5   **5.** $[(x, y, z)|x - 5y - 2z = 6\}$; Sec. 4.2, Ex. 4   **6.** $x = 110, y = 70$;

Sec. 4.3   **7. a.** x^3 **b.** 5^6 **c.** $5x$ **d.** $\dfrac{6y^2}{7}$; Sec. 5.1, Ex. 4   **8. a.** $16a^6$ **b.** $-\dfrac{8}{27}$ **c.** $\dfrac{64a^{15}}{b^9}$

d. $729x^3$ **e.** $\dfrac{a^4 c^8}{b^6}$; Sec. 5.2   **9. a.** $102.60   **b.** $12.60; Sec. 6.1, Ex. 9   **10. a.** $9x^2 - 6x + 1$

b. $\dfrac{1}{4}x^2 - 9$ **c.** $12x^2 - 16x - 35$; Sec. 5.4   **11. a.** $\dfrac{3x}{2}$ **b.** $x - 7$ **c.** $-\dfrac{1}{3y^2}$; Sec. 6.2, Ex. 1

12. $\dfrac{-x^2 + 23x + 38}{(x - 2)(x + 2)^2}$; Sec. 6.2   **13.** $3x^2 + 2x + 3 + \dfrac{-6x + 9}{x^2 - 1}$; Sec. 6.4, Ex. 6   **14. a.** $\dfrac{2a}{a - 1}$ **b.** $\dfrac{-3a - 18}{4a - 12}$ **c.** $\dfrac{y + x}{x^2 y^2}$; Sec. 6.3   **15.** -1;

Sec. 6.6, Ex. 4   **16.** $x^2 + 2x + 4$; Sec. 6.4   **17.** 6 mph; Sec. 6.7, Ex. 5   **18.** $8x + 4 + \dfrac{1}{x - 2}$; Sec. 6.5   **19. a.** 3   **b.** -3 **c.** -5

d. not a real number   **e.** $4x$; Sec. 7.1, Ex. 4   **20.** $-\dfrac{7}{5}$; Sec. 6.6   **21. a.** $\sqrt[4]{x^3}$ **b.** $\sqrt[6]{x}$ **c.** $\sqrt[6]{72}$; Sec. 7.2, Ex. 8   **22.** $k = \dfrac{1}{24}, y = \dfrac{1}{24}x$;

Sec. 6.8   **23. a.** $5\sqrt{3} + 3\sqrt{10}$ **b.** $\sqrt{35} + \sqrt{5} - \sqrt{42} - \sqrt{6}$ **c.** $21x - 7\sqrt{5x} + 15\sqrt{x} - 5\sqrt{5}$ **d.** $49 - 8\sqrt{3}$ **e.** $2x - 25$

f. $x + 22 + 10\sqrt{x - 3}$; Sec. 7.4, Ex. 4   **24. a.** 3   **b.** -3 **c.** $\dfrac{3}{8}$ **d.** x^3 **e.** $-5y^2$; Sec. 7.1   **25.** $\dfrac{\sqrt[4]{xy^3}}{3y^2}$; Sec. 7.5, Ex. 3   **26. a.** $a - a^{\frac{33}{4}}$

b. $x + 2x^{1/2} - 15$; Sec. 7.2   **27.** 3; Sec. 7.6, Ex. 4   **28. a.** 3   **b.** $2a$ **c.** $9ab^2\sqrt[3]{a^2}$; Sec. 7.3   **29.** $\dfrac{9 + i\sqrt{15}}{6}, \dfrac{9 - i\sqrt{15}}{6}$; Sec. 8.1, Ex. 8

30. a. $\dfrac{11\sqrt{5}}{12}$ **b.** $\dfrac{\sqrt[3]{3x}}{6}$; Sec. 7.4   **31.** $\dfrac{-1 + \sqrt{33}}{4}, \dfrac{-1 - \sqrt{33}}{4}$; Sec. 8.3, Ex. 2   **32.** $\dfrac{3\sqrt[3]{m^2 n}}{m^2 n^3}$; Sec. 7.5   **33.** $[0, 4]$; Sec. 8.4, Ex. 2

34. $4\sqrt{3}$ in.; Sec. 7.6   **35.** ; Sec. 8.5, Ex. 3   **36. a.** 1   **b.** i **c.** -1 **d.** $-i$; Sec. 7.7   **37. a.** $3x - 4$ **b.** $-x + 2$

c. $2x^2 - 5x + 3$ **d.** $\dfrac{x - 1}{2x - 3}$, where $x \neq \dfrac{3}{2}$; Sec. 9.1, Ex. 1

$F(x) = (x - 3)^2 + 1$

38. $\dfrac{-2 + \sqrt{5}}{2}, \dfrac{-2 - \sqrt{5}}{2}$; Sec. 8.1   **39.** $f^{-1}(x) = x - 3$; Sec. 9.2, Ex. 4

40. $\dfrac{3 - \sqrt{5}}{4}, \dfrac{3 + \sqrt{5}}{4}$; Sec. 8.3   **41. a.** 2   **b.** -1 **c.** $\dfrac{1}{2}$; Sec. 9.4, Ex. 3

42.

; Sec. 8.5

$y = -(x + 1)^2 + 1$

$(-1, 1)$

CHAPTER 10 CONIC SECTIONS

Mental Math **1.** upward **3.** to the left **5.** downward

Exercise Set 10.1 **1.** **3.** **5.** **7.**

9. **11.**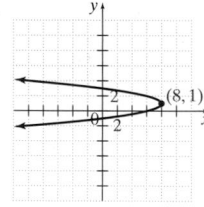

13. 5 units **15.** $\sqrt{41}$ units **17.** $\sqrt{10}$ units **19.** $\sqrt{5}$ units **21.** 13.88 units

23. 9 units **25.** $(4, -2)$ **27.** $\left(-5, \frac{5}{2}\right)$ **29.** $(3, 0)$ **31.** $\left(-\frac{1}{2}, \frac{1}{2}\right)$

33. $\left(\sqrt{2}, \frac{\sqrt{5}}{2}\right)$ **35.** $(6.2, -6.65)$

37. **39.** **41.** **43.** **45.**

47. 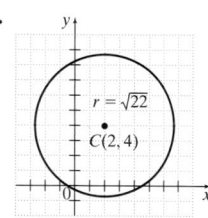 **49.** $(x - 2)^2 + (y - 3)^2 = 36$ **51.** $x^2 + y^2 = 3$

53. $(x + 5)^2 + (y - 4)^2 = 45$ **55.** answers may vary

57. **59.** **61.** **63.**

65. **67.** **69.** **71.** **73.**

75.   **77.**   **79.**   **81.**   **83.**

85.   **87.**   **89.**   **91.**

93.   **95.**   **97.**   **99.**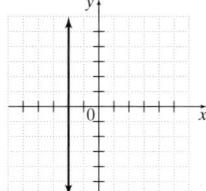

101. $\dfrac{\sqrt{10}}{4}$ **103.** $2\sqrt{5}$ **105. a.** 125 ft   **b.** 14 ft   **c.** 139 ft   **d.** $(0, 139)$ **e.** $x^2 + (y - 139)^2 = 125^2$

107. 20m   **109.** $y = -\dfrac{2}{125}x^2 + 40$ **111.**   **113.**

Mental Math   **1.** ellipse   **3.** hyperbola   **5.** hyperbola

Exercise Set 10.2   **1.**   **3.**   **5.**   **7.**

9.   **11.**   **13.**   **15.**   **17.**

19. 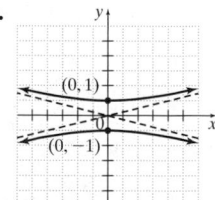   **21.** answers may vary   **23.** parabola   **25.** ellipse

27. hyperbola

29. circle

31. parabola

33. ellipse

35. hyperbola

37. parabola

39.

41.

43.

45.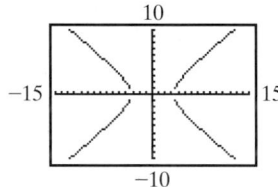

47. $(-\infty, 5)$ **49.** $[4, \infty)$ **51.** $-2x^3$ **53.** $-5x^4$

55. ellipses: C, E, H; circles: B, F; hyperbolas: A, D, G

57. A: 49, 7; B: 0, 0; C: 9, 3; D: 64, 8; E: 64, 8; F: 0, 0; G: 81, 9; H: 4, 2

59. A: $\frac{7}{6}$; B: 0; C: $\frac{3}{5}$; D: $\frac{8}{5}$ E: $\frac{8}{9}$; F: 0; G: $\frac{9}{4}$; H: $\frac{1}{6}$ **61.** equal to zero

63. answers may vary **65.** $\dfrac{x^2}{1.69 \cdot 10^{16}} + \dfrac{y^2}{1.5625 \cdot 10^{16}} = 1$

67. $9x^2 + 4y^2 = 36$

69.

71.

73.

Integrated Review

1.

2.

3.

4.

5.

6.

7.

8.

9.

10.

11.

12.

13.   **14.**   **15.**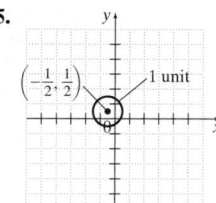

Exercise Set 10.3 **1.** $(3, -4), (-3, 4)$ **3.** $\left(\sqrt{2}, \sqrt{2}\right), \left(-\sqrt{2}, -\sqrt{2}\right)$ **5.** $(4, 0), (0, -2)$

7. $\left(-\sqrt{5}, -2\right), \left(-\sqrt{5}, 2\right), \left(\sqrt{5}, -2\right), \left(\sqrt{5}, 2\right)$ **9.** $\varnothing$ **11.** $(1, -2), (3, 6)$ **13.** $(2, 4), (-5, 25)$ **15.** $\varnothing$ **17.** $(1, -3)$

19. $(-1, -2), (-1, 2), (1, -2), (1, 2)$ **21.** $(0, -1)$ **23.** $(-1, 3), (1, 3)$ **25.** $\left(\sqrt{3}, 0\right), \left(-\sqrt{3}, 0\right)$ **27.** $\varnothing$ **29.** $(-6, 0), (6, 0), (0, -6)$

31.   **33.** 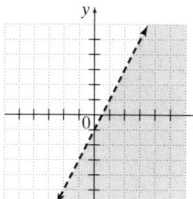   **35.** $(8x - 25)$ in.   **37.** $(4x^2 + 6x + 2)$ m   **39.** $0, 1, 2, 3,$ or 4 **41.** 9 and 7; 9 and -7; -9 and 7; -9 and -7 **43.** 15 cm by 19 cm   **45.** 15 thousand compact discs; price: $3.75   **47.**   **49.**

Exercise Set 10.4 **1.**   **3.**   **5.**   **7.**

9.   **11.**   **13.**   **15.**   **17.**

19.   **21.**   **23.**   **25.**   **27.**

29.   **31.**   **33.**   **35.**   **37.**

39. 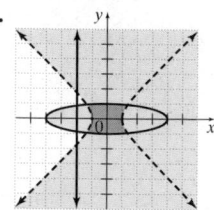   **41.** not a function   **43.** function   **45.** 1   **47.** $3a^2 - 2$ **49.** answers may vary

51.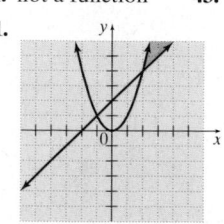

Chapter 10 Review   **1.** $\sqrt{197}$ units   **3.** $\sqrt{130}$ units   **5.** $7\sqrt{2}$ units   **7.** 16.60 units   **9.** $(-5, 5)$ **11.** $\left(-\dfrac{15}{2}, 1\right)$

13. $\left(\dfrac{1}{20}, -\dfrac{3}{16}\right)$ **15.** $\left(\sqrt{3}, -3\sqrt{6}\right)$ **17.** $(x + 4)^2 + (y - 4)^2 = 9$ **19.** $(x + 7)^2 + (y + 9)^2 = 11$

21.   **23.**   **25.**   **27.**   **29.**

31.   **33.** 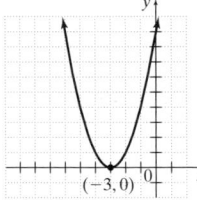   **35.** $(x - 5.6)^2 + (y + 2.4)^2 = 9.61$ **37.**

39.   **41.**   **43.**   **45.**   **47.**

49.   **51.** parabola   **53.** circle   **55.** circle   **57.** hyperbola

59. ellipse   **61.** hyperbola   **63.**   **65.** 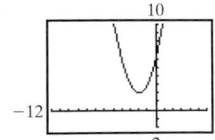   **67.** $(1, -2), (4, 4)$

69. $(-1, 1), (2, 4)$ **71.** $\left(2, 2\sqrt{2}\right), \left(2, -2\sqrt{2}\right)$ **73.** $(-1, 3), (-1, -3), (1, 3), (1, -3)$ **75.** $(1, 4)$ **77.** 15 ft by 10 ft

79.   **81.**   **83.**   **85.**   **87.**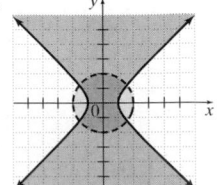

Chapter 10 Test   **1.** $2\sqrt{26}$ units   **2.** $\sqrt{95}$ units   **3.** $\left(-4, \dfrac{7}{2}\right)$

4.

5.

6.

7.

8.

9.

10.

11.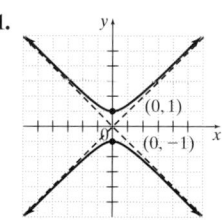

12. $(-5, -1), (-5, 1), (5, -1), (5, 1)$

13. $(6, 12), (1, 2)$

14.

15.

16.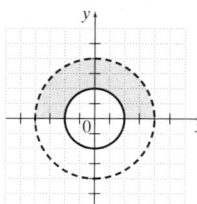

17. height: 10 ft; width: 30 ft

Chapter 10 Cumulative Review
1. $(4 \cdot 9)y = 36y$; Sec. 1.4, Ex. 7   **2.** $(-1, 7]$; Sec. 3.3

3. 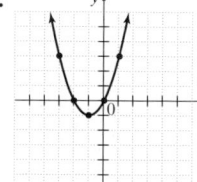 ; Sec. 2.3, Ex. 5   **4.** 3; Sec. 2.4   **5.** $\varnothing$; Sec. 4.1, Ex. 8   **6.** 150 mph, 175 mph; Sec. 4.3   **7. a.** $125x^6$ **b.** $\dfrac{8}{27}$ **c.** $\dfrac{9p^8}{q^{10}}$ **d.** $64y^2$ **e.** $\dfrac{y^{14}}{x^{35}z^7}$; Sec. 5.2, Ex. 2   **8. a.** 4^5 **b.** y^6 **c.** $8x$ **d.** $\dfrac{3a^4}{2}$; Sec. 5.7

9. $-\dfrac{1}{6}, 3$; Sec. 5.8, Ex. 5   **10. a.** $(3y + 5)(y + 3)$ **b.** $2a^3(2a + 5)(5a + 1)$ **c.** $(y - 1)(y - 7)$; Sec. 5.6   **11.** $\dfrac{12}{x - 1}$; Sec. 6.2, Ex. 6   **12.** $\dfrac{5a + 10}{3(a + 5)(a - 5)}$; Sec. 6.2

13. a. $\dfrac{1}{9xy^2}$ **b.** $\dfrac{x(x - 2)}{2(x + 2)}$ **c.** $\dfrac{x^2}{y^2}$; Sec. 6.3, Ex. 1   **14. a.** $\dfrac{ab}{b - a}$ **b.** $\dfrac{1}{2x + 1}$; Sec. 6.3   **15.** $2x - 5$; Sec. 6.4, Ex. 3   **16.** $\dfrac{4}{3}$; Sec. 6.6

17. 16; Sec. 6.5, Ex. 4   **18.** $k = 2$; $y = \dfrac{2}{x}$; Sec. 6.8   **19.** $-6, -1$; Sec. 6.6, Ex. 5   **20. a.** -2 **b.** 5   **c.** -6 **d.** $3x$ **e.** $12y$; Sec. 7.3

21. $2\dfrac{2}{9}$ hr; no; Sec. 6.7, Ex. 4   **22. a.** $2\sqrt{2}$ **b.** $\dfrac{2y^2\sqrt[3]{10}}{5}$ **c.** $2xy^2\sqrt[5]{x^2}$; Sec. 7.3   **23. a.** 1   **b.** -4 **c.** $\dfrac{2}{5}$ **d.** x^2 **e.** $-3x^3$; Sec. 7.1, Ex. 3   **24. a.** $2\sqrt{5} + 5\sqrt{3}$ **b.** $\sqrt{21} - \sqrt{3} - \sqrt{35} + \sqrt{5}$ **c.** $21 - 4\sqrt{5}$ **d.** -7; Sec. 7.4   **25. a.** $z - z^{17/3}$ **b.** $x^{2/3} - 3x^{1/3} - 10$; Sec. 7.2, Ex. 5   **26.** $\dfrac{\sqrt{3} - 3}{3}$; Sec. 7.5   **27. a.** 2   **b.** $\dfrac{5}{2}\sqrt{x}$ **c.** $14xy^2\sqrt[3]{x}$;   **d.** $4a^2b\sqrt[4]{2a}$; Sec. 7.3, Ex. 5   **28.** 6; Sec. 7.6   **29. a.** $\dfrac{5\sqrt{5}}{12}$ **b.** $\dfrac{5\sqrt[3]{7x}}{2}$; Sec. 7.4, Ex. 3   **30.** two complex but not real solutions; Sec. 8.3   **31.** $\dfrac{\sqrt{21xy}}{3y}$; Sec. 7.5, Ex. 2   **32.** 4; Sec. 6.6   **33.** 42; Sec. 7.6, Ex. 1

34. $[2, \infty)$; Sec. 8.4   **35. a.** $-i$ **b.** 1   **c.** -1 **d.** 1; Sec. 7.7, Ex. 6

36. 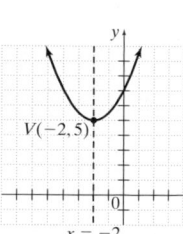   **37.** $-1 + \sqrt{5}, -1 - \sqrt{5}$; Sec. 8.1, Ex. 5   **38.** 13; Sec. 8.6   **39.** $2 + \sqrt{2}, 2 - \sqrt{2}$; Sec. 8.2, Ex. 3   **40.** $f^{-1}(x) = 2x - 1$; Sec. 9.2   **41.** $2, -2, i, -i$; Sec. 8.3, Ex. 3   **42. a.** $9x^2 - 21x + 12$ **b.** 90   **c.** $-3x^2 + 9x - 1$ **d.** -31; Sec. 9.1   **43.** $[-2, 3)$; Sec. 8.4, Ex. 6   **44.** ; Sec. 10.2

45. 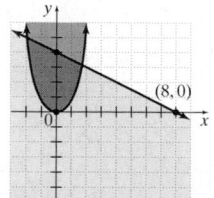 ; Sec. 8.5, Ex. 5   **46. a.** 1/3   **b.** 11/3   **c.** -2; Sec. 9.3   **47.** $(2, -16)$; Sec. 8.6, Ex. 4   **48.** ; Sec. 10.4   **49.** $\sqrt{2} \approx 1.414$; Sec. 10.1, Ex. 6   **50.** $(-6, 0), (0, 6)$; Sec. 10.3

CHAPTER 11 SEQUENCES, SERIES, AND THE BINOMIAL THEOREM

Exercise Set 11.1 **1.** $5, 6, 7, 8, 9$ **3.** $-1, 1, -1, 1, -1$ **5.** $\dfrac{1}{4}, \dfrac{1}{5}, \dfrac{1}{6}, \dfrac{1}{7}, \dfrac{1}{8}$ **7.** $2, 4, 6, 8, 10$ **9.** $-1, -4, -9, -16, -25$

11. $2, 4, 8, 16, 32$ **13.** $7, 9, 11, 13, 15$ **15.** $-1, 4, -9, 16, -25$ **17.** 75 **19.** 118 **21.** $\dfrac{6}{5}$ **23.** 729 **25.** $\dfrac{4}{7}$ **27.** $\dfrac{1}{8}$ **29.** -95

31. $-\dfrac{1}{25}$ **33.** $a_n = 4n - 1$ **35.** $a_n = -2^n$ **37.** $a_n = \dfrac{1}{3^n}$ **39.** 48 ft, 80 ft, and 112 ft **41.** $a_n = 0.10(2)^{n-1}$; $819.20

43. 2400 cases; 75 cases **45.** 50 sparrows in 2004: extinct in 2010

47. 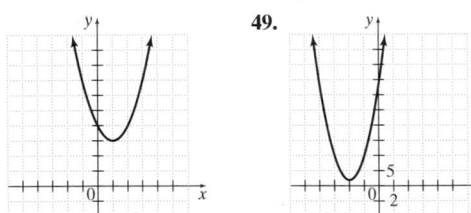 **49.** **51.** $\sqrt{13}$ units **53.** $\sqrt{41}$ units **55.** $1, 0.7071, 0.5774, 0.5, 0.4472$
57. $2, 2.25, 2.3704, 2.4414, 2.4883$

Exercise Set 11.2 **1.** $4, 6, 8, 10, 12$ **3.** $6, 4, 2, 0, -2$ **5.** $1, 3, 9, 27, 81$ **7.** $48, 24, 12, 6, 3$ **9.** 33 **11.** -875 **13.** -60 **15.** 96

17. -28 **19.** 1250 **21.** 31 **23.** 20 **25.** $a_1 = \dfrac{2}{3}; r = -2$ **27.** answers may vary **29.** $a_1 = 2; d = 2$ **31.** $a_1 = 5; r = 2$

33. $a_1 = \dfrac{1}{2}; r = \dfrac{1}{5}$ **35.** $a_1 = x; r = 5$ **37.** $a_1 = p; d = 4$ **39.** 19 **41.** $-\dfrac{8}{9}$ **43.** $\dfrac{17}{2}$ **45.** $\dfrac{8}{81}$ **47.** -19

49. $a_n = 4n + 50$; 130 seats **51.** $a_n = 6(3)^{n-1}$ **53.** $486, 162, 54, 18, 6$; $a_n = \dfrac{486}{3^{n-1}}$; 6 bounces **55.** $a_n = 4000 + 125(n - 1)$ or

$a_n = 3875 + 125n$; $5375 **57.** 25 g **59.** $\dfrac{11}{18}$ **61.** 40 **63.** $\dfrac{907}{495}$ **65.** $11,782.40, $5891.20, $2945.60, $1472.80 **67.** $19.652, 19.618,$

$19.584, 19.55$ **69.** answers may vary

Exercise Set 11.3 **1.** -2 **3.** 60 **5.** 20 **7.** $\dfrac{73}{168}$ **9.** $\dfrac{11}{36}$ **11.** 60 **13.** 74 **15.** 62 **17.** $\dfrac{241}{35}$ **19.** $\displaystyle\sum_{i=1}^{5}(2i - 1)$

21. $\displaystyle\sum_{i=1}^{4}4(3)^{i-1}$ **23.** $\displaystyle\sum_{i=1}^{6}(-3i + 15)$ **25.** $\displaystyle\sum_{i=1}^{4}\dfrac{4}{3^{i-2}}$ **27.** $\displaystyle\sum_{i=1}^{7}i^2$ **29.** -24 **31.** -13 **33.** 82 **35.** -20 **37.** -2

39. $1, 2, 3, \ldots, 10$; 55 trees **41.** $a_n = 6(2)^{n-1}$; 96 units **43.** $a_n = 50(2)^n$; n represents the number of 12-hour periods; 800 bacteria

45. 30 opossums; 68 opossums **47.** 6.25 lb; 93.75 lb **49.** 16.4 in.; 134.5 in. **51.** 10 **53.** $\dfrac{10}{27}$ **55.** 45 **57.** 90

59. a. $2 + 6 + 12 + 20 + 30 + 42 + 56$ **b.** $1 + 2 + 3 + 4 + 5 + 6 + 7 + 1 + 4 + 9 + 16 + 25 + 36 + 49$ **c.** Answers may vary.
d. True; answers may vary

Integrated Review **1.** $-2, -1, 0, 1, 2$ **2.** $\dfrac{7}{2}, \dfrac{7}{3}, \dfrac{7}{4}, \dfrac{7}{5}, \dfrac{7}{6}$ **3.** $1, 3, 9, 27, 81$ **4.** $-4, -1, 4, 11, 20$ **5.** 64 **6.** -14 **7.** $\dfrac{1}{40}$ **8.** $-\dfrac{1}{82}$

9. $7, 4, 1, -2, -5$ **10.** $-3, -15, -75, -375, -1875$ **11.** $45, 15, 5, \dfrac{5}{3}, \dfrac{5}{9}$ **12.** $-12, -2, 8, 18, 28$ **13.** 101 **14.** $\dfrac{243}{16}$ **15.** 384

16. 185 **17.** -10 **18.** $\dfrac{1}{5}$

Exercise Set 11.4 **1.** 36 **3.** 484 **5.** 63 **7.** 2.496 **9.** 55 **11.** 16 **13.** 24 **15.** $\dfrac{1}{9}$ **17.** -20 **19.** $\dfrac{16}{9}$ **21.** $\dfrac{4}{9}$ **23.** 185

25. $\dfrac{381}{64}$ **27.** $-\dfrac{33}{4}$, or -8.25 **29.** $-\dfrac{75}{2}$ **31.** $\dfrac{56}{9}$ **33.** $4000, 3950, 3900, 3850, 3800$; 3450 cars; 44,700 cars **35.** Firm A (Firm A, $265,000;

Firm B, $254,000) **37.** $39,930; $139,230 **39.** 20 min; 123 min **41.** 180 ft **43.** Player A, 45 points; Player B, 75 points **45.** $3050

47. $10,737,418.23 **49.** 720 **51.** 3 **53.** $x^2 + 10x + 25$ **57.** $\dfrac{8}{10} + \dfrac{8}{100} + \dfrac{8}{1000} + \cdots; \dfrac{8}{9}$ **59.** Answers may vary.

65. $8x^3 - 12x^2 + 6x - 1$

Exercise Set 11.5 **1.** $m^3 + 3m^2n + 3mn^2 + n^3$ **3.** $c^5 + 5c^4d + 10c^3d^2 + 10c^2d^3 + 5cd^4 + d^5$
5. $y^5 - 5y^4x + 10y^3x^2 - 10y^2x^3 + 5yx^4 - x^5$ **7.** Answers may vary. **9.** 8 **11.** 42 **13.** 360 **15.** 56

17. $a^7 + 7a^6b + 21a^5b^2 + 35a^4b^3 + 35a^3b^4 + 21a^2b^5 + 7ab^6 + b^7$ **19.** $a^5 + 10a^4b + 40a^3b^2 + 80a^2b^3 + 80ab^4 + 32b^5$

21. $q^9 + 9q^8r + 36q^7r^2 + 84q^6r^3 + 126q^5r^4 + 126q^4r^5 + 84q^3r^6 + 36q^2r^7 + 9qr^8 + r^9$

23. $1024a^5 + 1280a^4b + 640a^3b^2 + 160a^2b^3 + 20ab^4 + b^5$ **25.** $625a^4 - 1000a^3b + 600a^2b^2 - 160ab^3 + 16b^4$

27. $8a^3 + 36a^2b + 54ab^2 + 27b^3$ **29.** $x^5 + 10x^4 + 40x^3 + 80x^2 + 80x + 32$ **31.** $5cd^4$ **33.** d^7 **35.** $-40r^2s^3$

37. $6x^2y^2$　**39.** $30a^9b$　**41.** 　**43.** 　**45.** 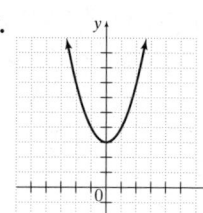　**47.** $x\sqrt[3]{x} + 5\sqrt{3}x^2 + 30x\sqrt{x} + 30\sqrt{3}x + 45\sqrt{x} + 9\sqrt{3}$

49. 126　**51.** 28

53. answers may vary

Chapter 11 Review

1. $-3, -12, -27, -48, -75$　**3.** $\dfrac{1}{100}$　**5.** $a_n = \dfrac{1}{6n}$　**7.** 144 ft, 176 ft, 208 ft

9. 450, 1350, 4050, 12,150, 36,450; 36,450 infected people in 2007　**11.** $-2, -\dfrac{4}{3}, -\dfrac{8}{9}, -\dfrac{16}{27}, -\dfrac{32}{81}$　**13.** 111　**15.** -83　**17.** $a_1 = 3; d = 5$

19. $a_n = \dfrac{3}{10^n}$　**21.** $a_1 = \dfrac{8}{3}, r = \dfrac{3}{2}$　**23.** $a_1 = 7x, r = -2$　**25.** 8, 6, 4.5, 3.4, 2.5, 1.9; good　**27.** $a_n = 2^{n-1}$, \$512, \$536,870,912

29. $a_n = 900 + (n-1)150$ or $a_n = 150n + 750$; \$1650/month　**31.** $1 + 3 + 5 + 7 + 9 = 25$　**33.** $\dfrac{1}{4} - \dfrac{1}{6} + \dfrac{1}{8} = \dfrac{5}{24}$　**35.** -4

37. -10　**39.** $\displaystyle\sum_{i=1}^{6} 3^{i-1}$　**41.** $\displaystyle\sum_{i=1}^{4} \dfrac{1}{4^i}$　**43.** $a_n = 20(2)^n$; n represents the number of 8-hour periods; 1280 yeast

45. Job A, \$48,300; Job B, \$46,600　**47.** 150　**49.** 900　**51.** -410　**53.** 936　**55.** 10　**57.** -25　**59.** \$30,418; \$99,868　**61.** \$58; \$553

63. 2696 mosquitoes　**65.** $\dfrac{5}{9}$　**67.** $x^5 + 5x^4z + 10x^3z^2 + 10x^2z^3 + 5xz^4 + z^5$　**69.** $16x^4 + 32x^3y + 24x^2y^2 + 8xy^3 + y^4$

71. $b^8 + 8b^7c + 28b^6c^2 + 56b^5c^3 + 70b^4c^4 + 56b^3c^5 + 28b^2c^6 + 8bc^7 + c^8$　**73.** $256m^4 - 256m^3n + 96m^2n^2 - 16mn^3 + n^4$　**75.** $35a^4b^3$

Chapter 11 Test

1. $-\dfrac{1}{5}, \dfrac{1}{6}, -\dfrac{1}{7}, \dfrac{1}{8}, -\dfrac{1}{9}$　**2.** 247　**3.** $a_n = \dfrac{2}{5}\left(\dfrac{1}{5}\right)^{n-1}$　**4.** $a_n = (-1)^n 9n$　**5.** 155　**6.** -330　**7.** $\dfrac{144}{5}$

8. 1　**9.** 10　**10.** -60　**11.** $a^6 - 6a^5b + 15a^4b^2 - 20a^3b^3 + 15a^2b^4 - 6ab^5 + b^6$　**12.** $32x^5 + 80x^4y + 80x^3y^2 + 40x^2y^3 + 10xy^4 + y^5$

13. 925 people; 250 people initially　**14.** $1 + 3 + 5 + 7 + 9 + 11 + 13 + 15$; 64 shrubs　**15.** 33.75 cm, 218.75 cm　**16.** 320 cm

17. 304 ft; 1600 ft　**18.** $\dfrac{14}{33}$

Chapter 11 Cumulative Review

1. a. -5　**b.** 3　**c.** $-\dfrac{1}{8}$　**d.** -4　**e.** $\dfrac{1}{4}$　**f.** undefined; Sec. 1.3, Ex. 5　**2. a.** $-a - 3$

b. $7x + 3$　**c.** $-7x - 23$; Sec. 1.4　**3.** \$2350; Sec. 1.6, Ex. 5　**4.** \$325; Sec. 1.6　**5. a.** $(-5, 2)$　**b.** $(1, 0)$; Sec. 4.5, Ex. 2

6. $f(x) = \dfrac{3}{2}x - \dfrac{13}{2}$; Sec. 3.5　**7. a.** $15x^7$　**b.** $-8x^4p^{12}$; Sec. 5.1, Ex. 2　**8.** $-5, -1, 1$; Sec. 5.8　**9.** $x^3 - 4x^2 - 3x + 11 + \dfrac{12}{x+2}$; Sec. 6.5,

Ex. 2　**10.** $\dfrac{34 - 9a}{15(a-2)}$; Sec. 6.2　**11. a.** $5\sqrt{2}$　**b.** $2\sqrt[3]{3}$　**c.** $\sqrt{26}$　**d.** $2\sqrt[4]{2}$; Sec. 7.3, Ex. 3　**12.** 3; Sec. 7.6　**13.** 10%; Sec. 8.1, Ex. 9

14. a. $\dfrac{\sqrt[3]{36x^2}}{3x}$　**b.** $3 + 2\sqrt{2}$; Sec. 7.5　**15.** 2, 7; Sec. 8.3, Ex. 4　**16.** $-\dfrac{7}{6}, -3$; Sec. 6.6　**17.** $\left(-\dfrac{7}{2}, -1\right)$; Sec. 8.4, Ex. 5

18. 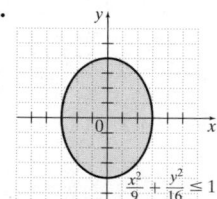 $x = -2$　$V(-2, -6)$; Sec. 8.5　**19.** $\dfrac{25}{4}$ ft; $\dfrac{5}{8}$ sec; Sec. 8.6, Ex. 5　**20.** $\left(-\dfrac{3}{2}, -\dfrac{81}{4}\right)$; Sec. 8.6　**21. a.** 25; 7　**b.** $x^2 + 6x + 9; x^2 + 3$;

Sec. 9.1, Ex. 2　**22.** $f^{-1}(x) = \dfrac{3 - x}{2}$ Sec. 9.2　**23.** $f^{-1} = \{(1, 0), (7, -2), (-6, 3), (4, 4)\}$; Sec. 9.2, Ex. 3

24. a. 7; 3　**b.** $x^2 + 2x - 1; x^2 - 1$; Sec. 9.1　**25. a.** 4　**b.** $\dfrac{3}{2}$　**c.** 6;　**d.** 1.431, Sec. 9.3, Ex. 4　**26. a.** 5

b. -3　**c.** $\dfrac{1}{32}$; Sec. 9.4　**27. a.** 2　**b.** -1　**c.** 3　**d.** 6; Sec. 9.4, Ex. 5　**28. a.** 3　**b.** 5/3　**c.** 8/3;

Sec. 9.3　**29. a.** $\log_{11} 30$　**b.** $\log_3 6$　**c.** $\log_2(x^2 + 2x)$; Sec. 9.5, Ex. 1　**30. a.** 5　**b.** -3　**c.** 1/5　**d.** 4; Sec. 9.6

31. \$2509.30; Sec. 9.6, Ex. 8　**32. a.** $\log_6 20$　**b.** $\log_8 3$　**c.** $\log_2 \dfrac{x^5}{(x-1)^2}$; Sec. 9.5　**33.** $\dfrac{\log 7}{\log 3} = 1.7712$; Sec. 9.7, Ex. 1　**34.** 34.7 yr.; Sec. 9.6

35. 18; Sec. 9.7, Ex. 2　**36.** 5/8; Sec. 9.7　**37.** ; Sec. 10.2, Ex. 4　**38.** $\sqrt{101}$ units; Sec. 10.1　**39.** $\left(2, \sqrt{2}\right)$; Sec. 10.3, Ex. 1

40. $(0, -6)$ $(6, 0)$; Sec. 10.3　**41.** ; Sec. 10.4, Ex. 1

$(-4, 5)$　$(4, 5)$

$(-4, -5)$ $(4, -5)$

$\dfrac{x^2}{16} - \dfrac{y^2}{25} = 1$

$\dfrac{x^2}{9} + \dfrac{y^2}{16} \leq 1$

42. ; Sec. 10.4   **43.** $0, 3, 8, 15, 24$; Sec. 11.1, Ex. 1   **44.** $2/3$; Sec. 11.1   **45.** 72; Sec. 11.2, Ex. 3   **46.** 6250; Sec. 11.2

47. a. $\dfrac{7}{2}$ **b.** 56; Sec. 11.3, Ex. 1   **48. a.** 40 **b.** 15; Sec. 11.3   **49.** 465; Sec. 11.4, Ex. 2

50. $15x^4y^2$; Sec. 11.5

APPENDIX A   EXERCISE SET

Sec. 1.5   **1.** $-\dfrac{8}{3}$ **2.** 6 **3.** 0 **4.** $\varnothing$ Sec. 3.2   **1.** $(-1, \infty)$ **2.** $\left(-\infty, \dfrac{3}{2}\right]$ **3.** 3 **4.** all real numbers or $(-\infty, \infty)$

Sec. 3.3   **1.** $[-1, 3]$ **2.** $(-1, 6)$ **3.** 5.1 **4.** $(-\infty, -4)$ **5.** $(-\infty, -3]$ Sec. 3.4   **1.** $\dfrac{5}{3}, -3$ **2.** $(-\infty, -2) \cup (2, \infty)$ **3.** 4 **4.** $\dfrac{1}{2}$

5. $[19, \infty)$ Sec. 3.5   **1.** $(-\infty, 4] \cup [18, \infty)$ **2.** $4, 18$ **3.** $(-3, 2)$ **4.** $-\dfrac{7}{9}, \dfrac{7}{9}$ **5.** 10 Sec. 5.8   **1.** $9, -\dfrac{1}{2}$ **2.** 0 **3.** $7, -\dfrac{21}{2}$

4. $\left(-\infty, \dfrac{1}{3}\right) \cup (3, \infty)$ **5.** $(-\infty, 2)$ **6.** $\left(-\infty, \dfrac{5}{4}\right)$ Sec. 6.6   **1.** $7, -2$ **2.** $-\dfrac{8}{5}, 2$ **3.** $(-\infty, 0]$ **4.** all real numbers or $(-\infty, \infty)$

5. $[-1, 5]$ **6.** $0, 5, -5$ Sec. 7.6   **1.** $\dfrac{1}{3}, -5$ **2.** $\dfrac{9}{5}$ **3.** $5, -\dfrac{1}{3}$ **4.** $\dfrac{25}{3}$ **5.** $(-\infty, -4]$ **6.** $(-\infty, -2) \cup (8, \infty)$ Sec. 8.2   **1.** $2 \pm \sqrt{17}$

2. $\dfrac{5 \pm \sqrt{17}}{2}$ **3.** $3, 2$ **4.** $(10, \infty)$ **5.** 3 **6.** $0, -\dfrac{22}{3}$ **7.** 2 **8.** $(-26, 34)$ Sec. 8.4   **1.** $5, -2$ **2.** $(-2, 5)$ **3.** -4

4. $(-\infty, -4] \cup (10, \infty)$ **5.** 23 **6.** $\dfrac{2 \pm \sqrt{10}}{5}$ **7.** $\varnothing$ **8.** 15 **9.** $(1, \infty)$ **10.** $(-\infty, 1]$ Sec. 10.7   **1.** 3 **2.** $\dfrac{\log 5 - 7 \log 2}{\log 2}$

3. $(-\infty, 0) \cup (9, \infty)$ **4.** $\dfrac{7}{6}$ **5.** $4, -1$ **6.** $\dfrac{-1 \pm \sqrt{10}}{3}$ **7.** $(2, 4]$ **8.** $\dfrac{61}{16}$ **9.** 1 **10.** $\left(-\infty, \dfrac{2}{3}\right) \cup (1, \infty)$

APPENDIX C   REVIEW OF VOLUME AND SURFACE AREA

1. $V = 72$ cu. in.; $SA = 108$ sq. in.   **3.** $V = 512$ cu. cm; $SA = 384$ sq. cm

5. $V = 4\pi$ cu. yd ≈ 12.56 cu. yd; $SA = \left(2\sqrt{13}\pi + 4\pi\right)$ sq yd ≈ 35.20 sq. yd

7. $V = \dfrac{500}{3}\pi$ cu. in. $\approx 523\dfrac{17}{21}$ cu. in.; $SA = 100\pi$ sq. in. $\approx 314\dfrac{2}{7}$ sq. in.   **9.** $V = 48$ cu. cm; $SA = 96$ sq. cm   **11.** $2\dfrac{10}{27}$ cu. in.   **13.** 26 sq. ft

15. $10\dfrac{5}{6}$ cu. in.   **17.** 960 cu. cm   **19.** 196π sq. in.   **21.** $7\dfrac{1}{2}$ cu. ft   **23.** $12\dfrac{4}{7}$ cu. cm

APPENDIX D   AN INTRODUCTION TO USING A GRAPHING UTILITY

Viewing Window and Interpreting Window Settings Exercise Set

1. yes   **3.** no   **5.** answers may vary   **7.** answers may vary   **9.** answers may vary   **11.** Xmin $= -12$ Ymin $= -12$
Xmax $= 12$ Ymax $= 12$
Xscl $= 3$ Yscl $= 3$

13. Xmin $= -9$ Ymin $= -12$ **15.** Xmin $= -10$ Ymin $= -25$ **17.** Xmin $= -10$ Ymin $= -30$ **19.** Xmin $= -20$ Ymin $= -30$
Xmax $= 9$ Ymax $= 12$ Xmax $= 10$ Ymax $= 25$ Xmax $= 10$ Ymax $= 30$ Xmax $= 30$ Ymax $= 50$
Xscl $= 1$ Yscl $= 2$ Xscl $= 2$ Yscl $= 5$ Xscl $= 1$ Yscl $= 3$ Xscl $= 5$ Yscl $= 10$

Graphing Equations and Square Viewing Window Exercise Set   **1.** Setting B   **3.** Setting B   **5.** Setting B

7.   **9.**   **11.**   **13.**

15.   **17.**   **19.**   **21.**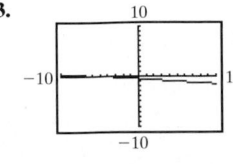

GRAPHING ANSWER SECTION

The Graphing Answer Section contains answers to exercises requiring graphical solutions, Spotlight on Decision Making exercises, and Chapter Projects.

CHAPTER 1 REAL NUMBERS, ALGEBRAIC EXPRESSIONS, AND EQUATIONS

1.2 Spotlight on Decision Making

Master mailing (5% discount): Wendy Alvarez, Clay Darden, Gabriella Evans, Ira Greenberg, Maude Issacson, Harold Jones, Avi Khalosef, Ricardo Martinez; Safe-driver discount (additional 10% discount): Wendy Alvarez, Gabriella Evans, Ira Greenberg, Maude Issacson, Harold Jones.

1.3 Spotlight on Decision Making

Hawaii; the possible bonus is $2000.

1.4 Spotlight on Decision Making

a. green **b.** dark blue **c.** green **d.** dark blue **e.** tan

1.6 Spotlight on Decision Making

12.6% of Marston households must cope with hunger. Answers may vary.

1.8 Spotlight on Decision Making

The patient would ingest 97.56 grams of fluoride. This amount is safe for a 50-pound child.

Chapter Project

1. *Wall Street Journal*: 17,854; *New York Times*: 12,191; *New York Daily News*: 30,010; *New York Post*: 89,909
All newspapers gained circulation by the amounts given.
2. The *New York Post* experienced the greatest change in circulation.

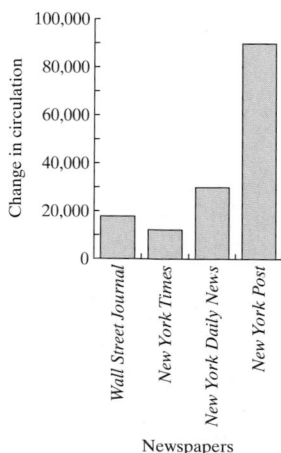

3. Total circulation in 2000: 4,008,345; total circulation in 2001: 4,158,309. Total circulation increased by 149,964.   **4.** Answers may vary. One possible answer is the increased interest in current events.   **5.** Approximately 0.56 newspapers per person.   **6.** Approximately 0.25 newspapers per person.   **7.** Answers may vary. One possible answer is that the figure found in Question 6 is more meaningful because New York City-based newspapers are likely to circulate in the entire metropolitan area. Neither figure could be meaningful since these newspapers may circulate nationally (or even internationally).   **8.** (Use 261 days per year.) *Wall Street Journal*: $1,780,605/day, $464,737,905/year; *New York Times*: $1,109,371/day, $289,545,831/year; *New York Daily News*: $367,236.5/day, $95,848,726.50/year; *New York Post*: $266,930/day, $69,668,730/year.   **9.** Answers may vary.

CHAPTER 2 GRAPHS AND FUNCTIONS

2.1 Spotlight on Decision Making

It appears that Felix will hit land along the coast of Mexico near the U.S. border. However, the gulf coast of Texas is not out of danger.

Exercise Set 2.1

1.

2.

3.

4.

5.

6.

7.

8.

9.

10.

41.

42.

57.

58.

59.

60.

61.

62.

63.

64.

65.

66.

67.

68.

69.

70.

71.

72.

73.

74.

75.

76.

77.

78.

79.

80.

81.

82.

83.

84.

85.

86.

91. a.

92. a.

103.

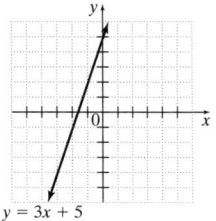

$y = 3x + 5$

104.

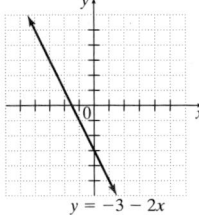

$y = -3 - 2x$

105.

106.

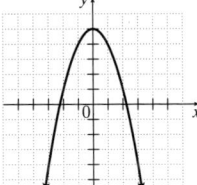

Exercise Set 2.2

115.

116.

117.

118.

119.

120.

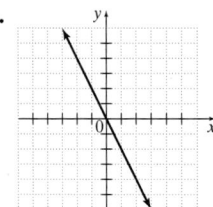

Exercise Set 2.3

1. $y = -2x$

2.

3.

4.

5.

6.

7.

8.

13.

14.

15.

16.

17.

18.

19.

20.

23.

24.

25.

26.

27.

28.

35.

36.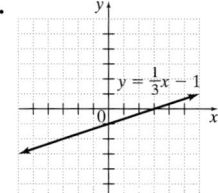
$y = \frac{1}{3}x - 1$

37.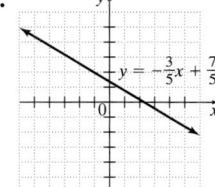
$y = -\frac{3}{5}x + \frac{7}{5}$

38.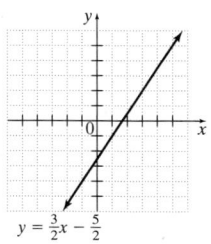
$y = \frac{3}{2}x - \frac{5}{2}$

39.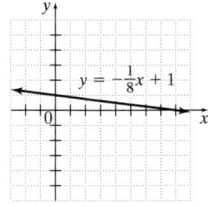
$y = -\frac{1}{8}x + 1$

40.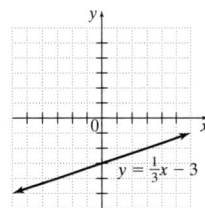
$y = \frac{1}{3}x - 3$

41.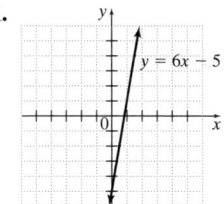
$y = 6x - 5$

42.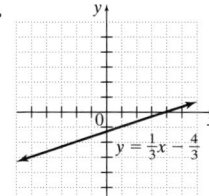
$y = \frac{1}{3}x - \frac{4}{3}$

43.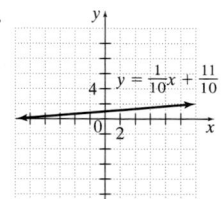
$y = \frac{1}{10}x + \frac{11}{10}$

44.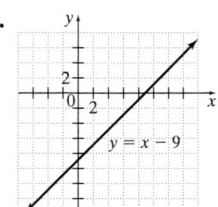
$y = x - 9$

45.
$y = \frac{3}{2}$

46.
$x = \frac{3}{2}$

47.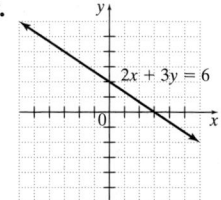
$2x + 3y = 6$

48.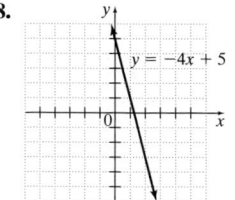
$y = -4x + 5$

49.
$x = -3$

50.
$y = 6$

51.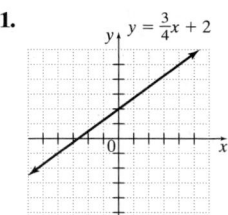
$y = \frac{3}{4}x + 2$

52.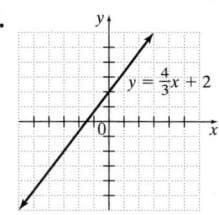
$y = \frac{4}{3}x + 2$

53.

54. $y = -x$

55. 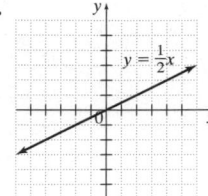 $y = \frac{1}{2}x$

56. $y = -2x$

57. 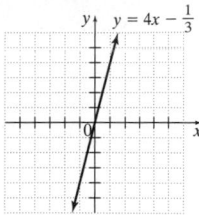 $y = 4x - \frac{1}{3}$

58. 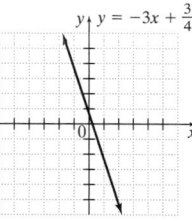 $y = -3x + \frac{3}{4}$

59. $x = -3$

60. $y = 3$

73. b.

Cost of driving (dollars)

$y = 0.2x + 24$

Distance driven (miles)

74. b.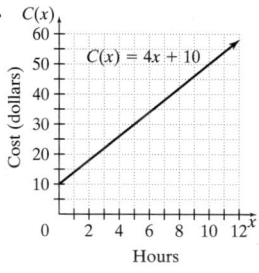

Cost (dollars)

$C(x) = 4x + 10$

Hours

2.4 Spotlight on Decision Making

Answers may vary.

Exercise Set 2.4

101.

102.

103. a.

b.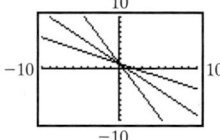

2.5 Spotlight on Decision Making

Answers may vary.

Exercise Set 2.5

7. $y = 5x$

8. 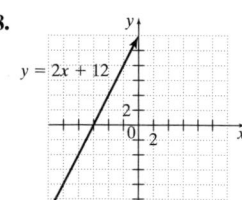 $y = 2x + 12$

9. 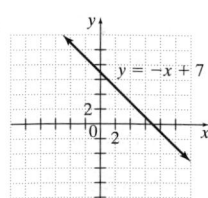 $y = -x + 7$

10. 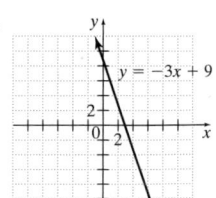 $y = -3x + 9$

11. 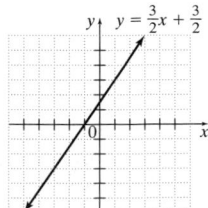 $y = \frac{3}{2}x + \frac{3}{2}$

12. 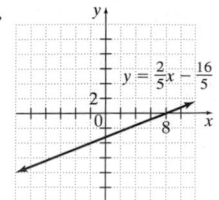 $y = \frac{2}{5}x - \frac{16}{5}$

83.

84.

85.

86.

Exercise Set 2.6

25.

26.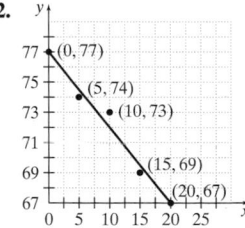

Chapter Project

1. **2.**

3. Answers may vary. One possibility: $(0, 77), (20, 67)$; $y = -\dfrac{1}{2}x + 77$ **4.** Slope: $m = -\dfrac{1}{2}$; the percentage of children who live with both parents declines by $\dfrac{1}{2}$ or 0.5 each year; yes **5.** 52; in the year 2030, 52% of U.S. children will live with both parents. **6.** Answers may vary. **8.** In the year 2035

Chapter 2 Review

1.

2.

7.

8.

9.

10.

11.

12.

13.

14.

39.

40.

41.

46.

47.

48.

49.

50.

51.

52.

53.

54. b.

70.

71.

72.

73.

92. a.

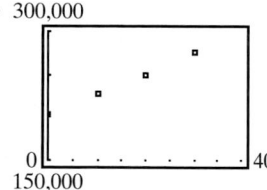

Chapter 2 Test

1.

3.

4.

5.

6.

27. a.   **b.**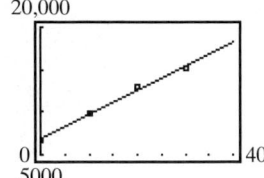

CHAPTER 3 EQUATIONS AND INEQUALITIES

3.1   Spotlight on Decision Making

10.2% of 39,400 is 4018.8, which is less than 4950. Therefore, choose (a). One example of other information is the average number of persons in a household, both in the United States and in Marston.

3.2   Spotlight on Decision Making

Temporary classrooms will be needed. The funding needs to be added to the budget within the next six years.

3.6   Spotlight on Decision Making

Extra tall: The point corresponding to 160 lb and 69 in. lies within the region designated as extra tall.

Exercise Set 3.6

1.   **2.**   **3.**   **4.**   **5.**

6.   **7.**   **8.**   **9.**   **10.**

11.   **12.**   **15.**   **16.**   **17.**

18.

19.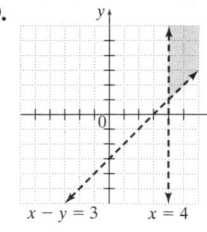
$x - y = 3$ $x = 4$

20.

21.

22.

23.

24.

25.

26.

27.

28.

29.

30.

31.

32.

33.

34.

35.

36.

37.

38.

39.

40.

41.

42.

43.

44.

45.

46.

69.

70.

71.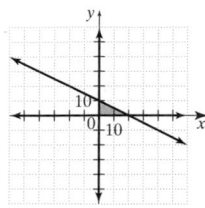

Chapter Project

1. $|x - 15,000| > 1800$; $|x - 50,000| > 6000$; $|x - 450,000| > 54,000$; $|x - 275,000| > 33,000$;
$|x - 90,000| > 10,800$; $|x - 30,000| > 3600$; $|x - 120,000| > 14,400$; $|x - 70,000| > 8400$;
$|x - 50,000| > 6000$; $|x - 140,000| > 16,800$; $|x - 80,000| > 9600$; $|x - 300,000| > 36,000$;
$|x - 400,000| > 48,000$; $|x - 50,000| > 6000$; $|x - 100,000| > 12,000$; $|x - 200,000| > 24,000$;
$|x - 250,000| > 30,000$; $|x - 100,000| > 12,000$

2. $13,200 < x < 16,800$; $44,000 < x < 56,000$; $396,000 < x < 504,000$; $242,000 < x < 308,000$;
$79,200 < x < 100,800$; $26,400 < x < 33,600$; $105,600 < x < 134,400$; $61,600 < x < 78,400$;
$44,000 < x < 56,000$; $123,200 < x < 156,800$; $70,400 < x < 89,600$; $264,000 < x < 336,000$;
$352,000 < x < 448,000$; $44,000 < x < 56,000$; $88,000 < x < 112,000$; $176,000 < x < 224,000$;
$220,000 < x < 280,000$; $88,000 < x < 112,000$

3.

Budget category	Budgeted amount	Minimum allowed	Actual expense	Maximum allowed	Within budget?	Amount over/ under budget
Immunization Programs	$15,000	$13,200	$14,800	$16,800	Yes	Under $200
Inspections	$50,000	$44,000	$41,900	$56,000	Yes	Under $8100
Fire Dept. Equipment	$450,000	$396,000	$375,000	$504,000	Yes	Under $75,000
Fire Dept. Salaries	$275,000	$242,000	$268,500	$308,000	Yes	Under $6500
Book/Periodical/Purchases	$90,000	$79,200	$107,300	$100,800	No	Over $17,300
Library Equipment	$30,000	$26,400	$29,000	$33,600	Yes	Under $1000
Library Salaries	$120,000	$105,600	$118,400	$134,400	Yes	Under $1600
Park Maintenance	$70,000	$61,600	$82,500	$78,400	No	Over $12,500
Playground Equipment	$50,000	$44,000	$45,000	$56,000	Yes	Under $5000
Parks & Rec. Salaries	$140,000	$123,200	$118,000	$156,800	Yes	Under $22,000
Summer Programs	$80,000	$70,400	$96,200	$89,600	No	Over $16,200
Police Dept. Equipment	$300,000	$264,000	$328,000	$336,000	No	Over $28,000
Police Dept. Salaries	$400,000	$352,000	$405,000	$448,000	No	Over $5000
Recycling	$50,000	$44,000	$48,100	$56,000	Yes	Under $1900
Sewage	$100,000	$88,000	$92,500	$112,000	Yes	Under $7500
Snow Removal & Road Salt	$200,000	$176,000	$268,300	$224,000	No	Over $68,300
Street Maintenance	$250,000	$220,000	$284,000	$280,000	No	Over $34,000
Water Treatment	$100,000	$88,000	$94,100	$112,000	Yes	Under $5900

Categories to be researched: inspections, fire dept. equipment, book/periodical purchases, park maintenance, parks and rec. salaries, summer programs, snow removal and road salt, and street maintenance

4. Answers may vary. Inspections: fewer violations reported than expected; fire dept. equipment: fewer repairs and replacements of equipment than anticipated; book/periodical purchases: rising costs of publications; park maintenance: higher usage than expected; parks and rec. salaries: open position left vacant; summer programs: more participation than anticipated; snow removal and road salt: more snowfall than anticipated; street maintenance: more pavement damage because of severe weather.   5. Answers may vary.   6. Answers may vary.

Chapter 3 Review

57.

58.

59.

60.

61.

62.

63.

64.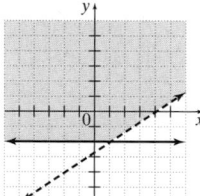

Chapter 3 Test

17.

18.

19.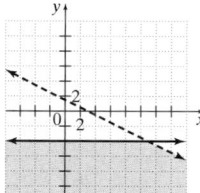

CHAPTER 4 SYSTEMS OF EQUATIONS

4.1   Spotlight on Decision Making

The cost of the bottled water for 52 weeks is $260. The cost of the filtered water for 52 weeks is $182.60. Choose the second option for the lower long-term cost.

Exercise Set 4.1

7.

8.

9.

10.

11.

12.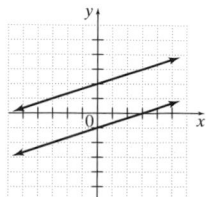

4.3   Spotlight on Decision Making

The AT&T plan is less expensive if usage is less than 200 minutes per month; the Qwest plan is less expensive when usage is greater than 200 minutes per month. Base the decision on anticipated usage. If long distance calls average 4 hours per month, choose the Qwest plan.

4.4   Spotlight on Decision Making

According to the model, 25 months are available to plan the changes. A consulting firm will not be needed.

Exercise Set 4.5

51.   **52.**   **53.**   **54.**

Chapter Project

1.

2. $A: y = -1.732x + 88.62; B: y = 0.577x - 18.004$

3. $(46.178, 8.640)$

4. Answers may vary.

Chapter 4 Review

1.   **2.**   **3.**   **4.**   **5.**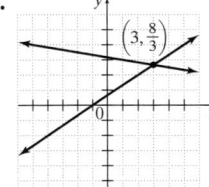

Chapter 4 Test

3.   **4.**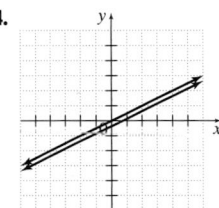

Chapter 4 Cumulative Review

25.   **28.**   **39.**   **40.**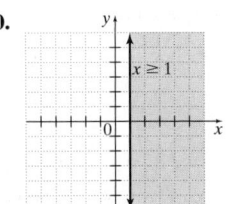

CHAPTER 5 EXPONENTS, POLYNOMIALS, AND POLYNOMIAL FUNCTIONS

5.1 Spotlight on Decision Making

Use a magnification setting of $43,000\times$.

5.4 Spotlight on Decision Making

Answers may vary.

Chapter Project

1. Icecaps and glaciers

2., 3.

	Water Volume (cubic Kilometers)	Percent
Atmosphere	1.3×10^4	9.56×10^{-4}
Average in stream channels	1.0×10^3	7.35×10^{-5}
Freshwater lakes	1.2×10^5	8.82×10^{-3}
Groundwater	8.3×10^6	6.10×10^{-1}
Icecaps and glaciers	2.9×10^7	2.13×10^0
Oceans	1.32×10^9	9.71×10^1
Saline lakes and inland seas	1.0×10^5	7.35×10^{-3}
Water in soil above ground water	6.7×10^5	4.93×10^{-2}
Total	1.36×10^9	100.00

4. 5.0225 cm   **5.** $y = 0.0020x^2 + 0.2662x + 2.4621$ **6. a.** 10.3671 cm   **b.** 20.7721 cm   **c.** 33.6771 cm   **d.** 49.0821 cm
7. Answers may vary.

Chapter 5 Cumulative Review

15.   **16.**   **19.**   **20.**   **25.**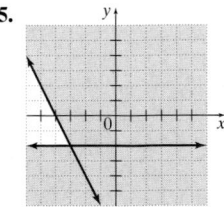

CHAPTER 6 RATIONAL EXPRESSIONS

6.1 Spotlight on Decision Making

4203; It dissipates 2049 watts of power, which is the closest of the three to 2000 watts.

Exercise Set 6.1

91.   **92.**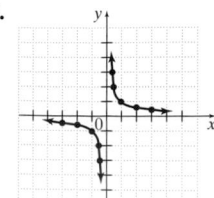

Exercise Set 6.6

72.   **73.**   **74.**   **75.**

6.7 Spotlight on Decision Making

Yes; the actual altitude is 7500 meters. Thus, the altimeter reading is about 1% off.

6.8 Spotlight on Decision Making

Yes; buy two gallons and one quart. This is enough paint for 1012.5 sq. ft. and will cost $48.95.

Exercise Set 6.8

72.   **73.**   **74.**   **75.**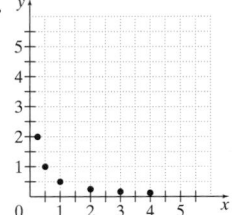

Chapter Project

1. $E = 0.015v^3$

2.

Wind Speed (miles per hour)	Electricity Generated Hourly (in watts)
15	50.625
17	73.695
19	102.885
21	138.915
23	182.505
25	234.375
27	295.245
29	365.835
31	446.865
33	539.055
35	643.125

a. 19 mph   **b.** 30 mph

3. ; increasing slowly at first, then rapidly   **4.** No; if the wind speed were 15 mph half the time and 25 mph the other half, the windmills would generate an average of 142.5 watts per hour.   **5.** 18.75 watts per hour; $\frac{1}{8}c$

Chapter 6 Cumulative Review

11.   **12.**   **15.**   **16.**   **19.**

20.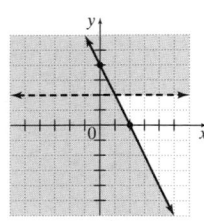

CHAPTER 7 RATIONAL EXPONENTS, RADICALS, AND COMPLEX NUMBERS

7.1 Spotlight on Decision Making

It is not possible for S/2001U1 to have an oxygen atmosphere, since the average speed of an oxygen molecule exceeds the moon's escape velocity of approximately 442 meters per second.

Exercise Set 7.1

85.   **86.**   **87.**   **88.**   **97.**

98.   **99.**   **100.**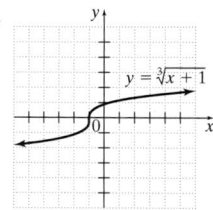

7.2 Spotlight on Decision Making

No; for the data since 1994, the model consistently underestimates usage by as much as 12%.

7.6 Spotlight on Decision Making

1: B; 2: C; 3: A; 4: C; 5: B

Chapter Project

Answers may vary.
Theoretical Times: 30 cm: 1.10 sec; 55 cm: 1.49 sec; 70 cm: 1.68 sec; Theoretical Lengths: 1 sec: 24.82 cm; 1.25 sec: 38.79 cm; 2 sec: 99.30 cm

Chapter 7 Review

24.

25.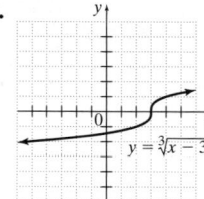

Chapter 7 Test

32.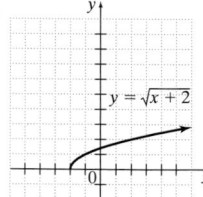

Chapter 7 Cumulative Review

11.

12.

15.

16.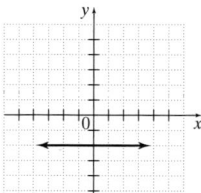

CHAPTER 8 QUADRATIC EQUATIONS AND FUNCTIONS

8.2 Spotlight on Decision Making

The RDA for a 40-pound child is 465.878 micrograms of vitamin A; this patient needs about 66 more micrograms each day. She will need to gain about 4.2 pounds before a daily intake of 500 micrograms is recommended.

Exercise Set 8.4

55.

56.

57.

58.

59.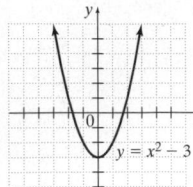

$y = x^2 - 3$

60.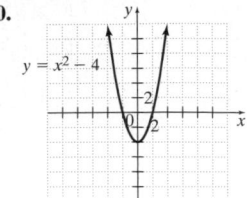

$y = x^2 - 4$

61.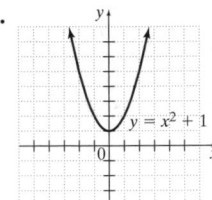

$y = x^2 + 1$

62.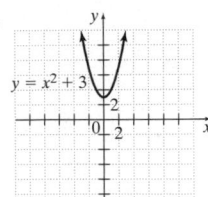

$y = x^2 + 3$

Exercise Set 8.5

1.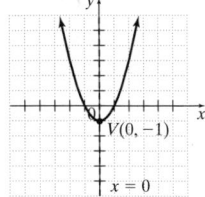

$V(0, -1)$
$x = 0$

2.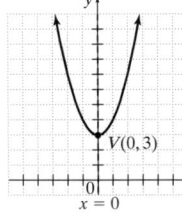

$V(0, 3)$
$x = 0$

3.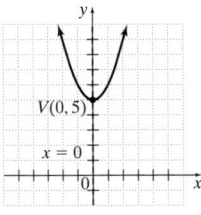

$V(0, 5)$
$x = 0$

4.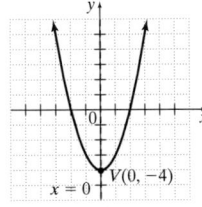

$V(0, -4)$
$x = 0$

5.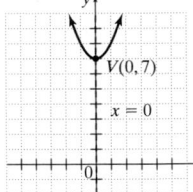

$V(0, 7)$
$x = 0$

6.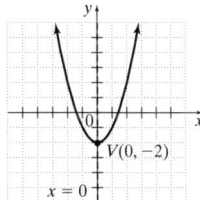

$V(0, -2)$
$x = 0$

7.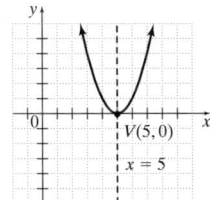

$V(5, 0)$
$x = 5$

8.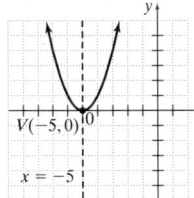

$V(-5, 0)$
$x = -5$

9.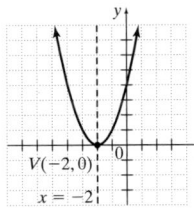

$V(-2, 0)$
$x = -2$

10.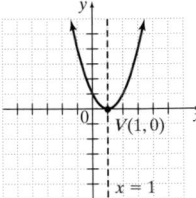

$V(1, 0)$
$x = 1$

11.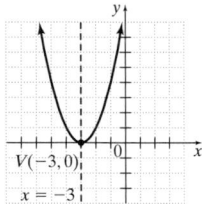

$V(-3, 0)$
$x = -3$

12.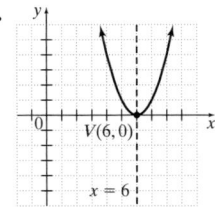

$V(6, 0)$
$x = 6$

13.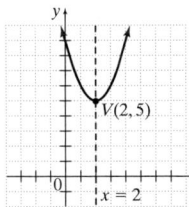

$V(2, 5)$
$x = 2$

14.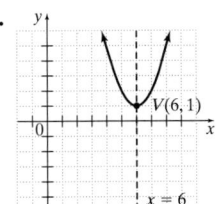

$V(6, 1)$
$x = 6$

15.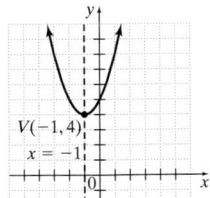

$V(-1, 4)$
$x = -1$

16.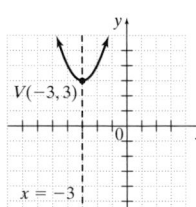

$V(-3, 3)$
$x = -3$

17.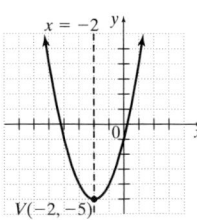

$x = -2$
$V(-2, -5)$

18.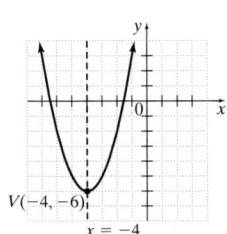

$V(-4, -6)$
$x = -4$

19.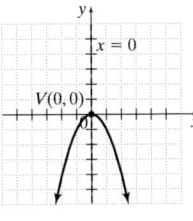

$x = 0$
$V(0, 0)$

20.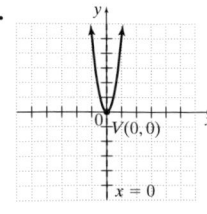

$V(0, 0)$
$x = 0$

21.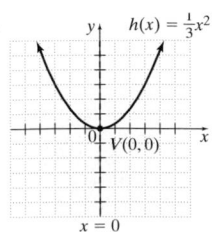

$h(x) = \frac{1}{3}x^2$
$V(0, 0)$
$x = 0$

22.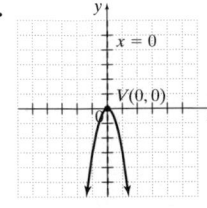

$x = 0$
$V(0, 0)$

23.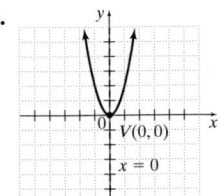

$V(0, 0)$
$x = 0$

24.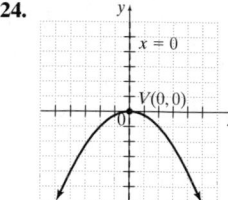

$x = 0$
$V(0, 0)$

25.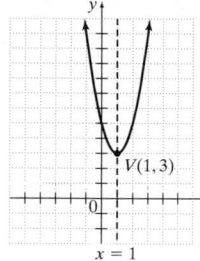

$V(1, 3)$
$x = 1$

26.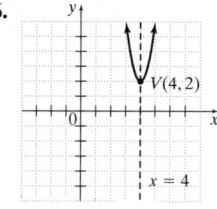

$V(4, 2)$
$x = 4$

27.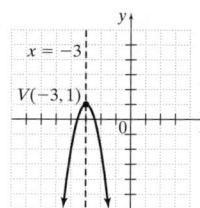

$x = -3$
$V(-3, 1)$

28.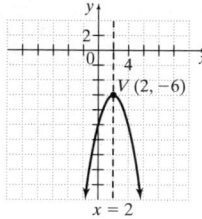

$V(2, -6)$
$x = 2$

29.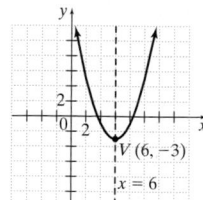

$V(6, -3)$
$x = 6$

30.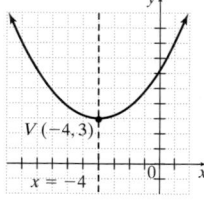

$V(-4, 3)$
$x = -4$

31.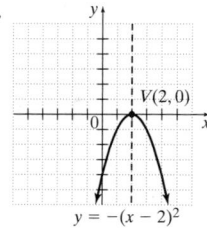

$V(2, 0)$
$y = -(x - 2)^2$

32.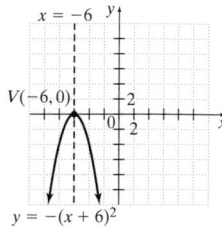

$x = -6$
$V(-6, 0)$
$y = -(x + 6)^2$

33.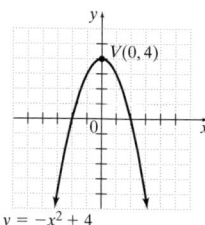

$V(0, 4)$
$y = -x^2 + 4$

34.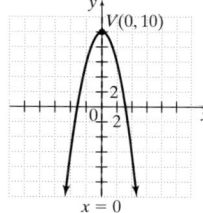

$V(0, 10)$
$x = 0$

35.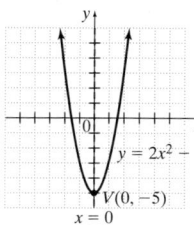

$y = 2x^2 - 5$
$V(0, -5)$
$x = 0$

36.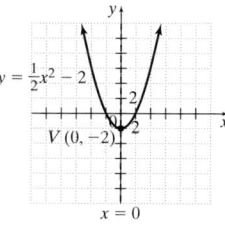

$y = \frac{1}{2}x^2 - 2$
$V(0, -2)$
$x = 0$

37.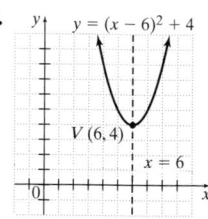

$y = (x - 6)^2 + 4$
$V(6, 4)$
$x = 6$

38.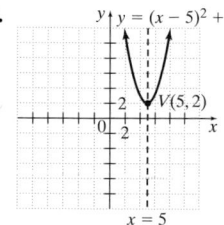

$y = (x - 5)^2 + 2$
$V(5, 2)$
$x = 5$

39.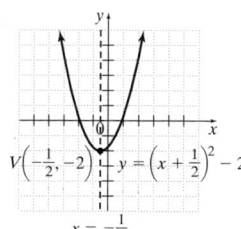

$V\left(-\frac{1}{2}, -2\right)$ $y = \left(x + \frac{1}{2}\right)^2 - 2$
$x = -\frac{1}{2}$

40.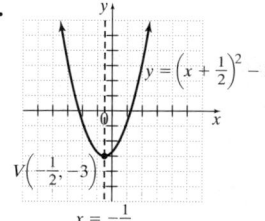

$y = \left(x + \frac{1}{2}\right)^2 - 3$
$V\left(-\frac{1}{2}, -3\right)$
$x = -\frac{1}{2}$

41.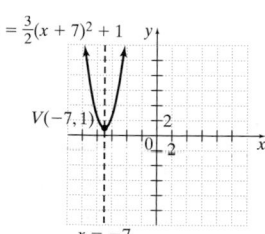

$y = \frac{3}{2}(x + 7)^2 + 1$
$V(-7, 1)$
$x = -7$

42.

$x = 1$
$V(1, -5)$
$y = -\frac{3}{2}(x - 1)^2 - 5$

43.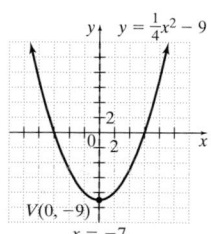

$y = \frac{1}{4}x^2 - 9$
$V(0, -9)$
$x = -7$

44.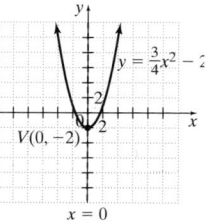

$y = \frac{3}{4}x^2 - 2$
$V(0, -2)$
$x = 0$

45.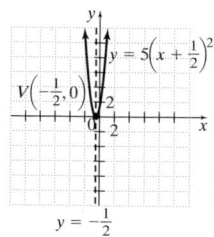

$y = 5\left(x + \frac{1}{2}\right)^2$
$V\left(-\frac{1}{2}, 0\right)$
$y = -\frac{1}{2}$

46.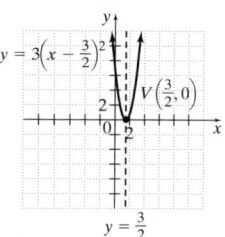

$y = 3\left(x - \frac{3}{2}\right)^2$
$V\left(\frac{3}{2}, 0\right)$
$y = \frac{3}{2}$

47.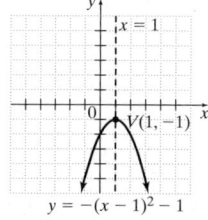

$x = 1$
$V(1, -1)$
$y = -(x - 1)^2 - 1$

48.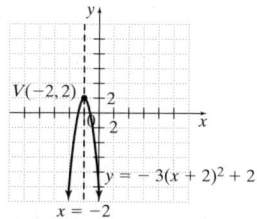

$V(-2, 2)$
$y = -3(x + 2)^2 + 2$
$x = -2$

49. $y = \sqrt{3}(x + 5)^2 + \frac{3}{4}$

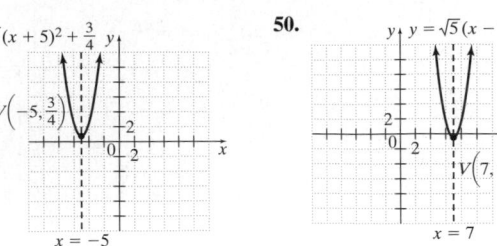

50. $y = \sqrt{5}(x - 7)^2 - \frac{1}{2}$

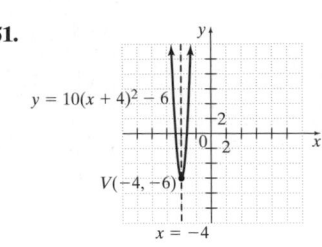

51. $y = 10(x + 4)^2 - 6$

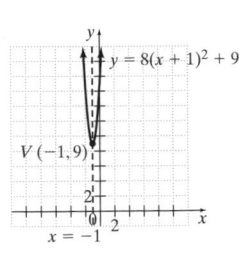

52. $y = 8(x + 1)^2 + 9$

53.

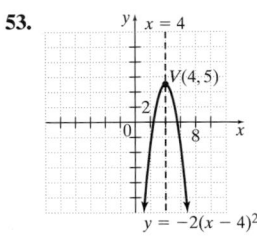

$y = -2(x - 4)^2 + 5$

54.

71.

72.

73.

74.

75.

76.

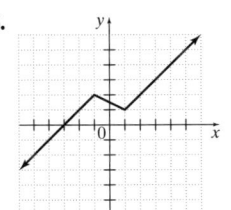

8.6 Spotlight on Decision Making

$12.50; The highest point on the graph, representing the most receipts, corresponds to a ticket price of $12.50

Exercise Set 8.6

13.

14.

15.

16.

17.

18.

19.

20.

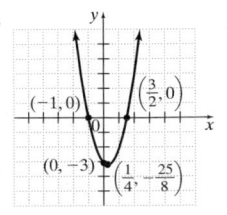

21. $y = x^2 + 8x + 15$

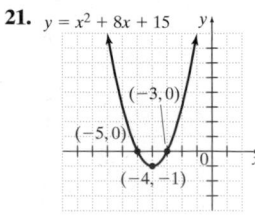

22. $y = (x + 5)^2 - 16$

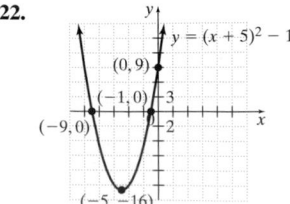

23. $y = x^2 - 6x + 5$

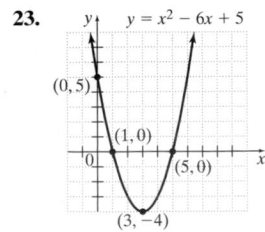

24. $y = x^2 - 4x + 3$

25.
$y = x^2 - 4x + 5$
$(0, 5)$
$(2, 1)$

26.
$y = x^2 - 6x + 11$
$(0, 11)$
$(3, 2)$

27.
$y = 2x^2 + 4x + 5$
$(0, 5)$
$(-1, 3)$

28.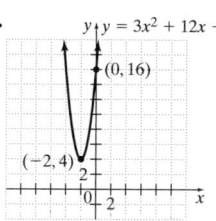
$y = 3x^2 + 12x + 16$
$(0, 16)$
$(-2, 4)$

29.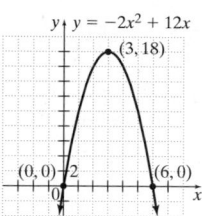
$y = -2x^2 + 12x$
$(3, 18)$
$(0, 0)$
$(6, 0)$

30.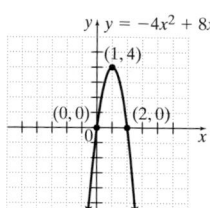
$y = -4x^2 + 8x$
$(1, 4)$
$(0, 0)$
$(2, 0)$

31.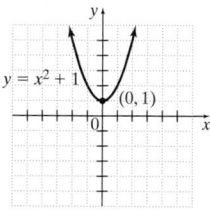
$y = x^2 + 1$
$(0, 1)$

32.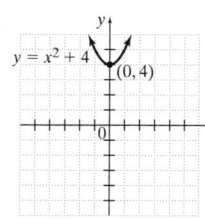
$y = x^2 + 4$
$(0, 4)$

33.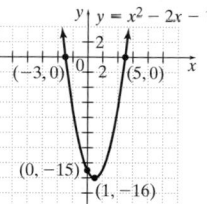
$y = x^2 - 2x - 15$
$(-3, 0)$
$(5, 0)$
$(0, -15)$
$(1, -16)$

34.
$y = x^2 - 4x + 3$
$(0, 3)$
$(1, 0)$
$(3, 0)$
$(2, -1)$

35.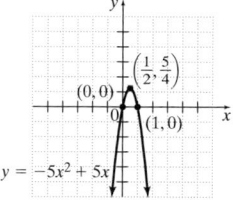
$(0, 0)$
$\left(\frac{1}{2}, \frac{5}{4}\right)$
$(1, 0)$
$y = -5x^2 + 5x$

36.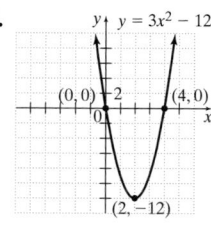
$y = 3x^2 - 12x$
$(0, 0)$
$(4, 0)$
$(2, -12)$

37.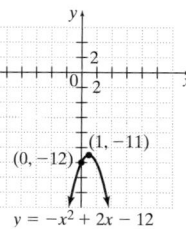
$(0, -12)$
$(1, -11)$
$y = -x^2 + 2x - 12$

38.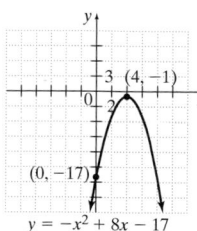
$(4, -1)$
$(0, -17)$
$y = -x^2 + 8x - 17$

39.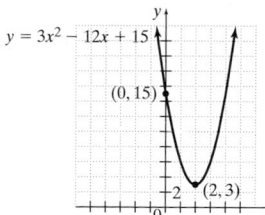
$y = 3x^2 - 12x + 15$
$(0, 15)$
$(2, 3)$

40.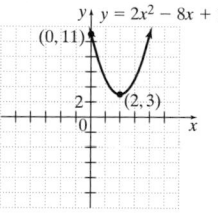
$y = 2x^2 - 8x + 11$
$(0, 11)$
$(2, 3)$

41.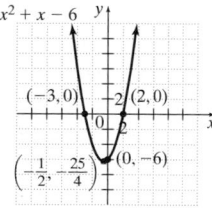
$y = x^2 + x - 6$
$(-3, 0)$
$(2, 0)$
$\left(-\frac{1}{2}, -\frac{25}{4}\right)$
$(0, -6)$

42.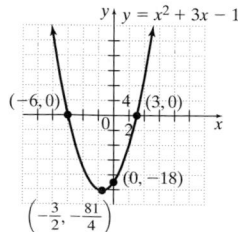
$y = x^2 + 3x - 18$
$(-6, 0)$
$(3, 0)$
$(0, -18)$
$\left(-\frac{3}{2}, -\frac{81}{4}\right)$

43.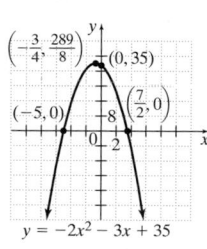
$\left(-\frac{3}{4}, \frac{289}{8}\right)$
$(0, 35)$
$(-5, 0)$
$\left(\frac{7}{2}, 0\right)$
$y = -2x^2 - 3x + 35$

44.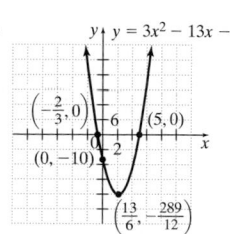
$y = 3x^2 - 13x - 10$
$\left(-\frac{2}{3}, 0\right)$
$(5, 0)$
$(0, -10)$
$\left(\frac{13}{6}, -\frac{289}{12}\right)$

55.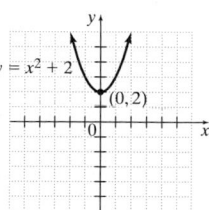
$y = x^2 + 2$
$(0, 2)$

56.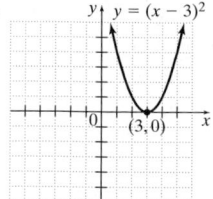
$y = (x - 3)^2$
$(3, 0)$

57.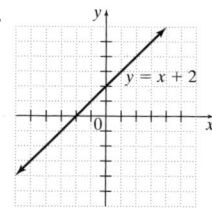
$y = x + 2$

58.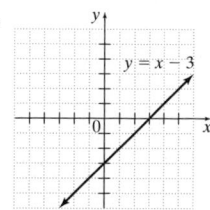
$y = x - 3$

59. 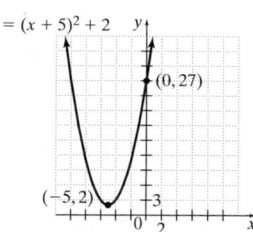 $y = (x + 5)^2 + 2$
$(0, 27)$
$(-5, 2)$

60.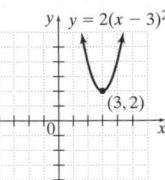
$y = 2(x - 3)^2 + 2$
$(3, 2)$

61.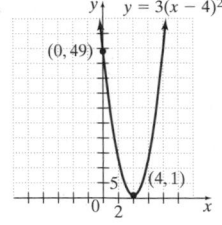
$y = 3(x - 4)^2 + 1$
$(0, 49)$
$(4, 1)$

62.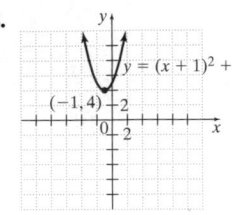
$y = (x + 1)^2 + 4$
$(-1, 4)$

63.
$\left(4, \frac{3}{2}\right)$
$\left(0, -\frac{29}{2}\right)$
$y = -(x - 4)^2 + \frac{3}{2}$

64.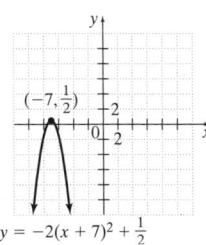
$\left(-7, \frac{1}{2}\right)$
$y = -2(x + 7)^2 + \frac{1}{2}$

65.

66.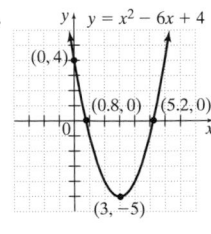
$y = x^2 - 6x + 4$
$(0, 4)$
$(0.8, 0)$ $(5.2, 0)$
$(3, -5)$

67.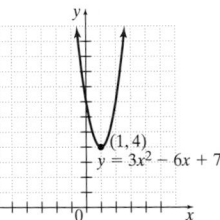
$(1, 4)$
$y = 3x^2 - 6x + 7$

68.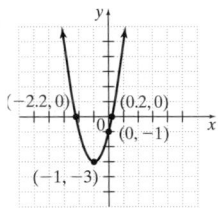
$(-2.2, 0)$ $(0.2, 0)$
$(0, -1)$
$(-1, -3)$

Chapter Project

1. $\begin{cases} 81a + 9b + c = 11 \\ 4761a + 69b + c = 33 \\ 10{,}201a + 101b + c = 66 \end{cases}$ **2.** $a = 0.007, b = -0.168, c = 11$ **3.** $y = 0.007x^2 - 0.168x + 11$ **4.** The model yields 15.095 pounds per person; answers may vary. **5.** 20.1 pounds per person **6.** 1988 **7.** 1912; 9.992 pounds per person **8.** Answers may vary. One possible answer: Poultry farmers could use this type of model to predict future demand for their products.

Chapter 8 Review

53.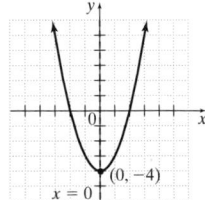
$(0, -4)$
$x = 0$

54.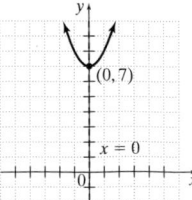
$(0, 7)$
$x = 0$

55.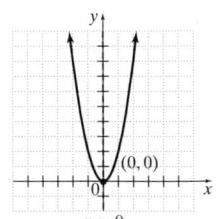
$(0, 0)$
$x = 0$

56.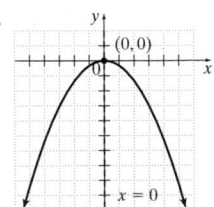
$(0, 0)$
$x = 0$

57.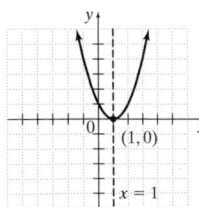
$(1, 0)$
$x = 1$

58.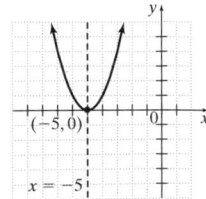
$(-5, 0)$
$x = -5$

59.
$(4, -2)$
$x = 4$

60.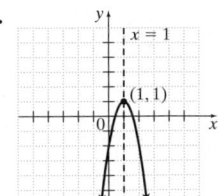
$x = 1$
$(1, 1)$

61.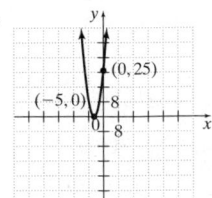
$(0, 25)$
$(-5, 0)$

62.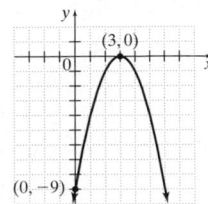
$(3, 0)$
$(0, -9)$

63.

64.

65.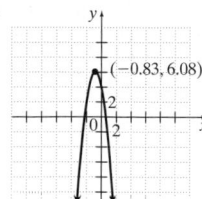

Chapter 8 Test

14.

15.

16.

17.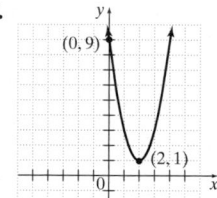

CHAPTER 9 EXPONENTIAL AND LOGARITHMIC FUNCTIONS

Exercise Set 9.2

23.

24.

25.

26.

27.

28.

29.

30.

61. c.

d.

62. c.

d.

65.

66.

67.   **68.**

Exercise Set 9.3

1.

2.

3.

4.

5.

6.

7.

8.

9.

10.

11.

12.

13.

14.

15.

16.

65.

66.

67.

68.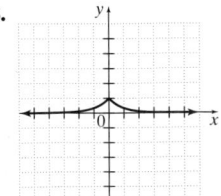

9.4   Spotlight on Decision Making

Logarithmic: the rate of growth is slowing.

Exercise Set 9.4

71.

72.

73.

74.

75.

76.

77.

78.

91. **92.**

93. **94.**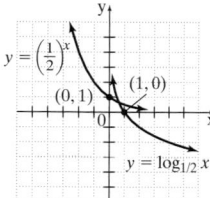

9.5 Spotlight on Decision Making

Answers may vary.

Exercise Set 9.5

63.

Integrated Review

16.

17.

18.

19.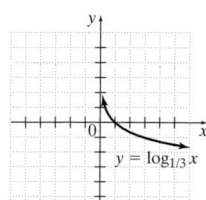

Exercise Set 9.6

79.

80.

81.

82.

83.

84.

85.

86.

87.

88.

89.

90.

91.

92.

93.

94.

95.

96.

97.

98.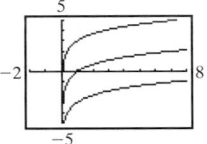

Exercise Set 9.7

75. a. 40,000

77. a. 80,000

79. a. 200,000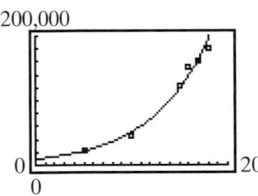

Chapter Project

answers may vary

Chapter 9 Review

29.

30.

31.

38.

39.

40.

41.

63.

64.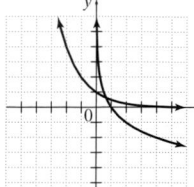

Chapter 9 Test

4.

23.

24.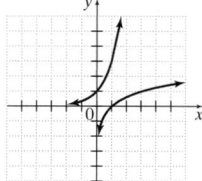

Chapter 9 Cumulative Review

3.

35.

42.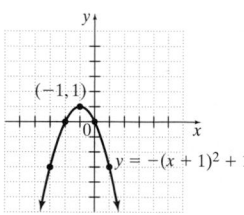

CHAPTER 10 CONIC SECTIONS

Exercise Set 10.1

1.

2.

3.

4.

5.

6.

7.

8.

9.

10.

11.

12.

37.

38.

39.

40.

41.

42.

43.

44.

45.

46.

47.

48.

56.

57.

58.

59.

60.

61.

62.

63.

64.

65.

66.

67.

68.

69.

70.

71.

72.

73.

74.

75.

76.

77.

78.

79.

80.

81.

82.

83.

84.

85.

86.

87.

88.

89.

90.

91.

92.

93.

94.

95.

96.

97.

98.

99.

110.

111.

112.

113.

Spotlight on Decision Making

a. hyperbola; one time **b.** ellipse; infinitely many times **c.** ellipse; infinitely many times **d.** ellipse; infinitely many times **e.** hyperbola; one time

Exercise Set 10.2

1.

2.

3.

4.

5.

6.

7.

8.

9.

10.

11.

12.

13.

14.

15.

16.

17.

18.

19.

20.

22.

23.

24.

25.

26.

27.

28.

29.

30.

31.

32.

33.

34.

35.

36.

37.

38.

39.

40.

41.

42.

43.

44.

45.

67.

68.

69.

70.

71.

72.

73.

74.

Integrated Review

1.

2.

3.

4.

5.

6.

7.

8.

9.

10.

11.

12.

13.

14.

15.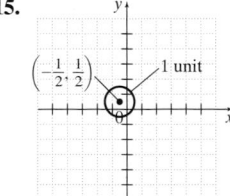

Exercise Set 10.3

31.

32.

33.

34.

47.

48.

49.

50.

10.4   Spotlight on Decision Making

The bridge must be redesigned; the clearance at the edge of the road is only 12.5 ft.

Exercise Set 10.4

1.

2.

3.

4.

5.

6.

7.

8.

9.

10.

11.

12.

13.

14.

15.

16.

17.

18.

19.

20.

21.

22.

23.

24.

25.

26.

27.

28.

29.

30.

31.

32.

33.

34.

35.   **36.**   **37.**   **38.**   **39.**

40.   **51.**

Chapter Project

1. Answers may vary.   **2.** Answers may vary.   **3.** Answers may vary.   **4.** Answers may vary.   **5.** Answers may vary.   **6.** Answers may vary.   **7.** Answers may vary.   **8.** These lines pass through the center of the ellipse.

Chapter 10 Review

21.   **22.**   **23.**   **24.**   **25.**

26.   **27.**   **28.**   **29.**

30.   **31.**   **32.**   **33.**

34.   **36.**   **37.**   **38.**

39.   **40.**   **41.**   **42.**

43.

44.

45.

46.

47.

48.

49.

50.

51.

52.

53.

54.

55.

56.

57.

58.

59.

60.

61.

62.

63.

64.

65.

66.

79.

80.

81.

82.

83.

84.

85.

86.

87.

88.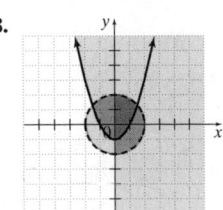

Chapter 10 Test

4.

5.

6.

7.

8.

9.

10.

11.

14.

15.

16.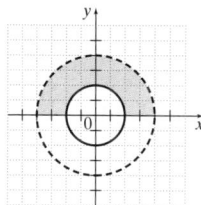

Chapter 10 Cumulative Review

3.

36.

44.

45.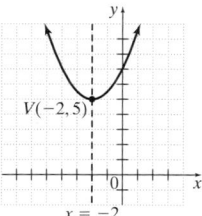

CHAPTER 11 SEQUENCES, SERIES, AND THE BINOMIAL THEOREM

11.1 Spotlight on Decision Making

Choose the second job. It will pay $15.15 per hour in the 5th year, whereas the first job will only pay $14.10 per hour. You may want to consider whether you will actually stay in that job 5 years; if you don't, the first job might be the better option.

Exercise Set 11.1

47.

48.

49.

50.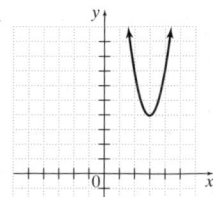

11.2 Spotlight on Decision Making

A culture dish that holds 5000 units will be large enough since 3417 units will be present after 7 hours.

11.4 Spotlight on Decision Making

The value of the Roth IRA will only be about $372,204 when the client is 65. You should suggest an additional investment.

Exercise Set 11.5

41. **42.** **43.** **44.** **45.**

46. **48.**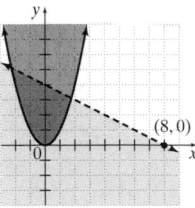

Chapter Project

1. geometric **2.** arithmetic **3.** $a_n = 3506(1.045)^{n-1}$ **4.** $a_n = 15{,}531 + 804.30(n - 1)$ or $a_n = 14{,}726.70 + 804.3n$ **5.** \$3828.64
6. \$18,748.20 **7.** \$14,999.33 **8.** \$76,601.40; S_7 is the total tuition for seven years beginning in the 1999–2000 academic year, but you are only interested in the total for the last four years of those years. Thus, you subtract S_3, the total tuition for three years beginning in the 1999–2000 academic year. **9.** Answers may vary.

Chapter 11 Cumulative Review

11. **13.** **18.** **42.**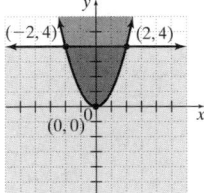

Exercise Set Appendix D

7. **8.** **9.** **10.**

11. **12.** **13.** **14.**

15. **16.** **17.** **18.**

19. **20.** **21.** **22.**

PHOTO CREDITS

Contents Reuters Media, Inc./Corbis/Bettmann, W.A. Harewood/Getty Images, Inc./Liaison, © Kevin Fleming/CORBIS, © Jason Hawkes/CORBIS, Index Stock Imagery, Inc., © Gregg Stott/Masterfile Corporation, Whitney, Frank/Getty Images, Inc./Image Bank, Mark Gibson/Mark and Audra Gibson Photography/© Gibson Stock Photography, © David R. Frazier/David R. Frazier Photolibrary, Inc. www.drfphoto.com, Telegraph Colour Library/FPG International LLC, © David Young Wolff/PhotoEdit Inc.

Chapter 1 Reuters Media, Inc./Corbis/Bettmann, (p.2) Sepp Seitz/Woodfin Camp & Associates, (p.3) © Rachel Epstein/PhotoEdit, (p.7) Boeing Commercial Airplane Group/Courtesy of The Boeing Company

Chapter 2 W.A. Harewood/Getty Images, Inc./Liaison, (p.122) © Bonn Sequenz/Imapress/The Image Works, (p.145) © Mark Richards/PhotoEdit, (p.151) Masterfile Corporation, (p.151) AGE Fotostock America, Inc., (p.178) Bob Daemmrich/Bob Daemmrich Photography, Inc., (p.184) Bill Bachmann/Photo Researchers, Inc., (p.190) © Mark Richards/PhotoEdit, (p.213) David Young Wolff/PhotoEdit Inc.

Chapter 3 © Kevin Fleming/CORBIS, (p.235) © Spencer Ainsley/The Image Works

Chapter 4 © Jason Hawkes/CORBIS, (p.284) © Lon C. Diehl/PhotoEdit Inc., (p.313) SuperStock, Inc., (p.314) © James Leynse/CORBIS/SABA Press Photos, Inc., (p.315) © Armando Arorizo/ZUMA/Corbis, (p.315) AP/Wide World Photos, (p.316) Pierre Tremblay/Masterfile Corporation, (p.333) T.A.Wiewandt/DRK Photo

Chapter 5 Index Stock Imagery, Inc., (p.354) Corbis Digital Stock, (p.355) Chris Butler/Science Photo Library, (p.355) © Rudi Von Briel/PhotoEdit Inc., (p.364) John Lemker/Animals Animals/Earth Scenes, (p.428) Getty Images, Inc./Photodisc

Chapter 6 © Gregg Stott/Masterfile Corporation, (p.450) Guirard, Greg/Getty Images Inc./Image Bank, (p.466) Pictor/Image State/International Stock Photography Ltd., (p.466) Kelly-Mooney Photography/Corbis/Bettmann, (p.485) Spike Mafford/Getty Images Inc./Photodisc, (p.485) Rob Melnychuk/Getty Images Inc./Photodisc, (p.491) Amy C. Etra/PhotoEdit, (p.492) John Serafin, (p.496) Michael Gadomski/Photo Researchers, Inc., (p.501) Richard A. Cooke III/Getty Images Inc./Stone Allstock, (p.505) © Ken Welsh/AGE Fotostock America, Inc.

Chapter 7 Whitney, Frank/Getty Images, Inc./Image Bank, (p.539) © Susan Findlay/Masterfile Corporation, (p.570) Steve Gottlieb/Getty Images Inc./Taxi

Chapter 8 Mark Gibson/Mark and Audra Gibson Photography/© Gibson Stock Photography, (p.602) Tony Freeman/PhotoEdit, (p.604) David Parker/Science Photo Library/Photo Researchers, Inc., (p.615) Dr. Chikaraishi/National Institute for Fusion Science, (p.616) © Shane Pedersen/Wildlight Photo Agency, (p.621) © ZEFA/Masterfile Corporation, (p.625) Tim Flach/Getty Images Inc./Stone Allstock, (p.625) Arthur S. Aubry Photography/Getty Images Inc./Photodisc, (p.669) AP/Wide World Photos

Chapter 9 © David R. Frazier/David R. Frazier Photolibrary, Inc. www.drfphoto.com, (p.700) © I and I/Masterfile Corporation, (p.701) Peter Arnold, Inc., (p.727) © Weiss/Sunset/Animals Animals/Earth Scenes

Chapter 10 Telegraph Colour Library/FPG International LLC

Chapter 11 © David Young Wolff/PhotoEdit Inc.

READ THIS LICENSE CAREFULLY BEFORE OPENING THIS PACKAGE. BY OPENING THIS PACKAGE, YOU ARE AGREEING TO THE TERMS AND CONDITIONS OF THIS LICENSE. IF YOU DO NOT AGREE, DO NOT OPEN THE PACKAGE. PROMPTLY RETURN THE UNOPENED PACKAGE AND ALL ACCOMPANYING ITEMS TO THE PLACE YOU OBTAINED THEM. THESE TERMS APPLY TO ALL LICENSED SOFTWARE ON THE DISK EXCEPT THAT THE TERMS FOR USE OF ANY SHAREWARE OR FREEWARE ON THE DISKETTES ARE AS SET FORTH IN THE ELECTRONIC LICENSE LOCATED ON THE DISK:

Single PC Site License

1. **GRANT OF LICENSE and OWNERSHIP:** The enclosed computer programs and any data ("Software") are licensed, not sold, to you by Pearson Education, Inc. publishing as Pearson Prentice Hall ("We" or the "Company") in consideration of your adoption of the accompanying Company textbooks and/or other materials, and your agreement to these terms. You own only the disk(s) but we and/or our licensors own the Software itself. This license allows instructors and students enrolled in the course using the Company textbook that accompanies this Software (the "Course") to use and display the enclosed copy of the Software on an unlimited number of computers, for academic use only, so long as you comply with the terms of this Agreement. You may make one copy for back up only. We reserve any rights not granted to you.

2. **USE RESTRICTIONS:** You may not sell or license copies of the Software or the Documentation to others. You may not transfer, distribute or make available the Software or the Documentation. You may not reverse engineer, disassemble, decompile, modify, adapt, translate or create derivative works based on the Software or the Documentation. You may be held legally responsible for any copying or copyright infringement that is caused by your failure to abide by the terms of these restrictions

3. **TERMINATION:** This license is effective until terminated. This license will terminate automatically without notice from the Company if you fail to comply with any provisions or limitations of this license. Upon termination, you shall destroy the Documentation and all copies of the Software. All provisions of this Agreement as to limitation and disclaimer of warranties, limitation of liability, remedies or damages, and our ownership rights shall survive termination.

4. **DISCLAIMER OF WARRANTY: THE COMPANY AND ITS LICENSORS MAKE NO WARRANTIES ABOUT THE SOFTWARE, WHICH IS PROVIDED "AS-IS." IF THE DISK IS DEFECTIVE IN MATERIALS OR WORKMANSHIP, YOUR ONLY REMEDY IS TO RETURN IT TO THE COMPANY WITHIN 30 DAYS FOR REPLACEMENT UNLESS THE COMPANY DETERMINES IN GOOD FAITH THAT THE DISK HAS BEEN MISUSED OR IMPROPERLY INSTALLED, REPAIRED, ALTERED OR DAMAGED. THE COMPANY DISCLAIMS ALL WARRANTIES, EXPRESS OR IMPLIED, INCLUDING WITHOUT LIMITATION, THE IMPLIED WARRANTIES OF MERCHANTABILITY AND FITNESS FOR A PARTICULAR PURPOSE. THE COMPANY DOES NOT WARRANT, GUARANTEE OR MAKE ANY REPRESENTATION REGARDING THE ACCURACY, RELIABILITY, CURRENTNESS, USE, OR RESULTS OF USE, OF THE SOFTWARE.**

5. **LIMITATION OF REMEDIES AND DAMAGES: IN NO EVENT, SHALL THE COMPANY OR ITS EMPLOYEES, AGENTS, LICENSORS OR CONTRACTORS BE LIABLE FOR ANY INCIDENTAL, INDIRECT, SPECIAL OR CONSEQUENTIAL DAMAGES ARISING OUT OF OR IN CONNECTION WITH THIS LICENSE OR THE SOFTWARE, INCLUDING, WITHOUT LIMITATION, LOSS OF USE, LOSS OF DATA, LOSS OF INCOME OR PROFIT, OR OTHER LOSSES SUSTAINED AS A RESULT OF INJURY TO ANY PERSON, OR LOSS OF OR DAMAGE TO PROPERTY, OR CLAIMS OF THIRD PARTIES, EVEN IF THE COMPANY OR AN AUTHORIZED REPRESENTATIVE OF THE COMPANY HAS BEEN ADVISED OF THE POSSIBILITY OF SUCH DAMAGES.** SOME JURISDICTIONS DO NOT ALLOW THE LIMITATION OF DAMAGES IN CERTAIN CIRCUMSTANCES, SO THE ABOVE LIMITATIONS MAY NOT ALWAYS APPLY.

6. **GENERAL:** THIS AGREEMENT SHALL BE CONSTRUED IN ACCORDANCE WITH THE LAWS OF THE UNITED STATES OF AMERICA AND THE STATE OF NEW YORK, APPLICABLE TO CONTRACTS MADE IN NEW YORK, AND SHALL BENEFIT THE COMPANY, ITS AFFILIATES AND ASSIGNEES. This Agreement is the complete and exclusive statement of the agreement between you and the Company and supersedes all proposals, prior agreements, oral or written, and any other communications between you and the company or any of its representatives relating to the subject matter. If you are a U.S. Government user, this Software is licensed with "restricted rights" as set forth in subparagraphs (a)-(d) of the Commercial Computer-Restricted Rights clause at FAR 52.227-19 or in subparagraphs (c)(1)(ii) of the Rights in Technical Data and Computer Software clause at DFARS 252.227-7013, and similar clauses, as applicable.

Should you have any questions concerning this agreement or if you wish to contact the Company for any reason, please contact in writing: Customer Service Pearson Prentice Hall, 200 Old Tappan Road, Old Tappan NJ 07675.

Minimum System Requirements

Windows

- Pentium II 300 MHz processor
- Windows 98 or later
- 64 MB RAM
- 800 x 600 resolution
- 8x or faster CD-ROM drive
- QuickTime 6.0 or later

Macintosh

- Power PC G3 233 MHz or better
- Mac OS 9.x or 10.x
- 64 MB RAM
- 800 x 600 resolution
- 8x or faster CD-ROM drive
- QuickTime 6.0 or later